Microscope Design

Volume 2:
Practice

Microscope Design

Volume 2: Practice

Dmitry N. Frolov

SPIE PRESS
Bellingham, Washington USA

Library of Congress Control Number: 2021950934

Published by
SPIE
P.O. Box 10
Bellingham, Washington 98227-0010 USA
Phone: +1 360.676.3290
Fax: +1 360.647.1445
Email: books@spie.org
Web: www.spie.org

Printed in the United States of America.
First printing.
For updates to this book, visit http://spie.org and type "PM350" in the search field.

SPIE.

This book is dedicated to my daughter Elena, my son Alexey, and my grandchildren, Violetta, Olesya, Dmitry, and Alexander, whom I love so much.

Table of Contents

Preface

This is the second volume of my book on microscopy. The first, *Microscope Design, Volume 1: Principles*, describes the old Soviet and modern Russian schools of optical instrument making, including the methodology of designing light microscopes and other optical devices. In this second volume, more attention is given to the practice of design. I also share my experiences and disclose some professional secrets.

I have always been inclined to share my professional experience with technical experts and researchers; it is beneficial to communicate with both microscope designers and users who can gain insight from my knowledge and expertise. In addition, the material in this book can be useful to students and teachers as well as microscope hobbyists who are passionate about making new discoveries and learning more about microscope design.

In this volume, I tried to systematize the practical experience that I gained over nearly 40 years of my professional life. Working at the Russian enterprise LOMO for almost 20 years allowed me to acquire practical design experience and introduced me to the production of light microscopes and other optical devices. The next 20 years of working at the Russian project Labor-Microscopes significantly expanded my horizons and brought me closer to researchers, people for whom a microscope is the main tool used in their daily activities. My dream has always been for developers and researchers to "speak the same language" so that these two groups of specialists can support each other and present each other with new tasks. I tried to realize this idea in this volume, where along with some theoretical aspects, the practice of microscope design is emphasized. In a broad sense, this practice is realized when you can trace the motivations and specific technical solutions of developers, as well as the perceptions of the results of this work by researchers. As a result, productivity, researcher efficiency, and the accuracy of information about an object under study can all be increased when working on a microscope.

This book presents specific engineering solutions, the constructive parameters of many real optical systems associated with a light microscope, and examples of the mechanical designs of such systems. Together with the presentation of the methodology for designing such systems, many examples of the quality assessment of the resulting images are also provided. At the same time, not only are graphs of the aberration correction of certain systems shown, but full-color micrographs of real microscopy objects made using the described systems are also presented. This allows readers to see the correlation between theoretical and real image quality. In my opinion, this is the practice of microscope design.

Thanks to the support from SPIE Press, I can now fully share my experience in the design and practical use of microscopy devices. I have no secrets from my readers; indeed, they can use this book for work, training, hobbies, or natural studies.

I am grateful to my Labor-Microscopes colleagues for the vast experience of their collaboration. Thanks to our clients, researchers, and friends who shared their invaluable experience with microscopes with me. Thanks also to SPIE Director of Publications and Platform Patrick Franzen for his attentive and gracious attitude toward my manuscript, as well as to SPIE Senior Editor Alexandra MacWade for her superior work on this publication. Thanks to my readers.

I hope that sharing my knowledge benefits readers. This is the main thing that a person desires: to help others and for there to be peace in the world, where all people live richly and are happy.

Dmitry N. Frolov
June 2023

Chapter 1
Synthesis of Optical Systems for Mass-produced Budget Microscopes

To begin this chapter, I would like to remember my "production" mentor, Ivanova Tatiana Aleksandrovna. After graduating from the Institute of Precision Mechanics and Optics, now called the University of Information Technology, I joined the Optical Design Bureau in the Department of Computational Optics of the optical–mechanical association LOMO in Russia. I was very fortunate to be a part of the optical design group for light microscopes. Tatiana Aleksandrovna was the main specialist in this group. She was a complicated and highly ambitious woman with whom I worked closely for just three years. Despite there being very few women professors in optics, she became a professor. Unfortunately, her health was undermined; working in a production plant can be a very difficult place for the implementation of innovative ideas and projects. She had a bright and short life.

1.1 Achromatic Correction Objectives

Currently, it is not possible to trace the original method of constructing circuit solutions for achromatic objectives used by Abbe. Analyzing the construction of classical achromats, Berek [1] proposed that the construction of small-magnification objectives be considered according to the type of the known Petzval objectives in the reverse course of the rays. The construction of high-magnification objective designs should be carried out in such a way that all surfaces of the system have the smallest possible values of the first two third-order aberration coefficients.

This condition is satisfied by constructions built on the principle of synthesis of frontal elements containing surfaces close to aplanatic, and elements forming the design of the Petzval objective. When constructing circuit solutions for achromatic objectives of the classical type, the system can be based on two thin components with optical φ_1, φ_2 powers separated by an air gap d [2]. Using ratios of paraxial optics, describing the angles and heights of the first paraxial beam [3],

$$\alpha_{i+1} = \alpha_i + h_i\varphi_i; \quad h_{i+1} = h_i - \alpha_{i+1}d_i;$$

given that $h_1 = \alpha_1 S_1 = S_1 V_{ob}$, we get

$$\varphi_1 = -\frac{(d - S_1)V + S_2'}{S_1 V d}, \quad \varphi_2 = \frac{d - S_1 V + S_2'}{S_2' d}, \tag{1.1}$$

where S_1 is the position of the plane of the object relative to the first component, and S_2' is the position of the image plane relative to the second component. We will keep in mind that for implementing the principle of interchangeability of the main nodes, microscopes with a finite optical length of the tube have several standardized parameters [4]:

- Objective height h: distance from the object plane to the objective reference (supporting) plane ($h_{ob} = 33$ or 37 mm and $h_{ob} = 45$ mm).
- Tube length 160 (170–205) mm: distance from the reference (supporting) plane of the objective lens to the reference (supporting) plane of the eyepiece (ocular).
- Removal of the reference plane of the eyepiece relative to its front focal plane ($h_{oc} = 13$ and 10 mm).
- A parametric series of linear magnifications of microscope objectives is used.

Using the example of objectives of small magnifications, we will follow the procedure for obtaining optical circuits based on a two-component system. If we accept the condition

$$-S_1 + d + S_{H'}' = h = 45\,\text{mm},$$

where $S_{H'}'$ is the position of the back main plane for the second component ($S_{H'}' \approx 5$ mm), and keep in mind that $S_2' \approx 147 + S_{H'}'$, we can get a set of overall solutions for the two-component system depending on the value of the frontal segment S_1 and the linear magnification V_{ob} of the objective.

We will keep in mind that the values of the relative aperture (diaphragm number) of the component are $\frac{D_i}{f_i'} = \frac{2S_i \sin\sigma_i}{f_i'}$, where f_i' is the focal length, $S_{IV} = \frac{1}{n}\Sigma\varphi$ is the Petzval sum, and $\Delta Z_p' = -\frac{1}{2}y'^2 S_{IV}$ is the corresponding curvature of the image.

An analysis of relations [Eq. (1.1)] shows that as the magnification V increases, the optical power of the first component goes into the region of positive values, S_{IV} increases, and the results lead to the classical design of the objective: an achromat of small magnification consisting of two positive components separated by an air gap.

When comparing sets of parameters for a fixed magnification depending on S_1, d, the results of the overall solution give very different values of the relative apertures of the components. The size of the relative apertures not only determines the aberration properties of the system, but also the sensitivity of its elements to manufacturing errors, i.e., the technological properties of the system. Thus, at the development stage, the prerequisites of the technological design of the objective's optical system are already established.

The area of magnification V_{ob}, leading to moderate values of the relative apertures, is $V = -4^x \div -6^x(-8^x)$. With magnification $V_{ob} = -10^x$, the focal length of the second component becomes negative, and the relative aperture of the first component increases to $A \approx 1:1.4$.

The natural way to reduce the relative apertures with increasing numerical aperture is to complicate the design of the objective, i.e., introduce a front lens built from surfaces close to aplanatic, and contribute to a magnification $V_{fr} \geq n$, where n is the refractive index of the lens material. To improve the manufacturability of achromats, the first surface of the frontal lens should be made flat and the second should be made close to the aplanatic surface of the second type [5]; thus, the ratio

$$r_2 \approx r_{apl} \approx \frac{n_2 S_2}{n_2 + 1},$$

where S_2 is the position of the plane of the object for the second surface.

According to this principle, old Abbe achromats as well as Soviet objectives with 8×/0.20, 20×/0.40, and 40×/0.65 achromatic objectives, which are analyzed in Volume 1 of this book (Chapter 1, Section 1.2.1), could be constructed. With a further increase in the numerical aperture, an aplanatic meniscus is introduced behind the frontal flat-convex lens, the first surface of which is close to the aplanatic surface of the first kind, the second to the aplanatic surface of the second kind. According to this scheme, the achromat 90×/1.25 oil was built.

Let us trace the relationship between the parameters of the frontal lens of the high-aperture objective of the microscope and the fourth monochromatic sum S_{IV}, which determines the curvature of the image. Figure 1.1 shows the simple frontal lens of an oil immersion objective, the material of which has a refractive index n_3, close to the refractive indices of immersion n_2 and the cover glass n_1. Let the surface of the lens facing the air be an aplanatic surface of the first kind. Then its radius of curvature is determined by the known relation

$$r_3 = r_{apl} = \frac{n_3 S_3}{n_3 + 1}, \tag{1.2}$$

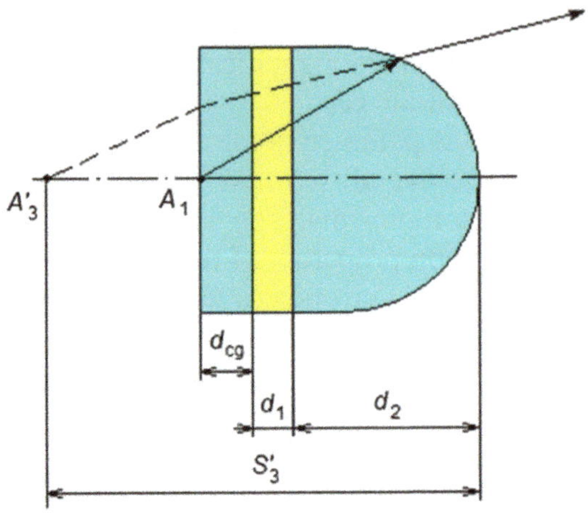

Figure 1.1 Simple frontal lens of an immersion objective.

where S_3 is the position of the plane of the object relative to the surface. Let us calculate the parameters of the first paraxial beam through the system of surfaces, bearing in mind that the plane of the object coincides with the first surface of the cover glass. For this system, we have

$$S_3 = -d_{cg}\frac{n_3}{n_1} - d_1\frac{n_3}{n_2} - d_2, \tag{1.3}$$

where d_{cg}, d_1, d_2 is the thickness of the cover glass, the immersion layer, and the lens, respectively.

Substituting Eq. (1.3) into Eq. (1.2),

$$d_2 = -\frac{1}{n_1 n_2 n_3}[n_3^2(d_{cg}n_2 + d_1 n_1) + r_3 n_1 n_2(n_3 + 1)]. \tag{1.4}$$

Let us express the radius of the aplanatic surface through the value $S_{IV} = -\frac{n_3-1}{n_3 r_3}$ and substitute it into Eq. (1.4). In addition, we take $n_1 = n_2$ (the refractive indices are $n_D = 1.515 \pm 0.002$ for the cover glass and $n_D = 1.515 \pm 0.001$ for the oil immersion). After simple transformations, we get

$$d_2 = \frac{1}{n_2 n_3^2 S_{IV}}[(n_3^2 - 1)n_2 - n_3^3(d_{cg} + d_1)S_{IV}]. \tag{1.5}$$

A magnification in such a system will be equal: $V_{fl} = n_3^2$. Depending on the S_{IV} value, the parameters of the frontal lens are as follows: $d_{cg} = 0.17$ mm, $d_1 = 0.3$ mm, and $n_3 = 1.5263$ (glass type LZ_K20). The results of the calculation of V_{fl}, r_3, d_2, and S'_3 are presented in Table 1.1.

Analyzing the data in Table 1.1 allows us to draw the following conclusions:

Table 1.1 The parameters of the frontal lens depending on the S_{IV} value.

n_3	V_{fl}	S_{IV}	r_3	d_2	S'_3	Use in Soviet objectives
1.5263	2.3296	0.4420	−0.78	0.93	−1.7773	Achromat 90×/1.25 oil Apochromat 90×/1.30 oil
		0.350	−0.9852	1.157	−2.4889	—
		0.3172	−1.087	1.3	−2.5417	Apochromat 60×/1.00 oil
		0.30	−1.1494	1.428	−2.9037	—
		0.25	−1.3793	1.809	−3.4845	
		0.20	−1.7241	2.38	−4.3556	
		0.15	−2.2988	3.331	−5.8075	

1. With a decrease in S_{IV}, the thickness of the lens increases.
2. With the increase in the thickness of the lens, the segment S'_3 increases.

The increase in the back segment limits the possibility of building a subsequent part of the objective from *thin components*. The height of a microscope's objective is standardized and equal to 45 mm (in older models, including the old sets of achromats and apochromats, it is 33 or 37 mm). If we keep in mind that the position of the image plane relative to the reference plane of the objectives with a tube length of 160 mm is $S' = 147$ mm, and make this segment equal to the back segment of the objective, we get $S'_3 < \frac{S' V_{fl}}{V_{ob}}$.

1.2 Investigation of Chromatic Aberrations of Objective Elements

The chromatic difference of positions in a system containing thin elements separated by an air gap is determined by the relation [6]:

$$\alpha'^2 \Delta S'_{F'C'} - = -\sum_1^n h_i^2 \frac{\varphi_i}{v_i}, \tag{1.6}$$

where α' is the angle of the first paraxial beam in the image space, h_i is the height of the first paraxial beam on this element, and v_i is the coefficient of dispersion of its material,

$$v_i = \frac{n_e - 1}{n_{F'} - n_{c'}}.$$

To fulfill the condition $\Delta S'_{F'-C'} = 0$, it is necessary to have elements with optical powers that are opposite in sign and made of materials with different dispersion coefficients. For a system of two thin lenses, between which the distance is *null*, the following relations are true:

$$h_1 = h_2, \quad \frac{\varphi_1}{v_1} + \frac{\varphi_2}{v_2} = 0, \quad \varphi_1 + \varphi_2 = \varphi_{equiv},$$

where φ_{equiv} is the equivalent optical power of the component. The result is

$$\begin{aligned} \varphi_1 &= -\varphi_{equiv}\frac{v_1}{v_2 - v_1}, \\ \varphi_2 &= \varphi_{equiv}\frac{v_2}{v_2 - v_1}. \end{aligned} \tag{1.7}$$

If we use glass type LZ_F1 (from a Russian glass catalog) as a flint lens material, $n_1 = 1.6169$ and $v = 36.70$; if we use crown glass type LZ_TKI4, $n_2 = 1.6155$ and $v = 60.34$; $\varphi_{equiv} = 0.02$ ($f' = 50$ mm), and then $\varphi_1 = 0.03105$.

In the simplest case of making optical elements, the construction of a component is determined by the following set of parameters:

$r_1 = \infty$	$d_1 = 0$	LZ_F1
$r_2 = 19.868$	$d_2 = 0$	1
$r_3 = \infty$	$d_3 = 0$	LZ_TK14
$r_4 = -12.056$		

Petzval's curvature of such a system is determined by the value $S_{IV} = \frac{1}{n}\sum\varphi \approx 0.0123$.

As a result of a change in the deflection of the positive and negative lenses, the radii of curvature of the inner surfaces of the component become equal, after which the component can be made as the glued one. This is a free parameter, α_4, that determines the ability to control the value of the spherical aberration of the component. As an initial value, it is advisable to choose

$$\alpha_4 = \alpha_{\min} = \frac{(2n_4 + 1)(V_2 + 1)}{2(n_2 + 2)},$$

where $V_2 = \frac{\alpha_3}{\alpha_5}$ is the linear magnification of the second element. The change in the selection of lens materials can cause coma aberration and higher values of spherical aberration and spherochromatism. This component—glued positive and negative lenses—is one of the basic elements for optical systems of microscope objectives. When developing objectives with an achromatic type of aberration correction, glasses with ordinary properties that are easily mass produced are used as materials.

Let us return to Eq. (1.7). The condition of equality $\Delta S'_{F'-C'}$ to null of the chromatic difference does not mean the elimination of the chromatic difference of the segments S' for any wavelength. The difference between the segments of the paraxial images of the axial point of the object for the wavelength λ and the main wavelength λ_0 with the corrected chromatic difference $\Delta S'_{F'-C'} = 0$ is called the secondary spectrum of the optical system:

$$\Delta S'_{SS} = S'_{\lambda} - S'_{\lambda_0}, \quad S'_{\lambda_1} - S'_{\lambda_2} = 0.$$

Let us determine the value of the secondary spectrum in the system of thin components. To do this, we construct Eq. (1.6) for arbitrary wavelengths λ, λ_0:

$$\Delta S'_{\lambda-\lambda_0} = -\sum_i^n h_i^2 \frac{\varphi_i}{v_i}, \qquad \text{where } v_i = \frac{n_{\lambda_0} - 1}{n_{\lambda} - n_{\lambda_0}}.$$

Multiply and divide the last expression by $n_{F'} - n_{C'}$:

$$\Delta S'_{\lambda-\lambda_0} = -\sum_i^n h_i^2 \frac{\varphi_2(n_{\lambda} - n_{\lambda_0})}{n_{\lambda_0} - 1} \cdot \frac{n_{F'} - n_{C'}}{n_{F'} - n_{C'}} = -\sum_i^n h_i^2 \frac{\varphi_i}{v_i} \cdot \frac{n_{\lambda} - n_{\lambda_0}}{n_{F'} - n_{C'}},$$

where the value $p_{\lambda_0,\,\lambda} = \frac{n_{\lambda} - n_{\lambda_0}}{n_{F'} - n_{C'}}$ is called the relative partial dispersion of the material. The value of the secondary spectrum of a system of thin components separated by an air gap is thus equal to [7]

$$\Delta S'_{\lambda-\lambda_0} = -\sum_i^n h_i^2 \frac{\varphi_i}{v_i} p_i. \tag{1.8}$$

Let us determine the value of the secondary spectrum of a system constructed of flint and crown lenses (the example above). For the optical material of the flint lens, choose glass with refractive index $n_e = 1.6169$ and dispersion coefficient $v_e = 36.70$ (glass type LZ_F1). For the optical material of the crown lens, choose glass type LZ_TK14 ($n_e = 1.6155$, $v_e = 60.34$). For glass type LZ_F1, $p_{F',e} = 0.5215$ (relative partial dispersion), and for glass type LZ_TKI4, $p_{F',e} = 0.5091$:

$$\Delta S'_{\lambda-\lambda_0} = -h^2 \left(\frac{\varphi_1}{v_1} p_1 + \frac{\varphi_2}{v_2} p_2 \right). \tag{1.9}$$

If normalization $h = f'$, $\alpha' = 1$ for an infinitely distant object,

$$\begin{aligned}\Delta S'_{\lambda-\lambda_0} &= -f'^2 \left(\frac{\varphi_1}{v_1} p_1 + \frac{\varphi_2}{v_2} p_2 \right) \\ &= -50^2 \left(-\frac{0.03105}{36.70} \cdot 0.5215 + \frac{0.05105}{60.34} \cdot 0.5091 \right) = 0.0262.\end{aligned}$$

If the component works with a relative aperture $A = 1{:}2.5$ ($\sin\sigma \approx 0.2$), then the value of the secondary spectrum in the wave measure will be

$$W_{\lambda,\lambda_0} = \frac{\Delta S'_{SS}\sin^2\sigma}{2\lambda} \approx 0.96\lambda.$$

Equation (1.9) should be converted by substituting the expression for φ_1, φ_2 from Eq. (1.7) and adopting $h^2 = S'^2$, $(\alpha' = 1)$:

$$\Delta S'_{\lambda-\lambda_0} = -\frac{p_2 - p_1}{\nu_2 - \nu_1} \cdot \frac{S'^2}{f'}. \tag{1.10}$$

It is obvious that to reduce the magnitude of the secondary spectrum, it is necessary to use optical materials with values of relative partial dispersions as close as possible to the largest possible difference in the dispersion coefficients. Optical materials that meet this condition belong to a group of materials with special properties, and their nomenclature is very limited.

The use of some combinations of glass types increases technological manufacturing difficulties. For example, the use of glass type LZ_OF4 ($n_e = 1.65419$, $\nu_e = 43.24$, $p_{F',e} = 0.5149$), applied as a special flint, in combination with glass type LZ_TK14 gives the difference $\Delta p_{F',e} = 0.00584$ (instead of $\Delta p_{F',e} = 0.0124$ for LZ_F1–LZ_TK14). However, the use of a new combination, LZ_OF4–LZ_TK14, leads to an increase in the optical powers of the elements. Thus, when the condition $\Delta S'_{F'-C'} = 0$, $\varphi_{\mathrm{LZ\_OF4}} = -0.05057$, $\varphi_{\mathrm{LZ\_TK14}} = 0.07057$ is fulfilled; i.e., the optical powers of the flint and crown elements increased by 1.629 and 1.382 times, respectively, compared with the combination LZ_F1–LZ_TK14. An increase in the optical forces of the positive and negative elements leads to an increase in the curvature of the surfaces and an increase in higher-order aberrations, which necessitates a complication of the component design. In turn, the latter circumstance, in combination with an increase in the difficulty of manufacturing parts from glass with special properties (reduced chemical stability of glasses, a decrease in grinding hardness, etc.), degrades the technological properties of the lens. This circumstance excludes or limits the use of *special* materials in objectives of high-volume production, which include achromatic lenses.

Consider the chromatic aberrations of a system of two thin elements separated by an air gap d. In this case, in Eq. (1.8), $h_2 = h_1 - \alpha_3 d$. Keeping this in mind, then $\alpha_3 = \alpha_1 + h_1\varphi_1 = h_1\varphi_1$ $(\alpha_1 = 0)$ and $h_2 = h_1(1 - \varphi_1 d)$. The condition for correcting the chromatic position difference in this case is determined by the expression

$$\frac{h_1^2\varphi_1}{\nu_1} + \frac{h_1^2(1 - \varphi_1 d)^2\varphi_2}{\nu_2} = 0. \tag{1.11}$$

Let us express the size of the air gap in fractions of the focal length: $d = k \cdot f'_1$; thus, Eq. (1.11) has the form

$$\frac{\varphi_1}{\varphi_2} = -\frac{v_1}{v_2}(1-k)^2. \tag{1.12}$$

This last expression can be transformed, bearing in mind that $\varphi_{equiv} = \varphi_1 + \varphi_2 - d\varphi_1\varphi_2 = \varphi_1 + (1-k)\varphi_2$, from where $\varphi_1 = \varphi_{equiv} - (1-k)\varphi_2$,

$$\frac{\varphi_{equiv} - (1-k)\varphi_2}{\varphi_2} = -\frac{v_1}{v_2}(1-k)^2,$$

$$\varphi_2 = \frac{\varphi_{equiv}}{(1-k)\cdot\left[1 - \frac{v_1}{v_2}(1-k)\right]}. \tag{1.13}$$

Accordingly, the optical power of the first element is

$$\varphi_1 = -\frac{\frac{v_1}{v_2}\varphi_{equiv}(1-k)}{1 - \frac{v_1}{v_2}(1-k)}. \tag{1.14}$$

Of interest is the definition of the S_{IV} of the system:

$$S_{IV} \approx \frac{1}{n}(\varphi_1 + \varphi_2), \quad S_{IV} \approx \frac{\varphi_{equiv}\left[1 - \frac{v_1}{v_2}(1-k)^2\right]}{n(1-k)\left[1 - \frac{v_1}{v_2}(1-k)\right]}. \tag{1.15}$$

Next, let us determine the value of the secondary spectrum of the system:

$$\Delta S'_{\lambda-\lambda_0} = -\left[h_1^2\frac{\varphi}{v_1}p_1 + h_2^2\frac{\varphi_2}{v_2}p_2\right].$$

Substitution in the last expression of values h_2, φ_1, φ_2 gives

$$\Delta S'_{\lambda-\lambda_0} = -h_1^2\frac{(p_2 - p_1)\cdot(1-k)\cdot\varphi_{equiv}}{[v_2 - v_1(1-k)]}.$$

For a system operating from infinity, the last expression is converted to the form

$$\Delta S'_{\lambda-\lambda_0} = -f'_{equiv}\frac{(p_2 - p_1)\cdot(1-k)}{v_2 - v_1(1-k)}, \tag{1.16}$$

which for $k = 0$ coincides with Eq. (1.10).

Table 1.2 Values φ_1, φ_2, S_{IV}, and $\Delta S'_{\lambda-\lambda_0}$ for various types of relative orientations of the elements of an objective's optical scheme.

No.	k	φ_1	φ_2	d	φ_{equiv}	S_{IV}	$\Delta S'_{\lambda-\lambda_0}$
1	−0.50	−0.20813	0.15208	2.4023		−0.0347	0.1758
2	−0.40	−0.11469	0.09621	3.4877		−0.0114	0.0969
3	−0.30	−0.07555	0.07350	3.9709		−0.0013	0.0638
4	−0.20	−0.05404	0.06170	3.7001		0.0047	0.0456
5	−0.10	−0.04043	0.05494	2.4734		0.0090	0.0342
6	0.0	−0.03105	0.05105	0.0	0.02	0.0124	0.0262
7	0.10	0.06169	−0.04632	1.6210		0.0095	0.0317
8	0.20	0.08343	−0.07929	2.3972		0.0026	0.0429
9	0.30	0.15254	−0.18934	1.9667		−0.0228	0.0784
10	0.40	−1.45973	2.46621	−0.2740		0.6230	−0.7500

Let us define the values φ_1, φ_2, S_{IV}, and $\Delta S'_{\lambda-\lambda_0}$ (see Table 1.2) depending on the k for various types of relative orientations of the elements: combinations *flint ahead* (LZ_F1–LZ_TK14) and *crown ahead* (LZ_TKI4–LZ_F1).

Analyzing the data in Table 1.2, which are the results of calculations, allows us to draw the following conclusions:

1. The flint ahead orientation is preferred because it gives smaller values of optical forces at the same S_{IV} values.
2. The use of a real air gap $d \neq 0$ in a system made up of thin components allows the curvature of the image to be corrected with a chromatic difference of positions.
3. A decrease in S_{IV} is accompanied by an increase in the secondary spectrum. With a corrected curvature ($S_{IV} = 0$), the secondary spectrum exceeds the volume $\Delta S'_{\lambda-\lambda_0}$ of the thin system by 2.5 times.

The flint ahead orientation gives rise to the *triplet* design, which can be used as the basis for creating achromats with reduced and corrected curvature of the image. Thus, as a result of the synthesis of a symmetrical system of two identical halves, in which an initial design works with magnification due to a change in the deflection of a positive lens, spherical aberration can be corrected. Moreover, additional correction parameters—the refractive indices of the flint and crown, as well as the coefficients of dispersion of materials—enable the correction of astigmatism and influence the higher orders of spherical aberration.

1.3 An Example of Creating a Set of Simple Achromatic Objectives

Further studies on the optimization of classical constructions of Abbe achromats led to the creation of a set of achromatic objectives for implementation in microscopes with a finite optical tube length and *infinity*. A distinctive feature of these objectives, in which the same level of aberration as the classical Abbe is achieved, is a reduced number of lenses, which lowers the cost of their manufacture. Figures 1.2–1.13 show the design parameters and graphs of aberration correction of these achromatic objectives. These objectives are designed to complete the mass segment microscopes.

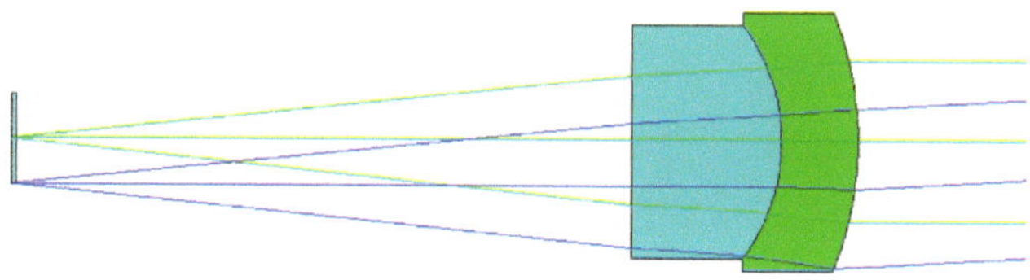

Lens: achro 5x0.10 ftl160

Object num aper 0.100000 Image height 9.000000 Primary wavln 0.546070

SRF	RADIUS	THICKNESS	APERTURE RADIUS	GLASS	SPECIAL
OBJ	0.000000	0.010000	1.781724	AIR	
AST	0.000000	0.170000	1.780000 A	LZ_K14	
2	0.000000	24.710000	1.790000	AIR	
3	0.000000	6.000000	4.270000	LZ_BK6	
4	-7.447000	3.000000	4.560000	LZ_TF3	
5	-13.490000	90.000000	5.050000	AIR	
6	214.800000	5.000000	6.930000	LZ_BK8	
7	-46.990000	0.300000	7.030000	AIR	
8	-46.880000	3.000000	7.020000	LZ_BF24	
9	-113.240000	157.804304 S	7.130000	AIR	
IMS	0.000000	0.000000	9.000000		

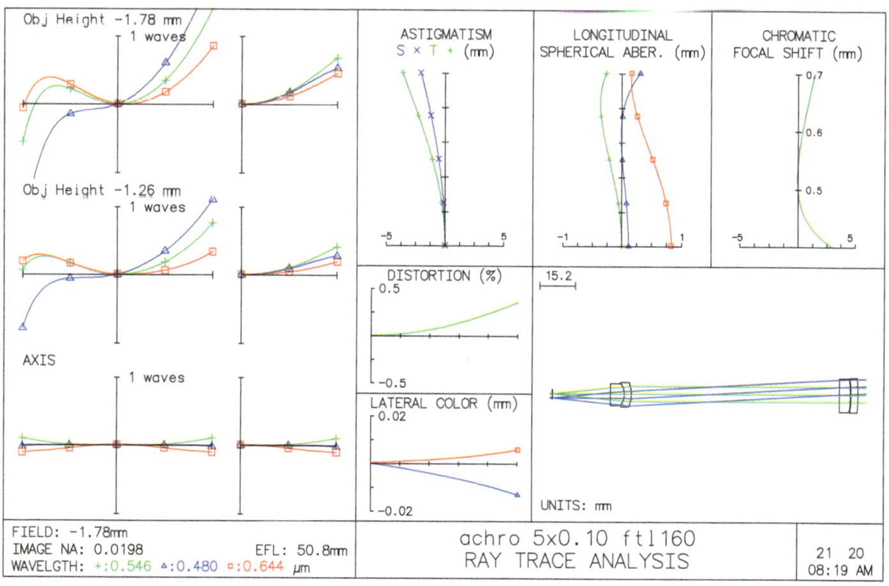

Figure 1.2 Optical design and graphs of aberration correction of a 5×/0.10 infinity achromatic objective ($F'_{ts} = 160$ mm). Here and in other drawings, the design parameters of the tube system are highlighted in gray.

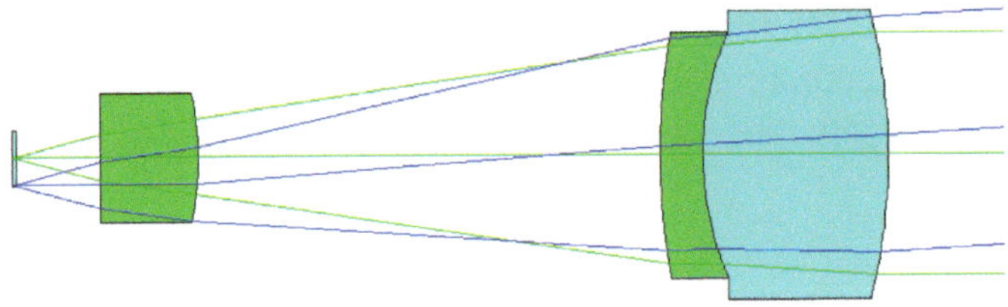

Lens: achro 10x0.25 ftl160

Object num aper 0.250000 Image height 9.000000 Primary wavln 0.546070

SRF	RADIUS	THICKNESS	APERTURE RADIUS	GLASS	SPECIAL
OBJ	0.000000	0.010000	0.929961	AIR	
AST	0.000000	0.170000	0.930000 A	LZ_K14	
2	0.000000	3.000000	0.960000	AIR	
3	0.000000	3.500000	1.740000	LZ_TF5	
4	-9.376000	16.160000	2.200000	AIR	
5	22.750000	1.500000	4.100000	LZ_TF5	
6	10.471000	6.500000	4.190000	LZ-K8	
7	-19.543000	90.000000	4.810000	AIR	
8	214.800000	5.000000	10.080000	LZ_BK8	
9	-46.990000	0.300000	10.150000	AIR	
10	-46.880000	3.000000	10.120000	LZ_BF24	
11	-113.240000	157.439906 S	10.240000	AIR	
IMS	0.000000	0.000000	9.000000		

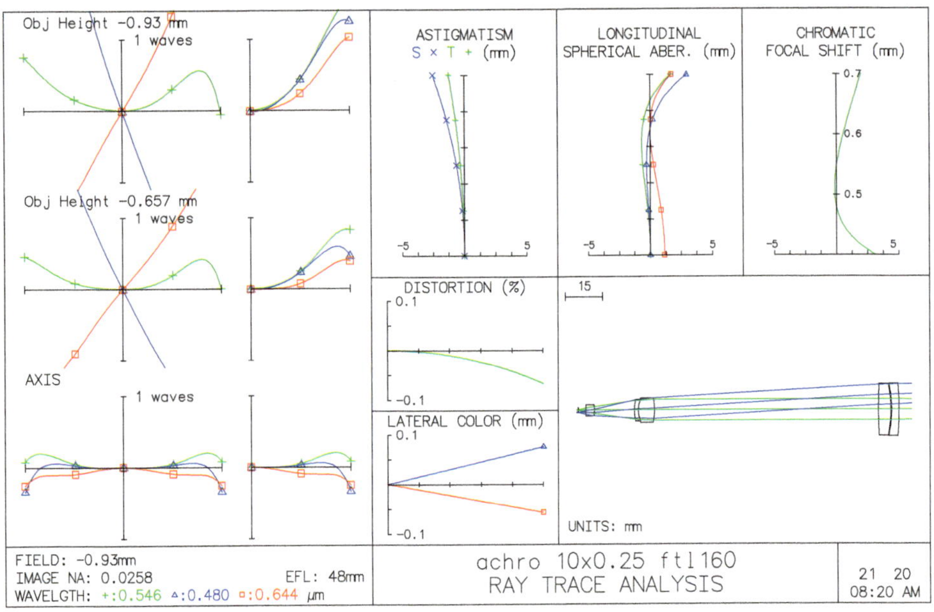

Figure 1.3 Optical design and graphs of aberration correction of a 10×/0.25 infinity achromatic objective ($F'_{ts} = 160$ mm; version 1).

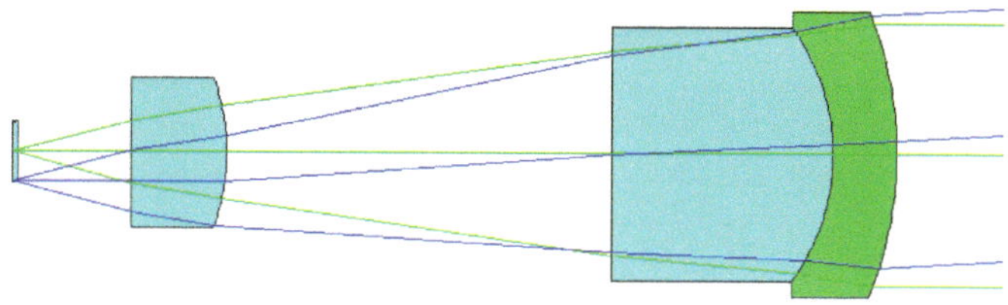

Lens: achro 10x0.25 ftl160

Object num aper 0.250000 Image height 9.000000 Primary wavln 0.546070

SRF	RADIUS	THICKNESS	APERTURE RADIUS	GLASS	SPECIAL
OBJ	0.000000	0.010000	0.908425	AIR	
AST	0.000000	0.170000	0.910000 A	LZ_K14	
2	0.000000	3.630000	0.940000	AIR	
3	0.000000	3.000000	1.870000	LZ_K8	
4	-6.607000	12.180000	2.310000	AIR	
5	0.000000	7.000000	3.160000	LZ_K8 P	
6	-6.607000	2.000000	3.920000	LZ_TF3	
7	-11.508000	90.000000	4.400000	AIR	
8	214.800000	5.000000	9.550000	LZ_BK8	
9	-46.990000	0.300000	9.630000	AIR	
10	-46.880000	3.000000	9.600000	LZ_BF24	
11	-113.240000	157.801570 S	9.720000	AIR	
IMS	0.000000	0.000000	9.000000		

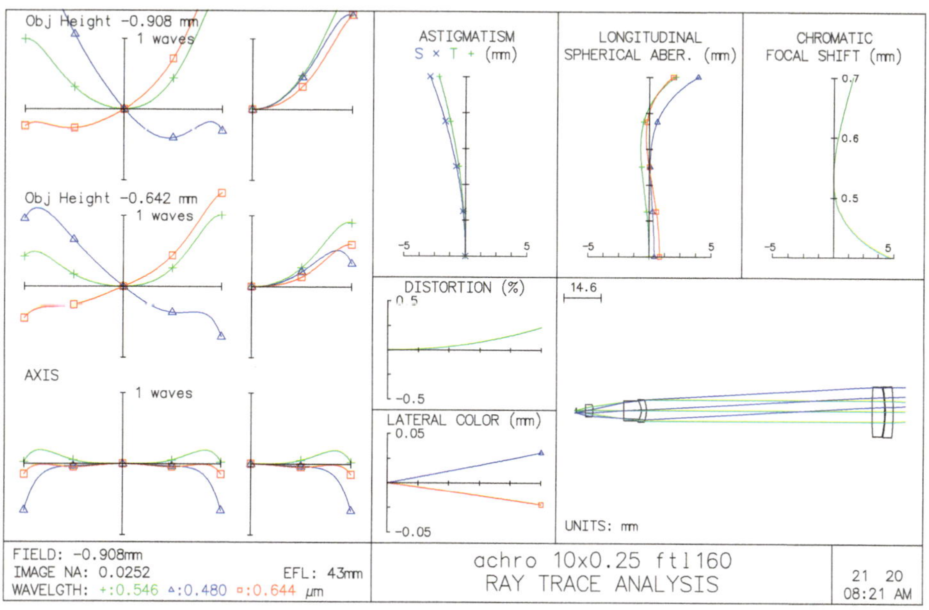

Figure 1.4 Optical design and graphs of aberration correction of a 10×/0.25 infinity achromatic objective ($F'_{ts} = 160$ mm; version 2).

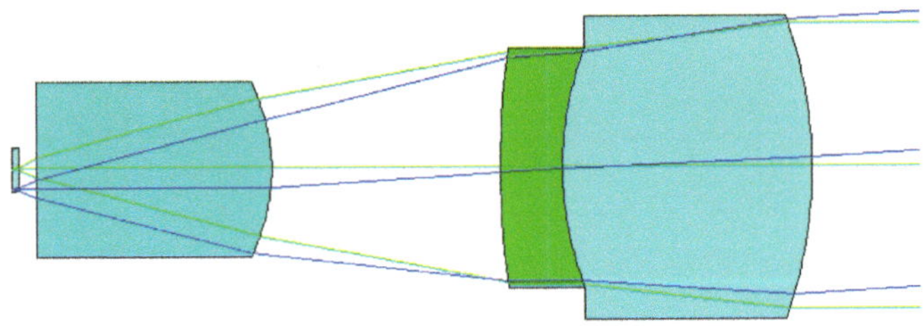

Lens: achro 20x0.40 ftl160

Object num aper 0.400000 Image height 9.000000 Primary wavln 0.546070

SRF	RADIUS	THICKNESS	APERTURE RADIUS		GLASS	SPECIAL
OBJ	0.000000	0.010000	0.470225		AIR	
AST	0.000000	0.170000	0.470000	A	LZ_K14	
2	0.000000	0.400000	0.520000		AIR	
3	0.000000	5.730000	0.690000		LZ_TK14	
4	-4.529000	5.500000	2.040000		AIR	
5	20.320000	1.500000	2.750000		LZ_TF5	
6	7.516000	6.000000	2.790000		LZ-K8	
7	-10.328000	90.000000	3.550000		AIR	
8	214.800000	5.000000	8.720000		LZ_BK8	
9	-46.990000	0.300000	8.810000		AIR	
10	-46.880000	3.000000	8.790000		LZ_BF24	
11	-113.240000	157.627954 S	8.900000		AIR	
IMS	0.000000	0.000000	9.000000			

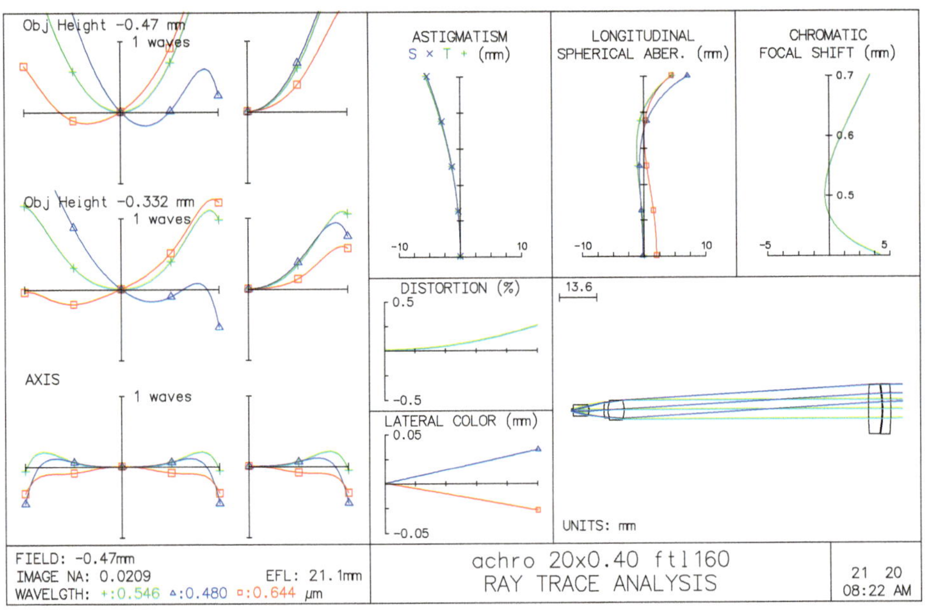

Figure 1.5 Optical design and graphs of aberration correction of a 20×/0.40 infinity achromatic objective ($F'_{ts} = 160$ mm; version 1).

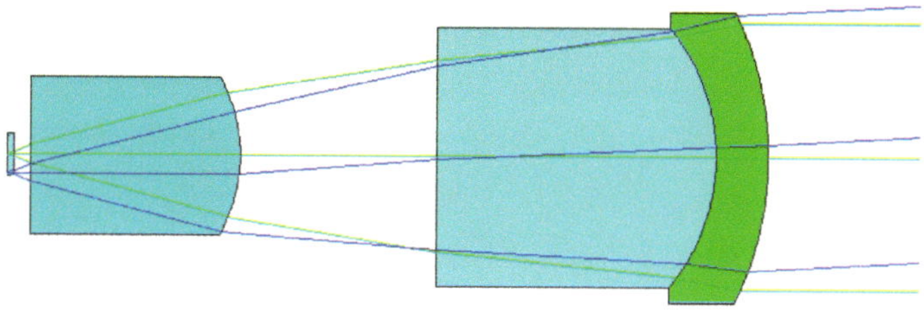

Lens: achro 20x0.40 ftl160

Object num aper 0.400000 Image height 9.000000 Primary wavln 0.546070

SRF	RADIUS	THICKNESS	APERTURE RADIUS	GLASS	SPECIAL
OBJ	0.000000	0.010000	0.457791	AIR	
AST	0.000000	0.170000	0.460000 A	LZ_K14	
2	0.000000	0.400000	0.510000	AIR	
3	0.000000	5.300000	0.680000	LZ_TK14	
4	-3.873000	4.900000	1.910000	AIR	
5	0.000000	7.000000	2.350000	LZ_K8	
6	-4.831000	1.300000	3.160000	LZ_TF5	
7	-7.980000	90.000000	3.550000	AIR	
8	214.800000	5.000000	8.680000	LZ_BK8	
9	-46.990000	0.300000	8.770000	AIR	
10	-46.880000	3.000000	8.750000	LZ_BF24	
11	-113.240000	157.770032 S	8.860000	AIR	
IMS	0.000000	0.000000	9.000000		

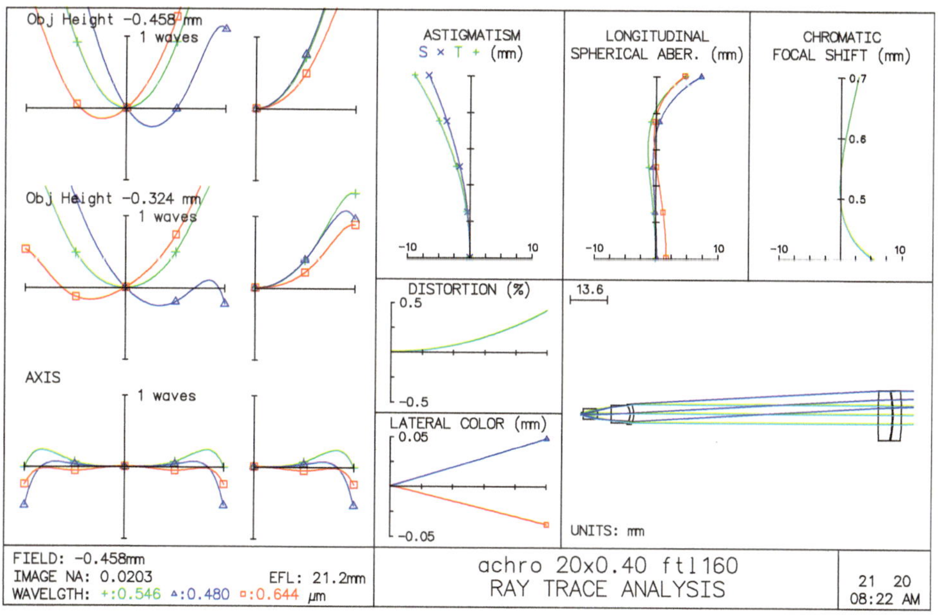

Figure 1.6 Optical design and graphs of aberration correction of a 20×/0.40 infinity achromatic objective ($F'_{ts} = 160$ mm; version 2).

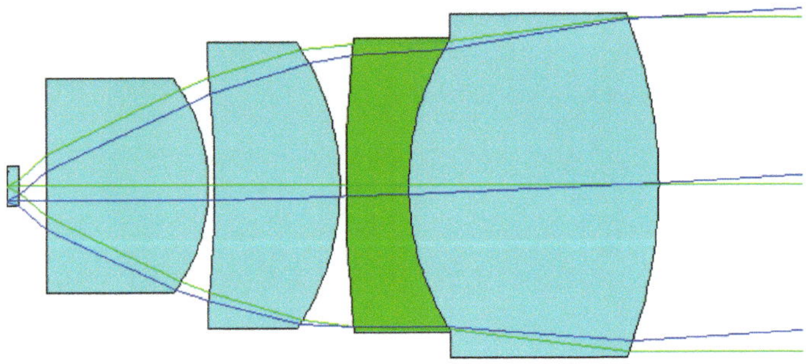

Lens: achro 40x0.65 d017 ftl160

Object num aper	0.650000	Image height	9.000000	Primary wavln	0.546070

SRF	RADIUS	THICKNESS	APERTURE RADIUS	GLASS	SPECIAL
OBJ	0.000000	0.010000	0.224889	AIR	
AST	0.000000	0.170000	0.230000 A	LZ_K14	
2	0.000000	0.450000	0.310000	AIR	
3	0.000000	2.590000	0.700000	LZ_K8	
4	-2.704000	0.100000	1.660000	AIR	
5	-15.631000	2.000000	1.850000	LZ_K8 P	
6	-3.908000	0.100000	2.220000	AIR	
7	20.650000	1.000000	2.280000	LZ_TF5	
8	4.285000	3.940000	2.260000	LZ-K8	
9	-7.244000	90.000000	2.660000	AIR	
10	214.800000	5.000000	7.920000	LZ_BK8	
11	-46.990000	0.300000	8.020000	AIR	
12	-46.880000	3.000000	8.010000	LZ_BF24	
13	-113.240000	157.994699 S	8.120000	AIR	
IMS	0.000000	0.000000	9.000000		

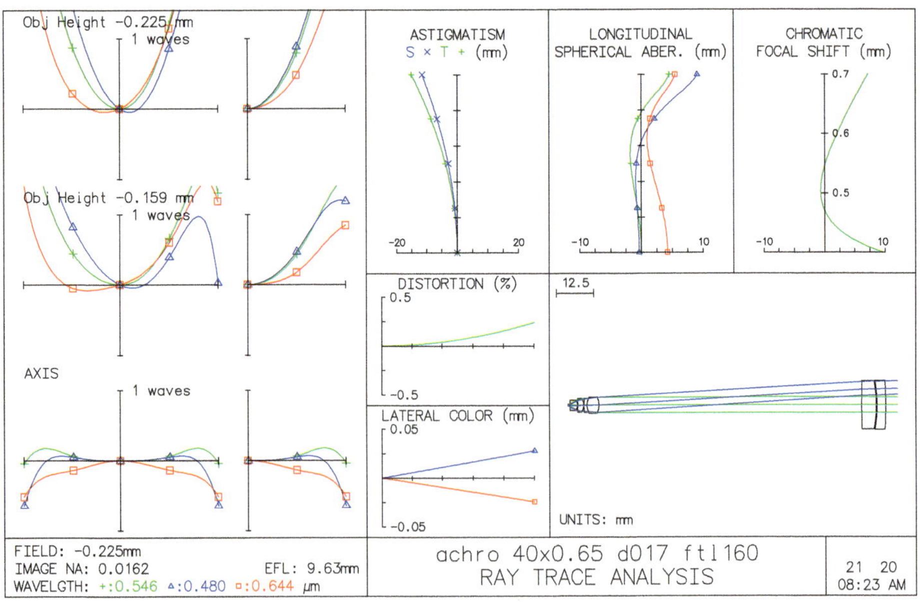

Figure 1.7 Optical design and graphs of aberration correction of a 40×/0.65 infinity achromatic objective ($F'_{ts} = 160$ mm; version 1).

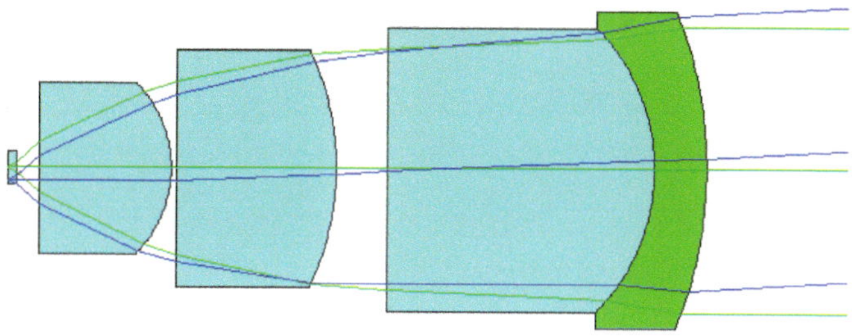

Lens: achro 40x0.65 d017 ftl160

Object num aper 0.650000 Image height 9.000000 Primary wavln 0.546070

SRF	RADIUS		THICKNESS		APERTURE RADIUS		GLASS		SPECIAL
OBJ	0.000000		0.010000		0.233229		AIR		
AST	0.000000		0.170000		0.240000	A	LZ_K14		
2	0.000000		0.448400		0.320000		AIR		
3	0.000000		2.560000		0.710000		LZ_K8		
4	-2.249000		0.100000		1.600000		AIR		
5	0.000000		3.100000		1.800000		LZ_K8	P	
6	-5.058000		1.000000		2.220000		AIR		
7	0.000000		5.200000		2.290000		LZ_K8	P	
8	-3.698000		1.000000		2.660000		LZ_TF5		
9	-7.870000		90.000000		2.970000		AIR		
10	214.800000		5.000000		8.110000		LZ_BK8		
11	-46.990000		0.300000		8.200000		AIR		
12	-46.880000		3.000000		8.180000		LZ_BF24		
13	-113.240000		157.803841	S	8.290000		AIR		
IMS	0.000000		0.000000		9.000000				

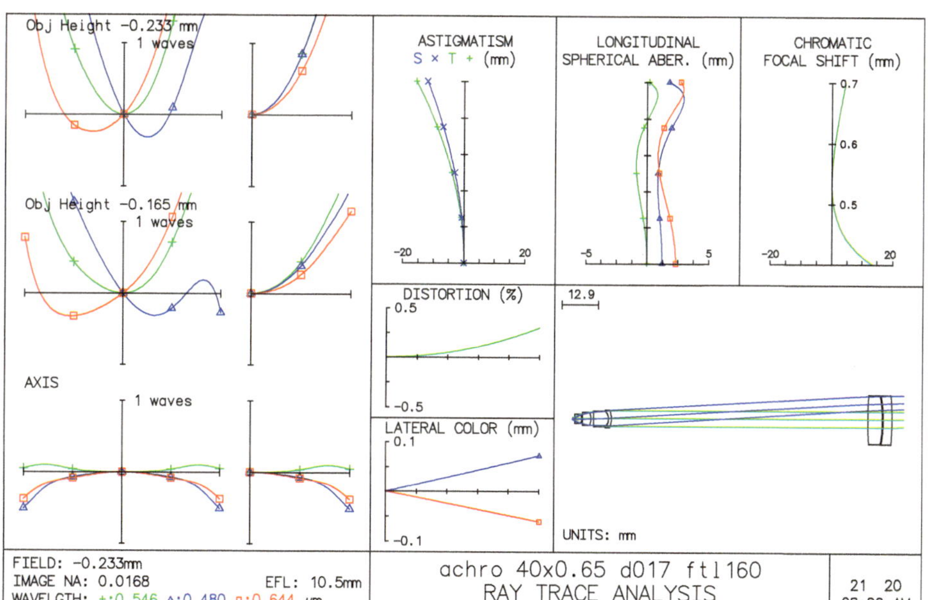

Figure 1.8 Optical design and graphs of aberration correction of a 40×/0.65 infinity achromatic objective ($F'_{ts} = 160$ mm; version 2).

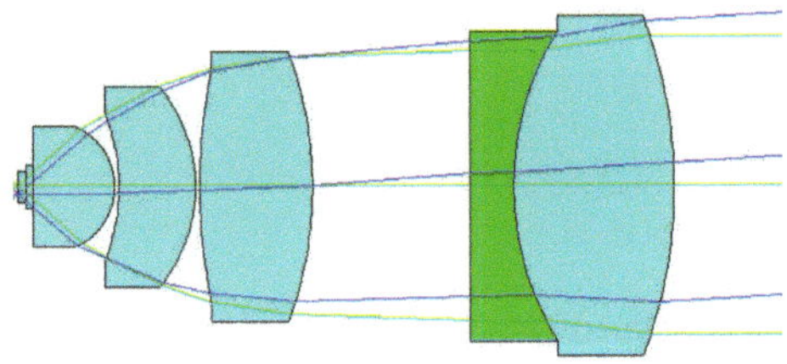

Lens: achro 50x1.00 oil ftl160

Object num aper 1.000000 Image height 9.000000 Primary wavln 0.546070

SRF	RADIUS		THICKNESS		APERTURE RADIUS		GLASS		SPECIAL
OBJ	0.000000		0.010000		0.179854		OIL		
AST	0.000000		0.170000		0.190000	A	LZ_K14		
2	0.000000		0.150000		0.340000		OIL		
3	0.000000		1.840000		0.470000		LZ_K8		
4	-1.393200		0.100000		1.300000		AIR		
5	-4.207000		1.700000		1.610000		LZ_K8	P	
6	-3.281000		0.100000		2.150000		AIR		
7	13.032000		2.500000		2.590000		LZ_K8	P	
8	-8.395000		3.540000		2.900000		AIR		
9	0.000000		1.000000		3.240000		LZ_TF5		
10	6.081000		3.600000		3.340000		LZ_K8	P	
11	-10.000000		90.000000		3.660000		AIR		
12	214.800000		5.000000		8.860000		LZ_BK8		
13	-46.990000		0.300000		8.950000		AIR		
14	-46.880000		3.000000		8.930000		LZ_BF24		
15	-113.240000		157.787062	S	9.040000		AIR		
IMS	0.000000		0.000000		9.000000				

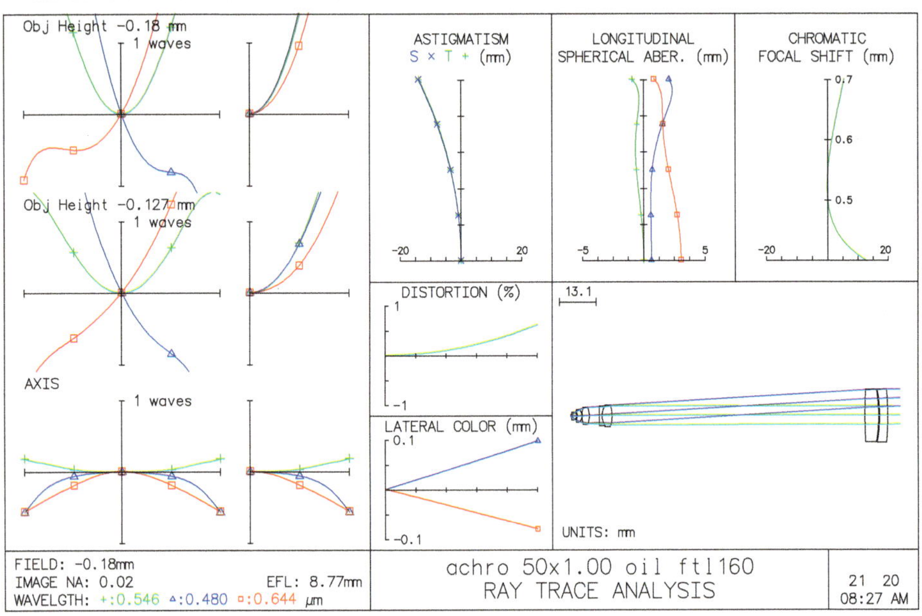

Figure 1.9 Optical design and graphs of aberration correction of a 50×/1.00 oil immersion infinity achromatic objective ($F'_{ts} = 160$ mm).

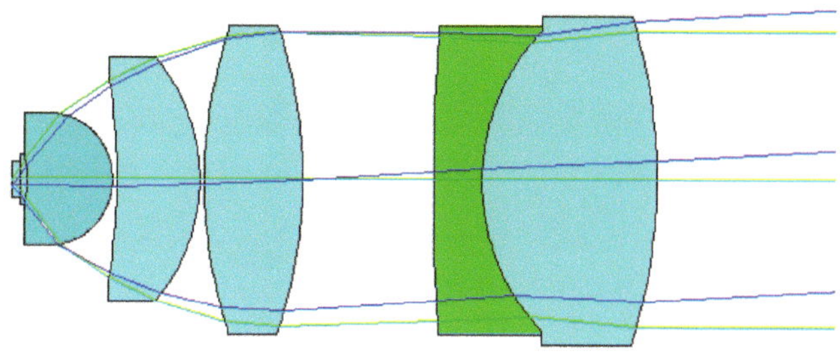

Lens: achro 63x1.15water d017 ftl160

Object num aper 1.150000 Image height 9.000000 Primary wavln 0.546070

SRF	RADIUS	THICKNESS	APERTURE RADIUS	GLASS	SPECIAL
OBJ	0.000000	0.010000	0.144200	WATER	
AST	0.000000	0.170000	0.160000 A	LZ_K14	
2	0.000000	0.150200	0.360000	WATER	
3	-1.940900	1.760000	0.500000	LZ_BK6	
4	-1.318300	0.100000	1.310000	AIR	
5	-11.668000	1.700000	1.970000	LZ_LK3	
6	-3.698000	0.100000	2.430000	AIR	
7	9.419000	2.000000	2.970000	LZ_LK3 P	
8	-9.419000	2.670000	3.080000	AIR	
9	42.660000	1.000000	3.070000	LZ_TF5	
10	4.285000	3.600000	3.000000	LZ_LK3 P	
11	-10.280000	90.000000	3.270000	AIR	
12	214.800000	5.000000	8.460000	LZ_BK8	
13	-46.990000	0.300000	8.550000	AIR	
14	-46.880000	3.000000	8.530000	LZ_BF24	
15	-113.240000	157.712483 S	8.640000	AIR	
IMS	0.000000	0.000000	9.000000		

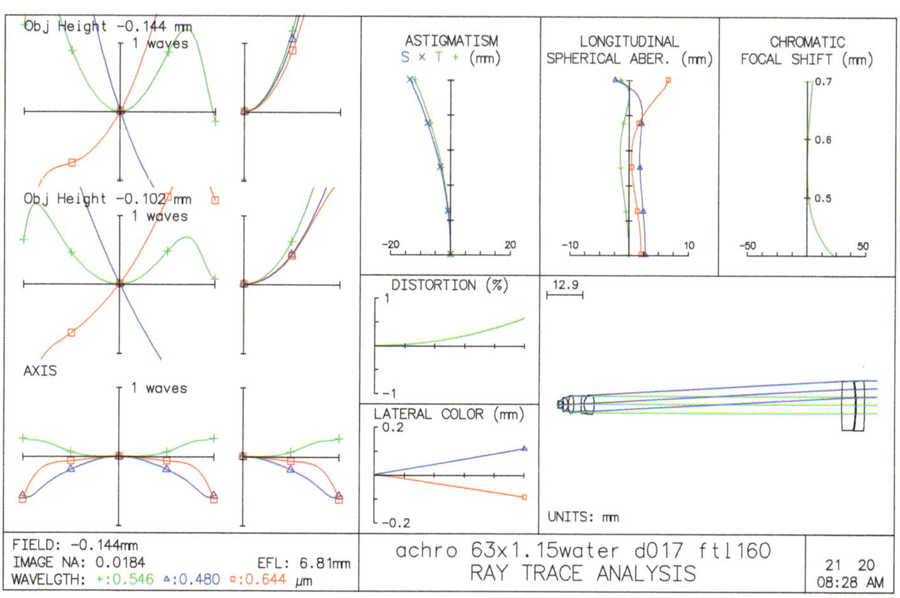

Figure 1.10 Optical design and graphs of aberration correction of a 63×/1.15 water immersion infinity achromatic objective ($F'_{ts} = 160$ mm).

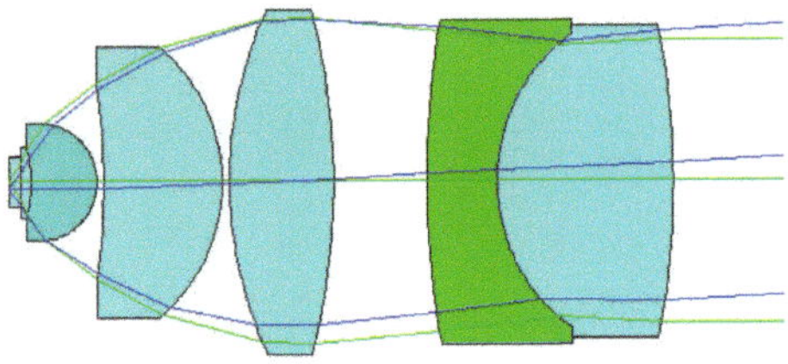

Lens: achro 100x1.20water d017 ftl160

Object num aper	1.200000	Image height	9.000000	Primary wavln	0.546070

SRF	RADIUS	THICKNESS		APERTURE RADIUS		GLASS		SPECIAL
OBJ	0.000000	0.010000		0.092234		WATER		
AST	0.000000	0.170000		0.120000	A	LZ_K14		
2	0.000000	0.150055		0.340000		WATER		
3	-1.770100	0.930000		0.500000		LZ_BK6		
4	-0.818200	0.100000		0.820000		AIR		
5	-9.773800	1.700000		1.380000		LZ_LK3		
6	-2.445700	0.100000		1.880000		AIR		
7	5.616000	1.500000		2.370000		LZ_LK3	P	
8	-8.999300	1.330000		2.400000		AIR		
9	13.130900	1.000000		2.250000		LZ_TF5		
10	2.499500	2.500000		2.040000		LZ_LK3	P	
11	-11.383200	90.000000		2.170000		AIR		
12	214.800000	5.000000		8.100000		LZ_BK8		
13	-46.990000	0.300000		8.220000		AIR		
14	-46.880000	3.000000		8.210000		LZ_BF24		
15	-113.240000	157.841504	S	8.330000		AIR		
IMS	0.000000	0.000000		9.000000				

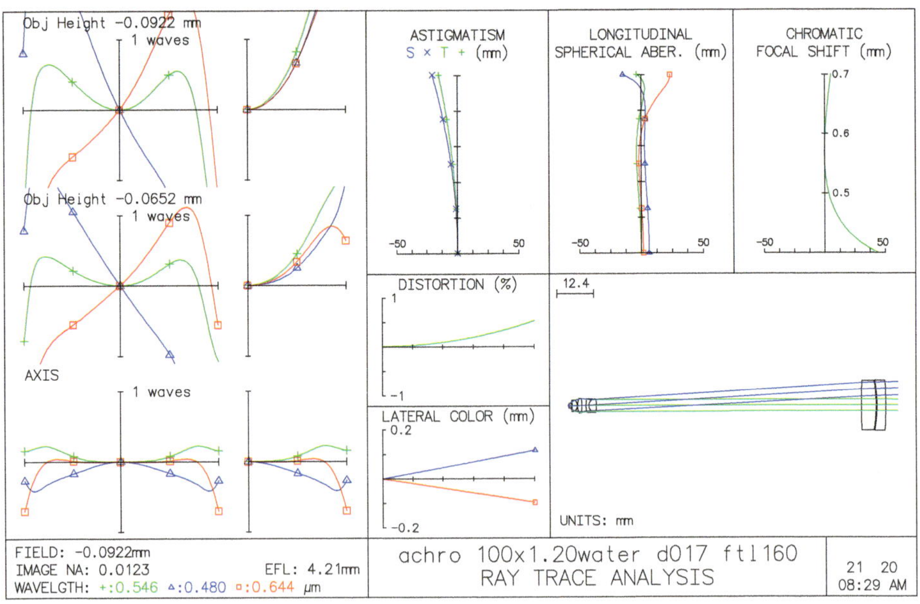

Figure 1.11 Optical design and graphs of aberration correction of a 100×/1.20 water immersion infinity achromatic objective ($F'_{ts} = 160$ mm).

Lens: 100x1.23water d017 tub160 h45 Efl 1.919

Object num aper 1.230000 Image height 9.000000 Primary wavln 0.546070

SRF	RADIUS	THICKNESS	APERTURE RADIUS	GLASS	SPECIAL
OBJ	0.000000	0.010000	0.090386	WATER	
AST	0.000000	0.170000	0.110000 A	K9	
2	0.000000	0.150200	0.350000	WATER P	
3	-2.360000	0.950000	0.550000	BAK2	
4	-0.828000	0.100000	0.830000	AIR	
5	-56.490000	1.700000	1.510000	H-QK3	
6	-2.710000	0.100000	1.950000	AIR	
7	10.516000	1.500000	2.290000	H-QK3 P	
8	-7.819000	1.300000	2.380000	AIR	
9	29.110000	1.000000	2.390000	H-ZF6	
10	3.250000	2.500000	2.330000	H-QK3 P	
11	-6.399000	185.668898 S	2.500000	AIR	
IMS	0.000000	0.000000	9.000000		

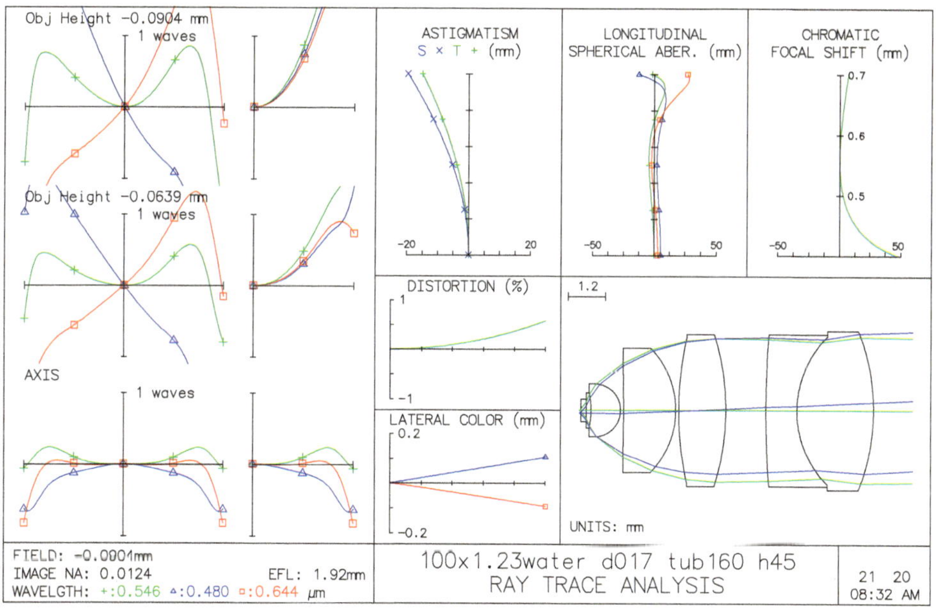

Figure 1.12 Optical design and graphs of aberration correction of a 100×/1.23 water immersion finite achromatic objective.

1.4 Studying the Properties of the Planar Elements

Among the various circuit solutions of microscope objectives with a flat image field, a large group consists of systems developed on the basis of a *planar* [8], the design of which was proposed by Paul Rudolph. It is one of the most common designs of photographic objectives.

It is advisable to begin studying the correction capabilities of these types of systems, as applied to microscope objectives, with the properties of a planar half. In contrast to the half (for example) of the triplet, here there are two

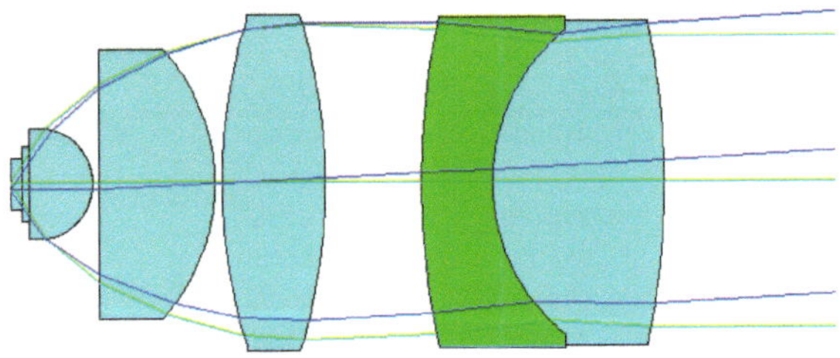

Lens: achro 100x1.25oil ftl160

Object num aper 1.250000 Image height 9.000000 Primary wavln 0.546070

SRF	RADIUS	THICKNESS		APERTURE RADIUS		GLASS		SPECIAL
OBJ	0.000000	0.010000		0.093742		OIL		
AST	0.000000	0.170000		0.110000	A	LZ_K14		
2	0.000000	0.115700		0.360000		OIL		
3	0.000000	0.930000		0.530000		K9		
4	-0.787000	0.100000		0.787000		AIR		
5	0.000000	1.700000		1.440000		LZ_LK3		
6	-2.729000	0.100000		1.910000		AIR		
7	7.727000	1.500000		2.300000		LZ_LK3	P	
8	-7.727000	1.410000		2.380000		AIR		
9	10.864000	1.040000		2.350000		LZ_TF5		
10	2.729000	2.500000		2.160000		LZ_LK3	P	
11	-10.864000	90.000000		2.300000		AIR		
12	214.800000	5.000000		7.990000		LZ_BK8		
13	-46.990000	0.300000		8.100000		AIR		
14	-46.880000	3.000000		8.090000		LZ_BF24		
15	-113.240000	157.838673	S	8.210000		AIR		
IMS	0.000000	0.000000		9.000000				

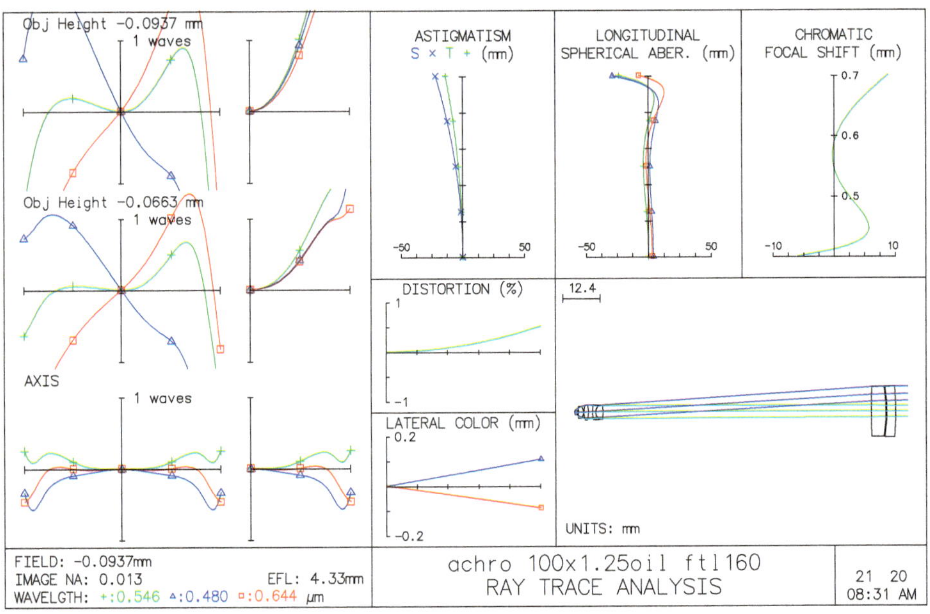

Figure 1.13 Optical design and graphs of aberration correction of a 100×/1.25 oil immersion infinity achromatic objective ($F'_{ts} = 160$ mm).

additional parameters: the thickness of the negative element and its deflection, determined by the angle α_2.

Consider the course of the first paraxial beam for an infinitely distant object through a system of two elements with optical forces that are opposite in sign. Denote the angle α_3, composed by a ray with an axis between the first and second element, by V (magnification in the second lens), and take $h_1 = 1$, $\alpha_5 = 1$ as the normalization condition. Denote the thickness of the first and second lenses by d_1, d_2, and the air $d_{air} = 0$. For simplicity, let us accept the condition of equality of the refractive indices for the main wavelength of the materials of the first and second lenses, $n_2 = n_4$. Thus, the expression for the first monochromatic sum of Seidel for the planar half can be represented as

$$\begin{aligned} S_1 = \frac{n}{(n-1)^2} \{ & h_1 V \left[\alpha_2^2(n+2) - \alpha_2 V(1+2n) + V^2 n\right] + \alpha_2^4 d_1 \\ & - \alpha_2 V d_1 \left[\alpha_2^2(n+2) - \alpha_2 V(1+2n) + V^2 n\right] \\ & + h_3(1-V) \cdot \left[\alpha_4^2(n+2) - \alpha_4(1+2n) \cdot (1+V) + n(1+V+V^2)\right] \\ & + \alpha_4^4 d_2 - \alpha_4 d_2 \left[\alpha_4^2(n+2) - \alpha_4(1+2n) + n\right] \} = 0. \end{aligned} \tag{1.17}$$

Consider solving the equation $S_1 = 0$ for $d_2 = 0$, $d_1 = 0$. As independent parameters, we choose the magnification V and the angle α_4, compiled by the aperture beam with the optical axis in the positive lens, setting the initial value α_4, which determines the minimum spherical aberration of the third order of the positive lens:

$$\alpha_4 = \alpha_{\min} = \frac{(2n+1)(V+1)}{2(n+2)}.$$

As a lens material, we choose glass type LZ_TKI7, which has a refractive index $n_e = 1.6305$, $v_e = 59.09$ for the main wavelength, close to the n_e of glass type LZ_OF4 (special flint) and glass type LZ_TF1 (heavy flint), $n_e = 1.6522$, $v_e = 33.62$, with which, if necessary, it can form an achromatic pair. Table 1.3 shows the results of solving the equation $S_I = 0$ for a series of three values of V and $\alpha_4 = \alpha_{\min}$, $\alpha_{\min} \pm 0.2$, and $\alpha_{\min} \pm 0.4$.

Analyzing the data in Table 1.3, which are the results of calculations, allows us to draw the following conclusions:

1. For all given values of the parameter V with $d_1 = 0$, there are real roots of the equation $S_1 = 0$.
2. When the positive lens is deflected by $\alpha_4 = \alpha_{min}$, the curvature of the surfaces of the negative lens is minimal.

Table 1.3 Studying the properties of the planar elements.

No.	V	α_4	α_2	r_1	r_2	r_3	r_4	S_{IV}
				$d_1 = 0$ (first root of the equation)				
		0.06944	−1.4947	−0.2587	−0.2818	2.0129	−0.7110	
		0.26944	−1.2023	−0.3216	−0.3582	0.9862	−1.1245	
1	−0.2	0.46944	−1.0853	−0.3563	−0.4017	0.6531	−2.6878	0.6133
		0.66944	−1.2022	−0.3216	−0.3582	0.4882	6.8891	
		0.86944	−1.4946	−0.2587	−0.2818	0.3898	1.5097	
		−0.0472	−1.3767	−0.2809	−0.3418	1.9589	−0.5848	
		0.15208	−1.1749	−0.3291	−0.4160	0.9730	−0.8384	
2	−0.4	0.35208	−1.0973	−0.3524	−0.4539	0.06473	−1.4803	0.6133
		0.55208	−1.1749	−0.3291	−0.416	0.4849	−6.3155	
		0.75208	−1.3766	−0.2809	−0.3418	0.3877	2.7865	
		0.0347	−1.2644	−0.3058	−0.4314	0.9603	−0.6683	
		0.2347	−1.2039	−0.3212	−0.4626	0.6416	−1.0214	
3	−0.6	0.4347	−1.2644	−0.3058	−0.4314	0.4817	−2.165	0.6133
		0.6347	−1.4255	−0.2713	−0.3657	0.3857	18.0774	
		0.8347	−1.6504	−0.2343	−0.3015	0.3215	1.7466	
		−0.08264	−1.3864	−0.2789	−0.4317	0.9478	−0.5556	
		0.11736	−1.3359	−0.2895	−0.4575	0.64	−0.7797	
4	−0.8	0.31736	−1.3864	−0.2789	−0.4317	0.04785	−1.3066	0.6133
		0.51736	−1.5234	−0.2538	−0.3744	0.3836	−4.0302	
		0.71736	−1.7188	−0.225	−0.3149	0.3201	3.7164	
		−0.4	−1.6421	−0.2355	−0.3759	1.8128	−0.3816	
		−0.2	−1.5214	−0.2542	−0.4258	0.9356	−0.4754	
5	−1.0	0.0	−1.4776	−0.2617	−0.4474	0.6305	−0.6305	0.6133
		0.2	−1.5214	−0.2542	−0.4258	0.4755	−0.9356	
		0.4	−1.6421	−0.2355	−0.3759	0.3816	−1.8128	
					$d_1 = 0.2$			
		−0.04792	−2.7138	−0.1425	−0.2417	3.0121	−0.9022	−0.5571
		0.15208	−1.3503	−0.2864	−0.4445	1.2358	−1.0648	0.1958
2	−0.4	0.35208	−1.2353	−0.3130	−0.4871	0.8072	−1.846	0.2471
		0.55208	−1.3503	−0.2864	−0.4445	0.6159	−8.021	0.1958
		0.75208	−2.7138	−0.1425	−0.2417	0.5981	4.2989	−0.5572
		0.0347	−1.4072	−0.2748	−0.4768	1.2305	−0.8564	0.1696
		0.2347	−1.3271	−0.2914	−0.5102	0.8119	−1.2924	0.2063
3	−0.6	0.4347	−1.4071	−0.2748	−0.4768	0.6173	−2.7743	0.1696
		0.6347	−1.6355	−0.2364	−0.4049	0.5118	23.9906	0.059
		0.8347	−2.0143	−0.192	−0.3295	0.4510	2.4503	0.1413
		0.08264	−1.5335	−0.2522	−0.4845	1.2384	−0.726	0.1094
		0.11736	−1.4689	−0.2633	−0.5115	0.8228	−1.0087	0.1406
4	−0.8	0.31736	−1.5335	−0.2522	−0.4845	0.6254	−1.7074	0.1094
		0.51736	−1.717	−0.2252	−0.4236	0.5153	−5.4141	0.0177
		0.71736	−2.0033	−0.1930	−0.3581	0.4484	5.2053	−0.1352
5	−1.0	−0.4	−1.8924	−0.2098	−0.4304	2.481	0.5223	−0.0482
		−0.2	−1.6848	−0.2295	−0.4825	1.2509	0.6357	0.0341
		0.0	−1.629	−0.2373	−0.5048	0.8359	−0.8359	0.0623
		0.2	−1.6848	−0.2295	−0.4825	0.6357	−1.2508	0.0341
		0.4	−1.8429	−0.2098	−0.4304	0.5223	−2.4810	−0.0481

(continued)

Table 1.3 (*Continued*) Studying the properties of the planar elements.

No.	V	α_4	α_2	r_1	r_2	r_3	r_4	S_{IV}
				$d_1 = 0$ (second root of the equation)				
1	−0.2	0.06944	1.26	0.3069	0.2797	2.0129	−0.711	
		0.26944	0.9675	0.3997	0.3547	0.9862	−1.1245	
		0.46944	0.8506	0.4546	0.3973	0.6531	−2.6878	0.6133
		0.66944	0.9675	0.3997	0.3547	0.4882	6.8891	
		0.86944	1.26	0.3069	0.2797	0.3898	1.5097	
2	−0.4	−0.04792	0.9097	0.4262	0.3355	1.9589	−0.5848	
		0.15208	0.7055	0.5481	0.4067	0.973	−0.8384	
		0.35208	0.6278	0.6159	0.4429	0.6473	−1.4803	0.6133
		0.55208	0.7054	0.5482	0.4067	0.4849	−6.3155	
		0.75208	0.9072	0.4263	0.3355	0.3877	2.7865	
3	−0.6	0.0347	0.5602	0.6903	0.4167	0.9603	−0.6683	
		0.2347	0.4997	0.7738	0.4456	0.6416	−1.0214	
		0.4347	0.5601	0.6903	0.4166	0.4817	−2.165	0.6133
		0.6347	0.7213	0.5361	0.355	0.3857	18.0774	
		0.8347	0.9462	0.4087	0.2943	0.3215	1.7466	
4	−0.8	−0.08264	0.4475	0.8641	0.4122	0.9478	−0.5556	
		0.11736	0.397	0.974	0.4356	0.64	−0.7797	
		0.31736	0.4475	0.8642	0.4122	0.4786	−1.3066	0.6133
		0.51736	0.5845	0.6616	0.3697	0.3836	−4.0302	
		0.71736	0.7798	0.4959	0.3044	0.3201	3.7164	
5	−1.0	−0.4	0.4684	0.8256	0.3575	1.8128	−0.3816	
		−0.2	0.3478	1.1119	0.4024	0.9356	−0.4754	
		0.0	0.3039	1.2722	0.4216	0.6305	−0.6305	0.6133
		0.2	0.3478	1.1119	0.4024	0.4755	−0.9356	
		0.4	0.4684	0.8256	0.3575	0.3816	−1.8128	
					$d_1 = 0.2$			
2	−0.4	−0.04792	2.0042	0.1929	0.103	0.1174	−0.3504	−0.3172
		0.15208	2.0909	0.1849	0.0963	0.5661	−0.4879	−0.4488
		0.35208	0.6435	0.601	0.3791	0.564	−1.2899	0.6088
		0.55208	2.0909	0.1849	0.0963	0.2821	−3.6744	−0.4488
		0.75208	2.0042	0.1929	0.103	0.2323	1.6696	−0.3172

3. With a change from α_4 to $\pm\Delta\alpha_4$ with respect to α_{min}, the deflection of the first lens changes by the same amount, that is,

$$\Delta r_1(+\Delta\alpha_4) = \Delta r_1(-\Delta\alpha_4),\ \Delta r_2(+\Delta\alpha_4) = \Delta r_2(-\Delta\alpha_4).$$

4. When the thickness $d_1 = 0$, S_{IV} for all roots is positive and equal to $S_{IV} = \frac{1}{n}\Sigma\varphi = 0.6133$.
5. Chromatic coefficient of a thin system $C_1 = -(\frac{\varphi_1}{\nu_1} + \frac{\varphi_2}{\nu_2})$, where in accordance with accepted normalization conditions $\varphi_1 = V$, $\varphi_2 = 1 - V$, and $\left(\varphi_2 = \frac{(\alpha_5 - \alpha_3)}{h_3}\right)$ equals null when $V \approx -0.55$ (glass type combination LZ_TF1–LZ_TK17; $\nu_{1e} = 33.62$, $\nu_{2e} = 59.09$) and equals null when

$V \approx -0.84$ (glass type combination LZ_OF4–LZ_TK17; $\nu_{1e} = 43.24$, $\nu_{2e} = 59.09$).

6. Of the two real roots of the quadratic equation $S_I = 0$ ($d_1 = 0$), one gives rise to the planar construction. The first lens has the shape of a meniscus facing concave surfaces to the object $\alpha_2 < -1$. The lens thickness plays the role of a correction parameter; i.e., using it leads to $h_2 \gg h_1$, and solving a fourth-degree equation $S_I = 0$ if $d_1 \neq 0$ creates systems in which $S_{IV} \approx 0$ (range of values $V = -0.4 \div -1.0$). The other root of the equation gives rise to the classical Petzval objective design, which can be considered a symmetrical-type system with a large air gap between the components. Here, the thickness of the first lens leads to the formation of systems that are not suitable for practical realization given the extremely steep surfaces of the elements.
7. Parameter V (magnification in the second lens) can be given positive values, which will lead to the construction of a telephoto objective that could be developed into a symmetrical-type system. The second root of the equation leads to the construction of an achromatic doublet with a transition to a symmetrical design like the Petzval objective, but with the reverse orientation of the components.

Let us explore the possibility of correcting chromatic aberrations in a planar half by introducing a *chromatic* radius alternately into the first and second lenses of a system. Slyusarev [6] and Ivanova and Kirillovsky [9] showed that the chromatic coefficients of the glued surface separating two media with similar values of refractive indices for the main wavelength are described with a sufficient degree of accuracy by the expressions

$$C_{gl.sur.} = (n-1)\left(\frac{1}{\nu_3} - \frac{1}{\nu_2}\right)\left(\alpha_2 - \frac{h_{gl.sur.}}{r_{gl.sur.}}\right),$$

$$C^*_{gl.sur.} = (n-1)\left(\frac{p_3}{\nu_3} - \frac{p_2}{\nu_2}\right)\left(\alpha_2 - \frac{h_{gl.sur.}}{r_{gl.sur.}}\right).$$

The condition for correcting the chromatic difference of positions in the planar half by introducing a chromatic radius into the negative meniscus and implementing positive lenses of the system from the same material (in this case glass type LZ_TKI7) will be expressed by the equation:

$$\Delta S'_{F'-C'} \approx h_1\left(-\frac{\alpha_2}{\nu_2}\right) + h_{gl.sur.}(n-1)\left(\frac{1}{\nu_3} - \frac{1}{\nu_2}\right)\left(\alpha_2 - \frac{h_{gl.sur.}}{r_{gl.sur.}}\right) + h_2\alpha_2,$$

where ν_2, ν_3 are the coefficients of the dispersion of materials of negative and positive elements, respectively. The value of the chromatic radius is determined by the expression

$$r_{gl.sur.} \approx \frac{h^2_{gl.sur.}(n-1)(\nu_2-\nu_3)}{h_{gl.sur.}(n-1)(\nu_2-\nu_3)\alpha_2+\alpha_2(h_2\nu_2-h_1\nu_3)-h_2\nu_2}. \tag{1.18}$$

By inserting the chromatic radius in the second component of the system (the meniscus and the first negative lens of the second component are made from the same glass type), the correction of the chromatism of the position is determined by the condition

$$\Delta S'_{F'-C'} \approx \frac{\alpha_2}{\nu_4}(h_2-h_1)-h_2\frac{\alpha_4}{\nu_4}+h_2\frac{\alpha_4-1}{\nu_5}+h_2(n-1)\left(\frac{1}{\nu_5}-\frac{1}{\nu_4}\right)\left(\alpha_4-\frac{h_2}{r_{gl.sur.}}\right)=0,$$

where ν_4 is the coefficient of dispersion of the meniscus material and the negative lens of the second component and ν_5 is the coefficient of dispersion of a positive lens; the value of the chromatic radius in this case will be expressed by the relation

$$r_{gl.sur.} \approx \frac{h_2^2(n-1)(\nu_4-\nu_5)}{h_2[(n-1)(\nu_4-\nu_5)\alpha_4+(\alpha_4-1)\nu_4]+[\alpha_2(h_2-h_1)-h_2\alpha_4]\nu_5}. \tag{1.19}$$

Let us find the value of the secondary spectrum in these planar halves. For the first option, we have

$$\Delta S'_{SS} \approx -h_1\alpha_1\frac{p_2}{\nu_2}+h_{gl.sur.}(n-1)\left(\frac{p_3}{\nu_3}-\frac{p_2}{\nu_2}\right)\left(\alpha_2-\frac{h_{gl.sur.}}{r_{gl.sur.}}\right)+h_2(\alpha_2-1)\frac{p_3}{\nu_3}.$$

For the second option, we have

$$\begin{aligned}\Delta S'_{SS} \approx{}& \frac{p_2}{\nu_2}[h_2(\alpha_2-V)-h_1\alpha_2]\\ &+h_2\left[(V-\alpha_4)\frac{p_4}{\nu_4}+(n-1)\left(\frac{p_5}{\nu_5}-\frac{p_4}{\nu_4}\right)\left(\alpha_4-\frac{h_2}{r_{gl.sur.}}\right)+(\alpha_4-1)\frac{p_5}{\nu_5}\right].\end{aligned}$$

By analyzing these two obtained methods for correcting the chromatism of the position in the half planar, we can conclude that it is preferable to introduce a chromatic radius into the negative meniscus. The curvature of the glued surface (and, consequently, the spherochromatic aberration) in this case is much less than in the second variant, and the value of the secondary spectrum remains almost unchanged.

1.5 Synthesis of Achromatic Objectives When the Image Curvature Is Reduced

The possibility of correcting the spherical aberration of the third order in the planar half, revealed by solving Eq. (1.17), is confirmed from the data obtained by studying the aberrations of the real system, which essentially is as follows. For each increase in V during deflection of a positive lens determined by given values of α_4, there was an angle α_2 at which the spherical aberration for the edge of the pupil was minimal (or equal to null). Studies have shown that the introduction of thickness $d_2 \neq 0$ and air gap $d_{\text{air}} \neq 0$ does not significantly affect the condition for correcting the spherical aberration of the planar half, a result of solving Eq. (1.17). Analyzing the results of these calculations allows us to draw the following conclusions.

1. At a selected value of V, the astigmatic difference $Z'_S - Z'_m$ depends on the angle α_4, which determines the deflection of a positive lens.
2. The value of η is negative for all solutions of systems.
3. A negative increment of $\Delta\alpha_4$ with respect to α_{min}, i.e., an increase in the radius of curvature of the third surface (the first surface of the positive lens), leads to a decrease in the astigmatic difference and a decrease in η.
4. With an increase of the magnification of the positive lens, S_{IV} decreases, moving to the region of negative values.
5. In the range of parameter values $V = -(0.45 \div 0.60)$ $S_{IV} \approx 0$, the astigmatic difference can be positive, negative, and equal to null depending on the deflection of the positive lens. A satisfactory correction of spherical aberration occurs for any value of α_4 corresponding to the condition for correcting astigmatism. In this case, the spherical aberration of higher orders has negative values.
6. The identified region of the parameter V is promising for building microscope objectives based on a half planar with a numerical aperture $\sin\sigma' \geq 0.20$.

For the subsequent calculations, the planar half from Table 1.4 (system 3b) was taken; its parameters were reduced to $F' = 30$ mm.

If we connect the two planar *halves* (Table 1.4, system 3b), we get a system with a linear magnification of $-1\times$; its design parameters are described by angles α. Then we can use the method described by Zverev and Ivanova [10], which essentially means that we can move from a symmetrical system having a linear magnification of $1\times$ to a *semi*-symmetrical design by adding a value $\Delta\alpha = \frac{1+V}{1-V}$ to all angles without a significant change in the quality of aberration correction. Thus, the initial objective system was obtained with a linear magnification of $-6.3\times$ and a numerical aperture of 0.20, the design parameters and aberration characteristics of which are shown in Table 1.5.

Table 1.4 Results of dimensional aberration calculation of planar halves ($F' = 30$ mm).

No.	α	d	r	$\Delta S'$	$\eta\%$	Z'_m	Z'_s	S_{IV}
	0			0	0			
	−1.38534	10	−13.9567	−0.1032	−1.1877			
1	0.2	1	−19.5551	−0.2086	−2.5899	−3.569	−2.054	0.00364
	0.46944	5	41.8327	−0.3214	−4.2817			
	1		−165.8617	−0.4927	−6.3948			
	0			0	0			
	−1.38534	10	−13.9567	−0.3684	−1.8835			
1a	−0.2	1	−19.5551	−0.8072	−4.1688	−6.811	−2.542	0.00351
	−0.66944	5	31.2703	−1.3553	−7.0493			
	1		418.1775	−2.0793	−10.8972			
	0			0	0			
	−1.38534	10	−13.9567	−0.3274	−0.8406			
1b	−0.2	1	−19.5551	−0.6533	−1.8347	−0.505	−0.35	0.00367
	0.26944	5	63.1705	−0.985	−3.034			
	1		−70.517	−1.3394	−4.521			
	0			0	0			
	−1.08534	10	−15.4678	0.0319	−1.3549			
2	−0.4	1	−24.056	0.0181	−3.0119	−3.529	−0.458	0.00485
	0.35208	5	40.715	−0.0957	−5.1135			
	1		−90.5064	−0.4129	−7.936			
	0			0	0			
	−1.4	10	−13.8106	0.0705	−1.7905			
2a	−0.4	1	−21.4333	0.0631	−3.975	−4.588	−1.711	0.00342
	0.55208	5	−31.2305	−0.0901	−6.7599			
	1		−389.3216	−0.5127	−10.5819			
	0			0	0			
	−1.3	10	−14.8729	−0.0915	−0.8712			
2b	−0.4	1	−23.0988	−0.1747	−1.9132	−1.554	−0.753	0.00437
	0.15208	5	61.6921	−0.2669	−3.1884			
	1		−45.7452	−0.4052	−4.8006			
	0			0	0			
	−1.47	10	−13.1529	0.535	−1.9871			
3	0.6	1	−22.7031	0.9245	−4.6362	−3.162	−1.17	0.00277
	0.2347	5	41.8981	0.8606	−8.5176			
	1		−65.4971	0.4375	−15.7204			
	0			0	0			
	−1.43534	10	−13.4705	0.0637	−1.2046			
3a	0.6	1	−23.3148	0.074	−2.6977	−1.433	−0.606	0.00301
	0.0347	5	62.3748	−0.4885	−4.6236			
	1		−43.2941	−0.4653	−7.2737			
	0			0	0			
	−1.68536	10	−11.4723	−0.0485	−1.3089			
3b	0.6	1	−19.6241	−0.0364	−2.9579	1.867	0.673	0.00037
	0.2	5	155.278	−0.0384	−5.1855			
	1		−32.5471	−0.3405	−8.6604			

Table 1.5 Design parameters and aberration characteristics of the initial objective system (6.3×/0.20).

α	d	r	n_e	Glass
−1.726	3	19.4333	1	Air
−0.52591	0.6	−24.4265	1.6305	LZ_TK17
−0.12582	6	10.6031	1	Air
0.95936	2.4	6.2398	1.6305	LZ_TK17
−0.726	6	−4.8005	1	Air
−2.41136	0.6	−9.4074	1.6305	LZ_TK17
−1.32619	3	−136.0768	1	Air
−0.92609		−15.0029	1.6305	LZ_TK17
0.274			1	Air

	Λ = 546.07 nm			y′ = 15 mm	
sinσ	$\Delta S'$	η%	$\Delta S'_{F'-C'}$	Z'_m	Z'_s
0	0	0	−23.466		
$0.2\sqrt{1/4}$	−0.791	0.0507	−24.050		
$0.2\sqrt{1/2}$	−1.5802	0.1264	−24.685	2.384	1.191
$0.2\sqrt{3/4}$	−2.4056	0.2091	−25.358		
0.2	−3.3146	0.3077	−26.058		

Note. $F' = 23.748$ mm, $S_F = -11.7389$ mm, $S'_{F'} = 5.3314$ mm, $S_1 = -15.5084$, $S' = 154.9446$ mm, and $V = -6.3\times$.

To correct the chromatic aberrations of the system, the material of the last lens was replaced with fluorite, the first lens was replaced with glass type LZ_K8 ($n_e = 1.5183$, $v_e = 63.83$; outer lens of the objective), and chromatic radii were inserted into the menisci. The first version of the 6.3×/0.20 objective was obtained by using special software to identify the influence of $\Delta\alpha_i$ parameters on the aberrations of the system and its subsequent optimization (Table 1.6).

Figure 1.14 shows the graphs of the transverse aberrations of the wide beam for $y' = 0$ (axial point) and $y' = 12.5$ mm (meridional and sagittal sections) in the plane of the best installation, spaced from the Gauss plane −0.58 mm. Wave aberrations for the line e do not exceed 0.05λ for the axial point of the object. In aberrations of wide oblique beams, with complete correction of coma, there is a positive spherical aberration characteristic of the planar construction and a small-value astigmatism. To reduce the aberrations of oblique beams, the entire set of calculations was repeated for the planar half. The new values allow for the creation of a system, the design parameters and aberration characteristics of which are shown in Table 1.7. It takes into account the standard cover glass $d = 0.17$ mm, the parameters of which are used in the calculation.

When correcting wave aberrations for the line e of the axial point of the object $W_{max} < 0.15\lambda$, the spherical aberration in oblique beams in the sagittal section is halved, the curvature of the image is almost completely corrected, and the scattering circle in the meridional section is somewhat reduced compared with the previous version, mainly due to reduction of the meridional

Table 1.6 Design parameters and aberration characteristics of the 6.3×/0.20 objective (first version).

r	d	n_e	Glass
17.9241	3	1	Air
−16.9477	0.6	1.5183	LZ_K8
9.2018	4	1	Air
−7.0607	2	1.6305	LZ_TK17
−4.5275	2.4	1.6542	LZ_OF4
40.2857	2	1	Air
−8.3664	4	1.6542	LZ_OF4
51.8746	0.6	1.6305	LZ_TK17
−15.3677	3	1	Air
		1.43497	Fluorite
		1	Air

	$\lambda = 546.07$ nm			$y' = 15$ mm		
sinσ	$\Delta S'$	η%	$\Delta S'_{F'-C'}$	Z'_m	Z'_s	LC
0	0	0	−0.6640			
$0.2\sqrt{1/4}$	−0.5599	−0.0138	−0.5708			
$0.2\sqrt{1/2}$	−0.8531	−0.0192	−0.4184	−0.9677	0.0894	0.169
$0.2\sqrt{3/4}$	−0.7884	−0.0168	−0.1811			
0.2	−0.2118	−0.0796	−0.1811			

Note. $S_1 = -15.5084$ mm, $S' = 151.1952$ mm, and $V = -6.3\times$.

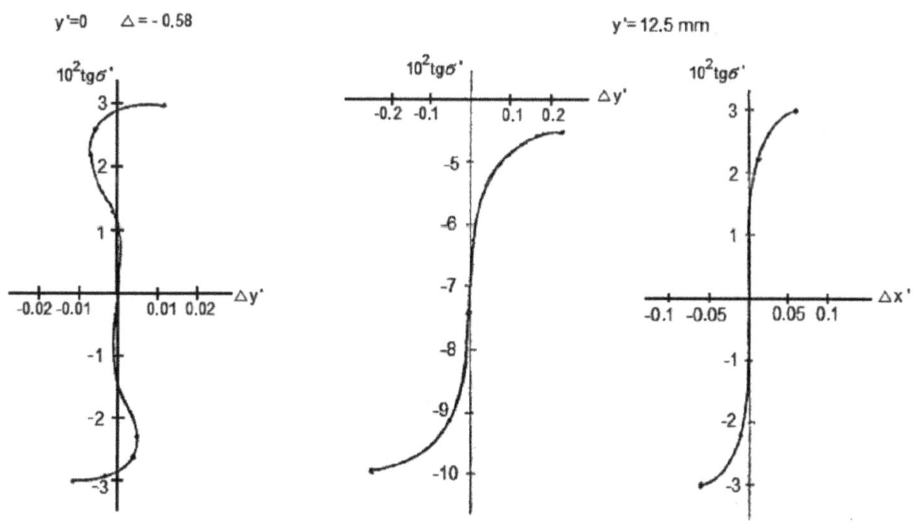

Figure 1.14 Planar, 6.3×/0.20 objective (first version).

curvature. The coma is almost absent. The spherochromatism of the system slightly exceeds the corresponding values in the previous version, since the chromatic difference of the positions in the original symmetrical system increases in proportion to the decrease in the parameter *V*. With proper

Table 1.7 Design parameters and aberration characteristics of the 6.3×/0.20 objective (final version).

r	d	n_e	Glass
∞	0.17	1.5168	LZ_K14
∞	15.864	1	Air
34.4022	3	1.5183	LZ_K8
−14.4272	0.6	1	Air
10.5372	4	1.6305	LZ_TK17
−30.1132	2	1.6542	LZ_OF4
7.1619	1.8	1	Air
−5.1069	2	1.6542	LZ_OF4
15.3714	4	1.6305	LZ_TK17
−9.1249	0.6	1	Air
80.7510	3	1.43497	Fluorite
−14.7074		1	Air

	$\lambda = 546.07$ nm				$y' = 15$ mm	
$\sin\sigma$	$\Delta S'$	η%	$\Delta S'_{F'-C'}$	Z'_m	Z'_s	LC
0	0.6	0	−1.0344			
$0.2\sqrt{1/4}$	0.0127	−0.026	−0.8642			
$0.2\sqrt{1/2}$	−0.3162	−0.041	−0.6325	−0.113	0.250	0.258
$0.2\sqrt{3/4}$	−0.3156	−0.043	−0.3252			
0.2	0.1161	−0.031	0.0984			

Note. $S_1 = -0.01$ mm, $S' = 157.76$ mm, and $V = -6.3\times$.

selection of optical materials in the system, the plan apochromatic type of aberration correction with corrected lateral color (LC; chromatism of magnification) and a high level of correction of aberrations of the inclined beam (including coma, astigmatism, and curvature of the image) can be achieved. The main disadvantage of the planar design is the spherical aberration of a wide oblique beam; however, this can be reduced by introducing some vignetting.

Thus, the planar design is promising for creating microscope objectives with improved aberration correction along the field. The advantages of the planar include a long working distance (LWD) of the objective, since the first half of the original symmetrical system is an inverted telephoto objective in the reverse direction of the rays. It should be noted that during the transition from a symmetrical structure to an asymmetrical one, by adding some value of the first paraxial beam to the initial angles, the correction load in both halves does not change equally. A comparative analysis of the aberrations of the two halves draws attention to underscore the fact that the spherical aberration of the first half is positive and contains almost no higher orders. The spherical aberration of the second half is negative, has significant positive higher orders, and its course causes the residual spherical aberration of the entire lens. The curvature of the image in the first half is negative, but smaller than in a thin system working with the same magnification; in the second half, it has a positive value. The study of aberrations in the planar system and its individual halves is of fundamental importance for creating microscope objectives with reduced image curvature. If we consider the first half of the transformed planar as part of the frontal system, then, due to the increased frontal segment (half of the planar is an inverted telephoto lens), the radius of curvature of the frontal convex lens can be increased. Accordingly, the S_{IV} of the objective is reduced.

Studies have shown that subsequent objective systems can be made as a single lens, and the correction capabilities of the transformed planar halves make it possible to achieve constancy of LC (chromatic of magnifications) in the whole set of achromats with the value $\frac{\Delta y'_{F'-C'}}{y'_e} \approx 0.7 \div 0.8\%$. The optical design of a set of objectives demonstrates that systems of achromatic objectives can be constructed based on a planar. In the optical structures of these objectives, there is only one lens more than in the classical Abbe achromats. But these objectives differ from the reduced image curvature and constancy of LC (or chromatic of magnification) found in the whole set of classical objectives.

1.5.1 An example of creating a set of achromatic objectives when the image curvature and LC are reduced

Figures 1.15–1.20 present the design parameters and graphs of aberration correction of these achromatic objectives. These objectives are designed to complete the finite length of the tube of mass segment microscopes. We classified these objectives as LC reduction (LCR) semi–plan achromats.

Unfortunately, it must be stated that this OX-100 objective failed. It can be argued that the design based on half of the planar stopped working effectively for such a large value of the linear magnification of the objective. Indeed, if we study the distribution of the power load among the components, we can see that the third and fourth components (along the rays from the object from left to right) have very small radii of curvature values, including for the glued surfaces. The high degree of influence of technological aberrations on the resulting image quality was confirmed in the manufacture of prototypes of this lens. In addition, in this design of the objective, it was not possible to achieve an LC value (chromatism of magnification) equal to 0.7–0.8% (as is achieved in other objectives in the kit). Therefore, despite the high image quality produced by the prototypes of this objective, organizing its serial production was determined to be inexpedient.

Therefore, the development of another objective with a linear magnification of 100×, a numerical aperture of 1.25 oil immersion, and a correction level of 0.8% for LC was performed. This new objective used a different concept for building an optical system, which made it possible to achieve the required quality of LC correction and minimize technological aberrations in production. This objective was recommended for production and use in mass segment microscopes in conjunction with other objectives in the set. Figure 1.21 shows the design parameters and graphs of aberration correction of this objective.

Lens: OX10-1 10x0.30 d017 tub160 h45 Efl 16.27

Object num aper 0.300000 Image height 9.000000 Primary wavln 0.546070

SRF	RADIUS	THICKNESS	APERTURE RADIUS	GLASS	SPECIAL
OBJ	0.000000	0.010000	0.918531	AIR	
AST	0.000000	0.170000	0.920000 A	LZ_K14	
2	0.000000	7.525000	0.950000	AIR	
3	143.880000	1.500000	3.330000	LZ_TF4	
4	9.376000	4.000000	3.700000	LZ_BF25	
5	-9.120000	14.000000	4.190000	AIR	
6	11.376000	4.000000	5.000000	LZ_TK21	
7	26.550000	1.500000	4.560000	LZ_TF4 P	
8	9.506000	2.500000	4.340000	AIR	
9	23.120000	5.000000	4.760000	LZ_LK3	
10	-31.190000	151.620000	5.050000	AIR	
IMS	0.000000	0.000000	9.000000		

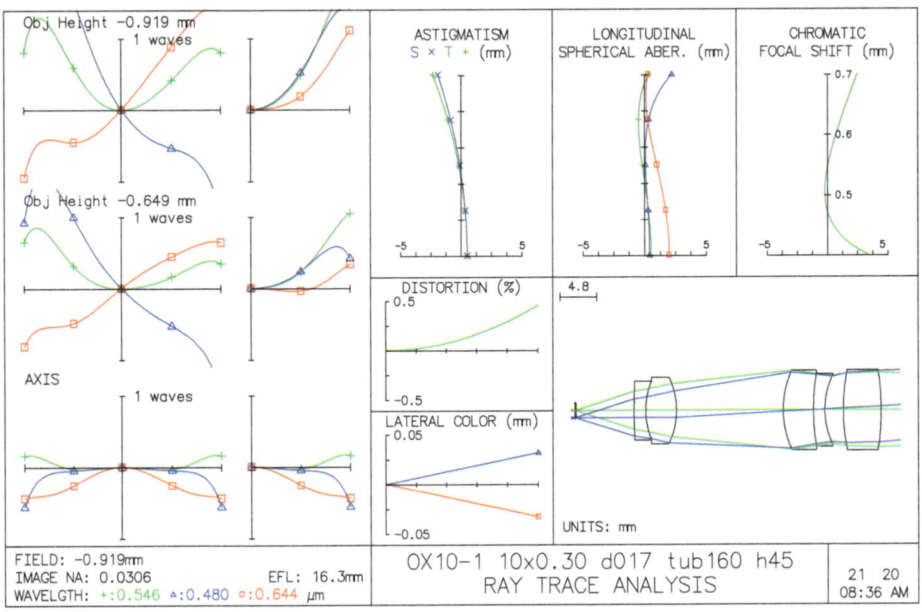

Figure 1.15 Optical design and graphs of aberration correction of a 10×/0.30 finite LCR semi–plan achromatic objective.

Figure 1.22 shows a photograph of a set of these objectives; design parameters and graphs of aberration correction of these objectives are shown in Figs. 1.15–1.17, 1.19, and 1.21. Table 1.8 shows the main technical parameters of these budget objectives (we called them *second-generation* objectives).

Figures 1.23–1.27 show photographs of real biological objects made using these objectives (the preparations were made by Anatoly Mikhaltsov, a respected microscopy expert from the Russian city of Omsk). These objectives were manufactured for more than 20 years at the Russian enterprise LOMO.

These photographs allow us to compare the quality of the images obtained using these objectives with the level of their aberration correction; this level of

Lens: OX25-1 25x0.50 d017 tub160 h45 Efl 7.07

Object num aper 0.500000 Image height 9.000000 Primary wavln 0.546070

SRF	RADIUS	THICKNESS	APERTURE RADIUS		GLASS		SPECIAL
OBJ	0.000000	0.010000	0.360541		AIR		
AST	0.000000	0.170000	0.370000	A	LZ_K14		
2	0.000000	1.495500	0.430000		AIR		
3	0.000000	5.950000	1.290000		LZ_TK21		
4	-6.668000	0.100000	2.960000		AIR		
5	22.490000	1.300000	3.170000		LZ_TF4		
6	13.996000	2.000000	3.250000		LZ_LK3		
7	-14.689000	0.100000	3.390000		AIR		
8	11.614000	2.000000	3.430000		LZ_TK14		
9	-48.420000	1.300000	3.310000		LZ_TF4	P	
10	8.551000	1.800000	3.150000		AIR		
11	35.970000	2.500000	3.370000		LZ_TK21	P	
12	-13.868000	173.270000	3.590000		AIR		
IMS	0.000000	0.000000	9.000000				

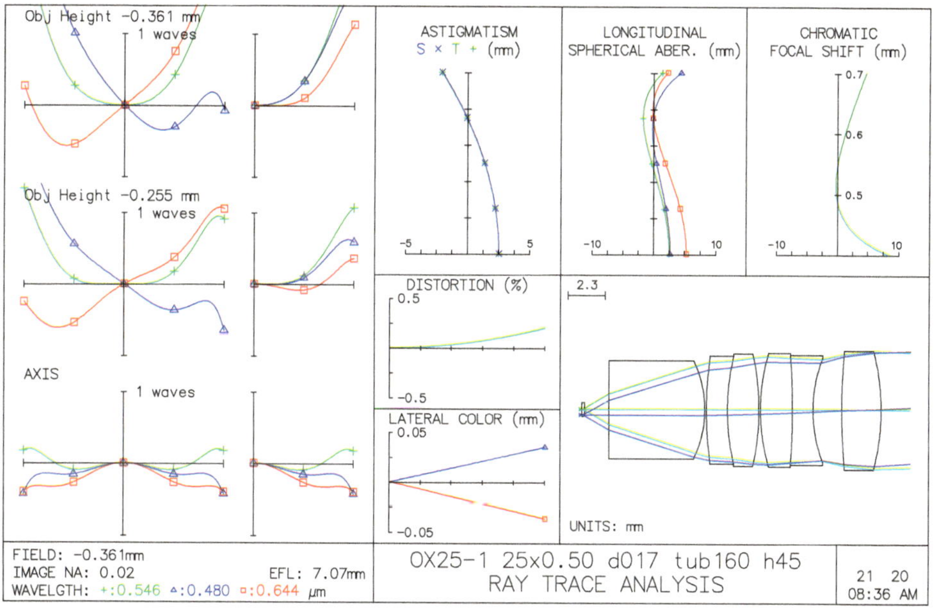

Figure 1.16 Optical design and graphs of aberration correction of a 25×/0.50 finite LCR semi–plan achromatic objective.

aberration correction is quite sufficient to complete modern models of low-cost microscopes with a mass segment of these objectives.

1.6 Synthesis of Plan Achromatic Objectives

Creating objectives with an aberration correction such as a plan achromat, whose designs would be quite simple for use in mass segment microscopes, is a

Lens: OX40-2 40x0.65 d017 tub160 h45 — Efl 4.6

Object num aper 0.650000 Image height 9.000000 Primary wavln 0.546070

SRF	RADIUS	THICKNESS	APERTURE RADIUS	GLASS	SPECIAL
OBJ	0.000000	0.010000	0.226693	AIR	
AST	0.000000	0.170000	0.230000 A	LZ_K14	
2	0.000000	0.456000	0.310000	AIR	
3	0.000000	3.700000	0.710000	LZ_TK12	
4	-3.251000	0.100000	2.060000	AIR	
5	18.239000	1.200000	2.340000	LZ_TF4	
6	13.305000	1.700000	2.460000	LZ_LK3	
7	-7.870000	0.100000	2.620000	AIR	
8	8.017000	1.800000	2.610000	LZ_TK14	
9	-20.140000	1.200000	2.560000	LZ_TF4 P	
10	5.888000	1.800000	2.470000	AIR	
11	18.967000	1.700000	2.980000	LZ_TK21	
12	-11.402000	178.060000	3.120000	AIR	
IMS	0.000000	0.000000	9.000000		

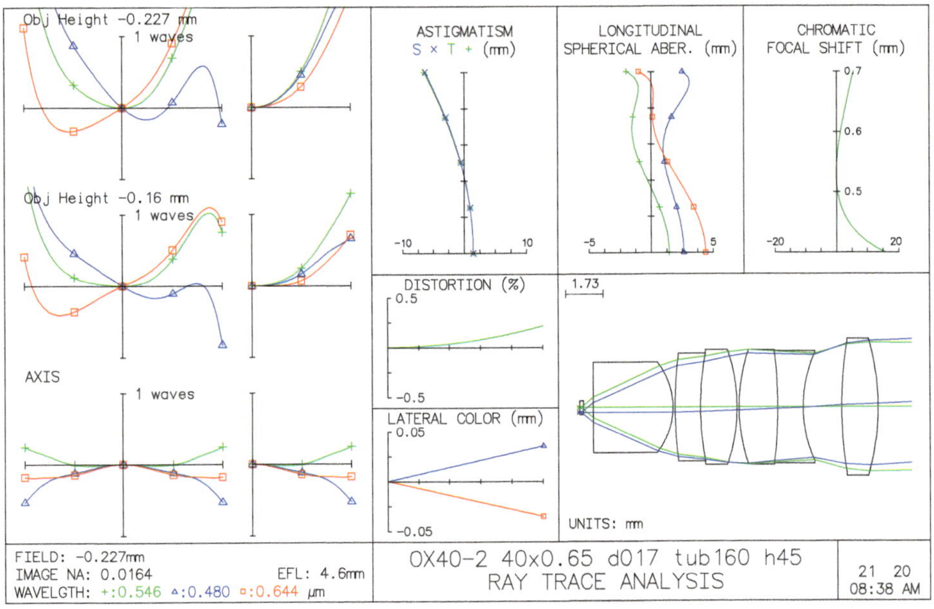

Figure 1.17 Optical design and graphs of aberration correction of a 40×/0.65 finite LCR semi-plan achromatic objective.

rather complicated task. Indeed, plan achromats, in which the curvature in the image of the object is corrected, are needed mainly for microprojection of the image onto a digital receiver. Unlike the observer's eye (visual microscopy), this receiver is not capable of certain *optical effects*, such as accommodation, convergence, and others that (when observed through a microscope with eyes) in some way smooth the images, adjusting them under the features of the visual apparatus of a particular researcher. Moreover, the mental activity of a person working behind a microscope makes some

Lens: OX40B-2 40x0.75water d0 tub160 Efl 4.58

Object num aper 0.750000 Image height 9.000000 Primary wavln 0.546070

SRF	RADIUS	THICKNESS	APERTURE RADIUS	GLASS	SPECIAL
OBJ	0.000000	1.0000e-20	0.222098	AIR	
AST	0.000000	0.299500	0.230000 A	WATER	
2	0.000000	3.000000	0.430000	LZ_K8	
3	-2.270000	0.200000	1.700000	AIR	
4	48.190000	1.200000	2.050000	LZ_TF4	
5	18.880000	2.500000	2.280000	LZ_LK3	
6	-5.675000	0.200000	2.660000	AIR	
7	7.870000	2.000000	2.850000	LZ_TK14	
8	-18.967000	2.300000	2.820000	LZ_TF4 P	
9	6.792000	2.400000	2.770000	AIR	
10	30.270000	2.200000	3.460000	LZ_TK21	
11	-11.508000	175.700000	3.670000	AIR	
IMS	0.000000	0.000000	9.000000		

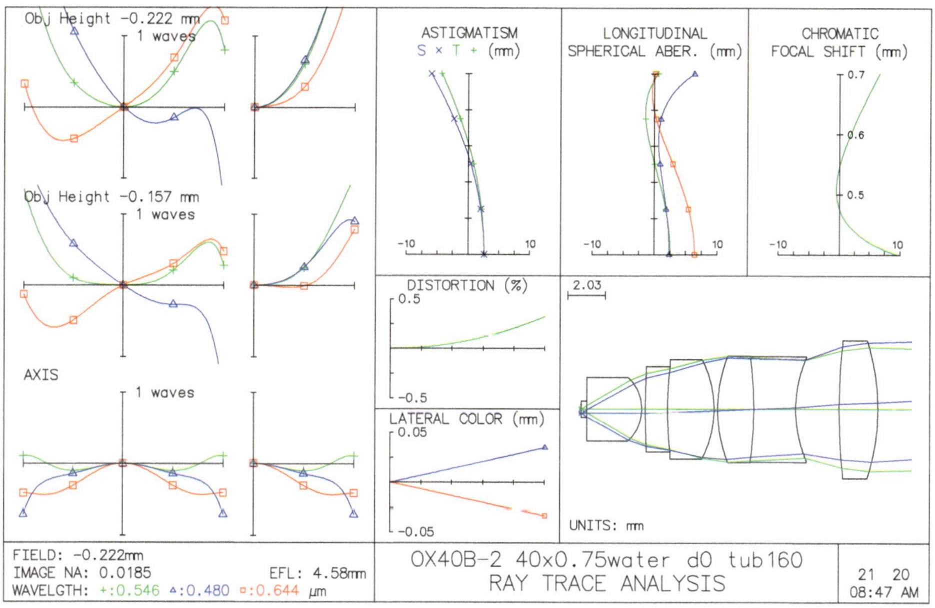

Figure 1.18 Optical design and graphs of aberration correction of a 40×/0.75 water immersion finite LCR semi–plan achromatic objective.

adjustments in the study of images, allowing the researcher to see exactly what he expects. Digital image receivers are devoid of these and other features in image processing; therefore, the optics of the microscope, of the objective, in the first place, must be of a higher quality than that intended for visual observation. In addition, the spectral characteristics of digital image receivers differ from those of the human eye. To construct optical systems of objectives with a flat image field, i.e., plan achromats, certain components (lenses) should be used in their optical schemes that contribute to this type of

Lens: ox50 50x1.0oil tub160 h45 Efl 3.77

Object num aper 1.000000 Image height 9.000000 Primary wavln 0.546070

SRF	RADIUS	THICKNESS	APERTURE RADIUS	GLASS	SPECIAL
OBJ	0.000000	0.010000	0.182032	OIL	
AST	0.000000	0.170000	0.190000 A	LZ_K14	
2	0.000000	0.151000	0.340000	OIL	
3	0.000000	2.700000	0.470000	LZ_K8	
4	-1.940900	0.050000	1.790000	AIR	
5	0.000000	1.500000	2.390000	LZ_TK21	
6	-5.943000	0.100000	2.630000	AIR	
7	67.760000	1.000000	2.770000	LZ_TF4	
8	9.462000	1.700000	2.930000	LZ_LK3	
9	-6.166000	0.100000	3.020000	AIR	
10	10.520000	1.800000	3.070000	LZ_LK3 P	
11	-6.516000	1.200000	3.030000	LZ_TF1	
12	7.980000	4.400000	2.990000	AIR	
13	618.000000	1.700000	3.970000	LZ_TK21 P	
14	-12.823000	175.436694 S	4.120000	AIR	
IMS	0.000000	0.000000	9.000000		

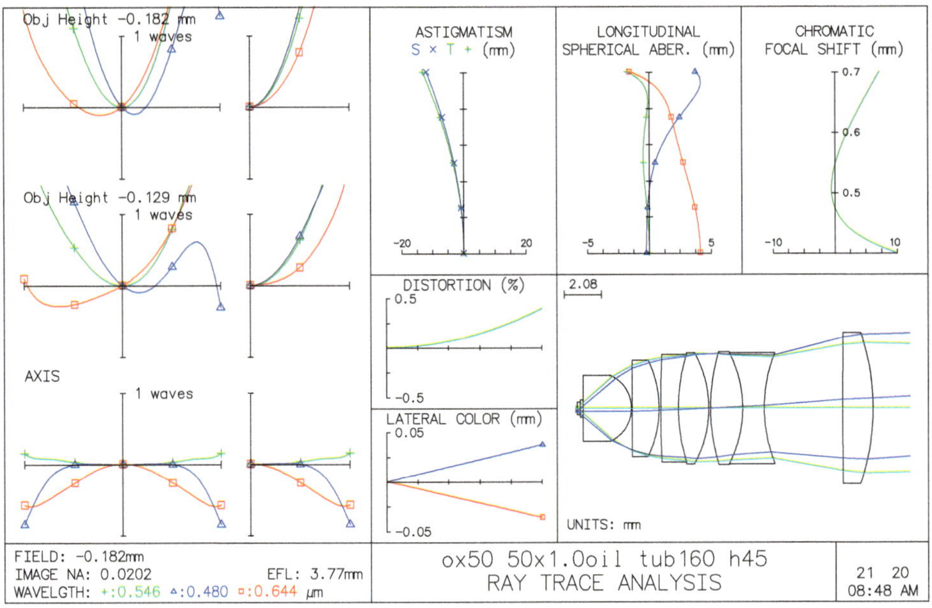

Figure 1.19 Optical design and graphs of aberration correction of a 50×/1.00 oil immersion finite LCR semi–plan achromatic objective.

aberration correction. The principal lens elements that are used to obtain objectives with a flat image field are menisci and biconcave lenses. Consider the properties of the meniscus, which is usually used in the *plan* objective as a frontal component (from the object from left to right). Let us consider these components.

Lens: ox100 100x1.25oil tub160 h45 — Efl 1.95

Object num aper 1.250000 Image height 9.000000 Primary wavln 0.546070

SRF	RADIUS	THICKNESS	APERTURE RADIUS	GLASS	SPECIAL
OBJ	0.000000	0.010000	0.090545	OIL	
AST	0.000000	0.170000	0.110000 A	LZ_K14	
2	0.000000	0.120000	0.360000	OIL	
3	0.000000	0.930000	0.530000	LZ_K8	
4	-0.772700	0.050000	0.770000	AIR	
5	-10.093000	1.000000	1.290000	LZ-TK12	
6	-2.070000	0.100000	1.520000	AIR	
7	8.166000	1.000000	1.800000	LZ_TF5	
8	3.499000	1.700000	1.870000	LZ_TK14	
9	-11.117000	0.050000	1.990000	AIR	
10	7.112000	1.500000	2.040000	LZ_STK3	
11	-6.730000	1.200000	1.980000	LZ_TF5 P	
12	4.246000	2.450000	1.890000	AIR	
13	-348.300000	1.700000	2.540000	LZ_BF16	
14	-6.792000	180.020000	2.760000	AIR	
IMS	0.000000	0.000000	9.000000		

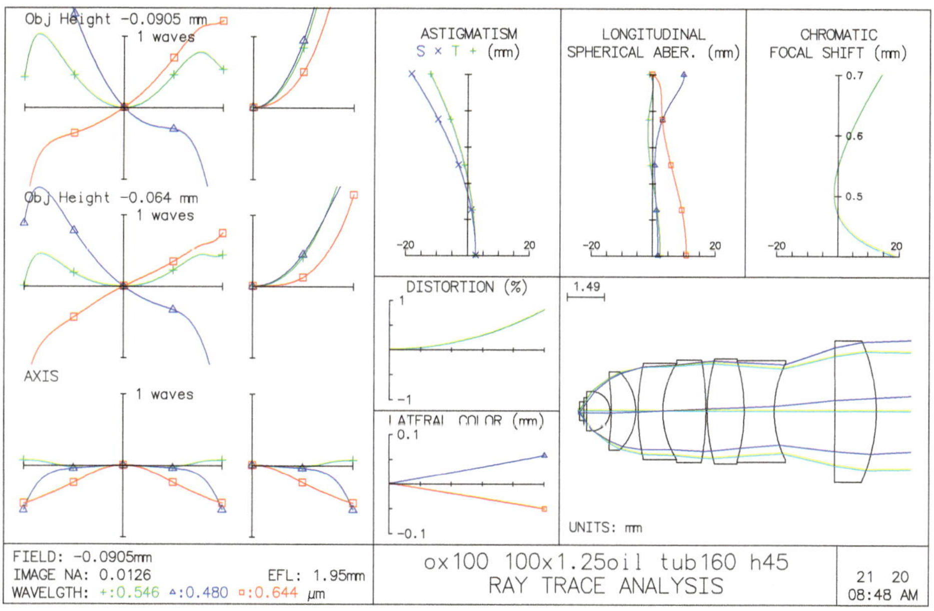

Figure 1.20 Optical design and graphs of aberration correction of a 100×/1.25 oil immersion finite LCR semi–plan achromatic objective.

1.6.1 Frontal part in the form of a thick meniscus

Thick menisci are often used as frontal components in objectives with a flat field [9], making it possible to correct the Petzval coefficient. With their use, the image of the axial point of the object after refraction on the first surface coincides with the aplanatic point of the third type of the second surface. The

Lens: ox100-1 100x1.25 oil tub 160mm — Efl 1.78

Object num aper 1.250000 — Image height 9.000000 — Primary wavln 0.546070

SRF	RADIUS	THICKNESS	APERTURE RADIUS	GLASS	SPECIAL
OBJ	0.000000	0.010000	0.089940	OIL	
AST	0.000000	0.170000	0.100000 A	LZ_K14	
2	0.000000	0.100000	0.350000	OIL P	
3	0.000000	1.000000	0.500000	LZ_K8	
4	-0.831800	0.050000	0.830000	AIR	
5	0.000000	1.700000	1.470000	LZ_LK3	
6	-2.938000	0.050000	1.960000	AIR	
7	59.430000	1.000000	2.260000	LZ_BF16	
8	4.875000	2.600000	2.580000	LZ_LK3	
9	-3.945000	0.050000	2.500000	AIR	
10	60.260000	1.000000	2.840000	LZ_TF3	
11	4.207000	2.900000	2.830000	LZ_LK3 P	
12	-8.954000	13.820000	3.020000	AIR	
13	5.861000	3.340000	3.750000	LZ_TF5	
14	4.446000	164.663743 S	2.830000	AIR	
IMS	0.000000	-0.457430	8.986926 S		

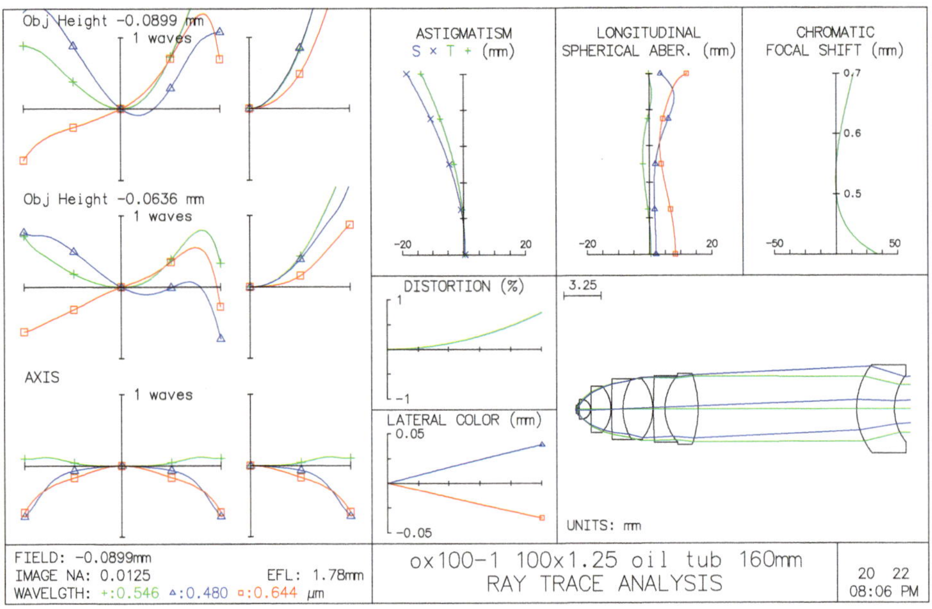

Figure 1.21 Optical design and graphs of aberration correction of a 100×/1.25 oil immersion finite LCR semi–plan achromatic objective (successful version).

second surface of the meniscus does not introduce spherical aberration, coma, and astigmatism, while the first surface, if the object is not located at its center of curvature, introduces these aberrations. Coma and spherical aberration practically do not reach large values, since the working distance of objectives, especially high-aperture objectives, is small. Astigmatism introduced by the first surface can reach significant values; because of the condition for

Figure 1.22 Photograph of the set of objectives in which the image curvature and LC are reduced.

Table 1.8 Main technical parameters and basic optical layout of second-generation objectives.

Magnification	*NA*	*WD* (mm)	*F′* (mm)	*R* (μm)	*DF* (μm)	*FOV* on object (mm)	*FOV* on image (mm)	The principal optical layout	Optical design at figure no.
10×	0.30	7.2	16.27	1.12	3.73	1.8	18		1.15
25×	0.50	1.2	7.07	0.67	1.33	0.72	18	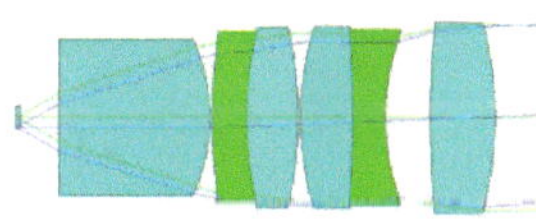	1.16
40×	0.65	0.35	4.6	0.48	0.69	0.45	18		1.17
50×	1.0 oil	0.13	3.77	0.33	0.33	0.36	18	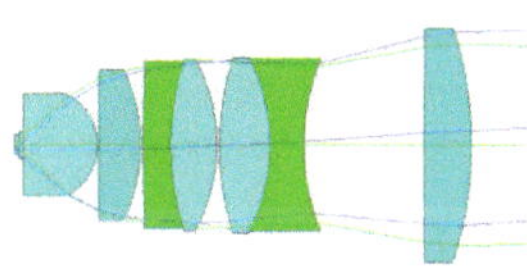	1.19
100×	1.25 oil	0.10	1.78	0.26	0.20	0.18	18	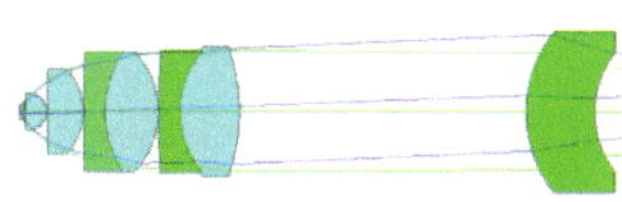	1.21

Note. R, resolution; *DF*, depth of focus; *FOV*, field of view.

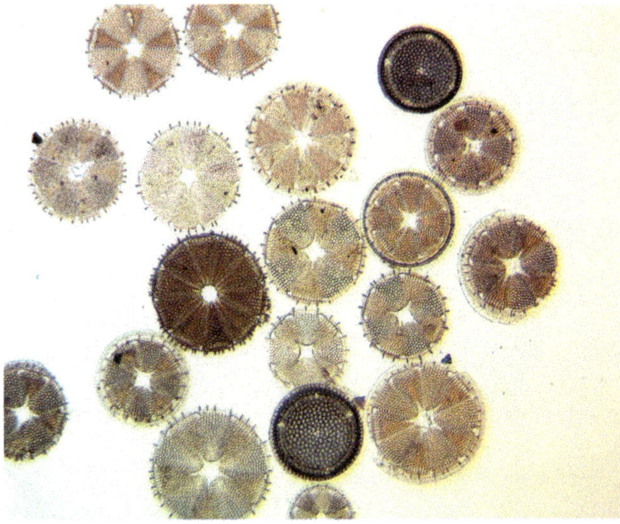

Figure 1.23 Photograph of real objects, diatoms, using a 10×/0.30 finite LCR semi–plan achromatic objective.

Figure 1.24 Photograph of a real object, *Pteridium aquilinum*, using a 25×/0.50 finite LCR semi–plan achromatic objective.

correcting the Petzval coefficient, it has significant curvature. Figure 1.28 shows an image of a "thick" frontal meniscus.

The object is located at the point S at a distance S_1 from the first surface, and its image after refraction on the first surface coincides with the aplanatic point of the second surface. Letters A_1, A_2, and A_3 designate aplanatic points of the first surface (first, second, and third type). The magnitude of the linear

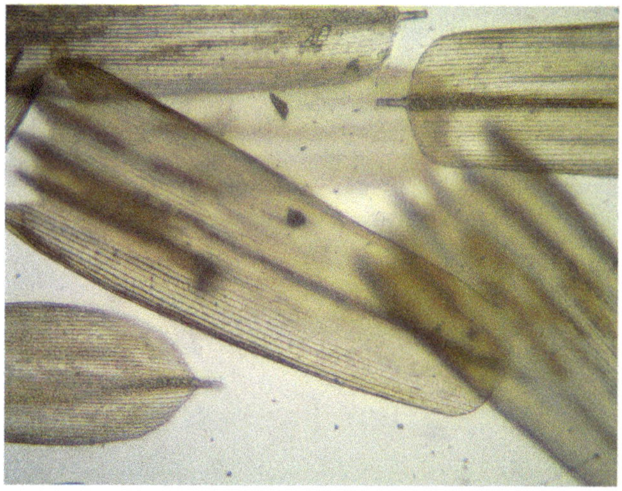

Figure 1.25 Photograph of real objects, wings, using a 40×/0.65 finite LCR semi–plan achromatic objective.

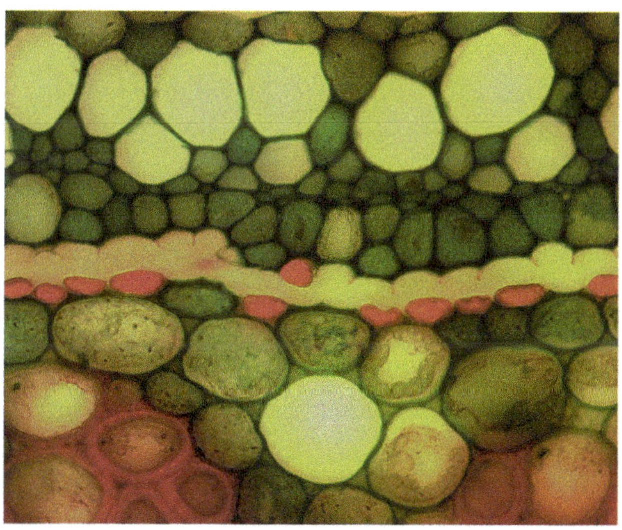

Figure 1.26 Photograph of a real object, *Pteridium aquilinum*, using a 50×/1.00 oil immersion finite LCR semi–plan achromatic objective.

field in microscope objectives, especially strong ones, is small, and the meniscus astigmatism is quite accurately determined by third-order aberrations.

The location of the object in the aplanatic points of the first surface is impractical. At the point A_1, there is no working distance, since $S_1 = 0$; at point A_2, astigmatism is close to the maximum value; and at point A_3, the aperture for the subsequent part does not decrease, since in this case the linear magnification in the meniscus is 1. When designing objectives with a medium

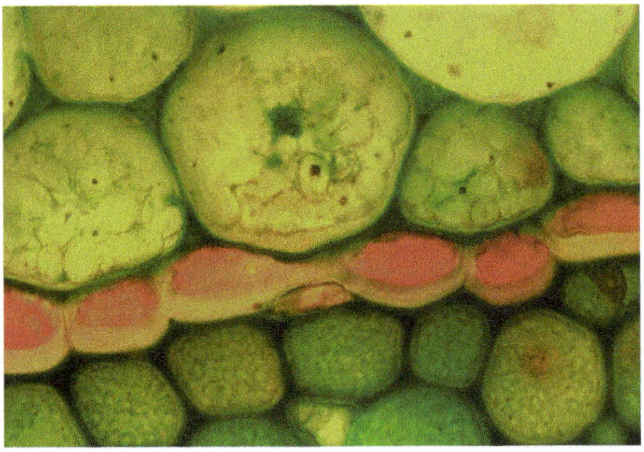

Figure 1.27 Photograph of a real object, *Pteridium aquilinum*, using a 100×/1.25 oil immersion finite LCR semi–plan achromatic objective.

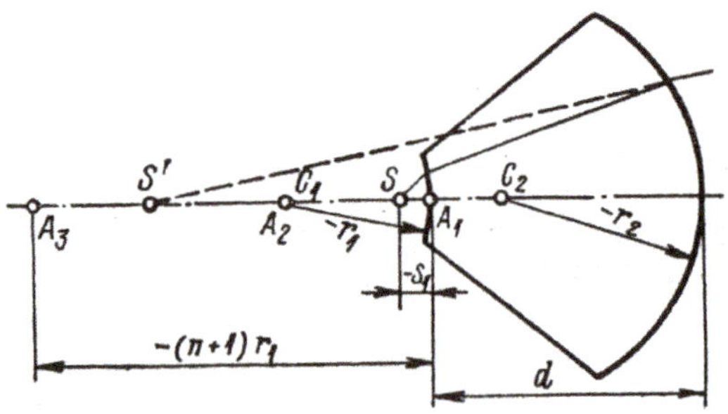

Figure 1.28 "Thick" frontal meniscus.

or high numerical aperture, it is advisable to position the object near the first aplanatic point A_1. If $S_1 = (0.1 \div 0.3) \cdot r_1$, then the magnification in the frontal meniscus is within $V = (2 \div 3)^x$ and the astigmatism is twice less than the maximum value. Optical materials used in the front meniscus should have high values of the refractive indices and coefficients of the average dispersion to achieve the highest values of linear magnification and the best correction of chromatic aberrations.

The first surface of such a meniscus plays the role of a power element, and the second surface and thickness are the projection parameters that determine the possibility of aplanatism and correction of the curvature of the image in the objective as a whole. Indeed, if we bear in mind that the increase in the optical system is determined by the ratio $V = \frac{n_1\alpha_1}{n'\alpha'}$ [9], if $\alpha' = 1$, $n' = 1$ and we build a second surface of the front lens according to the principle of aplanatism, then the parameters of the first paraxial beam for this lens are determined by the relations

$$\alpha_1 = V/n_1,$$
$$\alpha_2 = n_2, \qquad h_1 = S_1 V/n_1,$$
$$\alpha_3 = 1, \qquad h_2 = (n_2 + 1)r_2.$$

Accordingly, the structural elements of the front lens are determined by the relations

$$r_1 = \frac{S_1 V}{n_1} \cdot \frac{n_2 - n_1}{n_2^2 - V}, \tag{1.20}$$

$$r_2 = \frac{(n_2 - 1) \cdot n_1 r_1}{(n_2 - n_1) - n_1 n_2 r_1 S_{IV}}, \tag{1.21}$$

$$d = \frac{S_1 V/n_1 - r_2(n_2 + 1)}{n_2}. \tag{1.22}$$

Obviously, when $V = n_2$ and $n_1 = 1$, the first surface becomes concentric to the axial point of the object and the lens is converted into a classical aplanatic element based on which achromats and apochromats are built. Such an element, as is well known, is characterized by residual astigmatism, which is determined by the properties of aplanatic surfaces of the first type; Grammatin [5] and Ivanova and Kirillovsky [9] studied the properties of frontal lenses, the basic elements of plan objectives with high numerical apertures. They demonstrated that for small values of the segment S_1 ($|S_1| < 1$ mm), useful magnifications lie in the range $n_2 < V < 3$. In this area, spherical aberration η and astigmatism have values that can be compensated for in the subsequent components of the objective, and the Petzval sum can be negative and have the value needed to correct the curvature of the image after the objective.

At the same time, the range of work of the menisci can be extended by introducing the segment S_1 as the parameter. An analysis of Eqs. (1.20–1.22) shows that for a lens in the air ($n_1 = 1$), the values of parameters of structural elements suitable for practical implementation depend on S_1, n_2, V, and S_{IV}. Thus, r_1 is directly proportional to S_1, and the value of r_2 depends on r_1 and S_{IV}; the denominator of the fraction in Eq. (1.21) becomes 0 with $r_1 = \frac{n_2 - 1}{n_2 S_{IV}}$, and thus $r_2 = \infty$ and Eq. (1.22) loses its meaning.

Table 1.9 shows the design parameters of a set of frontal lenses depending on S_1 and V for two values of the refractive indices of glass types LZ_K8 and LZ_CTK19; in addition, we will conduct calculations based on two values of S_{IV}.

It should be noted that for building frontal systems, lenses for which the absolute value of the first radius r_1 exceeds 0.8 mm and the thickness d value is $0.5 \leq d \leq 5$ mm can have practical applications.

Analyzing the data in Table 1.9 allows us to draw the following conclusions:

Table 1.9 Design parameters of a set of frontal lenses depending on S_1 and V for refractive indices of glass types LZ_K8 and LZ_CTK19.

No.	S_1	V	r_1	S_{IV}	r_2	d
			$n_D = 1.5163$ (LZ_K8)			
1	−1.0	0.8	−0.2755	0.050	−0.2647	—
				−0.050	−0.3583	0.0669
		1.0	−0.3974	0.050	−0.3755	—
				−0.050	−0.4220	0.0408
		1.6	−1.1815	0.050	−1.0068	0.616
				−0.050	−1.4295	1.317
		2.0	−3.4512	0.050	−2.2904	2.482
				−0.050	−6.9959	10.291
2	−2.0	0.8	−0.5510	0.050	−0.5098	—
				−0.050	−0.4745	—
		1.0	−0.7948	0.050	−0.7118	—
				−0.050	−0.9000	0.1745
		1.6	−2.3630	0.050	−1.7543	0.801
				−0.050	−3.6186	3.895
		2.0	−6.9024	0.050	−3.4273	3.050
				−0.050	508.7048	—
3	−4.0	0.8	−1.1020	0.050	−0.9485	—
				−0.050	−1.3148	0.0715
		1.0	−1.5896	0.050	−1.2888	—
				−0.050	−2.0735	0.803
		1.6	−4.7260	0.050	−2.7899	0.4090
				−0.050	−15.4433	21.407
		2.0	−13.8046	0.050	−4.5603	2.2918
				−0.050	−13.1914	16.6151
4	−6.0	0.8	−1.653	0.050	−6.8109	8.1371
				−0.050	−2.1827	0.4566
		1.0	−2.384	0.050	−1.7660	—
				−0.050	−3.6678	2.1297
		1.6	−7.089	0.050	−3.4732	—
				−0.050	172.6462	280.175
		2.0	−20.707	0.050	−2.0862	—
				−0.050	10.1471	12.9465
			$n_D = 1.7440$ (LZ_CTK19)			
1	−1.0	0.8	−0.2655	0.050	−0.2574	—
				−0.050	−0.2740	—
		1.0	−0.3644	0.050	−0.3494	—
				−0.050	−0.3807	0.0256
		1.6	−0.8258	0.050	−0.7529	0.2672
				−0.050	−0.9143	0.5211
		2.0	−1.4287	0.050	−1.2236	0.7787
				−0.050	−1.7161	1.5530

(continued)

Table 1.9 (*Continued*) Design parameters of a set of frontal lenses depending on S_1 and V for refractive indices of glass types LZ_K8 and LZ_CTK19.

No.	S_1	V	r_1	S_{IV}	r_2	d
2	−2.0	0.8	−0.5310	0.050	−0.4998	—
				−0.050	−0.5659	—
		1.0	−0.7288	0.050	−0.6714	—
				−0.050	−0.7974	0.1078
		1.6	−1.6516	0.050	−1.3838	0.3424
				−0.050	−2.0480	1.3874
		2.0	−2.8574	0.050	−2.1405	1.0743
				−0.050	−4.2965	4.4665
3	−4.0	0.8	−1.0620	0.050	−0.9444	—
				−0.050	−1.2129	0.0735
		1.0	−1.4576	0.050	−1.2450	—
				−0.050	−1.7580	0.472
		1.6	−3.3032	0.050	−2.3814	0.0772
				−0.050	−5.3895	4.810
		2.0	−5.7148	0.050	−3.4225	1.2473
				−0.050	−17.3048	23.558
4	−6.0	0.8	−1.5932	0.050	−1.3420	—
				−0.050	−1.9588	0.329
		1.0	−2.1866	0.050	−1.7405	—
				−0.050	−2.9402	1.874
		1.6	−4.9548	0.050	−3.1344	—
				−0.050	−11.8192	13.092
		2.0	−8.5723	0.050	−4.2761	—
				−0.050		—

1. With an increase in the value of the segment S_1, the value of V decreases, and then the values of r_1, r_2, and d used for realization can be determined.
2. With an increase in the refractive index, the values of V, which determine the required structural elements, also increase.
3. There is a fairly large set of elements in which S_{IV} can have negative values for the segment S_1 that are acceptable for a constructive implementation of the values of r_1, r_2, and d. These elements can work as front lenses in small- and medium-magnification objectives.
4. In addition, elements whose design parameters are determined by $|S_1| > 1$ and region $0.8 < V < n_2$ can work as the second element of a plan objective; the second element is located after the front lens, having $V > n$, $S_{IV} > 0$. Such a construction of the frontal system can give the best solutions for high numerical aperture objectives.

1.6.2 Last part of the objective in the form of a meniscus

In the last part of the objective, finite-thickness menisci, proposed by Boegehold [11], are also often used. The use of this element not only allows the Petzval sum to be reduced, but it also enables the compactness of the overall dimensions of optical systems with plan objectives, whose height is standardized.

Figure 1.29 shows a schematic diagram of the middle part (conventionally indicated by the combined main planes $H_K H'_K$) and the last part of the microscope objective as a meniscus.

The position of the plane of the object of the middle system s_K is known from the calculation of the frontal part of the objective; the position of the image plane is defined as $s'_K \approx s_K \cdot V_K$, where V_K is the magnification in the middle part. When analyzing the aberration properties of the meniscus, magnification in the meniscus V_M and its front segment value s_M can be used as variable parameters.

If we assume that the position of the last surface of the meniscus coincides with the reference plane of the objective, and the objective is designed to work, for example, with the finite length of the tube, then we can get the ratio of the radii of curvature of the surfaces through the angles and heights of the first paraxial ray:

$$\alpha_1 = 1/V_M$$

$$h_1 = -s'_M/V_M$$

$$\alpha_2 = (-s'_M + s_M V_M)/V_M d$$

$$h_2 = -s_M$$

$$\alpha_3 = 1.$$

The meniscus thickness d can be used as a correction parameter. The following considerations are true:

1. For the selected value of magnification in V_M, there is a region of constructive solutions, defined by the segment s_M, in which the radii of curvature of the meniscus have acceptable values for practical implementation.

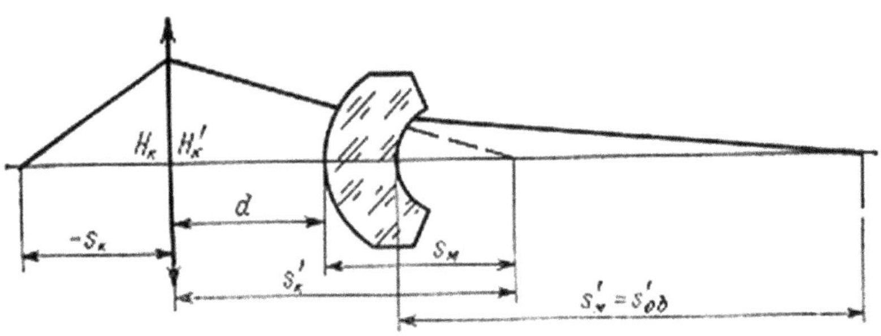

Figure 1.29 Schematic diagram of the last part of the microscope objective.

2. Corrective capabilities of menisci to compensation for residual aberrations of frontal systems are rather limited. But the value of LC (chromatism of magnification) can be significant and the opposite of the LC of the frontal part.
3. Increasing the value of segment s_M at a selected value of V_M is not only advisable to enhance the compensation properties of the meniscus, but it also reduces the relative apertures of the components of the middle part of the objective.

To determine the chromatic aberrations of the meniscus, the chromatic coefficients C_1 and C_2 are calculated for the outer surfaces of the meniscus:

$$C_1 = (\alpha_1 - \alpha_2)/\nu; \quad C_2 = (\alpha_2 - 1)/\nu.$$

To enhance the correctional qualities of the latter part of the objective, made in the form of a meniscus, we can make a meniscus by gluing together two (or three) lenses. Consider a double-glued lens. If, when introduced into the meniscus of the *achromatization radius* $r_{gl.sur}$, the crown and flint lenses are made of glass types that have near values of the refractive indices for the main wavelength, then the chromatic coefficient of the glued surface separating two media will be determined by the expression [6, 9]

$$C_{gl.sur.} = (n - 1)(1/\nu_3 - 1/\nu_2)(\alpha_2 - h_{gl.sur.}/r_{gl.sur.}),$$

where $h_{gl.sur.}$ is the height of the first paraxial ray on the glued surface, $h_{gl.sur.} = h_1 - \alpha_2 d_1$, and d_1 is the thickness of the first lens, selected for design reasons; ν_2, ν_3 are coefficients of dispersion of materials. To obtain chromatic compensation of the frontal part, large values of the chromatic coefficient $C_{gl.sur.}$ must be present in the last part, which leads to the appearance of a negative chromatic difference of positions and increases the correction load for the middle part of the objective. Studying the properties of individual elements of objectives reveals that constructing the last part of a plan objective is a process of optimizing the parameters of the middle part and the meniscus of finite thickness.

1.6.3 An example of creating plan achromatic (semi–plan achromatic) objectives

By the way, the optical system of the objective presented in Fig. 1.21 is based on the classical Abbe scheme with the addition of a finite-thickness meniscus. Similar results can be achieved using the modernized scheme of classical achromatic objectives (when a single positive lens is used instead of the first double-glued lens) with the addition of such a meniscus. This approach allows either the values of monochromatic and chromatic aberrations of off-axis beams, including curvature of the image and LC (chromatism of magnification), to be significantly reduced, or it completely corrects them [LC free (LCF)].

Studies on the optimization of objectives, designed to complete the mass segment microscopes, led to the creation of sets of plan achromatic and semi–plan achromatic objectives for implementation in microscopes with a finite optical tube length and infinity. Figures 1.30–1.59 present the principal layouts, design parameters, and graphs of aberration correction of these objectives.

Figure 1.60 shows a photograph of a set of objectives; design parameters and graphs of aberration correction of these objectives are shown in Figs. 1.36, 1.41, 1.45, 1.49, and 1.57. These are the last generation of objectives with an aberration correction of the LCF plan achromat. Although the 100× oil immersion objective does not have the same high quality of curvature aberration correction, it looks practically like a plan objective (the image quality is similar to that shown in Fig. 1.27). Table 1.10 shows the main technical parameters of these budget objectives (we called them *third-generation* objectives).

These objectives were tested by obtaining images of real objects. Figures 1.61–1.63 show photographs of real objects taken using these objectives. Some testing photographs were taken by Sergey Dushenbiev, a respected expert from the Russian city of Volgograd. Figure 1.61 shows photographs taken when testing a 10×/0.30 objective. Figure 1.62 shows photographs taken when testing a 20×/0.65 objective. Figure 1.63 shows photographs taken when testing a 40×/0.70 objective. Figures 1.64–1.110 are photographs of other biological objects obtained using a series of LCF plan achromatic objectives.

1.6.4 An example of creating plan achromatic (semi–plan achromatic) objectives for studying under a microscope using special methods

Objects examined under a microscope are not always prepared in the standard fashion described in Volume 1 of this book (Chapter 1, Fig. 1.35). A brief classification of the most common types of microscopic objects is shown in Fig. 2.8 (Volume 1, Chapter 2). Naturally, this scheme is not complete, especially because nuances such as the location of an object in a particular environment, the features of its manufacture, and the methods and techniques used to study them are not taken into account. The studied objects are often not covered with a coverslip; sometimes they are located in the depths of some liquid or immersion, for example, water or oil. Moreover, there is often a need to sufficiently penetrate into the depths of a liquid or immersion as well as view a microscopic object in its thickness. To study such objects, special types of microscopes are developed and used. These microscopes must be equipped with special objectives.

1.6.4.1 Budget objectives for inverted biological microscopes

The optical system, which is the basis for the design of an inverted microscope, is presented in Fig. 2.21 (Volume 1, Chapter 2). Such a microscope not only differs from the usual so-called *direct* one in terms of the

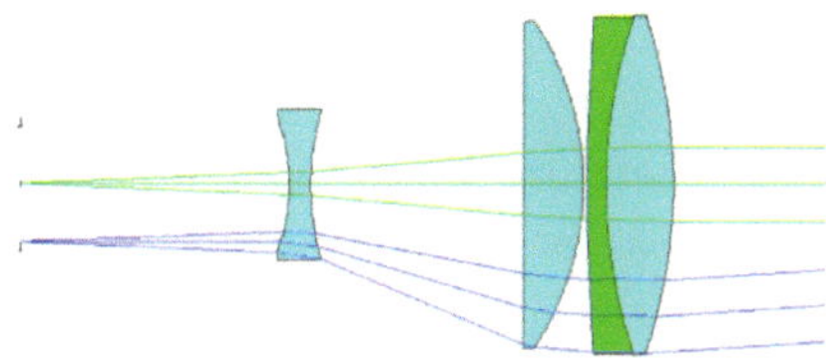

Lens: plan 2.5x0.04 Ftl 160

Object num aper 0.040000 Image height 10.000000 Primary wavln 0.546070

SRF	RADIUS		THICKNESS		APERTURE RADIUS		GLASS		SPECIAL
OBJ	0.000000		1.0000e-20		4.002093		AIR		
AST	0.000000		18.940000		4.000000	A	AIR		
2	-13.213000		1.500000		4.690000		LZ_LK3		
3	18.836000		15.020000		5.160000		AIR		
4	0.000000		4.200000		10.990000		LZ_LK3	P	
5	-19.861000		0.300000		11.190000		AIR		
6	144.210000		1.300000		11.500000		LZ_TF5		
7	34.510000		4.800000		11.510000		LZ_LK3	P	
8	-34.510000		90.000000		11.580000		AIR		
9	214.800000		5.000000		5.770000		LZ_BK8		
10	-46.990000		0.300000		5.040000		AIR		
11	-46.880000		3.000000		5.000000		LZ_BF24		
12	-113.240000		157.745776	S	4.880000		AIR		
IMS	0.000000		0.000000		10.000000				

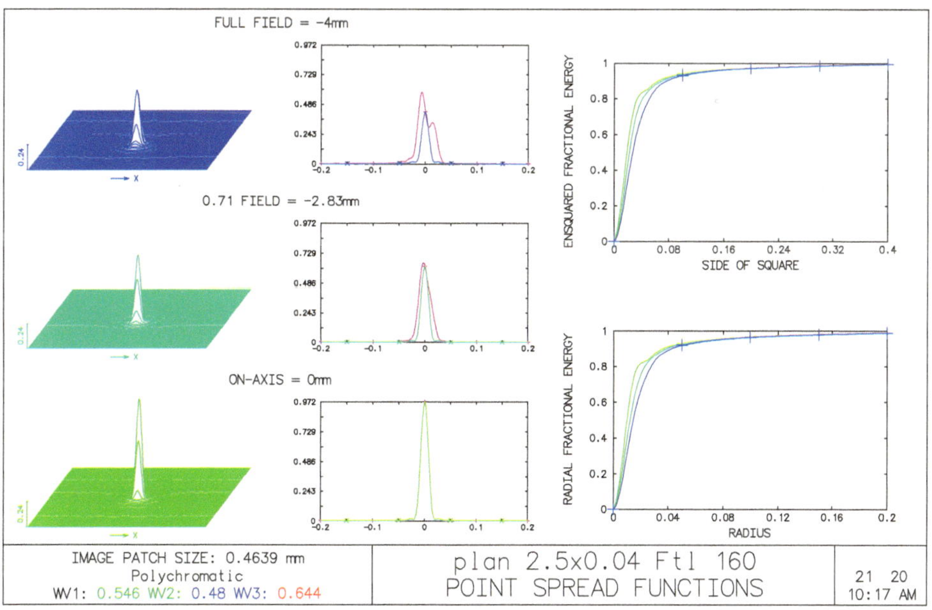

Figure 1.30 Optical design and graphs of aberration correction of a 2.5×/0.04 infinity LCF plan achromatic objective ($F'_{ts} = 160$ mm).

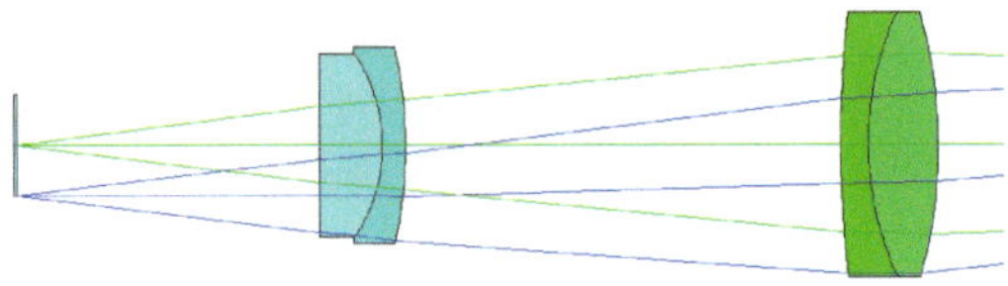

Lens: plan 4x0.13 tub160 h45				Efl	31.5
Object num aper	0.130000	Image height	10.000000	Primary wavln	0.546070
SRF	**RADIUS**	**THICKNESS**	**APERTURE RADIUS**	**GLASS**	**SPECIAL**
OBJ	0.000000	0.010000	2.498916	AIR	
AST	0.000000	0.170000	2.510000 A	K9	
2	0.000000	15.290000	2.520000	AIR	
3	0.000000	3.200000	4.530000	K9 P	
4	-7.800000	1.200000	4.670000	LZ_CTK12	
5	-20.460000	22.000000	4.990000	AIR	
6	55.150000	1.400000	6.870000	H-ZF3	
7	15.394000	3.500000	6.860000	BAF3	
8	-24.500000	148.265879 S	6.890000	AIR	
IMS	0.000000	0.000000	10.000000		

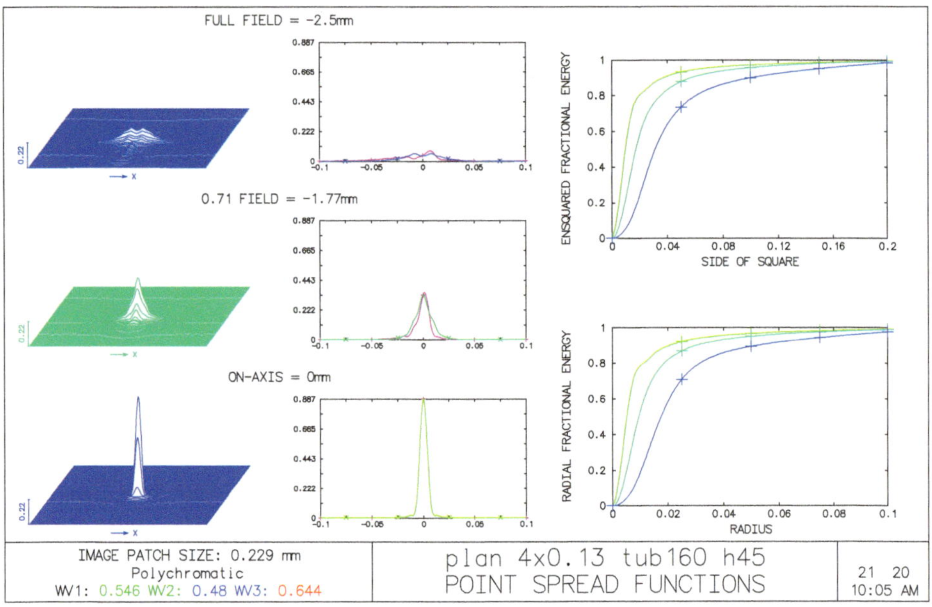

Figure 1.31 Optical design and graphs of aberration correction of a 4×/0.13 finite LCF plan achromatic objective.

arrangement of the main functional units and accessories, such as objectives, a stage, and a lighting system; the main difference is the studied objects, or rather, their location relative to the illuminating transmitted light of the system (condenser) and the objective. In an ordinary *upright* microscope, the illumination system ensures that light passes through the object being examined from below directly in front of the object, and the objective looks at

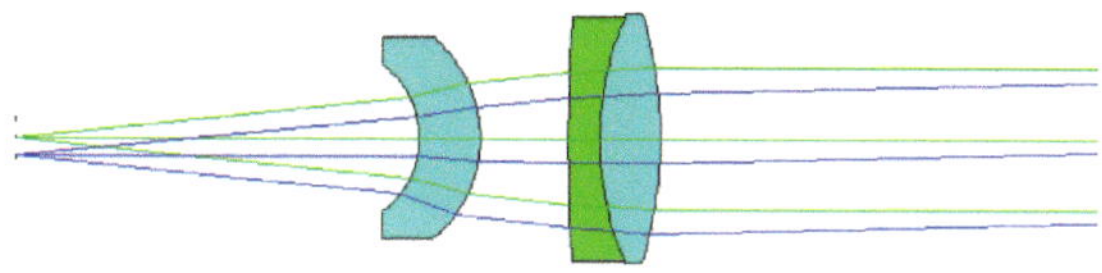

Lens: Plan 5x0.10 ft1200

Object num aper	0.100000	Image height	10.000000	Primary wavln	0.546070
SRF	**RADIUS**	**THICKNESS**	**APERTURE RADIUS**	**GLASS**	**SPECIAL**
OBJ	0.000000	1.0000e-20	2.000199	AIR	
AST	0.000000	23.006800	2.000199 AS	AIR	
2	-5.160000	3.500000	4.120000	H-LAK3	
3	-7.200000	5.000000	5.640000	AIR	
4	112.040000	1.800000	6.690000	ZF3	
5	16.900000	3.500000	6.850000	H-ZK6	
6	-23.550000	200.000000	6.940000	AIR	
7	268.500000	6.250000	11.300000	LZ_BK8	
8	-58.738000	0.375000	11.380000	AIR	
9	-58.600000	3.750000	11.350000	LZ_BF24	
10	-141.550000	197.297328 S	11.470000	AIR	
IMS	0.000000	0.000000	10.000000 S		

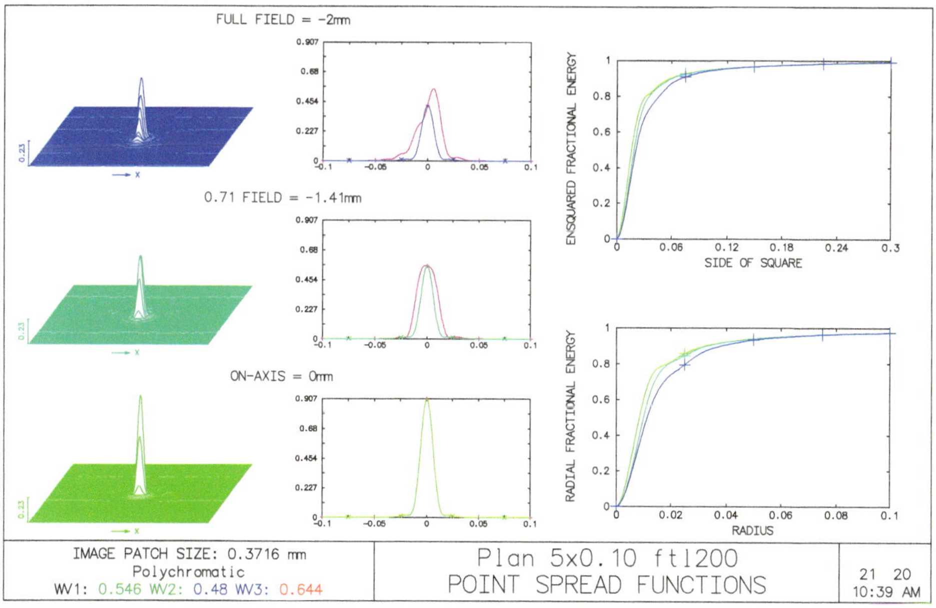

Figure 1.32 Optical design and graphs of aberration correction of a 5×/0.1 infinity LCF plan achromatic objective ($F'_{ts} = 200$ mm).

the object from above. On the contrary, in an inverted microscope, the objective is located under the object being studied, and the lighting system (transmitted light) is located above it. Figure 1.111 shows an optical diagram and a photograph of a Russian inverted microscope, the Biolam-P, with which the explanation of the principle of operation of an inverted microscope becomes more understandable.

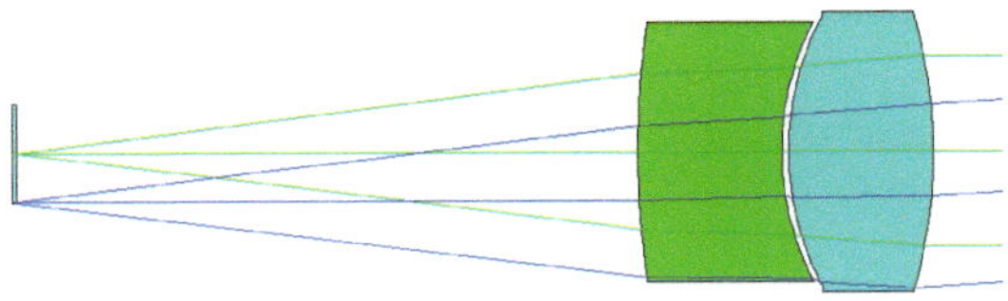

Lens: ocx 5x0.12 d=0-0.17 Ftl 160

Object num aper 0.120000 Image height 10.000000 Primary wavln 0.546070

SRF	RADIUS	THICKNESS	APERTURE RADIUS	GLASS	SPECIAL
OBJ	0.000000	0.010000	2.000460	AIR	
AST	0.000000	0.170000	2.000000 A	LZ_K14	
2	0.000000	26.091200	2.010000	AIR	
3	35.650000	6.000000	5.210000	LZ_TF2	
4	10.990000	0.300000	5.290000	AIR	
5	11.342000	6.000000	5.470000	LZ_BK6	
6	-20.230000	90.000000	5.700000	AIR	
7	214.800000	5.000000	7.620000	LZ_BK8	
8	-46.990000	0.300000	7.730000	AIR	
9	-46.880000	3.000000	7.720000	LZ_BF24	
10	-113.240000	157.837338 S	7.840000	AIR	
IMS	0.000000	0.000000	10.000000		

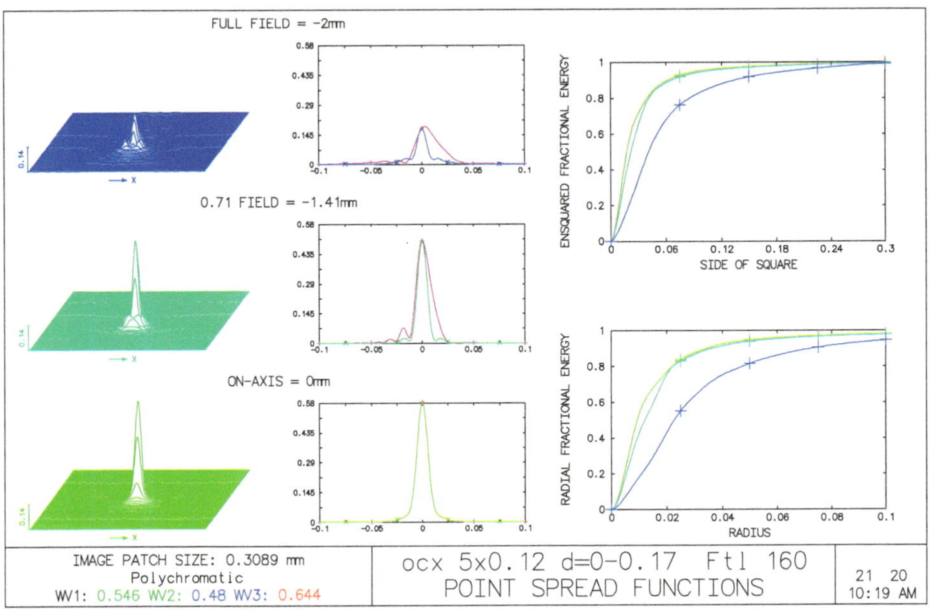

Figure 1.33 Optical design and graphs of aberration correction of a 5×/0.12 infinity LCF plan achromatic objective ($F'_{ts} = 160$ mm).

The constructive implementation of such a microscope, as well as its principle of operation, is determined by the specific features of the object under study; the object is located in a special bowl and is quite thick and loose. The light source 1 is projected by the collector 14 into the plane of the aperture diaphragm 16, and with the help of a lens 12, it uniformly illuminates the test

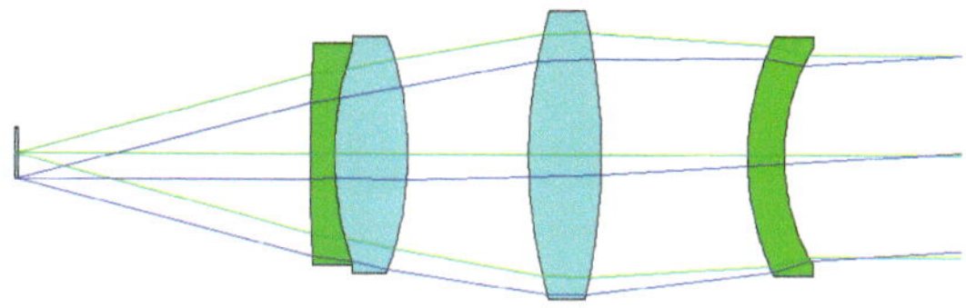

Lens: ocx 10x0.25 d=0-0.17 Ftl 160

Object num aper 0.250000 Image height 10.000000 Primary wavln 0.546070

SRF	RADIUS	THICKNESS	APERTURE RADIUS	GLASS	SPECIAL
OBJ	0.000000	0.010000	1.000125	AIR	
AST	0.000000	0.170000	1.000000 A	LZ_K14	
2	0.000000	11.997400	1.030000	AIR	
3	112.760000	1.000000	4.150000	LZ_TF5	
4	13.183000	3.000000	4.360000	LZ_LK3	
5	-13.183000	5.000000	4.670000	AIR	
6	19.498000	3.000000	5.700000	LZ_LK3 P	
7	-24.320000	6.000000	5.690000	AIR	
8	10.520000	1.500000	4.720000	LZ_TF5 P	
9	8.395000	90.000000	4.280000	AIR	
10	214.800000	5.000000	9.340000	LZ_BK8	
11	-46.990000	0.300000	9.440000	AIR	
12	-46.880000	3.000000	9.420000	LZ_BF24	
13	-113.240000	157.866743 S	9.540000	AIR	
IMS	0.000000	0.000000	10.000000		

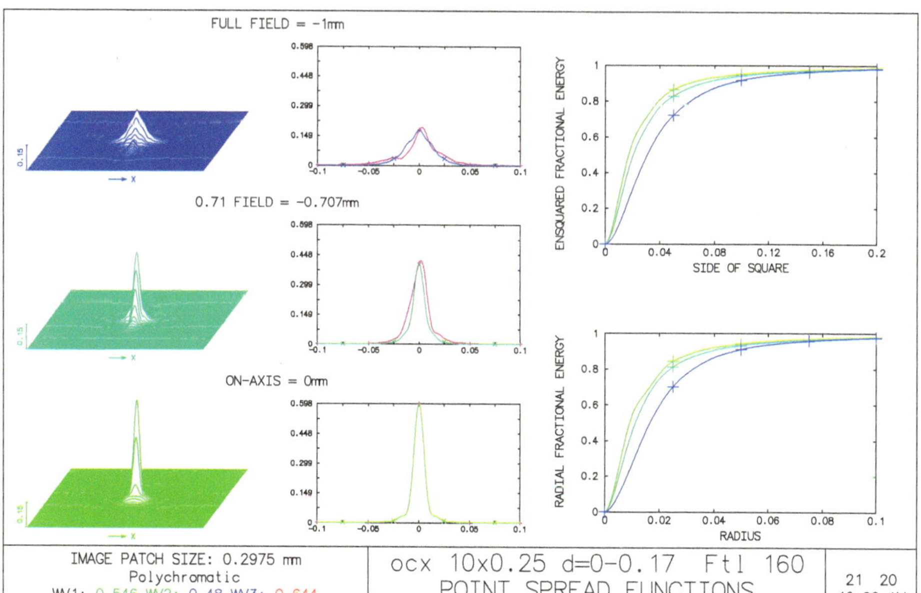

Figure 1.34 Optical design and graphs of aberration correction for a 10×/0.25 infinity LCF plan achromatic objective ($F'_{ts} = 160$ mm).

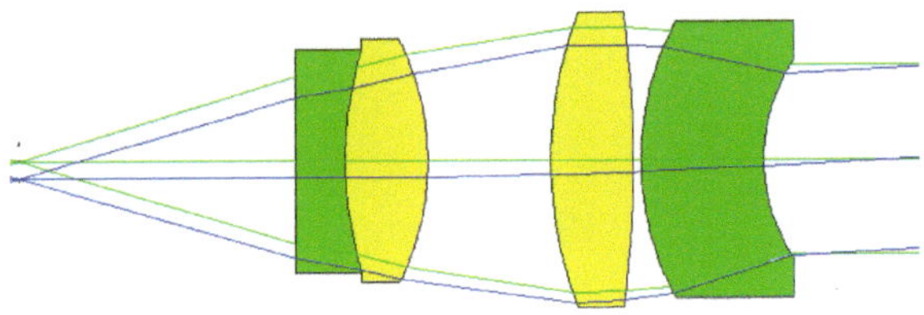

Lens: LD plan 10x0.28 d0 ft1200

Object num aper 0.280000 Object height -1.000000 Primary wavln 0.546070

SRF	RADIUS	THICKNESS	APERTURE RADIUS	GLASS	SPECIAL
OBJ	0.000000	1.0000e-20	1.000000	AIR	
AST	0.000000	16.825500	1.000000 AS	AIR	
2	0.000000	3.000000	5.910000	H-ZF6	
3	21.760000	5.000000	6.560000	H-FK61	
4	-15.672000	7.500000	7.130000	AIR	
5	23.710000	5.000000	8.640000	H-FK61 P	
6	-66.000000	0.500000	8.530000	AIR	
7	16.200000	7.500000	8.110000	H-ZF6 P	
8	10.233000	200.000000	5.850000	AIR	
9	268.500000	6.250000	15.390000	LZ_BK8	
10	-58.738000	0.375000	15.440000	AIR	
11	-58.600000	3.750000	15.390000	LZ_BF24	
12	-141.550000	197.288748 S	15.520000	AIR	
IMS	0.000000	0.000000	10.089890 S		

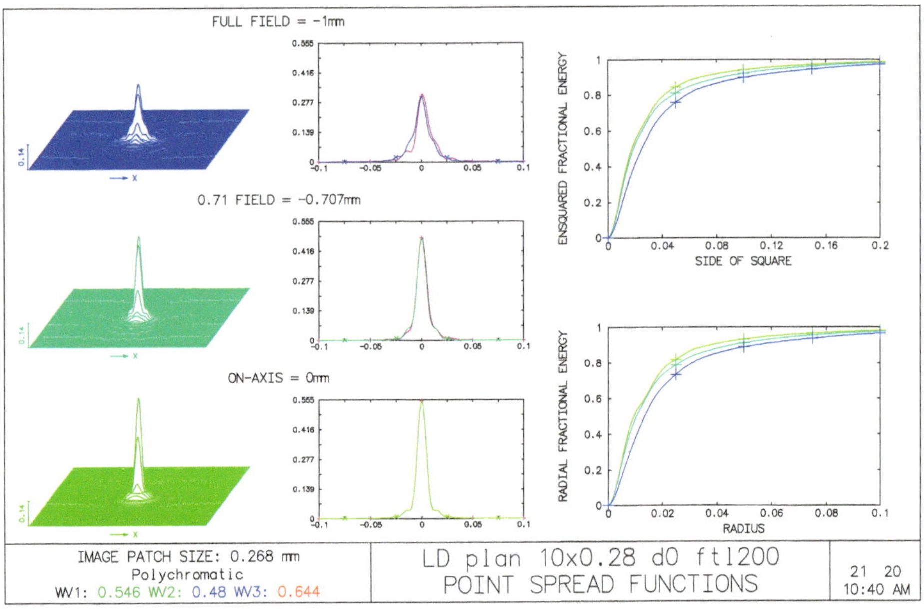

Figure 1.35 Optical design and graphs of aberration correction of a 10×/0.28 infinity LCF plan achromatic objective ($F'_{ts} = 200$ mm).

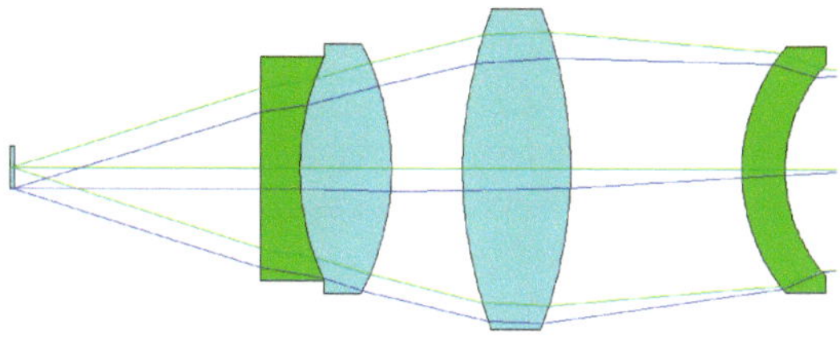

Lens: Plan 10x0.30 tub160 h45 Efl 15.05

Object num aper 0.300000 Image height 10.000000 Primary wavln 0.546070

SRF	RADIUS	THICKNESS	APERTURE RADIUS	GLASS	SPECIAL
OBJ	0.000000	0.010000	0.971539	AIR	
AST	0.000000	0.170000	0.970000 A	K9	
2	0.000000	11.495000	1.010000	AIR	
3	0.000000	1.800000	4.620000	H-ZF7LA	
4	12.330000	4.200000	5.110000	H-QK3	
5	-12.330000	3.300000	5.680000	AIR	
6	20.270000	5.000000	7.220000	H-QK3	
7	-20.270000	8.000000	7.290000	AIR	
8	8.700000	2.000000	5.610000	H-ZF7LA P	
9	7.080000	159.020000	4.810000	AIR	
IMS	0.000000	0.000000	10.000000		

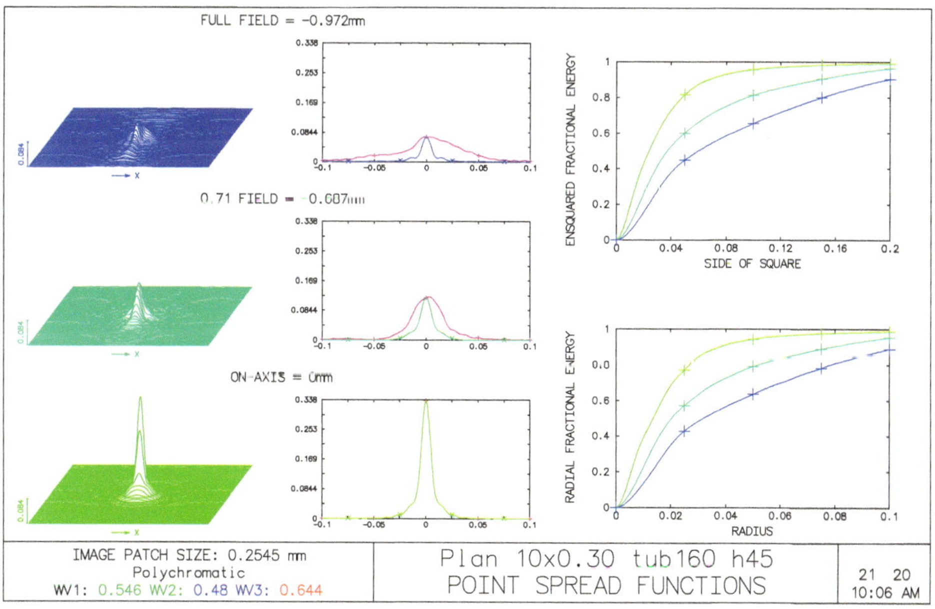

Figure 1.36 Optical design and graphs of aberration correction of a 10×/0.30 finite LCF plan achromatic objective.

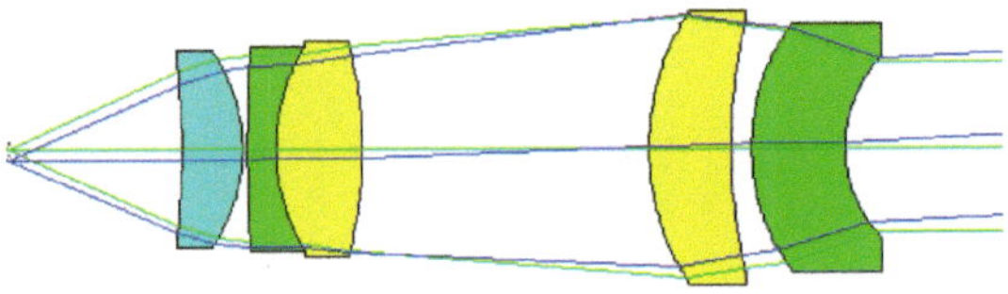

Lens: LD Plan 20x0.40 d0 ft1200

Object num aper	0.400000	Image height	10.000000	Primary wavln	0.546070
SRF	**RADIUS**	**THICKNESS**	**APERTURE RADIUS**	**GLASS**	**SPECIAL**
OBJ	0.000000	1.0000e-20	0.500528	AIR	
AST	0.000000	8.490000	0.500528 AS	AIR	
2	-34.130000	3.000000	4.100000	ZK6	
3	-7.800000	0.250000	4.630000	AIR	
4	76.058000	1.400000	4.780000	ZF3	
5	9.032000	4.200000	4.820000	H-FK61	
6	-19.269000	14.000000	5.070000	AIR	
7	11.495000	4.000000	6.470000	H-FK61 P	
8	26.410000	1.000000	6.070000	AIR	
9	9.720000	4.500000	5.840000	ZF3 P	
10	6.087000	200.000000	4.300000	AIR	
11	268.500000	6.250000	14.160000	LZ_BK8	
12	-58.738000	0.375000	14.220000	AIR	
13	-58.600000	3.750000	14.170000	LZ_BF24	
14	-141.550000	197.406777 S	14.300000	AIR	
IMS	0.000000	0.000000	10.000000 S		

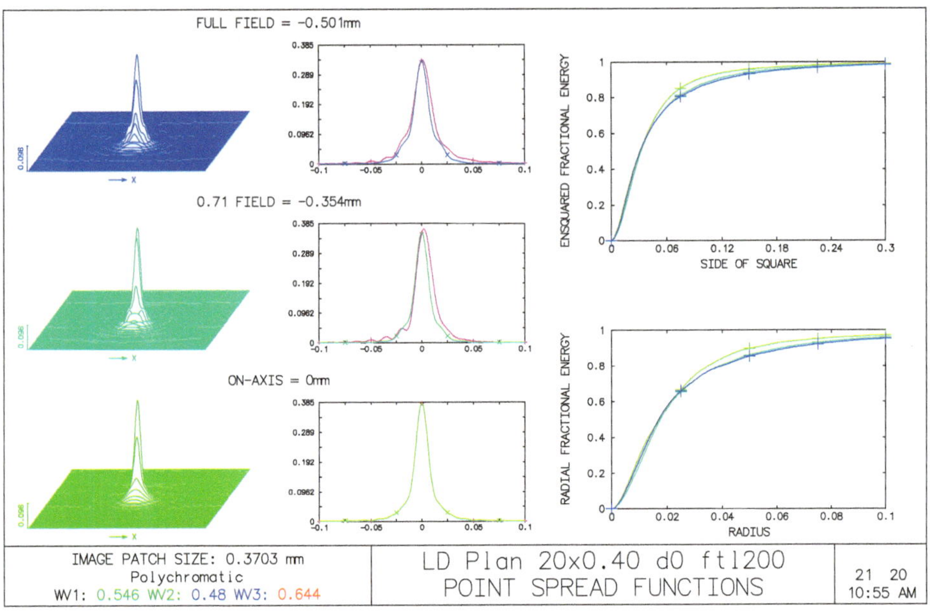

Figure 1.37 Optical design and graphs of aberration correction of a 20×/0.40 infinity LCF long working distance (LD) plan achromatic objective ($F'_{ts} = 200$ mm).

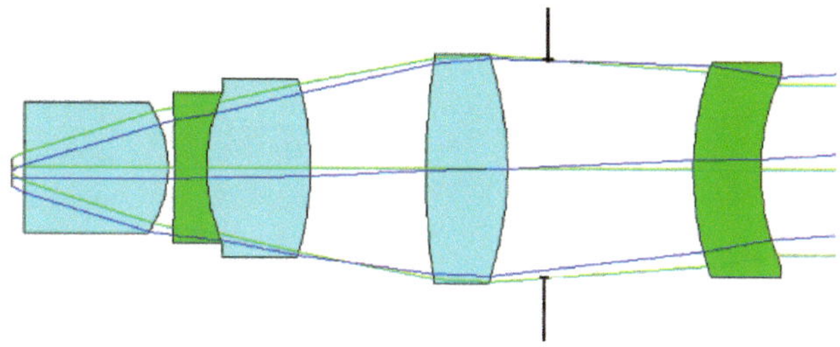

Lens: ocx 20x0.45 d0 ftl160

Object num aper	0.450000	Image height	10.000000	Primary wavln	0.546070

SRF	RADIUS	THICKNESS	APERTURE RADIUS	GLASS	SPECIAL
OBJ	0.000000	1.206350	0.496655	AIR	
AST	0.000000	6.200000	1.100000 A	LZ_K8	
2	-5.035000	0.300000	2.770000	AIR	
3	-75.160000	1.400000	2.930000	LZ_TF5	
4	7.870000	4.500000	3.150000	LZ_K8	
5	-12.647000	5.000000	3.780000	AIR	
6	30.550000	3.500000	4.630000	LZ_K8 P	
7	-15.136000	8.000000	4.810000	AIR	
8	13.583000	2.900000	4.520000	LZ_TF5	
9	9.290000	90.000000	4.020000	AIR	
10	214.800000	5.000000	9.670000	LZ_BK8	
11	-46.990000	0.300000	9.760000	AIR	
12	-46.880000	3.000000	9.740000	LZ_BF24	
13	-113.240000	157.832953 S	9.860000	AIR	
IMS	0.000000	0.000000	10.000000		

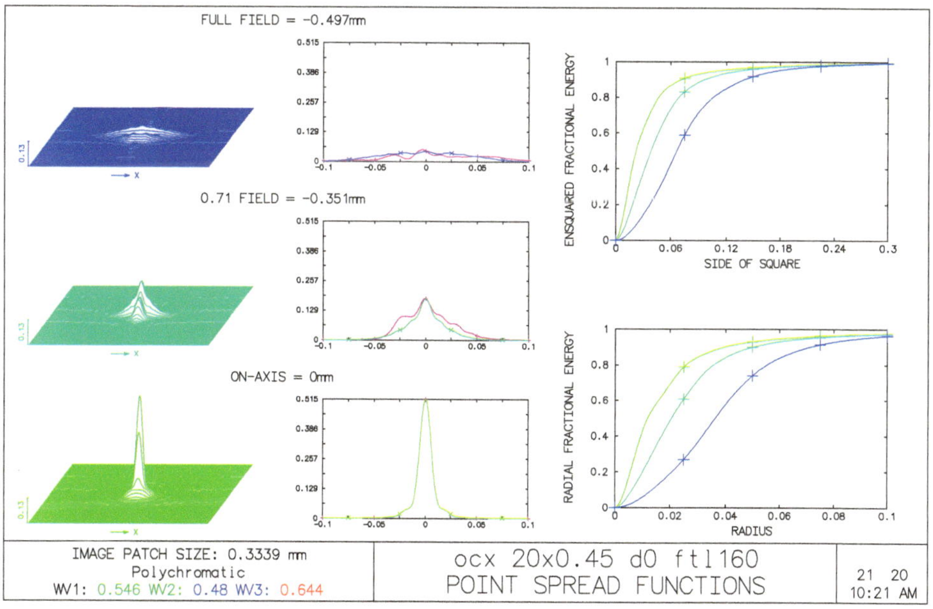

Figure 1.38 Optical design and graphs of aberration correction of a 20×/0.45 infinity LCF semi–plan achromatic objective ($F'_{ts} = 160$ mm).

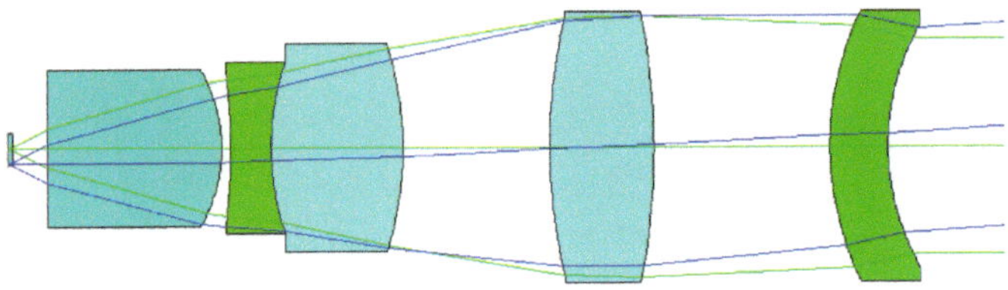

Lens: ocx 20x0.45 d=0.17 Ftl 160

Object num aper 0.450000 Image height 10.000000 Primary wavln 0.546070

SRF	RADIUS	THICKNESS	APERTURE RADIUS	GLASS	SPECIAL
OBJ	0.000000	0.010000	0.501369	AIR	
AST	0.000000	0.170000	0.500000 A	LZ_K14	
2	0.000000	1.200000	0.560000	AIR	
3	0.000000	6.000000	1.160000	LZ_CTK12	
4	-5.297000	0.300000	2.630000	AIR	
5	-13.742000	1.400000	2.690000	LZ_TF5	
6	8.630000	4.500000	2.990000	LZ_LK3	
7	-8.630000	5.000000	3.730000	AIR	
8	13.742000	3.500000	4.670000	LZ_LK3 P	
9	-42.660000	6.000000	4.700000	AIR	
10	7.551000	2.000000	4.550000	LZ_TF5 P	
11	5.970000	90.000000	3.930000	AIR	
12	214.800000	5.000000	9.730000	LZ_BK8	
13	-46.990000	0.300000	9.830000	AIR	
14	-46.880000	3.000000	9.810000	LZ_BF24	
15	-113.240000	157.835329 S	10.010000	AIR	
IMS	0.000000	0.000000	10.000000		

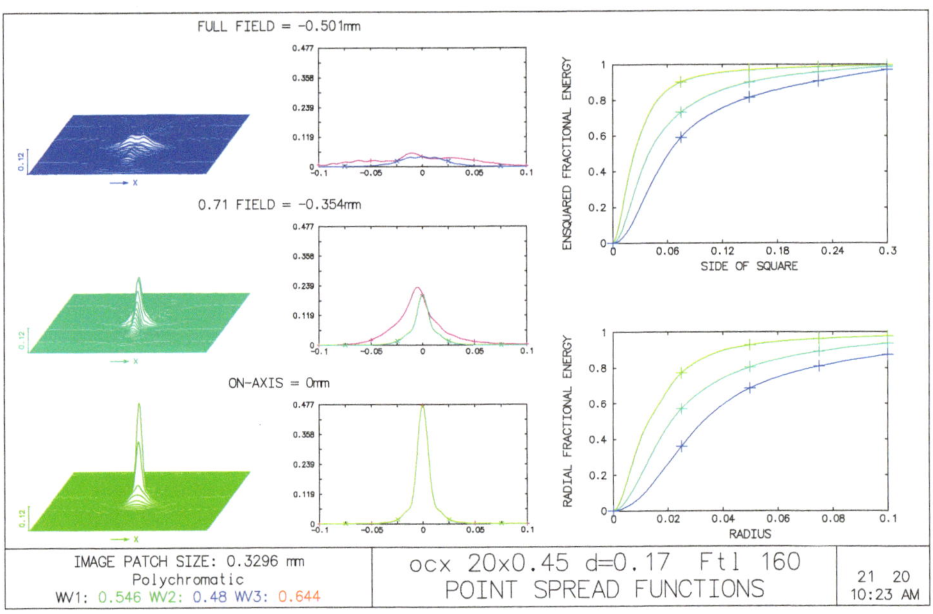

Figure 1.39 Optical design and graphs of aberration correction of a 20×/0.45 infinity LCF semi–plan achromatic objective ($F'_{ts} = 160$ mm).

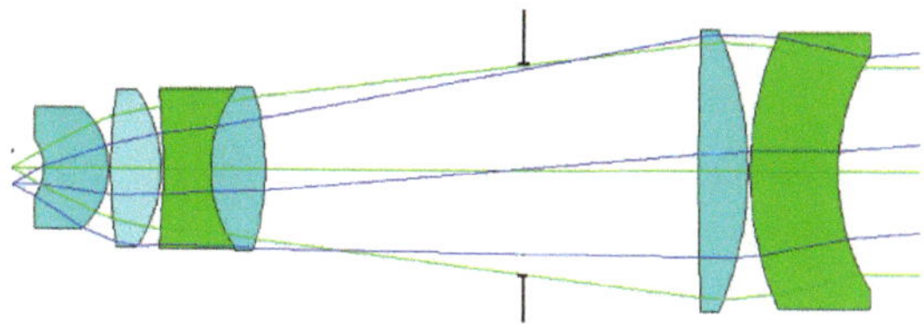

Lens: plan 20x0.50 d=0 Ftl 160

Object num aper 0.500000 Image height 12.500000 Primary wavln 0.546070

SRF	RADIUS	THICKNESS	APERTURE RADIUS	GLASS	SPECIAL
OBJ	0.000000	1.0000e-20	0.625925	AIR	
AST	0.000000	1.248300	0.625925 AS	AIR	
2	-1.836500	2.600000	1.110000	LZ_CTK12	
3	-3.133000	0.100000	2.350000	AIR	
4	20.510000	2.000000	2.890000	LZ_LK3	
5	-5.297000	0.100000	3.060000	AIR	
6	-31.840000	1.900000	3.030000	LZ_TF2	
7	5.272000	2.200000	3.090000	LZ_LK3 P	
8	-10.593000	17.240000	3.170000	AIR	
9	177.830000	2.000000	5.370000	LZ_LK3 P	
10	-13.335000	0.100000	5.460000	AIR	
11	13.122000	3.500000	5.340000	LZ_TF5	
12	8.710000	100.000000	4.560000	AIR	
13	214.800000	5.000000	12.290000	LZ_BK8	
14	-46.990000	0.300000	12.380000	AIR	
15	-46.880000	3.000000	12.350000	LZ_BF24	
16	-113.240000	157.841160 S	12.520000	AIR	
IMS	0.000000	0.000000	12.500000 S		

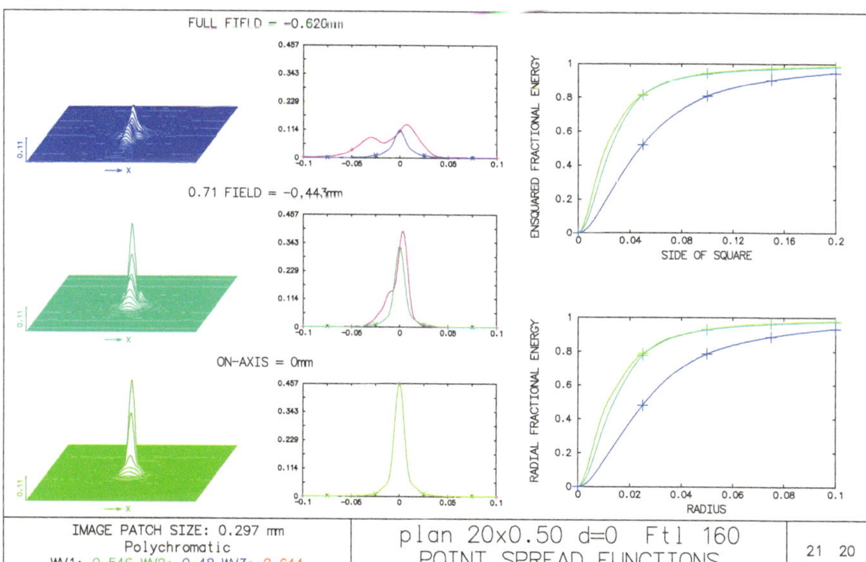

Figure 1.40 Optical design and graphs of aberration correction of a 20×/0.50 infinity LCF plan achromatic objective ($F'_{ts} = 160$ mm).

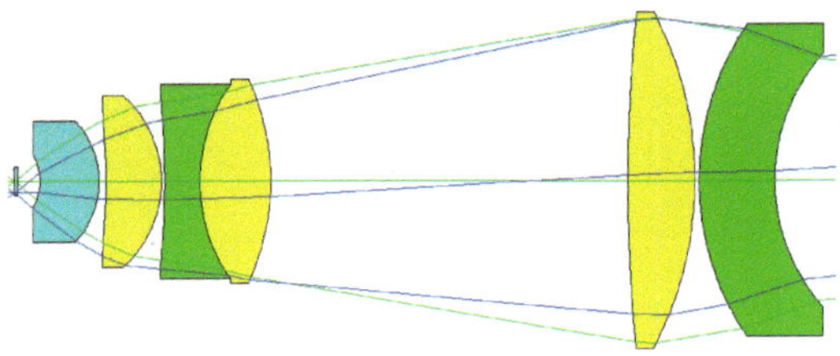

Lens: Plan 20x0.65 d017 tub 160 h45 — Efl 8.37

Object num aper 0.650000 — Image height 14.000000 — Primary wavln 0.546070

SRF	RADIUS	THICKNESS	APERTURE RADIUS	GLASS	SPECIAL
OBJ	0.000000	0.010000	0.700679	AIR	
AST	0.000000	0.170000	0.690000 A	K9	
2	0.000000	1.004100	0.770000	AIR	
3	-2.260000	2.680000	1.280000	H-LAK3	
4	-3.746000	0.300000	2.750000	AIR	
5	-32.880000	2.500000	3.540000	H-FK61	
6	-4.847000	0.140000	3.860000	AIR	
7	-39.800000	1.600000	3.980000	ZF2	
8	7.060000	3.200000	4.240000	H-FK61 P	
9	-10.233000	16.000000	4.390000	AIR	
10	66.620000	3.000000	7.170000	H-FK61 P	
11	-15.232000	0.200000	7.250000	AIR	
12	11.495000	3.400000	6.740000	ZF6	
13	7.900000	160.807441 S	5.510000	AIR	
IMS	0.000000	0.000000	14.000000 S		

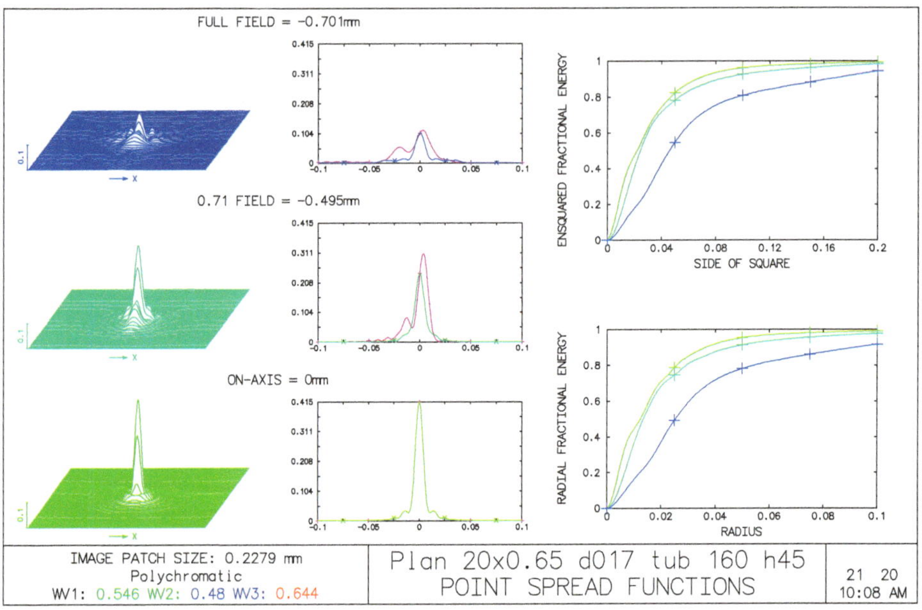

Figure 1.41 Optical design and graphs of aberration correction of a 20×/0.65 finite LCF plan achromatic objective.

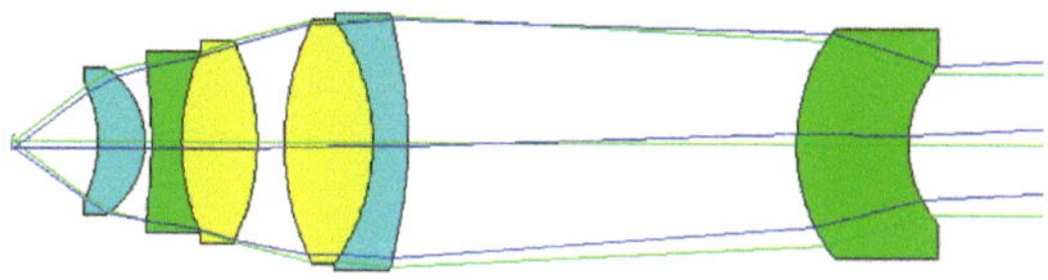

Lens: LD Plan 40x0.60 d0 ft1200

Object num aper 0.600000 Image height 10.000000 Primary wavln 0.546070

SRF	RADIUS	THICKNESS	APERTURE RADIUS	GLASS	SPECIAL
OBJ	0.000000	1.0000e-20	0.250282	AIR	
AST	0.000000	3.661000	0.250282 AS	AIR	
2	-5.580000	2.000000	2.540000	H-LAK61	
3	-3.704000	0.300000	3.120000	AIR	
4	-23.390000	1.300000	3.450000	H-ZF6	
5	9.480000	3.300000	3.870000	H-FK61	
6	-9.480000	1.200000	4.310000	AIR	
7	12.057000	3.800000	5.200000	H-FK61 P	
8	-9.000000	1.500000	5.210000	H-LAK61 P	
9	-19.500000	17.000000	5.460000	AIR	
10	7.930000	4.800000	4.870000	H-ZF6 P	
11	5.090000	200.000000	3.390000	AIR	
12	268.500000	6.250000	13.340000	LZ_BK8	
13	-58.738000	0.375000	13.410000	AIR	
14	-58.600000	3.750000	13.370000	LZ_BF24	
15	-141.550000	197.116357 S	13.490000	AIR	
IMS	0.000000	0.000000	10.000000 S		

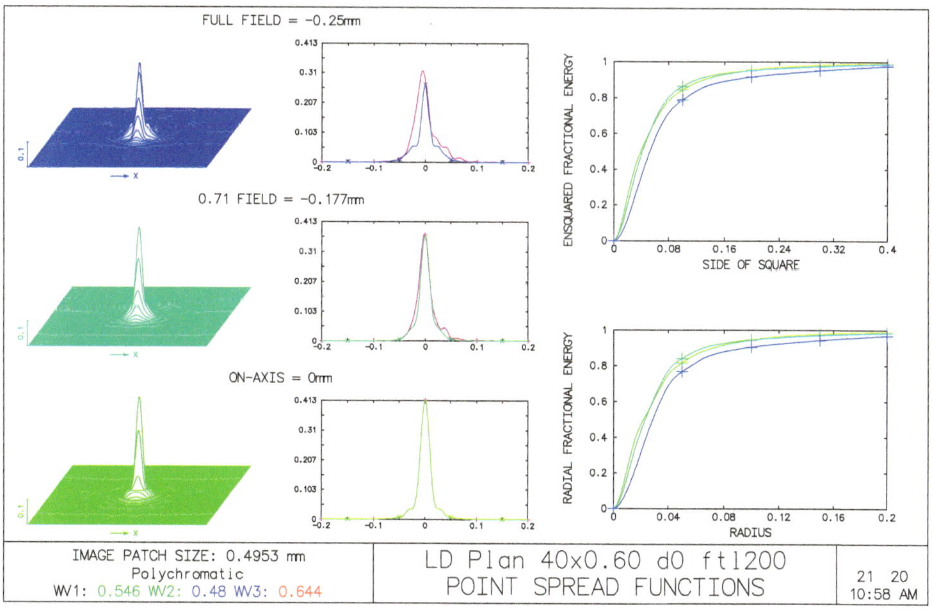

Figure 1.42 Optical design and graphs of aberration correction of a 40×/0.60 infinity LCF LD plan achromatic objective ($F'_{ts} = 200$ mm).

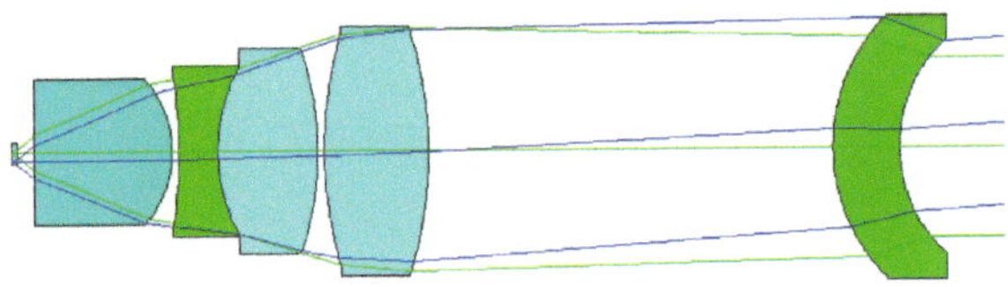

Lens: ocx 40x0.65 d=0.17 Ftl 160

Object num aper	0.650000	Image height	10.000000	Primary wavln	0.546070

SRF	RADIUS	THICKNESS	APERTURE RADIUS	GLASS	SPECIAL
OBJ	0.000000	0.010000	0.251967	AIR	
AST	0.000000	0.170000	0.230000 A	LZ_K14	
2	0.000000	0.500000	0.310000	AIR	
3	0.000000	4.200000	0.740000	LZ_CTK12	
4	-3.076000	0.200000	2.130000	AIR	
5	-15.171000	1.200000	2.230000	LZ_TF5	
6	5.035000	3.000000	2.480000	LZ_LK3	
7	-9.120000	0.200000	3.000000	AIR	
8	11.722000	3.100000	3.400000	LZ_LK3 P	
9	-11.722000	12.200000	3.630000	AIR	
10	5.297000	2.000000	3.840000	LZ_TF5 P	
11	4.207000	90.000000	3.130000	AIR	
12	214.800000	5.000000	8.110000	LZ_BK8	
13	-46.990000	0.300000	8.200000	AIR	
14	-46.880000	3.000000	8.190000	LZ_BF24	
15	-113.240000	158.064793 S	8.290000	AIR	
IMS	0.000000	0.000000	10.000000		

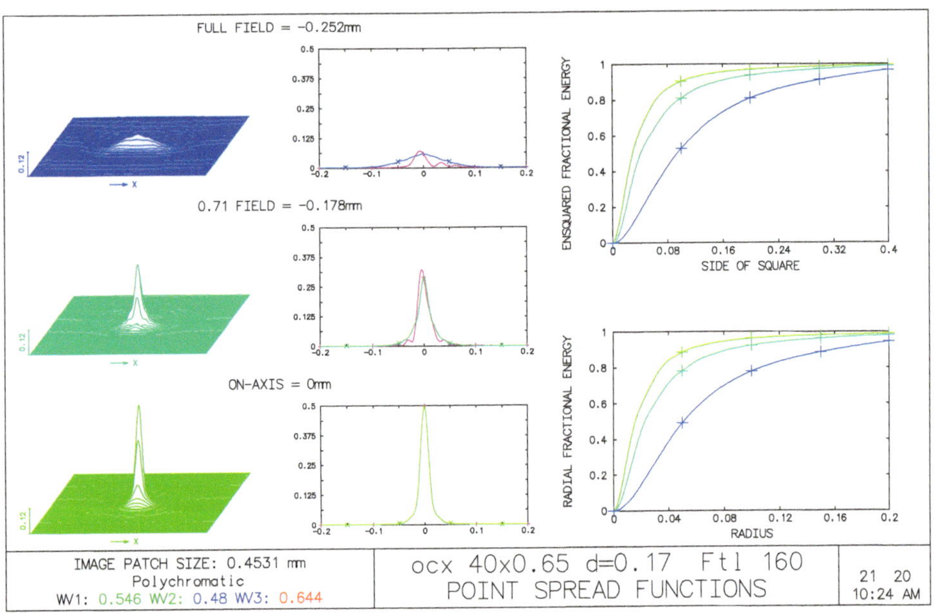

Figure 1.43 Optical design and graphs of aberration correction of a 40×/0.65 infinity LCF semi-plan achromatic objective ($F'_{ts} = 160$ mm).

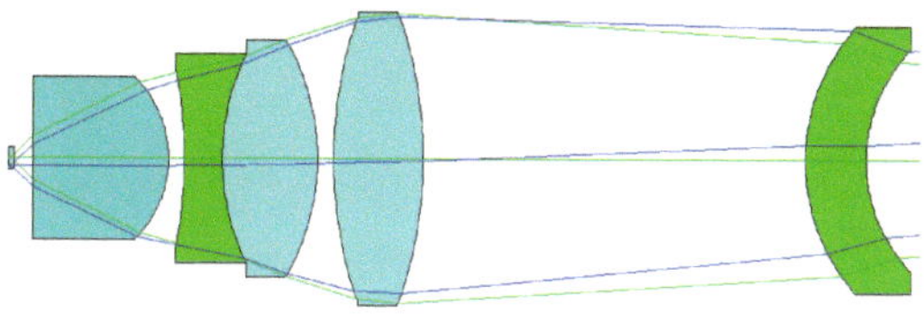

Lens: Plan 40x0.70 tub160 h45 Efl 4.49

Object num aper 0.700000 Image height 10.000000 Primary wavln 0.546070

SRF	RADIUS	THICKNESS	APERTURE RADIUS		GLASS		SPECIAL
OBJ	0.000000	0.010000	0.252959		AIR		
AST	0.000000	0.170000	0.260000	A	K9		
2	0.000000	0.599000	0.350000		AIR		
3	0.000000	4.300000	0.940000		H-ZK6		
4	-3.400000	0.500000	2.490000		AIR		
5	-14.054000	1.200000	2.750000		H-ZF6		
6	7.127000	3.000000	3.200000		H-QK3		
7	-7.127000	0.500000	3.640000		AIR		
8	13.090000	2.800000	4.370000		H-QK3	P	
9	-13.090000	12.000000	4.490000		AIR		
10	6.066000	1.900000	4.080000		H-ZF6	P	
11	4.800000	168.020000	3.410000		AIR		
IMS	0.000000	0.000000	10.000000				

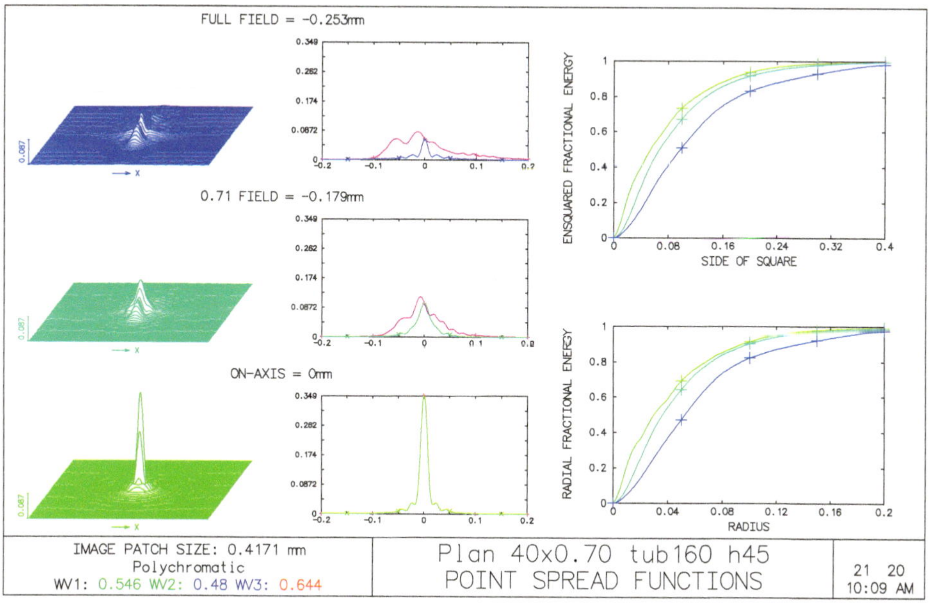

Figure 1.44 Optical design and graphs of aberration correction of a 40×/0.70 finite LCF plan achromatic objective.

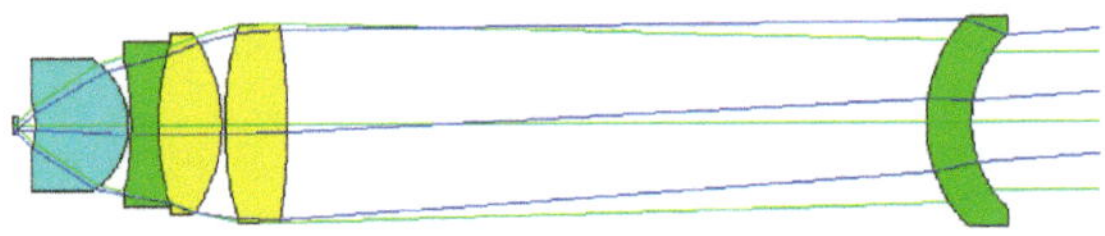

Lens: Plan 40x0.70 d=0.17 f180

Object num aper 0.700000 Image height 11.000000 Primary wavln 0.546070

SRF	RADIUS	THICKNESS	APERTURE RADIUS	GLASS	SPECIAL
OBJ	0.000000	0.010000	0.275281	AIR	
AST	0.000000	0.170000	0.285083 AS	H-K9	
2	0.000000	0.853500	0.370000	AIR	
3	-3.000000	4.420000	1.030000	H-LAK3	
4	-3.500000	0.150000	2.980000	AIR	
5	-15.076000	1.300000	3.280000	H-ZF7LA	
6	12.840000	3.000000	3.760000	H-FK61	
7	-7.000000	0.150000	4.130000	AIR	
8	15.489000	3.000000	4.510000	H-FK61 P	
9	-28.278000	30.000000	4.550000	AIR	
10	6.792000	2.100000	4.790000	H-ZF7LA P	
11	5.536000	180.000000	4.040000	AIR	
12	241.650000	5.630000	15.749988	LZ_BK8	
13	-52.860000	0.340000	15.842065	AIR	
14	-52.740000	3.380000	15.794750	LZ_BF24	
15	-127.400000	177.592047 S	15.902217	AIR	
IMS	0.000000	0.000000	11.000000 S		

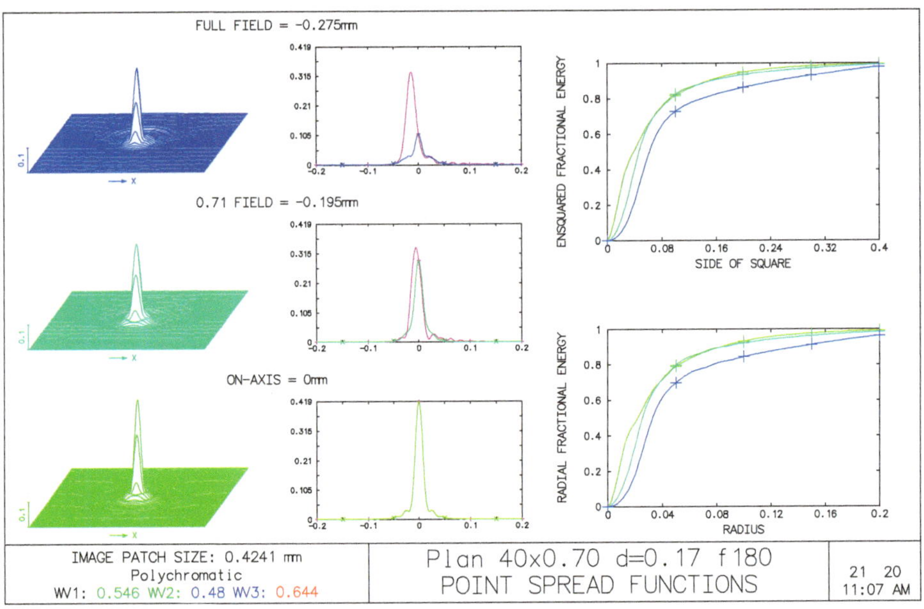

Figure 1.45 Optical design and graphs of aberration correction of a 40×/0.70 infinity LCF plan achromatic objective ($F'_{ts} = 180$ mm).

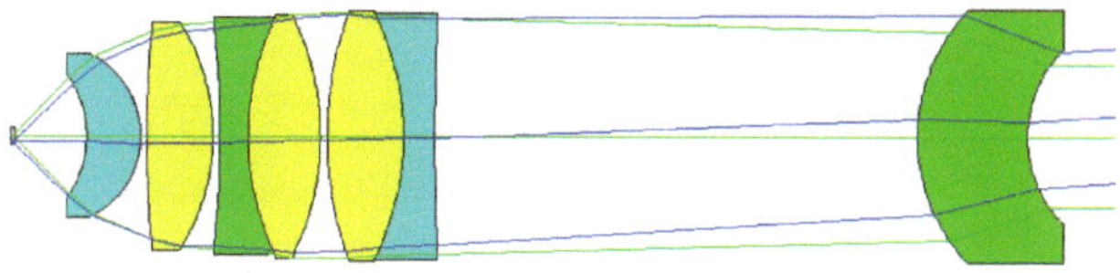

Lens: LD plan 40x0.70 d=0.17 ftl180

Object num aper 0.700000	Image height 11.000000	Primary wavln 0.546070			
SRF	**RADIUS**	**THICKNESS**	**APERTURE RADIUS**	**GLASS**	**SPECIAL**
OBJ	0.000000	0.010000	0.276361	AIR	
AST	0.000000	0.170000	0.280000 A	H-K9	
2	0.000000	3.300000	0.370000	AIR	
3	-4.650000	2.400000	2.740000	LZ_CTK12	
4	-4.000000	0.300000	3.610000	AIR	
5	54.700000	3.000000	4.740000	H-FK61	
6	-9.120000	0.300000	5.030000	AIR	
7	-49.660000	1.300000	5.070000	H-ZF5	
8	11.760000	3.300000	5.260000	H-FK61	
9	-11.760000	0.300000	5.400000	AIR	
10	16.200000	3.500000	5.540000	H-FK61 P	
11	-11.760000	1.400000	5.440000	LZ_CTK12 P	
12	144.660000	22.000000	5.400000	AIR	
13	7.960000	5.000000	5.590000	H-ZF7LA	
14	5.297000	180.000000	3.840000	AIR	
15	241.650000	5.630000	15.749988	LZ_BK8	
16	-52.860000	0.340000	15.842065	AIR	
17	-52.740000	3.380000	15.794750	LZ_BF24	
18	-127.400000	177.518943 S	15.902217	AIR	
IMS	0.000000	0.000000	11.000000 S		

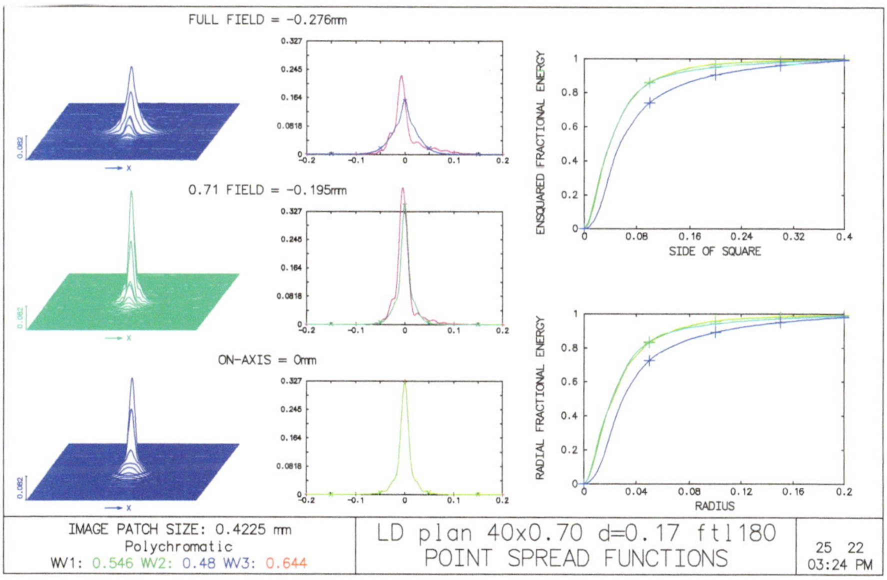

Figure 1.46 Optical design and graphs of aberration correction of a 40×/0.70 infinity LCF LD plan achromatic objective ($F'_{ts} = 180$ mm).

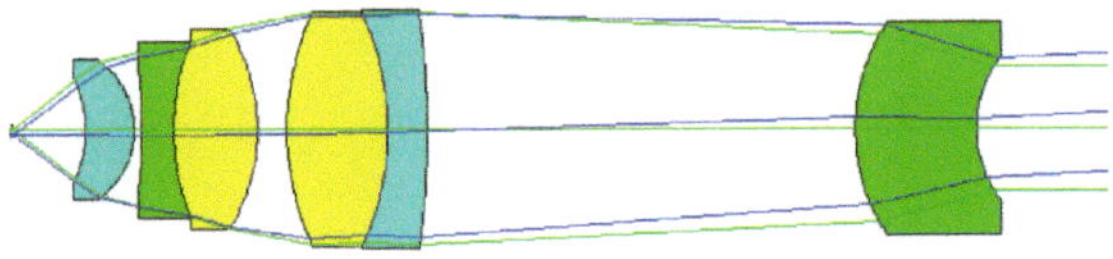

Lens: LD Plan 50x0.60 d0 ft1200

Object num aper	0.600000	Image height	10.000000	Primary wavln	0.546070

SRF	RADIUS	THICKNESS		APERTURE RADIUS		GLASS		SPECIAL
OBJ	0.000000	1.0000e-20		0.199864		AIR		
AST	0.000000	3.000000		0.199864	AS	AIR		
2	-4.200000	1.950000		2.050000		H-LAK61		
3	-3.200000	0.300000		2.710000		AIR		
4	-32.190000	1.300000		3.070000		H-ZF7LA		
5	9.370000	3.300000		3.410000		H-FK61		
6	-7.400000	1.200000		3.880000		AIR		
7	10.411100	4.000000		4.560000		H-FK61	P	
8	-10.411100	1.500000		4.510000		H-LAK61	P	
9	-32.520000	17.000000		4.610000		AIR		
10	7.350000	4.800000		4.140000		H-ZF7LA	P	
11	4.500000	200.000000		2.770000		AIR		
12	268.500000	6.250000		12.600000		LZ_BK8		
13	-58.738000	0.375000		12.670000		AIR		
14	-58.600000	3.750000		12.630000		LZ_BF24		
15	-141.550000	197.455075	S	12.750000		AIR		
IMS	0.000000	0.000000		10.000000	S			

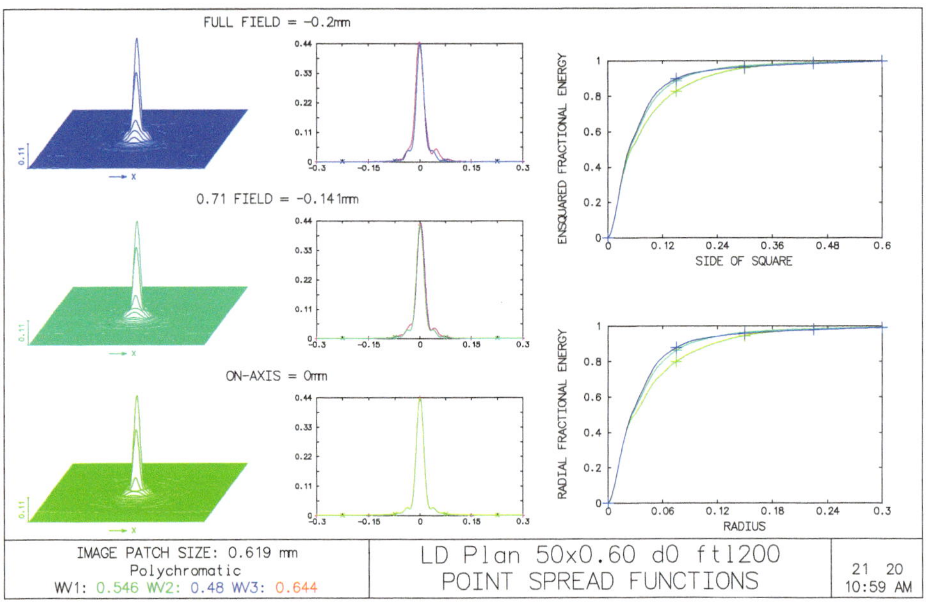

Figure 1.47 Optical design and graphs of aberration correction of a 50×/0.60 infinity LCF LD plan achromatic objective ($F'_{ts} = 200$ mm).

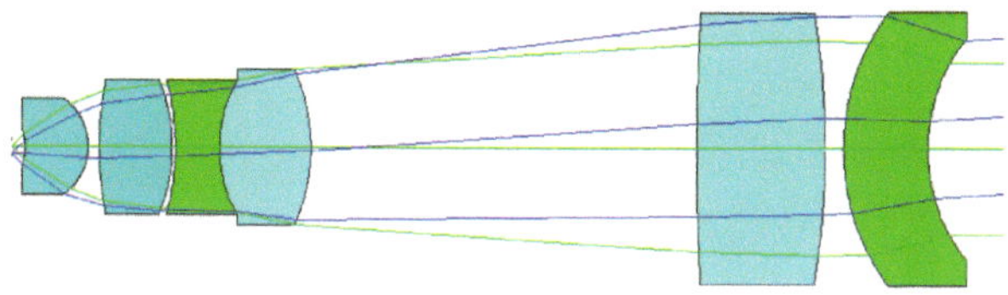

Lens: plan 50x0.70 d=0 Ftl 160

Object num aper	0.700000	Image height	10.000000	Primary wavln	0.546070

SRF	RADIUS	THICKNESS	APERTURE RADIUS	GLASS	SPECIAL
OBJ	0.000000	1.0000e-20	0.200195	AIR	
AST	0.000000	0.381860	0.200195 AS	AIR	
2	-1.195300	1.720000	0.470000	LZ_CTK19	
3	-1.522500	0.300000	1.270000	AIR	
4	8.752300	2.000000	1.630000	LZ_CTK19 P	
5	-4.554500	0.100000	1.750000	AIR	
6	-6.384600	1.200000	1.720000	LZ_TF5	
7	3.570700	2.500000	1.770000	LZ_LK3	
8	-4.805800	10.500000	2.050000	AIR	
9	55.584700	3.500000	3.340000	LZ_LK3 P	
10	-22.810500	0.500000	3.540000	AIR	
11	5.815500	2.300000	3.570000	LZ_TF5 P	
12	4.511400	90.000000	2.910000	AIR	
13	214.800000	5.000000	8.580000	LZ_BK8	
14	-46.990000	0.300000	8.680000	AIR	
15	-46.880000	3.000000	8.670000	LZ_BF24	
16	-113.240000	157.829763 S	8.790000	AIR	
IMS	0.000000	0.000000	10.000000		

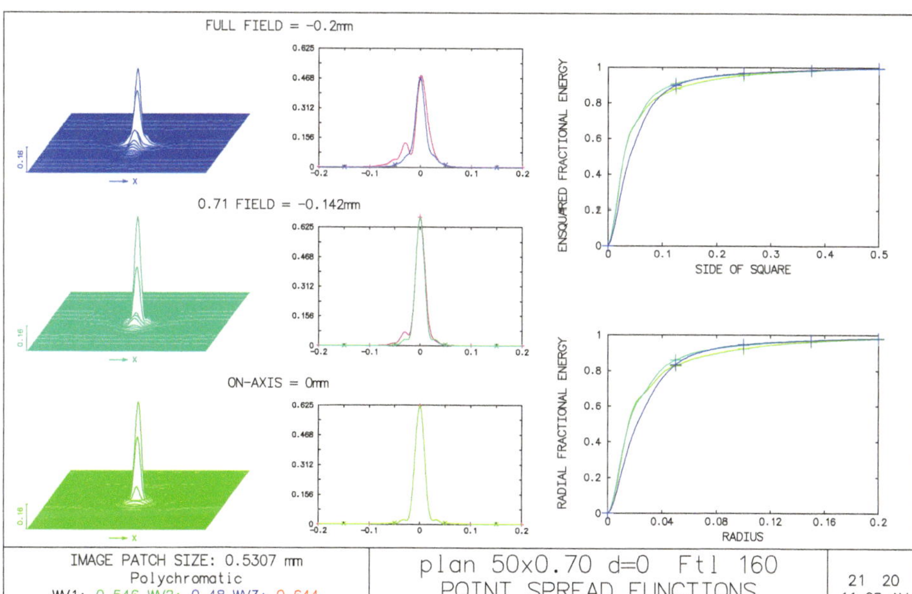

Figure 1.48 Optical design and graphs of aberration correction of a 50×/0.70 infinity LCF plan achromatic objective ($F'_{ts} = 160$ mm).

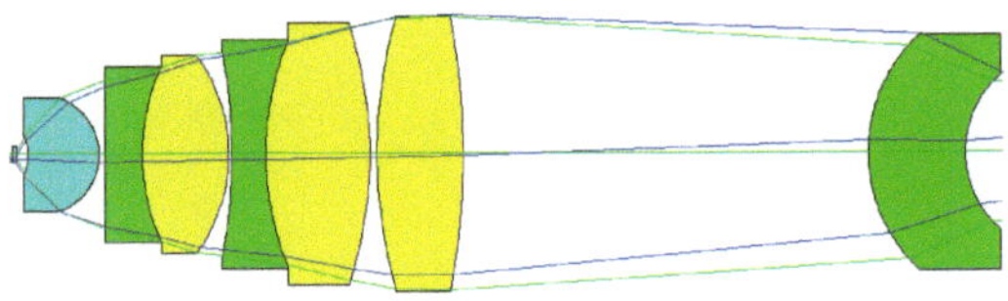

Lens: plan 60x0.85 d017 tub160 h45 Efl 2.79

Object num aper 0.850000 Image height 14.000000 Primary wavln 0.546070

SRF	RADIUS	THICKNESS	APERTURE RADIUS	GLASS	SPECIAL
OBJ	0.000000	0.010000	0.231813	AIR	
AST	0.000000	0.170000	0.240000 A	K9	
2	0.000000	0.399900	0.360000	AIR	
3	-1.506000	2.470000	0.710000	H-LAK3	
4	-2.001000	0.200000	1.920000	AIR	
5	0.000000	1.300000	2.410000	ZF1	
6	6.897000	3.000000	2.780000	H-FK61	
7	-5.390000	0.110000	3.190000	AIR	
8	-17.701000	1.200000	3.280000	ZF3	
9	9.600000	3.600000	3.620000	H-FK61 P	
10	-13.305000	0.240000	4.200000	AIR	
11	15.672000	3.000000	4.560000	H-FK61 P	
12	-24.210000	14.000000	4.640000	AIR	
13	5.209000	3.400000	4.120000	H-ZF6	
14	3.307000	161.870000	2.720000	AIR	
IMS	0.000000	0.000000	14.000000		

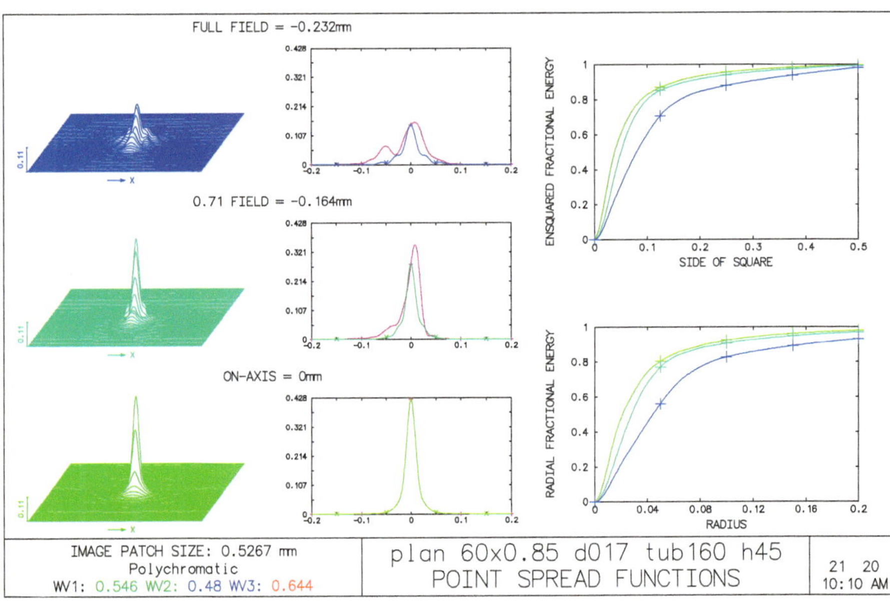

Figure 1.49 Optical design and graphs of aberration correction of a 60×/0.85 finite LCF plan achromatic objective.

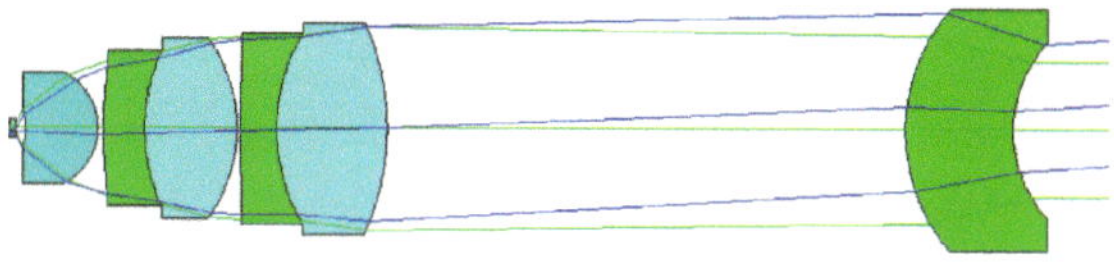

Lens: ocx 63x0.85 d=0.17 Ftl 160

Object num aper 0.850000 Image height 10.000000 Primary wavln 0.546070

SRF	RADIUS	THICKNESS	APERTURE RADIUS	GLASS	SPECIAL
OBJ	0.000000	0.010000	0.158977	AIR	
AST	0.000000	0.170000	0.170000 A	LZ_K14	
2	0.000000	0.350000	0.290000	AIR	
3	-2.377000	2.330000	0.690000	LZ_CTK19	
4	-1.977000	0.200000	1.770000	AIR	
5	25.700000	1.300000	2.210000	LZ_TF3	
6	5.834000	3.000000	2.450000	LZ_LK3	
7	-4.786000	0.100000	2.850000	AIR	
8	101.160000	1.200000	2.940000	LZ_TF3 P	
9	5.834000	3.600000	3.040000	LZ_LK3 P	
10	-8.166000	16.900000	3.360000	AIR	
11	5.834000	3.440000	3.840000	LZ_TF5	
12	4.130000	90.000000	2.790000	AIR	
13	214.800000	5.000000	8.160000	LZ_BK8	
14	-46.990000	0.300000	8.260000	AIR	
15	-46.880000	3.000000	8.240000	LZ_BF24	
16	-113.240000	157.903083 S	8.360000	AIR	
IMS	0.000000	0.000000	10.000000		

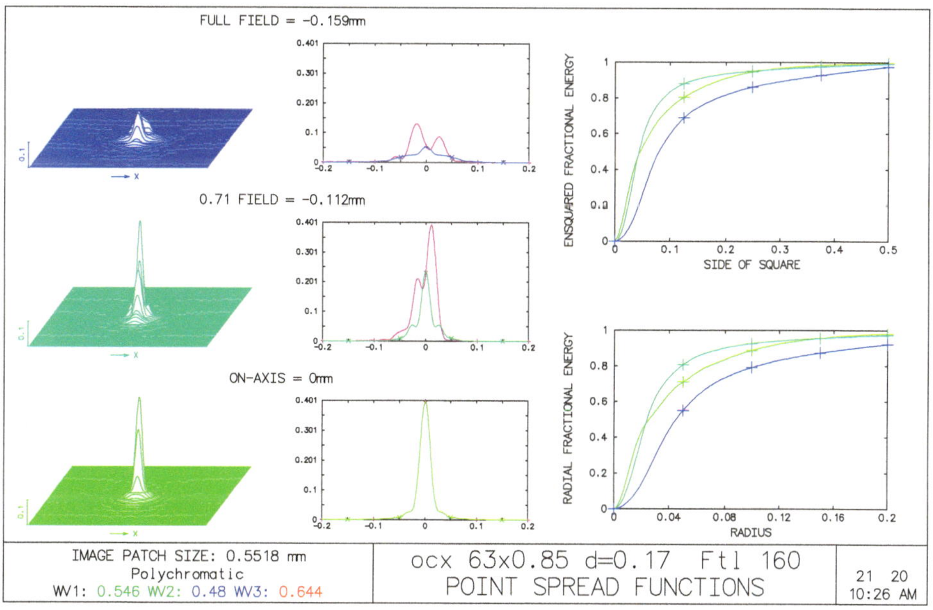

Figure 1.50 Optical design and graphs of aberration correction of a 63×/0.85 infinity LCF semi–plan achromatic objective ($F'_{ts} = 160$ mm).

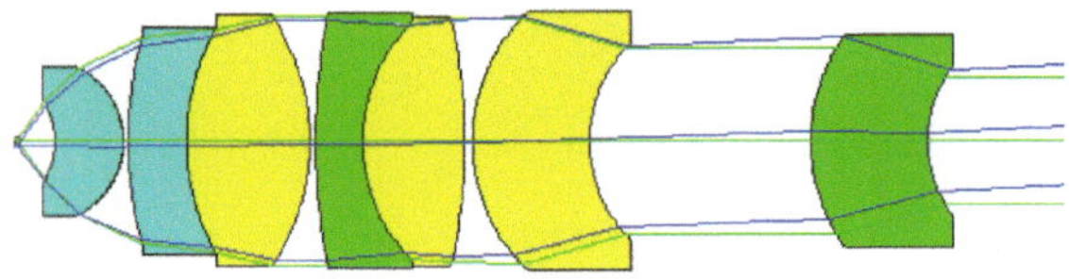

Lens: LD Plan 100x0.80 d0 ft1200

Object num aper	0.800000	Image height	10.000000	Primary wavln	0.546070

SRF	RADIUS	THICKNESS	APERTURE RADIUS	GLASS	SPECIAL
OBJ	0.000000	1.0000e-20	0.099999	AIR	
AST	0.000000	1.000000	0.099999 AS	AIR	
2	-1.900000	1.800000	1.030000	H-LAK3	
3	-2.000000	0.130000	1.860000	AIR	
4	9.546000	1.500000	2.610000	LZ_CTK12	
5	5.495000	3.200000	2.820000	H-FK61	
6	-5.495000	0.130000	3.130000	AIR	
7	15.000000	1.200000	3.170000	H-ZF7LA	
8	4.200000	2.600000	3.000000	H-FK61 P	
9	-13.270000	0.200000	3.110000	AIR	
10	4.368000	3.000000	3.220000	H-FK61 P	
11	3.370000	5.700000	2.450000	AIR	
12	4.368000	3.000000	2.670000	H-ZF7LA P	
13	3.000000	200.000000	1.850000	AIR	
14	268.500000	6.250000	12.400000	LZ_BK8	
15	-58.738000	0.375000	12.470000	AIR	
16	-58.600000	3.750000	12.440000	LZ_BF24	
17	-141.550000	197.282718 S	12.560000	AIR	
IMS	0.000000	0.000000	10.000000 S		

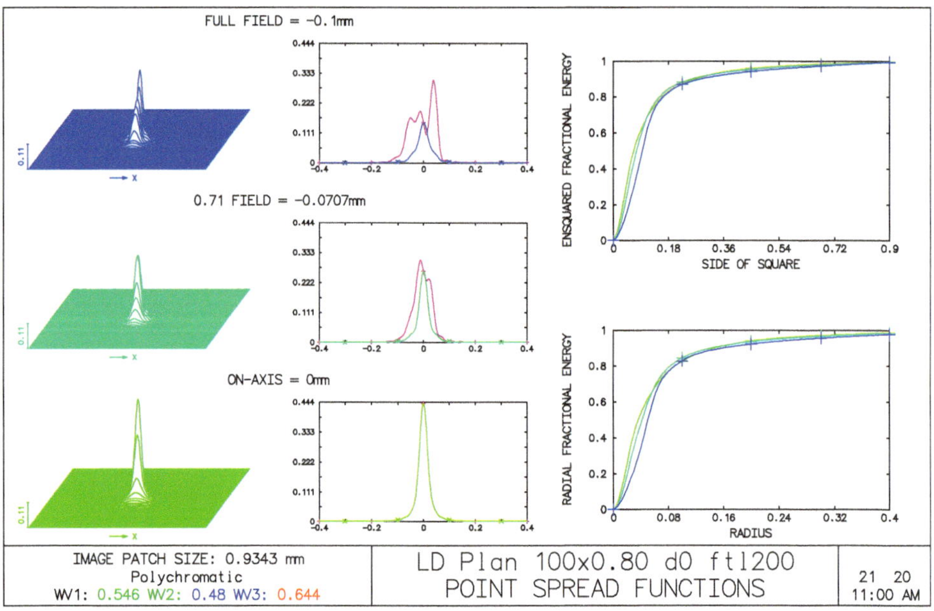

Figure 1.51 Optical design and graphs of aberration correction of a 100×/0.80 infinity LD plan achromatic objective ($F'_{ts} = 200$ mm).

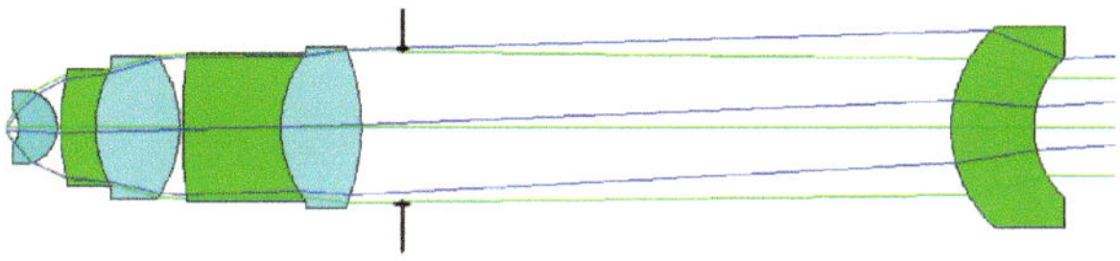

Lens: Plan 100x0.85 d=0 f160

Object num aper 0.850000 Image height 10.000000 Primary wavln 0.546070

SRF	RADIUS	THICKNESS	APERTURE RADIUS	GLASS	SPECIAL
OBJ	0.000000	1.0000e-20	0.099517	AIR	
AST	0.000000	0.391000	0.100000 A	AIR	
2	-0.831800	1.100000	0.480000	LZ_CTK19	
3	-1.056800	0.100000	1.020000	AIR	
4	5.598000	1.000000	1.500000	LZ_TF5	
5	3.311000	2.400000	1.630000	LZ_LK3	
6	-3.311000	0.100000	1.980000	AIR	
7	18.880000	2.790000	2.050000	LZ_TF5 P	
8	3.311000	2.300000	2.050000	LZ_LK3 P	
9	-5.728000	16.760000	2.230000	AIR	
10	3.767000	2.330000	2.780000	LZ_TF5 P	
11	2.630000	90.000000	1.960000	AIR	
12	214.800000	5.000000	7.040000	LZ_BK8	
13	-46.990000	0.300000	7.150000	AIR	
14	-46.880000	3.000000	7.130000	LZ_BF24	
15	-113.240000	157.820984 S	7.240000	AIR	
IMS	0.000000	0.000000	10.000000		

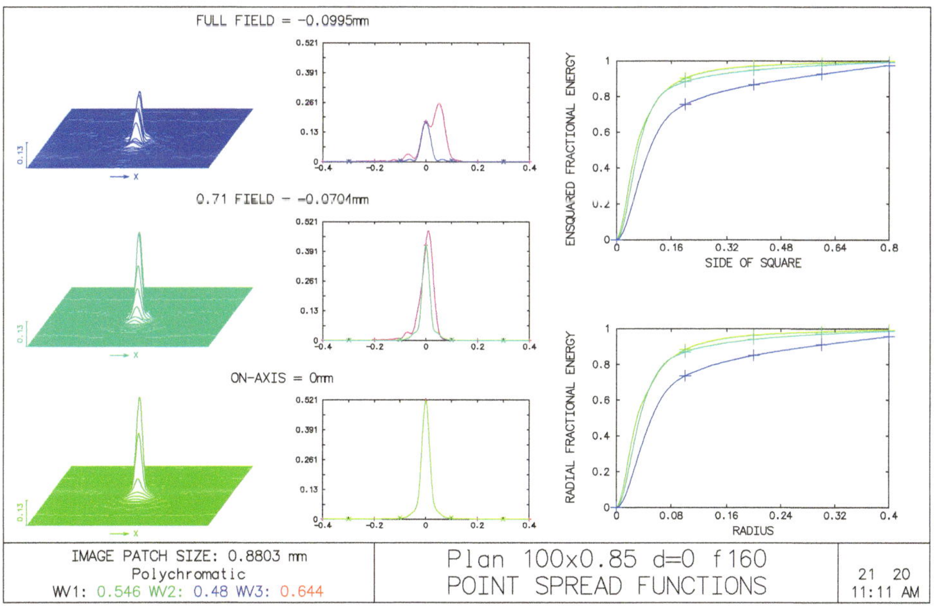

Figure 1.52 Optical design and graphs of aberration correction of a 100×/0.85 infinity LCF plan achromatic objective ($F'_{ts} = 160$ mm).

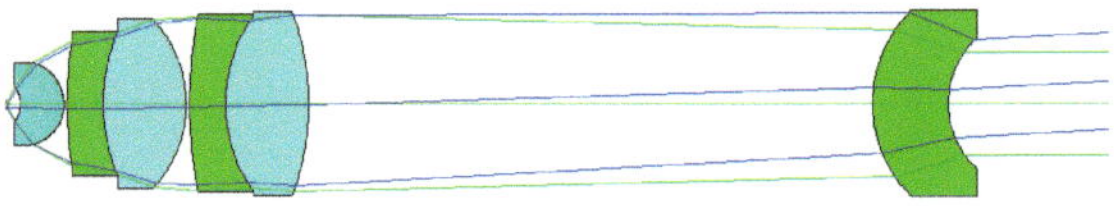

Lens: Plan 100x0.90 d=0 ftl200

Object num aper	0.900000	Image height	10.000000	Primary wavln	0.546070

SRF	RADIUS	THICKNESS	APERTURE RADIUS	GLASS	SPECIAL
OBJ	0.000000	1.0000e-20	0.099939	AIR	
AST	0.000000	0.499890	0.099939 AS	AIR	
2	-1.000000	1.600000	0.640000	H-LAK3	
3	-1.450000	0.120000	1.440000	AIR	
4	13.550000	1.300000	2.190000	H-ZF7LA	
5	6.169000	3.000000	2.490000	H-QK3	
6	-4.140000	0.120000	2.930000	AIR	
7	16.762000	1.300000	3.080000	H-ZF7LA P	
8	5.000000	3.000000	3.020000	H-QK3 P	
9	-8.750000	20.000000	3.190000	AIR	
10	4.400000	2.700000	3.240000	H-ZF6	
11	3.000000	200.000000	2.270000	AIR	
12	268.500000	6.250000	12.400000	LZ_BK8	
13	-58.738000	0.375000	12.470000	AIR	
14	-58.600000	3.750000	12.440000	LZ_BF24	
15	-141.550000	197.327025 S	12.560000	AIR	
IMS	0.000000	0.000000	10.000000 S		

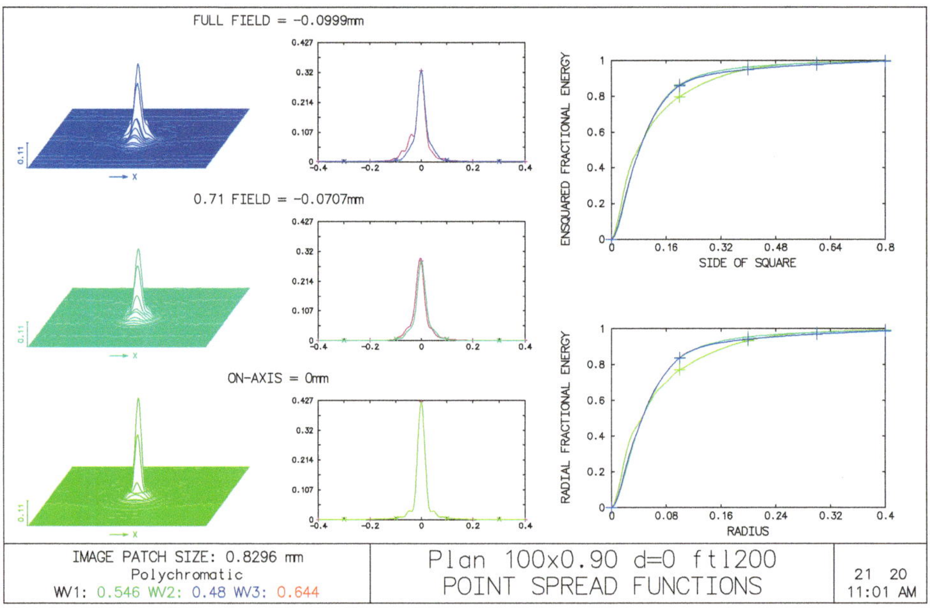

Figure 1.53 Optical design and graphs of aberration correction of a 100×/0.90 infinity LCF plan achromatic objective ($F'_{ts} = 200$ mm).

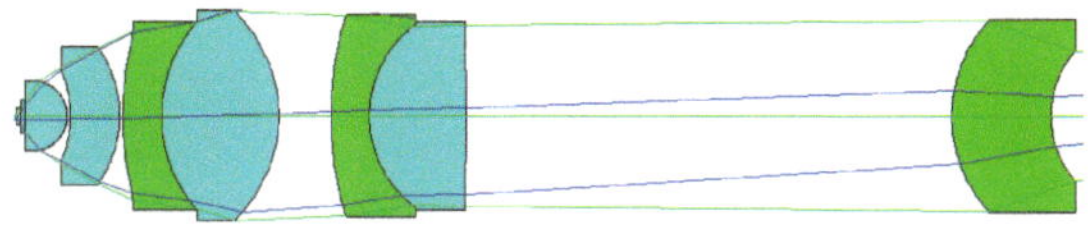

Lens: ocx 100x1.20water d017 tub160 Efl 1.78

Object num aper 1.200000 Image height 9.000000 Primary wavln 0.546070

SRF	RADIUS	THICKNESS	APERTURE RADIUS	GLASS	SPECIAL
OBJ	0.000000	0.010000	0.091724	WATER	
AST	0.000000	0.170000	0.100000 A	K9	
2	0.000000	0.121200	0.260000	WATER P	
3	0.000000	1.300000	0.500000	H-ZK1	
4	-1.090000	0.100000	1.060000	AIR	
5	-3.600000	1.500000	1.450000	LZ_CTK12	
6	-3.600000	0.100000	2.090000	AIR	
7	10.351000	1.200000	2.590000	H-ZF2	
8	4.338000	3.600000	2.850000	H-QK3	
9	-4.338000	1.600000	3.190000	AIR	
10	10.351000	1.200000	3.070000	H-ZF3	
11	3.448000	3.000000	2.780000	H-QK3 P	
12	-100.350000	15.000000	2.840000	AIR	
13	4.322000	3.000000	2.920000	H-ZF6	
14	3.000000	162.995758 S	2.000000	AIR	
IMS	0.000000	0.000000	9.000000		

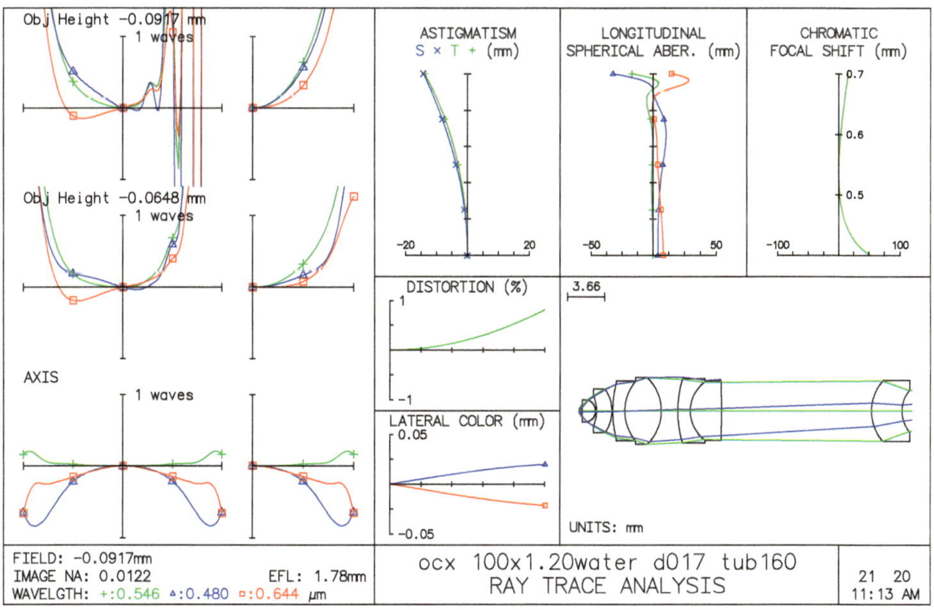

Figure 1.54 Optical design and graphs of aberration correction of a 100×/1.20 water immersion finite LCF semi–plan achromatic objective.

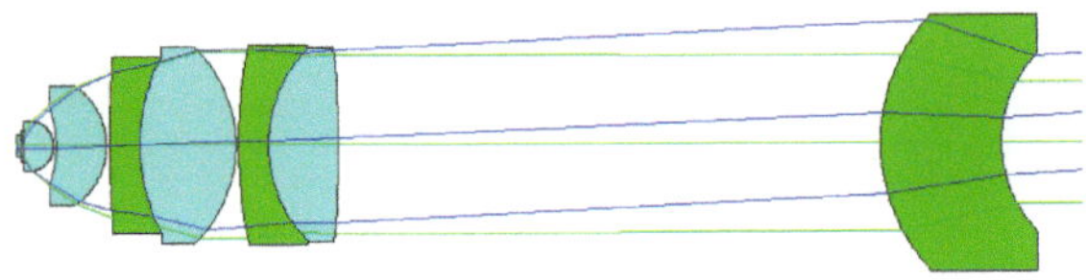

Lens: ocx 100x1.25oil ftl160

Object num aper	1.250000	Image height	10.000000	Primary wavln	0.546070

SRF	RADIUS	THICKNESS	APERTURE RADIUS	GLASS	SPECIAL
OBJ	0.000000	0.009960	0.099874	OIL	
AST	0.000000	0.170000	0.110000 A	LZ_K14	
2	0.000000	0.100000	0.360000	OIL	
3	0.000000	1.000000	0.510000	LZ_K8	
4	-0.831800	0.100000	0.830000	AIR	
5	-3.631000	1.700000	1.300000	LZ_LK3	
6	-2.291000	0.100000	1.930000	AIR	
7	37.240000	1.000000	2.470000	LZ_TF1	
8	5.943000	3.200000	2.840000	LZ_LK3 P	
9	-4.365000	0.100000	3.190000	AIR	
10	16.069000	1.000000	3.230000	LZ_TF3	
11	4.365000	2.300000	3.090000	LZ_LK3 P	
12	-29.310000	17.900000	3.140000	AIR	
13	5.943000	4.000000	4.130000	LZ_TF5	
14	4.130000	90.000000	2.880000	AIR	
15	214.800000	5.000000	8.100000	LZ_BK8	
16	-46.990000	0.300000	8.200000	AIR	
17	-46.880000	3.000000	8.190000	LZ_BF24	
18	-113.240000	157.550529 S	8.300000	AIR	
IMS	0.000000	0.000000	10.000000		

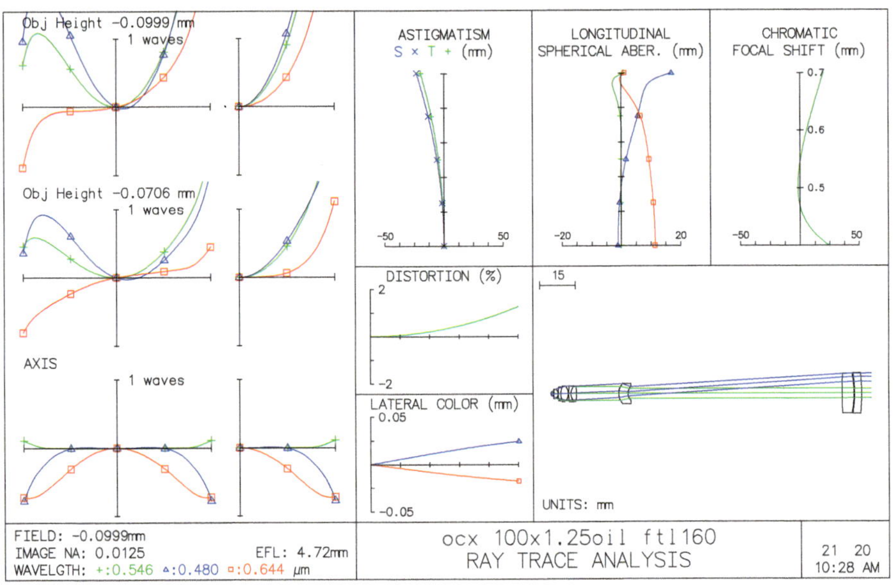

Figure 1.55 Optical design and graphs of aberration correction of a 100×/1.25 oil immersion infinity LCF semi-plan achromatic objective ($F'_{ts} = 160$ mm).

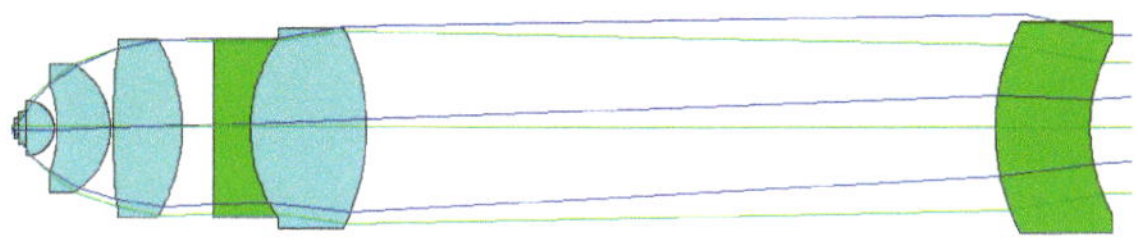

Lens: ocx 100x1.30 oil tub160 h45 — Efl 1.77

Object num aper 1.300000 — Image height 9.000000 — Primary wavln 0.546070

SRF	RADIUS	THICKNESS	APERTURE RADIUS	GLASS	SPECIAL
OBJ	0.000000	0.010000	0.089093	OIL	
AST	0.000000	0.170000	0.110000 A	K9	
2	0.000000	0.211000	0.310000	OIL P	
3	0.000000	0.910000	0.510000	K9 P	
4	-0.850000	0.100000	0.850000	AIR	
5	-4.120000	1.700000	1.430000	K9 P	
6	-2.290000	0.100000	1.990000	AIR	
7	21.180000	2.200000	2.500000	H-QK3	
8	-5.297000	1.000000	2.760000	AIR	
9	143.700000	1.200000	2.780000	H-ZF7LA	
10	4.800000	3.700000	2.780000	H-QK3 P	
11	-7.000000	20.000000	3.110000	AIR	
12	7.000000	3.000000	3.280000	H-ZF6	
13	5.500000	161.455160 S	2.680000	AIR	
IMS	0.000000	0.000000	9.000000		

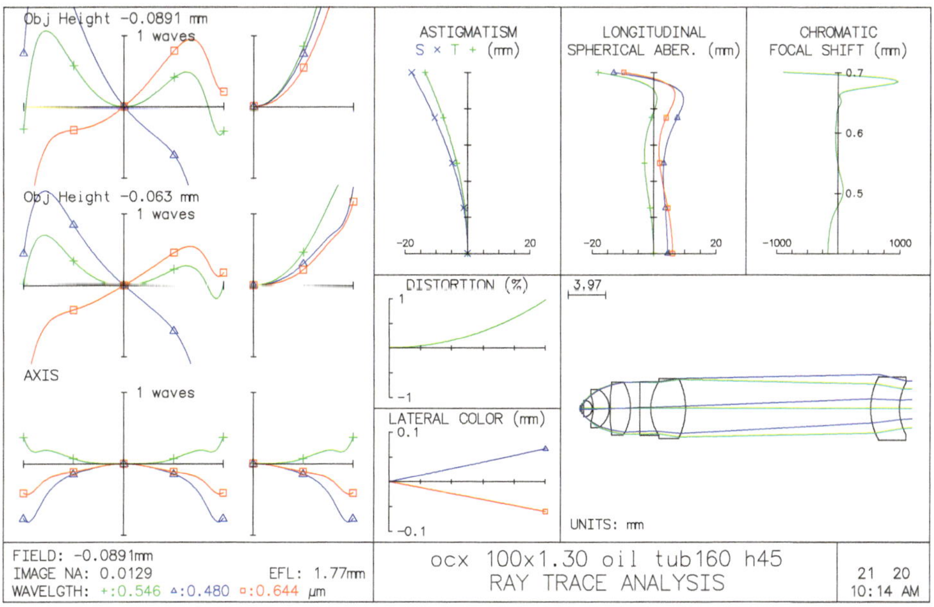

Figure 1.56 Optical design and graphs of aberration correction of a 100×/1.30 oil immersion finite LCR semi–plan achromatic objective.

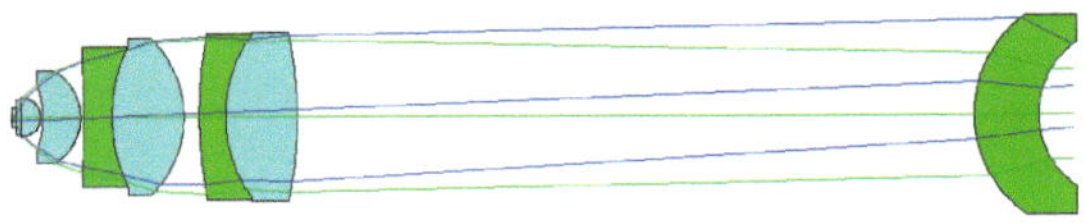

Lens: S-Plan 100x1.30 oil tub160 h45				Efl	1.69
Object num aper	1.300000	Image height	10.000000	Primary wavln	0.546070
SRF	**RADIUS**	**THICKNESS**	**APERTURE RADIUS**	**GLASS**	**SPECIAL**
OBJ	0.000000	0.010000	0.100521	OIL	
AST	0.000000	0.170000	0.110000 A	K9	
2	0.000000	0.210600	0.400000	OIL P	
3	0.000000	0.910000	0.750000	K9 P	
4	-0.850000	0.100000	0.850000	AIR	
5	-4.120000	1.700000	1.440000	K9 P	
6	-2.290000	0.120000	2.010000	AIR	
7	-50.920000	1.300000	2.510000	ZF1	
8	7.000000	3.200000	3.050000	H-QK3	
9	-4.600000	0.700000	3.410000	AIR	
10	18.710000	1.200000	3.580000	ZF6	
11	5.750000	3.200000	3.480000	H-QK3 P	
12	-17.320000	30.000000	3.620000	AIR	
13	5.160000	2.880000	4.310000	H-ZF6	
14	3.746000	149.221753 S	3.120000	AIR	
IMS	0.000000	0.000000	10.000000		

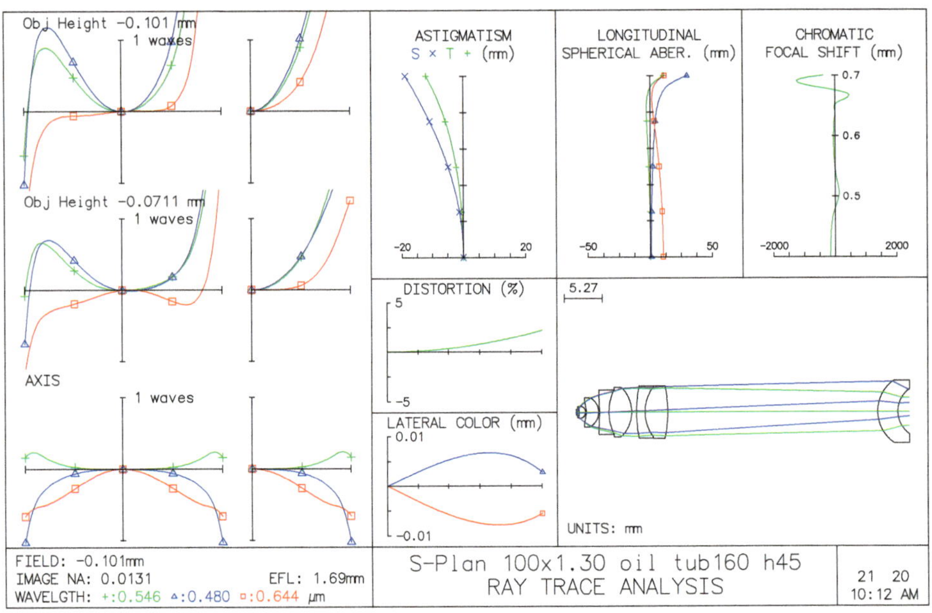

Figure 1.57 Optical design and graphs of aberration correction of a 100×/1.30 oil immersion finite LCF semi–plan achromatic objective.

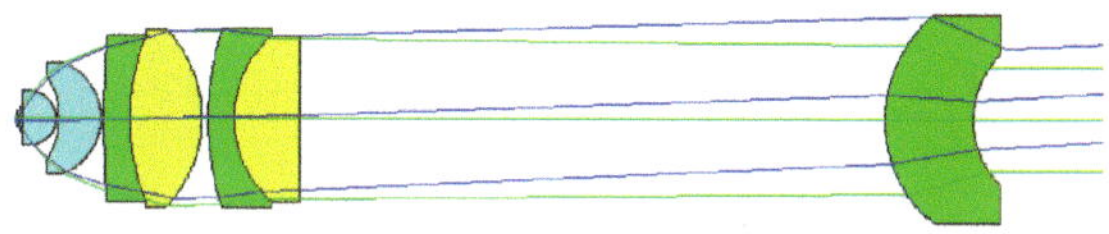

Lens: S_Plan 100x1.30 oil ftl180

Object num aper 1.300000 Image height 10.000000 Primary wavln 0.546070

SRF	RADIUS	THICKNESS	APERTURE RADIUS	GLASS	SPECIAL
OBJ	0.000000	0.010000	0.100343	OIL	
AST	0.000000	0.170000	0.116948 AS	K9	
2	0.000000	0.132300	0.350000	OIL	
3	0.000000	1.500000	0.570000	K9 P	
4	-1.220000	0.130000	1.220000	AIR	
5	-3.597000	1.900000	1.890000	K9 P	
6	-2.675000	0.140000	2.500000	AIR	
7	39.990000	1.200000	3.360000	ZF3	
8	9.817000	3.300000	3.710000	H-FK61	
9	-5.702000	0.300000	3.990000	AIR	
10	12.942000	1.200000	4.010000	ZF3 P	
11	4.966000	3.000000	3.720000	H-FK61 P	
12	0.000000	27.000000	3.730000	AIR	
13	6.081000	4.000000	4.300000	H-ZF6	
14	4.177000	180.000000	2.970000	AIR	
15	241.650000	5.630000	12.000000	LZ_BK8	
16	-52.860000	0.340000	12.000000	AIR	
17	-52.740000	3.380000	12.000000	LZ_BF24	
18	-127.400000	177.554185 S	12.000000	AIR	
IMS	0.000000	0.000000	10.000000 S		

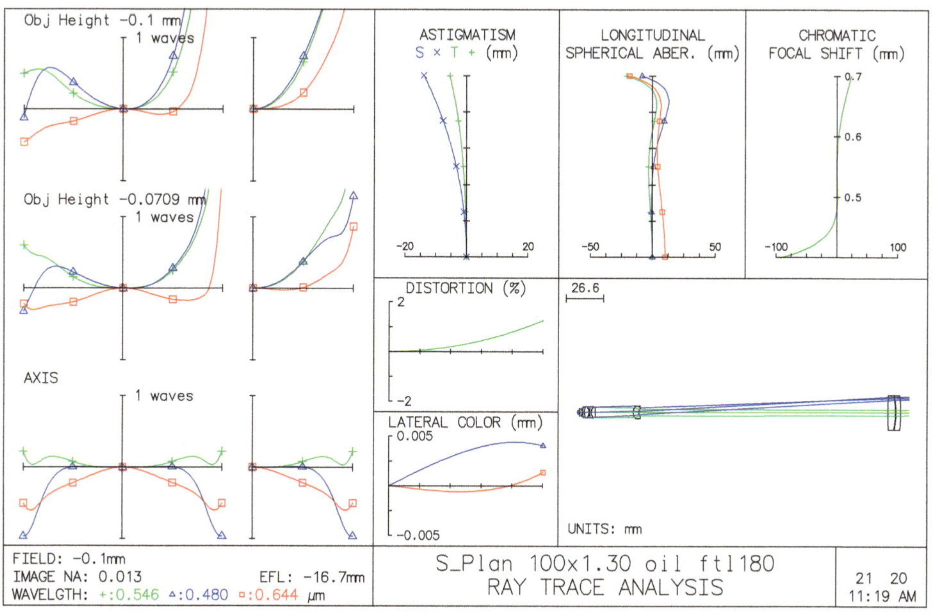

Figure 1.58 Optical design and graphs of aberration correction of a 100×/1.30 oil immersion infinity LCF semi–plan achromatic objective ($F'_{ts} = 180$ mm).

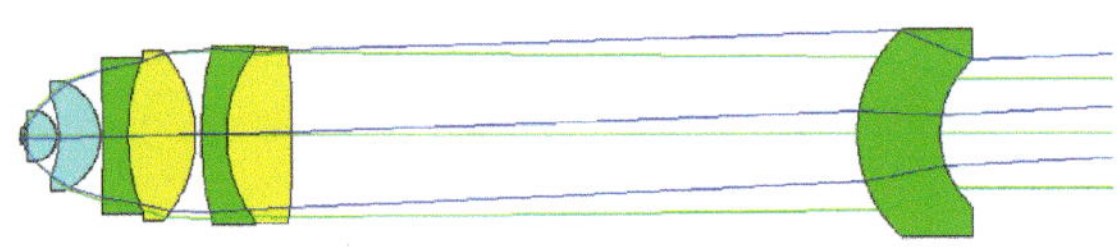

Lens: S_Plan 100x1.30 oil ftl200

Object num aper 1.300000 Image height 10.000000 Primary wavln 0.546070

SRF	RADIUS	THICKNESS	APERTURE RADIUS	GLASS	SPECIAL
OBJ	0.000000	0.010000	0.100024	OIL	
AST	0.000000	0.170000	0.116629 AS	K9	
2	0.000000	0.180160	0.390000	OIL	
3	0.000000	1.450000	0.610000	K9 P	
4	-1.200000	0.130000	1.200000	AIR	
5	-4.370000	2.000000	1.930000	K9 P	
6	-2.729000	0.130000	2.530000	AIR	
7	-102.330000	1.200000	3.200000	ZF3	
8	9.461000	3.300000	3.640000	H-FK61	
9	-5.747000	0.300000	3.970000	AIR	
10	16.982000	1.200000	4.150000	ZF3 P	
11	6.535000	3.200000	4.030000	H-FK61 P	
12	-47.986000	27.000000	4.100000	AIR	
13	6.535000	4.000000	4.850000	H-ZF6	
14	4.635000	200.000000	3.450000	AIR	
15	268.500000	6.250000	11.300000	LZ_BK8	
16	-58.738000	0.375000	11.380000	AIR	
17	-58.600000	3.750000	11.350000	LZ_BF24	
18	-141.550000	197.275739 S	11.470000	AIR	
IMS	0.000000	0.000000	10.000000 S		

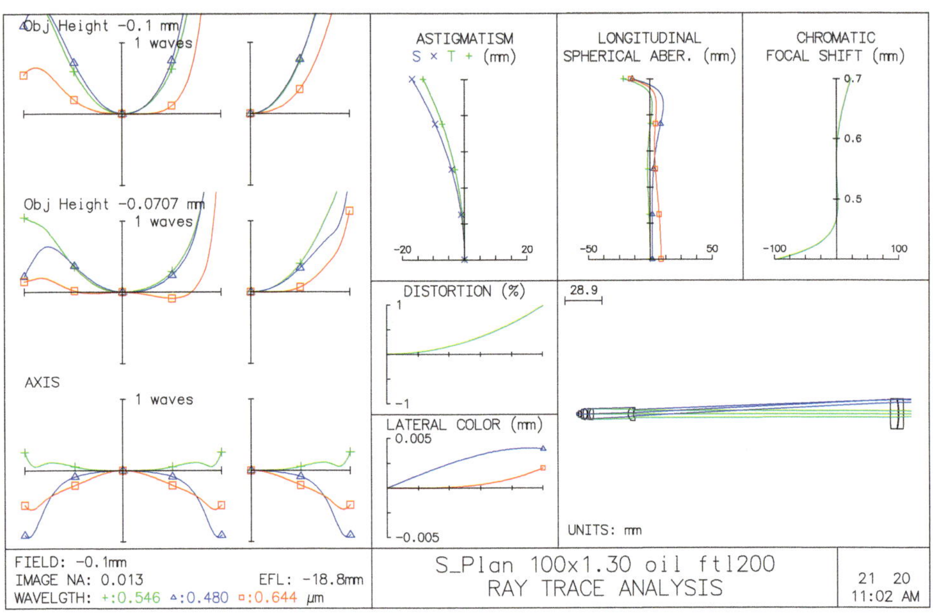

Figure 1.59 Optical design and graphs of aberration correction of a 100×/1.30 oil immersion infinity LCF semi–plan achromatic objective ($F'_{ts} = 200$ mm).

Figure 1.60 Photograph of a set of modern LCF plan achromatic objectives.

Table 1.10 Main technical parameters and basic optical layout of third-generation objectives.

Magnification	*NA*	*WD* (mm)	*F′* (mm)	*R* (μm)	*DF* (μm)	*FOV* on object (mm)	*FOV* on image (mm)	The principal optical layout	Optical design at figure no.
10×	0.30	11.2	15.05	1.12	3.73	2.0	20	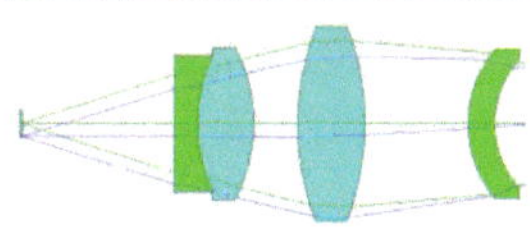	1.36
20×	0.65	0.51	8.37	0.52	0.79	1.4	28	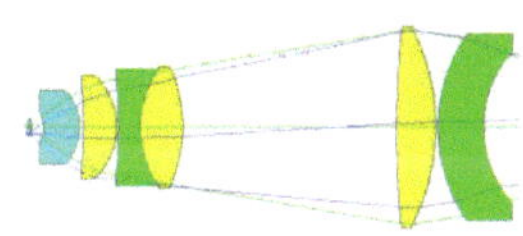	1.41
40×	0.70	0.45	4.49	0.48	0.69	0.5	20	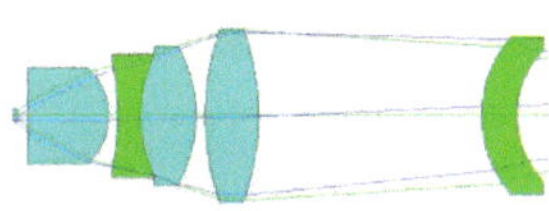	1.45
60×	0.85	0.18	2.79	0.38	0.43	0.47	28	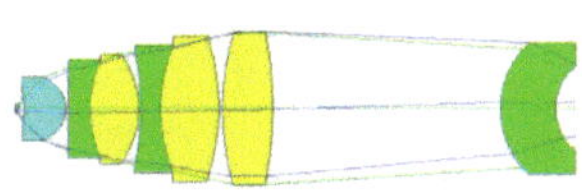	1.49
100×	1.3 oil	0.18	1.69	0.26	0.20	0.2	20	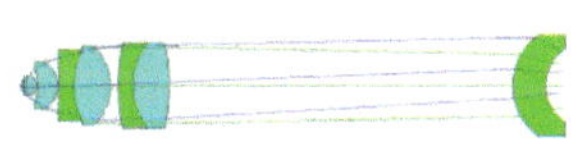	1.57

object 5. When working with objects placed in laboratory glassware, the wall height of which does not exceed 45 mm, a special lens 4 is introduced into the beam, which increases numerical aperture of the lighting system. The optical

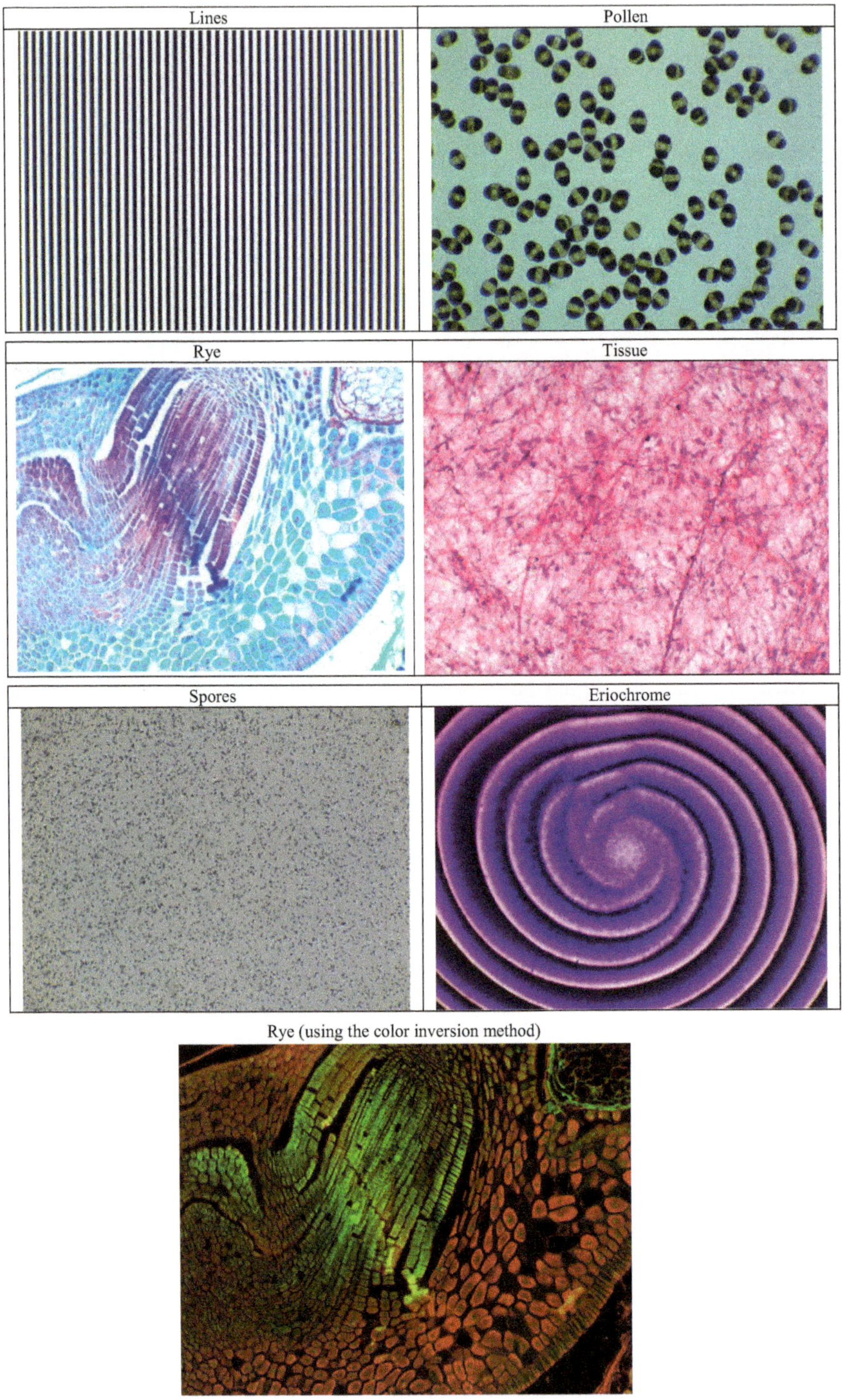

Figure 1.61 Photographs of real testing objects using a 10×/0.30 LCF plan achromatic objective.

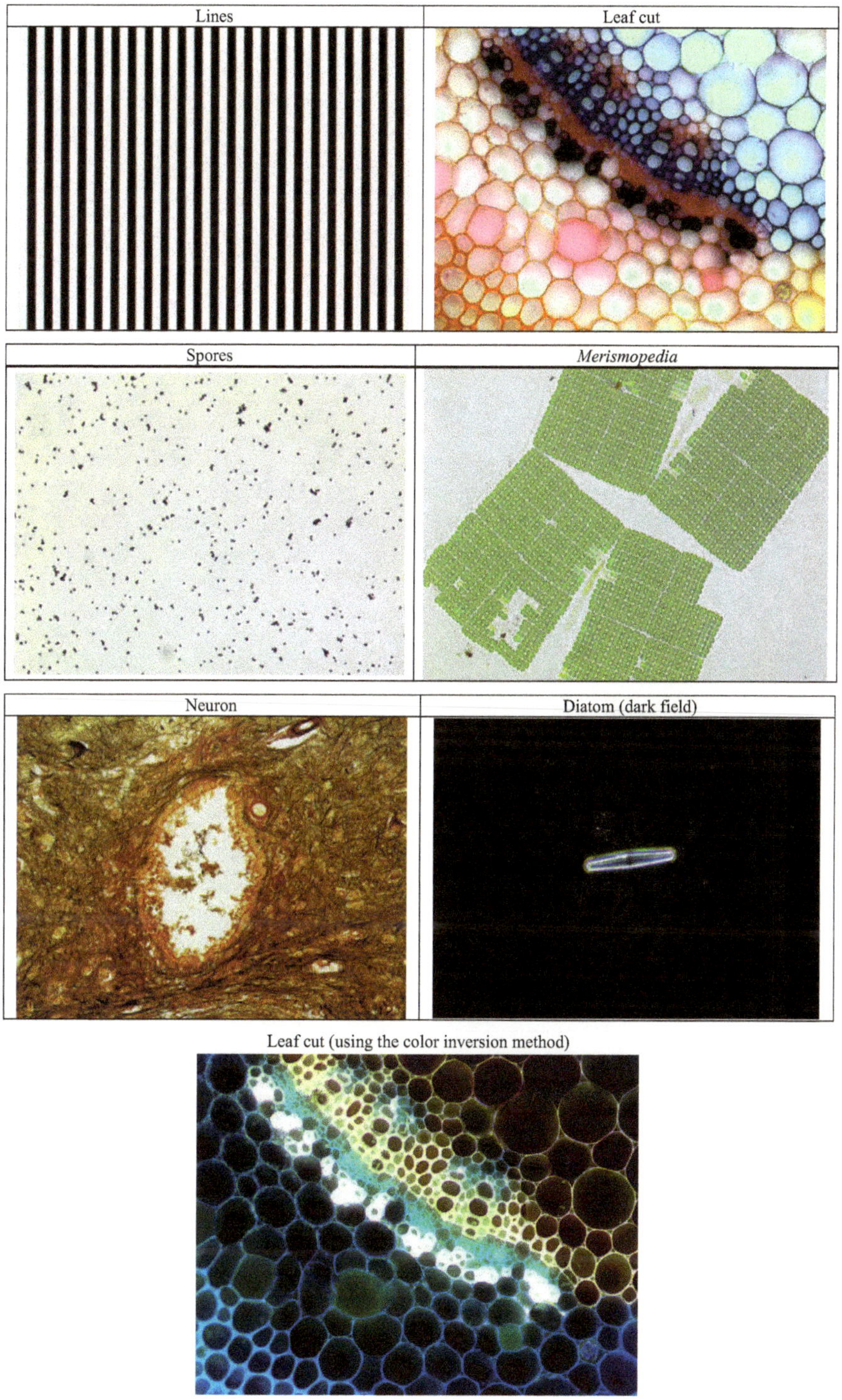

Figure 1.62 Photographs of real testing objects using a 20×/0.65 LCF plan achromatic objective.

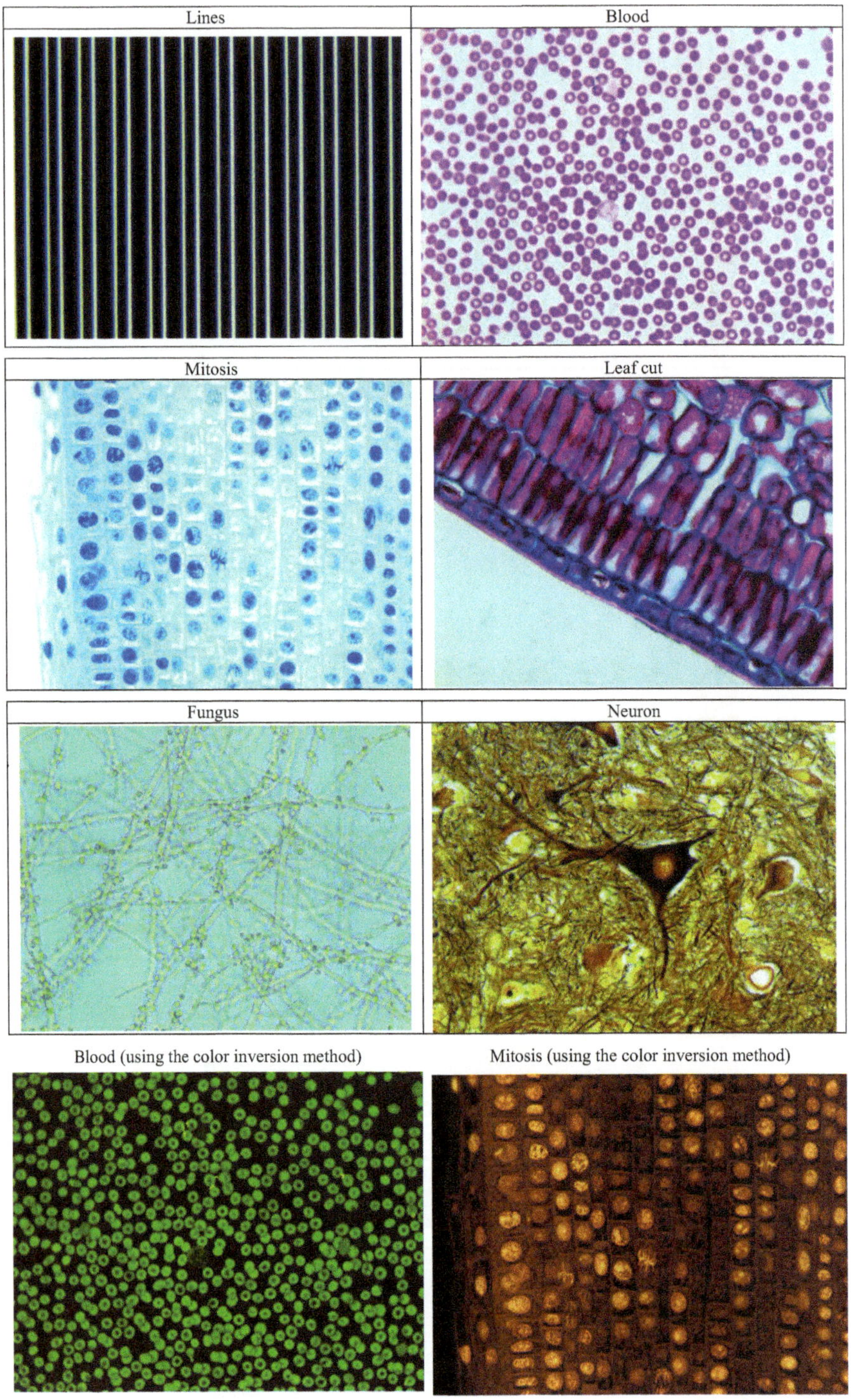

Figure 1.63 Photographs of real testing objects using a 40×/0.70 LCF plan achromatic objective.

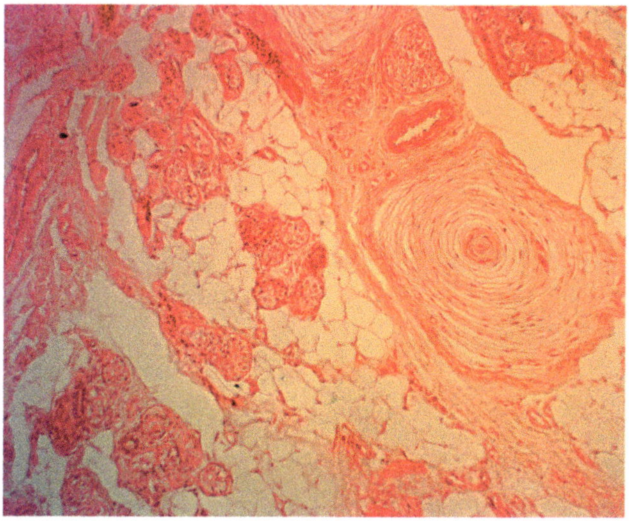

Figure 1.64 Photograph of a real object using a 10×/0.30 LCF plan achromatic objective.

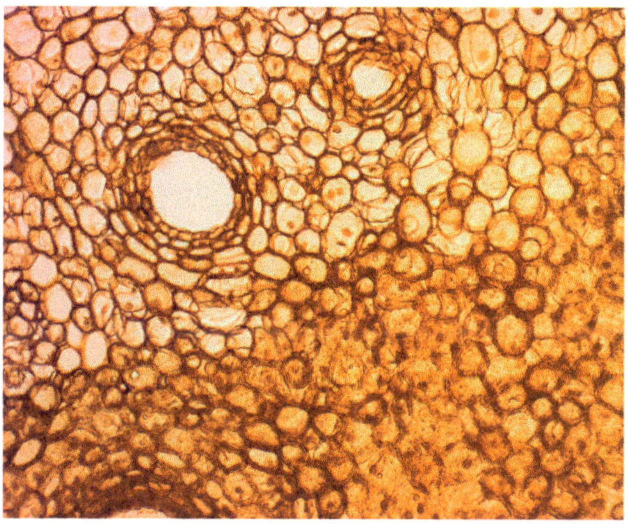

Figure 1.65 Photograph of a real object using a 10×/0.30 LCF plan achromatic objective.

scheme also uses a heat-shielding filter 15 and frosted glass 2, which is introduced into the course of the rays when using small-magnification objectives. The image of the object is obtained by projection by the objective 10, lenses 7 and 11, and mirrors 6 and 8 into the plane of the field diaphragm of the eyepieces installed in the visual head 13.

An interesting feature of this microscope is the use of an original engineering solution that makes it possible to implement research methods such as phase contrast and interference contrast. In this case, ordinary *bright*

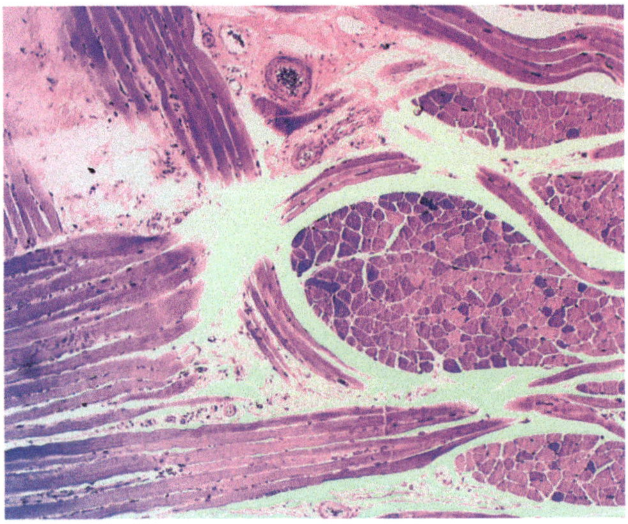

Figure 1.66 Photograph of a real object using a 10×/0.30 LCF plan achromatic objective.

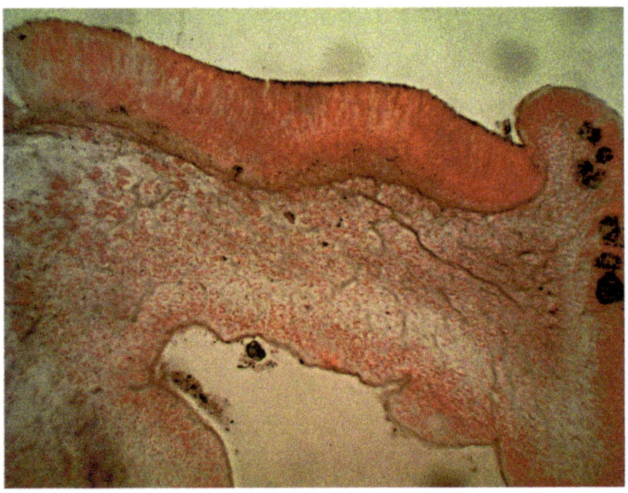

Figure 1.67 Photograph of a real object using a 10×/0.30 LCF plan achromatic objective.

field objectives are used (without phase rings inside the objectives); the system of lenses 7 and 11 also serves to optically transfer the plane of the exit pupil of the objectives to a special remote plane where the phase ring 9 is located (it is deposited on a plane-parallel plate). In a similar manner, the differential-interference contrast method is also implemented when a special prism is located in this plane. The object can be placed in a special liquid medium (water, saline, etc.). The researcher's task is not only to fix the image of the object in a static state, but also to observe a change in the shape and morphology of the object under the influence of some external factors (in accordance with the scientific research methodology).

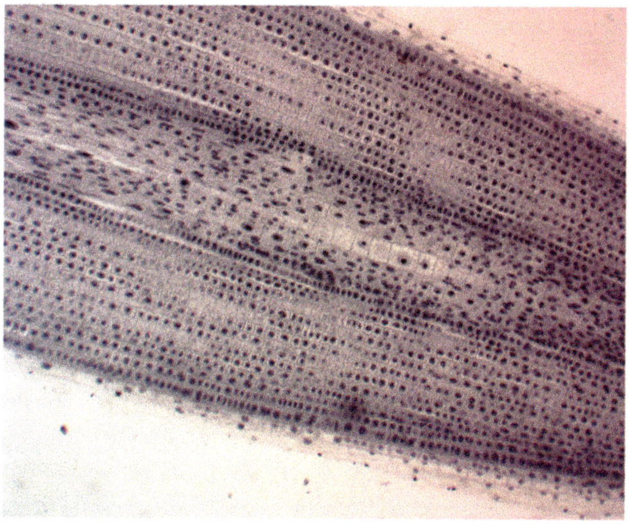

Figure 1.68 Photograph of a real object using a 10×/0.30 LCF plan achromatic objective.

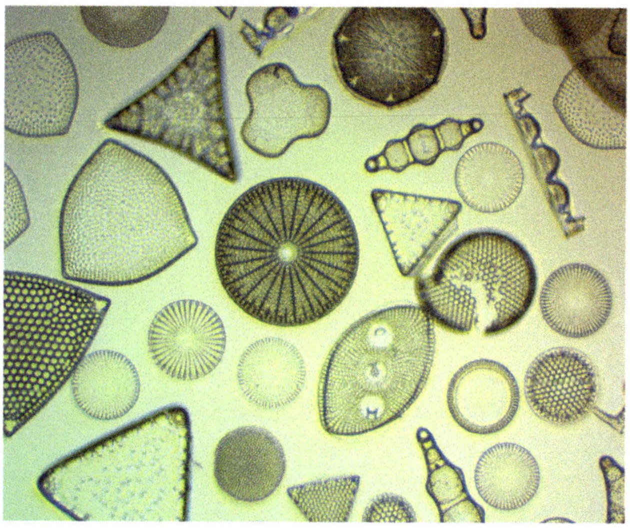

Figure 1.69 Photograph of a real object using a 10×/0.30 LCF plan achromatic objective.

For the builder of the optical and mechanical constructions of an inverted microscope, the initial requirements are the characteristics of special laboratory glassware, such as the thickness of its bottom and the optical and dispersion parameters of the optical material from which it is made. The optical and dispersion parameters of the fluid used are also important. These parameters are taken into account when calculating the optical system of a microscope, primarily in the optical design of lenses. So, for example, as the initial parameter for design, the nominal value of the thickness of the bottom of the

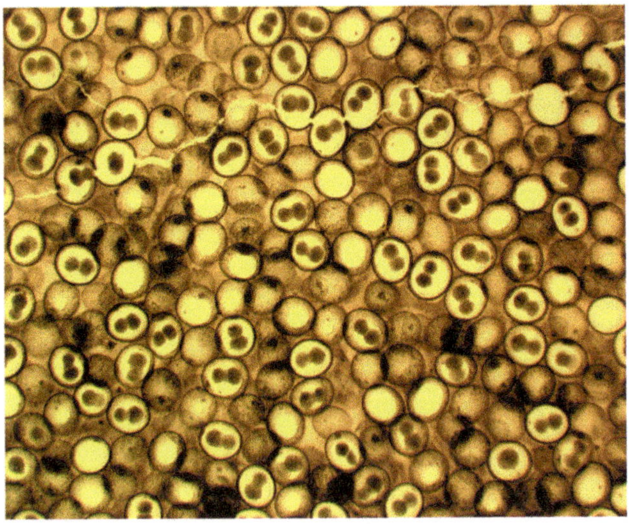

Figure 1.70 Photograph of a real object using a 10×/0.30 LCF plan achromatic objective.

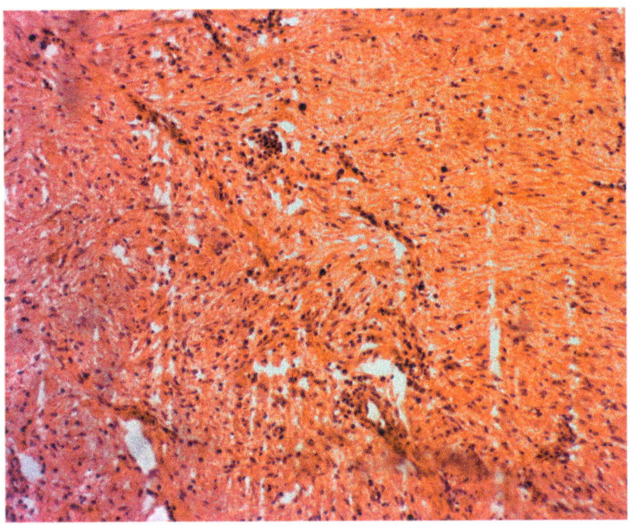

Figure 1.71 Photograph of a real object using a 10×/0.30 LCF plan achromatic objective.

laboratory glassware is determined to be 1.5 mm; in practice, this means that this value can vary over a wide range, for example, from 0.8 to 4 mm. A common glass slide is often used, the thickness of which is 1.1–1.2 mm. The refractive index of the optical material for dishes is determined in accordance with the parameters of the Soviet optical glass type LZ_LK5 or its analogues; refractive index $n_e = 1.479903$, dispersion coefficient $\nu_e = 65.d47$. It is taken into account that the object under study is located in a liquid medium, the refractive index and dispersion coefficient of which are determined to be

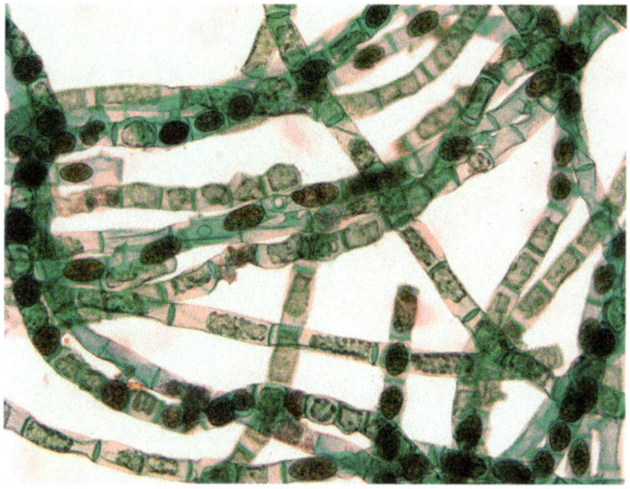

Figure 1.72 Photograph of a real object using a 10×/0.30 LCF plan achromatic objective.

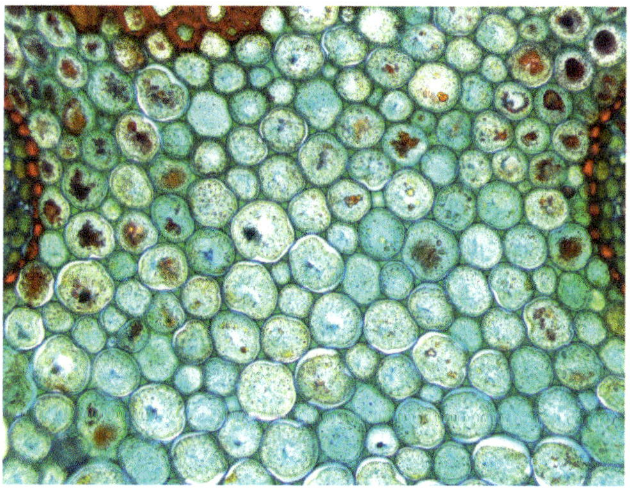

Figure 1.73 Photograph of a real object using a 20×/0.65 LCF plan achromatic objective.

$n_e = 1.336143$ and $\nu_e = 56.02$, respectively. Objectives having an increased free working distance should be used. A feature of the mechanical design for some objectives may be the use of a so-called *correction frame*, a description and principle of operation of which are shown in Fig. 1.17 (Volume 1, Chapter 1).

Biolam-P is designed to work with objectives on the so-called *finite tube length* of 160 mm with an eyepiece height of 13 mm and a linear eyepiece field of 18 mm; 2.5×, 4×, 10×, 20×, and 40× objectives are included with this microscope. Figures 1.112–1.114 show the design parameters and graphs of aberration correction of some of these objectives. Note that the lenses in these objectives are unified for various designs: for working without cover

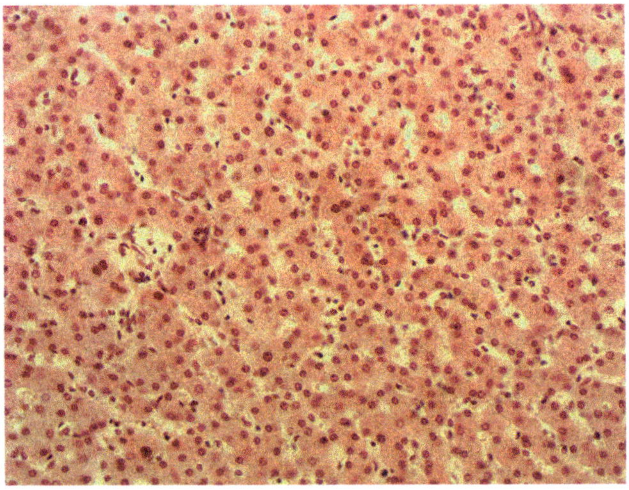

Figure 1.74 Photograph of a real object using a 20×/0.65 LCF plan achromatic objective.

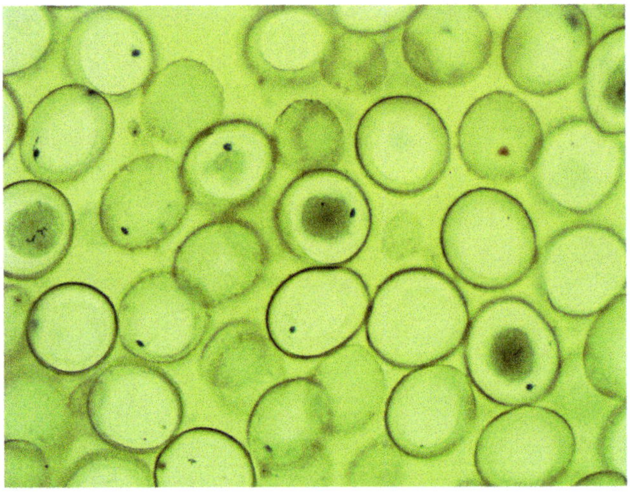

Figure 1.75 Photograph of a real object using a 20×/0.65 LCF plan achromatic objective.

glass, for working with standard cover glass ($d = 0.17$ mm), and for inverted microscopes.

Figures 1.115–1.126 are photographs of real objects taken with these and other objectives that complete the Biolam-P inverted microscope. The application of some contrasting methods, such as a dark field and phase contrast, is shown. Images of standard objects from below were obtained through a slide and a cover glass. Figures 1.127–1.132 show the design parameters and graphs of aberration correction of some objectives designed to operate in inverted microscopes for an infinite length of the tube.

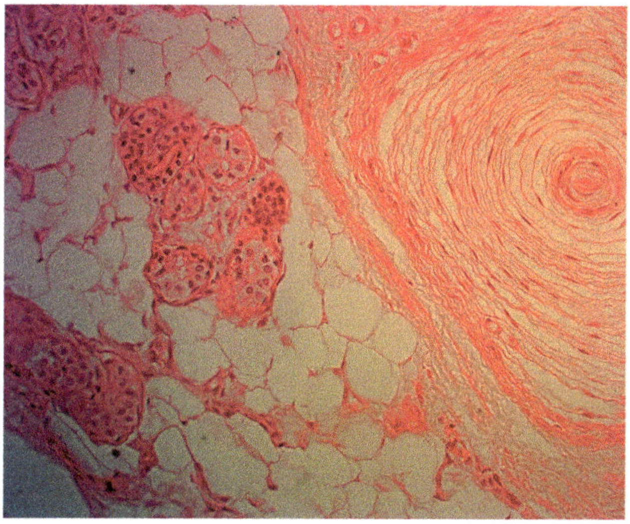

Figure 1.76 Photograph of a real object using a 20×/0.65 LCF plan achromatic objective.

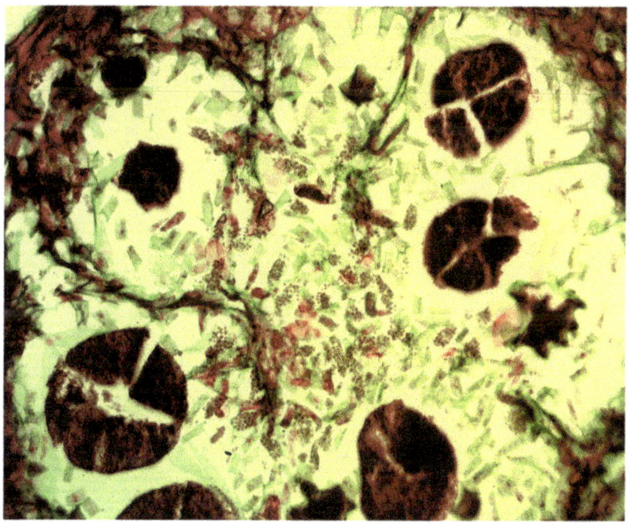

Figure 1.77 Photograph of a real object using a 20×/0.65 LCF plan achromatic objective.

1.6.4.2 Budget objectives for inverted (and upright) metallography microscopes

In general, metallographic inverted microscopes are structurally similar to biological ones. Figure 1.133 shows a photograph of a Russian budget metallographic microscope, the Metam-RV, and its principal optical layout. This is a rather interesting microscope. Although the design of the device is

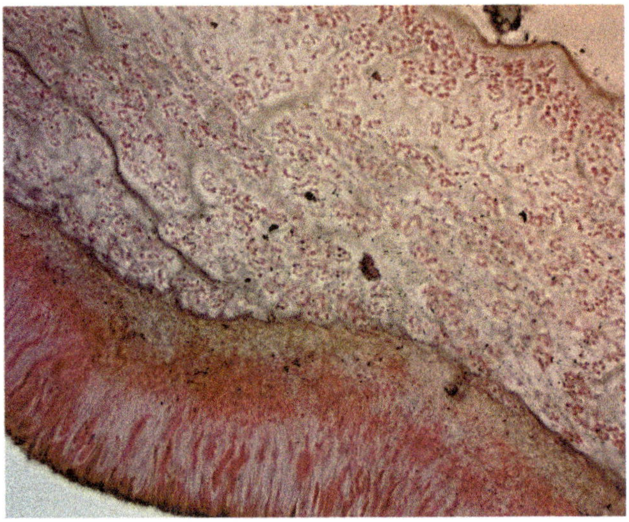

Figure 1.78 Photograph of a real object using a 20×/0.65 LCF plan achromatic objective.

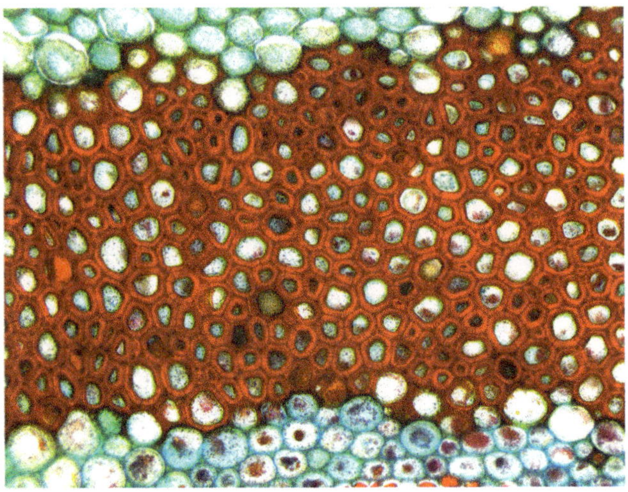

Figure 1.79 Photograph of a real object using a 20×/0.65 LCF plan achromatic objective.

quite compact, dark field, polarization, and Nomarski differential contrast methods are implemented here.

When observing in a bright field, the rays from light source 1 pass through lens 2, heat filter 3, lighting lens 5, and iris diaphragm 6, are reflected from a plane-parallel translucent reflector 7, and sent through objective 9 to object 11. Rays reflected from the surface of the object pass through the objective and the reflector again, fall on the mirror 18, and are reduced by the lens 17 to the focal plane of the eyepiece 13, where a real inverse and enlarged image of

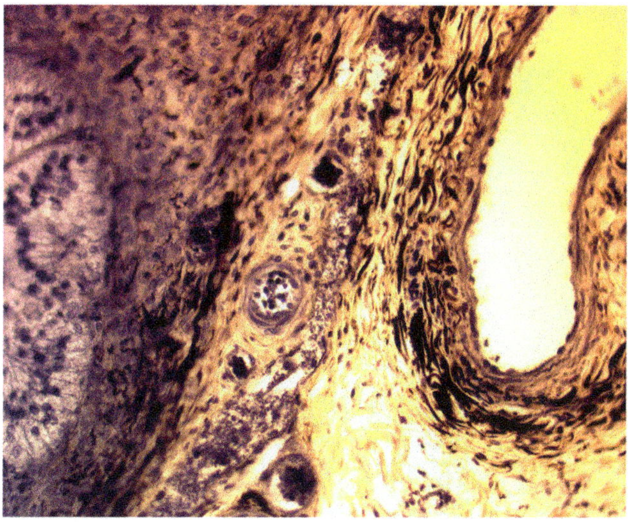

Figure 1.80 Photograph of a real object using a 20×/0.65 LCF plan achromatic objective.

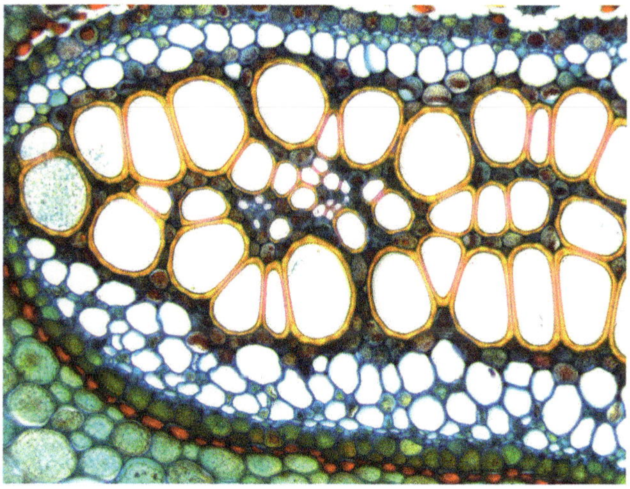

Figure 1.81 Photograph of a real object using a 20×/0.65 LCF plan achromatic objective.

the object is created. Using a prism 15 changes the direction of the optical axis of the microscope. The prism block 14 divides the beam into the binocular head. When observing in a dark field, the reflector 7 and the lens 5 are turned off from the path of the rays and the diaphragm 19 is introduced, the central zone of which is shielded. The light, passing through the diaphragm 19, is reflected from the annular reflector 8 and falls on a parabolic condenser 10, which collects a beam of rays on the object. Rays diffusely reflected from the irregularities of the object fall into the objective. In the microscope's field of view, the irregularities of the object are shown bright against a general dark

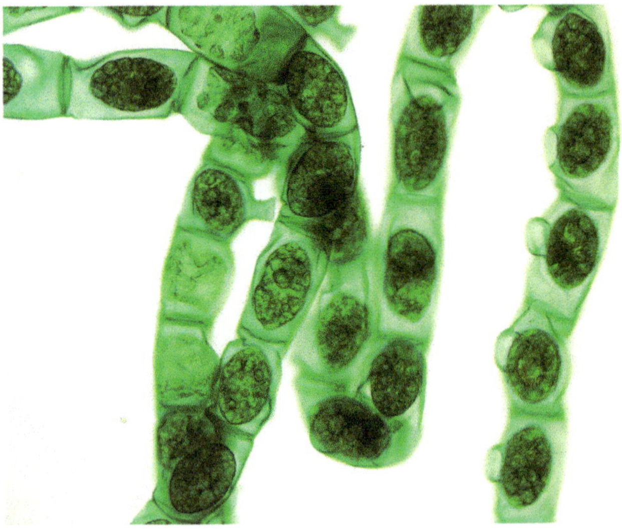

Figure 1.82 Photograph of a real object using a 20×/0.65 LCF plan achromatic objective.

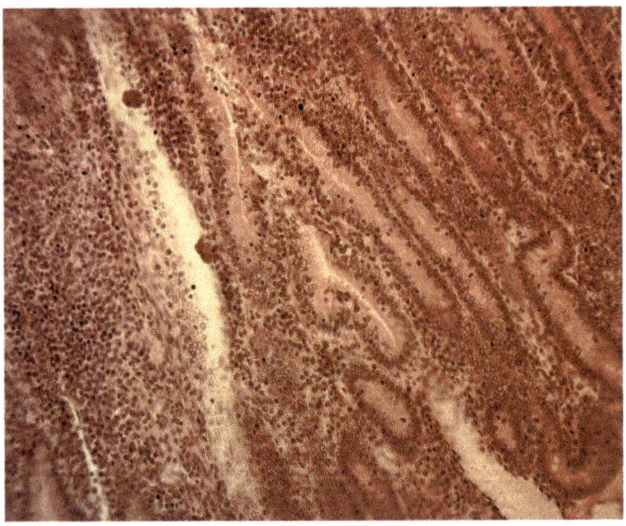

Figure 1.83 Photograph of a real object using a 20×/0.65 LCF plan achromatic objective.

background. To obtain uniform illumination of the object under study in the bright field, an illumination lens 5 with a frosted surface is introduced into the rays, and a filter 4 is introduced to increase the contrast of the image of the object. To obtain uniform illumination in the dark field, a stub is introduced into the rays. When observing in polarized light, the input of the rays is introduced by a reflector 7, an analyzer 16, and a polarizer 20.

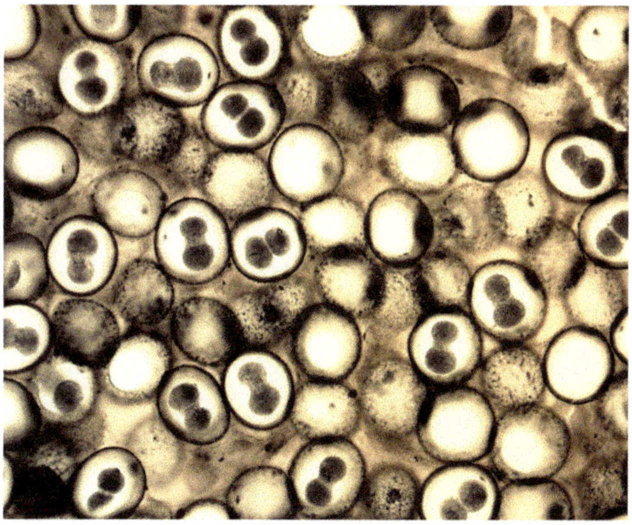

Figure 1.84 Photograph of a real object using a 20×/0.65 LCF plan achromatic objective.

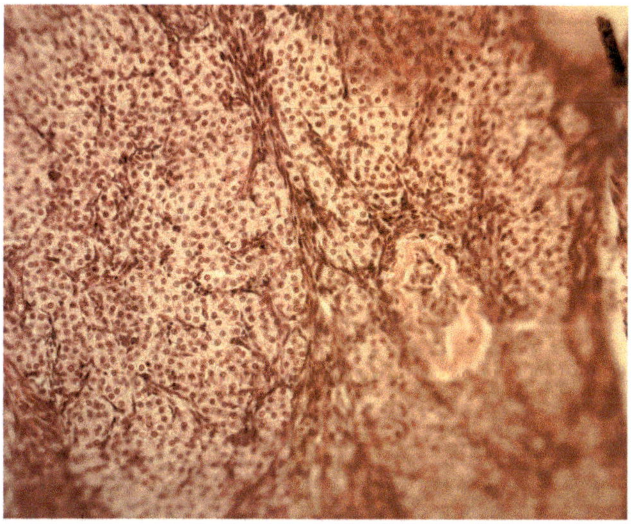

Figure 1.85 Photograph of a real object using a 20×/0.65 LCF plan achromatic objective.

When observing using the differential interference contrast method, a reflector 7, an analyzer 16, and a prism 12 are introduced into the course of the rays. The linearly polarized light exiting the polarizer is reflected from the reflector 7, hits a birefringent prism 12 oriented in space in such a way that the angle between the plane of polarization of the polarizer 20 and the axis of the prism 12 is 45 deg, and is split into two beams passing through the prism. Rays exiting prism 12, polarized in mutually perpendicular planes and having equal intensities, pass through objective 9 and fall on object 11, upon

Figure 1.86 Photograph of a real object using a 20×/0.65 LCF plan achromatic objective.

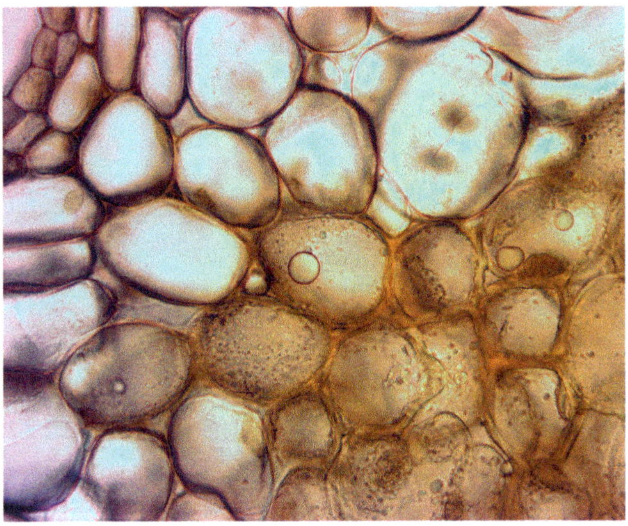

Figure 1.87 Photograph of a real object using a 40×/0.70 LCF plan achromatic objective.

reflection from which a phase difference of these rays arises due to surface irregularities.

Reflecting from the object and again passing through the objective and the prism, the rays are connected in the localization plane of the prism, which is aligned with the rear focal plane of the objective. When entering the analyzer 16, the plane of polarization of which is located at an angle 45 deg to the axis of the prism, interference of the rays is achieved. A double image of the object

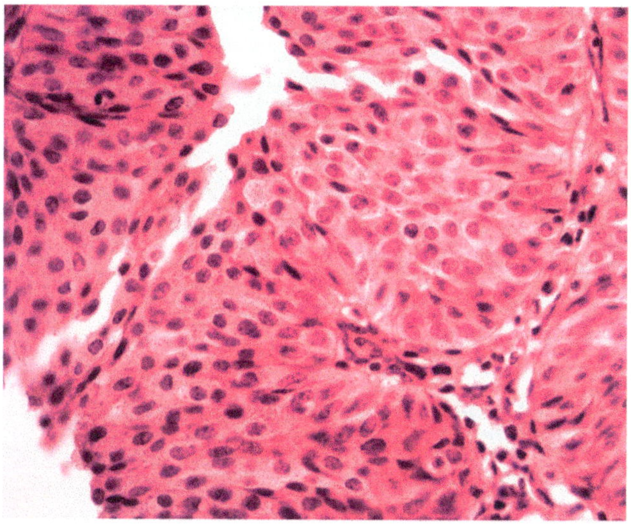

Figure 1.88 Photograph of a real object using a 40×/0.70 LCF plan achromatic objective.

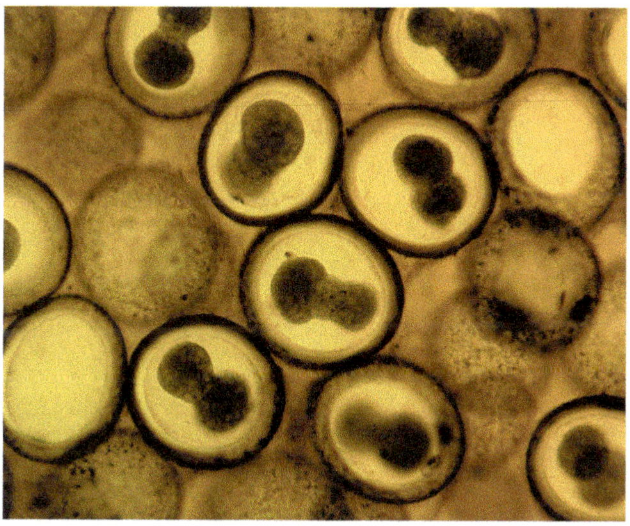

Figure 1.89 Photograph of a real object using a 40×/0.70 LCF plan achromatic objective.

is obtained; however, the bifurcation is so insignificant (close, but to the resolution limit of the lens) that it is practically invisible and the object is perceived in relief.

By the way, a fairly common question that many researchers have is, how different are the images on metallographic microscopes when they are made using a variety of contrast methods? The answer can be found when working on this microscope. Figure 1.134 shows the difference in images made by the

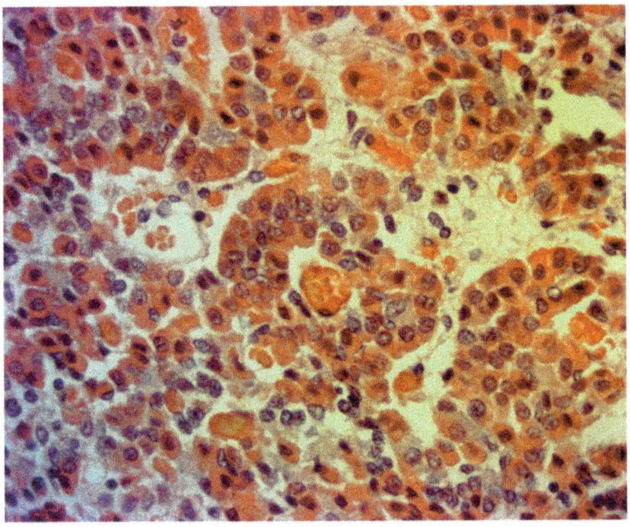

Figure 1.90 Photograph of a real object using a 40×/0.70 LCF plan achromatic objective.

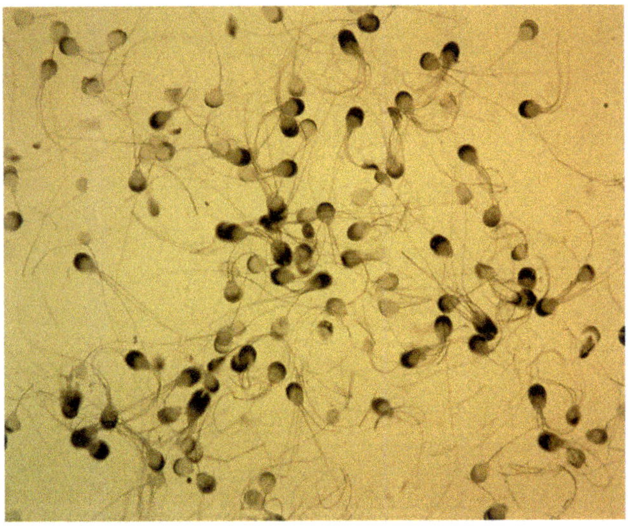

Figure 1.91 Photograph of a real object using a 40×/0.70 LCF plan achromatic objective.

bright field, dark field, polarization, and Nomarski differential contrast methods. These photographs make it clear that even for budget microscopy, in the study of *metallographic objects*, polarization and differential contrast methods are important; the ability to implement these contrasting methods should be mandatory when designing budget metallographic microscopes.

It is interesting to use objectives in the Metam-RV microscope, particularly an objective (bright field) that has a linear magnification of

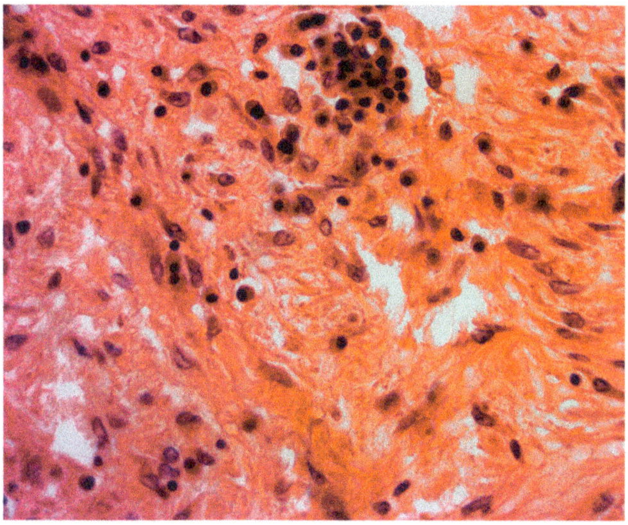

Figure 1.92 Photograph of a real object using a 40×/0.70 LCF plan achromatic objective.

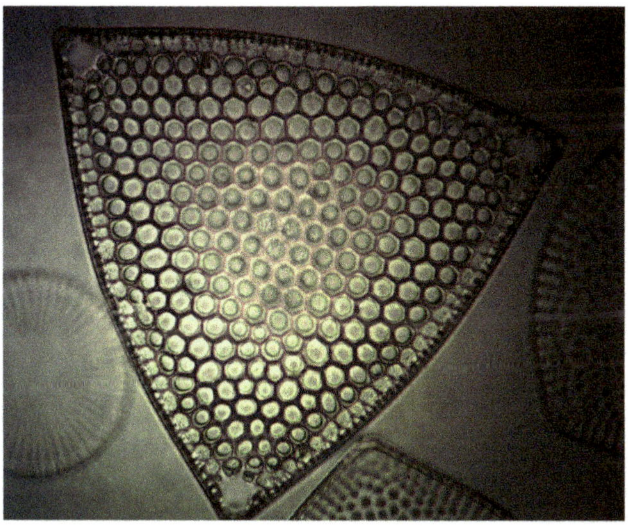

Figure 1.93 Photograph of a real object using a 40×/0.70 LCF plan achromatic objective.

50×, a numerical aperture of 0.85, and plan apochromatic aberration correction. However, today these objectives are obsolete and require replacement with new ones. In contrast to inverted biological microscopes, in (inverted) metallographic microscopes, in almost all cases, opaque or translucent samples are examined. For this, the so-called *reflected light illuminator* is used. For inverted microscopes, it is very important that an increased working distance is ensured between the object being examined and

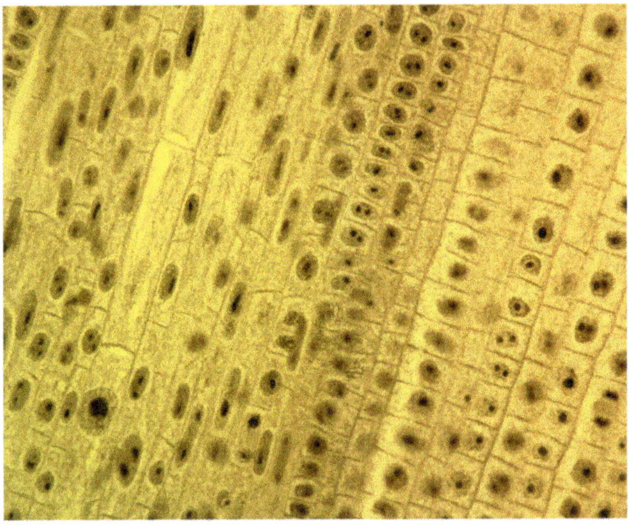

Figure 1.94 Photograph of a real object using a 40×/0.70 LCF plan achromatic objective.

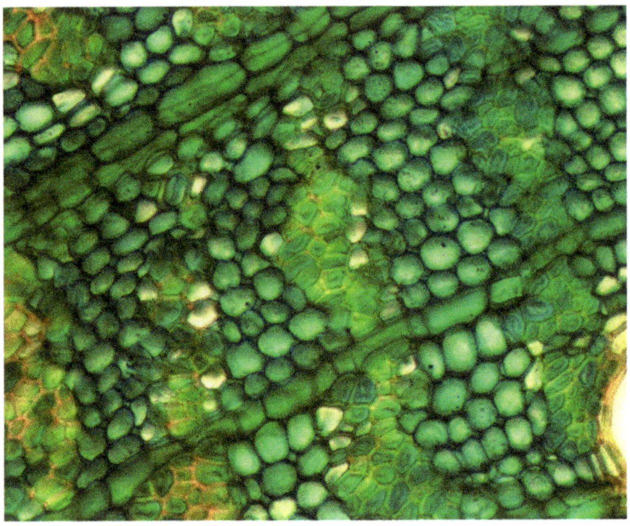

Figure 1.95 Photograph of a real object using a 40×/0.70 LCF plan achromatic objective.

the first surface of the objective (frame or, if it protrudes from the frame, lens, which sometimes happens). This is a problem for metallographic inverted microscopes, because, as a rule (unlike, for example, biological inverted microscopes), in-depth viewing of the specimen being studied does not occur since its surface is flat. Therefore, the parfocal property of objectives and their exchange during microscopy are often used. The surface of the object must be in the same plane for all objectives in the kit. When changing the objective in

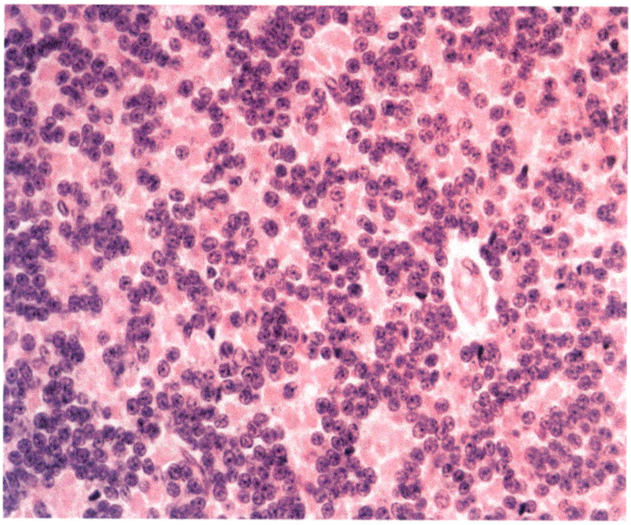

Figure 1.96 Photograph of a real object using a 40×/0.70 LCF plan achromatic objective.

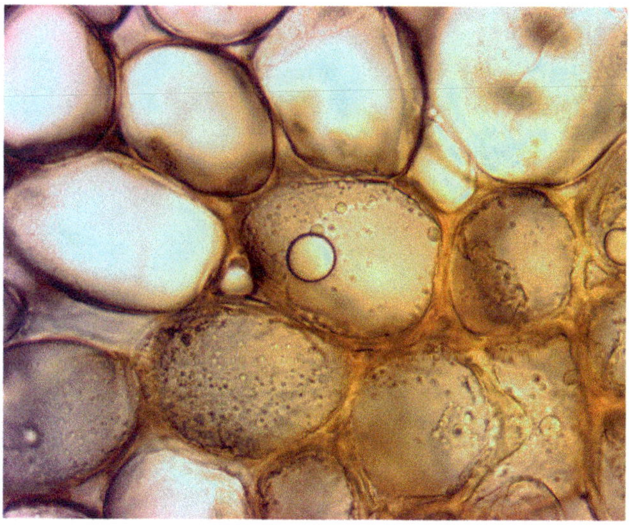

Figure 1.97 Photograph of a real object using a 60×/0.85 LCF plan achromatic objective.

modern inverted metallographic objectives, such parfocality is achieved; when switching the revolving device (into which the objectives are screwed), the objectives do not touch and do not cling to the lower plane of the subject table. However, basically, changing objectives without problems is achieved by using objectives with increased working distances. The accepted name of objectives with an increased working distance is long working distance (LWD, LD, or LLD); the value of the working distance in such objectives is at least

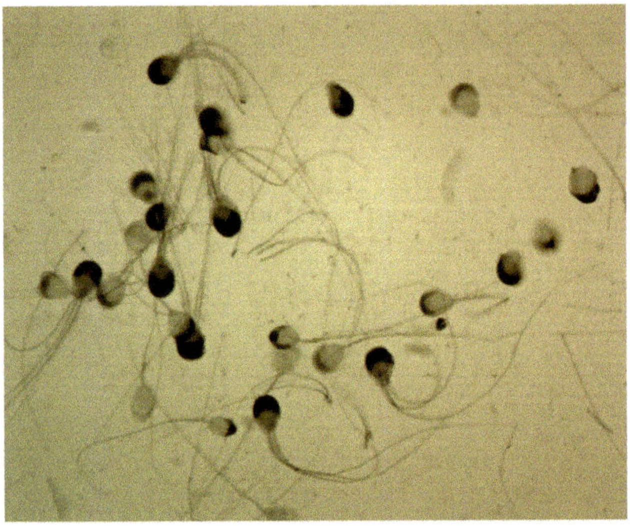

Figure 1.98 Photograph of a real object using a 60×/0.85 LCF plan achromatic objective.

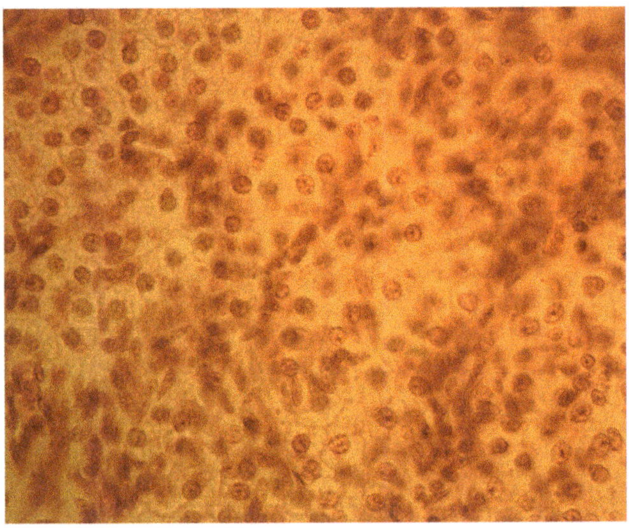

Figure 1.99 Photograph of a real object using a 60×/0.85 LCF plan achromatic objective.

half of its focal length. Figures 1.135–1.137 show the design parameters and graphs of aberration correction of modern budget objectives intended for use in metallographic microscopes. The optical systems of these objectives used the original optical glass (type LZ_OK4), characterized by the following optical parameters: refractive index $n_e = 1.4485$ and dispersion coefficient $\nu_e = 91.53$. The absence of internal stresses was achieved with this optical material, making it usable, including in polarization studies.

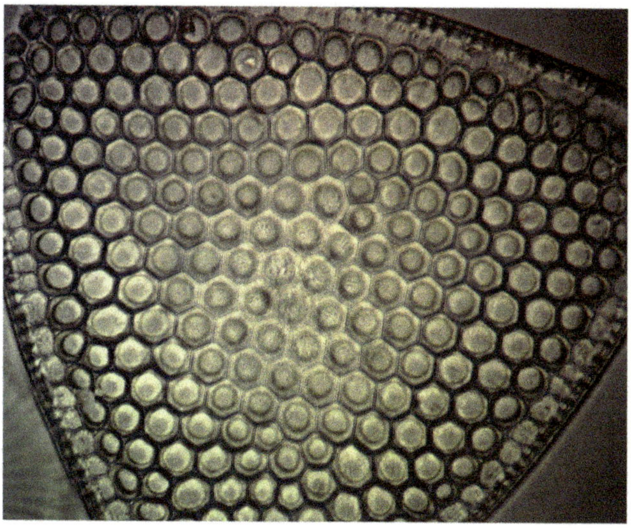

Figure 1.100 Photograph of a real object using a 60×/0.85 LCF plan achromatic objective.

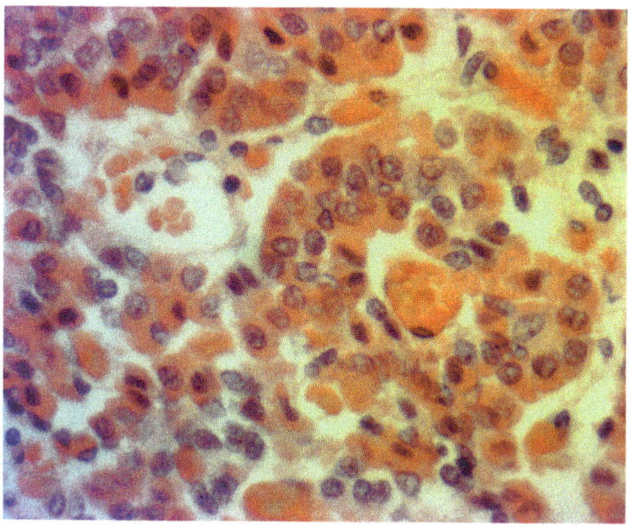

Figure 1.101 Photograph of a real object using a 60×/0.85 LCF plan achromatic objective.

Making a "changeling" microscope, designed for research in reflected light, could be useful for designing a budget model of a metallographic microscope. Figure 1.138 shows handmade and computer simulation drawings of the basic design of such a microscope, proposed about 20 years ago.

Two stages are permanent parts and, at the same time, part of the mechanical construction of the microscope. The main *working* element of the mechanical structure is a rotary assembly rotated clockwise. This element

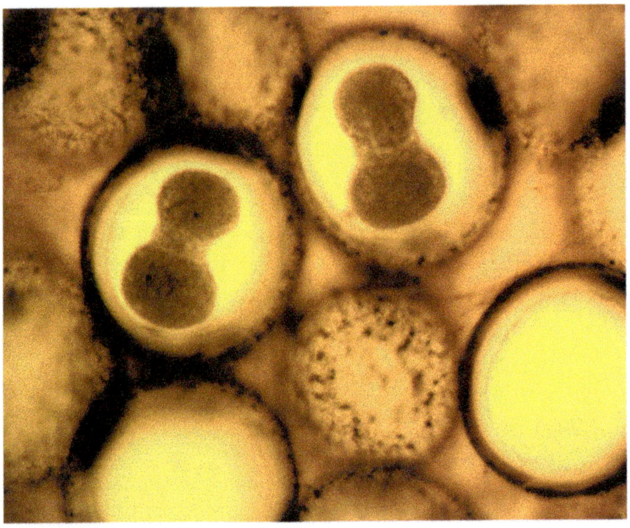

Figure 1.102 Photograph of a real object using a 60×/0.85 LCF plan achromatic objective.

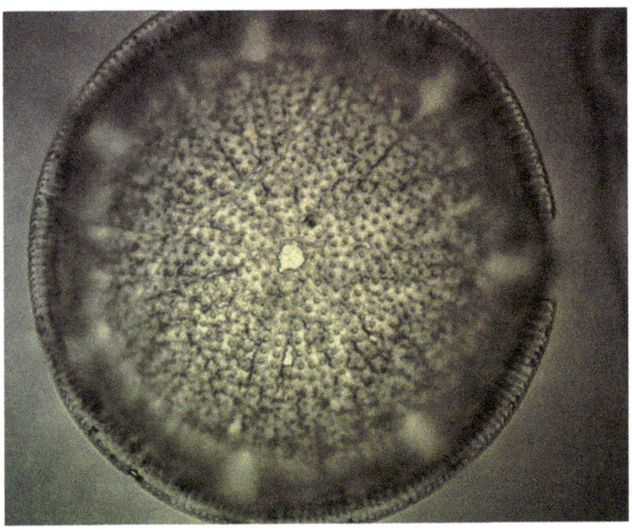

Figure 1.103 Photograph of a real object using a 60×/0.85 LCF plan achromatic objective.

contains a lighting system, a mirror system that allows both bright field and dark field lighting options to be implemented, polarizing filters, and a nosepiece with objectives. In an inverted microscope variant, the objective is directed upward and the object being studied is located on the upper stage; in the upright metallographic microscope variant, the objective is directed downward and the object is located on the lower stage. On the front side of the *rotary element* may be a visualization system containing a visual head (also

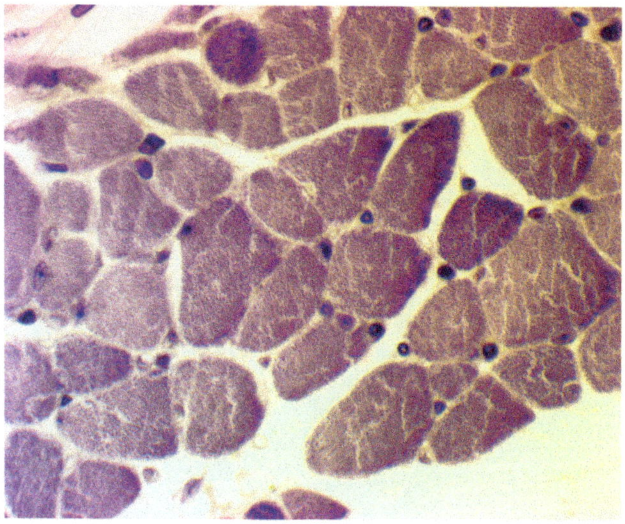

Figure 1.104 Photograph of a real object using a 60×/0.85 LCF plan achromatic objective.

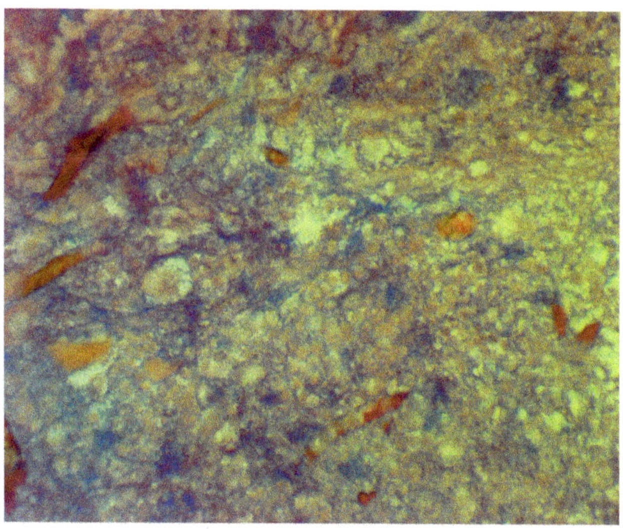

Figure 1.105 Photograph of a real object using a 60×/0.85 LCF plan achromatic objective.

having the ability to turn) or a digital image receiver. In addition, a technological groove is made in the direction of the front side, allowing you to set and move perpendicular to the optical axis of the objective, i.e., the Wollaston prism system for implementing the differential contrast method.

1.6.4.3 Budget objectives for polarizing microscopes

The most complete consumer qualities of a polarizing microscope are manifested when studying objects from the ore and mining industry. Studies

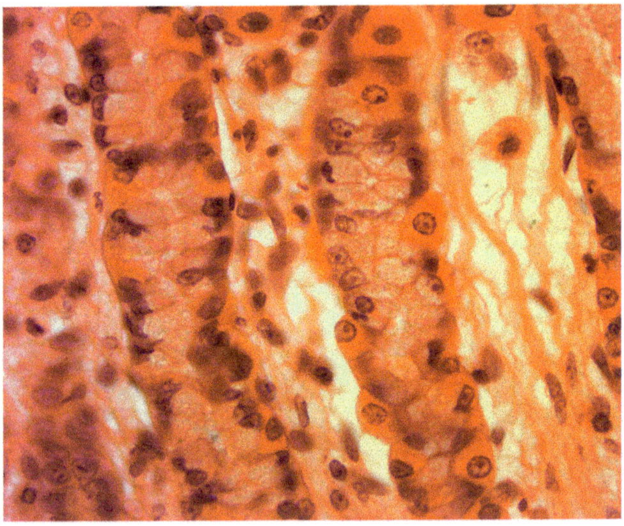

Figure 1.106 Photograph of a real object using a 60×/0.85 LCF plan achromatic objective.

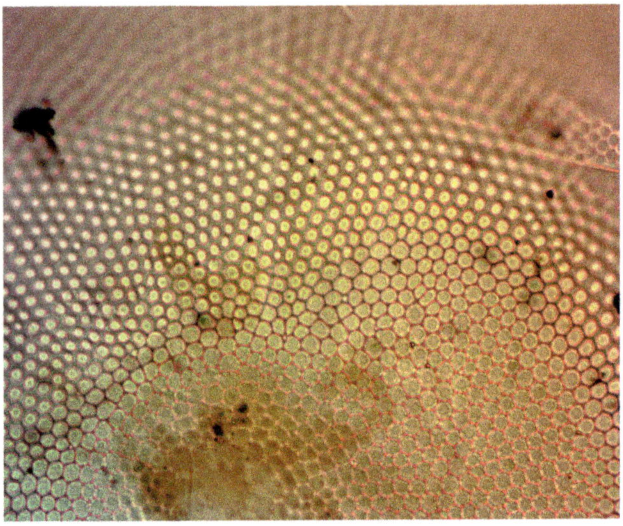

Figure 1.107 Photograph of a real object using a 100×/1.30 oil immersion LCF semi–plan achromatic objective.

of such samples are more difficult to carry out than others, especially in the field of crystallography, petrography, and chemistry. It is necessary to use special techniques. To implement these techniques, certain conditions are required for the microscope's optics and objectives. Orthoscopic and conoscopic observations use different physical and geometric laws of light propagation. Thus, in orthoscopy, the illumination of the object occurs with a

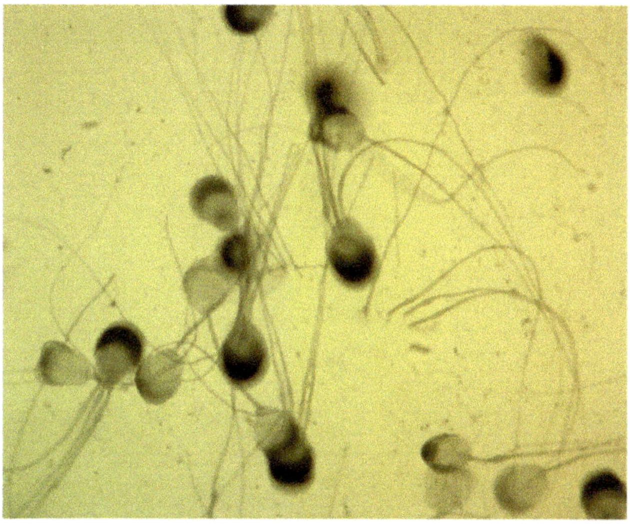

Figure 1.108 Photograph of a real object using a 100×/1.30 oil immersion LCF semi–plan achromatic objective.

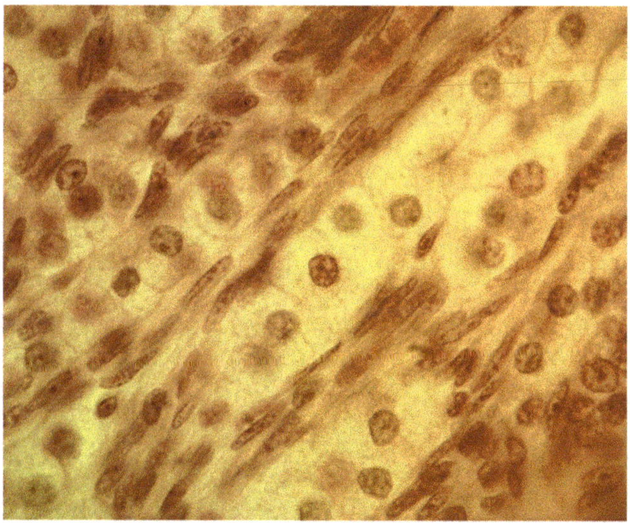

Figure 1.109 Photograph of a real object using a 100×/1.30 oil immersion LCF semi–plan achromatic objective.

small numerical aperture; the rays coming from the light source fall on the object at small angles. For orthoscopy, small- and medium-magnification objectives with a small numerical aperture are usually used. Observations made using conoscopy methods in a strongly converging beam of polarized light are more specific. The images of the objects themselves are not studied, but rather some colored axial images are examined, which represent

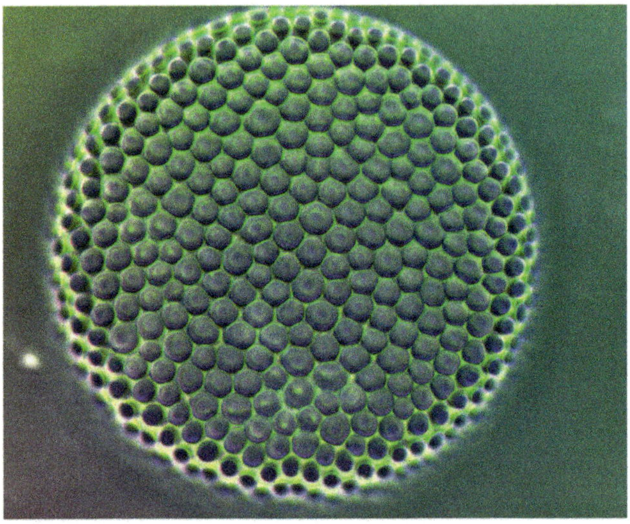

Figure 1.110 Photograph of a real object using a 100×/1.30 oil immersion LCF semi–plan achromatic objective (using the color inversion method).

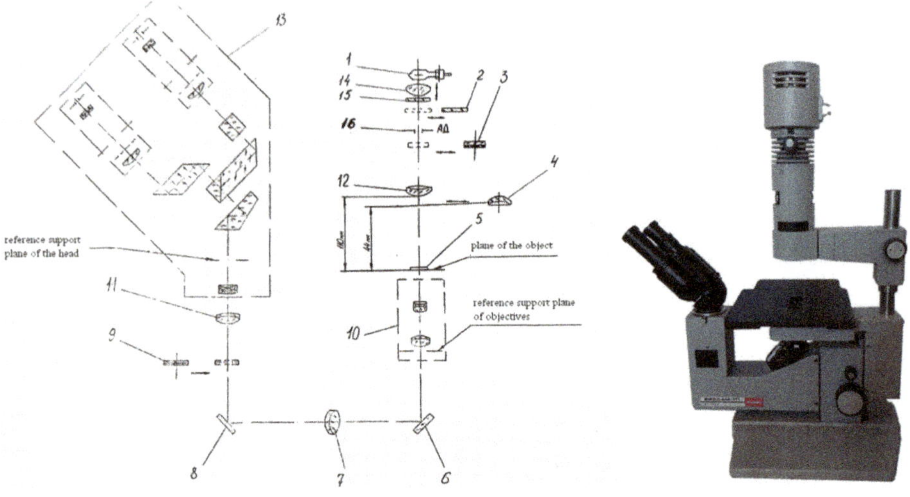

Figure 1.111 Optical diagram and a photograph of a Biolam-P microscope.

interference patterns. For example, in the study of minerals, the conoscopy method makes it possible to determine important parameters of a mineral, such as shape, index of refraction, angle of quenching, positive or negative strong or weak birefringence, number of optical axes, crystallographic structure, etc. Recently, in connection with the emergence of techniques for studying finely dispersed crystalline formations, there has been an increased demand for polarization microscopes. In addition, methods have been found

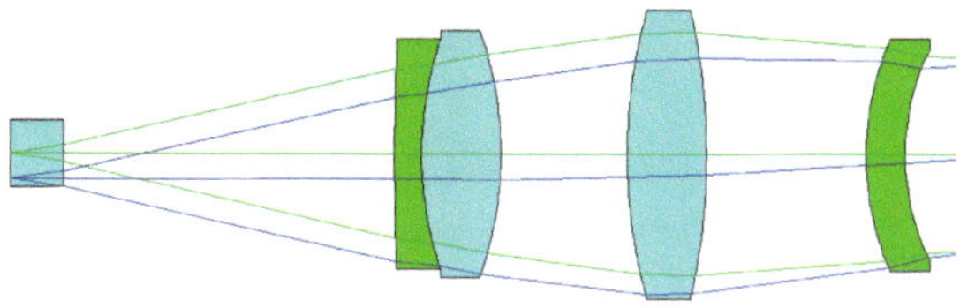

Lens: plan 10x0.22 d=2 Finity

Object num aper 0.220000 Image height 9.000000 Primary wavln 0.546070

SRF	RADIUS	THICKNESS	APERTURE RADIUS	GLASS	SPECIAL
OBJ	0.000000	0.010000	0.936504	FISIO	
AST	0.000000	2.000000	0.938174 AS	LZ_LK5	
2	0.000000	12.510000	1.240000	AIR	
3	112.760000	1.000000	4.080000	LZ_TF5	
4	13.183000	3.000000	4.260000	LZ_LK3	
5	-13.183000	4.760000	4.550000	AIR	
6	19.498000	3.000000	5.320000	LZ_LK3 P	
7	-24.320000	6.050000	5.300000	AIR	
8	10.520000	1.500000	4.290000	LZ_TF5 P	
9	8.395000	158.170000	3.880000	AIR	
IMS	0.000000	0.000000	8.978712 S		

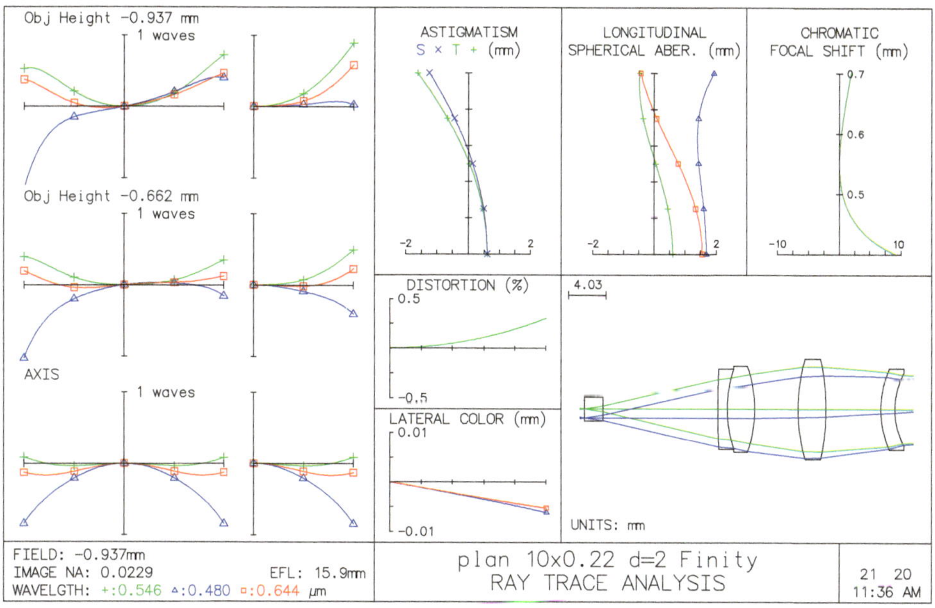

Figure 1.112 Optical design and graphs of aberration correction of a 10×/0.22 finite LCF plan achromatic objective (cover glass; $d = 2.0$ mm).

that make it possible to determine the refractive index of 1-μm or smaller particles. For conoscopic studies, medium- to large-magnification lenses with a high numerical aperture (from 0.5 and above) are used. And for some methods, immersion objectives are needed. In addition to common methods, special observation methods, such as focal shielding, phase contrast, fluorescence

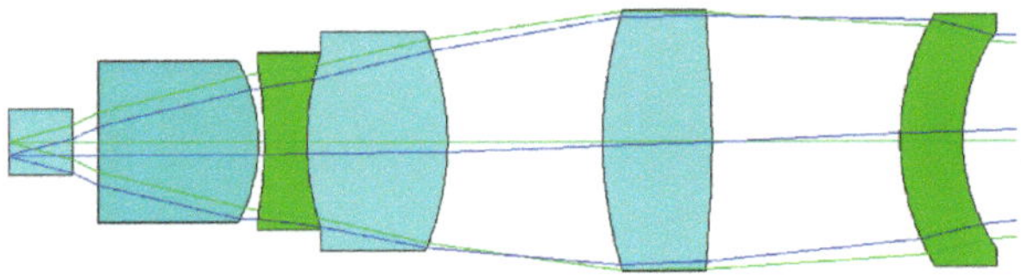

Lens: ocx 20x0.40 d=2 Finity

Object num aper	0.400000	Image height	9.000000	Primary wavln	0.546070

SRF	RADIUS	THICKNESS	APERTURE RADIUS	GLASS	SPECIAL
OBJ	0.000000	0.010000	0.432009	FISIO	
AST	0.000000	2.000000	0.435146 AS	LZ_LK5	
2	0.000000	0.861000	1.040000	AIR	
3	0.000000	5.150000	1.420000	LZ_CTK12	
4	-5.297000	0.200000	2.520000	AIR	
5	-13.742000	1.400000	2.550000	LZ_TF5	
6	8.630000	4.500000	2.780000	LZ_LK3	
7	-8.630000	5.000000	3.440000	AIR	
8	13.742000	3.500000	4.090000	LZ_LK3 P	
9	-42.660000	6.000000	4.090000	AIR	
10	7.551000	2.000000	3.940000	LZ_TF5 P	
11	5.970000	161.380000	3.410000	AIR	
IMS	0.000000	0.000000	8.995998 S		

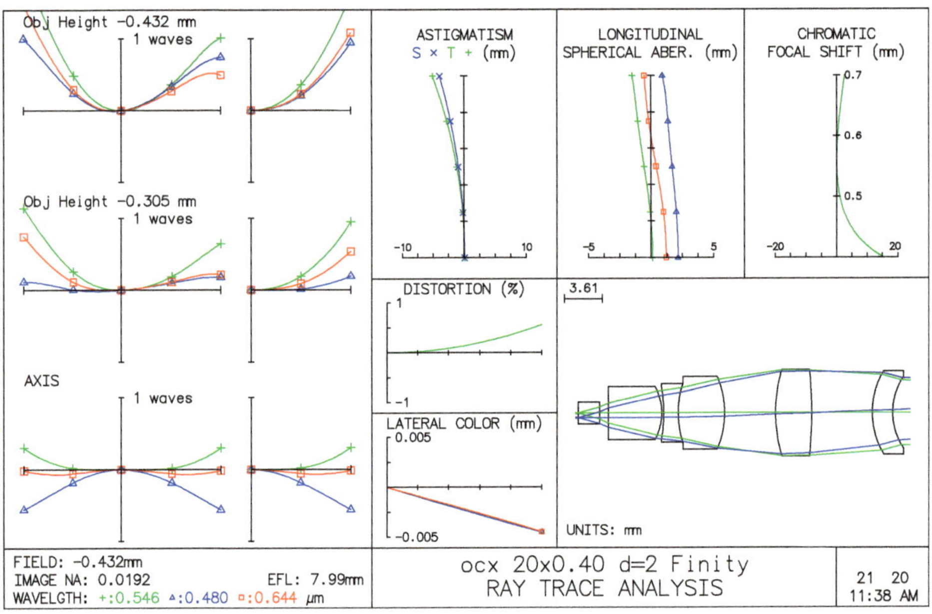

Figure 1.113 Optical design and graphs of aberration correction of a 20×/0.40 finite LCF S-plan achromatic (cover glass; $d = 2.0$ mm).

analysis, and others, can be used on polarization microscopes. For example, the focal shielding method can be used for immersion objects when comparing their refractive indices with the surrounding immersion medium.

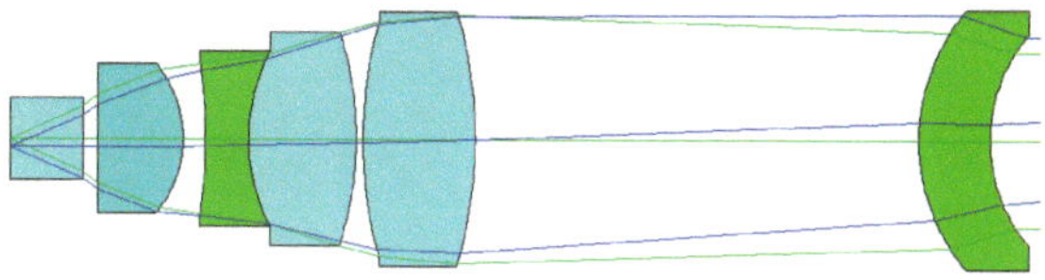

Lens: ocx 40x0.60 d=2 Finity

Object num aper 0.600000 Image height 9.000000 Primary wavln 0.546070

SRF	RADIUS	THICKNESS	APERTURE RADIUS	GLASS	SPECIAL
OBJ	0.000000	0.010000	0.215224	FISIO	
AST	0.000000	2.000000	0.220249 AS	LZ_LK5	
2	0.000000	0.399100	1.110000	AIR	
3	0.000000	2.380000	1.410000	LZ_CTK12	
4	-3.076000	0.620000	2.020000	AIR	
5	-15.171000	1.200000	2.140000	LZ_TF5	
6	5.035000	3.000000	2.350000	LZ_LK3	
7	-9.120000	0.190000	2.870000	AIR	
8	11.722000	3.100000	3.210000	LZ_LK3 P	
9	-11.722000	12.200000	3.420000	AIR	
10	5.297000	2.000000	3.470000	LZ_TF5 P	
11	4.207000	164.901157 S	2.840000	AIR	
IMS	0.000000	0.000000	9.000000 S		

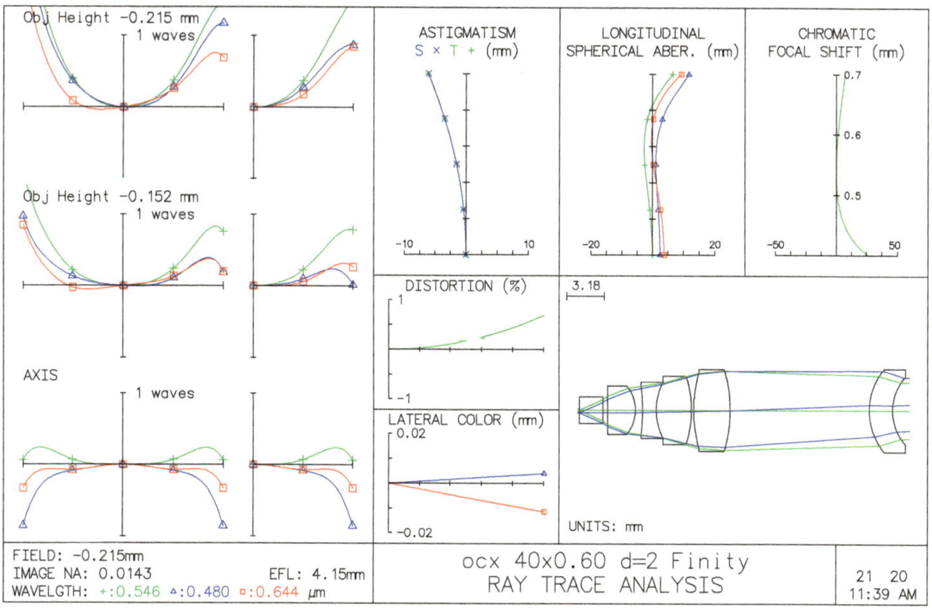

Figure 1.114 Optical design and graphs of aberration correction of a 40×/0.60 finite LCF S-plan achromatic objective (cover glass; $d = 2.0$ mm).

Figure 1.139 shows a drawing of a budget Russian microscope, the Polam-R, which also has some interesting distinguishing features. It is intended mainly for the study of opaque objects in reflected ordinary and

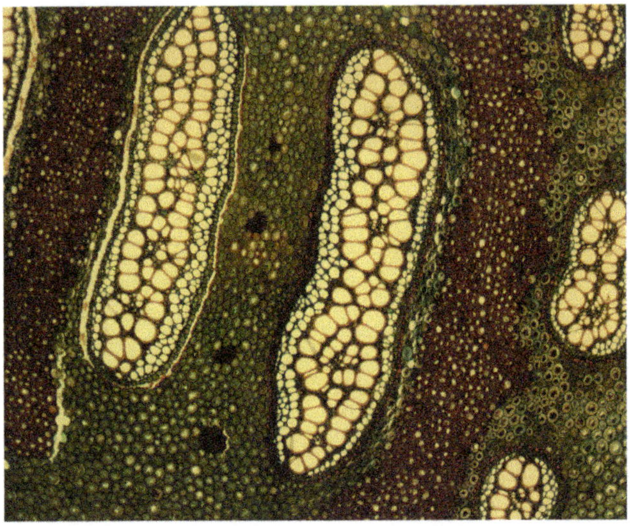

Figure 1.115 Photograph of a real object using a 4×/0.12 LCF plan achromatic objective.

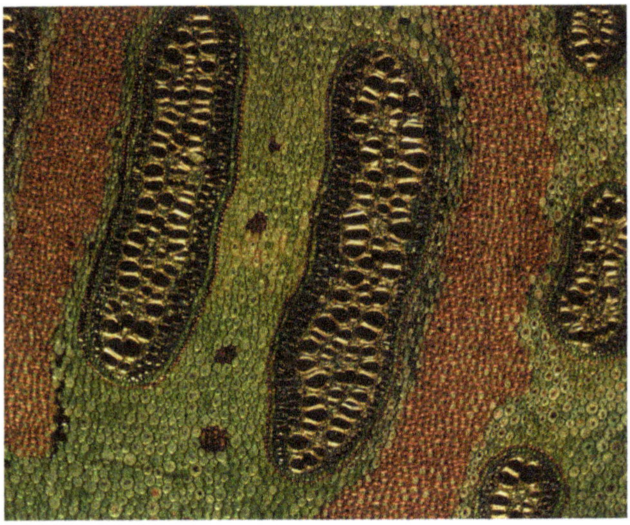

Figure 1.116 Photograph of a real object using a 4×/0.12 LCF plan achromatic objective (using the dark field method).

polarized light. It is used primarily for work in the field of mineralography, carbon petrography, metallography, etc. The main feature of this microscope is that immersion objectives of small- and medium-magnification are used in it, e.g., an objective having a linear magnification of 11× and a numerical aperture (in oil immersion) of 0.25 or an objective with a 30× linear magnification and a 0.65 oil numerical aperture. The illuminator OI-12,

Figure 1.117 Photograph of a real object using a 10×/0.22 LCF plan achromatic objective.

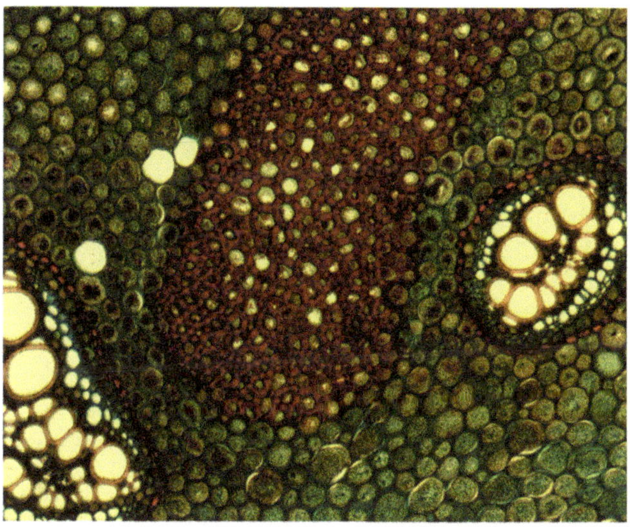

Figure 1.118 Photograph of a real object using a 10×/0.22 LCF plan achromatic objective.

shown in Fig. 1.105 (Volume 1, Chapter 1), also includes these objectives. A photograph of these objectives is also shown in Fig. 1.139.

Objectives for polarization research must meet specific requirements that are not imposed on conventional objectives, for example, for biological research. One of these basic requirements is isotropy. That is, on the one hand, the optical materials used must comply with the requirement that there be no internal stresses in the glass as well as no defects or other inclusions that can cause a deviation from isotropy. On the other hand, as a rule, the lenses used

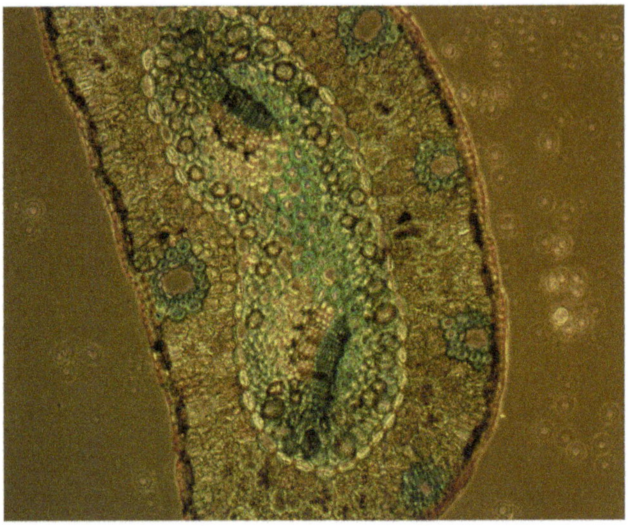

Figure 1.119 Photograph of a real object using a 10×/0.22 LCF plan achromatic objective (using the phase contrast method).

Figure 1.120 Photograph of a real object using a 20×/0.40 LCF S-plan achromatic objective.

in polarizing microscopes (optical glass) are influenced by the mechanical effects of the frames. In this case, birefringence may appear. Of course, we must carefully mount the lenses into the frames. For example, special glue should be selected that evenly distributes the bonding forces along the diameter of the lens. When calculating the optical system of a polarizing

Figure 1.121 Photograph of a real object using a 20×/0.40 LCF S-plan achromatic objective (using the phase contrast method).

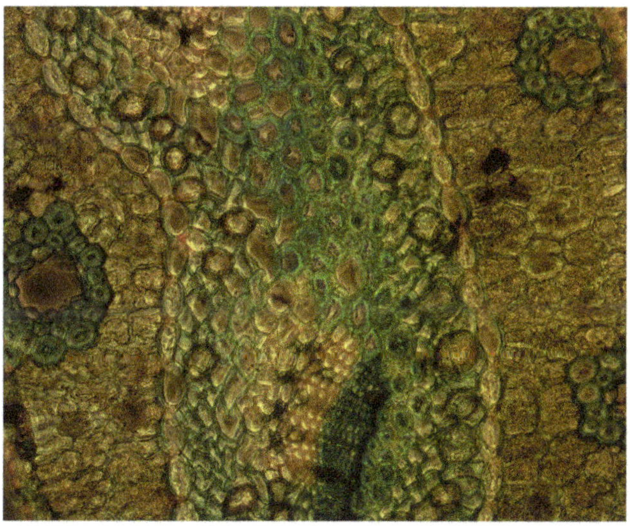

Figure 1.122 Photograph of a real object using a 20×/0.40 LCF S-plan achromatic objective (using the phase contrast method).

objective, special attention should be paid to glass, which is used as positive crown lenses. The requirement for isotropy in modern objectives is supplemented by the need for more perfect aberration correction than in old achromatic objectives, which used ordinary silicate glass such as *light crown* and *crown*. With old achromatic objectives, there was no way to correct (or reduce) the aberration of the secondary spectrum; i.e., it was not possible

Figure 1.123 Photograph of a real object using a 40×/0.60 LCF S-plan achromatic objective.

Figure 1.124 Photograph of a real object using a 40×/0.60 LCF S-plan achromatic objective.

to get an objective to work as part of a polarizing microscope with apochromatic or semi-apochromatic correction of aberrations.

To improve the image quality of a polarizing microscope, a higher aberration correction level compared with achromats must be achieved. Here, the effect of different aberrations on the resulting image quality should be

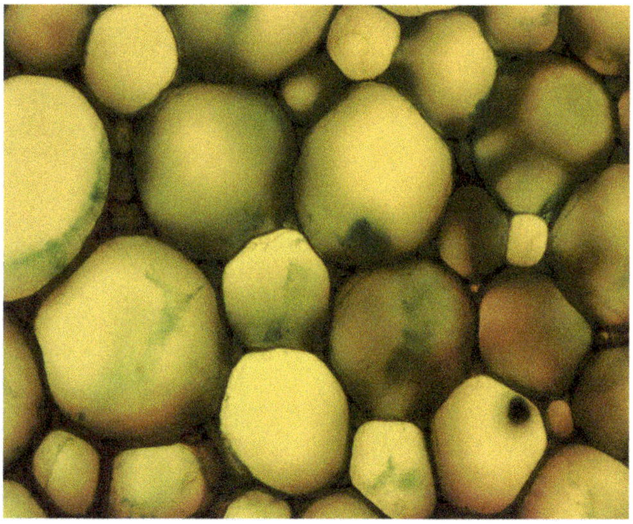

Figure 1.125 Photograph of a real object using a 40×/0.60 LCF S-plan achromatic objective.

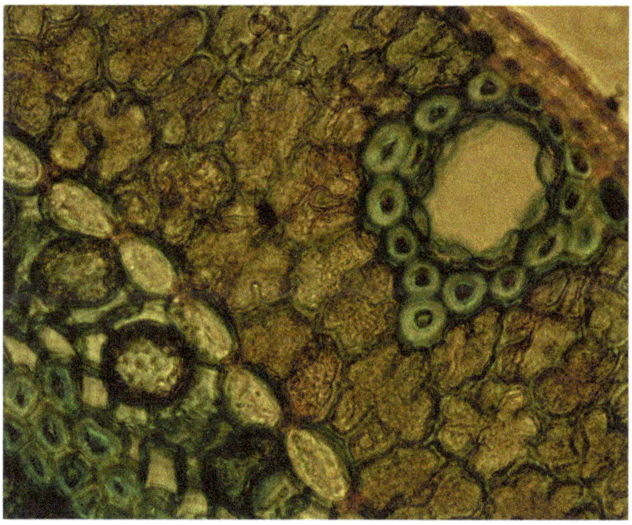

Figure 1.126 Photograph of a real object using a 40×/0.60 LCF S-plan achromatic objective (using the phase contrast method).

enumerated. First, for a more reliable study of polarization preparations, a high level of correction of a chromatic aberration such as secondary chromatism (LC) is required. Often, various inclusions and artifacts are examined with a microscope, the morphology of which allows the most accurate identification of the object under study. Second, it is desirable to

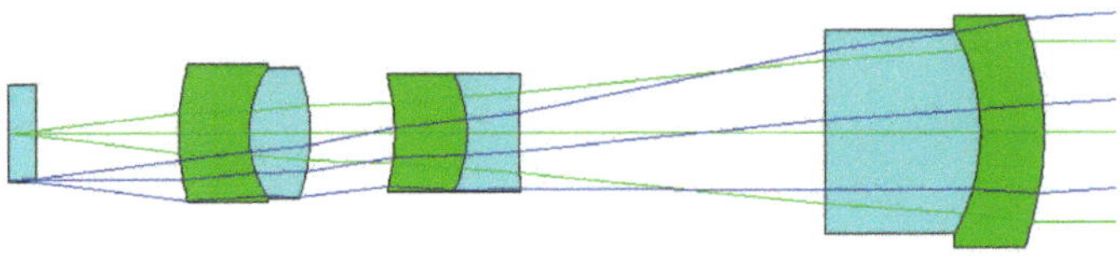

Lens: Plan 5x0.12 d1.2 ftl160

Object num aper	0.120000	Image height	10.000000	Primary wavln	0.546070

SRF	RADIUS	THICKNESS	APERTURE RADIUS	GLASS	SPECIAL
OBJ	0.000000	0.010000	2.000140	FISIO	
AST	0.000000	1.200000	1.990000 A	LZ_LK5	
2	0.000000	6.346400	2.090000	AIR	
3	10.544000	3.100000	2.910000	LZ_TF5	
4	5.012000	2.700000	2.710000	LZ_LK3	
5	-7.798000	3.700000	2.780000	AIR	
6	-8.913000	3.100000	2.340000	LZ_TF5 P	
7	-5.943000	2.300000	2.500000	LZ_K8	
8	54.330000	13.400000	2.460000	AIR	
9	-247.200000	6.800000	3.550000	LZ_K8 P	
10	-8.750000	2.800000	4.320000	LZ_TF5 P	
11	-14.289000	90.000000	4.900000	AIR	
12	214.800000	5.000000	10.610000	LZ_BK8	
13	-46.990000	0.300000	10.690000	AIR	
14	-46.880000	3.000000	10.660000	LZ_BF24	
15	-113.240000	157.832612 S	10.790000	AIR	
IMS	0.000000	0.000000	10.000000		

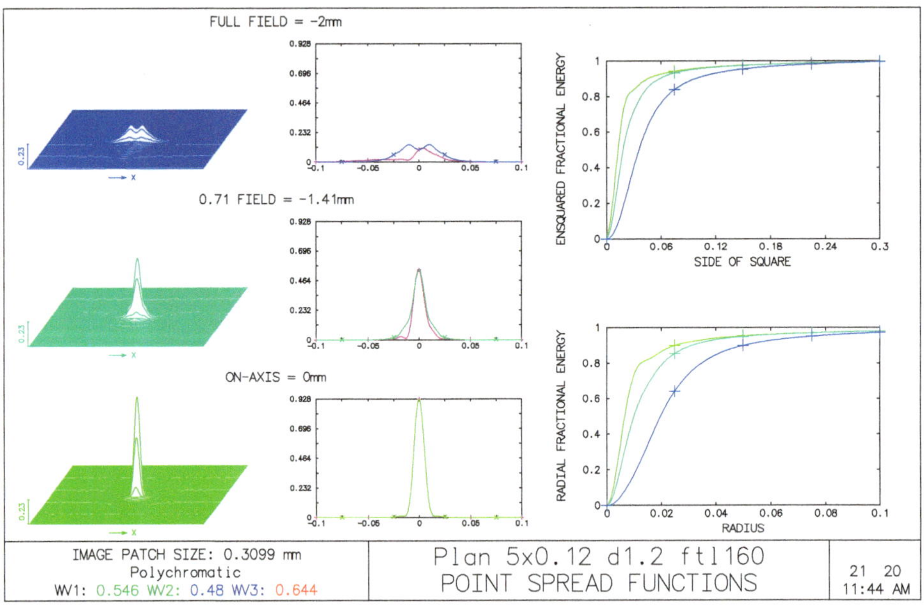

Figure 1.127 Optical design and graphs of aberration correction of a 5×/0.12 infinity LCF plan achromatic objective ($d = 1.2$ mm; $F'_{ts} = 160$ mm).

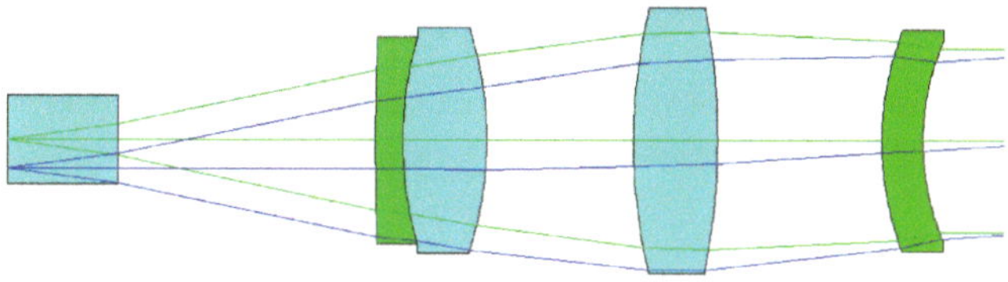

Lens: Plan 10x0.20 d4.0 ftl160

Object num aper 0.200000 Image height 10.000000 Primary wavln 0.546070

SRF	RADIUS	THICKNESS	APERTURE RADIUS	GLASS	SPECIAL
OBJ	0.000000	0.010000	1.005312	FISIO	
AST	0.000000	4.000000	1.010000 A	LZ_LK5	
2	0.000000	9.307500	1.550000	AIR	
3	112.720000	1.000000	3.460000	LZ_TF5	
4	13.183000	3.000000	3.610000	LZ_LK3	
5	-13.183000	5.270000	3.940000	AIR	
6	19.498000	3.000000	4.660000	LZ_LK3 P	
7	-24.320000	5.920000	4.650000	AIR	
8	10.520000	1.500000	3.870000	LZ_TF5 P	
9	8.395000	90.000000	3.520000	AIR P	
10	214.800000	5.000000	8.550000	LZ_BK8	
11	-46.990000	0.300000	8.650000	AIR	
12	-46.880000	3.000000	8.630000	LZ_BF24	
13	-113.240000	157.829169 S	8.760000	AIR	
IMS	0.000000	0.000000	10.000000		

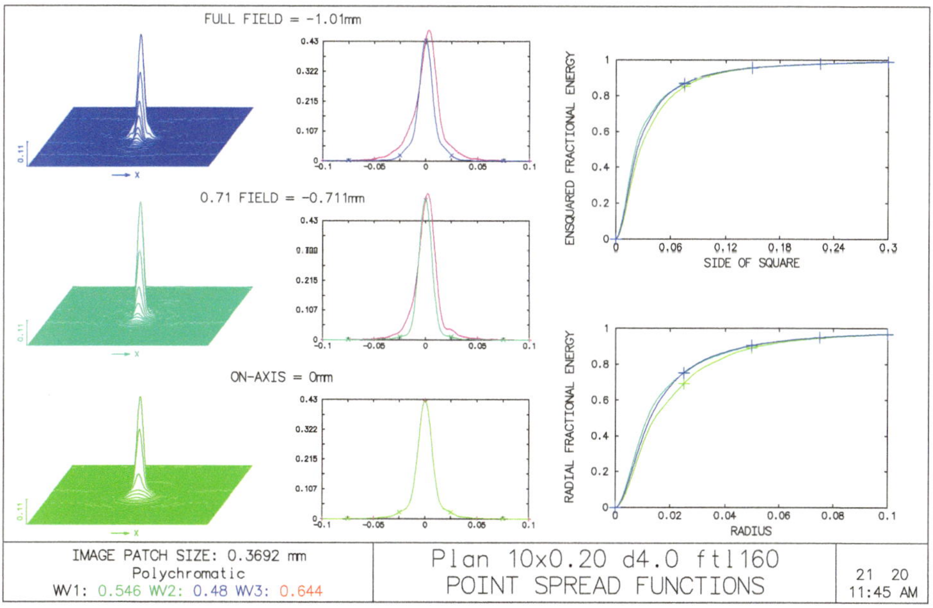

Figure 1.128 Optical design and graphs of aberration correction of a 10×/0.20 infinity LCF plan achromatic objective ($d = 4.0$ mm; $F'_{ts} = 160$ mm).

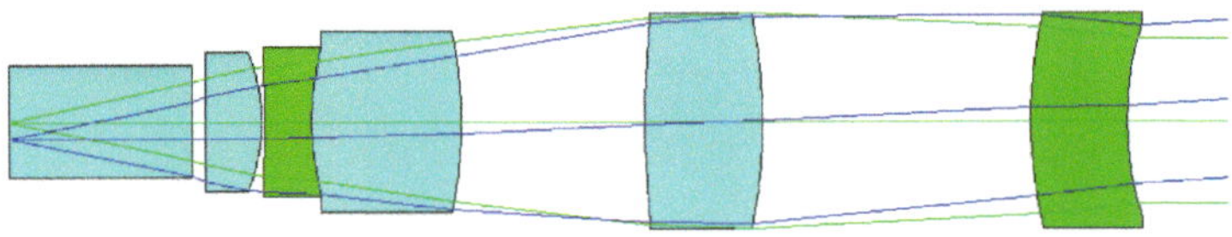

Lens: Plan 20x0.3 d5.5 ftl160

Object num aper	0.300000	Image height	10.000000	Primary wavln	0.546070

SRF	RADIUS		THICKNESS		APERTURE RADIUS		GLASS		SPECIAL
OBJ	0.000000		0.010000		0.500659		FISIO		
AST	0.000000		5.500000		0.500000	A	LZ_LK5		
2	0.000000		0.419800		1.640000		AIR		
3	0.000000		1.700000		1.770000		LZ_K8		
4	-5.035000		0.100000		2.030000		AIR		
5	-75.160000		1.400000		2.070000		LZ_TF5		
6	7.870000		4.500000		2.180000		LZ_K8	P	
7	-12.647000		5.500000		2.630000		AIR		
8	30.550000		3.500000		3.040000		LZ_K8	P	
9	-15.136000		8.000000		3.150000		AIR		
10	13.583000		2.900000		3.150000		LZ_TF5	P	
11	9.290000		90.000000		2.850000		AIR	P	
12	214.800000		5.000000		8.520000		LZ_BK8		
13	-46.990000		0.300000		8.630000		AIR		
14	-46.880000		3.000000		8.610000		LZ_BF24		
15	-113.240000		157.819283	S	8.740000		AIR		
IMS	0.000000		0.000000		10.000000				

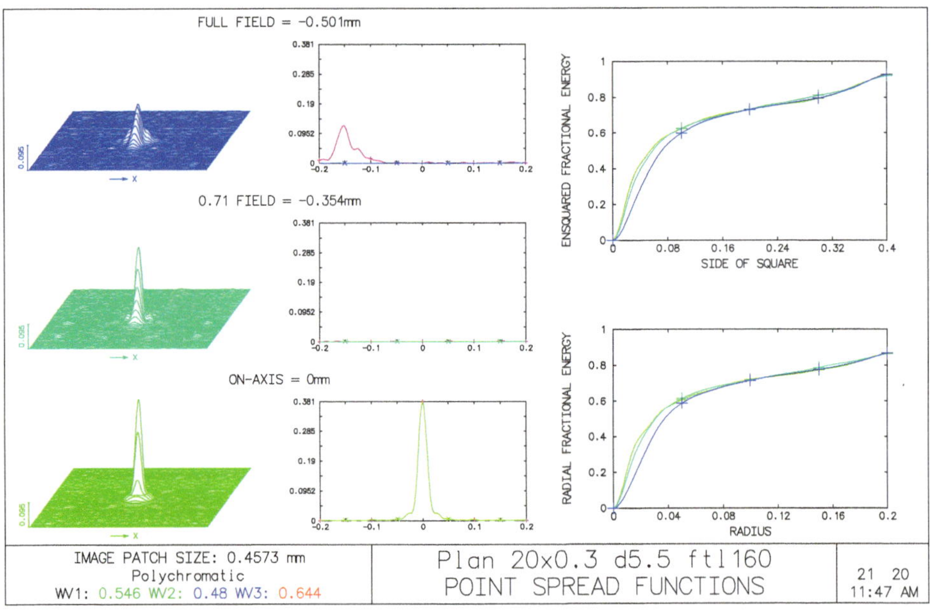

Figure 1.129 Optical design and graphs of aberration correction of a 20×/0.30 infinity LCF plan achromatic objective ($d = 5.5$ mm; $F'_{ts} = 160$ mm).

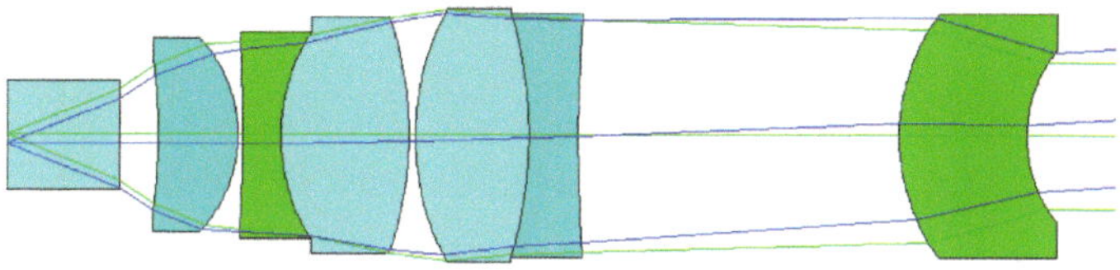

Lens: Plan 40x0.55 d3.5 ftl160

Object num aper	0.550000	Image height	10.000000	Primary wavln	0.546070

SRF	RADIUS	THICKNESS	APERTURE RADIUS	GLASS	SPECIAL
OBJ	0.000000	0.010000	0.252388	FISIO	
AST	0.000000	3.500000	0.260000 A	LZ_LK5	
2	0.000000	1.206900	1.660000	AIR	
3	-17.947000	2.500000	2.350000	LZ_CTK19	
4	-4.207000	0.150000	2.940000	AIR	
5	-48.420000	1.200000	3.010000	LZ_TF5	
6	5.728000	4.000000	3.130000	LZ_K8	
7	-10.666000	0.200000	3.580000	AIR	
8	7.980000	3.500000	3.860000	LZ_K8 P	
9	-12.794000	1.500000	3.710000	LZ_CTK19 P	
10	54.580000	10.000000	3.650000	AIR	
11	6.223000	4.000000	3.710000	LZ_TF5 P	
12	4.093000	90.000000	2.600000	AIR P	
13	214.800000	5.000000	8.070000	LZ_BK8	
14	-46.990000	0.300000	8.180000	AIR	
15	-46.880000	3.000000	8.160000	LZ_BF24	
16	-113.240000	157.841152 S	8.280000	AIR	
IMS	0.000000	0.000000	10.000000		

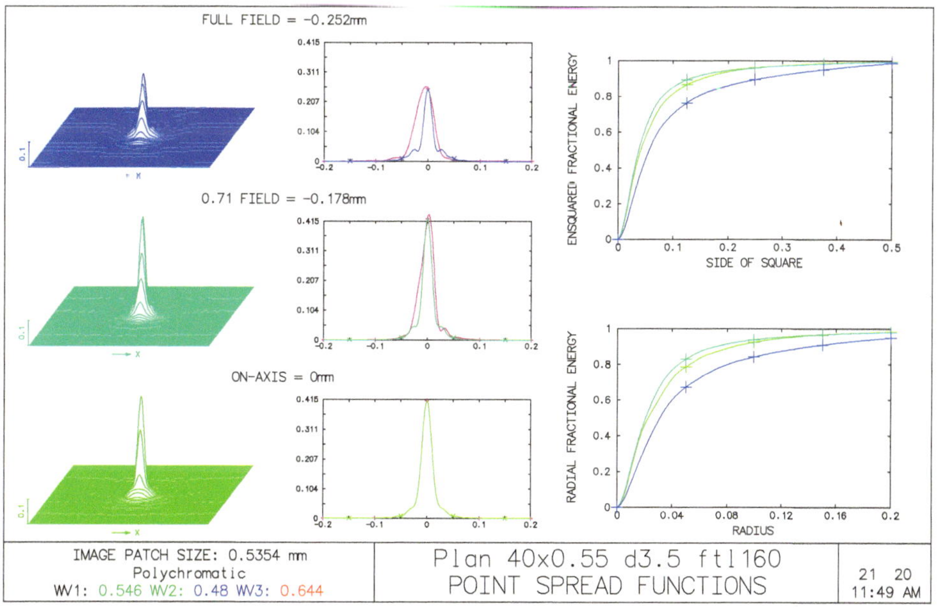

Figure 1.130 Optical design and graphs of aberration correction of a 40×/0.55 infinity LCF plan achromatic objective ($d = 3.5$ mm; $F'_{ts} = 160$ mm).

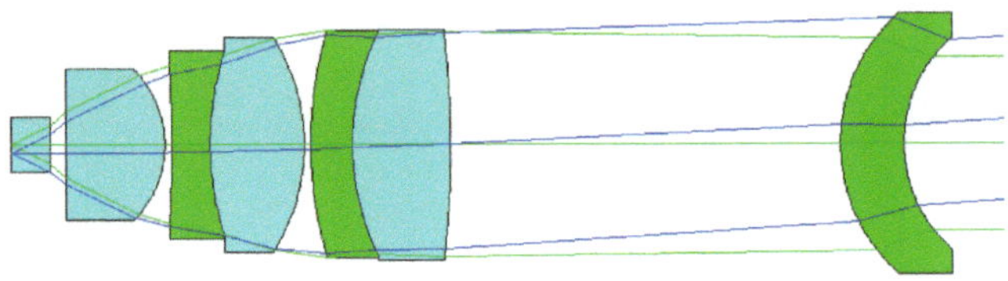

Lens: S_Plan 40x0.65 d1.2 ftl160

Object num aper 0.650000 Image height 10.000000 Primary wavln 0.546070

SRF	RADIUS	THICKNESS	APERTURE RADIUS	GLASS	SPECIAL
OBJ	0.000000	0.010000	0.254956	FISIO	
AST	0.000000	1.200000	0.260000 A	LZ_LK5	
2	0.000000	0.496400	0.850000	AIR	
3	0.000000	3.160000	1.270000	LZ_K8	
4	-3.192000	0.220000	2.310000	AIR	
5	-40.090000	1.200000	2.580000	LZ_TF5	
6	9.036000	3.000000	2.870000	LZ_LK3	
7	-5.970000	0.200000	3.280000	AIR	
8	11.588000	1.300000	3.470000	LZ_TF5 P	
9	7.551000	3.100000	3.400000	LZ_LK3 P	
10	-32.890000	12.200000	3.520000	AIR	
11	5.224000	2.000000	3.980000	LZ_TF5 P	
12	4.207000	90.000000	3.250000	AIR P	
13	214.800000	5.000000	8.510000	LZ_BK8	
14	-46.990000	0.300000	8.610000	AIR	
15	-46.880000	3.000000	8.590000	LZ_BF24	
16	-113.240000	157.757721 S	8.700000	AIR	
IMS	0.000000	0.000000	10.000000		

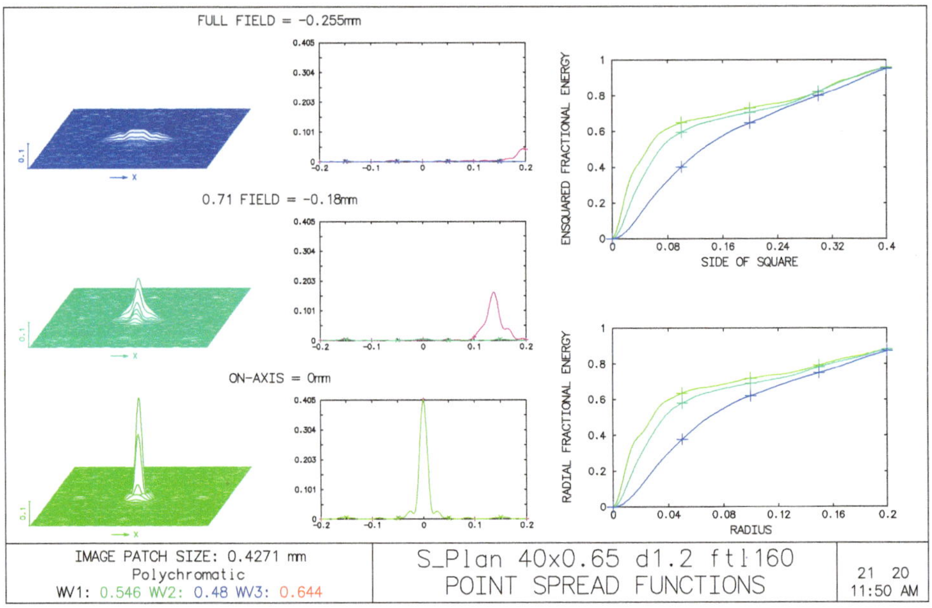

Figure 1.131 Optical design and graphs of aberration correction of a 40×/0.65 infinity LCF semi–plan achromatic objective ($d = 1.2$ mm; $F'_{ts} = 160$ mm).

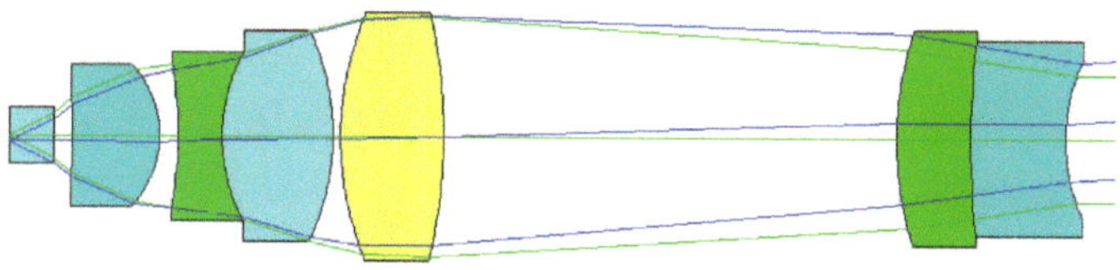

Lens: Plan 63x0.65 d1.2 ftl160

Object num aper 0.650000 Image height 10.000000 Primary wavln 0.546070

SRF	RADIUS	THICKNESS	APERTURE RADIUS	GLASS	SPECIAL
OBJ	0.000000	0.010000	0.159796	FISIO	
AST	0.000000	1.200000	0.160000 A	LZ_LK5	
2	0.000000	0.496900	0.750000	AIR	
3	-23.442000	2.350000	1.150000	LZ_CTK19	
4	-2.656000	0.490000	1.860000	AIR	
5	-11.940700	1.200000	1.970000	LZ_TF5	
6	4.572000	3.000000	2.220000	LZ_LK3	
7	-6.223000	0.210000	2.770000	AIR	
8	8.205000	2.800000	3.180000	CAF2	
9	-14.061000	12.200000	3.270000	AIR	
10	8.995000	2.000000	2.820000	LZ_TF5 P	
11	23.442000	2.600000	2.550000	LZ_CTK19 P	
12	4.922000	90.000000	2.070000	AIR P	
13	214.800000	5.000000	7.560000	LZ_BK8	
14	-46.990000	0.300000	7.680000	AIR	
15	-46.880000	3.000000	7.660000	LZ_BF24	
16	-113.240000	157.753179 S	7.780000	AIR	
IMS	0.000000	0.000000	10.000000		

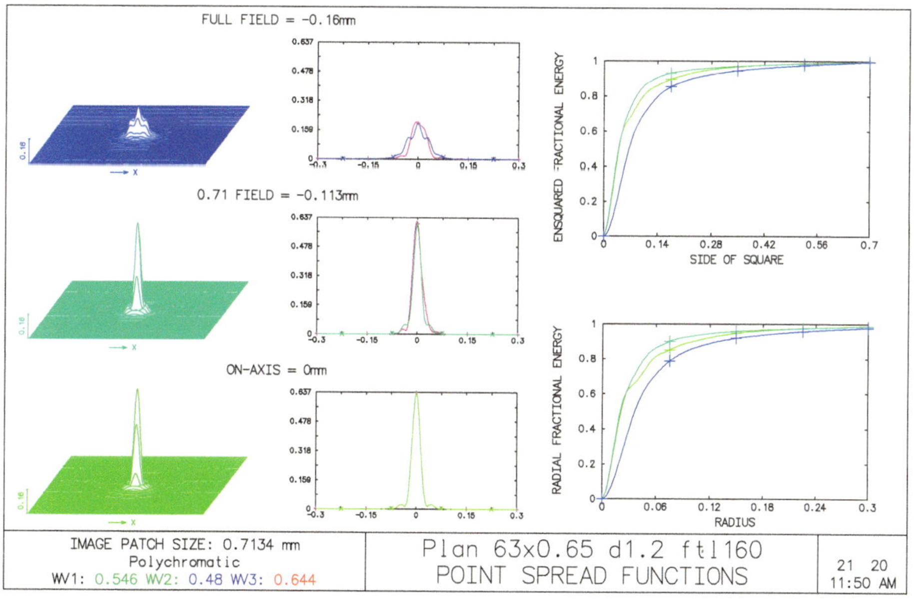

Figure 1.132 Optical design and graphs of aberration correction of a 63×/0.65 infinity LCF semi–plan achromatic objective ($d = 1.2$ mm; $F'_{ts} = 160$ mm).

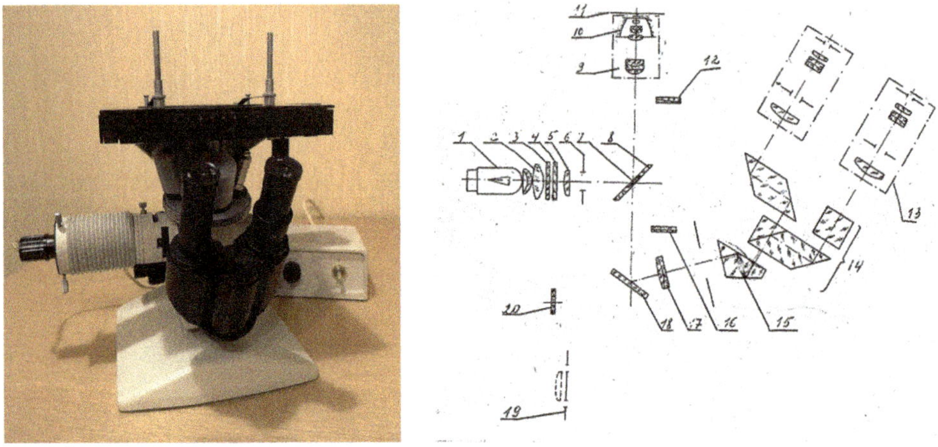

Figure 1.133 Photograph of the Metam-RV metallurgical microscope and its principal optical layout.

reduce an aberration such as the secondary spectrum. Microscopic polarizing objects have a rich spectrum of natural colors and shades. When studying and comparing these spectra and shades, identification of one or another type of object is carried out. Correct color rendering and the absence (or reduction) of chromatic aberration are important parameters in a polarizing microscope, both in orthoscopic and conoscopic studies.

A well-known tool for reducing the secondary spectrum in an optical system is the use of fluorite crystal lenses. However, fluorite cannot be recommended for polarizing objectives because it almost always differs in anomalous birefringence. Modern types of optical glass make it possible to replace fluorite with similar materials in terms of their optical and mechanical properties. In modern catalogs of companies that manufacture optical glass, some brands of fluorophosphate glasses are presented that have unique properties, such as relative partial dispersion. Some brands in their optical parameters are analogues of fluorite.

In optical schemes of modern objectives for budget polarizing microscopes, one can use either ordinary silicate glass grades or optical materials that have advantages such as relative partial dispersion; during testing, the optical materials used in the lenses (and other optics of a polarizing microscope) must meet the requirements for isotropy. Thus, all the objectives presented in this chapter can be applied to polarization methods if the corresponding requirements are met.

If we discuss even broader categories, the creation of objectives (and other optics of a light microscope) that can be used universally in most known research methods using a polarizing microscope can be considered a feat of engineering. Optical systems should enable the use of techniques such as

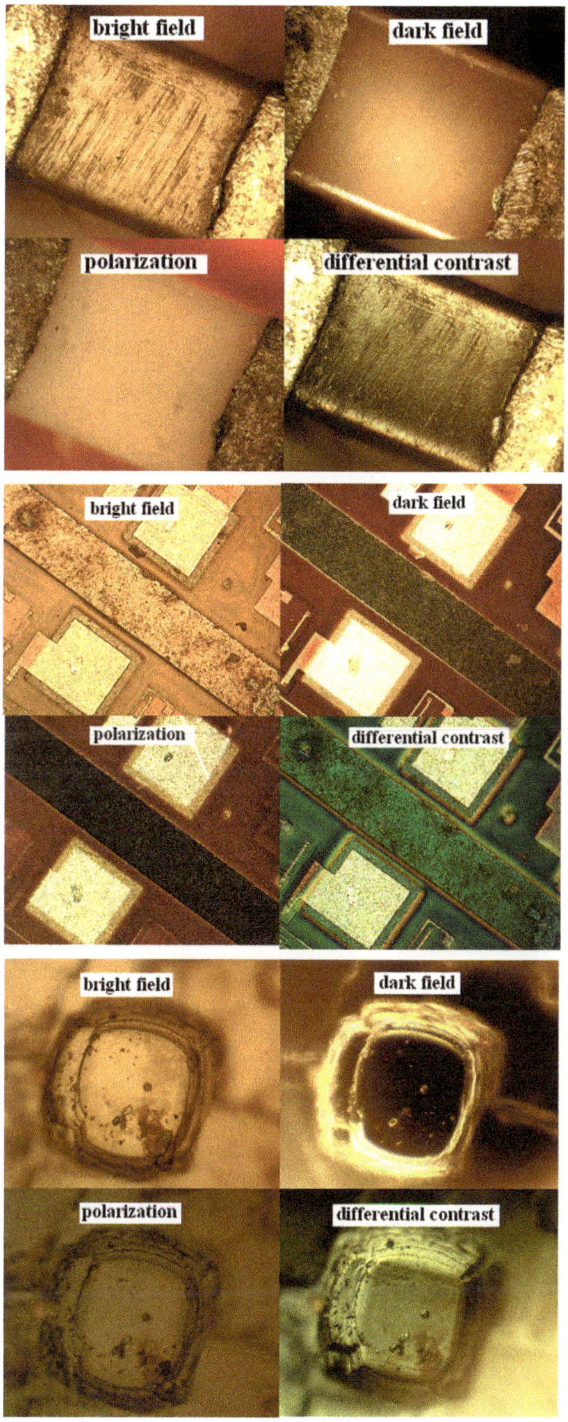

Figure 1.134 Comparison of images using different contrast methods.

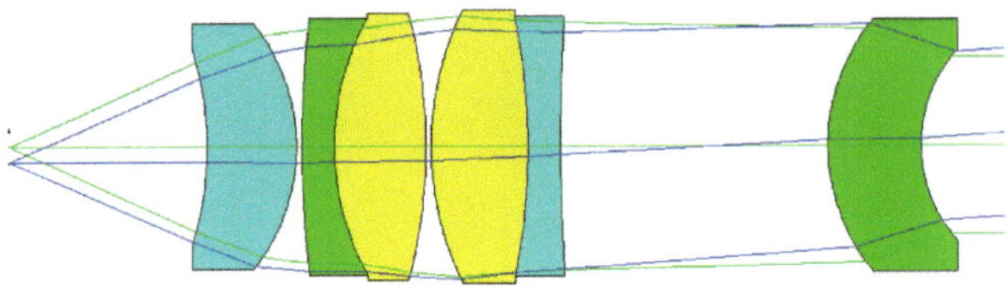

Lens: Plan 20x0.40 LD Ftl160

Object num aper	0.400000	Image height	11.000000	Primary wavln	0.546070

SRF	RADIUS	THICKNESS	APERTURE RADIUS	GLASS	SPECIAL
OBJ	0.000000	1.0000e-20	0.549970	AIR	
AST	0.000000	7.500000	0.549970 AS	AIR	
2	-11.623100	3.400000	3.570000	LZ_CTK19	
3	-6.875100	0.200000	4.470000	AIR	
4	37.490500	1.200000	4.670000	LZ_TF5	
5	9.485100	3.400000	4.660000	LZ_OK4	
6	-18.233400	0.200000	4.850000	AIR	
7	10.864300	3.600000	4.960000	LZ_OK4 P	
8	-24.115900	1.200000	4.730000	LZ_CTK19 P	
9	89.851000	10.000000	4.610000	AIR	
10	6.933400	3.500000	4.570000	LZ_TF5 P	
11	5.097400	100.000000	3.490000	AIR	
12	214.800000	5.000000	10.270000	LZ_BK8	
13	-46.990000	0.300000	10.360000	AIR	
14	-46.880000	3.000000	10.340000	LZ_BF24	
15	-113.240000	157.857380 S	10.480000	AIR	
IMS	0.000000	0.000000	11.000000 S		

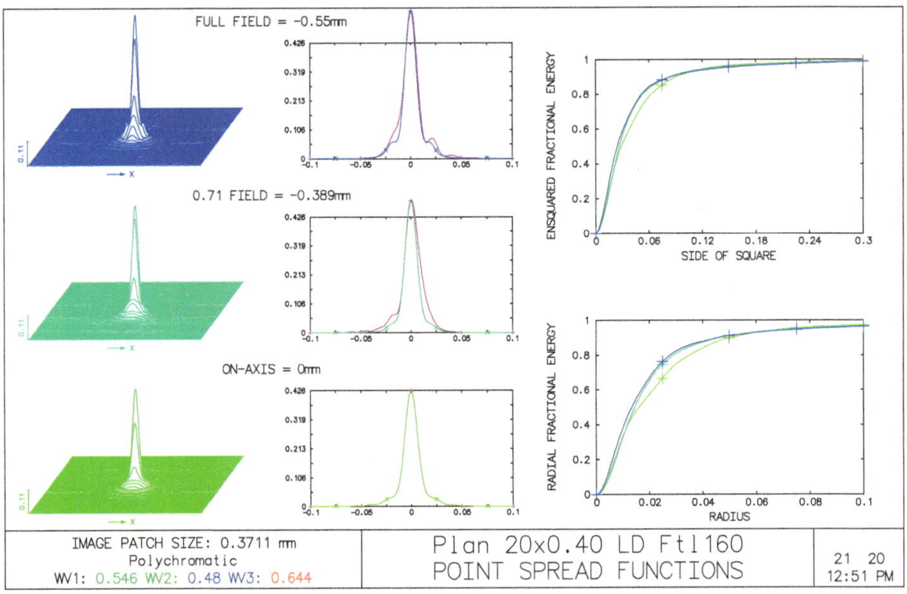

Figure 1.135 Optical design and graphs of aberration correction of a 20×/0.40 infinity LCF LD plan achromatic objective ($F'_{ts} = 160$ mm).

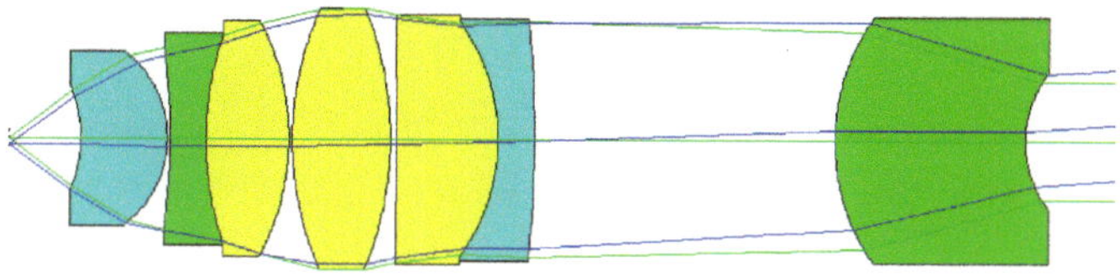

Lens: Plan 50x0.60 LD Ftl160

Object num aper 0.600000 Image height 11.000000 Primary wavln 0.546070

SRF	RADIUS	THICKNESS	APERTURE RADIUS	GLASS	SPECIAL
OBJ	0.000000	1.0000e-20	0.220112	AIR	
AST	0.000000	2.400000	0.220112 AS	AIR	
2	-4.433000	2.900000	1.750000	LZ_CTK19	
3	-3.514000	0.100000	2.840000	AIR	
4	-22.777600	1.200000	3.110000	LZ_TF5	
5	10.171400	2.800000	3.450000	LZ_OK4	
6	-7.788800	0.100000	3.820000	AIR	
7	10.092300	3.300000	4.230000	LZ_OK4 P	
8	-10.872100	0.200000	4.240000	AIR	
9	-215.802200	3.400000	4.050000	LZ_OK4 P	
10	-6.318900	1.200000	3.820000	LZ_CTK19 P	
11	-37.554200	10.000000	3.930000	AIR	
12	6.751800	6.400000	3.990000	LZ_TF5 P	
13	3.527300	100.000000	2.230000	AIR	
14	214.800000	5.000000	9.110000	LZ_BK8	
15	-46.990000	0.300000	9.230000	AIR	
16	-46.880000	3.000000	9.210000	LZ_BF24	
17	-113.240000	157.853249 S	9.350000	AIR	
IMS	0.000000	0.000000	11.000000 S		

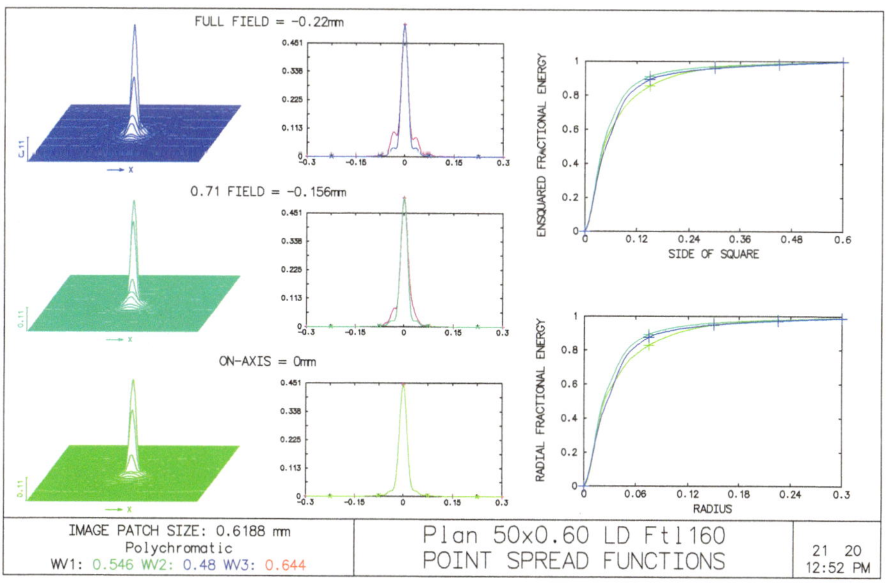

Figure 1.136 Optical design and graphs of aberration correction of a 50×/0.60 infinity LCF LD plan achromatic ($F'_{ts} = 160$ mm).

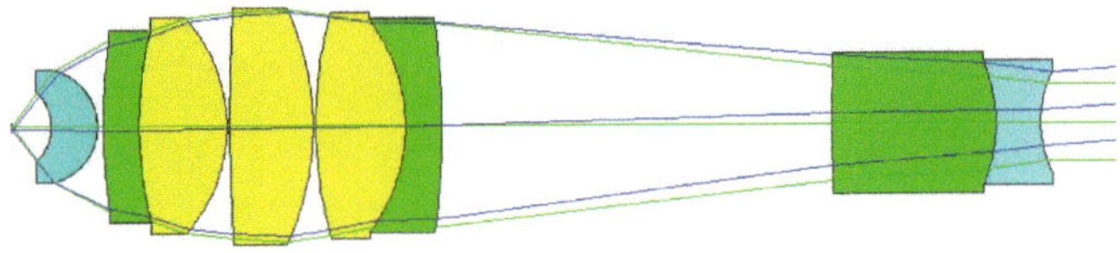

Lens: Plan 100x0.80 LD Ftl160

Object num aper	0.800000	Image height	11.000000	Primary wavln	0.546070

SRF	RADIUS	THICKNESS	APERTURE RADIUS	GLASS	SPECIAL
OBJ	0.000000	1.0000e-20	0.109864	AIR	
AST	0.000000	1.310000	0.100000 A	AIR	
2	-1.524700	1.500000	1.150000	LZ_CTK19	
3	-1.885300	0.200000	1.840000	AIR	
4	21.661000	1.200000	2.810000	LZ_TF5	
5	11.827900	3.000000	3.110000	LZ_OK4	
6	-5.090400	0.100000	3.510000	AIR	
7	78.613000	2.800000	3.710000	LZ_OK4 P	
8	-8.692200	0.100000	3.850000	AIR	
9	13.253500	3.000000	3.670000	LZ_OK4 P	
10	-5.656300	1.200000	3.450000	LZ_TF5 P	
11	-26.699500	13.000000	3.470000	AIR	
12	196.300200	5.500000	2.290000	LZ_TF5 P	
13	-5.268600	1.500000	2.020000	LZ_K8	
14	3.615800	100.000000	1.710000	AIR	
15	214.800000	5.000000	8.730000	LZ_BK8	
16	-46.990000	0.300000	8.850000	AIR	
17	-46.880000	3.000000	8.830000	LZ_BF24	
18	-113.240000	158.482417 S	8.970000	AIR	
IMS	0.000000	0.000000	11.000000 S		

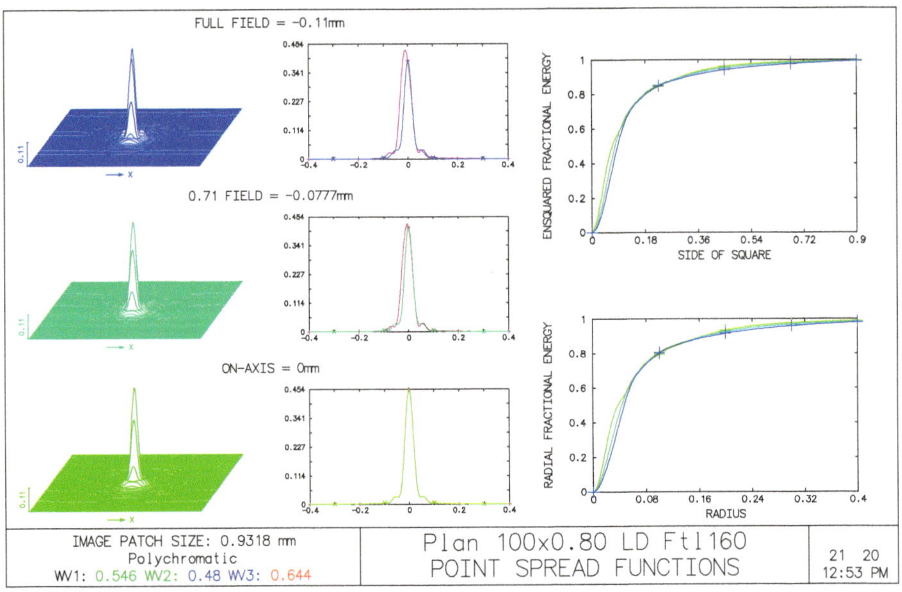

Figure 1.137 Optical design and graphs of aberration correction of a 100×/0.80 infinity LCF LD plan achromatic ($F'_{ts} = 160$ mm).

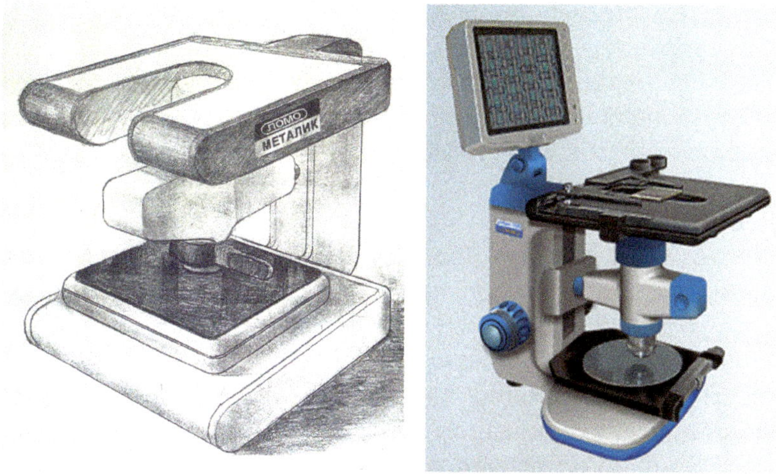

Figure 1.138 Drawing of the basic layout of a "changeling" microscope.

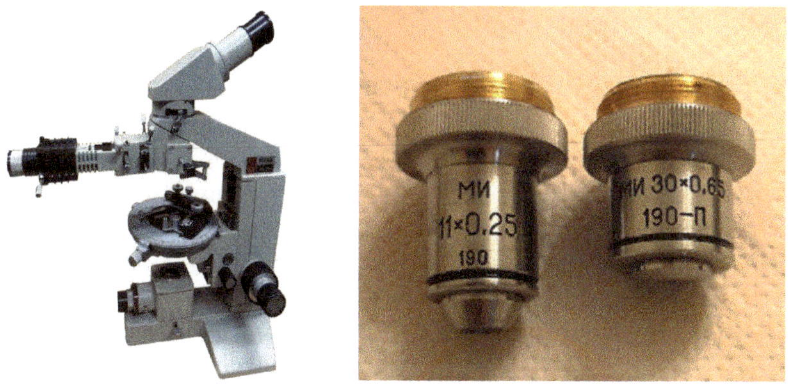

Figure 1.139 Drawing of a Polam-R microscope and photograph of oil immersion objectives.

orthoscopy, conoscopy, phase contrast, focal screening, fluorescence, and others. For example, some modern ones implement polarization and fluorescence studies simultaneously. Therefore, the optical systems of these objectives should not only be free of stresses and have isotropy, but also work in the near UV region, where luminescence excitation usually occurs. Optical glass should have good transmission throughout the entire spectral range. In addition, the requirements for optical adhesives and coatings on lenses should also take all of these factors into account. To implement the methods of focal screening and phase contrast, the optical systems of the lenses should make it possible to integrate iris aperture diaphragms, as well as plane-parallel phase plates and wedges for differential contrast, into the optical scheme. If the equivalent focus plane in the objectives is located in the air gap, then such integration is possible.

By the way, it is precisely for polarization studies that an objective that can work with various types of immersions (water, glycerin, oil, etc.) can be used most effectively. Finally, to achieve high contrast in conoscopy and, especially, in orthoscopy, attention must be paid to reducing the influence of spurious illumination that occurs when reflections from planar *mirror* fragments of the object under study. Optical lens systems should also reduce the likelihood of a false birefringence effect.

When working with some objectives, the use of an immersion is expected. For example, with an immersion oil, the refractive index is close to 1.518 for a wavelength of 546 nm; the Abbe number is about 48. In the proposed set below, optical systems were created, including the *twin* objectives, in which for each objective working without immersion there is a corresponding objective with the same linear magnification and numerical aperture operating in an oil immersion. Such objectives are necessary for the study of both polished minerals those of weakly reflecting objects, for example, coals. The fine structure of charcoal grinding, which is poorly observed when viewed with *dry* objectives, is of higher contrast with the use of immersion.

Another big problem is that the presence of a significant amount of scattered light and reflexes from the lens surface reduces image clarity. Grammatin et al. [12] and Arlievsky [13] have studied the possibility of reducing scattered light in objectives operating on reflection. They concluded that the most effective way to reduce scattered light is to perform an optical design based on the use of lenses with a certain configuration. For objectives operating in a polarizing microscope, these conclusions are also valid. The use of immersion in objectives (especially small magnifications) is also one way to reduce scattered light. In addition, with the use of immersion, there are several other qualitative changes in the image due to the enhancement in the immersion of a number of optical contrasts: reflectivity, the color of a mineral in white light, the effects of anisotropy, the strength and color of internal reflexes, etc. Figures 1.140–1.142 show the design parameters and graphs of aberration correction of budget oil immersion objectives intended for use in mass segment polarizing microscopes. Figures 1.143–1.149 are photographs of a real sample taken on a polarizing microscope with a 5× linear magnification objective at various rotational positions of the stage. Figure 1.150 is a photograph of a real sample taken on a polarizing microscope with 2×, 4×, 20×, and 40× linear magnification objectives.

1.7 Topical Reasoning about Optics for Budget Microscopes

This may seem strange, but despite the advances made in motorization, for example, the sensory and *stepping* mechanisms used in many consumer devices, the light microscope of the mass-market segment is still a very primitive *optical–mechanical* device that uses technologies from the 19th

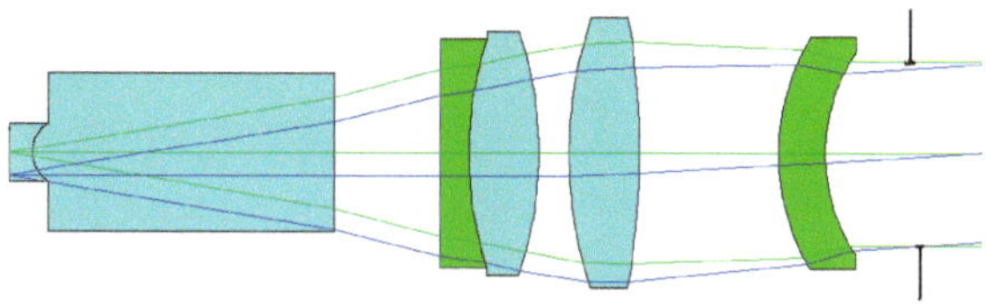

Lens: Plan 10x0.25 oil f160

Object num aper 0.250000 Image height 10.000000 Primary wavln 0.546070

SRF	RADIUS	THICKNESS	APERTURE RADIUS	GLASS	SPECIAL
OBJ	0.000000	1.0000e-20	0.966923	AIR	
AST	0.000000	0.992200	1.000000 A	OIL	
2	1.499700	13.000000	1.240000	LZ_K8	
3	0.000000	4.500000	3.320000	AIR	
4	0.000000	1.200000	4.510000	LZ_TF5	
5	15.560000	3.000000	4.790000	LZ_K8 P	
6	-15.560000	1.300000	5.110000	AIR	
7	18.030000	3.000000	5.650000	LZ_K8 P	
8	-32.810000	6.000000	5.640000	AIR	
9	9.290000	2.000000	4.860000	LZ_TF5 P	
10	7.656000	90.000000	4.260000	AIR	
11	214.800000	5.000000	9.160000	LZ_BK8	
12	-46.990000	0.300000	9.260000	AIR	
13	-46.880000	3.000000	9.240000	LZ_BF24	
14	-113.240000	157.832179 S	9.370000	AIR	
IMS	0.000000	0.000000	10.000000		

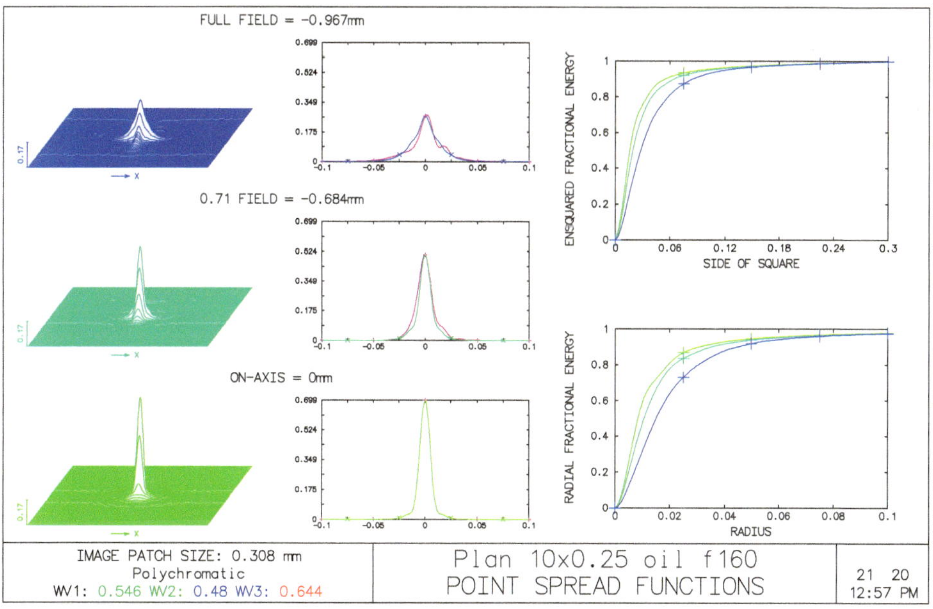

Figure 1.140 Optical design and graphs of aberration correction of a 10×/0.25 oil immersion infinity LCF plan achromatic objective ($F'_{ts} = 160$ mm).

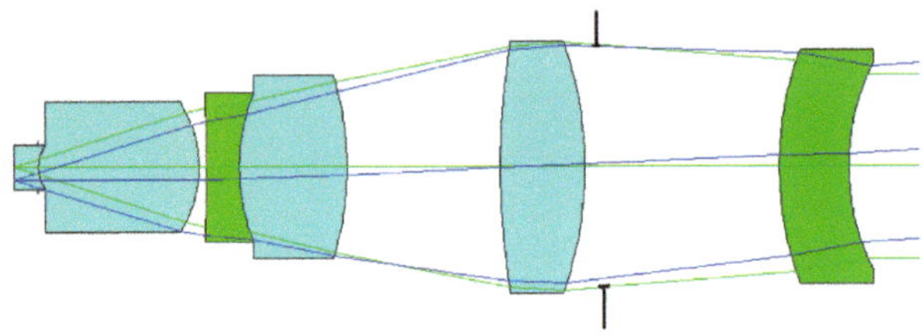

Lens: ocx 20x0.45 oil ftl160

Object num aper 0.450000 Image height 10.000000 Primary wavln 0.546070

SRF	RADIUS	THICKNESS	APERTURE RADIUS	GLASS	SPECIAL
OBJ	0.000000	1.0000e-20	0.521993	AIR	
AST	0.000000	0.996100	0.521993 AS	OIL	
2	1.499700	6.730000	0.910000	LZ_K8	
3	-5.035000	0.300000	2.660000	AIR	
4	-75.160000	1.400000	2.820000	LZ_TF5	
5	7.870000	4.500000	3.050000	LZ_K8	
6	-12.647000	6.410000	3.720000	AIR	
7	30.550000	3.500000	4.920000	LZ_K8 P	
8	-15.136000	8.000000	5.130000	AIR	
9	13.583000	2.900000	4.760000	LZ_TF5	
10	9.290000	90.000000	4.210000	AIR	
11	214.800000	5.000000	10.220000	LZ_BK8	
12	-46.990000	0.300000	10.310000	AIR	
13	-46.880000	3.000000	10.280000	LZ_BF24	
14	-113.240000	157.832557 S	10.400000	AIR	
IMS	0.000000	0.000000	10.000000		

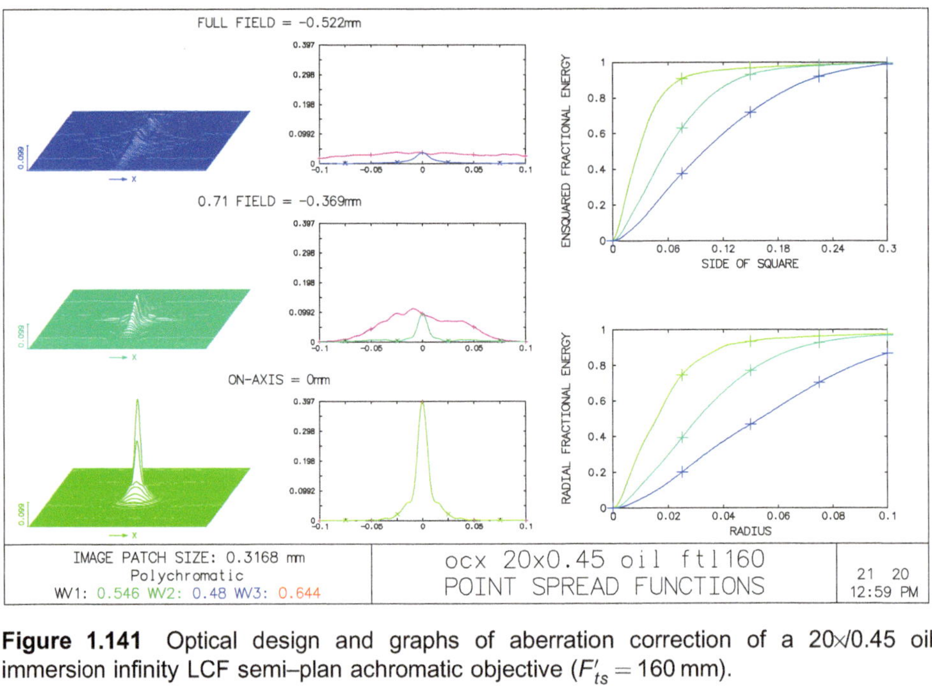

Figure 1.141 Optical design and graphs of aberration correction of a 20×/0.45 oil immersion infinity LCF semi–plan achromatic objective ($F'_{ts} = 160$ mm).

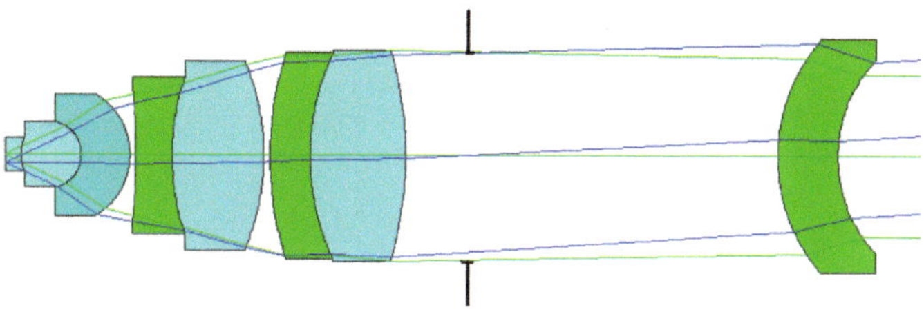

Lens: ocx 40x0.65 oil ftl160

Object num aper 0.650000 Image height 10.000000 Primary wavln 0.546070

SRF	RADIUS	THICKNESS		APERTURE RADIUS		GLASS		SPECIAL
OBJ	0.000000	1.0000e-20		0.249086		AIR		
AST	0.000000	0.519670		0.249086	AS	OIL		
2	1.499700	2.000000		0.540000		LZ_K8		
3	-1.056800	1.600000		1.040000		LZ_CTK19		
4	-2.270000	0.200000		1.960000		AIR		
5	-20.700000	1.200000		2.220000		LZ_TF5		
6	7.727000	3.000000		2.510000		LZ_LK3		
7	-7.727000	0.200000		3.040000		AIR		
8	10.814000	1.300000		3.310000		LZ_TF5	P	
9	7.727000	3.100000		3.270000		LZ_LK3	P	
10	-12.359000	12.200000		3.390000		AIR		
11	5.598000	2.000000		3.750000		LZ_TF5		
12	4.529000	90.000000		3.120000		AIR		
13	214.800000	5.000000		8.620000		LZ_BK8		
14	-46.990000	0.300000		8.730000		AIR		
15	-46.880000	3.000000		8.710000		LZ_BF24		
16	-113.240000	157.831710	S	8.830000		AIR		
IMS	0.000000	0.000000		10.000000				

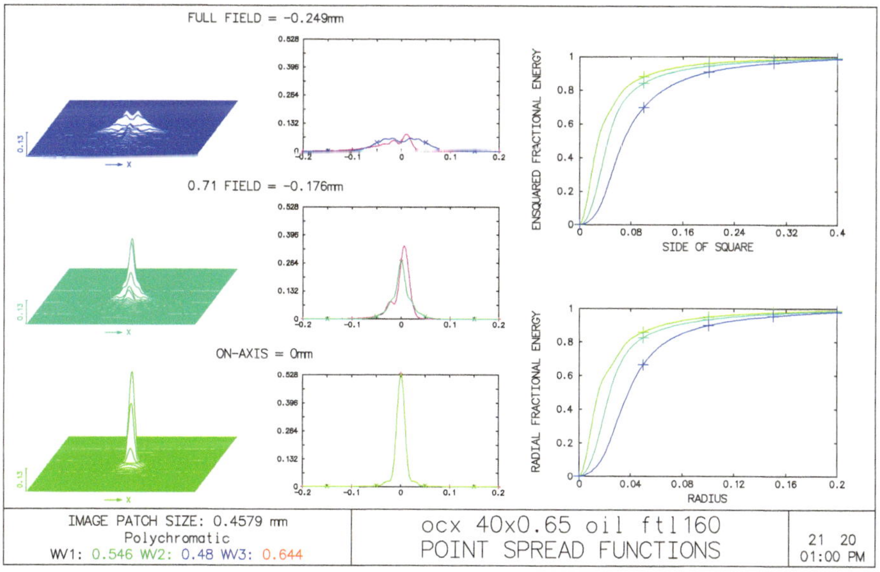

Figure 1.142 Optical design and graphs of aberration correction of a 40×/0.65 oil immersion infinity LCF semi–plan achromatic objective ($F'_{ts} = 160$ mm).

Figure 1.143 Photograph of a real polarizing object using a 5× objective at position 1.

Figure 1.144 Photograph of a real polarizing object using a 5× objective at position 2.

century. For example, why are researchers forced to use a mechanical focusing system, rotating the thumbs of the drives with their hands, and look through the eyepiece when digital image receivers are developing so rapidly? Compared with any modern digital camera—not to mention smartphones and other digital gadgets—a modern budget microscope seems like a medieval monster that cannot adapt to the realities of today.

Figure 1.145 Photograph of a real polarizing object using a 5× objective at position 3.

Figure 1.146 Photograph of a real polarizing object using a 5× objective at position 4.

1.7.1 The design features of objectives for microscopy of mass production

What is the explanation for this significant obstacle in the progress of budget microscopy? The answer is obviously optics, the optical system of a microscope, the basis of which is the objective. In budget microscopy, which occupies at least 75% of the market, *first-generation* objectives, proposed by

Figure 1.147 Photograph of a real polarizing object using a 5× objective at position 5.

Figure 1.148 Photograph of a real polarizing object using a 5× objective at position 6.

Abbe many decades ago, are used. These are achromatic objectives, the aberration correction of which is far from perfect; moreover, they are full of compromises. Attempts to modernize the optical systems of Abbe objectives have not been undertaken for more than 100 years since these systems seemed to be as simple and cheap to manufacture as possible. However, in Russia, optical designs were made and objectives were produced in which, with a simpler optical system, it was possible to maintain the quality of aberration

Figure 1.149 Photograph of a real polarizing object using a 5× objective at position 7.

Figure 1.150 Photograph of a real polarizing object using 2×, 4×, 20×, and 40× objectives.

correction of Abbe objectives. A description of the construction of optical systems and the design parameters of the simplest objectives for use in mass-produced microscopes is presented in [14]. Optical designs and graphs of aberration correction of some of these objectives are shown in Figs. 1.2–1.13. It can be seen that to improve the technological parameters of these objectives, for example, lenses with flat surfaces or the same radii are often used. Principal optical schemes and sketches of mechanical design are presented in Fig. 1.151.

It is believed that optical design, mechanical design, and manufacturing of objectives for the mass segment of light microscopes are simpler tasks than, for example, mastering complex objectives for research microscopes. This is partly true if we do not consider the most important criterion for the consumer of objectives of the mass segment: their low cost. The cost of complex objectives does not have a significant value in the overall price structure, as research microscopes can be extremely expensive; their number in the microscope market is not very large.

If we consider the final cost of objectives for microscopes of the mass segment (it should be low) as a complex characteristic combining all the costs of developing and manufacturing each objective, we can conclude that solving the problem of mass production of such objectives is not a trivial task. Let us list the main technological requirements that such objectives should meet.

1. The number of optical (and mechanical, respectively) elements in each objective should be minimal.
2. All optical details of the objectives in the kit should be manufactured using a single technology, which can be called *grouped*, i.e., a group of parts can be made by using the same forming tool. An example of this is the so-called *ball technology* used for producing the front lenses of objectives in which the required shape of the ball is achieved by rubbing a large number of glass blanks against each other when the

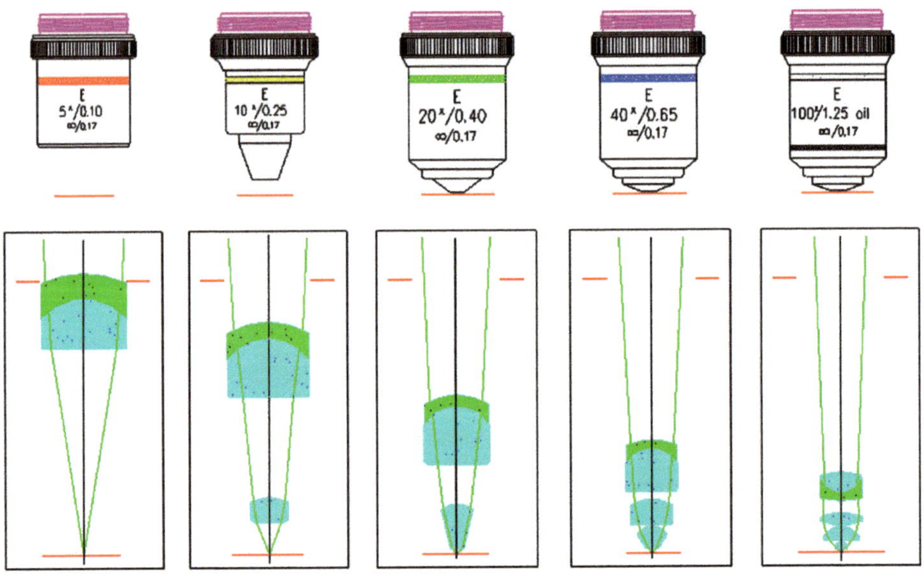

Figure 1.151 Optical layouts and sketches of the mechanical design of the simplest objectives.

drum rotates, inside which they are all placed. For other types of lenses, there are different standard technologies that are well mastered in the mass production of objectives. An additional requirement is that the brands of optical materials used in such lenses should be well mastered for the mass production of objectives (these should be the usual and cheapest brands of silicate glasses; for example, the use of crystals is not allowed). Such optical materials should have the best technological parameters, such as low hardness on grinding, low coefficient of temperature expansion, low hygroscopic, and others. Additional requirements for lenses for objectives of the mass segment may involve the formation of lenses. For example, it is necessary to avoid the critical magnitudes of the ratio of the lens diameters to the values of their radii and their thickness along the axis, the thin edge of the lens, and so on.

3. All mechanical parts of the objectives in the kit should be manufactured by *group methods*; for example, the same type of parts should be made of the same alloys and metals, preferably of the standard size. The use of high-performance machining machines with numerical program control has become a common practice in the manufacture of parts of this kind.
4. The use of a unified, complex method of assembling objectives for mass-produced microscopes is the main requirement that should be envisaged at the optical design stage and confirmed at all other design and manufacturing stages. For example, a so-called *filled-type* way of assembling the lens should be designed in which individual components are sequentially integrated ("filled") into the common housing. These components (as a rule) are lenses glued into mechanical frames. Note that the design of the component allows for the possibility of subsequent disassembly (dismantling) of the component.
5. The advent of gluing the lens into the frame with the help of superglue, which can polymerize under the influence of UV radiation for several minutes, significantly increased the productivity in the manufacture of components but eliminated the possibility of subsequent dismantling. The preferred method of inserting the lens into the frame is the so-called *collimation* method, which enables high accuracy of the component as a whole, using lenses and frames made with sufficiently wide manufacturing tolerances. The previously used technology of integration into a common lens housing without a frame can now be recognized as not modern, as in this case the accuracy of manufacturing both the lenses themselves and the case is significantly increased. The subsequent operation of adjusting the components (lens in the frame) is much simpler in comparison with others.

6. The techniques and methods of alignment of objectives should be maximally unified (preferably formalized). The technology, which is currently used for alignment of objectives for mass-produced microscopes, provides the constructive possibility of moving one of the components within small limits along and perpendicular to the equivalent optical axis of the objective. Thus, the high image quality of the assembled objective is achieved; moving the component along the optical axis eliminates residual spherical aberration, and moving perpendicular to the optical axis eliminates coma. These aberrations are revealed during assembly, arising from deviations of the real values of the design parameters of the components from nominal ones. In addition, for example, one of the components can be rotated relative to the optical axis to reduce aberrations such as astigmatism or field asymmetry. This method of objective alignment cannot be considered very high performance, because to eliminate spherical aberration, for example, multiple assembling/disassembling of the objective must be performed. An alternative to this method of assembling and aligning the objective is a technique that eliminates spherical aberration without disassembling the objective. To do this, some frames of neighboring components can be made with a small eccentric, and the design of the common housing should allow these components to be turned relative to each other without disassembling the objective. Another useful tool for assembling and adjusting objectives of the mass segment can be the use of adaptive-selective assembly (ASA) [15] technology, which, to a large extent, allows the assembly to be formalized. A description of the application of ASA technology to the sphere of optical instrumentation is given in [15]. This technology is currently not used for assembling microscope objectives; however, it provides for a comprehensive approach to assembly without involving highly qualified specialists.
7. The set of tools necessary for testing and attestation of manufactured objectives for mass-produced microscopes may differ from traditional ones, which are now widely used in optical enterprises. To assess the image quality of the finished objective, the so-called *starry sky* or a simpler method is used, such as the Abbe plate or a standard regular biological object such as a diatom. The assembly and final alignment can also be performed using these tools. As a rule, testing and attestation of objectives is carried out by the most experienced specialist who has thoroughly studied and is able to apply all of the above requirements in practice. Nevertheless, we note that this specialist very rarely knows the degree of aberration correction that was specified in the optical design. Thus, the assembly of the objective, as well as its attestation, is carried out with a large share of the subjective world view, which will obviously change when these

operations are performed by another person. This is a problem in mass production when lenses from one optical design have differences in image quality. Therefore, the problem of ensuring high reproducibility in the assembly and certification of objectives for microscopes of the mass segment is relevant. The use of ASA is only one way to solve this problem. Another tool can be a high degree of formalization of the assembly process, or rather alignment of the objective, for which it is necessary to use specially designed stands and other auxiliary equipment. Such a stand can become the main structural unit from a set of instruments for attestation of the objective. One of the algorithms for attestation is the comparison of the image obtained after the real objective with the theoretical one, which is specified in the optical design. However, the algorithms for attestation may be different if they achieve a significant result: an increase in the degree of formalization of the process with a reduction in production time.

If we analyze the above requirements when manufacturing objectives for the assembly of mass-produced microscopes, we can agree that the task of designing and manufacturing them is complex—very complex—and involves many production and technological factors. Factors associated with the moral and psychological aspects of the people working in this area, including their qualifications and the degree of their creative interest in the work, should also not be overlooked. As an example, Fig. 1.152 is an assembly drawing of an objective with a linear magnification of 40× and a numerical aperture of 0.65. You can see that the manufacturing tolerances are wide and that the design of the objective is fairly simple. This is an illustration of the technological level in the manufacture of budget objectives for mass-produced microscopes.

However, these budget objectives (conditionally *first-generation +*) did not find wide application because some new financial investments were required for their development. And compared with the classical Abbe, the

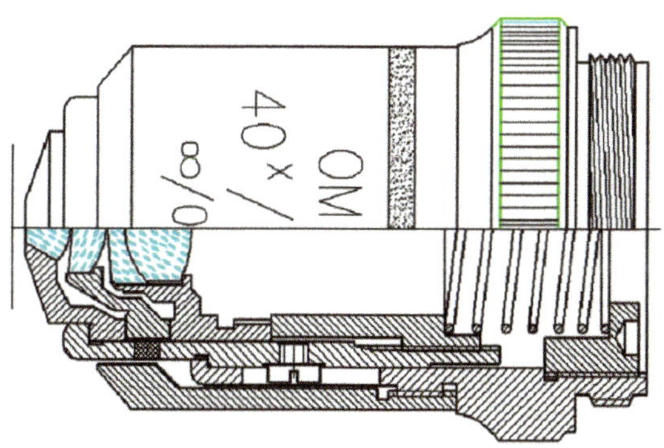

Figure 1.152 Assembly drawing illustrating the technological level in the manufacture of budget objectives for mass-produced microscopes.

objectives did not provide advantages in image quality. Thus, our engineers designed budget objectives, the image quality of which was improved compared with the classical Abbe objectives. A set of these objectives was mastered in series production; they were used in several budget biological microscopes of the mass segment. We can conditionally call these objectives the second-generation objectives. These objectives are distinguished by high-quality correction of aberrations, but such aberrations as image curvature and lateral chromaticity remain incompletely corrected (for example, when the microscope is equipped, the residual value of the lateral chromaticity of 0.8% for all objectives is compensated in the visual channel by a special eyepiece).

The optical designs and graphs of aberration correction of some of these objectives are shown in Figs. 1.15–1.20. Although these objectives were designed more than 30 years ago, they can be used in modern models of budget microscopes of the mass segment; the objectives have fairly simple optical designs. The value of the residual aberrations of the curvature of the image and the lateral chromaticity (LC) are not so big. Figure 1.153 shows the quality of the image obtained using objectives with a linear magnification of 25× and a numerical aperture of 0.50, and a linear magnification of 40× and a numerical aperture of 0.65, with direct projection on a digital receiver. Photographs of the lines obtained with their help are characteristic of images through optics that are not fully adjusted in terms of image curvature and LC.

If we analyze these two images, we can see that the fragments of the structure of the object located in the center of the field look very sharp and do not have parasite color. But the images of the fragments of the object located on the periphery of the linear field have obvious distortions in shape and color. In general, this image quality is due to the lenses' optical design in which the aberration correction did not involve full correction of aberrations such as image curvature and lateral chromaticity. On the other hand, attempts at the end of the last century to introduce into mass production third-generation budget objectives in which these aberrations (curvature of the image and lateral chromaticity) would be completely corrected were also not successful. These attempts failed because at the time, the technological

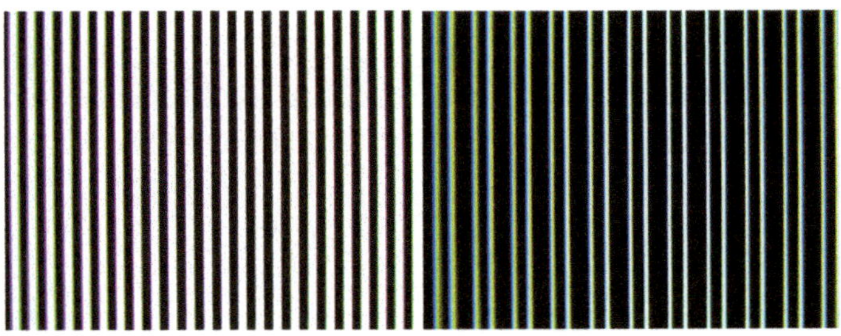

Figure 1.153 Images of lines after using 25×/0.50 and 40×/0.65 second-generation objectives.

equipment of optical plants in the mass production segment was inadequate. Some optical elements turned out to be too expensive for the price.

Today, the technological situation has changed, and the demand for budget third-generation objectives is expected to massively rise. Researchers who use simple microscopes are less comfortable with the image quality of first- and second-generation mass-market objectives, primarily due to the rising trend in the use of digital imaging systems. An increasing number of microscopes, even those that are used primarily in schools, are equipped with digital cameras; this is now standard for light microscopy. Accordingly, the elemental base of light microscopes, primarily objectives, is being improved.

What are the distinguishing characteristics of third-generation budget microscope objectives? Today, the market presents many different models of light microscopes made by several manufacturers. Some microscopes use the finite length of the tube (160 mm for the objectives) while others use the infinite length of the optical tube; the parametric series of focal lengths for the objectives vary among manufacturers. But regardless of the manufacturer, the dimensions, or the design of the microscopes and objectives, the differences are clearly seen in the quality of the optical images. This quality varies due to fundamental differences in the concept of aberration correction quality of the objectives.

Studies of many engineering solutions (through the analysis of patents, articles, and real microscopes) have shown that until now in the segment of mass-produced microscopes, many manufacturers (even recognized leaders in the field of optical instrumentation) use a compensation method of aberration correction when, for example, the additional focusing (tube) system was designed to have compensatory values of some aberrations. In this case, the optical design of all objectives used in the microscope (often used only in a specific model) takes the influence of *compensatory* aberrations of the tube system into account.

This design is insufficient for researchers because, in this case, the so-called *free choice* of selecting objectives from different microscope models is impossible. Indeed, in real use, researchers who have a stand microscope would like to choose the best objectives from their available arsenal, and not necessarily ones from the same manufacturer. This is impossible with this design because the resulting image quality from the combination of the selected objective and the microscope (from different manufacturers) will be unsatisfactory in most cases. In the optical system of a microscope, it is the objective that is of paramount importance. But must researchers buy a new tripod (microscope) every time they use an interesting objective if the manufacturers of the objective and the microscope are different? Of course, the aberration correction of microscope objectives, including those in the budget segment of the market, should be independent of any additional components that exist in the microscope's optical system. It is directly in the optical system of the objective itself that aberrations such as curvature of the image and lateral chromaticity must be corrected.

Many microscopists use older microscopes, which for them have become good and reliable associates of their studies. These individuals are usually experienced professionals who are well versed in the nuances of working on a microscope, and they are often not ready to give up their time-tested instruments. But there is a need for modern objectives that possess enhanced consumer qualities. First, we are talking about objectives for microscopes using the *classical* finite optical length of the tube (160 mm). On the other hand, the use of a finite length of the tube, as a rule, means the absence of any additional optical elements after the objective; the image is built directly by the objective itself, and the image quality is determined only by the degree of its aberration correction. This is undoubtedly an advantage in constructing a microscope with a finite length of the tube; on the other hand, it is obvious that the aberration correction of all objectives used in such a microscope must be independent.

For microscopes using a finite length of the tube, we produced a set of such third-generation objectives, the optical design of which takes into account all the consumer qualities of modern lenses for budget microscopes. All technological requirements were also taken into account. We manufactured and offered a complete set of LCF-type plan achromatic objectives that achieve an aberration correction. The kit included 4×, 10×, 20×, 40×, 60×, 100× water immersion, and 100× oil immersion objectives.

We specifically included a 100× water immersion objective in the kit because water immersion has significant advantages over oil immersion. Perhaps water immersion objectives will soon replace oil immersion objectives. Unlike oil, water is not toxic, is not an aggressive medium, is harmless to the researcher, and does not affect the front lens of the objective. Working with water immersion objectives is also much more convenient and cheaper than working with oil immersion objectives; in most cases, simple tap water can be used (there is no need to buy expensive oil). In contrast to oil immersion objectives, it is much easier to care for water immersion objectives because the front lens does not need to be wiped with special solutions to remove residual oil. With a water immersion objective, even if the water is dried after use, it will dissolve when the objective is reused. The resolving power values of both objectives are practically the same. These advantages allowed us to offer a 100× water immersion objective for use in mass-produced budget microscopes.

An example of these objectives is shown in Fig. 1.60; design parameters and graphs of aberration correction of these objectives are shown in Figs. 1.36, 1.41, 1.45, 1.49, and 1.57, and in Table 1.10. These are the *latest-generation* LCF plan achromatic aberration correction objectives for microscopes using a finite length of the tube.

The distinctive properties of new objectives are a high degree of unification and universality. Objectives can be manufactured for different instruments, including biological light microscopes of transmitted or reflected light; luminescent, polarizing microscopes; phase and interference contrast devices; and dark field microscopes. Manufacturing objectives with iris aperture diaphragms is also possible for specific studies, for example, in a dark field or on polarization microscopes. These objectives have the following main characteristics:

1. Objectives are designed for use in microscopes of the optical finite length of the tube (160 mm).
2. Objectives comply with the international standard DIN (parfocal distance, 45 mm).
3. A linear field in the image space is 20 mm.
4. The type of optical correction is a plan achromat.
5. Corrected lateral chromaticity (classification LCF).
6. In comparison with analogues, increased numerical apertures.
7. The absence of reflexes and scattered light to work in reflected light.
8. Optimized for microphotography and video projection.

Another distinctive feature of the new objectives is that they can be used in microscopic studies in *reflected light in a bright field*. This is possible due to the absence of reflexes and scattered light when the object under study is illuminated by the incident light through the objective. It is no secret that many analogues suffer from such a drawback; in the transmitted light the image quality is satisfactory, whereas in the reflected light it looks milky. An additional distinguishing feature of these new objectives is that they can be used in measuring microscopes. Objectives do not distort the shape of the object under study, instead building the image strictly congruent to the original structure. The resulting image is not burdened with aberrations and other distortions and thus, it can be measured and examined. The results of the measurements will give an adequate idea of the investigated object.

By analyzing the design parameters of these objectives, one can see that many of the specified technological requirements for budget objectives for use in mass-market microscopes are fulfilled. So, for example, optical materials manufactured with well-mastered lens grinding and polishing processes are used. Tolerances for the manufacture of lenses are quite wide. The radii of some lenses are made equal. All this makes it possible to keep the costs of manufacturing these objectives minimal. At the same time, the aberration correction quality of these objectives is high. Figure 1.154 shows the quality of the image obtained with the use of these third-generation objectives with a linear magnification of 20× and a numerical aperture of 0.65, and a linear magnification of 40× and a numerical aperture of 0.70, with direct projection

Figure 1.154 Images of lines after using 20×/0.65 and 40×/0.70 third-generation objectives.

onto the digital receiver. Photographs of the lines obtained with their help are characteristic of images through optics that are corrected with respect to image curvature and lateral chromaticity.

These comparative photographs were taken by Sergey Dushenbiev. Obviously, the images of an object obtained with third-generation objectives show significant advantages in quality compared with the ones obtained with second-generation objectives. This is further confirmed by direct comparison of images of the same real object taken by two 40× objectives; these are presented in Fig. 1.155 (left, second-generation objective; right, third-generation objective).

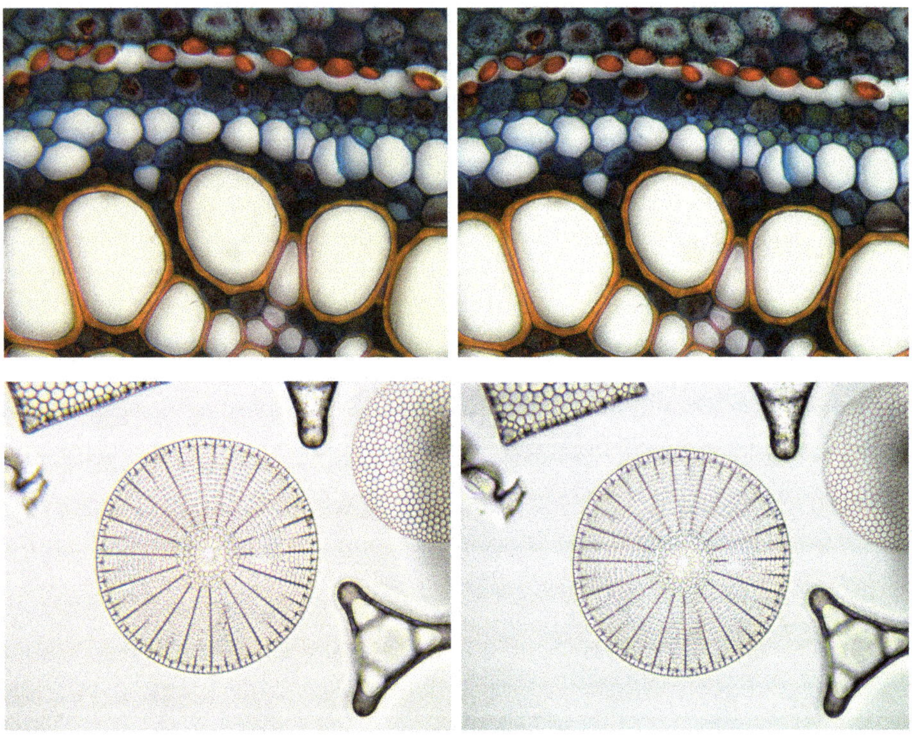

Figure 1.155 Images taken with 40× second-generation (left) and third-generation (right) objectives for comparison.

Let us again turn to Figs. 1.17 and 1.45, which show the aberration characteristics of 40× second- and third-generation objectives, respectively, for the finite length of the tube. Figure 1.156 shows graphs in the same format.

If we analyze the graphs of aberration correction presented in Fig. 1.156, we can see that the aberration correction for the axial (aperture) beams of

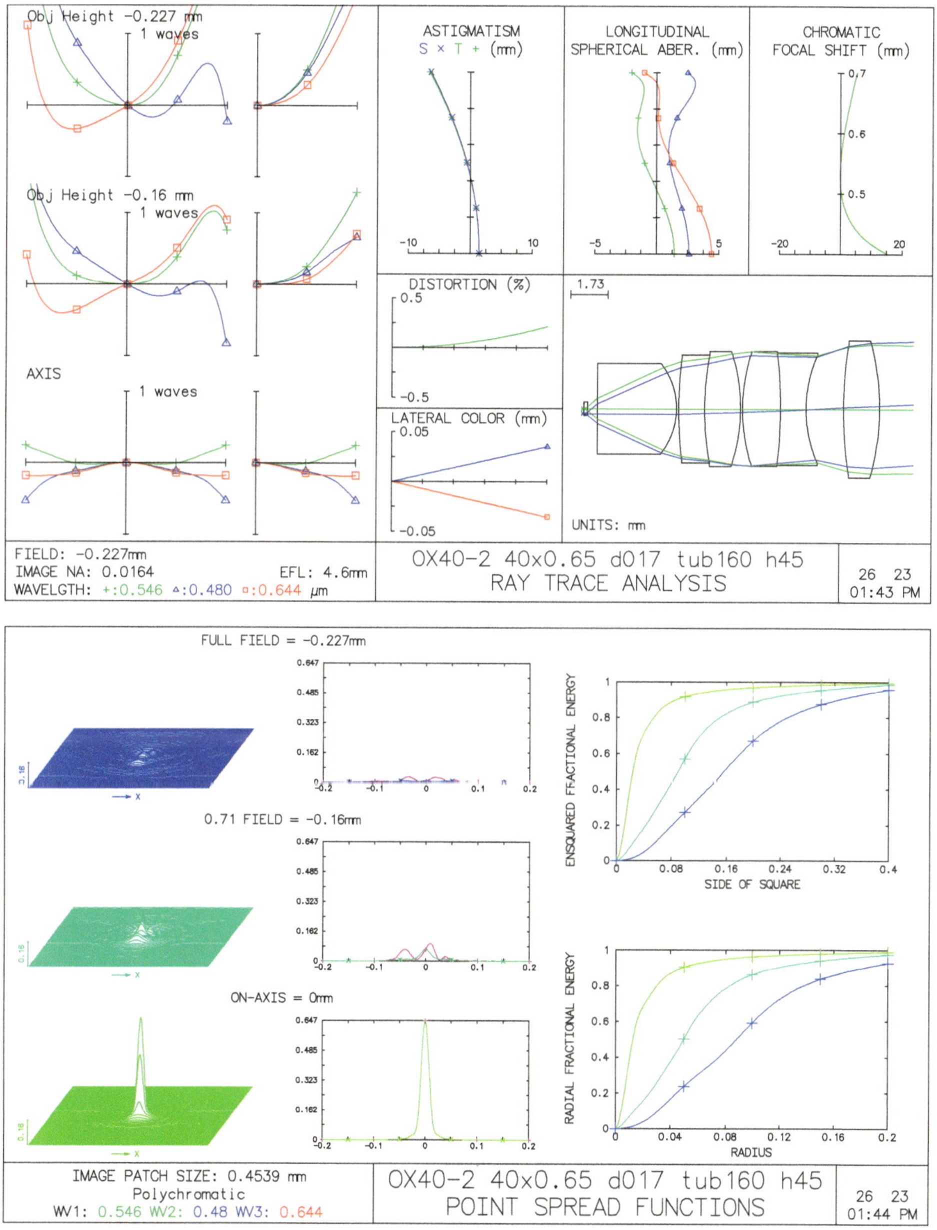

Figure 1.156 Comparison of aberration correction of 40× second- and third-generation objectives.

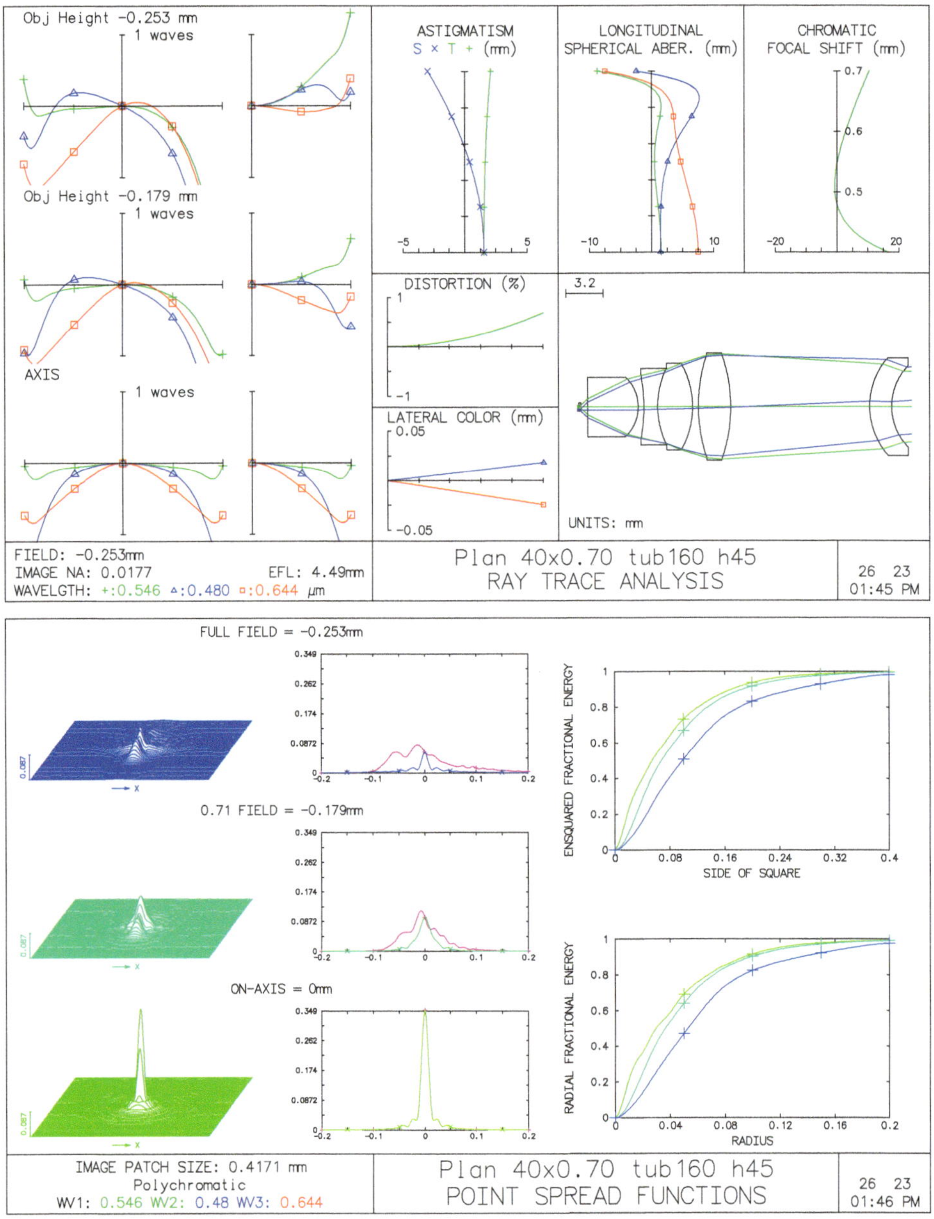

Figure 1.156 (*Continued*) Comparison of aberration correction of 40× second- and third-generation objectives.

both objectives is almost the same. However, the correction of aberrations such as image curvature and lateral chromaticity is different. In the case of the second-generation objective, the curvature of the image is only reduced (approximately twice as compared with the classical Abbe objectives); with the third-generation objective, this aberration is completely corrected. A similar situation is seen in the degree of lateral chromaticity correction: in the first case this value is 0.8%, whereas in the second case it is practically corrected. As a result, you can compare the image quality of these two objectives. It is much better with the third-generation objective. This is not only confirmed theoretically, but also by obtaining real images.

Let us focus on the description of 20× and 60× objectives. Objectives with such linear magnifications, as a rule, are not included in the basic microscope set. They can be purchased by the client separately. Therefore, we isolated these two objectives from the general set, increasing their numerical apertures and expanding the effective linear field. The linear field for these objectives is 28 mm (in the image plane). The optical designs of these objectives use original circuit solutions (Figs. 1.41 and 1.49) as well as modern fluorophosphate optical glass with a refractive index value of about 1.5 and an Abbe number of about 81.5. Although the use of this optical glass slightly increases the cost of these objectives, it provides many advantages (in front of analogues).

1.7.2 Some thoughts on the design of phase contrast and differential contrast in budget transmitted light microscopes

Consider the fundamental features of the construction of devices in which some of the methods that are commonly called *contrasting methods* are implemented in microscopy. More precisely, consider a condenser–objective pair. It is known that when building, for example, phase contrast devices, a plate with light rings, an interference prism, a polarizer, or other device that is part of a system for phase shift in a light beam is placed in the plane of the entrance pupil of the condenser. In the plane of the exit pupil of the objective, which is interfaced with the plane of the entrance pupil of the condenser, there is a phase plate, or an interference prism, analyzer, etc. Of fundamental importance for the optical calculation of such systems is the location of equivalent focal planes in the optical systems of the condenser and objective. Figure 1.157 shows two possible options when building some types of contrast devices. We assume that the use of a microscope lighting system constructed according to Köhler is preferable when constructing contrast devices.

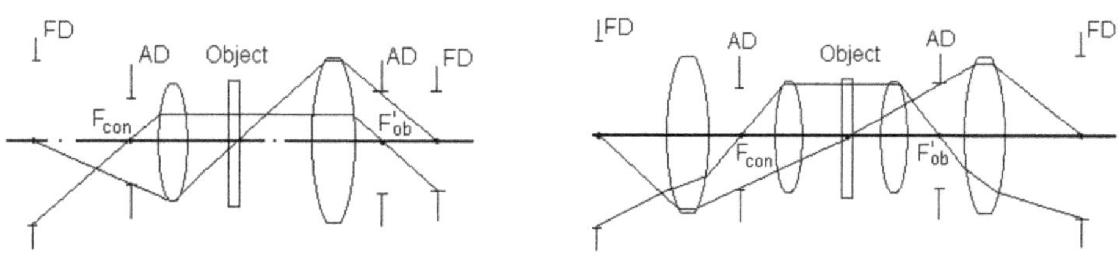

Figure 1.157 Two options for arranging equivalent focal planes in the condenser and objective.

If the focal planes of the condenser and the objective are *out extended* (as shown in Fig. 1.157, left), the optical design technique is obvious; it is in these planes that the contrasting elements should be located. In the case when a condenser and an objective are used in which equivalent focal planes are located inside their optical systems (Fig. 1.157, right), the optical design should be made in a slightly different way. In this case, in the overall calculation of the condenser–objective optical system, it is necessary to take into account a requirement such as the conjugation of the real intersection planes of the so-called *reference* rays with the optical axis. In some cases, these planes do not coincide with the planes of equivalent focal planes. This is characteristic of short-focus condensers and objectives that have strong telephoto elements in circuit. The numerical apertures in these systems are also quite high. Thus, some *negligence* or errors in the optical design of such systems can make it impossible to get the highest image quality and the maximum effect of the contrast device. When calculating contrasting devices based on a single condenser and a whole set of different objectives, combinations of different cases of arrangement of equivalent focal planes are obtained, which must also be taken into account. In the calculation technique for phase contrast devices, for example, it is necessary to take into account the optical aberration values in the pupils, or more precisely, in the planes where the *lighting rings* are supposed to be placed in the condenser and the *phase rings* are supposed to be placed in the objective. In practice, this can be done like this. For a condenser in the reverse path of rays from the object through the front and other lenses, a calculation is made *from infinity* to a finite distance (developers of optical systems call these systems *type 0-1*). The plane in which the beam crosses the optical axis (or the most suitable place in the air gap between the lenses) can preferably be used to position the light diaphragm (light ring) in it. At the same time, the rest of the condenser (if there are still lenses or mirrors) is not taken into account. It is similar for the objectives. For objectives with small linear magnifications, if a type 0-1 system is calculated, the plane of intersection of the beam coincides with the equivalent focal plane (plane of the exit pupil of the objective). However, for such objectives it is often difficult to place the exit pupil plane in predetermined parfocal dimensions, for example, 45 mm. This explains the small number, for example, of objectives for phase contrast with linear magnifications of 1×, 2.5×, or 4×. However, this problem can be solved using the relative complication of the optical design of such objectives. It should be noted that the placement of the light ring in the condenser or phase ring in the objective may not be strictly in the planes of intersection of the beam with the optical axis; it can be some other plane (near the plane of the axis crossing the beam). The important thing is that they are conjugate for the condenser and objective.

Figures 2.60–2.65 (Volume 1, Chapter 2), for example, show the high quality of phase contrast (comparable to those developed at other optical companies) for the device used in the old Soviet microscopes (a photograph of the device itself is shown in Volume 1, Chapter 2, Fig. 2.59). Consider a set of objectives designed to implement the phase contrast method in the Russian microscope the MikMed-2. Figure 1.158 presents the design parameters and graphs of aberration correction of the 10×/0.25 objective. According to the proposed classification, it can be called a 10×/0.25 finite LCF plan achromatic objective. This objective (and others for this microscope) was designed for use in a microscope with a finite tube length of 160 mm and an eyepiece height of 13 mm.

Lens: plan 10x0.25 d=017 Finity Efl 15.82

Object num aper 0.250000 Image height 9.000000 Primary wavln 0.546070

SRF	RADIUS	THICKNESS	APERTURE RADIUS	GLASS	SPECIAL
OBJ	0.000000	0.010000	0.924544	AIR	
AST	0.000000	0.170000	0.930000 A	LZ_K14	
2	0.000000	13.813000	0.960000	AIR	
3	112.760000	1.000000	4.550000	LZ_TF5	
4	13.183000	3.000000	4.770000	LZ_LK3	
5	-13.183000	4.470000	5.030000	AIR	
6	19.498000	3.000000	5.970000	LZ_LK3 P	
7	-24.320000	6.320000	5.960000	AIR	
8	10.520000	1.500000	4.770000	LZ_TF5 P	
9	8.395000	158.720000	4.310000	AIR	
IMS	0.000000	0.000000	9.000000		

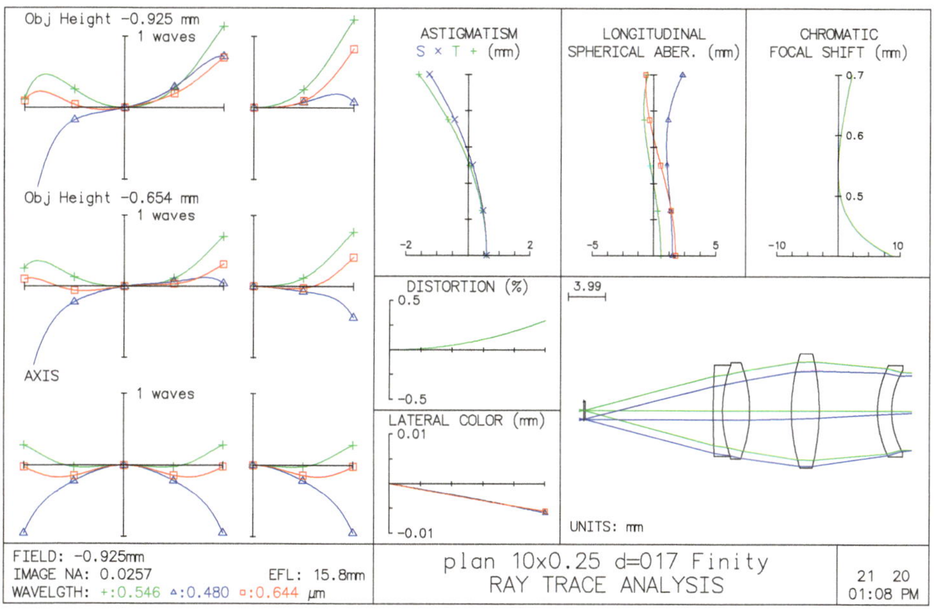

Figure 1.158 Optical design and graphs of aberration correction of a 10×/0.25 finite LCF plan achromatic objective.

To design this objective as an element of a phase contrast device, we will calculate the beam from infinity and determine the position of the objective's equivalent focal length. The input parameters for the calculation will be the following values:

1. The image size is 8 mm, determined from the ratio for the diameter of the exit pupil $Ø = 2F' \cdot NA$, where F' is the focal length of the objective and NA is the input numerical aperture.
2. The height of the beam on the first surface is 0.9 mm, based on half of the linear field of the objective in the space of objects (this calculation is shown in Fig. 1.159).

Thus, in the rear equivalent focal plane of this objective at a distance of 5.36 mm from the center of the sphere (of the second surface of the last lens following the rule of optical calculation from left to right), we can arrange the plate on

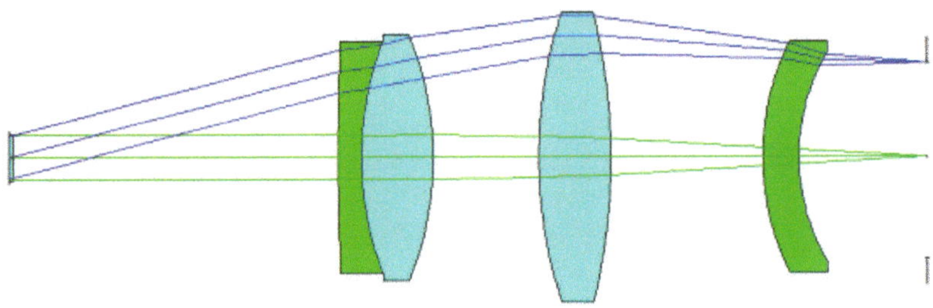

Lens: plan 10x0.25 d=017 Finity pupil

Ent beam radius 0.900000 Image height 4.000000

SRF	RADIUS		THICKNESS		APERTURE RADIUS		GLASS	
OBJ	0.000000		1.0000e+08		2.5292e+07		AIR	
1	0.000000		0.010000		0.902529	S	AIR	
AST	0.000000		0.170000		0.930000	A	LZ_K14	
3	0.000000		13.813000		0.960000		AIR	
4	112.760000		1.000000		4.550000		LZ_TF5	
5	13.183000		3.000000		4.770000		LZ_LK3	
6	-13.183000		4.470000		5.030000		AIR	
7	19.498000		3.000000		5.970000		LZ_LK3	P
8	-24.320000		6.320000		5.960000		AIR	
9	10.520000		1.500000		4.770000		LZ_TF5	P
10	8.395000		5.359359		4.310000		AIR	
11	0.000000		-1.9509e-15	S	4.000000	S	AIR	
IMS	0.000000		0.000000		4.000000	S		

Figure 1.159 Optical design and drawings of enter and exit pupils of the 10×/0.25 finite LCF plan achromatic objective.

one of the surfaces of which the phase ring (or system of rings) is applied. In another case, the Wollaston prism can be placed in this plane to obtain the effect of interference contrast. In the practice of optical design in both cases, we can assume that this is a plate, for example, 2-mm thick (from two glued plates, inside of which a phase ring is applied) made of quartz glass or other optical material. As an example, as shown in Fig. 1.160, we take the optical material of an LZ_TF1 plate and show that its integration into the

Lens: plan 10x0.25 d=017 Finity — Efl 15.82

Object num aper 0.250000 — Image height 9.000000 — Primary wavln 0.546070

SRF	RADIUS	THICKNESS	APERTURE RADIUS		GLASS		SPECIAL
OBJ	0.000000	0.010000	0.924544		AIR		
AST	0.000000	0.170000	0.930000	A	LZ_K14		
2	0.000000	13.813000	0.960000		AIR		
3	112.760000	1.000000	4.550000		LZ_TF5		
4	13.183000	3.000000	4.770000		LZ_LK3		
5	-13.183000	4.470000	5.030000		AIR		
6	19.498000	3.000000	5.970000		LZ_LK3	P	
7	-24.320000	6.320000	5.960000		AIR		
8	10.520000	1.500000	4.770000		LZ_TF5	P	
9	8.395000	4.360000	4.310000		AIR		
10	0.000000	1.000000	4.168401		LZ_TF1		
11	0.000000	1.000000	4.116964		LZ_TF1	P	
12	0.000000	152.340000	4.090215		AIR		
IMS	0.000000	0.000000	9.000000				

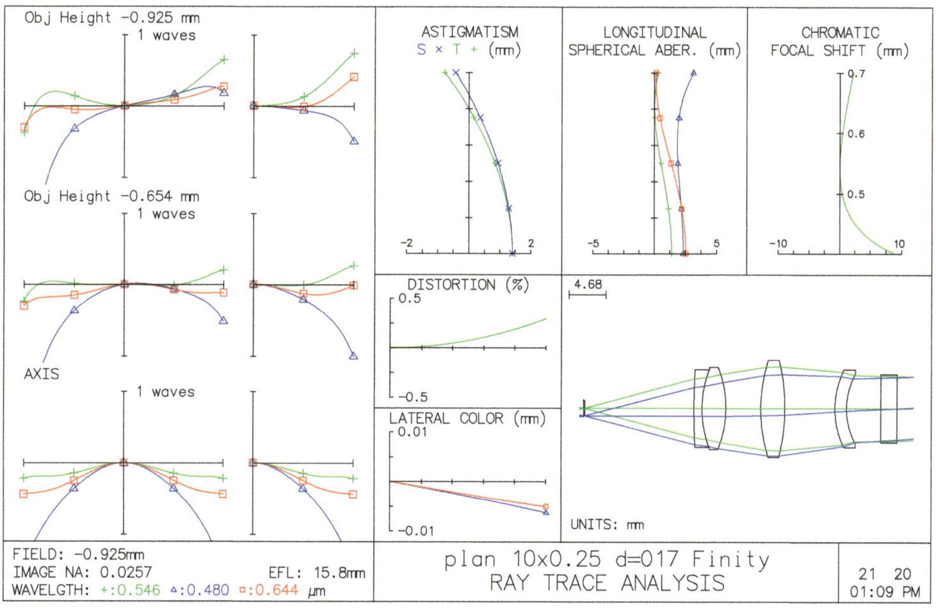

Figure 1.160 Optical design and graphs of aberration correction of a 10×/0.25 phase plate finite LCF plan achromatic objective.

optical circuit of the objective does not lead to a significant decrease in image quality.

Let us make a similar calculation for another objective with a linear magnification of 20× and a numerical aperture of 0.40. The design parameters and graphs of aberration correction of the bright field version of this objective are presented in Fig. 1.161. This objective was included in the set for the phase contrast device for the specified microscope. Its distinctive feature is that the plane of intersection of the *parallel beam* with the optical axis does not coincide with the location of the equivalent focal plane of the objective; the

Lens: om27 20x0.4 finity h45 Efl 8.42

Object num aper 0.400000 Image height 9.000000 Primary wavln 0.546070

SRF	RADIUS	THICKNESS	APERTURE RADIUS	GLASS	SPECIAL
OBJ	0.000000	0.010000	0.416394	AIR	
AST	0.000000	0.170000	0.450000 A	LZ_K14	
2	0.000000	1.759000	0.500000	AIR	
3	0.000000	2.000000	1.280000	LZ_K14 P	
4	-4.000000	1.200000	1.720000	AIR	
5	0.000000	1.500000	2.000000	LZ_TF1	
6	6.500000	1.600000	2.200000	LZ_K14 P	
7	-7.470000	7.400000	2.330000	AIR	
8	53.280000	1.000000	3.360000	LZ_TF1 P	
9	8.550000	1.800000	3.510000	LZ_K14 P	
10	-10.510000	173.560000	3.570000	AIR	
IMS	0.000000	0.000000	9.000000		

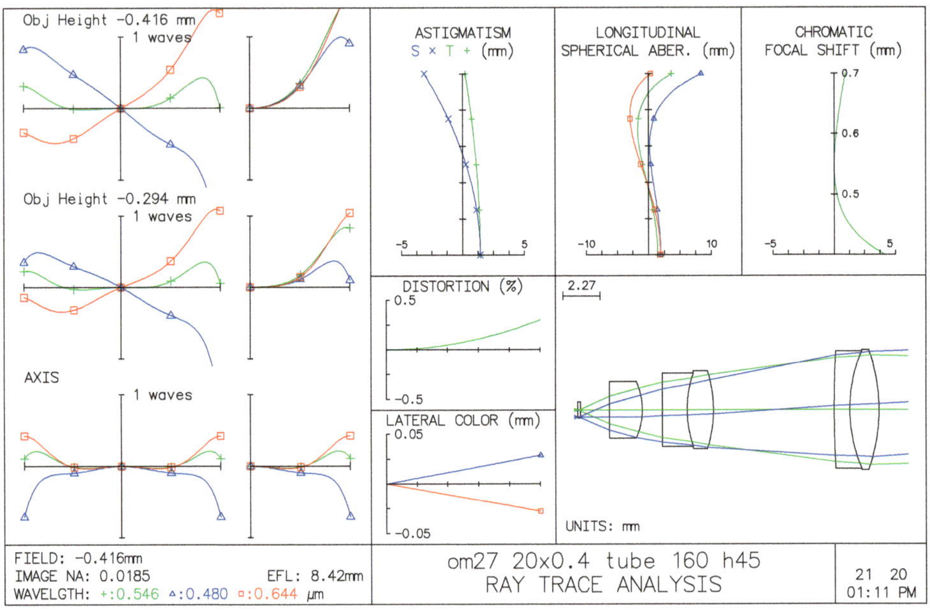

Figure 1.161 Optical design and graphs of aberration correction of a 20×/0.40 finite achromatic objective.

position difference is shown by the value X in Fig. 1.162. Figure 1.163 shows a drawing of the result of the calculation of the parallel beam (green) and the location of its intersection with the optical axis. Figure 1.164 shows the design parameters and graphs of aberration correction of a 20×/0.40 objective used for phase contrast.

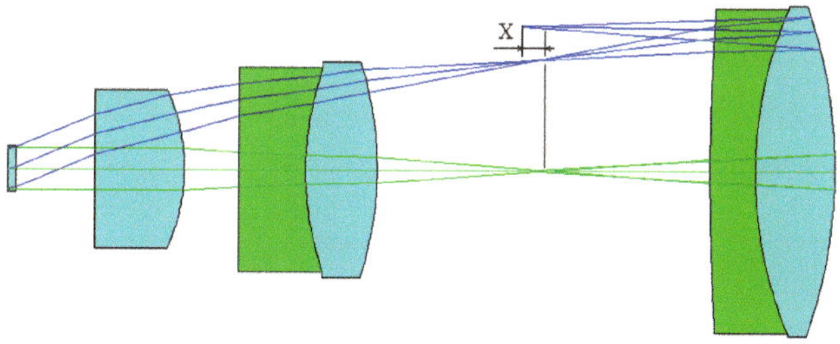

Figure 1.162 Mismatch of the beam crossing plane with the optical axis and the equivalent focal plane of a 20×/0.40 finite achromatic objective.

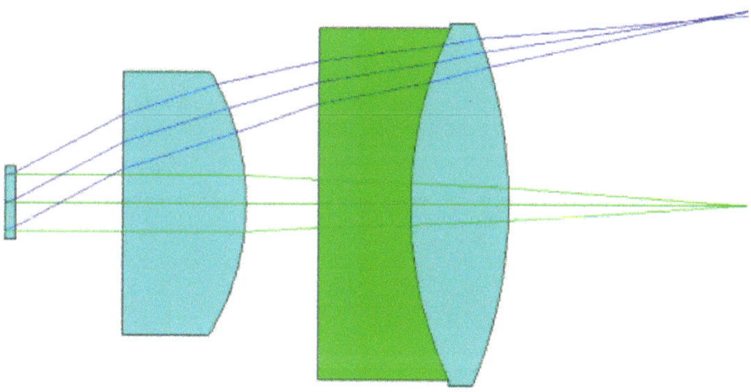

Lens: om27 20x0.4 finity h45 pupil

Ent beam radius 0.450000 Image height 3.370000 Primary wavln 0.546070

SRF	RADIUS	THICKNESS	APERTURE RADIUS	GLASS	SPECIAL
OBJ	0.000000	1.0000e+08	5.2594e+07	AIR	
1	0.000000	0.010000	0.455259 S	AIR	
AST	0.000000	0.170000	0.520000 A	LZ_K14	
3	0.000000	1.759000	0.580000	AIR	
4	0.000000	2.000000	1.610000	LZ_K14 P	
5	-4.000000	1.200000	2.100000	AIR	
6	0.000000	1.500000	2.520000	LZ_TF1	
7	6.500000	1.600000	2.810000	LZ_K14 P	
8	-7.470000	3.900000	2.890000	AIR	
9	0.000000	-0.008294 S	3.370000	AIR	
IMS	0.000000	0.000000	3.370000		

Figure 1.163 Optical design and drawings of the intersection of the parallel beam with the optical axis of a 20×/0.40 finite achromatic objective.

Lens: om27 20x0.4 finity h45+phase — Efl 8.04

Object num aper 0.400000 — Image height 9.000000 — Primary wavln 0.546070

SRF	RADIUS	THICKNESS	APERTURE RADIUS	GLASS	SPECIAL
OBJ	0.000000	0.010000	0.397129	AIR	
AST	0.000000	0.170000	0.410000 A	LZ_K14	
2	0.000000	1.835000	0.450000	AIR	
3	0.000000	2.000000	1.250000	LZ_K14 P	
4	-4.000000	1.200000	1.700000	AIR	
5	0.000000	1.500000	1.980000	LZ_TF1	
6	6.500000	1.600000	2.190000	LZ_K14 P	
7	-7.470000	2.900000	2.320000	AIR	
8	0.000000	1.000000	2.530000	LZ_TF1 P	
9	0.000000	1.000000	2.690000	LZ_TF1 P	
10	0.000000	2.500000	2.690000	AIR	
11	53.280000	1.000000	3.170000	LZ_TF1 P	
12	8.550000	1.800000	3.310000	LZ_K14 P	
13	-10.510000	173.480000	3.390000	AIR	
IMS	0.000000	0.000000	9.000000		

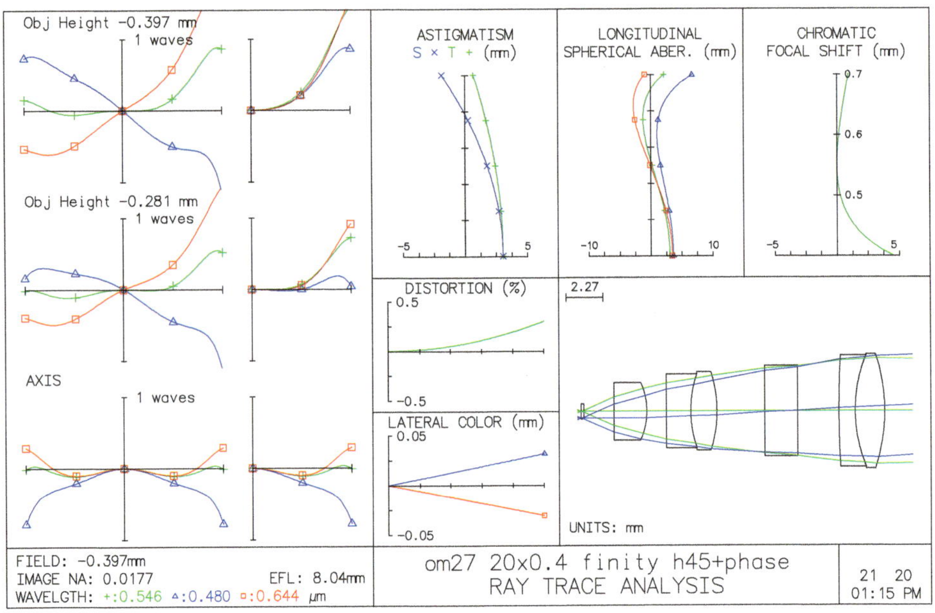

Figure 1.164 Optical design and graphs of aberration correction of a 20×/0.40 phase plate finite achromatic objective.

The next objective from the set for the phase contrast device for the MikMed-2 microscope is one with a linear magnification of 40× and a numerical aperture of 0.65. The design parameters and graphs of aberration correction of the bright field version of this objective are shown in Fig. 1.165. A distinctive feature of this objective is the presence in the optical system of a lens with a powerful telephoto element. In this case, the plane of intersection

Lens: ocx 40x0.65 d=0.17 Finity				Efl	4.045
Object num aper	0.650000	Image height	9.000000	Primary wavln	0.546070
SRF	**RADIUS**	**THICKNESS**	**APERTURE RADIUS**	**GLASS**	**SPECIAL**
OBJ	0.000000	0.010000	0.209058	AIR	
AST	0.000000	0.170000	0.230000 A	LZ_K14	
2	0.000000	0.385600	0.220000	AIR	
3	0.000000	4.530000	0.630000	LZ_CTK12	
4	-3.076000	0.100000	2.150000	AIR	
5	-15.171000	1.200000	2.230000	LZ_TF5	
6	5.035000	3.000000	2.470000	LZ_LK3	
7	-9.120000	0.350000	2.990000	AIR	
8	11.722000	3.100000	3.420000	LZ_LK3 P	
9	-11.722000	12.200000	3.630000	AIR	
10	5.297000	2.000000	3.630000	LZ_TF5 P	
11	4.207000	164.950000	2.960000	AIR	
IMS	0.000000	0.000000	9.000000		

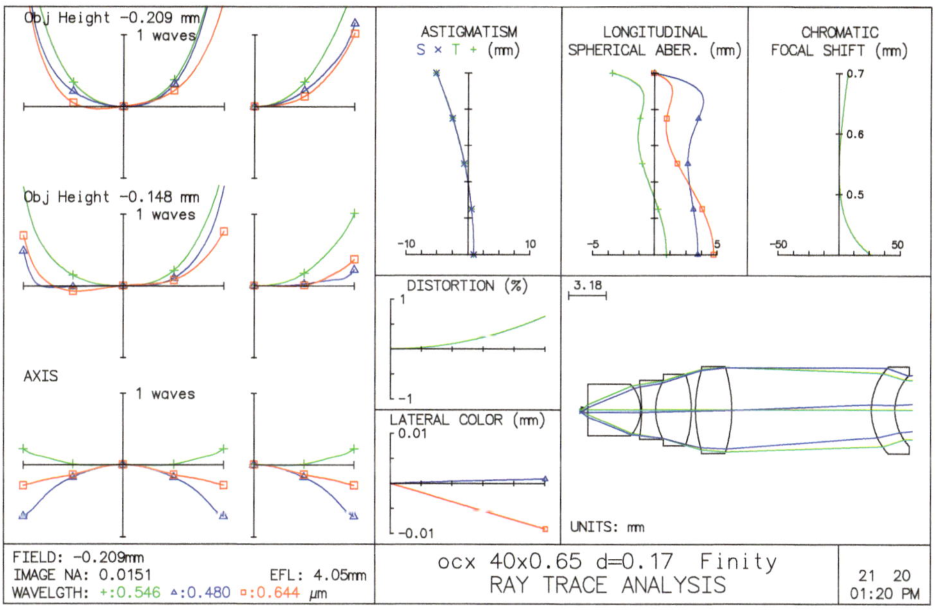

Figure 1.165 Optical design and graphs of aberration correction of a 40×/0.65 finite LCF semi–plan achromatic objective.

of the parallel beam with the optical axis does not coincide with the location of the equivalent focal plane of the objective, as shown in Fig. 1.166. Figure 1.167 shows a drawing of the result of the calculation of the parallel beam (green) and the location of its intersection with the optical axis. Figure 1.168 shows the design parameters and graphs of aberration correction of the 40×/0.65 objective used for phase contrast.

The last objective in the set for the phase contrast device is one with a linear magnification of 100× and a numerical aperture of 1.25 oil immersion. The design parameters and graphs of aberration correction of the bright field

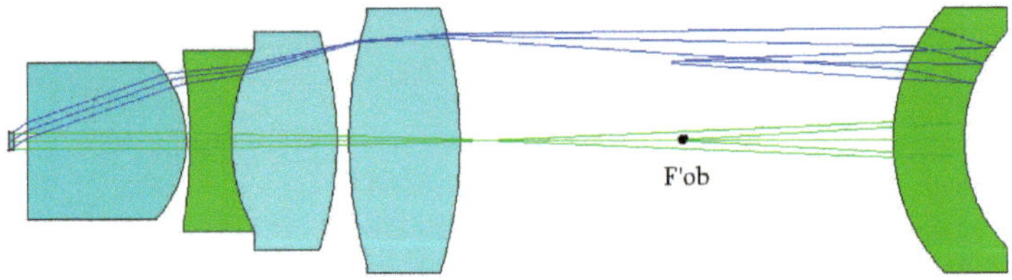

Lens:ocx 40x0.65 d=0.17	Finity pupil			Efl	4.045
Ent beam radius	0.225000 Image height		2.600000 Primary wavln		0.546070
SRF	RADIUS	THICKNESS	APERTURE RADIUS	GLASS	SPECIAL
OBJ	0.000000	1.0000e+08	6.4269e+07	AIR	
1	0.000000	0.010000	0.231427 S	AIR	
AST	0.000000	0.170000	0.230000 A	LZ_K14	
3	0.000000	0.385600	0.220000	AIR	
4	0.000000	4.530000	0.630000	LZ_CTK12	
5	-3.076000	0.100000	2.150000	AIR	
6	-15.171000	1.200000	2.230000	LZ_TF5	
7	5.035000	3.000000	2.470000	LZ_LK3	
8	-9.120000	0.350000	2.990000	AIR	
9	11.722000	3.100000	3.420000	LZ_LK3 P	
10	-11.722000	12.200000	3.630000	AIR	
11	5.297000	2.000000	3.630000	LZ_TF5 P	
12	4.207000	-8.254138	2.960000	AIR	
13	0.000000	-1.9962e-15 S	2.600000 S	AIR	
IMS	0.000000	0.000000	9.000000		

Figure 1.166 Mismatch of the beam crossing plane with the optical axis and the equivalent focal plane of a 40×/0.65 finite LCF plan achromatic objective.

version of this objective are presented in Fig. 1.169. In this case, the plane of intersection of the parallel beam with the optical axis practically coincides with the location of the equivalent focal plane of the objective, as shown in Fig. 1.170. Here, the phase plate is applied directly to the lens (in this case, a flint lens is selected). Theoretically, the aberration correction does not change. The graphs of aberration correction are similar to those shown in Fig. 1.169; however, in practice, an objective made in a variant for phase contrast has a significantly lower contrast and more spherical aberration caused by the influence of a phase ring.

Figure 1.171 shows photographs of the MikMed-2 microscope and the KF-4M phase contrast device. It is interesting to note that if you turn on the 10× objective on the phase contrast device and use the light diaphragm in the condenser designed to work with the 100× objective, you can get a dark field effect. This is shown in Fig. 1.172. Figures 1.173–1.179 show photographs of real objects (diatoms) taken using the discussed phase contrast objectives with a MikMed-2 microscope.

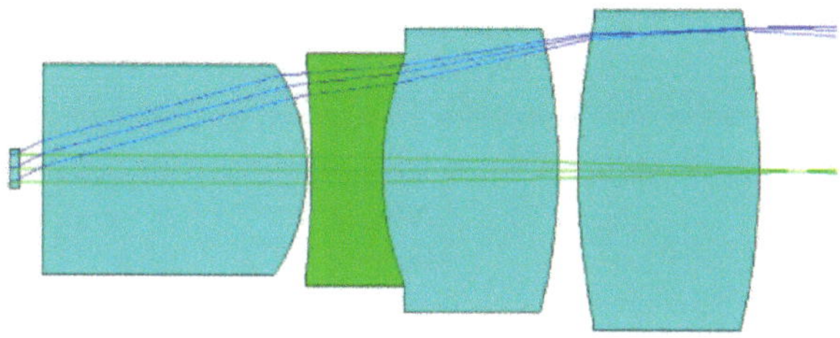

Lens:ocx 40x0.65 d=0.17 Finity pupil

Ent beam radius 0.225000 Image height 2.600000

SRF	RADIUS	THICKNESS	APERTURE RADIUS		GLASS	
OBJ	0.000000	1.0000e+08	4.9692e+07		AIR	
1	0.000000	0.010000	0.229969	S	AIR	
AST	0.000000	0.170000	0.260000	A	LZ_K14	
3	0.000000	0.385600	0.320000		AIR	
4	0.000000	4.530000	0.530000		LZ_CTK12	
5	-3.076000	0.100000	1.740000		AIR	
6	-15.171000	1.200000	1.780000		LZ_TF5	
7	5.035000	3.000000	1.920000		LZ_LK3	
8	-9.120000	0.350000	2.350000		AIR	
9	11.722000	3.100000	2.480000		LZ_LK3	P
10	-11.722000	12.200000	2.650000		AIR	
IMS	0.000000	0.000000	2.822888	S		

Figure 1.167 Optical design and drawings of the intersection of the parallel beam with the optical axis of a 40×/0.65 finite LCF semi–plan achromatic objective.

The same considerations for phase contrast devices can be applied to the construction of interference contrast devices. The Wollaston prisms are placed in the corresponding planes in the optical systems of the condenser and objective. It should be noted that implementing contrasting techniques such as phase and differential interference contrast can be achieved by using special additional conjugating optical systems that provide optical transfer of planes, i.e., their removal beyond the limits of their own condenser and objective systems. In this case, there is no need to place light rings in the condenser or phase rings in the objective; they can be placed in other areas of the optical system of the microscope, for example, on special flat plates.

There are two examples of phase and interference contrast devices that are among the most successful to have been implemented into mass-produced budget microscopes. German microscopes remained faithful to the traditions of high-quality engineering. Carl Zeiss Jena budget microscopes had what can be called peculiar standards for constructing contrast systems, including phase

Lens: ocx 40x0.65 d=0.17 Finity+plate Efl 4.06

Object num aper 0.650000 Image height 9.000000 Primary wavln 0.546070

SRF	RADIUS	THICKNESS	APERTURE RADIUS	GLASS	SPECIAL
OBJ	0.000000	0.010000	0.207043	AIR	
AST	0.000000	0.170000	0.230000 A	LZ_K14	
2	0.000000	0.385600	0.300000	AIR	
3	0.000000	4.530000	0.630000	LZ_CTK12	
4	-3.076000	0.100000	2.150000	AIR	
5	-15.171000	1.200000	2.230000	LZ_TF5	
6	5.035000	3.000000	2.470000	LZ_LK3	
7	-9.120000	0.350000	2.990000	AIR	
8	11.722000	3.100000	3.420000	LZ_LK3 P	
9	-11.722000	0.300000	3.630000	AIR	
10	0.000000	0.500000	3.630000	LZ_TF1	
11	0.000000	1.000000	3.630000	LZ_TF1 P	
12	0.000000	10.400000	3.630000	AIR	
13	5.297000	2.000000	3.650000	LZ_TF5 P	
14	4.207000	164.950000	2.970000	AIR	
IMS	0.000000	0.000000	9.000000		

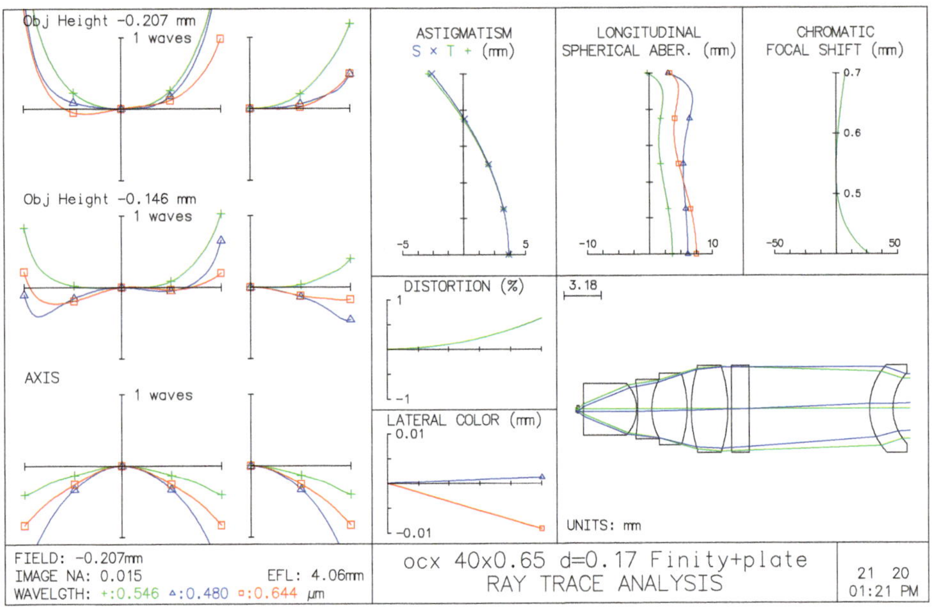

Figure 1.168 Optical design and graphs of aberration correction of a 40×/0.65 phase plate finite LCF semi–plan achromatic objective.

and interference contrast. Figure 1.180 shows photographs of the Interphaco microscopes of different generations.

Figure 1.181 is a photograph of the Polish BIOLAR PZO microscope, which was also unique in the budget microscope sector. This microscope used a fairly wide range of accessories for contrasting, including a device for

Lens: 100x1.25 oil finity				Efl	1.99
Object num aper	1.250000	Image height	9.000000	Primary wavln	0.546070
SRF	**RADIUS**	**THICKNESS**	**APERTURE RADIUS**	**GLASS**	**SPECIAL**
OBJ	0.000000	0.010000	0.094990	OIL	
AST	0.000000	0.170000	0.130000 A	LZ_K14	
2	0.000000	0.117500	0.380000	OIL	
3	0.000000	0.930000	0.540000	LZ_K20	
4	-0.780000	0.050000	0.780000	AIR	
5	-7.880000	1.000000	1.270000	LZ_K8	
6	-2.000000	0.100000	1.500000	AIR	
7	-34.020000	1.000000	1.750000	LZ_TF1	
8	5.290000	1.750000	2.080000	LZ_K2	
9	-4.020000	2.000000	2.270000	AIR	
10	101.860000	1.000000	2.610000	LZ_TF4	
11	5.649000	1.600000	2.710000	LZ_K8 P	
12	-7.447000	182.280000	2.770000	AIR	
IMS	0.000000	0.000000	9.000000		

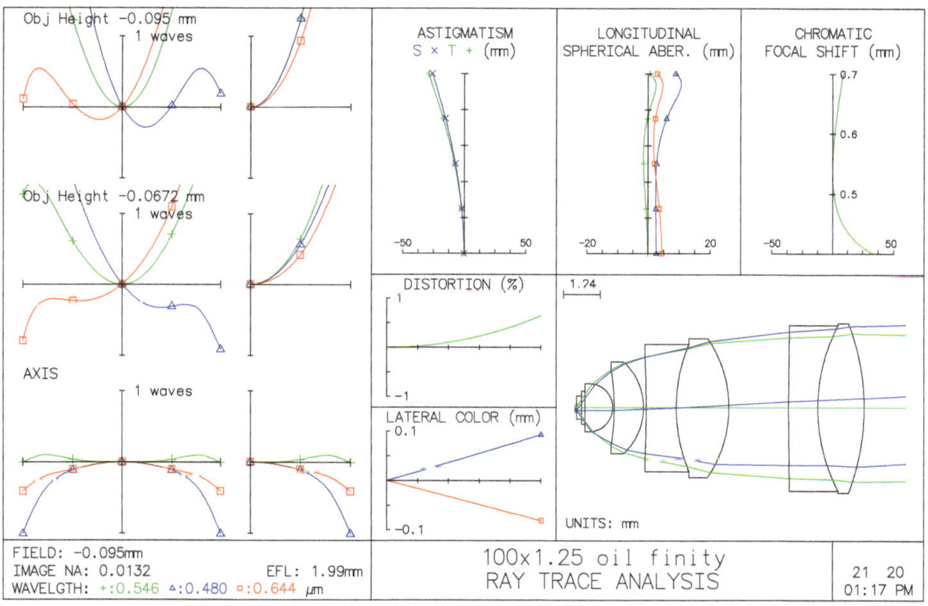

Figure 1.169 Optical design and graphs of aberration correction of a 100×/1.25 oil immersion finite achromatic objective.

positive phase contrast and one for negative phase contrast. A unique feature of this microscope was its implementation of interference contrast according to Pluta.

It is interesting to note that the technical descriptions of these two types of microscopes are excellent independent textbooks that provide not only the theory of some methods of contrasting in light microscopy, but also practical recommendations and a detailed description of the principles for setting up

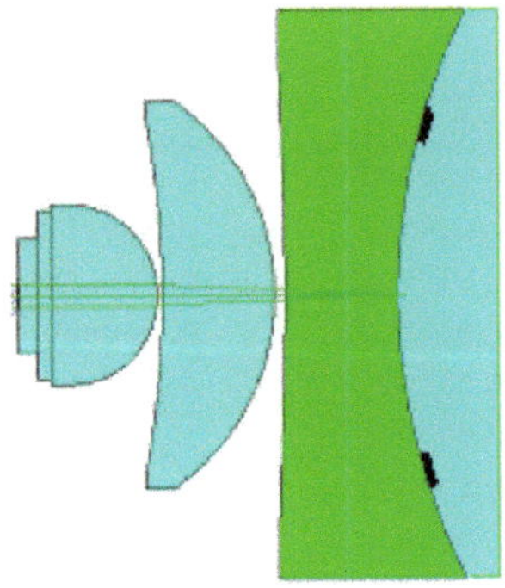

Lens: 100x1.25 oil finity

Ent beam radius 0.090000 Image height 2.100000

SRF	RADIUS	THICKNESS	APERTURE RADIUS	GLASS
OBJ	0.000000	1.0000e+08	1.0369e+08	OIL
1	0.000000	0.010000	0.105736 S	AIR
AST	0.000000	0.170000	0.140000 A	LZ_K14
3	0.000000	0.117500	0.490000	OIL
4	0.000000	0.930000	0.730000	LZ_K20
5	-0.780000	0.050000	0.780000	AIR
6	-7.880000	1.000000	1.510000	LZ_K8
7	-2.000000	0.100000	1.670000	AIR
8	-34.020000	1.000000	2.040000	LZ_TF1
9	5.290000	0.063551 S	2.450000	LZ_K2
IMS	0.000000	0.000000	2.100000 S	

Figure 1.170 Optical design and drawings of the location of the intersection of the parallel beam with the optical axis for a 100×/1.25 oil immersion finite achromatic objective.

Figure 1.171 Photographs of the MikMed-2 microscope and the KF-4M phase contrast device.

Figure 1.172 Photograph of a real object using a 10×/0.25 phase LCF plan achromatic objective (dark field option).

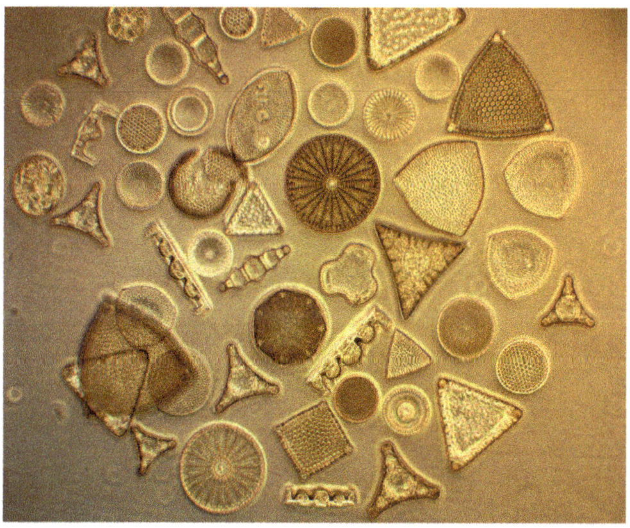

Figure 1.173 Photograph of a real object using a 10×/0.25 Ph LCF plan achromatic objective.

microscopes, i.e., extensive information on practice. Here is a quote from the BIOLAR microscope user manual [16]:

> "Using an ordinary biological microscope, you can only observe objects that, in relation to their environment, absorb light to varying degrees and, therefore, differ in some natural contrast. Such objects are often called amplitude, since they cause a change in the amplitude of the light passing through them. However, in nature there are such micro-objects

Figure 1.174 Photograph of a real object using a 20×/0.40 Ph achromatic objective.

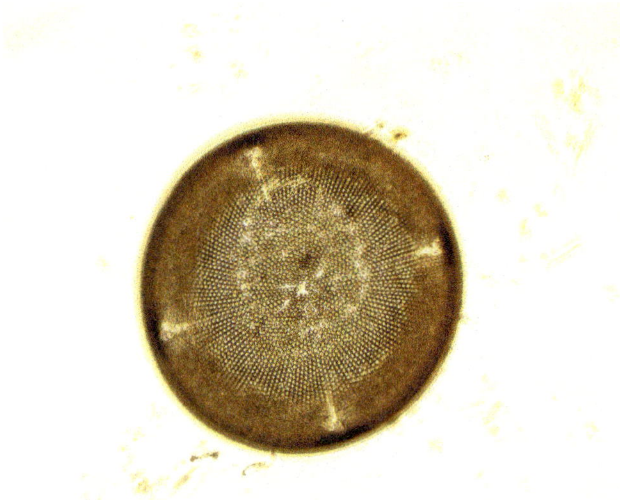

Figure 1.175 Photograph of a real object using a 20×/0.40 Ph achromatic objective.

and structures that do not show any difference in the absorption of light and differ from their environment only in the refractive index or thickness.

Such objects are usually called phase objects, since they cause only a change in the phase of the transmitted light wave. In an ordinary microscope, in the passing light, they are completely invisible or almost invisible, because the human eye reacts only to changes in the power of

Figure 1.176 Photograph of a real object using a 40×/0.65 Ph LCF semi–plan achromatic objective.

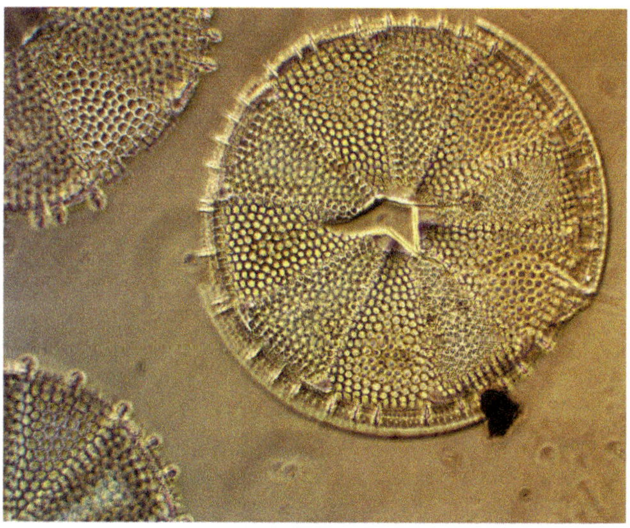

Figure 1.177 Photograph of a real object using a 40×/0.65 Ph LCF semi–plan achromatic objective.

light, i.e., to changes in the amplitude of the light wave (as you know, the light intensity is proportional to the square of the amplitude).

In order to be able to observe phase objects using an ordinary biological microscope, it is necessary to color them. The staining of the preparations is however difficult and entails various undesirable effects. In particular, living

Figure 1.178 Photograph of a real object using a 100×/1.25 Ph oil immersion achromatic objective.

Figure 1.179 Photograph of a real object using a 100×/1.25 Ph oil immersion achromatic objective.

cells and tissues cannot be stained, since staining is usually associated with their killing.

In this regard, the possibility of studying completely transparent and non-light-absorbing living microorganisms using an ordinary microscope is largely limited.

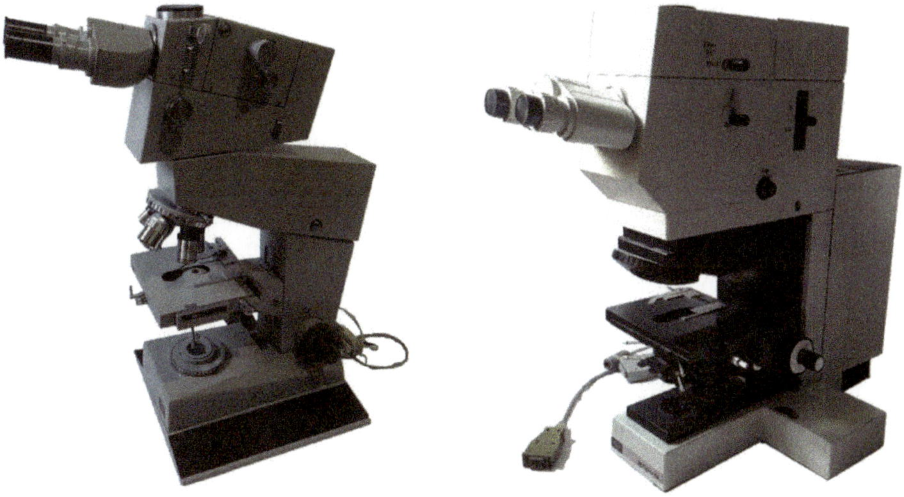

Figure 1.180 Interphaco microscopes made by Carl Zeiss Jena.

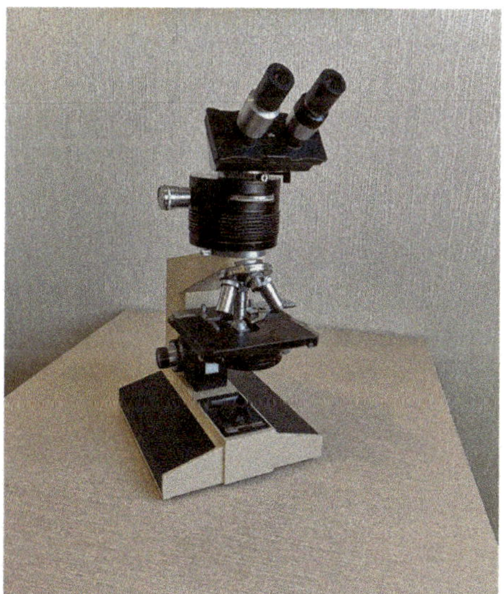

Figure 1.181 Photograph of the BIOLAR PZO microscope.

Currently, phase-contrast microscopes are used, almost without exception, to observe phase objects. The weak side of these microscopes (besides some defects like the halo, the small contrast of closely packed and adjacent structures, as well as the poor reproduction of stretched objects with relatively large transverse dimensions) is that they mainly serve for qualitative and descriptive studies. Phase contrast allows you to quickly assess whether the refractive index of a given investigated object is larger or smaller in relation to the coefficient of the environment, and even makes possible, in some cases, the exact determination of the value of this coefficient (by selecting the appropriate immersion fluid).

But this, in essence, ends with the possibility of quantitative studies of this method (apart from the possibility of measuring linear and planimetric quantities). Significantly more opportunities for quantitative studies are presented by the interference microscope as a whole, and in particular the BIOLAR PI polarization-interference microscope or the UPI polarization-interference device with the BIOLAR biological microscope.

This microscope is used to observe all kinds of micro-objects, both phase (transparent) and amplitude (absorbing light). Some measurements are possible with its help: differences in the optical path (phase shift), gradient of the difference in the optical path of thickness, refractive index, birefringence, contact angle, type of microsurface, concentration of substances, dry matter content in cells, light transmission, and other physical quantities.

The microscope allows for qualitative and quantitative studies using the following interference methods: the strip method, the differential and uniform field method with a large image splitting.

The polarization-interference microscope finds application, first of all, in biological and medical sciences (especially in cytology, histology, morphology, biochemistry, microbiology), as well as in physical chemistry, crystallography, and mineralogy; in the textile industry, in the technology of thin layers and films, as well as in other fields of science and technology."

These microscopes, given as examples, currently have no analogues in the sector of budget microscopy and are highly appreciated by professionals. Figures 1.182–1.186 show photographs of real objects obtained by applying

Figure 1.182 Photograph of a real object.

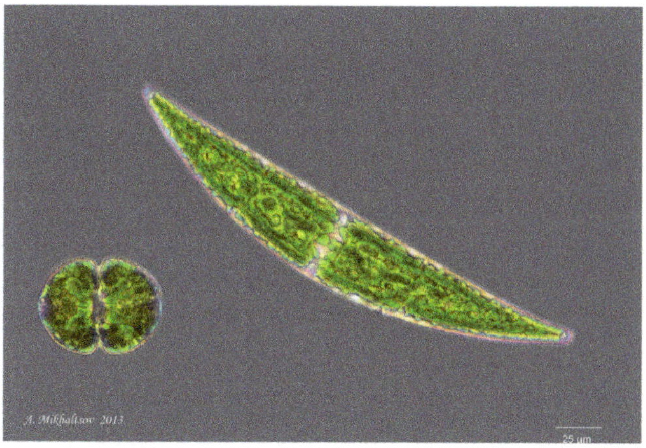

Figure 1.183 Photograph of a real object.

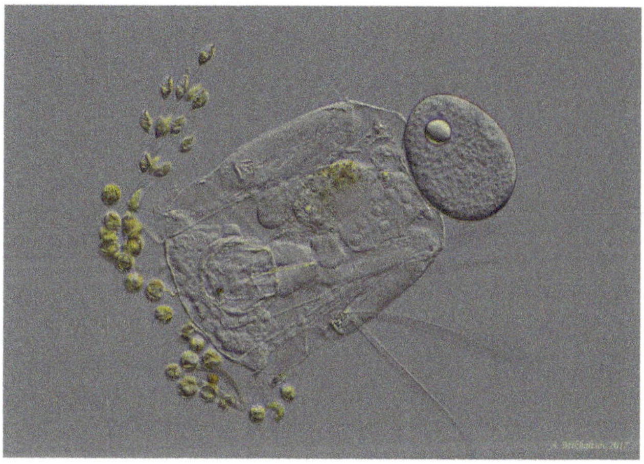

Figure 1.184 Photograph of a real object.

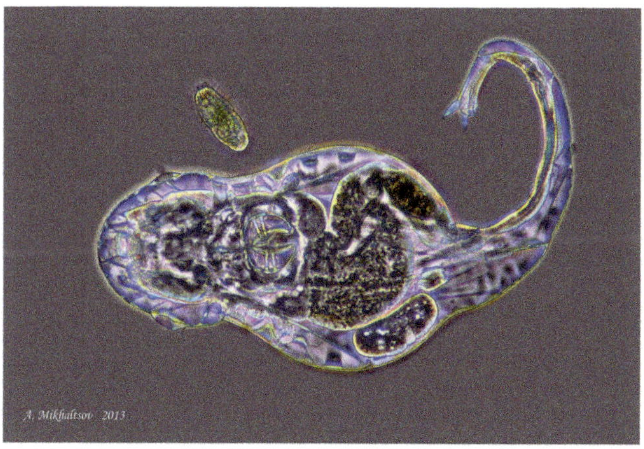

Figure 1.185 Photograph of a real object.

Figure 1.186 Photograph of a real object.

interference contrast methods. These photographs were taken by Anatoly Mikhaltsov.

We can also use some nonstandard engineering solutions in the construction of phase contrast systems. For example, if we use *liquid crystal* technology, we can get topological drawings of light rings of various configurations, for example, on a plate located in the front focal plane of the condenser (or close to this plane) or directly on one of the condenser lenses. Elementary electrical circuits allow the process of building the desired topology of the picture to be controlled. Such a technical solution may allow the traditional methods of mechanical changing of light rings in capacitors to be abandoned; mechanical constructions of phase contrast capacitor designs can be greatly simplified. In addition, a more precise adjustment ("by place") of the size of the light ring of the condenser to the size of the phase ring in the objective (including the opportunity to use an objective from a different phase contrast set) may be possible. Implementing some special effects associated with the redistribution of light energy in the plane of the pupil is also possible. With the development of this liquid crystal phase contrast technology, it may also be possible to manufacture the actual phase rings in the objectives, thereby replacing the relatively expensive and time-consuming technology that is used today. Moreover, liquid crystal technology can be implemented in variants of the "out place remote" phase contrast. Such technology can be particularly valuable in microscope systems where direct access of the operator to the microscope is difficult. An example of such a microscope is one designed to work underwater, where the operator's work is carried out remotely and the number of mechanisms moved relative to each other is minimal.

LED technology can also significantly change both the fundamental lighting systems of light microscopes in general and as applied to phase contrast devices. In addition, these technologies can be successfully applied to build the so-called *dark field* lighting in transmitted light microscopes. LED technology in lighting systems can give some impetus to the development of new techniques and practice in microscopy. For example, the effects of strobe, modulation, colorimetry, and others.

1.7.3 Independent correction terminology

The terminology used for independent objective correction should not address the fact that microscope objectives are designed to work with the infinite length of the tube. Terms such as *infinite color system* (ICS) or *infinite color correction system* (ICCS) mean that the objectives are designed to work under the conditions of an infinite length of the microscope tube, when the aperture rays propagate parallel to the optical axis after the objective and then are focused by a special tube system. The term *independent correction* should only refer to an objective quality parameter, such as aberration correction, regardless of whether it is intended for operation for the finite or infinite optical length of the tube. Nevertheless, the infinite length of the tube is gaining a more stable position in the construction of optical systems of modern microscopes, including mass-produced budget microscopes. Therefore, we developed a set of budget objectives *for infinity* for potential use in such microscopes. However, these objectives (as well as the objectives considered above for the finite length of the tube) also have high correction quality with respect to image curvature and lateral chromaticity. Table 1.11 shows the main technical parameters of a set of third-generation budget objectives for mass-produced microscopes of an infinite length of tube. All objectives are designed for a 45-mm parfocal distance standard and use the additional tube system $F' = 200$ mm. The calculated linear field in the image space was chosen equal to $2y' = 20$ mm, but the real field reaches 22 mm without a noticeable deterioration in quality.

Some objectives are designed for use without a cover glass and are mainly intended to work as part of metallographic microscopes. Some objectives use a 0.17-mm cover glass while others can work equally well with or without a cover glass. The use of an infinite length of the tube for such objectives has become advantageous, since when focusing on an object, the optical length of the tube after the objective does not change and there are no additional defocus aberrations in the image area.

It is believed that the difference between using a 0.17-mm-thick cover glass and not using one cannot be detected when using objectives of weak magnifications. To what extent can the estimated image quality deteriorate if, for example, an objective designed for working with a cover glass is used without it? Conversely, to what extent can it deteriorate if an objective designed for working without a cover glass ($d = 0$) is used with a 0.17-mm

Table 1.11 Main technical parameters and basic optical layout of a set of third-generation budget objectives for mass-produced microscopes of an infinite length of tube.

Magnification	NA	WD (mm)	F' (mm)	R (μm)	DF (μm)	FOV on object (mm)	FOV on image (mm)	Optical design at figure no.	The principal optical layout
5	0.10	20	40	3.33	33	4.4	22	1.32	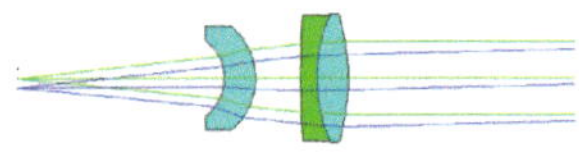
10×	0.28	15.5	20.0	1.2	4.25	2.2	22	1.35	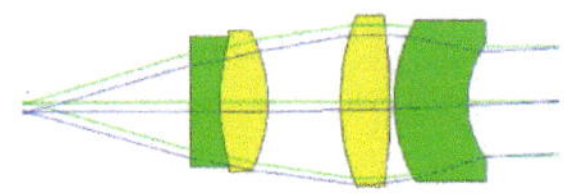
20×	0.40	8.0	10.0	0.84	2.1	1.1	22	1.37	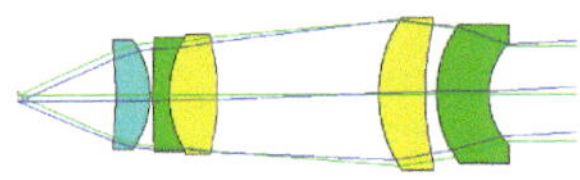
40×	0.60	2.9	5	0.56	0.93	0.55	22	1.42	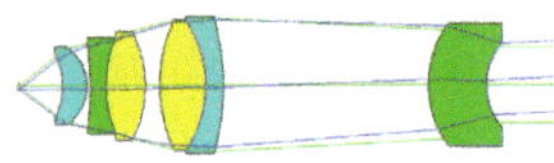
50×	0.60	2.4	4.0	0.56	0.92	0.55	22	1.47	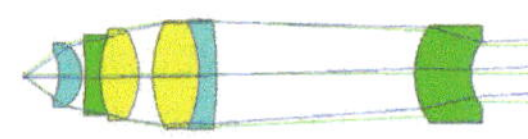
100×	0.80	0.52	2.0	0.42	0.52	0.22	22	1.51	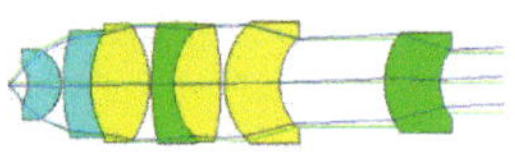
100×	0.90	0.15	2.0	0.37	0.41	0.22	22	1.53	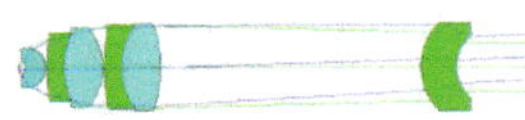
100×	1.30 oil	0.12	2.0	0.26	0.20	0.22	22	1.59	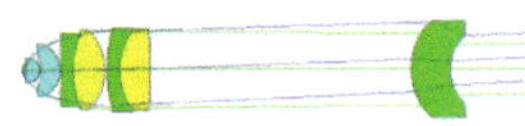

one? It has been reported [17] that for small-magnification objectives (up to 16–20×, or more precisely for objectives with numerical apertures to about 0.40), the loss in image quality is insignificant; there is practically no difference in the study of an object with and without a coverslip. As an argument, formulas have been given for calculating the spherical aberration that arises when using an objective in the not quite normal mode, concluding that the magnitude of this spherical aberration is small, so there should not be milkiness in the image. Nevertheless, my practical studies [18] have shown that objectives designed and assembled in a traditional way (in performance only at $d = 0.17$ mm or at $d = 0$) noticeably lose their image quality when operating in not quite normal mode.

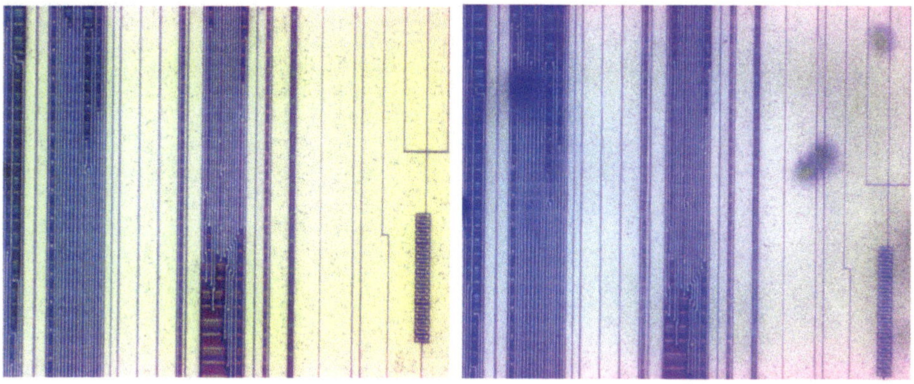

Figure 1.187 Object obtained without and with a cover glass using a 10×/0.25 standard objective.

The photograph in Fig. 1.187 confirms this, illustrating a standard 10×/0.25 objective designed for use without a coverslip but used with a cover glass. First, my colleagues and I took a photograph of the chip without a cover glass, and then covered the chip with a cover glass and took another photograph. There was an obvious difference in the quality of the image. It is interesting to note that in the second image, characteristic spots in the cover glass layer appeared in the field of view.

Using the information obtained during our research, we made an optical design of three objectives with linear magnifications of 10×, 20×, and 40× that can work equally well with objects covered or not by a cover glass. This idea is similar to the one that is often used in high-aperture objectives for microscopes, when a so-called *correction frame* is used to correct the deviation from the nominal value in the value of the cover glass. In this case, one or more of the objective components can move within small limits along the optical axis. In our case, we performed a *mathematical approximation* of the nominal thickness of the cover glass, taking (for optical design) $d = 0.085$ mm. In this case, the residual spherical aberration is twice less, the objective has high image quality in the range of cover glass thickness from 0 to 0.17 mm, and no additional movements of the components are required. Another useful quality of these objectives is that they provide a large working distance from the object to the first lens of the objective. This is very useful because the microscopist often uses this procedure when investigating an object: first a dry objective followed by an immersion objective. But if, for example, another part of the object needs to be investigated after the immersion, it is inconvenient to wipe the immersion from the object. Thus, it is very convenient to have in the arsenal of dry objectives one with a large working distance, allowing the researcher to view through the immersion without

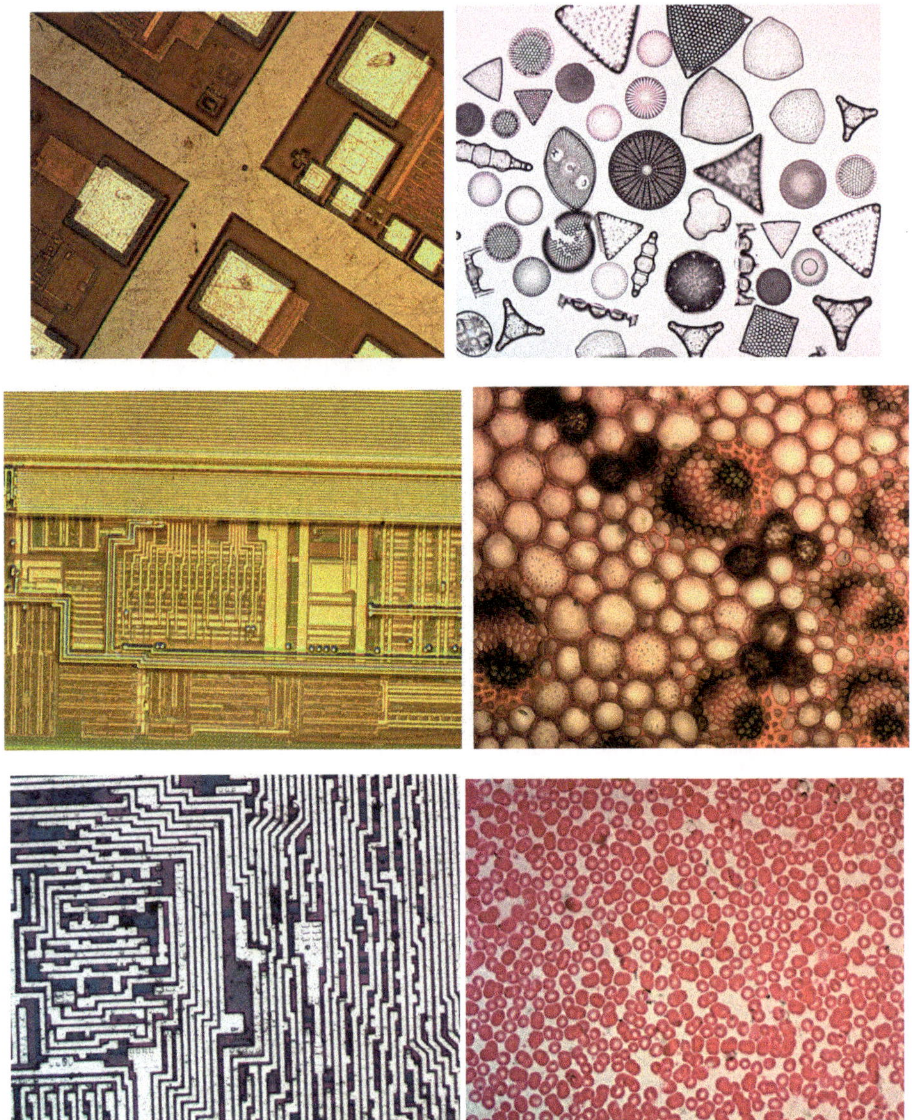

Figure 1.188 Photographs of metallographic objects not covered by a cover glass and biological objects covered with a 0.17-mm cover glass using the objectives shown in Fig. 1.189.

touching it. Such objectives allow this to be done, which has been confirmed by tests and microscopy practice.

Figure 1.188 shows photographs of metallographic objects not covered by a cover glass and biological objects covered with a 0.17-mm cover glass, where, as examples, 10×/0.28, 20×/0.40, and 40×/0.60 objectives are used (parameters of optical design and aberration corrections are shown in Figs. 1.35, 1.37, and

Figure 1.189 Photograph of the set of objectives ($d = 0$–0.17 mm).

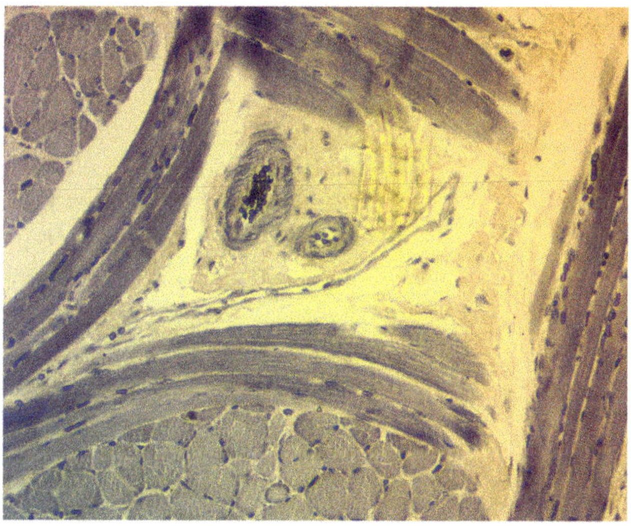

Figure 1.190 Photograph of an object made through oil immersion with a 20×/0.40 dry objective.

1.42, respectively). Figure 1.189 shows a photograph of the set of these objectives.

Figure 1.190 shows a photograph of a biological object made through oil immersion (from the air without touching the immersion) with a 20×/0.40 objective ($d = 0$–0.17 mm). This photograph confirms the high image quality. [By the way, a 40×/0.70 objective, the parameters of which are presented in Fig. 1.46, has similar properties. It is also designed to work as part of microscopes using the infinite length of the tube, but unlike the objectives considered here, it uses an additional tube system ($F' = 180$ mm).]

1.8 Topical Reasoning about the Parts of Mass-produced Microscopes

It would, of course, be convenient to have a universal microscope in which objectives for both the finite and infinite length of the tube could be used. In this case, when getting a new interesting objective, microscopists would not have to wonder whether it was necessary to adapt their existing microscope to work with this objective. Let us briefly consider the optical system of the projection branch of a light (budget) microscope. We will not consider a monocular microscope or a microscope equipped with a so-called *lensless* visual head. In other cases, as a rule, this system consists of four blocks: the actual objective, some additional lenses (tube system), a head with prisms, and eyepieces. In any combination on the finite or infinite length of the tube, the image is constructed jointly by the objective, the tube system, and the head prisms in the so-called *intermediate image plane*, i.e., the plane of the eyepiece's field diaphragm. The differences between the finite and infinite length of the tube are as follows:

1. In the finite case, the image of the object after the objective is formed in converging aperture beams of rays (designers of optical systems call these "1–1"-type systems). The tube system in this case also operates according to the 1–1 principle (that is, from the finite distance to the finite one) and projects the image obtained after the objective into the plane of the field diaphragm of the eyepiece.
2. In the infinite case, the object is located strictly in the front focal plane of the objective, and its image after the objective is formed by a parallel aperture beam of rays (at infinity). Designers of optical systems call these "1–0"-type systems. In this case, the tube system will operate as a 0–1-type system and build the image obtained after the objective in its focal plane, which coincides with the plane of the field diaphragm of the eyepiece.

1.8.1 Design of tube systems

Here, let us briefly discuss the design of visual heads as well as some intermediate optical systems, which are often located between the objective and the visual binocular head. This applies to the so-called *tube systems*, which are one or more lenses. If a microscope is designed for a finite tube length of 160 mm, such a lens system pushes the image plane obtained after the objective by the required amount. This is because the value of the so-called *posterior paraxial segment* of the objective is not enough to build an image after the prisms used inside the binocular head. Prisms are needed to provide the required rotation of the rays so that the researcher can comfortably work at the microscope. As a rule, binocular nozzles provide optimal observation conditions for ergonomics; the inclination of the binocular tubes of the horizon, as a rule, is 30–45 deg, and the possible values of the displacement of the ocular tubes in accordance with the eye base of a particular observer are 45–75 mm. The dimensions of the prisms, in turn, are due to these and other ergonomic requirements, ensuring the researcher's comfort.

It is known that the thickness of the glass, even if it is a parallel plate, introduces spherical aberration as well as image shift along the optical axis. But most importantly, the thickness of the glass introduces significant chromatic aberration due to the dispersions of optical materials. When calculating the optical scheme of a light microscope, the optical design of the so-called *projection system*, i.e., the objective and the tube system, receives the most attention. Some optical designers prefer to calculate objectives and tube systems separately; in this case, the quality of the aberration correction of the objective is obtained as is. The task of the tube optical system is to obtain an image after the last lens in the plane of the field diaphragm of the eyepiece without introducing any additional aberrations. In this case, the quality of the aberration correction of the tube system should be ideal. An example of such a tube system for a microscope with a finite tube length of 160 mm is shown in Volume 1 (Chapter 1, Fig. 1.53). As a rule, the design of such tube systems is made as needed, based on the actual construction conditions of a particular microscope. This condition also applies to the calculations to set the plane of the entrance (exit) pupil.

A slightly different situation is if we make the optical design of the objective and the tube system separately for a microscope with an infinite optical length of the tube. This explains why so many optical designs of objectives presented in this book are made using virtual tube systems that are not used in microscopy practice. With this kind of optical design, the focus is on the design of the objective as well as on observing the telecentric path of the rays in the microscope's projection system. For example, if the tube system (tube lens) has a focal length $F'_{ts} = 200$ mm, then the estimated distance between the objective and the tube system will be 200 mm. According to the optical design of the tube system, the position for the entrance pupil will be equal to its focal length dimension. The design is considered ideal if the planes of the exit pupil for the objective and the entrance pupil for the tube system coincide. In a real situation this is an approximate value because the positions of the exit pupil planes are different for different objectives. In short-focus lenses, they are usually located deep inside the optical system of the objective; for long-focus objectives, they are usually not far from the reference plane. It should be noted that there is a technical possibility of aligning the positions of the pupil planes in the set of objectives; however, their optical systems will be complicated, which is not always acceptable for the budget version of the microscope. In some cases, the ideal optical design of the projection system is ignored, and the virtual location of the entrance pupil for the tube system is taken in a different plane, for example, at a distance of 160 mm from the first lens of the tube system (with its focal length $F' = 200$ mm). An important parameter for the tube system is called the *relative hole*; it directly depends on the value of the output aperture (after the tube system) and the location of the entrance pupil. In turn, the value of the output numerical aperture can be calculated based on the values of the input numerical aperture and the focal length of the objective, as well as the focal length of the tube system. The value of the output numerical aperture of the tube system will be equal to

$$NA' = NA_{ob} \cdot \frac{F'_{ob}}{F'_{ts}}.$$

Naturally, since the same tube system is used for the entire set of objectives, the calculated value of its output aperture should correspond to the lens with the largest relative aperture (in the set of objectives for microscopes, these are usually small-magnification objectives, some of which have an increased output numerical aperture). Such objectives, for example, can have a linear magnification (the ratio of the focal length of the tube system to the focal length of the lens) of $-4\times$ and an input numerical aperture $NA_{ob} = 0.2$, or a linear magnification of $-10\times$ and an input numerical aperture $NA_{ob} = 0.4$. Thus, the optical design of the tube system should be based on the value of the output numerical aperture $F'_{ts} = 0.04$.

Figures 1.191–1.193 show examples of optical designs of tube systems taken from open sources. Figure 1.191 shows the design parameters and graphs of aberration correction of a tube system calculated for focal length $F' = 180$ mm and position of the entrance pupil $S_p = -180$ mm. The same is shown in Figs. 1.192 and 1.193 but with position of the entrance pupil $S_p = -100$ and -50 mm, respectively.

By the way, by measuring the diameter of the lenses of this tube system in a real head, we can determine for what value the position of the entrance pupil was made in the optical design. The examined tube system has a high image quality, which does not deteriorate (but improves) when the value of the position of the entrance pupil decreases. However, it is not intended as an attachment to work in conjunction with a prism visual head, since for this distance from the last surface of the last lens to the image plane is not enough. This confirms that this tube system is only intended to carry out the optical designs of the objectives as a focusing tube system with an ideal aberration correction.

Figure 1.194 shows the design parameters and graphs of aberration correction of a tube system calculated for focal length $F' = 200$ mm and position of the entrance pupil $S_p = -160$ mm (the tube system is not calculated for position of the entrance pupil $S_p = -200$ mm). The same is shown in Figs. 1.195 and 1.196 but with position of the entrance pupil $S_p = -100$ and -50 mm, respectively.

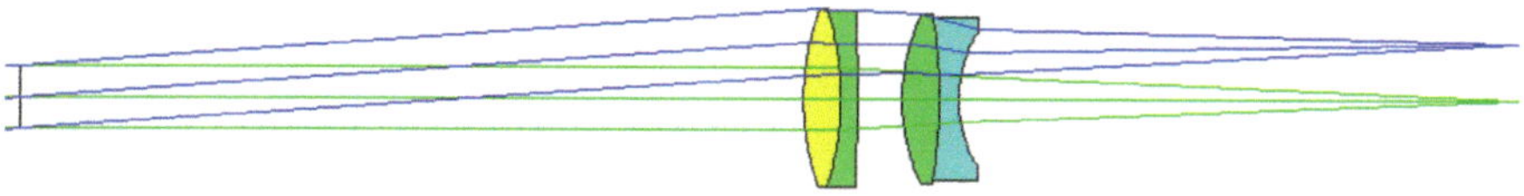

Lens: Ftl 180 NA=0.04 2y'=25 Sp=-180 Efl 179.99

Image num aper 0.040000 Image height 12.500000 Primary wavln 0.546074

SRF	RADIUS	THICKNESS	APERTURE RADIUS	GLASS	SPECIAL
OBJ	0.000000	1.0000e+20	6.9449e+18	AIR	
AST	0.000000	180.000000	7.199481 AS	AIR	
2	60.435700	8.500000	19.900000	S-FPL51	
3	-67.232800	3.800000	19.750000	S-NBH8	
4	-640.476000	10.290000	19.630000	AIR	
5	44.058600	8.500000	18.840000	S-BAH28	
6	-113.886300	4.400000	18.200000	S-NBM51	
7	28.037100	129.540916 S	15.480000	AIR	
IMS	0.000000	0.000000	12.500000 S		

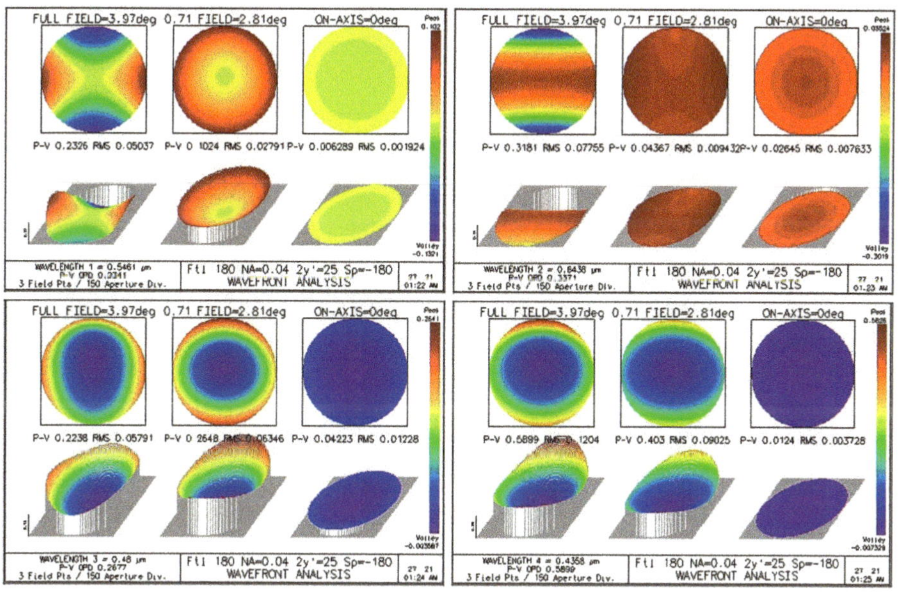

Figure 1.191 Design parameters and graphs of aberration correction of a tube system ($F' = 180$ mm; position of the entrance pupil $S_p = -180$ mm).

As the graphs of aberration correction show, the examined tube system ($F'_{ts} = 200$ mm) is close to ideal at the calculated positions of the entrance pupil $S_p = -160$ and -100 mm, but the quality deteriorates when the plane of the entrance pupil approaches the image.

Nevertheless, this tube system can work together with the prism system of the visual head. Although some values of monochromatic and chromatic aberrations are added, in general, the quality of the aberration correction

Lens: Ftl 180 NA=0.04 2y'=25 Sp=-100 Efl 179.99

Image num aper 0.040000 Image height 12.500000 Primary wavln 0.546074

SRF	RADIUS	THICKNESS	APERTURE RADIUS	GLASS	SPECIAL
OBJ	0.000000	1.0000e+20	6.9449e+18	AIR	
AST	0.000000	100.000000	7.199481 AS	AIR	
2	60.435700	8.500000	14.250000	S-FPL51	
3	-67.232800	3.800000	14.060000	S-NBH8	
4	-640.476000	10.290000	14.050000	AIR	
5	44.058600	8.500000	13.820000	S-BAH28	
6	-113.886300	4.400000	13.000000	S-NBM51	
7	28.037100	129.540916 S	11.700000	AIR	
IMS	0.000000	0.000000	12.500000 S		

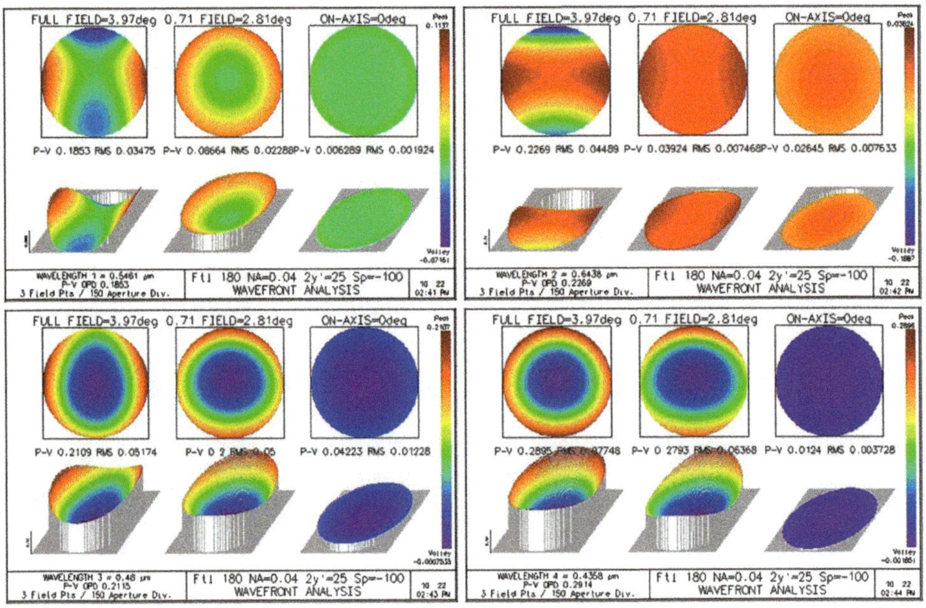

Figure 1.192 Design parameters and graphs of aberration correction of a tube system ($F' = 180$ mm; position of the entrance pupil $S_p = -100$ mm).

remains satisfactory for budget microscopes. Figure 1.197 shows the design parameters and graphs of aberration correction with this combination (as an example, a prism system from one prism head was selected). If the basic unified parameters shown in Fig. 1.50 (Volume 1, Chapter 1) are fulfilled when designing the budget microscope (in particular, the size of 41.5 mm), we can assign the position plane of the entrance pupil at a distance of -50 mm from the first lens of the tube system.

This is very surprising, but practically no mirror visual attachments are used in budget light microscopes. This is due to the highly complicated nature of manufacturing mirrors (as compared with manufacturing prisms) and the difficulty of adjusting such a head. In addition, mirror elements often cause

Lens: Ftl 180 NA=0.04 2y'=25 Sp=-50 Efl 179.99

Image num aper 0.040000 Image height 12.500000 Primary wavln 0.546074

SRF	RADIUS	THICKNESS	APERTURE RADIUS	GLASS	SPECIAL
OBJ	0.000000	1.0000e+20	6.9449e+18	AIR	
AST	0.000000	50.000000	7.199481 AS	AIR	
2	60.435700	8.500000	10.730000	S-FPL51	
3	-67.232800	3.800000	10.640000	S-NBH8	
4	-640.476000	10.290000	10.680000	AIR	
5	44.058600	8.500000	10.720000	S-BAH28	
6	-113.886300	4.400000	10.020000	S-NBM51	
7	28.037100	129.540916 S	9.270000	AIR	
IMS	0.000000	0.000000	12.500000 S		

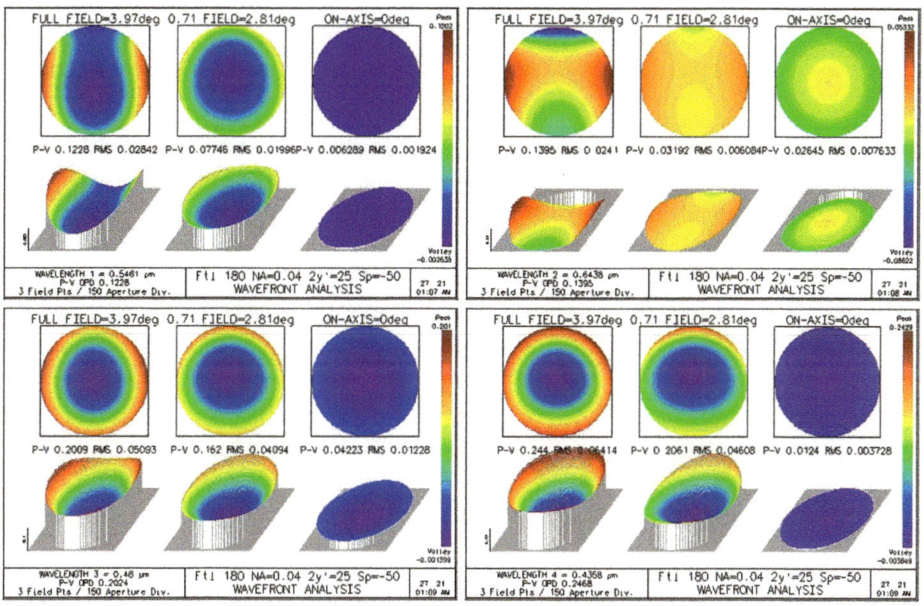

Figure 1.193 Design parameters and graphs of aberration correction of a tube system ($F' = 180$ mm; position of the entrance pupil $S_p = -50$ mm).

glare and double images, leading to depolarization. However, using mirror heads can be a comprehensive solution to many visual head problems. In this case, the optical elements of the head do not introduce their own aberrations; image quality is determined solely by the objective's aberration correction if it is a microscope with a finite length of the tube, or by the objective and tube focusing system (tube lens) if it is a microscope with an infinite length of tube. Mirror visual heads are significantly less expensive than optical glass. Compared with the prism head, the mirror head has practically no weight. The development of optical and alignment technologies will most likely lead to more frequent use of mirror visual heads and other attachments.

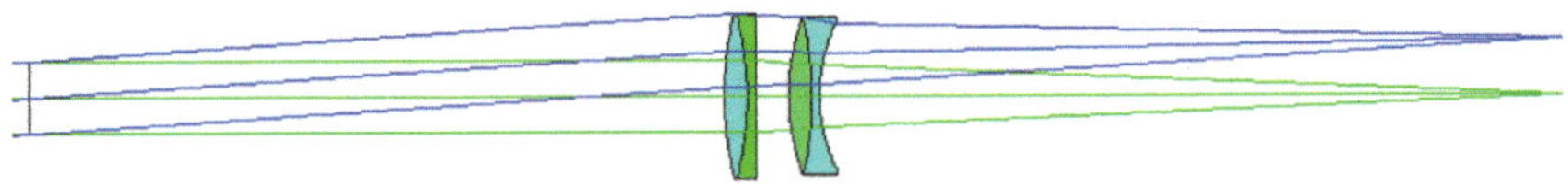

Lens: Ft1200 NA=0.04 2y'=25 Sp=-160 Efl 200.97

Image num aper 0.040000 Image height 12.500000 Primary wavln 0.546074

SRF	RADIUS	THICKNESS	APERTURE RADIUS	GLASS	SPECIAL
OBJ	0.000000	1.0000e+20	6.2200e+18	AIR	
AST	0.000000	160.000000	8.038590 AS	AIR	
2	75.043000	5.100000	18.130000	S-BSM10	
3	-75.043000	2.000000	18.090000	S-LAM7	
4	1.6006e+03	7.500000	17.950000	AIR	
5	50.260000	5.100000	17.400000	BAH26	
6	-84.541000	1.800000	17.360000	S-NBM51	
7	36.911000	169.236753 S	15.980000	AIR	
IMS	0.000000	0.000000	12.500000		

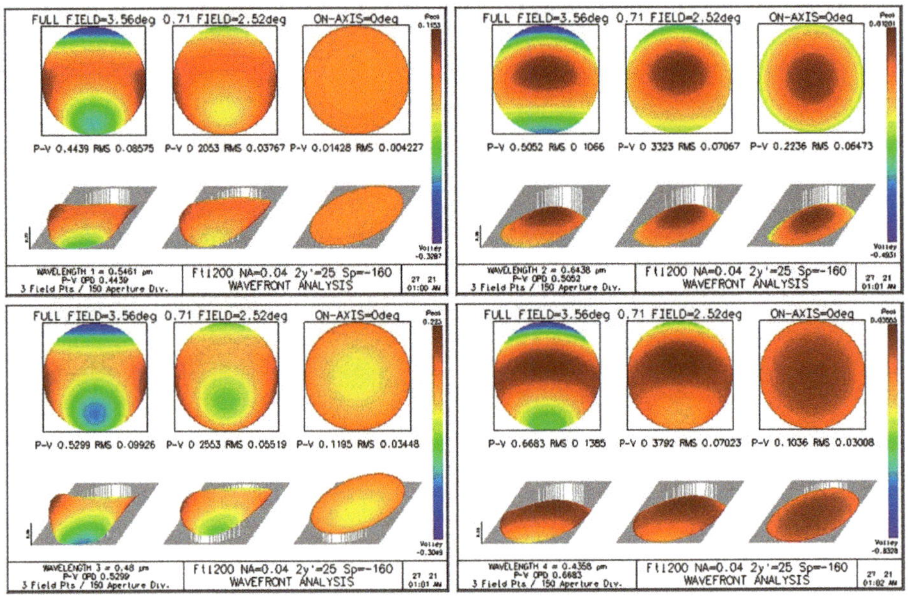

Figure 1.194 Design parameters and graphs of aberration correction of a tube system ($F' = 200$ mm; position of the entrance pupil $S_p = -160$ mm).

Figures 1.198–1.201 show examples of optical designs of some tube systems that can be used in light microscopes; tube systems differ in parameters such as focal length, position of the entrance pupil, numerical aperture, and the linear field in the image space.

It was interesting to investigate the possibility of using objectives for photographic cameras as tube systems for a microscope. For this study, photographic objectives were chosen whose equivalent focal lengths are close to the focal lengths of tube systems used in light microscopes.

Lens: Ft1200 NA=0.04 2y'=25 Sp=-100 Efl 200.97

Image num aper 0.040000 Image height 12.500000 Primary wavln 0.546074

SRF	RADIUS	THICKNESS	APERTURE RADIUS	GLASS	SPECIAL
OBJ	0.000000	1.0000e+20	6.2200e+18	AIR	
AST	0.000000	100.000000	8.038590 AS	AIR	
2	75.043000	5.100000	14.340000	S-BSM10	
3	-75.043000	2.000000	14.260000	S-LAM7	
4	1.6006e+03	7.500000	14.190000	AIR	
5	50.260000	5.100000	13.930000	BAH26	
6	-84.541000	1.800000	13.660000	S-NBM51	
7	36.911000	169.236753 S	12.920000	AIR	
IMS	0.000000	0.000000	12.500000		

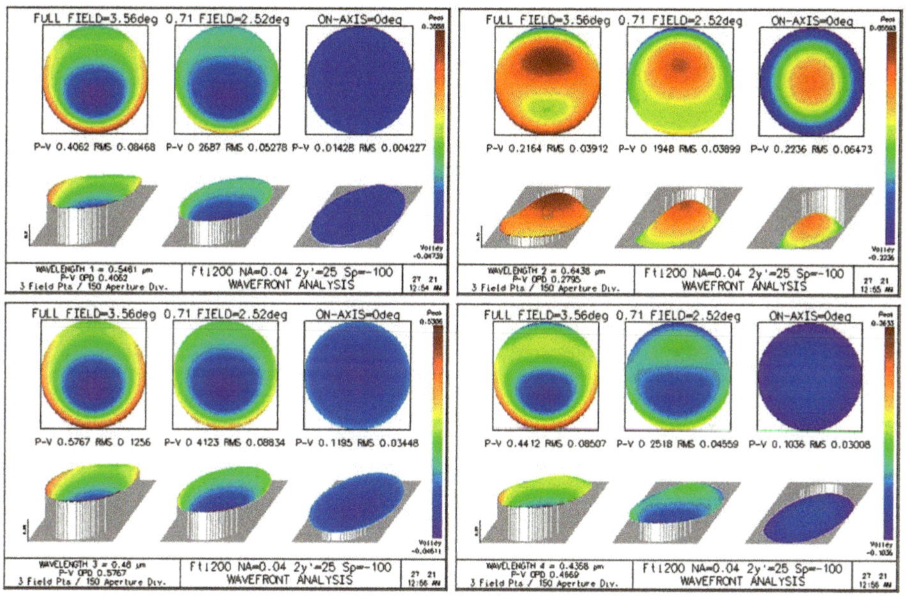

Figure 1.195 Design parameters and graphs of aberration correction of a tube system ($F' = 200$ mm; position of the entrance pupil $S_p = -100$ mm).

Figures 1.202–1.206 (comparative tables) show the design parameters of the selected objectives and the aberration correction graphs in the normal mode as photographic objectives (top) and in the tube system of a microscope (bottom). In this case, the working conditions were chosen as follows: the position of the entrance pupil corresponds to the equivalent focal length of the objective, the output numerical aperture is 0.03, and the linear field in the image space is 25 mm.

The Helios 104 objective can be used very well as a tube system with the indicated numerical aperture and linear field parameters. The image quality is slightly degraded at the periphery.

Lens: Ft1200 NA=0.04 2y'=25 Sp=-50 Efl 200.97

Image num aper 0.040000 Image height 12.500000 Primary wavln 0.546074

SRF	RADIUS	THICKNESS	APERTURE RADIUS	GLASS	SPECIAL
OBJ	0.000000	1.0000e+20	6.2200e+18	AIR	
AST	0.000000	50.000000	8.038590 AS	AIR	
2	75.043000	5.100000	11.200000	S-BSM10	
3	-75.043000	2.000000	11.130000	S-LAM7	
4	1.6006e+03	7.500000	11.110000	AIR	
5	50.260000	5.100000	11.040000	BAH26	
6	-84.541000	1.800000	10.740000	S-NBM51	
7	36.911000	169.236753 S	10.340000	AIR	
IMS	0.000000	0.000000	12.500000		

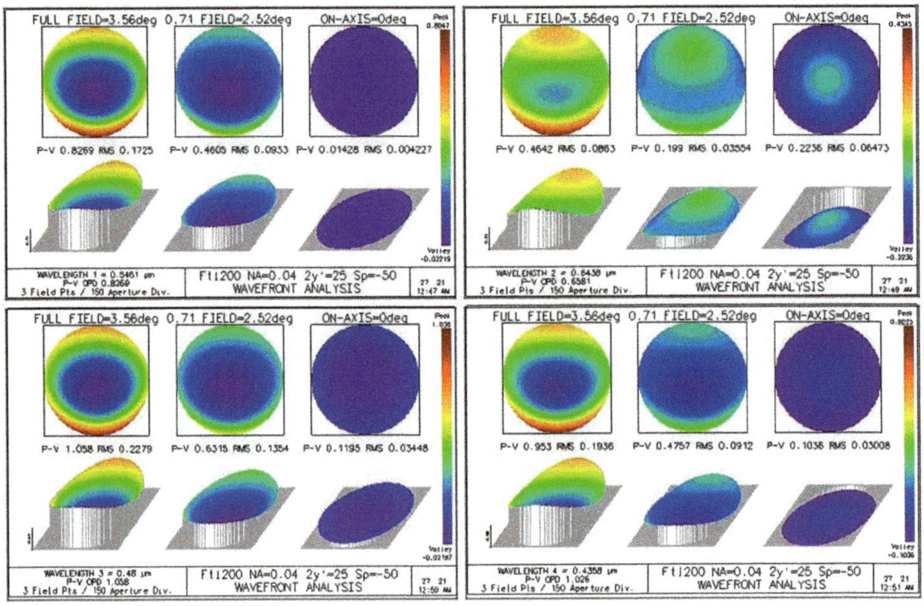

Figure 1.196 Design parameters and graphs of aberration correction of a tube system ($F' = 200$ mm; position of the entrance pupil $S_p = -50$ mm).

The Industar 37 objective can well be used as a tube system with the indicated numerical aperture and linear field parameters. The image quality is slightly degraded at the periphery.

The Industar 51 objective can be used as a tube system with the indicated numerical aperture parameters. The maximum linear field can be about 20 mm. The image quality is reduced at the periphery.

The Tair 56 objective can be used as a tube system with the indicated numerical aperture parameters. The maximum linear field can be about 20 mm. The image quality is reduced at the periphery.

The Jupiter 6 objective can be used as a tube system with the indicated numerical aperture parameters. The maximum linear field can even be

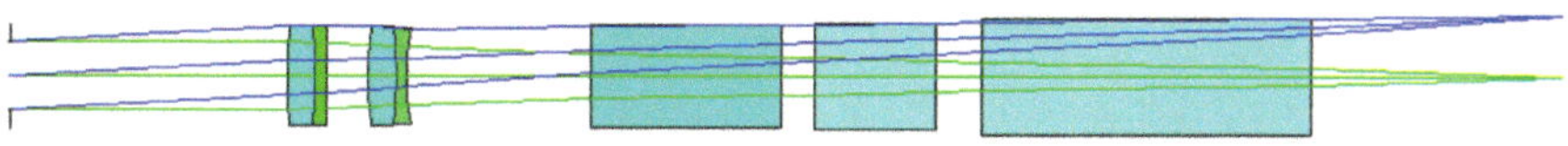

Lens: Ft1200 NA=0.03 2y'=22 Sp=-50 Efl 200.97

Image num aper 0.030000 Image height 11.000000 Primary wavln 0.546074

SRF	RADIUS		THICKNESS		APERTURE RADIUS		GLASS		SPECIAL
OBJ	0.000000		1.0000e+20		5.4736e+18		AIR		
AST	0.000000		50.000000		6.028943	AS	AIR		
2	75.043000		5.100000		8.790000		S-BSM10		
3	-75.043000		2.000000		8.750000		S-LAM7		
4	1.6006e+03		7.500000		8.740000		AIR		
5	50.260000		5.100000		8.740000		BAH26		
6	-84.541000		1.800000		8.470000		S-NBM51		
7	36.911000		33.408000		8.230000		AIR		
8	0.000000		34.600000		8.760000		LZ_BK10		
9	0.000000		6.000000		9.130000		AIR		
10	0.000000		22.000000		9.230000		LZ_K8		
11	0.000000		8.000000		9.470000		AIR		
12	0.000000		59.500000		9.600000		LZ_K8	P	
13	0.000000		46.130315	S	10.250000		AIR		
IMS	0.000000		0.000000		11.000000	S			

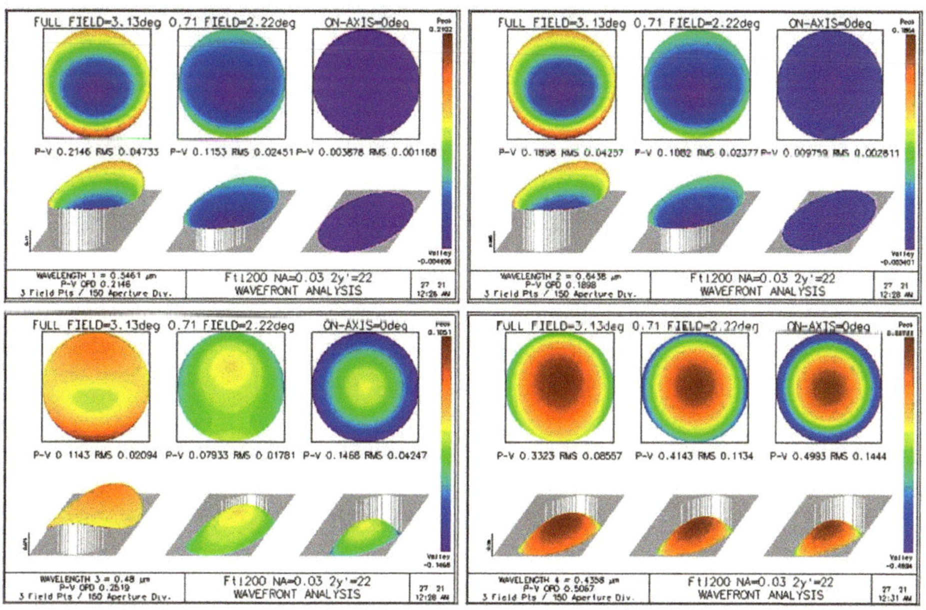

Figure 1.197 Design parameters and graphs of aberration correction of a tube system ($F' = 200$ mm) with prisms of the visual head (position of the entrance pupil $S_p = -50$ mm).

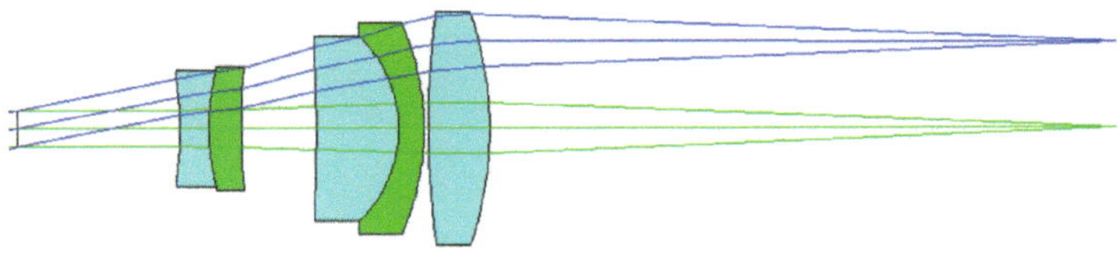

Lens: Ftl 110 NA=0.04 2y'=41.5mm

Image num aper 0.040000 Image height 20.750000 Primary wavln 0.546074

SRF	RADIUS	THICKNESS	APERTURE RADIUS	GLASS	SPECIAL
OBJ	0.000000	1.0000e+20	1.8859e+19	AIR	
AST	0.000000	0.001000	4.401078 AS	BK7 C	
2	0.000000	40.000000	4.401202 S	AIR	
3	-72.811700	7.000000	11.930000	LZ_K8	
4	53.180000	8.000000	13.720000	LZ_TF5	
5	144.787100	18.500000	14.550000	AIR	
6	-399.783800	20.000000	19.710000	LZ_K8 P	
7	-30.091900	6.000000	22.000000	LZ_TF5 P	
8	-61.912800	1.000000	25.150000	AIR	
9	211.140500	15.000000	26.850000	LZ_K8 P	
10	-81.951100	151.999796 S	27.640000	AIR	
IMS	0.000000	0.000000	20.750000 S		

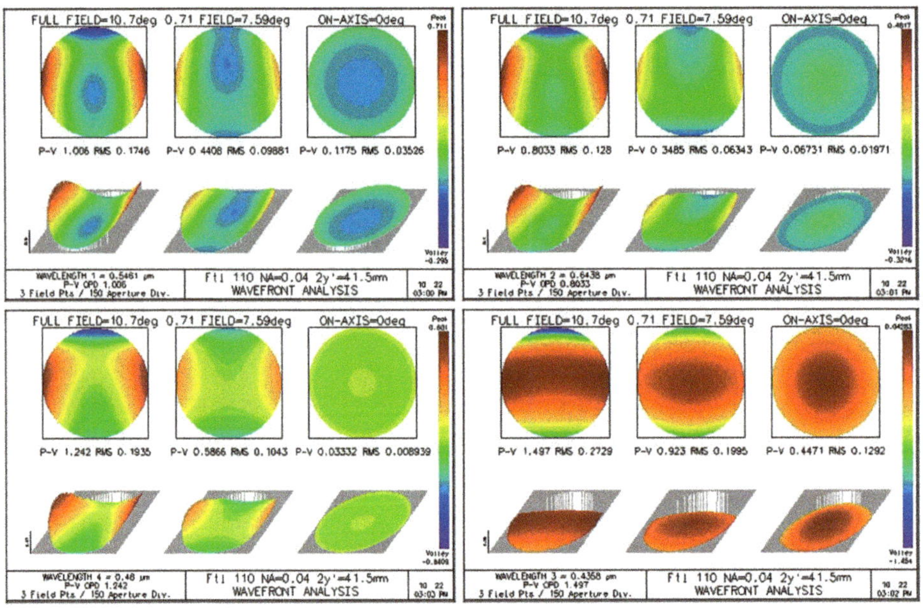

Figure 1.198 Design parameters and graphs of aberration correction of a tube system ($F' = 110$ mm; position of the entrance pupil $S_p = -40$ mm; $NA' = 0.04$; $2y' = 41.5$ mm).

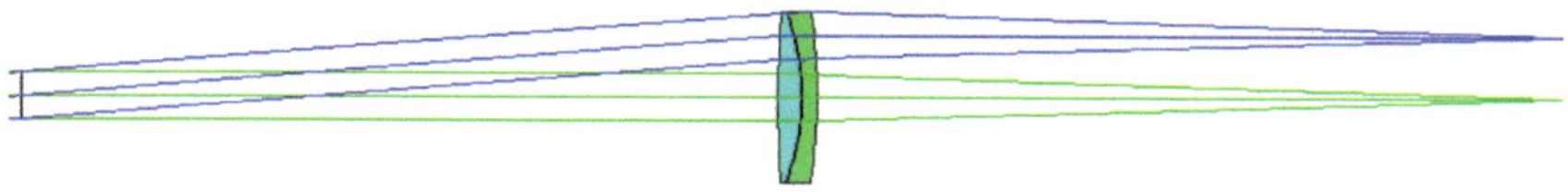

Lens: Ftl 160 NA=0.03 2y'=25mm Efl 160.0

Image num aper 0.030000 Image height 12.500000 Primary wavln 0.546074

SRF	RADIUS	THICKNESS	APERTURE RADIUS	GLASS	SPECIAL
OBJ	0.000000	1.0000e+20	7.8109e+18	AIR	
AST	0.000000	0.001000	4.801010 AS	BK7 C	
2	0.000000	160.000000	4.801061 S	AIR	
3	214.800000	5.000000	17.390000	LZ_BK8	
4	-46.990000	0.300000	17.410000	AIR	
5	-46.880000	3.000000	17.350000	LZ_BF24	
6	-113.240000	157.833626 S	17.560000	AIR	
IMS	0.000000	0.000000	12.500000 S		

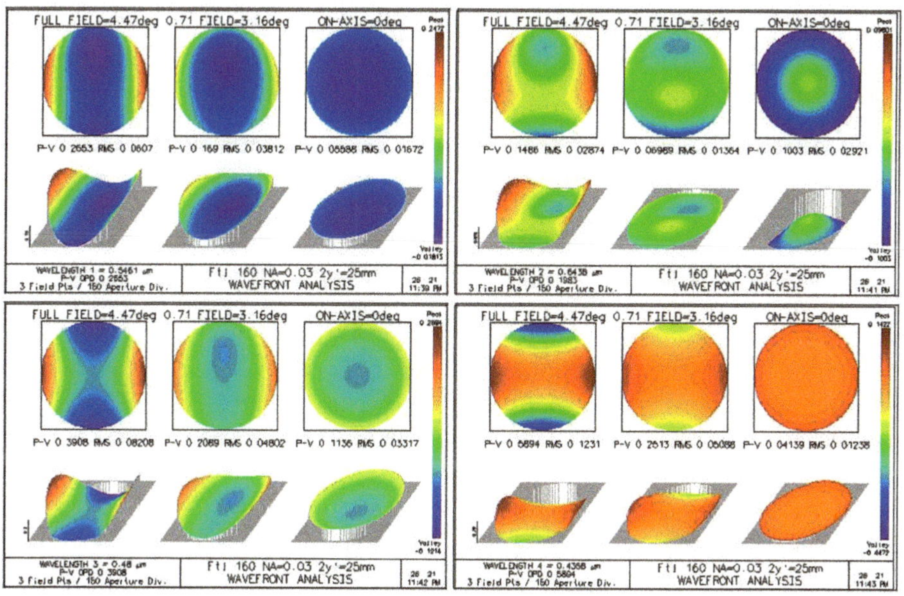

Figure 1.199 Design parameters and graphs of aberration correction of a tube system ($F' = 160$ mm; position of the entrance pupil $S_p = -160$ mm; $NA' = 0.03$; $2y' = 25$ mm).

increased to a value of, for example, 28 mm without a noticeable decrease in image quality at the periphery.

In general, it turned out to be possible to use standard photographic objectives with 150–300-mm focal lengths as tube systems of a microscope. At the same time, the values of the required light diameters were somewhat less than those with normal use; however, when working as tube systems, vignetting does not occur, despite a change from the initial position of the entrance pupil. As tube systems, these photographic objectives provide an

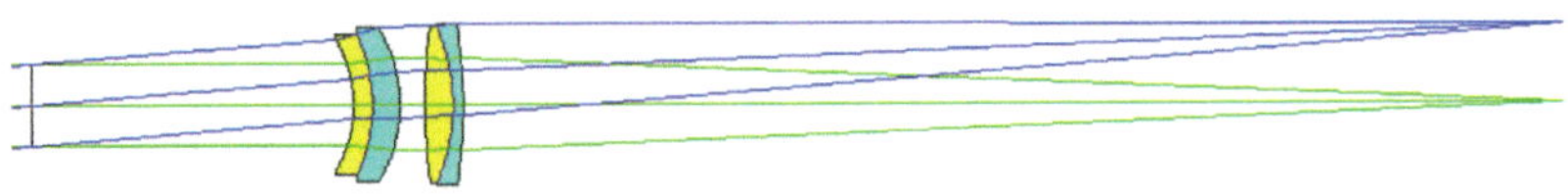

Lens: Ftl 180 NA=0.04 2y'=28mm Sp=-60 Efl 180.03

Image num aper 0.040000 Image height 14.000000 Primary wavln 0.546074

SRF	RADIUS	THICKNESS	APERTURE RADIUS	GLASS	SPECIAL
OBJ	0.000000	1.0000e+09	7.7763e+07	AIR	
AST	0.000000	60.000000	7.201353 AS	AIR	
2	-22.390000	3.500000	11.620000	H-FK71	
3	-28.689000	5.000000	12.540000	H-ZLAF50E	
4	-25.800000	4.500000	13.660000	AIR	
5	121.110000	5.000000	13.910000	H-FK71 P	
6	-34.350000	1.800000	13.890000	H-ZLAF50E P	
7	-89.000000	200.056342 S	14.150000	AIR	
IMS	0.000000	0.000000	14.000000 S		

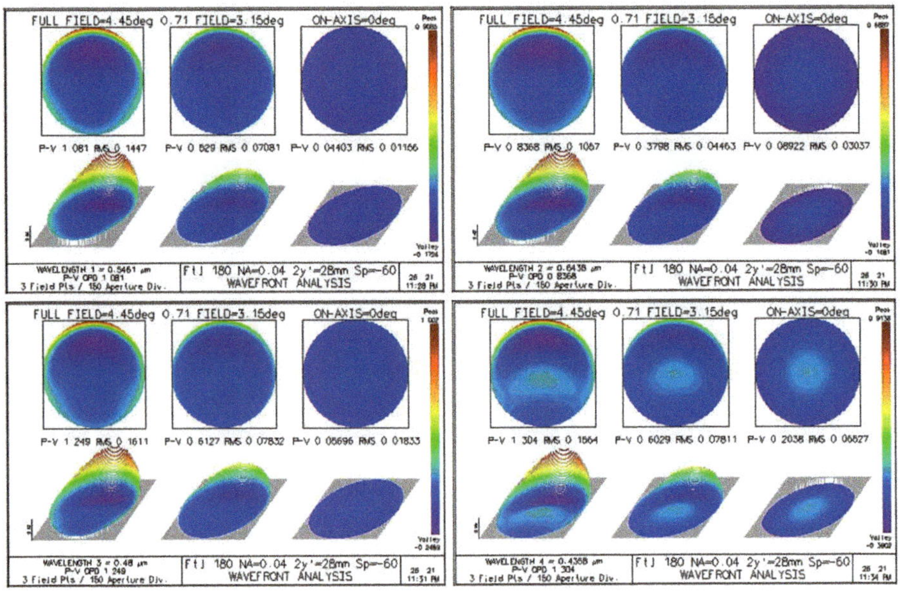

Figure 1.200 Design parameters and graphs of aberration correction of a tube system ($F' = 180$ mm; position of the entrance pupil $S_p = -60$ mm; $NA' = 0.04$; $2y' = 28$ mm).

output numerical aperture of 0.03, demonstrating the level of aberration correction adopted in microscopy, despite the fact that the aberration correction in the (initial) photographic objective systems is very different. It can be assumed that the use of zoom photographic objectives (in the indicated range of equivalent focal lengths) is also possible.

Figure 1.207 shows an example of standard zoom photographic objectives that were used as tube systems in one of the experiments. When using these zoom photographic objectives in a microscope, noticeable distortions and a

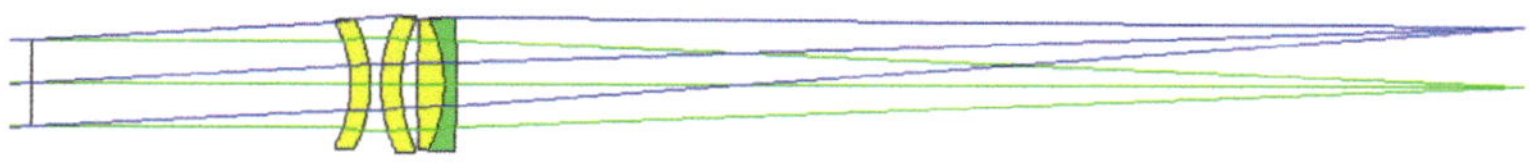

Lens: Ftl 200 NA=0.04 2y'=22mm Sp=-60 Efl 200.05

Image num aper 0.040000 Image height 11.000000 Primary wavln 0.546074

SRF	RADIUS		THICKNESS		APERTURE RADIUS		GLASS		SPECIAL
OBJ	0.000000		1.0000e+09		5.4985e+07		AIR		
AST	0.000000		0.001000		8.002222	AS	BK7	C	
2	0.000000		60.000000		8.000000		AIR		
3	-22.780000		3.700000		11.140000		H_FK61		
4	-24.210000		2.000000		11.900000		AIR		
5	27.980000		3.700000		12.330000		H_FK61	P	
6	26.750000		3.000000		11.870000		AIR		
7	217.000000		5.000000		11.900000		H_FK61	P	
8	-25.800000		2.300000		11.940000		LZ_KF4		
9	-140.000000		200.049424	S	12.060000		AIR		
IMS	0.000000		0.000000		11.000000	S			

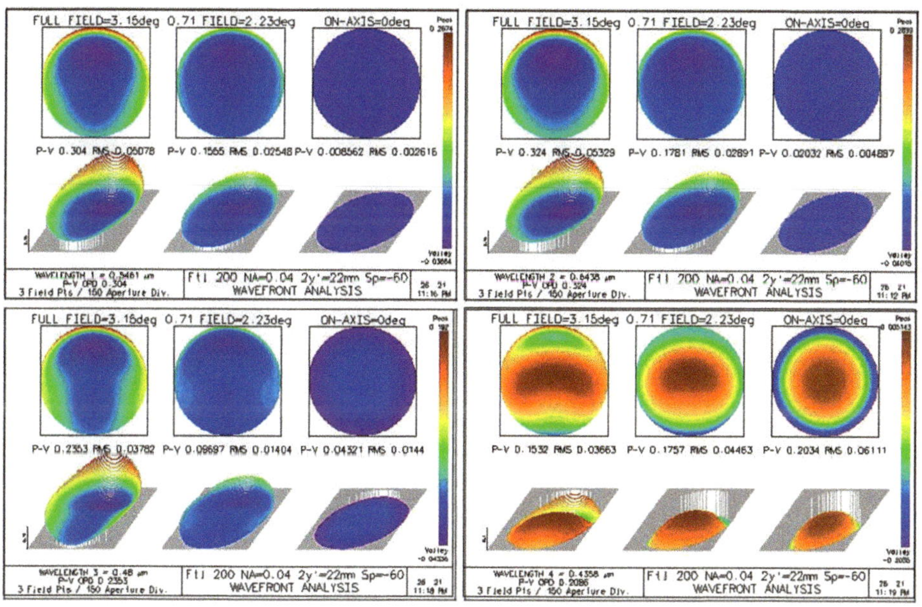

Figure 1.201 Design parameters and graphs of aberration correction of a tube system ($F' = 200$ mm; position of the entrance pupil $S_p = -60$ mm; $NA' = 0.04$; $2y' = 22$ mm).

significant deterioration in image quality were not detected. At the same time, optics of a rather large diameter is a significant advantage, since it allows (if necessary) the use of such objectives in the mode of the so-called *telecentric ray path*.

Quite often we can find a constructive solution when using a conventional digital camera (usually the so-called *body*) with a standard lens with a fixed or zoom focal length to obtain photographs of microscopic objects. A microscope objective is attached to the lens through the adapter ring, allowing for high resolution. In this version, it turns out that the standard lens of a digital camera works as a tube system, and the sensor plane replaces the intermediate image plane (as it would be in a conventional microscope). Such a design really works; however, it cannot fully replace a microscope,

Lens: gelios 104 Efl 174.34

Image num aper 0.172000 Field angle 15.000000 Primary wavln 0.546074

SRF	RADIUS		THICKNESS		APERTURE RADIUS		GLASS		SPECIAL
OBJ	0.000000		1.0000e+20		2.6795e+19		AIR		
1	131.830000		15.840000		44.000000		LZ_TK16		
2	984.000000		0.140000		41.000000		AIR		
3	69.820000		30.000000		36.000000		LZ_TK20		
4	-8.4720e+03		6.000000		30.000000		LZ_LF5		
5	43.550000		20.400000		20.300000		AIR		
AST	0.000000		14.400000		17.900000	A	AIR		
7	-62.290000		7.480000		18.000000		LZ_TF1		
8	78.520000		34.340000		20.100000		LZ_BF13		
9	-82.790000		0.400000		30.000000		AIR		
10	184.080000		35.560000		33.000000		LZ_BF13	P	
11	-288.400000		86.727156	S	37.000000		AIR		
IMS	0.000000		0.000000		46.714657	S			

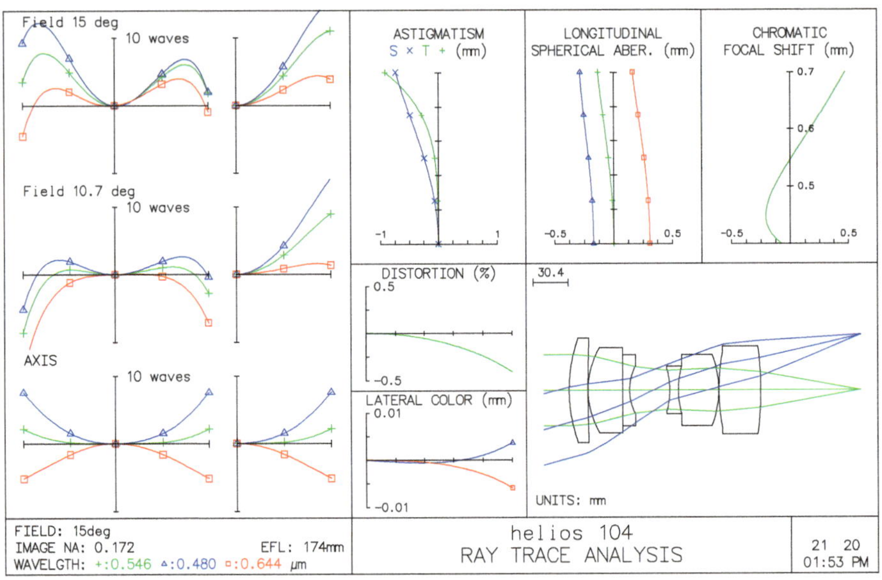

Figure 1.202 Standard design (top) and design for use as a microscope tube system (bottom) of the Helios 104 photographic camera objective.

Lens: gelios 104 as a tube system Efl 174.34

Image num aper 0.030000 Image height 12.500000 Primary wavln 0.546074

SRF	RADIUS	THICKNESS		APERTURE RADIUS		GLASS		SPECIAL
OBJ	0.000000	1.0000e+20		7.1698e+18		AIR		
AST	0.000000	170.000000		5.230000	A	AIR		
2	131.830000	15.840000		17.510000		LZ_TK16		
3	984.000000	0.140000		17.420000		AIR		
4	69.820000	30.000000		17.420000		LZ_TK20		
5	-8.4720e+03	6.000000		14.710000		LZ_LF5		
6	43.550000	34.800000		13.870000		AIR		
7	-62.290000	7.480000		14.860000		LZ_TF1		
8	78.520000	34.340000		16.120000		LZ_BF13		
9	-82.790000	0.400000		19.650000		AIR		
10	184.080000	35.560000		19.790000		LZ_BF13	P	
11	-288.400000	86.727156	S	19.120000		AIR		
IMS	0.000000	0.000000		12.500000	S			

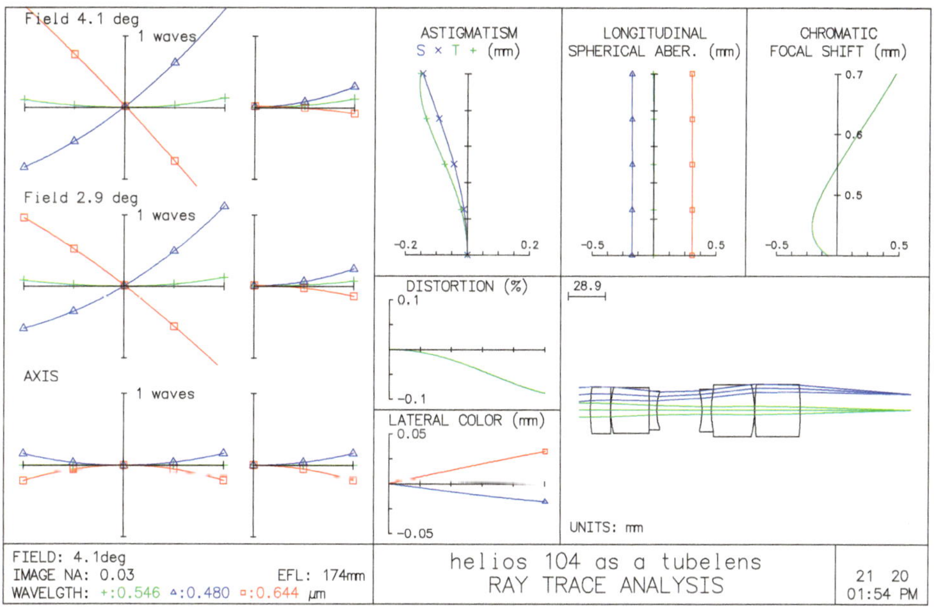

Figure 1.202 (*Continued*) Standard design (top) and design for use as a microscope tube system (bottom) of the Helios 104 photographic camera objective.

especially when using medium- and large-magnification lenses, since the resolution of the sensor of a digital receiver, in this case, is not fully used. Some problems associated with the appearance of vignetting due to inconsistency of the entrance and exit pupils of the elements of the optical system are also possible.

1.8.2 Design of the visual head

Using this reasoning, we can suggest a constructive solution for the head. For this, the head must be divided into two independent components. The first is a lens tube system, which is attached to the outside of the mechanical part of the head. The second is the prism system, which is fixed inside the mechanical part of the head. It should be noted that in most microscopes, lenses of the tube system (which consists of two lenses) are spaced apart: the first lens is located outside and is fixed to the mechanical part of the head, and the second lens is inside the head and is located between the prisms. Thus, for the system of a finite tube, the goal is to place both lenses of the tube system outside the

Lens: industar 37 Efl 302.23

Image num aper 0.055000 Field angle 26.300000 Primary wavln 0.546074

SRF	RADIUS	THICKNESS	APERTURE RADIUS	GLASS	SPECIAL
OBJ	0.000000	1.0000e+20	4.9423e+19	AIR	
1	82.410000	12.000000	33.350000	LZ_TK16	
2	0.000000	12.500000	32.500000	AIR	
3	-193.200000	6.000000	29.750000	LZ_LF5	
4	74.470000	11.700000	28.250000	AIR	
AST	0.000000	8.000000	28.600000 A	AIR	
6	-608.100000	4.200000	29.200000	LZ_OF1	
7	75.860000	15.900000	29.800000	LZ_TK20	
8	-121.060000	266.413468 S	30.300000	AIR	
IMS	0.000000	0.000000	149.372155 S		

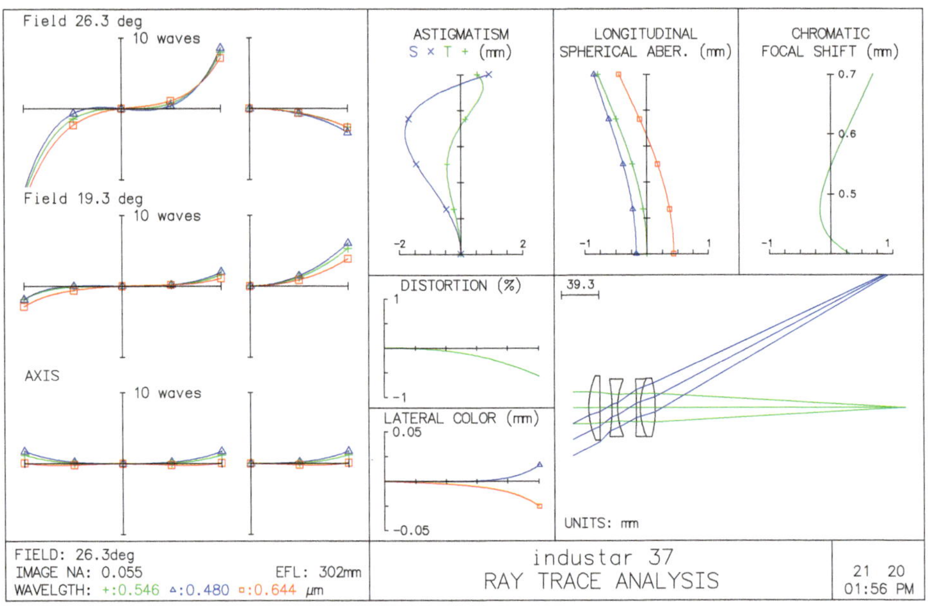

Figure 1.203 Standard design (top) and design for use as a microscope tube system (bottom) of the Industar 37 photographic camera objective.

Lens: industar 37 as a tube system Efl 302.23

Image num aper 0.030000 Image height 12.500000 Primary wavln 0.546074

SRF	RADIUS	THICKNESS	APERTURE RADIUS	GLASS	SPECIAL
OBJ	0.000000	1.0000e+20	4.1359e+18	AIR	
AST	0.000000	200.000000	9.070000 A	AIR	
2	82.410000	12.000000	17.420000	LZ_TK16	
3	0.000000	12.500000	16.850000	AIR	
4	-193.200000	6.000000	15.780000	LZ_LF5	
5	74.470000	19.700000	15.550000	AIR	
6	-608.100000	4.200000	16.970000	LZ_OF1	
7	75.860000	15.900000	17.370000	LZ_TK20	
8	-121.060000	266.413468 S	17.930000	AIR	
IMS	0.000000	0.000000	12.500000 S		

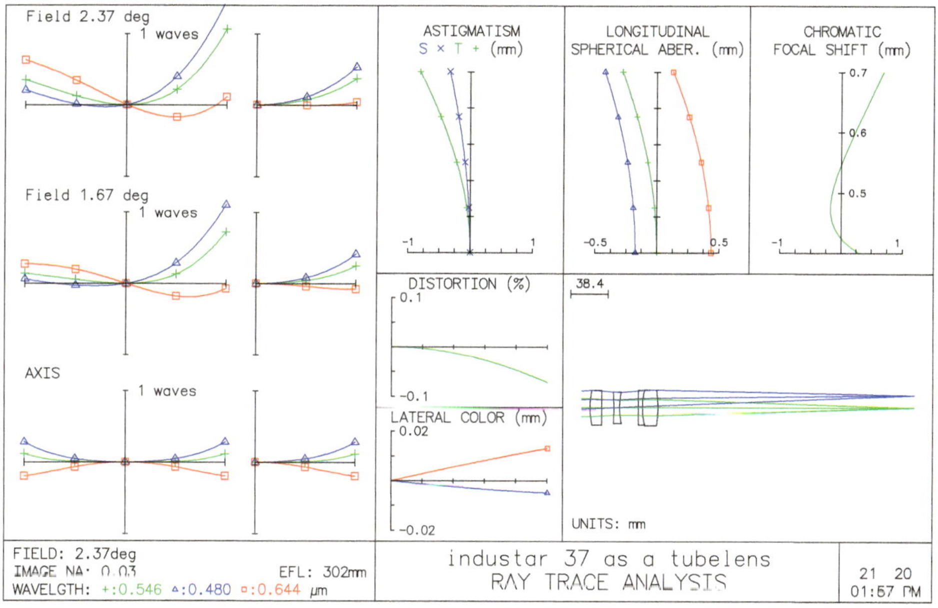

Figure 1.203 (*Continued*) Standard design (top) and design for use as a microscope tube system (bottom) of the Industar 37 photographic camera objective.

mechanical part of the head. For an infinite length of the tube, this is generally observed, and the lenses of the tube system are located outside. Figure 1.208 shows examples of such a constructive design of a visual head.

Let us consider some specific examples of the design of these types of heads. These heads were used in Soviet microscopes MikMed-2 and Lumam RPO-11. One of the heads has a strongly protruding lens composition; this version of the head is designed to work as part of a microscope designed for a finite tube length of 160 mm. The second head is designed to work as part of a microscope for an infinite length of the tube. Figures 1.209 and 1.210 show the design parameters and the level of aberration correction of these heads.

It is only interesting to compare these heads in terms of the possibility of unifying their optical–mechanical structures; image quality in the infinite visual head cannot be called high because there are uncorrected chromatic aberrations of position and magnification (LC). The head included in the MikMed-2 microscope has high image quality; secondary chromatism (LC) is completely corrected. This head is well suited for use with LCF objectives, including plan achromatic objectives. However, it is important to remember that in the factory, the MikMed-2 was equipped with objectives, a set of which is shown in Fig. 1.22. LC is not fully corrected in these objectives; the residual value for each objective is 0.8%. To compensate for this residual LC, so-called *compensation eyepieces* were used in this microscope, in which the

Lens: industar 51 Efl 211.15

Image num aper 0.110000 Field angle 28.000000 Primary wavln 0.546074

SRF	RADIUS	THICKNESS	APERTURE RADIUS	GLASS	SPECIAL
OBJ	0.000000	1.0000e+20	5.3171e+19	AIR	
1	57.680000	8.400000	23.350000	LZ_TK16	
2	0.000000	9.300000	22.750000	AIR	
3	-129.420000	4.000000	20.750000	LZ_LF5	
4	52.970000	8.100000	19.750000	AIR	
AST	0.000000	4.100000	20.000000 A	AIR	
6	-407.400000	3.000000	20.250000	LZ_OF1	
7	54.080000	11.200000	20.750000	LZ_TK20	
8	-83.560000	184.299401 S	20.800000	AIR	
IMS	0.000000	0.000000	112.268002 S		

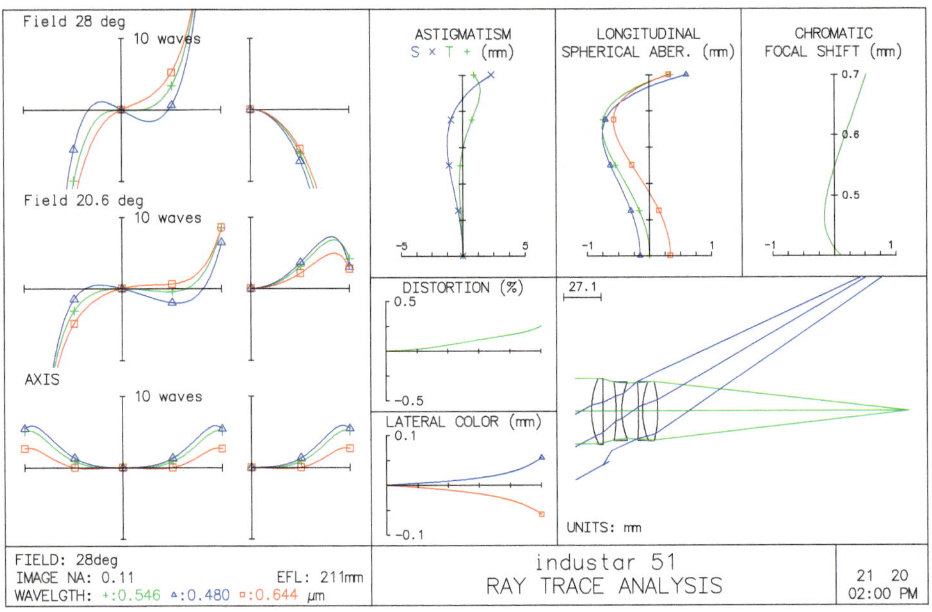

Figure 1.204 Standard design (top) and design for use as a microscope tube system (bottom) of the Industar 51 photographic camera objective.

Lens: industar 51 asa tube system — Efl 211.15

Image num aper 0.030000 Image height 12.500000 Primary wavln 0.546074

SRF	RADIUS	THICKNESS	APERTURE RADIUS	GLASS	SPECIAL
OBJ	0.000000	1.0000e+20	5.9201e+18	AIR	
AST	0.000000	200.000000	6.334362 AS	AIR	
2	57.680000	8.400000	18.390000	LZ_TK16	
3	0.000000	9.300000	17.910000	AIR	
4	-129.420000	4.000000	16.710000	LZ_LF5	
5	52.970000	12.200000	16.380000	AIR	
6	-407.400000	3.000000	17.480000	LZ_OF1	
7	54.080000	11.200000	18.080000	LZ_TK20	
8	-83.560000	184.299401 S	18.480000	AIR	
IMS	0.000000	0.000000	12.500000 S		

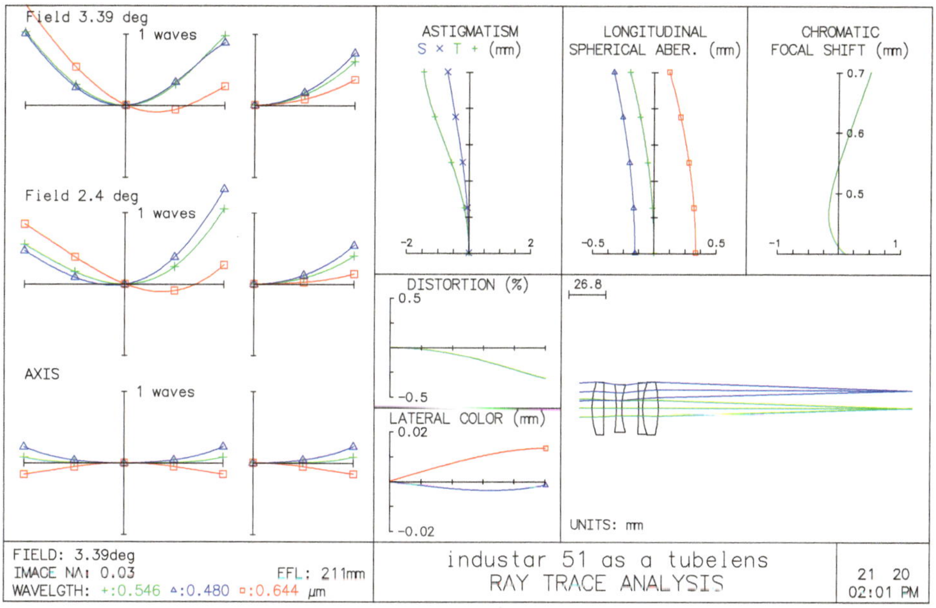

Figure 1.204 (*Continued*) Standard design (top) and design for use as a microscope tube system (bottom) of the Industar 51 photographic camera objective.

compensation value of LC is –0.8%. If we turn to the reasoning, which is illustrated in Table 2.3 (Volume 1, Chapter 2), we can see that in the MikMed-2 microscope, the LC correction (in accordance with option 1) is not optimal. Option 2 is considered the more acceptable option (given that these objectives are used in this microscope) when the LC is compensated in the optical system of the head. An example of the optical design of such a head is shown in Fig. 1.211, which presents the design parameters and graphs of aberration correction of such a head. The LC compensation value of about –1% was achieved in the head itself (there is no need for compensation eyepieces in this case).

In this version of the head, the prism blocks are unified with the head from the Lumam RPO-11 microscope.

Figures 1.212 and 1.213 show variants of joint work of the compensation head and two objectives, the OX-25-1 and the OX-100-1 (design parameters and graphs of aberration correction of these are presented in Figs. 1.16 and 1.21, respectively). This arrangement makes it possible to almost completely compensate for secondary chromatism (LC), as shown by the graphs of aberrations.

A variant of the other (most common) arrangement, when the aberration of secondary chromatism is compensated by the eyepiece itself, is shown in

Lens: tair 56 Efl 150.0

Image num aper 0.178000 Field angle 5.000000 Primary wavln 0.546074

SRF	RADIUS	THICKNESS	APERTURE RADIUS	GLASS	SPECIAL
OBJ	0.000000	1.0000e+20	8.7489e+18	AIR	
1	186.210000	16.400000	26.800000	LZ_TK9	
2	-84.140000	4.400000	26.250000	LZ_TF3	
3	2.3550e+03	0.100000	25.950000	AIR	
4	78.340000	10.100000	25.750000	LZ_TK9 P	
5	229.600000	5.800000	24.600000	AIR	
AST	0.000000	17.400000	23.650000 A	AIR	
7	26.000000	10.600000	18.850000	LZ_K8	
8	20.180000	88.070682 S	14.860000	AIR	
IMS	0.000000	0.000000	13.123440 S		

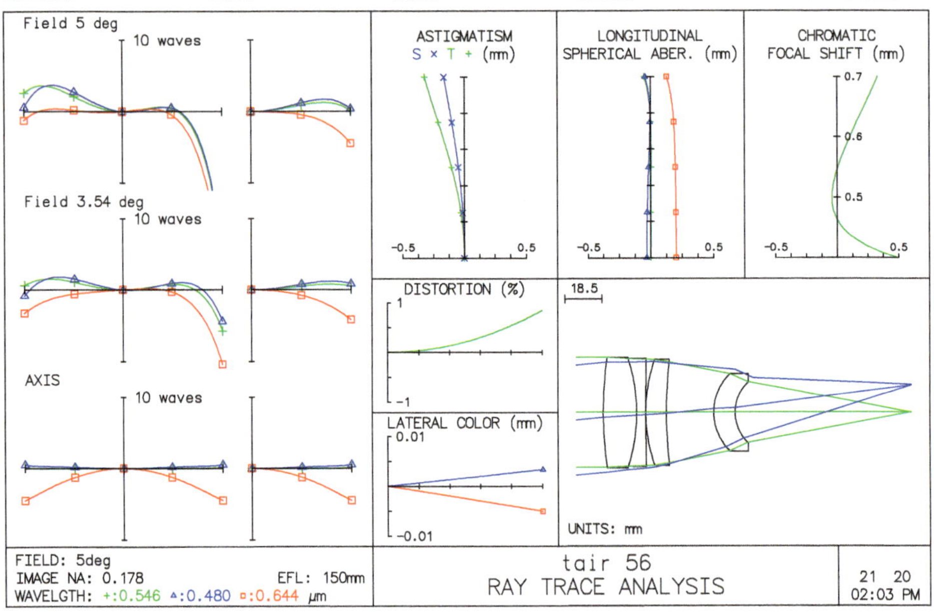

Figure 1.205 Standard design (top) and design for use as a microscope tube system (bottom) of the Tair 56 photographic camera objective.

Lens: tair 56 as tube system — Efl 150.0

Image num aper 0.030000 — Image height 12.500000 — Primary wavln 0.546074

SRF	RADIUS		THICKNESS		APERTURE RADIUS		GLASS		SPECIAL
OBJ	0.000000		1.0000e+20		8.3332e+18		AIR		
AST	0.000000		150.000000		4.500048	AS	AIR		
2	186.210000		16.400000		16.950000		LZ_TK9		
3	-84.140000		4.400000		17.180000		LZ_TF3		
4	2.3550e+03		0.100000		17.340000		AIR		
5	78.340000		10.100000		17.450000		LZ_TK9	P	
6	229.600000		23.200000		16.960000		AIR		
7	26.000000		10.600000		15.760000		LZ_K8		
8	20.180000		88.070682	S	13.030000		AIR		
IMS	0.000000		0.000000		12.500000	S			

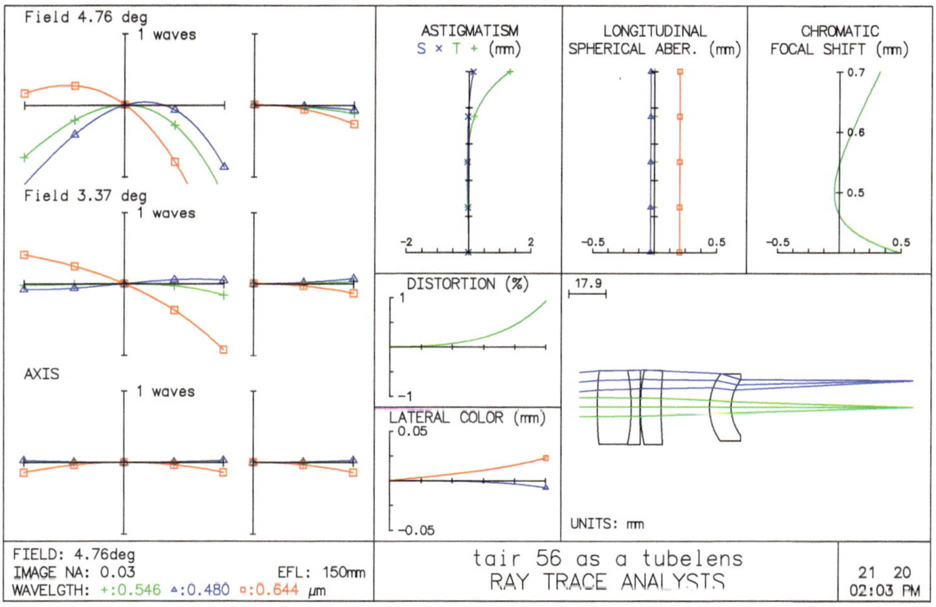

Figure 1.205 (*Continued*) Standard design (top) and design for use as a microscope tube system (bottom) of the Tair 56 photographic camera objective.

Fig. 1.33 (Volume 1, Chapter 1), where the combination of the OM-21 objective and the 7×/18 eyepiece is used as an example. There, the aberration plots also show full LC compensation. Unfortunately, the graphs of aberration correction of both variants of the possible layout of the projection optical system of a light microscope do not reveal the advantages of one of the layouts; both graphs show almost complete compensation for secondary chromatism (LC). In practice, the advantage of the compensation option *through the tube system* is obvious. Thus, what can be confirmed again and again is that the theory and practice of optical and mechanical design must be supplemented by the practice of real work and research using a microscope.

It is possible to unify the visual head of a budget microscope, making it possible to use (change) different tube systems. These can be tube lens systems for operating a microscope with objectives for a finite length of the tube (magnification of the head, $M = 1\times$) or for an infinite length of the tube. At the same time, the microscope can use the set of tube systems for infinity with different focal lengths (for example, $F'_{ts} = 160$, 180, and 200 mm). Figures 1.214–1.217 show examples of the optical design of unified heads.

Lens: upiter 6 Efl 180.36

Image num aper 0.179000 Field angle 7.000000 Primary wavln 0.546074

SRF	RADIUS		THICKNESS		APERTURE RADIUS		GLASS		SPECIAL
OBJ	0.000000		1.0000e+20		1.2278e+19		AIR		
1	88.810000		8.600000		34.000000		LZ_TK20		
2	441.860000		2.280000		33.300000		AIR		
3	55.910000		10.000000		30.300000		LZ_TK20	P	
4	77.660000		10.800000		27.700000		LZ_K1		
5	-959.700000		14.500000		25.600000		LZ_TF2		
6	37.020000		37.700000		19.400000		AIR		
AST	0.000000		2.100000		16.000000	A	AIR		
8	186.800000		3.800000		15.400000		LZ_TF7		
9	-2.3678e+03		83.262707	S	15.500000		AIR		
IMS	0.000000		0.000000		22.145944	S			

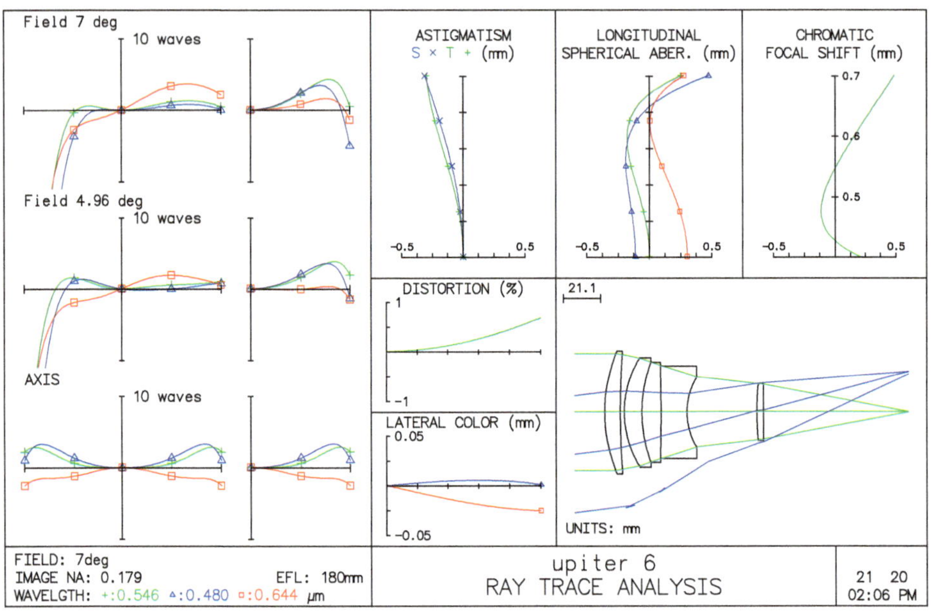

Figure 1.206 Standard design (top) and design for use as a microscope tube system (bottom) of the Jupiter 6 photographic camera objective.

Lens: upiter 6 as a tube system　　Efl 180.36

Image num aper 0.030000　Image height 12.500000　Primary wavln 0.546074

SRF	RADIUS	THICKNESS	APERTURE RADIUS	GLASS	SPECIAL
OBJ	0.000000	1.0000e+20	6.9304e+18	AIR	
AST	0.000000	180.000000	5.410927 AS	AIR	
2	88.810000	8.600000	17.990000	LZ_TK20	
3	441.860000	2.280000	17.730000	AIR	
4	55.910000	10.000000	17.570000	LZ_TK20 P	
5	77.660000	10.800000	16.270000	LZ_K1	
6	-959.700000	14.500000	15.010000	LZ_TK2	
7	37.020000	39.800000	12.910000	AIR	
8	186.800000	3.800000	14.380000	LZ_TF7	
9	-2.3678e+03	83.262707 S	14.250000	AIR	
IMS	0.000000	0.000000	12.500000 S		

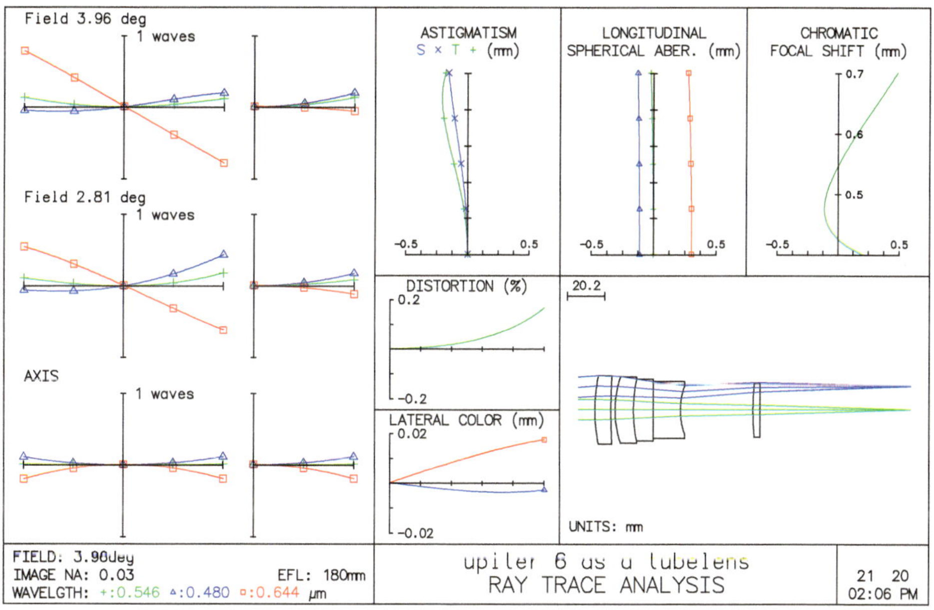

Figure 1.206 (*Continued*) Standard design (top) and design for use as a microscope tube system (bottom) of the Jupiter 6 photographic camera objective.

Let us look at the option of working with a set of third-generation finite LCF plan achromatic objectives (shown in Fig. 1.60) in conjunction with a unified head (Fig. 1.214). It should be noted that the aberration correction remains practically unchanged if you make the optical design separately first for the objectives, then for the objectives and the tube system, and then for the objectives, tube system, and head prisms. However, we need to pay attention to the location of the exit pupils in order to select the eyepiece for visual observation and the adapter for projecting the image onto a digital receiver in

Figure 1.207 Standard zoom photographic objectives.

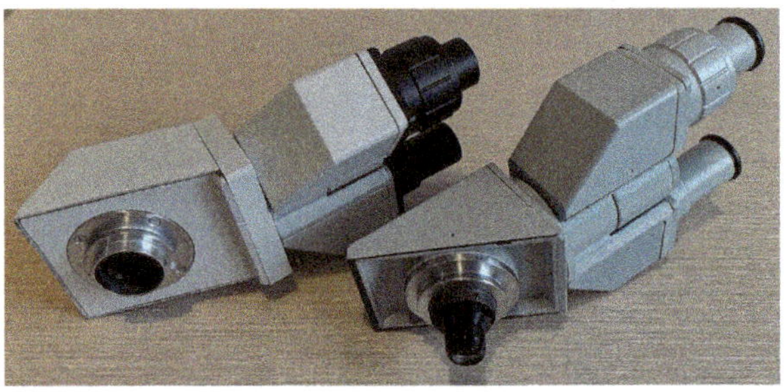

Figure 1.208 Examples of visual head designs for budget microscopes.

the future. As an example, Table 1.12 shows the calculated values for the location of the exit pupils after the objectives, after the tube system, and after the head, respectively.

Let us now consider the option of working with a set of third-generation infinity LCF plan achromatic objectives in conjunction with a unified head (Fig. 1.215). Table 1.13 shows the calculated values of the location of the exit pupils after the objective, after the tube system, and after the head, respectively.

Finally, let us consider the option of working with a set of third-generation infinity LCF plan achromatic objectives in conjunction with a unified head (Fig. 1.217). Table 1.14 shows the calculated values of the location of the exit pupils after the objective, after the tube system, and after the head, respectively.

When analyzing Tables 1.12–1.14, we can conclude that if the calculations are carried out for a standard microscope (the distance between the reference plane of the objective and the reference plane of the head is 41.5 mm), then the location of the exit pupils is different. For the variant of the finite optical tube length of 160 mm, the pupil location values are in the range of about 460–650 mm from the last surface of the head prism; we can arbitrarily call

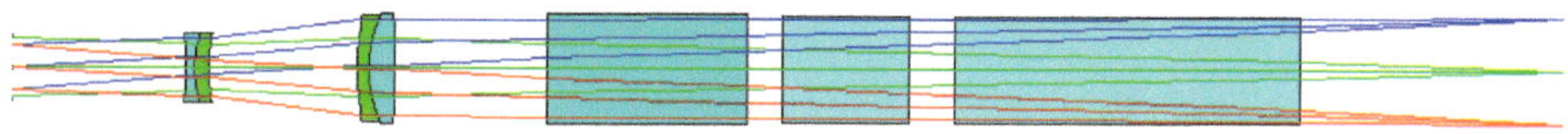

Lens:head MicMed-2 Lateral color free

Object num aper	0.030000	Image height	9.000000	Primary wavln	0.546070

SRF	RADIUS	THICKNESS	APERTURE RADIUS	GLASS	SPECIAL
OBJ	0.000000	-160.500000	8.894082	AIR	
AST	0.000000	30.000000	4.817168 AS	AIR	
2	-97.720000	1.500000	5.630000	LZ_BK8	
3	17.660000	2.500000	5.720000	LZ_TF3	
4	32.060000	25.500000	5.720000	AIR	
5	47.100000	2.000000	9.110000	LZ_TF2	
6	21.580000	4.500000	9.100000	LZ_TK12	
7	-116.410000	26.330000	9.160000	AIR	
8	0.000000	34.600000	9.130000	LZ_BK10	
9	0.000000	6.000000	9.110000	AIR	
10	0.000000	22.000000	9.110000	LZ_K8	
11	0.000000	8.000000	9.090000	AIR	
12	0.000000	59.500000	9.090000	LZ_K8 P	
13	0.000000	46.131708 S	9.060000	AIR	
IMS	0.000000	0.000000	9.000000 S		

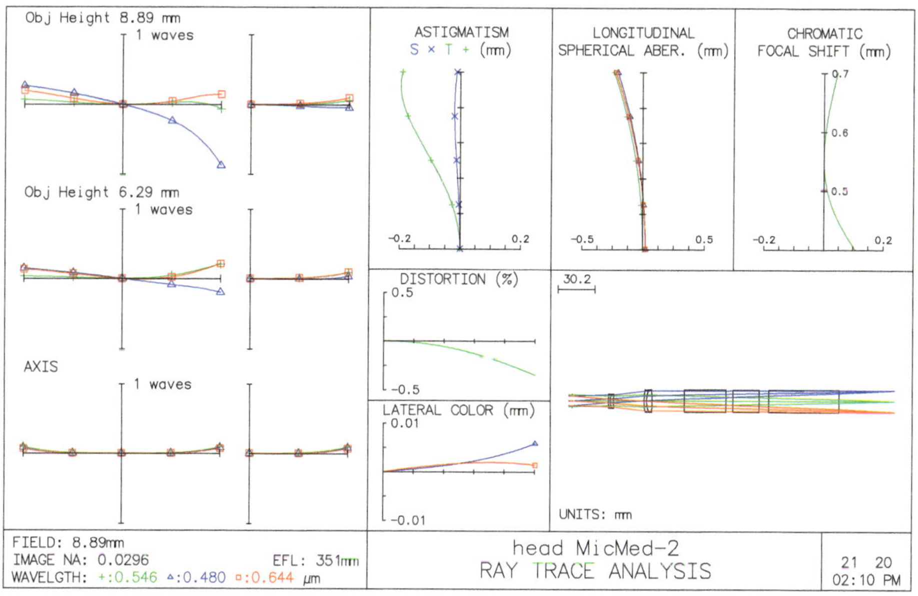

Figure 1.209 Optical design and graphs of aberration correction of a finite visual head for an LCF MikMed-2 microscope (magnification, 1×).

this the *distant position of the exit pupil.* For a microscope variant with an infinite optical tube length, the pupil location values are in the range of about 100–250 mm from the last surface of the head prism; we can arbitrarily call this the *near position of the exit pupil.* These values should be taken into

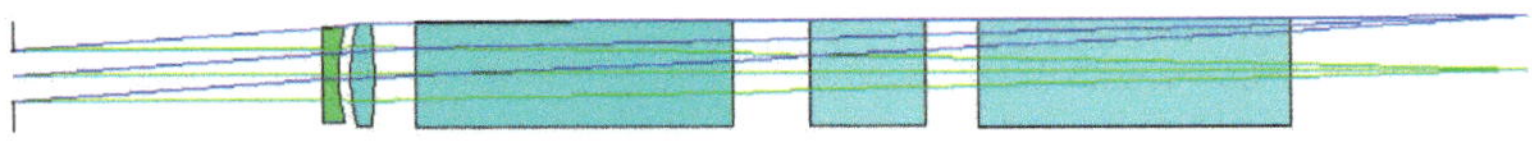

Lens: head Lumam F160 NA=0.03 2y'=20

Image num aper 0.030000 Image height 10.000000 Primary wavln 0.546070

SRF	RADIUS	THICKNESS	APERTURE RADIUS	GLASS	SPECIAL
OBJ	0.000000	1.0000e+20	6.2411e+18	AIR	
AST	0.000000	0.001000	10.000000 A	BK7 C	
2	0.000000	60.000000	10.000000	AIR	
3	-485.300000	3.000000	8.550000	LZ_LF5	
4	31.190000	2.300000	8.750000	AIR	
5	34.750000	4.500000	9.370000	LZ_TK2	
6	-69.820000	7.800000	9.500000	AIR	
7	0.000000	60.400000	9.530000	LZ_BK10	
8	0.000000	14.300000	9.640000	AIR	
9	0.000000	22.000000	9.680000	LZ_K8	
10	0.000000	10.000000	9.720000	AIR	
11	0.000000	59.500000	9.750000	LZ_K8 P	
12	0.000000	44.779220 S	9.870000	AIR	
IMS	0.000000	0.000000	10.000000 S		

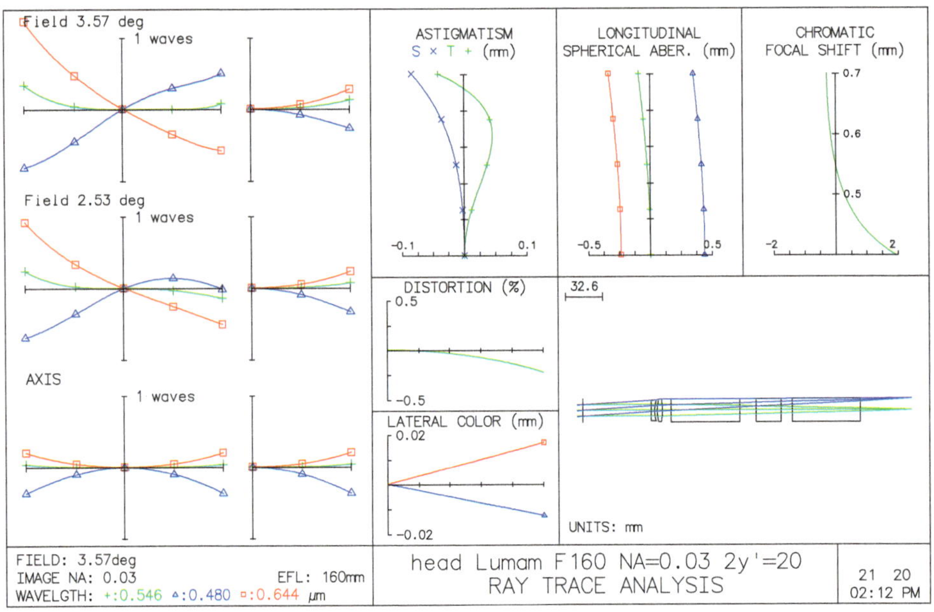

Figure 1.210 Optical design and graphs of aberration correction of an infinite visual head of a Lumam RPO-11 microscope (focal length, 160 mm).

account when constructing optical systems of a light microscope in its specific design, i.e., when calculating image visualization systems.

These heads use the same prism systems as the heads from MikMed-2 microscopes (noncompensation option). However, for the finite length of the tube, the strong protrusion of the lens composition was eliminated; the

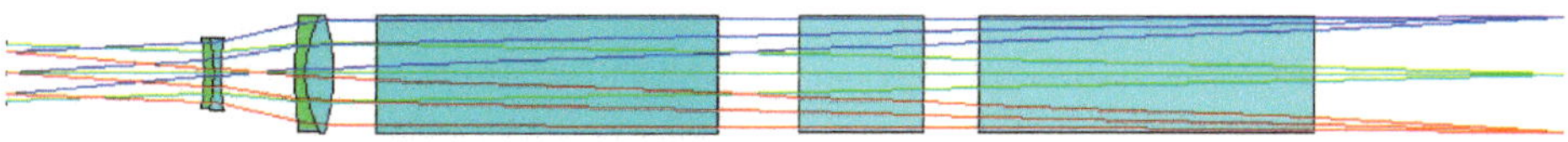

Lens:head MicMed-2 LC =-1% compensate

Object num aper 0.030000 Image height 9.000000 Primary wavln 0.587560

SRF	RADIUS	THICKNESS	APERTURE RADIUS	GLASS	SPECIAL
OBJ	0.000000	-154.300000	8.894525	AIR	
AST	0.000000	35.000000	4.631084 AS	AIR	
2	-49.790000	1.800000	5.630000	LZ_F1	
3	-20.600000	1.200000	5.710000	LZ_TK16	
4	29.910000	13.320000	5.870000	AIR	
5	79.280000	1.800000	8.950000	LZ_LF5	
6	17.330000	5.000000	9.360000	LZ_TK2	
7	-34.300000	7.800000	9.450000	AIR	
8	0.000000	60.400000	9.430000	LZ_BK10	
9	0.000000	14.300000	9.340000	AIR	
10	0.000000	22.000000	9.310000	LZ_K8	
11	0.000000	10.000000	9.280000	AIR	
12	0.000000	59.500000	9.250000	LZ_K8 P	
13	0.000000	44.560381 S	9.160000	AIR	
IMS	0.000000	0.000000	9.000000 S		

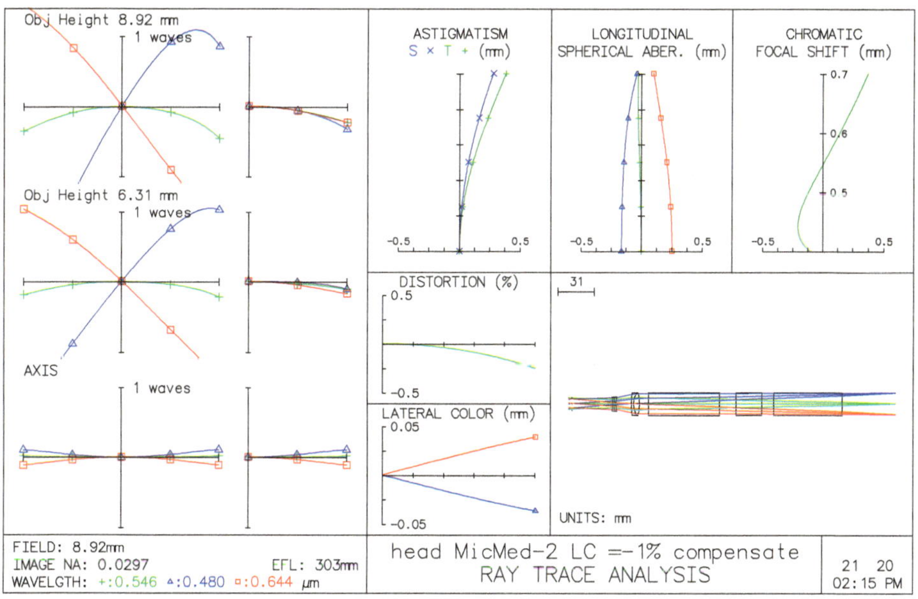

Figure 1.211 Optical design and graphs of aberration correction of a finite visual head (magnification, 1×; LC compensating about 1%; position of the exit pupil $S'_p = -280$ mm).

location of the first lens of the head may be at the same level as the mechanical supporting plane of the nozzle. For the infinite length of the tube, the protrusion of the lens composition is also absent. The mechanical design of a unified head can compensate for the replacement of lens components made

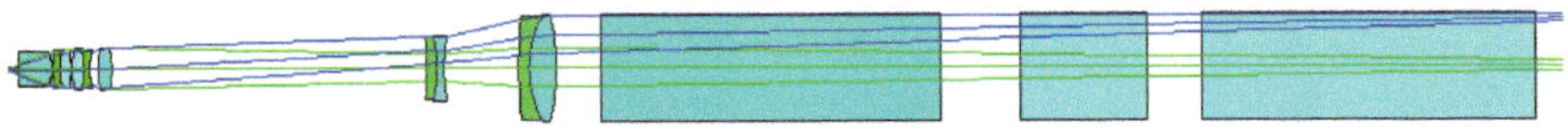

Lens: OX25-1 tub160+ head compens

Object num aper 0.500000 Image height 9.000000 Primary wavln 0.546070

SRF	RADIUS	THICKNESS	APERTURE RADIUS	GLASS	SPECIAL
OBJ	0.000000	0.010000	0.356746	AIR	
AST	0.000000	0.170000	0.362519 AS	LZ_K14	
2	0.000000	1.495000	0.420000	AIR	
3	0.000000	5.950000	1.280000	LZ_TK21	
4	-6.668000	0.100000	2.950000	AIR	
5	22.490000	1.300000	3.160000	LZ_TF4	
6	13.996000	2.000000	3.250000	LZ_LK3	
7	-14.689000	0.100000	3.380000	AIR	
8	11.614000	2.000000	3.420000	LZ_TK14	
9	-48.420000	1.300000	3.300000	LZ_TF4 P	
10	8.551000	1.800000	3.150000	AIR	
11	35.970000	2.500000	3.370000	LZ_TK21 P	
12	-13.868000	56.690000	3.590000	AIR	
13	-49.790000	1.800000	5.380000	LZ_F1	
14	-20.600000	1.200000	5.460000	LZ_TK16	
15	29.910000	13.320000	5.610000	AIR	
16	79.280000	1.800000	8.630000	LZ_LF5	
17	17.330000	5.000000	9.010000	LZ_TK2	
18	-34.300000	7.800000	9.140000	AIR	
19	0.000000	60.400000	9.140000	LZ_BK10	
20	0.000000	14.300000	9.190000	AIR	
21	0.000000	22.000000	9.200000	LZ_K8	
22	0.000000	10.000000	9.220000	AIR	
23	0.000000	59.500000	9.230000	LZ_K8 P	
24	0.000000	44.576755 S	9.270000	AIR	
IMS	0.000000	0.000000	9.000000 S		

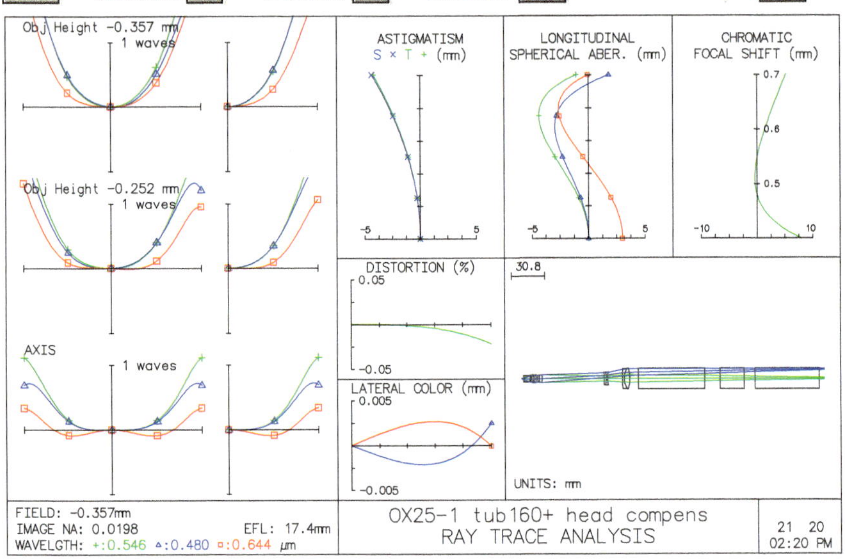

Figure 1.212 Optical design and graphs of aberration correction of an OX-25-1 finite LCR semi–plan achromatic objective and an LC compensation visual head (position of the exit pupil $S'_p = -400$ mm).

either in the form of separate optical–mechanical units or as part of a common unit using a revolving device. Figure 1.218 shows an example of the mechanical design of a unified head; a similar design can be made for the trinocular head for using a digital image pickup.

Safety requirements for binocular visual devices to ensure stereoscopic image perception

We assume that when someone is observing through a binocular microscope, they can sense the depth of the space perceived by their vision and, based on their visual sensation, can estimate the (relative) distance to the object under

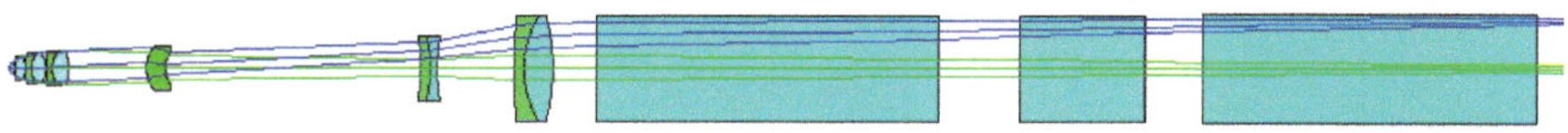

Lens: ox100-1+ head compens Efl 4.39

Object num aper 1.250000 Image height 9.000000 Primary wavln 0.546070

SRF	RADIUS		THICKNESS		APERTURE RADIUS		GLASS		SPECIAL
OBJ	0.000000		0.010000		0.089147		OIL		
AST	0.000000		0.170000		0.100000	A	LZ_K14		
2	0.000000		0.100000		0.350000		OIL	P	
3	0.000000		1.000000		0.500000		LZ_K8		
4	-0.831800		0.050000		0.830000		AIR		
5	0.000000		1.700000		1.470000		LZ_LK3		
6	-2.938000		0.050000		1.960000		AIR		
7	59.430000		1.000000		2.260000		LZ_BF16		
8	4.875000		2.600000		2.580000		LZ_LK3		
9	-3.945000		0.050000		2.500000		AIR		
10	60.260000		1.000000		2.840000		LZ_TF3		
11	4.207000		2.900000		2.830000		LZ_LK3	P	
12	-8.954000		13.820000		3.020000		AIR		
13	5.861000		3.340000		3.750000		LZ_TF5		
14	4.446000		45.250000		2.830000		AIR		
15	-49.790000		1.800000		5.380000		LZ_F1		
16	-20.600000		1.200000		5.460000		LZ_TK16		
17	29.910000		13.320000		5.610000		AIR		
18	79.280000		1.800000		8.630000		LZ_LF5		
19	17.330000		5.000000		9.010000		LZ_TK2		
20	-34.300000		7.800000		9.140000		AIR		
21	0.000000		60.400000		9.140000		LZ_BK10		
22	0.000000		14.300000		9.190000		AIR		
23	0.000000		22.000000		9.200000		LZ_K8		
24	0.000000		10.000000		9.220000		AIR		
25	0.000000		59.500000		9.230000		LZ_K8	P	
26	0.000000		44.565241	S	9.270000		AIR		
IMS	0.000000		0.000000		9.000000				

Figure 1.213 Optical design and graphs of aberration correction of an OX-100-1 finite LCR semi–plan achromatic objective and an LC compensation visual head (position of the exit pupil $S'_p = -400$ mm).

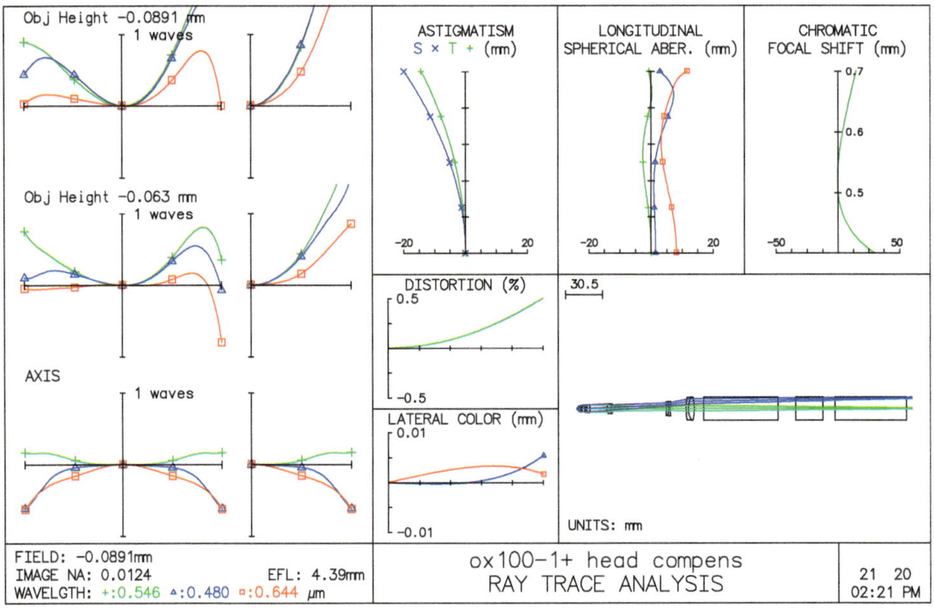

Figure 1.213 (*Continued*) Optical design and graphs of aberration correction of an OX-100-1 finite LCR semi–plan achromatic objective and an LC compensation visual head (position of the exit pupil $S'_p = -400$ mm).

study; that is, use the so-called *spatial vision* or *stereoscopic effect*. These phenomena inevitably occur when two eyes simultaneously observe the same fragment of the object under study. With this observation, the necessary types of work for each of the observer's eyes are accommodation and convergence. The eyes are accommodated by changing the optical power of the lens to obtain a sharp image of the object under study; the volume of accommodation for each of the eyes may be different. Each of the eyes also converges, turning around the centers of the eyes at a certain angle so that both eyes converge at some single point on the object under study. Figure 1.219 illustrates these considerations.

If the distance between the centers of the eyes is designated as b, then for normal stereoscopic perception, the following condition must be fulfilled: $-a \cdot \phi = b = \text{const}$. We assume that the volume of accommodation for people with normal vision is ~4 diopter, although this ability is highly dependent on the age and condition of the observer; in all people, the ability to accommodate decreases over the years. The limits of accommodation in people with normal vision begin from infinity and are limited to the nearest point T at a distance depending on age from 100 to 250 mm. If the subject in question is at infinity, the axes of the eyes are parallel and accommodation is 0.

It is known that the smallest angular stereoscopic parallax still perceptible by the eye is 10″; the average limit of angular stereoscopic parallax is

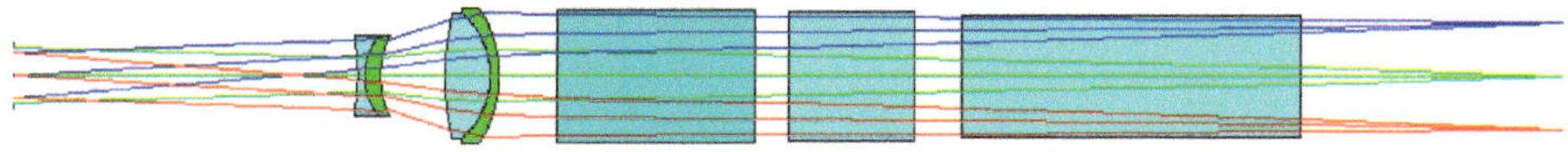

Lens: Finity Unif head M=1.0x 2y'=18

Object num aper 0.030000 Object height 9.000000 Primary wavln 0.546070

SRF	RADIUS		THICKNESS		APERTURE RADIUS		GLASS		SPECIAL
OBJ	0.000000		-160.000000		9.000000		AIR		
AST	0.000000		60.000000		4.802161	AS	AIR		
2	-19.170000		1.500000		6.360000		LZ_LK3		
3	11.345000		2.800000		7.020000		LZ_TF5		
4	17.450000		7.500000		6.990000		AIR		
5	61.110000		10.000000		9.330000		LZ_TK12		
6	-12.370000		2.600000		9.970000		LZ_TF5	P	
7	-20.520000		11.500000		11.090000		AIR		
8	0.000000		34.600000		10.890000		LZ_BK10		
9	0.000000		6.000000		10.580000		AIR		
10	0.000000		22.000000		10.490000		LZ_K8		
11	0.000000		8.000000		10.290000		AIR		
12	0.000000		59.500000		10.180000		LZ_K8	P	
13	0.000000		46.140370	S	9.630000		AIR		
IMS	0.000000		0.000000		9.000000				

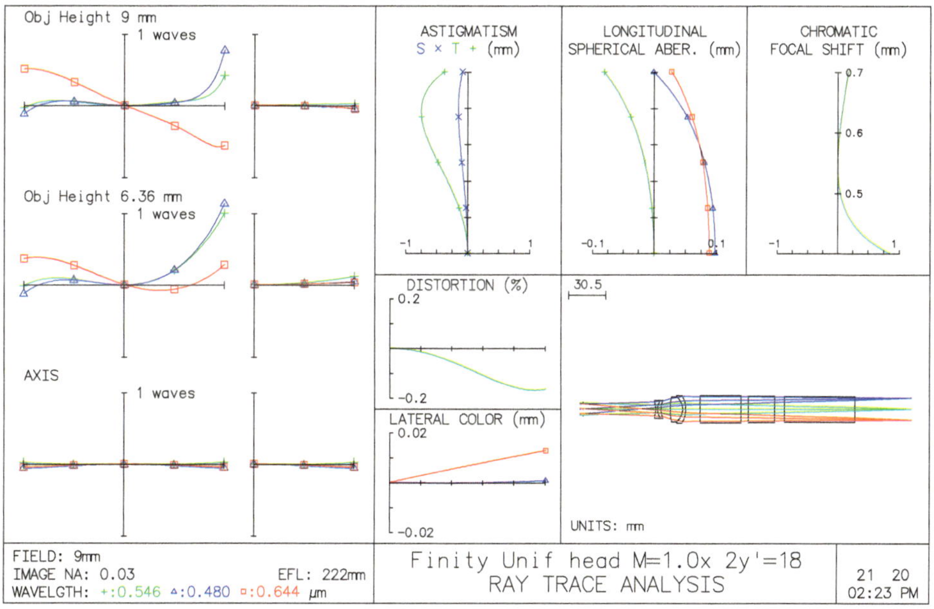

Figure 1.214 Optical design and graphs of aberration correction of a finite (160-mm) unified visual head (magnification, 1×; position of the exit pupil $S'_p = -526$ mm).

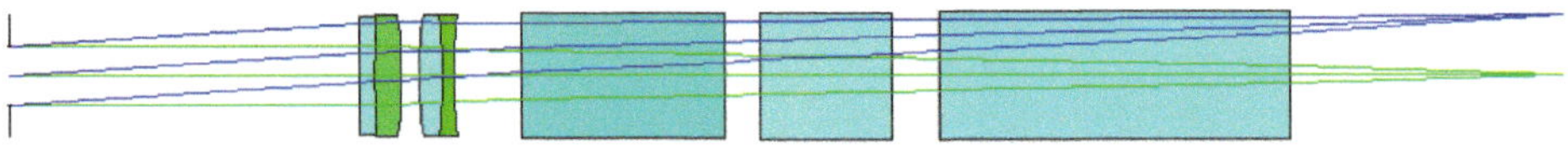

Lens: head Ftl F160 NA=0.03 2y'=20 Efl 160.0

Image num aper 0.030000 Image height 10.000000 Primary wavln 0.546070

SRF	RADIUS	THICKNESS		APERTURE RADIUS		GLASS		SPECIAL
OBJ	0.000000	1.0000e+20		6.2500e+18		AIR		
AST	0.000000	0.001000		10.000000	A	BK7	C	
2	0.000000	60.000000		10.000000		AIR		
3	0.000000	2.600000		8.790000		LZ_LK3		
4	267.300000	4.600000		9.680000		LZ_F2		
5	-87.600000	3.000000		10.120000		AIR		
6	66.650000	4.000000		10.110000		LZ_LK3	P	
7	-102.100000	2.000000		10.050000		LZ_TF2		
8	92.460000	11.500000		10.090000		AIR		
9	0.000000	34.600000		10.160000		LZ_BK10		
10	0.000000	6.000000		10.290000		AIR		
11	0.000000	22.000000		10.400000		LZ_K8		
12	0.000000	8.000000		10.470000		AIR		
13	0.000000	59.500000		10.530000		LZ_K8	P	
14	0.000000	46.138310	S	10.730000		AIR		
IMS	0.000000	0.000000		10.000000	S			

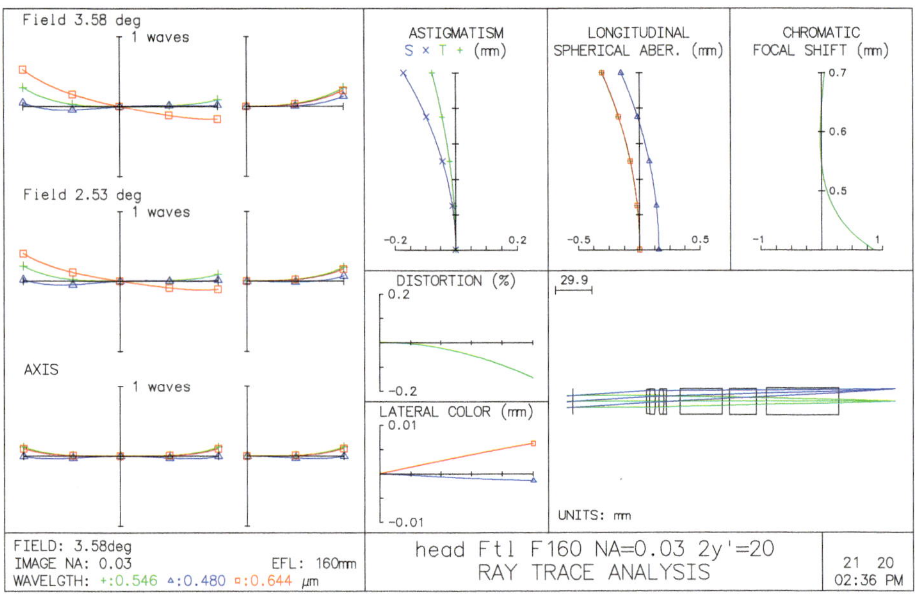

Figure 1.215 Optical design and graphs of aberration correction of an infinite unified visual head ($F' = 160$ mm; position of the exit pupil $S'_p = -205$ mm).

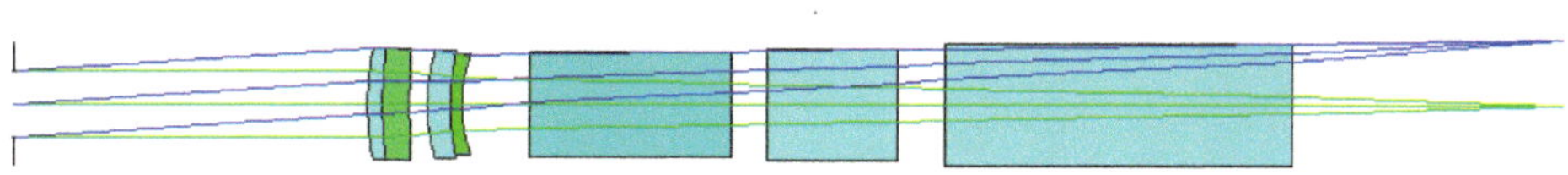

Lens: head Ftl F180 NA=0.03 2y'=22 Efl 179.99

Image num aper 0.030000 Image height 11.000000 Primary wavln 0.546070

SRF	RADIUS		THICKNESS		APERTURE RADIUS		GLASS		SPECIAL
OBJ	0.000000		1.0000e+20		6.1115e+18		AIR		
AST	0.000000		0.001000		10.000000	A	BK7	C	
2	0.000000		60.000000		10.000000		AIR		
3	45.070000		2.600000		9.120000		LZ_LK3		
4	82.900000		4.600000		9.070000		LZ_F2		
5	172.770000		3.000000		8.920000		AIR		
6	28.630000		4.000000		8.830000		LZ_LK3	P	
7	43.290000		2.000000		8.420000		LZ_TF2		
8	24.260000		11.500000		8.090000		AIR		
9	0.000000		34.600000		8.290000		LZ_BK10		
10	0.000000		6.000000		8.730000		AIR		
11	0.000000		22.000000		8.850000		LZ_K8		
12	0.000000		8.000000		9.140000		AIR		
13	0.000000		59.500000		9.300000		LZ_K8	P	
14	0.000000		46.132357	S	10.090000		AIR		
IMS	0.000000		0.000000		11.000000	S			

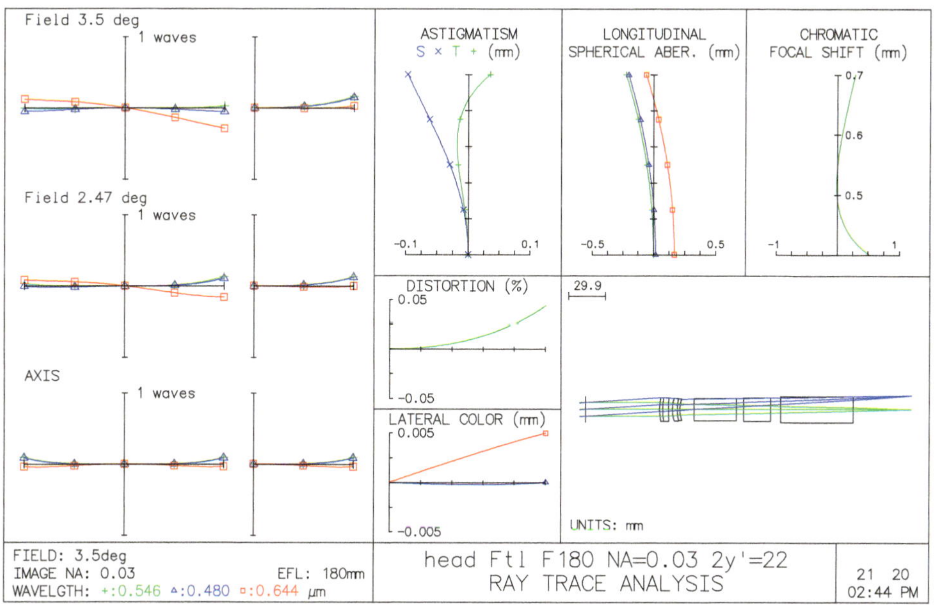

Figure 1.216 Optical design and graphs of aberration correction of an infinite unified visual head ($F' = 180$ mm; position of the exit pupil $S'_p = -175$ mm).

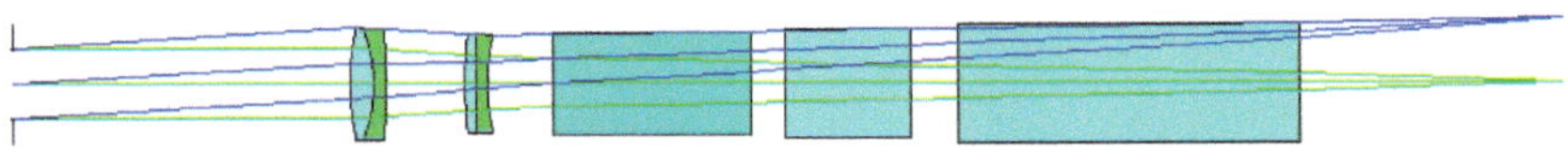

Lens: head Ftl F200 NA=0.03 2y'=22 Efl 200.0

Image num aper 0.030000 Image height 11.000000 Primary wavln 0.546070

SRF	RADIUS	THICKNESS	APERTURE RADIUS	GLASS	SPECIAL
OBJ	0.000000	1.0000e+20	5.5000e+18	AIR	
AST	0.000000	0.001000	10.000000 A	BK7 C	
2	0.000000	60.000000	10.000000	AIR	
3	42.120000	4.200000	9.360000	LZ_LK3	
4	-23.530000	2.000000	9.310000	LZ_KF4	
5	-259.640000	14.000000	9.210000	AIR	
6	77.600000	2.000000	8.310000	LZ_LK3 P	
7	-331.400000	2.000000	8.200000	LZ_KF4 P	
8	31.670000	11.500000	7.970000	AIR	
9	0.000000	34.600000	8.190000	LZ_BK10	
10	0.000000	6.000000	8.650000	AIR	
11	0.000000	22.000000	8.770000	LZ_K8	
12	0.000000	8.000000	9.080000	AIR	
13	0.000000	59.500000	9.240000	LZ_K8 P	
14	0.000000	46.139528 S	10.060000	AIR	
IMS	0.000000	0.000000	11.000000 S		

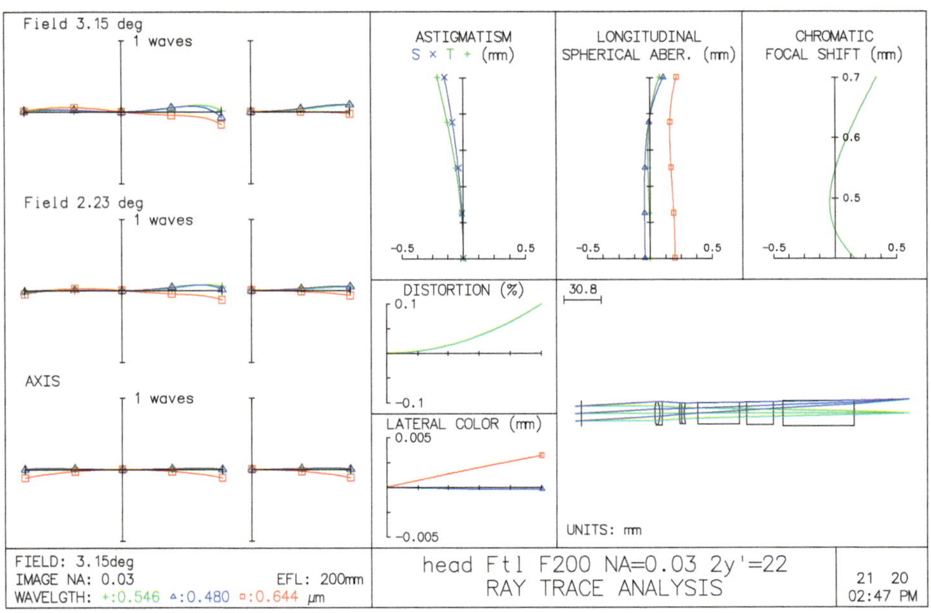

Figure 1.217 Optical design and graphs of aberration correction of an infinite unified visual head ($F' = 200$ mm; position of the exit pupil $S'_p = -172$ mm).

Table 1.12 Estimated exit pupil placement (finite length of the tube).

Objective	Figure	Position of the exit pupil from objective reference plane S'_p (mm)	Position of the exit pupil after tube system (without prisms) S'_p (mm)	Position of the exit pupil after prisms (with the tube system) S'_p (mm)
10×/0.30	1.36	−4.85	−365	−465
20×/0.65	1.41	−17.24	−530	−630
40×/0.70	1.45	−25.94	−710	−810
60×/0.85	1.49	−18.26	−548	−648
100×/1.30 oil	1.57	−18.43	−552	−652

Table 1.13 Estimated exit pupil placement (infinite length of the tube system; $F' = 160$ mm).

Objective	Figure	Position of the exit pupil from objective reference plane S'_p (mm)	Position of the exit pupil after tube system (without prisms) S'_p (mm)	Position of the exit pupil after prisms (with the tube system) S'_p (mm)
5×/0.12	1.33	+22.95	−32	−132
10×/0.25	1.34	−7.7	−81	−181
20×/0.45	1.39	−19.9	−107	−207
40×/0.65	1.43	−26.5	−127	−227
100×/1.25 oil	1.55	−25.0	−123	−223

Table 1.14 Estimated exit pupil placement (infinite length of the tube system; $F' = 200$ mm).

Objective	Figure	Position of the exit pupil from objective reference plane S'_p (mm)	Position of the exit pupil after tube system (without prisms) S'_p (mm)	Position of the exit pupil after prisms (with the tube system) S'_p (mm)
5×/0.10	1.32	+46.7	−12	−112
10×/0.28	1.35	+7.4	−44	−144
20×/0.40	1.37	−9.62	−61	−161
50×/0.60	1.47	−14.03	−66	−166
100×/1.30 oil	1.59	−18.01	−71	−171

30–40′. These values determine the requirements for the maximum allowable amount of nonparallelism of the optical axes of the binocular visual head; nonparallelism over the permissible limit causes the image to double and, accordingly, additional accommodation by the eyes, which makes the observer more tired. The limiting parallactic angles can be determined approximately, based on the diameter of the pupil of the eye (~2 mm) and the limits of the distance of the best vision (100–250 mm):

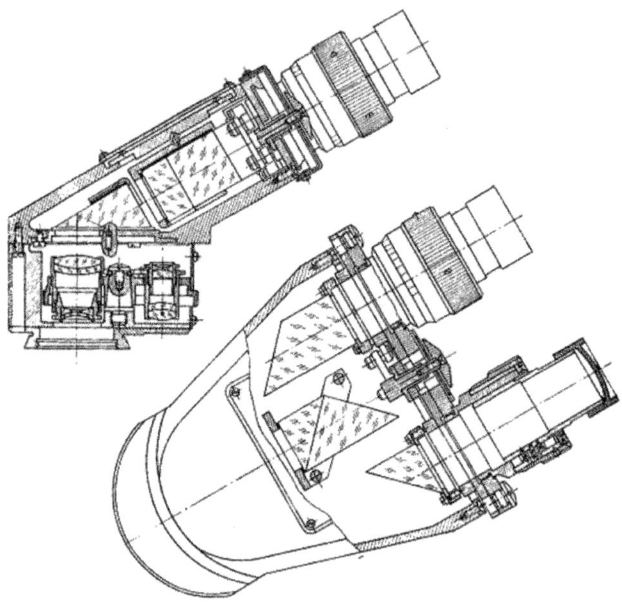

Figure 1.218 Example of the mechanical design of a unified head.

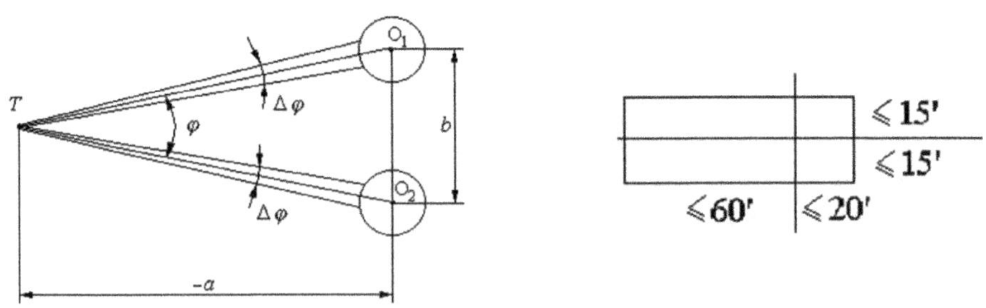

Figure 1.219 Observation through a binocular visual head.

$$\Delta\varphi = \frac{\varnothing_{eye.pupil}}{2a}\,\text{(from drawing)},$$

$$\Delta\varphi = \frac{1}{250} = 0.004\,\text{radian} \approx 13.3'\ \text{for one eye},$$

$$\Delta\varphi = \frac{1}{100} = 0.010\,\text{radian} \approx 33.3'\ \text{for one eye}.$$

$$\text{For two eyes}: \Delta\varphi \approx 2^{7'} \div 6^{7'}.$$

To assess the permissible parallactic stereoscopic angle, we use the concept of *prism diopters* Δ, where $1\Delta = \frac{1\text{cm}}{1\text{m}}$.

Based on the physiology of the eye, 0.5Δ per eye is painlessly overcome by the muscular apparatus of the eye:

$$0.5\Delta = \frac{0.5\,\text{cm}}{1\,\text{m}} = \frac{5\,\text{mm}}{1000\,\text{mm}} = 0.005\,\text{radian} \approx 17.24'.$$

Accordingly, in both eyes: $17.24' \times 2 \approx 35'$.

We will take into account the special working conditions associated with using binocular microscopes (large values of linear magnifications, location and small diameters of exit pupils, resolution of objects, large linear fields of view, etc.). Even with a theoretically ideal microscope optical system, prolonged use causes fatigue and discomfort. You can set the following requirements for the values of permissible deviations from the parallelism of the rays emerging from the eyepieces:

- The convergence of beams of rays in a horizontal plane of not more than 20′,
- Beam divergence in the vertical plane of not more than 15′, and
- Beam divergence in the horizontal plane of not more than 60′.

1.8.3 Design of eyepieces

It is the eye—or, more precisely, its optical characteristics—that should determine the parameters of the optical and mechanical design of the eyepiece. Indeed, the whole theory of light microscopy, which has not yet been questioned, is based on the fact that the microscope is an observational device; i.e., it is intended for observation by eyes. At the same time, in light microscopy theory, eye parameters such as resolution, spectral sensitivity (there is no point in correcting in the eyepiece, for example, an aberration such as the secondary spectrum), the eye's base, the location of the pupil, and some others are taken into consideration. However, in the optical design of eyepieces (and in a broader sense of light microscopes), parameters such as the intrinsic aberrations of the observer's eye, the eye's ability to accommodate and converge, and the degree of tension or relaxation of the eye when observing at different distances are not always considered. As a rule, the initial numerical aperture of the projection system (the rear numerical aperture of the objective or the output aperture of the tube system), as well as the linear field in the so-called intermediate image plane on the microscope, is taken as the initial data for calculating the eyepiece. The diameter of the exit pupil for the eyepiece depends on these parameters; it is believed that the eye is able to adapt to this size (changing aberrations and lighting parameters are not taken into account).

It should be noted that there is some inconsistency between the same parameters for different parts of a light microscope. For example, a projection system is characterized by relatively high numerical apertures and small angular field values. In contrast, the ocular system is characterized by small numerical apertures (the value of the diameter of the exit pupil is small) and a significant angular field. Especially large values of the angular field are achieved in eyepieces intended for astronomical research. However, in microscopy they can be significant, for example, in short-focus optical systems of eyepieces. This inconsistency sometimes leads to some inaccuracies in optical designs. For example, the value of secondary chromatism (LC) for the projection system (objective) within the entire linear field is constant (the angular field is small); the LC value within the linear field of the eyepiece varies by no less than 10%, and this difference in values increases in proportion to the equivalent focal length of the eyepiece. Even for short-focus eyepieces in which the optical design is based on the conditions of independent optical correction (the LC value is close to zero), this difference does not very critically affect the resulting image quality. On the contrary, if compensation eyepieces are used in the microscope system to compensate for the LC of the objective, then full compensation within full field does not occur. Sufficiently complete compensation in such cases can occur only in some zones (either in the center of the field or in peripheral zones), but not uniformly throughout the field. That is, the negative impact on the resulting image quality in such a system is very significant.

Similar considerations can be made regarding compensation for astigmatism and distortion. It is believed that the distortion in the objectives cannot be specifically corrected (unless these are original objectives for measuring purposes); it is also believed that correction of distortion in the eyepieces is not always required, especially when it comes to bundling budget microscopes. However, the magnitude of the distortion in the eyepieces (unless this aberration is specifically corrected), as a rule, depends on the value equivalent to the focal distance of the eyepiece as well as on the layout of its optical design.

It should be noted that in some cases, nonspherical surfaces, such as parabolic surfaces, are used to compensate for distortion and correct astigmatism in the optical systems of the eyepiece (especially short-focus ones). It is believed that the use of one parabolic surface allows one to *simplify* the optical system on one lens. In this case, it is possible to increase the relative aperture of the optical system. However, one can only partially agree with the argument that the optical system, when using nonspherical surfaces, is simplified; the manufacture of such surfaces (if glass or crystals are used as optical materials) is a rather time-consuming process. If plastic is used as optical materials, the manufacturing process of such surfaces, for example, by casting or prototyping, is not difficult. In this case, it is possible to recommend optical calculation of eyepieces in the optical system in which nonspherical surfaces are used. In general, all surfaces of all lenses can be made nonspherical, which corresponds to the concept of innovative technology, the so-called *free-form optics*. The construction of optical systems for eyepieces (as well as lighting systems) using this technology is justified and can bring maximum effect. By the way, in ordinary microscope objectives (if we are talking about mass-produced budget objectives), the advantages of this technology do not seem obvious; the practice of designing an objective using nonspherical surfaces did not reveal significant advantages as compared with the use of traditional spherical lens surfaces. At the same time, using free-form optics technology for mirror or mirror-lens optical elements in microscopy could be revolutionary, since it significantly reduces the central screening zone in such systems and redistributes the energy of light beams.

Figures 1.220–1.237 show optical design options for eyepieces, designed primarily to equip budget light microscopes.

Given that the value of a microscope's full magnification during visual observation is the product of the magnification in the objective and the magnification in the eyepiece, we consider the possible combinations when choosing an objective and an eyepiece. According to the theory of useful microscope magnification, the total magnification should not be less than 500×/*NA* or more than 1000×/*NA* (*NA* is the numerical aperture of the

Lens: eyepiece 5/20 LC=0% Efl 49.99

Image num aper 0.030000 Image height 10.000000 Primary wavln 0.546070

SRF	RADIUS	THICKNESS	APERTURE RADIUS	GLASS	SPECIAL
OBJ	0.000000	1.0000e+09	2.0001e+08	AIR	
AST	0.000000	10.850000	1.499935 AS	AIR	
2	107.890000	1.500000	3.750000	LZ_TF5	
3	35.970000	5.500000	3.920000	LZ_TK14	
4	-29.110000	49.500000	4.510000	AIR	
5	-70.150000	4.000000	9.860000	LZ_LK3	
6	-24.830000	-13.149352 S	10.260000	AIR	
IMS	0.000000	0.000000	10.000000 S		

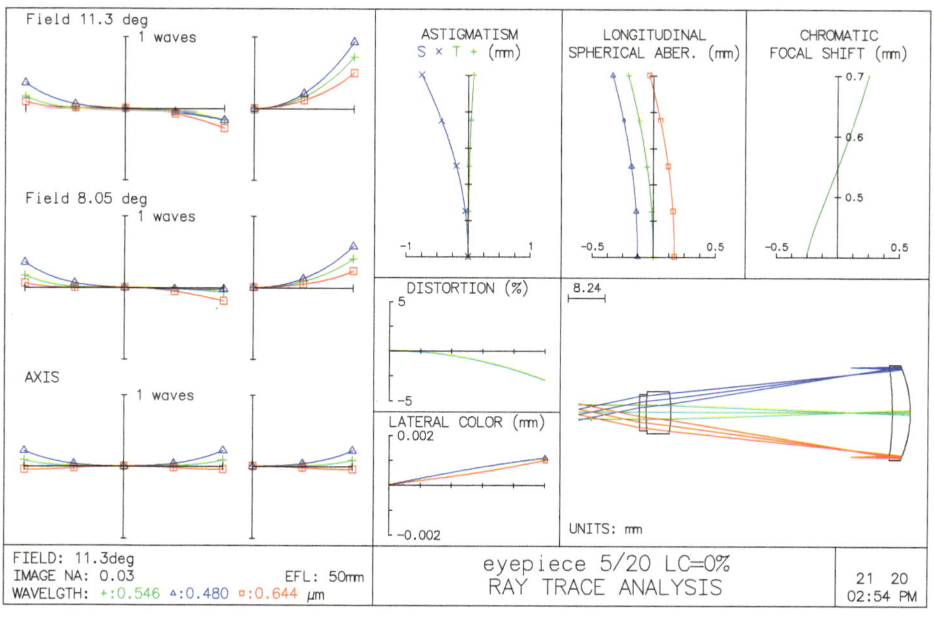

Figure 1.220 Design parameters and graphs of aberration correction of a 5×/20 eyepiece (LC = 0).

Lens: eyepiece 5/25 LC, dis=0% Efl 49.98

Image num aper 0.030000 Image height 12.500000 Primary wavln 0.546070

SRF	RADIUS	THICKNESS	APERTURE RADIUS	GLASS	SPECIAL
OBJ	0.000000	1.0000e+09	2.5010e+08	AIR	
AST	0.000000	22.000000	1.499403 AS	AIR	
2	18.800000	6.000000	7.380000	LZ_LK3	
3	-37.370000	0.600000	7.440000	AIR	
4	31.750000	7.000000	7.270000	LZ_K8	
5	12.820000	6.000000	6.260000	AIR	
6	-8.686000	2.500000	6.390000	LZ_F4	
7	30.470000	8.000000	9.230000	LZ_LK3 P	
8	-16.700000	0.600000	10.650000	AIR	
9	103.900000	7.000000	12.680000	LZ_CTK12	
10	-32.380000	24.972723 S	13.240000	AIR	
IMS	0.000000	0.000000	12.500000 S		

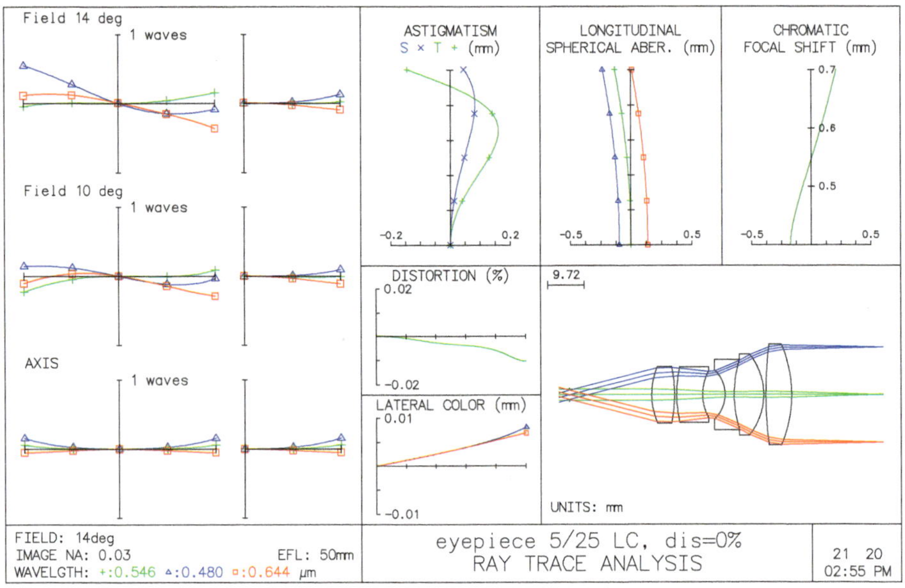

Figure 1.221 Design parameters and graphs of aberration correction of a 5×/25 eyepiece [LC = 0; distortion (DIS) = 0].

objective in the space of an object). Table 1.15 illustrates such combinations for standard objectives (having conventional and enlarged numerical apertures) and standard eyepieces, which are most often used for visual observation with a light microscope.

If we follow the rule of useful microscope magnification, it turns out that standard eyepieces with magnifications of 10× and 12.5× can be used in combination with objectives with conventional numerical apertures; however, a 12.5× eyepiece is preferred. Therefore, the optical design of such an eyepiece with a linear field of 22 mm (in the plane of the field diaphragm) was made; this is not difficult and can be performed according to the classical scheme shown in

Lens: Ocular 8x28 LC=-0.35% Efl 31.22

Image num aper 0.030000 Image height 14.000000 Primary wavln 0.546070

SRF	RADIUS	THICKNESS	APERTURE RADIUS	GLASS	SPECIAL
OBJ	0.000000	1.0000e+09	4.4849e+08	AIR	
AST	0.000000	6.000000	0.780000 A	AIR	
2	60.400000	2.000000	3.740000	LZ_LF7	
3	8.200000	6.000000	4.620000	LZ_TK2	
4	-14.600000	31.000000	5.570000	AIR	
5	-50.800000	7.300000	12.530000	LZ_TK2 P	
6	-21.000000	-22.589701 S	13.600000	AIR	
IMS	0.000000	0.000000	14.000000		

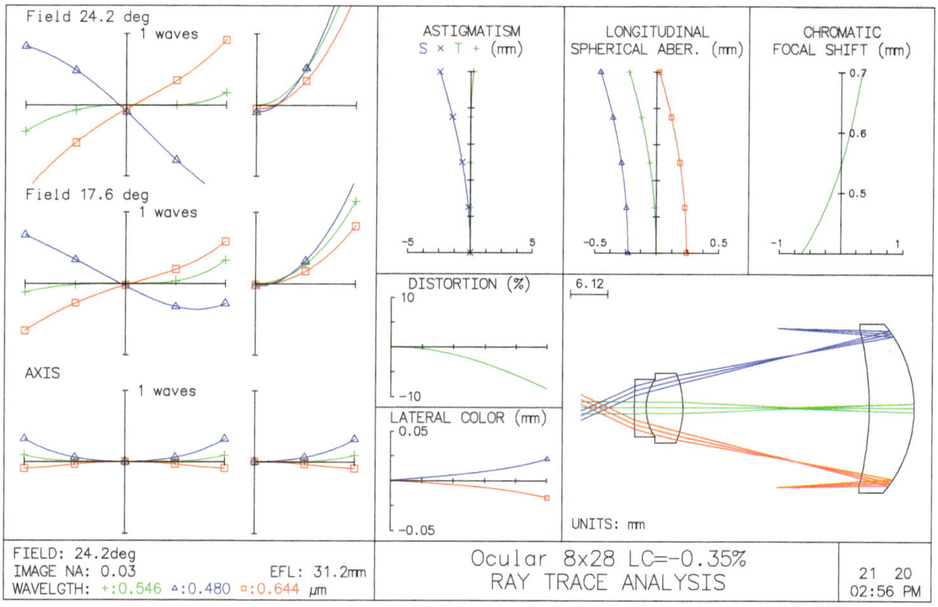

Figure 1.222 Design parameters and graphs of aberration correction of an 8×/28 eyepiece (LC = −0.35%).

Fig. 1.231. When using such an eyepiece, we do not limit the field in comparison, for example, with the 10×/22 eyepiece. A 12.5×/22 eyepiece is recommended to be used as the main and universal eyepiece for assembling budget light microscopes using objectives with standard numerical apertures.

If we consider the options for completing eyepieces with objectives that have increased numerical apertures, we can find a much wider range of possibilities. At the same time, one must take into account the researcher's natural desire to achieve the maximum possible value of full magnification on a microscope, when the theoretical image quality is not yet deteriorating. In this case, no one eyepiece can be recommended as universal for the entire set of objectives. The 10× and even 12.5× eyepieces are not preferred; the most commonly used eyepieces are those with linear magnifications of 15–25×. For

Lens: Eyepiece 10/18 LC=-0.6% Efl 24.9

Image num aper 0.030000 Image height 9.000000 Primary wavln 0.546070

SRF	RADIUS	THICKNESS	APERTURE RADIUS	GLASS	SPECIAL
OBJ	0.000000	1.0000e+09	3.6138e+08	AIR	
AST	0.000000	15.330000	0.747139 AS	AIR	
2	-43.250000	5.500000	7.000000	LZ_TK14	
3	-15.346000	0.500000	8.190000	AIR	
4	192.310000	6.000000	8.640000	LZ_TK14 P	
5	-14.060000	2.000000	8.840000	LZ_TF4	
6	-194.090000	6.500000	9.440000	AIR	
7	26.550000	5.500000	11.130000	LZ_K8	
8	0.000000	14.406163 S	11.000000	AIR	
IMS	0.000000	0.000000	9.000000 S		

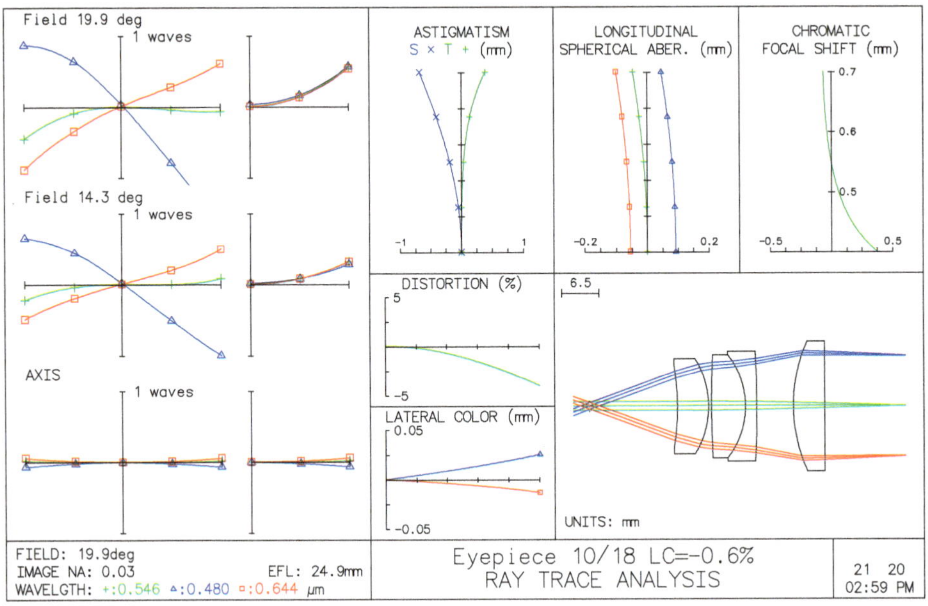

Figure 1.223 Design parameters and graphs of aberration correction of a 10×/18 eyepiece (LC = −0.6%).

small-magnification objectives, eyepieces with magnifications of 50× and even 100× are required, which were previously used only in astronomy. Moreover, the input numerical apertures for such eyepieces can be significantly increased and reach values of 0.08–0.1; perhaps this is not required in budget microscopes, but it can be used in the construction of research microscopes.

As an example of an optical design in which the objective, the tube system with the head's prisms, and the eyepiece work together, we used a 10×/0.28 infinity LCF plan achromatic objective (shown in Fig. 1.35), a unified head with the tube system ($F'_{ts} = 200$ mm; shown in Fig. 1.217), and a simple 10×/18 eyepiece (shown in Fig. 1.224). Figure 1.238 shows this design (image at infinity).

Lens: Eyepiece 10/18 LC=0% Efl 24.98

Image num aper 0.030000 Image height 9.000000 Primary wavln 0.546070

SRF	RADIUS	THICKNESS	APERTURE RADIUS	GLASS	SPECIAL
OBJ	0.000000	1.0000e+09	3.6031e+08	AIR	
AST	0.000000	14.490000	0.749353 AS	AIR	
2	-46.880000	6.000000	6.070000	LZ_TK16	
3	-15.740000	1.000000	7.350000	AIR	
4	0.000000	7.000000	7.740000	LZ_TK16 P	
5	-14.791000	2.000000	8.140000	LZ_TF4	
6	-108.140000	6.000000	8.640000	AIR	
7	26.550000	5.500000	9.970000	LZ_K8	
8	0.000000	14.403014 S	9.840000	AIR	
IMS	0.000000	0.000000	9.000000 S		

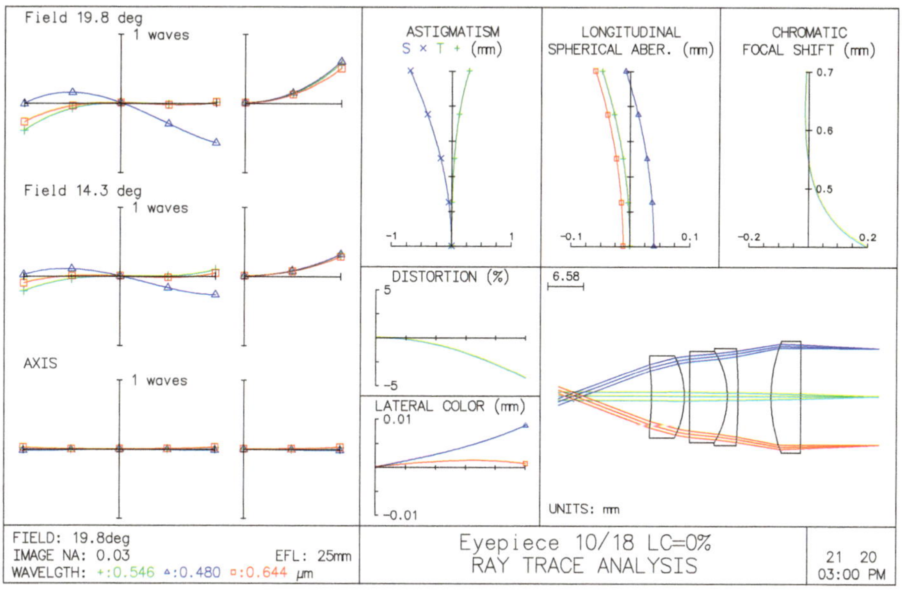

Figure 1.224 Design parameters and graphs of aberration correction of a 10×/18 eyepiece (LC = 0).

Modernization of the mechanical design of the eyepiece is also possible. More precisely, modernization involves integrating an ocular tube with the mechanical design. It does not apply to aberration correction of eyepieces, but rather to their overall performance. Older eyepiece models were designed to work at a height of 13 mm; they were integrated into the optical system of the microscope. The height of later generations of eyepieces was 10 mm, which was due to the implementation of the 20-mm ocular field in budget microscopes (with the standard value of the mechanical size of the ocular tube of 23.2 mm). In these cases, the image plane after the projection system of the microscope (objective or objective plus the tube system) is below the so-called *stand* plane of the eyepiece. The engineering solution is apparent

Lens: Eyepiece 10/20 LC=0% dis=0% Efl 24.99

Image num aper 0.040000 Image height 10.000000 Primary wavln 0.546070

SRF	RADIUS		THICKNESS		APERTURE RADIUS		GLASS		SPECIAL
OBJ	0.000000		1.0000e+09		4.0015e+08		AIR		
AST	0.000000		15.300000		0.999618	AS	AIR		
2	-205.100000		2.500000		7.090000		LZ_TF5		
3	81.660000		2.500000		7.790000		AIR		
4	-64.750000		9.500000		8.630000		LZ_CTK19		
5	-24.040000		0.300000		11.270000		AIR		A
6	-296.500000		2.000000		12.000000		LZ_TF5	P	
7	42.660000		7.500000		12.710000		LZ_CTK19	P	
8	-29.240000		5.400000		13.140000		AIR		
9	20.840000		6.000000		12.340000		LZ_CTK19	P	
10	57.020000		2.000000		11.320000		LZ_TF4		
11	15.031000		15.157529	S	9.830000		AIR		
IMS	0.000000		0.000000		10.000000	S			

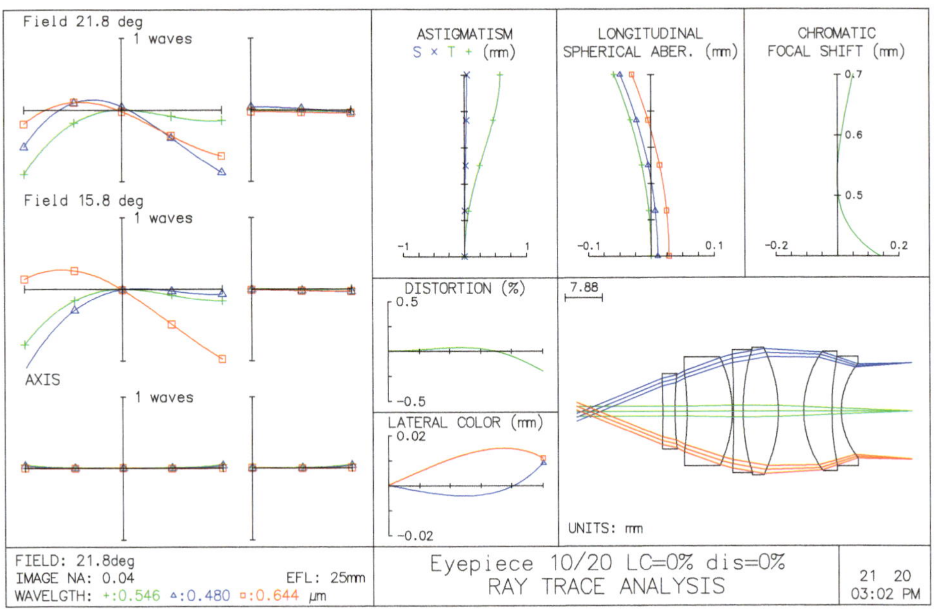

Figure 1.225 Design parameters and graphs of aberration correction of a 10×/20 eyepiece (LC = 0; DIS = 0; R4, parabolic).

when the designs of the ocular tube and the eyepiece are made in the manner illustrated in Fig. 1.239. In this case, the image plane after the projection system of the microscope is located above the stand plane of the eyepiece. This gives certain advantages, for example, the possibility of implementing a larger linear field (more than 20 mm) using a standard ocular tube with an internal diameter of 23.2 mm, and also makes it convenient to use and replace various scales, grids, crosshairs, etc., in special eyepieces.

Lens: Eyepiece 10/22 LC=0%			Zoom 1 of 1	Efl 24.858316	
Image num aper 0.030000		Image height 11.000000		Primary wavln 0.546070	
SRF	**RADIUS**	**THICKNESS**	**APERTURE RADIUS**	**GLASS**	**SPECIAL**
OBJ	0.000000	1.0000e+09	4.4251e+08	AIR	
AST	0.000000	23.580000	0.745749 AS	AIR	
2	-62.520000	6.800000	11.320000	LZ_CTK3	
3	-23.660000	0.300000	12.750000	AIR	
4	324.300000	3.200000	13.510000	LZ_TF4	
5	40.180000	9.500000	13.970000	LZ_CTK3 P	
6	-64.420000	0.300000	14.530000	AIR	
7	26.180000	14.500000	14.680000	LZ_CTK3 P	
8	-26.180000	3.200000	13.120000	LZ_TF4 P	
9	30.130000	8.783994 S	11.340000	AIR	
IMS	0.000000	0.000000	11.000000 S		

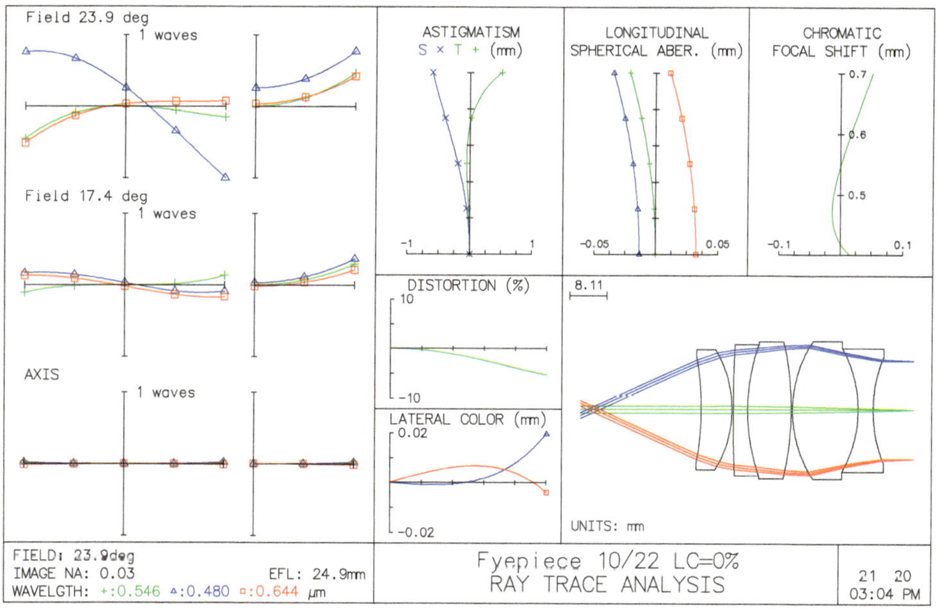

Figure 1.226 Design parameters and graphs of aberration correction of a 10×/22 eyepiece (LC = 0).

An interesting solution may be to unify (using, for example, various adapter bushings and simple modernization of the ocular tubes of the microscope), making it possible to use older 13-mm eyepieces as well as 10-mm and new ones for which the stand plane is higher. These engineering solutions can be used to manufacture budget microscopes. For example, objectives, heads, and oculars can be sold separately for microscopes used by researchers. New optics is characterized by high image quality that is ensured by effective correction of aberrations such as image curvature and lateral chromaticity. As a result, the image quality can be significantly improved

Lens: Eyepiece 10/22 LC,DIS=0%							Efl 25.0	
Image num aper	0.030000	Image height	11.000000	Primary wavln	0.546070			

SRF	RADIUS		THICKNESS		APERTURE RADIUS		GLASS		SPECIAL
OBJ	0.000000		1.0000e+09		4.3995e+08		AIR		
AST	0.000000		17.420000		0.750086	AS	AIR		
2	-233.500000		2.500000		8.350000		LZ_TF4		
3	58.580000		9.500000		9.210000		LZ_CTK19		
4	-16.770000		0.300000		10.590000		AIR		A
5	-173.000000		2.500000		10.580000		LZ_TF4	P	
6	22.160000		7.500000		10.700000		LZ_CTK19	P	
7	-310.700000		19.781867	S	10.800000		AIR		
IMS	0.000000		0.000000		11.000000	S			

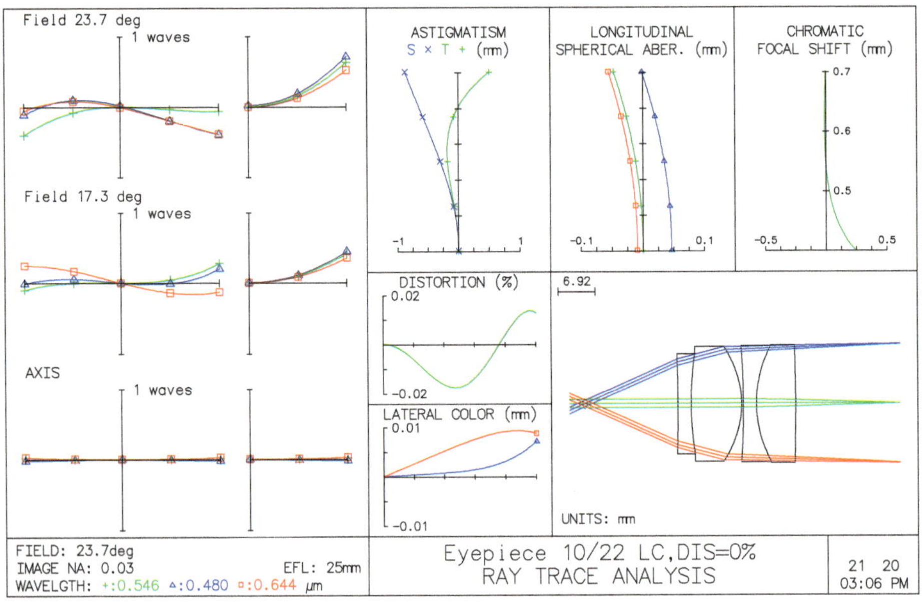

Figure 1.227 Design parameters and graphs of aberration correction of a 10×/22 eyepiece (LC = 0; DIS = 0; R3, parabolic).

compared with that of most budget objectives on the market today. In the near future, we can expect a new generation of objectives to be used in budget light microscopes. In turn, the inclusion of new optics in the design of budget microscopes will spur the use of modern positioning devices for object placement as well as image processing based on digital technologies. Separately, lighting systems built using LED technologies could present a real revolution in light microscopy.

1.8.4 Design of digital receiver projection systems

A common question for optical microscope designers is, how can the size of an image obtained in the intermediate image plane on a microscope be

Lens: Eyepiece 10/25 LC=0% Efl 24.89

Image num aper 0.040000 Image height 12.500000 Primary wavln 0.546070

SRF	RADIUS	THICKNESS	APERTURE RADIUS	GLASS	SPECIAL
OBJ	0.000000	1.0000e+09	5.0212e+08	AIR	
AST	0.000000	17.394400	0.995779 AS	AIR	
2	-11.641000	3.500000	8.390000	LZ_CTK19	
3	-14.388000	0.300000	10.340000	AIR	
4	-159.220000	2.000000	12.590000	LZ_TF5	
5	83.560000	7.500000	13.760000	LZ_CTK19 P	
6	-25.470000	0.300000	14.490000	AIR	A
7	-200.400000	2.500000	15.100000	LZ_TF5 P	
8	47.530000	10.500000	15.810000	LZ_CTK19 P	
9	-47.530000	0.300000	16.430000	AIR	
10	18.880000	6.000000	15.830000	LZ_CTK19 P	
11	33.570000	2.000000	15.230000	LZ_TF5 P	
12	14.322000	17.013811 S	12.270000	AIR	
IMS	0.000000	0.000000	12.500000 S		

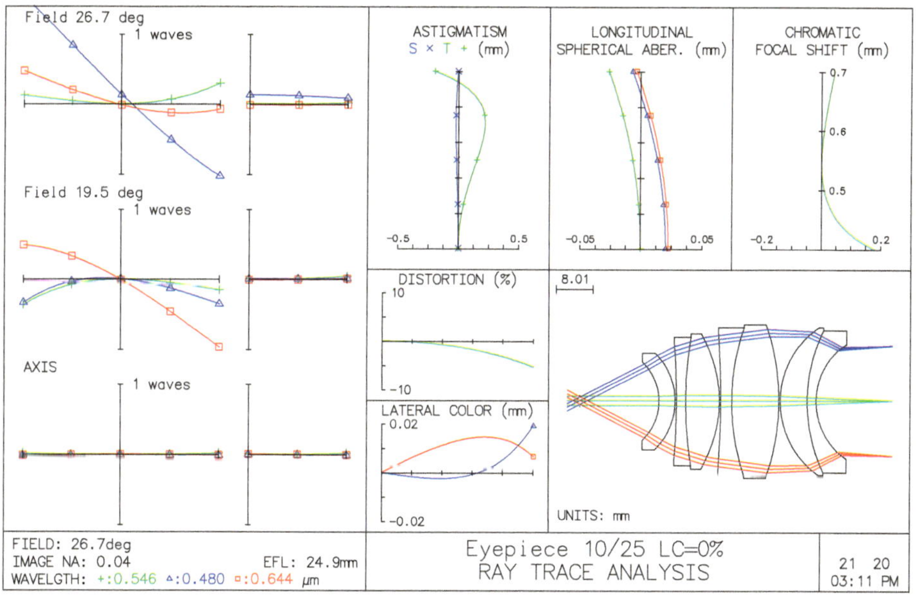

Figure 1.228 Design parameters and graphs of aberration correction of a 10×/25 eyepiece (LC = 0; R5, parabolic).

correlated with that of a digital image receiver? The fact is that until now, *round* optics were widely used in light microscopes; the researcher observes a circle-shaped image through the eyepieces. This is not entirely optimal if we keep in mind the capabilities of a person's so-called *peripheral vision*; i.e., not all useful information about the object under investigation falls into the brain of a human. But until now, if we spoke about visual observation with a microscope, this was the case. On the contrary, both film photography and

Lens: ocular 10/28 LC=0%				Efl	25.01
Image num aper	0.040000	Image height	14.000000	Primary wavln	0.546070

SRF	RADIUS	THICKNESS	APERTURE RADIUS	GLASS	SPECIAL
OBJ	0.000000	1.0000e+09	5.5968e+08	AIR	
AST	0.000000	16.000000	1.000578 AS	AIR	
2	-54.800000	2.500000	10.580000	LZ_TF5	
3	101.000000	2.000000	12.430000	AIR	
4	230.000000	9.000000	14.950000	LZ_CTK19	
5	-25.600000	0.500000	16.250000	AIR	
6	97.000000	2.000000	18.400000	LZ_TF5 P	
7	34.700000	8.500000	18.720000	LZ_CTK19 P	
8	-80.700000	0.500000	18.770000	AIR	
9	30.800000	8.500000	18.240000	LZ_CTK19 P	
10	-114.000000	2.000000	17.880000	LZ_TF4	
11	27.400000	16.033667 S	15.230000	AIR	
IMS	0.000000	0.000000	14.000000		

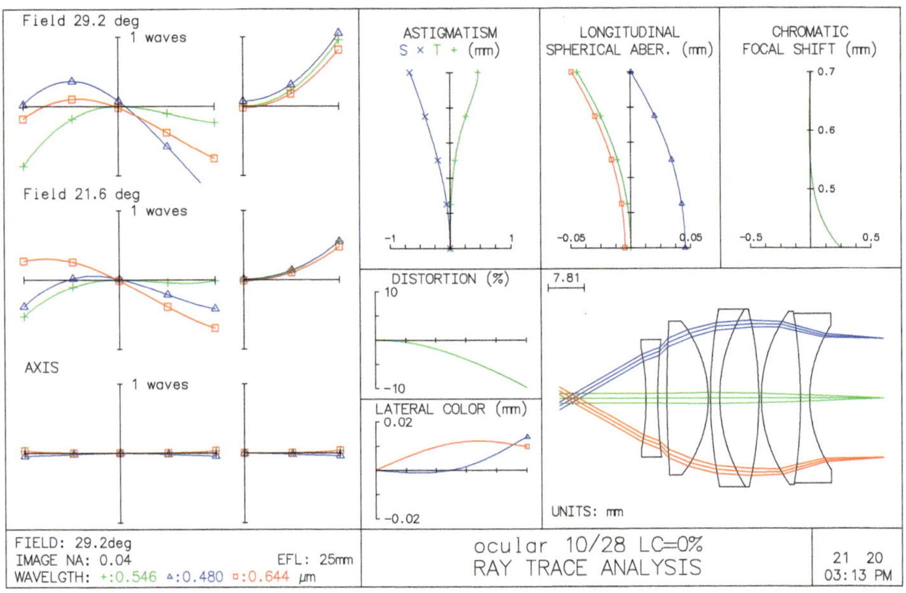

Figure 1.229 Design parameters and graphs of aberration correction of a 10×/28 eyepiece (LC = 0).

image receivers (analog and digital), the circle shape is almost never used. As a rule, an image is constructed in a rectangle having an appropriate aspect ratio. Without going into details, we use the standard format of an old 24×36-mm film camera as an example (as well as the modern so-called full-format camera). Figure 1.240 shows the options for constructing an image on such a receiver.

The image can be completely entered into the rectangle according to its smaller sides ($d = 24$ mm); in this case, there is no loss of information about

Lens: ocular 10/28 LC and DIS =0% Efl 24.86

Image num aper 0.040000 Image height 14.000000 Primary wavln 0.546070

SRF	RADIUS		THICKNESS		APERTURE RADIUS		GLASS		SPECIAL
OBJ	0.000000		1.0000e+09		5.6319e+08		AIR		
AST	0.000000		15.850000		0.994336	AS	AIR		
2	0.000000		2.500000		9.930000		LZ_TF5		
3	40.500000		2.000000		11.110000		AIR		
4	96.000000		9.000000		12.920000		LZ_CTK19		
5	-16.000000		0.500000		13.810000		AIR		A
6	-21.560000		2.000000		13.980000		LZ_TF5	P	
7	-66.600000		8.500000		16.110000		LZ_CTK19	P	
8	-22.530000		0.500000		17.160000		AIR		
9	26.800000		8.500000		16.340000		LZ_CTK19	P	
10	76.300000		2.000000		14.860000		LZ_TF5	P	
11	18.260000		15.854186	S	12.890000		AIR		
IMS	0.000000		0.000000		14.000000	S			

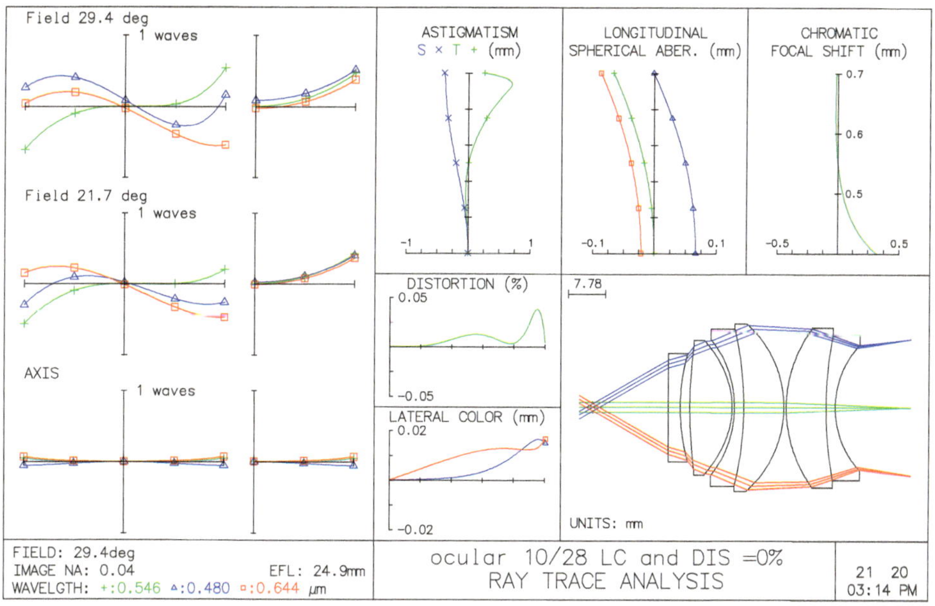

Figure 1.230 Design parameters and graphs of aberration correction of a 10×/28 eyepiece (LC = 0; DIS = 0; R4, parabolic).

some fragments of the image (i.e., all information about the object under study, the image of which is formed by the optics of the microscope, is transmitted to the receiver). But in this case, only 52% of the receiver area is used. If we fit the image along the longer side of the rectangle ($d = 36$ mm), there will be some loss of information about the object under study since there are four corners and two segments where the receiver does not receive the image. This is not a big loss, given that approximately 85% of the information about the object under investigation falls on the receiver. If the image

Lens: Eyepiece 12.5/22 LC=-0.6% Efl 20.16

Image num aper 0.030000 Image height 11.000000 Primary wavln 0.546070

SRF	RADIUS	THICKNESS	APERTURE RADIUS	GLASS	SPECIAL
OBJ	0.000000	1.0000e+09	5.4568e+08	AIR	
AST	0.000000	16.270000	0.604752 AS	AIR	
2	0.000000	5.750000	10.420000	LZ_CTK12	
3	-20.140000	0.300000	11.180000	AIR	
4	30.900000	7.800000	11.830000	LZ_CTK12 P	
5	-17.539000	2.500000	11.740000	LZ_TF10	
6	239.900000	13.243417 S	11.450000	AIR	
IMS	0.000000	0.000000	11.000000 S		

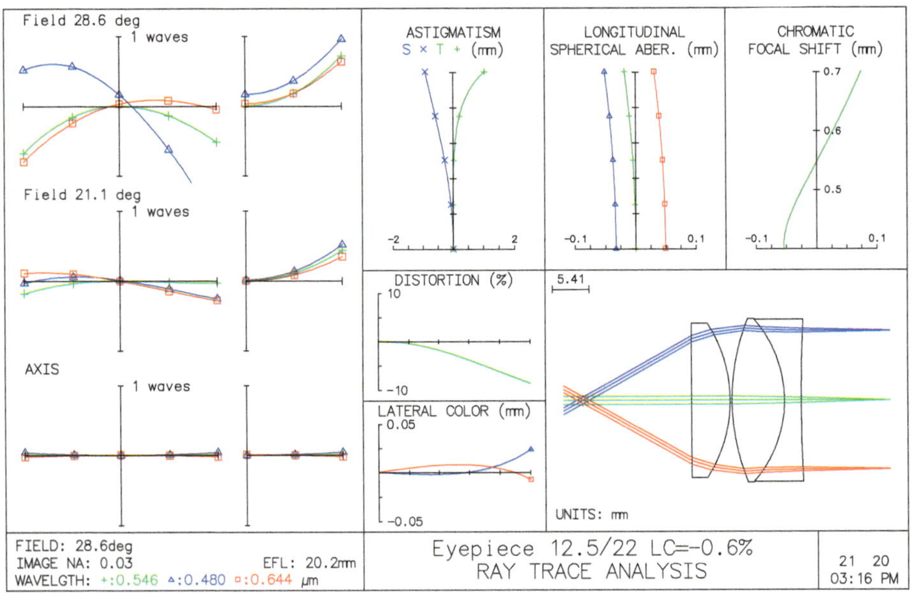

Figure 1.231 Design parameters and graphs of aberration correction of a 12.5×/22 eyepiece (LC = −0.6%).

describes the rectangle, then the receiver does not receive the image in four segments; only 60% of the information about the object under investigation is received by the receiver. One could probably easily explain why optical designers of traditional photographic objectives took into account the linear field value of 43.27 mm (i.e., the third option). Indeed, if a photographer took a photograph using a natural landscape, for example, they could afford to cut off part of the sky and part of the earth (as well as a couple of trees on the sides) in the image; the plot value of the photograph and the information content would not be changed. Figure 1.241 illustrates a certain middle option (approximately 75% of the information about the object under investigation is transmitted to a digital receiver) of constructing an image (a microscopic object) on a digital receiver.

Lens: Eyepiece 15/16 LC=0% Efl 16.68

Image num aper 0.030000 Image height 8.000000 Primary wavln 0.546070

SRF	RADIUS	THICKNESS	APERTURE RADIUS	GLASS	SPECIAL
OBJ	0.000000	1.0000e+09	4.7972e+08	AIR	
AST	0.000000	15.900000	0.500293 AS	AIR	
2	-41.690000	4.500000	8.210000	LZ_CTK3	
3	-15.776000	0.200000	9.120000	AIR	
4	216.300000	2.100000	9.750000	LZ_TF4	
5	26.790000	6.300000	10.090000	LZ_CTK3 P	
6	-42.950000	0.200000	10.430000	AIR	
7	17.620000	9.700000	10.550000	LZ_CTK3 P	
8	-17.620000	2.100000	9.670000	LZ_TF4 P	
9	20.090000	5.940296 S	8.220000	AIR	
IMS	0.000000	0.000000	8.000000 S		

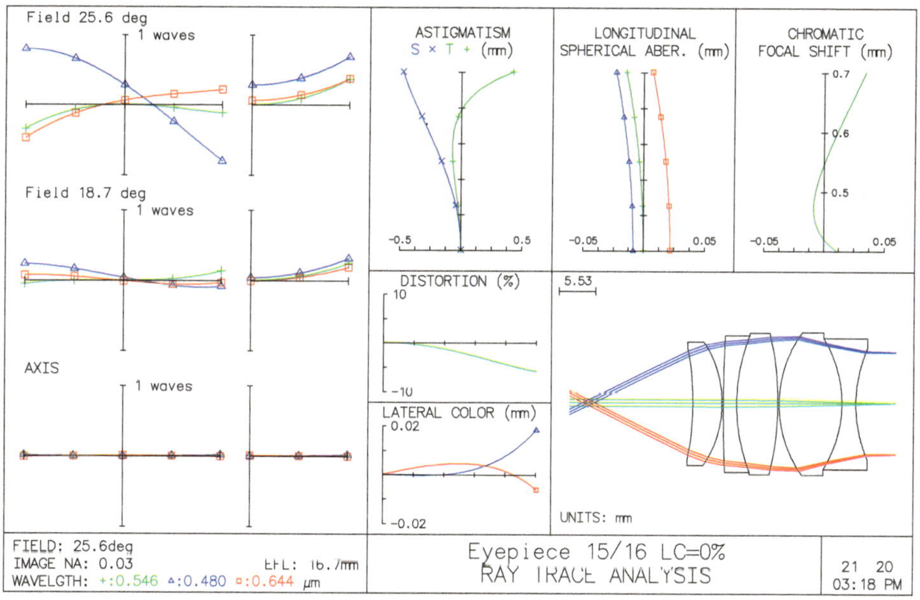

Figure 1.232 Design parameters and graphs of aberration correction of a 15×/16 eyepiece (LC = 0).

1.8.4.1 Optimal receiver geometry

Such a photograph as described above cannot even be an example of beauty and aesthetics; it looks very unusual, and the presence of the black corners in the image emphasizes that it is not complete. However, if we study Fig. 1.241, we can agree that it reflects exactly what the observer sees through the eyepieces, i.e., a circle whose diameter depends on the class of the microscope. For budget microscopes, the diameter of the intermediate image on the microscope can be, for example, 18, 20, or 22 mm. If the diameter is larger, the microscope is of a higher class. This is a complex issue, but for budget

Lens: Eyepiece 15/20 LC,DIS=0% Efl 16.67

Image num aper 0.040000 Image height 10.000000 Primary wavln 0.546070

SRF	RADIUS		THICKNESS		APERTURE RADIUS		GLASS		SPECIAL
OBJ	0.000000		1.0000e+09		5.9989e+08		AIR		
AST	0.000000		14.320000		0.666793	AS	AIR		
2	-9.078000		3.000000		7.160000		LZ_TF5		
3	-11.110000		0.300000		8.900000		AIR		
4	-467.000000		7.400000		11.560000		LZ_CTK19		
5	-15.200000		0.300000		12.350000		AIR		A
6	-30.000000		1.500000		12.300000		LZ_TF5	P	
7	36.370000		7.500000		13.510000		LZ_CTK19	P	
8	-26.720000		0.300000		13.710000		AIR		
9	21.130000		7.000000		12.710000		LZ_CTK19	P	
10	-48.000000		2.200000		12.290000		LZ_TF5	P	
11	25.370000		4.000000		10.140000		AIR		
12	-34.870000		1.900000		10.010000		LZ_TF5	P	
13	75.800000		3.999209	S	9.870000		AIR		
IMS	0.000000		0.000000		10.000000	S			

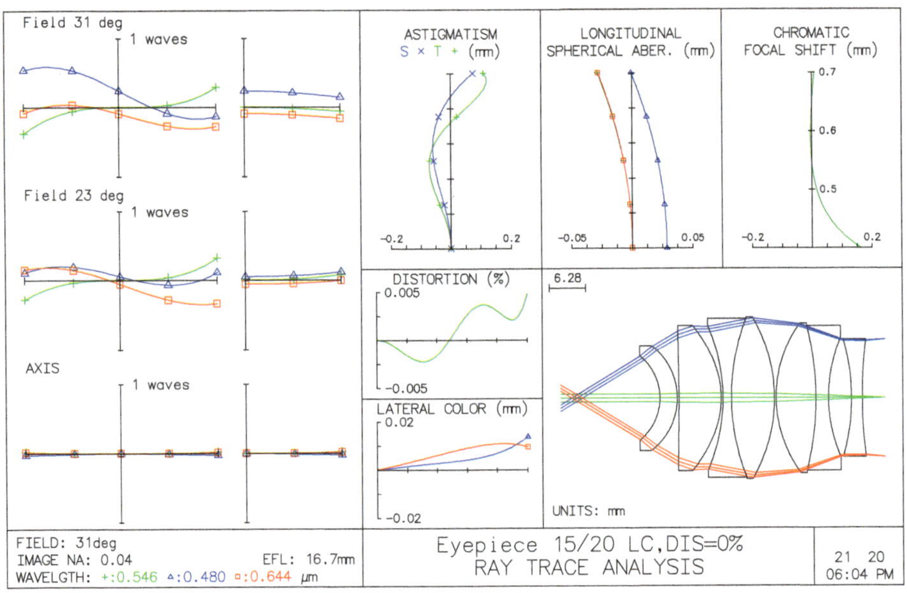

Figure 1.233 Design parameters and graphs of aberration correction of a 15×/20 eyepiece (LC = 0; DIS = 0; R4, parabolic).

microscopes, especially mass-produced ones, it does not make sense to use expensive wide-field optics (including objectives) and a full-format camera, which cost about four to five times more than simpler cameras whose sensor size is smaller. In this case, it makes sense to favor expediency and practicality; for mass-produced budget light microscopes, we can use the experience of photographers and select the option in which the image describes the rectangle of a digital receiver.

Lens: Eyepiece 16/13 LC=0% Efl 15.82

Image num aper 0.030000 Image height 6.500000 Primary wavln 0.546070

SRF	RADIUS		THICKNESS		APERTURE RADIUS		GLASS		SPECIAL
OBJ	0.000000		1.0000e+09		4.1084e+08		AIR		
AST	0.000000		13.780000		0.474635	AS	AIR		
2	0.000000		2.400000		6.520000		LZ_TK23		
3	-13.183000		0.100000		6.680000		AIR		
4	31.770000		3.100000		6.910000		LZ_TK23	P	
5	-13.772000		1.300000		6.890000		LZ_TF5		
6	0.000000		1.300000		6.980000		AIR		
7	35.340000		2.500000		7.070000		LZ_LK3		
8	0.000000		10.420850	S	7.010000		AIR		
IMS	0.000000		0.000000		6.500000	S			

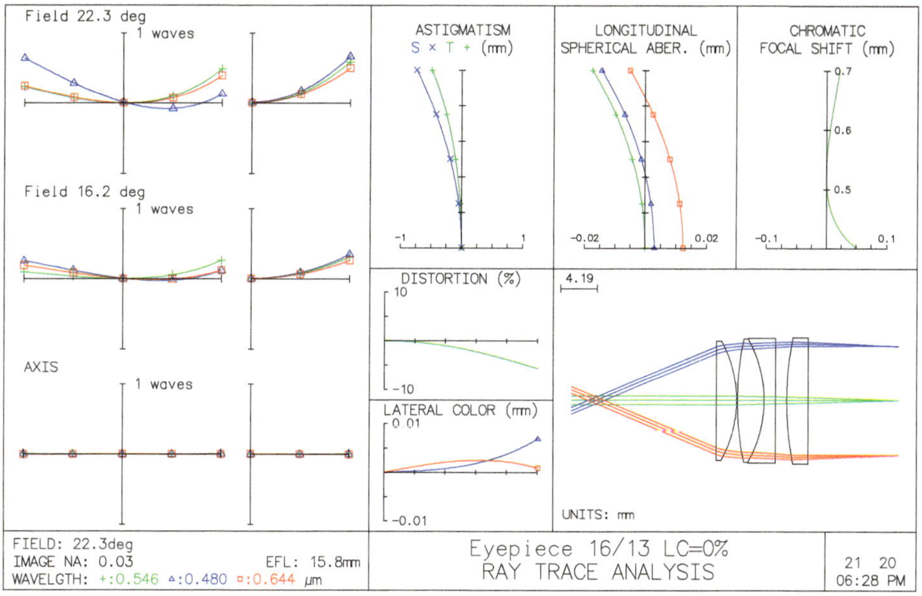

Figure 1.234 Design parameters and graphs of aberration correction of a 16×/13 eyepiece (LC = 0).

It probably makes little sense to list in detail the models of modern digital image receivers and their technical parameters. Digital receivers are constantly being improved in at least three areas:

1. Reduction in the size of a single pixel,
2. Increase in the total size of the matrix, and
3. Reduction of *electronic noise*.

Today, as a rule, these three parameters do not improve simultaneously, i.e., the problem is not solved comprehensively. This might be a temporary inhibition of progress until the invention and use of any new, relatively

Lens: Eyepiece 25x 2y'=7.6 LC,DIS=min   Efl 10.13

Image num aper 0.060000   Image height 3.800000   Primary wavln 0.546070

SRF	RADIUS	THICKNESS	APERTURE RADIUS	GLASS	SPECIAL
OBJ	0.000000	1.0000e+09	3.7531e+08	AIR	
AST	0.000000	16.040000	0.607503 AS	AIR	
2	-100.000000	3.000000	6.540000	LZ_TF5	
3	15.460000	6.400000	7.800000	LZ_K8	
4	-15.460000	0.600000	8.520000	AIR	
5	19.800000	5.600000	10.460000	LZ_K8 P	
6	-33.800000	0.600000	10.430000	AIR	
7	16.740000	6.800000	9.240000	LZ_CTK19	
8	-320.000000	2.000000	7.630000	LZ_TF4	
9	18.000000	8.000000	6.320000	AIR	
10	-7.400000	2.600000	4.030000	LZ_K8 P	
11	80.000000	-3.942645 S	4.220000	AIR	
IMS	0.000000	0.000000	3.800000 S		

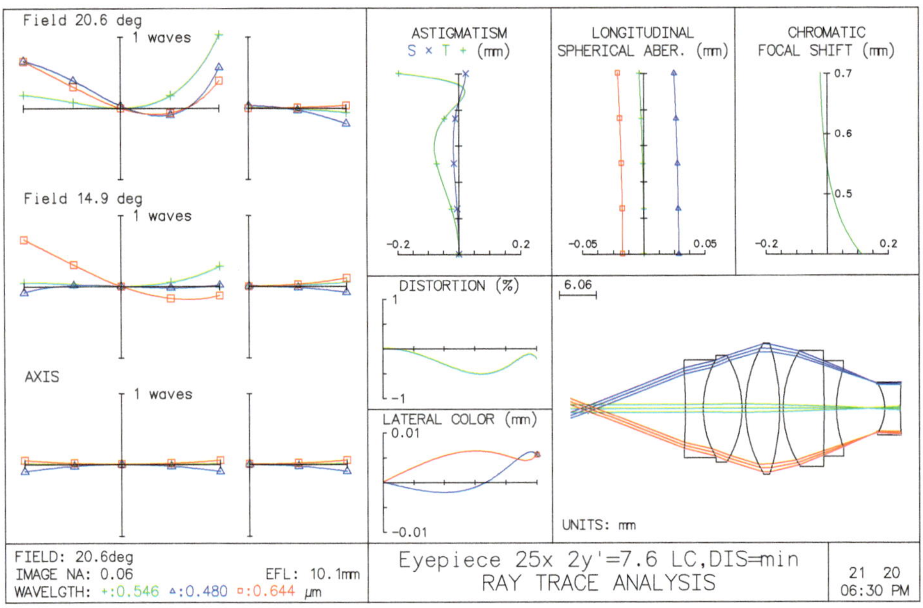

Figure 1.235 Design parameters and graphs of aberration correction of a 25×/7.6 eyepiece (LC, DIS min).

speaking, composite materials as the elemental base instead of the silicon used today. You might think that for budget light microscopy the most in-demand trend is reduction in the size of a single pixel. But because light microscopy theory implies that an image is less than a quarter of the wavelength used, this will not be possible to obtain. This circumstance is a natural limiter; overcoming it does not make practical sense when it comes to the use of a digital receiver in light microscopy.

Lens: Eyepiece 50x 2y'=3.8 LC,DIS=min Efl 5.06

Image num aper 0.080000 Image height 1.900000 Primary wavln 0.546070

SRF	RADIUS	THICKNESS	APERTURE RADIUS	GLASS	SPECIAL
OBJ	0.000000	1.0000e+09	3.7543e+08	AIR	
AST	0.000000	8.020000	0.404866 AS	AIR	
2	-50.000000	1.500000	3.380000	LZ_TF5	
3	7.730000	3.200000	4.030000	LZ_K8	
4	-7.730000	0.300000	4.340000	AIR	
5	9.900000	3.000000	5.390000	LZ_K8 P	
6	-16.900000	0.300000	5.370000	AIR	
7	8.370000	3.400000	4.700000	LZ_CTK19	
8	-160.000000	1.000000	3.890000	LZ_TF4	
9	9.000000	4.000000	3.200000	AIR	
10	-3.700000	1.300000	2.030000	LZ_K8 P	
11	40.000000	-1.995850 S	2.150000	AIR	
IMS	0.000000	0.000000	1.900000 S		

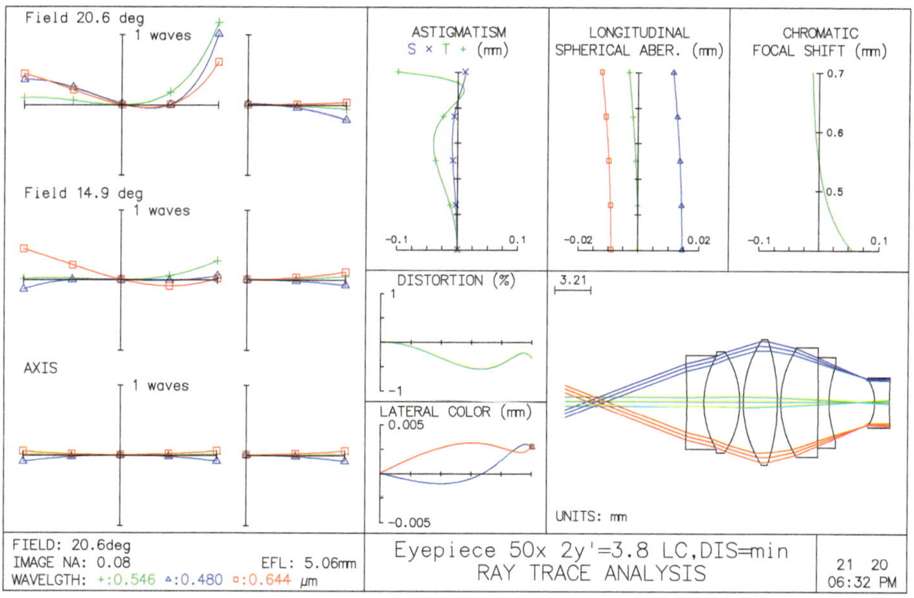

Figure 1.236 Design parameters and graphs of aberration correction of a 50×/3.8 eyepiece (LC, DIS min).

We will proceed from the fact that if a person observes an image in a microscope with their eyes through an eyepiece, they receive some information about the object under study. In this context, it is logical to ask a question: What additional benefits can the use of a so-called *visualization system* based on a digital image receiver provide? Of course, the possibility of documentation is an argument in favor of using these systems; a device called a *drawing device* has long been successfully used for these purposes. But replacing the drawing apparatus with a modern digital

Lens: Eyepiece 100x 2y'=2.0 LC,DIS=0% Efl 2.5

Image num aper 0.100000 Image height 1.000000 Primary wavln 0.546070

SRF	RADIUS		THICKNESS		APERTURE RADIUS		GLASS		SPECIAL
OBJ	0.000000		1.0000e+09		4.0015e+08		AIR		
AST	0.000000		7.000000		0.249909	AS	AIR		
2	15.710000		1.500000		3.180000		LZ_TF5		
3	5.640000		3.200000		3.460000		LZ_LK3		
4	-7.910000		0.300000		3.800000		AIR		
5	7.070000		2.800000		4.450000		SILICA		
6	-53.800000		0.300000		4.330000		AIR		
7	4.425000		3.400000		3.790000		SILICA	P	
8	6.077000		4.000000		2.740000		AIR		
9	-3.562000		1.300000		1.130000		LZ_TK12		
10	1.385000		1.300000		0.940000		LZ_TF5	P	
11	2.768000		-2.152475	S	0.780000		AIR		
IMS	0.000000		0.000000		1.000000	S			

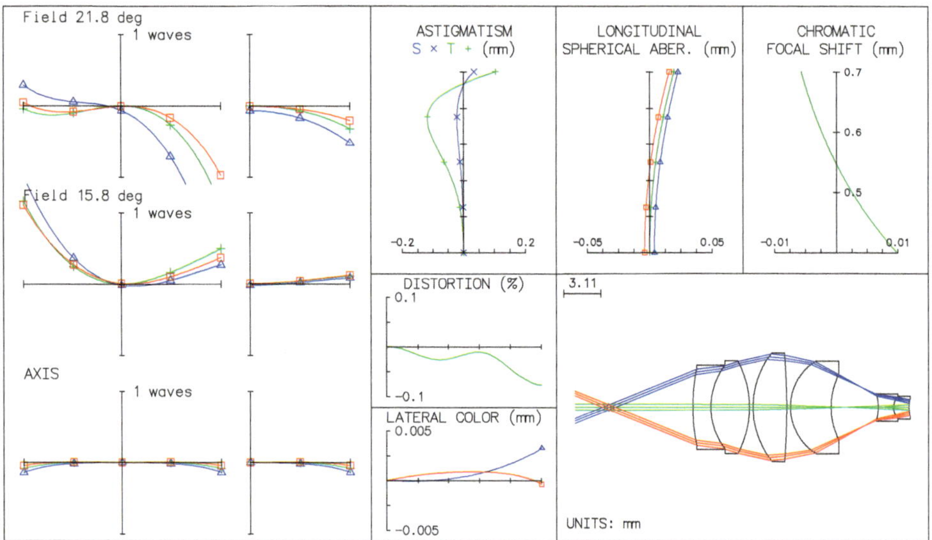

Figure 1.237 Design parameters and graphs of aberration correction of a 100×/2.0 eyepiece (LC = 0; DIS = 0).

gadget (one might say, a *fashionable* gadget) is just an option that does not affect the basic functions of the microscope; it cannot change the technical essence and principles of the microscope or improve its basic technical parameters. Descriptions of various types of visualization systems are given in Volume 1 (Chapter 2, Section 2.6), and we will try to expand on these considerations with practical examples. Table 1.16 shows the main geometric parameters of the sensors of some cameras used in modern budget microscopy.

Table 1.15 The combination of standard objective and eyepiece linear magnifications using a light microscope.

	Total magnification for objectives					
	4×/0.10 4×/0.20	10×/0.25 10×/0.60	20×/0.40 20×/0.80	40×/0.65 40×/0.95	60×/0.85 60×/1.40	100×/0.85 100×/1.40
Eye-pieces	$50 \leq M \leq 100$ $100 \leq M \leq 200$	$125 \leq M \leq 250$ $300 \leq M \leq 600$	$200 \leq M \leq 400$ $400 \leq M \leq 800$	$325 \leq M \leq 650$ $475 \leq M \leq 950$	$425 \leq M \leq 850$ $700 \leq M \leq 1400$	$425 \leq M \leq 850$ $700 \leq M \leq 1400$
10×	40×	100×	200×	400×	600×	1000×
12.5×	50×	125×	250×	500×	750×	1250×
15×	60×	150×	300×	600×	900×	1500×
16×	64×	160×	320×	640×	960×	1600×
20×	80×	200×	400×	800×	1200×	2000×
25×	100×	250×	500×	1000×	1500×	2500×
50×	200×	500×	1000×	2000×	3000×	5000×
100×	400×	1000×	2000×	4000×	6000×	10,000×

Note. M, useful magnification of a microscope.

Consider the technical parameters of two cameras that have approximately the same total resolution:

1. An 18-megapixel (MP) camera: pixel size, 1.25×1.25 μm; diagonal size of the matrix, ø 7.68 mm.
2. A 20-MP camera: pixel size, 2.4×2.4 μm; diagonal size of the matrix, ø 15.73 mm.

With nearly the same (we will assume this) resolution of the cameras, they nevertheless differ significantly in their technical parameters. Namely, the size of a single pixel in a 20-MP camera is two times larger than in an 18-MP camera; however, the sizes of the diagonals of the matrix also differ by half. The logical question is, which camera is more suitable for use as an imaging system for a light microscope?

To answer this question, we will use the following arguments from the practice of microscopy. Suppose we want to project an image after a 10×/0.40 objective, the output aperture of which is maximum in the set of standard microscope objectives. The value of the output numerical aperture in the image space for this lens is $NA' = 0.04$. Accordingly, the resolution in the space of images (after this objective) is equal to 8.33 μm. Let the linear field in the image space on the microscope be 22 mm (which is the usual value for budget microscopes). We assume that the linear fields of the images observed in the microscope through the eyepieces and using the visualization system should be the same, which allows for more accurate identification of details and nuances in the images. Then, to pair the linear field of the microscope and

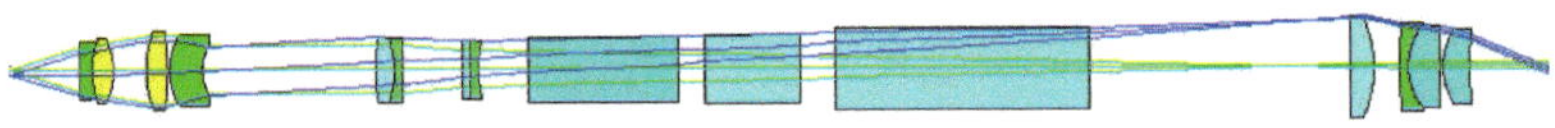

Lens: LD plan 10x0.28+head+ocular — Efl -2.47

Object num aper 0.280000 — Object height -1.000000 — Primary wavln 0.546070

SRF	RADIUS	THICKNESS	APERTURE RADIUS	GLASS	SPECIAL
OBJ	0.000000	1.0000e-20	1.000000	AIR	
AST	0.000000	16.825500	1.000000 AS	AIR	
2	0.000000	3.000000	5.910000	H-ZF6	
3	21.760000	5.000000	6.560000	H-FK61	
4	-15.672000	7.500000	7.130000	AIR	
5	23.710000	5.000000	8.640000	H-FK61 P	
6	-66.000000	0.500000	8.530000	AIR	
7	16.200000	7.500000	8.110000	H-ZF6 P	
8	10.233000	41.175000	5.850000	AIR	
9	42.120000	4.200000	7.290000	LZ_LK3	
10	-23.530000	2.000000	7.230000	LZ_KF4	
11	-259.640000	14.000000	7.180000	AIR	
12	77.600000	2.000000	6.620000	LZ_LK3 P	
13	-331.400000	2.000000	6.530000	LZ_KF4 P	
14	31.670000	11.500000	6.390000	AIR	
15	0.000000	34.600000	6.670000	LZ_BK10	
16	0.000000	6.000000	7.250000	AIR	
17	0.000000	22.000000	7.410000	LZ_K8	
18	0.000000	8.000000	7.790000	AIR	
19	0.000000	59.500000	8.000000	LZ_K8 P	
20	0.000000	60.540000	9.030000	AIR	
21	0.000000	5.500000	11.170000	LZ_K8 P	
22	-26.550000	6.000000	11.330000	AIR	
23	108.140000	2.000000	9.790000	LZ_TF4	
24	14.791000	7.000000	9.150000	LZ_TK16	
25	0.000000	1.000000	8.800000	AIR	
26	15.740000	6.000000	8.310000	LZ_TK16 P	
27	46.880000	17.770000	7.000000	AIR	
28	0.000000	1.6124e+05 S	0.723934 S	AIR	
IMS	0.000000	0.000000	6.5116e+04 S		

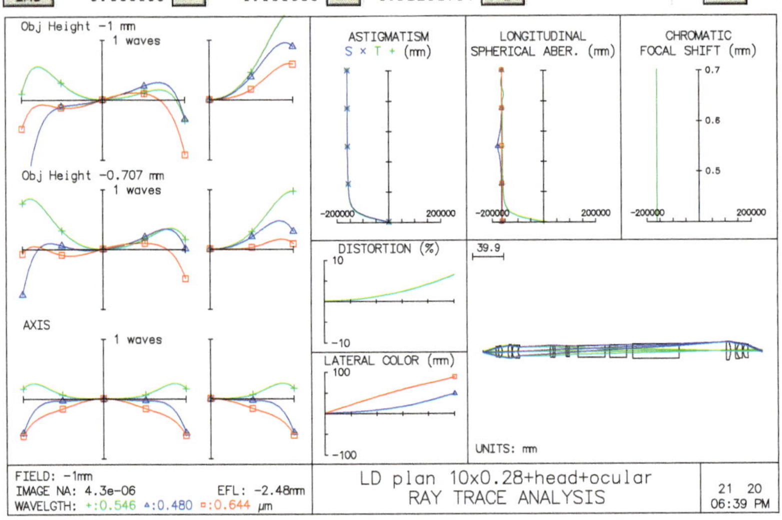

Figure 1.238 Parameters of design and graphs of aberration correction of a 10×/0.28 LCF plan achromatic objective with an infinite tube system ($F'_{ts} = 200$ mm) and a 10×/18 eyepiece.

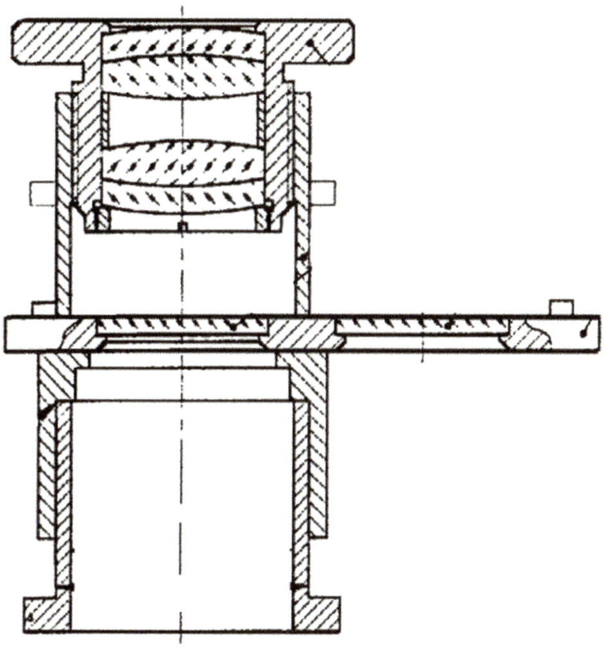

Figure 1.239 Example design of an eyepiece for a budget microscope.

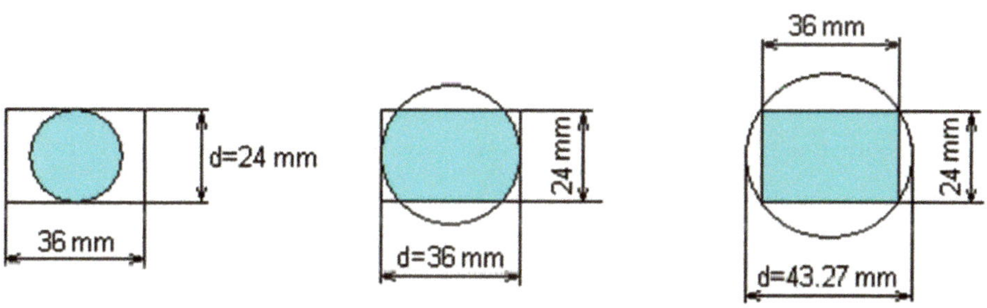

Figure 1.240 Examples of image building on a receiver.

the size of the diagonal of the matrix, an additional optical system *projective* is required and the value of its own linear increase is less than one.

- For an 18-MP camera whose matrix diagonal is 7.68 mm, the linear magnification of the projective will be $M = 0.35\times$. In this case, the output numerical aperture after the projection will be $NA' = 0.114$. This will provide a resolution in the plane of the matrix of approximately 3 μm.
- For a 20-MP camera, the linear magnification of projective will be $M = 0.715\times$. The numerical aperture after the projection will be

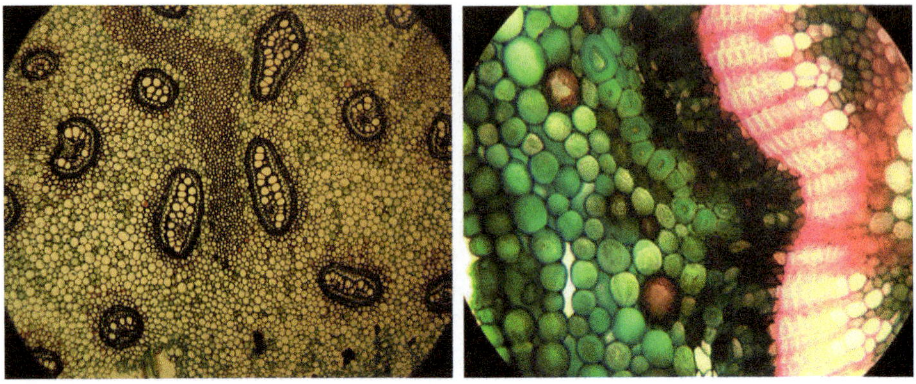

Figure 1.241 Photographs in which approximately 75% of the information about the object under investigation is transmitted to a digital receiver.

Table 1.16 Comparison of geometric parameters of some digital camera sensors.

Resolution, sensor dimensions (mm)	Diagonal of a matrix, mm	The size of a single pixel, μm
20 MP, 1 inch (13.06 × 8.76)	15.73	2.4 × 2.4
18 MP, 1/2.3 inch (6.14 × 4.61)	7.68	1.25×1.25
14 MP, 1/2.3 inch (5.73 × 4.60)	7.35	1.4 × 1.4
5.1 MP, 1/2.5 inch (5.70 × 4.28)	7.13	2.2 × 2.2

Note. MP, megapixel.

$NA' = 0.057$. This will provide a resolution in the plane of the matrix of about 6 μm.

We assume that for confident image reception, at least 2 pixels and the distance between them are required; i.e., the line of the sensitive layer will be approximately as follows:

- For the 18-MP camera, $1.25 + 1.25 + 1.25 = 3.75$ μm, which is comparable to the resolution in the image after the 10×/0.40 objective and 0.35× projection (it is about 3 μm).
- For a 20-MP camera, $2.4 + 2.4 + 2.4 = 7.2$ μm, which is comparable to the resolution in the image after the 10×/0.40 objective and 0.715× projection (it is about 6 μm).

These considerations show that in this case, the use of a 20-MP camera is preferable since it has a larger field in the sensor plane, despite the fact that the pixel size of the 18-MP camera is smaller. Strictly speaking, a 20-MP camera should have less electronic noise, more developed pixel-manufacturing technologies, and most likely a simpler optical design of the optical projective.

The effect of digital zoom when transferring an image to a monitor screen (where we can see the real image) will be the same for both cameras.

The microscope kit has several objectives. If we consider the parameters of the objective of the middle magnification, for example, the 40×/0.85 objective, then the following is obtained: the output aperture of the 40×/0.85 objective in the image space will be approximately $NA' = 0.02$ (i.e., twice smaller than for the 10×/0.40 objective), and the resolution in the space of images will be approximately 16.5 μm. In this case, there is no practical sense in using an 18-MP camera, since even a 20-MP camera has a significant margin in resolution (the resolution in the image after the 40×/0.85 objective and the 0.715× projective is about 11 μm). When using, for example, a 100×/1.40 oil immersion objective, the 20-MP camera has an even more significant reserve of resolution. To make the fullest possible use of a camera parameter such as the minimum size of a single pixel, we can ignore the condition for the correspondence of linear fields in the visual channel (when observed through the eyepiece) and in the visualization system. We can use the 18-MP camera, which has a minimum pixel size when examined with a microscope, for example, with a 40×/0.85 objective and 0.2× projection. This will increase the value of the output aperture after the projection and, therefore, will use the full resolution of the camera.

1.8.4.2 Increasing information content

Today, developers of optical elements of microscopes are focusing their efforts on solving optical problems that increase the information content of systems. We assume that increasing the linear field parameter will become a main goal in the design of modern light microscopes, as the ability to view a large linear field with high information content can increase productivity and optimize the microscopic process. A person's ability to adequately analyze an image observed by their eyes through the eyepieces of a microscope are somewhat limited. Studies have shown that for comfortable and long-term work when observing through eyepieces, a researcher needs to use a linear field on a microscope, which does not exceed a value of 18–20 mm. If the linear field significantly exceeds these values, fatigue quickly occurs; the person loses the ability to adequately assess the picture. Therefore, the most advanced models of light microscopes, although calculated for a linear field of 25–26.5 mm, can ensure the effective work of the researcher only for a fairly short time. Therefore, such microscopes (when observed through the eyepiece) are used to solve complex research problems. In addition, as a rule, such microscopes are equipped with a modern imaging system that allows the image to be projected onto a digital receiver.

Thus, when projecting images of an object under study onto a digital receiver, it is the visualization system that can significantly increase the linear field and, therefore, the amount of information about the object. Moreover, unlike a person, such systems do not get tired when covering large linear fields, adequately display information, and are not influenced by moods or emotional states; they lack phenomena inherent in the human eye such as accommodation, convergence, and others that can alter the structure and quality of the observed image. It must be said that digital receivers have spurred this progress in microscopy, since their development is ahead of that of wide-field microscope optics; somehow, imperceptibly, there was a qualitative change in the development of digital image receivers. On the one hand, modern receivers can significantly increase resolution by reducing the linear dimensions of a single pixel and eliminate the influence of electronic noise. But most importantly is the emergence of full-sized matrices for digital cameras, which can now receive 24×36-mm images. New wide-field lenses are required to work with such receivers. Obviously, the next goal in the field of digital receivers will be to achieve the image size of the next photographic format of 60×90 mm with a diagonal of approximately 110 mm. The creation of optics for digital receivers of this size may become an actual trend for optical designers, including microscope developers.

1.8.4.3 What can software not fix?

Image processing using specialized software is also a modern trend in creating visualization systems for microscopes, including budget ones. It is believed that such software can significantly improve the quality of the photograph by adjusting colors, saturation, contrast, and other image parameters. Some software developers even claim that their products can correct certain aberrations, including, for example, the often-mentioned LC. Digital image processing and a high-quality software program can be very good additions to a digital camera as well as the optical system of a microscope. But the assertion that such software can correct some imperfections and errors in the optical system of a microscope is most likely inaccurate. For example, in systems for automated scanning of the surface of a sample, a software program is used that allows several images to be stitched together. However, this stitching process cannot replace high-quality optics, which could cover the entire field (scan) as a whole. The fact is that in the vast majority of modern lenses, there are differences in image quality in the center of the linear field and on its periphery. These can be imperceptible during visual observation, when a person's visual apparatus, including visual functions such as accommodation, convergence, and others, is working with their ability to analyze intellectually. But it is very noticeable with digital projection. When several images are digitally stitched, they are formed as a whole. In this case, a mosaic picture is always made up of fragments in which the image quality in the central zones differs from that on the periphery of the fragment. Moreover, the deterioration of image quality at the periphery of fragments can be caused not only by the curvature in the image (this, by the way, can be fixed using the *multifocus* function), but by aberrations such as astigmatism, distortion, secondary chromatism, and others. It is impossible to eliminate such borders of image fragments in a single image using a software program.

1.8.4.4 Examples of visualization system designs

In accordance with the considerations described in Volume 1 (Chapter 2, Section 2.6), we will use the intermediate image plane on the microscope as the object plane for the adapter. In this case, the value of the input aperture for the adapter will correspond to the value of the output numerical aperture of the *highest exit aperture* objective from the microscope kit. To calculate the adapter for projecting the image onto the sensor of a digital receiver, we first use information about the positions of the exit pupils after the projection system. If this is only an objective, we need to take into account the position of its exit pupil in its optical system. If these calculations are carried out for the tube system, we assume that the prisms (of head) are turned off from the path of the rays when the image is projected onto a digital receiver. Figures 1.242 and 1.243 show the optical designs (from the unified head) of the finite (160 mm) and infinite tube systems (focal length, $F'_{ts} = 200$ mm, for which the position of the exit pupils is the most unfavorable).

Aberration correction graphs show that the image quality is almost the same in both cases. Therefore, we will consider the possibility of designing one unified adapter for all variants of the tube systems considered. For the primary optical design, we take the value of the position of the entrance pupil for the tube system equal to $S_p = -250$ mm. We will consider a virtual objective with a linear magnification of 4× and an input numerical aperture of 0.13, which is sometimes found in sets of budget microscopes. The value of its output numerical aperture is $NA' = 0.0325$; we will use this value for optical design. If we keep in mind that in budget microscopy, cameras with a sensor diagonal of approximately 7 mm (for example, for cameras listed in Table 1.6) are mainly used, it becomes relevant to use an adapter with its own linear magnification of 0.35×. Moreover, we assume that the linear field in the intermediate plane of the image of the budget microscope is 20 mm; the output numerical aperture of the adapter should be $NA' = 0.114$ (this value corresponds to a resolution of about 3 μm on the receiver sensor). We assume that the aberration correction of the adapter should be such that the values of transverse spherical aberration for all wavelengths and within the entire linear field of the receiver sensor do not exceed this value, i.e., 3 μm. This approach is recommended as a criterion for determining the level of aberration correction of systems projecting an image after the intermediate plane of the microscope onto the sensor of a digital receiver. Figure 1.244 shows an example of an optical design (design parameters and graphs of aberration correction) of a unified adapter with a linear magnification of 0.35×.

Lens: Finity tubus lens M=1.0x 2y'=18

Object num aper 0.030000 Object height 9.000000 Primary wavln 0.546070

SRF	RADIUS	THICKNESS	APERTURE RADIUS	GLASS	SPECIAL
OBJ	0.000000	-160.000000	9.000000	AIR	
AST	0.000000	60.000000	4.802161 AS	AIR	
2	-19.170000	1.500000	6.360000	LZ_LK3	
3	11.345000	2.800000	7.020000	LZ_TF5	
4	17.450000	7.500000	6.990000	AIR	
5	61.110000	10.000000	9.330000	LZ_TK12	
6	-12.370000	2.600000	9.970000	LZ_TF5 P	
7	-20.520000	147.338797 S	11.090000	AIR	
8	0.000000	0.000000	9.000000	AIR	
IMS	0.000000	0.000000	9.000000		

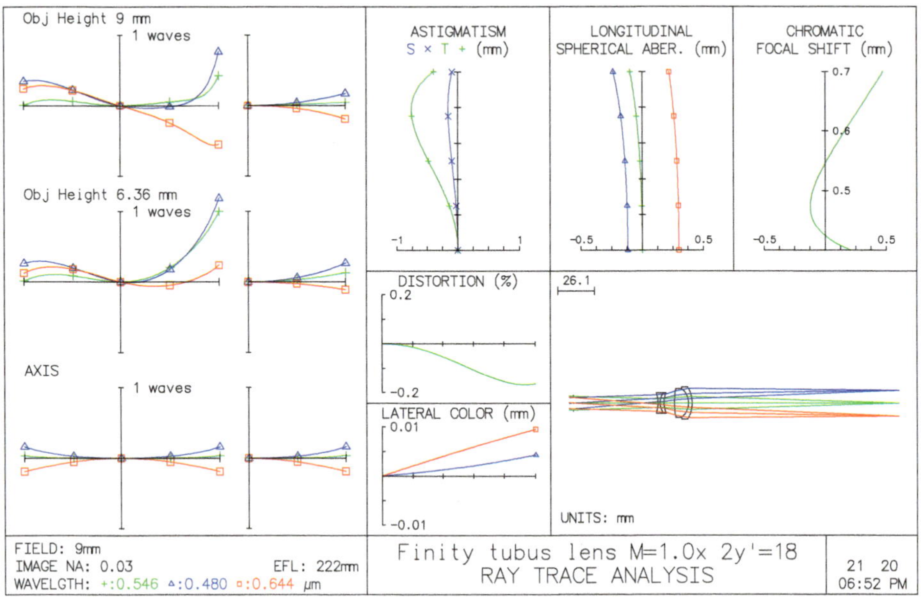

Figure 1.242 Parameters of design and graphs of aberration correction of a finite tube system.

In the first optical design, the position of the entrance pupil from the first lens of the adapter was $S_p = -250$ mm. Consider the graphs of aberration correction of the simplest microscope, with a so-called *straight tube*, which consists of a 20×/0.65 LCF plan achromatic objective (shown in Fig. 1.41) and a unified adapter with its own linear magnification of 0.35×. This is shown in Fig. 1.245.

This is an unfavorable case of the construction of a microscope that projects an image onto the sensor of a digital receiver: the positions of the exit pupils after the previous lenses are too close to the adapter optical system (*nearest pupil position*). However, as this example shows, the image quality has an acceptable level of aberration correction. Another unfavorable case of the

Lens: TS F200 NA=0.03 2y'=22				Efl 200.00	
Image num aper 0.030000	Image height 11.000000		Primary wavln 0.546070		
SRF	**RADIUS**	**THICKNESS**	**APERTURE RADIUS**	**GLASS**	**SPECIAL**
OBJ	0.000000	1.0000e+20	5.5000e+18	AIR	
AST	0.000000	0.001000	10.000000 A	BK7 C	
2	0.000000	60.000000	10.000000	AIR	
3	42.120000	4.200000	9.360000	LZ_LK3	
4	-23.530000	2.000000	9.310000	LZ_KF4	
5	-259.640000	14.000000	9.210000	AIR	
6	77.600000	2.000000	8.310000	LZ_LK3 P	
7	-331.400000	2.000000	8.200000	LZ_KF4 P	
8	31.670000	147.337955 S	7.970000	AIR	
IMS	0.000000	0.000000	11.000000 S		

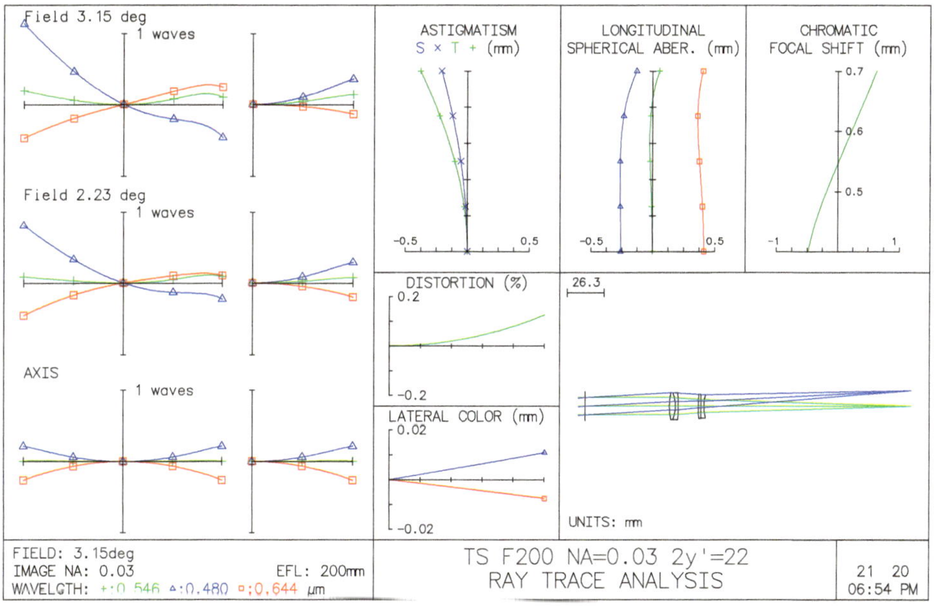

Figure 1.243 Parameters of design and graphs of aberration correction of an infinite tube system ($F'_{ts} = 200$ mm).

layout of the microscope for projecting an image onto the sensor of a digital receiver is a more complex microscope using a finite tube system (for using a visual head); this is a case of the *distant pupil position*. In this case (in accordance with the data shown in Table 1.12), the position of the exit pupils is too far from the adapter. Figure 1.246 shows the design parameters and graphs of aberration correction of a finite tube system and the unified adapter, which has a linear magnification of 0.35×.

Figure 1.247 shows the design parameters and graphs of aberration correction of a 20×/0.65 LCF plan achromatic objective in conjunction with a finite tube system and a unified adapter with a linear magnification of 0.35×.

Lens: Uni adapter 0.35x 2y'=7 mm

Image num aper 0.114000 Image height 3.520000 Primary wavln 0.546070

SRF	RADIUS	THICKNESS		APERTURE RADIUS		GLASS		SPECIAL
OBJ	0.000000	1.0000e-20		10.065697		AIR		
AST	0.000000	47.660000		10.065697	AS	AIR		
2	46.220000	8.500000		12.050000		H_LAK3		
3	-100.500000	0.500000		11.480000		AIR		
4	22.490000	8.000000		10.520000		LZ_CTK19		
5	89.500000	4.000000		8.400000		AIR		
6	-536.600000	4.000000		6.190000		LZ_CTK19	P	
7	22.630000	4.000000		4.780000		LZ_TF5		
8	5.976000	7.000000		3.370000		AIR		
9	7.292000	4.300000		2.990000		H_FK71		
10	227.200000	2.000000		2.360000		LZ_TF5	P	
11	6.823000	5.200000		2.070000		AIR		
12	63.500000	5.200000		3.780000		H_FK71	P	
13	-9.855000	0.800000		4.810000		AIR		
14	28.630000	6.200000		5.170000		LZ_TK12		
15	-28.630000	24.991946	S	5.290000		AIR		
IMS	0.000000	0.000000		3.520000	S			

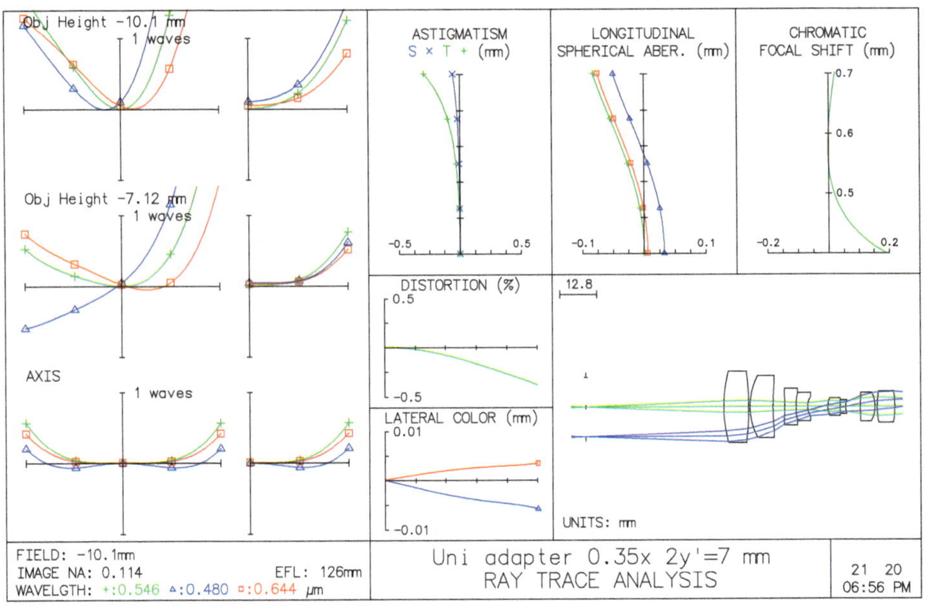

Figure 1.244 Parameters of design and graphs of aberration correction of a unified 0.35× adapter.

As this example shows, the image quality has an acceptable level of aberration correction. Figure 1.248 shows the design parameters and graphs of aberration correction of an infinite tube system ($F'_{ts} = 200$ mm) and a unified adapter with a linear magnification of 0.35×.

To confirm the effectiveness of the unified adapter, consider the microscope layout that includes a 10×/0.28 infinity LCF plan achromatic objective (shown in Fig. 1.35), infinite tube system ($F'_{ts} = 200$ mm) and a unified 0.35× adapter. This is shown in Fig. 1.249.

Adapters (as well as eyepieces) can perform some functions of compensation systems. The aberration correction of an adapter can be corrective for the selected objective according to the type of aberration, such as spherical aberration, astigmatism, LC, and other chromatic aberrations. However, the most interesting is the use of an adapter to compensate for aberration of

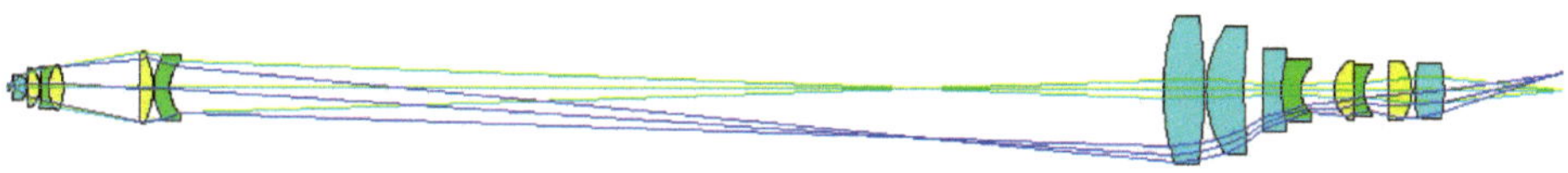

Lens: Plan 20x0.65tub160+adpter 0.35x

Object num aper 0.650000 Object height 0.500000 Primary wavln 0.546070

SRF	RADIUS		THICKNESS		APERTURE RADIUS		GLASS		SPECIAL
OBJ	0.000000		0.010000		0.500000		AIR		
AST	0.000000		0.170000		0.508553	AS	K9		
2	0.000000		1.004100		0.590000		AIR		
3	-2.260000		2.680000		1.170000		H-LAK3		
4	-3.746000		0.300000		2.610000		AIR		
5	-32.880000		2.500000		3.320000		H-FK61		
6	-4.847000		0.140000		3.710000		AIR		
7	-39.800000		1.600000		3.880000		ZF2		
8	7.060000		3.200000		4.200000		H-FK61	P	
9	-10.233000		16.000000		4.380000		AIR		
10	66.620000		3.000000		7.180000		H-FK61	P	
11	-15.232000		0.200000		7.250000		AIR		
12	11.495000		3.400000		6.720000		ZF6		
13	7.900000		208.390000		5.470000		AIR		
14	46.220000		8.500000		14.900000		H_LAK3		
15	-100.500000		0.500000		14.430000		AIR		
16	22.490000		8.000000		12.840000		LZ_CTK19		
17	89.500000		4.000000		10.780000		AIR		
18	-536.600000		4.000000		8.240000		LZ_CTK19	P	
19	22.630000		4.000000		6.410000		LZ_TF5		
20	5.976000		7.000000		4.530000		AIR		
21	7.292000		4.300000		5.800000		H_FK71		
22	227.200000		2.000000		5.390000		LZ_TF5	P	
23	6.823000		5.200000		4.540000		AIR		
24	63.500000		5.200000		5.580000		H_FK71	P	
25	-9.855000		0.800000		6.040000		AIR		
26	28.630000		6.200000		5.630000		LZ_TK12		
27	-28.630000		25.001418		4.860000		AIR		
28	0.000000		0.000000	S	3.494390	S	AIR		
IMS	0.000000		0.000000		3.494390	S			

Figure 1.245 Parameters of design and graphs of aberration correction of a 20×/0.65 LCF plan achromatic objective and a unified 0.35× adapter.

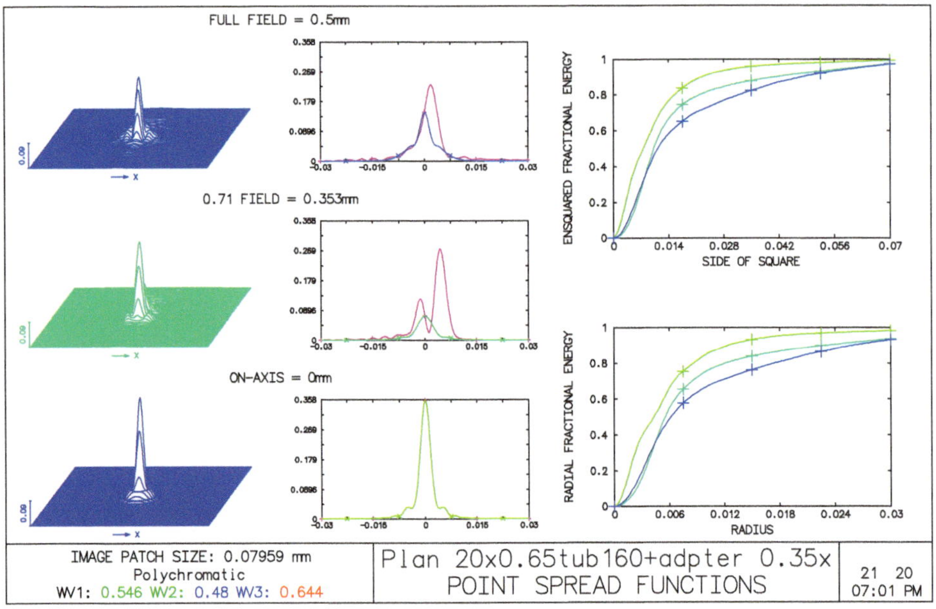

Figure 1.245 *(Continued)* Parameters of design and graphs of aberration correction of a 20×/0.65 LCF plan achromatic objective and a unified 0.35× adapter.

distortion. This aberration is not specifically corrected in most objectives, even ones of a high class, unless they are special objectives for measuring. Correction of distortion in the optical systems of budget objectives may not be justified due to the complexity of their optical systems. If accurate measurements are required on a microscope using such objectives, a compensation adapter in combination with this objective can provide a complete correction of distortion in such an optical system. As an example, let us take the same 20×/0.65 objective whose own distortion value is approximately 1.7%. Figure 1.250 shows the design parameters and graphs of aberration correction of the 20×/0.65 lens and a special adapter (its design is made for the diagonal size of the matrix equal to 1 inch) under the conditions of the microscope layout according to the straight tube principle.

Some possible developments for making accurate measurements using a light microscope may include building a microscope based on the principle of telecentricity for all the microscope's components. Such a construction involves the use of an infinite optical length of the tube in a microscope. The studied object is located strictly in the front focal plane of the objective, and the tube system is positioned so that its equivalent front focal plane coincides with (or is located close to) the equivalent rear focal plane of the objective. The image of the investigated object is built in the rear focal plane of the tube system; the front focal plane of the eyepiece (in case of visual observation) or the adapter for projection onto a digital receiver coincide with the rear focal

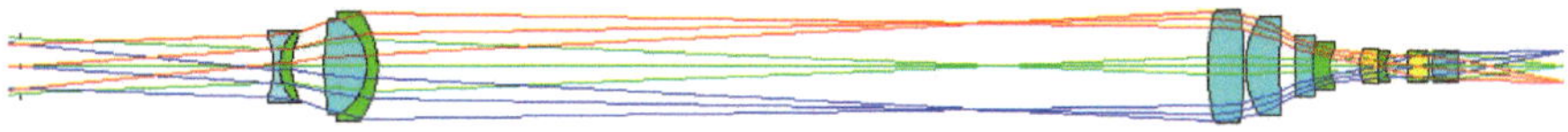

Lens: Finity TS+ Uni adapter 0.35x

Image num aper 0.114000 Image height 3.520000 Primary wavln 0.546070

SRF	RADIUS		THICKNESS		APERTURE RADIUS		GLASS		SPECIAL
OBJ	0.000000		-160.000000		10.077169		AIR		
AST	0.000000		60.000000		6.376370	AS	AIR		
2	-19.170000		1.500000		7.310000		LZ_LK3		
3	11.345000		2.800000		8.310000		LZ_TF5		
4	17.450000		7.500000		8.250000		AIR		
5	61.110000		10.000000		10.700000		LZ_TK12		
6	-12.370000		2.600000		11.060000		LZ_TF5	P	
7	-20.520000		194.930000		12.600000		AIR		
8	46.220000		8.500000		13.070000		H_LAK3		
9	-100.500000		0.500000		12.540000		AIR		
10	22.490000		8.000000		11.390000		LZ_CTK19		
11	89.500000		4.000000		9.270000		AIR		
12	-536.600000		4.000000		6.960000		LZ_CTK19	P	
13	22.630000		4.000000		5.420000		LZ_TF5		
14	5.976000		7.000000		3.860000		AIR		
15	7.292000		4.300000		3.950000		H_FK71		
16	227.200000		2.000000		3.370000		LZ_TF5	P	
17	6.823000		5.200000		2.960000		AIR		
18	63.500000		5.200000		3.400000		H_FK71	P	
19	-9.855000		0.800000		3.660000		AIR		
20	28.630000		6.200000		3.520000		LZ_TK12		
21	-28.630000		25.000361		3.600000		AIR		
22	0.000000		-1.5570e-14	S	3.520000	S	AIR		
IMS	0.000000		0.000000		3.520000	S			

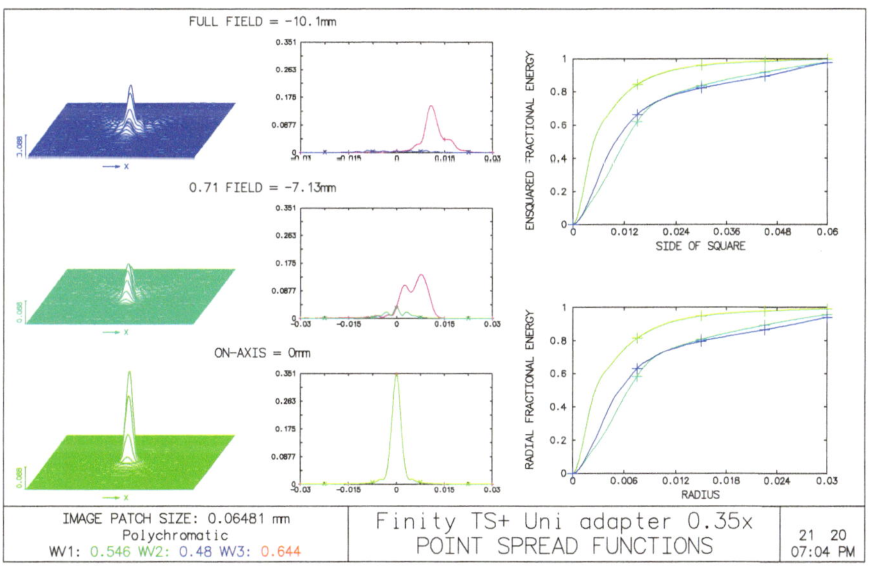

Figure 1.246 Parameters of design and graphs of aberration correction of a finite tube system and a unified 0.35× adapter.

plane of the tube system. We will call such a microscope a *telecentric microscope*; its parameters and advantages over conventional microscopes correspond to those of the well-known *telecentric objectives* over conventional objectives. One of the consequences of using telecentric systems is the

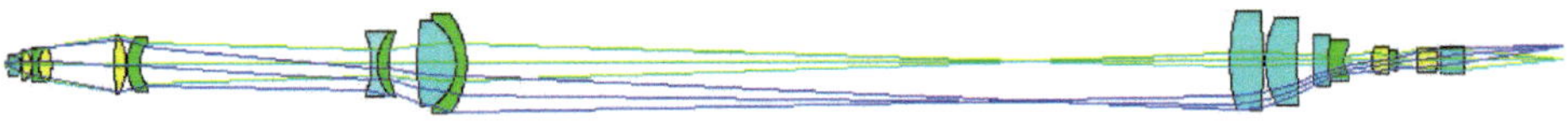

Lens:Plan 20x0.65tub160+TL+ Uniadpter

Object num aper 0.650000 Object height 0.500000 Primary wavln 0.546070

SRF	RADIUS	THICKNESS		APERTURE RADIUS		GLASS		SPECIAL
OBJ	0.000000	0.010000		0.500000		AIR		
AST	0.000000	0.170000		0.508553	AS	K9		
2	0.000000	1.004100		0.590000		AIR		
3	-2.260000	2.680000		1.170000		H-LAK3		
4	-3.746000	0.300000		2.610000		AIR		
5	-32.880000	2.500000		3.320000		H-FK61		
6	-4.847000	0.140000		3.710000		AIR		
7	-39.800000	1.600000		3.880000		ZF2		
8	7.060000	3.200000		4.200000		H-FK61	P	
9	-10.233000	16.000000		4.380000		AIR		
10	66.620000	3.000000		7.180000		H-FK61	P	
11	-15.232000	0.200000		7.250000		AIR		
12	11.495000	3.400000		6.720000		ZF6		
13	7.900000	60.802600		5.470000		AIR		
14	-19.170000	1.500000		7.090000		LZ_LK3		
15	11.345000	2.800000		7.990000		LZ_TF5		
16	17.450000	7.500000		7.940000		AIR		
17	61.110000	10.000000		10.410000		LZ_TK12		
18	-12.370000	2.600000		10.840000		LZ_TF5	P	
19	-20.520000	194.930000		12.280000		AIR		
20	46.220000	8.500000		12.500000		H_LAK3		
21	-100.500000	0.500000		11.930000		AIR		
22	22.490000	8.000000		10.870000		LZ_CTK19		
23	89.500000	4.000000		8.730000		AIR		
24	-536.600000	4.000000		6.470000		LZ_CTK19	P	
25	22.630000	4.000000		5.000000		LZ_TF5		
26	5.976000	7.000000		3.530000		AIR		
27	7.292000	4.300000		3.240000		H_FK71		
28	227.200000	2.000000		2.600000		LZ_TF5	P	
29	6.823000	5.200000		2.250000		AIR		
30	63.500000	5.200000		2.460000		H_FK71	P	
31	-9.855000	0.800000		3.000000		AIR		
32	28.630000	6.200000		3.180000		LZ_TK12		
33	-28.630000	25.000361		3.420000		AIR		
34	0.000000	0.000591	S	3.489630	S	AIR		
IMS	0.000000	0.000000		3.489630	S			

Figure 1.247 Parameters of design and graphs of aberration correction of a 20×/0.65 LCF plan achromatic objective, a finite tube system, and a unified 0.35× adapter.

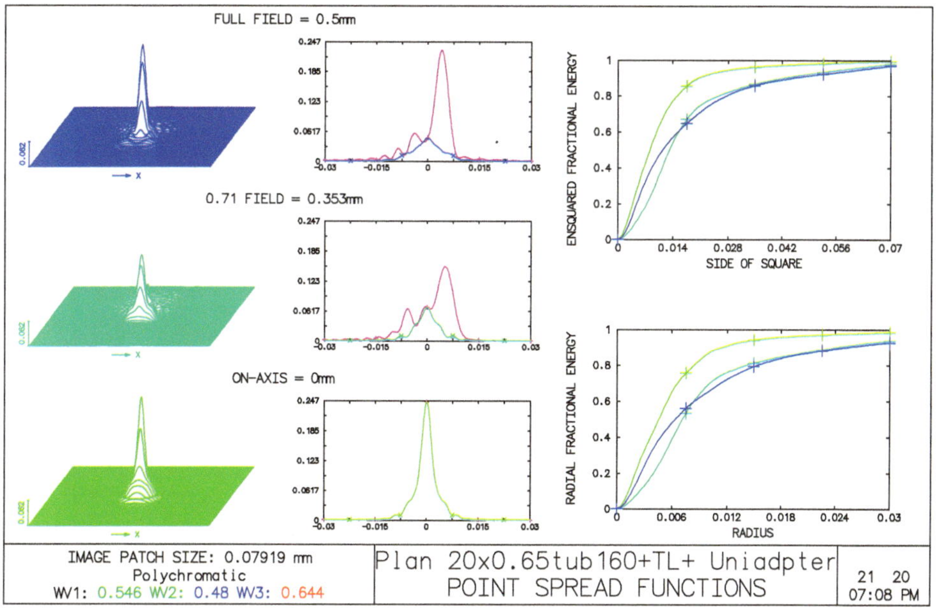

Figure 1.247 (*Continued*) Parameters of design and graphs of aberration correction of a 20×/0.65 LCF plan achromatic objective, a finite tube system, and a unified 0.35× adapter.

complete correction of distortion aberration; in this case, the distortion can be corrected in some cases in each of the components used or can be compensated *inside the system*. Figure 1.251 shows the design parameters and graphs of aberration correction of an adapter with a linear magnification of 0.73× in which the distortion aberration is completely corrected. Figure 1.252 shows an example of the optical design of a telecentric microscope consisting of an infinity 10×/0.28 LCF plan achromatic objective, a tube system ($F'_{ts} = 200$ mm), and a 0.73× adapter.

Let us also choose the value of the adapter's own linear increase equal to $M = 0.6\times$. Thus, the value of the input numerical aperture for the adapter will be $NA = 0.04$, and the value of its output aperture will be $NA' = 0.067$ (it determines the value of the resolution of the adapter in the sensor plane, equal to about 5 μm). For example, we will use a sensor measuring 1 inch (approximately 16 mm diagonally) as an image receiver. Figure 1.253 shows the design parameters and graphs of aberration correction of an adapter having a linear magnification of 0.6× and (with some margin) an output numerical aperture equal to $NA' = 0.07$ for projection onto a 1-inch sensor. The aberration correction of the adapter provides some compensation value for distortion because most microscope objectives (if they are not measuring) have a residual positive distortion value.

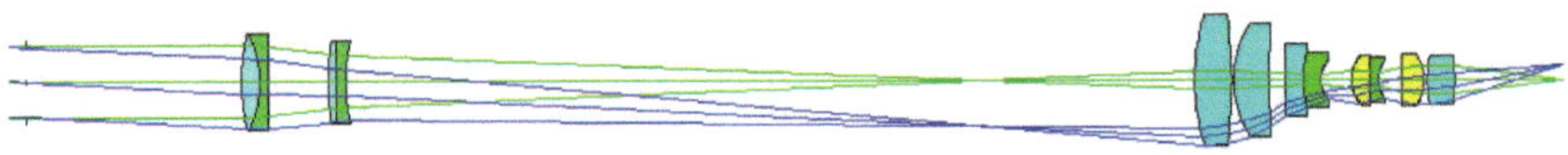

Lens: TS 200+Uni adapter 0.35x 2y'=7			Efl -69.95	
Image num aper 0.114000	Image height 3.520000		Primary wavln 0.546070	

SRF	RADIUS	THICKNESS	APERTURE RADIUS	GLASS	SPECIAL
OBJ	0.000000	1.0000e+08	5.0319e+06	AIR	
AST	0.000000	50.000000	7.974763 AS	AIR	
2	42.120000	4.200000	10.560000	LZ_LK3	
3	-23.530000	2.000000	10.540000	LZ_KF4	
4	-259.640000	14.000000	10.370000	AIR	
5	77.600000	2.000000	9.180000	LZ_LK3 P	
6	-331.400000	2.000000	9.050000	LZ_KF4 P	
7	31.670000	194.930000	8.750000	AIR	
8	46.220000	8.500000	14.530000	H_LAK3	
9	-100.500000	0.500000	14.050000	AIR	
10	22.490000	8.000000	12.550000	LZ_CTK19	
11	89.500000	4.000000	10.480000	AIR	
12	-536.600000	4.000000	8.010000	LZ_CTK19 P	
13	22.630000	4.000000	6.250000	LZ_TF5	
14	5.976000	7.000000	4.440000	AIR	
15	7.292000	4.300000	5.540000	H_FK71	
16	227.200000	2.000000	5.090000	LZ_TF5 P	
17	6.823000	5.200000	4.350000	AIR	
18	63.500000	5.200000	5.410000	H_FK71 P	
19	-9.855000	0.800000	5.870000	AIR	
20	28.630000	6.200000	5.520000	LZ_TK12	
21	-28.630000	25.000307	4.790000	AIR	
22	0.000000	5.4537e-14 S	3.520000 S	AIR	
IMS	0.000000	0.000000	3.520000 S		

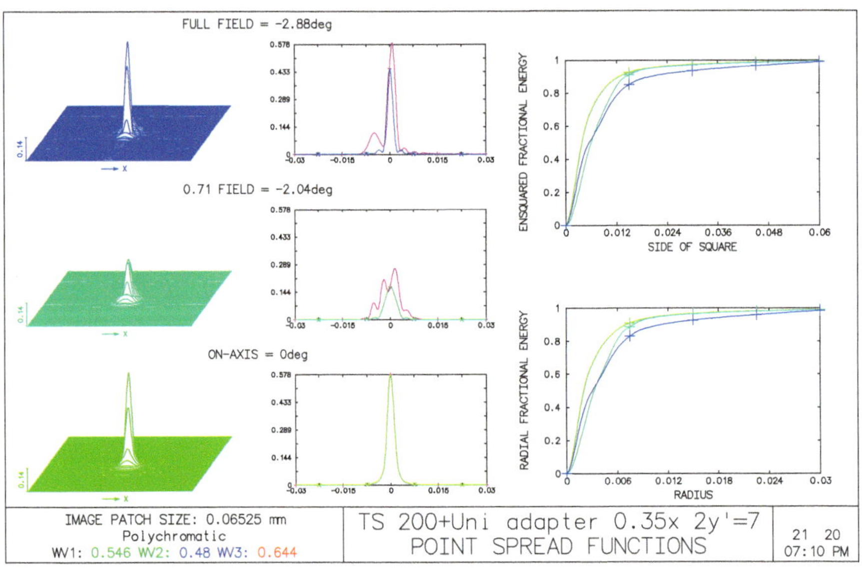

Figure 1.248 Parameters of design and graphs of aberration correction of an infinite tube system ($F'_{ts} = 200$ mm) with a unified 0.35× adapter.

In this calculation, it is assumed that the linear field on the microscope (after the lens) is 28 mm (which is *taken with a margin* since it is not found in budget microscopes); however, this adapter can also be used when projecting images onto other sensors with smaller sensor sizes. So, for example, if we use a matrix with a sensor size of 2/3 inch with a diagonal of about 11 mm, the size of the projected linear field of the microscope will be about 18 mm. This value is typical for most mass-produced budget microscopes.

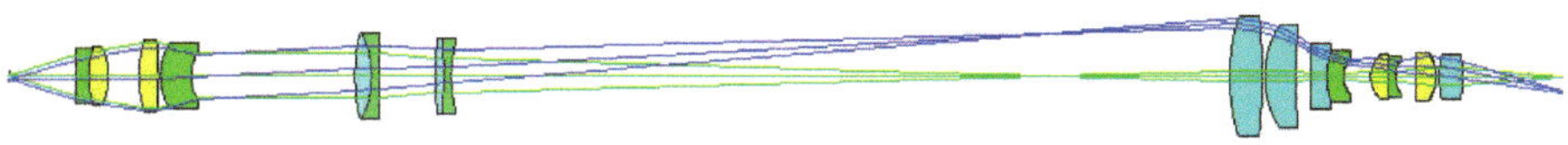

Lens:LD plan 10x0.28+TS+adapter 0.35x

Object num aper 0.280000 Object height -1.000000 Primary wavln 0.546070

SRF	RADIUS	THICKNESS		APERTURE RADIUS		GLASS		SPECIAL
OBJ	0.000000	1.0000e-20		1.000000		AIR		
AST	0.000000	16.825500		1.000000	AS	AIR		
2	0.000000	3.000000		5.910000		H-ZF6		
3	21.760000	5.000000		6.560000		H-FK61		
4	-15.672000	7.500000		7.130000		AIR		
5	23.710000	5.000000		8.640000		H-FK61	P	
6	-66.000000	0.500000		8.530000		AIR		
7	16.200000	7.500000		8.110000		H-ZF6	P	
8	10.233000	41.175000		5.850000		AIR		
9	42.120000	4.200000		10.560000		LZ_LK3		
10	-23.530000	2.000000		10.540000		LZ_KF4		
11	-259.640000	14.000000		10.370000		AIR		
12	77.600000	2.000000		9.180000		LZ_LK3	P	
13	-331.400000	2.000000		9.050000		LZ_KF4	P	
14	31.670000	194.930000		8.750000		AIR		
15	46.220000	8.500000		14.530000		H_LAK3		
16	-100.500000	0.500000		14.050000		AIR		
17	22.490000	8.000000		12.550000		LZ_CTK19		
18	89.500000	4.000000		10.480000		AIR		
19	-536.600000	4.000000		8.010000		LZ_CTK19	P	
20	22.630000	4.000000		6.250000		LZ_TF5		
21	5.976000	7.000000		4.440000		AIR		
22	7.292000	4.300000		5.540000		H_FK71		
23	227.200000	2.000000		5.090000		LZ_TF5	P	
24	6.823000	5.200000		4.350000		AIR		
25	63.500000	5.200000		5.410000		H_FK71	P	
26	-9.855000	0.800000		5.870000		AIR		
27	28.630000	6.200000		5.520000		LZ_TK12		
28	-28.630000	24.999473		4.790000		AIR		
29	0.000000	3.7604e-14	S	3.528156	S	AIR		
IMS	0.000000	0.000000		3.528156	S			

Figure 1.249 Parameters of design and graphs of aberration correction of a 10×/0.28 infinity LCF plan achromatic objective with an infinity tube system ($F'_{ts} = 200$ mm) and a unified 0.35× adapter.

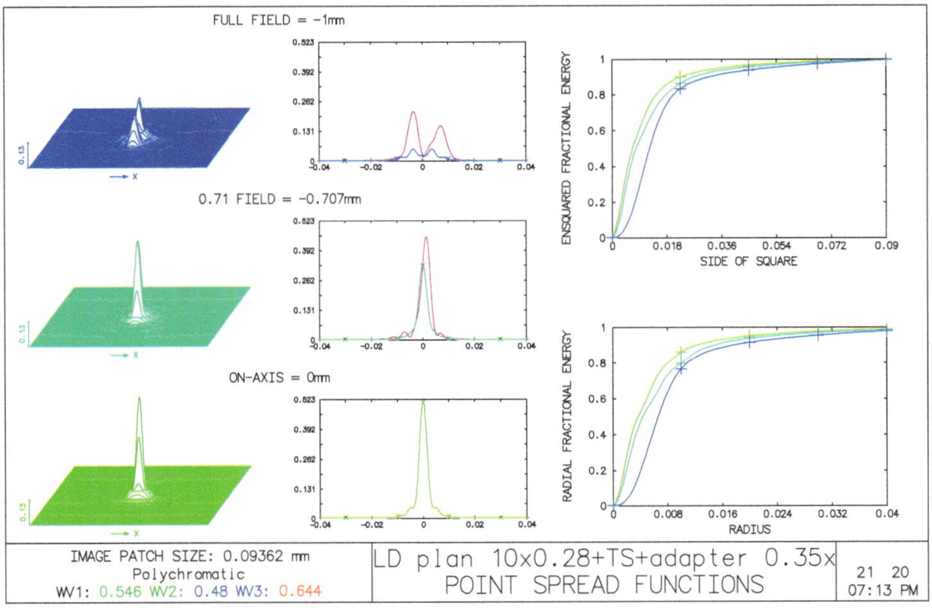

Figure 1.249 (*Continued*) Parameters of design and graphs of aberration correction of a 10×/0.28 infinity LCF plan achromatic objective with an infinity tube system ($F'_{ts} = 200$ mm) and a unified 0.35× adapter.

Without thinking about the optical principles and design of a light microscope, many researchers take photographs of images obtained directly after a regular eyepiece. A person looks through the eyepiece, positions the object of interest, and then, instead of the eye, *attaches* a photographic camera with a standard objective to the eyepiece, and takes a micrograph. Some progress with this image visualization technique has been made with the use of smartphones; it turned out to be convenient and practical. The image quality obtained after the eyepiece accumulates all the flaws and aberrations of the optical system of the microscope. In this case, the technical level of the gadget used can affect the image quality on the sensor of a digital receiver (camera or smartphone). Of course, the accuracy of the positioning of the gadget is of great importance; it is necessary to exclude factors such as skew (tilt of the gadget), hand shake, and others. This technique is often called *ocular imaging*. Figure 1.254 shows examples of ocular imaging (using a smartphone) in which the combination of the projection system of a light microscope and the optics of a smartphone turned out to be successful.

However, these photographs were taken using an objective with a linear magnification of 10× and a numerical aperture of 0.30 in combination with a 10× eyepiece. The objective is an aberration correction LCF plan achromat; secondary chromatism in the eyepiece is also corrected. (It is even slightly more corrected because a blue border is observed around the circumference.

By the way, if a yellow border were observed around the circumference, it would be clear that the eyepiece used is compensatory; by the thickness of the rim you can judge the strength of an eyepiece's LC compensation ability.) An analysis of these photographs shows that the resolution of the smartphone's optics is enough to get fairly high-quality photographs. Indeed, with the value

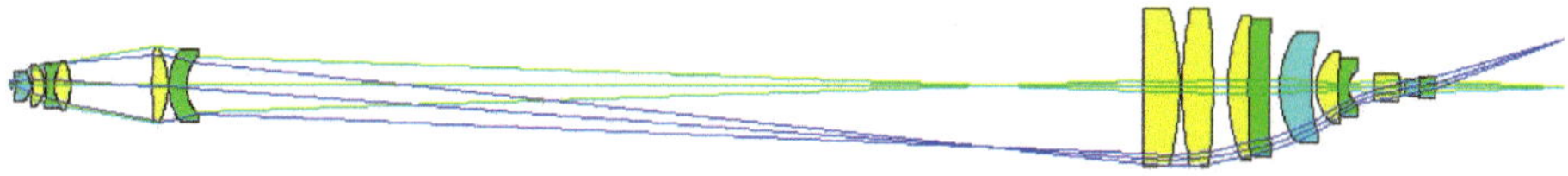

Lens:Plan 20x0.65tub160+adpter 2y'=18

Object num aper 0.650000 Image height 9.000000 Primary wavln 0.546070

SRF	RADIUS	THICKNESS		APERTURE RADIUS		GLASS		SPECIAL
OBJ	0.000000	0.010000		0.577179		AIR		
AST	0.000000	0.170000		0.690000	A	K9		
2	0.000000	1.004100		0.770000		AIR		
3	-2.260000	2.680000		1.280000		H-LAK3		
4	-3.746000	0.300000		2.750000		AIR		
5	-32.880000	2.500000		3.540000		H-FK61		
6	-4.847000	0.140000		3.860000		AIR		
7	-39.800000	1.600000		3.980000		ZF2		
8	7.060000	3.200000		4.240000		H-FK61	P	
9	-10.233000	16.000000		4.390000		AIR		
10	66.620000	3.000000		7.170000		H-FK61	P	
11	-15.232000	0.200000		7.250000		AIR		
12	11.495000	3.400000		6.740000		ZF6		
13	7.900000	185.800000		5.510000		AIR		
14	0.000000	7.000000		14.630000		H_FK71		
15	-57.823200	0.500000		14.980000		AIR		
16	68.675700	6.000000		14.900000		H_FK71	P	
17	-131.166700	3.000000		14.600000		AIR		
18	34.807300	4.000000		13.460000		H_FK71	P	
19	165.534000	4.000000		13.010000		LZ_TF5		
20	216.190700	2.000000		12.160000		AIR		
21	18.132200	6.000000		10.420000		LZ_CTK12		
22	27.483600	1.000000		8.470000		AIR		
23	8.608400	4.300000		6.710000		H_FK71	P	
24	51.059800	2.000000		5.640000		LZ_TF1		
25	3.665400	5.200000		3.220000		AIR		
26	-14.737700	5.200000		2.600000		H_FK71	P	
27	-5.818700	0.800000		2.200000		AIR		
28	34.850900	3.000000		1.760000		H_LAK3		
29	-53.001500	3.000000		1.560000		LZ_TF5		
30	0.000000	25.005133	S	2.010000		AIR		
IMS	0.000000	0.000000		9.000000	S			

Figure 1.250 Parameters of design and graphs of aberration correction of a 20×/0.65 LCF plan achromatic objective and an adapter for a 1-inch receiver.

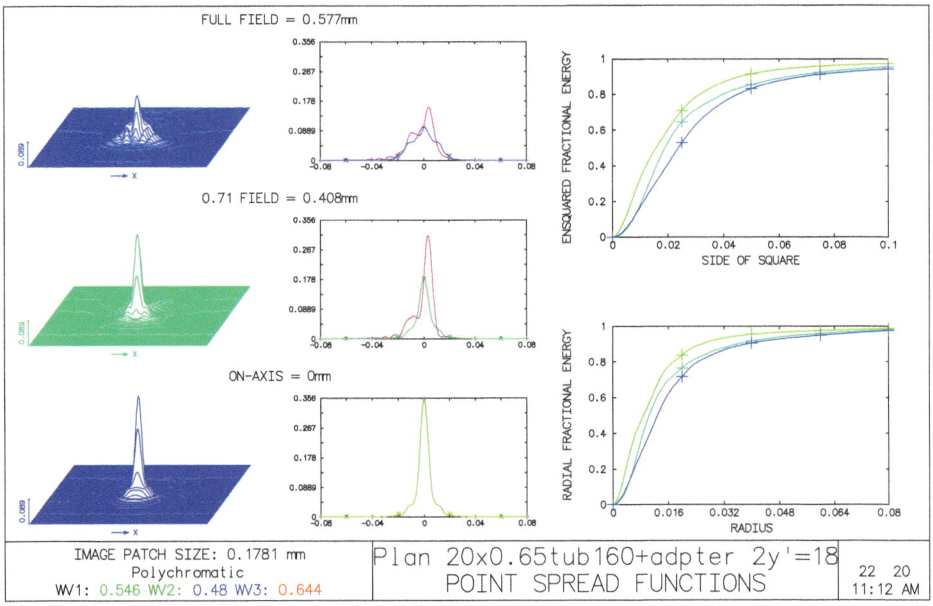

Figure 1.250 (*Continued*) Parameters of design and graphs of aberration correction of a 20×/0.65 LCF plan achromatic objective and an adapter for a 1-inch receiver.

of the output numerical aperture of the projection system (in particular, the microscope objective) equal to $NA' = 0.03$, the resulting "aperture" diameter on the eyepiece is approximately 1.5 mm. If we take the focal length of the smartphone's objective as $F' = 5$ mm, then the output numerical aperture on the sensor of the digital receiver is $NA' = 0.075$, which corresponds to the resolution on the sensor, about 4.5 μm.

For verification, we can use one of the optical objective designs, pinhole, which was developed for covert surveillance cameras; its focal length is 6 mm, and the linear field in the image space (digital receiver sensor) is 3.8 mm. Figure 1.255 shows the design parameters and graphs of aberration correction of this objective.

By combining the 10×/18 eyepiece (shown in Fig. 1.223) and this pinhole objective into one design, we can get a system for projecting the image from the intermediate plane onto the sensor of a digital receiver (we assume that we modeled the technique of ocular imaging using a smartphone). Figure 1.256 shows the design parameters and graphs of aberration correction of such an optical system.

In this variant of the combination, nearly the entire linear field (approximately 16 mm) is used. If we use a longer-focus system as an eyepiece (for example, the 5×/25 eyepiece shown in Fig. 1.221), we can capture a larger linear field in the intermediate plane of the microscope. This variant is shown in Fig. 1.257.

Lens: adapter 0.73 for 1' matrix			Efl 57.63		
Image num aper 0.040000	Image height 8.500000		Primary wavln 0.546070		
SRF	**RADIUS**	**THICKNESS**	**APERTURE RADIUS**	**GLASS**	**SPECIAL**
OBJ	0.000000	1.0000e-20	11.643597	AIR	
AST	0.000000	40.000000	0.630000 A	AIR	
2	0.000000	8.000000	12.810000	H-FK71	
3	-33.407100	0.300000	12.920000	AIR	
4	75.132100	13.000000	12.260000	H-FK71 P	
5	-27.555700	5.000000	10.590000	H-LAK3	
6	336.592600	0.300000	10.110000	AIR	
7	60.473500	7.000000	10.020000	H-FK71 P	
8	78.686800	7.000000	9.240000	H-ZF7LA	
9	121.128400	30.000000	8.450000	AIR	
10	62.025200	3.000000	3.840000	LZ_CTK12	
11	-52.394100	1.000000	3.520000	AIR	
12	7.027200	4.300000	3.070000	H-FK71 P	
13	113.341500	2.000000	1.960000	ZF1	
14	6.972900	5.200000	1.430000	AIR	
15	-4.833800	5.200000	2.260000	H-FK71 P	
16	-8.749600	0.800000	3.930000	AIR	
17	28.703000	6.000000	4.500000	H-LAK3 P	
18	0.000000	25.004585	4.950000	AIR	
19	0.000000	0.000000	8.500891 S	AIR	
IMS	0.000000	0.000000	8.500891 S		

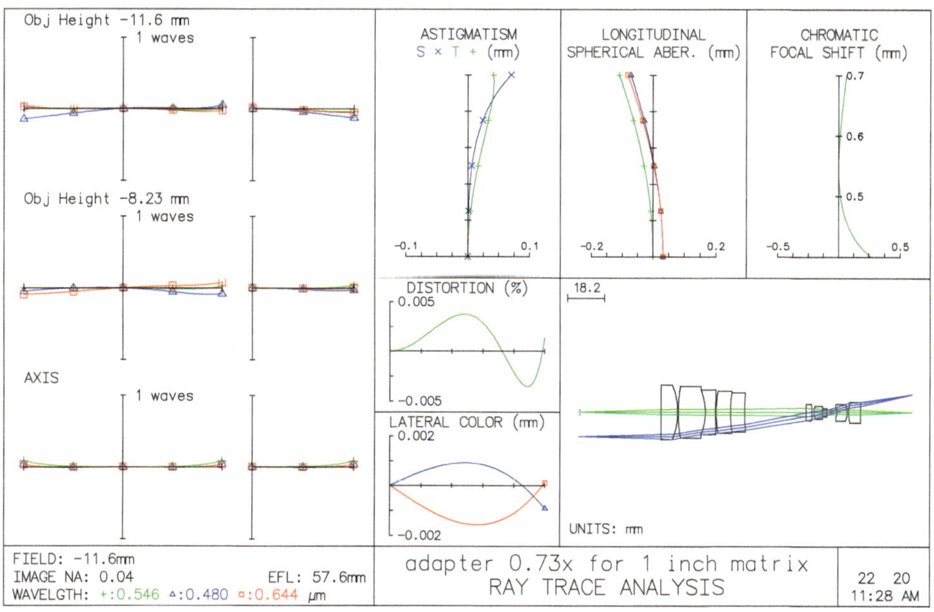

Figure 1.251 Parameters of design and graphs of aberration correction of a 0.73× adapter with LC in which distortion aberration is completely corrected.

However, the big problem with this arrangement is the inability to ensure the stability of the position of the sensor of the digital receiver on the optical axis of the eyepiece. Some designs were developed as technical solutions to this problem; one of these is presented in Fig. 1.258.

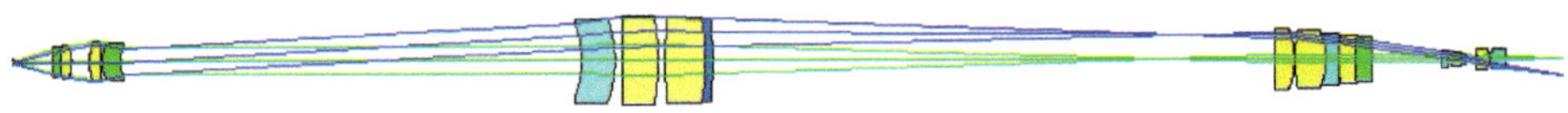

Lens:plan 10x0.28+telecent micriscope

Object num aper 0.280000 Object height -1.000000 Primary wavln 0.546070

SRF	RADIUS	THICKNESS	APERTURE RADIUS	GLASS	SPECIAL
OBJ	0.000000	1.0000e-20	1.000000	AIR	
AST	0.000000	16.825500	1.000000 A	AIR	
2	0.000000	3.000000	5.910000	H-ZF6	
3	21.760000	5.000000	6.560000	H-FK61	
4	-15.672000	7.500000	7.130000	AIR	
5	23.710000	5.000000	8.640000	H-FK61 P	
6	-66.000000	0.500000	8.530000	AIR	
7	16.200000	7.500000	8.110000	H-ZF6 P	
8	10.233000	200.000000	5.850000	AIR	
9	-43.930000	14.000000	15.230000	H-QK3	
10	-51.200000	3.000000	17.330000	AIR	
11	667.000000	15.000000	17.680000	CAF2	
12	-105.900000	3.000000	18.090000	AIR	
13	194.900000	17.000000	17.920000	CAF2 P	
14	-118.000000	3.000000	17.180000	H-ZLAF50B	
15	-754.000000	240.000000	17.150000	AIR	
16	0.000000	8.000000	12.810000	H-FK71	
17	-33.407100	0.300000	12.920000	AIR	
18	75.132100	13.000000	12.260000	H-FK71 P	
19	-27.555700	5.000000	10.590000	H-LAK3	
20	336.592600	0.300000	10.110000	AIR	
21	60.473500	7.000000	10.020000	H-FK71 P	
22	78.686800	7.000000	9.240000	H-ZF7LA	
23	121.128400	30.000000	8.450000	AIR	
24	62.025200	3.000000	3.840000	LZ_CTK12	
25	-52.394100	1.000000	3.520000	AIR	
26	7.027200	4.300000	3.070000	H-FK71 P	
27	113.341500	2.000000	1.960000	ZF1	
28	6.972900	5.200000	1.430000	AIR	
29	-4.833800	5.200000	2.260000	H-FK71 P	
30	-8.749600	0.800000	3.930000	AIR	
31	28.703000	6.000000	4.500000	H-LAK3 P	
32	0.000000	24.966192 S	4.208040	AIR	
33	0.000000	0.000000	7.355021 S	AIR	
IMS	0.000000	0.000000	7.355021 S		

Figure 1.252 Parameters of design and graphs of aberration correction of a 10×/0.28 LCF LD plan achromatic objective, tube lens ($F' = 200$ mm), and 0.73× adapter (telecentric microscope).

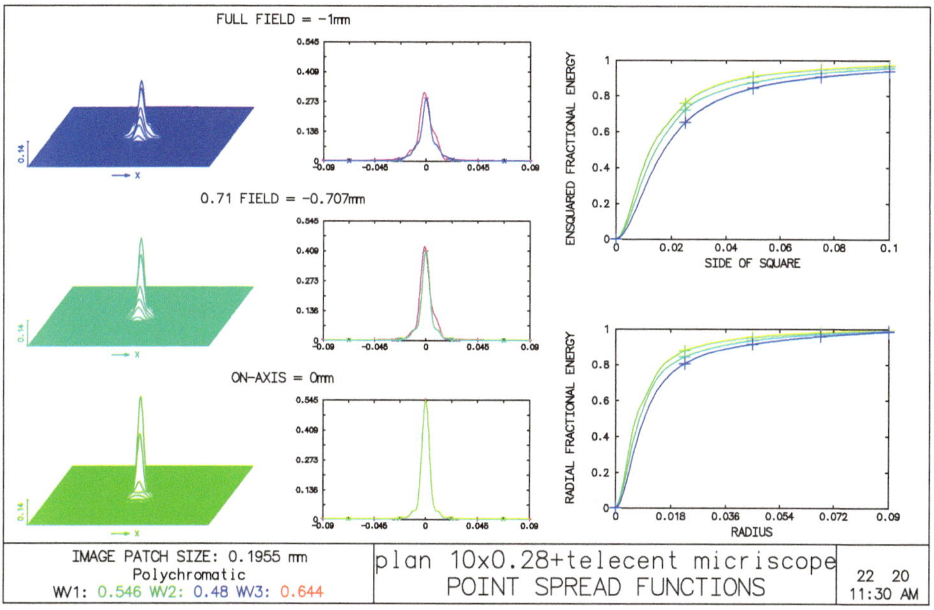

Figure 1.252 (*Continued*) Parameters of design and graphs of aberration correction of a 10×/0.28 LCF LD plan achromatic objective, tube lens ($F' = 200$ mm), and 0.73× adapter (telecentric microscope).

But such a constructive solution (like many others) does not provide the necessary high level of accuracy of positioning the digital receiver sensor on the optical axis. Moreover, when using microscope objectives of a medium or large magnification, as opposed to a small magnification (4–10×), the image quality is not very high. Two factors affecting image quality can be noted. In terms of the mechanical parameter, the first is the difficulty of developing reliable positioning of a digital camera or smartphone on the optical axis of the microscope (eyepiece) and, as a result, the second is the difficulty of accurately focusing the lens of the camera or smartphone on the object under study (through all previous optical systems). In terms of the optical parameter, very often the aberration correction of the optical systems of digital cameras and smartphones is not high enough; there is, for example, aberration such as distortion. Figure 1.259 shows an example of a photograph using a microscope objective with a linear magnification of 40× and a budget smartphone. In the photograph on the left, the image is not sharp; in the photograph on the right, the influence of distortion is obvious.

Nevertheless, given the considerations outlined above, it would be advisable to use the fundamental solution of ocular imaging in order to construct an optical projection system from the intermediate image plane on a microscope onto a digital receiver sensor. This is a convenient and practical

Lens: adapter 0.6 — Efl 40.89

Image num aper 0.070000 — Image height 8.100000 — Primary wavln 0.546070

SRF	RADIUS		THICKNESS		APERTURE RADIUS		GLASS		SPECIAL
OBJ	0.000000		1.0000e-20		13.500361		AIR		
AST	0.000000		25.000000		13.500361	AS	AIR		
2	174.944400		8.500000		14.790000		H-FK71		
3	-30.225300		0.500000		14.800000		AIR		
4	106.601800		6.000000		13.320000		H-FK71	P	
5	-152.475500		4.000000		12.230000		AIR		
6	-30.575800		4.000000		11.300000		H-FK71	P	
7	168.586900		4.000000		10.560000		LZ_TF5		
8	-87.050500		2.000000		10.230000		AIR		
9	22.965700		6.000000		8.970000		LZ_CTK12		
10	67.598800		1.000000		7.400000		AIR		
11	9.429500		4.300000		5.970000		H-FK71	P	
12	-27.672800		2.000000		4.910000		LZ_TF1		
13	4.318900		5.200000		3.170000		AIR		
14	-14.468600		5.200000		2.860000		LZ_LK3		
15	-6.703000		0.800000		3.040000		AIR		
16	24.445900		3.000000		3.210000		H_LAK3		
17	-25.073900		3.000000		3.460000		LZ_TF5	P	
18	0.000000		24.998382		3.770000		AIR		
19	0.000000		6.3385e-15	S	8.100000	S	AIR		
IMS	0.000000		0.000000		8.100000	S			

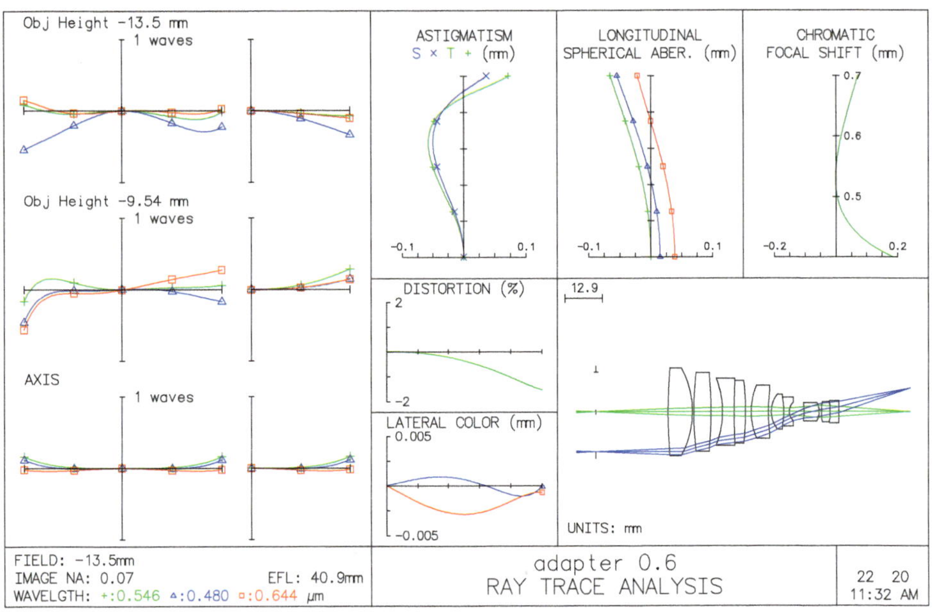

Figure 1.253 Parameters of design and graphs of aberration correction of a 0.6× adapter.

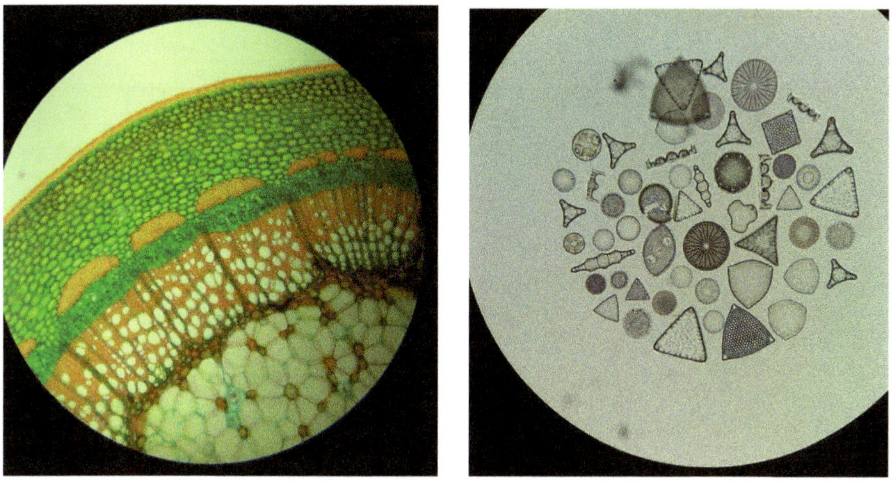

Figure 1.254 Examples of ocular imaging.

option because it allows the use of an eyepiece in combination with a specific set of objectives. The eyepiece may be compensatory or have independent aberration correction. Such a design will allow the use of a measuring grid or scale, pointer, or special marker. Thus, an optical design was developed for a special adapter, which (after the eyepiece) projects the image onto the sensor. We assume that budget digital cameras use the so-called *C-mount* standard; therefore, the distance from the adapter lens to the sensor plane should not be less than 20 mm. The value of the linear increase of such a system for ocular imaging will be determined by the ratio of the focal length of the eyepiece to the focal length of the adapter; the smaller the focal length of the adapter, the shorter the focus of the eyepiece can be. The light microscope mainly uses eyepieces with linear magnifications of 10× (focal length, 25 mm) and 5× (focal length, 50 mm). The use of shorter-focus eyepieces for ocular imaging does not seem appropriate due to the incomplete use of the linear field in the intermediate plane of the microscope image. To obtain a value of 20–25 mm from the last surface of the adapter lens, it was possible to achieve a focal length of the adapter of 15 mm; it works on the principle of a telephoto objective with a remote focal plane. In this embodiment, an acceptable level of aberration correction can be achieved.

Figure 1.260 shows the design parameters and the level of aberration correction of the adapter ($F' = 15$ mm) for ocular imaging. This adapter is designed to compensate for this type of aberration, such as distortion. Figure 1.261 shows the design parameters and graphs of aberration correction of the combination of a 10×/18 eyepiece (shown in Fig. 1.224) and an adapter ($F' = 15$ mm). Figure 1.262 shows the design parameters and graphs of aberration correction of the combination of a 10×/18 compensation eyepiece (shown in Fig. 1.223) and an adapter ($F' = 15$ mm).

Lens: pinhole F=6 F/6 2y'=1/4 inch			Efl 5.97		
Image num aper 0.070000		Image height 1.900000		Primary wavln 0.587560	
SRF	**RADIUS**	**THICKNESS**	**APERTURE RADIUS**	**GLASS**	**SPECIAL**
OBJ	0.000000	1.0000e+20	3.1817e+19	AIR	
AST	0.000000	1.500000	0.418021 AS	AIR	
2	3.404000	1.500000	0.895270	LZ_TF5	
3	3.048000	0.200000	0.997422	AIR	
4	24.890000	1.400000	1.070758	LZ_CTK19	
5	-1.870700	1.000000	1.339390	LZ_TF5 P	
6	-4.446000	4.459662	1.534586	AIR	
7	0.000000	-7.9302e-16 S	1.900000 S	AIR	
IMS	0.000000	0.000000	1.900000 S		

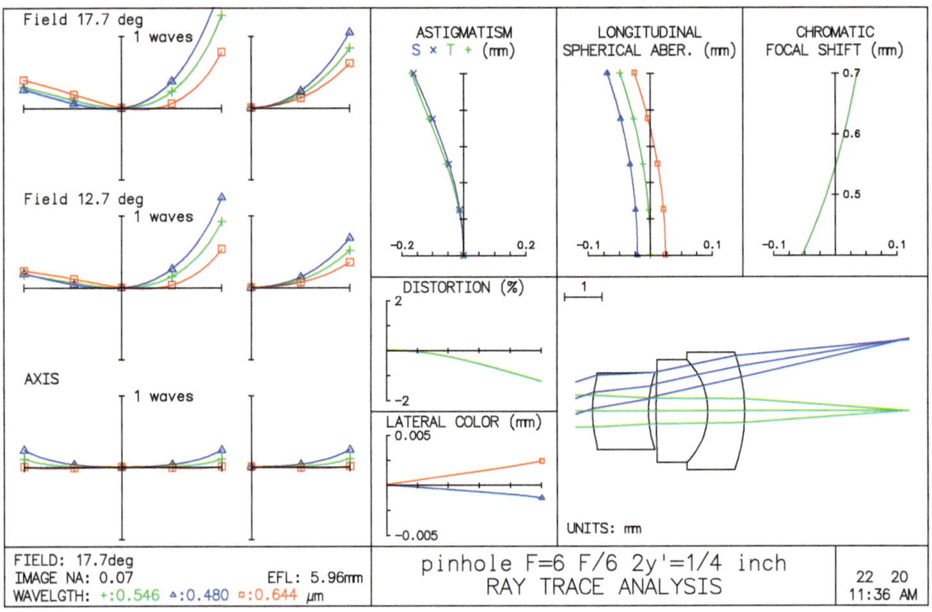

Figure 1.255 Parameters of design and graphs of aberration correction of a pinhole objective ($F' = 6$ mm).

However, in both variants, only half of the linear field in the intermediate plane of the image on the microscope is used. This is due to the use of a low-end budget microscope in which achromatic objectives are used. The aberration of image curvature for images after such objectives is significant; using only half of the field obtained (the total field is usually 18–20 mm) on the microscope allows you to level this, and the image on the sensor of the digital receiver appears flat and sharp throughout the field. If we use a higher-end budget microscope equipped with, for example, an LCF plan achromatic objective, it is advisable to use a longer-focus eyepiece, for example, a 5× eyepiece in combination with an adapter ($F' = 15$ mm). Figure 1.263 shows the design parameters and graphs of aberration correction of the combination of a 5×/25 eyepiece (shown in Fig. 1.221) and an adapter ($F' = 15$ mm).

Lens: ocul 10x comp +adapter F'=6

Image num aper 0.070000 Image height 1.900000 Primary wavln 0.546070

SRF	RADIUS	THICKNESS		APERTURE RADIUS		GLASS		SPECIAL
OBJ	0.000000	1.0000e-20		7.934819		AIR		
AST	0.000000	14.406000		7.934819	AS	AIR		
2	0.000000	5.500000		8.080000		LZ_K8		
3	-26.550000	6.500000		8.130000		AIR		
4	194.090000	2.000000		6.970000		LZ_TF4		
5	14.060000	6.000000		6.590000		LZ_TK14		
6	-192.310000	0.500000		6.110000		AIR		
7	15.346000	5.500000		5.810000		LZ_TK14	P	
8	43.250000	15.990000		4.600000		AIR	P	
9	3.404000	1.500000		1.160000		LZ_TF5		
10	3.048000	0.200000		1.140000		AIR		
11	24.890000	1.400000		1.140000		LZ_CTK19		
12	-1.870700	1.000000		1.400000		LZ_TF5	P	
13	-4.446000	4.448469		1.600000		AIR		
14	0.000000	4.6234e-12	S	1.900000	S	AIR		
IMS	0.000000	0.000000		1.900000	S			

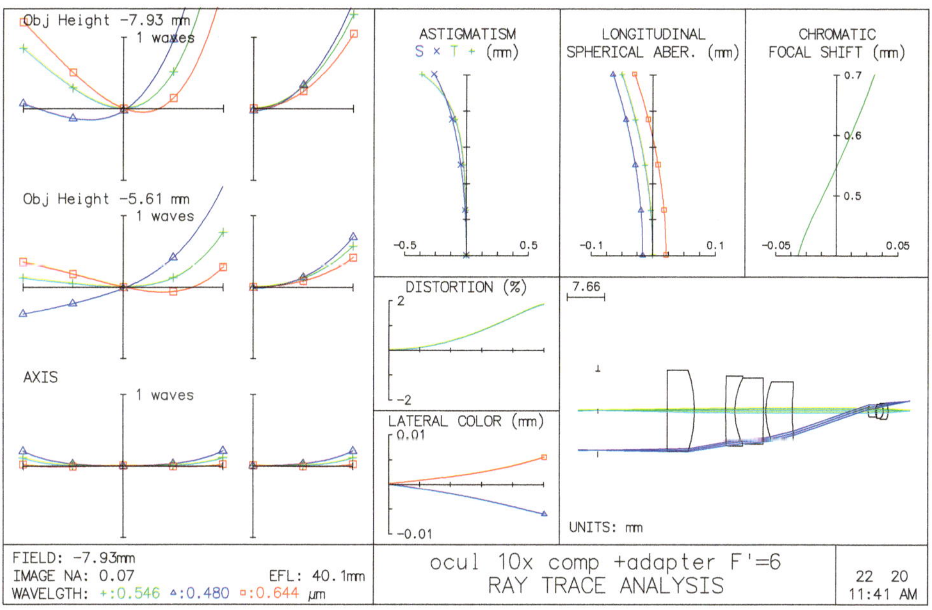

Figure 1.256 Parameters of design and graphs of aberration correction of a 10×/18 eyepiece and a pinhole objective ($F' = 6$ mm).

There is a concept in which a digital image receiver can completely replace the human eye in studies using a light microscope. At one Russian firm, a developer suggested abandoning the use of observation through eyepieces in a *digital* light microscope. This was the first time in budget microscopes that such a technical solution was implemented. Figure 1.264 shows a photograph of the prototype layout of one of the first budget microscopes without a visual

Lens: ocul 5x LC=0+adapter F'=6

Image num aper	0.070000	Image height	1.900000	Primary wavln	0.546070

SRF	RADIUS	THICKNESS	APERTURE RADIUS	GLASS	SPECIAL
OBJ	0.000000	1.0000e-20	15.924096	AIR	
AST	0.000000	24.973000	15.924096 AS	AIR	
2	32.380000	7.000000	16.380000	LZ_CTK12	
3	-103.900000	0.600000	16.090000	AIR	
4	16.700000	8.000000	12.100000	LZ_LK3	
5	-30.470000	2.500000	11.790000	LZ_F4	
6	8.686000	6.000000	7.150000	AIR	
7	-12.820000	7.000000	7.140000	LZ_K8	
8	-31.750000	0.600000	8.270000	AIR	
9	37.370000	6.000000	8.460000	LZ_LK3 P	
10	-18.800000	23.480000	8.380000	AIR	
11	3.404000	1.500000	0.940000	LZ_TF5	
12	3.048000	0.200000	1.000000	AIR	
13	24.890000	1.400000	1.020000	LZ_CTK19	
14	-1.870700	1.000000	1.200000	LZ_TF5 P	
15	-4.446000	0.000000	1.440000	AIR	
16	0.000000	4.448455 S	1.533989 S	AIR	
IMS	0.000000	0.000000	1.900000 S		

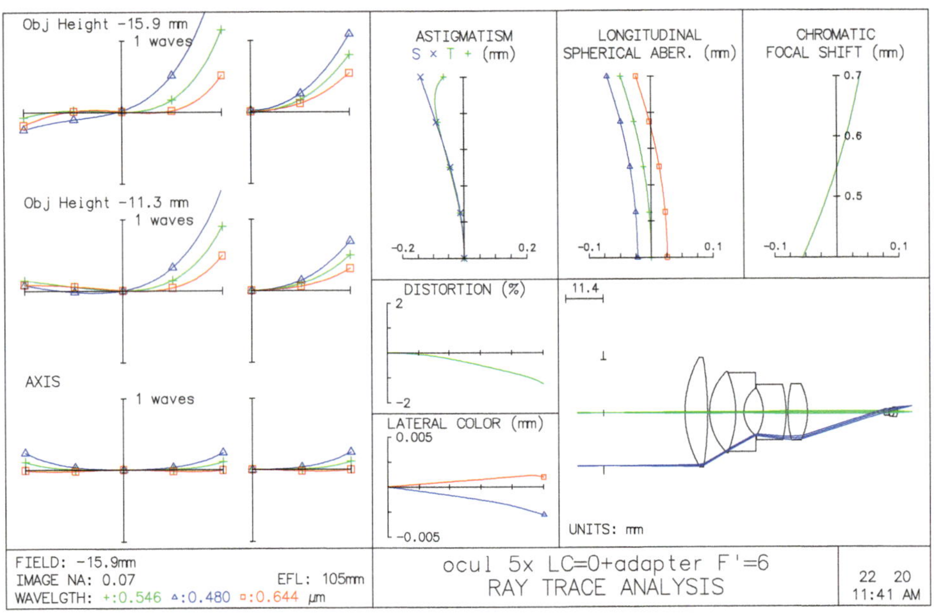

Figure 1.257 Parameters of design and graphs of aberration correction of a 5×/25 eyepiece and a pinhole objective ($F' = 6$ mm).

channel, as well as a technical drawing showing its main parts and assemblies. According to the principle of construction, it can be characterized as a telecentric microscope.

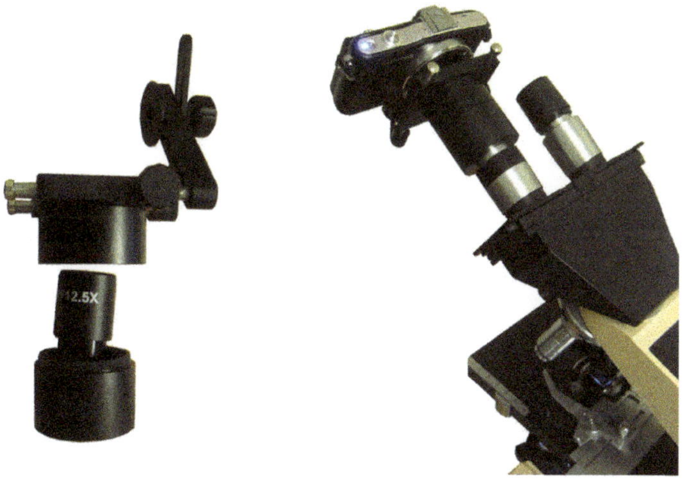

Figure 1.258 An example of a mechanical combination of an eyepiece and a digital camera.

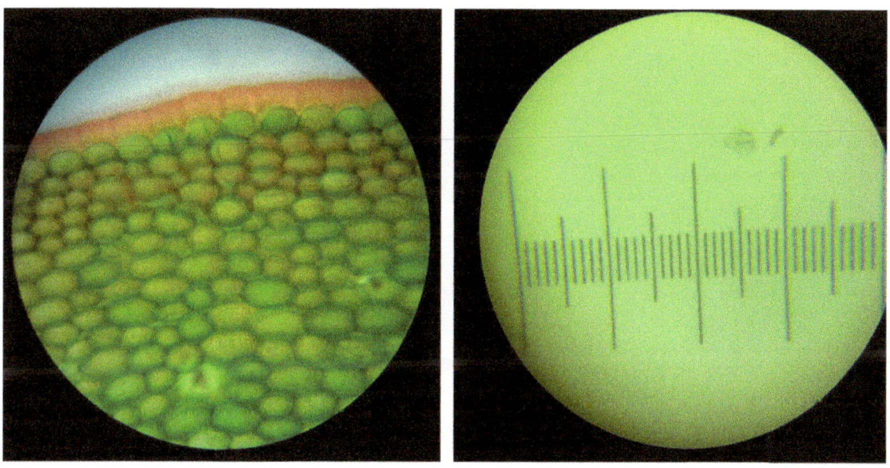

Figure 1.259 Examples of imperfections in image quality with ocular imaging by a smartphone.

The LED reflected light illuminator 5 is mounted on the axis of the rotary mechanism 9, which is rigidly fixed on the guide rail 14 of the focusing mechanism. The mechanism of rotation of the illuminator allows it to rotate relative to the horizontal plane and be installed in two fixed positions for working in direct and inverted mode (such a technical solution is presented in Fig. 1.138). The illuminator is rotated to a fixed position using the handle 6. An LED with an elliptical collector and a beam splitter plate are installed in the illuminator case. There are two threaded holes for mounting objectives along the axis of the beam splitter plate on the top and bottom in the

Lens: adpter F=15 NA'=0.05 2y'=6.2 Efl 15.01

Image num aper 0.050000 Image height 3.100000 Primary wavln 0.546070

SRF	RADIUS		THICKNESS		APERTURE RADIUS		GLASS		SPECIAL
OBJ	0.000000		1.0000e+20		2.0646e+19		AIR		
AST	0.000000		0.001000		0.750733	AS	BK7	C	
2	0.000000		2.520000		0.750733		AIR		
3	-5.550000		4.000000		1.260000		SILICA		
4	6.900000		1.000000		2.240000		AIR		
5	15.070000		6.000000		2.760000		LZ_CTK19		
6	-11.080000		0.800000		3.780000		AIR		
7	69.200000		1.500000		3.920000		LZ_TF5		
8	8.670000		3.200000		3.980000		LZ_CTK19	P	
9	10.650000		0.600000		4.070000		AIR		
10	11.170000		3.000000		4.350000		SILICA	P	
11	-10.250000		25.019668	S	4.470000		AIR		
IMS	0.000000		0.000000		3.100000	S			

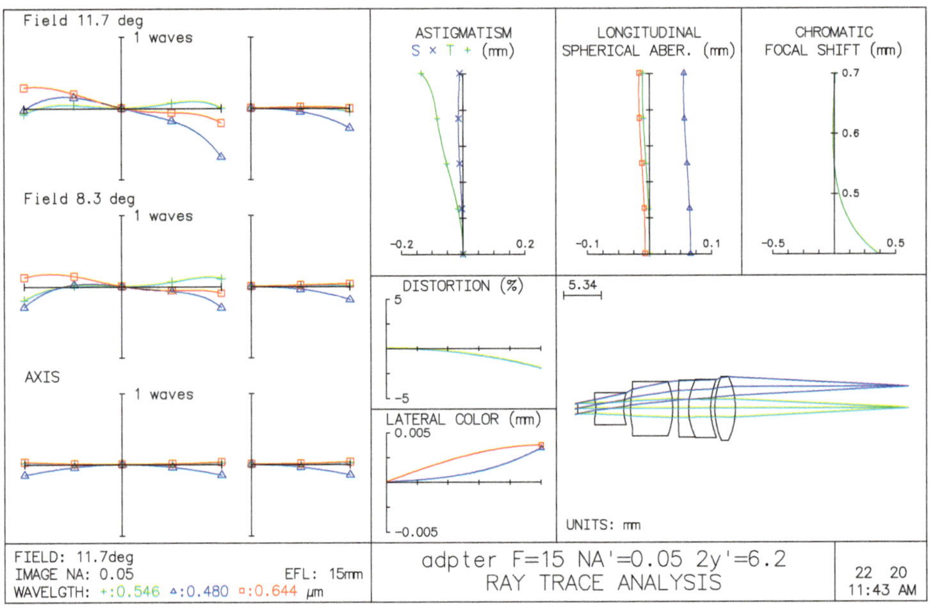

Figure 1.260 Parameters of design and graphs of aberration correction of an adapter ($F' = 15$ mm).

illuminator body. A removable objective 8 is mounted in one of the holes; a projection objective is mounted in the other hole, on the housing of which the digital camera 4 is rigidly fixed. A distinctive feature of such a microscope is the possibility of using large-magnification objectives, including those with high numerical apertures, for example, $NA = 0.95$ in air or $NA = 1.40$ in oil immersion. In fact, the idea of a telecentric microscope was implemented using this prototype. Figure 1.265 presents a principal optical diagram of such a microscope.

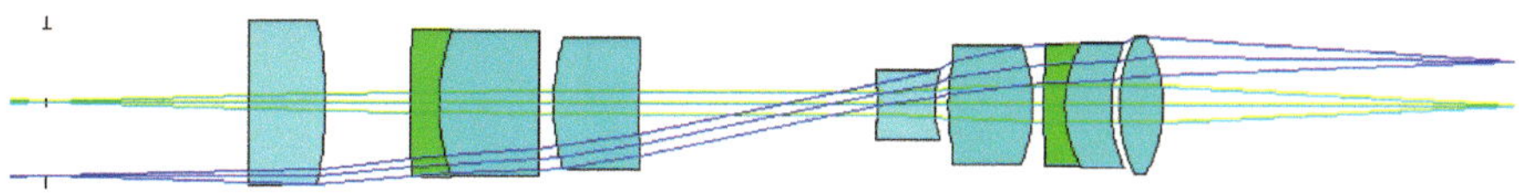

Lens:ocul 10x LC=0+adapter F'=15 LC=0

Object num aper	0.030000	Image height	3.100000	Primary wavln	0.546070

SRF	RADIUS	THICKNESS	APERTURE RADIUS	GLASS	SPECIAL
OBJ	0.000000	1.0000e-20	5.157089	AIR	
AST	0.000000	14.403000	5.157089 AS	AIR	
2	0.000000	5.500000	5.620000	LZ_K8	
3	-26.550000	6.000000	5.720000	AIR	
4	108.140000	2.000000	5.150000	LZ_TF4	
5	14.791000	7.000000	4.960000	LZ_TK16	
6	0.000000	1.000000	4.670000	AIR	
7	15.740000	6.000000	4.540000	LZ_TK16 P	
8	46.880000	17.022400	3.670000	AIR P	
9	-5.550000	4.000000	1.390000	SILICA	
10	6.900000	1.000000	2.430000	AIR	
11	15.070000	6.000000	2.980000	LZ_CTK19	
12	-11.080000	0.800000	4.040000	AIR	
13	69.200000	1.500000	4.190000	LZ_TF5	
14	8.670000	3.200000	4.250000	LZ_CTK19 P	
15	10.650000	0.600000	4.340000	AIR	
16	11.170000	3.000000	4.660000	SILICA P	
17	-10.250000	25.019867	4.750000	AIR	
18	0.000000	-5.4238e-05 S	3.100003 S	AIR	
IMS	0.000000	0.000000	3.100000 S		

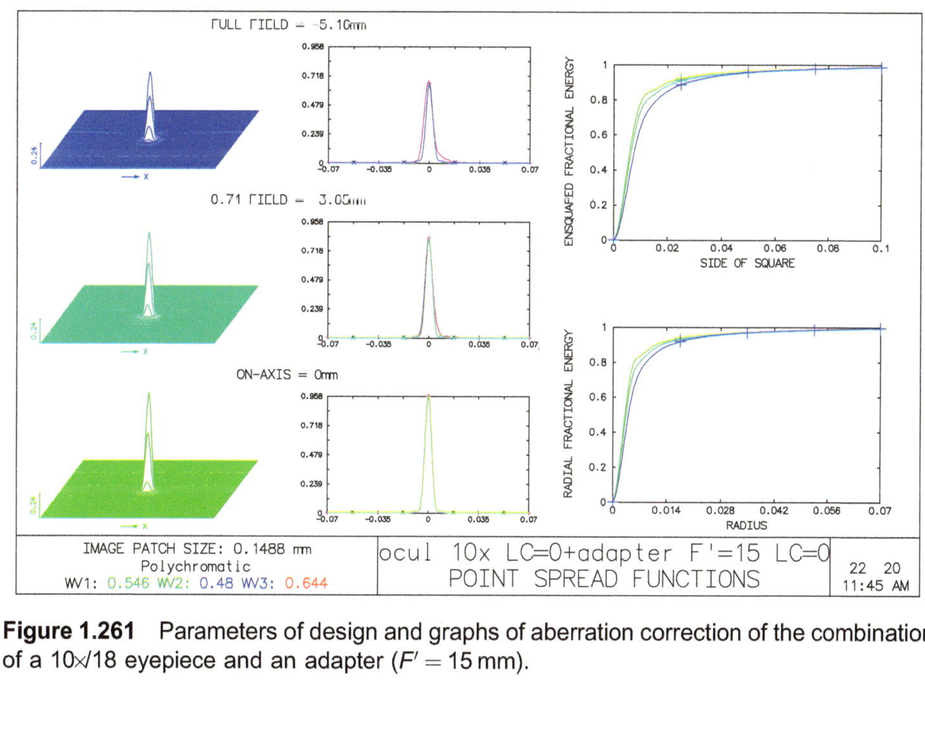

Figure 1.261 Parameters of design and graphs of aberration correction of the combination of a 10×/18 eyepiece and an adapter ($F' = 15$ mm).

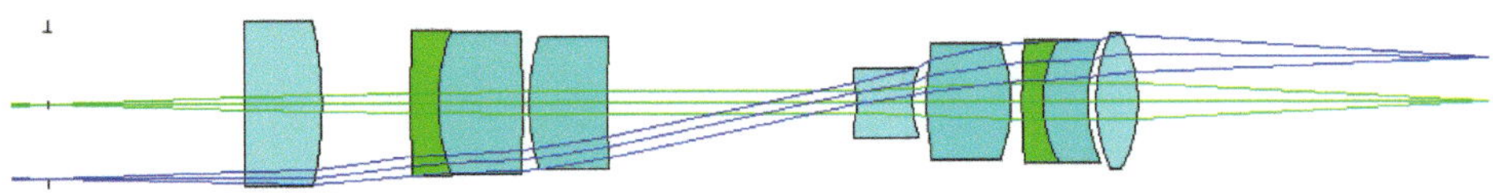

Lens:ocul 10x comp+adapter F'=15 LC=0

Object num aper 0.030000 Image height 3.100000 Primary wavln 0.546070

SRF	RADIUS		THICKNESS		APERTURE RADIUS		GLASS		SPECIAL
OBJ	0.000000		1.0000e-20		5.141846		AIR		
AST	0.000000		14.406000		5.141846	AS	AIR		
2	0.000000		5.500000		5.610000		LZ_K8		
3	-26.550000		6.500000		5.710000		AIR		
4	194.090000		2.000000		5.100000		LZ_TF4		
5	14.060000		6.000000		4.930000		LZ_TK14		
6	-192.310000		0.500000		4.750000		AIR		
7	15.346000		5.500000		4.660000		LZ_TK14	P	
8	43.250000		17.867000		3.850000		AIR	P	
9	-5.550000		4.000000		1.390000		SILICA		
10	6.900000		1.000000		2.430000		AIR		
11	15.070000		6.000000		2.980000		LZ_CTK19		
12	-11.080000		0.800000		4.040000		AIR		
13	69.200000		1.500000		4.190000		LZ_TF5		
14	8.670000		3.200000		4.250000		LZ_CTK19	P	
15	10.650000		0.600000		4.340000		AIR		
16	11.170000		3.000000		4.660000		SILICA	P	
17	-10.250000		25.019867		4.750000		AIR		
18	0.000000		-4.4603e-15	S	3.100000	S	AIR		
IMS	0.000000		0.000000		3.100000	S			

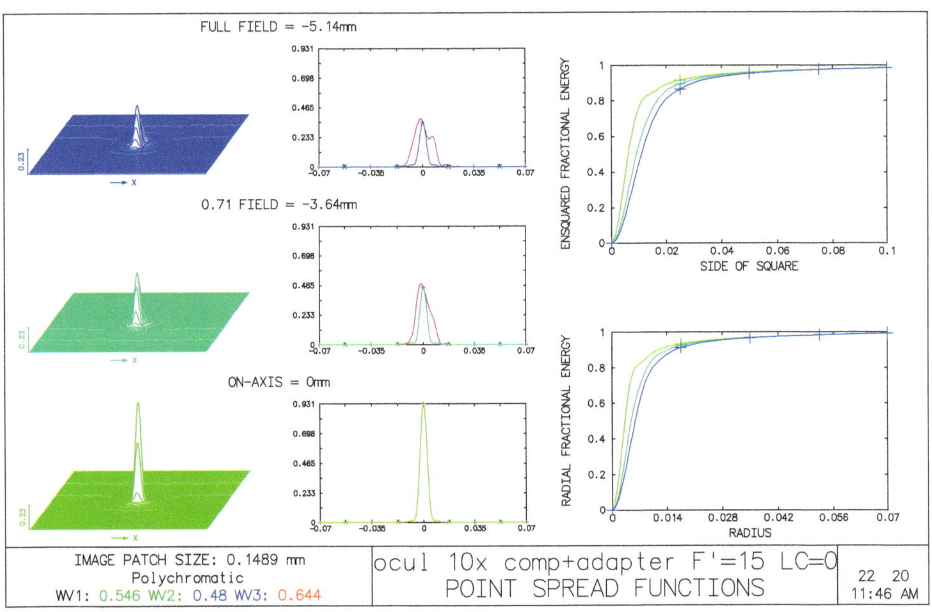

Figure 1.262 Parameters of design and graphs of aberration correction of the combination of a 10×/18 compensation eyepiece and an adapter ($F' = 15$ mm).

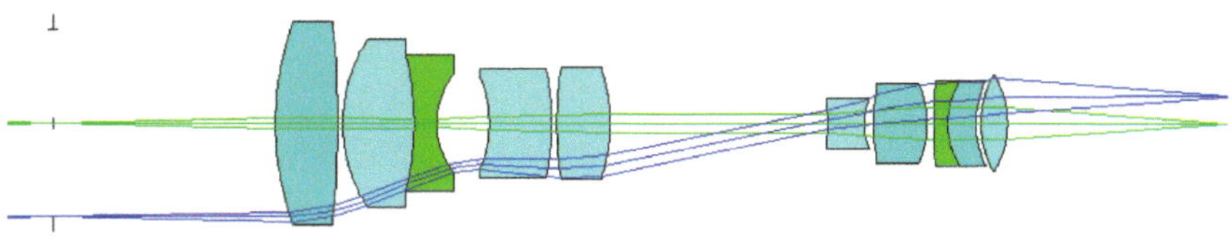

Lens: ocul 5x LC=0+adapter F'=15 LC=0

Image num aper 0.070000 Image height 3.100000 Primary wavln 0.546070

SRF	RADIUS	THICKNESS	APERTURE RADIUS	GLASS	SPECIAL
OBJ	0.000000	1.0000e-20	10.318970	AIR	
AST	0.000000	24.973000	10.318970 AS	AIR	
2	32.380000	7.000000	11.110000	LZ_CTK12	
3	-103.900000	0.600000	10.500000	AIR	
4	16.700000	8.000000	9.270000	LZ_LK3	
5	-30.470000	2.500000	7.480000	LZ_F4	
6	8.686000	6.000000	5.570000	AIR	
7	-12.820000	7.000000	5.320000	LZ_K8	
8	-31.750000	0.600000	6.050000	AIR	
9	37.370000	6.000000	6.130000	LZ_LK3 P	
10	-18.800000	24.510000	6.020000	AIR	
11	-5.550000	4.000000	1.590000	SILICA	
12	6.900000	1.000000	2.730000	AIR	
13	15.070000	6.000000	3.360000	LZ_CTK19	
14	-11.080000	0.800000	4.460000	AIR	
15	69.200000	1.500000	4.620000	LZ_TF5	
16	8.670000	3.200000	4.680000	LZ_CTK19 P	
17	10.650000	0.600000	4.760000	AIR	
18	11.170000	3.000000	5.190000	SILICA P	
19	-10.250000	25.019867	5.230000	AIR	
20	0.000000	-8.4426e-05 S	3.100006 S	AIR	
IMS	0.000000	0.000000	3.100000 S		

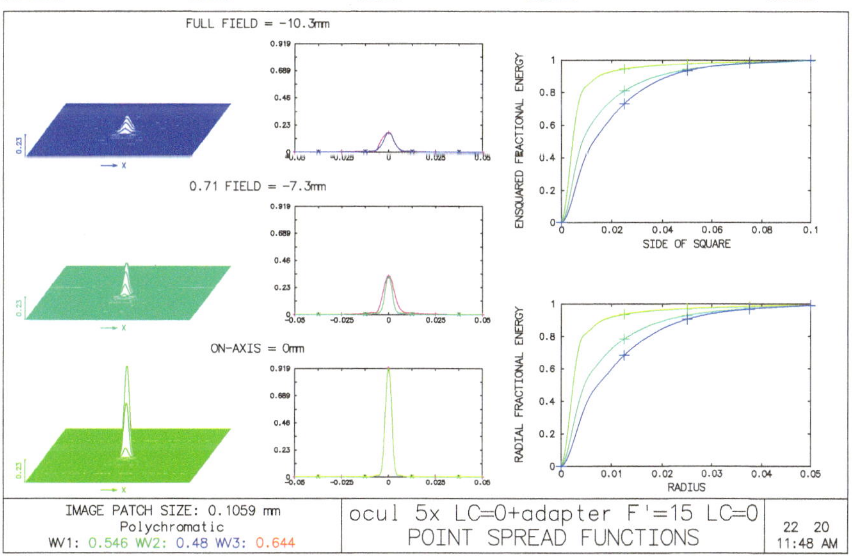

Figure 1.263 Parameters of design and graphs of aberration correction of the combination of a 5×/25 eyepiece and an adapter ($F' = 15$ mm).

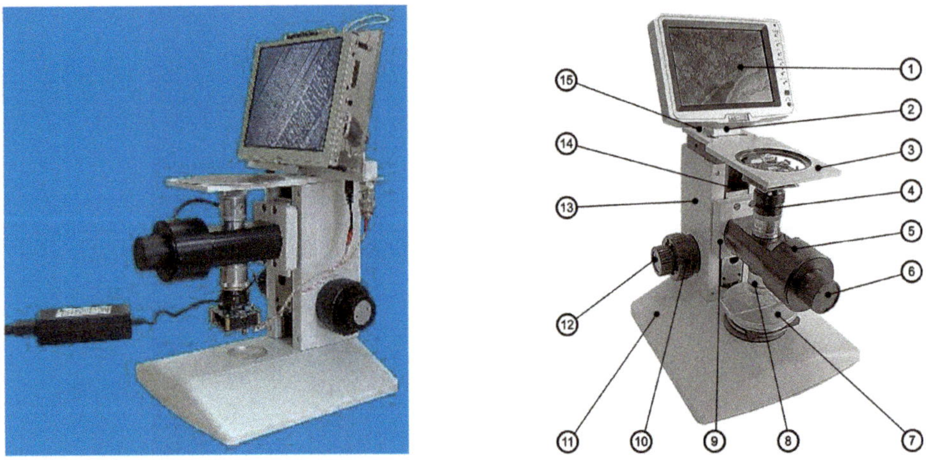

Figure 1.264 Photograph and technical drawing of a model of a digital telecentric microscope without a visual channel. 1, liquid crystal display; 2, mechanism of display rotation; 3, upper subject stage; 4, a camera with a projection objective; 5, LED illuminator of reflected light; 6, the handle of rotation of the illuminator; 7, lower subject stage; 8, interchangeable objective; 9, the mechanism of rotation of the illuminator; 10, coarse focus handle; 11, a stand; 12, handle fine focus; 13, case; 14, guide rail of focusing mechanism; 15, the handle of the rotation mechanism.

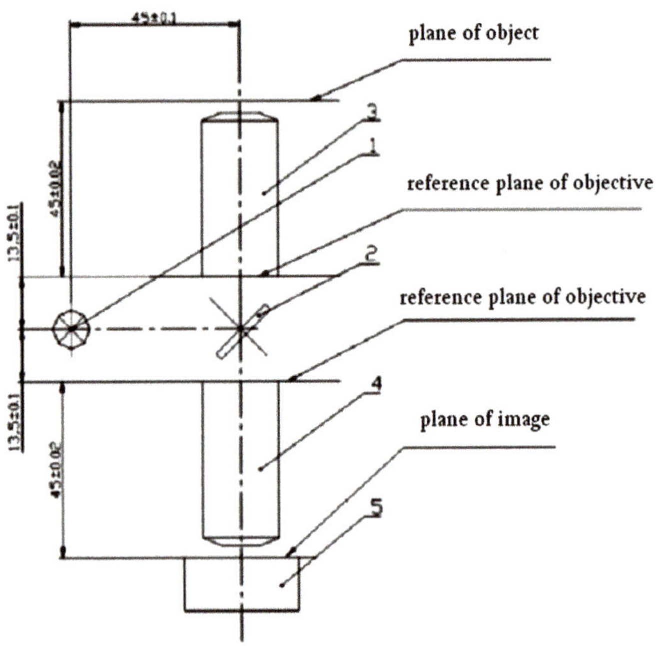

Figure 1.265 Principal optical diagram of a telecentric microscope.

The lighting system provides normally incident illumination of the object through the objective. Rays of light from the light source 1, the LEDs, are sent to the beam splitter plate 2 and, reflected from it, illuminate with the interchangeable objective 3 the investigated surface of the object located in the focal plane of this objective. The rays of light reflected from the surface of the object and carrying its image are transmitted by the objective 3 to infinity. Next, the image of the surface of the object is projected onto the matrix of the digital camera 5, located in the rear focal plane of the projection objective 4. As a projection objective, a wide-field completely color free (CCF) 2.5×/0.05 plan apochromatic objective is used. Figures 1.266–1.269 show parameters of design and graphs of aberration correction of combinations of 10×, 20×, 50×, and 100× objectives with a 2.5× objective as a telecentric microscope.

This digital microscope uses a black and white matrix as the image receiver. A method of synthesizing a color image is used in which an illuminator uses three separate radiation sources: red, green, and blue. By turning them on alternately with pulses, we can synthesize a frame of three black and white images obtained in three primary colors. Although such a lighting method leads to a decrease in the frame rate by about three times, there will be three times as many dots in the image, which is more important for a microscope (as we know, there are three to four times fewer dots in a color matrix of the same size than in a black and white matrix). Therefore, the choice of a black and white camera as an image receiver is advisable. To build such a three-color illuminator, it is advisable to use three white LEDs, highlighting the necessary parts of the spectrum by optical methods. This can be done, for example, by using color filters. The principal optical system of such an illuminator is shown in Fig. 1.270.

Specialized software was developed for this microscope that allows the user to receive images from the camera via USB, show them on the display, change the main characteristics of the image (scale, brightness, contrast, saturation, etc.), and write and read data. The program also lets the user control the parameters of the camera, including changing the automatic optimal parameter values (these values are selected by the processor that controls the sensor), as well as set the exposure time, gain, and color ratio. When working with a black and white matrix and a three-color illuminator in its manual control mode, the original special module allows black and white images to be sequentially registered in three primary colors from which a color image can then be synthesized. In addition, to use a three-color illuminator in the dynamic mode, some modernization of the black and white camera and control drivers was required to synchronize the operation of the illuminator and the camera and identify frames by color.

Based on the idea of excluding the visual channel in a light microscope, a whole line of such microscopes called the MICROVISOR® (μVIZO®) was built. However, serial production uses the *simplified principle* in which

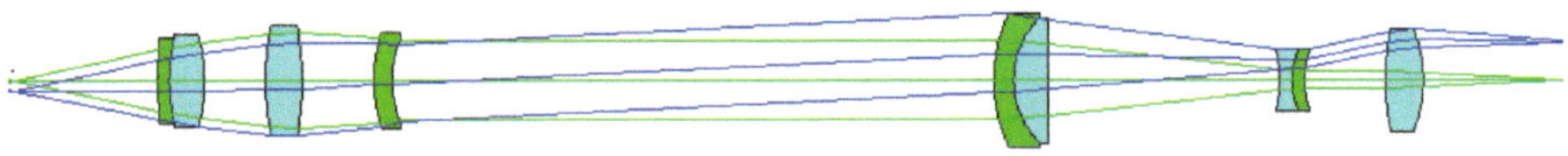

Lens: plan 10x0.25+plapo 2.5x0.05

Object num aper 0.200000 Image height 3.000000 Primary wavln 0.546070

SRF	RADIUS	THICKNESS	APERTURE RADIUS	GLASS	SPECIAL
OBJ	0.000000	1.0000e-20	0.767784	AIR	
AST	0.000000	12.120000	0.767784 AS	AIR	
2	112.760000	1.000000	3.250000	LZ_TF5	
3	13.183000	3.000000	3.390000	LZ_LK3	
4	-13.183000	5.000000	3.740000	AIR	
5	19.498000	3.000000	4.440000	LZ_LK3 P	
6	-24.320000	6.000000	4.440000	AIR	
7	10.520000	1.500000	3.770000	LZ_TF5 P	
8	8.395000	5.610000	3.430000	AIR	
9	0.000000	7.900000	3.200000	AIR	
10	0.000000	35.000000	3.580000	AIR	
11	0.000000	1.000000	5.270000	AIR	
12	12.246000	1.600000	5.380000	LZ_TF1	
13	7.047000	2.800000	4.990000	LZ_K8	
14	258.200000	18.920000	4.920000	AIR	
15	-8.790000	1.000000	2.470000	LZ_K8 P	
16	6.026000	1.000000	2.500000	LZ_TF1 P	
17	7.047000	6.700000	2.480000	AIR	
18	21.680000	3.100000	3.970000	LZ_LK3 P	
19	-11.117000	11.563421	4.130000	AIR	
20	0.000000	2.1252e-15 S	3.000000 S	AIR	
IMS	0.000000	0.000000	3.000000 S		

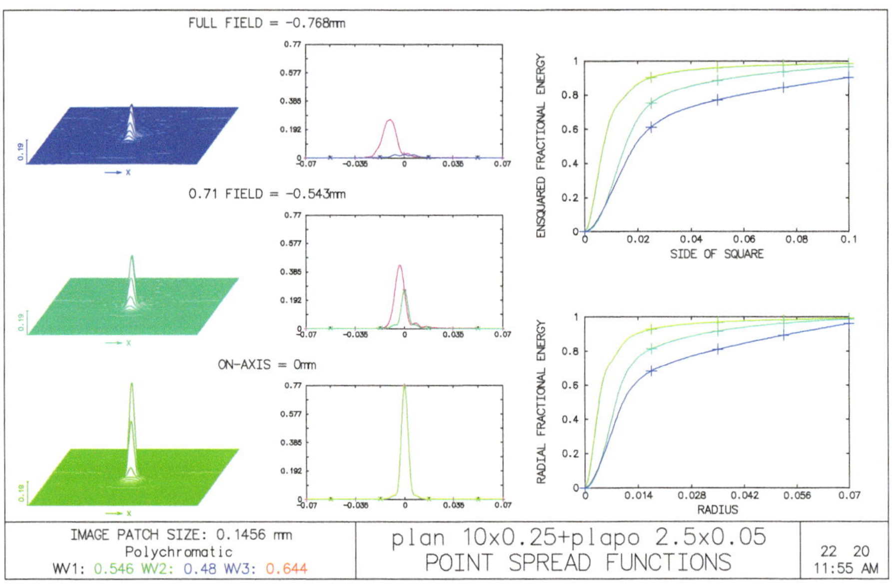

Figure 1.266 Parameters of design and graphs of aberration correction of a 10×/0.25 LCF plan achromatic objective combined with a CCF 2.5×/0.05 plan apochromatic objective.

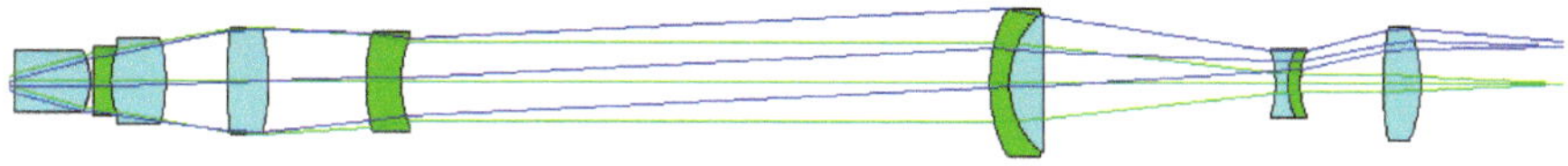

Lens: ocx 20x+plapo 2.5x

Object num aper 0.400000 Image height 3.000000 Primary wavln 0.546070

SRF	RADIUS	THICKNESS	APERTURE RADIUS		GLASS		SPECIAL
OBJ	0.000000	1.206350	0.381246		AIR		
1	0.000000	0.000000	0.907740	S	AIR		
AST	0.000000	6.200000	0.907740	AS	LZ_K8		
3	-5.035000	0.300000	2.430000		AIR		
4	-75.160000	1.400000	2.560000		LZ_TF5		
5	7.870000	4.500000	2.740000		LZ_K8		
6	-12.647000	5.000000	3.360000		AIR		
7	30.550000	3.500000	4.070000		LZ_K8	P	
8	-15.136000	8.000000	4.230000		AIR		
9	13.583000	2.900000	3.960000		LZ_TF5		
10	9.290000	12.000000	3.520000		AIR		
11	0.000000	35.000000	4.070000		AIR		
12	0.000000	1.000000	5.760000		AIR		
13	12.246000	1.600000	5.880000		LZ_TF1		
14	7.047000	2.800000	5.410000		LZ_K8		
15	258.200000	18.920000	5.370000		AIR		
16	-8.790000	1.000000	2.660000		LZ_K8	P	
17	6.026000	1.000000	2.690000		LZ_TF1	P	
18	7.047000	6.700000	2.670000		AIR		
19	21.680000	3.100000	4.280000		LZ_LK3		
20	-11.117000	11.563421	4.440000		AIR		
21	0.000000	0.002814 S	3.000227	S	AIR		
IMS	0.000000	0.000000	3.000000	S			

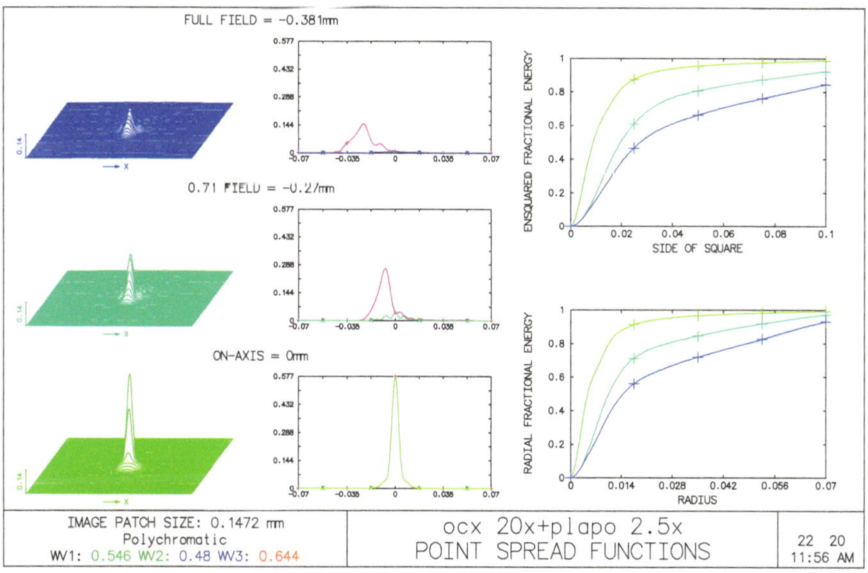

Figure 1.267 Parameters of design and graphs of aberration correction of a 20×/40 LCF semi–plan achromatic objective combined with a CCF 2.5×/0.05 plan apochromatic objective.

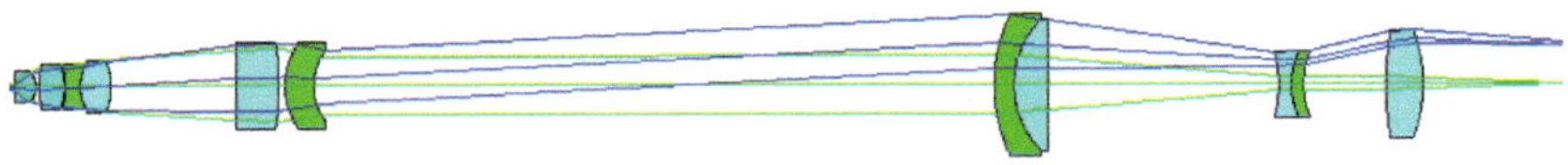

Lens: plan 50x+ Plapo 2.5x

Object num aper	0.700000	Image height	3.000000	Primary wavln	0.546070

SRF	RADIUS	THICKNESS	APERTURE RADIUS	GLASS	SPECIAL
OBJ	0.000000	1.0000e-20	0.153673	AIR	
AST	0.000000	0.381860	0.153673 AS	AIR	
2	-1.195300	1.720000	0.440000	LZ_CTK19	
3	-1.522500	0.300000	1.240000	AIR	
4	8.752300	2.000000	1.590000	LZ_CTK19 P	
5	-4.554500	0.100000	1.730000	AIR	
6	-6.384600	1.200000	1.700000	LZ_TF5	
7	3.570700	2.500000	1.770000	LZ_LK3	
8	-4.805800	10.500000	2.050000	AIR	
9	55.584700	3.500000	3.220000	LZ_LK3 P	
10	-22.810500	0.500000	3.390000	AIR	
11	5.815500	2.300000	3.410000	LZ_TF5 P	
12	4.511400	20.000000	2.780000	AIR	
13	0.000000	35.000000	3.710000	AIR	
14	0.000000	1.000000	5.420000	AIR	
15	12.246000	1.600000	5.540000	LZ_TF1	
16	7.047000	2.800000	5.120000	LZ_K8	
17	258.200000	18.920000	5.060000	AIR	
18	-8.790000	1.000000	2.550000	LZ_K8 P	
19	6.026000	1.000000	2.580000	LZ_TF1 P	
20	7.047000	6.700000	2.560000	AIR	
21	21.680000	3.100000	4.110000	LZ_LK3 P	
22	-11.117000	11.563421	4.270000	AIR	
23	0.000000	0.002326 S	3.000198 S	AIR	
IMS	0.000000	0.000000	3.000000 S		

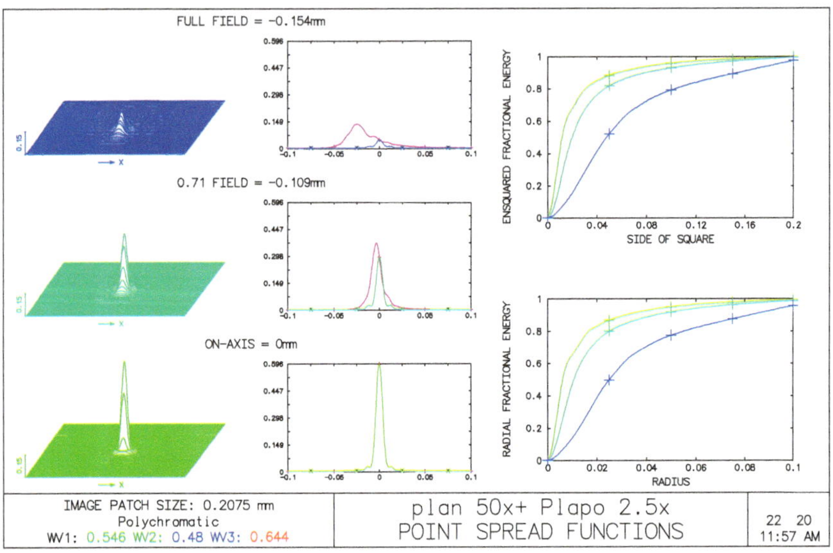

Figure 1.268 Parameters of design and graphs of aberration correction of a 50×/70 LCF plan achromatic objective combined with a CCF 2.5×/0.05 plan apochromatic objective.

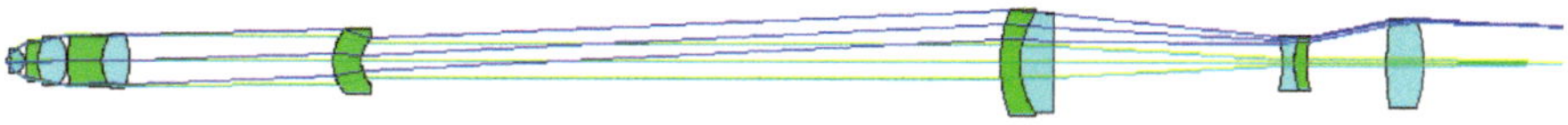

Lens: Plan 100x+plapo 2.5x

Object num aper 0.850000 Image height 3.000000 Primary wavln 0.546070

SRF	RADIUS	THICKNESS	APERTURE RADIUS	GLASS	SPECIAL
OBJ	0.000000	1.0000e-20	0.076389	AIR	
AST	0.000000	0.391000	0.076389 AS	AIR	
2	-0.831800	1.100000	0.470000	LZ_CTK19	
3	-1.056800	0.100000	1.010000	AIR	
4	5.598000	1.000000	1.500000	LZ_TF5	
5	3.311000	2.400000	1.630000	LZ_LK3	
6	-3.311000	0.100000	1.980000	AIR	
7	18.880000	2.790000	2.050000	LZ_TF5 P	
8	3.311000	2.300000	2.040000	LZ_LK3 P	
9	-5.728000	16.760000	2.210000	AIR	
10	3.767000	2.330000	2.550000	LZ_TF5 P	
11	2.630000	16.670000	1.810000	AIR	
12	0.000000	35.000000	2.550000	AIR	
13	0.000000	1.000000	4.180000	AIR	
14	12.246000	1.600000	4.260000	LZ_TF1	
15	7.047000	2.800000	4.020000	LZ_K8	
16	258.200000	18.920000	3.910000	AIR	
17	-8.790000	1.000000	2.100000	LZ_K8 P	
18	6.026000	1.000000	2.130000	LZ_TF1 P	
19	7.047000	6.700000	2.120000	AIR	
20	21.680000	3.100000	3.420000	LZ_LK3 P	
21	-11.117000	11.563421	3.590000	AIR	
22	0.000000	0.000985 S	3.000071 S	AIR	
IMS	0.000000	0.000000	3.000000 S		

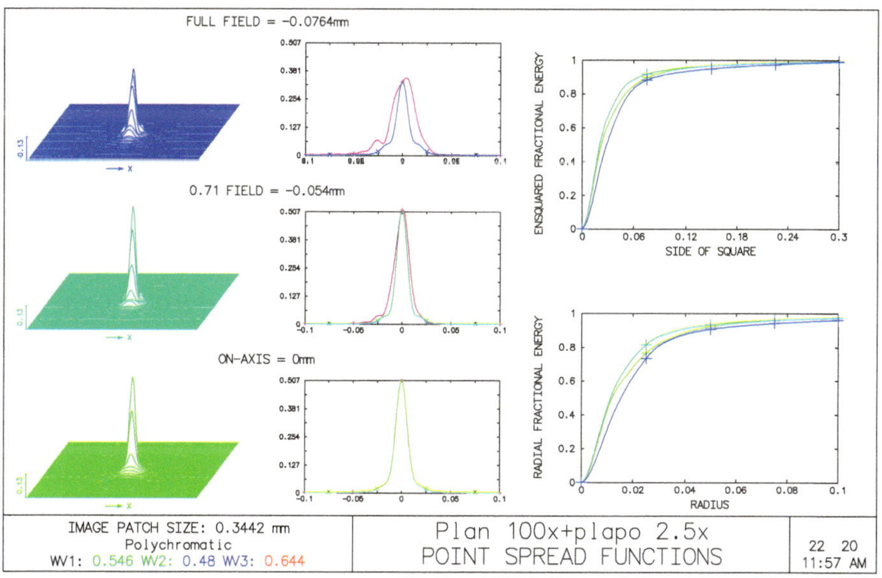

Figure 1.269 Parameters of design and graphs of aberration correction of a 100×/85 LCF plan achromatic objective combined with a CCF 2.5×/0.05 plan apochromatic objective.

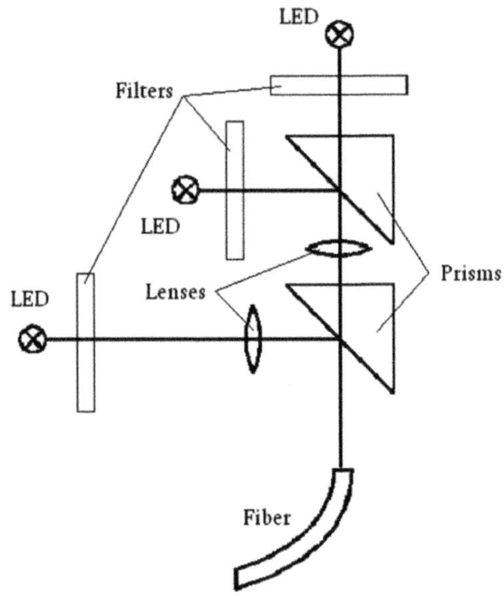

Figure 1.270 Optical system of three-LED illuminator.

the visual observation system in conventional budget microscopes is replaced by a digital visualization system. But according to the results of our experience with such devices, we began to be pessimistic about such a basic technical solution. In our company Labor-Microscopes, we proposed a more balanced and optimal combination of a visual observation system and a visualization system (MicroScreener; www.labor-microscopes.ru). Figure 1.271 shows photographs of modern models of digital microscopes such as the MICROVISOR and MicroScreener, which are offered in budget microscopy.

Today, a complete rejection of the visual channel in budget microscopy does not appear to be entirely justified and is not an optimal technical solution, especially when it comes to models like the MICROVISOR. However, such considerations relate mainly to image quality and consumer parameters (namely, resolution). If we compare using visual observation techniques through the eyepieces with using the indicated visualization system, effectiveness and quality are still higher with visual observation through the eyepieces; the described visualization system can be used mainly as an auxiliary one. However, this MICROVISOR model uses modern objectives with plan achromatic aberration correction with corrected LC. Figure 1.272 shows a photograph of this objective kit. Medium- and high-magnification objectives use the optical designs shown in Figs. 1.135–1.137; small linear magnification objectives are unified with objectives designed to complete inverted biological microscopes. Figures 1.273–1.290 show photographs of real objects taken with some of these objectives.

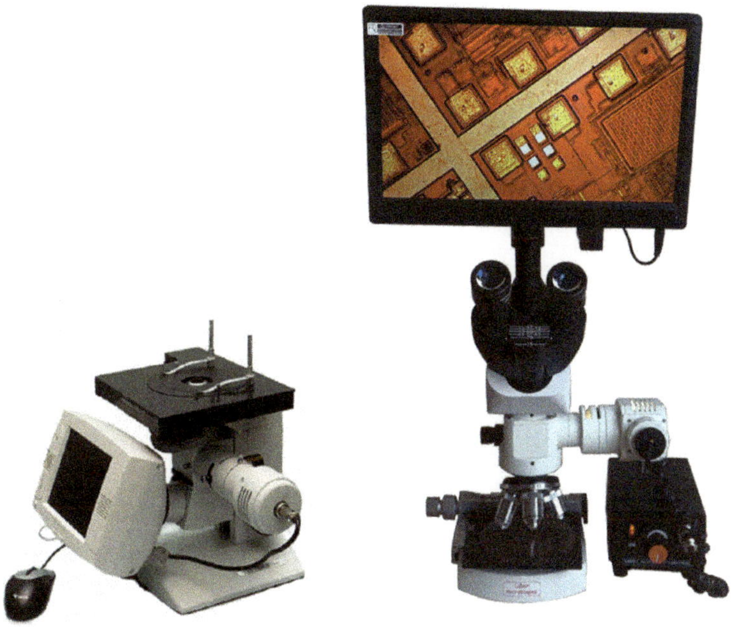

Figure 1.271 Photographs of the MICROVISOR and MicroScreener digital microscopes.

Figure 1.272 Set of metallographic LCF plan achromatic objectives for an infinite length of the tube.

The MicroScreener appears to offer a preferable technical solution than the MICROVISOR because the MicroScreener user may not have to think about matching the parameters of the visual channel with the digital one; there is an optimal balance of optical dimensions and aberration correction. The system does not introduce shading, vignetting, or other image artifacts associated with the need to coordinate the location of the entrance and exit pupils of individual elements of the optical system.

Figure 1.273 Photograph of a real object using a 5×/0.12 infinity LCF plan achromatic objective (bright field reflected light).

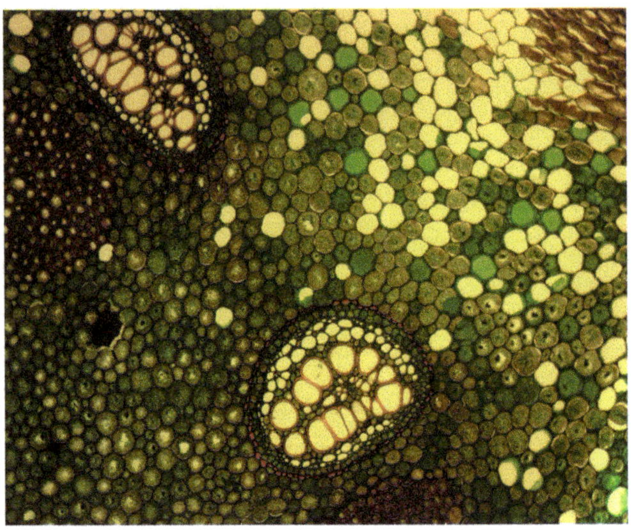

Figure 1.274 Photograph of a real object using a 5×/0.12 infinity LCF plan achromatic objective (bright field transmitted light).

1.8.5 Design of focusing mechanisms

This technical solution does not directly apply to the problem of constructing a unified microscope that allows the use of objectives designed for different lengths of the tube. But it should be noted that at least three standards have been used to date for the parfocal distance of microscope objectives. Some budget microscope manufacturers use 33–37 mm as the parfocal distance of

Figure 1.275 Photograph of a real object using a 5×/0.12 infinity LCF plan achromatic objective (bright field reflected light).

Figure 1.276 Photograph of a real object using a 5×/0.12 infinity LCF plan achromatic objective (dark field reflected light).

the objectives (the old Japanese standard), whereas other manufacturers prefer 45 mm as the height of the objectives (the international standard proposed by the Germans). The most interesting objectives can be designed for a parfocal height of 60 or even 75 mm. And this does not include objectives with a parfocal distance of 95 mm or promising objectives that could possibly have a height of 120 mm. The idea of unification is to *lengthen* the focusing

Figure 1.277 Photograph of a real object using a 10×/0.20 infinity LCF LD plan achromatic objective (dark field reflected light).

Figure 1.278 Photograph of a real object using a 10×/0.20 infinity LCF LD plan achromatic objective (bright field transmitted light).

mechanism of budget microscopes, i.e., to provide the possibility of using objectives with a height of 33–60 mm (preferably 75 or even 95 mm). The mechanical layout and design of the microscope stand allow this to be done; it can be achieved by using a longer focusing bar than usual.

Figure 1.291 shows photographs of an old Soviet Lumam-R8 microscope, the stand design of which made it possible to work with objectives with

Figure 1.279 Photograph of a real object using a 10×/0.20 infinity LCF LD plan achromatic objective (dark field reflected light).

Figure 1.280 Photograph of a real object using a 10×/0.20 infinity LCF LD plan achromatic objective (dark field reflected light).

different parfocal distances. Although the mounting diameter for mounting objectives is 27 mm, it is possible to change the nosepiece in order to use modern objectives with a mounting diameter of 30 or even 32 mm. In addition, if we use, for example, unified visual attachments (discussed above in Section 1.8.2, Design of the visual head), then such a microscope stand can become a prototype for a unified microscope that can be used practically for

Figure 1.281 Photograph of a real object using a 10×/0.20 infinity LCF LD plan achromatic objective (bright field reflected light).

Figure 1.282 Photograph of a real object using a 10×/0.20 infinity LCF LD plan achromatic objective (dark field reflected light).

all occasions. It must be said that this is not the only example of a microscope in which objectives of various parfocal heights can be used; the MikMed-1 (Biolam), MikMed-2, Min-8, and some other microscopes also have this design feature.

However, it should be noted that the operating conditions of the microscope lighting system at various positions of the stage (condenser) can

Figure 1.283 Photograph of a real object using a 20×/0.35 infinity LCF LD plan achromatic objective (bright field transmitted light).

Figure 1.284 Photograph of a real object using a 20×/0.35 infinity LCF LD plan achromatic objective (bright field transmitted light).

be different when using objectives with different parfocal heights. This happens if the condenser operates in converging beams of rays (in space from the collector); if the microscope implements the Köhler principle of illumination, strict correspondence will be achieved only at one of the positions of the condenser (for example, if the microscope is focused using an

Figure 1.285 Photograph of a real object using a 20×/0.35 infinity LCF LD plan achromatic objective (bright field reflected light).

Figure 1.286 Photograph of a real object using a 20×/0.35 infinity LCF LD plan achromatic objective (dark field reflected light).

objective with a parfocal height of 45 mm). In other cases, strict compliance with this principle will not be respected. However, as the practice of microscopy shows, this does not lead to a significant decrease in image quality, especially if you keep in mind that this is a budget microscope.

Figure 1.287 Photograph of a real object using a 50×/0.60 infinity LCF LD plan achromatic objective (bright field reflected light).

Figure 1.288 Photograph of a real object using a 50×/0.60 infinity LCF LD plan achromatic objective (dark field reflected light).

1.8.6 Design of illumination systems

Why do light microscopes have such complex lighting systems? This logical question does not have a very clear answer. A common opinion is that microscope lighting systems are becoming increasingly complex. Remembering older microscopes, in which a mirror and a simple two-lens condenser

Figure 1.289 Photograph of a real object using a 100×/0.75 infinity LCF LD plan achromatic objective (bright field reflected light).

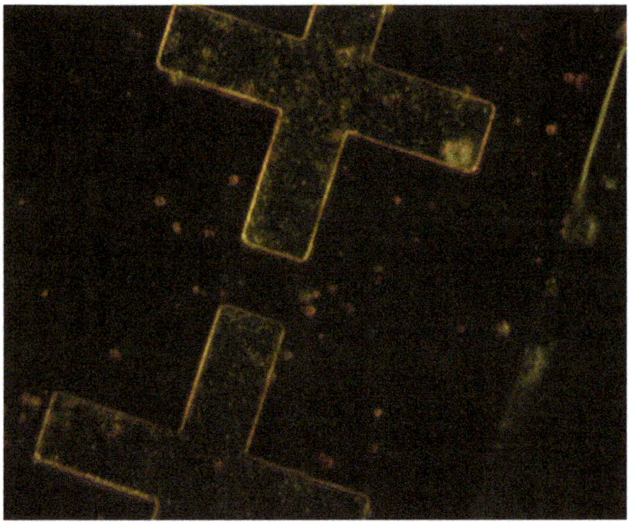

Figure 1.290 Photograph of a real object using a 100×/0.75 infinity LCF LD plan achromatic objective (dark field reflected light).

ensured satisfactory lighting, a researcher may wonder, why complicate them? If the answer can be as simple as the question, it might sound something like this: developers are once again trying to at least slightly improve the so-called Lagrange–Helmholtz invariant to solve the main contradiction of the microscope, i.e., a large linear field and a low numerical aperture, or a small linear field and a high numerical aperture. For objectives, the projection

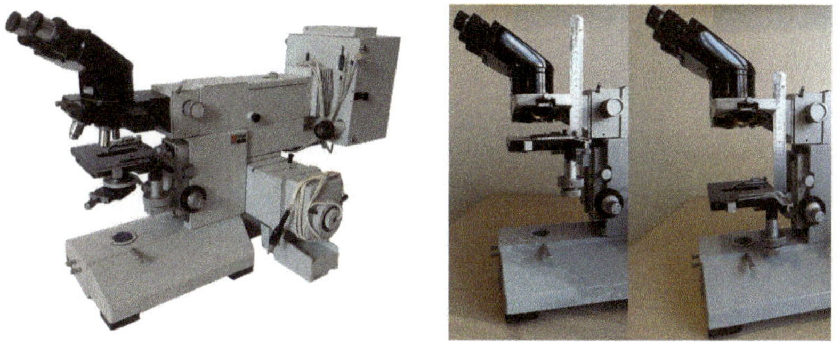

Figure 1.291 Photographs of an old Soviet Lumam-R8 microscope with a longer than usual focusing bar.

system of the microscope, these attempts have not been successful. In light microscopes, objectives of various linear magnifications (linear fields on the object) and numerical apertures are used. For the lighting system, developers frequently think that there is an original and beautiful engineering solution that will allow for a unified lighting system that can be used for all occasions. Of course, this is a correct and fair aspiration when it comes to mass-produced budget microscopes.

1.8.6.1 Complex lighting systems

Microscopists working with old microscopes know that, as a rule, lamps with a conventional tungsten spiral as the glow body were used as a light source. In terms of the projection (using the illuminator) of this thread on an object, using the critical lighting method, it provided terrible illumination on a microscope; for some reason, many researchers thought it was acceptable for the thread to be projected onto the plane of the pupil of the lens using the Köhler method. The image of this shapeless and colored thread, which trembled even from the researcher's breath, was used in microscopes on a greatly enlarged, oppressive scale. Halogen lamps, mercury, and other modern light sources were not only much brighter (which was required with the introduction of thick glass objectives, for example, plan achromats), but stable and convenient. And even a characteristic of the radiation source such as *color temperature* has since become satisfactory for researchers. However, the main disadvantages of light sources remained: they still had very small linear dimensions of the luminous body. The image scale of the lighting system should be significant. Table 1.17 shows examples of the linear dimensions of the luminous body of some halogen lamps used in light microscopes.

It is obvious that the degree of resolution by the objective and the reliability of research depend on the quality of an object's illumination. At the same time, both critical and Köhler-based lighting methods suggest the presence of (at least) collector and condenser systems. Various modernized

Table 1.17 Examples of halogen lamps used in light microscopes and linear dimensions of the luminous body.

Name	Linear dimensions of the luminous body (mm)	Notes
KGM 6-20	1.8×1.8	To ensure the numerical aperture of the condenser $NA=0.90$, a linear magnification in the lighting system $V=13$–14x is required.
KGM 6-30	1.75×1.75	To ensure the numerical aperture of the condenser $NA=0.90$, a linear magnification in the lighting system $V=13$–14x is required.
KGM 9-70	2.6×2.5	To ensure the numerical aperture of the condenser $NA=0.90$, a linear magnification in the lighting system $V=10$–11x is required.
KGM 12-50	2.5×2.5	To ensure the numerical aperture of the condenser $NA=0.90$, a linear magnification in the lighting system $V=10$–11x is required.
KGM 12-100	3.0×3.0	To ensure the numerical aperture of the condenser $NA=0.90$, a linear magnification in the lighting system $V=8$–9x is required.

circuit solutions are known to improve the quality of lighting; such schemes typically include intermediate matching systems. In the practice of constructing light microscopes, as well as in specialized literature on the design of lighting systems, there are many examples of original lighting systems. However, almost all of them have a disadvantage: in some there are no intermediate image planes, others do not allow aggregation, and others do not provide telecentric lighting or have limitations on other indicators of functionality.

Our focus is on the principal optical schemes for constructing such systems; gabarits and aberration correction parameters may be the subject of a separate study. There is some discussion among developers and users of light microscopes about which is better: critical or Köhler-based lighting methods. But it makes little sense to give preference to one of them. A lot depends on the object of microscopy and the general observation conditions. The most important differences and advantages, as always, are contained in the nuances. The logical result of such discussions and studies can only be real images of real microscopic objects.

Figure 1.292 shows a principal optical diagram of a Lumam-R8 microscope (photographs of which are shown in Fig. 1.291). Here we describe the optical system of this microscope. In this microscope, there are two systems of illumination, reflected light (it is the main one for this type of microscope) and transmitted light, constructed according to the classical two-component scheme. A description of lighting systems is a good illustration of their different options.

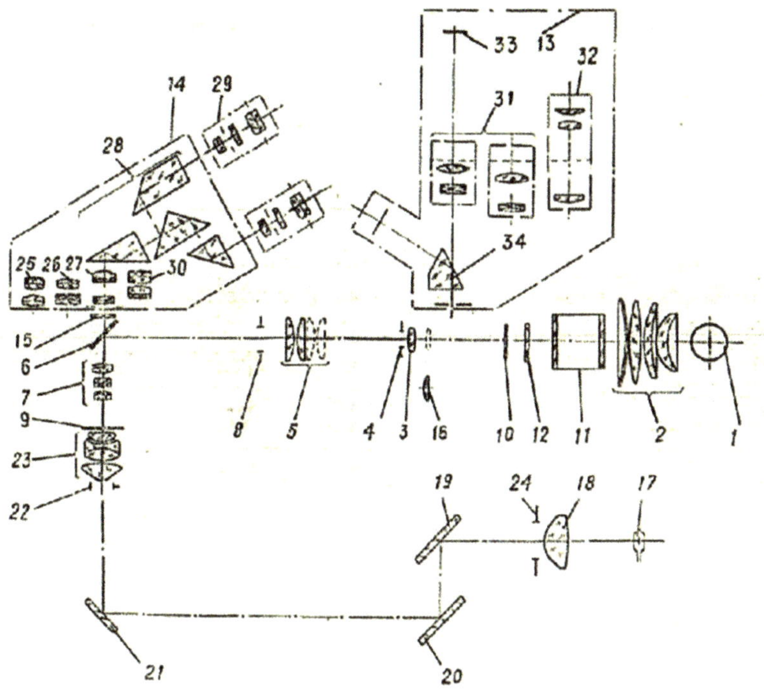

Figure 1.292 Principal optical layout of a Lumam-R8 microscope.

The principle of the Lumam-R8 microscope is based on the use of the phenomenon of fluorescence of objects, which occurs under the action of rays of a certain spectral composition. The study of objects in the light of fluorescence is possible while using the reflected light illuminator or the transmitted light illuminator. In some cases, it makes sense to use mixed lighting.

The *reflected light illuminator* operates as follows. The light source 1 by the collector 2 and lens 3 is projected into the plane of the aperture diaphragm 4, and then the lens system 5 and the beam splitter plate 6 are transferred to the plane of the exit pupil of the objective 7. The field diaphragm 8 by the objective 7 is projected into the plane of the object 9 from the total radiation of the source, the light that excites fluorescence, which is highlighted by means of interchangeable filters 10. To protect the filters from being heated by a 250-W mercury lamp, the system has a heat filter 12 and a special cuvette 11, which is filled with distilled water or a copper sulfate solution. The beam splitter plates 6 are interchangeable. One of the plates is coated, reflecting light mainly in the spectral range of 340 to 370 nm and transmitting light in the spectral range of 420 to 500 nm. This plate is used in the study of blue–blue luminescence excited by the long-wave portion of the UV spectrum. The other plate is coated with light reflecting mainly in the spectral range of 520 to 545 nm and transmitting light in the spectral range of 610 to 700 nm. This plate is used in the study of red luminescence by the excited green region of the spectrum.

To study the yellow–green–red luminescence excited by the blue-violet region of the spectrum, one of three plates is selected that is coated with reflective light in the spectral ranges of 400 to 440 nm, 450 to 480 nm, and 480 to 500 nm and transmitting light, respectively, in the spectral ranges of 500 to 700 nm, 520 to 700 nm, and 540 to 700 nm. The image of the object 9 in the light of fluorescence by the objective 7 is projected into the front focal plane of the eyepiece of the monocular head 13 or the binocular head 14. To eliminate the excitation scattered in the observation system, the so-called *blocking* filters 15 are used. Using this microscope, it is also possible to observe objects illuminated by the dark field method, for which lens 16 is turned on in the course of the rays; the lens system 5 is set to the position shown by the dotted line. Instead of a beam splitter plate 6, a special dark field mirror is installed; special objectives are installed in the turret (these objectives are shown in Volume 1, Chapter 1, Fig. 1.88).

When using a *transmitted light illuminator*, the light source 17 by the collector 18 and the mirror system 19, 20, and 21 is projected onto the plane by the aperture of the diaphragm 22 of the condenser 23. The image of the field diaphragm 24 by the mirror system 19, 20, and 21 and the condenser 23 is projected into the plane of the object 9. To excite luminescence with UV rays, light filters made of UFS6 glass with a thickness of 3 and 5 mm are used. To excite luminescence with blue-violet rays, FS1 glass filters with a thickness of 1, 2, 4, and 6 mm, CCI5 glass filters with a thickness of 2 and 4 mm, or an interference filter that emits excitation light from the total radiation of the source in the spectral range of 450 to 480 nm is used. To excite luminescence with green beams, a 3-mm-thick filter made of 3C11 glass is used. An ultrahigh-pressure mercury lamp is used as a source of illumination for the reflected light illuminator (the size of the luminous area is approximately 0.2–0.3 mm); in transmitted light, a KGM 9-70 halogen lamp is used.

As a comparison to the engineering solutions discussed above, let us consider another option for constructing the optical system of a microscope. Figure 1.293 shows a schematic optical diagram and a photograph of a Biolam-I microscope. This microscope is built on the same stand as the Lumam-R8 fluorescent microscope; however, the lighting systems of both transmitted and reflected light are constructed differently. The optical system of the Biolam-I microscope allows binocular observation and photographs to be taken of the studied objects during microscopy.

The *transmitted light illuminator* operates as follows. The light source 1 by the collector 2, the mirror system 3, 4, and 5, the *pancratical* (zoom) system 6, and the achromatic lens 7 is projected into the plane of the aperture

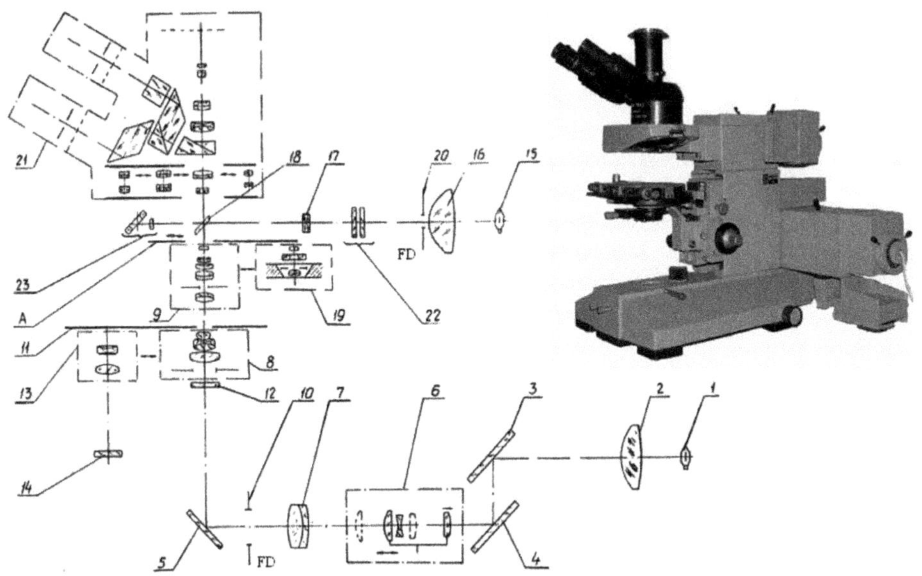

Figure 1.293 Schematic optical diagram and a photograph of a Biolam-I microscope.

diaphragm of the condenser 8 (type KON-1K; information about this condenser is presented in Volume 1, Chapter 1, Figs. 1.38 and 1.39), coupled to the output plane pupil of the objective 9. The field diaphragm 10 with a mirror 5 and a condenser 8 is projected into the plane of the studied object 11. When examining biological objects, interchangeable light filters can be introduced into the microscope lighting system 12. To illuminate large fields (when low numerical aperture 2.5× and 10× objectives), a condenser 13 with a numerical aperture of 0.22 and frosted glass 14 is used.

The *reflected light illuminator* operates as follows. The light source 15 by the collector 16, the lens 17, and the beam splitter plate 18 is projected into the plane of the exit pupil of the objective 9 (epi objective 19). The field diaphragm 20 with a lens 17, a beam splitter plate 18, and an objective is projected into the plane of the test object 11. The light reflected from the object passes through the objective and a beam splitter plate 18 and is projected into the front focal plane of the eyepiece of the binocular head 21. Interchangeable filters are used to increase the contrast of the observed object 22 installed in the lighting system after the collector 16. To work in a dark field, an annular diaphragm and a mirror are in the course of the rays, which is a reflector of the dark field 23. In this case, the rays from the light source 15 pass through the lens 17 with a wide beam. Then, the central part of the beam is shielded by an annular diaphragm, and the edge rays of the beam are guided by an annular mirror into the mirror condenser of the objective 19 and then onto the object.

This is considered one of the best of the old Soviet microscopes in terms of image quality (high-quality lighting and the use of plan apochromatic objectives). However, despite the use of a very progressive lighting system for transmitted light, based on the so-called pancratical (zoom) collector, it was not possible to achieve the universality of a condenser. As a result, the microscope used two types of capacitors that must be periodically changed during the microscopy process; each time the condenser was changed, the lighting needed to be adjusted. To make the condenser universal, i.e., to allow it to be used without major reconfigurations (without changing it) when working with the entire set of objectives, from a small increase (a large linear field at the object and a moderate numerical aperture) to a large linear increase (a small field at the object at the maximum numerical apertures), developers created many condenser designs. Some of them can be called universal. One example is a condenser in which three types of frontal lenses (different apertures and linear fields) can occupy a place on the optical axis using a revolving device. Another example is the so-called pancratical (zoom) condenser, which provides the required parameters of the linear field and numerical apertures by moving a group of lenses inside the mechanical structure. Figure 1.294 shows photographs of these condensers.

However, use of these condensers is not widespread in mass-produced budget microscopes due to the relative complexity of their designs. In microscopes of this class, the standard Abbe condenser is often used; the size of the front lens is increased mechanically in comparison with the classical one because it is believed that an acceptable result can be achieved by illuminating a large field at the object. But in practice this is not so. Achieving high uniformity causes problems, and the method of lighting large fields is characterized by the use of a light spot scattered randomly on the front lens. Even for budget microscopes, and even for less serious microscopic studies, such a condenser is not always suitable.

If we proceed from the concept of building a modern microscope, as described in Volume 1 (Chapter 2, Section 2.5), then the system of the so-called telecentric lighting device can become the basis for its optical design. Let us briefly recall the structure. In the front focal plane of the collector there is a light source. The front focal plane of each subsequent component coincides with the rear focal plane of the previous component, and the plane of the object under study is located in the rear focal plane of the condenser (the description of the optical system is given *from left to right*, as is customary in optical design practice). You can trace the logic of constructing basic optical schemes for telecentric lighting systems, for example, depending on the number of components. Figure 1.295 shows examples of telecentric lighting systems.

The first system uses only two components: a collector and a condenser. The light source 1 by the collector 2 and the condenser 4 is projected into the

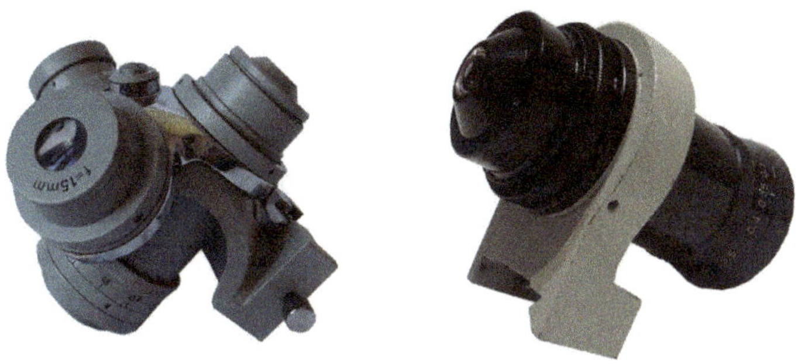

Figure 1.294 Universal condensers.

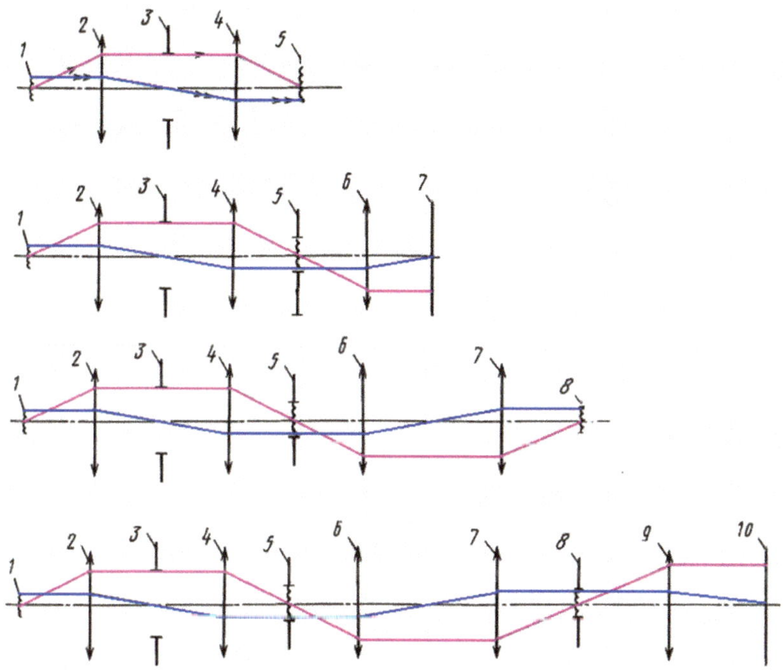

Figure 1.295 Examples of telecentric lighting systems.

plane of the studied object 5. The field diaphragm of the collector, conjugated with the rear focal plane of the collector and the front focal plane of the condenser, is projected into the plane of the exit pupil of the objective. This is a critical lighting method. The second system uses three components: a collector, a condenser, and a matching system located between them. The image of the light source 1 by the collector 2 and matching system 4 is built in the plane 5, the aperture diaphragm of the condenser 6, and then in the exit pupil of the objective. The field diaphragm 3, conjugated with the rear focal plane of the collector and the front focal plane of the matching system, is projected by the matching system and the condenser into the plane of the object under study 7. This is a Köhler-based lighting method.

The third system uses four components: a collector, a condenser, and two matching systems located between them. The image of the light source 1 by the collector 2 and matching system 4 is built in the plane 5, and then by the matching system 6 and the capacitor 7 in the plane of the object under study 8. Field aperture 3 through matching systems 4 and 6 is paired with the front focal plane of the condenser 7 and beyond with the exit pupil of the objective. This lighting method, although it refers formally to the critical method, is probably its modernization; it can be used successfully if the light source is sufficiently evenly luminous. The fourth system uses five components, three of which are coordinating. The image of the light source 1 by the collector 2 and matching systems 4, 6, and 7 is built in the front focal plane 8 of the condenser 9, and then in the plane of the exit pupil of the objective; however, in the optical scheme of the illuminator, there is a real image of the light source (plane 5, in which it is convenient to position the iris aperture diaphragm). Field diaphragm 3 due to matching systems 4, 6, and 7 and a condenser 9 is projected into the plane of the investigated object 10; however, in the optical scheme of the illuminator, there is a real image of the field diaphragm 3. It is located between components 6 and 7, where you can place the iris field diaphragm.

The number of components can be increased up to 6 or 7 units, for example, when building an optical system is required for an inverted metallographic microscope in which the distance from the light source to the plane of the object under study is significant. The optical design of the telecentric lighting system can be easily formalized and can be carried out using a primitive software algorithm; for example, the input parameters for the calculation are the size of the luminous body of the light source and the overall length of the system, and the output is the numerical aperture after the condenser and the size of the linear field on the object.

Let us consider an example of the overall calculation of a three-component lighting optical system with a telecentric main beam path that implements the Köhler-based lighting method. This system can be used in optical instruments in which high uniformity of lighting, high resolution, and high contrast are required. These are systems of photolithography and microscopy. Consider the example of microscopy. Figure 1.296 shows a drawing of such a system with the designations of the main planes, the positions of the field and aperture diaphragms, the planes of the location of the *filament* of the light source, and the object under study.

The light source (with a filament body size X) is located in the front focus of the collector system (with focal length $F1$); the material iris field diaphragm (whose size is Y) is located in its back focus. In turn, the iris field diaphragm is

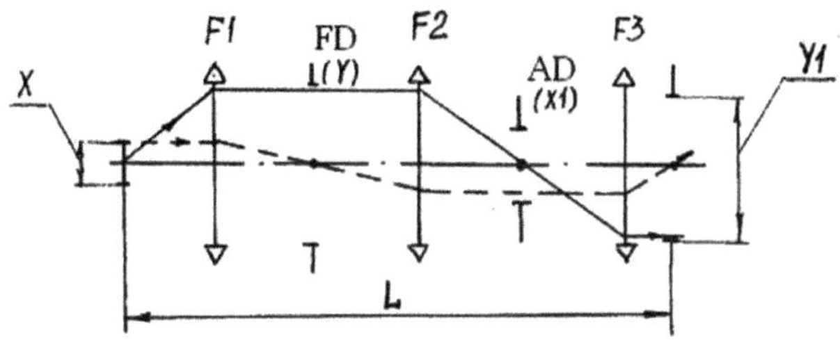

Figure 1.296 A three-components telecentric lighting system (similar to the second system shown in Fig. 1.295).

located in the front focus of the conjugating tube system (with focal length $F2$). The material iris aperture diaphragm (with an opening diameter of $X1$) is located in the back focus of the conjugating system. The iris aperture diaphragm is located in the front focus of the condenser system (with focal length $F3$). The illuminated object (the size of which is equal to $Y1$) is located in the back focus of the condenser system. From the relationships of geometric optics, we get

$$\begin{cases} X1 = X\dfrac{F2}{F1} \\ Y1 = Y\dfrac{F3}{F2} \\ L = 2 \cdot F1 + 2 \cdot F2 + 2 \cdot F3 \end{cases}$$

where L is the total length of the system.

Obviously, having given the values of the parameters X, Y, $X1$, and $Y1$ and solving the system of equations with respect to the parameter L, one can determine the focal lengths of the individual components of the lighting system. As an example, we will calculate a three-component lighting system with the following initial data (corresponding to existing and currently used accessories in microscopes):

- The size of the body of the glow of the light source $X = 2$ mm
- The field diaphragm diameter $Y = 40$ mm
- The aperture diaphragm diameter $X1 = 18$ mm
- The size of the illuminated object $Y1 = 2.5$ mm
- The total length of the system $L = 360$ mm

For convenience, the software program Mathcad was used.

$$F1=1 \quad F2=1 \quad F3=1 \quad X1=18 \quad X=2.0 \quad Y1=2.5 \quad Y=40 \quad L=360$$

Given

$$X1 \cdot F1 - XF2 = 0$$

$$Y1 \cdot F2 - Y \cdot F3 = \mathrm{L}$$

$$2 \cdot F1 + 2 \cdot F2 + 2 \cdot F3 = \mathrm{L}$$

$$\mathrm{vec} : = \mathrm{Find}(F1, F2, F3)$$

$$\text{Solution:} \quad \mathrm{vec}\begin{pmatrix} 17.041 \\ 153.373 \\ 9.586 \end{pmatrix}.$$

Given the initial values of $X1$, X, $Y1$, and Y and a given length of the system $L = 360$ mm, we obtain the following values of the focal lengths for the collector, the conjugate system, and the condenser: $F1 = 17.041$ mm, $F2 = 153.373$ mm, $F3 = 9.586$ mm, respectively. The obtained values correspond to values that are actually possible for the calculation of the lighting system. At the same time, the linear dependence of the calculated parameters on the total length of the lighting system is obvious; a change in the focal lengths of individual components is directly proportional to a change in length.

Thus, for the accepted value $L = 250$ mm, and all other things being equal, we obtain:

$$F1=1 \quad F2=1 \quad F3=1 \quad X1=18 \quad X=2.0 \quad Y1=2.5 \quad Y=40 \quad L=250$$

Given

$$X1 \cdot F1 - XF2 = 0$$

$$Y1 \cdot F2 - Y \cdot F3 = 0$$

$$2 \cdot F1 + 2 \cdot F2 + 2 \cdot F3 = L$$

$$\mathrm{vec} : = \mathrm{Find}(F1, F2, F3)$$

$$\text{Solution:} \quad \mathrm{vec}\begin{pmatrix} 11.834 \\ 106.509 \\ 6.657 \end{pmatrix}.$$

For the accepted value $L = 160$ mm, with other constant conditions, we obtain:

$$F1 = 1 \quad F2 = 1 \quad F3 = 1 \quad X1 = 18 \quad X = 2.0 \quad Y1 = 2.5 \quad Y = 40 \quad L = 160$$

Given

$$X1 \cdot F1 - XF2 = 0$$

$$Y1 \cdot F2 - Y \cdot F3 = 0$$

$$2 \cdot F1 + 2 \cdot F2 + 2 \cdot F3 = L$$

$$\text{vec} := \text{Find}(F1, F2, F3)$$

$$\text{Solution:} \quad \text{vec} \begin{pmatrix} 7.574 \\ 68.166 \\ 4.26 \end{pmatrix}.$$

Let us analyze the results obtained for the possibility of calculating components having a finite thickness along the axis and an acceptable relative aperture. In the first case, at $L = 360$ mm, the obtained values of the focal lengths of the thin components of the collector and the matching system are close to the real possible values. Thus,

the relative aperture of collector in this case equals $\frac{\varnothing_{FDcol}}{f'_{\text{col}}} = \frac{40}{17.04}$, i.e., 1/0.43;

the relative aperture of the conjugating system is $\frac{\varnothing_{FDcol}}{f'_{tl}} = \frac{40}{153.37}$, i.e., 1/3.83;

and the relative opening of the condenser is $\frac{\varnothing_{ADcon}}{f'_{\text{con}}} = \frac{18}{9.586}$, i.e., 1/0.53.

In the second case, with $L = 250$ mm, the creation of a lighting system is not possible because the relative openings of the collector and condenser systems are 1/0.30 and 1/0.37, respectively. The creation of a lighting system is also not possible in the third case, with $L = 160$ mm. Let us assume that an acceptable relative collector opening can take the value $\frac{\varnothing_{FDcol}}{f'_{\text{col}}} = 1 : 0.6$, which follows from the experience of calculating real optical systems of microscopes. To calculate the lighting system with the real possible values of the foci of the components $F1$, $F2$, and $F3$, we will adjust the initial parameters. We will bear in mind that it is necessary to change the relative aperture of the collector system by a factor of 1.4 (with respect to its value 1/0.43); as a result, it is necessary to decrease the initial value of the diameter of the collector field diaphragm *FD* (to a value of 28 mm, respectively).

For the value $Y = 28$ mm, the diameter of the field diaphragm of the collector, with other constant initial data, we will calculate the focal lengths of the components with the value of the total length of the system $L = 360$ mm:

$F1 = 1 \quad F2 = 1 \quad F3 = 1 \quad X1 = 18 \quad X = 2.0 \quad Y1 = 2.5 \quad Y = 28 \quad L = 360$

Given

$$X1 \cdot F1 - XF2 = 0$$

$$Y1 \cdot F2 - Y \cdot F3 = 0$$

$$2 \cdot F1 + 2 \cdot F2 + 2 \cdot F3 = L$$

$\text{vec} := \text{Find}(F1, F2, F3)$

$$\text{Solution:} \quad \text{vec}\begin{pmatrix} 16.661 \\ 149.95 \\ 13.388 \end{pmatrix}.$$

The following focal lengths of the components were obtained:

- Collector, $F' = 16.661$ mm (with a relative aperture of 1/0.6)
- Conjugating tube system, $F' = 149.95$ mm (with a relative aperture 1/5.36)
- Condenser, $F' = 13.388$ mm (with a relative aperture of 1/0.74)

At a given value of the diameter of the aperture diaphragm *AD*, its numerical aperture is $NA_{\text{con}} = \frac{\varnothing_{ADcon}}{2f'_{\text{con}}} = \frac{18}{2 \cdot 13.39} = 0.67$. This is sufficient when used to work on a general-purpose microscope in conjunction with objectives having their own numerical aperture up to 0.90–0.95, if we keep in mind that in order to increase the contrast when observing with a microscope, the optimal ratio of the numerical apertures of the condenser and objective should be 2/3. However, for working with high-aperture immersion objectives without loss of resolution on a microscope, such a value of the numerical aperture of the condenser is not sufficient.

To continue the research, we will take $NA_{con} = 0.9$, and then the diameter of the aperture diaphragm will be equal: $\varnothing_{ADcon} = NA \cdot 2f'_{\text{con}} = 0.9 \cdot 2 \cdot 13.39 = 24.1$ mm. However, for working with high-aperture immersion objectives without loss of resolution on a microscope, such a value of the numerical aperture of the condenser is not sufficient. To continue the research, we will take $NA_{con} = 0.9$, and then the diameter of the aperture diaphragm will be equal. We will change the source data to calculate the following:

- Diameter of the condenser's aperture diaphragm, $X1 = 24$ mm
- The size of the light source, $X = 2.0$ mm
- Illuminated field at the object under study, $Y1 = 2.5$ mm
- Diameter of the collector's field diaphragm *FD*, $Y = 28$ mm

In this case, we take the total length of the system to be $L = 360$ mm. After making calculations using the Mathcad software program, we obtain the following values of the focal lengths of the components:

$$F1=1 \quad F2=1 \quad F3=1 \quad X1=18 \quad X=2.0 \quad Y1=2.5 \quad Y=28 \quad L=360$$

Given

$$X1 \cdot F1 - XF2 = 0$$

$$Y1 \cdot F2 - Y \cdot F3 = 0$$

$$2 \cdot F1 + 2 \cdot F2 + 2 \cdot F3 = L$$

$$\text{vec} := \text{Find}(F1, F2, F3)$$

$$\text{Solution:} \quad \text{vec} \begin{pmatrix} 12.792 \\ 153.503 \\ 13.706 \end{pmatrix}.$$

In this case, the collector focal length of ~12.8 mm is not enough. Therefore, the size of the source must be larger. Let the source body glow size be $X = 4$ mm (instead of 2 mm). We calculate the focal lengths of the components:

$$F1=1 \quad F2=1 \quad F3=1 \quad X1=18 \quad X=2.0 \quad Y1=2.5 \quad Y=28 \quad L=360$$

Given

$$X1 \cdot F1 - XF2 = 0$$

$$Y1 \cdot F2 - Y \cdot F3 = 0$$

$$2 \cdot F1 + 2 \cdot F2 + 2 \cdot F3 = L$$

$$\text{vec} := \text{Find}(F1, F2, F3)$$

$$\text{Solution:} \quad \text{vec} \begin{pmatrix} 23.886 \\ 143.318 \\ 12.796 \end{pmatrix}.$$

With the system length $L = 360$ mm, the following focal lengths of the components were obtained: a collector $F' = 23.886$ mm, a conjugating tube system $F' = 143.318$ mm, and a condenser $F' = 12.796$ mm. In this case, the relative aperture of the collector is large, namely, 1/0.85. Therefore, we set the system length $L = 250$ mm:

$$F1 = 1 \quad F2 = 1 \quad F3 = 1 \quad X1 = 18 \quad X = 2.0 \quad Y1 = 2.5 \quad Y = 28 \quad L = 250$$

Given

$$X1 \cdot F1 - XF2 = 0$$

$$Y1 \cdot F2 - Y \cdot F3 = 0$$

$$2 \cdot F1 + 2 \cdot F2 + 2 \cdot F3 = L$$

$$\text{vec} := \text{Find}(F1, F2, F3)$$

$$\text{Solution:} \quad \text{vec} \begin{pmatrix} 16.588 \\ 99.526 \\ 8.886 \end{pmatrix}.$$

With the system length $L = 250$ mm, the following focal lengths of the components were obtained: a collector $F' = 16.588$ mm, a conjugating tube system $F' = 99.526$ mm, and a condenser $F' = 8.886$ mm. In the case of $L = 250$ mm, the real possible values of the focal lengths of the components of the microscope lighting system were obtained: a collector $F' \approx 16.6$ mm, a conjugating tube system $F' \approx 99.5$ mm, and a condenser $F' \approx 8.9$ mm for real possible relative holes of the components. The relative aperture of the collector is 1/0.59, and the numerical aperture of the condenser is $NA_{\text{con}} = \frac{\varnothing_{ADcon}}{2f'_{\text{con}}} = \frac{24}{2 \cdot 8.9} = 1.35$.

That is, it is possible to build a lighting system with the following characteristics:

- The size of the light source, 4 mm
- The size of the illuminated field, 2.5 mm
- Diameter of the condenser's aperture diaphragm, 24 mm
- Diameter of the collector's field diaphragm, 28 mm
- Numerical aperture of the condenser, $NA = 1.35$ (it may be to the oil immersion condenser)

For the final calculation, we will use a real halogen light source with a filament body size (diagonal) of 3.6 mm; using the same initial data, we will calculate the focal lengths of the components with the value of the total length of the system $L = 250$ mm:

$F1 = 1 \quad F2 = 1 \quad F3 = 1 \quad X1 = 18 \quad X = 2.0 \quad Y1 = 2.5 \quad Y = 28 \quad L = 250$

Given

$$X1 \cdot F1 - XF2 = 0$$

$$Y1 \cdot F2 - Y \cdot F3 = 0$$

$$2 \cdot F1 + 2 \cdot F2 + 2 \cdot F3 = L$$

$$\text{vec} := \text{Find}(F1, F2, F3)$$

$$\text{Solution:} \qquad \text{vec} \begin{pmatrix} 15.13 \\ 100.865 \\ 9.006 \end{pmatrix}.$$

With the system length $L = 250$ mm, the following focal lengths of the components were obtained: a collector $F' = 15.13$ mm, a conjugating tube system $F' = 100.865$ mm, and a condenser $F' = 9.006$ mm. In this case, the relative aperture of the collector, 1/0.54, is not suitable. We determine the focal lengths of the components with an increased length of the system $L = 275$ mm and the unchanged source data:

$F1 = 1 \quad F2 = 1 \quad F3 = 1 \quad X1 = 18 \quad X = 2.0 \quad Y1 = 2.5 \quad Y = 28 \quad L = 275$

Given

$$X1 \cdot F1 - XF2 = 0$$

$$Y1 \cdot F2 - Y \cdot F3 = 0$$

$$2 \cdot F1 + 2 \cdot F2 + 2 \cdot F3 = L$$

$$\text{vec} := \text{Find}(F1, F2, F3)$$

$$\text{Solution:} \qquad \text{vec} \begin{pmatrix} 16.643 \\ 110.951 \\ 9.906 \end{pmatrix}.$$

At $L = 275$ mm, the following focal lengths of the components were obtained:

- Collector, $F' = 16.643$ mm (with a relative aperture of 1/0.6)
- Conjugating tube system, $F' = 110.951$ mm (with a relative aperture of 1/4)
- Condenser, $F' = 9.906$ mm (with a relative aperture of 1/0.4)

In this case, when conducting an aberration calculation, we can work with these data.

Such an illuminator is universal in terms of optical design; it allows similar optical circuits to be built for illuminators of transmitted and reflected light, in which the objective itself acts as a condenser. In the corresponding planes conjugated with the front focal plane of the condenser (objective in the case of a reflected light illuminator), various additional elements can be placed for implementation in a microscope, including contrast methods. Figure 1.297 shows a technical drawing of the Polam-LPO, a budget Russian microscope from the unified line, the so-called *common series* (biological, metallographic, polarizing, and others), in which a telecentric lighting system was successfully implemented for the first time.

In this microscope (as in other common series unified microscopes), five-component telecentric systems are used in illuminators of transmitted and reflected light. The knobs for adjusting the field and aperture diaphragms are located in convenient places, improving the ergonomic parameters of the microscope. The transverse dimensions of the illuminators are balanced with the overall dimensions of the microscope.

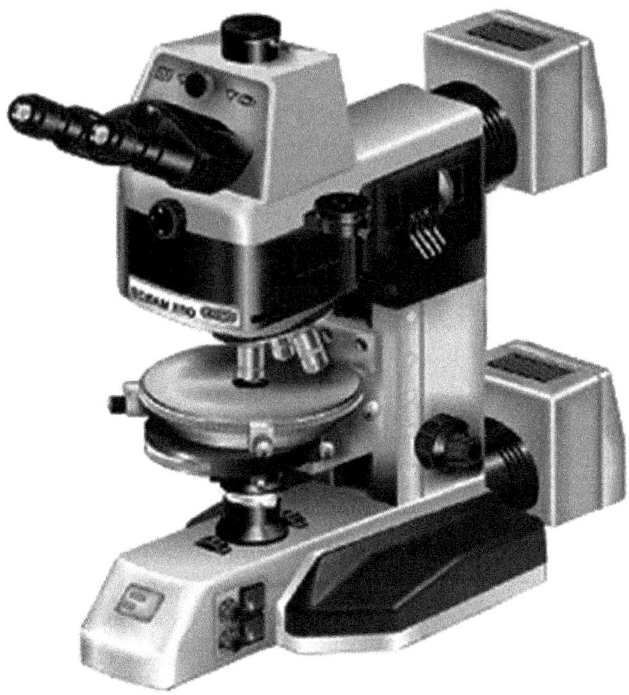

Figure 1.297 Technical drawing of a Russian Polam-LPO microscope.

1.8.6.2 Simplified lighting systems

Figure 1.298 shows an example of an optical–mechanical design of a simple transmitted light illumination system for a budget microscope. Of course, developers are constantly striving to optimize and simplify the applied lighting systems of a light microscope. One interesting engineering solution is to use a lamp combined with a reflector as a light source without using a lens collector. In this case, it is possible to make the optical design in such a way as to obtain a uniformly illuminated circle in the plane of the entrance pupil of the condenser (in which the influence of central shielding is not felt). In this case, the numerical aperture of such a modified collector system is sufficient to obtain high-quality lighting. Another interesting engineering solution, which can significantly simplify the optical system of the illuminator, is the use of polyfocal single lenses. The principle of operation of polyfocal single lenses is based on the possibility of changing the radii of curvature and thickness in the center of the lens from a special optical material, with a radial effect on the edge of the lens. When the lens "squeezes," its parameters such as focal length and optical power, respectively, change. Such a lens can be used as a kind of pancratical (zoom) element of the illuminator.

The ability to use modern LED technology provides an exceptional value when building simplified lighting systems. A detailed discussion of this is in Volume 1 (Chapter 2, Section 2.15). Here, let us discuss options for the practical use of LED technology and show that it can lead to a significant simplification of not only the lighting system as such, but also the microscope as a whole, without loss of image quality. We will proceed from the fact that LED technologies made possible a fundamental technical solution that was not previously available, namely, the ability to directly increase the size of the luminous body of the light source. At the same time, it became possible to significantly improve the uniformity of light over the area of the source due, among other things, to the development of technologies for the production of transparent plastics.

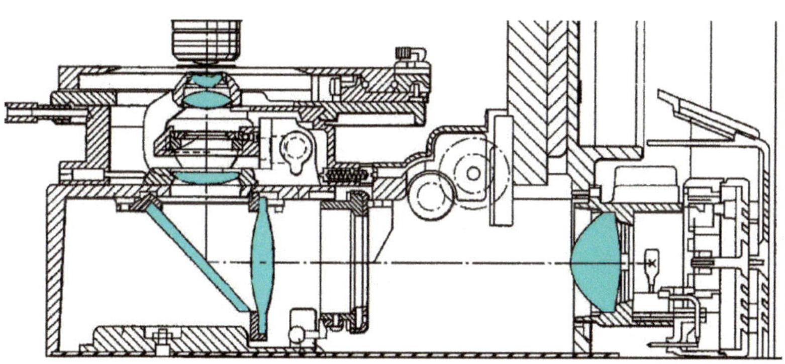

Figure 1.298 Design of a simple transmitted light system for a budget microscope.

1.8.6.2.1 Light panels with LEDs arranged in a circle

The optical design of a lighting system can be simplified by using an LED panel, which includes many single LEDs located inside the panel area at the minimum possible distance apart from each other. The outer plane of the panel is made of transparent plastic, which allows a high degree of uniformity of illumination to be achieved throughout the entire surface of the panel (single LEDs are practically not visible). With a critical method of illumination in a microscope, such an LED panel can be located in the so-called plane of the field diaphragm. However, the collector in such a system is absent. The image of the LED panel by means of a capacitor is built in the plane of the object under study. In this case, the optical design can be carried out in such a way that it becomes possible to fulfill the requirements of lighting a large field (for lenses with small linear magnifications) and to obtain a high numerical aperture for lenses with medium and large linear magnifications. In this case, the entrance pupil of the condenser is filled with chaotic light from the LED panel. The mechanical design of the condenser in this case can be simplified; there is no need to use an iris diaphragm. In fact, the system may not have any iris diaphragms. Figures 1.299–1.301 are photographs of real samples using this engineering solution.

With a Köhler-based lighting method, the LED panel can be located in front of the condenser (we can assume that it is located in the equivalent front focal plane of the condenser, i.e., in the plane of the entrance pupil). Because the main goal in a light microscope is to evenly fill the exit pupil of the objective with light, an engineering solution that allows you to place the light

Figure 1.299 Photograph of a real object using a standard 10× objective (LED panel in a field diaphragm).

Figure 1.300 Photograph of a real object using a standard 20× objective (LED panel in a field diaphragm).

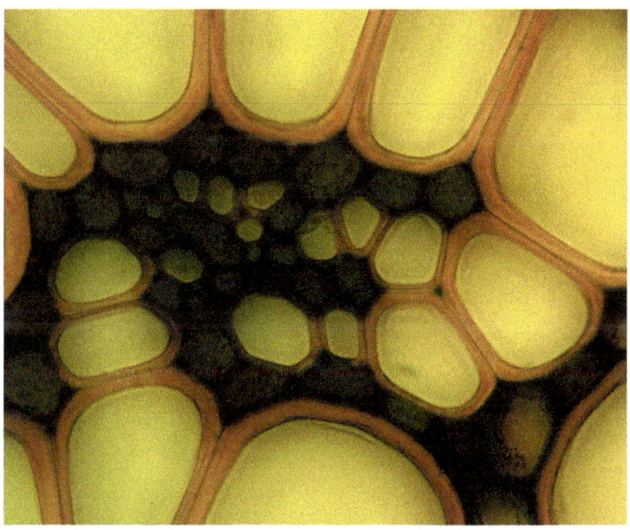

Figure 1.301 Photograph of a real object using a standard 40× objective (LED panel in a field diaphragm).

source directly on the plane of the entrance pupil of the condenser is valid because these two planes are conjugated in the microscope. In this case, the diameter of the entrance pupil (any of the mechanical frames located nearby) of the condenser will work as an aperture diaphragm; it is possible to make an optical design such that the illumination of large fields (working with low-magnification objectives) is provided with chaotic scattered light after the front lens of the condenser. If we assume that the average size of the entrance pupil of the condenser is 25–30 mm, the use of an LED panel of such a diameter is absolutely valid. Figures 1.302–1.304 are photographs of the same object obtained using an engineering solution in which the LED panel is located in the plane of the entrance pupil of the condenser.

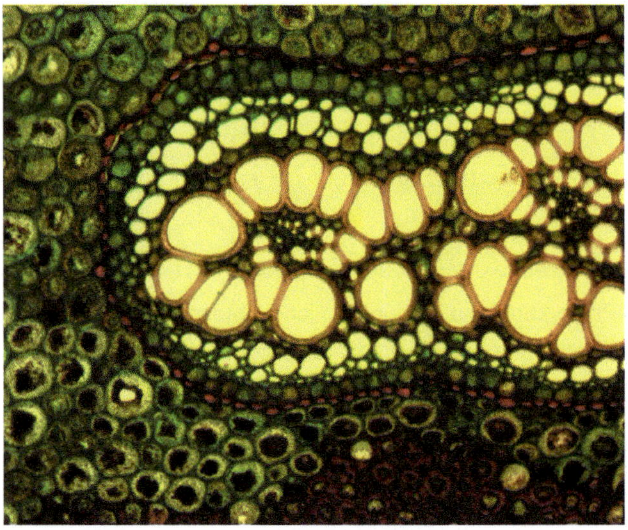

Figure 1.302 Photograph of a real object using a standard 10× objective (LED panel in the entrance pupil of the condenser).

Figure 1.303 Photograph of a real object using a standard 20× objective (LED panel in the entrance pupil of the condenser).

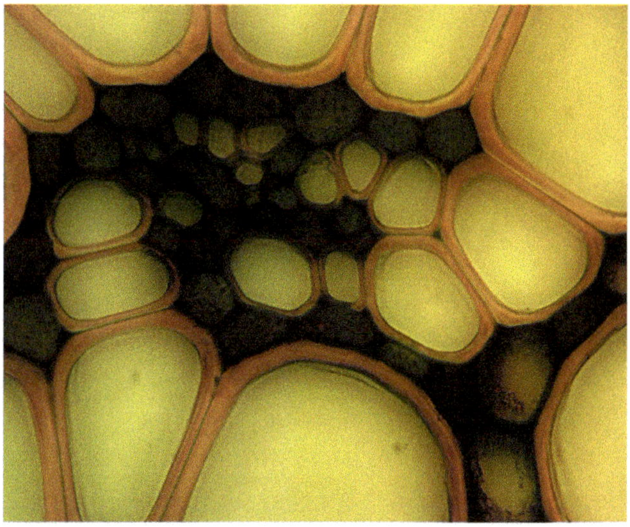

Figure 1.304 Photograph of a real object using a standard 40× objective (LED panel in the entrance pupil of the condenser).

Because electrical connections are possible in such an LED panel, contrast systems can be built. The most common can be used to adjust color temperature as well as the spectral range of the emitted light. These can be controlled (on/off) for each LED (from the set included in the panel) separately or in combination using, for example, specialized algorithms. It is possible to use such an LED panel as a component for a contrast system in a microscope; indeed, it is possible to form a different configuration of the on/off LEDs, enabling microscopy methods such as dark field, oblique lighting, and certain types of phase and contrast methods. Particularly interesting results can be obtained by combining the capabilities of liquid crystal technologies (whose potential uses are discussed above) and LED technologies. For example, to implement a phase contrast methodology on a microscope, liquid crystal technology can be used to build phase rings on plates or lenses located in the plane of the exit pupil of the objective, and LED technology can be used to build a profile of light rings on a luminous LED panel placed in the plane of the input pupil of the condenser. Such an illuminator can become the basis for the rapid transformation of bright field objectives into dark field objectives; the photographs in Fig. 1.305 illustrate this idea. On the left is a circular illuminator based on the so-called fiber illumination, and on the right is one based on LEDs. However, such an engineering solution is possible only for objectives with a small linear magnification because the numerical aperture of the emitted rays is small.

Figure 1.305 Examples of circular illuminators.

1.8.6.2.2 Light panels based on LED strips

In this embodiment, many single LEDs are also used; they are located only around the circumference of a certain plastic disk. Although such an engineering solution does not allow certain special contrasting methods to be implemented in a microscope, it makes it possible to significantly simplify a microscope's design while preserving image quality as compared with classical illumination. Figure 1.306 shows the design of such an illuminator.

The illuminator is a circular disk made of light-conducting plastic. On one of the surfaces, a reflective coating can be applied, and on the other, a translucent coating with a series of microlenses can be applied. The combination of these is a lens raster. For example, the LED strip can be located on the ends of this disk. The light rays emitted by the LEDs are randomly reflected from both surfaces of the disk, providing not only uniform illumination on one of the surfaces (above), but also a certain numerical aperture of the emerging rays. Through calculation and experimental methods, the so-called *lens raster pitch* can be selected to provide an acceptable balance between the resolution of the illuminator and the contrast in the light rays. Experiments on the use of such an illuminator with a conventional light microscope show high functionality and an acceptable aperture of the lighting beams. We can refuse to use many parts of a conventional light microscope, such as a flashlight with a light source, a collector system, intermediate matching lenses, and a condenser. Such an illuminator can serve as a stage while the investigated object is located directly on the working surface of the illuminator. It can also be used in the engineering solutions described above. Figure 1.307 shows a photograph of a

Figure 1.306 The design of an illuminator.

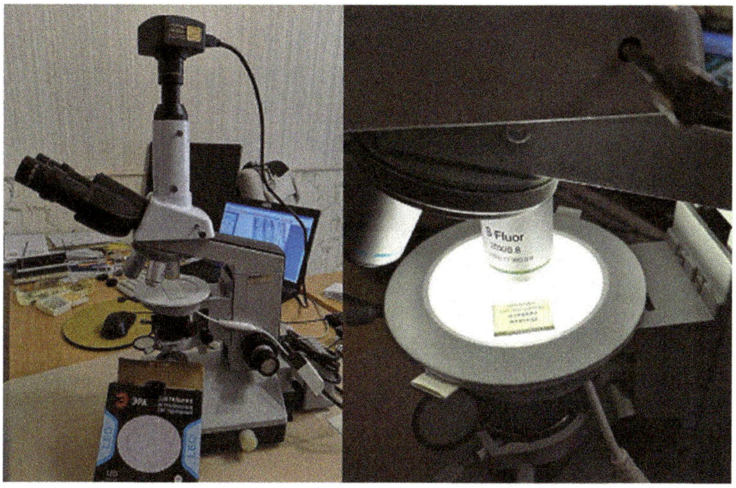

Figure 1.307 An LED illuminator is used as an object stage.

microscope in which such an illuminator is used as an object stage. Figures 1.308–1.324 show images of real objects using such an illuminator (it was mounted on a microscope instead of a stage) with objectives of various linear magnifications and numerical apertures.

Such an illuminator, slightly modified (it is necessary to have a central hole of the required diameter), can become the basis for an engineering solution for the unification of objectives that use a special lighting mirror to

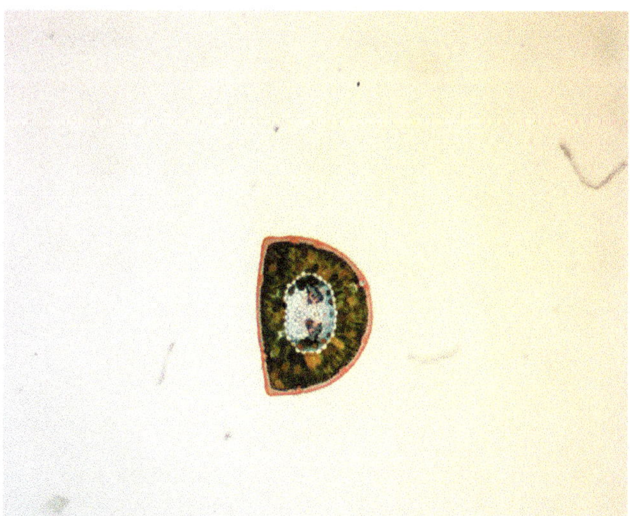

Figure 1.308 Photograph of a real object using a standard 4× objective (LED panel instead of a stage).

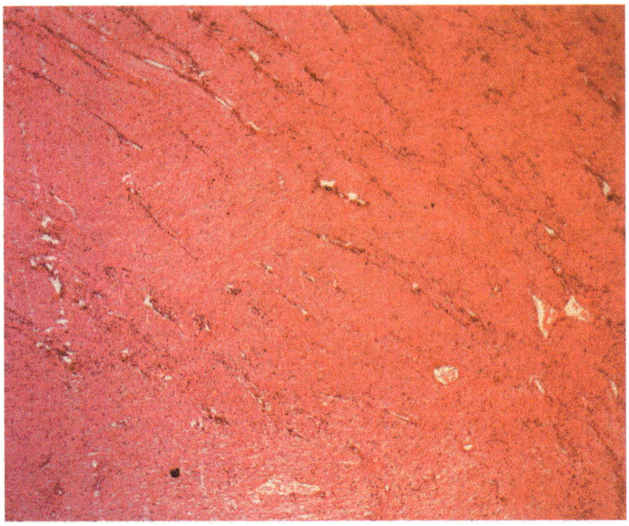

Figure 1.309 Photograph of a real object using a standard 4× objective (LED panel instead of a stage).

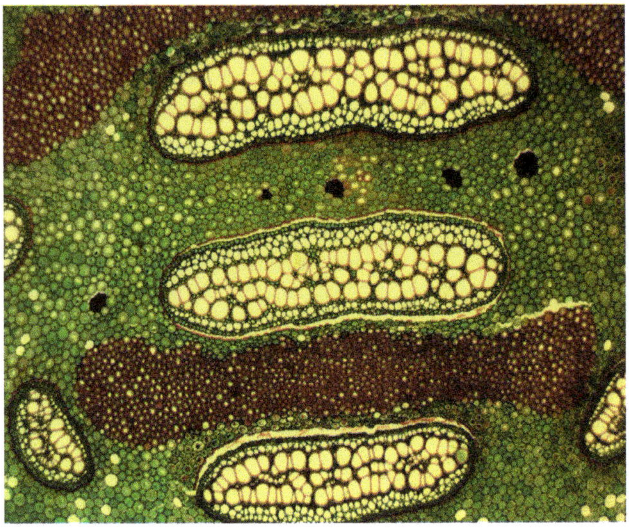

Figure 1.310 Photograph of a real object using a standard 4× objective (LED panel instead of a stage).

work in bright and dark fields. Figure 1.325 is a photograph of an objective that illustrates this.

Traditionally, to integrate such a mirror into the optical system of the lens (to make it a dark field objective), its diameter is increased where the objective is screwed to the nosepiece (or another special mount is used, as shown in Fig. 1.325). This leads to an increase in the dimensions of the revolving device.

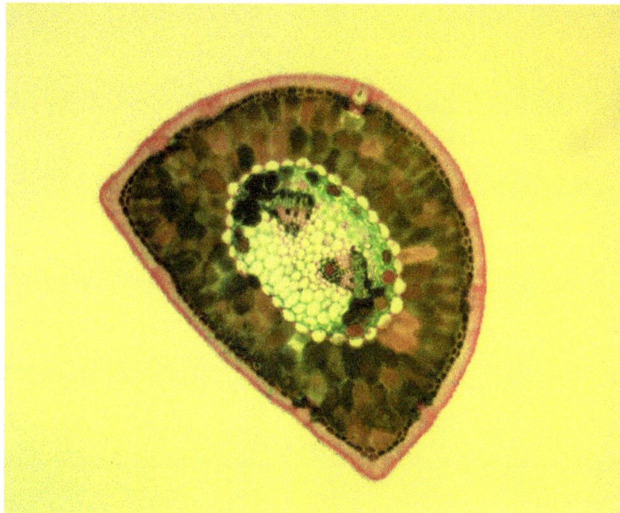

Figure 1.311 Photograph of a real object using a standard 10× objective (LED panel instead of a stage).

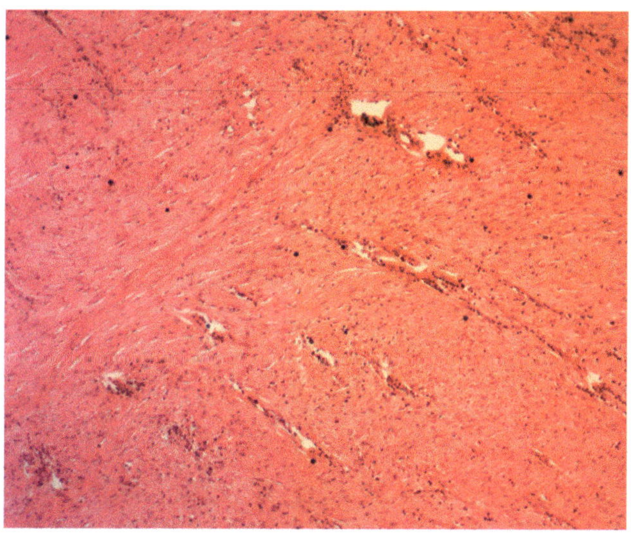

Figure 1.312 Photograph of a real object using a standard 10× objective (LED panel instead of a stage).

A so-called *Beck plate* (or similar reflector) is also used in the lighting system of metallographic microscopes, for example, to increase the dimensions of the illuminator itself. Bright field objectives initially have a smaller diameter where the thread of the mount to the nosepiece is located. The goal of using such an illuminator is to “dress” it on a bright field objective but use only the

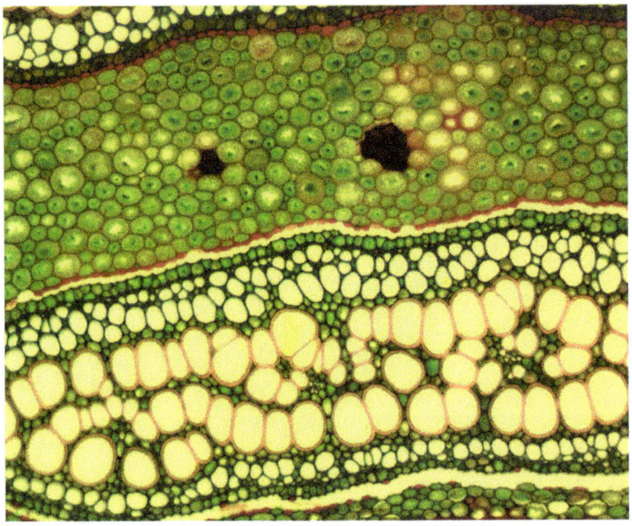

Figure 1.313 Photograph of a real object using a standard 10× objective (LED panel instead of a stage).

Figure 1.314 Photograph of a real object using a standard 20× objective (LED panel instead of a stage).

space near the front (nose) of the objective. In this case, the use of a special mirror is also required (since it provides the necessary numerical aperture of illumination), but its own longitudinal dimensions are small. Thus, it is possible to build an optical–mechanical unit consisting of the described illuminator and mirror; the addition of this unit to a conventional bright field objective allows it to work as a dark field objective.

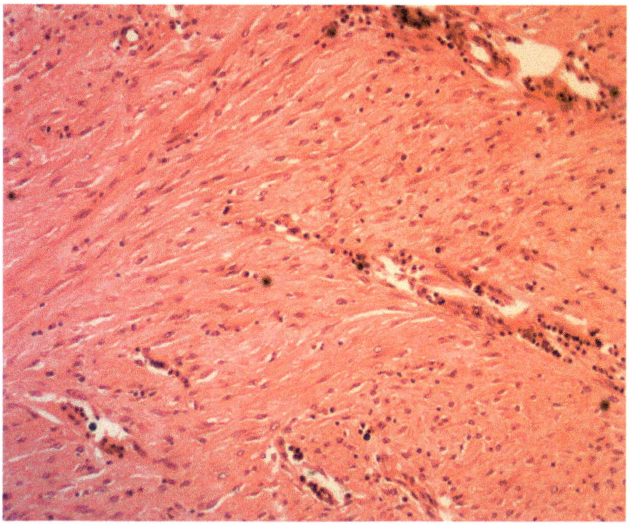

Figure 1.315 Photograph of a real object using a standard 20× objective (LED panel instead of a stage).

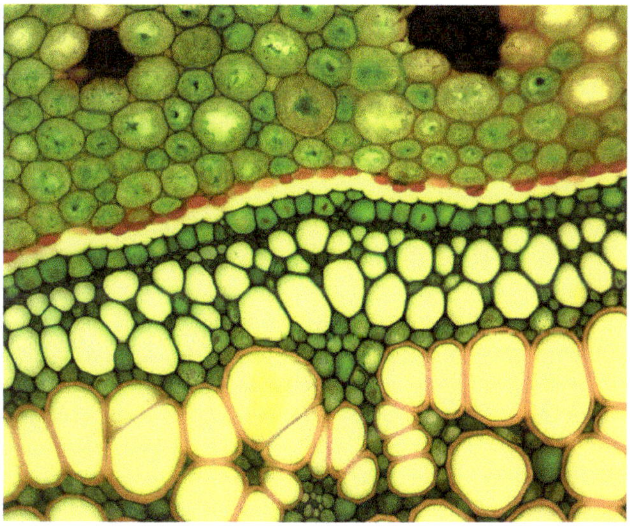

Figure 1.316 Photograph of a real object using a standard 20× objective (LED panel instead of a stage).

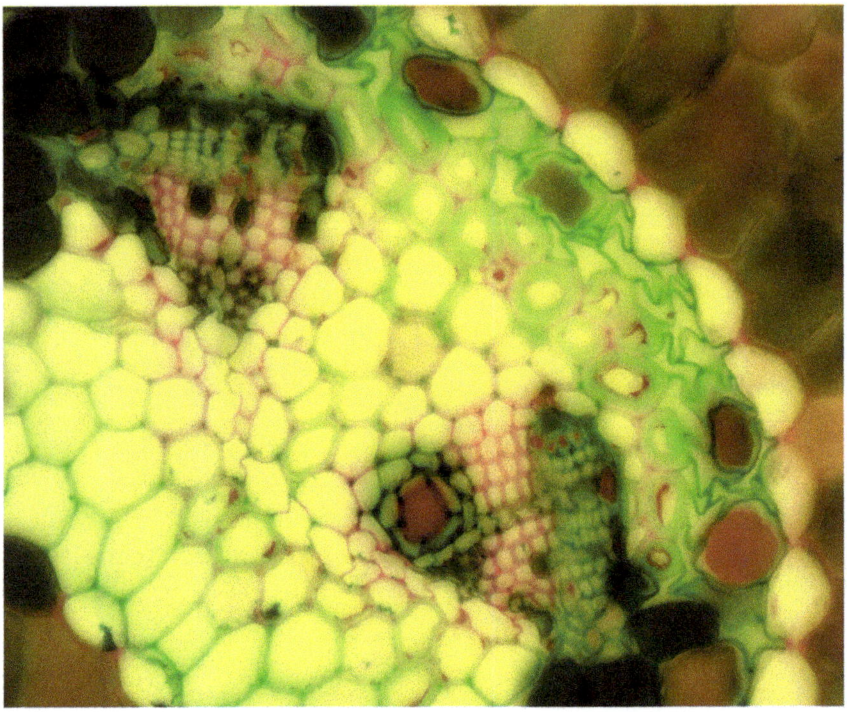

Figure 1.317 Photograph of a real object using a standard 40× objective (LED panel instead of a stage).

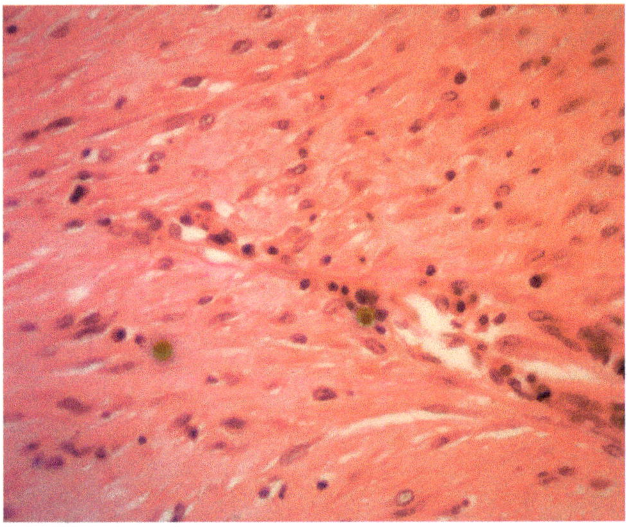

Figure 1.318 Photograph of a real object using a standard 40× objective (LED panel instead of a stage).

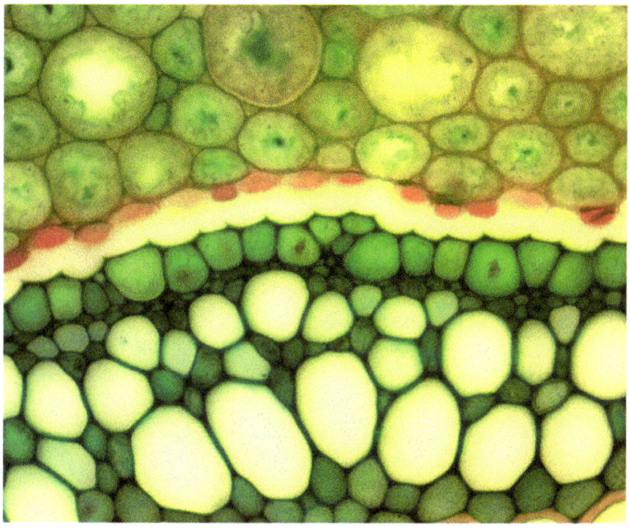

Figure 1.319 Photograph of a real object using a standard 40× objective (LED panel instead of a stage).

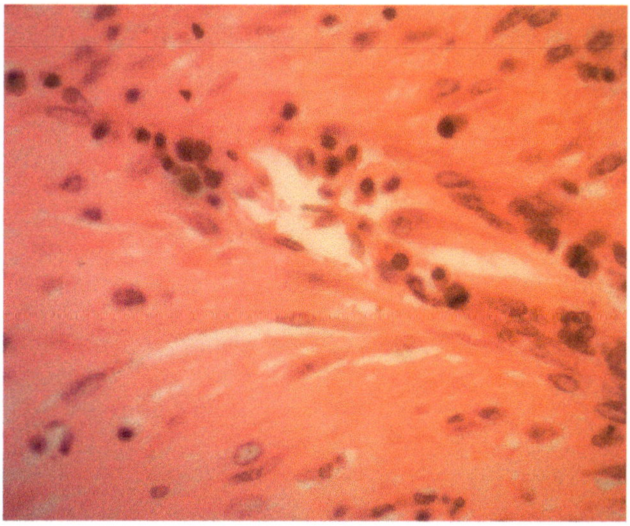

Figure 1.320 Photograph of a real object using a standard 60× objective (LED panel instead of a stage).

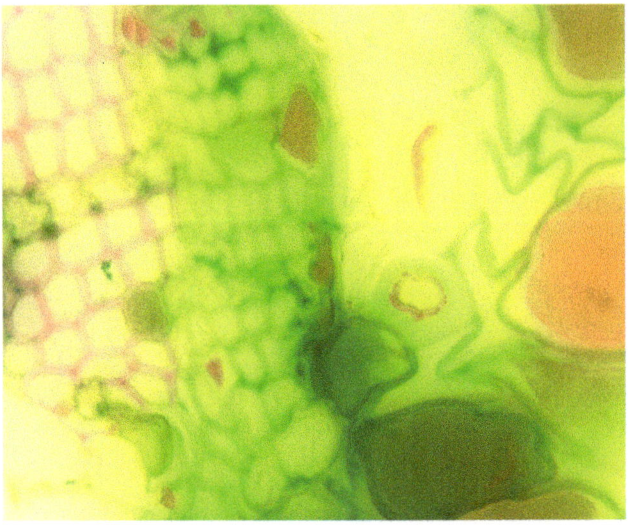

Figure 1.321 Photograph of a real object using a 100×/1.20 water immersion objective (LED panel instead of a stage).

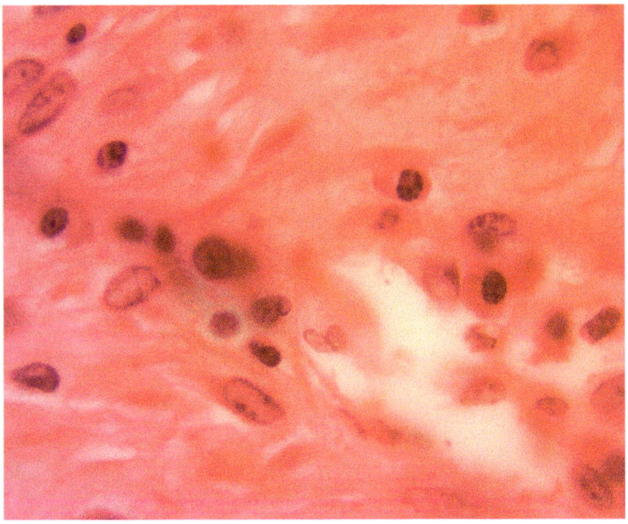

Figure 1.322 Photograph of a real object using a 100×/1.20 water immersion objective (LED panel instead of a stage).

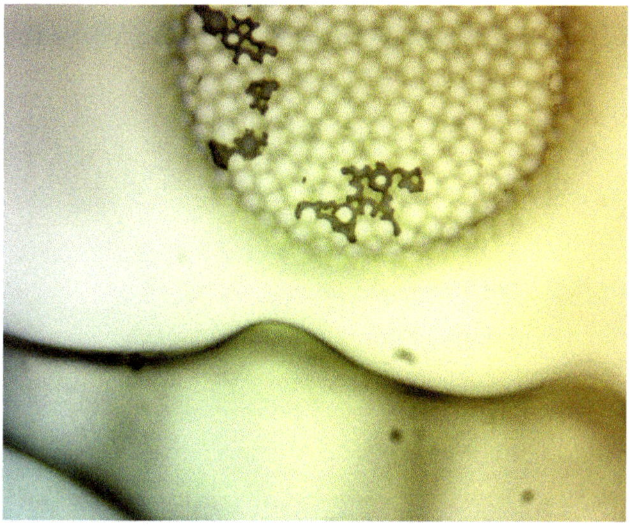

Figure 1.323 Photograph of a real object using a 100×/1.35 oil immersion objective (LED panel instead of a stage).

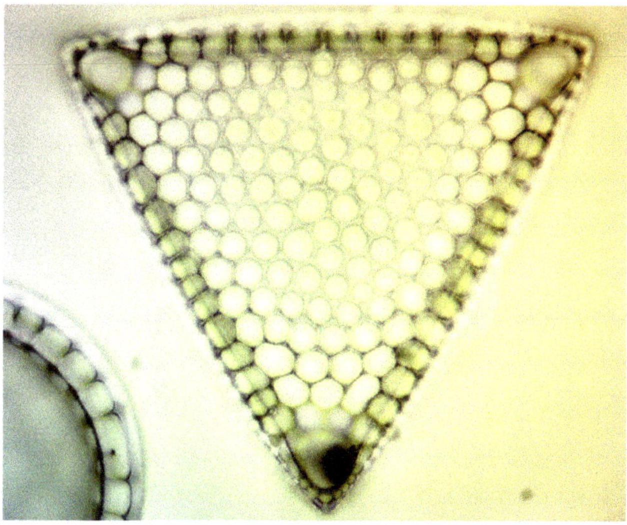

Figure 1.324 Photograph of a real object using a 100×/1.35 oil immersion objective (LED panel instead of a stage).

Figure 1.325 Example of bright–dark field objective building.

References

[1] M. Berek, *Fundamentals of Practical Optics*, Gostekhizdat, Moscow (1933).

[2] T. A. Ivanova, R. M. Larina, and L. A. Bogdanova, "Microscope objectives of small magnifications with improved parameters," *J. Opt. Technol.* No. 3 (1986).

[3] V. N. Churilovsky, *Theory of Optical Instruments*, Mashinostroenie, Leningrad (1966).

[4] Research Report No. NG2-455-81, "Investigation of the possibility of creating a set of improved achromatic objectives for large-scale microscopes of the Biolam C and R series," LOMO, Saint Petersburg (1981).

[5] A. P. Grammatin, "Some differential properties of aplanatic surfaces and the use of these properties to evaluate higher order aberrations. Modern methods of calculation and design of optical systems," *GOI Proceedings* **37**(167) (1970).

[6] G. G. Slyusarev, *Optical System Calculation Methods*, Mashinostroenie, Leningrad (1969).

[7] T. A. Ivanova, "Calculation of three-lens glued components with a reduced secondary spectrum," *J. Opt. Technol.* No. 9 (1979).

[8] M. M. Rusinov, *Technical Optics*, Mashinostroenie, Leningrad (1979).

[9] T. A. Ivanova and V. K. Kirillovsky, *Design and Control in the Optics of Microscopes*, Mashinostroenie, Leningrad (1984).

[10] V. A. Zverev and T. A. Ivanova, "The method of converting optical systems of a symmetric type," *J. Opt. Technol.* No. 6 (1985).

[11] H. Boegehold, *Das Optische System des Mikroskops*, Verlag Technik, Berlin (1958).

[12] A. P. Grammatin, L. S. Agroskin, and R. M. Larina, "Field illumination in reflected light microscopes," *J. Opt. Technol.* No. 3 (1969).
[13] A. G. Arlievsky, "Investigation of first-order reflexes in the objectives of microscopes of reflected light," *J. Opt. Technol.* No. 4 (1979).
[14] D. N. Frolov, O. A. Vinogradova, V. N. Frolov, and P. S. Vakulov, "Optical design and unification of optical systems of objectives for microscopes," *Proc. SPIE* **10690**, 1069021 (2018) [https://doi.org/10.1117/12.2311799].
[15] O. A. Vinogradova, A. D. Pavliy, A. D. Frolov, and D. N. Frolov, "The concept of constructing system for adaptive-selective assembly of optical devices," *Proc. SPIE* **7849**, 78492E (2010) [https://doi.org/10.1117/12.870322].
[16] User manual, "BIOLAR polarization and interference microscope," Polskie Zakłady Optyczne, WKC Warszawa (1976).
[17] Optical Society of America, *Handbook of Optics, Volume 1: Fundamentals, Techniques, and Design*, 2nd ed., McGraw-Hill, New York (1994).
[18] D. N. Frolov, O. A. Vinogradova, L. Wen, and F. Jing, "Optical design of budget objectives for mass production microscopes," *Proc. SPIE* **10814**, 1081415 (2018) [https://doi.org/10.1117/12.2500425].

Chapter 2
Synthesis of Optical Systems for Medium Series Budget Microscopes

The classification of light microscopes into mass-produced, *medium series*, and other categories is conditional. The production of optical, optical–mechanical, and optical–electronic products is primarily based on market demand. It is not always true that there is less demand for more expensive and complex products. It is often the case that modules and accessories are designed to be interchangeable, allowing them to be used in products of various price points and technical specifications. This applies to microscopes, in which most of the nodes and accessories are unified and interchangeable. The optimal choice of the configuration of a particular microscope for solving specific research problems and the effectiveness of the research itself depend on the qualifications of the engineer completing the microscope as well as the experience of a researcher working with the microscope.

Often, the use of a medium series mass-produced device can successfully solve a research problem without the need for more expensive equipment. For example, using objectives of a different class on the tripods of ordinary microscopes can improve the consumer qualities of these standard instruments. The same applies to the retrofitting of conventional microscopes with additional accessories, such as those using contrast methods for research (polarization, dark field, phase contrast, and others), as well as equipping imaging systems with digital receivers. Many researchers follow this path, gradually equipping their microscope with various additional accessories and complicating its original basic configuration; the sequence of the microscope's improvement by the trial-and-error method makes it possible to obtain the optimal configuration for solving a research problem. In this context, the main task for microscope engineers and developers is to ensure the possibility of such modernization and customization of standard microscopes with additional equipment. Another task is to develop and offer a wide range of standard and original components and devices for initiating such creative customization by researchers working with a microscope. In this sense, a fairly clear separation can be made between mass-produced and medium series class microscopes. In the first case, the microscope is built to perform highly specialized tasks, for example, in biology or metallography. In the second case, the microscope is a device capable of some transformation and adjustment to the solution of original research problems; that is, it has more pronounced signs of universality. Of course, both classes of microscopes must meet the requirements of universality, maybe only to varying degrees.

Experienced engineers and researchers sometimes use a microscope (as a rule, medium series instruments) not quite for its intended purpose, that is, to solve nonstandard research tasks. For example, in biology, the polarization technique (and a full-fledged polarization microscope) can be very effectively used because some microorganisms have inclusions with special birefringence properties. In a transmitted study technique, these inclusions can be effectively detected. A polarizing microscope can be used very effectively for metallography and forensic science when working, for example, in reflected light. To solve many research problems, a combination within the same configuration, for example, polarization contrast and fluorescence techniques, can be very effective.

At the same time, objectives remain the main elements of a light microscope that can most effectively transform a microscope to solve an original research problem. Engineers must provide the researcher, first and foremost, with a wide range of objectives with various technical and consumer parameters. The researcher should be able to choose objectives from a wide range of products and compare research results when using different objectives.

2.1 An Example of Creating LCF Plan MicroFluar Objectives with Enlarged Numerical Apertures and 27-mm Linear Fields

Over many years, microscopy has developed some traditions regarding the values of linear magnifications, numerical apertures, and linear fields of objectives, depending on the class of the microscope models in which objectives are used (for basic equipment). There are even some regulatory documents describing what linear magnification values need to be applied when designing objectives for various tasks. With respect to numerical apertures and linear fields, although there are no such clear recommendations, there are some conventions when higher-class objectives, higher numerical apertures, and larger linear fields are implemented on a microscope. Naturally, objectives that implement higher numerical apertures and linear fields have more complex than usual optical structures. In addition, optical materials can be used in these objectives, which differ from ordinary objectives in their refractive and dispersive properties. The use of optical materials enables a higher level of quality and consumer properties of the microscope. In addition, these kinds of objectives use glass type H-FK61 as an optical material for producing many positive lenses; this glass is characterized by the parameters $n_d = 1.497$, $\gamma_d = 81.6$, and $n_{F-C} = 0.00609$, and allows for a significant reduction of chromatic aberrations, particularly of the secondary spectrum.

Here, it is worth remarking on the name of this type of aberration correction. Some manufacturers use "fluor," "fluotar," and similar words to name their objectives. One may assume that these objectives contain fluorite as part of their optical systems. However, there is no connection between this name and the type of optical correction of an objective; objectives can have achromatic, semi-apochromatic, or apochromatic aberration correction. At the same time, it would be logical to establish a connection between the terms fluor, fluotar, and others with the type of aberration correction of the objective (the degree of correction of chromatic aberrations, in particular, in the secondary spectrum). This is also logical because an apochromatic correction of chromatic aberrations without the use of fluorite in the composition of an objective's optical system can be achieved. The use of optical glass type H-FK61 can significantly reduce the value of the secondary spectrum and achieve semi-apochromatic aberration correction in objectives. We will indicate the type of aberration correction for objectives whose aberration of the secondary spectrum is reduced as "MicroFluars" regardless of whether their optical systems contain fluorite or not.

When we talk about modern objectives for polarization research, there are at least two parameters that we would like to improve. Essentially, modern objectives refer to budget objectives that can be used in the laboratory and on polarizing microscopes. First, the linear ocular field of these microscopes (and objectives) should be increased. Second, the numerical aperture of these objectives must also be increased. Both of these parameters significantly affect the quality of the studies being performed on the microscope. Correcting chromatic aberration is very important in polarization microscopy [1]. Polarized objects display different colors for different parts and fragments, and proper identification of these requires good color reproduction without distortion. The primary chromatic aberration is always corrected. Until recently, the secondary chromatic aberration was not corrected in polarizing objectives (especially in budget objectives). Of course, this is a very big disadvantage, particularly if surveillance on a large field is attempted. A 27-mm ocular linear field provides an enhanced quality of visual observation on a polarizing microscope. Therefore, my colleagues and I developed objectives that have a secondary aberration chromaticity corrected [lateral color free (LCF)]. We also began to reduce the distortion aberration, which can interfere with measurements of image fragments. Correction of curvature of the image is important because polarized objects are usually flat. It is necessary to avoid the appearance of non-sharpness and reduce the contrast in the image during projection through the objective, especially in the case of large-magnification objectives. Also, it is very beneficial to correct astigmatism and coma in inclined beams.

Figures 2.1–2.7 show the layouts and design parameters of objectives with enlarged numerical apertures and 27-mm linear fields in the image space. The optical designs of these objectives enable a high quality of aberration correction. The lateral color (LC; chromatism of magnification) is corrected; we call this type of aberration correction *LCF*. The secondary spectrum is reduced. We call this set of objectives the *Seventh Series LCF plan MicroFluars*; these are universal objectives. Their advantage is that they can be used, including in polarizing microscopes, where special conditions for birefringence must be observed.

Starting from a 10× objective and up to a 60× objective, we can analyze the optical designs of objectives and trace the logic of the construction of optical

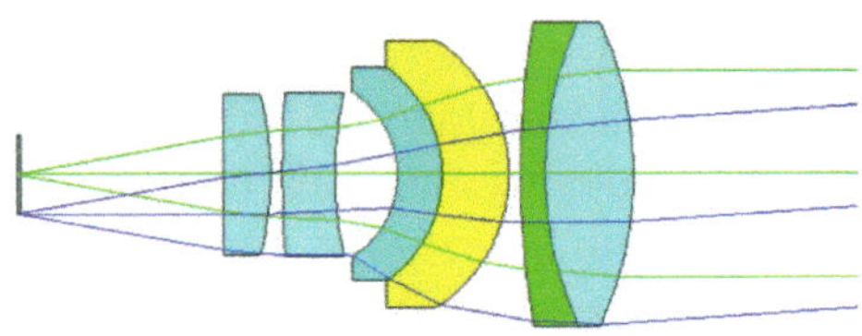

Lens: PlanFluar 5x0.17 d017 ftl200

Object num aper 0.170000 Image height 13.500000 Primary wavln 0.546070

SRF	RADIUS	THICKNESS	APERTURE RADIUS	GLASS	SPECIAL
OBJ	0.000000	0.010000	2.700296	AIR	
AST	0.000000	0.170000	2.702021 AS	K9	
2	0.000000	14.000000	2.630000	AIR	
3	0.000000	3.100000	5.050000	K9 P	
4	-20.600000	0.800000	5.320000	AIR	
5	80.180000	3.500000	5.380000	K9 P	
6	23.140000	4.300000	5.380000	AIR	
7	-6.300000	3.000000	5.440000	H-LAK61	
8	-8.395000	4.500000	6.980000	H-FK61	
9	-10.351000	0.800000	8.790000	AIR	
10	59.460000	1.600000	9.990000	H-ZF7LA	
11	23.140000	6.000000	10.040000	K9 P	
12	-23.140000	200.000000	10.200000	AIR	
13	238.880000	3.700000	16.770000	LZ_K8	
14	56.360000	2.000000	16.870000	AIR	
15	56.750000	3.700000	17.300000	LZ_TF1	
16	53.680000	13.000000	17.250000	H-FK61	
17	-56.640000	3.700000	17.420000	LZ_CTK12	
18	-103.020000	197.356583 S	17.700000	AIR	
IMS	0.000000	0.000000	13.500000 S		

Figure 2.1 Optical design and graphs of aberration correction of a 5×/0.17 infinity LCF plan MicroFluar objective ($F'_{ts} = 200$ mm).

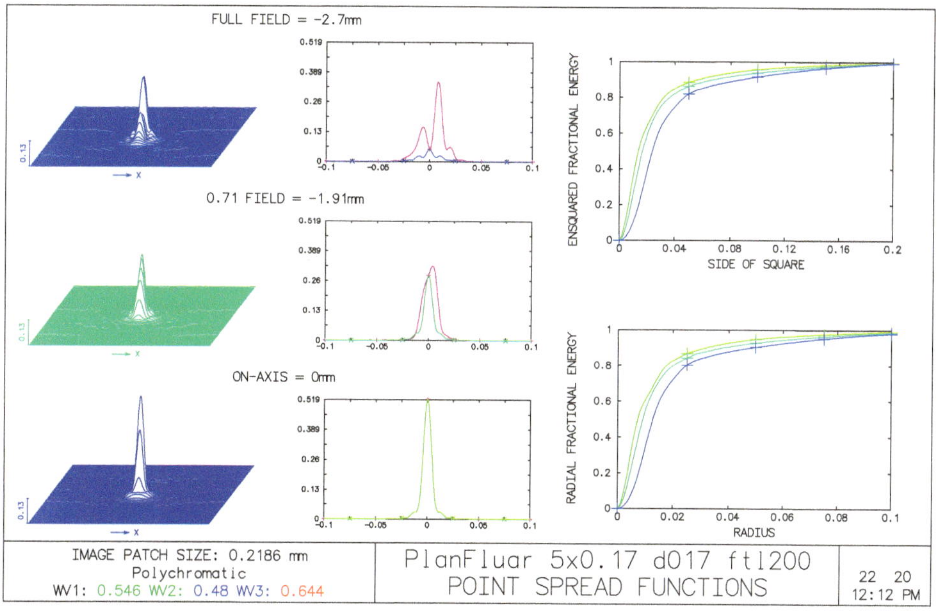

Figure 2.1 (*Continued*) Optical design and graphs of aberration correction of a 5×/0.17 infinity LCF plan MicroFluar objective ($F'_{ts} = 200$ mm).

circuits. The optical circuits of these objectives include a thick meniscus located in the front and a positive component. Depending on the objective's magnification and numerical aperture, this positive component can be made in the form of one, two, or three single positive lenses. So, for the 10× objective, this additional component is not needed because the corrective capabilities of the first meniscus are enough. For 20× and 40× objectives, one additional positive component is used, and for the 50× and 60× objectives, there are two. Then, in accordance with this theory, the optical circuit includes a component in the form of a negative double-cemented lens, which, together with the previous component, is half the design of the planar. Further, along the path of the ray, another component of the middle part of the objective is still present. The single meniscus of finite thickness closes the optical design. This meniscus was turned by a concavity to the image space.

These objectives are built based on synthesis using the composition of at least two types of methods: the half of the planar and the thick meniscus techniques. The half planar technique does not work for an objective with a linear magnification of 100×, so its optical design can be constructed using other principles and basic elements of the optical scheme. As a frontal element, a double-glued lens, rather than a single, can be used. It is the same for the last meniscus; it can be double glued. Figure 2.7 shows the design parameters and graphs of aberration correction of a 100×/1.37 oil immersion objective.

In this objective, a component is used that acts as an "ungluing" of a triple gluing, in which the second surface of the first lens and the first surface of the next lens are rather close together (surface numbers 11 and 12); one can say that an additional correction parameter appears, the *air lens*, the deflection of which allows the smoothing of spherochromatic aberrations that are quite difficult to correct in the objective. This component is also used in other objectives. Table 2.1 shows the main technical parameters of these objectives.

In the optical design of the next objective for some lenses, optical glass type H-FK71 was used, which has outstanding dispersion properties and is characterized by the following optical parameters: $n_d = 1.4565$, $\gamma_d = 90.27$,

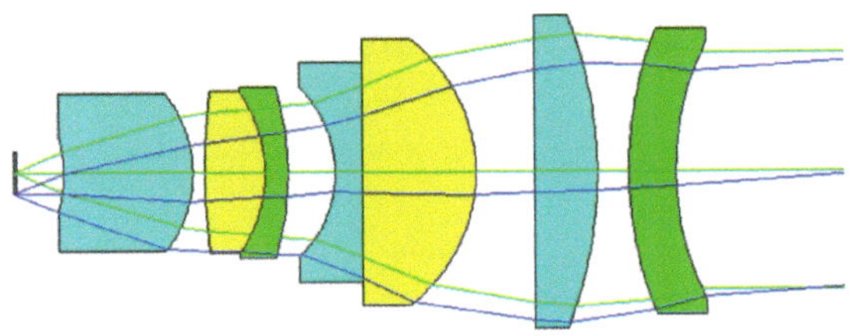

Lens: planFluar 10x0.37 d017 ft1200

Object num aper 0.370000 Image height 13.500000 Primary wavln 0.546070

SRF	RADIUS		THICKNESS		APERTURE RADIUS		GLASS		SPECIAL
OBJ	0.000000		0.010000		1.351438		AIR		
AST	0.000000		0.170000		1.355421	AS	K9		
2	0.000000		3.131400		1.340000		AIR		
3	-8.511000		8.300000		2.440000		H-LAK61		
4	-7.686000		0.800000		4.900000		AIR		
5	44.730000		3.800000		5.050000		H-FK61		
6	-8.801000		1.400000		5.050000		H-ZF7LA		
7	-19.790000		3.000000		5.310000		AIR		
8	-7.269000		1.700000		5.280000		H-LAK61	P	
9	0.000000		7.200000		6.840000		H-FK61		
10	-10.411100		3.700000		8.290000		AIR		
11	0.000000		4.000000		9.540000		H-LAK61	P	
12	-25.800000		2.000000		9.730000		AIR		
13	23.600000		3.000000		8.880000		H-ZF7LA	P	
14	17.385000		200.000000		8.030000		AIR		
15	238.880000		3.700000		19.480000		LZ_K8		
16	56.360000		2.000000		19.570000		AIR		
17	56.750000		3.700000		20.070000		LZ_TF1		
18	53.680000		13.000000		20.000000		H-FK61		
19	-56.640000		3.700000		20.100000		LZ_CTK12		
20	-103.020000		197.351929	S	20.430000		AIR		
IMS	0.000000		0.000000		13.500000	S			

Figure 2.2 Optical design and graphs of aberration correction of a 10×/0.37 infinity LCF plan MicroFluar objective ($F'_{ts} = 200$ mm).

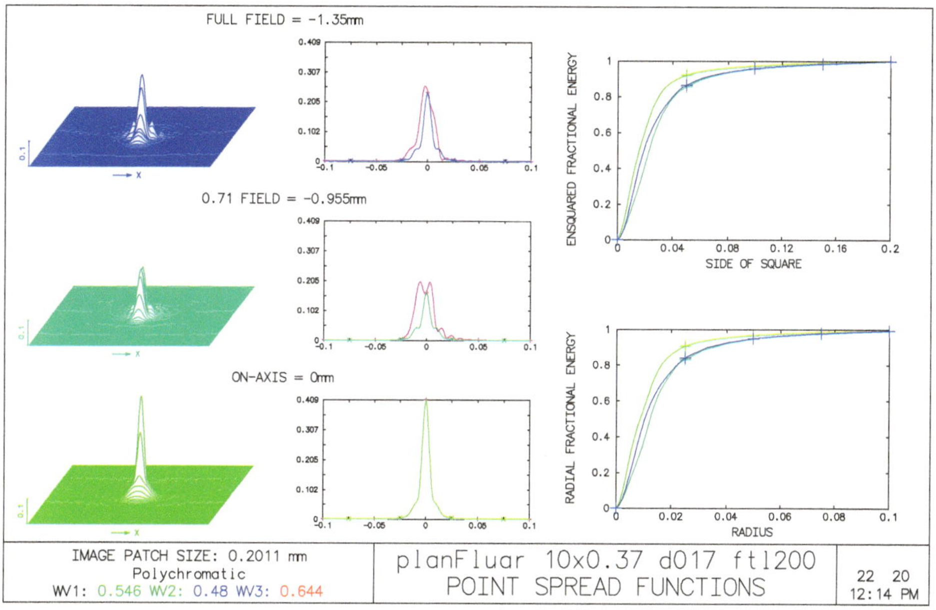

Figure 2.2 (*Continued*) Optical design and graphs of aberration correction of a 10×/0.37 infinity LCF plan MicroFluar objective ($F'_{ts} = 200$ mm).

and $n_{F-C} = 0.00506$. The use of this optical material made it possible to increase the value of the numerical aperture and light power of the objective with a linear magnification of 20×, as well as increase the value of its working distance. Figure 2.8 shows the design parameters and graphs of aberration correction of this objective.

It should be noted that the basic version of the optical design of this objective is presented here. The optical system includes the penultimate element: a single lens located near the plane of the exit pupil of the objective. Its first surface is made in the form of a plane, and the curvature of the second surface is minimal. This lens (in another embodiment of the optical design of the objective) can be made in the form of the so-called kinoform element, or as a composition of a large number of microspherical lenses. By the way, microspheres (or the so-called *nanospheres*) can also be applied on the surface of an 'object's glass and cover glass.

2.2 An Example of Creating LCF Plan MicroFluar Objectives with Long Working Distances

Objectives designed for use in the laboratory and in research models of metallographic and inspection microscopes are mainly used to control microelectronic products. For some tasks, it is very important to ensure a significant value of the so-called free working distance [long working distance

(LWD)] from the object under study to the frame of the front lens of the objective. In this case, the light diameters of the lenses of the frontal part of the objective significantly increase (compared with conventional objectives), causing some aberrations that are very difficult to reduce. A set of such objectives was designed using glass type H-FK71; it can be said that this glass became the basis for the optical structures of these objectives.

In metallographic microscopy, most objects are opaque and translucent. As a rule, such objects are studied by the reflection of light from their surfaces. These objects include metals, alloys, minerals, and ores, as well as some

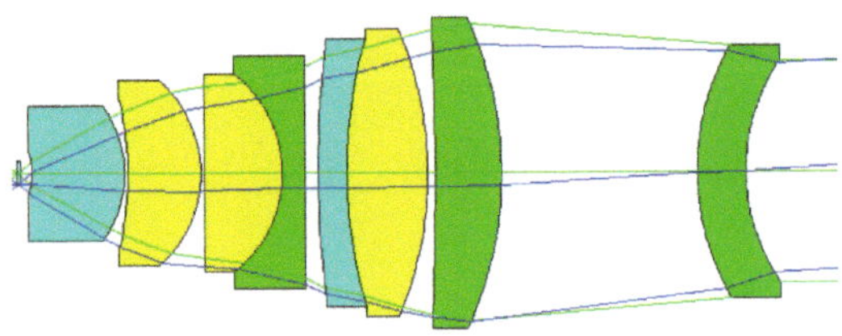

Lens: PlanFluar 20x0.67 d017 ftl200

Object num aper 0.670000 Image height 13.500000 Primary wavln 0.546070

SRF	RADIUS	THICKNESS	APERTURE RADIUS	GLASS	SPECIAL
OBJ	0.000000	0.010000	0.674896	AIR	
AST	0.000000	0.170000	0.683921 AS	K9	
2	0.000000	0.747400	0.740000	AIR	
3	-4.200000	5.800000	1.240000	H-LAK61	
4	-6.700000	0.200000	4.050000	AIR	
5	-19.000000	4.600000	4.550000	H-FK61	
6	-7.200000	0.200000	5.590000	AIR	
7	0.000000	4.800000	5.860000	H-FK61 P	
8	-7.600000	1.400000	5.980000	H-ZF7LA	
9	0.000000	0.800000	7.030000	AIR	
10	58.920000	1.800000	7.630000	H-LAK61 P	
11	29.000000	5.000000	8.090000	H-FK61 P	
12	-21.880000	0.500000	8.670000	AIR	
13	-195.500000	4.000000	9.120000	H-ZF7LA P	
14	-22.390000	12.000000	9.410000	AIR	
15	14.940000	3.000000	7.640000	H-ZF7LA P	
16	11.581000	200.000000	6.750000	AIR	
17	238.880000	3.700000	19.640000	LZ_K8	
18	56.360000	2.000000	19.740000	AIR	
19	56.750000	3.700000	20.240000	LZ_TF1	
20	53.680000	13.000000	20.160000	H-FK61	
21	-56.640000	3.700000	20.260000	LZ_CTK12	
22	-103.020000	197.332985 S	20.590000	AIR	
IMS	0.000000	0.000000	13.500000 S		

Figure 2.3 Optical design and graphs of aberration correction of a 20×/0.67 infinity LCF plan MicroFluar objective ($F'_{ts} = 200$ mm).

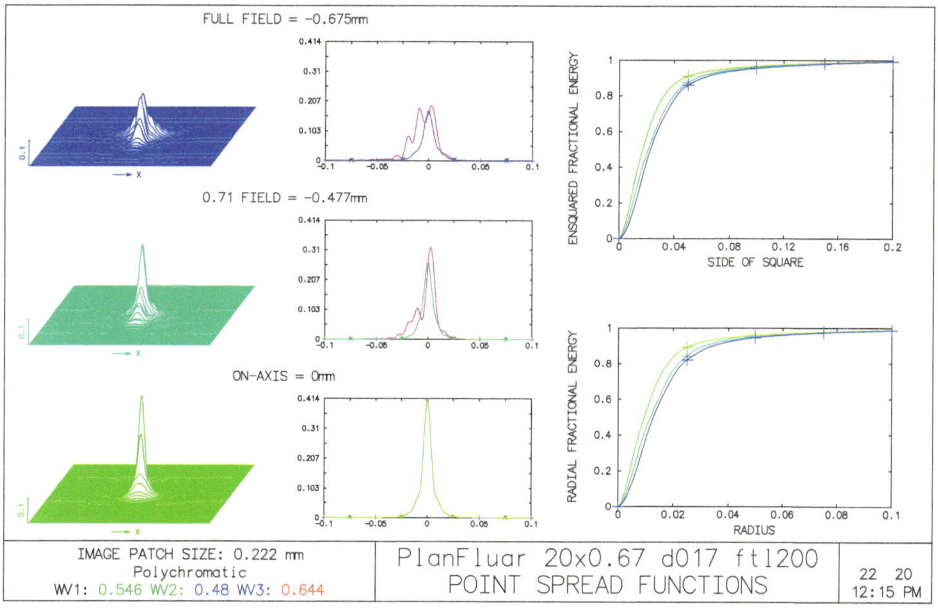

Figure 2.3 (*Continued*) Optical design and graphs of aberration correction of a 20×/0.67 infinity LCF plan MicroFluar objective ($F'_{ts} = 200$ mm).

biological preparations. Illumination of these objects in metallographic microscopes can occur through an objective, which in this case plays the role of a condenser. In this case, the probability of parasitic and scattered light is high. But parasitic and scattered light can also be present in the light passing through an object. Studies have established that scattered light, formed as a result of the reflection of the light flux from the working surfaces of the objective lens, has a significant effect on the image quality of objects. The individual components of this reflected light flux, called *first-order reflexes*, can create significant illumination of the observer's field of illumination, thereby reducing the contrast of the image of the object. To estimate the loss of contrast due to first-order reflexes, it is necessary to know the relation of the illumination to the illumination of the main image. To determine the illumination [2], we suggested that the effect of an optical system be considered, including the aperture and field diaphragms of the illuminator, reflecting the surface of the lens and the parts enclosed between the illuminator, the reflecting surface and the image plane of the object. The reflective images of the aperture and field diaphragms constructed by this system, as well as the material diaphragm and the rim of the last surface of the objective, form the light tube of the reflected stream, which creates a glare of the image. In this case, the illumination is determined by the diameter of one of these diaphragms, visible at the smallest angle from the axial point of the image.

Figures 2.9–2.15 show the layouts, design parameters, and graphs of aberration correction of LCF plan MicroFluar objectives with LWDs. A high degree of aberration correction of the LCF type for a 25-mm and larger linear field in the image space for these objectives is achieved.

Figure 2.16 shows a photograph of a set of objectives; the layouts, design parameters, and graphs of aberration correction of these objectives are shown in Figs. 2.9 and 2.11–2.14. These are the last-generation objectives for metallographic and microelectronic research. They are designed to work without a cover glass and have an LCF LD plan MicroFluar aberration correction. These objectives are distinguished by increased free working

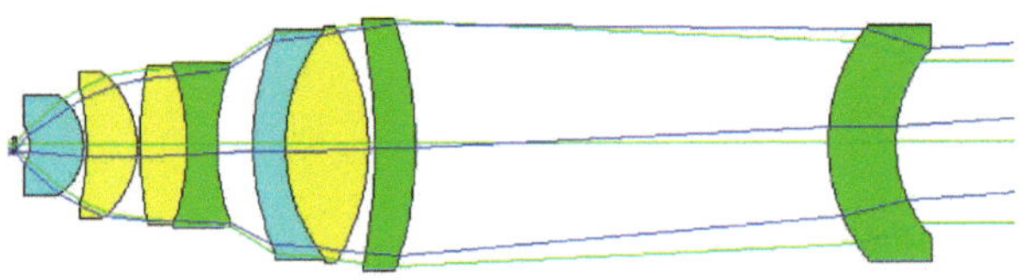

Lens: PlanFluar 40x0.77 d017 ft1200

Object num aper 0.770000 Image height 13.500000 Primary wavln 0.546070

SRF	RADIUS		THICKNESS		APERTURE RADIUS		GLASS		SPECIAL
OBJ	0.000000		0.010000		0.337371		AIR		
AST	0.000000		0.170000		0.349439	AS	K9		
2	0.000000		0.656800		0.430000		AIR		
3	-1.820000		2.700000		0.920000		H-LAK61		
4	-2.830000		0.140000		2.390000		AIR		
5	-12.953000		2.500000		3.000000		H-FK61		
6	-4.368000		0.150000		3.500000		AIR		
7	18.000000		2.300000		3.750000		H-FK61	P	
8	-11.169000		1.500000		3.760000		H-ZF7LA		
9	11.169000		1.800000		3.950000		AIR		
10	12.953000		1.600000		5.260000		H-LAK61	P	
11	8.705000		4.000000		5.460000		H-FK61	P	
12	-10.627000		0.300000		5.580000		AIR		
13	-30.000000		2.000000		5.720000		H-ZF7LA	P	
14	-18.000000		20.000000		5.940000		AIR		
15	8.801000		3.300000		5.580000		H-ZF7LA	P	
16	6.625000		200.000000		4.500000		AIR		
17	238.880000		3.700000		17.200000		LZ_K8		
18	56.360000		2.000000		17.300000		AIR		
19	56.750000		3.700000		17.730000		LZ_TF1		
20	53.680000		13.000000		17.680000		H-FK61		
21	-56.640000		3.700000		17.830000		LZ_CTK12		
22	-103.020000		197.422855	S	18.120000		AIR		
IMS	0.000000		0.000000		13.500000	S			

Figure 2.4 Optical design and graphs of aberration correction of a 40×/0.77 infinity LCF plan MicroFluar objective ($F'_{ts} = 200$ mm).

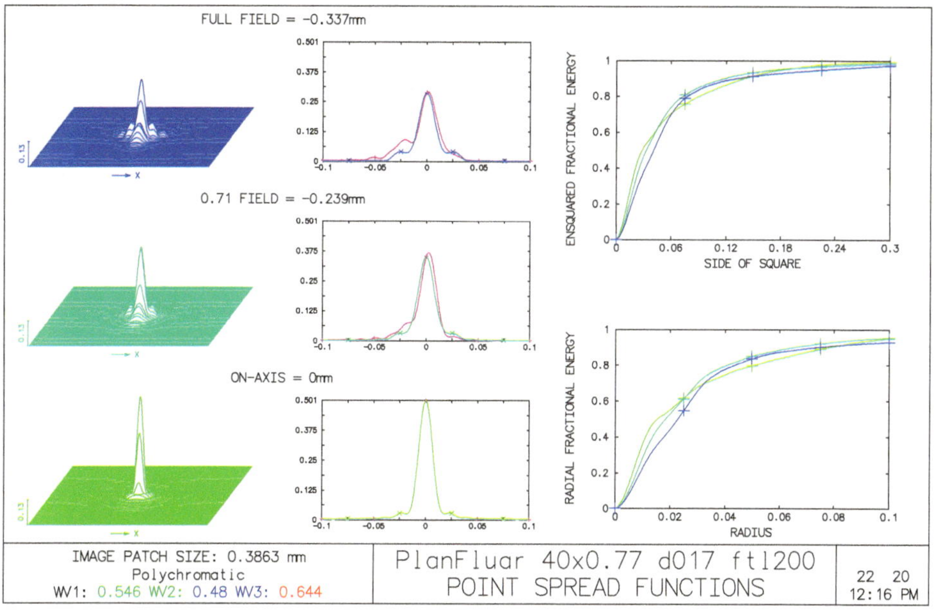

Figure 2.4 (*Continued*) Optical design and graphs of aberration correction of a 40×/0.77 infinity LCF plan MicroFluar objective ($F'_{ts} = 200$ mm).

distances from the frame of the frontal lens to the object under study. Table 2.2 shows the main technical parameters of these kinds of budget objectives.

Objectives with large linear magnifications (20×, 50×, 100×, and 250×) are designed so that the effective working linear field in the space of objects is increased, compared with that of typical budget light microscopes, by 1.5 times or more. For a 50× objective, the linear field is 0.84 mm on the object and 42 mm in the image space; this allows the corresponding tube system to use the so-called *full-format matrix* when projecting an image onto a digital receiver. Figures 2.17–2.29 are photographs of real objects taken using these objectives.

These design parameters, aberration characteristics, and photographs of real objects demonstrate that such objectives can be recommended for completing modern models of metallographic and inspection microscopes for use in working laboratories as well as for research.

2.3 Optical Materials with Special Dispersive Properties

After considering techniques that use a combination of objectives of various configurations, it is advisable when designing a microscope objective to consider one of the most powerful parameters in the optical systems of light microscopes: the type of optical material to use for the lenses. Easily produced and relatively cheap optical materials should be used in the optical systems of budget microscopes, whereas optical materials with specific dispersive properties should be used in higher classes of microscopes. Using these types of optical materials can significantly improve the consumer properties of objectives, as they may increase the numerical apertures, linear fields, and light power of the lenses. However, these advantages are due not only to the use of nonstandard optical materials in advanced objectives but also to the use of original optical designs.

Nonstandard optical materials include types of optical glasses and optical crystals that have special dispersive properties that can reduce the secondary

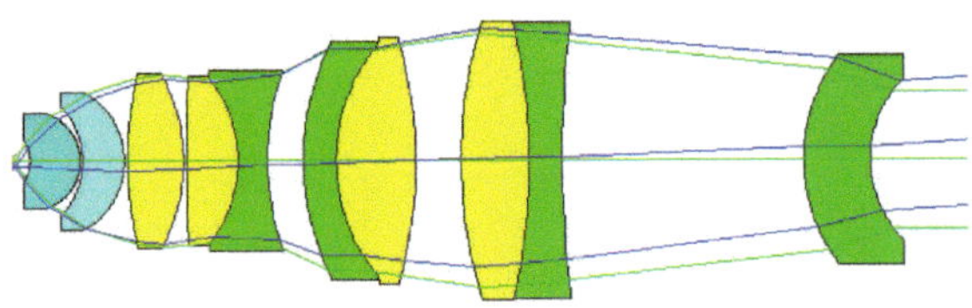

Lens: PlanFluar 50x0.87 d=0 ftl200

Object num aper 0.870000 Image height 13.000000 Primary wavln 0.546070

SRF	RADIUS	THICKNESS	APERTURE RADIUS	GLASS	SPECIAL
OBJ	0.000000	1.0000e-20	0.260158	AIR	
AST	0.000000	0.770000	0.260158 AS	AIR	
2	-1.650000	2.550000	0.980000	H-LAK61	
3	-2.390000	0.130000	2.290000	AIR	
4	-3.800000	2.100000	2.670000	SILICA	
5	-3.800000	0.150000	3.380000	AIR	
6	16.749000	2.800000	4.140000	H-FK61	
7	-9.000000	0.150000	4.270000	AIR	
8	69.000000	2.600000	4.150000	H-FK61 P	
9	-6.400000	1.500000	4.080000	H-ZF7LA	
10	13.140000	1.800000	4.420000	AIR	
11	12.840000	1.600000	5.780000	H-ZF7LA P	
12	8.868000	4.000000	5.810000	H-FK61 P	
13	-20.750000	2.300000	6.020000	AIR	
14	22.570000	3.400000	6.840000	H-ZF7LA P	
15	-29.720000	1.600000	6.770000	H-FK61 P	
16	54.800000	12.000000	6.470000	AIR	
17	8.511000	3.300000	5.150000	H-ZF7LA P	
18	5.600000	200.000000	3.950000	AIR	
19	238.880000	3.700000	17.200000	LZ_K8	
20	56.360000	2.000000	17.500000	AIR	
21	56.750000	3.700000	17.730000	LZ_TF1	
22	53.680000	13.000000	17.680000	H-FK61	
23	-56.640000	3.700000	17.830000	LZ_CTK12	
24	-103.020000	197.330917 S	18.120000	AIR	
IMS	0.000000	0.000000	13.000000 S		

Figure 2.5 Optical design and graphs of aberration correction of a 50×/0.87 infinity LCF plan MicroFluar objective ($F'_{ts} = 200$ mm).

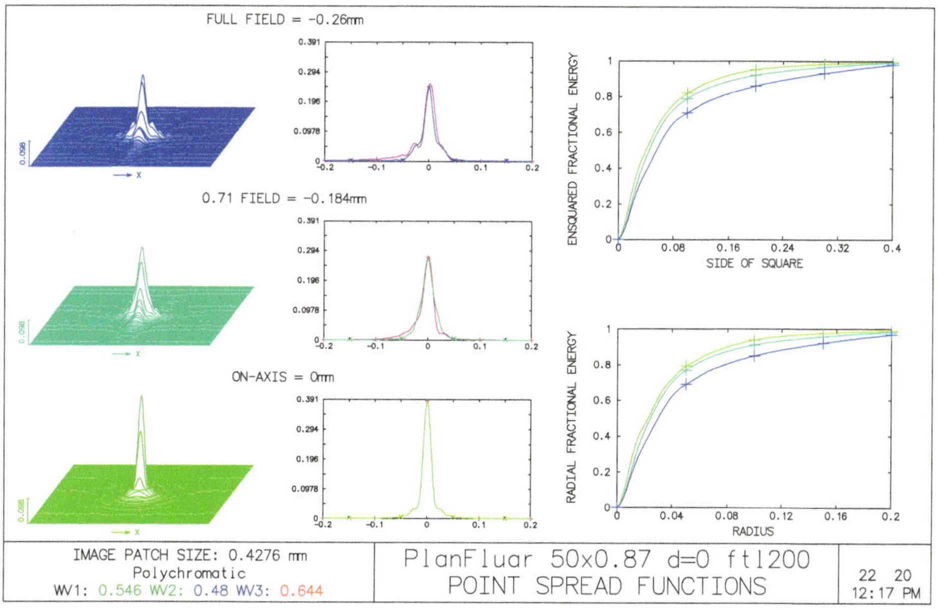

Figure 2.5 (*Continued*) Optical design and graphs of aberration correction of a 50×/0.87 infinity LCF plan MicroFluar objective ($F'_{ts} = 200$ mm).

spectrum in optical systems. This is characterized by the graphical dependence of the relative partial dispersion on the coefficient of dispersion: $p_{\lambda 1,\, \lambda 2} = f(\nu_\lambda)$. The refractive index n_e for the wavelength $\lambda_e = 0.546$ μm, located near the maximum sensitivity of the human eye, is the main refractive index, the average variance $n_{F'} - n_{C'}$ is the main dispersion, and the coefficient of dispersion $\nu_e = \frac{n_e - 1}{n_{F'} - n_{C'}}$ is the Abbe number (in its modern interpretation). If we plot on the diagram the relative partial dispersions of optical glasses calculated for the same wavelengths, for example $p_{F',g} = \frac{n_g - n_{F'}}{n_{F'} - n_{C'}}$, depending on the dispersion coefficient ν_e, a pattern observed by Abbe is found: the coordinates of most glasses are located near a straight line called *normal*; these graphs are shown in all optical textbooks and catalogs of optical glass. To fix the normal straight line, we can choose, for example, optical glass types K18 and F13.

Materials that differ in ν_e by more than 3 units from glasses with a normal course of dispersion with equal values of relative private dispersions $p_{F',g}$ are considered special. The nomenclature of optical materials with a special dispersive quality is very limited. Information on several Soviet optical glasses as well as glasses from the catalog of the German company Schott are provided in Table 2.3; data are given relative to the spectral line *d*, which is chosen as the main line. The value of $\Delta p_{g,F} \cdot 10^4$, which characterizes the special properties of glass relative to the normal straight line constructed for glass types K7 ($n_d = 1.511121$ and $\nu_d = 60.41$) and F2 ($n_d = 1.620040$ and $\nu_d = 36.37$), is shown in column 9.

The advantages of glasses from the Schott catalog are demonstrated by the reduction of the optical forces of elements when these glasses are combined, for example, with fluorite, as well as by the reduction of the secondary spectrum. Reduction of the optical forces of the elements leads to a simplification of the component structures and a decrease in the higher orders of spherical aberration and spherochromatism. Objectives and other microscope

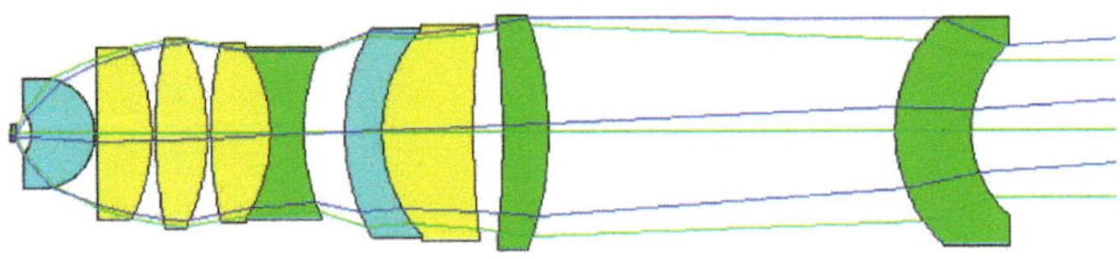

Lens: PlanFluar 60x0.87 d017 ft1200

Object num aper 0.870000 Image height 13.500000 Primary wavln 0.546070

SRF	RADIUS		THICKNESS		APERTURE RADIUS		GLASS		SPECIAL
OBJ	0.000000		0.010000		0.224760		AIR		
AST	0.000000		0.170000		0.242405	AS	K9		
2	0.000000		0.690000		0.350000		AIR		
3	-1.450000		2.740000		0.950000		H-LAK61		
4	-2.360000		0.130000		2.330000		AIR		
5	-227.450000		2.400000		3.220000		H-FK61		
6	-7.900000		0.150000		3.670000		AIR		
7	19.000000		2.300000		3.980000		H-FK61	P	
8	-9.000000		0.150000		4.040000		AIR		
9	17.000000		2.600000		3.830000		H-FK61	P	
10	-6.647000		1.500000		3.660000		H-ZF7LA		
11	8.670000		1.800000		3.600000		AIR		
12	9.157000		1.600000		4.460000		H-LAK61	P	
13	6.457000		4.000000		4.360000		H-FK61	P	
14	42.250000		1.300000		4.600000		AIR		
15	-42.250000		2.000000		4.760000		H-ZF7LA	P	
16	-13.900000		15.000000		4.980000		AIR		
17	6.647000		3.300000		4.860000		H-ZF7LA	P	
18	4.760000		200.000000		3.620000		AIR		
19	238.880000		3.700000		16.850000		LZ_K8		
20	56.360000		2.000000		16.970000		AIR		
21	56.750000		3.700000		17.400000		LZ_TF1		
22	53.680000		13.000000		17.360000		H-FK61		
23	-56.640000		3.700000		17.530000		LZ_CTK12		
24	-103.020000		197.321111	S	17.820000		AIR		
IMS	0.000000		0.000000		13.500000	S			

Figure 2.6 Optical design and graphs of aberration correction of a 60×/0.87 infinity LCF plan MicroFluar objective ($F'_{ts} = 200$ mm).

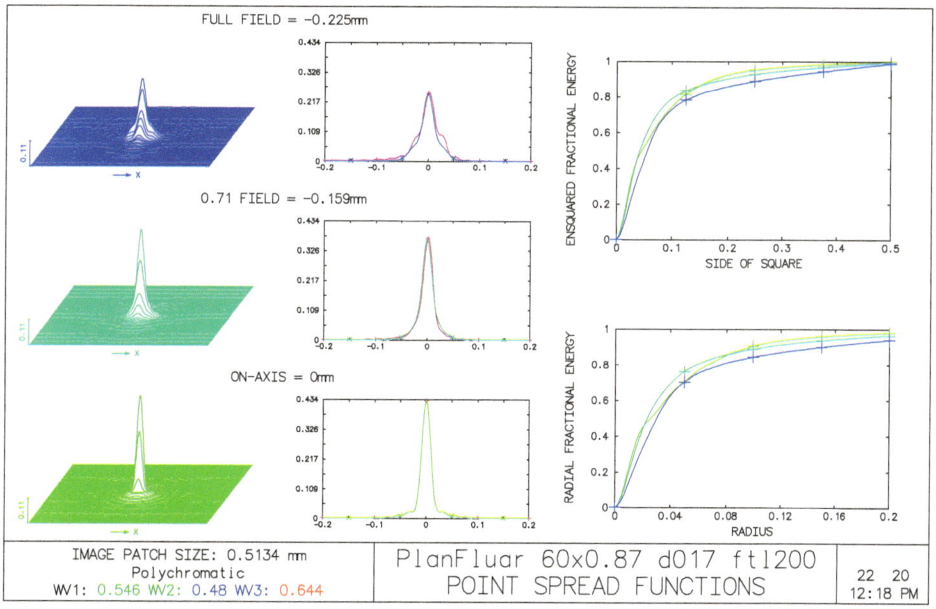

Figure 2.6 (*Continued*) Optical design and graphs of aberration correction of a 60×/0.87 infinity LCF plan MicroFluar objective ($F'_{ts} = 200$ mm).

lenses designed from glasses from the Schott catalog are distinguished by greater compactness and better technological properties with the same basic circuit solutions.

In principle, the optical materials listed in blue in Table 2.3, and especially in their combination with fluorite (the optical materials listed in red), are sufficient to create objectives with improved consumer properties to complete laboratory and research models of light microscopes. Quartz is also a very interesting optical material (also listed in red); like fluorite, it has outstanding dispersion properties. As an *achromatic pair*, quartz and fluorite can be used in optical systems (objectives and other optical components of light microscopes) if a wide range of achromatization is required, for example, when creating poly-apochromats or multi-apochromats. The unique dispersion properties of these two materials are complemented by a wide transmission spectrum (especially in the near UV region); this property is not present in ordinary silicate glass types. Strange as it may seem, quartz is a highly technological and well-developed material in the optical industry.

Throughout the long history of microscopy, optical crystals, especially natural and subsequently synthetic fluorite, have been used in the optical components of microscopes designed for nontrivial studies. As a material of natural origin, fluorite has many flaws that do not allow for the mass production of lenses. These are mainly its technological characteristics; lenses with high requirements for deviation tolerances from the nominal values of

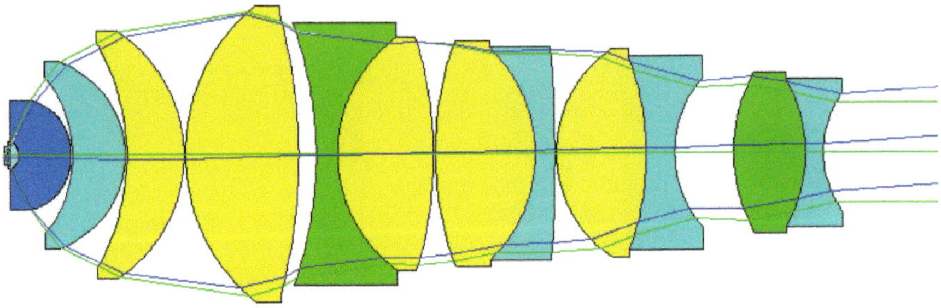

Lens: PlanFluar 100x1.37oil ftl200

Object num aper	1.370000	Image height	13.500000	Primary wavln	0.546070
SRF	**RADIUS**	**THICKNESS**	**APERTURE RADIUS**	**GLASS**	**SPECIAL**
OBJ	0.000000	0.010000	0.134909	OIL	
AST	0.000000	0.170000	0.155898 AS	K9	
2	0.000000	0.127800	0.500000	OIL	
3	0.000000	0.500000	0.760000	K9	
4	-0.828000	3.000000	0.760000	H-ZLAF3	
5	-3.076000	0.100000	3.010000	AIR	
6	-6.329000	3.000000	4.240000	H-LAK61	
7	-5.297000	0.100000	5.100000	AIR	
8	-12.057000	3.200000	6.170000	H-FK61	
9	-8.000000	0.100000	6.760000	AIR	
10	10.233000	6.400000	8.100000	H-FK61 P	
11	-29.000000	1.000000	7.710000	AIR	
12	-20.240000	1.200000	7.180000	H-ZF5	
13	7.900000	5.300000	6.360000	H-FK61 P	
14	-21.180000	0.100000	6.410000	AIR	
15	17.602000	5.600000	6.180000	H-FK61 P	
16	-7.900000	1.100000	5.820000	H-LAK61 P	
17	-83.110000	0.100000	5.850000	AIR	
18	6.792000	5.000000	5.720000	H-FK61 P	
19	-15.394000	1.600000	5.310000	H-LAK61 P	
20	6.066000	3.300000	4.070000	AIR	
21	9.000000	4.000000	4.380000	H-ZF7LA	
22	-8.000000	1.000000	4.060000	LZ_CTK12	
23	5.495000	200.000000	3.270000	AIR	
24	238.880000	3.700000	18.690000	LZ_K8	
25	56.360000	2.000000	18.840000	AIR	
26	56.750000	3.700000	19.340000	LZ_TF1	
27	53.680000	13.000000	19.410000	H-FK61	
28	-56.640000	3.700000	19.480000	LZ_CTK12	
29	-103.020000	197.262910 S	19.820000	AIR	
IMS	0.000000	0.000000	13.500000 S		

Figure 2.7 Optical design and graphs of aberration correction of a 100×/1.37 infinity LCF plan MicroFluar objective ($F'_{ts} = 200$ mm).

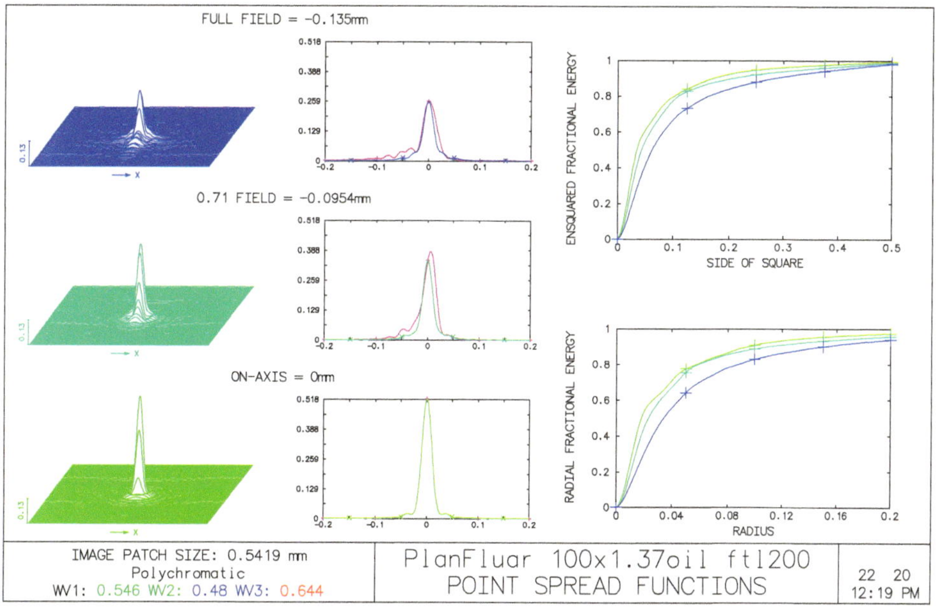

Figure 2.7 (*Continued*) Optical design and graphs of aberration correction of a 100×/1.37 infinity LCF plan MicroFluar objective ($F'_{ts} = 200$ mm).

Table 2.1 Main technical parameters and basic optical layout of the Seventh Series objectives.

Magnification	*NA*	*WD* (mm)	*F′* (mm)	*R* (μm)	*DF* (μm)	*FOV* on object (mm)	*FOV* on image (mm)	Optical design at figure no.	The principal optical layout
5×	0.17	13.7	40.0	1.96	11.5	5.4	27	2.1	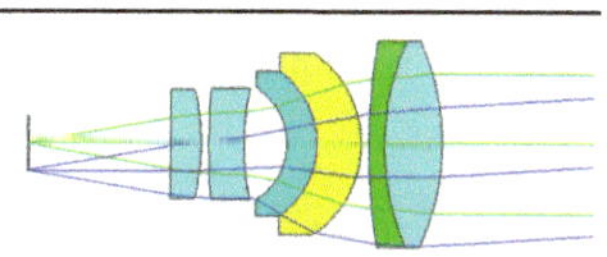
10×	0.37	2.60	20.0	0.90	2.43	2.7	27	2.2	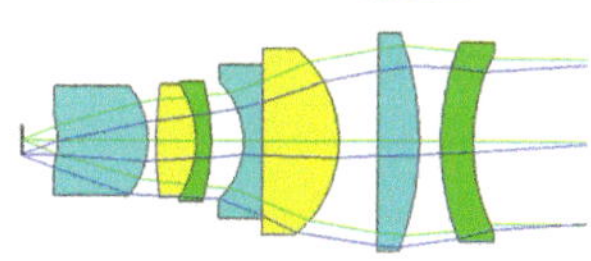
20×	0.67	0.48	10.0	0.50	0.74	1.35	27	2.3	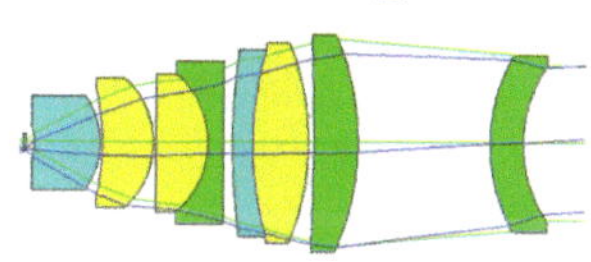
40×	0.77	0.41	5.0	0.43	0.56	0.68	27	2.4	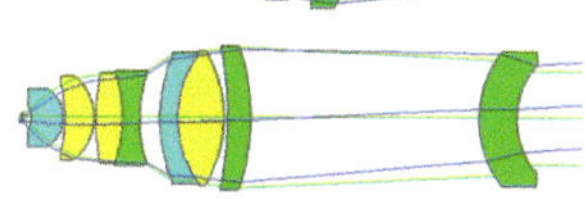

(*continued*)

Table 2.1 (*Continued*) Main technical parameters and basic optical layout of the Seventh Series objectives.

Magnification	*NA*	*WD* (mm)	*F′* (mm)	*R* (μm)	*DF* (μm)	*FOV* on object (mm)	*FOV* on image (mm)	Optical design at figure no.	The principal optical layout
50×	0.87	0.40	4.0	0.38	0.44	0.54	27	2.5	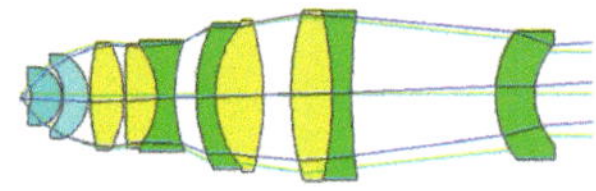
60×	0.87	0.30	3.3	0.38	0.44	0.45	27	2.6	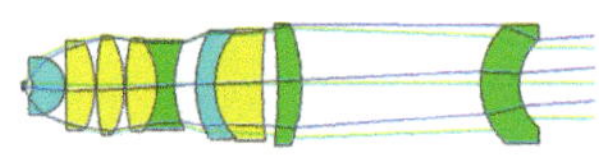
100×	1.37 oil	0.11	2.0	0.24	0.18	0.27	27	2.7	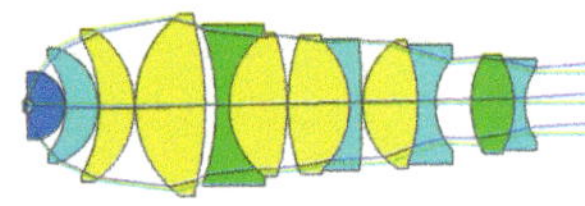

the design parameters are quite laborious and expensive to manufacture. At the same time, as compared with other objectives, the advantages of objectives with fluorite lenses are obvious, and the quality of the images obtained with them is substantially higher. The development of a substitute for fluorite, that is, an optical material more suitable for the conditions of mass production, has become an important goal for the optical industry.

The optical powers of the positive and negative elements of the doublets composed of the corresponding materials, found from the condition $C_1 = 0$: $\varphi_1 = \frac{\nu_1}{\nu_1 - \nu_2}$, $\varphi_2 = -\frac{\nu_2}{\nu_1 - \nu_2}$, and the chromatic coefficient $C_2^* = -\frac{p_2 - p_1}{\nu_2 - \nu_1}$, characterizing the secondary spectrum of the doublet for the *g*, *F* spectral lines, are shown in Table 2.4.

2.3.1 Optical materials: Analogues of fluorite

International researchers continue to work on developing a fluorite substitute. In the catalogs of leading German, American, Japanese, and Russian firms specializing in the manufacture of optical glass, some glass types are presented whose refractive indices and dispersion parameters are close to those of fluorite. These are very important achievements, as such glasses are also free from their own internal voltages (characteristic of optical crystals), which allows them to be used for the optical design of polarization optics. In Russia, work on the development of a complete analogue of fluorite has been carried out [3].

The technology of introducing fluorine into oxide glasses has been investigated; such an introduction gives these glasses several valuable properties that are unattainable for pure oxide glasses. Of particular interest is the development and study of oxygen-free fluorite glasses. The development of these glasses can solve the problem of creating optical systems with low-light losses, allowing materials with a wide range of transparency, including UV, visible, and IR ranges, to be obtained. Introducing rare-earth and d-elements as activators in fluorite glasses provides them with several new, unique properties that cannot be described within the framework of existing

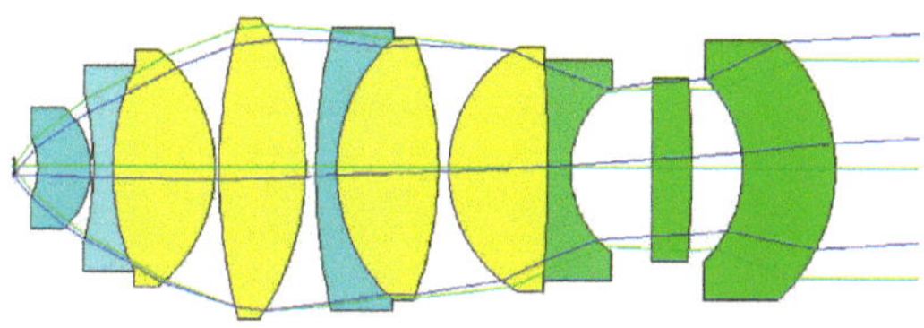

Lens: PlanFluor 20x0.80 ftl200 h=60

Object num aper 0.800000 Image height 12.500000 Primary wavln 0.546070

SRF	RADIUS	THICKNESS	APERTURE RADIUS	GLASS	SPECIAL
OBJ	0.000000	0.010000	0.625003	AIR	
AST	0.000000	0.170000	0.638336 AS	K9	
2	0.000000	1.700000	0.740000	AIR	
3	-5.160000	4.000000	2.290000	H-LAF50A	
4	-5.890000	0.200000	4.460000	AIR	
5	-23.390000	1.500000	5.560000	K9 P	
6	18.232000	7.700000	7.530000	H-FK71	
7	-10.850000	0.300000	8.610000	AIR	
8	39.370000	6.600000	10.740000	H-FK71 P	
9	-19.350000	0.800000	10.960000	AIR	
10	44.000000	1.600000	10.320000	H-LAF50A P	
11	12.750000	7.700000	9.530000	H-FK71 P	
12	-25.966000	0.800000	9.560000	AIR	
13	10.233000	7.500000	8.960000	H-FK71 P	
14	-117.800000	1.700000	8.040000	ZBAF4	
15	7.503000	6.000000	6.040000	AIR	
16	0.000000	3.000000	6.450000	H-ZF7LA	
17	-85.500000	3.900000	6.660000	AIR	
18	-9.160000	7.000000	6.760000	H-ZF7LA P	
19	-12.700000	200.000000	9.520000	AIR	
20	238.880000	3.700000	22.670000	LZ_K8	
21	56.360000	2.000000	22.740000	AIR	
22	56.750000	3.700000	23.340000	LZ_TF1	
23	53.680000	13.000000	23.210000	H-FK61	
24	-56.640000	3.700000	23.250000	LZ_CTK12	
25	-103.020000	197.341868 S	23.630000	AIR	
IMS	0.000000	0.000000	12.500000 S		

Figure 2.8 Optical design and graphs of aberration correction of a 20×/0.80 infinity LCF plan achromatic objective ($F'_{ts} = 200$ mm; $h_{ob} = 60$ mm).

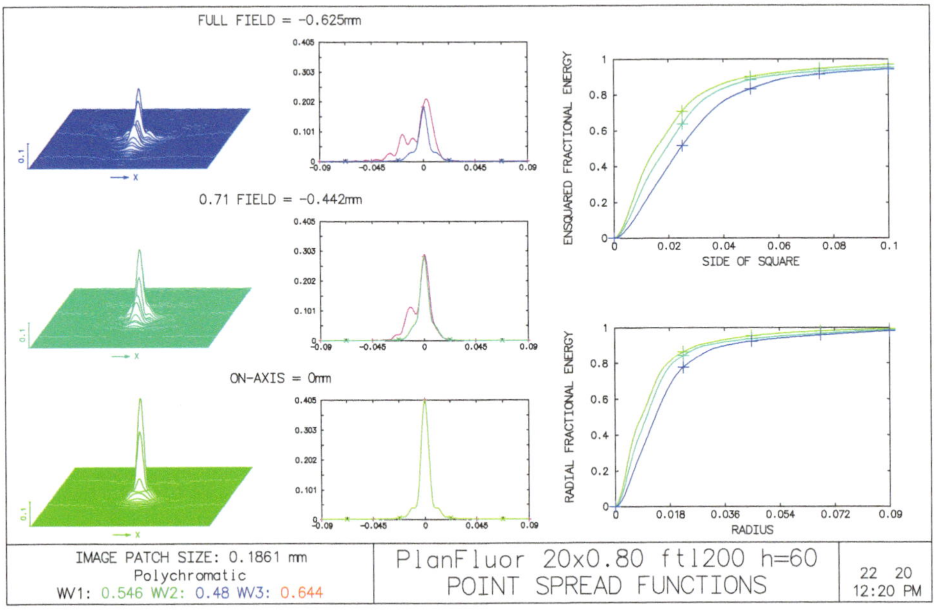

Figure 2.8 (*Continued*) Optical design and graphs of aberration correction of a 20×/0.80 infinity LCF plan achromatic objective ($F'_{ts} = 200$ mm; $h_{ob} = 60$ mm).

ideas about the behavior of activators, both in oxygen-containing glasses and in fluoride monocrystals.

The first fluoroberyllate glasses were developed based on technology and formulations suitable for practical use, but production was difficult due to their high toxicity. Fluoroaluminate glasses, first obtained in 1949, are an independent group of oxygen-free halide glasses. However, their extremely high tendency to crystallize did not allow them to find practical application. The discovery of fluorozirconate glasses in 1974 stimulated interest in fluoride-based glasses, primarily as a material for the manufacture of optical fibers, and made it possible to return to fluoroaluminate glasses, which have several advantages when used in fiber optics, including those with high mechanical strength, chemical resistance, and heat resistance. As compared with fluorozirconate glasses, fluoroaluminate glasses have a lower refractive index, approaching that of calcium fluoride (CaF_2, fluorite) crystals and fluoroberyllate glasses.

Factors that significantly complicate the search and study of new glasses in fluoride systems as a whole should be noted. These include the absence of sufficiently stable glasses in binary and even in trinary systems; the strong influence of experimental conditions (the formulation of the mixture, the dynamics of heating, etc.) on the processes of glass formation and crystallization; the effect of the atmosphere on the melt at all stages of the glassmaking process; the effects of temperature and cooking time, as well as

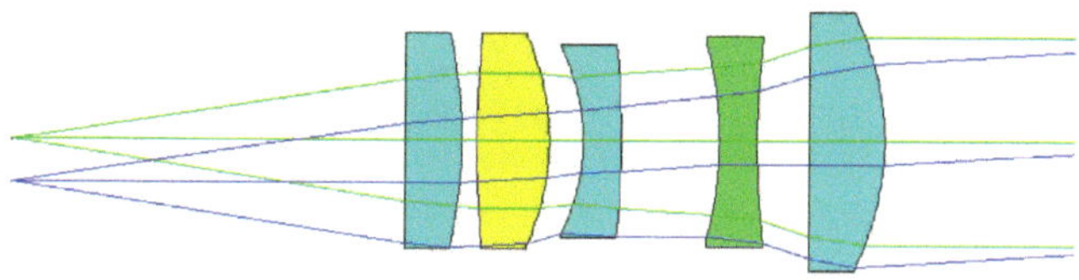

Lens: LD PlanFluar 5x0.15 d=0 ftl180

Object num aper 0.150000 Image height 12.500000 Primary wavln 0.546070

SRF	RADIUS	THICKNESS	APERTURE RADIUS	GLASS	SPECIAL
OBJ	0.000000	1.0000e-20	2.499930	AIR	
AST	0.000000	21.004500	2.499930 AS	AIR	
2	0.000000	3.000000	5.680000	H-LAK3	
3	-24.500000	0.800000	5.880000	AIR	
4	59.320000	3.800000	5.820000	H-FK71	
5	-13.090000	1.800000	5.700000	AIR	
6	-10.516000	2.000000	5.090000	H-LAK4L	
7	-56.640000	5.200000	5.210000	AIR	
8	-19.140000	2.000000	5.210000	ZF6	
9	43.250000	2.800000	5.580000	AIR	
10	0.000000	4.000000	6.380000	H-LAK3 P	
11	-14.510000	180.000000	6.800000	AIR	
12	214.992000	3.330000	16.563528	LZ_K8	
13	50.724000	1.800000	16.628257	AIR	
14	51.075000	3.330000	16.987214	LZ_TF1	
15	48.312000	11.700000	16.951954	H-FK61	
16	-50.976000	3.330000	17.236565	LZ_CTK12	
17	-92.718000	177.609948 S	17.438747	AIR	
IMS	0.000000	0.000000	12.500000 S		

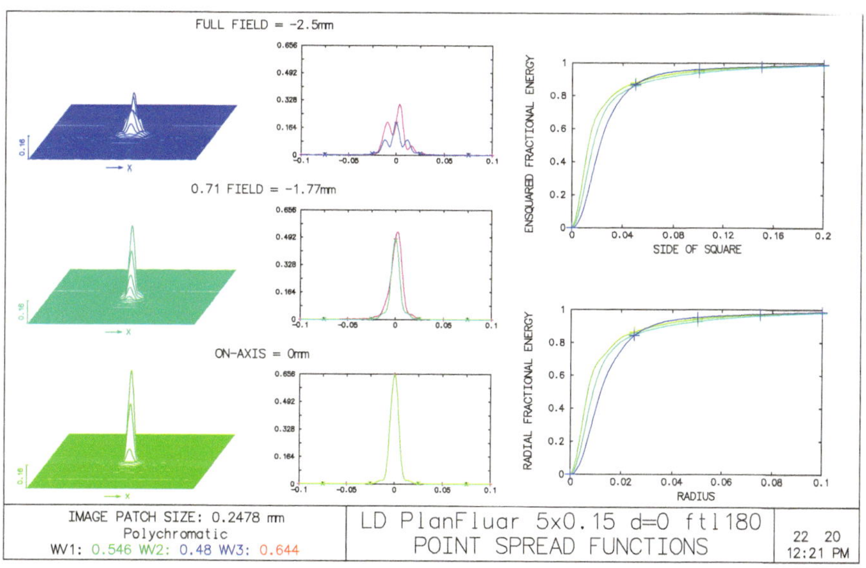

Figure 2.9 Optical design and graphs of aberration correction of a 5×/0.15 infinity LCF LD plan MicroFluar objective ($F'_{tl} = 180$ mm).

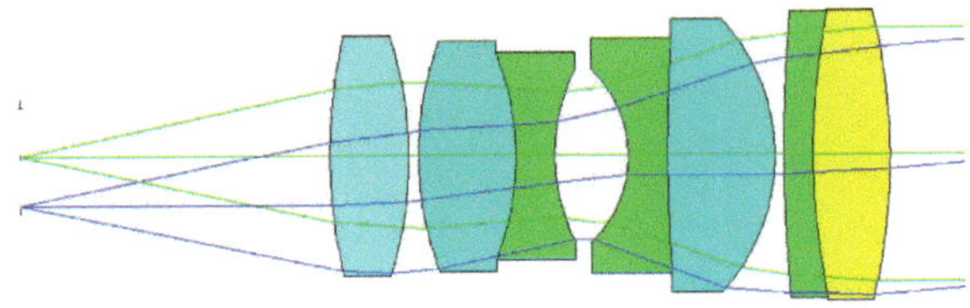

Lens: PlanFluar 5x0.20 d=0 Ftl160

Object num aper 0.200000 Image height 12.500000 Primary wavln 0.546070

SRF	RADIUS	THICKNESS	APERTURE RADIUS	GLASS	SPECIAL
OBJ	0.000000	-1.0000e-08	2.490819	AIR	
AST	0.000000	16.135000	2.500000 A	AIR	
2	25.700000	4.100000	5.930000	LZ_K8	
3	-19.187000	0.500000	6.040000	AIR	
4	15.922000	5.000000	5.790000	LZ_TK17	
5	-13.900000	2.000000	5.190000	LZ_OF4	
6	9.683000	3.800000	4.380000	AIR	
7	-5.970000	2.000000	4.380000	LZ_OF4 P	
8	98.170000	5.500000	5.900000	LZ_TK17 P	
9	-10.023000	0.500000	6.880000	AIR	
10	80.170000	1.500000	7.180000	LZ_OF4 P	
11	35.320000	4.000000	7.210000	LZ_OK1	
12	-30.340000	90.000000	7.290000	AIR	
13	214.800000	5.000000	12.760000	LZ_BK8	
14	-46.990000	0.300000	12.840000	AIR	
15	-46.880000	3.000000	12.800000	LZ_BF24	
16	-113.240000	157.827540 S	12.980000	AIR	
IMS	0.000000	0.000000	12.500000		

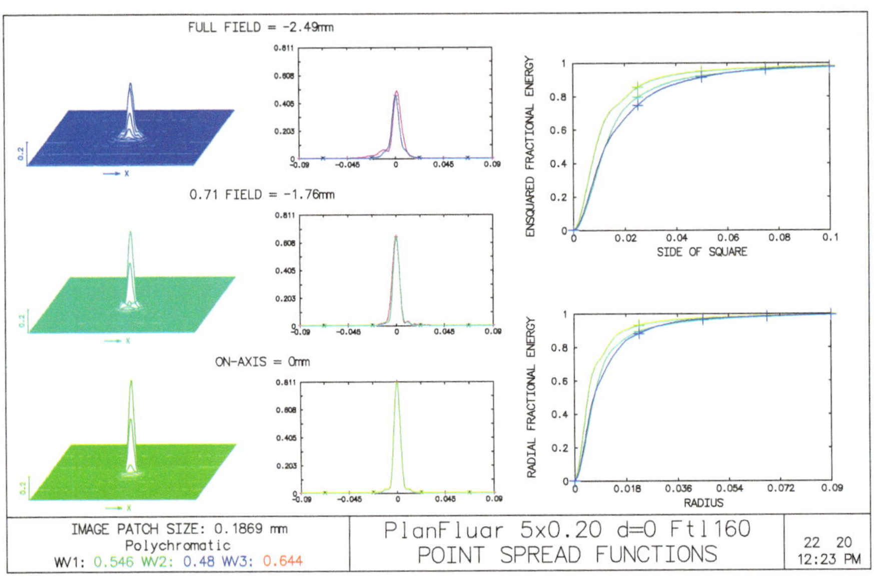

Figure 2.10 Optical design and graphs of aberration correction of a 5×/0.20 infinity LCF LD plan MicroFluar objective ($F'_{tl} = 160$ mm).

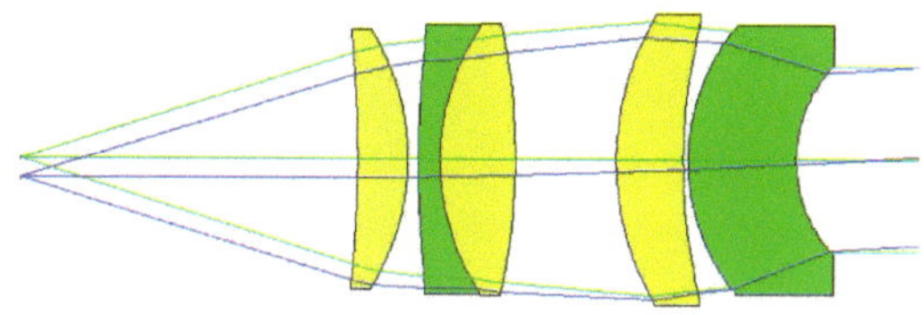

Lens: LD planFluar 10x0.30 d=0 ftl180

Object num aper 0.300000 Image height 12.500000 Primary wavln 0.546070

SRF	RADIUS	THICKNESS		APERTURE RADIUS		GLASS		SPECIAL
OBJ	0.000000	1.0000e-20		1.249540		AIR		
AST	0.000000	19.977000		1.249540	AS	AIR		
2	-68.137000	3.000000		7.400000		H-FK71		
3	-13.900000	0.600000		7.670000		AIR		
4	68.137000	1.300000		7.930000		ZF2		
5	13.900000	4.500000		7.920000		H-FK71	P	
6	-36.130000	6.000000		8.040000		AIR		
7	16.200000	4.000000		8.510000		H-FK71	P	
8	31.340000	0.300000		8.100000		AIR		
9	11.870000	6.500000		7.870000		ZF6		
10	8.050000	180.000000		5.520000		AIR		
11	214.992000	3.330000		17.838852		LZ_K8		
12	50.724000	1.800000		17.896832		AIR		
13	51.075000	3.330000		18.273580		LZ_TF1		
14	48.312000	11.700000		18.225136		H-FK61		
15	-50.976000	3.330000		18.490307		LZ_CTK12		
16	-92.718000	177.594224	S	18.697095		AIR		
IMS	0.000000	0.000000		12.500000	S			

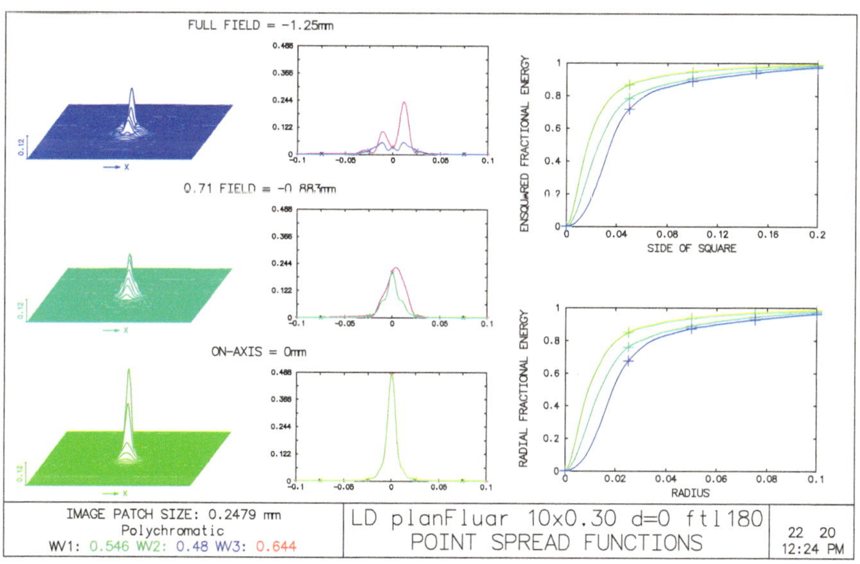

Figure 2.11 Optical design and graphs of aberration correction of a 10×/0.30 infinity LCF LD plan MicroFluar objective ($F'_{tl} = 180$ mm).

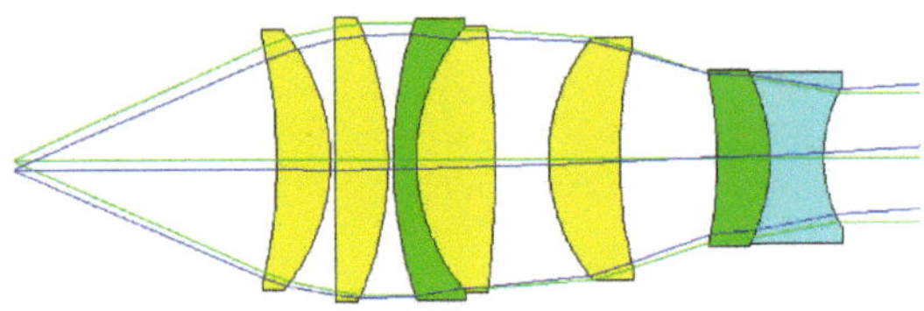

Lens: LD PlanFluar 20x0.40 d=0 ftl180

Object num aper	0.400000	Image height	15.000000	Primary wavln	0.546070

SRF	RADIUS	THICKNESS	APERTURE RADIUS	GLASS	SPECIAL
OBJ	0.000000	1.0000e-20	0.749766	AIR	
AST	0.000000	15.022900	0.749766 AS	AIR	
2	-29.000000	3.000000	6.940000	H-FK71	
3	-11.092000	0.300000	7.340000	AIR	
4	0.000000	3.000000	7.890000	H-FK71	
5	-18.270000	0.300000	8.030000	AIR	
6	23.710000	1.300000	7.880000	ZF6	
7	11.092000	4.500000	7.400000	H-FK71 P	
8	-55.590000	3.000000	7.350000	AIR	
9	10.233000	4.000000	6.720000	H-FK71 P	
10	25.500000	5.500000	6.080000	AIR	
11	-23.140000	3.000000	4.910000	ZF6	
12	-10.233000	3.000000	4.790000	H-K9L	
13	7.600000	180.000000	3.980000	AIR	
14	214.992000	3.330000	18.610000	LZ_K8	
15	50.724000	1.800000	18.760000	AIR	
16	51.075000	3.330000	19.260000	LZ_TF1	
17	48.312000	11.700000	19.220000	H-FK61	
18	-50.976000	3.330000	19.330000	LZ_CTK12	
19	-92.718000	177.610949 S	19.710000	AIR	
IMS	0.000000	0.000000	15.000000 S		

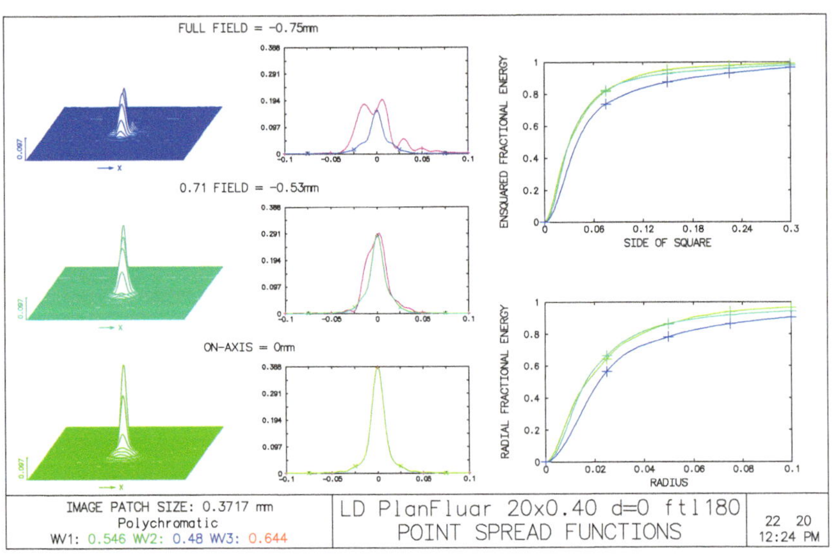

Figure 2.12 Optical design and graphs of aberration correction of a 20×/0.40 infinity LCF LD plan MicroFluar objective ($F'_{tl} = 180$ mm).

the conditions and rate of cooling of the melt; the low quality and scarcity of some raw materials (fluoride and fluorinating additives); fluoride toxicity; the low amount of basic physic-chemical information available as compared with oxides (state diagrams, properties, and compositions); the aggressive influence of melts on construction materials; and the high volatility of fluorides.

A special type of glass, *special crowns*, is already used in the optical industry. These glasses are designed based on a fluorophosphate substance and are of particular interest for the optical design of objectives and other optics for light microscopes. Such optical systems are represented by both

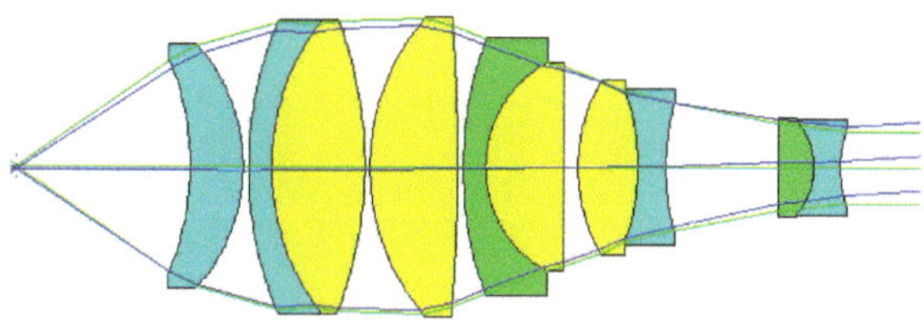

Lens: LD PlanFluar 50x0.55 d=0 f180

Object num aper 0.550000 Image height 20.000000 Primary wavln 0.546070

SRF	RADIUS	THICKNESS	APERTURE RADIUS	GLASS	SPECIAL
OBJ	0.000000	1.0000e-20	0.399910	AIR	
AST	0.000000	10.000700	0.399910 AS	AIR	
2	-15.280000	3.000000	6.140000	H-ZLAF55D	
3	-10.000000	0.300000	6.950000	AIR	
4	20.530000	1.300000	8.310000	H-ZLAF55D P	
5	13.090000	5.300000	8.160000	H-FK71	
6	-22.390000	0.300000	8.280000	AIR	
7	12.560000	5.000000	8.360000	H-FK71 P	
8	-97.840000	0.300000	7.990000	AIR	
9	20.000000	1.300000	7.160000	ZF2	
10	6.516000	4.500000	5.790000	H-FK71 P	
11	0.000000	0.800000	5.580000	AIR	
12	9.360000	3.400000	4.890000	H-FK71 P	
13	-12.560000	1.600000	4.400000	H-ZLAF55D P	
14	12.560000	6.500000	3.760000	AIR	
15	0.000000	2.000000	3.000000	H-ZF7LA	
16	-4.000000	1.500000	2.960000	H-LAK3	
17	6.730000	180.000000	2.560000	AIR	
18	214.992000	3.330000	22.070000	LZ_K8	
19	50.724000	1.800000	22.340000	AIR	
20	51.075000	3.330000	23.060000	LZ_TF1	
21	48.312000	11.700000	23.030000	H-FK61	
22	-50.976000	3.330000	23.040000	LZ_CTK12	
23	-92.718000	177.657959 S	23.610000	AIR	
IMS	0.000000	0.000000	20.000000 S		

Figure 2.13 Optical design and graphs of aberration correction of a 50×/0.55 infinity LCF LD plan MicroFluar objective (F'_{tl} = 180 mm).

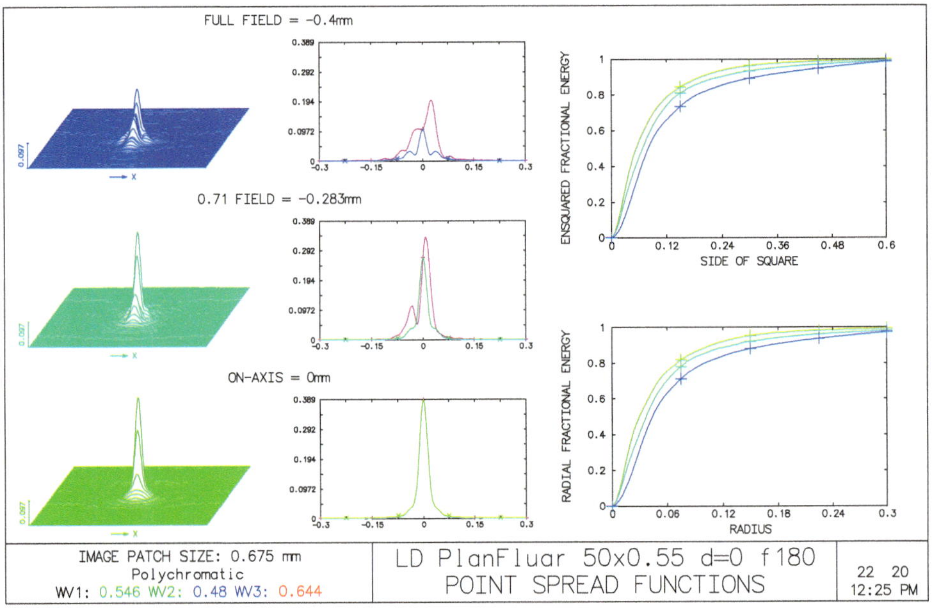

Figure 2.13 (*Continued*) Optical design and graphs of aberration correction of a 50×/0.55 infinity LCF LD plan MicroFluar objective (F'_{tl} = 180 mm).

large lenses with diameters up to 50–60 mm (microscope tube and adapter lenses) and micro lenses with diameters from 0.5 to 20–25 mm (in objectives). This is an approximate range of linear dimensions of lenses, which should ideally be made of fluorite or its equivalent (substitute). CaF_2 is a very non–thermo-stable material, which, being a cubic crystal, easily splits even when grinding or polishing on the cleavage planes, which generally complicates the technology; moreover, it is a scarce and expensive material. Fluoroaluminate glasses with the addition of $Ba(PO_3)_2$ can have optical constants very close to fluorite but without these and other shortcomings. Of course, these glasses also have their drawbacks; for example, they are characterized by an increased tendency to crystallize.

The goal was developed to assess the feasibility of synthesizing fluoroaluminate glass so that it would have optical parameters as similar as possible to those of natural fluorite. We developed a set of such glasses; the main characteristics of these are presented in Tables 2.5 and 2.6.

It can be seen in Table 2.5 that the glass that is the most similar to CaF_2 in terms of its optical parameters is F10, the constants of which coincide with the optical constants of fluorite with an accuracy of $0 \div (-3)\times 10^{-4}$ (red text). F10 glass is characterized by a high reproducibility of results. Deviations of the refractive index and the average dispersion of glasses do not exceed $\pm 2\times 10^{-4}$ for n_D and $\pm 2\times 10^{-5}$ for $n_F - n_C$.

One of the goals in the industrial development of such glass is for it to be used in objectives and other optical components in research-grade polarization microscopes. At the same time, the degree of aberration correction in laboratory and research microscopes should be at the level of plan apochromats, or even better, at the level of plan poly-apochromats. The use of optical materials that are isotropic is of fundamental importance for conducting research on a polarizing microscope. If used in optical glass, isotropic materials, which have refractive and dispersive characteristics similar

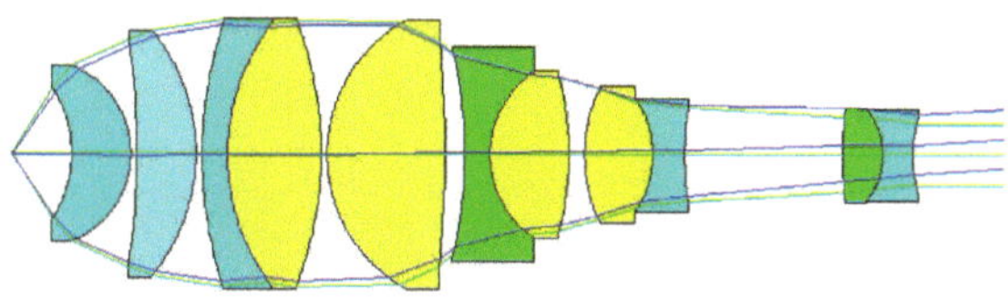

Lens: LD PlanFluar 100x0.85 d=0 f180

Object num aper 0.850000 Image height 15.000000 Primary wavln 0.546070

SRF	RADIUS	THICKNESS	APERTURE RADIUS	GLASS	SPECIAL
OBJ	0.000000	1.0000e-20	0.149951	AIR	
AST	0.000000	3.000120	0.149951 AS	AIR	
2	-6.000000	3.000000	3.350000	H-ZLAF55D	
3	-4.650000	0.300000	4.370000	AIR	
4	-39.800000	3.000000	5.690000	SILICA	
5	-9.375000	0.300000	6.090000	AIR	
6	20.460000	1.300000	6.700000	H-ZLAF55D P	
7	10.516000	4.800000	6.540000	H-FK71	
8	-17.660000	0.300000	6.690000	AIR	
9	7.490000	6.000000	6.720000	H-FK71 P	
10	-58.100000	1.000000	6.000000	AIR	
11	-22.180000	1.300000	5.380000	ZF3	
12	4.950000	4.000000	4.190000	H-FK71 P	
13	-15.700000	0.800000	4.080000	AIR	
14	7.490000	3.400000	3.460000	H-FK71 P	
15	-4.950000	1.600000	2.890000	H-ZLAF55D P	
16	15.078000	8.000000	2.640000	AIR	
17	18.883000	2.000000	2.580000	H-ZF7LA	
18	-3.569000	1.500000	2.530000	H-LAK3	
19	6.516000	180.000000	2.200000	AIR	
20	214.992000	3.330000	16.620000	LZ_K8	
21	50.724000	1.800000	16.770000	AIR	
22	51.075000	3.330000	17.220000	LZ_TF1	
23	48.312000	11.700000	17.200000	H-FK61	
24	-50.976000	3.330000	17.360000	LZ_CTK12	
25	-92.718000	177.670528 S	17.700000	AIR	
IMS	0.000000	0.000000	15.000000 S		

Figure 2.14 Optical design and graphs of aberration correction of a 100×/0.85 infinity LCF LD plan MicroFluar objective ($F'_{tl} = 180$ mm).

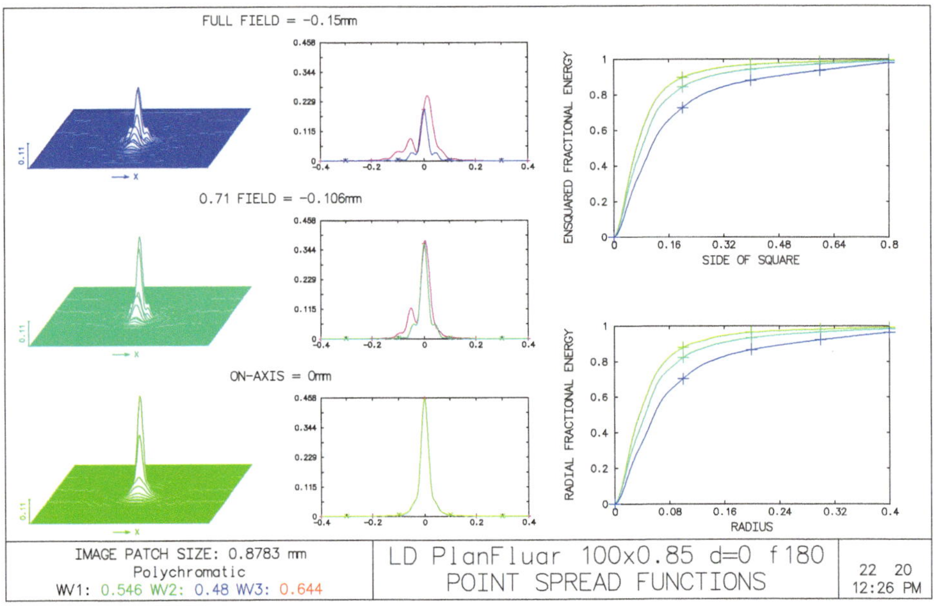

Figure 2.14 (*Continued*) Optical design and graphs of aberration correction of a 100×/0.85 infinity LCF LD plan MicroFluar objective (F'_{tl} = 180 mm).

to fluorite, can provide the required level of aberration correction for use in research polarization microscopes. Naturally, objectives for polarization studies can also be used in biological (metallographic) microscopes, thus providing a wide unification of complex objectives and other optical systems of light microscopes.

2.3.2 Other optical materials needed to advance microscope optics

Modern optical glass catalogs from the world's leading manufacturers include a wide variety of glass types with a range of refractive and dispersive properties. However, as practical experience shows, only a small glass nomenclature is used in microscopy. On the one hand, for mass-produced microscopes, the list of optical materials is rather limited; it is determined by the requirements for high technology and low cost in the manufacture of lenses. On the other hand, when designing laboratory and research microscopes, optical materials that have nonstandard refractive and dispersive properties (even to the detriment of the requirement of high processability) are required. The availability of such optical materials is quite limited. Moreover, the most interesting optical materials are not even offered. Here is a list of some optical materials, whose use can help further the progress of microscope optics:

- Lithium fluoride (LiF). The development of this crystal in the mass production of lenses and other optical components for microscopy can bring significant progress to the field. This material is a kind of continuation of fluorite; it has even more pronounced and unique dispersive properties. Lithium fluoride has a broad spectral transmittance range of 0.15 to 6 μm. The refractive index in this range varies

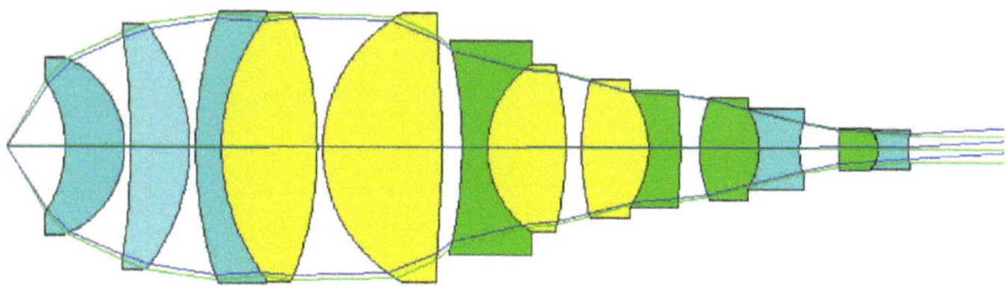

Lens: LD PlanFluar 250x0.85 d=0 f180

Object num aper 0.850000 Image height 20.000000 Primary wavln 0.546070

SRF	RADIUS	THICKNESS		APERTURE RADIUS		GLASS		SPECIAL
OBJ	0.000000	1.0000e-20		0.079976		AIR		
AST	0.000000	3.000278		0.079976	AS	AIR		
2	-6.000000	3.000000		3.310000		H-ZLAF55D		
3	-4.650000	0.300000		4.350000		AIR		
4	-39.800000	3.000000		5.680000		SILICA		
5	-9.375000	0.300000		6.080000		AIR		
6	20.460000	1.300000		6.680000		H-ZLAF55D	P	
7	10.516000	4.800000		6.520000		H-FK71		
8	-17.660000	0.300000		6.670000		AIR		
9	7.490000	6.000000		6.650000		H-FK71	P	
10	-58.100000	1.000000		5.870000		AIR		
11	-22.180000	1.300000		5.270000		ZF3		
12	4.950000	4.000000		4.130000		H-FK71	P	
13	-15.700000	0.800000		4.000000		AIR		
14	13.250000	3.400000		3.490000		H-FK71	P	
15	-4.820000	1.600000		2.950000		H-ZF7LA		
16	-53.680000	1.000000		2.830000		AIR		
17	6.220000	3.000000		2.610000		H-ZF7LA	P	
18	-3.877000	2.000000		2.120000		H-ZLAF55D	P	
19	3.877000	2.000000		1.390000		AIR		
20	-11.041000	2.000000		1.140000		H-ZF7LA		
21	-1.320000	1.500000		1.090000		H-LAK53A		
22	3.000000	180.000000		0.910000		AIR		
23	214.992000	3.330000		21.320000		LZ_K8		
24	50.724000	1.800000		21.640000		AIR		
25	51.075000	3.330000		22.370000		LZ_TF1		
26	48.312000	11.700000		22.350000		H-FK61		
27	-50.976000	3.330000		22.400000		LZ_CTK12		
28	-92.718000	177.729822	S	22.290000		AIR		
IMS	0.000000	0.000000		20.000000	S			

Figure 2.15 Optical design and graphs of aberration correction of a 250×/0.85 infinity LCF LD plan MicroFluar objective ($F'_{tl} = 180$ mm).

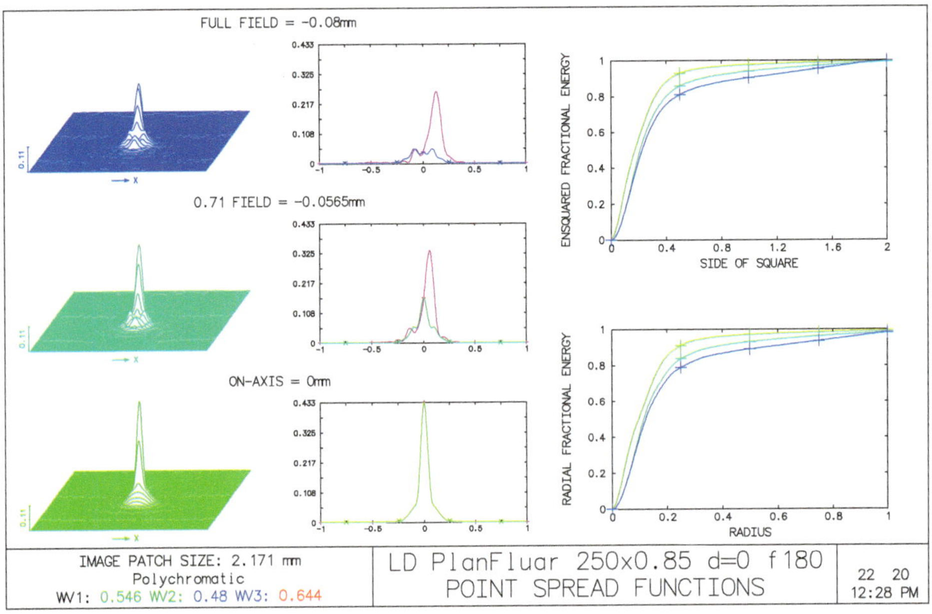

Figure 2.15 (*Continued*) Optical design and graphs of aberration correction of a 250×/0.85 infinity LCF LD plan MicroFluar objective ($F'_{tl} = 180$ mm).

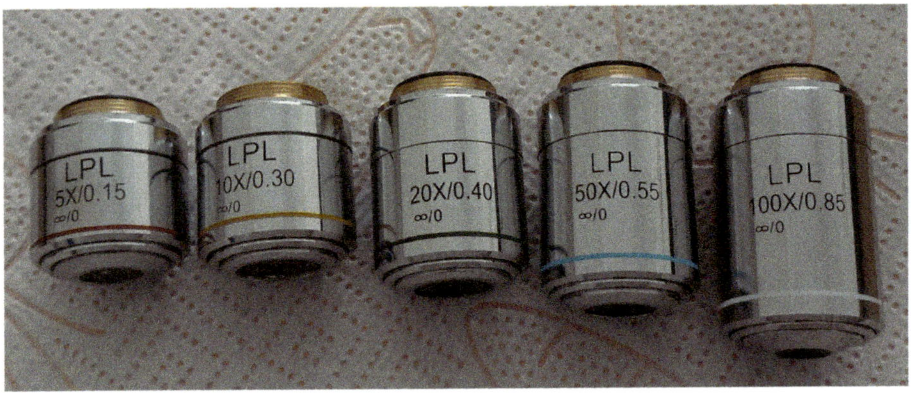

Figure 2.16 Photograph of a set of modern LCF LD plan MicroFluar metallographic objectives.

from approximately 1.3 to 1.4. This crystal is optically isotropic and medium hard. However, this material is rather unstable when operating in a humid atmosphere and at high radiation. Although it can withstand temperatures of about 500°C (in a dry atmosphere), it is quite sensitive to thermal shock. Due to its relatively low refractive index, lithium fluoride (in many cases) can be used without an anti-reflex coating. Because of its unique dispersive properties, it can also be

Table 2.2 Main technical parameters and basic optical layout of third-generation LCF LD plan MicroFluar objectives.

Magnification	*NA*	*WD* (mm)	*F'* (mm)	*R* (μm)	*DF* (μm)	*FOV* on object (mm)	*FOV* on image (mm)	Optical design at figure no.	The principal optical layout
5×	0.15	20.5	36	2.22	14.8	5	25	2.9	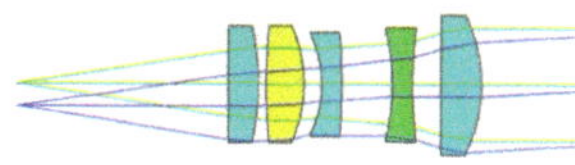
10×	0.30	18.2	18	1.11	3.7	2.5	25	2.11	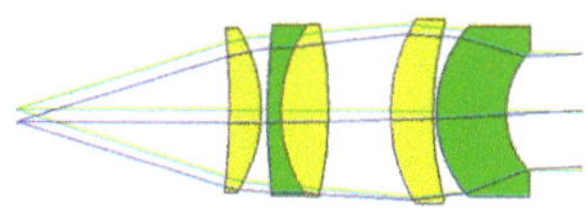
20×	0.40	12.9	9	0.83	2.08	1.5	30	2.12	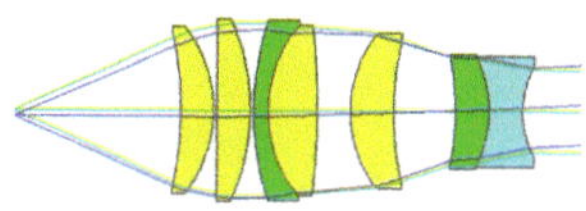
50×	0.55	7.6	3.6	0.61	1.1	0.8	40	2.13	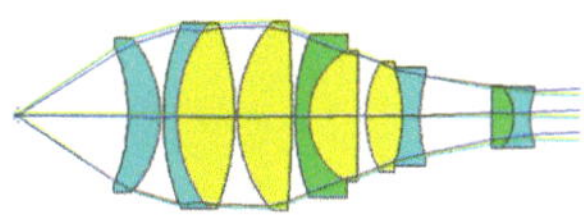
100×	0.85	1.8	1.8	0.39	0.46	0.3	30	2.14	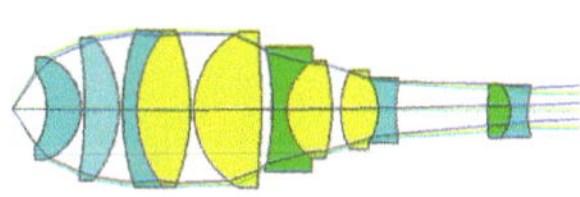
250×	0.85	1.8	1.8	0.39	0.46	0.16	40	2.15	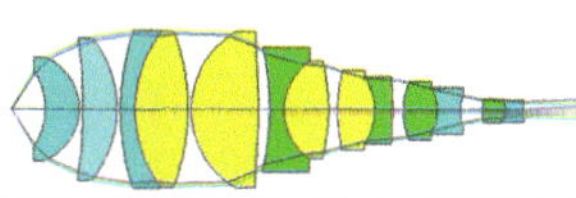

used in complex optical systems with super-apochromatic, poly-apochromatic, and multi-apochromatic aberration corrections. It is interesting to note its potential use in the manufacture of spherical, aspherical, and cylindrical lenses, as well as prisms.

It should be said that researchers at the Saint Petersburg State Institute of Technology in Russia have been working on developing a replacement for natural lithium fluoride, much like the work that has been done on the synthesis of glass similar to fluorite. Research has shown the fundamental possibility of creating such glass; its industrial development promises substantial simplification and cost reduction of newly developed, wide-spectrum dioptric optics, including optics for light microscopes.

- Sapphire (Al_2O_3). This material is widely used in protective optical windows, sight glasses, prisms, and optical lenses in the range of 190 to 4500 nm. Sapphire has a unique combination of a wide range of

Figure 2.17 Photograph of a real object using a 5×/0.15 LCF LD plan MicroFluar objective.

Figure 2.18 Photograph of a real object using a 10×/0.30 LCF LD plan MicroFluar objective.

transmission—from the UV region to the IR part of the spectrum—and high transmittance, as well as durability. Sapphire is interesting for optical design because it has a high refractive index combined with a high Abbe number. Thus, $n_d = 1.768207$, $\nu_d = 71.8$. This material is convenient to use in the front lenses of high-aperture lenses, especially in immersion lenses; it has a very high hardness, which ensures that the deviation tolerances from the nominal values for the thickness and centering of the lens are very tight.

Figure 2.19 Photograph of a real object using a 10×/0.30 LCF LD plan MicroFluar objective (using the Nomarski contrast method).

Figure 2.20 Photograph of a real object using a 20×/0.40 LCF LD plan MicroFluar objective.

- Magnesium fluoride (MgF_2). Due to transparency in a wide spectral range (especially in the vacuum UV spectral range of about 121 nm), combined with mechanical and thermal strength, chemical and radiation resistance, natural birefringence, and favorable thermophysical characteristics, single-crystal magnesium fluoride is widely used in various optical devices. Unlike other fluoride crystals, MgF_2 crystals are nonblocking.
- Yttrium aluminum garnet (YAG). YAG is an optical material suitable for use in UV and IR optics. Products from YAG can be used as

Figure 2.21 Photograph of a real object using a 20×/0.40 LCF LD plan MicroFluar objective (using the Nomarski contrast method).

Figure 2.22 Photograph of a real object using a 20×/0.40 LCF LD plan MicroFluar objective (using the Nomarski contrast method).

optical elements in a wide spectral range of 250 to 5000 nm. Although the mechanical and chemical properties of YAG are close to sapphire, YAG does not have birefringence and it is somewhat simpler to process than sapphire. Because of its high strength, damage threshold, refractive index, and thermal conductivity, YAG can be used at high temperatures and for high-power laser focusing systems.

Figure 2.23 Photograph of a real object using a 20×/0.40 LCF LD plan MicroFluar objective (using the Nomarski contrast method).

Figure 2.24 Photograph of a real object using a 50×/0.55 LCF LD plan MicroFluar objective.

2.3.3 Immersion liquids

In light microscopy, 20–25% of objectives are designed to be used with immersion liquids. In every microscope, regardless of its price segment, there is an immersion objective; as a rule, for mass-produced microscopes, it is an oil immersion objective with a linear magnification of 100× and a numerical aperture of 1.25–1.30. In laboratory and research models of light microscopes,

Figure 2.25 Photograph of a real object using a 50×/0.55 LCF LD plan MicroFluar objective (using the Nomarski contrast method).

Figure 2.26 Photograph of a real object using a 50×/0.55 LCF LD plan MicroFluar objective (using the Nomarski contrast method).

oil immersion objectives with a numerical aperture of up to 1.49 are used. Along with oil immersion objectives, glycerin immersion and water immersion objectives are also used. To solve special problems in microscopy, so-called *high-refraction* immersions are also used, for example, iodine-methylene and other (modern) types. Such immersions make it possible to realize a numerical aperture of up to 1.75, which is used for the latest microscopic examination

Figure 2.27 Photograph of a real object using a 50×/0.55 LCF LD plan MicroFluar objective (using the Nomarski contrast method).

Figure 2.28 Photograph of a real object using a 100×/0.85 LCF LD plan MicroFluar objective.

techniques. Let us compare the main optical constants of certain types of immersions that are used in light microscopy. Table 2.7 presents such data for different immersions.

In some objectives in which the numerical aperture of 1.70–1.75 is reached, sapphire is used as the optical material for the frontal lens. This is logical because, in addition to having a fairly high refractive index, sapphire has several other unique characteristics; its value of average dispersion is

Figure 2.29 Photograph of a real object using a 100×/0.85 LCF LD plan MicroFluar objective (using the Nomarski contrast method).

Table 2.3 Data on certain types of optical glasses.

No.	Glass type 1	n_d 2	ν_d 3	n_F 4	n_C 5	n_g 6	$n_F - n_C$ 7	$\frac{n_g - n_F}{n_F - n_C}$ 8	$\Delta p_{g,F} \cdot 10^4$ 9
1	LZ-K18	1.519179	60.37	1.525150	1.516550	1.529806	0.00860	0.5414	0
2	F13	1.620048	36.35	1.632096	1.615086	1.642054	0.01706	0.5837	0
3	LZ-OK1	1.522262	76.35	1.527015	1.520175	1.530707	0.00684	0.5398	265
4	LZ-OF1	1.529492	51.81	1.536620	1.526400	1.542251	0.01022	0.5510	−55
5	LZ-OF4	1.650632	43.46	1.661105	1.646135	1.669508	0.01497	0.5613	−98
6	LZ-CTK3	1.659502	57.34	1.667504	1.656004	1.673767	0.0115	0.5446	−21
7	LZ-CTK19	1.744132	50.43	1.754423	1.739667	1.762531	0.01476	0.5493	−96
8	LZ-TF4	1.740024	28.16	1.758714	1.732434	1.774548	0.02628	0.6025	44
9	Silica	1.581461	67.72	1.463137	1.456367	1.466623	0.00677	0.5149	−138
10	CaF_2	1.433879	95.32	1.437043	1.432491	1.439494	0.004557	0.5384	585
11	K7	1.511121	60.41	1.505781	1.497641	1.510186	0.00814	0.5412	0
12	F2	1.620040	36.37	1.632083	1.615033	1.642016	0.01705	0.5826	0
13	K51	1.505180	59.55	1.511064	1.502581	1.515644	0.008483	0.5399	−37
14	KzF5	1.521300	52.74	1.528184	1.518299	1.533609	0.009885	0.5488	−62
15	KzFS2	1.557809	53.85	1.564982	1.554623	1.570608	0.010359	0.5431	−101
16	KzFS1	1.613099	44.34	1.62276	1.60894	1.63048	0.013826	0.5581	−111
17	TiF6	1.616501	30.97	1.63070	1.61080	1.64308	0.019904	0.6082	304
18	LakN9	1.691002	54.71	1.69979	1.68716	1.70667	0.012631	0.5447	−70
19	Lak24	1.696800	56.18	1.70542	1.69301	1.71214	0.012404	0.5415	−78
20	SF13	1.740771	27.60	1.75988	1.73304	1.77621	0.026841	0.6082	109
21	Lak28	1.744287	50.77	1.75451	1.73985	1.76257	0.014659	0.5498	−86
22	SF14	1.761820	26.53	1.78229	1.75358	1.79998	0.028718	0.6124	133
23	SF11	1.784720	25.76	1.80645	1.77599	1.82518	0.028718	0.6147	143
24	LaSF5	1.880692	41.01	1.89577	1.87430	1.90793	0.021476	0.5658	−90

Table 2.4 Optical powers of the positive and negative elements of the doublets composed of the corresponding optical materials.

No.	Glass type		φ_1	φ_2	C_2^*
1	LZ-K18	LZ-F13	2.5133	−1.5131	0.00176
2	CaF_2	LZ-OF1	2.1908	−1.1908	0.00029
3	CaF_2	LZ-OF4	1.8380	−0.8380	0.00044
4	CaF_2	LZ-CTK3	2.5097	−1.5097	0.00016
5	CaF_2	Silica	3.4536	−2.4536	−0.00085
6	LZ-OF4	LZ-TF4	2.8405	−1.8408	0.00269
7	LZ-CTK3	LZ-TF4	1.9650	−0.9650	0.00198
8	LZ-K7	LZ-F2	2.5128	−1.5128	0.00172
9	CaF_2	KzF5	2.2386	−1.2386	0.00024
10	CaF_2	KzFS2	2.2985	−1.2985	0.00011
11	CaF_2	KzFS1	1.8698	−0.8698	0.00039
12	K7	LaSF5	3.1139	−2.1139	0.00127
13	KzFS1	SF11	2.3864	−1.3864	0.00305
14	LakN9	SF11	1.8898	−0.8898	0.00242

Table 2.5 Properties of developed fluorophosphate glasses.

Glass	ρ, g/cm^3	t_g, °C	T_{cr}, °C	T_l, °C	n_C	n_D	n_F	ν
Fluorite	3.400	—	—	—	1.4322	1.4338	1.4370	95.35
1176	3.510	410	530	640	1.4431	1.4445	1.4478	93.13
F1	3.501	395	525	635	1.4405	1.4421	1.4450	97.40
F2	3.517	400	550	645	1.4398	1.4415	1.4446	91.48
F6	3.508	415	560	655	1.4416	1.4426	1.4462	95.76
F9	3.526	400	560	660	1.4392	1.4411	1.4439	94.34
F10	3.433	415	565	680	1.4322	1.4336	1.4367	95.26
F11	3.456	410	560	670	1.4306	1.4319	1.4351	96.42

$\Delta n_C - n_F = 0.0107$, its Abbe number is $\nu_d = \frac{n_d - 1}{n_F - n_C} = 71.8$, and its relative partial dispersion value is $p_{F,g} = \frac{n_g - n_F}{n_F - n_C} = 0.526$.

Let us take as an analogy the principle of constructing simple oil immersion objectives for which nearly equal values of the refractive indexes are used (for the main wavelength) in the optical design of the immersion, cover glass, and frontal lens. Then, for the case of using sapphire as the optical material of the frontal lens, we need to select the appropriate type of immersion, namely, modern. This is done in some objectives in which the numerical aperture of 1.70–1.75 is reached. However, if we apply a classification to the immersion, the same as for the other optical materials,

Table 2.6 Physical and chemical characteristics of glass type F10.

No.	Characteristics	Value
1	Refractive index:	
	$-n_C$	$1.4322 \pm 2 \times 10^{-4}$
	$-n_D$	$1.4336 \pm 2 \times 10^{-4}$
	$-n_F$	$1.4367 \pm 2 \times 10^{-4}$
2	Average dispersion: $(n_F - n_C) \times 10^{-5}$	450
3	Dispersion coefficient: ν	95.26
4	Density: ρ, g/cm^3	$3.433 \pm 2 \times 10^{-3}$
5	Thermal expansion coefficient: $\alpha \times 10^7$, K^1	
	from 20 to 120	154
	from 20 to 200	154
	from 20 to 300	163
6	Temperature, °C at	
	$\lg_\eta = 12$	370
	$\lg_\eta = 11{,}5$	372
	$\lg_\eta = 10$	403
	$\lg_\eta = 9$	421
7	Glass transition temperature: t_g, °C	385
8	Crystallization ability, °C	510–620
9	The increment of the specific optical density in the region of the maximum $\lambda = 480$ nm at γ-radiation dose 1×10^5 rad ΔD, cm^{-1}	0.68

Table 2.7 Dispersion properties of immersion liquids.

Type of immersion	n_d	n_C	n_F	$\varkappa$ $n_C - n_F$	ν_d	$p_{F,g} = \frac{n_g - n_F}{n_F - n_C}$
Water	1.33304	1.3312	1.3372	0.006	55.51	0.513
Glycerol	1.474	1.4717	1.4795	0.0078	60.77	0.564
Cedar oil	1.51501	1.5117	1.5224	0.0107	48.0	0.559
Gelatin	1.53821	1.53467	1.5465	0.0118	45.61	0.569
Iodine-methylene	1.74174	1.73135	1.7684	0.0371	20.0	0.686
Modern	1.7845	1.7762	1.8045	0.0283	27.72	0.548

then (conventionally) such an immersion will be of the *heavy flint* type if we take into account its dispersion parameters. It should be noted that when a heavy flint is used in the frontal part (in the area of maximum numerical aperture), the optical design of the high-aperture objective is greatly complicated; even a small thickness of the layer of such an immersion (approximately 0.1–0.12 mm) introduces significant chromatic aberrations that, if corrected, require a very complex optical design of the objective. But the biggest drawback of this immersion is its very low transmission ability, especially in the left border of the visible region. Thus, the transmittance is limited by the spectral line of 404 nm; with an immersion layer thickness of 0.1 mm, the transmittance is 63%. This characteristic does not allow the use of such an immersion, for example, in fluorescent studies in the near UV region. In addition, because the design of the objective is rather complicated, its own transmission is also reduced. Therefore, some high-aperture lenses in which the numerical aperture of 1.70–1.75 is reached can work only in the visible spectral range.

At the same time, by using water as an immersion, it is possible to achieve similar values of resolution, as with the use of a modern, high-refractive immersion; the area of research (the value of the main wavelength) must be shifted to the left border of the visible range using, for example, the wavelength of 441.6 nm, in which one of the lasers operates. In addition, this spectral range is still visible to the human eye and is available for most (even budget) digital image receivers. Thus, the resolution at a numerical aperture of 1.3 in the water immersion will be equal to 0.17 μm. The same resolution value is achieved using a modern immersion at a wavelength of, for example, 589 nm with a numerical aperture of 1.75. Of course, based on this argument, water is perhaps the most practical and technologically advanced immersion for use in light microscopes. Moreover, its transparency is rather high and extends to the near UV range of up to 190-nm wavelengths (in special preparation). Glycerin is another recommended immersion. A good combination is a mixture of water and glycerin in certain proportions.

2.4 Using Fluorite in Budget Objectives

Our discussion is focused on low-cost objectives that are designed to complete mass-produced light microscopes. The use of fluorite in the optical systems of objectives (and other optical components of microscopes) has so far significantly complicated and drastically increased the cost of products; however, its indisputable advantages (or those of its direct analogue) make it necessary to find technological solutions to reduce the cost of its use. Modern glass-processing technologies make it possible to grind and polish fluorite at nearly the same cost as grinding and polishing ordinary silicate glass. It is only necessary to select the optimal operating modes of the machine. If the process (for example, the assignment and provision of the optimal operation modes of the tool during lens shaping) is implemented by a smart machine with numerical control (by special software), then the cost of such processing can be quite low. Another issue is the cost of the fluorite itself. It is considered to be a rather expensive material; however, in budget objectives, the diameters of the lenses are quite small and the accuracy of shaping is relatively low. If we use a fluorite lens in a budget objective, the objective will be about 30–50% more expensive as compared with the use of ordinary silicate glass only. But if we use a direct glass equivalent of fluorite (for example, glass F10; see Table 2.5), the cost will increase by only 5–10%.

Throughout this book, budget objectives containing fluorite lenses are referred to as MicroFluars. As a rule, the use of fluorite in objectives can significantly reduce an aberration such as the secondary spectrum; that is, this type of aberration correction, like a semi-apochromatic or apochromatic, can be achieved in budget objectives without complicating the objectives' optical design. In addition, the use of fluorite significantly helps increase the numerical apertures of objectives while ensuring achromatic aberration correction without significantly complicating the objectives' optical structures. In other cases, the use of fluorite can help increase the values of the free working distance in some objectives.

By design, these objectives hardly differ from the achromats, semi–plan achromats, or plan achromats designed for use in mass-produced microscopes, except that they contain fluorite lenses. Figures 2.30–2.61 present the principal layouts, design parameters, and graphs of aberration correction of some of these objectives. Figure 2.62 is a photograph of this objective; it has been used in microscopes manufactured by the Russian enterprise LOMO for about 30 years. Figures 2.63–2.65 are photographs of real objects taken using this objective.

Figures 2.66–2.68 present the principal layouts, design parameters, and graphs of aberration correction of more objectives made with fluorite.

2.4.1 Optical designs of MicroFluar microscope objectives: Compromises in the aberration correction

MicroFluar objectives are mainly intended for use in fluorescence microscopes, whose main parameters are their light power and an increased numerical aperture of the objective. Based on this, we can consider some compromises, mainly concerning the quality of the aberration correction, in the optical design of such objectives. The following two types of aberrations, the correction of which can be different in objectives of different classes and purposes, can be considered:

1. Field curvature. This aberration occurs mainly when using short-focus objectives with steep spherical refracting surfaces. The shorter the focal distance of the objectives, the higher the curvature of the image transmitted by it is. When we study an object (using short-focus objectives) from the center to the periphery of the field, the field curvature aberration requires refocusing the microscope. The resulting image is sharp only in the center of the field or in one of its circular segments, and not in other areas. A visual observer feels the most discomfort when a microscope is focused exactly on the field edge and a

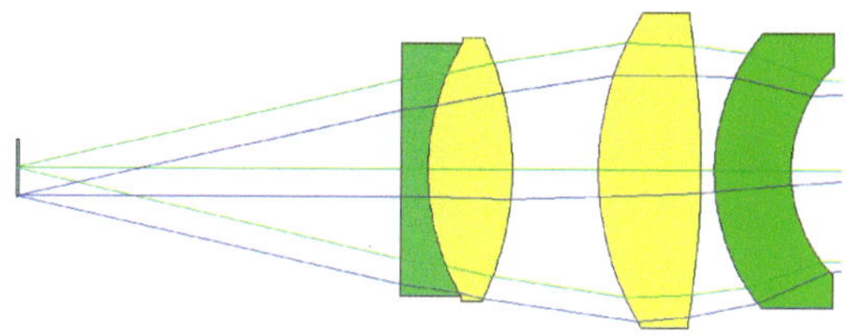

Lens: PlanFluar 6.3x0.22 d017 tub160 Efl 21.9

Object num aper 0.220000 Image height 10.000000 Primary wavln 0.546070

SRF	RADIUS		THICKNESS		APERTURE RADIUS		GLASS		SPECIAL
OBJ	0.000000		0.010000		1.591507		AIR		
AST	0.000000		0.170000		1.590000	A	H-K9		
2	0.000000		22.490000		1.610000		AIR		
3	0.000000		1.600000		6.680000		LZ_BF21		
4	13.900000		5.000000		7.180000		CAF2		
5	-16.762000		5.000000		7.490000		AIR		
6	16.762000		6.000000		8.920000		CAF2	P	
7	-54.850000		0.800000		8.680000		AIR		
8	12.134000		4.500000		7.790000		LZ_TF5		
9	8.336000		149.457230	S	5.910000		AIR		
IMS	0.000000		0.000000		10.000000				

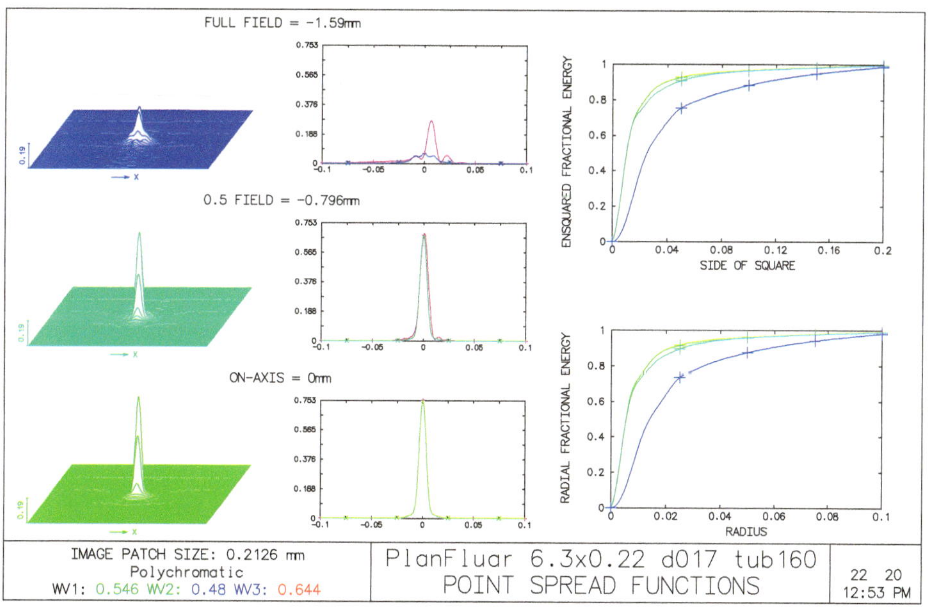

Figure 2.30 Optical design and graphs of aberration correction of a 6.3×/0.22 finite LCF plan MicroFluar objective.

considerable part of its center is non-sharp. The same is true in the case of microphotography. We should also note that the structure and form of the image curvature of a studied object are not distorted; i.e., when you refocus the microscope and achieve sharpness in any area of the field

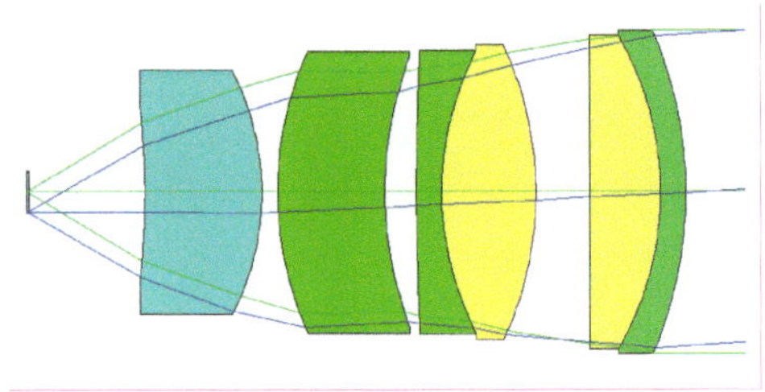

Lens: OF 10x0.50 d=0.17 Ftl 160

Object num aper 0.500000 Image height 10.000000 Primary wavln 0.546070

SRF	RADIUS	THICKNESS	APERTURE RADIUS	GLASS	SPECIAL
OBJ	0.000000	0.010000	1.004126	AIR	
AST	0.000000	0.170000	1.010000 A	LZ_K14	
2	0.000000	5.965000	1.070000	AIR	
3	-40.270000	6.000000	4.370000	LZ_CTK19	
4	-13.032000	0.800000	6.060000	AIR	
5	16.596000	5.500000	6.990000	LZ_TF8	
6	18.365000	1.600000	6.690000	AIR	
7	129.420000	1.300000	6.790000	LZ_TF8 P	
8	15.031000	4.800000	7.030000	CAF2	
9	-16.943000	2.800000	7.310000	AIR	
10	0.000000	3.600000	7.720000	CAF2 P	
11	-15.031000	1.300000	7.810000	LZ_BF11	
12	-20.230000	90.000000	8.050000	AIR	
13	214.800000	5.000000	13.750000	LZ_BK8	
14	-46.990000	0.300000	13.790000	AIR	
15	-46.880000	3.000000	13.740000	LZ_BF24	
16	-113.240000	157.842565 S	13.890000	AIR	
IMS	0.000000	0.000000	10.000000		

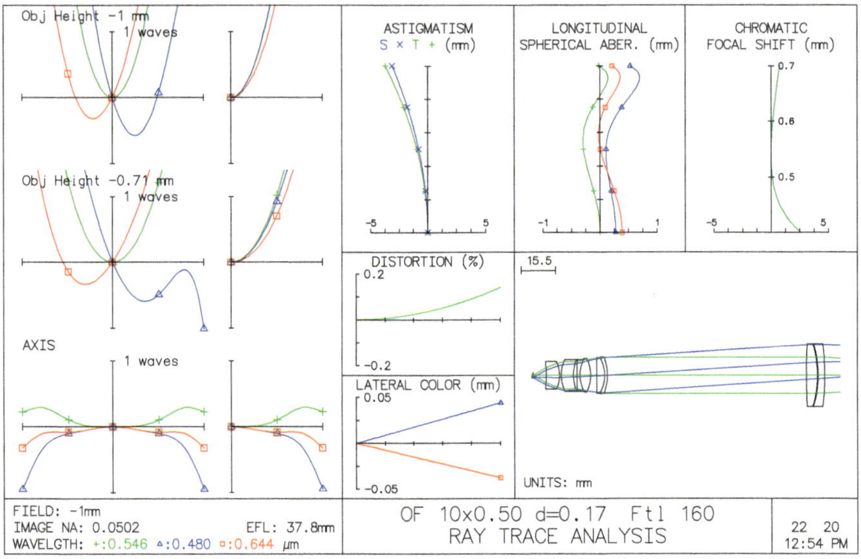

Figure 2.31 Optical design and graphs of aberration correction of a 10×/0.50 infinity LCR MicroFluar objective ($F'_{ts} = 160$ mm).

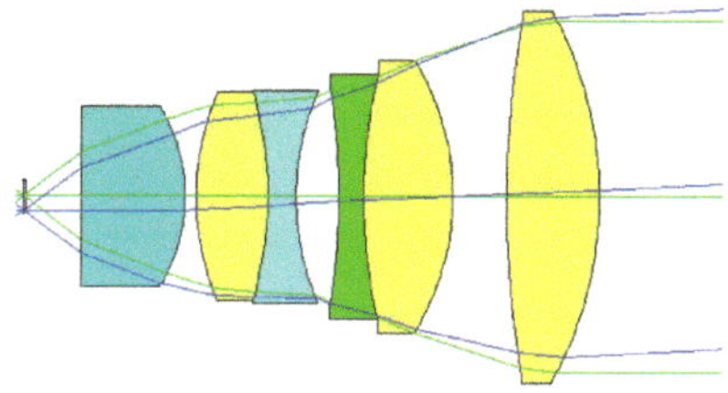

Lens: OF 10x0.60 d=0.17 ft1200

Object num aper 0.600000 | Image height 10.000000 | Primary wavln 0.546070

SRF	RADIUS	THICKNESS	APERTURE RADIUS	GLASS	SPECIAL
OBJ	0.000000	0.010000	0.989169	AIR	
AST	0.000000	0.170000	1.010000 A	K9	
2	0.000000	3.900000	1.080000	AIR	
3	0.000000	7.200000	4.010000	H-LAK2	
4	-11.426000	0.800000	6.080000	AIR	
5	17.320000	5.000000	6.990000	CAF2	
6	-24.500000	2.000000	7.020000	K9 P	
7	17.320000	3.000000	7.150000	AIR	
8	-44.140000	1.700000	7.410000	ZF2	
9	34.130000	6.300000	8.260000	CAF2	
10	-16.762000	3.700000	9.190000	AIR	
11	68.570000	6.400000	12.120000	CAF2	
12	-25.350000	200.000000	12.520000	AIR	
13	-36.100000	14.000000	22.070000	K9	
14	-43.560000	3.000000	25.750000	AIR	
15	377.100000	15.000000	26.580000	CAF2	
16	-81.500000	3.000000	26.820000	AIR	
17	221.300000	17.000000	25.720000	CAF2	
18	-72.000000	3.000000	24.380000	H-LAK2 P	
19	0.000000	199.991812 S	24.080000	AIR	
IMS	0.000000	0.000000	10.000000		

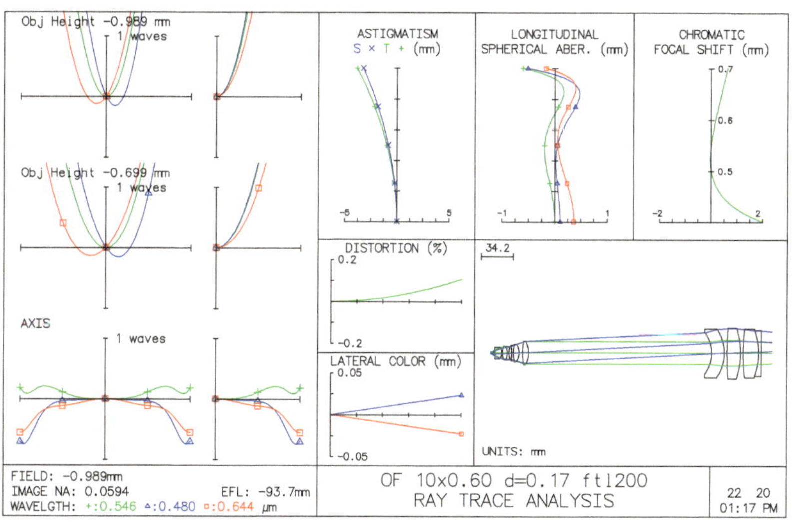

Figure 2.32 Optical design and graphs of aberration correction of a 10×/0.60 infinity LCR MicroFluar objective ($F'_{ts} = 200$ mm).

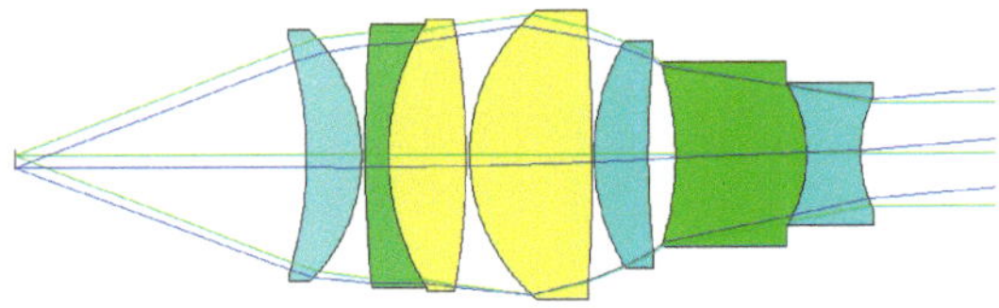

Lens: PlanFluar LD 20x0.35 d0 ftl160

Object num aper	0.350000	Image height	12.500000	Primary wavln	0.546070

SRF	RADIUS	THICKNESS		APERTURE RADIUS		GLASS		SPECIAL
OBJ	0.000000	1.0000e-20		0.625298		AIR		
AST	0.000000	16.298800		6.350000	A	AIR		
2	-21.180000	3.000000		6.350000		LZ_LK3		
3	-9.506000	0.200000		6.760000		AIR		
4	76.210000	1.300000		7.050000		LZ_BF28		
5	13.062000	4.300000		7.100000		CAF2		
6	-41.110000	0.200000		7.290000		AIR		
7	10.093000	6.700000		7.760000		CAF2	P	
8	-103.280000	0.100000		7.010000		AIR		
9	11.588000	3.000000		6.110000		LZ_CTK3		
10	52.000000	1.500000		5.350000		AIR		
11	-16.032000	7.300000		5.010000		LZ_TF5		
12	-7.244000	3.000000		3.840000		LZ_CTK19		
13	6.730000	90.000000		3.020000		AIR		
14	214.800000	5.000000		9.910000		LZ-BK8		
15	-46.990000	0.300000		10.030000		AIR		
16	-46.880000	3.000000		10.010000		LZ-BF24		
17	-113.240000	157.880817	S	10.160000		AIR		
IMS	0.000000	0.000000		12.500000				

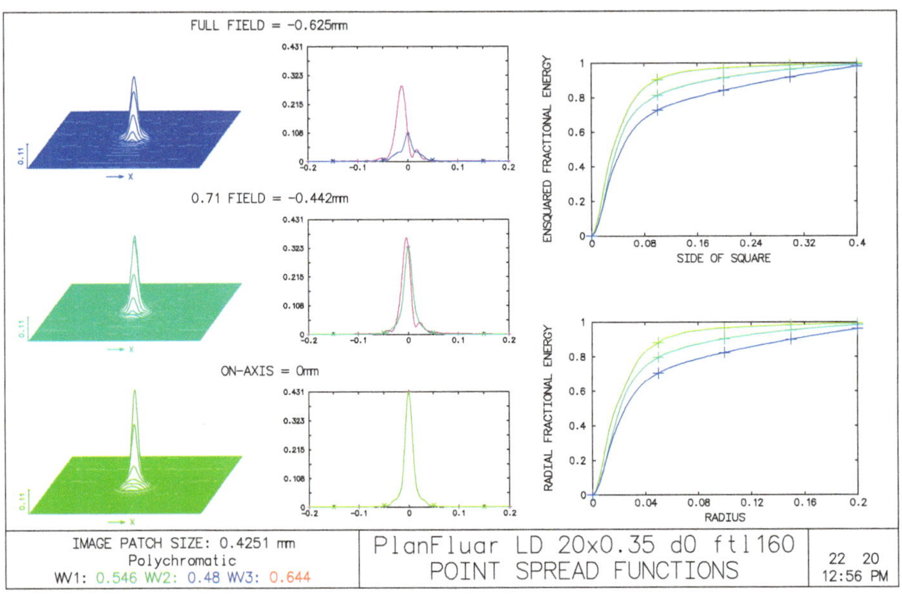

Figure 2.33 Optical design and graphs of aberration correction of a 20×/0.35 infinity LCF LD plan MicroFluar objective ($F'_{ts} = 160$ mm).

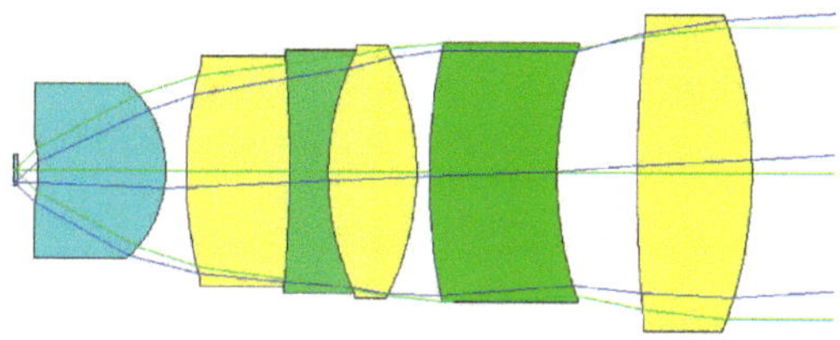

Lens: OF 20x0.70 d=0.17 Ftl 160

Object num aper 0.700000 Image height 10.000000 Primary wavln 0.546070

SRF	RADIUS	THICKNESS		APERTURE RADIUS		GLASS		SPECIAL
OBJ	0.000000	0.010000		0.497875		AIR		
AST	0.000000	0.170000		0.520000	A	LZ_K14		
2	0.000000	0.812200		0.610000		AIR		
3	-5.970000	5.000000		1.260000		LZ_CTK19		
4	-4.446000	0.800000		3.350000		AIR		
5	15.524000	4.100000		4.100000		CAF2		
6	-49.430000	1.500000		4.400000		LZ_BF21		
7	10.423000	3.500000		4.650000		CAF2	P	
8	-10.423000	0.500000		4.850000		AIR		
9	23.770000	5.000000		4.960000		LZ_TF3		
10	14.421000	3.130000		4.860000		AIR		
11	56.360000	4.500000		5.560000		CAF2		
12	-16.904000	90.000000		6.050000		AIR		
13	214.800000	5.000000		11.790000		LZ_BK8		
14	-46.990000	0.300000		11.860000		AIR		
15	-46.880000	3.000000		11.820000		LZ_BF24		
16	-113.240000	157.844790	S	11.960000		AIR		
IMS	0.000000	0.000000		10.000000				

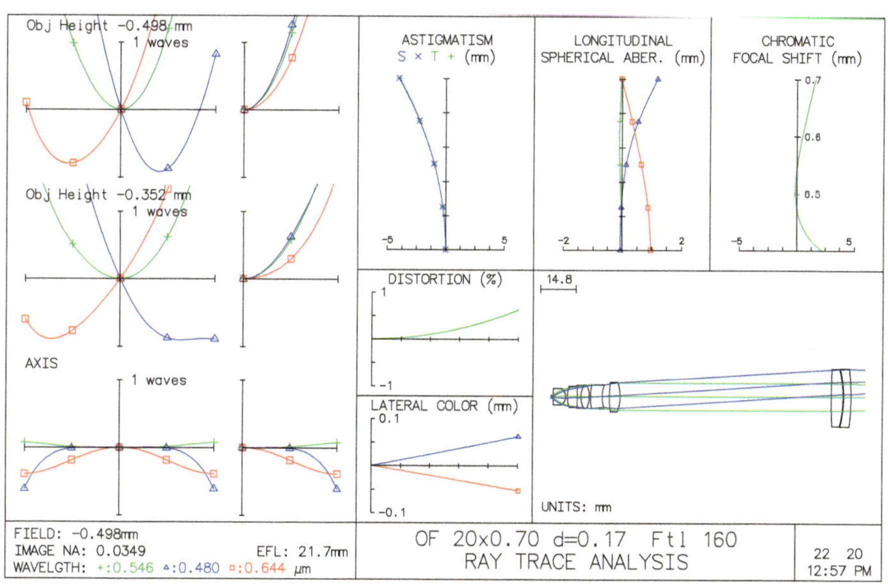

Figure 2.34 Optical design and graphs of aberration correction of a 20×/0.70 infinity LCR MicroFluar objective (F'_{ts} = 160 mm).

(even in the periphery), the information on the segment of the object under consideration remains reliable.

2. LC aberration. When dealing with medium and high magnification, we consider the color of the image components of an object's structure. The impact of LC on image quality in visual microscopy has been a subject of discussions among developers and researchers for 50 years. Old achromatic and apochromatic objectives used in many microscopes do not correct LC aberration. A color border visible for the whole linear field appears at the fragment edges of structural images. We should also

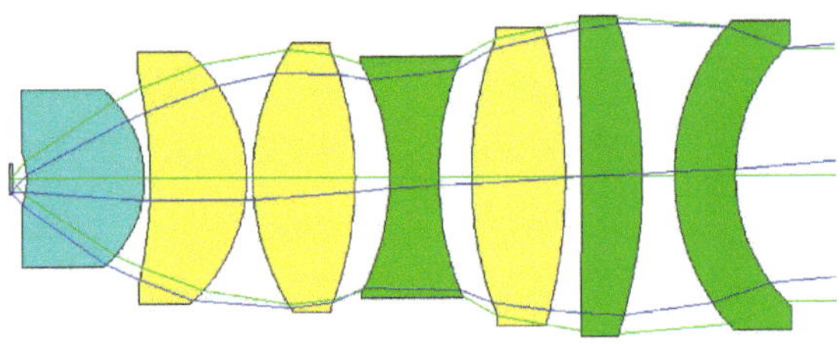

Lens: PlanFluar 20x0.75 d017 f160

Object num aper 0.750000 Object height -0.616700 Primary wavln 0.546070

SRF	RADIUS	THICKNESS	APERTURE RADIUS	GLASS	SPECIAL
OBJ	0.000000	0.010000	0.616700	AIR	
AST	0.000000	0.170000	0.630000 A	LZ_K14	
2	0.000000	0.700800	0.730000	AIR	
3	-3.631000	5.900000	1.260000	LZ_CTK19	
4	-5.445000	0.300000	4.250000	AIR	
5	-22.180000	4.800000	5.040000	CAF2	
6	-7.727000	0.300000	6.000000	AIR	
7	11.588000	5.000000	6.450000	CAF2 P	
8	-15.996000	1.700000	6.200000	AIR	
9	-11.588000	2.400000	5.530000	LZ_TF5	
10	14.997000	1.600000	5.750000	AIR	
11	18.793000	4.600000	6.640000	CAF2 P	
12	-25.350000	0.800000	7.110000	AIR	
13	-510.500000	3.000000	7.420000	LZ_TF5 P	
14	-24.210000	1.600000	7.620000	AIR	
15	10.914000	3.000000	7.360000	LZ_TF5 P	
16	8.551000	100.000000	6.280000	AIR	
17	677.600000	4.000000	13.670000	LZ_TK13	
18	-30.970000	1.500000	13.690000	LZ_F1	
19	53.210000	5.400000	13.970000	AIR	
20	63.530000	4.800000	15.300000	LZ_TF1	
21	-107.400000	168.883943 S	15.390000	AIR	
IMS	0.000000	0.000000	12.500000		

Figure 2.35 Optical design and graphs of aberration correction of a 20×/0.75 infinity LCR plan MicroFluar objective ($F'_{ts} = 160$ mm).

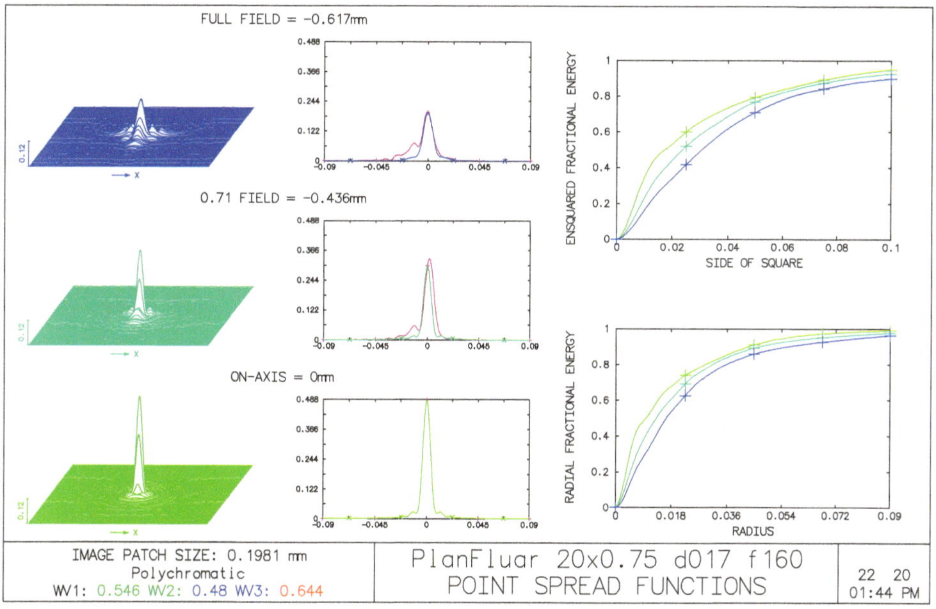

Figure 2.35 (*Continued*) Optical design and graphs of aberration correction of a 20×/0.75 infinity LCR plan MicroFluar objective (F'_{ts} = 160 mm).

note that the presence of LC in the image of an object distorts its structure and form. Refocusing the microscope, when we move from the center to the periphery, does not correct such distortions [4].

2.4.1.1 A description of the model specimen

Let us consider an object to be studied with a biological microscope. In a typical case, it is a specimen, i.e., a thin slice of matter or a smear glued with one of its sides to the object plate and its opposite side glued up with the cover glass. We should note that to properly describe the technical aspects of microscopy, we should assume that the object is a composition of all these components. Evidently, an object prepared in this way (object plate, glue, specimen, and cover glass) within the linear field from the center to the periphery will not be uniformly plain (there will be a kind of general wedging and local hills and holes in the surface layer of the biological culture under consideration). Figure 1.35 in the first chapter of Volume 1 of this book shows the configuration of a real biological object.

Microscopists are aware that different fragments of an object to be studied are usually located at different heights from the object plate (and cover glass). If we compare the spread in the location of a fragment of the biological culture with the focal depth of the objectives used, we can determine if they are equivalent for medium- and low-magnification objectives. But for a high

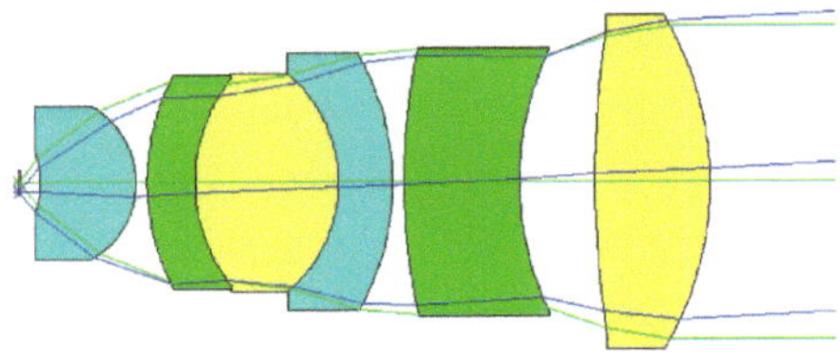

Lens: OF 20x0.80 d=0.17 ft1200

Object num aper 0.800000 | Image height 10.000000 | Primary wavln 0.546070

SRF	RADIUS	THICKNESS	APERTURE RADIUS	GLASS	SPECIAL
OBJ	0.000000	0.010000	0.499987	AIR	
AST	0.000000	0.170000	0.510000 A	K9	
2	0.000000	1.003900	0.620000	AIR	
3	-4.546000	5.000000	1.580000	H-LAK2	
4	-4.368000	0.580000	3.910000	AIR	
5	11.250000	2.500000	5.440000	ZF6	
6	8.395000	7.400000	5.250000	CAF2	
7	-7.060000	2.800000	5.550000	H-LAK2 P	
8	-12.953000	0.600000	6.540000	AIR	
9	31.220000	6.000000	6.820000	ZF3	
10	14.791000	4.000000	6.680000	AIR	
11	53.270000	6.000000	7.870000	CAF2	
12	-16.200000	200.000000	8.500000	AIR	
13	-36.100000	14.000000	18.410000	K9	
14	-43.560000	3.000000	21.470000	AIR	
15	377.100000	15.000000	22.060000	CAF2	
16	-81.500000	3.000000	22.350000	AIR	
17	221.300000	17.000000	21.710000	CAF2	
18	-72.000000	3.000000	20.510000	H-LAK2 P	
19	0.000000	200.009934 S	20.330000	AIR	
IMS	0.000000	0.000000	10.000000		

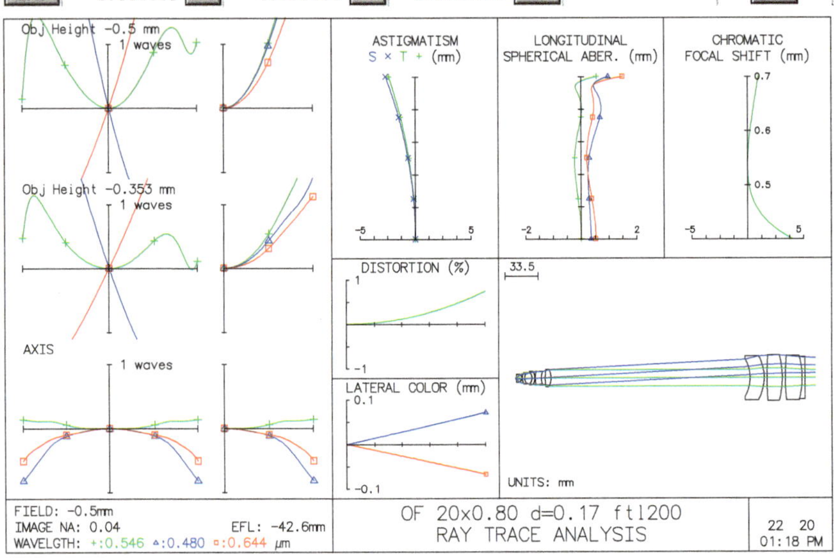

Figure 2.36 Optical design and graphs of aberration correction of a 20×/0.80 infinity LCR MicroFluar objective ($F'_{ts} = 200$ mm).

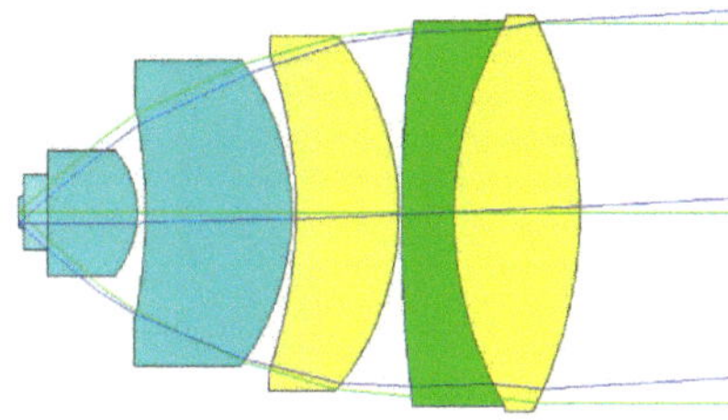

Lens: achro 25x1.0 oil ftl160

Object num aper 1.000000 Image height 9.000000 Primary wavln 0.546070

SRF	RADIUS	THICKNESS		APERTURE RADIUS		GLASS		SPECIAL
OBJ	0.000000	0.010000		0.358084		OIL		
AST	0.000000	0.170000		0.370000	A	LZ_K14		
2	0.000000	0.842800		0.520000		OIL		
3	0.000000	3.100000		1.260000		LZ_BK6		
4	-3.251000	0.300000		2.090000		AIR		
5	-17.140000	5.000000		3.540000		LZ_CTK19		
6	-8.472000	0.150000		5.110000		AIR		
7	-16.181000	3.500000		5.320000		CAF2		
8	-9.462000	0.100000		5.950000		AIR		
9	52.840000	1.800000		6.340000		LZ_TF5		
10	12.618000	4.300000		6.450000		CAF2	P	
11	-14.859000	90.000000		6.630000		AIR		
12	214.800000	5.000000		11.970000		LZ_BK8		
13	-46.990000	0.300000		12.030000		AIR		
14	-46.880000	3.000000		11.990000		LZ_BF24		
15	-113.240000	157.880284	S	12.120000		AIR		
IMS	0.000000	0.000000		9.000000				

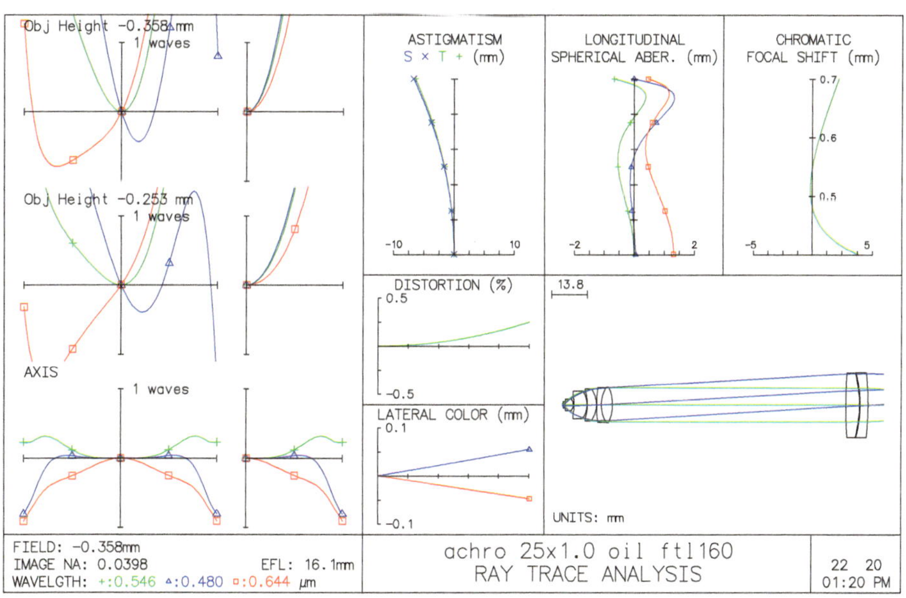

Figure 2.37 Optical design and graphs of aberration correction of a 25×/1.00 oil immersion infinity LCR achromatic objective ($F'_{ts} = 160$ mm).

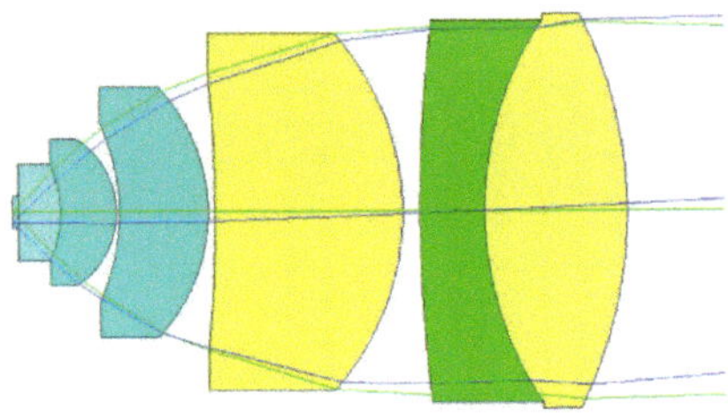

Lens: OF 32x0.95water d017 tub160 h45			Efl	5.55	
Object num aper	0.950000	Image height 9.000000	Primary wavln	0.546070	
SRF	**RADIUS**	**THICKNESS**	**APERTURE RADIUS**	**GLASS**	**SPECIAL**
OBJ	0.000000	0.010000	0.280722	WATER	
AST	0.000000	0.170000	0.290000 A	LZ_K14	
2	0.000000	1.190000	0.430000	WATER	
3	-3.196600	1.600000	1.340000	LZ_BK6	
4	-2.335200	0.100000	2.020000	AIR	
5	-5.564300	2.500000	2.460000	LZ_CTK19	
6	-4.938500	0.200000	3.430000	AIR	
7	-37.516000	5.300000	3.910000	CAF2	
8	-7.240500	0.500000	4.920000	AIR	
9	43.437000	1.800000	5.210000	LZ_TF2	
10	9.276000	4.000000	5.260000	CAF2 P	
11	-12.000000	174.609347 S	5.420000	AIR	
IMS	0.000000	0.000000	9.000000		

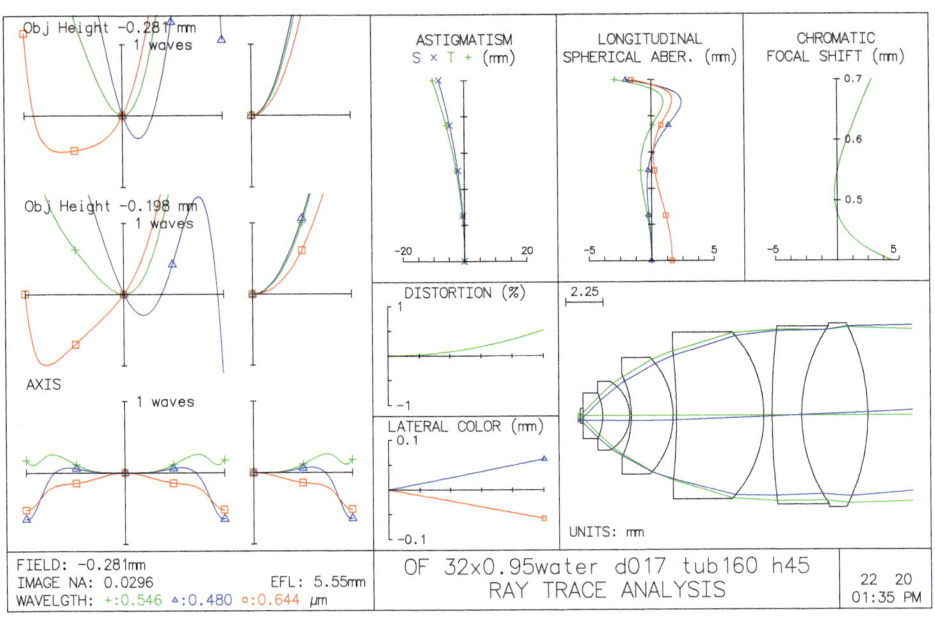

Figure 2.38 Optical design and graphs of aberration correction of a 32×/0.95 water immersion finite LCR MicroFluar objective.

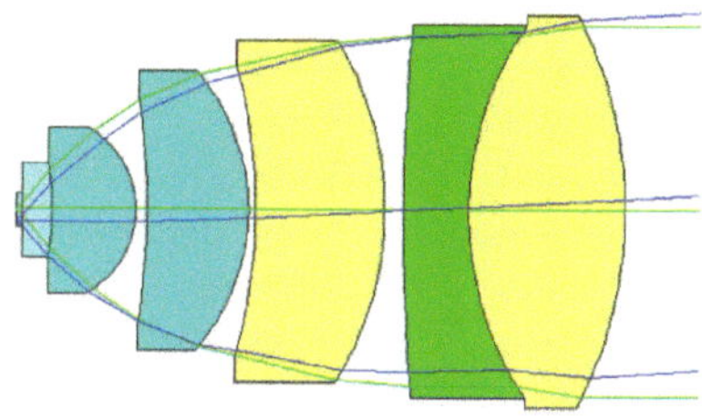

Lens: OF 32x1.0water d017 ftl160

Object num aper	1.000000	Image height	9.000000	Primary wavln	0.546070

SRF	RADIUS	THICKNESS	APERTURE RADIUS	GLASS	SPECIAL
OBJ	0.000000	0.010000	0.281077	WATER	
AST	0.000000	0.170000	0.290000 A	LZ_K14	
2	0.000000	0.840000	0.440000	WATER	
3	-7.178000	2.290000	1.270000	LZ_BK6	
4	-2.512000	0.300000	2.230000	AIR	
5	-21.680000	2.790000	3.100000	LZ_TK14	
6	-5.598000	0.200000	3.740000	AIR	
7	-14.060000	3.500000	3.920000	CAF2	
8	-8.710000	0.500000	4.590000	AIR	
9	47.750000	1.800000	4.880000	LZ_TF2	
10	8.710000	4.300000	4.990000	CAF2 P	
11	-11.220000	90.000000	5.270000	AIR	
12	214.800000	5.000000	10.740000	LZ_BK8	
13	-46.990000	0.300000	10.810000	AIR	
14	-46.880000	3.000000	10.780000	LZ_BF24	
15	-113.240000	157.891730 S	10.900000	AIR	
IMS	0.000000	0.000000	9.000000		

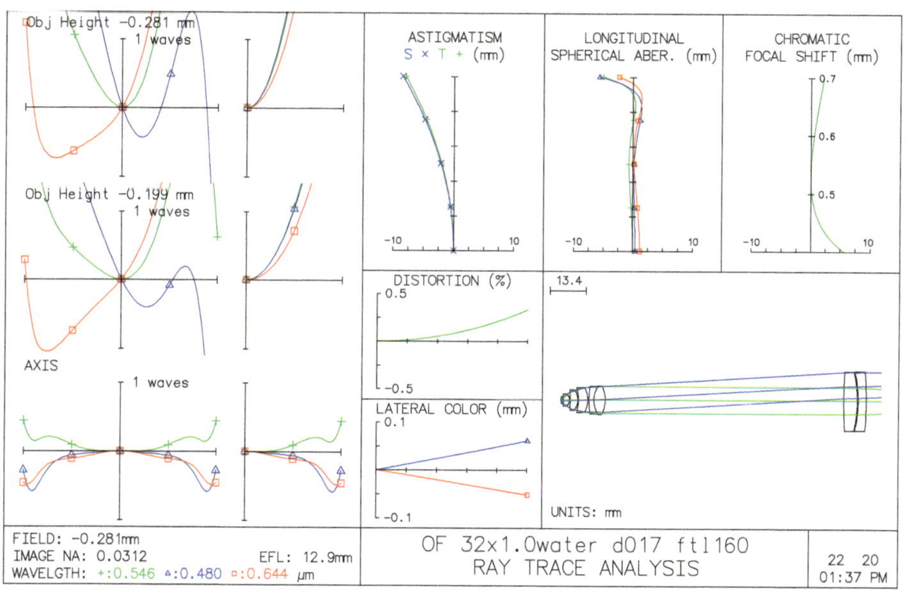

Figure 2.39 Optical design and graphs of aberration correction of a 32×/1.00 water immersion infinity LCR MicroFluar objective ($F'_{ts} = 160$ mm).

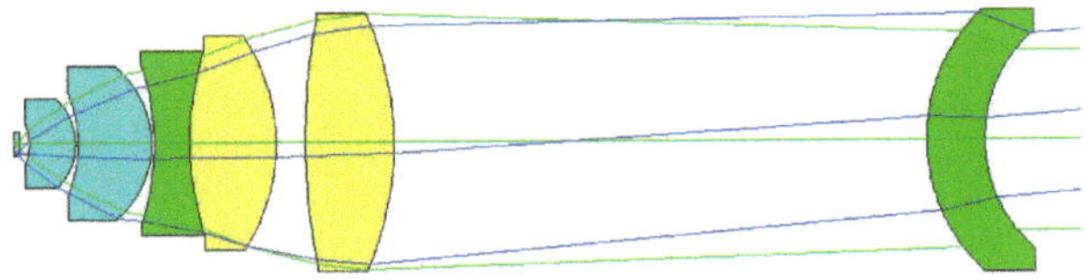

Lens: PlanFluor 40x0.75 d017 ftl160

Object num aper 0.500000 Image height 12.500000 Primary wavln 0.546070

SRF	RADIUS	THICKNESS	APERTURE RADIUS	GLASS	SPECIAL
OBJ	0.000000	0.010000	0.311421	AIR	
AST	0.000000	0.170000	0.317195 AS	LZ_K14	
2	0.000000	0.350000	0.410000	AIR	
3	-1.644400	1.600000	0.650000	LZ_CTK19	
4	-2.014000	0.100000	1.490000	AIR	
5	-4.285000	2.500000	1.720000	LZ_CTK19 P	
6	-3.251000	0.100000	2.550000	AIR	
7	-8.166000	1.200000	2.640000	LZ_TF5	
8	9.863000	3.000000	3.070000	CAF2	
9	-6.138000	1.000000	3.540000	AIR	
10	21.380000	3.000000	4.100000	CAF2 P	
11	-10.139000	18.100000	4.270000	AIR	
12	5.861000	1.970000	4.340000	LZ_TF5 P	
13	4.699000	90.000000	3.610000	AIR	
14	214.800000	5.000000	10.730000	LZ-BK8	
15	-46.990000	0.300000	10.840000	AIR	
16	-46.880000	3.000000	10.820000	LZ-BF24	
17	-113.240000	157.825327 S	10.990000	AIR	
IMS	0.000000	0.000000	12.500000 S		

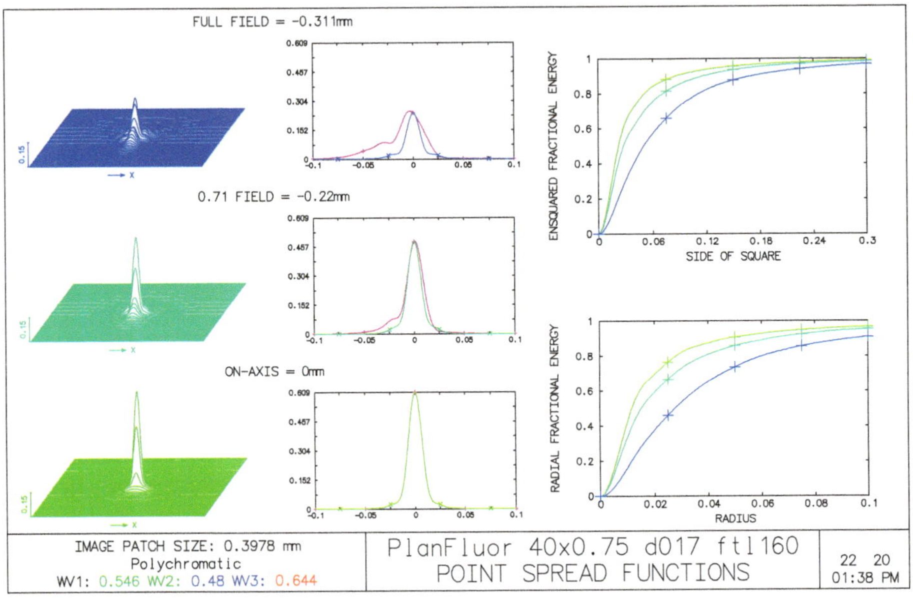

Figure 2.40 Optical design and graphs of aberration correction of a 40×/0.75 infinity LCF plan MicroFluar objective ($F'_{ts} = 160$ mm).

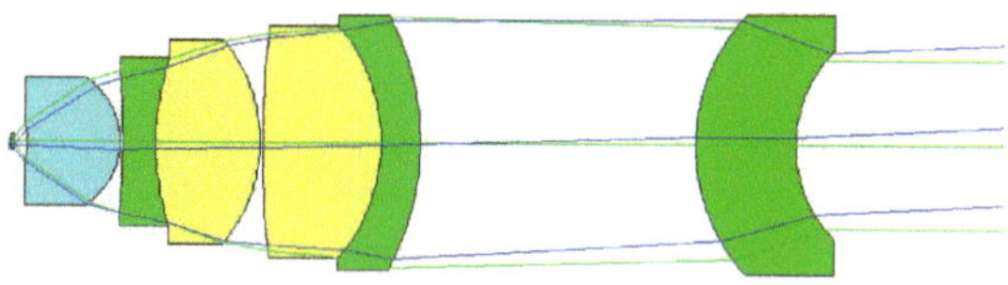

Lens: OF 40x0.80 d=0.17 ft1200

Object num aper 0.800000 Image height 10.000000 Primary wavln 0.546070

SRF	RADIUS	THICKNESS	APERTURE RADIUS	GLASS	SPECIAL
OBJ	0.000000	0.010000	0.248993	AIR	
AST	0.000000	0.170000	0.260000 A	K9	
2	0.000000	0.600000	0.370000	AIR	
3	-4.600000	4.500000	1.020000	LZ_CTK12	
4	-3.520000	0.130000	3.040000	AIR	
5	-47.240000	1.600000	3.520000	ZF6	
6	12.700000	5.000000	3.960000	CAF2	
7	-7.420000	0.130000	4.800000	AIR	
8	48.800000	5.700000	5.140000	CAF2 P	
9	-8.336000	1.900000	5.460000	ZF6 P	
10	-12.415000	13.300000	6.010000	AIR	
11	8.850000	4.900000	6.100000	ZF6	
12	6.281000	200.000000	4.460000	AIR	
13	-36.100000	14.000000	13.710000	K9	
14	-43.560000	3.000000	16.000000	AIR	
15	377.100000	15.000000	16.330000	CAF2	
16	-81.500000	3.000000	16.680000	AIR	
17	221.300000	17.000000	16.390000	CAF2	
18	-72.000000	3.000000	15.490000	LZ_CTK12 P	
19	0.000000	199.936678 S	15.410000	AIR	
IMS	0.000000	0.000000	10.000000		

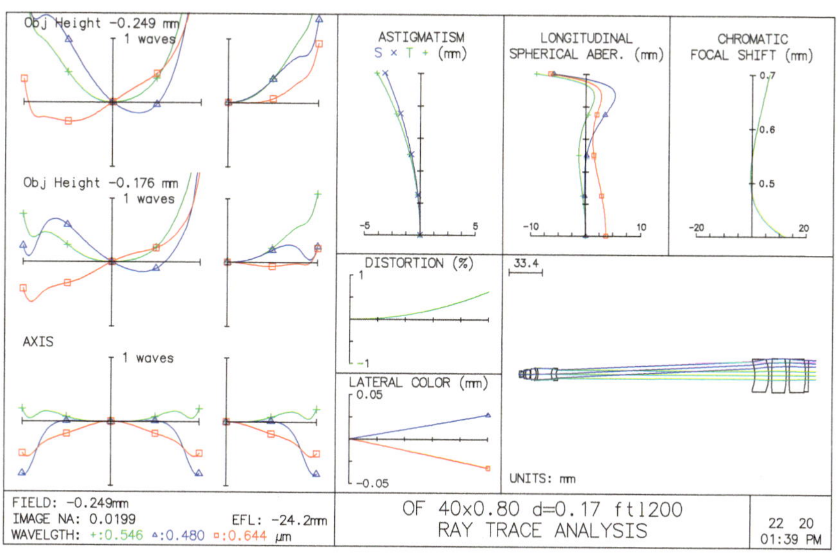

Figure 2.41 Optical design and graphs of aberration correction of a 40×/0.80 infinity LCR semi–plan MicroFluar objective ($F'_{ts} = 200$ mm).

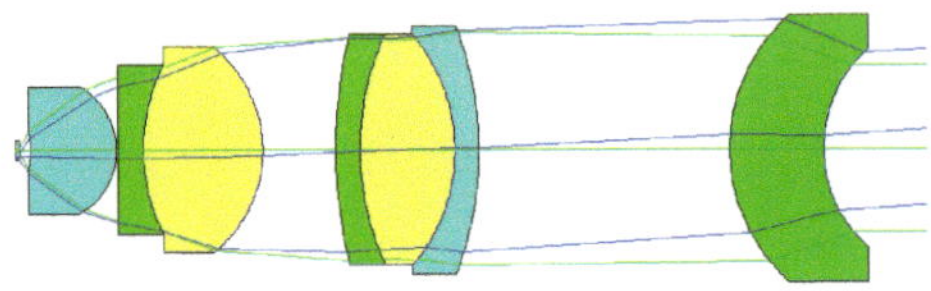

Lens: OF 40x0.85 d=0.17 Ftl 160

Object num aper 0.850000 Image height 10.000000 Primary wavln 0.546070

SRF	RADIUS	THICKNESS	APERTURE RADIUS	GLASS	SPECIAL
OBJ	0.000000	0.010000	0.251746	AIR	
AST	0.000000	0.170000	0.270000 A	LZ_K14	
2	0.000000	0.450000	0.380000	AIR	
3	-3.837000	3.600000	0.920000	LZ_CTK19	
4	-2.992000	0.100000	2.610000	AIR	
5	0.000000	1.000000	3.130000	LZ_TF3	
6	7.691000	5.000000	3.450000	CAF2	
7	-5.346000	3.000000	4.200000	AIR	
8	19.724000	1.000000	4.730000	LZ_TF1	
9	11.143000	4.000000	4.690000	CAF2	
10	-7.145000	1.000000	4.730000	LZ_CTK19 P	
11	-13.964000	10.500000	5.070000	AIR	
12	7.278000	3.900000	5.450000	LZ_TF5	
13	5.346000	90.000000	4.060000	AIR	
14	214.800000	5.000000	8.990000	LZ_BK8	
15	-46.990000	0.300000	9.080000	AIR	
16	-46.880000	3.000000	9.050000	LZ_BF24	
17	-113.240000	157.346787 S	9.160000	AIR	
IMS	0.000000	0.000000	10.000000		

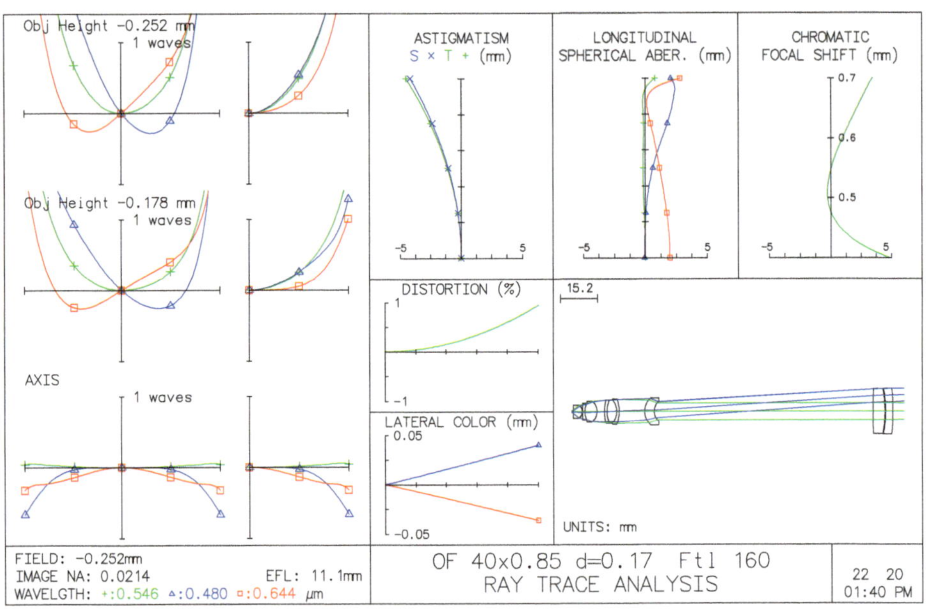

Figure 2.42 Optical design and graphs of aberration correction of a 40×/0.85 infinity LCR semi–plan MicroFluar objective (F'_{ts} = 160 mm).

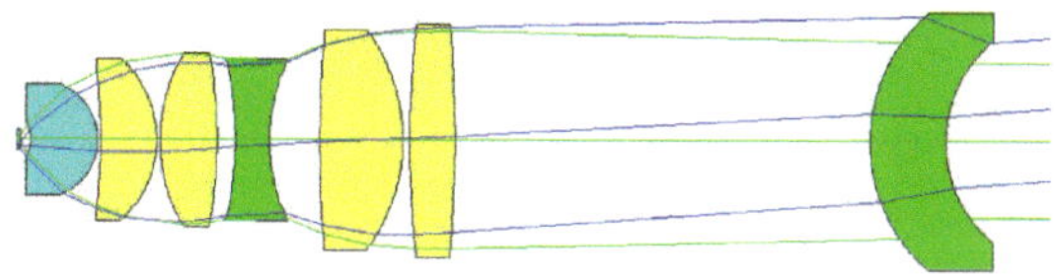

Lens: PlanFluar 40x0.85 d017 f160

Object num aper 0.850000 Object height -0.304000 Primary wavln 0.546070

SRF	RADIUS	THICKNESS	APERTURE RADIUS	GLASS	SPECIAL
OBJ	0.000000	0.010000	0.304000	AIR	
AST	0.000000	0.170000	0.320000 A	LZ_K14	
2	0.000000	0.449800	0.440000	AIR	
3	-1.644400	2.960000	0.810000	LZ_CTK19	
4	-2.559000	0.150000	2.390000	AIR	
5	-22.180000	2.470000	3.130000	CAF2	
6	-4.699000	0.140000	3.530000	AIR	
7	7.482000	2.500000	3.800000	CAF2 P	
8	-20.890000	0.800000	3.690000	AIR	
9	-15.560000	1.500000	3.540000	LZ_TF5	
10	8.433000	2.200000	3.530000	AIR	
11	51.050000	3.600000	4.270000	CAF2 P	
12	-8.091000	0.300000	4.790000	AIR	
13	45.920000	2.000000	5.020000	CAF2 P	
14	-60.530000	18.300000	5.080000	AIR	
15	7.482000	3.280000	5.590000	LZ_TF5 P	
16	5.728000	100.000000	4.400000	AIR	
17	677.600000	4.000000	11.940000	LZ_TK13	
18	-30.970000	1.500000	12.000000	LZ_F1	
19	53.210000	5.400000	12.230000	AIR	
20	63.530000	4.800000	13.440000	LZ_TF1	
21	-107.400000	168.947776 S	13.550000	AIR	
IMS	0.000000	0.000000	12.500000		

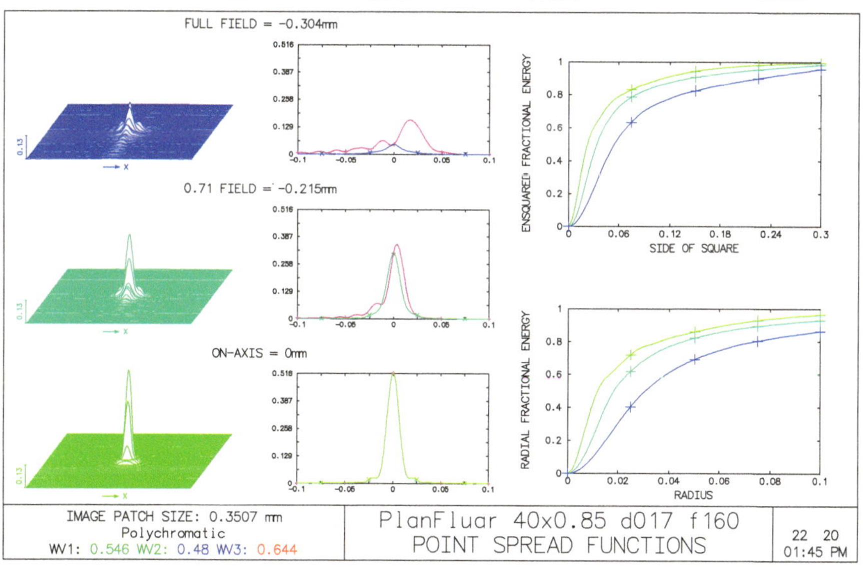

Figure 2.43 Optical design and graphs of aberration correction of a 40×/0.85 infinity LCR plan MicroFluar objective ($F'_{ts} = 160$ mm).

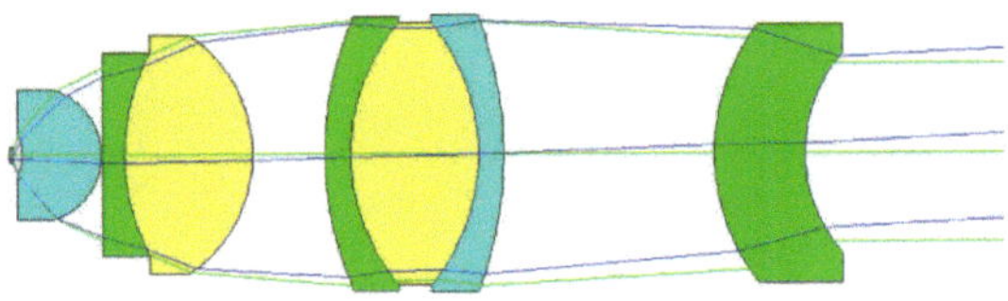

Lens: OF 40x0.95 d=0.17 ftl200

Object num aper	0.950000	Image height	10.000000	Primary wavln	0.546070

SRF	RADIUS	THICKNESS	APERTURE RADIUS		GLASS		SPECIAL
OBJ	0.000000	0.010000	0.255037		AIR		
AST	0.000000	0.170000	0.280000	A	K9		
2	0.000000	0.504600	0.420000		AIR		
3	-3.000000	4.500000	1.200000		H-LAK3		
4	-3.644000	0.110000	3.520000		AIR		
5	0.000000	1.200000	4.810000		ZF1		
6	11.370000	7.200000	5.490000		CAF2		
7	-7.600000	4.000000	6.450000		AIR		
8	18.270000	1.400000	7.410000		ZF1	P	
9	10.920000	7.200000	7.140000		CAF2	P	
10	-10.920000	1.400000	7.140000		H-LAK3	P	
11	-20.460000	11.800000	7.530000		AIR		
12	11.250000	5.000000	6.990000		ZF6		
13	8.000000	200.000000	5.420000		AIR		
14	-36.100000	14.000000	15.280000		K9		
15	-43.560000	3.000000	17.740000		AIR		
16	377.100000	15.000000	18.170000		CAF2		
17	-81.500000	3.000000	18.500000		AIR		
18	221.300000	17.000000	18.130000		CAF2		
19	-72.000000	3.000000	17.150000		LZ_CTK12		
20	0.000000	199.946741 S	17.050000		AIR		
IMS	0.000000	0.000000	10.000000				

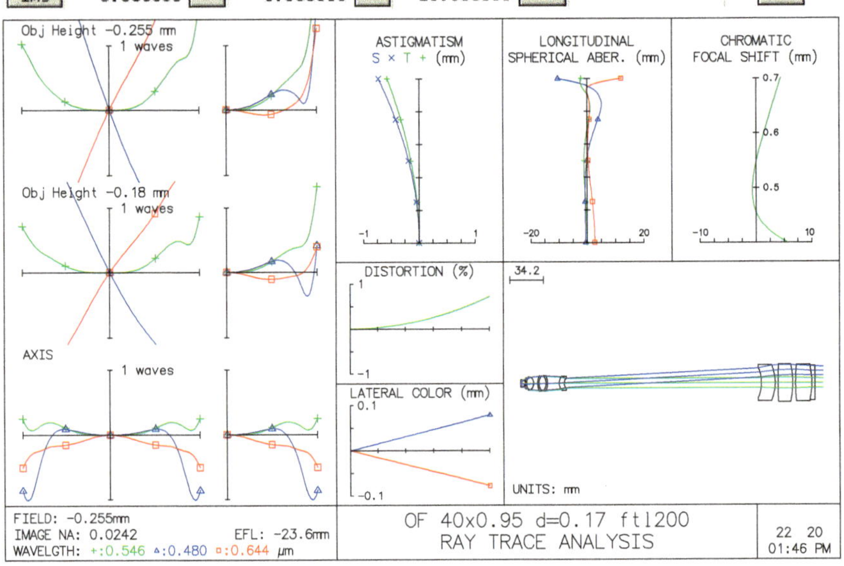

Figure 2.44 Optical design and graphs of aberration correction of a 40×/0.95 infinity LCR semi–plan MicroFluar objective ($F'_{ts} = 200$ mm).

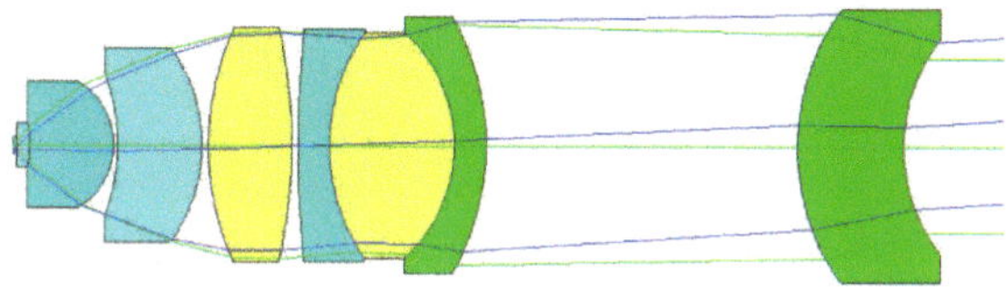

Lens: OF 40x1.0 water d017 ftl160

Object num aper	1.000000	Image height	10.000000	Primary wavln	0.546070

SRF	RADIUS	THICKNESS	APERTURE RADIUS	GLASS	SPECIAL
OBJ	0.000000	0.010000	0.252276	WATER	
AST	0.000000	0.170000	0.260000 A	LZ_K14	
2	0.000000	0.600000	0.410000	WATER	
3	-6.000000	4.000000	1.000000	LZ_CTK19	
4	-3.339800	0.200000	2.900000	AIR	
5	-9.535100	4.000000	3.360000	SILICA	
6	-7.415000	0.300000	4.500000	AIR	
7	12.654500	4.000000	5.300000	CAF2	
8	-22.817500	0.300000	5.410000	AIR	
9	48.955000	1.500000	5.380000	LZ_CTK19 P	
10	8.972500	6.000000	5.190000	CAF2 P	
11	-6.840300	1.500000	5.220000	LZ_TF5	
12	-11.574900	15.000000	5.940000	AIR	
13	10.646900	5.000000	6.300000	LZ_TF5 P	
14	7.973300	90.000000	4.960000	AIR	
15	214.800000	5.000000	10.560000	LZ_BK8	
16	-46.990000	0.300000	10.630000	AIR	
17	-46.880000	3.000000	10.600000	LZ_BF24	
18	-113.240000	157.633342 S	10.730000	AIR	
IMS	0.000000	0.000000	10.000000		

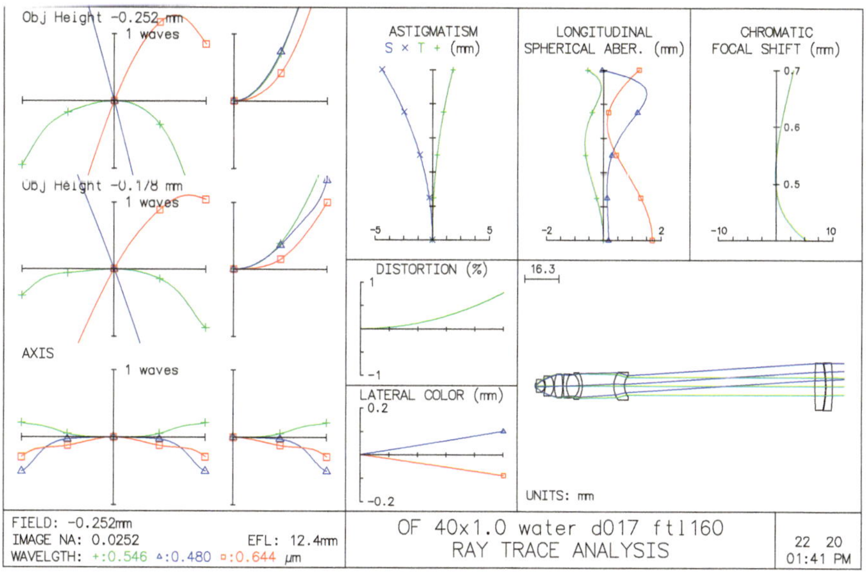

Figure 2.45 Optical design and graphs of aberration correction of a 40×/1.00 water immersion infinity LCR semi–plan MicroFluar objective ($F'_{ts} = 160$ mm).

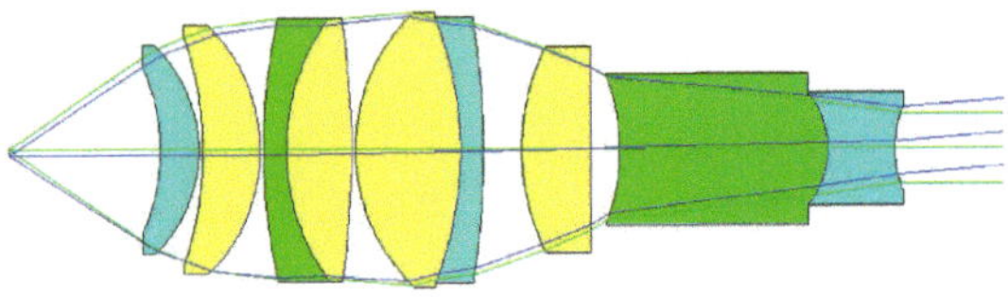

Lens: PlanFluar LD 50x0.55 d0 ftl160

Object num aper 0.550000 Image height 12.500000 Primary wavln 0.546070

SRF	RADIUS	THICKNESS	APERTURE RADIUS	GLASS	SPECIAL
OBJ	0.000000	1.0000e-20	0.249362	AIR	
AST	0.000000	7.996600	4.230000 A	AIR	
2	-9.908000	2.000000	4.230000	LZ_CTK12	
3	-7.244000	0.200000	5.280000	AIR	
4	-17.701000	3.000000	5.830000	CAF2	
5	-8.551000	0.200000	6.280000	AIR	
6	29.920000	1.200000	6.670000	LZ_TF2	
7	11.117000	3.400000	6.590000	CAF2 P	
8	-42.660000	0.200000	6.650000	AIR	
9	9.462000	5.500000	6.970000	CAF2 P	
10	-17.061000	1.200000	6.720000	LZ_CTK19	
11	-39.630000	2.000000	6.500000	AIR	
12	11.075600	3.700000	5.220000	CAF2 P	
13	0.000000	1.505000	4.280000	AIR	
14	-10.209000	11.000000	3.810000	LZ_TF5	
15	-4.285000	3.500000	2.830000	LZ_CTK12 P	
16	5.272000	90.000000	2.130000	AIR	
17	214.800000	5.000000	9.800000	LZ-BK8	
18	-46.990000	0.300000	9.920000	AIR	
19	-46.880000	3.000000	9.900000	LZ-BF24	
20	-113.240000	157.720518 S	10.060000	AIR	
IMS	0.000000	0.000000	12.500000		

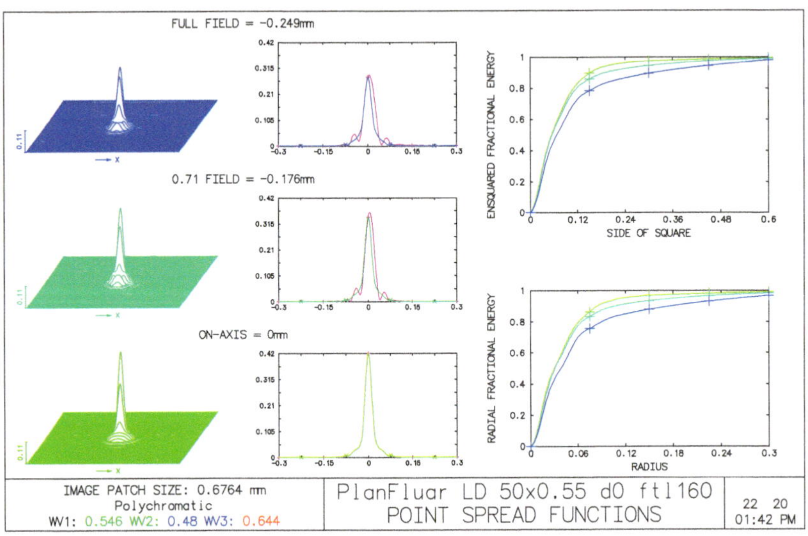

Figure 2.46 Optical design and graphs of aberration correction of a 50×/0.55 infinity LCF LD plan MicroFluar objective ($F'_{ts} = 160$ mm).

magnification and high-aperture objectives, this spread is 5–10 times more than its focal depth. This situation requires that the object be reviewed in depth (by refocusing the microscope). When you achieve exact focusing on a fragment, you may expect that images of other fragments (even those located in the vicinity of the first fragment but at a different depth) would not be sharp. This is the point of the widespread method of visual microscopy, in

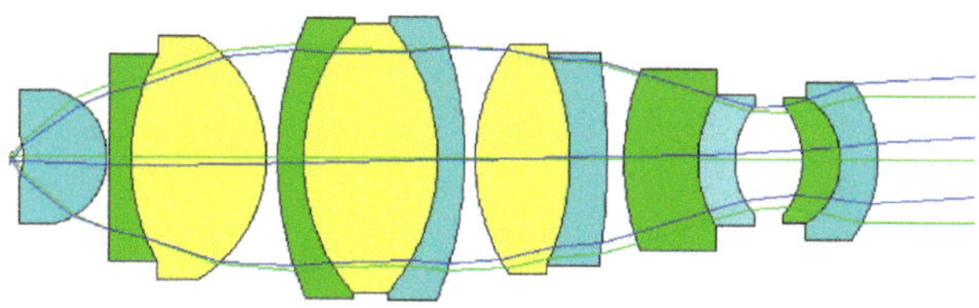

Lens: PlanFluar 50x0.85 d=0 F200

Object num aper	0.850000	Image height	12.500000	Primary wavln	0.546070

SRF	RADIUS	THICKNESS		APERTURE RADIUS		GLASS		SPECIAL
OBJ	0.000000	1.0000e-20		0.248413		AIR		
AST	0.000000	0.625000		0.250000	A	AIR		
2	-3.000000	4.500000		1.200000		H-LAK3		
3	-3.644000	0.110000		3.520000		AIR		
4	0.000000	1.200000		4.810000		ZF1		
5	11.370000	7.200000		5.490000		CAF2		
6	-7.600000	0.600000		6.450000		AIR		
7	18.270000	1.400000		7.410000		ZF1	P	
8	10.920000	7.200000		7.140000		CAF2	P	
9	-10.920000	1.400000		7.140000		H-LAK3	P	
10	-20.460000	0.600000		7.530000		AIR		
11	10.802000	5.100000		6.060000		CAF2	P	
12	-13.305000	2.000000		5.630000		H-LAK3	P	
13	-35.320000	0.900000		5.410000		AIR		
14	13.305000	4.000000		4.810000		H-ZF6		
15	6.630000	2.000000		3.500000		H-K9		
16	4.368000	3.600000		2.920000		AIR		
17	-4.650000	2.000000		2.940000		H-ZF6	P	
18	-4.000000	2.000000		3.360000		H-LAK3	P	
19	-7.047000	200.000000		4.180000		AIR		
20	-36.100000	14.000000		15.280000		K9		
21	-43.560000	3.000000		17.740000		AIR		
22	377.100000	15.000000		18.170000		CAF2		
23	-81.500000	3.000000		18.500000		AIR		
24	221.300000	17.000000		18.130000		CAF2		
25	-72.000000	3.000000		17.150000		H-LAK2		
26	0.000000	199.929182	S	17.050000		AIR		
IMS	0.000000	0.000000		12.500000				

Figure 2.47 Optical design and graphs of aberration correction of a 50×/0.85 infinity LCF plan MicroFluar objective ($F'_{ts} = 200$ mm).

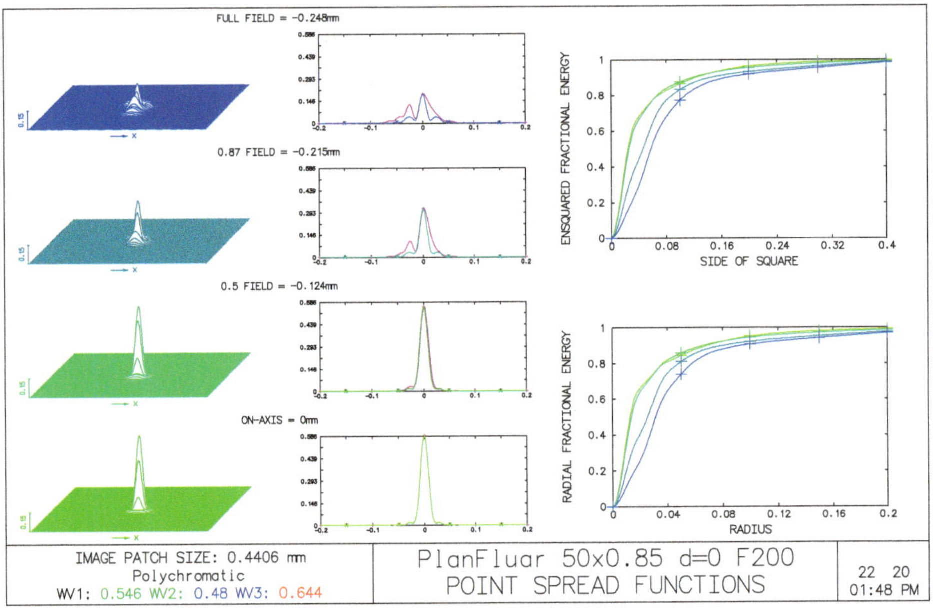

Figure 2.47 (*Continued*) Optical design and graphs of aberration correction of a 50×/0.85 infinity LCF plan MicroFluar objective ($F'_{ts} = 200$ mm).

which the researcher selects the location depth for the most interesting fragment of the biological culture. In addition, there is always a kind of wedge in the process of manufacturing the object plate; e.g., if the left edge is sharp, the right edge can be non-sharp. The lighting and the mechanical systems for microscope positioning also have an adverse effect. Figure 2.69 is a microphotograph of a real biological specimen made using a 100×/1.2 water immersion objective (in a dark field). This photograph confirms our conclusions.

2.4.1.2 Compromises in curvature correction in visual microscopy

It is interesting to consider the expediency of correcting curvature aberration or, specifically, the optimal degree for correcting it in short-focus objectives with achromatic and apochromatic adjustments used in biological microscopes for visual studies. These objectives usually have a high magnification (40–100×) and numerical apertures of 0.95 for the air, 1.20 for water immersion, and 1.40 for oil immersion. Such objectives have considerable image curvature, and for visual microscopy studies, researchers using these objectives have to refocus two to four times when studying an image from the center to the periphery over the field. This takes into account the natural ability of the eyes to accommodate. Let us consider that for practical microscopy, high-magnification and high-aperture objectives are not used as panoramic objectives. The area of an object that requires further investigation

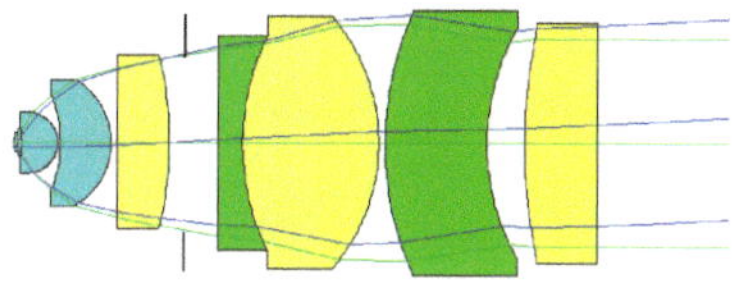

Lens: OF 50x1.20water d017 ftl200

Object num aper	1.200000	Image height	10.000000	Primary wavln	0.546070

SRF	RADIUS	THICKNESS	APERTURE RADIUS	GLASS	SPECIAL
OBJ	0.000000	0.010000	0.200137	WATER	
AST	0.000000	0.170000	0.220000 A	K9	
2	0.000000	0.179400	0.440000	WATER P	
3	-2.379000	1.700000	0.630000	K9 P	
4	-1.400000	0.120000	1.400000	AIR	
5	-6.000000	2.400000	2.200000	ZK1	
6	-3.323000	0.300000	2.910000	AIR	
7	0.000000	2.500000	3.570000	CAF2	
8	-16.684000	2.300000	4.030000	AIR	
9	0.000000	1.200000	4.660000	ZF6	
10	10.920000	6.500000	4.960000	CAF2 P	
11	-8.750000	0.300000	5.860000	AIR	
12	14.510000	4.900000	6.110000	ZF6 P	
13	10.700000	1.800000	5.320000	AIR	
14	25.560000	3.500000	5.460000	CAF2 P	
15	0.000000	200.000000	5.570000	AIR	
16	-36.100000	14.000000	15.670000	K9	
17	-43.560000	3.000000	18.300000	AIR	
18	377.100000	15.000000	18.760000	CAF2	
19	-81.500000	3.000000	19.090000	AIR	
20	221.300000	17.000000	18.690000	CAF2	
21	-72.000000	3.000000	17.670000	LZ_CTK12	
22	0.000000	200.045250 S	17.560000	AIR	
IMS	0.000000	0.000000	10.000000		

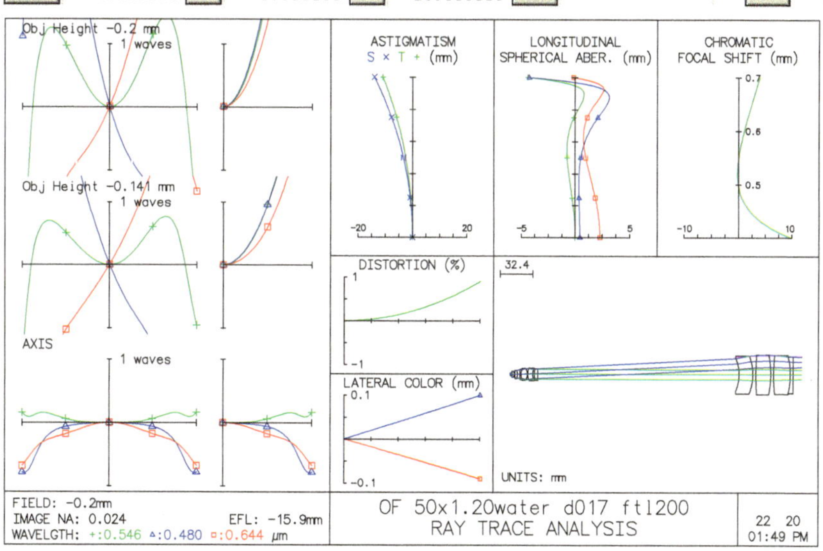

Figure 2.48 Optical design and graphs of aberration correction of a 50×/1.20 water immersion infinity LCR MicroFluar objective ($F'_{ts} = 200$ mm).

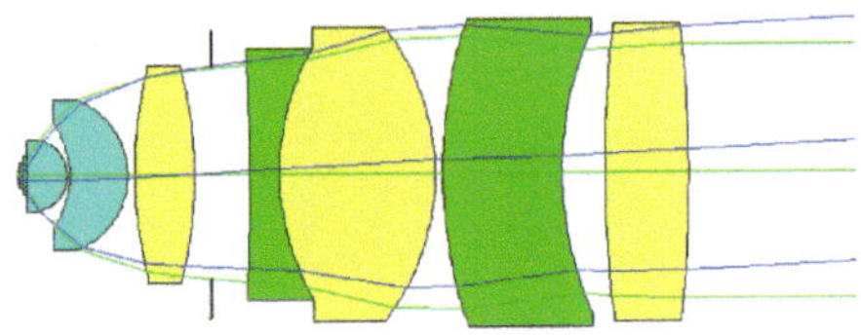

Lens: OF 50x1.30 oil ftl200

Object num aper	1.300000	Image height	10.000000	Primary wavln	0.546070

SRF	RADIUS	THICKNESS	APERTURE RADIUS	GLASS	SPECIAL
OBJ	0.000000	0.010000	0.198859	OIL	
AST	0.000000	0.170000	0.210000 A	K9	
2	0.000000	0.151100	0.500000	OIL P	
3	0.000000	1.700000	0.750000	K9 P	
4	-1.450000	0.100000	1.440000	AIR	
5	-3.736000	2.400000	2.150000	H-LAK3	
6	-3.280300	0.300000	3.040000	AIR	
7	17.061000	2.500000	4.170000	CAF2	
8	-17.061000	2.300000	4.420000	AIR	
9	-84.720000	1.200000	4.810000	ZF6	
10	10.281000	6.500000	5.100000	CAF2 P	
11	-9.461000	0.300000	5.960000	AIR	
12	18.883000	5.000000	6.240000	ZF6 P	
13	14.000000	1.800000	5.700000	AIR	
14	50.289000	3.500000	5.850000	CAF2 P	
15	-50.289000	200.000000	6.050000	AIR	
16	-36.100000	14.000000	16.620000	K9	
17	-43.560000	3.000000	19.430000	AIR	
18	377.100000	15.000000	19.960000	CAF2	
19	-81.500000	3.000000	20.290000	AIR	
20	221.300000	17.000000	19.820000	CAF2	
21	-72.000000	3.000000	18.760000	LZ_CTK12	
22	0.000000	200.079485 S	18.640000	AIR	
IMS	0.000000	0.000000	10.000000		

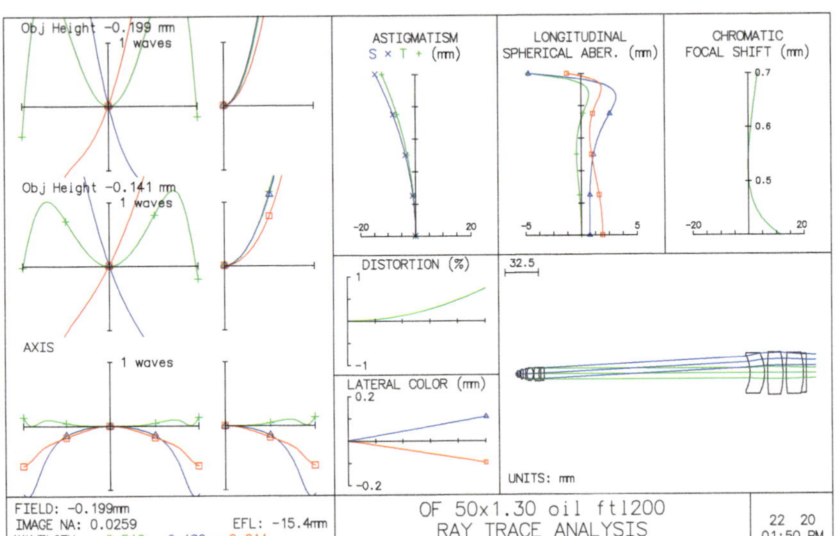

Figure 2.49 Optical design and graphs of aberration correction of a 50×/1.30 oil immersion infinity LCR MicroFluar objective ($F'_{ts} = 200$ mm).

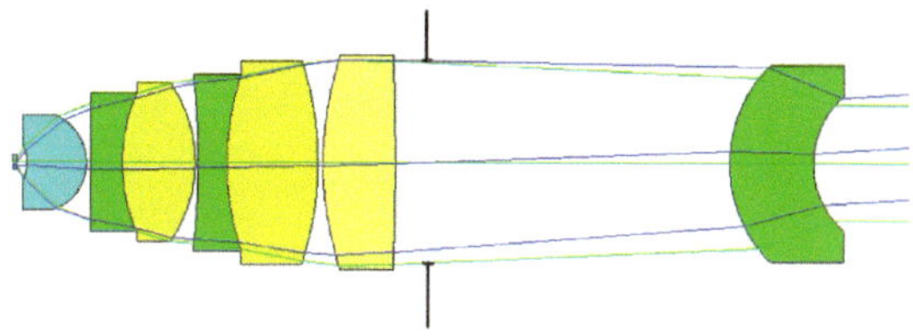

Lens: planFluar 60x0.85 d017 Ftl164

Object num aper	0.850000	Image height	10.000000	Primary wavln	0.546070

SRF	RADIUS	THICKNESS	APERTURE RADIUS	GLASS	SPECIAL
OBJ	0.000000	0.010000	0.166646	AIR	
AST	0.000000	0.170000	0.170000 A	K9	
2	0.000000	0.400000	0.280000	AIR	
3	-1.506000	2.470000	0.670000	H-LAK3	
4	-2.001000	0.200000	1.900000	AIR	
5	0.000000	1.300000	2.410000	ZF1	
6	6.897000	3.000000	2.790000	H-FK61	
7	-5.390000	0.110000	3.190000	AIR	
8	-31.742000	1.200000	3.300000	H-ZF5	
9	10.850000	3.800000	3.530000	H-FK61 P	
10	-12.465000	0.200000	4.060000	AIR	
11	13.550000	3.000000	4.260000	CAF2	
12	-105.700000	14.000000	4.290000	AIR	
13	5.453000	3.460000	3.960000	H-ZF6	
14	3.550000	164.000000	2.700000	AIR	
15	220.328000	5.128700	11.570000	BK8_LZOS	
16	-48.199500	0.307700	11.630000	AIR	
17	-48.086700	3.077200	11.590000	BF24_LZOS	
18	-116.154700	161.165658 S	11.710000	AIR	
IMS	0.000000	0.000000	10.000000		

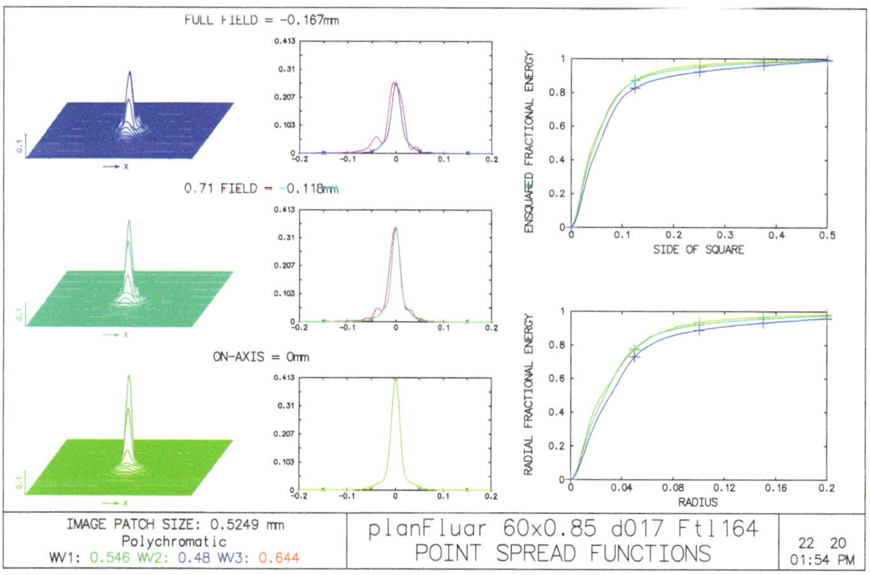

Figure 2.50 Optical design and graphs of aberration correction of a 60×/0.85 infinity LCF plan MicroFluar objective ($F'_{ts} = 164$ mm).

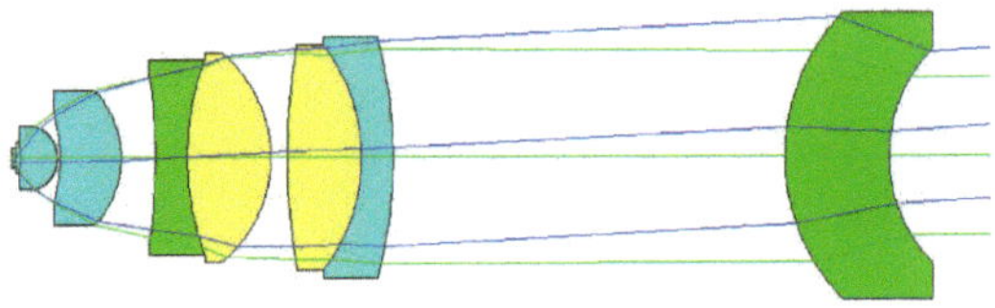

Lens: OF 63x1.15 water d017 ftl160

Object num aper	1.150000	Image height	10.000000	Primary wavln	0.546070

SRF	RADIUS	THICKNESS	APERTURE RADIUS		GLASS		SPECIAL
OBJ	0.000000	0.010000	0.158213		WATER		
AST	0.000000	0.170000	0.170000	A	LZ-K14		
2	0.000000	0.150400	0.370000		WATER	P	
3	-3.499000	1.450000	0.550000		LZ-BK6		
4	-1.180300	0.100000	1.170000		AIR		
5	-5.623000	2.300000	1.670000		LZ-CTK19		
6	-3.499000	1.300000	2.460000		AIR		
7	-21.880000	1.200000	3.070000		LZ-TF5		
8	9.772000	3.200000	3.540000		CAF2		
9	-4.786000	0.600000	3.800000		AIR		
10	22.700000	2.800000	4.080000		CAF2	P	
11	-6.577000	1.200000	4.090000		LZ-CTK19	P	
12	-16.331000	15.000000	4.380000		AIR		
13	7.278000	3.950000	5.210000		LZ-TF5	P	
14	5.445000	90.000000	3.910000		AIR		
15	214.800000	5.000000	8.990000		LZ-BK8		
16	-46.990000	0.300000	9.080000		AIR		
17	-46.880000	3.000000	9.060000		LZ-BF24		
18	-113.240000	157.819563 S	9.170000		AIR		
IMS	0.000000	0.000000	10.000000				

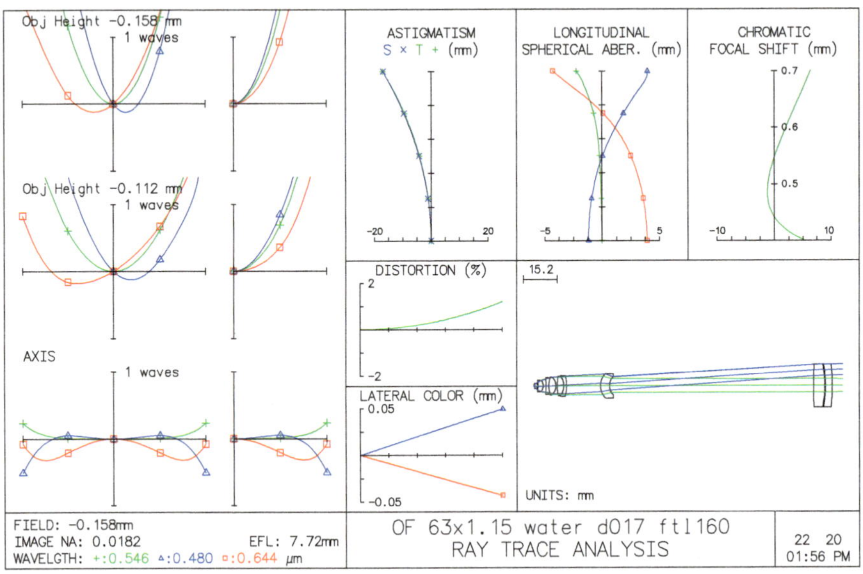

Figure 2.51 Optical design and graphs of aberration correction of a 63×/1.15 water immersion infinity LCR semi–plan MicroFluar objective ($F'_{ts} = 160$ mm).

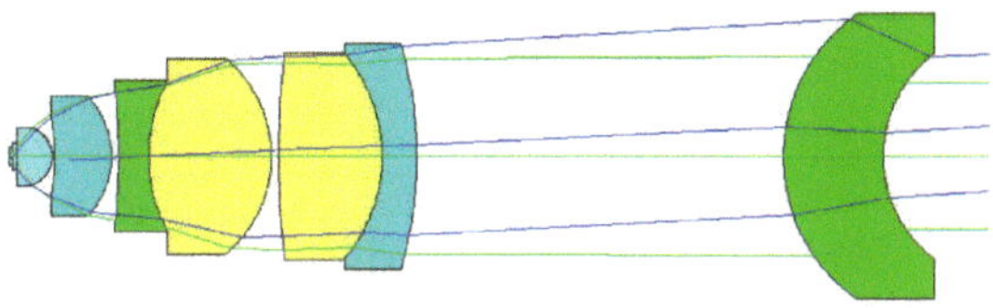

Lens: OF 63x1.15oil ftl160

Object num aper	1.150000	Image height	10.000000	Primary wavln	0.546070

SRF	RADIUS	THICKNESS	APERTURE RADIUS	GLASS	SPECIAL
OBJ	0.000000	0.010000	0.158733	OIL	
AST	0.000000	0.170000	0.170000 A	LZ_K14	
2	0.000000	0.150000	0.370000	OIL	
3	0.000000	1.450000	0.540000	LZ_K8	
4	-1.180300	0.100000	1.150000	AIR	
5	-9.550000	2.300000	1.700000	LZ_CTK19	
6	-3.436000	0.300000	2.390000	AIR	
7	-23.170000	1.200000	2.640000	LZ_TF5	
8	6.368000	5.000000	3.010000	CAF2	
9	-4.966000	0.300000	3.900000	AIR	
10	34.670000	4.200000	4.080000	CAF2 P	
11	-6.194000	1.400000	4.120000	LZ_CTK19 P	
12	-15.560000	15.000000	4.490000	AIR	
13	6.918000	4.000000	5.680000	LZ_TF5	
14	5.129000	90.000000	4.130000	AIR	
15	214.800000	5.000000	7.000000	LZ_BK8	
16	-46.990000	0.300000	7.040000	AIR	
17	-46.880000	3.000000	7.020000	LZ_BF24	
18	-113.240000	157.448725 S	7.080000	AIR	
IMS	0.000000	0.000000	10.000000		

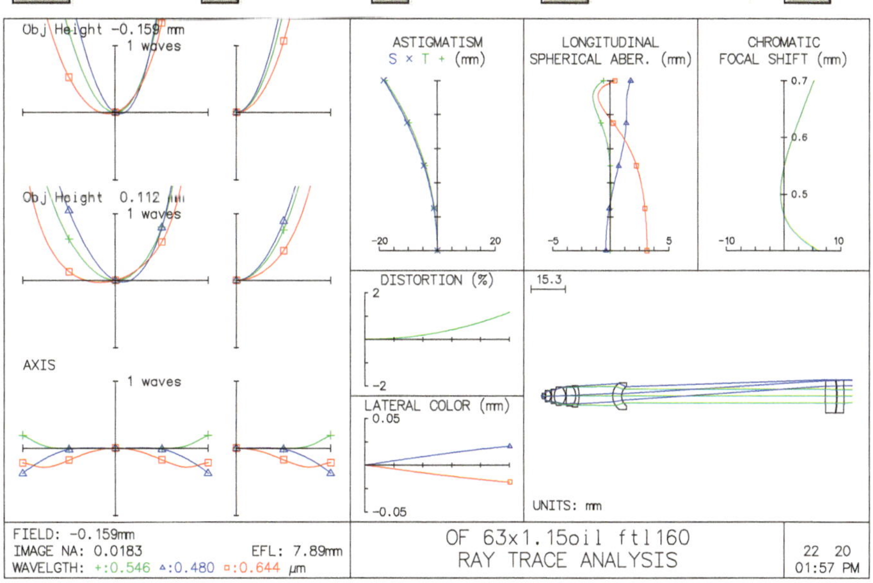

Figure 2.52 Optical design and graphs of aberration correction of a 63×/1.15 oil immersion infinity LCR MicroFluar objective (F'_{ts} = 160 mm).

is selected using low- and medium-magnification objectives (usually with corrected field curvature aberration), and then high-aperture and high-magnification objectives are used for finer investigations. The inherent motivation of a researcher to bring a fragment of the object under consideration to the central area of the visible linear field not only provides reassurance but also maximum reliability of the study. If a curvature

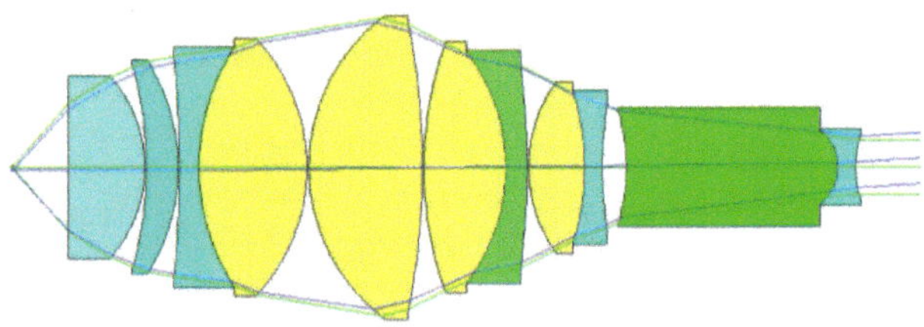

Lens:LD PlanFluar 100x0.75 d=0 ftl200

Object num aper	0.750000	Image height	12.500000	Primary wavln	0.546070

SRF	RADIUS	THICKNESS	APERTURE RADIUS	GLASS	SPECIAL
OBJ	0.000000	1.0000e-20	0.124893	AIR	
AST	0.000000	3.196000	0.124893 AS	AIR	
2	0.000000	4.440000	4.490000	H-K9L	
3	-7.503000	0.100000	5.130000	AIR	
4	-20.750000	1.800000	5.680000	H-LAK3	
5	-11.092000	0.100000	5.980000	AIR	
6	-60.900000	1.100000	6.240000	H-LAK3	
7	12.190000	6.200000	6.730000	CAF2	
8	-10.411100	0.100000	7.200000	AIR	
9	10.627000	6.400000	8.510000	CAF2 P	
10	-41.582000	0.100000	8.170000	AIR	
11	20.110000	4.700000	7.040000	CAF2 P	
12	-10.411100	1.200000	6.500000	ZBAF4	
13	-44.140000	0.100000	5.840000	AIR	
14	8.128000	3.200000	4.810000	CAF2 P	
15	-13.450000	1.000000	4.290000	H-LAK3 P	
16	23.230000	1.400000	3.600000	AIR	
17	-10.950000	12.000000	3.300000	H-ZF6	
18	-2.618000	1.000000	2.090000	H-LAK3 P	
19	5.401000	200.000000	1.820000	AIR	
20	-36.100000	14.000000	15.950000	H-K9L P	
21	-43.560000	3.000000	18.310000	AIR	
22	377.100000	15.000000	18.810000	CAF2	
23	-81.500000	3.000000	19.370000	AIR	
24	221.300000	17.000000	19.270000	CAF2	
25	-72.000000	3.000000	18.660000	LZ_CTK12	
26	0.000000	199.473986 S	18.650000	AIR	
IMS	0.000000	0.000000	12.500000		

Figure 2.53 Optical design and graphs of aberration correction of a 100×/0.75 infinity LCF LD plan MicroFluar objective ($F'_{ts} = 200$ mm).

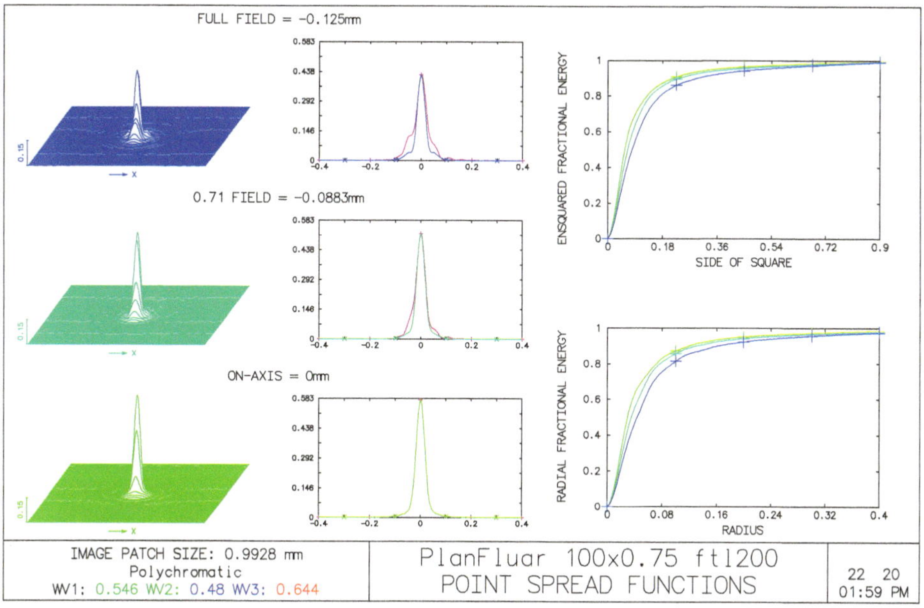

Figure 2.53 (*Continued*) Optical design and graphs of aberration correction of a 100×/0.75 infinity LCF LD plan MicroFluar objective ($F'_{ts} = 200$ mm).

aberration has not been corrected in the objective, there is a technical opportunity to compensate for it in the field of the intermediate image of the microscope. To do this, the proper compensator must be added to the microscope's design. However, in this case, each objective from the microscope kit must have its own compensator, which is impractical. A well-known compensation method for curvature aberration after the plane of the intermediate image uses eyepieces (although it was used in Leitz's old microscopes [5], it did not become popular due to the same reason: each objective needs its own eyepiece). Moreover, there is no experimental evidence or positive reviews by microscopists on the efficiency of this method, suggesting that it should not be used in the design of new microscopes. Of course, there is an opportunity to correct curvature aberration in the optical systems of objectives themselves. But doing so complicates their optical systems. Thus, we can make these intermediate conclusions:

- In principle, it is technically feasible to fully compensate for curvature aberration in biological microscopes when using standard achromatic and apochromatic objectives of high magnifications and maximum numerical apertures. However, it cannot be considered expedient for visual microscopy due to the curvature in the image of the studied objects. Moreover, most microscopy studies do not require the use of high-aperture objectives with corrected field curvature aberration (plan

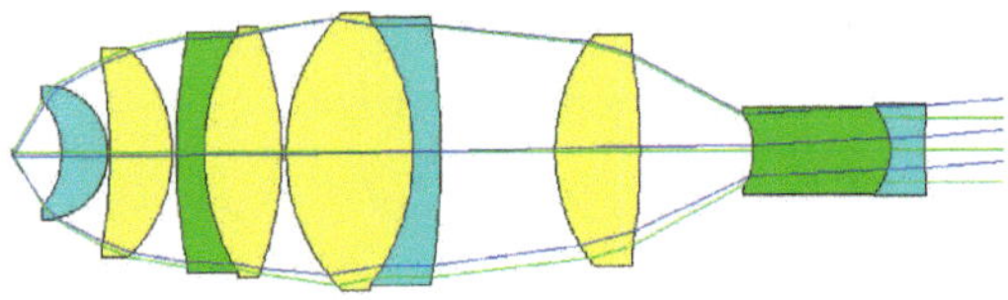

Lens: PlanFluor LD 100x0.85 d0 ftl160

Object num aper 0.850000 Image height 12.500000 Primary wavln 0.546070

SRF	RADIUS	THICKNESS	APERTURE RADIUS	GLASS	SPECIAL
OBJ	0.000000	1.0000e-20	0.125000	AIR	
AST	0.000000	2.200058	2.290000 A	AIR	
2	-3.486800	2.000000	2.290000	LZ_TK17	
3	-2.848800	0.100000	2.820000	AIR	
4	-21.385100	2.700000	3.920000	CAF2	
5	-6.122700	0.200000	4.420000	AIR	
6	28.000100	1.200000	4.990000	LZ_TF10	
7	9.407600	3.400000	5.070000	CAF2 P	
8	-13.941200	0.200000	5.280000	AIR	
9	8.343800	5.500000	5.860000	CAF2 P	
10	-9.438600	1.200000	5.640000	H-ZLAF50A	
11	-37.444600	5.000000	5.650000	AIR	
12	7.936400	3.700000	4.880000	CAF2 P	
13	-25.492500	5.000000	4.390000	AIR	
14	-3.052000	6.000000	1.600000	LZ_TF10 P	
15	-2.775200	1.500000	1.850000	H-ZLAF50A P	
16	56.730000	90.000000	1.920000	AIR	
17	214.800000	5.000000	9.010000	LZ-BK8	
18	-46.990000	0.300000	9.150000	AIR	
19	-46.880000	3.000000	9.140000	LZ-BF24	
20	-113.240000	157.836495 S	9.290000	AIR	
IMS	0.000000	0.000000	12.500000		

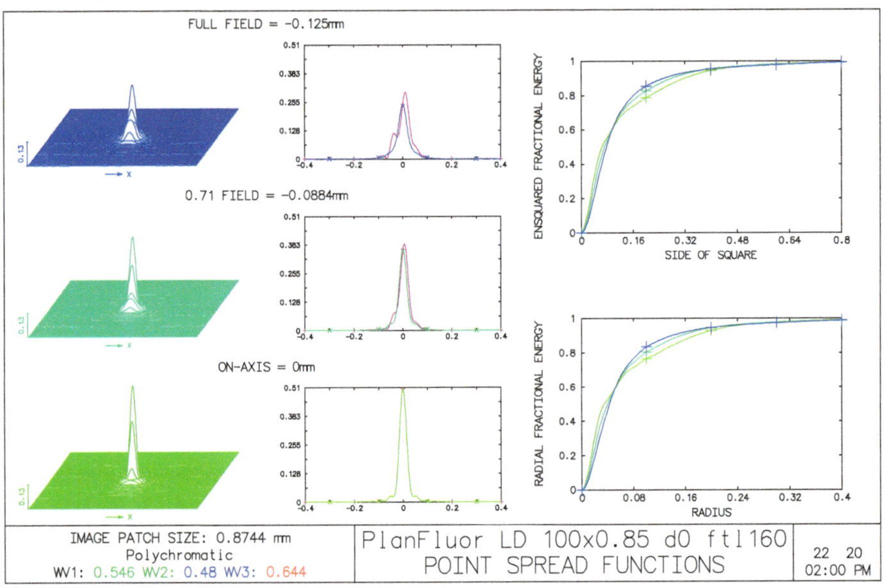

Figure 2.54 Optical design and graphs of aberration correction of a 100×/0.85 infinity LCF LD plan MicroFluar objective ($F'_{ts} = 160$ mm).

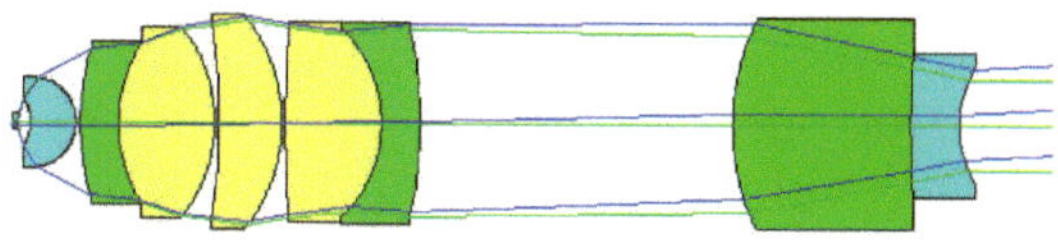

Lens: PlanFlu 100x0.95 d=0.17 f180

Object num aper 0.950000 Image height 10.000000 Primary wavln 0.546074

SRF	RADIUS	THICKNESS	APERTURE RADIUS	GLASS	SPECIAL
OBJ	0.000000	0.010000	0.099978	AIR	
AST	0.000000	0.170000	0.150000 A	K9	
2	0.000000	0.500000	0.290000	AIR	
3	-1.459600	1.770000	0.890000	LZ_CTK19	
4	-1.664700	0.200000	1.660000	AIR	
5	9.564800	1.500000	2.780000	LZ_TF5	
6	5.439300	3.800000	3.050000	CAF2	
7	-5.439300	0.150000	3.590000	AIR	
8	-27.583300	2.500000	3.810000	CAF2 P	
9	-6.075600	0.150000	4.010000	AIR	
10	73.124300	3.800000	3.750000	CAF2 P	
11	-4.687600	1.500000	3.500000	LZ_TF5 P	
12	-16.423300	12.200000	3.780000	AIR	
13	8.705600	7.000000	3.950000	LZ_TF5 P	
14	36.341100	1.900000	2.690000	LZ_CTK12	
15	4.022400	180.000000	2.180000	AIR	
16	-32.490000	12.600000	15.750000	K9	
17	-39.204000	2.700000	18.070000	AIR	
18	339.390000	13.500000	18.610000	CAF2	
19	-73.350000	2.700000	19.130000	AIR	
20	199.170000	15.300000	19.010000	CAF2	
21	-64.800000	2.700000	18.410000	H-LAK2	
22	0.000000	179.924480 S	18.400000	AIR	
IMS	0.000000	0.000000	12.500000		

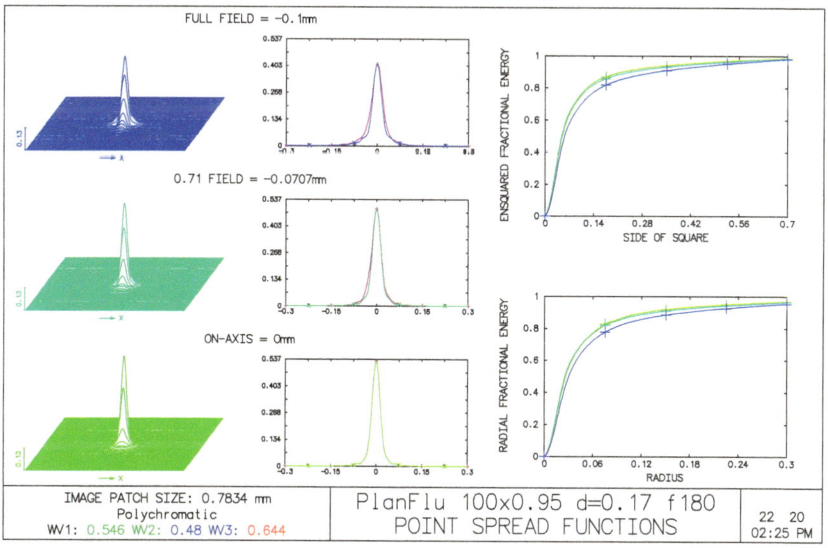

Figure 2.55 Optical design and graphs of aberration correction of a 100×/0.95 infinity LCF plan MicroFluar objective ($F'_{ts} = 180$ mm).

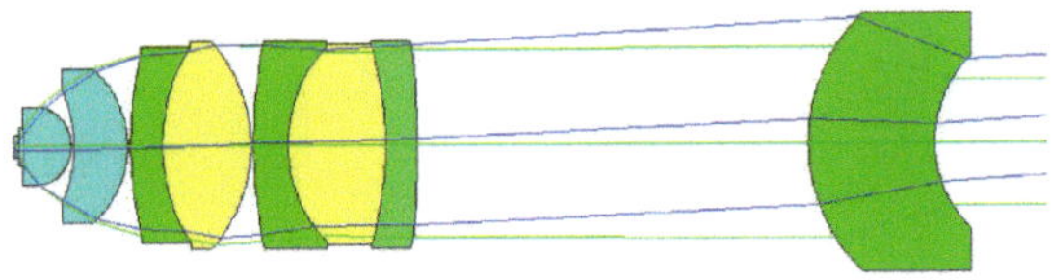

Lens: OF 100x1.20 water d017 ftl160

Object num aper 1.200000 Image height 10.000000 Primary wavln 0.546070

SRF	RADIUS		THICKNESS		APERTURE RADIUS		GLASS		SPECIAL
OBJ	0.000000		0.010000		0.100233		WATER		
AST	0.000000		0.170000		0.120000	A	LZ_K14		
2	0.000000		0.150100		0.340000		WATER	P	
3	-4.207000		1.500000		0.570000		LZ_BK6		
4	-1.202300		0.100000		1.200000		AIR		
5	-3.981000		1.700000		1.720000		LZ_CTK19		
6	-3.048000		0.100000		2.340000		AIR		
7	11.858000		1.000000		2.860000		LZ_TF5		
8	5.420000		2.800000		2.960000		CAF2		
9	-4.613000		0.100000		3.140000		AIR		
10	15.382000		1.100000		3.130000		LZ_TF1		
11	4.169000		3.100000		2.960000		CAF2	P	
12	-10.280000		1.000000		3.030000		LZ_OF4		
13	-25.470000		12.350000		3.130000		AIR		
14	5.346000		4.000000		3.920000		LZ_TF5	P	
15	3.565000		160.000000		2.600000		AIR		
16	214.800000		5.000000		10.800000		LZ_BK8		
17	-46.990000		0.300000		10.880000		AIR		
18	-46.880000		3.000000		10.850000		LZ_BF24		
19	-113.240000		157.512717	S	10.980000		AIR		
IMS	0.000000		0.000000		10.000000				

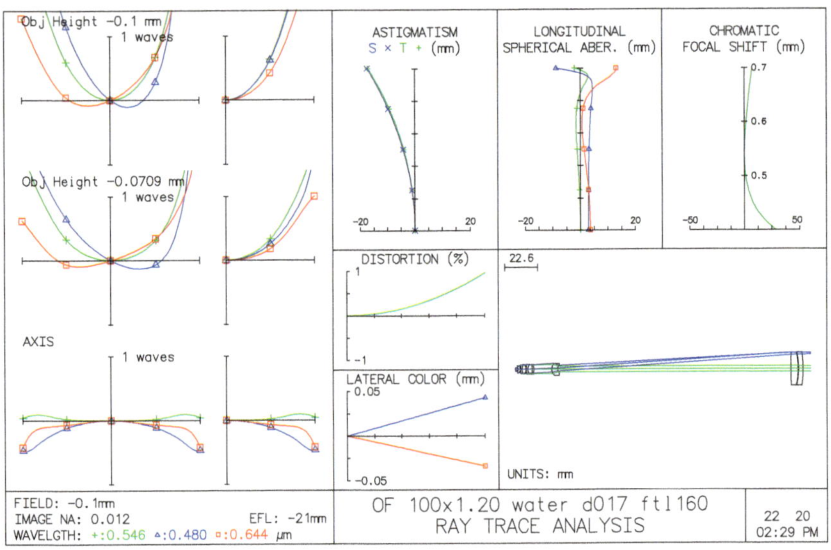

Figure 2.56 Optical design and graphs of aberration correction of a 100×/1.20 water immersion infinity LCR semi–plan MicroFluar objective ($F'_{ts} = 160$ mm).

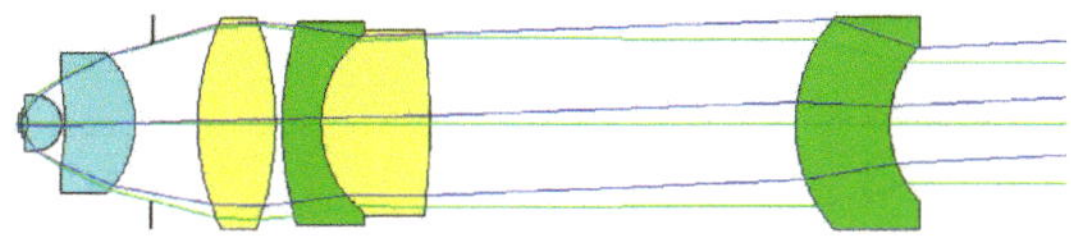

Lens: OF 100x1.20water d017 ftl200

Object num aper 1.200000 Image height 10.000000 Primary wavln 0.546070

SRF	RADIUS		THICKNESS		APERTURE RADIUS		GLASS		SPECIAL
OBJ	0.000000		0.010000		0.101193		WATER		
AST	0.000000		0.170000		0.120000	A	K9		
2	0.000000		0.200000		0.340000		WATER	P	
3	-2.000000		1.400000		0.580000		BAK2		
4	-1.109000		0.130000		1.110000		AIR		
5	-9.000000		2.900000		1.800000		ZK1		
6	-4.000000		2.600000		2.800000		AIR		
7	9.817000		3.200000		4.200000		CAF2		
8	-12.210000		0.300000		4.260000		AIR		
9	14.894000		1.600000		4.100000		ZF6		
10	4.500000		4.500000		3.590000		CAF2	P	
11	-36.130000		15.000000		3.730000		AIR		
12	6.400000		3.900000		4.290000		ZF6	P	
13	4.600000		200.000000		3.120000		AIR		
14	-36.100000		14.000000		12.110000		K9		
15	-43.560000		3.000000		14.160000		AIR		
16	377.100000		15.000000		14.450000		CAF2		
17	-81.500000		3.000000		14.780000		AIR		
18	221.300000		17.000000		14.570000		CAF2		
19	-72.000000		3.000000		13.790000		LZ_CTK12		
20	0.000000		200.051658	S	13.730000		AIR		
IMS	0.000000		0.000000		10.000000				

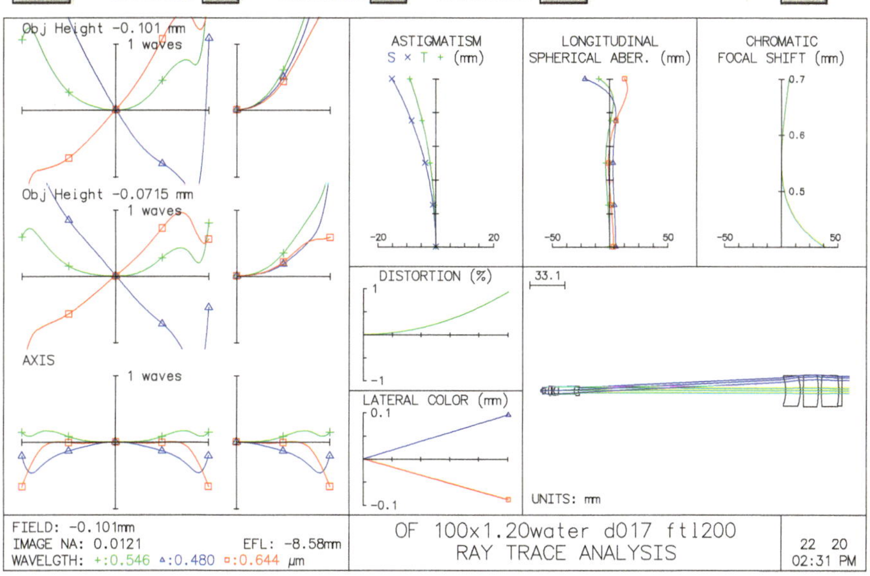

Figure 2.57 Optical design and graphs of aberration correction of a 100×/1.20 water immersion infinity LCR semi–plan MicroFluar objective ($F'_{ts} = 200$ mm).

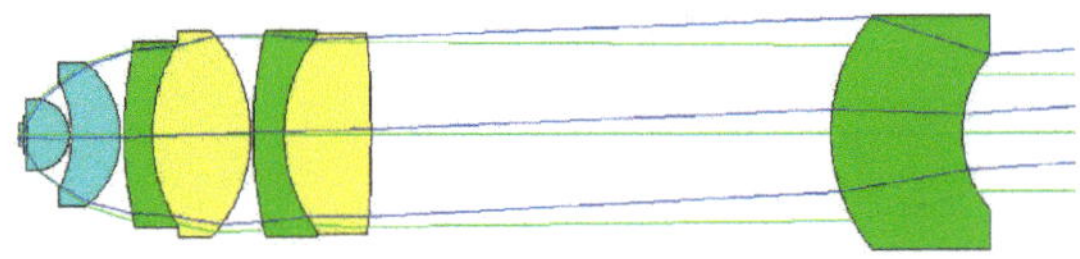

Lens:S_plan 100x1.25water d017 ftl160

Object num aper 1.250000 Image height 10.000000 Primary wavln 0.546070

SRF	RADIUS	THICKNESS	APERTURE RADIUS	GLASS	SPECIAL
OBJ	0.000000	0.010000	0.101381	WATER	
AST	0.000000	0.170000	0.130000 A	LZ_K14	
2	0.000000	0.150000	0.380000	WATER	
3	-4.207000	1.500000	0.650000	LZ_BK6	
4	-1.202300	0.090000	1.200000	AIR	
5	-3.981000	1.700000	1.820000	LZ_CTK19	
6	-3.048000	0.142800	2.440000	AIR	
7	13.243000	1.000000	3.040000	LZ_TF5	
8	5.808000	3.500000	3.170000	CAF2	
9	-5.058000	0.100000	3.490000	AIR	
10	14.421000	1.100000	3.510000	LZ_TF5 P	
11	5.395000	3.000000	3.320000	CAF2 P	
12	-43.350000	16.000000	3.400000	AIR	
13	6.194000	4.600000	3.970000	LZ_TF5 P	
14	4.055000	90.000000	2.660000	AIR	
15	214.800000	5.000000	7.680000	LZ_BK8	
16	-46.990000	0.300000	7.780000	AIR	
17	-46.880000	3.000000	7.760000	LZ_BF24	
18	-113.240000	157.877678 S	7.870000	AIR	
IMS	0.000000	0.000000	10.000000		

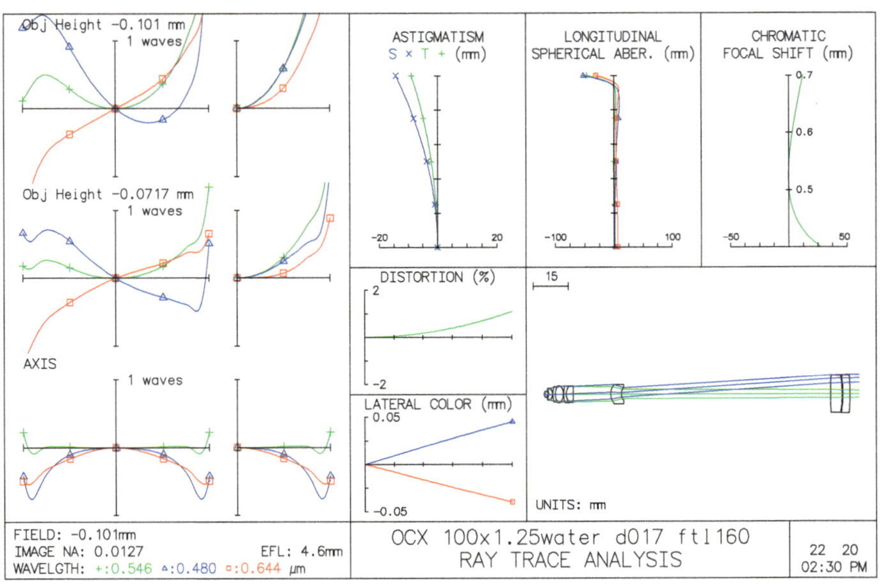

Figure 2.58 Optical design and graphs of aberration correction of a 100×/1.25 water immersion infinity LCR semi–plan MicroFluar objective ($F'_{ts} = 160$ mm).

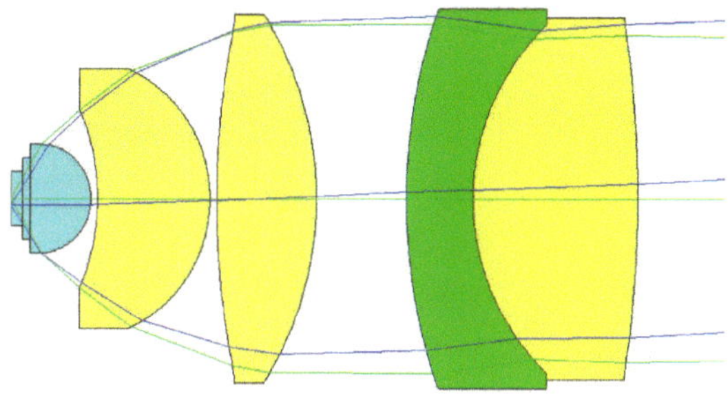

Lens: Semi Apo 100x1.30 oil finity — Efl 1.88

Object num aper 1.300000 — Image height 9.000000 — Primary wavln 0.546070

SRF	RADIUS		THICKNESS		APERTURE RADIUS		GLASS		SPECIAL
OBJ	0.000000		0.010000		0.090183		OIL		
AST	0.000000		0.170000		0.120000	A	LZ_K14		
2	0.000000		0.115700		0.390000		OIL		
3	0.000000		0.930000		0.590000		LZ_K8		
4	-0.787000		0.100000		0.790000		AIR		
5	-3.254300		1.700000		1.300000		CAF2		
6	-2.054800		0.100000		1.880000		AIR		
7	12.186700		1.500000		2.580000		CAF2	P	
8	-4.925600		1.350000		2.670000		AIR		
9	8.106700		1.000000		2.750000		LZ_TF5		
10	3.416700		2.500000		2.530000		CAF2	P	
11	-16.402100		182.521007	S	2.620000		AIR		
IMS	0.000000		0.000000		9.000000	S			

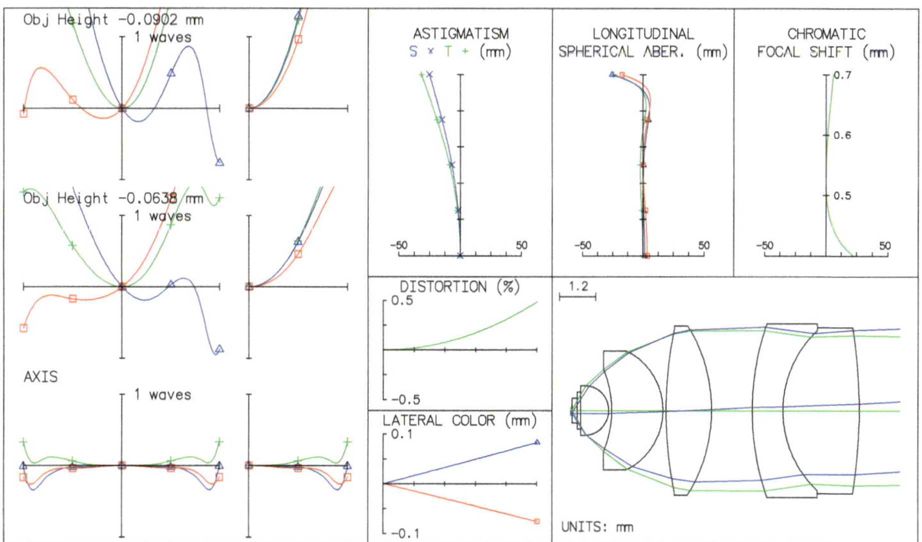

Figure 2.59 Optical design and graphs of aberration correction of a 100×/1.30 oil immersion finite semi-apochromatic objective.

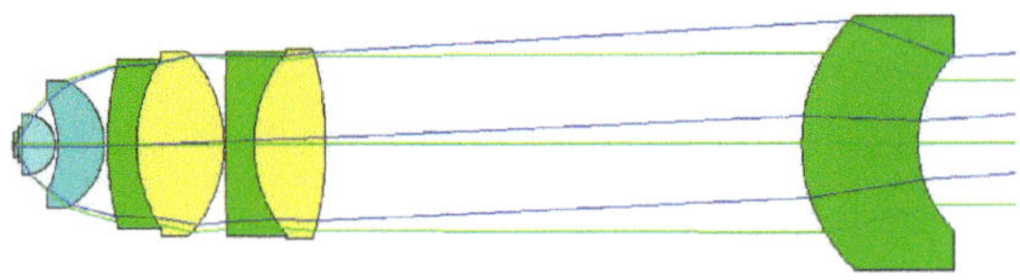

Lens: OF 100x1.30 oil ftl160

Object num aper	1.300000	Image height	10.000000	Primary wavln	0.546070

SRF	RADIUS	THICKNESS	APERTURE RADIUS	GLASS	SPECIAL
OBJ	0.000000	0.009960	0.100230	OIL	
AST	0.000000	0.170000	0.120000 A	LZ_K14	
2	0.000000	0.100000	0.400000	OIL	
3	0.000000	1.180000	0.570000	LZ_K8	
4	-1.000000	0.100000	1.000000	AIR	
5	-3.404000	1.600000	1.590000	LZ_CTK19	
6	-2.421000	0.100000	2.140000	AIR	
7	11.402000	1.000000	2.730000	LZ_TF5	
8	5.321000	3.000000	2.850000	CAF2	
9	-4.529000	0.050000	3.090000	AIR	
10	77.270000	1.000000	3.090000	LZ_TF2	
11	4.699000	2.500000	3.050000	CAF2 P	
12	-11.614000	16.500000	3.160000	AIR	
13	5.888000	4.000000	4.250000	LZ_TF5	
14	4.130000	90.000000	2.960000	AIR	
15	214.800000	5.000000	7.510000	LZ_BK8	
16	-46.990000	0.300000	7.600000	AIR	
17	-46.880000	3.000000	7.580000	LZ_BF24	
18	-113.240000	157.835880 S	7.680000	AIR	
IMS	0.000000	0.000000	10.000000		

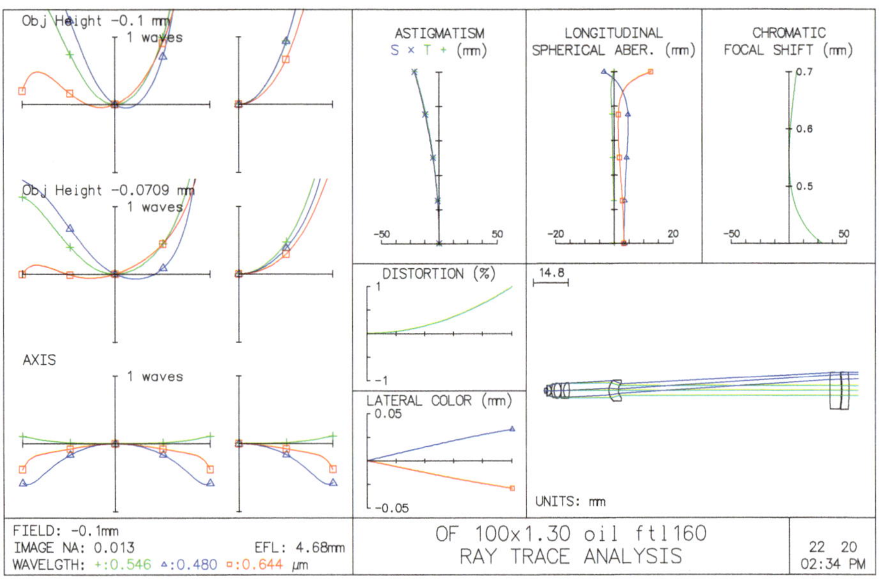

Figure 2.60 Optical design and graphs of aberration correction of a 100×/1.30 oil immersion infinity LCR semi–plan MicroFluar objective ($F'_{ts} = 160$ mm).

achromats and plan apochromats). We should take into account the peculiarities of studied objects, the methods of their preparation, and the application of real-world microscopy practice.

- The value of the numerical aperture (not the linear magnification) is the main criterion when considering expediency and proper correction of the field curvature aberration in objectives in biological microscopes.

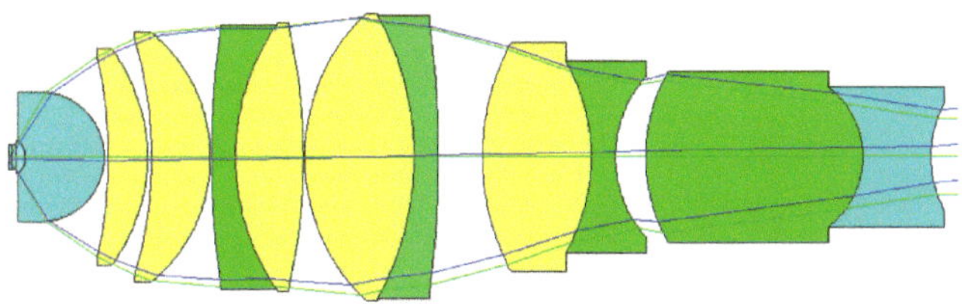

Lens: PlanFluar 100x1.32 oil Ftl160

Object num aper	1.320000	Image height	12.500000	Primary wavln	0.546070

SRF	RADIUS	THICKNESS	APERTURE RADIUS	GLASS	SPECIAL
OBJ	0.000000	1.0000e-20	0.125034	OIL	
AST	0.000000	0.279680	0.125034 AS	OIL	
2	0.000000	0.500000	0.620000	K9	
3	-1.096500	4.100000	0.830000	LZ_CTK19	
4	-3.221000	0.200000	3.220000	AIR	
5	-22.080000	2.000000	4.970000	CAF2	
6	-8.790000	0.200000	5.370000	AIR	
7	-20.560000	3.000000	5.890000	CAF2 P	
8	-7.870000	0.050000	6.220000	AIR	
9	46.670000	1.200000	6.590000	LZ_TF4	
10	11.535000	3.400000	6.590000	CAF2	
11	-29.790000	0.050000	6.670000	AIR	
12	9.817000	5.500000	7.150000	CAF2 P	
13	-14.421000	1.200000	7.020000	LZ_OF4	
14	-54.700000	2.000000	6.740000	AIR	
15	12.589000	5.500000	5.710000	CAF2 P	
16	-9.162000	1.200000	4.770000	LZ_TF1	
17	6.310000	1.500000	4.020000	AIR	
18	9.162000	11.000000	4.280000	LZ_TF4 P	
19	-4.487000	3.500000	3.470000	LZ_CTK12	
20	6.310000	90.000000	2.660000	AIR	
21	214.800000	5.000000	9.850000	LZ_BK8	
22	-46.990000	0.300000	9.980000	AIR	
23	-46.880000	3.000000	9.970000	LZ_BF24	
24	-113.240000	157.824933 S	10.130000	AIR	
IMS	0.000000	0.000000	12.500000		

Figure 2.61 Optical design and graphs of aberration correction of a 100×/1.32 oil immersion infinity LCF plan MicroFluar objective ($F'_{ts} = 160$ mm).

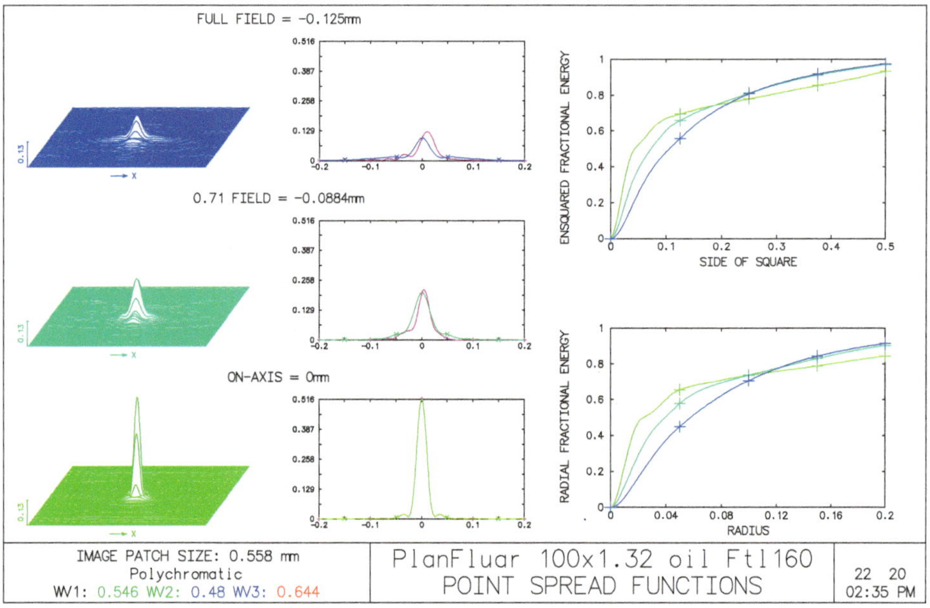

Figure 2.61 (*Continued*) Optical design and graphs of aberration correction of a 100×/1.32 oil immersion infinity LCF plan MicroFluar objective ($F'_{ts} = 160$ mm).

Figure 2.62 Photograph of a 100×/1.32 oil immersion LCF plan MicroFluar objective ($F'_{ts} = 160$ mm).

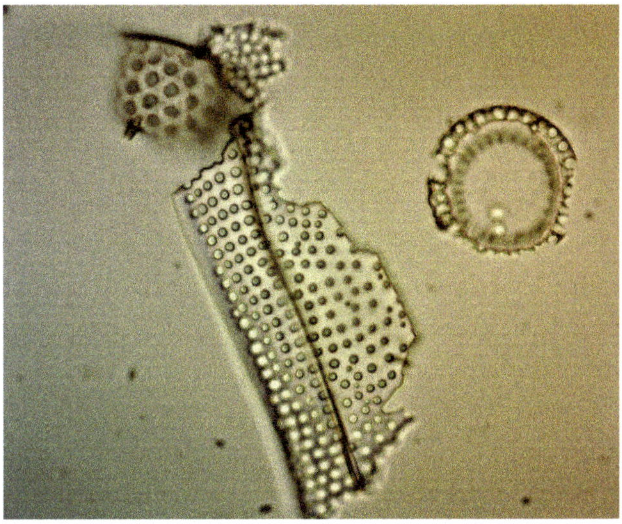

Figure 2.63 Photograph of a real object using a 100×/1.32 oil immersion LCF plan MicroFluar objective.

Figure 2.64 Photograph of a real object using a 100×/1.32 oil immersion LCF plan MicroFluar objective.

2.4.1.3 Compromise in LC correction in visual microscopy

Let us consider the image coloration of the structural elements of objects under study when they are projected by medium- and high-magnification objectives. The impact of LC on image quality in a microscope, as opposed to the impact of other types of chromatic aberrations (e.g., chromatic focal shift, secondary spectrum), is a poorly studied subject, although the nature of its

Figure 2.65 Photograph of a real object using a 100×/1.32 oil immersion LCF plan MicroFluar objective.

origin is the same: different refraction powers at the lens surface for different colors of the operational spectral range. The nature of chromatic focal shift and secondary spectrum correction in the optical system of a particular objective (or microscope) is not of much interest; these are corrected or not, depending on the class of the objective and the complexity of its optical design. It is known that to reduce the secondary spectrum, materials that have the closest possible relative partial dispersion values to the largest possible difference in the dispersion coefficients must be used. However, let us note that full correction of the secondary spectrum in objectives designed for visual microscopy is not always expedient. The theoretical light sensitivity of the eye is reduced in the spectral range of the near-UV (NUV) range. This is why complete aberration correction for wavelengths in this spectral range is unnecessary for visual observation in light microscopes. The other case is LC aberration, which can remain uncorrected even with the correction of other chromatic aberrations and even when very complex optical designs for objectives are used. Examples of this are old objectives with apochromatic aberrations that are still used today by many microscopists.

It is evident that if it is not corrected intentionally, LC aberration is associated with an objective's linear magnification: the higher the magnification, the higher the LC. This is why LC aberration is insignificant in low- and even medium-magnification objectives; it has a significant impact in high-magnification objectives. At the edges of image fragments, we can see an iridescent color border over all the contours. This is especially visible if we use a regular structure, scale, or reticle as the studied object. The periphery of elements of the regular structure will have more color if we use the objectives that have a higher linear magnification. The LC value of the objective within the whole linear field, from the center to the periphery, remains practically the same. Figure 2.70 is a photograph of an object with a regular structure that shows the coloration of elements over the periphery.

LC aberration in the image of a studied object distorts its structure and form; when the microscope is refocused from the center of the field to any area of the periphery, information about the fragment of the object being studied will not be accurate. Thus, we can conclude that for visual observation, this aberration must be fully or at least partly (up to the degree not comprehensible by the observer) corrected by the microscope. We should

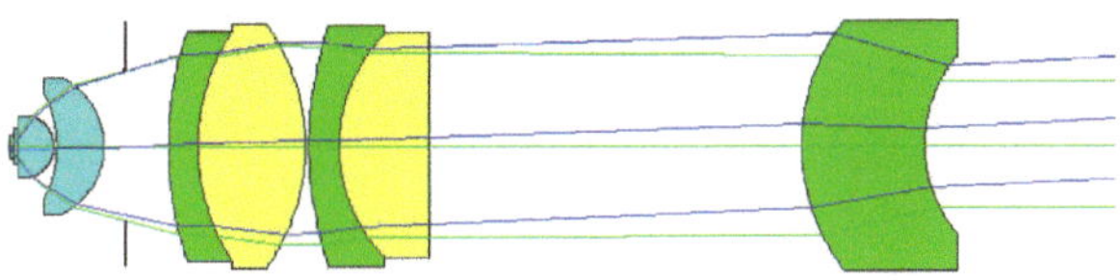

Lens: OF 100/1.35 oil

Object num aper	1.350000	Image height	10.000000	Primary wavln	0.546070

SRF	RADIUS	THICKNESS	APERTURE RADIUS	GLASS	SPECIAL
OBJ	0.000000	0.010000	0.101950	OIL	
AST	0.000000	0.170000	0.121426 AS	K9	
2	0.000000	0.130750	0.450000	OIL P	
3	0.000000	1.500000	0.710000	K9 P	
4	-1.200000	0.130000	1.200000	AIR	
5	-4.200000	1.900000	2.050000	H-LAK2	
6	-2.890000	2.600000	2.650000	AIR	
7	13.450000	1.200000	4.390000	ZF6	
8	8.000000	4.300000	4.440000	CAF2	
9	-8.000000	0.130000	4.720000	AIR	
10	14.100000	1.200000	4.670000	ZF6	
11	6.066000	3.600000	4.320000	CAF2 P	
12	0.000000	15.000000	4.360000	AIR	
13	7.625000	5.000000	4.850000	ZF6 P	
14	5.297000	200.000000	3.440000	AIR	
15	-36.100000	14.000000	13.030000	K9	
16	-43.560000	3.000000	15.230000	AIR	
17	377.100000	15.000000	15.570000	CAF2	
18	-81.500000	3.000000	15.920000	AIR	
19	221.300000	17.000000	15.670000	CAF2	
20	-72.000000	3.000000	14.840000	H-LAK2	
21	0.000000	199.978573 S	14.770000	AIR	
IMS	0.000000	0.000000	10.000000 S		

Figure 2.66 Optical design and graphs of aberration correction of a 100×/1.35 oil immersion infinity LCR semi–plan MicroFluar objective ($F'_{ts} = 200$ mm).

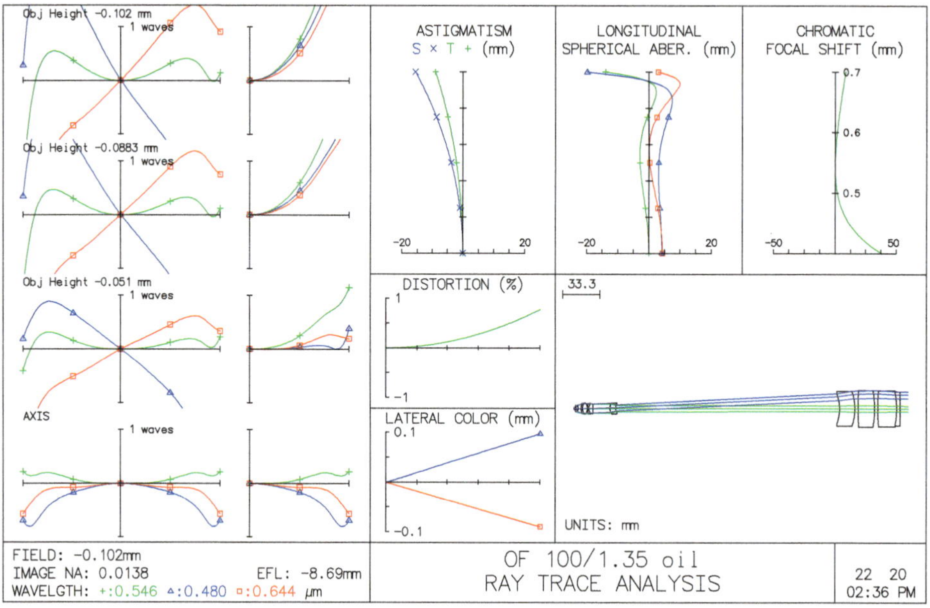

Figure 2.66 (*Continued*) Optical design and graphs of aberration correction of a 100×/1.35 oil immersion infinity LCR semi–plan MicroFluar objective ($F'_{ts} = 200$ mm).

also note that an observer's eye does not accommodate for the properties of colored artifacts.

It is technically feasible to compensate for LC aberration in the plane of the intermediate image in a microscope if the aberration is not intentionally corrected in the objective. To achieve this, a proper compensator, called a *focusing tube system*, must be added to the microscope's design. However, in this case, each lens in the microscope kit would need to use its own compensator, which would be impractical. There is another method that uses special eyepieces to compensate for LC aberration after the intermediate image plane. This method was widely incorporated in the 1950s by nearly all leading microscope manufacturers [6]. For the purposes of unification, all objectives in these kits are designed to have an LC aberration of 2%. A considerable issue with this method is the presence of a yellow border in the eyepiece field.

The value of eyepiece LC aberration correction changes in another way (contrary to the objective): from the center to the periphery; i.e., there is no full compensation. Thus, the methods described above are suboptimal. A logical solution would be to correct the LC aberration directly in the objective together with the use of non-compensating focusing systems and eyepieces. But such correction complicates the optical systems of objectives.

Let us note a possible compromise in LC aberration correction for visual microscopy that can serve as a prerequisite for designing achromatic and

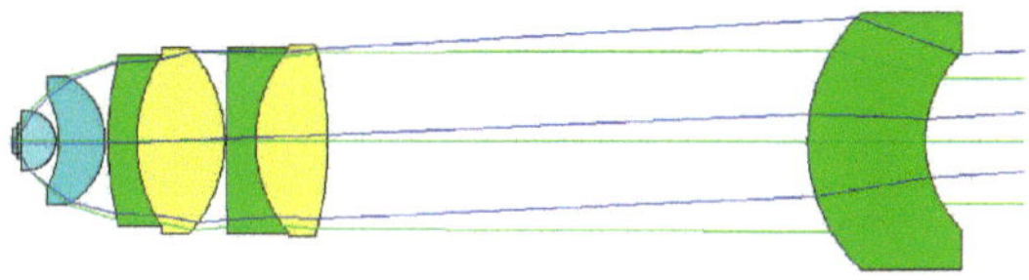

Lens: 100x1.40 oil tub160 h45				Efl	1.71
Object num aper	1.400000	Image height	10.000000	Primary wavln	0.546070
SRF	**RADIUS**	**THICKNESS**	**APERTURE RADIUS**	**GLASS**	**SPECIAL**
OBJ	0.000000	0.010000	0.099107	OIL_IMM	
AST	0.000000	0.170000	0.120000 A	K9	
2	0.000000	0.099900	0.540000	OIL_IMM P	
3	0.000000	1.300000	0.780000	LZ_BK6	
4	-1.116900	0.100000	1.120000	AIR	
5	-2.630000	1.600000	1.770000	LZ_CTK19	
6	-2.443000	0.100000	2.340000	AIR	
7	8.204000	1.000000	3.610000	LZ_TF5	
8	5.200000	3.500000	3.630000	CAF2	
9	-5.200000	0.050000	3.740000	AIR	
10	45.600000	1.000000	3.680000	LZ_TF2	
11	4.699000	3.000000	3.560000	CAF2 P	
12	-11.350000	18.200000	3.660000	AIR	
13	6.166000	4.450000	4.840000	LZ_TF5 P	
14	4.169000	160.376708 S	3.230000	AIR	
IMS	0.000000	0.000000	10.000000		

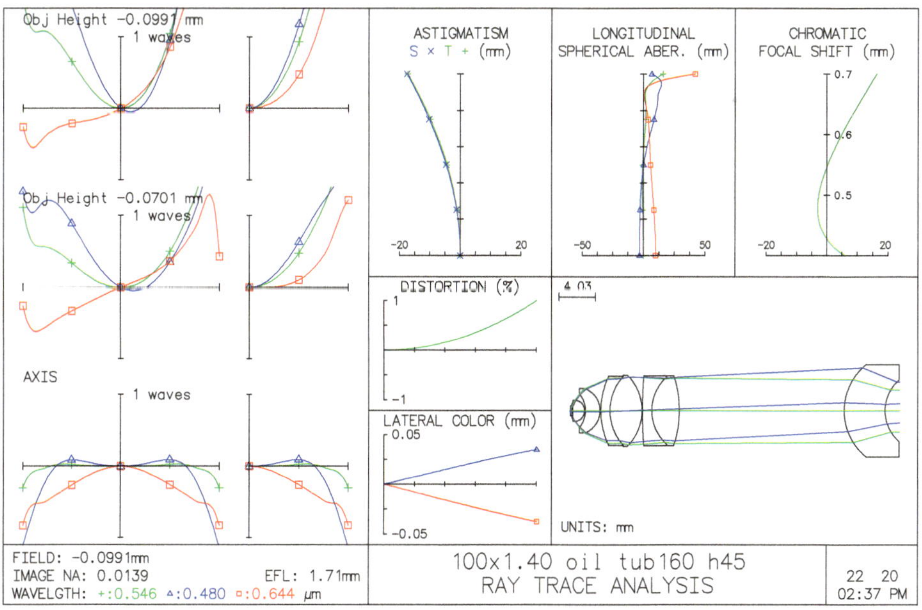

Figure 2.67 Optical design and graphs of aberration correction of a 100×/1.40 oil immersion finite LCR semi–plan MicroFluar objective.

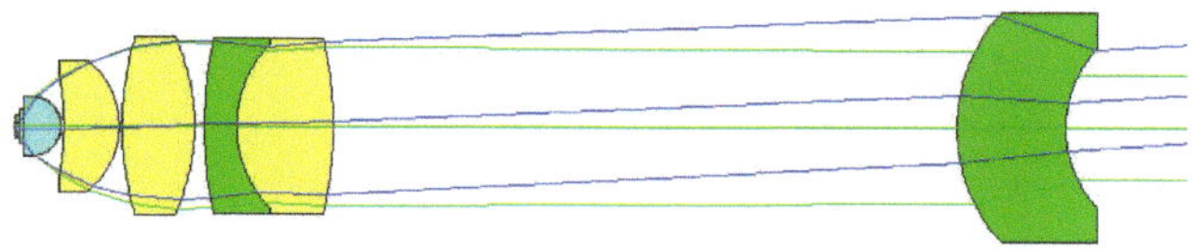

Lens: OF 125x1.25oil ftl160

Object num aper	1.250000	Image height	10.000000	Primary wavln	0.546070

SRF	RADIUS		THICKNESS		APERTURE RADIUS		GLASS		SPECIAL
OBJ	0.000000		0.010000		0.079839		OIL		
AST	0.000000		0.170000		0.090000	A	LZ_K14		
2	0.000000		0.116940		0.340000		OIL		
3	0.000000		1.150000		0.510000		K9		
4	-0.895400		0.100000		0.900000		AIR		
5	-7.015000		1.700000		1.480000		CAF2		
6	-2.466000		0.100000		1.980000		AIR		
7	8.790000		2.300000		2.500000		CAF2	P	
8	-6.081000		0.300000		2.670000		AIR		
9	11.912000		1.000000		2.650000		LZ_TF5		
10	3.467000		3.000000		2.470000		CAF2	P	
11	-11.912000		19.500000		2.630000		AIR		
12	5.012000		3.400000		3.450000		LZ_TF5	P	
13	3.436000		90.000000		2.380000		AIR		
14	214.800000		5.000000		7.280000		LZ_BK8		
15	-46.990000		0.300000		7.380000		AIR		
16	-46.880000		3.000000		7.370000		LZ_BF24		
17	-113.240000		157.817116	S	7.470000		AIR		
IMS	0.000000		0.000000		10.000000				

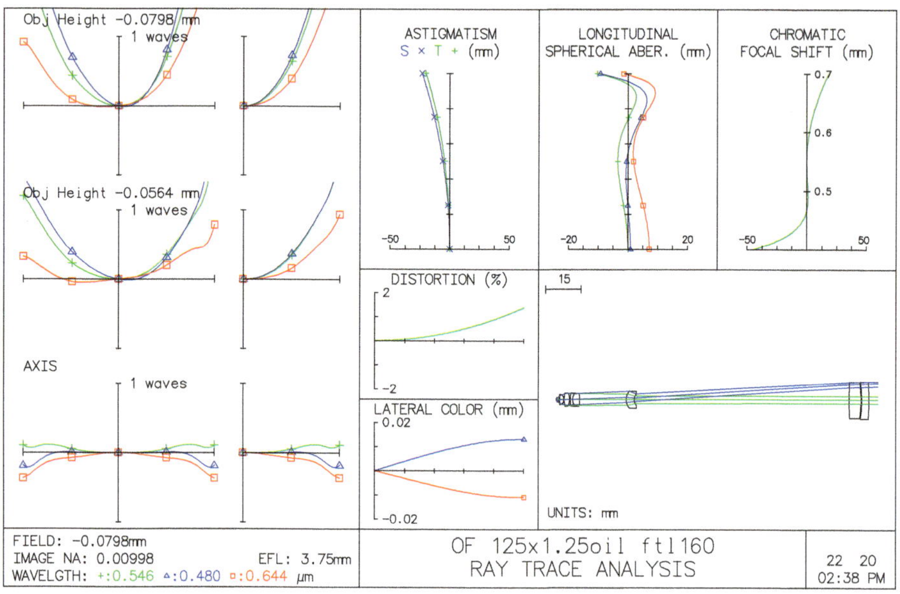

Figure 2.68 Optical design and graphs of aberration correction of a 125×/1.25 oil immersion infinity LCF semi–plan MicroFluar objective (F'_{ts} = 160 mm).

Figure 2.69 Microphotograph of a real biological object made with a 100×/1.2 water immersion objective.

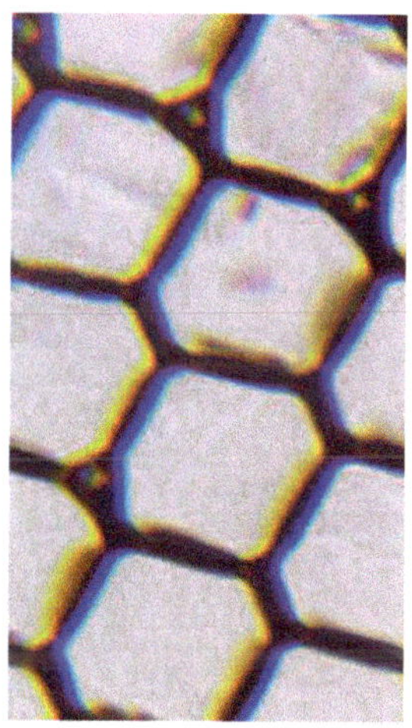

Figure 2.70 Photograph of an object with a regular structure.

apochromatic objectives. According to studies by Russian researchers, for high-magnification objectives, the residual LC of 0.5% will not have a considerable impact on the reliability of results for visual observation, even without the use of a compensating eyepiece.

2.4.1.4 Modernizing the design of old Zeiss achromats and apochromats

The optical designs of old Zeiss achromats and apochromats allow the chromatic focal shift to be corrected and use optical materials with specific relative partial dispersion values (fluorites and other crystals). These designs also allow the secondary spectrum aberration to be reduced in achromats and corrected in apochromats. In these classical designs, there are no methods to fully correct curvature aberration nor LC. Thus, when using such objectives, the images of fragments of studied objects always have non-sharpness and coloration at the periphery of the field of vision.

The initial solution to this problem was to add to the original achromat or apochromat design a negative long-focus (near the afocal) concave–convex lens that turned with its concavity to the image space and was made out of heavy flint optical glass. This solution develops Bogehold's idea [7]. The combined properties of this concave–convex lens not only allow the reduction or full correction of the LC aberration to the required level in the objective, but also a considerable reduction or full correction of the curvature aberration in the image. This component is used in several modern high-magnification objectives [8]. Moreover, it is so universal that achromatic and apochromatic objectives have been optically designed to achieve much better aberration correction as compared with old achromats and apochromats. This is mainly due to the considerable reduction in field curvature and LC aberration. As shown in Table 2.8, this optical design is not complicated with an increased number of lenses.

Based on the above, we can develop technical requirements for designing high-aperture achromatic and apochromatic objectives for biological microscopes for visual studies. Different requirements for curvature aberration correction must be applied for objectives of different magnifications. For a 10× objective, the aberration must be corrected more carefully than for a typical achromat. With this objective, the researcher selects an interesting area of the object under study and then continues the work using a higher-magnification objective. A 20× objective can have a high numerical aperture but also certain compromises for field curvature and LC aberration correction. Such an objective has a very large aperture and is quite useful for the fluorescence method. The same objectives with 50× magnification are used for water and oil immersion methods. A 40× objective must be corrected

Table 2.8 Principal optical layouts of old and new objectives.

Old classical design	New design
100×/1.25 oil achromat 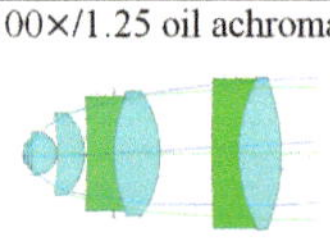	100×/1.30 oil plan achromat 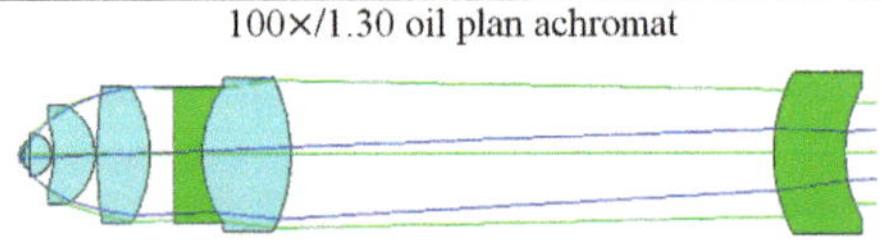
40×/0.95 apochromat 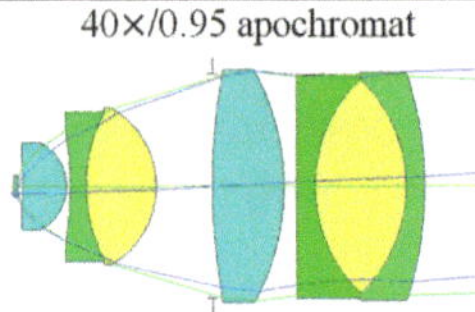	40×/0.95 plan MicroFluar 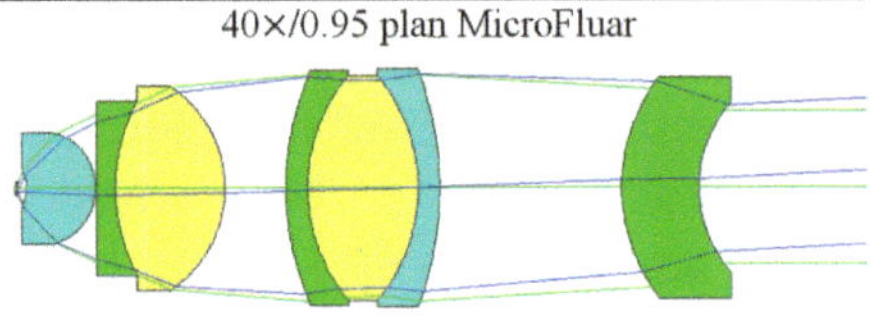

more carefully for curvature aberration. Use of this objective could become widespread, as it has a relatively high numerical aperture and does not use immersion. When maximum linear magnifications (100× immersion objectives) are achieved, the curvature aberration must be reduced by a factor of two (relative to classical achromats and Abbe apochromats).

The optical systems of objectives must retain the advantages of simple designs. These objectives must be manufactured in large quantities at a reasonable cost. A small quantity of lenses in optical designs allows objectives to be manufactured according to DIN or other standards. These designs provide high optical transmission and a small amount of diffused light. They also enable high reproducibility and consistent image quality for mass production.

2.4.1.5 A description of fluorescence objectives

Optical components intended for use in fluorescence microscopes must meet certain requirements that those for conventional biological and metallographic microscopes do not. However, there are specific requirements for optical components in almost every specialized research method that uses a light microscope. Recall that in polarization optics, the main requirement is the absence of intrinsic stresses in the lenses, both before and after mounting them in mechanical frames (by the way, this is useful for all types of optics, although it significantly increases the complexity of manufacturing such parts). In optics for use in fluorescence analysis, the main requirement is the lack of intrinsic fluorescence in the optical details. A source of intrinsic fluorescence in lenses can be, for example, an insufficiently high level of glass-melting technology when inclusions from foreign materials appear. Although the process of glass melting is its own science and practice, some brands of optical materials are more prone than others to their own fluorescence. The practice of developing special optical components (MicroFluar lenses, for example) has demonstrated the necessity of limiting the use of certain kinds of glass (for example, heavy flint) in the optical design. Moreover, before manufacturing lenses, it is necessary to check the glass blanks from which the lenses are to be made for the absence of intrinsic fluorescence. Optical glue can also be the source of intrinsic fluorescence. There are special types of adhesives recommended for bonding lenses that are used in fluorescence studies. In high-aperture immersion objectives, special types of immersions in which there is no intrinsic fluorescence should be used. Although widely used in other studies, oil immersion is often unsuitable for fluorescence; glycerin and water (as well as a mixture of the two) are the best immersions for fluorescence analysis. In light microscopy, the so-called *ordinary* or *Stokes* method of fluorescence is mainly used in which antireflection coatings shift the boundary of the maximum transmission of optics for fluorescence toward the NUV (for example, wavelengths of

365 and 405 nm). Therefore, it is precisely in this spectral region that it is necessary to thoroughly check for the absence of intrinsic fluorescence in optical components.

Given these considerations, two sets of objectives with increased numerical apertures and increased light power can be proposed for use in medium series budget fluorescent microscopes. Objectives in these kits may be referred to as MicroFluars or plan MicroFluars. These objectives are characterized by reduced aberration values, such as image curvature and LC. The first set consists of five objectives designed for operation in a fluorescent microscope with an infinite tube length; the focal length of the additional tube system is $F'_{ts} = 160$ mm.

Figure 2.71 shows photographs of a budget Russian fluorescence microscope, the Lumam, and a set of objectives designed for it. The design parameters and graphs of aberration correction of these objectives are presented in Figs. 2.31, 2.34, 2.42, 2.56, and 2.60. These objectives have been manufactured for almost 30 years at LOMO. These objectives can have significant advantages over their analogues. Table 2.9 lists the main technical parameters of the objectives shown in Fig. 2.71.

The Lumam microscope had two options for visual heads depending on the objectives used; the first head design was discussed in Volume 1 of this book (Chapter 1, Fig. 1.210; Lumam RPO-11). Let us consider the optical designs of MicroFluar objectives combined with the visual head (the second head design) of the Lumam RPO-12 microscope. This head has its own LC compensation value of −0.6% and an increased calculated value of the output numerical aperture. Figure 2.72 shows the design parameters and graphs of aberration correction of this head, and Figs. 2.73–2.77 show the design parameters and graphs of aberration correction of the combined design of this head with the indicated MicroFluar objectives.

The second set also consists of five objectives that are designed for use in a fluorescence microscope with an infinite tube length; the focal length of the additional tube system is $F'_{ts} = 200$ mm. These objectives have simpler optical

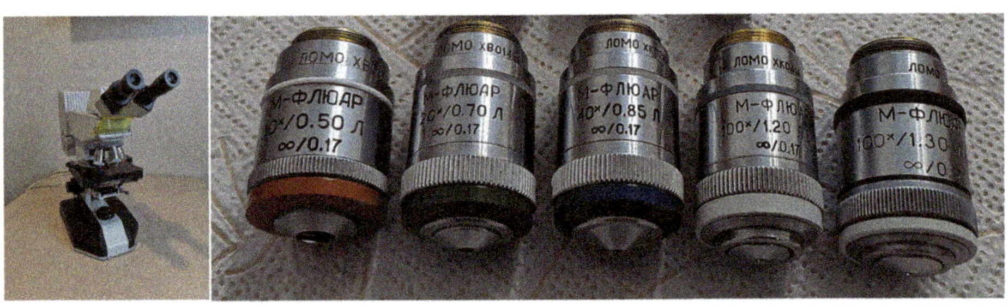

Figure 2.71 Photographs of the Lumam microscope and its MicroFluar objectives.

Table 2.9 Main technical parameters and basic optical layout of the medium series fluorescent MicroFluar objectives (version 1).

Magnification	*NA*	*WD* (mm)	*F'* (mm)	*R* (μm)	*DF* (μm)	*FOV* on object (mm)	*FOV* on image (mm)	Optical design at figure no.	The principal optical layout
10×	0.50	5.0	16	0.67	1.33	2.0	20	2.31	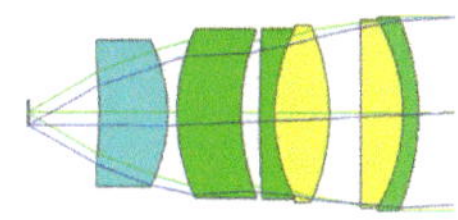
20×	0.70	0.42	8	0.48	0.68	1.0	20	2.34	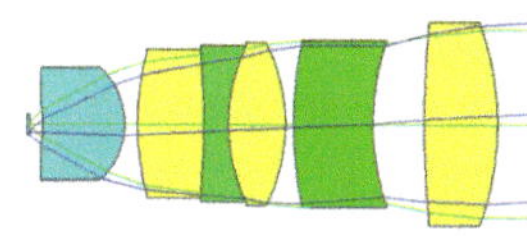
40×	0.85	0.14	4	0.39	0.46	0.5	20	2.42	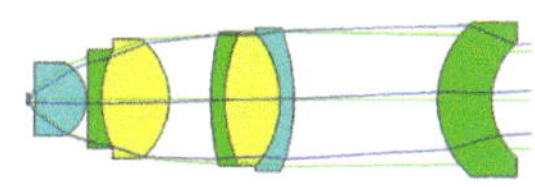
100×	1.20 water	0.12	1.6	0.28	0.23	0.1	20	2.56	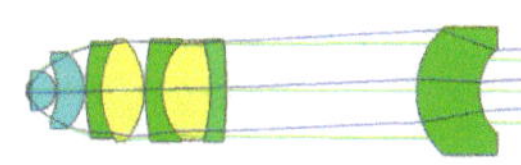
100×	1.30 oil	0.11	1.6	0.26	0.20	0.1	20	2.60	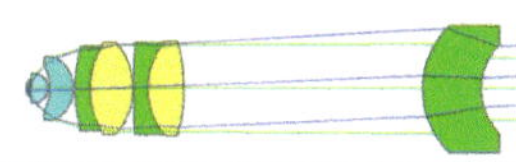

systems than some of the objectives in the first kit. However, the numerical apertures are increased compared with previous objectives. In this kit's design, the original method for correcting LC in a microscope for visual observation is proposed. If the intrinsic LC of the objective is about 1.5–1.7%, then partial compensation will be useful with a focusing (tube) system that has an LC compensation value of about 0.5–0.7%. An eyepiece with an LC compensation value of about 0.5–0.7% can also be used. The resulting residual LC, with a value of no more than 0.5%, does not substantially affect the quality and reliability of the studies.

Figure 2.78 is a photograph of these budget objectives, which are currently produced by the Russian firm Labor-Microscopes. The design parameters and graphs of aberration correction of these objectives are presented in Figs. 2.32, 2.36, 2.44, 2.57, and 2.66. Table 2.10 shows the main technical parameters of the objectives shown in Fig. 2.78.

Let us begin with the fact that the most important element in manufacturing microscope lenses is their optical design, as it determines the main consumer qualities and parameters of the objective. Because mechanical designs can vary significantly for the same optical design of an objective, it is not practical to describe them in detail here. However, it may be necessary to

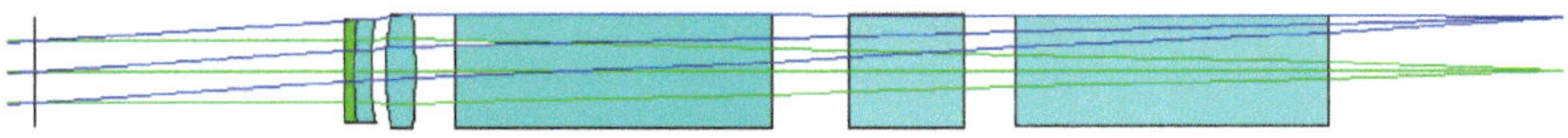

Lens:head F160 compens NA=0.0352y'=20 Efl 159.94

Image num aper 0.035000 Image height 10.000000 Primary wavln 0.546070

SRF	RADIUS		THICKNESS		APERTURE RADIUS		GLASS		SPECIAL
OBJ	0.000000		1.0000e+20		6.2524e+18		AIR		
AST	0.000000		0.001000		10.000000	A	BK7	C	
2	0.000000		60.000000		10.000000		AIR		
3	0.000000		2.200000		9.360000		LZ_TF4		
4	95.060000		3.200000		9.450000		LZ_CTK19		
5	43.750000		3.000000		9.590000		AIR		
6	49.770000		5.600000		10.340000		LZ_CTK19	P	
7	-104.710000		7.800000		10.510000		AIR		
8	0.000000		60.400000		10.480000		LZ_BK10		
9	0.000000		14.300000		10.360000		AIR		
10	0.000000		22.000000		10.320000		LZ_K8		
11	0.000000		10.000000		10.290000		AIR		
12	0.000000		59.500000		10.260000		LZ_K8	P	
13	0.000000		44.566161	S	10.160000		AIR		
IMS	0.000000		0.000000		10.000000	S			

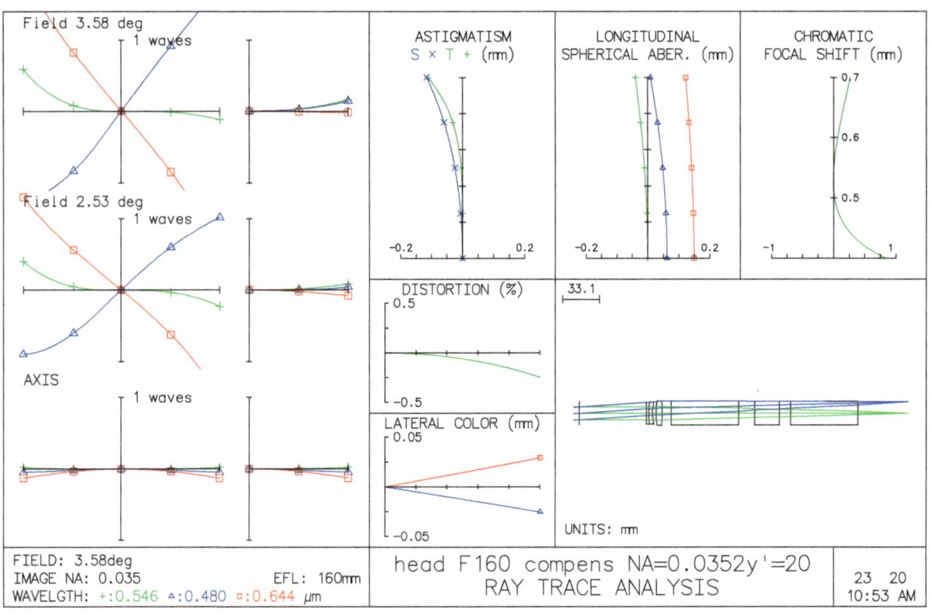

Figure 2.72 Optical design and graphs of aberration correction of an infinity (LC compensate) visual head (F'_{ts} = 160 mm).

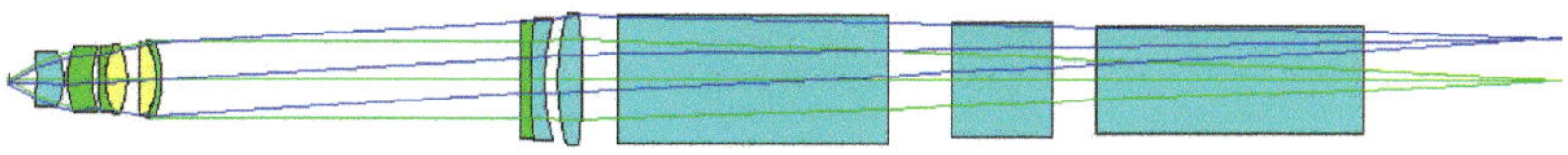

Lens: OF 10x0.50 +comp head F'=160 mm

Object num aper 0.500000 Image height 9.000000 Primary wavln 0.546070

SRF	RADIUS	THICKNESS	APERTURE RADIUS	GLASS	SPECIAL
OBJ	0.000000	0.010000	0.904274	AIR	
AST	0.000000	0.170000	0.910047 AS	LZ_K14	
2	0.000000	5.965100	0.970000	AIR	
3	-40.270000	6.000000	4.280000	LZ_CTK19	
4	-13.032000	0.800000	5.980000	AIR	
5	16.596000	5.500000	6.900000	LZ_TF8	
6	18.365000	1.600000	6.630000	AIR	
7	129.420000	1.300000	6.730000	LZ_TF8 P	
8	15.031000	4.800000	6.980000	CAF2	
9	-16.943000	2.800000	7.270000	AIR	
10	0.000000	3.600000	7.710000	CAF2 P	
11	-15.031000	1.300000	7.810000	LZ_BF11	
12	-20.230000	80.000000	8.050000	AIR	
13	0.000000	2.200000	12.610000	LZ_TF4	
14	95.060000	3.200000	12.710000	LZ_CTK19	
15	43.750000	3.000000	12.860000	AIR	
16	49.770000	5.600000	13.840000	LZ_CTK19 P	
17	-104.710000	7.800000	13.950000	AIR	
18	0.000000	60.400000	13.700000	LZ_BK10	
19	0.000000	14.300000	12.600000	AIR	
20	0.000000	22.000000	12.190000	LZ_K8	
21	0.000000	10.000000	11.780000	AIR	
22	0.000000	59.500000	11.490000	LZ_K8 P	
23	0.000000	44.566000	10.370000	AIR	
24	0.000000	-0.000816 S	9.000066 S	AIR	
IMS	0.000000	0.000000	9.000000 S		

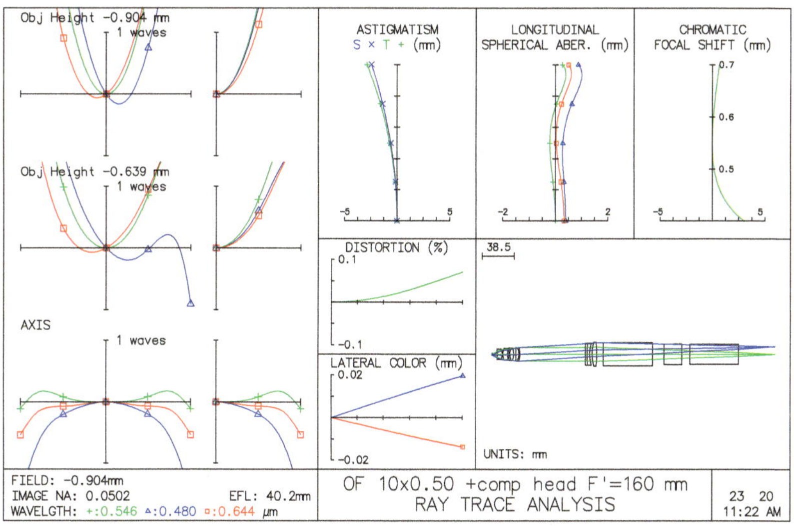

Figure 2.73 Optical design and graphs of aberration correction of a 10×/0.50 infinity LCR MicroFluar objective combined with a (LC compensate) visual head ($F'_{ts} = 160$ mm).

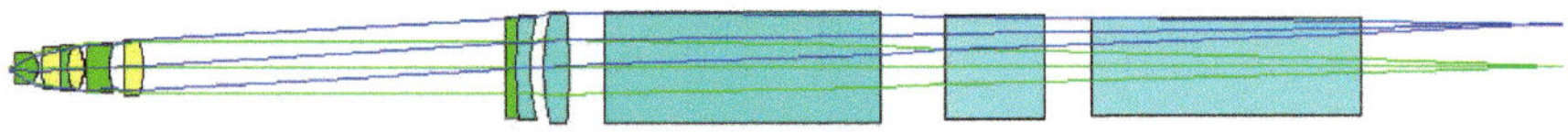

Lens: OF 20x0.70+comp head F'=160 mm

Object num aper 0.700000 Image height 9.000000 Primary wavln 0.546070

SRF	RADIUS	THICKNESS	APERTURE RADIUS	GLASS	SPECIAL
OBJ	0.000000	0.010000	0.448354	AIR	
AST	0.000000	0.170000	0.458156 AS	LZ_K14	
2	0.000000	0.812200	0.540000	AIR	
3	-5.970000	5.000000	1.220000	LZ_CTK19	
4	-4.446000	0.800000	3.310000	AIR	
5	15.524000	4.100000	4.070000	CAF2	
6	-49.430000	1.500000	4.380000	LZ_BF21	
7	10.423000	3.500000	4.640000	CAF2 P	
8	-10.423000	0.500000	4.850000	AIR	
9	23.770000	5.000000	4.960000	LZ_TF3	
10	14.421000	3.130000	4.850000	AIR	
11	56.360000	4.500000	5.540000	CAF2	
12	-16.904000	80.000000	6.010000	AIR	
13	0.000000	2.200000	10.550000	LZ_TF4	
14	95.060000	3.200000	10.640000	LZ_CTK19	
15	43.750000	3.000000	10.770000	AIR	
16	49.770000	5.600000	11.580000	LZ_CTK19 P	
17	-104.710000	7.800000	11.720000	AIR	
18	0.000000	60.400000	11.590000	LZ_BK10	
19	0.000000	14.300000	10.980000	AIR	
20	0.000000	22.000000	10.760000	LZ_K8	
21	0.000000	10.000000	10.530000	AIR	
22	0.000000	59.500000	10.370000	LZ_K8 P	
23	0.000000	44.566000	9.750000	AIR	
24	0.000000	0.011312 S	9.000335 S	AIR	
IMS	0.000000	0.000000	9.000000 S		

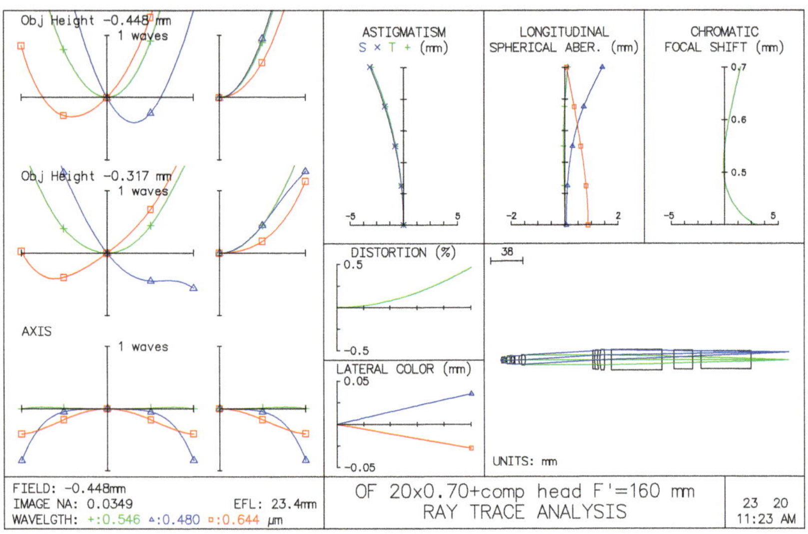

Figure 2.74 Optical design and graphs of aberration correction of a 20×/0.70 infinity LCR MicroFluar objective combined with a (LC compensate) visual head (F'_{ts} = 160 mm).

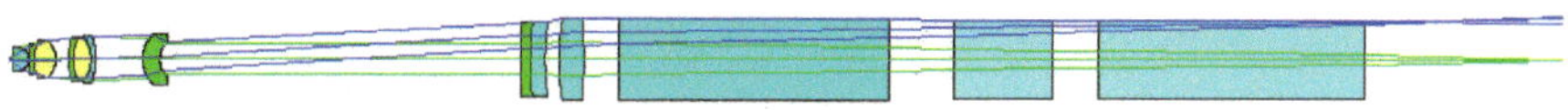

Lens: OF 40x0.85+comp head F'160 mm

Object num aper 0.850000 Image height 9.000000 Primary wavln 0.546070

SRF	RADIUS	THICKNESS	APERTURE RADIUS	GLASS	SPECIAL
OBJ	0.000000	0.010000	0.226461	AIR	
AST	0.000000	0.170000	0.242597 AS	LZ_K14	
2	0.000000	0.449700	0.360000	AIR	
3	-3.837000	3.600000	0.910000	LZ_CTK19	
4	-2.992000	0.100000	2.600000	AIR	
5	0.000000	1.000000	3.110000	LZ_TF3	
6	7.691000	5.000000	3.450000	CAF2	
7	-5.346000	3.000000	4.200000	AIR	
8	19.724000	1.000000	4.720000	LZ_TF1	
9	11.143000	4.000000	4.680000	CAF2	
10	-7.145000	1.000000	4.720000	LZ_CTK19 P	
11	-13.964000	10.500000	5.040000	AIR	
12	7.278000	3.900000	5.400000	LZ_TF5	
13	5.346000	80.000000	4.010000	AIR	
14	0.000000	2.200000	8.010000	LZ_TF4	
15	95.060000	3.200000	8.080000	LZ_CTK19	
16	43.750000	3.000000	8.190000	AIR	
17	49.770000	5.600000	8.780000	LZ_CTK19 P	
18	-104.710000	7.800000	8.930000	AIR	
19	0.000000	60.400000	8.900000	LZ_BK10	
20	0.000000	14.300000	8.730000	AIR	
21	0.000000	22.000000	8.670000	LZ_K8	
22	0.000000	10.000000	8.610000	AIR	
23	0.000000	59.500000	8.570000	LZ_K8 P	
24	0.000000	44.566000	8.400000	AIR	
25	0.000000	-0.012787 S	9.000762 S	AIR	
IMS	0.000000	0.000000	9.000000 S		

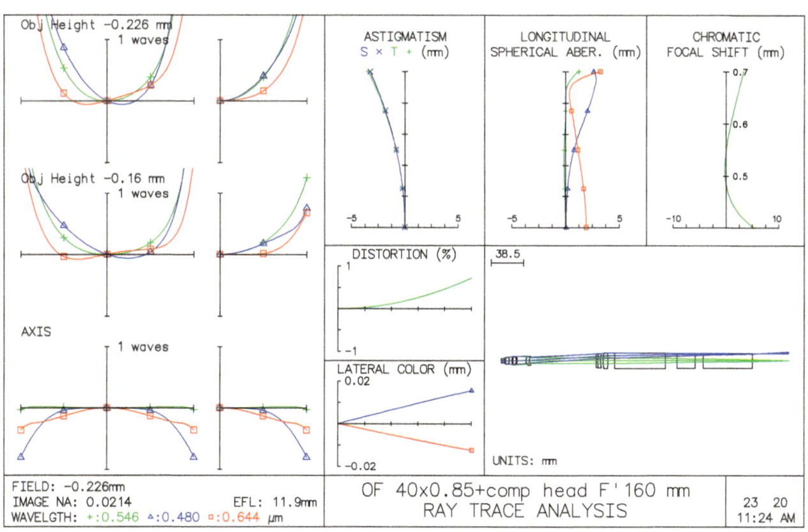

Figure 2.75 Optical design and graphs of aberration correction of a 40×/0.85 infinity LCR MicroFluar objective combined with a (LC compensate) visual head ($F'_{ts} = 160$ mm).

Lens: OF100x1.2water+comp head f160mm

Object num aper	1.200000	Image height	9.000000	Primary wavln	0.546070

SRF	RADIUS	THICKNESS		APERTURE RADIUS		GLASS		SPECIAL
OBJ	0.000000	0.010000		0.090277		WATER		
AST	0.000000	0.170000		0.110831	AS	LZ_K14		
2	0.000000	0.150057		0.340000		WATER	P	
3	-4.207000	1.500000		0.570000		LZ_BK6		
4	-1.202300	0.100000		1.200000		AIR		
5	-3.981000	1.700000		1.720000		LZ_CTK19		
6	-3.048000	0.100000		2.340000		AIR		
7	11.858000	1.000000		2.860000		LZ_TF5		
8	5.420000	2.800000		2.960000		CAF2		
9	-4.613000	0.100000		3.140000		AIR		
10	15.382000	1.100000		3.130000		LZ_TF1		
11	4.169000	3.100000		2.960000		CAF2	P	
12	-10.280000	1.000000		3.030000		LZ_OF4		
13	-25.470000	12.350000		3.130000		AIR		
14	5.346000	4.000000		3.910000		LZ_TF5	P	
15	3.565000	80.000000		2.600000		AIR		
16	0.000000	2.200000		6.480000		LZ_TF4		
17	95.060000	3.200000		6.540000		LZ_CTK19		
18	43.750000	3.000000		6.640000		AIR		
19	49.770000	5.600000		7.140000		LZ_CTK19	P	
20	-104.710000	7.800000		7.300000		AIR		
21	0.000000	60.400000		7.330000		LZ_BK10		
22	0.000000	14.300000		7.460000		AIR		
23	0.000000	22.000000		7.510000		LZ_K8		
24	0.000000	10.000000		7.560000		AIR		
25	0.000000	59.500000		7.620000		LZ_K8	P	
26	0.000000	44.566000		8.910000		AIR		
27	0.000000	-7.0446e-05	S	9.000003	S	AIR		
IMS	0.000000	0.000000		9.000000	S			

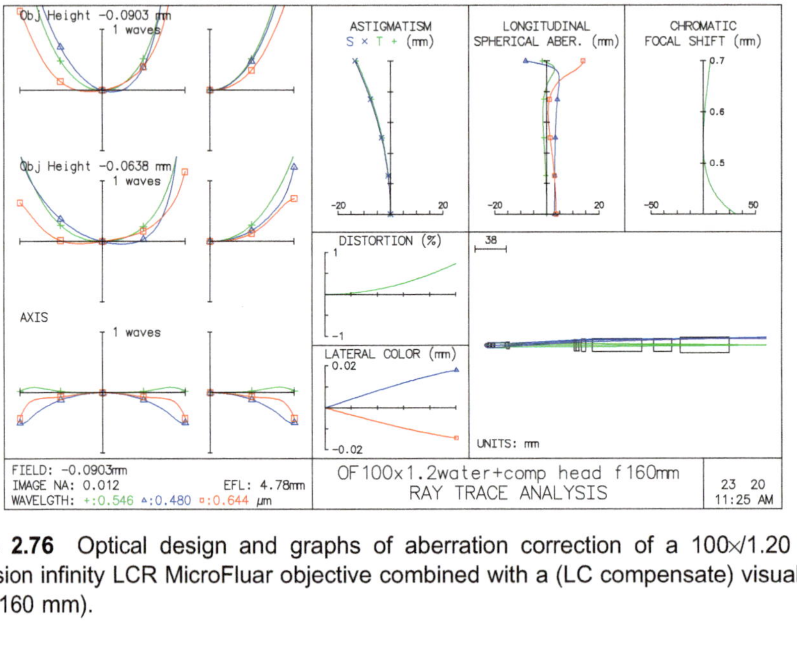

Figure 2.76 Optical design and graphs of aberration correction of a 100×/1.20 water immersion infinity LCR MicroFluar objective combined with a (LC compensate) visual head ($F'_{ts} = 160$ mm).

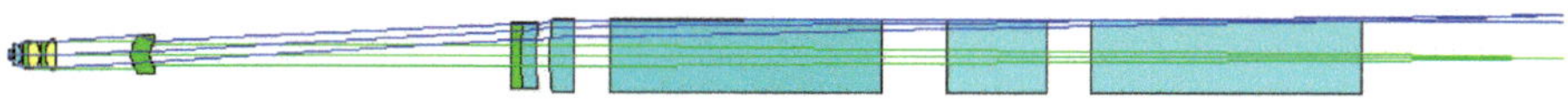

Lens: OF 100x1.30+comp head F'160 mm

Object num aper 1.300000 Image height 9.000000 Primary wavln 0.546070

SRF	RADIUS	THICKNESS	APERTURE RADIUS	GLASS	SPECIAL
OBJ	0.000000	0.009960	0.090261	OIL	
AST	0.000000	0.170000	0.106799 AS	LZ_K14	
2	0.000000	0.100000	0.390000	OIL	
3	0.000000	1.180000	0.560000	LZ_K8	
4	-1.000000	0.100000	1.000000	AIR	
5	-3.404000	1.600000	1.590000	LZ_CTK19	
6	-2.421000	0.100000	2.140000	AIR	
7	11.402000	1.000000	2.730000	LZ_TF5	
8	5.321000	3.000000	2.840000	CAF2	
9	-4.529000	0.050000	3.080000	AIR	
10	77.270000	1.000000	3.070000	LZ_TF2	
11	4.699000	2.500000	3.030000	CAF2 P	
12	-11.614000	16.500000	3.140000	AIR	
13	5.888000	4.000000	4.110000	LZ_TF5	
14	4.130000	80.000000	2.870000	AIR	
15	0.000000	2.200000	6.740000	LZ_TF4	
16	95.060000	3.200000	6.800000	LZ_CTK19	
17	43.750000	3.000000	6.900000	AIR	
18	49.770000	5.600000	7.420000	LZ_CTK19 P	
19	-104.710000	7.800000	7.570000	AIR	
20	0.000000	60.400000	7.580000	LZ_BK10	
21	0.000000	14.300000	7.650000	AIR	
22	0.000000	22.000000	7.670000	LZ_K8	
23	0.000000	10.000000	7.700000	AIR	
24	0.000000	59.500000	7.720000	LZ_K8 P	
25	0.000000	44.566000	7.780000	AIR	
26	0.000000	0.002412 S	9.000018 S	AIR	
IMS	0.000000	0.000000	9.000000 S		

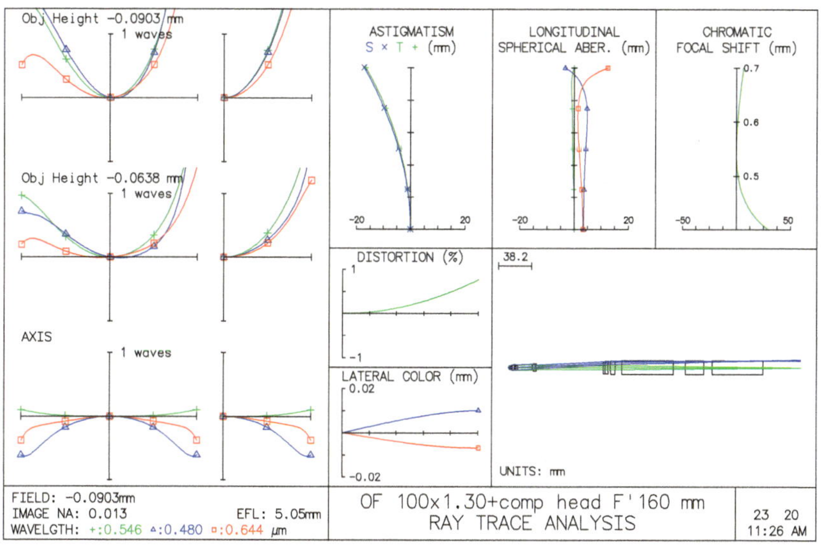

Figure 2.77 Optical design and graphs of aberration correction of a 100×/1.30 oil immersion infinity LCR MicroFluar objective combined with a (LC compensate) visual head ($F'_{ts} = 160$ mm).

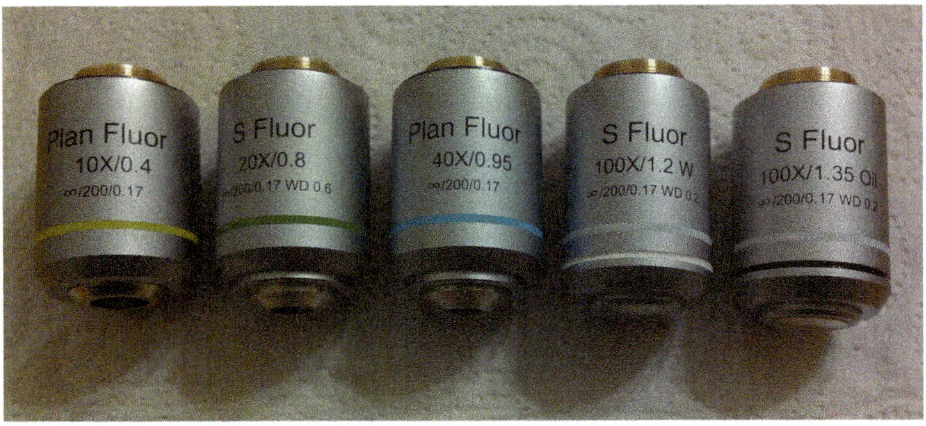

Figure 2.78 Photograph of modern MicroFluar objectives labeled "Fluor" (version 2).

carefully consider one of the elements of the optical design, namely, the design documentation for the manufacture of optical elements of these objectives. In any case, and this is confirmed by the manufacture of a large number of real objectives, if we produce objectives in accordance with the drawings below, the image quality will correspond to the level determined by the optical design. The drawings are presented here in the order listed in Table 2.10.

Table 2.10 Main technical parameters and basic optical layout of medium series fluorescent MicroFluar objectives (version 2).

Magnification	*NA*	*WD* (mm)	*F*′ (mm)	*R* (μm)	*DF* (μm)	*FOV* on object (mm)	*FOV* on image (mm)	Optical design at figure no.	The principal optical layout
10×	0.40 screw 0.8"	3.77	20.0	0.84	2.10	2.0	20	—	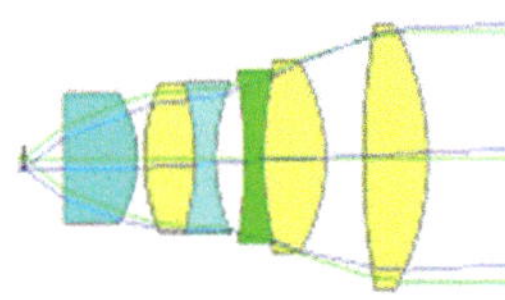
10×	0.60 screw 25 mm	3.6	20.0	0.56	0.93	2.0	20	2.32	
20×	0.80	0.60	10.0	0.42	0.52	1.0	20	2.36	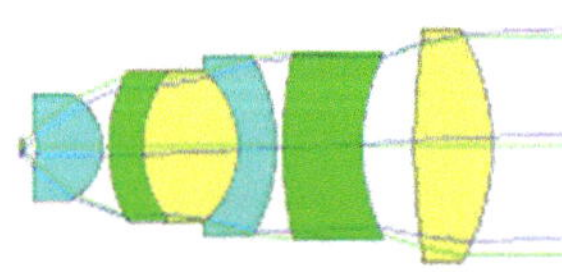
40×	0.95	0.17	5.0	0.35	0.37	0.5	20	2.44	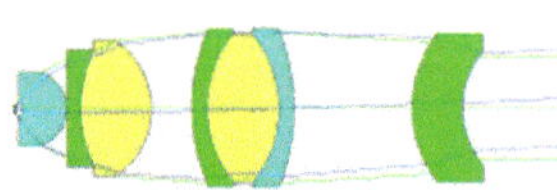
100×	1.20 water	0.12	1.6	0.28	0.23	0.1	20	2.56	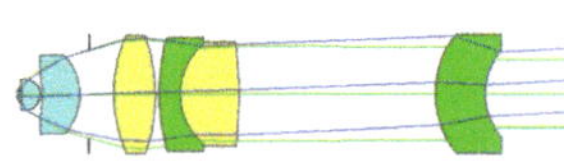
100×	1.35 oil	0.16	2.0	0.25	0.18	0.20	20	2.66	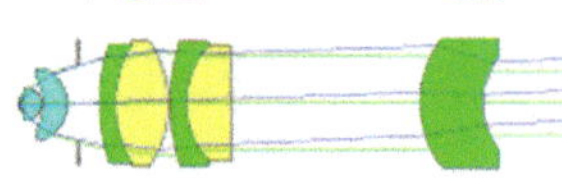

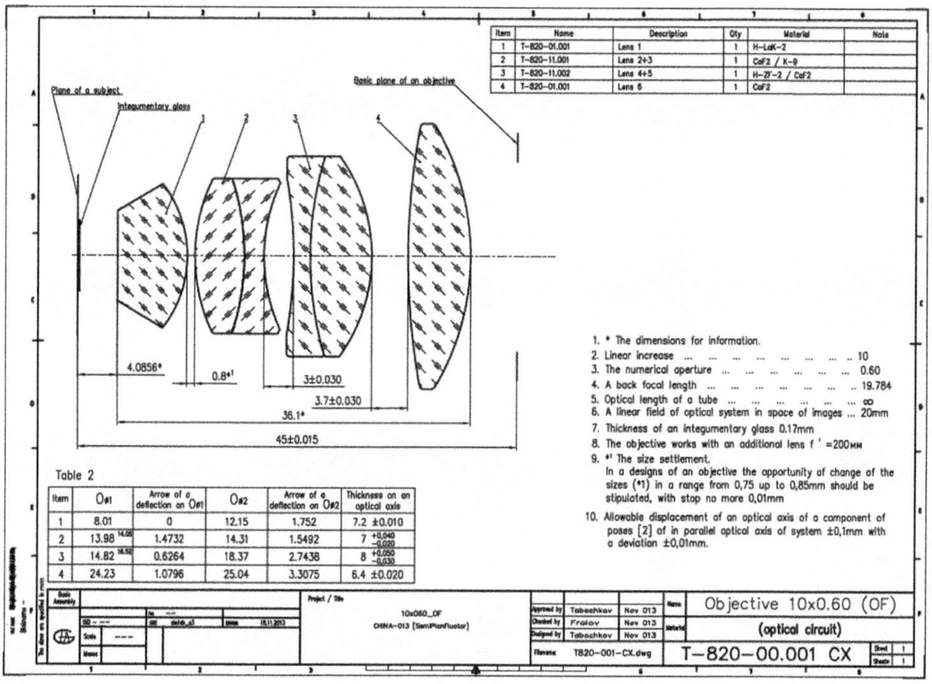

Figure 2.79 A drawing of the principal optical layout of a 10×/0.60 infinity LCR MicroFluar objective.

Thus, Fig. 2.79 shows a drawing of the principal optical layout of the objective with a linear magnification of 10× and a numerical aperture of 0.60. Figures 2.80–2.87 are drawings of the lenses of this objective.

Let us consider the drawing of the principal optical layout of this objective. This drawing, like all subsequent ones, can provide an experienced engineer with a lot of information or become an excellent study guide for students. By studying the drawings of optical parts, it is possible, for example, to make preliminary conclusions about the possible technological processes that could be used for manufacturing lenses. These technological processes include, for example, rounding, shaping, and centering, as well as the so-called *blocks* made during the grinding and polishing of lens surfaces. By using the principal optical layout drawing, a basic optical system can be made and some conclusions can be drawn regarding, among other things, the future mechanical design. Again, although the actual mechanical design may be different, some of its basic parameters are determined by the optical design. Let us illustrate this statement with the design example shown in Fig. 2.79. We will discuss here only one of the important parameters of the 10× objective: the diameters of lenses. The diameters of lenses determine whether the objective can be attached to a nosepiece device, the thread diameters of which can vary in microscope models made by different manufacturers. The more light power

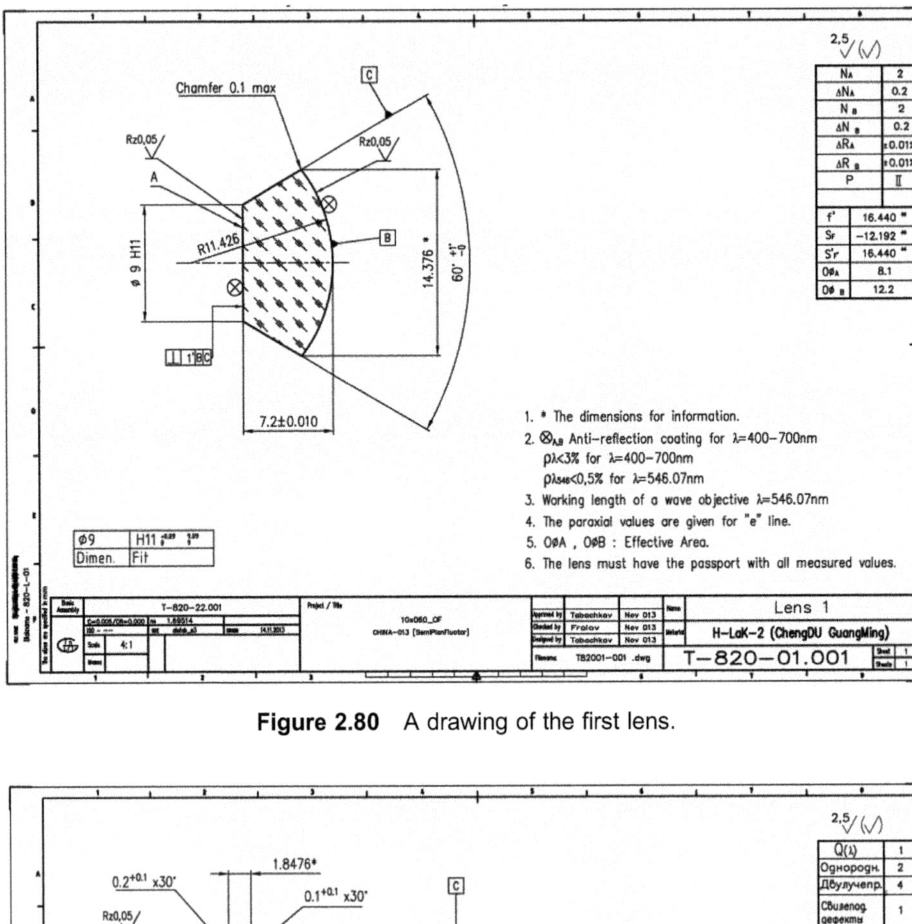

Figure 2.80 A drawing of the first lens.

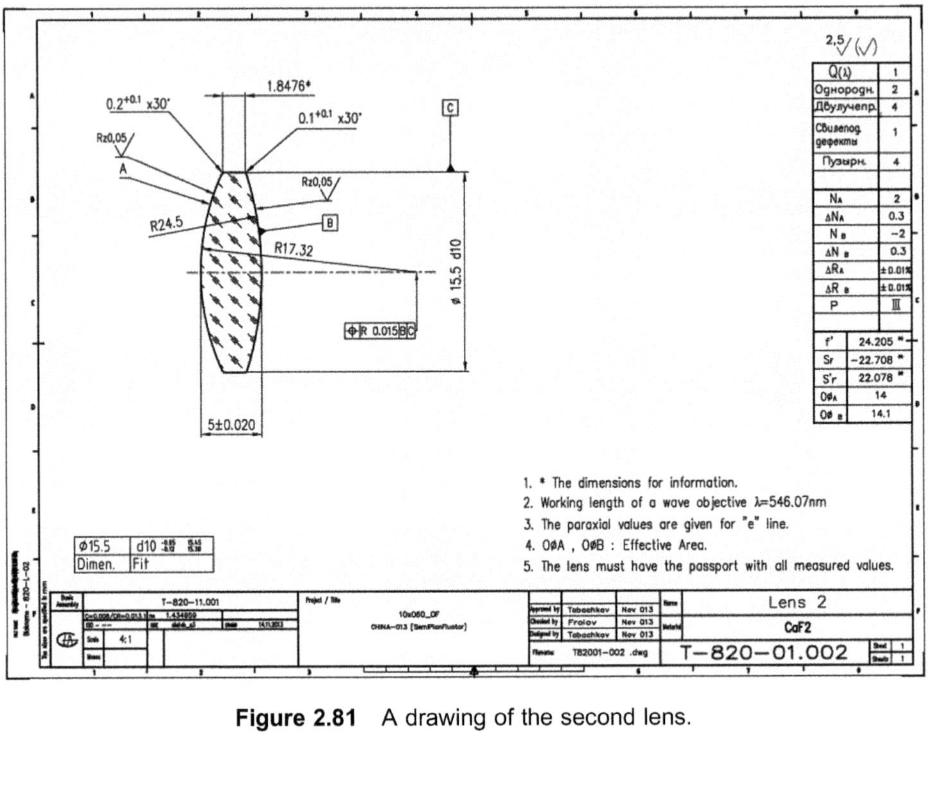

Figure 2.81 A drawing of the second lens.

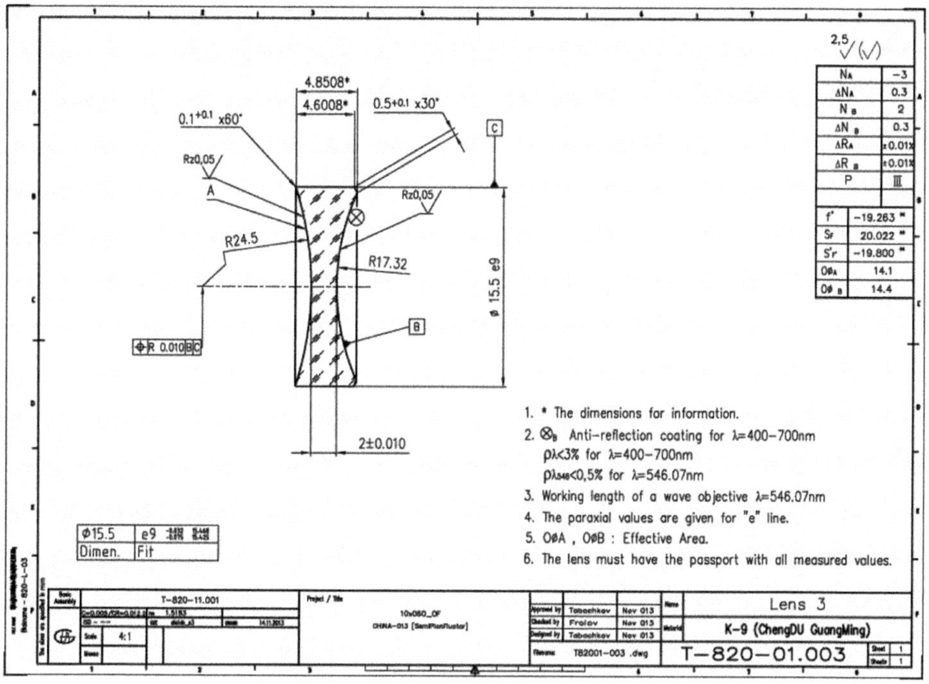

Figure 2.82 A drawing of the third lens.

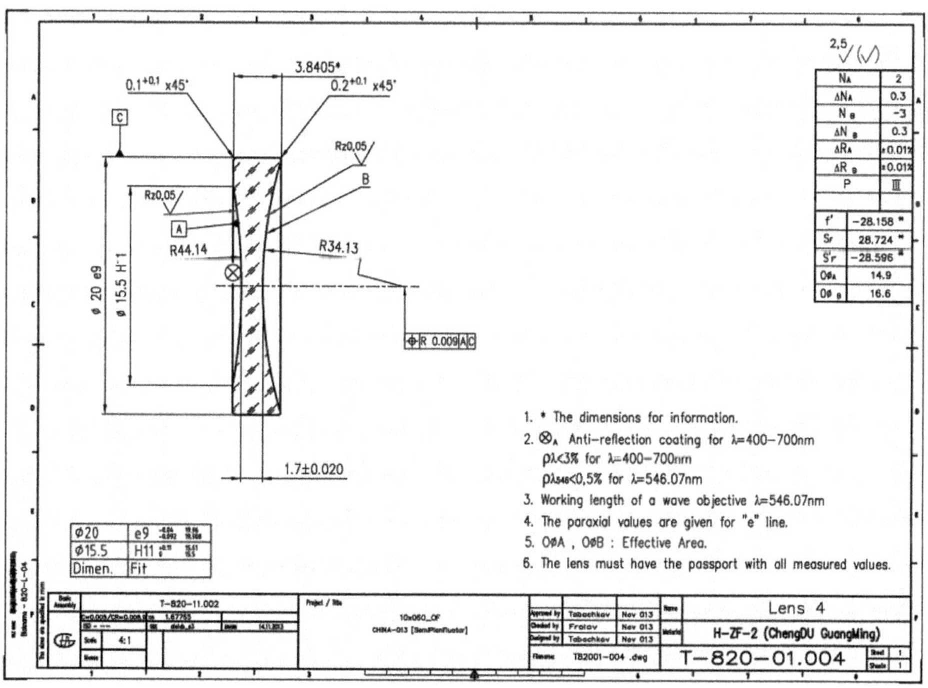

Figure 2.83 A drawing of the fourth lens.

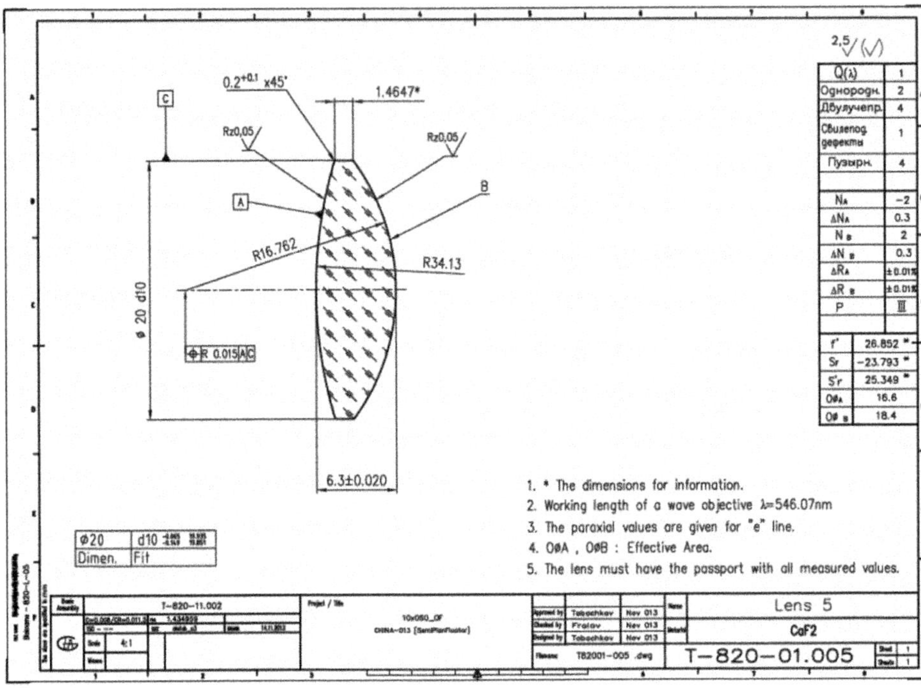

Figure 2.84 A drawing of the fifth lens.

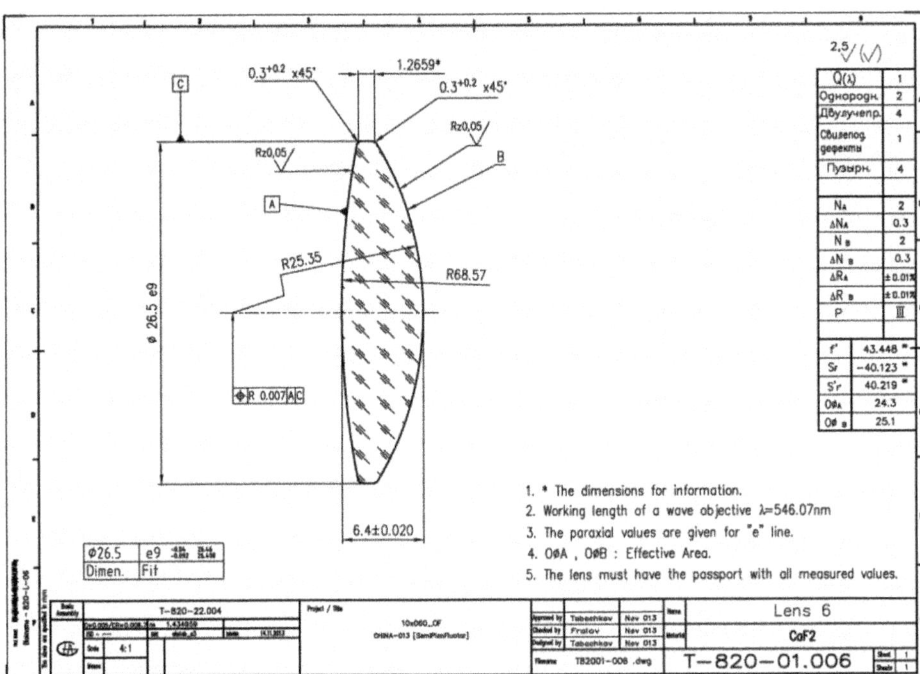

Figure 2.85 A drawing of the sixth lens.

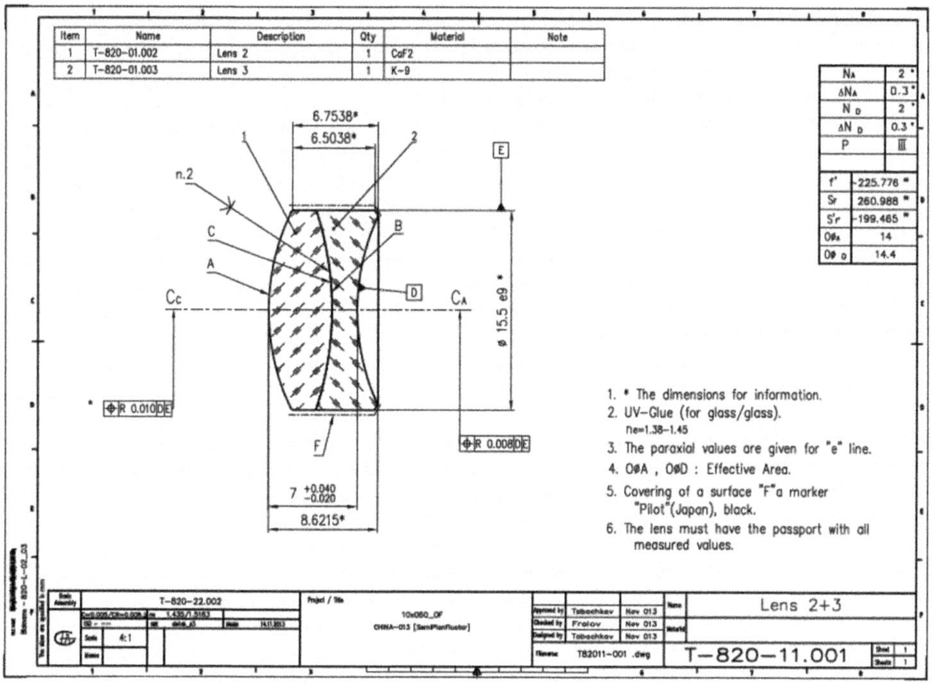

Figure 2.86 A drawing of the second and third lenses.

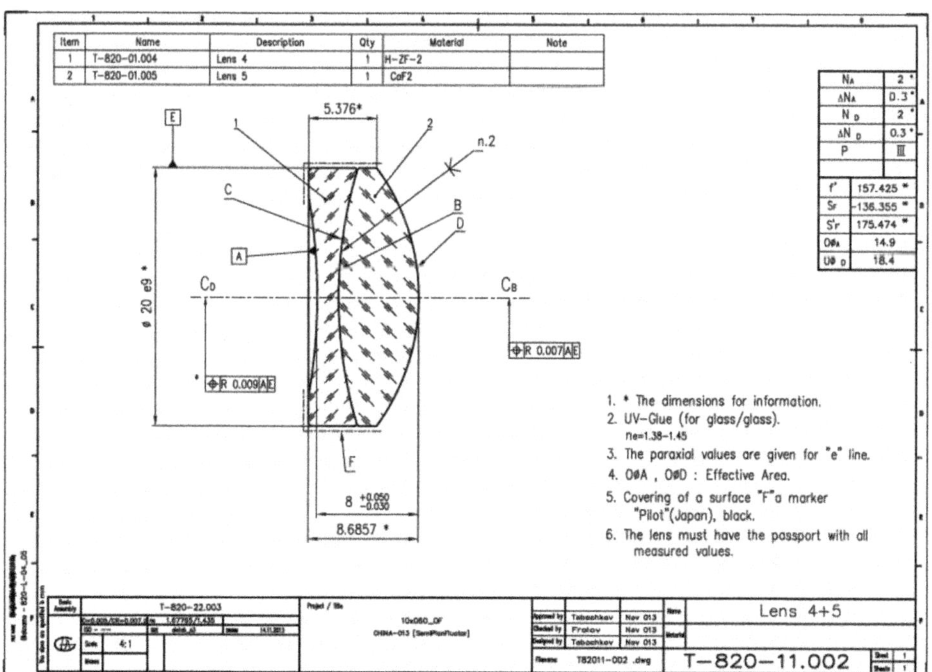

Figure 2.87 A drawing of the fourth and fifth lenses.

an objective has, the larger the lens diameters can be in the optical design; in these cases, larger diameters of the connecting thread (to the microscope nosepiece) must be used. Currently, mass-produced microscopes use a conventional thread with a diameter of 20.32 mm. However, in medium series microscope models, thread diameters of 25, 26, and 27 mm are increasingly found. In our case, the optical design allows us to achieve a numerical aperture of 0.60 in the 10× objective; however, in this case, the required thread diameter of the objective must be at least 25 mm. To be able to use conventional threads with a diameter of 20.32 mm, it is necessary to reduce the numerical aperture to 0.40.

Figure 2.88 shows a drawing of a principal optical layout of an objective with a linear magnification of 20× and a numerical aperture of 0.80. Figures 2.89–2.96 show drawings of the lenses of this objective.

From this drawing, some conclusions can be made about the level of complexity of the lens manufacturing, mechanical design, and assembly of a 20×/0.80 objective. Consider, for example, the values of the air gaps between the lenses as well as the allowable deviations from the nominal values (tolerances). But first, let us determine the influence of the first air gap, which is used as a *correction* gap when assembling the objective according to image quality or spherical aberration. By adjusting this air gap, it is possible to correct the spherical aberration that occurs when the lens is assembled due to

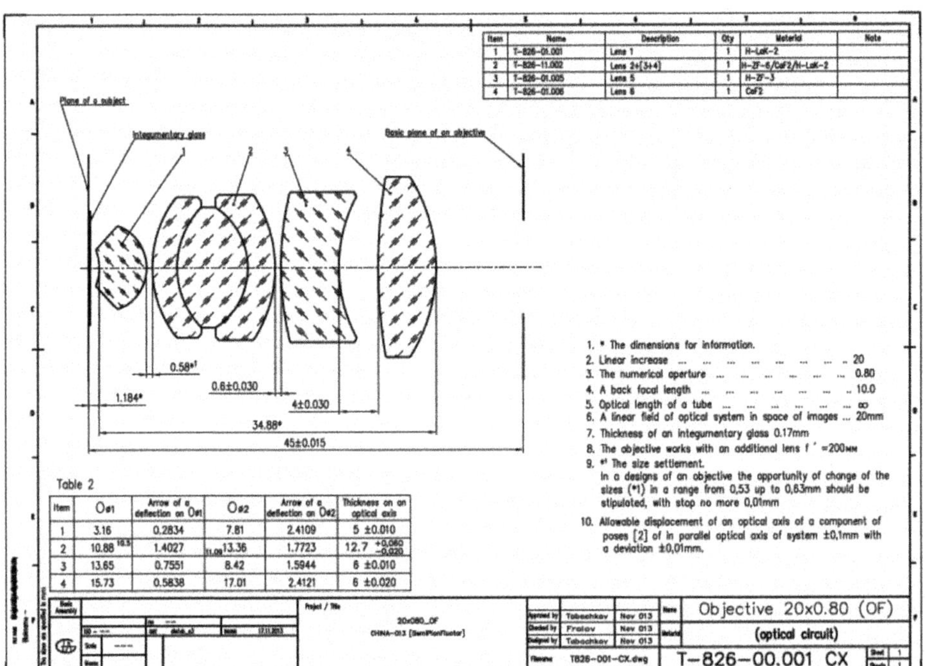

Figure 2.88 The principal optical layout of a 20×/0.80 infinity LCR MicroFluar objective.

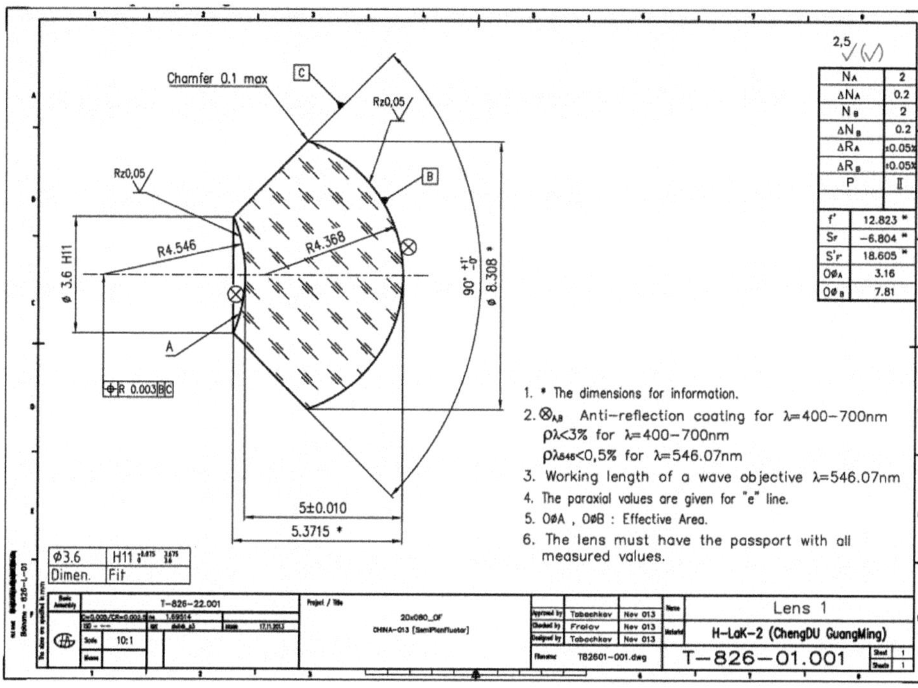

Figure 2.89 A drawing of the first lens.

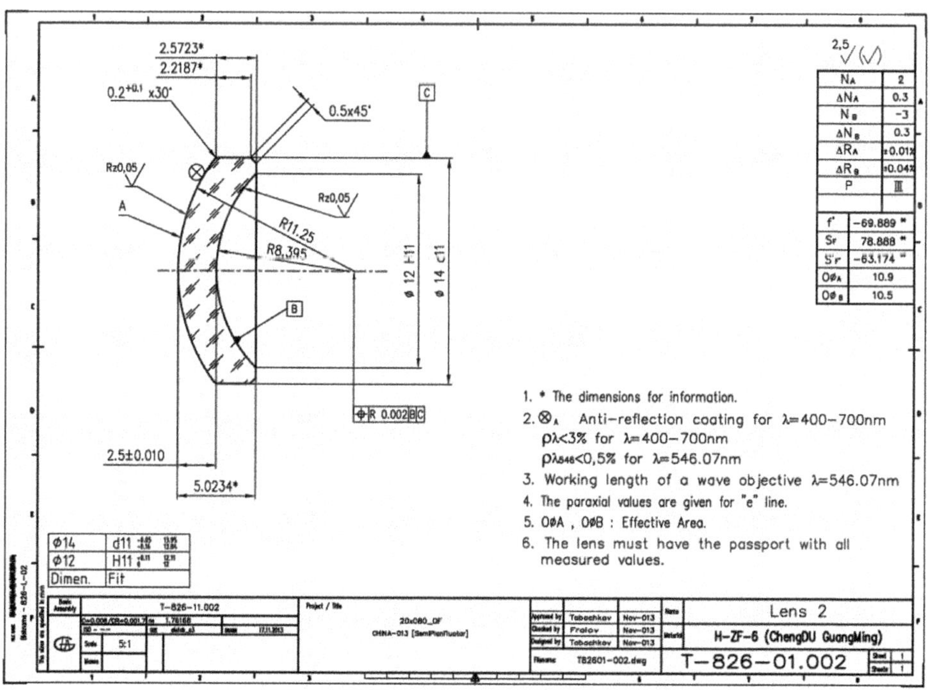

Figure 2.90 A drawing of the second lens.

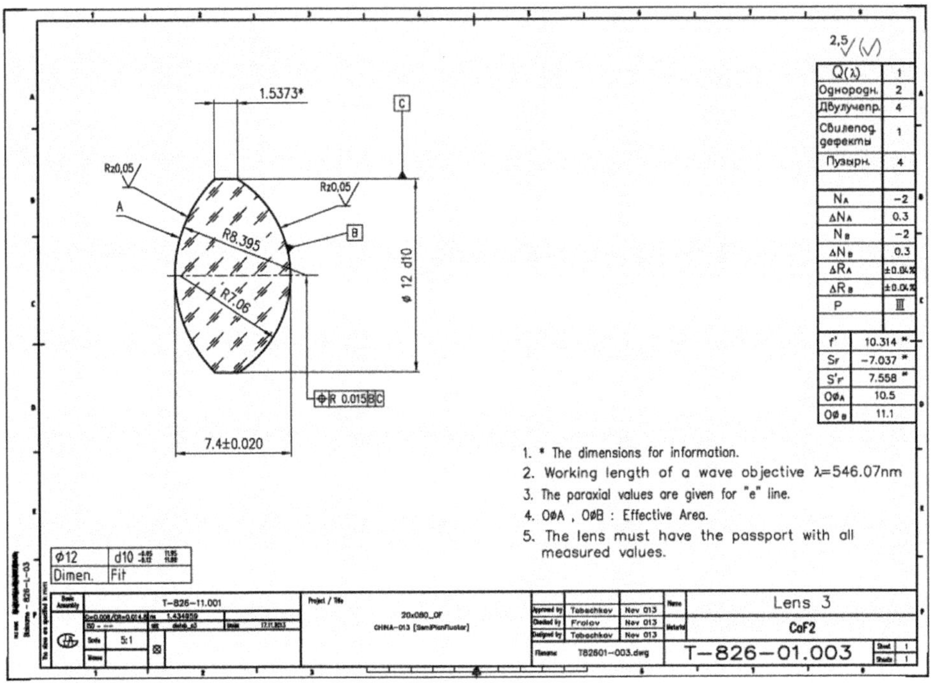

Figure 2.91 A drawing of the third lens.

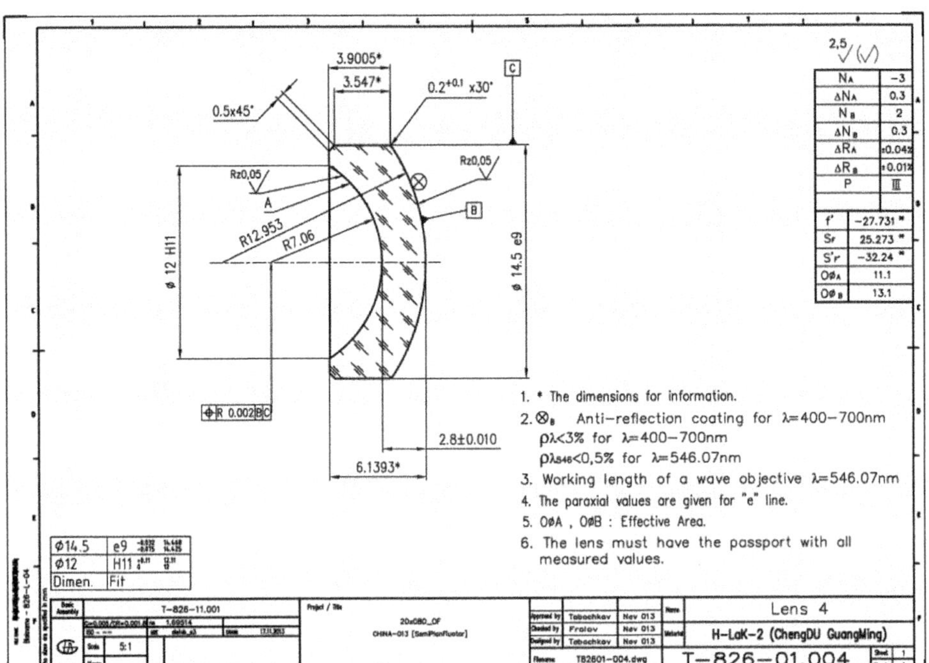

Figure 2.92 A drawing of the fourth lens.

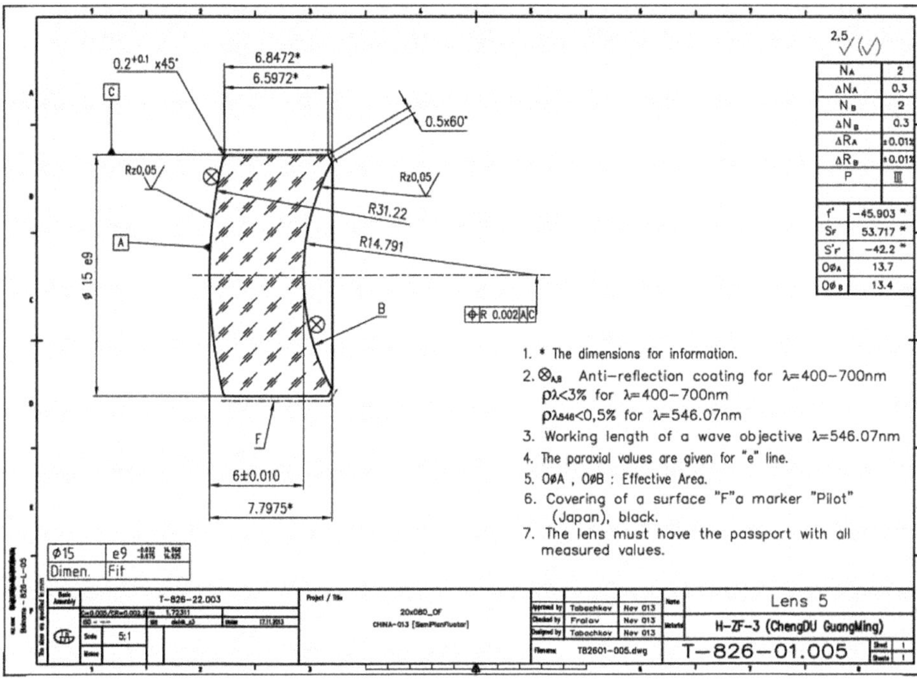

Figure 2.93 A drawing of the fifth lens.

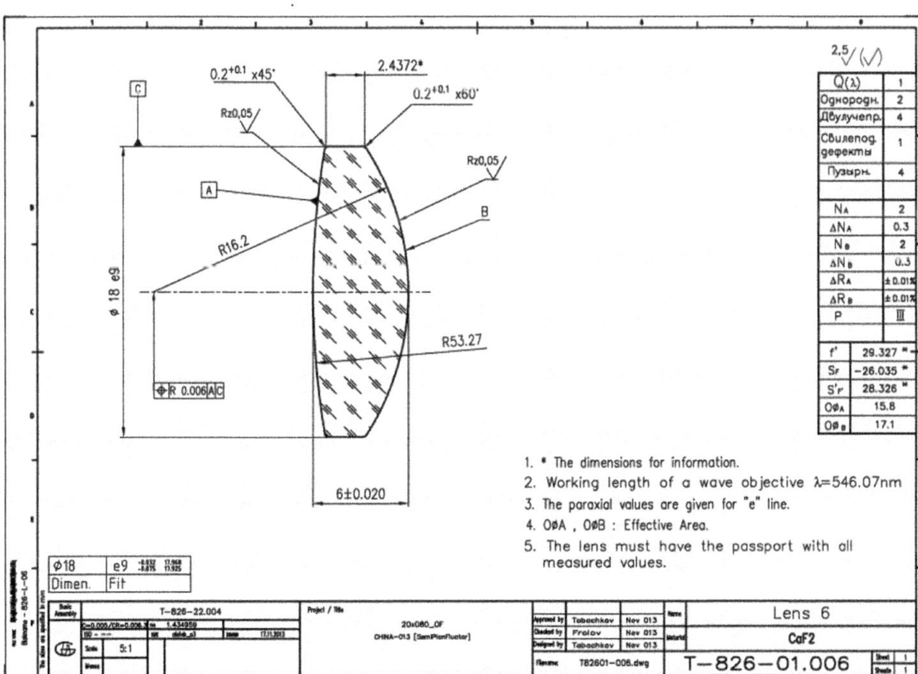

Figure 2.94 A drawing of the sixth lens.

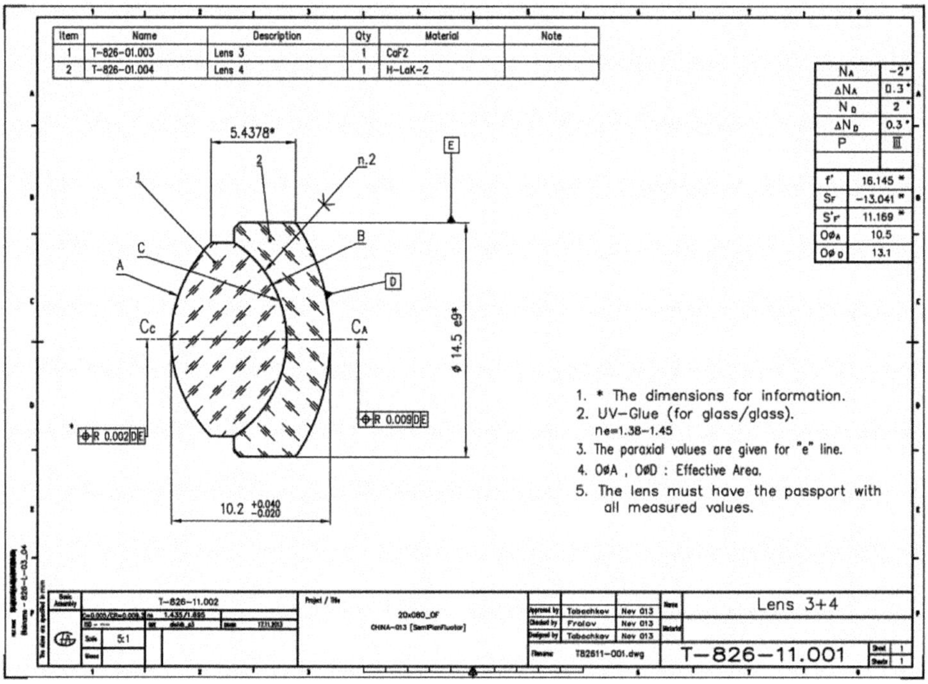

Figure 2.95 A drawing of the third and fourth lenses.

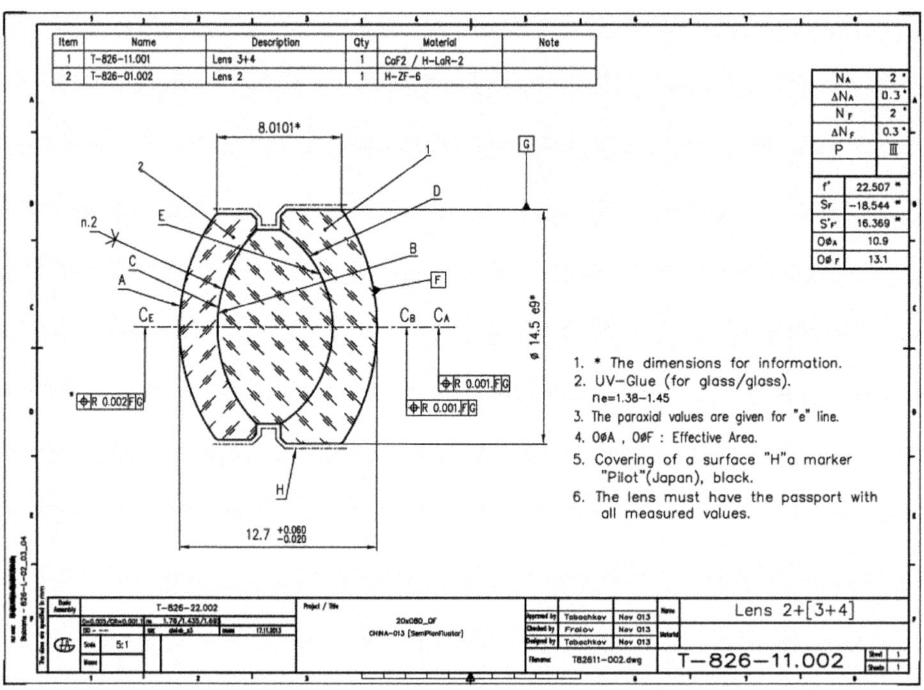

Figure 2.96 A drawing of the second, third, and fourth lenses.

the deviation of the values, for example, in lens thickness from the ones in the optical design. This is an objective with a rather high-input numerical aperture. Therefore, a slight change in the indicated air gap will cause the appearance of spherical aberration; it can be used as compensation. The option to implement this change in the air gap should be included in the mechanical design of the lens. There may be a variety of engineering solutions that allow you to change this air gap within small limits (no more than ± 0.05 mm for this objective).

It can be assumed that the mechanical design will use the so-called *bulk method* of objective assembly. In this case, each lens will be placed in its frame (most likely glued) and then centered using the so-called *autocollimation method*. In this case, the centering operation involves achieving a high degree of alignment of the optical axis of the lens (determined by the centers of its curvature) with the mechanical axis of the frame into which the lens is glued. To achieve the required values of the air gap between the lenses (lenses in the frames), the so-called *vertex segments* are measured, that is, the distances from the top of one of the lens radii to the edge of the frame into which this lens is glued. Centering, measuring the vertex segments, and achieving the required values of frame thickness can be performed by using the *one installation method* on a special machine. The result will be a so-called *component*, which is a lens mounted in a frame of a specific thickness; the optical axis of the lens, in this case, is combined with the mechanical axis of the frame.

Thus, it should be possible to exchange (during the objective assembly) the first air gap between the first and second lens with the air gap between the first and second component (lens in the frame). As a rule, such a change is ensured by the fact that the initially chosen distance between these components is less than the nominal distance (for example, $d = 0.5$ mm for this objective). Then, by using so-called *gasket rings* (which are made of copper and can have thicknesses between 0.002 and 0.01 mm), the value of this air gap is changed, achieving the minimum values of the spherical aberration of the objective. There are other ways to change the air gap within small limits, for example, by using an eccentric frame.

Figure 2.97 shows a drawing of the principal optical layout of the objective with a linear magnification of 40× and a numerical aperture of 0.95. Figures 2.98–2.107 show drawings of the lenses of this objective.

An attentive person will immediately see that for this lens, the nominal value of the air gap between the first and second components is 0.11 mm. In item 9 of the explanatory inscription in Fig. 2.97, it is written that it is possible to change the value of this air gap in the range of 0.06 to 0.16 mm. Indeed, the numerical aperture of this objective is close to the maximum possible for systems operating in the air, that is, not using immersion. To assemble this objective according to spherical aberration, it is necessary to use spacer rings

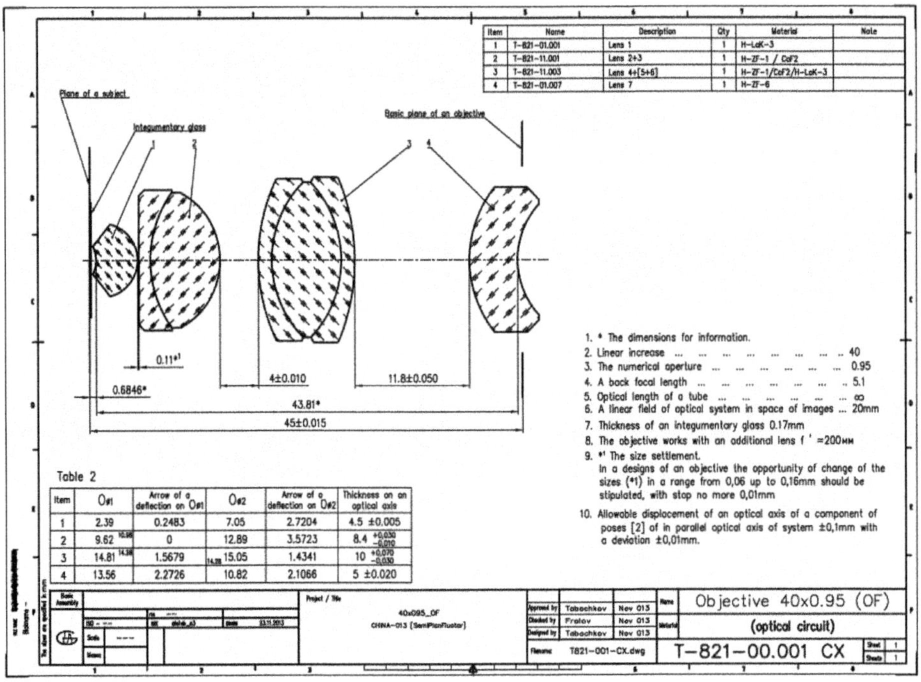

Figure 2.97 The principal optical layout of a 40×/0.95 infinity LCR MicroFluar objective.

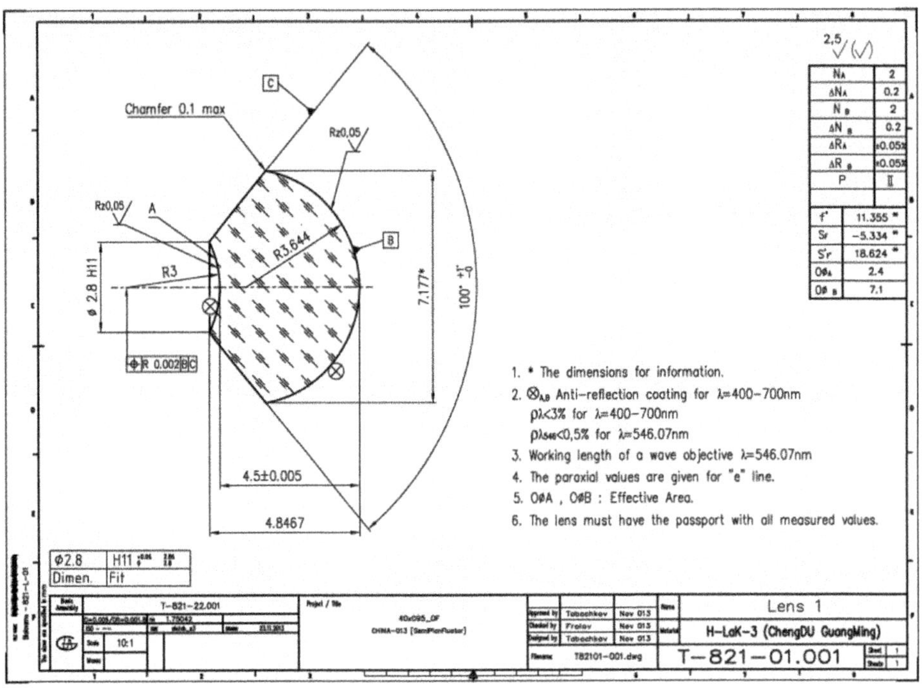

Figure 2.98 A drawing of the first lens.

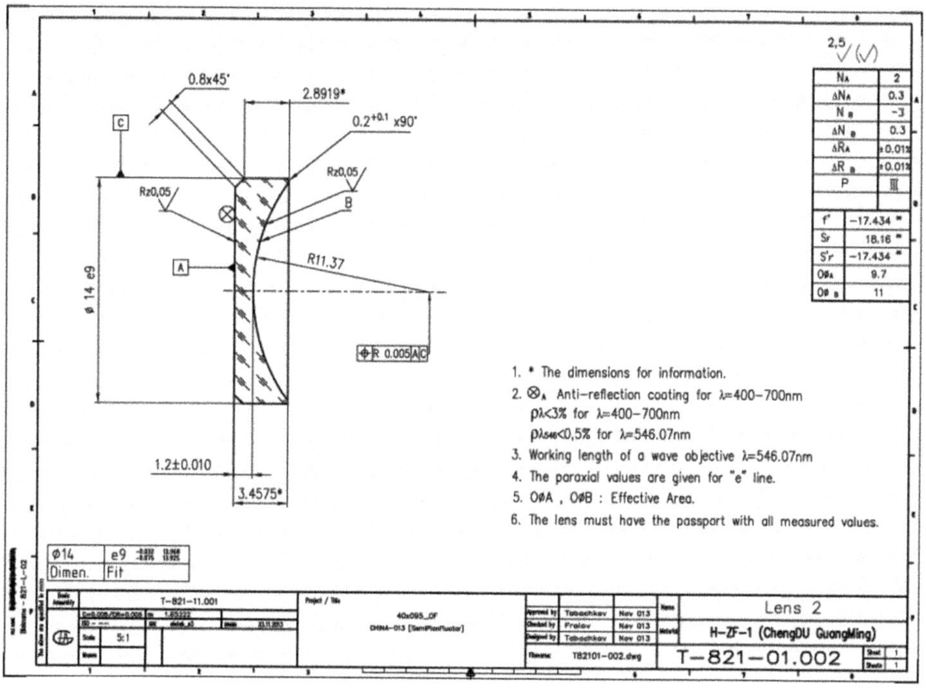

Figure 2.99 A drawing of the second lens.

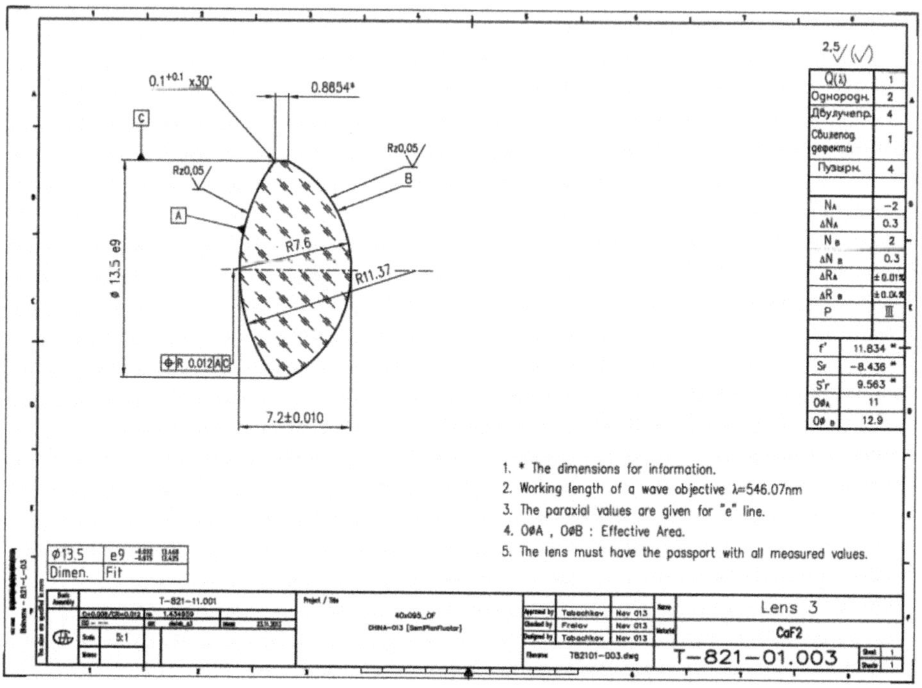

Figure 2.100 A drawing of the third lens.

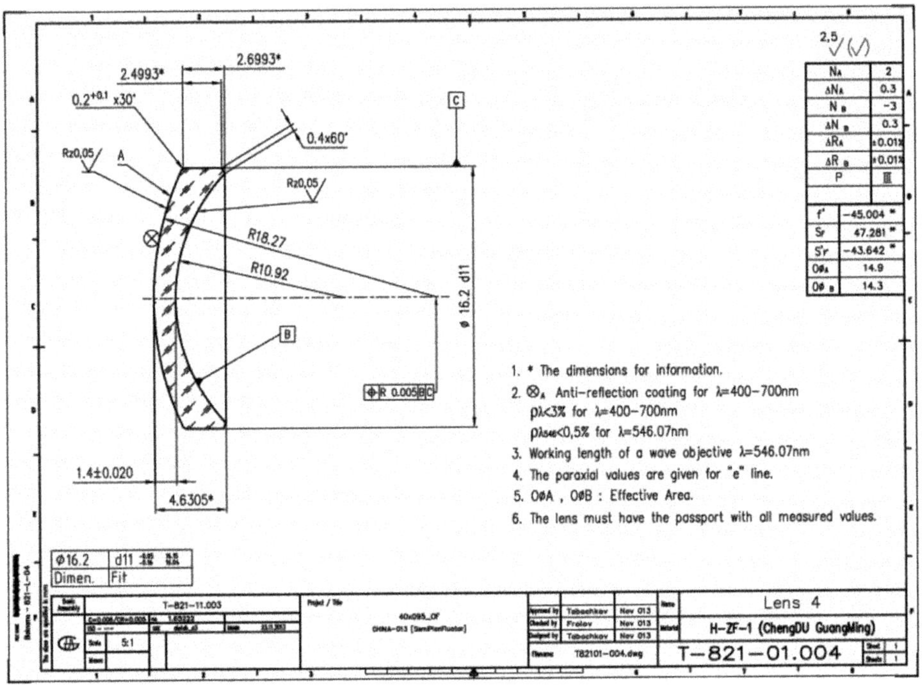

Figure 2.101 A drawing of the fourth lens.

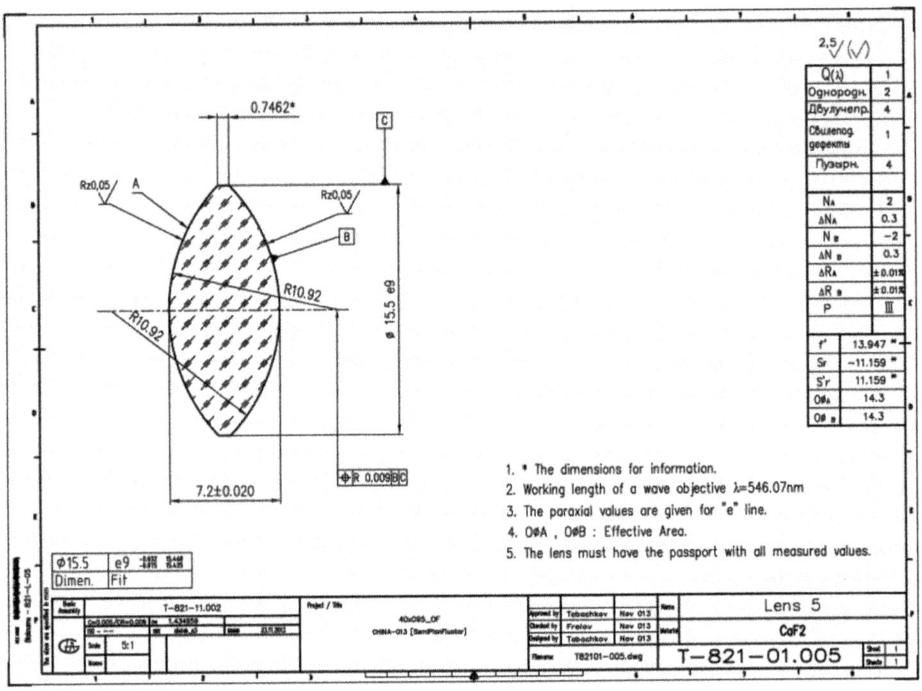

Figure 2.102 A drawing of the fifth lens.

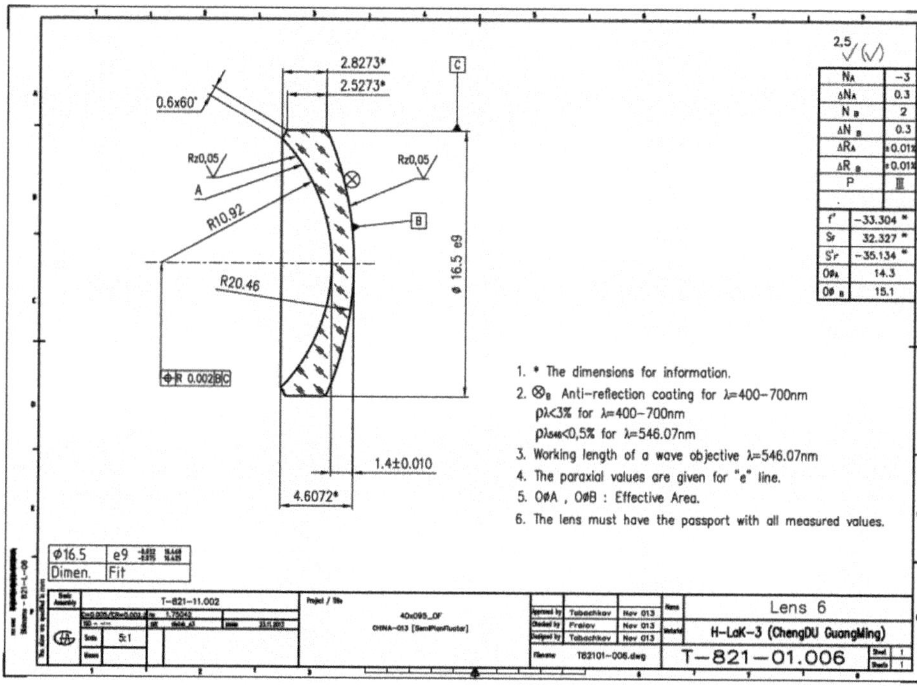

Figure 2.103 A drawing of the sixth lens.

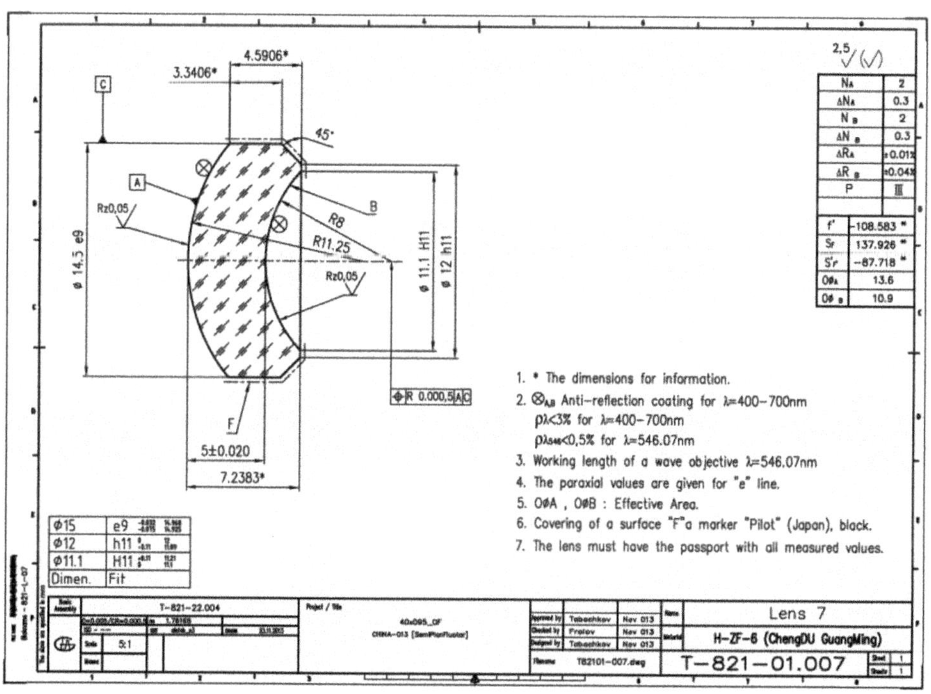

Figure 2.104 A drawing of the seventh lens.

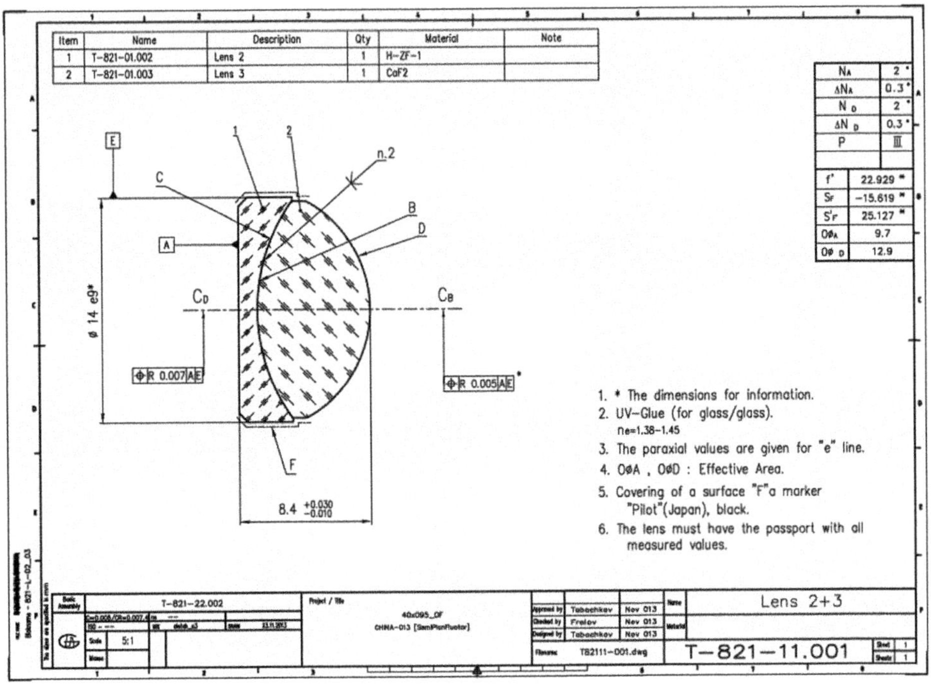

Figure 2.105 A drawing of the second and third lenses.

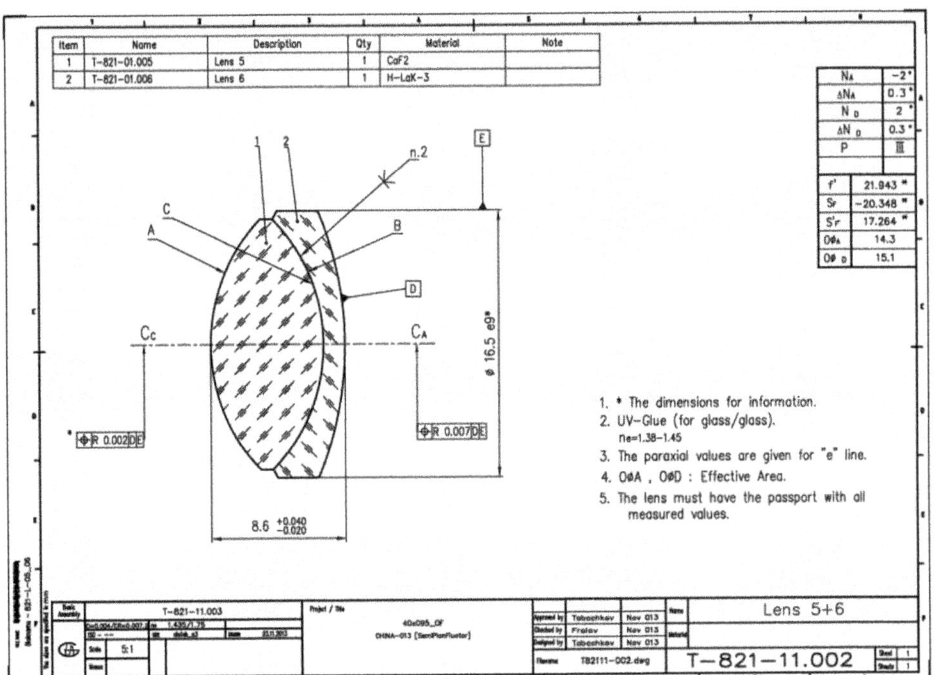

Figure 2.106 A drawing of the fifth and sixth lenses.

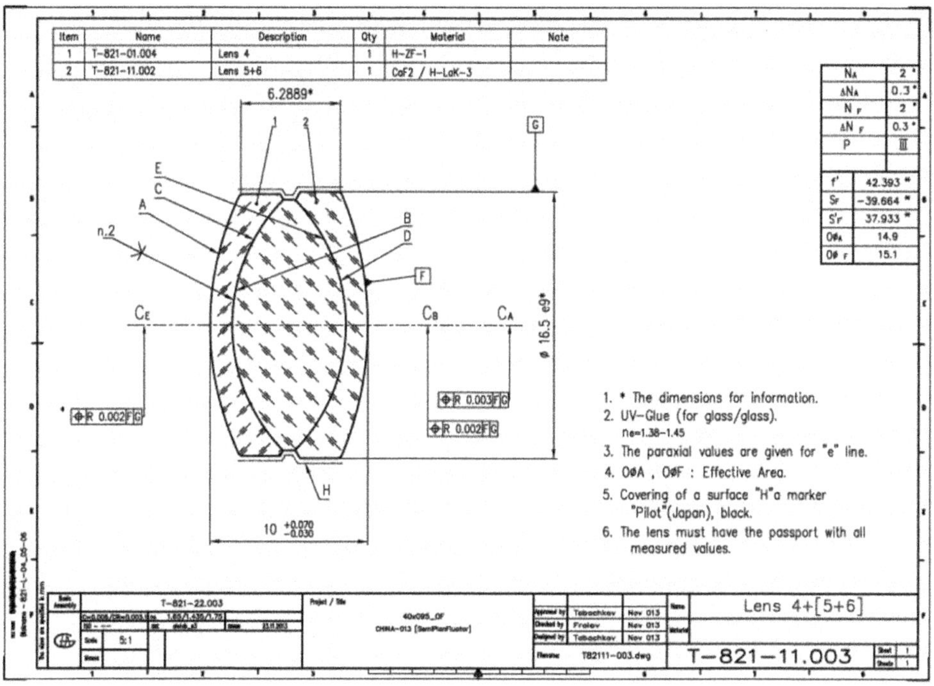

Figure 2.107 A drawing of the fourth, fifth, and sixth lenses.

between the first and second components that have a nominal thickness of no more than 0.003–0.005 mm.

But there is one more parameter that must be provided when assembling an objective. This is related to centering the objective as a whole. Indeed, if we talk about individual objective components, each of them is centered before the objective is assembled. These components are poured into the common body of the objective, the inner diameter of which, as a whole, corresponds to the values of the external diameters of the components' frames. However, if the inner diameter of the case and the outer diameters of the component frames were strictly the same, it would be impossible to install the components in the body. Thus, the so-called *guaranteed clearance* must be maintained between the component frames and the body. The so-called coma aberration appears due to the presence of such clearances. This aberration is not related to spherical aberration; however, these two aberrations have a poor correlation with each other during lens assembly. Unfortunately, spherical aberration is rather poorly visible when there is a coma in the objective system; only with the elimination of coma does it become possible to estimate the magnitude of spherical aberration.

How can coma aberration arising from the installation of components in a common body be eliminated? For this (as the optical design prescribes), the mechanical design of the objective should allow for one of the components to be displaced perpendicular to the optical axis of the objective by a certain amount. In the 40×/0.95 objective under consideration, such a component is the second (along the ray from left to right) component. The mechanical design can make component displacement easy; the gap between the common objective's body and the frame of the selected component is slightly larger than for the other components. Four (as a rule) holes are made in the body; the worker can shift the component perpendicular to the optical axis using a special device.

Figure 2.108 shows a drawing of a principal optical layout of a water immersion objective with a linear magnification of 100× and a numerical aperture of 1.20. Figures 2.109–2.115 show drawings of the lenses of this objective.

There are some nuances to be aware of when choosing a component to displace perpendicular to the optical axis of the lens. For instance, it makes little sense to move components containing single lenses or those glued from three lenses off the axis. For such a displacement, it is better to choose a component that contains double-glued lenses. As a rule, the optical powers of these lenses differ not only in sign (one of them is positive, the other is negative) but also in the absolute value of the optical power, which creates a certain *shoulder* and makes the offset operation more efficient. That is, in this

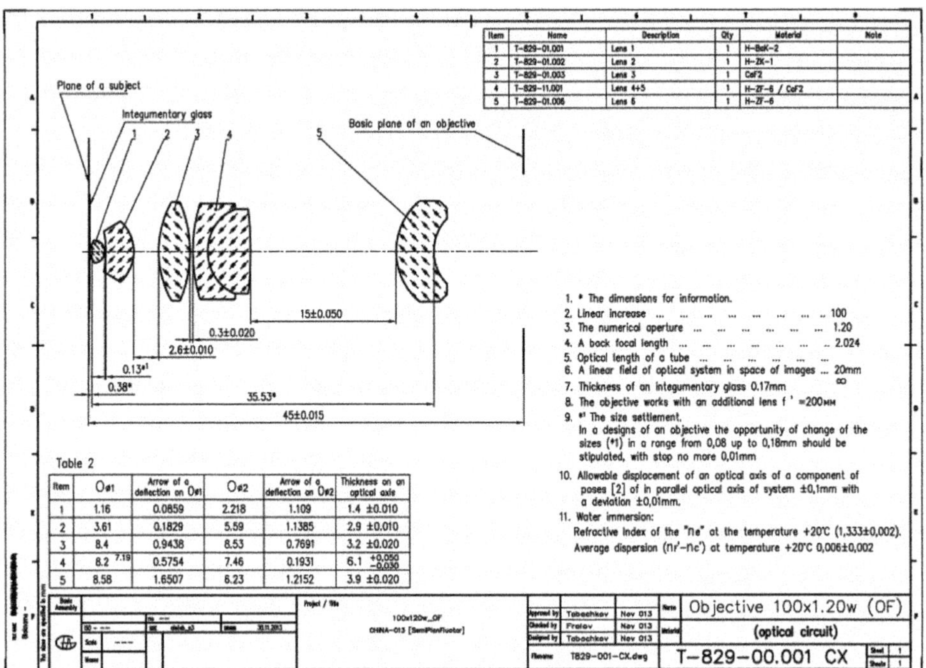

Figure 2.108 The principal optical layout of a 100×/1.20 water immersion infinity LCR MicroFluar objective.

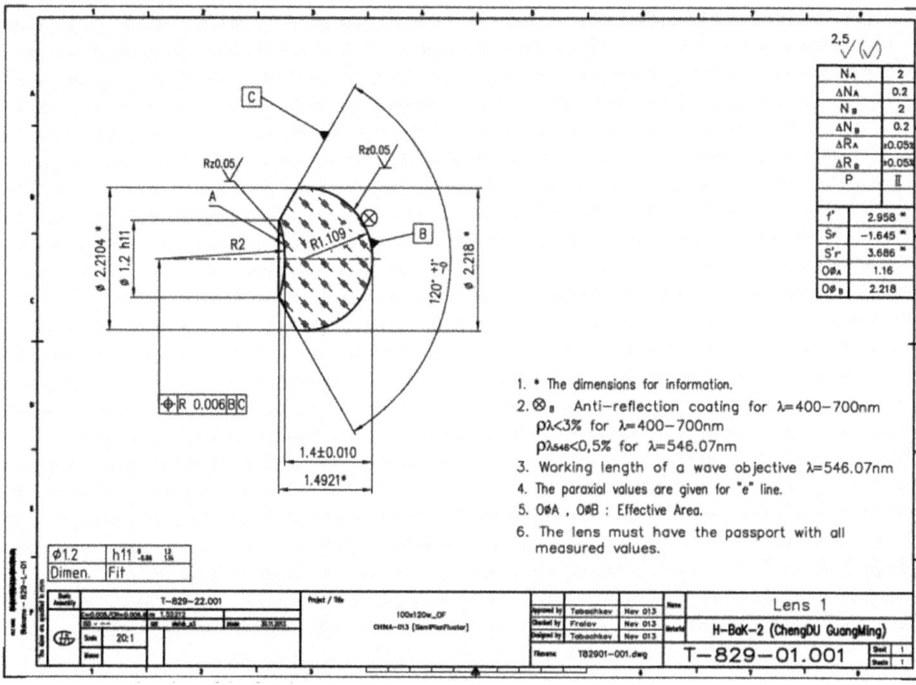

Figure 2.109 A drawing of the first lens.

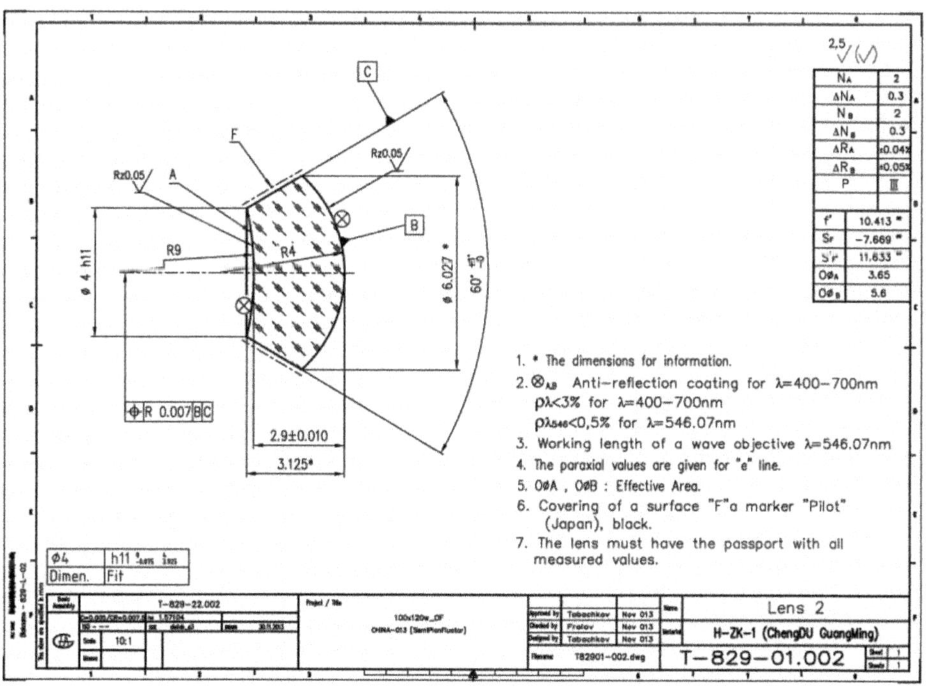

Figure 2.110 A drawing of the second lens.

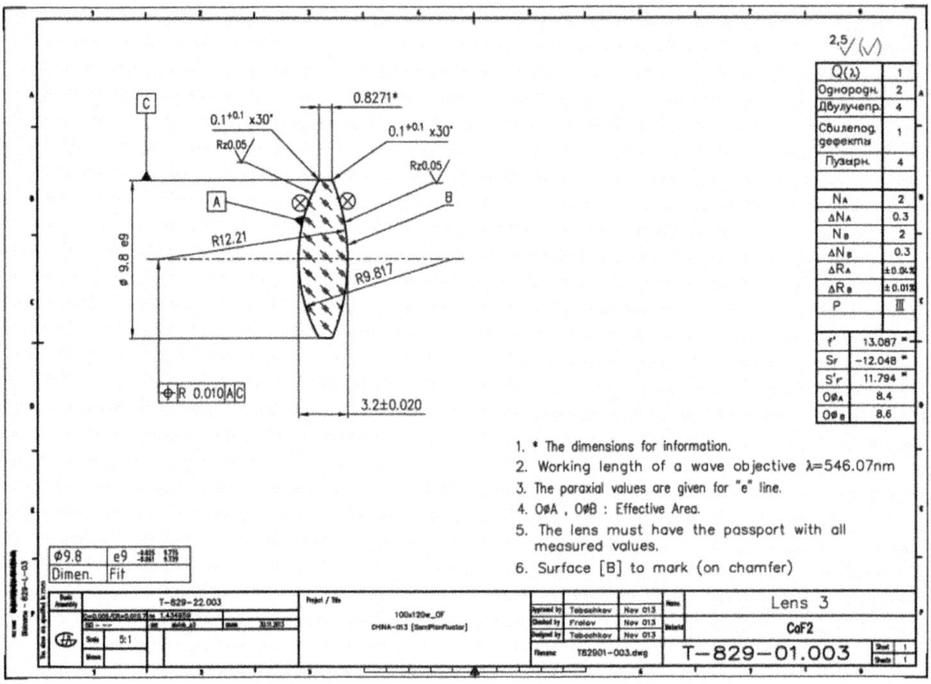

Figure 2.111 A drawing of the third lens.

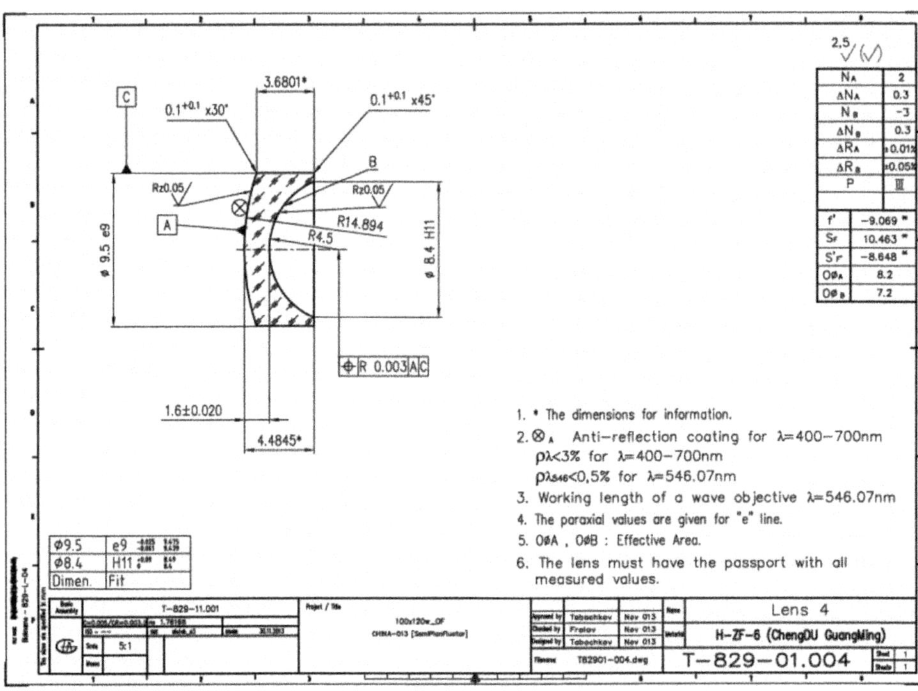

Figure 2.112 A drawing of the fourth lens.

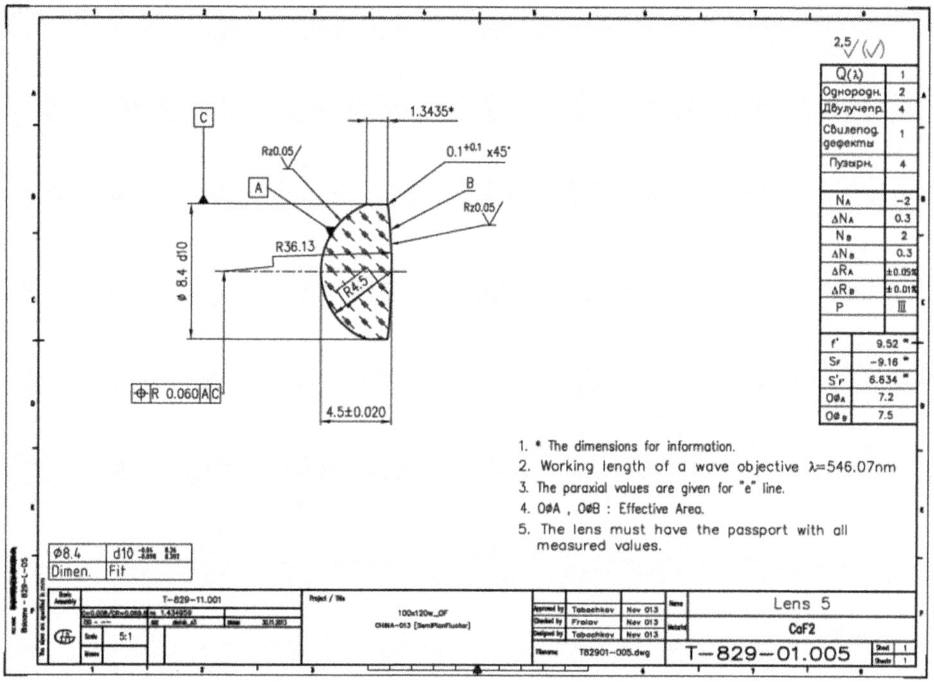

Figure 2.113 A drawing of the fifth lens.

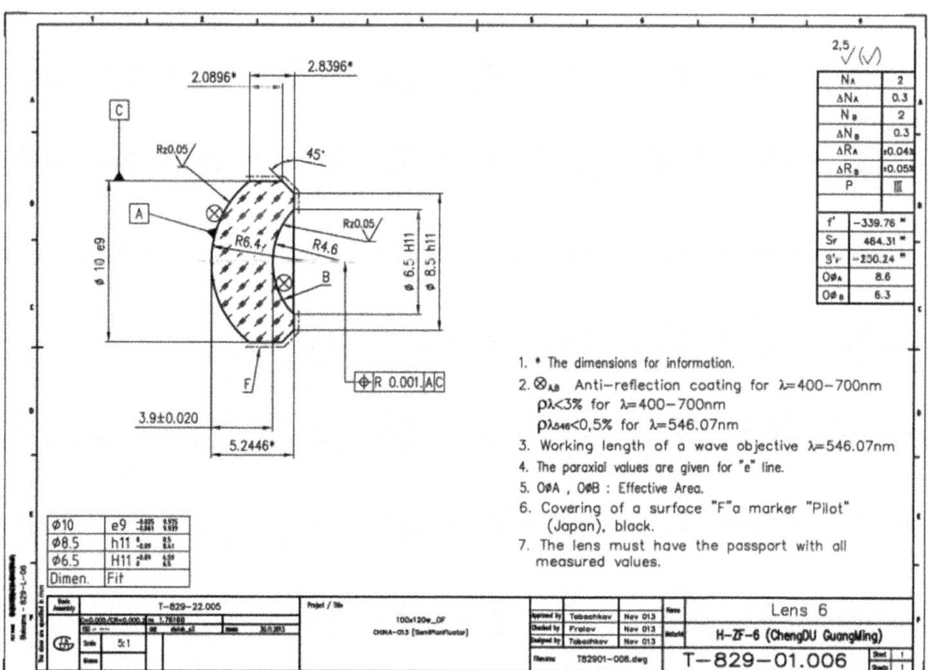

Figure 2.114 A drawing of the sixth lens.

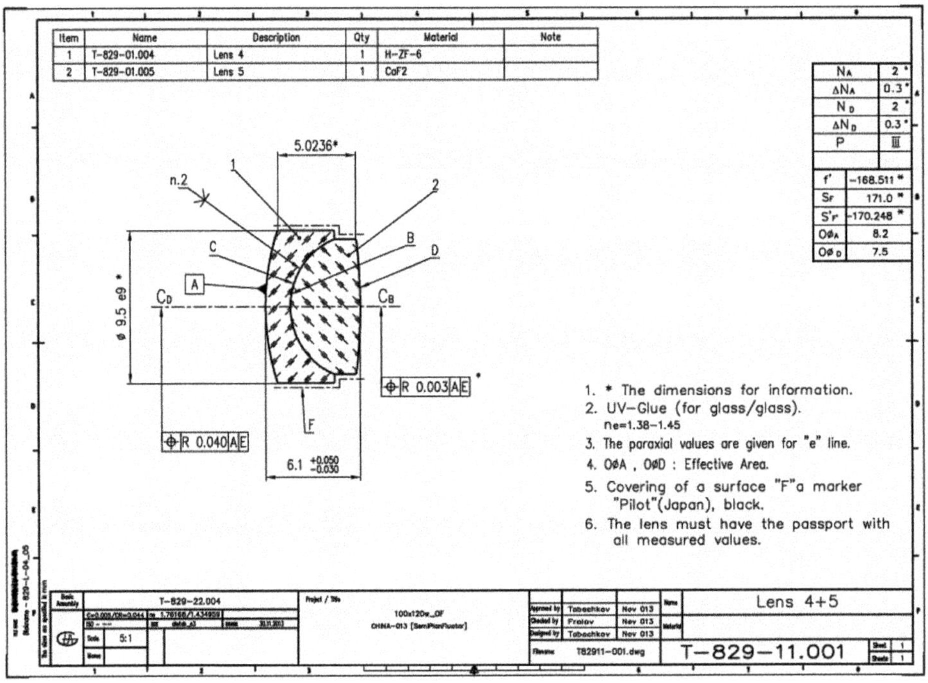

Figure 2.115 A drawing of the fourth and fifth lenses.

case, a slight displacement of the selected component from the optical axis of the objective will be required in order to achieve coma correction. Thus, for displacement in this lens, you can select the fourth component.

Another notable feature of the optical design (and, accordingly, the future mechanical design) is the presence of a significant air gap between the second and third components. In the middle of this gap is the so-called *back focal plane* of the objective. This provides a very big advantage for the objective because the following additional accessories can be located here, which can significantly enhance the objective's consumer qualities:

- We can place here the material iris aperture diaphragm. The principal assembly drawing of such a diaphragm is shown in Volume 1 of this book (Chapter 1, Fig. 1.45), although the designs may be different. This allows the objective to be used when implementing the dark field method in transmitted light. Using such an aperture in an eccentric frame makes it possible to implement the oblique vision method or create a cartoon effect in the pupil of the objective.
- Here we can arrange a plate with the so-called *phase rings* deposited on it (we can make these plates removable for implementing various phase contrast methods).

- An interference prism can be placed here (as is done in some objectives of the BIOLAR microscope shown in Chapter 1, Fig. 1.158). These prisms can also be interchangeable.

Let us return to what was discussed in Volume 1 of this book (Chapter 1, Section 1.2.3) about some features of water immersion objectives; the optical and mechanical designs of the front lens of the objective should ensure retention of a drop of water; thus, if the first surface of the frontal lens has an irregular shape, water can slide, preventing accurate focusing. Water itself has interesting refractive and dispersion properties; it has a unique light transmission, including in the near UV range. In the practice of microscopy, images obtained using a water immersion objective look clearer as compared with, for example, those obtained using an oil immersion objective.

Figure 2.116 shows a drawing of a principal optical layout of an oil immersion objective with a linear magnification of 100× and a numerical aperture of 1.35. Figures 2.117–2.125 show drawings of the lenses of this objective.

The design of these objectives takes into account the experience gained from using the objectives shown in Fig. 2.71. Thus, compared with those objectives, the aberration curvature correction in the image after the 10× (if it is made with a numerical aperture of 0.40, such as the version shown in the

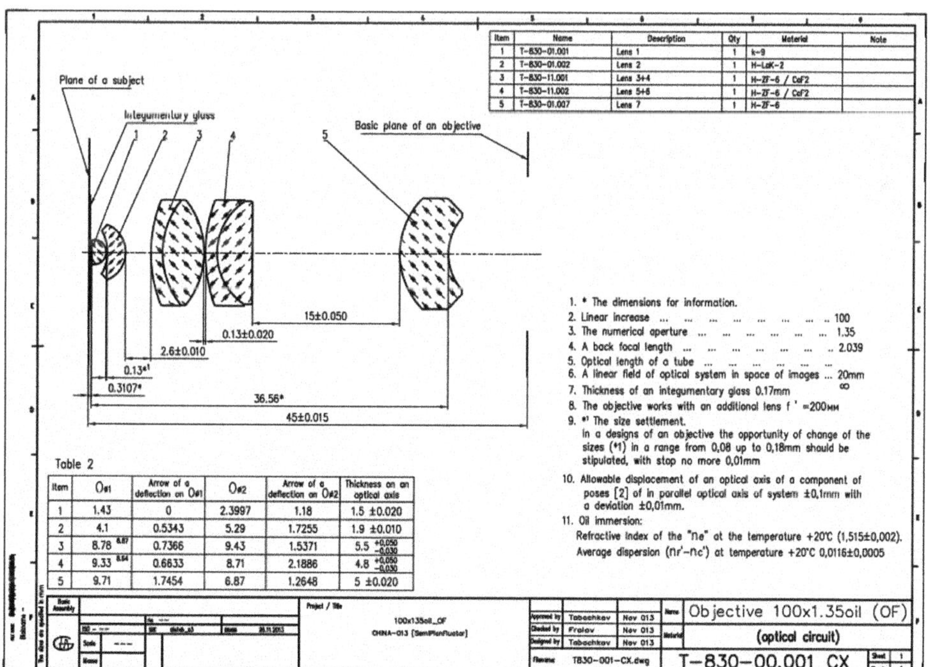

Figure 2.116 The principal optical layout of a 100×/1.35 oil immersion infinity LCR MicroFluar objective.

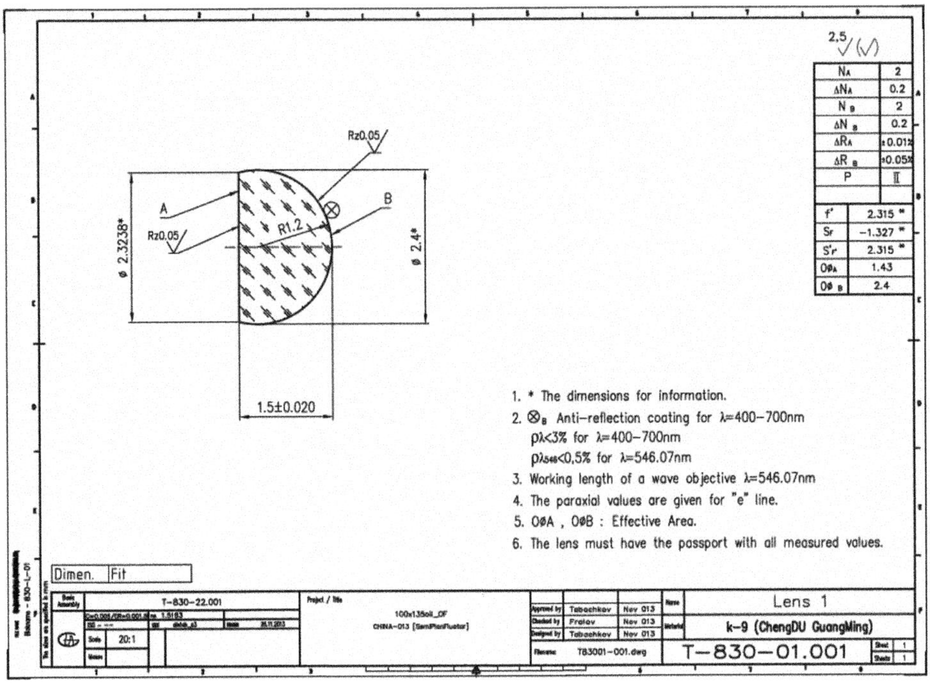

Figure 2.117 A drawing of the first lens.

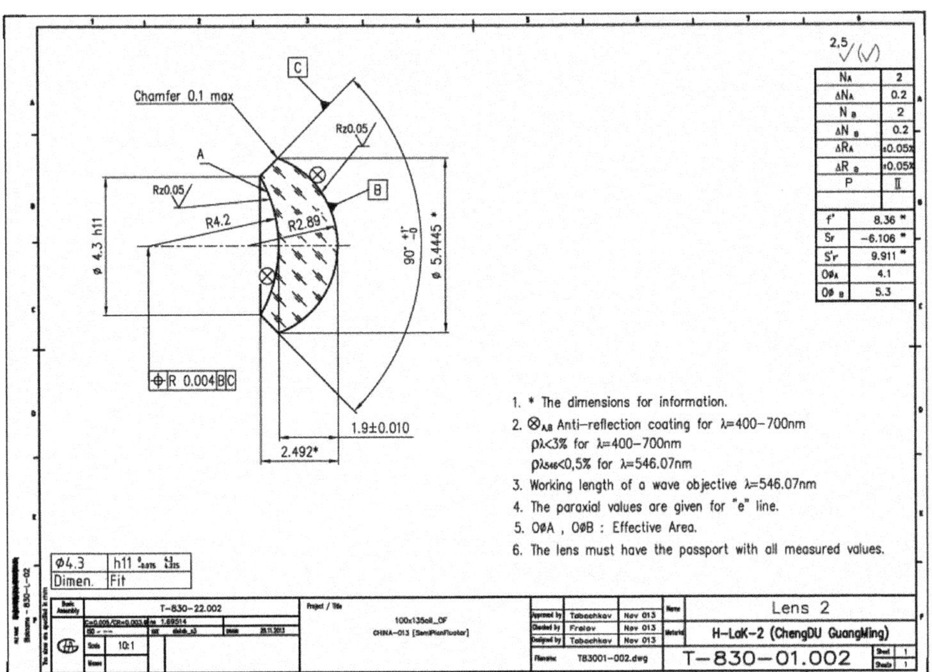

Figure 2.118 A drawing of the second lens.

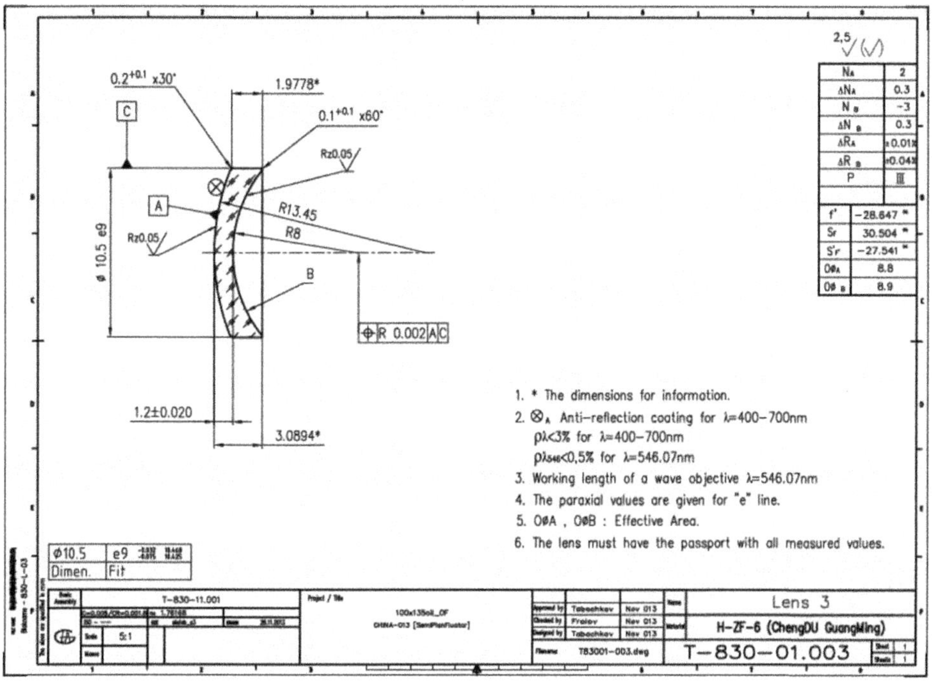

Figure 2.119 A drawing of the third lens.

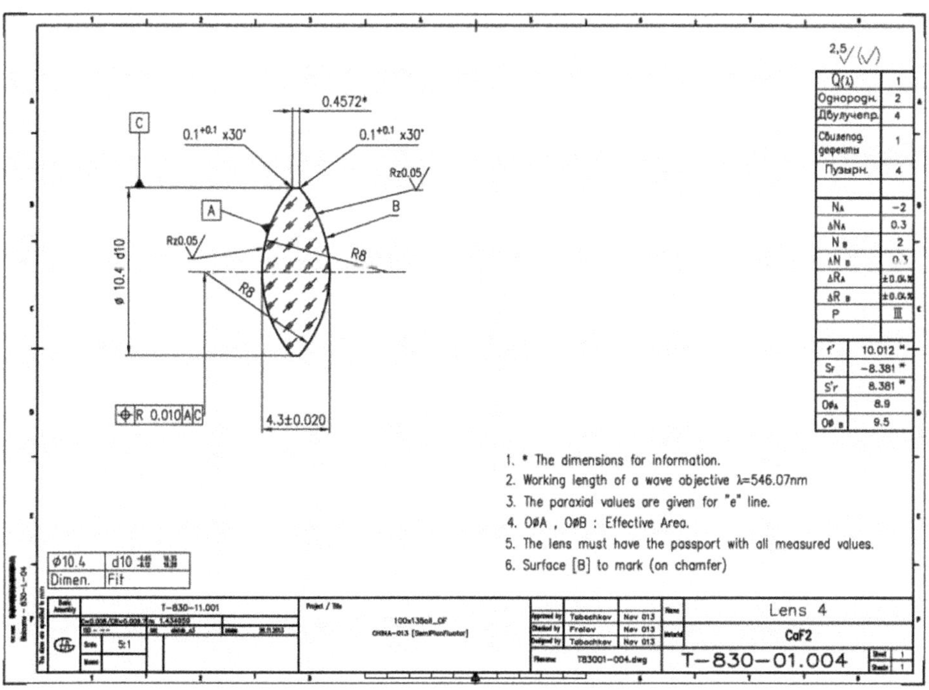

Figure 2.120 A drawing of the fourth lens.

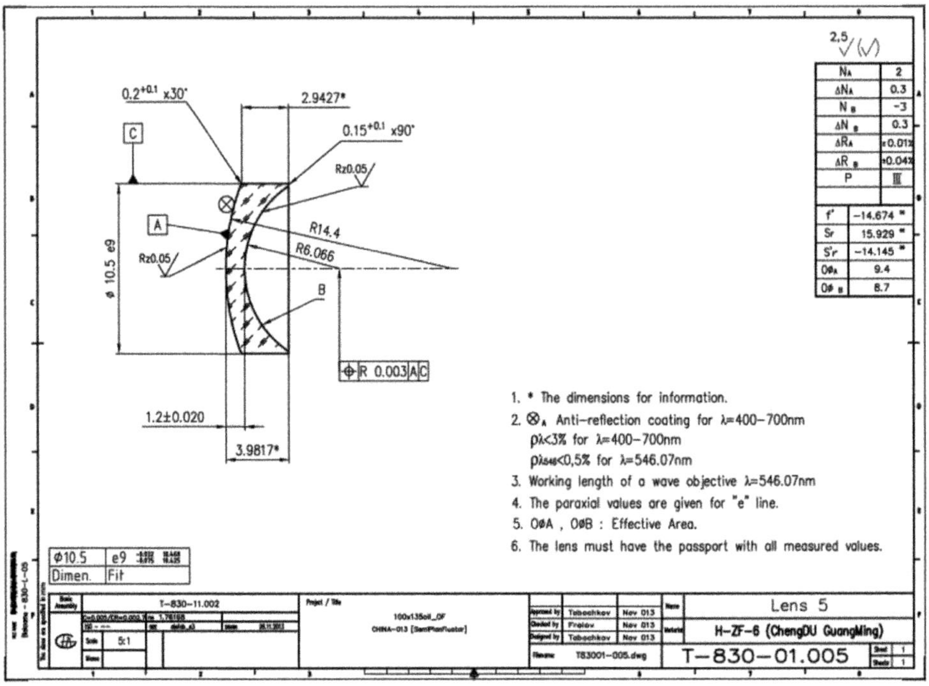

Figure 2.121 A drawing of the fifth lens.

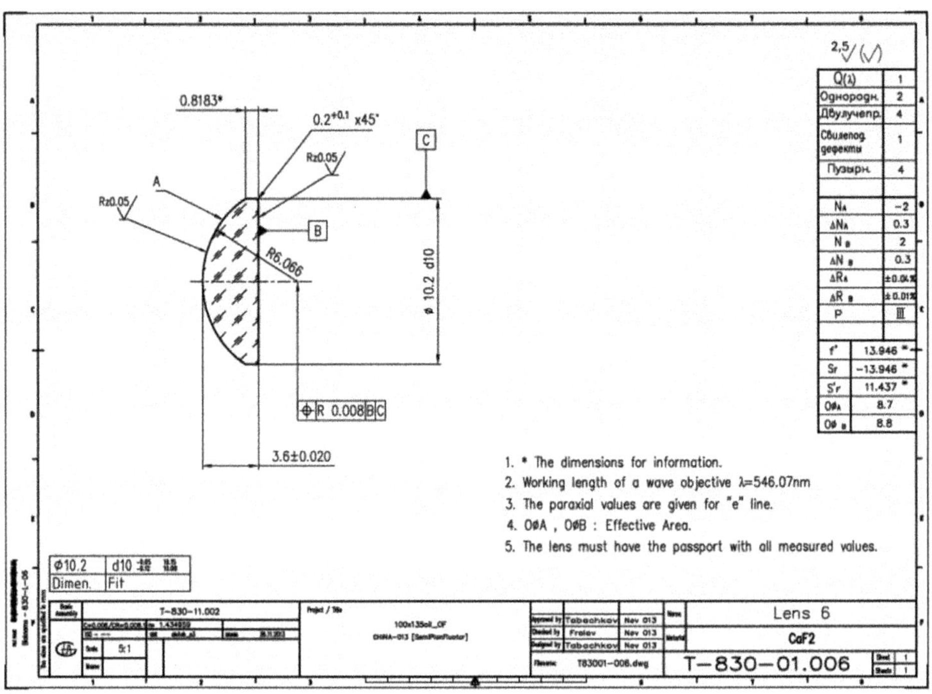

Figure 2.122 A drawing of the sixth lens.

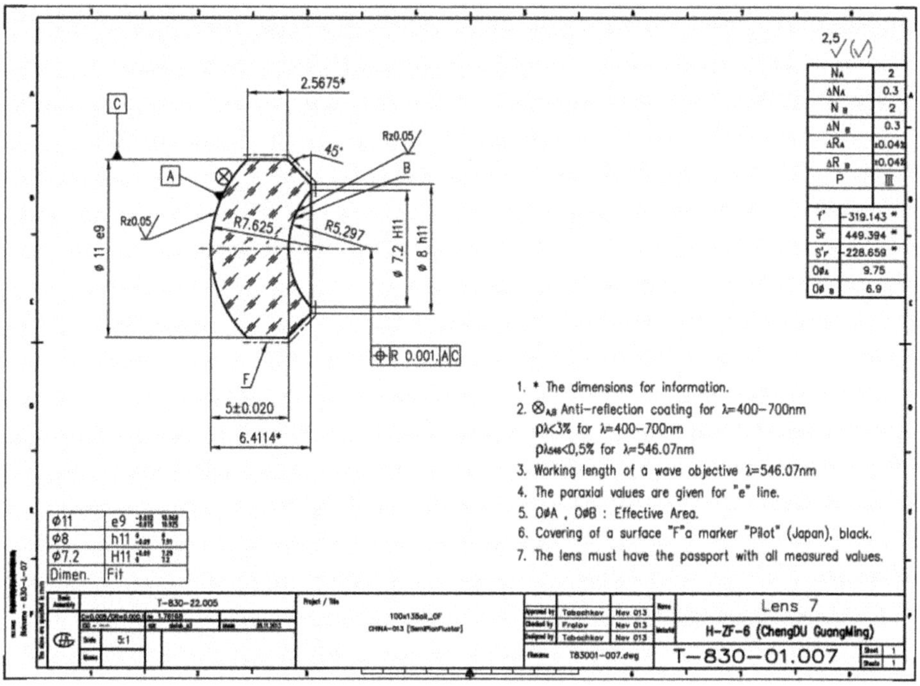

Figure 2.123 A drawing of the seventh lens.

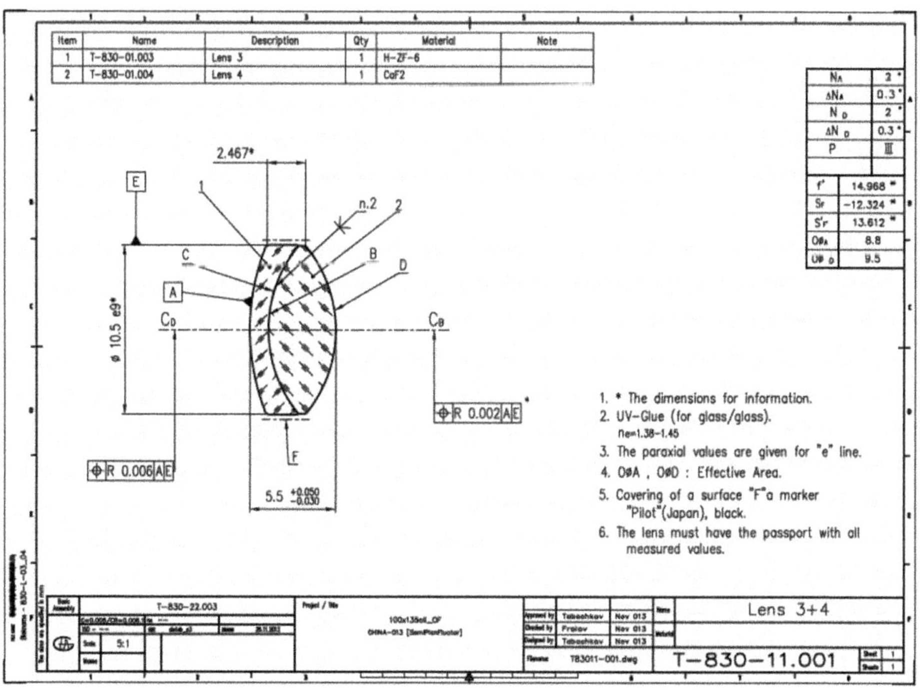

Figure 2.124 A drawing of the third and fourth lenses.

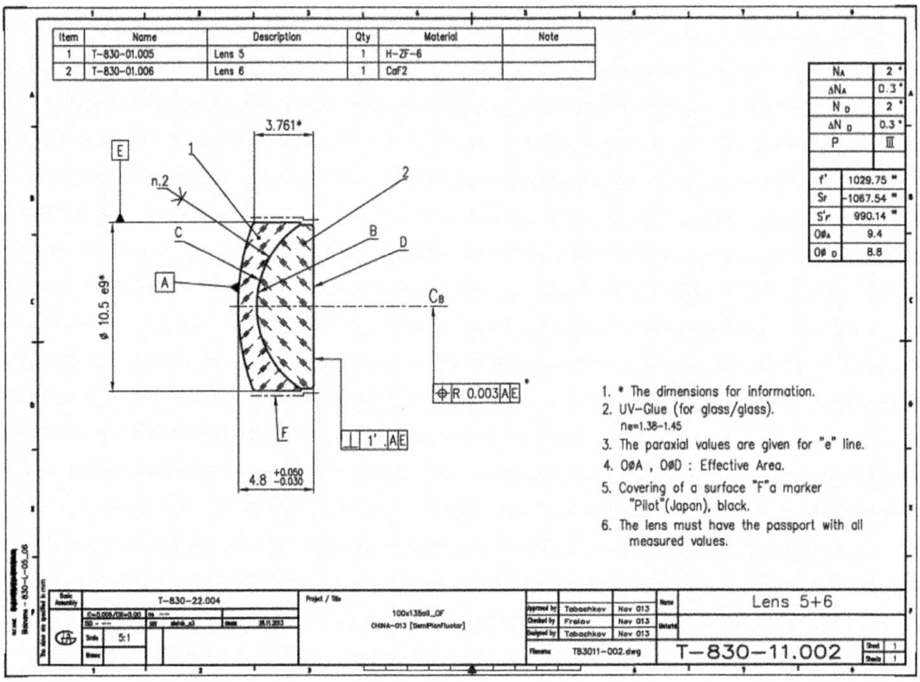

Figure 2.125 A drawing of the fifth and sixth lenses.

photograph) and 40× objectives is significantly improved; for 20×, 40×, and 100× oil immersion objectives, the numerical aperture is significantly increased. In addition, the optical systems of these objectives allow the implementation of microscopy techniques such as phase and interference contrast; an iris aperture diaphragm can also be added to such objectives for using dark field methods as well as for photometric studies.

Let us consider the optical designs of these objectives in conjunction with a visual head that has its own LC compensation value of −0.7% and an increased calculated value of the output numerical aperture. Figure 2.126 shows the design parameters and graphs of aberration correction of this head, and Figs. 2.127–2.131 show the design parameters and graphs of aberration correction of this head combined with the indicated MicroFluar objectives.

If we follow the rule of *useful magnification* on a microscope (discussed in Chapter 1, Table 1.15), only two types of eyepieces should be combined when using MicroFluar objectives with increased numerical apertures: 12.5× (high-magnification objectives) and 20× (medium-magnification objectives). For these objectives, an eyepiece with a linear magnification of 12.5× turned out to be universal. Therefore, the optical design of such an eyepiece with a linear field of 22 mm (in the plane of the field diaphragm, i.e., an intermediate image on a microscope) was made; this eyepiece is not difficult to make and can have a very simple optical design (an example is shown in Fig. 1.231 in Chapter 1).

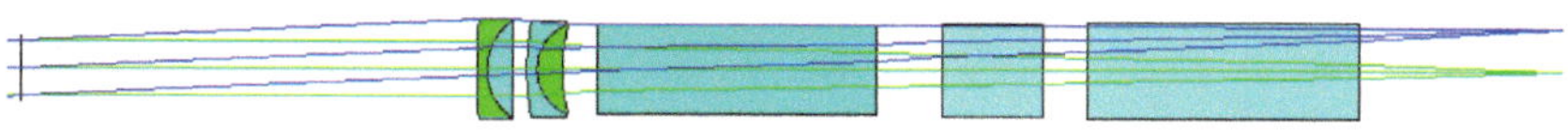

Lens: head comp F200 LC=-0.7% Efl 199.19

Image num aper 0.030000 Image height 9.000000 Primary wavln 0.546070

SRF	RADIUS	THICKNESS	APERTURE RADIUS	GLASS	SPECIAL
OBJ	0.000000	1.0000e+20	4.5184e+18	AIR	
AST	0.000000	0.001000	5.975539 AS	BK7 C	
2	0.000000	100.000000	7.000000	AIR	
3	113.230000	2.200000	10.500000	LZ_TF4	
4	12.498000	5.600000	10.350000	LZ_CTK19	
5	-587.000000	3.000000	10.350000	AIR	
6	49.190000	2.200000	10.240000	LZ_CTK19 P	
7	10.069000	5.600000	9.430000	LZ_TF4 P	
8	36.810000	7.800000	9.400000	AIR	
9	0.000000	60.400000	9.410000	LZ_BK10	
10	0.000000	14.300000	9.600000	AIR	
11	0.000000	22.000000	9.700000	LZ_K8	
12	0.000000	10.000000	9.800000	AIR	
13	0.000000	59.500000	9.860000	LZ_K8 P	
14	0.000000	44.517000	10.130000	AIR	
15	0.000000	0.000345 S	9.000001 S	AIR	
IMS	0.000000	0.000000	9.000000 S		

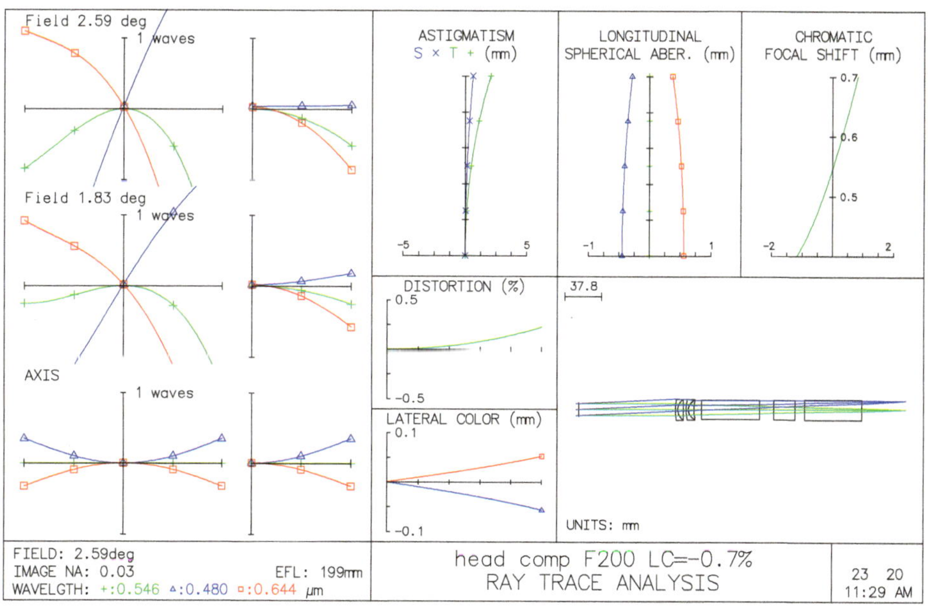

Figure 2.126 Optical design and graphs of aberration correction of an infinity (LC compensate) visual head ($F'_{ts} = 200$ mm).

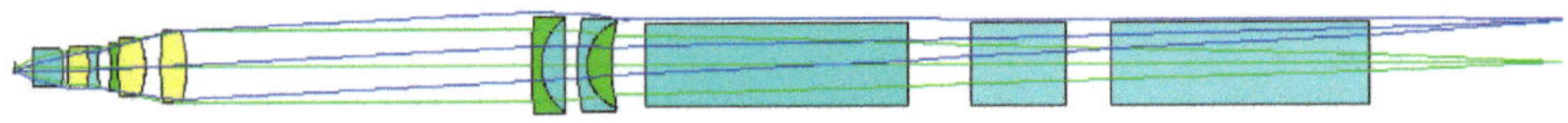

Lens: OF 10x0.40+comp head F200

Object num aper	0.400000	Image height	9.000000	Primary wavln	0.546070

SRF	RADIUS	THICKNESS	APERTURE RADIUS	GLASS	SPECIAL
OBJ	0.000000	0.010000	0.893927	AIR	
AST	0.000000	0.170000	0.898292 AS	K9	
2	0.000000	3.900000	0.940000	AIR	
3	0.000000	7.200000	2.640000	H-LAK2	
4	-11.426000	0.800000	4.200000	AIR	
5	17.320000	5.000000	4.520000	CAF2	
6	-24.500000	2.000000	4.580000	K9 P	
7	17.320000	3.000000	4.660000	AIR	
8	-44.140000	1.700000	5.050000	ZF2	
9	34.130000	6.300000	5.440000	CAF2	
10	-16.762000	3.700000	6.440000	AIR	
11	68.570000	6.400000	7.620000	CAF2	
12	-25.350000	80.000000	8.160000	AIR	
13	113.230000	2.200000	10.510000	LZ_TF4	
14	12.498000	5.600000	10.350000	LZ_CTK19	
15	-587.000000	3.000000	10.350000	AIR	
16	49.190000	2.200000	10.220000	LZ_CTK19 P	
17	10.069000	5.600000	9.390000	LZ_TF4 P	
18	36.810000	7.800000	9.350000	AIR	
19	0.000000	60.400000	9.330000	LZ_BK10	
20	0.000000	14.300000	9.250000	AIR	
21	0.000000	22.000000	9.250000	LZ_K8	
22	0.000000	10.000000	9.250000	AIR	
23	0.000000	59.500000	9.250000	LZ_K8 P	
24	0.000000	44.517000	9.250000	AIR	
25	0.000000	-0.003534 S	9.000260 S	AIR	
IMS	0.000000	0.000000	9.000000 S		

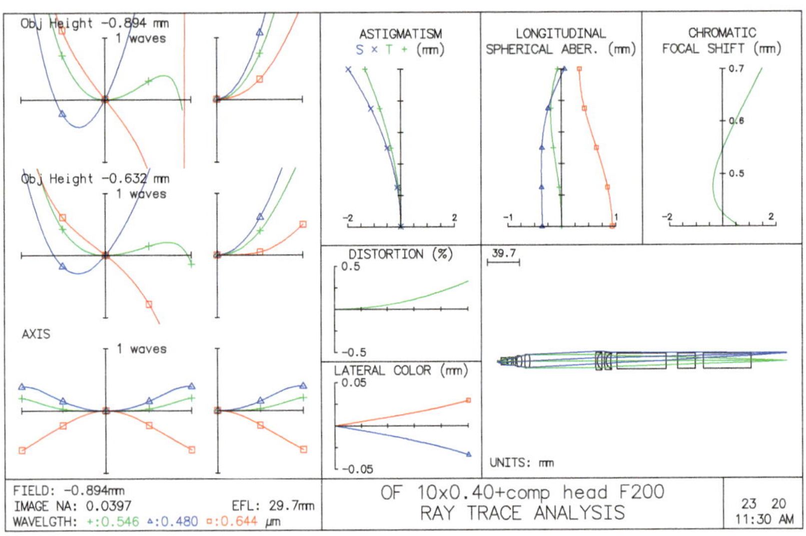

Figure 2.127 Optical design and graphs of aberration correction of a 10×/0.40 infinity LCR MicroFluar objective combined with a (LC compensate) visual head ($F'_{ts} = 200$ mm).

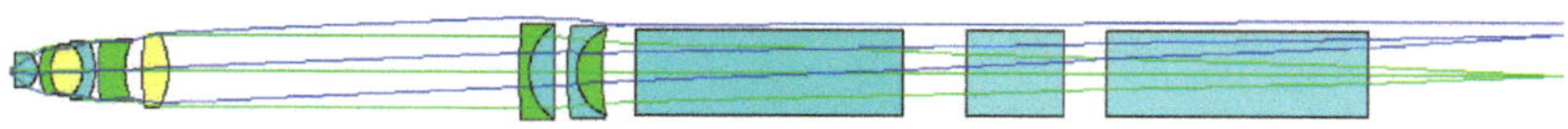

Lens: OF 20x0.80+comp head F200

Object num aper	0.800000	Image height	9.000000	Primary wavln	0.546070

SRF	RADIUS	THICKNESS	APERTURE RADIUS	GLASS	SPECIAL
OBJ	0.000000	0.010000	0.451810	AIR	
AST	0.000000	0.170000	0.465144 AS	K9	
2	0.000000	1.003900	0.570000	AIR	
3	-4.546000	5.000000	1.540000	H-LAK2	
4	-4.368000	0.580000	3.880000	AIR	
5	11.250000	2.500000	5.420000	ZF6	
6	8.395000	7.400000	5.240000	CAF2	
7	-7.060000	2.800000	5.550000	H-LAK2 P	
8	-12.953000	0.600000	6.540000	AIR	
9	31.220000	6.000000	6.820000	ZF3	
10	14.791000	4.000000	6.620000	AIR	
11	53.270000	6.000000	7.670000	CAF2	
12	-16.200000	80.000000	8.270000	AIR	
13	113.230000	2.200000	10.510000	LZ_TF4	
14	12.498000	5.600000	10.350000	LZ_CTK19	
15	-587.000000	3.000000	10.350000	AIR	
16	49.190000	2.200000	10.220000	LZ_CTK19 P	
17	10.069000	5.600000	9.400000	LZ_TF4 P	
18	36.810000	7.800000	9.360000	AIR	
19	0.000000	60.400000	9.340000	LZ_BK10	
20	0.000000	14.300000	9.290000	AIR	
21	0.000000	22.000000	9.300000	LZ_K8	
22	0.000000	10.000000	9.310000	AIR	
23	0.000000	59.500000	9.320000	LZ_K8 P	
24	0.000000	44.531441 S	9.350000	AIR	
25	0.000000	0.000000	9.000000 S	AIR	
IMS	0.000000	0.000000	9.000000 S		

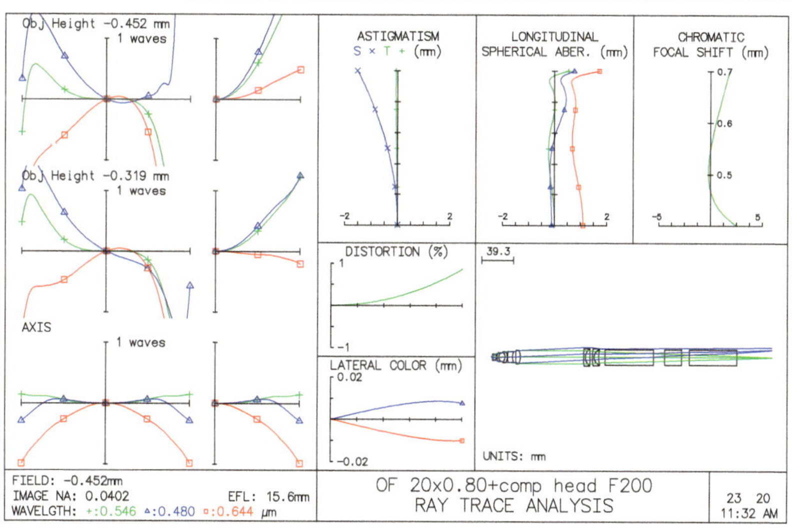

Figure 2.128 Optical design and graphs of aberration correction of a 20×/0.80 infinity LCR MicroFluar objective combined with a (LC compensate) visual head ($F'_{ts} = 200$ mm).

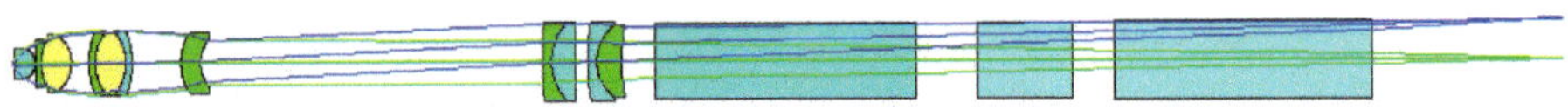

Lens: OF 40x0.95+comp head F200

Object num aper 0.950000 Image height 9.000000 Primary wavln 0.546070

SRF	RADIUS	THICKNESS	APERTURE RADIUS	GLASS	SPECIAL
OBJ	0.000000	0.010000	0.230526	AIR	
AST	0.000000	0.170000	0.260950 AS	K9	
2	0.000000	0.504600	0.400000	AIR	
3	-3.000000	4.500000	1.190000	H-LAK3	
4	-3.644000	0.110000	3.520000	AIR	
5	0.000000	1.200000	4.810000	ZF1	
6	11.370000	7.200000	5.490000	CAF2	
7	-7.600000	4.000000	6.450000	AIR	
8	18.270000	1.400000	7.390000	ZF1 P	
9	10.920000	7.200000	7.130000	CAF2 P	
10	-10.920000	1.400000	7.120000	H-LAK3 P	
11	-20.460000	11.800000	7.500000	AIR	
12	11.250000	5.000000	6.930000	ZF6	
13	8.000000	80.000000	5.370000	AIR	
14	113.230000	2.200000	8.940000	LZ_TF4	
15	12.498000	5.600000	8.900000	LZ_CTK19	
16	-587.000000	3.000000	8.880000	AIR	
17	49.190000	2.200000	8.760000	LZ_CTK19 P	
18	10.069000	5.600000	8.220000	LZ_TF4 P	
19	36.810000	7.800000	8.030000	AIR	
20	0.000000	60.400000	8.080000	LZ_BK10	
21	0.000000	14.300000	8.370000	AIR	
22	0.000000	22.000000	8.480000	LZ_K8	
23	0.000000	10.000000	8.590000	AIR	
24	0.000000	59.500000	8.660000	LZ_K8 P	
25	0.000000	44.517000	8.960000	AIR	
26	0.000000	-0.048226 S	9.005201 S	AIR	
IMS	0.000000	0.000000	9.000000 S		

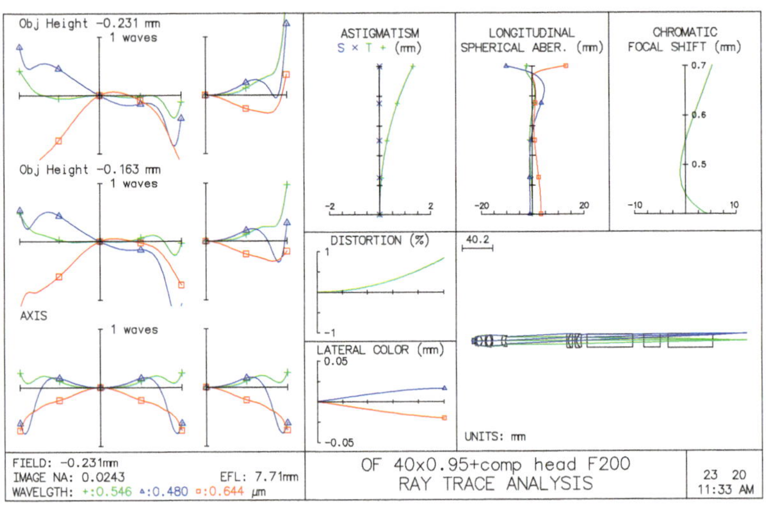

Figure 2.129 Optical design and graphs of aberration correction of a 40×/0.95 infinity LCR MicroFluar objective combined with a (LC compensate) visual head ($F'_{ts} = 200$ mm).

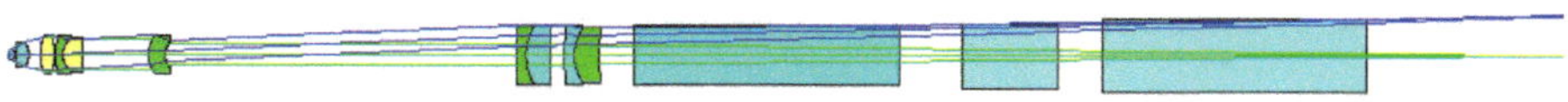

Lens: OF 100x1.20water+comp head F200

Object num aper 1.200000 Image height 9.000000 Primary wavln 0.546070

SRF	RADIUS	THICKNESS	APERTURE RADIUS	GLASS	SPECIAL
OBJ	0.000000	0.010000	0.091426	WATER	
AST	0.000000	0.170000	0.111980 AS	K9	
2	0.000000	0.200000	0.330000	WATER P	
3	-2.000000	1.400000	0.570000	BAK2	
4	-1.109000	0.130000	1.110000	AIR	
5	-9.000000	2.900000	1.800000	ZK1	
6	-4.000000	2.600000	2.790000	AIR	
7	9.817000	3.200000	4.170000	CAF2	
8	-12.210000	0.300000	4.240000	AIR	
9	14.894000	1.600000	4.070000	ZF6	
10	4.500000	4.500000	3.570000	CAF2 P	
11	-36.130000	15.000000	3.710000	AIR	
12	6.400000	3.900000	4.200000	ZF6 P	
13	4.600000	80.000000	3.050000	AIR	
14	113.230000	2.200000	6.350000	LZ_TF4	
15	12.498000	5.600000	6.350000	LZ_CTK19	
16	-587.000000	3.000000	6.340000	AIR	
17	49.190000	2.200000	6.300000	LZ_CTK19 P	
18	10.069000	5.600000	6.070000	LZ_TF4 P	
19	36.810000	7.800000	5.830000	AIR	
20	0.000000	60.400000	5.940000	LZ_BK10	
21	0.000000	14.300000	6.530000	AIR	
22	0.000000	22.000000	6.750000	LZ_K8	
23	0.000000	10.000000	6.970000	AIR	
24	0.000000	59.500000	7.120000	LZ_K8 P	
25	0.000000	44.517000	7.740000	AIR	
26	0.000000	0.055837 S	8.999937 S	AIR	
IMS	0.000000	0.000000	9.000000 S		

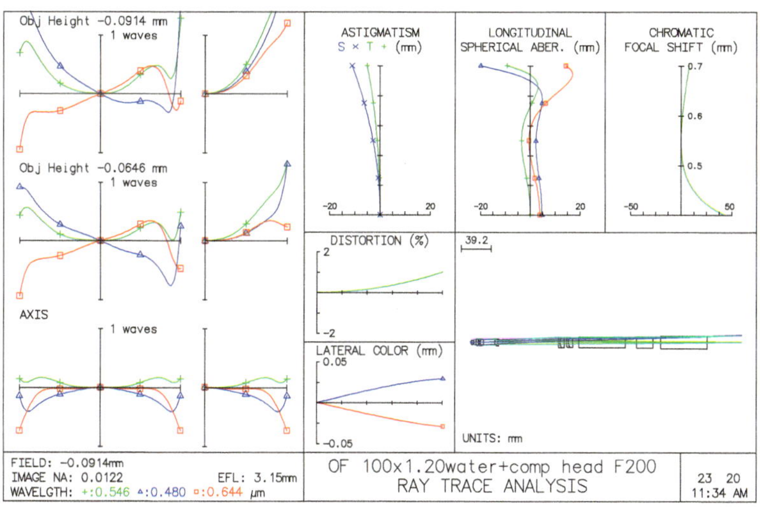

Figure 2.130 Optical design and graphs of aberration correction of a 100×/1.20 water immersion infinity LCR MicroFluar objective combined with a (LC compensate) visual head ($F'_{ts} = 200$ mm).

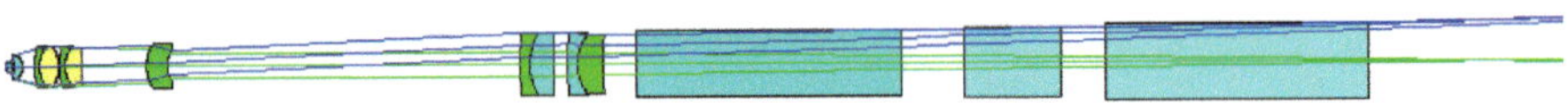

Lens: OF 100x1.35oil+comp head F200

Object num aper	1.350000	Image height	9.000000	Primary wavln	0.546070

SRF	RADIUS		THICKNESS		APERTURE RADIUS		GLASS		SPECIAL
OBJ	0.000000		0.010000		0.092139		OIL		
AST	0.000000		0.170000		0.111616	AS	K9		
2	0.000000		0.130750		0.450000		OIL	P	
3	0.000000		1.500000		0.700000		K9	P	
4	-1.200000		0.130000		1.200000		AIR		
5	-4.200000		1.900000		2.050000		H-LAK2		
6	-2.890000		2.600000		2.650000		AIR		
7	13.450000		1.200000		4.370000		ZF6		
8	8.000000		4.300000		4.420000		CAF2		
9	-8.000000		0.130000		4.700000		AIR		
10	14.400000		1.200000		4.640000		ZF6		
11	6.066000		3.600000		4.300000		CAF2	P	
12	0.000000		15.000000		4.330000		AIR		
13	7.625000		5.000000		4.770000		ZF6	P	
14	5.297000		80.000000		3.380000		AIR		
15	113.230000		2.200000		6.890000		LZ_TF4		
16	12.498000		5.600000		6.890000		LZ_CTK19		
17	-587.000000		3.000000		6.880000		AIR		
18	49.190000		2.200000		6.840000		LZ_CTK19	P	
19	10.069000		5.600000		6.560000		LZ_TF4	P	
20	36.810000		7.800000		6.320000		AIR		
21	0.000000		60.400000		6.430000		LZ_BK10		
22	0.000000		14.300000		7.040000		AIR		
23	0.000000		22.000000		7.270000		LZ_K8		
24	0.000000		10.000000		7.500000		AIR		
25	0.000000		59.500000		7.660000		LZ_K8	P	
26	0.000000		44.517000		8.280000		AIR		
27	0.000000		-0.016663	S	9.000988	S	AIR		
IMS	0.000000		0.000000		9.000000	S			

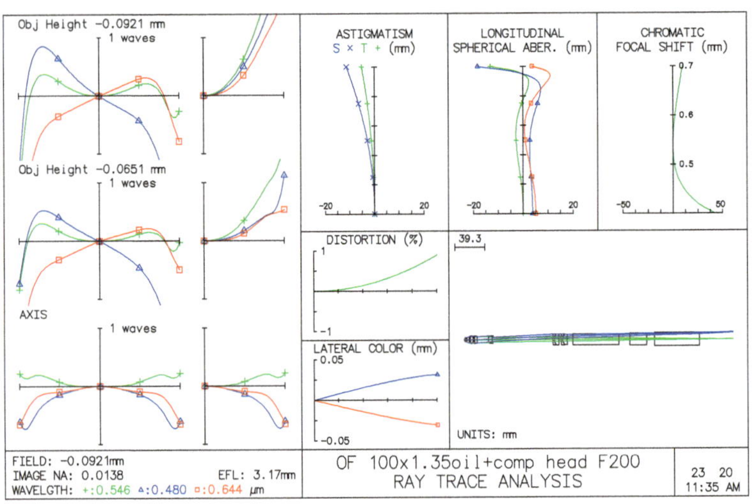

Figure 2.131 Optical design and graphs of aberration correction of a 100×/1.35 oil immersion infinity LCR MicroFluar objective combined with a (LC compensate) visual head ($F'_{ts} = 200$ mm).

Using such an eyepiece does not limit the field compared with the linear field of the 10× eyepiece, which is now the most commonly used eyepiece in light microscopy for visual observation. By the way, the 12.5× eyepiece could become universal in lenses configured with standard numerical apertures (not enlarged, as the lenses in MicroFluar objectives) for all types of light microscopes for visual observation.

My colleagues and I once developed an optical design for a 20× wide-angle eyepiece with a linear field (in the plane of the field diaphragm) of 22 mm. Its optical system turned out to be quite complex. Moreover, the appropriateness of using such an eyepiece is questionable, as its angular field (about 100 deg) is very large and may not have a correlation with the physiological capabilities of the human eye. In other words, a designed and manufactured 20×/22 mm eyepiece may not satisfy the requirement for comfortable visual observation with a light microscope.

We conducted a study using a conventional biological light microscope; the 12.5× eyepiece was also used. MicroFluar objectives of medium magnification were compared with lenses that had similar linear magnifications but standard numerical apertures. We found that the real (visible through the eyepiece) resolution of 10×/0.40 and 20×/0.80 objectives significantly exceeded that of standard objectives (10×/0.25 and 20×/0.40), although this did not seem to be correlated with the so-called useful magnification condition for visual observation with a light microscope. We also conducted another experiment in which a 12.5× eyepiece was used. In this experiment, we examined image quality and resolution when using a 10×/0.40 MicroFluar-type objective. Maximum visual resolution when observing through the eyepiece was achieved when the numerical apertures of the condenser and the objective were (approximately) equal. By decreasing the aperture of the condenser, a noticeable decrease in resolution was observed. With an increase in the aperture of the condenser (to values larger than the aperture of the objective), changes in resolution were not visually detected. Thus, in some cases, we can disregard the condition of useful magnification on a microscope (at the lower end) and recommend the use of objectives in combination with a 12.5× eyepiece.

In light microscopy, although it is not necessary, some standard magnification values should be adhered to. Typically, a 10× eyepiece is used as the base eyepiece. If we use a 12.5× eyepiece in combination with objectives that have standard 10×, 20×, 40×, and 100× magnifications, the total magnification on the microscope will be 1.25 times different than if we were using a regular 10× eyepiece. This situation calls for an original engineering solution, particularly because different microscope manufacturers use different parametric series of focal lengths of their objectives and other optical components of microscopes. Simply put, different manufacturers use different focal lengths for tube systems. This creates a great deal of inconvenience for researchers, as it is quite difficult to find the right combination of objectives, tube systems, and eyepieces. However, this provides us with an opportunity to find an original engineering solution. Indeed, we can mix objectives and other optical components from different manufacturers in one microscope, including in the optical design.

If we use objectives such as the MicroFluar, described earlier as version 2 combined with a compensation head for the tube system ($F' = 160\,\text{mm}$), then the values of their own linear magnifications will decrease by 1.25 times (in proportion to the ratios $F' = 160/F' = 200$); that is, instead of a 10× objective, for example, we will have an 8× objective. This can be very relevant for fluorescence microscopy because, for example, the light power of the objective, proportional to the fourth power of its output aperture, rises by almost 2.5 times. Naturally, the resolution in the plane of the intermediate image on the microscope in this case will increase 1.25 times. Using an eyepiece with a linear magnification of 12.5× compensates for the loss of full magnification on a microscope, returning it to a standard and familiar value. The only technical difficulty in implementing such an engineering solution may be the increase in the dimensions of the prisms in the visual head; for objectives of small and medium linear magnifications when using prisms from a standard head with an eyepiece tube diameter of 23.2 mm, significant vignetting may occur. However, this problem can be easily solved if standard prisms are used from a head with an eyepiece tube diameter of 30 mm.

Let us consider some optical designs that implement such an engineering solution. Figure 2.132 shows the design parameters and graphs of aberration correction of such a visual head ($F' = 160\,\text{mm}$) that has its own LC compensation value of −0.8% and an increased calculated value of the output numerical aperture ($NA' = 0.04$). Figures 2.133–2.137 show the design parameters and graphs of aberration correction of the joint design of this head, the version 2 MicroFluar objectives, as well as an eyepiece with its own linear magnification of 12.5× (the parameters of which are shown in Chapter 1, Fig. 1.208).

In the examples above, we used the MicroFluar objectives as is, i.e., without changing the light diameters of the lenses. However, when implementing the considered engineering solution, there is an additional opportunity to increase the size of the linear field; its value can be increased by at least 1.25 times. Doing so requires that the light diameters of the lenses of the objectives be increased, which is possible if implemented in the original mechanical design. A similar solution was implemented in old Soviet microscopes, for example, in the JenaMed and JenaVal series, by the engineers of Carl Zeiss Jena. Although production of these microscopes was discontinued long ago, their optical image quality is still unrivaled in the field of budget microscopes.

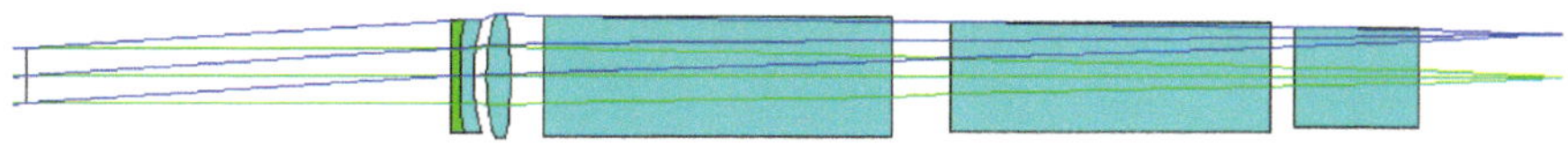

Lens:head F160 compens NA=0.04 D=30mm				Efl 159.97	
Image num aper 0.040000	Image height 10.000000		Primary wavln 0.546070		
SRF	RADIUS	THICKNESS	APERTURE RADIUS	GLASS	SPECIAL
OBJ	0.000000	1.0000e+20	6.2511e+18	AIR	
AST	0.000000	0.001000	6.398849 AS	BK7 C	
2	0.000000	100.000000	6.398890 S	AIR	
3	0.000000	2.200000	12.660000	LZ_TF4	
4	105.000000	3.200000	12.770000	LZ_CTK19	
5	43.750000	3.000000	12.930000	AIR	
6	49.770000	5.600000	13.920000	LZ_CTK19 P	
7	-104.710000	7.800000	14.040000	AIR	
8	0.000000	82.400000	13.830000	LZ_CTK19 P	
9	0.000000	14.000000	12.700000	AIR	
10	0.000000	75.000000	12.380000	LZ_CTK19 P	
11	0.000000	5.500000	11.370000	AIR	
12	0.000000	30.000000	11.240000	LZ_CTK19 P	
13	0.000000	34.290098 S	10.840000	AIR	
IMS	0.000000	0.000000	10.000000 S		

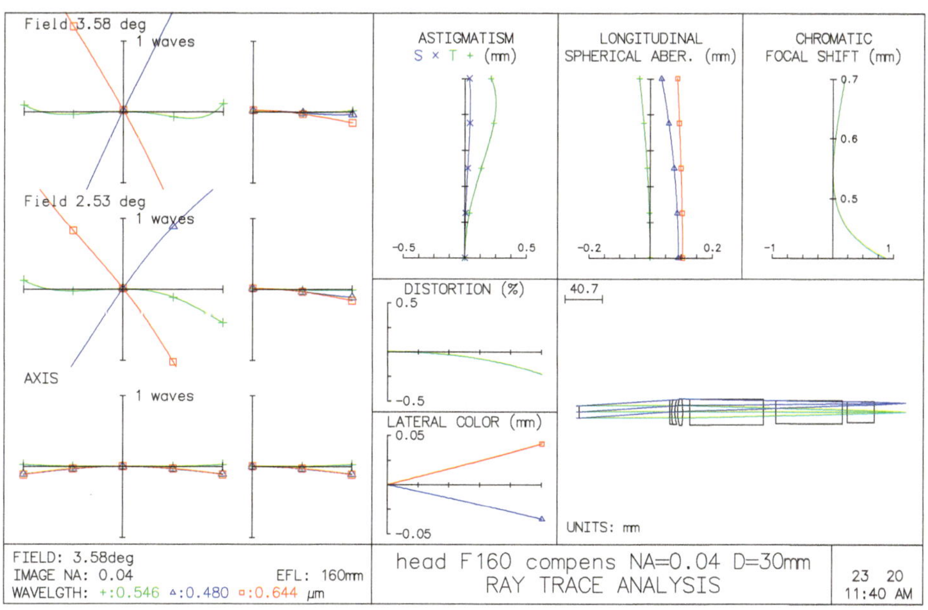

Figure 2.132 Optical design and graphs of aberration correction of an infinity (LC compensate) visual head ($F'_{ts} = 160$ mm; increased prism diameters).

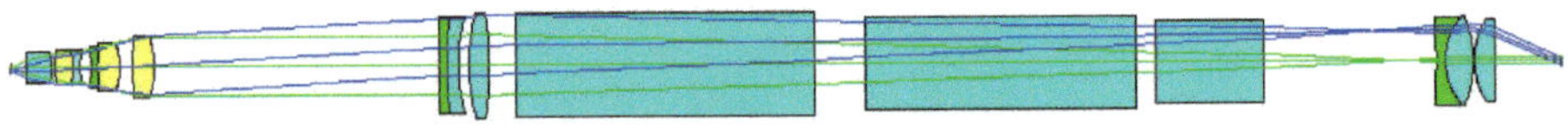

Lens: OF10x0.4+compheadF160+okul12.5x | Efl -2.49

Object num aper 0.400000 | Object height -1.000000 | Primary wavln 0.546070

SRF	RADIUS	THICKNESS	APERTURE RADIUS	GLASS	SPECIAL
AST	0.000000	0.170000	1.004364 AS	K9	
2	0.000000	3.900000	1.030000	AIR	
3	0.000000	7.200000	2.620000	LZ_CTK12	
4	-11.426000	0.800000	4.180000	AIR	
5	17.320000	5.000000	4.500000	CAF2	
6	-24.500000	2.000000	4.570000	K9 P	
7	17.320000	3.000000	4.650000	AIR	
8	-44.140000	1.700000	5.050000	ZF2	
9	34.130000	6.300000	5.440000	CAF2	
10	-16.762000	3.700000	6.440000	AIR	
11	68.570000	6.400000	7.800000	CAF2	
12	-25.350000	80.000000	8.430000	AIR	
13	0.000000	2.200000	12.660000	LZ_TF4	
14	105.000000	3.200000	12.770000	LZ_CTK19	
15	43.750000	3.000000	12.930000	AIR	
16	49.770000	5.600000	13.920000	LZ_CTK19 P	
17	-104.710000	7.800000	14.040000	AIR	
18	0.000000	82.400000	13.830000	LZ_CTK19 P	
19	0.000000	14.000000	12.700000	AIR	
20	0.000000	75.000000	12.380000	LZ_CTK19 P	
21	0.000000	5.500000	11.370000	AIR	
22	0.000000	30.000000	11.240000	LZ_CTK19 P	
23	0.000000	47.533300	10.840000	AIR	
24	-239.900000	2.500000	11.450000	LZ_TF10	
25	17.539000	7.800000	11.740000	LZ_CTK12	
26	-30.900000	0.300000	11.830000	AIR	
27	20.140000	5.750000	11.180000	LZ_CTK12	
28	0.000000	17.720000	10.420000	AIR	
29	0.000000	1.7763e+05 S	1.432924 S	AIR	
IMS	0.000000	0.000000	7.1253e+04 S		

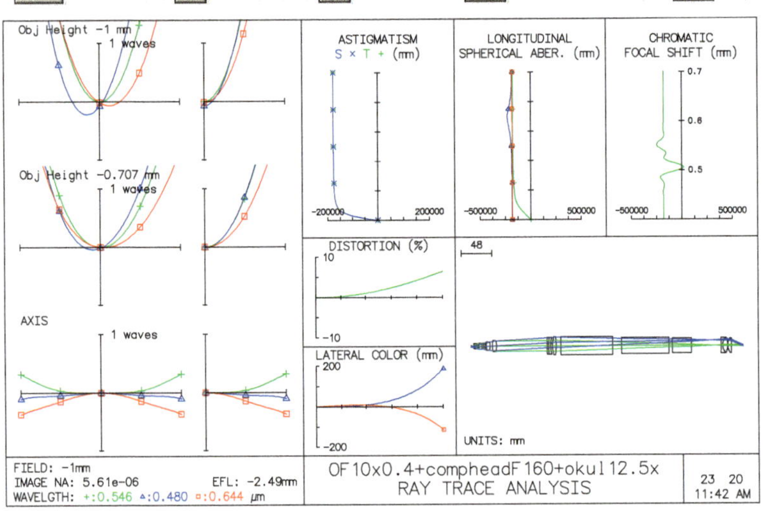

Figure 2.133 Optical design and graphs of aberration correction of a 10×/0.40 infinity LCR MicroFluar objective combined with a visual head (F'_{ts} = 160 mm) and a 12.5× eyepiece.

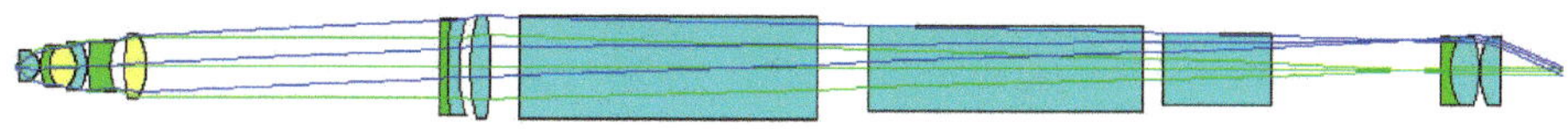

Lens: OF20x0.8+compheadF160+okul12.5x				Efl	-1.26
Object num aper	0.800000	Object height	-0.500000	Primary wavln	0.546070
SRF	**RADIUS**	**THICKNESS**	**APERTURE RADIUS**	**GLASS**	**SPECIAL**
OBJ	0.000000	0.010000	0.500000	AIR	
AST	0.000000	0.170000	0.513333 AS	K9	
2	0.000000	1.003900	0.620000	AIR	
3	-4.546000	5.000000	1.580000	LZ_CTK12	
4	-4.368000	0.580000	3.910000	AIR	
5	11.250000	2.500000	5.440000	ZF6	
6	8.395000	7.400000	5.250000	CAF2	
7	-7.060000	2.800000	5.550000	LZ_CTK12 P	
8	-12.953000	0.600000	6.540000	AIR	
9	31.220000	6.000000	6.820000	ZF3	
10	14.791000	4.000000	6.680000	AIR	
11	53.270000	6.000000	7.870000	CAF2	
12	-16.200000	80.000000	8.500000	AIR	
13	0.000000	2.200000	12.600000	LZ_TF4	
14	105.000000	3.200000	12.690000	LZ_CTK19	
15	43.750000	3.000000	12.810000	AIR	
16	49.770000	5.600000	13.720000	LZ_CTK19 P	
17	-104.710000	7.800000	13.820000	AIR	
18	0.000000	82.400000	13.520000	LZ_CTK19 P	
19	0.000000	14.000000	11.900000	AIR	
20	0.000000	75.000000	11.420000	LZ_CTK19 P	
21	0.000000	5.500000	9.950000	AIR	
22	0.000000	30.000000	9.760000	LZ_CTK19 P	
23	0.000000	47.533300	9.170000	AIR	
24	-239.900000	2.500000	8.670000	LZ_TF10	
25	17.539000	7.800000	8.940000	LZ_CTK12	
26	-30.900000	0.300000	9.340000	AIR	
27	20.140000	5.750000	9.160000	LZ_CTK12	
28	0.000000	17.720000	8.340000	AIR	
29	0.000000	-4.3664e+04 S	1.991604 S	AIR	
IMS	0.000000	0.000000	1.7326e+04 S		

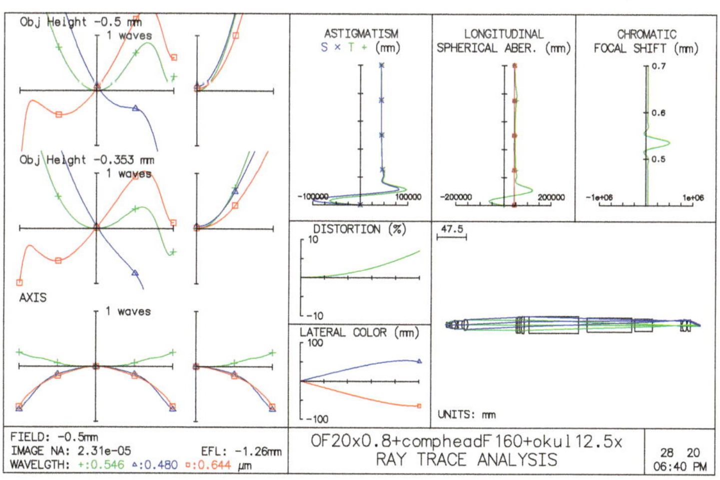

Figure 2.134 Optical design and graphs of aberration correction of a 20×/0.80 infinity LCR MicroFluar objective combined with a visual head ($F'_{ts} = 160$ mm) and a 12.5× eyepiece.

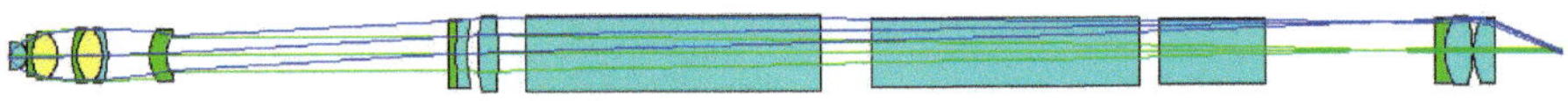

Lens:OF40x0.95+compheadF160+ocul12.5x Efl -0.64

Object num aper 0.950000 Object height -0.250000 Primary wavln 0.546070

SRF	RADIUS	THICKNESS	APERTURE RADIUS	GLASS	SPECIAL
OBJ	0.000000	0.010000	0.250000	AIR	
AST	0.000000	0.170000	0.280424 AS	K9	
2	0.000000	0.504600	0.420000	AIR	
3	-3.000000	4.500000	1.200000	H-LAK3	
4	-3.644000	0.110000	3.520000	AIR	
5	0.000000	1.200000	4.810000	ZF1	
6	11.370000	7.200000	5.490000	CAF2	
7	-7.600000	4.000000	6.450000	AIR	
8	18.270000	1.400000	7.400000	ZF1 P	
9	10.920000	7.200000	7.140000	CAF2 P	
10	-10.920000	1.400000	7.140000	H-LAK3 P	
11	-20.460000	11.800000	7.520000	AIR	
12	11.250000	5.000000	6.980000	ZF6	
13	8.000000	80.000000	5.410000	AIR	
14	0.000000	2.200000	9.280000	LZ_TF4	
15	105.000000	3.200000	9.360000	LZ_CTK19	
16	43.750000	3.000000	9.460000	AIR	
17	49.770000	5.600000	10.140000	LZ_CTK19 P	
18	-104.710000	7.800000	10.280000	AIR	
19	0.000000	82.400000	10.160000	LZ_CTK19 P	
20	0.000000	14.000000	9.530000	AIR	
21	0.000000	75.000000	9.340000	LZ_CTK19 P	
22	0.000000	5.500000	8.760000	AIR	
23	0.000000	30.000000	8.680000	LZ_CTK19 P	
24	0.000000	47.533300	8.450000	AIR	
25	-239.900000	2.500000	8.700000	LZ_TF10	
26	17.539000	7.800000	8.910000	LZ_CTK12	
27	-30.900000	0.300000	9.270000	AIR	
28	20.140000	5.750000	9.030000	LZ_CTK12	
29	0.000000	17.720000	8.170000	AIR	
30	0.000000	1.3059e+04 S	2.284443 S	AIR	
IMS	0.000000	0.000000	5.0785e+03 S		

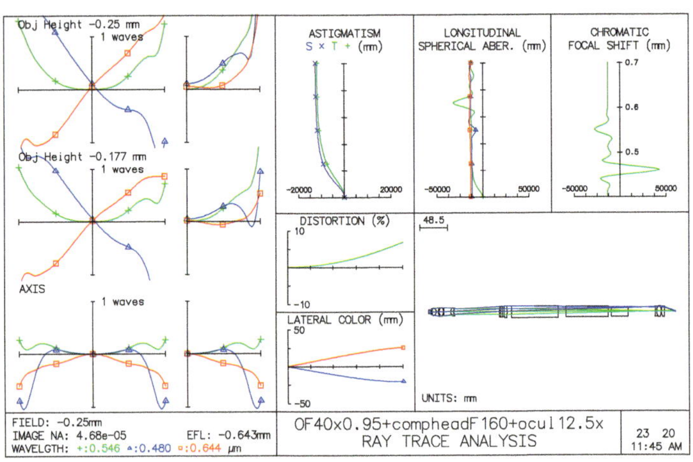

Figure 2.135 Optical design and graphs of aberration correction of a 40×/0.95 infinity LCR MicroFluar objective combined with a visual head ($F'_{ts} = 160$ mm) and a 12.5× eyepiece.

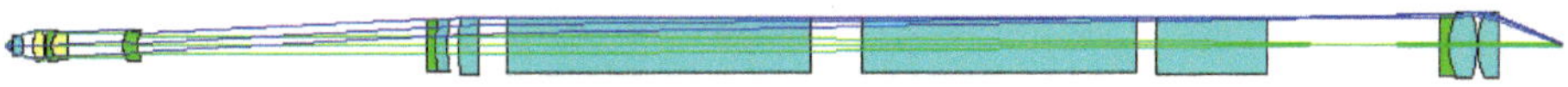

Lens:100x1.2wate+compheadF160+oku12.5 Efl -0.255

Object num aper 1.200000 Object height -0.100000 Primary wavln 0.546070

SRF	RADIUS	THICKNESS		APERTURE RADIUS		GLASS		SPECIAL
OBJ	0.000000	0.010000		0.100000		WATER		
AST	0.000000	0.170000		0.120554	AS	K9		
2	0.000000	0.200000		0.340000		WATER	P	
3	-2.000000	1.400000		0.580000		BAK2		
4	-1.109000	0.130000		1.110000		AIR		
5	-9.000000	2.900000		1.800000		ZK1		
6	-4.000000	2.600000		2.800000		AIR		
7	9.817000	3.200000		4.200000		CAF2		
8	-12.210000	0.300000		4.260000		AIR		
9	14.894000	1.600000		4.100000		ZF6		
10	4.500000	4.500000		3.590000		CAF2	P	
11	-36.130000	15.000000		3.730000		AIR		
12	6.400000	3.900000		4.290000		ZF6	P	
13	4.600000	80.000000		3.120000		AIR		
14	0.000000	2.200000		6.710000		LZ_TF4		
15	105.000000	3.200000		6.770000		LZ_CTK19		
16	43.750000	3.000000		6.870000		AIR		
17	49.770000	5.600000		7.370000		LZ_CTK19	P	
18	-104.710000	7.800000		7.510000		AIR		
19	0.000000	82.400000		7.500000		LZ_CTK19	P	
20	0.000000	14.000000		7.450000		AIR		
21	0.000000	75.000000		7.440000		LZ_CTK19	P	
22	0.000000	5.500000		7.390000		AIR		
23	0.000000	30.000000		7.380000		LZ_CTK19	P	
24	0.000000	47.533300		7.510000		AIR		
25	-239.900000	2.500000		8.220000		LZ_TF10		
26	17.539000	7.800000		8.370000		LZ_CTK12		
27	-30.900000	0.300000		8.710000		AIR		
28	20.140000	5.750000		8.450000		LZ_CTK12		
29	0.000000	17.720000		7.560000		AIR		
30	0.000000	-1.1287e+04	S	1.005851	S	AIR		
IMS	0.000000	0.000000		4.4261e+03	S			

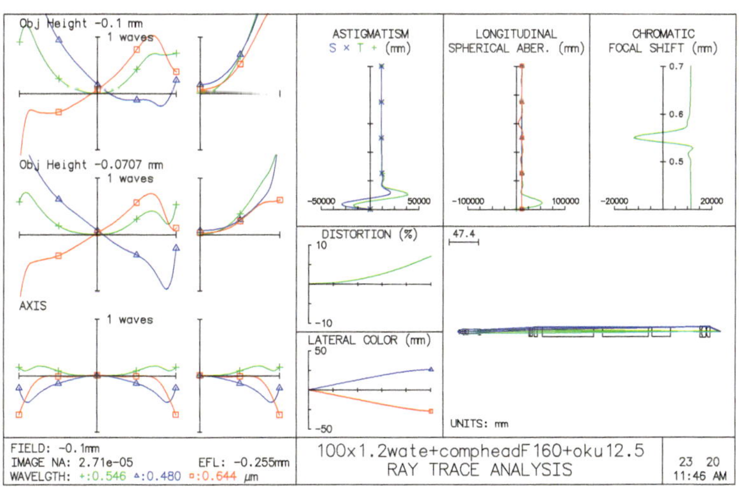

Figure 2.136 Optical design and graphs of aberration correction of a 100×/1.20 water immersion infinity LCR MicroFluar objective combined with a visual head (F'_{ts} = 160 mm) and a 12.5× eyepiece.

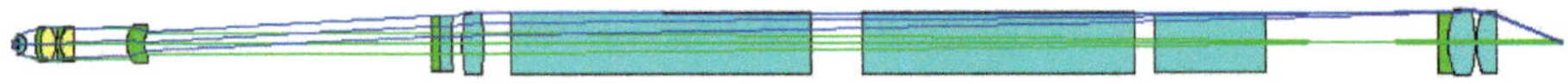

Lens:100x1.35oil+compheadF160+oku12.5 Efl -0.26

Object num aper 1.350000 Object height -0.100000 Primary wavln 0.546070

SRF	RADIUS	THICKNESS		APERTURE RADIUS		GLASS		SPECIAL
OBJ	0.000000	0.010000		0.100000		OIL		
AST	0.000000	0.170000		0.120000	A	K9		
2	0.000000	0.130750		0.450000		OIL	P	
3	0.000000	1.500000		0.710000		K9	P	
4	-1.200000	0.130000		1.200000		AIR		
5	-4.200000	1.900000		2.050000		LZ_CTK12		
6	-2.890000	2.600000		2.650000		AIR		
7	13.450000	1.200000		4.390000		ZF6		
8	8.000000	4.300000		4.440000		CAF2		
9	-8.000000	0.130000		4.720000		AIR		
10	14.400000	1.200000		4.670000		ZF6		
11	6.066000	3.600000		4.320000		CAF2	P	
12	0.000000	15.000000		4.360000		AIR		
13	7.625000	5.000000		4.850000		ZF6	P	
14	5.297000	80.000000		3.440000		AIR		
15	0.000000	2.200000		7.280000		LZ_TF4		
16	105.000000	3.200000		7.350000		LZ_CTK19		
17	43.750000	3.000000		7.450000		AIR		
18	49.770000	5.600000		8.000000		LZ_CTK19	P	
19	-104.710000	7.800000		8.150000		AIR		
20	0.000000	82.400000		8.140000		LZ_CTK19	P	
21	0.000000	14.000000		8.050000		AIR		
22	0.000000	75.000000		8.030000		LZ_CTK19	P	
23	0.000000	5.500000		7.950000		AIR		
24	0.000000	30.000000		7.940000		LZ_CTK19	P	
25	0.000000	47.533300		7.910000		AIR		
26	-239.900000	2.500000		8.090000		LZ_TF10		
27	17.539000	7.800000		8.200000		LZ_CTK12		
28	-30.900000	0.300000		8.500000		AIR		
29	20.140000	5.750000		8.200000		LZ_CTK12		
30	0.000000	17.720000		7.270000		AIR		
31	0.000000	3.7778e+04	S	1.064895	S	AIR		
IMS	0.000000	0.000000		1.4702e+04	S			

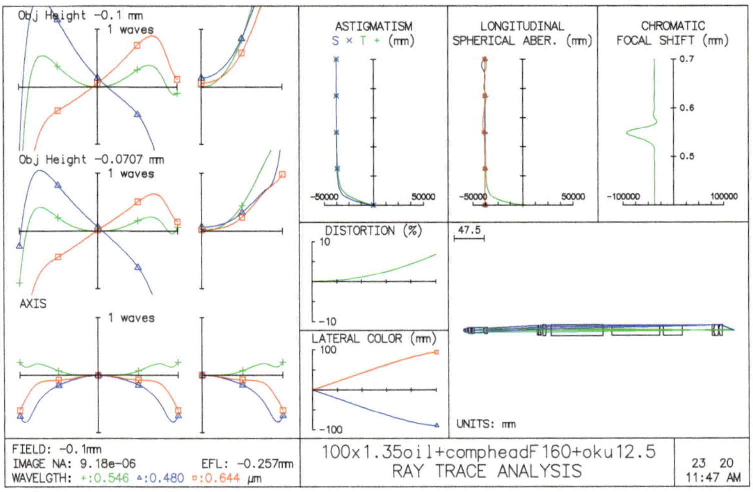

Figure 2.137 Optical design and graphs of aberration correction of a 100×/1.35 oil immersion infinity LCR MicroFluar objective combined with a visual head ($F'_{ts} = 160$ mm) and a 12.5× eyepiece.

Thus, for visual microscopy, when observation is performed through the eyepiece, some compromises in image quality are possible (and even sometimes necessary) for the sake of, for example, increasing the light power of optics. However, for microphotography, that is, when projecting images onto a digital receiver sensor, the image quality after optical systems should be as good as possible. Following the methodology for constructing digital receiver projection systems described in Chapter 1, Section 1.8.4, we can propose an adapter whose optical design is made specifically for use with MicroFluar-series objectives (in particular, version 2). Figures 2.138

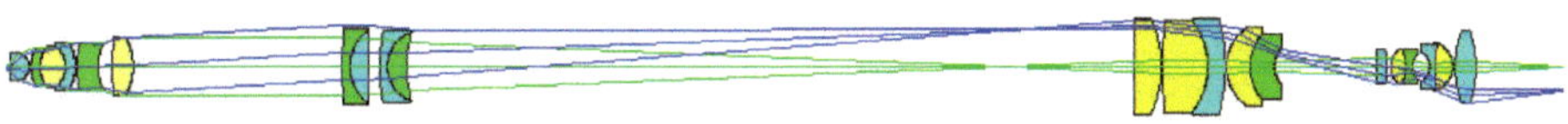

Lens:OF 20x0.80+ comp TL+adapter0.65x

Object num aper 0.800000 Object height -0.500000 Primary wavln 0.546070

SRF	RADIUS	THICKNESS	APERTURE RADIUS	GLASS	SPECIAL
OBJ	0.000000	0.010000	0.500000	AIR	
AST	0.000000	0.170000	0.513333 AS	K9	
2	0.000000	1.003900	0.620000	AIR	
3	-4.546000	5.000000	1.580000	LZ_CTK12	
4	-4.368000	0.580000	3.910000	AIR	
5	11.250000	2.500000	5.440000	ZF6	
6	8.395000	7.400000	5.250000	CAF2	
7	-7.060000	2.800000	5.550000	LZ_CTK12 P	
8	-12.953000	0.600000	6.540000	AIR	
9	31.220000	6.000000	6.820000	ZF3	
10	14.791000	4.000000	6.630000	AIR	
11	53.270000	6.000000	7.690000	CAF2	
12	-16.200000	60.000000	8.290000	AIR	
13	113.230000	2.200000	10.500000	LZ_TF4	
14	12.498000	5.600000	10.350000	LZ_CTK19	
15	-587.000000	3.000000	10.350000	AIR	
16	49.190000	2.200000	10.240000	LZ_CTK19 P	
17	10.069000	5.600000	9.430000	LZ_TF4 P	
18	36.810000	208.780000	9.400000	AIR	
19	0.000000	8.000000	12.840000	H-FK71	
20	-30.979700	0.300000	13.180000	AIR	
21	118.418900	13.000000	13.100000	H_FK71	
22	-19.618700	5.000000	12.590000	H-LAK3	
23	-47.557000	0.300000	13.380000	AIR	
24	12.974200	7.000000	10.970000	H_FK71	
25	13.829200	7.000000	9.010000	H-ZF7LA	
26	6.829200	30.000000	4.800000	AIR	
27	159.816300	3.000000	4.420000	LZ_CTK12	
28	-17.233600	1.000000	4.740000	AIR	
29	5.735400	4.300000	4.950000	H_FK71	
30	-12.024400	2.000000	4.580000	ZF1	
31	4.571700	5.200000	3.540000	AIR	
32	-4.260800	1.500000	4.040000	H-LAK4L	
33	41.496200	5.200000	6.300000	H_FK71	
34	-6.996700	0.800000	6.840000	AIR	
35	30.865900	6.000000	9.640000	H_LAK3	
36	-30.093900	25.000000	9.990000	AIR	
37	0.000000	0.010879 S	6.477644 S	AIR	
IMS	0.000000	0.000000	6.476387 S		

Figure 2.138 Optical design and graphs of aberration correction of a 20×/0.80 infinity LCR MicroFluar objective combined with a compensate tube system and a 0.65× compensate adapter.

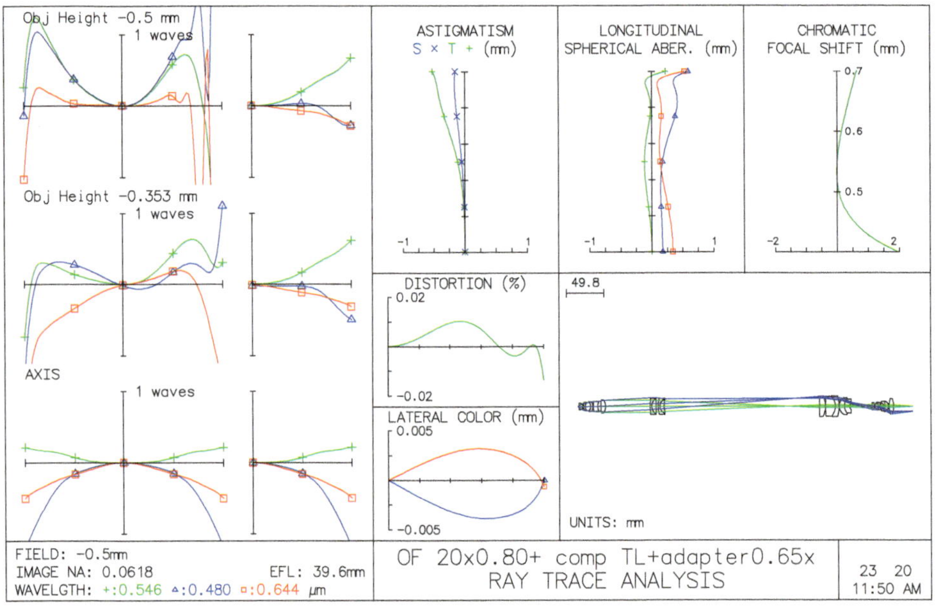

Figure 2.138 (*Continued*) Optical design and graphs of aberration correction of a 20×/0.80 infinity LCR MicroFluar objective combined with a compensate tube system and a 0.65× compensate adapter.

and 2.139 show the design parameters and the quality of the aberration correction of two examples of MicroFluar objectives combined with a compensation tube system (LC = −0.7%) and an adapter with a linear magnification of 0.65× for projecting an image onto a digital receiver with a 2/3 inch sensor size. In both cases, the aberration graphs show a high resulting level of aberration correction, including LC and distortion.

Figures 2.140–2.157 are photographs of some biological objects made using MicroFluar-series objectives (version 2).

2.4.2 Optical designs of some nonstandard objectives

If special research methods are essentially not found in budget microscopes (with the exception of contrasting methods), they can be found in medium series microscopes. Special research methods refers to the study of objects, which are covered or not with coverslips, with the use of nonstandard microscopy techniques. These special methods allow researchers to achieve results that are impossible under normal conditions. At the same time, specialized elements in the base of the microscope, special lighting and image-registering devices, and special optics should be used here. Attention must be paid to the optical designs of objectives for such microscopes, as the objective is the main part of the microscope, regardless of its purpose and the research methods to be used. Moreover, it is the objective's design that determines whether a particular technique on a light microscope can be implemented.

2.4.2.1 LWD water immersion objectives

The likelihood of researchers using water immersion in microscopic studies is difficult to overstate. Water is, perhaps, the only immersion that allows the study of organisms to be conducted safely on a microscope. The advantages of high-aperture water immersion objectives compared with, for example, oil

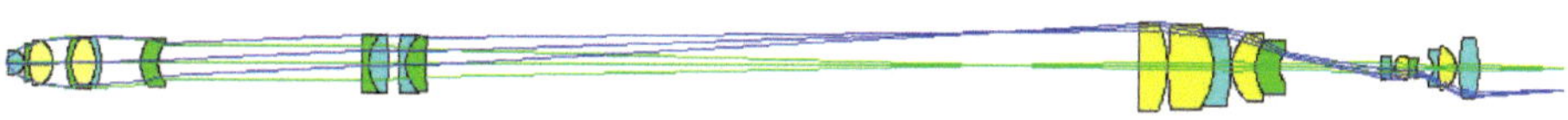

Lens: OF 40x0.95+comp TL+adater0.65x

Object num aper 0.950000 Object height -0.250000 Primary wavln 0.546070

SRF	RADIUS	THICKNESS	APERTURE RADIUS	GLASS	SPECIAL
OBJ	0.000000	0.010000	0.250000	AIR	
AST	0.000000	0.170000	0.280424 AS	K9	
2	0.000000	0.504600	0.420000	AIR	
3	-3.000000	4.500000	1.200000	H-LAK3	
4	-3.644000	0.110000	3.520000	AIR	
5	0.000000	1.200000	4.810000	ZF1	
6	11.370000	7.200000	5.490000	CAF2	
7	-7.600000	4.000000	6.450000	AIR	
8	18.270000	1.400000	7.410000	ZF1 P	
9	10.920000	7.200000	7.140000	CAF2 P	
10	-10.920000	1.400000	7.140000	H-LAK3 P	
11	-20.460000	11.800000	7.530000	AIR	
12	11.250000	5.000000	6.990000	ZF6	
13	8.000000	60.000000	5.420000	AIR	
14	113.230000	2.200000	8.300000	LZ_TF4	
15	12.498000	5.600000	8.290000	LZ_CTK19	
16	-587.000000	3.000000	8.280000	AIR	
17	49.190000	2.200000	8.200000	LZ_CTK19 P	
18	10.069000	5.600000	7.770000	LZ_TF4 P	
19	36.810000	208.780000	7.550000	AIR	
20	0.000000	8.000000	12.610000	H-FK71	
21	30.979700	0.300000	12.830000	AIR	
22	118.418900	13.000000	12.260000	H_FK71	
23	-19.618700	5.000000	11.020000	H-LAK3	
24	-47.557000	0.300000	11.070000	AIR	
25	12.974200	7.000000	9.960000	H_FK71	
26	13.829200	7.000000	7.840000	H-ZF7LA	
27	6.829200	30.000000	4.690000	AIR	
28	159.816300	3.000000	3.100000	LZ_CTK12	
29	-17.233600	1.000000	3.420000	AIR	
30	5.735400	4.300000	3.610000	H_FK71	
31	-12.024400	2.000000	3.100000	ZF1	
32	4.571700	5.200000	2.650000	AIR	
33	-4.260800	1.500000	3.330000	H-LAK4L	
34	41.496200	5.200000	5.280000	H_FK71	
35	-6.996700	0.800000	6.140000	AIR	
36	30.865900	6.000000	8.340000	H_LAK3	
37	-30.093900	25.000000	8.620000	AIR	
38	0.000000	-0.015625 S	6.349881 S	AIR	
IMS	0.000000	0.000000	6.348180 S		

Figure 2.139 Optical design and graphs of aberration correction of a 40×/0.95 infinity LCR MicroFluar objective combined with a compensate tube system and a 0.65× compensate adapter.

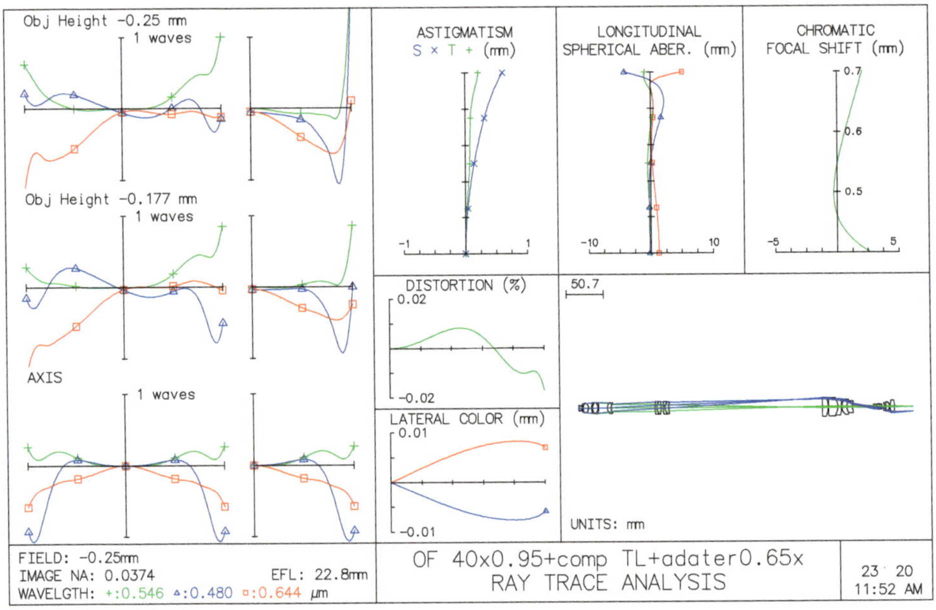

Figure 2.139 (*Continued*) Optical design and graphs of aberration correction of a 40×/0.95 infinity LCR MicroFluar objective combined with a compensate tube system and a 0.65× compensate adapter.

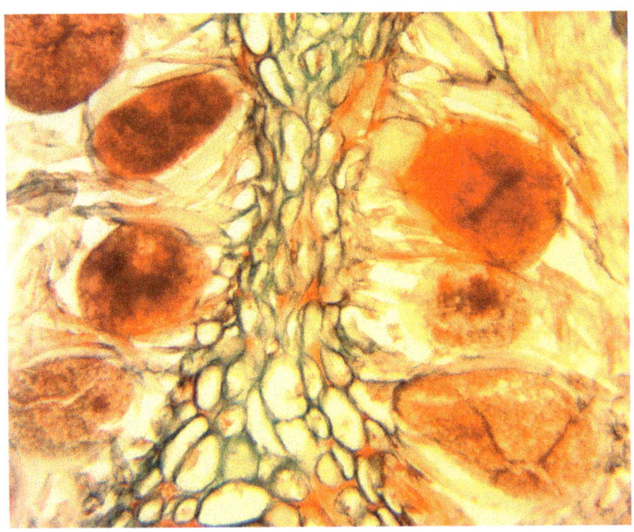

Figure 2.140 Photograph of a real object using a 10×/0.40 LCR MicroFluar objective.

immersion objectives are obvious. Water is not aggressive and does not provide a viscous environment; there is no toxicity. When using water immersion, it is easy to clean the frontal lens; the objective is less prone to

Figure 2.141 Photograph of a real object using a 10×/0.40 LCR MicroFluar objective.

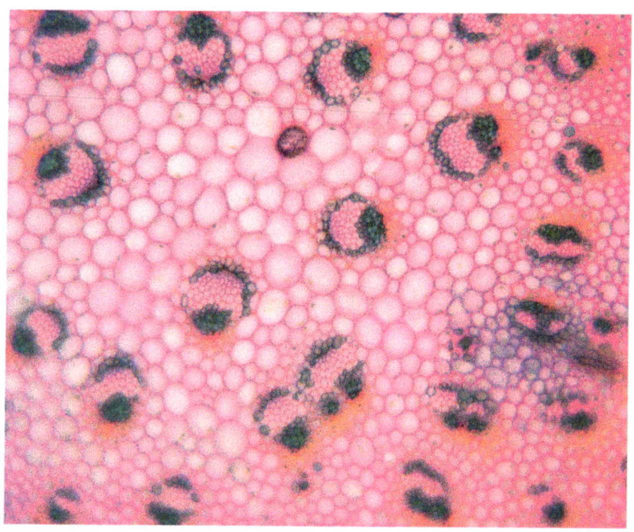

Figure 2.142 Photograph of a real object using a 10×/0.40 LCR MicroFluar objective.

destruction and thus more reliable and durable. All these conditions are especially important for fluorescent light microscopy. However, designing water immersion objectives raises theoretical and practical challenges. The characteristics of water determine the necessary shape of the lens with which water will be in contact, which in turn imposes limitations on the corrective aberration capabilities of the lens and the objective as a whole. The refractive index and dispersion characteristics of water as an optical material are not the

Figure 2.143 Photograph of a real object using a 10×/0.40 LCR MicroFluar objective.

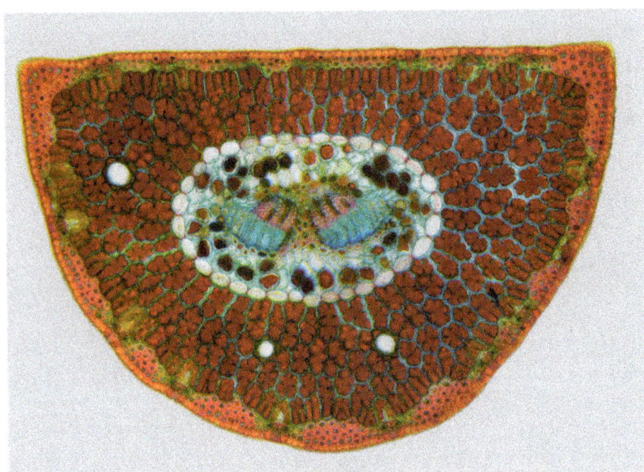

Figure 2.144 Photograph of a real object, *Pinus*, using a 20×/0.80 LCR MicroFluar objective (taken by Anatoly Mikhaltsov).

same as those of other immersion media. The frontal lenses of water immersion objectives should be made of optical materials that are resistant to the influence of a humid atmosphere. Designing water immersion objectives, especially those for luminescent analysis, is a time-consuming task that requires developers who have special knowledge as well as practical experience.

Figures 2.158–2.161 show the design parameters and graphs of aberration correction of these objectives. These objectives are designed to work in

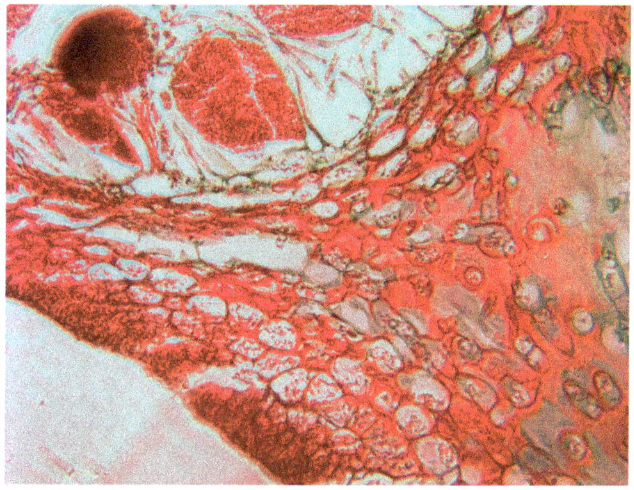

Figure 2.145 Photograph of a real object using a 20×/0.80 LCR MicroFluar objective.

Figure 2.146 Photograph of a real object using a 20×/0.80 LCR MicroFluar objective.

microscopes with an infinite tube length; the focal length of the additional tube system is $F' = 160$ mm. Table 2.11 shows the main technical parameters of special water immersion objectives that were designed to accommodate special requirements.

With these high numerical apertures, it is possible to fulfill the basic requirement: to obtain a large working distance for all objectives in the water environment. The advantage that these objectives provide, as compared with

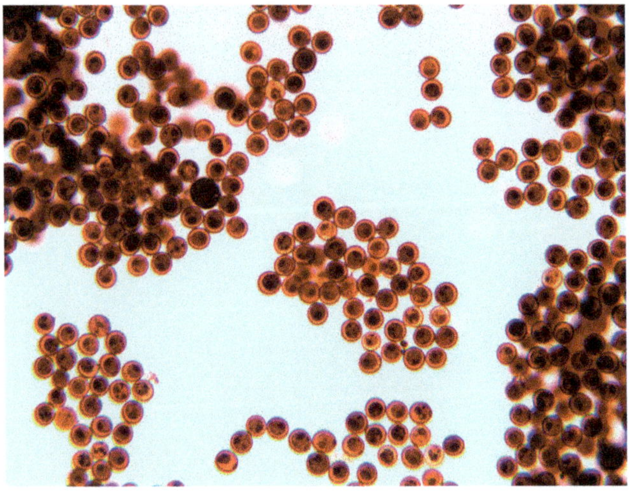

Figure 2.147 Photograph of a real object using a 20×/0.80 LCR MicroFluar objective.

Figure 2.148 Photograph of a real object using a 40×/0.95 LCR semi–plan MicroFluar objective.

other water immersion objectives, is their potential use for microscopic techniques, primarily luminescent analysis, based on micromanipulators, microinjection devices, and others. Such objectives can also be used in confocal ophthalmic microscopes, genetic engineering systems, etc. The refractive indexes and dispersion parameters of water are close to the parameters of the physiological fluid that is the basis of the environment for most living organisms.

Figure 2.149 Photograph of a real object using a 40×/0.95 LCR semi–plan MicroFluar objective.

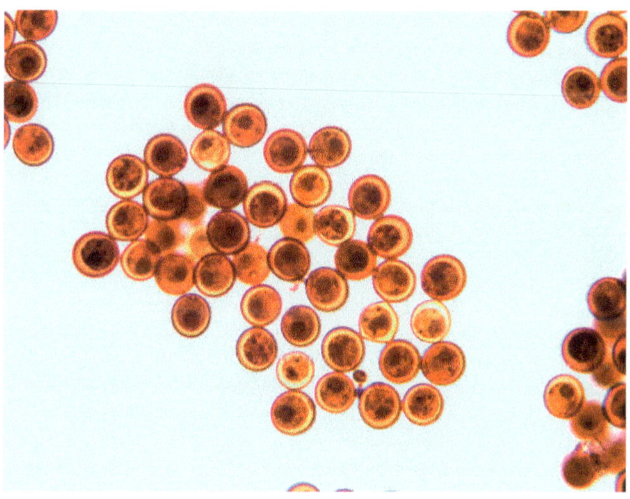

Figure 2.150 Photograph of a real object using a 40×/0.95 LCR semi–plan MicroFluar objective.

2.4.2.2 LWD dualchromat objectives

Another example of objectives that can be used for the above tasks are those designed primarily for use in inverted microscopes (although such objectives can also be used in so-called *upright* microscopes). These objectives have several design features that distinguish them from conventional objectives for light microscopes:

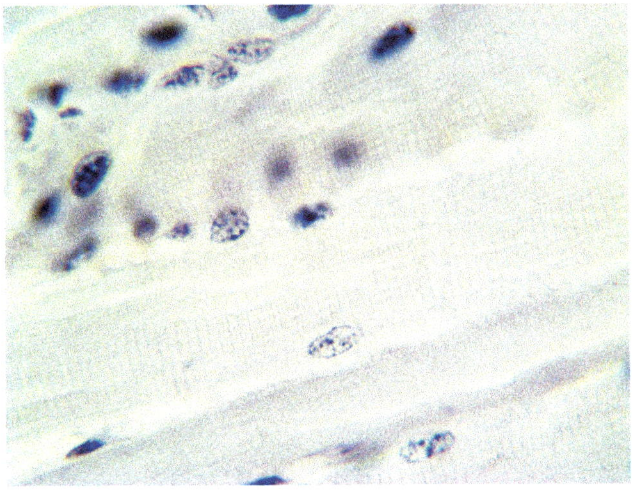

Figure 2.151 Photograph of a real object using a 100×/1.20 water immersion LCR semi–plan MicroFluar objective.

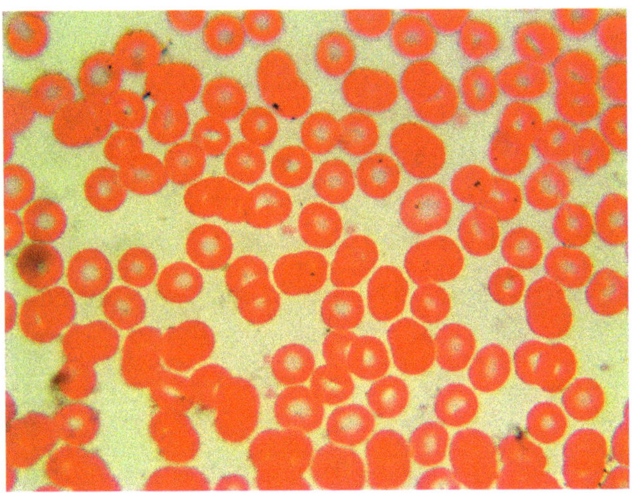

Figure 2.152 Photograph of a real object using a 100×/1.20 water immersion LCR semi–plan MicroFluar objective.

- A minimum number of lenses is required to obtain maximum transmission in the spectral region from 350 to 1800 nm. This does not mean that objective achromatization is performed for such a wide spectral range, but rather that objectives can be used to implement fluorescence techniques, including *anti-Stokes* methods, at the wavelength of the studied near-IR fluorescence. The optical materials and antireflection coatings used in the objectives provide a high-transmission requirement in the indicated spectral range.

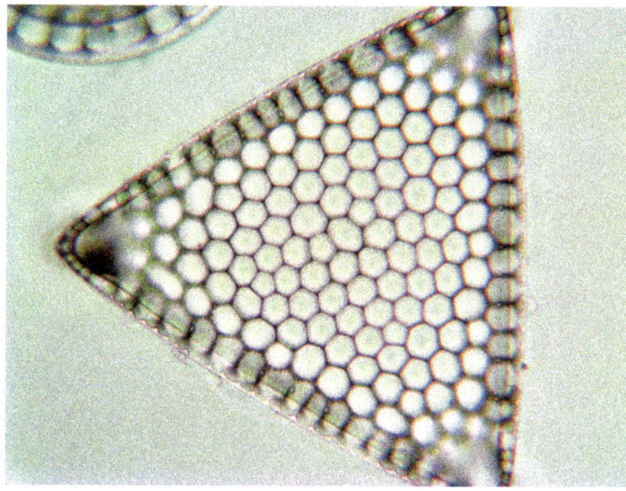

Figure 2.153 Photograph of a real object using a 100×/1.20 water immersion LCR semi–plan MicroFluar objective.

Figure 2.154 Photograph of a real object using a 100×/1.20 water immersion LCR semi–plan MicroFluar objective.

- An increased (long) free working distance makes it possible to use various kinds of manipulators, contrasting devices, and shielding devices (for example, color and interference filters, polarizer, screens, diaphragms, etc.).
- The aberration correction of these objectives is designed to account for the influence of the thickness of the bottom of the laboratory glassware that is typically used in inverted microscopes. Moreover, this type of

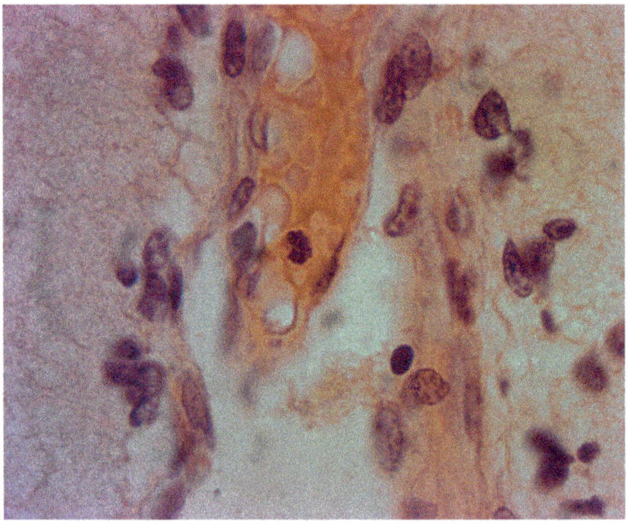

Figure 2.155 Photograph of a real object using a 100×/1.20 water immersion LCR semi–plan MicroFluar objective.

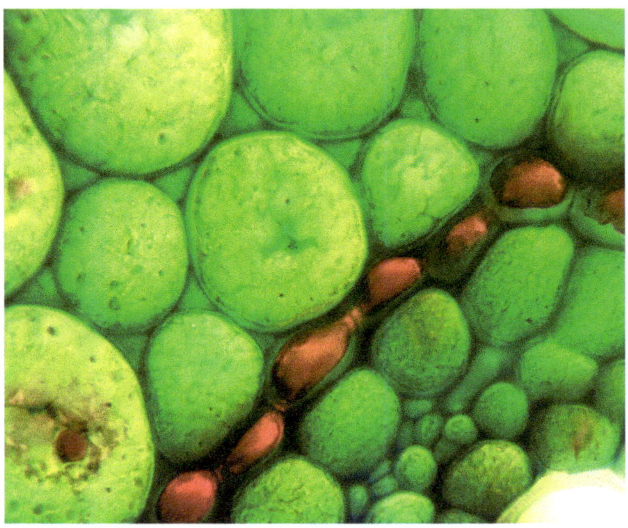

Figure 2.156 Photograph of a real object using a 100×/1.35 oil immersion LCR semi–plan MicroFluar objective.

aberration correction can be characterized as *dualchromat*, as described in Volume 1 (Chapter 2, Section 2.10.4). In the optical design for these objectives, the working (main) wavelength is 1480 ± 50 nm, and the auxiliary wavelength is 546.07 nm.

- The intrinsic height of each objective is no more than 25 mm, which makes it possible to integrate additional devices, such as beam-splitting

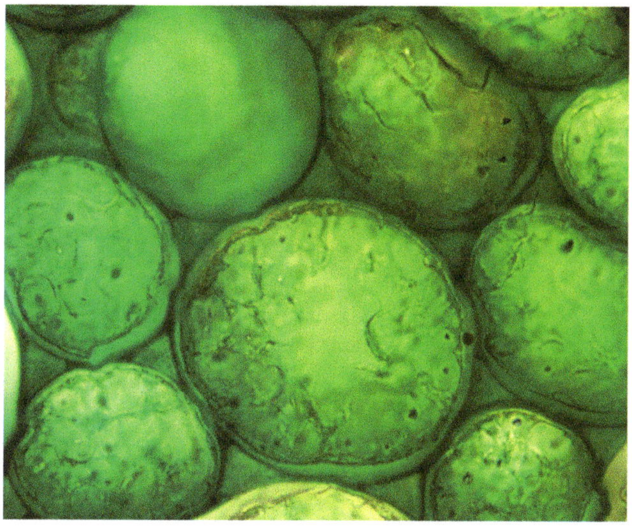

Figure 2.157 Photograph of a real object using a 100×/1.35 oil immersion LCR semi–plan MicroFluar objective.

prisms and mirrors, interference and diffraction elements, LED lighting systems, laser emitters, and other elements, into the mechanical design of the objectives. In some cases, these additional elements allow original microscopic research techniques to be implemented.

Figures 2.162–2.165 show the design parameters and graphs of aberration correction of LWD dualchromat objectives. These objectives are designed to work as part of microscopes with an infinite tube length; the focal length of the additional tube system is $F' = 160\,\text{mm}$. Table 2.12 shows the main technical parameters of these objectives.

An example of how these objectives are used is the ZILOS-tk® device in which 40× objectives have been incorporated for more than 15 years. Figure 2.166 shows the design of these devices (using a 40× objective). These objectives can also be used for *two-photon microscopy*, which allows images of living tissue to be obtained with a depth of up to 1 mm.

2.4.2.3 Correction frame objectives for inverted biological microscopes

Some objectives designed for use in inverted microscopes allow for a so-called *correction frame*. The basic mechanical design of a correction frame is shown in Volume 1, Fig. 1.17. Some objectives are able to correct aberrations that occur when the actual thickness of the cover glass deviates from its expected value; as a rule, a corrective frame allows objectives to be used with coverslips that are between 0.10- and 0.20-mm thick. However, for inverted microscopes, the glass thickness (the bottom thickness of laboratory glassware) is assumed to be 1.2 mm, and the permissible limits for its change

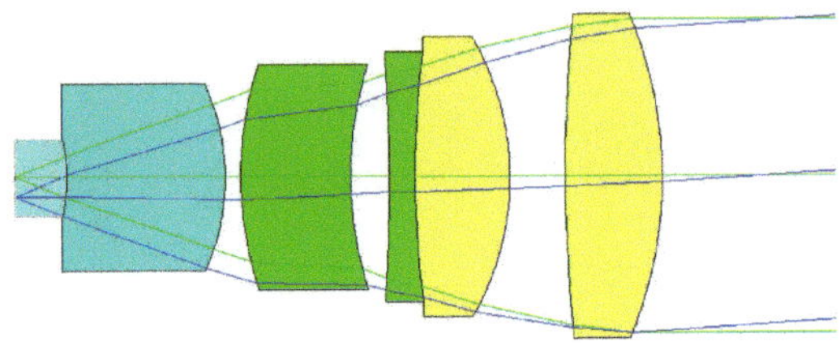

Lens: OF 10x0.50 LD water ftl 160

Object num aper	0.500000	Image height	10.000000	Primary wavln	0.546070

SRF	RADIUS	THICKNESS	APERTURE RADIUS	GLASS	SPECIAL
OBJ	0.000000	1.0000e-20	0.970813	FISIOL	
AST	0.000000	2.700000	0.970813 AS	FISIOL P	
2	-8.485500	8.000000	2.000000	LZ_CTK12	
3	-10.847900	0.800000	4.650000	AIR	
4	17.761100	5.500000	5.460000	LZ-TF5	
5	17.959200	2.000000	5.560000	AIR	
6	-79.951000	1.300000	5.830000	LZ-TF5 P	
7	45.328000	4.800000	6.220000	CAF2	
8	-13.700000	2.800000	6.930000	AIR	
9	74.683000	4.900000	7.770000	CAF2 P	
10	-20.551000	90.000000	8.000000	AIR P	
11	214.800000	5.000000	13.930000	LZ-BK8	
12	-46.990000	0.300000	13.970000	AIR	
13	-46.880000	3.000000	13.930000	LZ-BF24	
14	-113.240000	157.854180 S	14.070000	AIR	
IMS	0.000000	0.000000	10.000000 S		

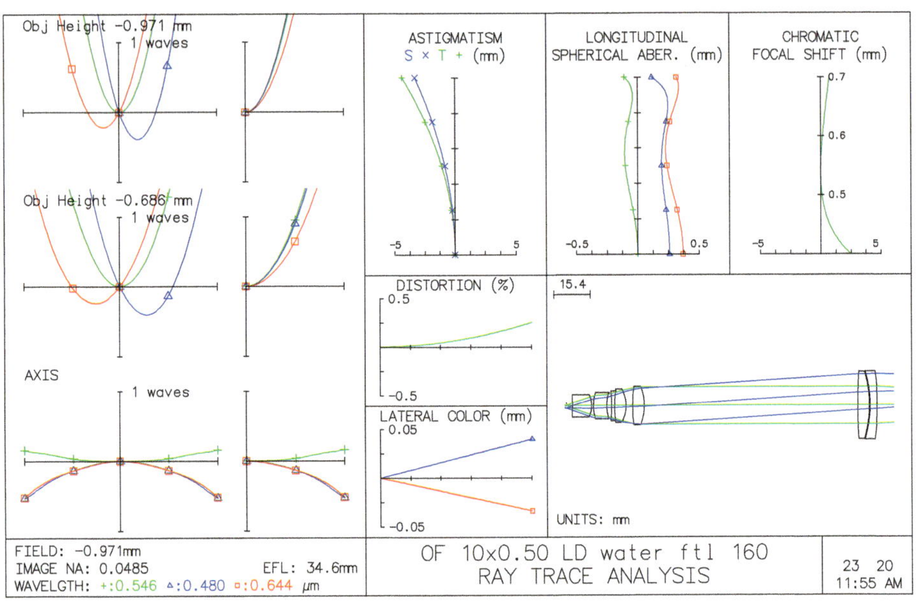

Figure 2.158 Optical design and graphs of aberration correction of a 10×/0.50 water immersion infinity LCR LD MicroFluar objective.

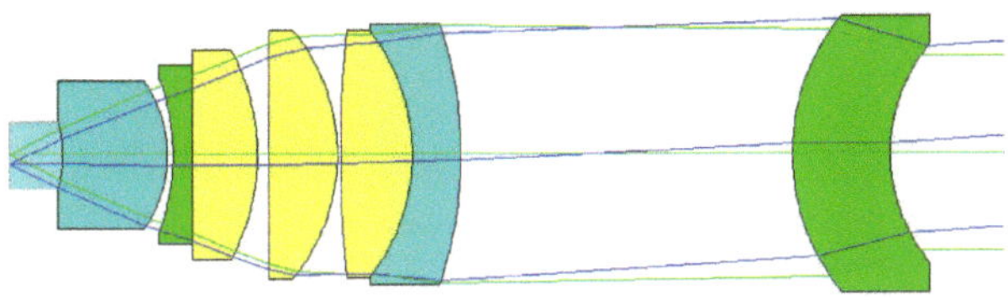

Lens: OF 20x0.60 LD water ftl 160

Object num aper 0.600000 Image height 10.000000 Primary wavln 0.546070

SRF	RADIUS	THICKNESS	APERTURE RADIUS	GLASS	SPECIAL
OBJ	0.000000	1.0000e-20	0.509884	FISIOL	
AST	0.000000	2.700000	0.509884 AS	FISIOL P	
2	-7.656000	5.380000	1.760000	LZ_CTK12	
3	-6.400000	0.300000	3.710000	AIR	
4	-10.940000	1.000000	3.920000	LZ-TF5	
5	0.000000	3.330000	4.480000	CAF2	
6	-10.940000	0.600000	5.280000	AIR	
7	0.000000	3.520000	5.960000	CAF2	
8	-9.550000	0.160000	6.240000	AIR	
9	66.070000	3.700000	6.210000	CAF2 P	
10	-9.817000	2.500000	6.160000	LZ_CTK19	
11	-20.320000	17.000000	6.550000	AIR	
12	10.940000	4.940000	6.940000	LZ-TF5 P	
13	8.460000	90.000000	5.520000	AIR	
14	214.800000	4.940000	10.920000	LZ-BK8	
15	-46.990000	0.300000	10.990000	AIR	
16	-46.880000	3.000000	10.960000	LZ-BF24	
17	-113.240000	157.378060 S	11.090000	AIR	
IMS	0.000000	0.000000	10.000000 S		

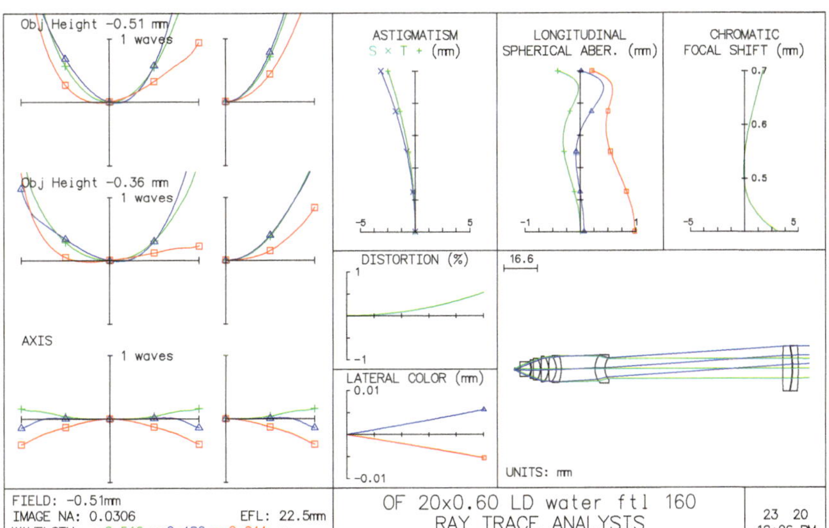

Figure 2.159 Optical design and graphs of aberration correction of a 20×/0.60 water immersion infinity LCR LD MicroFluar objective.

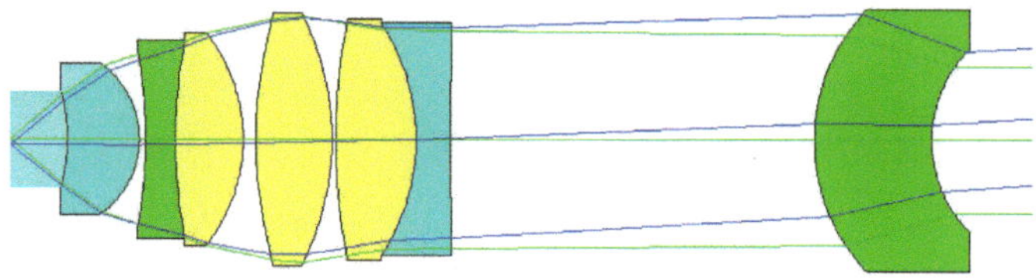

Lens: OF 40x0.85 LD water ftl 160

Object num aper 0.850000 Image height 10.000000 Primary wavln 0.546070

SRF	RADIUS	THICKNESS	APERTURE RADIUS	GLASS	SPECIAL
OBJ	0.000000	1.0000e-20	0.247093	FISIOL	
AST	0.000000	2.700000	0.247093 AS	FISIOL P	
2	-7.656000	3.450000	2.210000	LZ_CTK12	
3	-4.093000	0.300000	3.480000	AIR	
4	-23.230000	1.400000	4.010000	LZ-TF5	
5	23.230000	3.200000	4.520000	CAF2	
6	-7.656000	0.600000	4.880000	AIR	
7	19.588000	3.600000	5.640000	CAF2	
8	-12.474000	0.160000	5.750000	AIR	
9	27.425000	3.700000	5.510000	CAF2 P	
10	-9.550000	1.600000	5.290000	LZ_CTK19	
11	216.810000	17.000000	5.310000	AIR	
12	8.460000	5.500000	5.990000	LZ-TF5 P	
13	5.830000	90.000000	4.170000	AIR	
14	214.800000	4.940000	9.250000	LZ-BK8	
15	-46.990000	0.300000	9.330000	AIR	
16	-46.880000	3.000000	9.310000	LZ-BF24	
17	-113.240000	157.804261 S	9.420000	AIR	
IMS	0.000000	0.000000	10.000000 S		

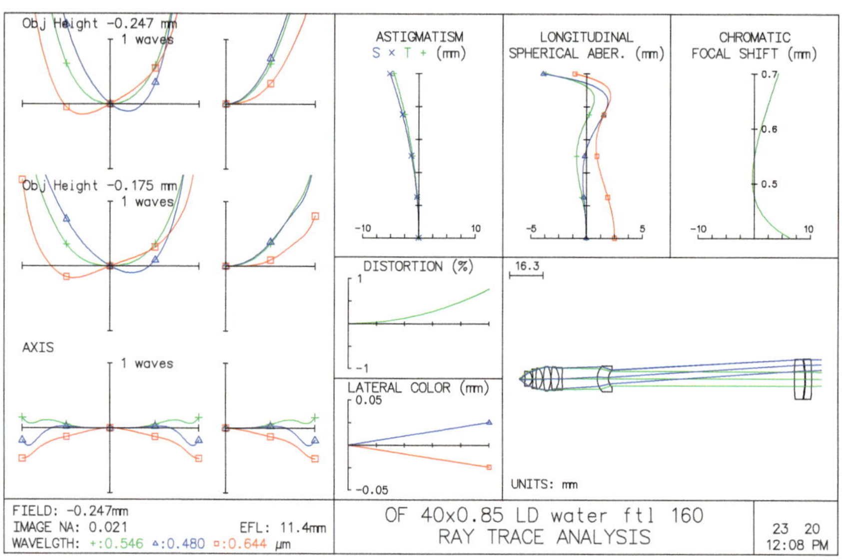

Figure 2.160 Optical design and graphs of aberration correction of a 40×/0.85 water immersion infinity LCR LD MicroFluar objective.

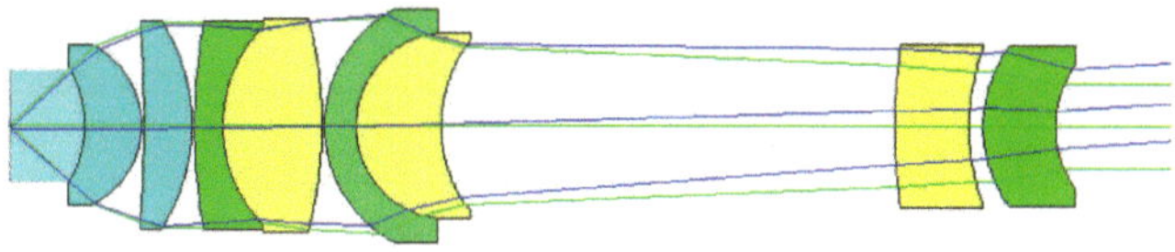

Lens: OF 100x0.90 LD water ftl 160

Object num aper 0.900000 Image height 10.000000 Primary wavln 0.546070

SRF	RADIUS	THICKNESS	APERTURE RADIUS	GLASS	SPECIAL
OBJ	0.000000	1.0000e-20	0.101500	FISIOL	
AST	0.000000	2.700063	0.101500 AS	FISIOL P	
2	-3.809000	2.000000	2.030000	LZ_CTK12	
3	-3.046900	0.100000	2.750000	AIR	
4	-36.848300	1.700000	3.380000	LZ-CTK19	
5	-6.488300	0.100000	3.580000	AIR	
6	20.362400	1.000000	3.570000	LZ-TF5	
7	4.793800	3.500000	3.430000	CAF2	
8	-14.465000	0.100000	3.680000	AIR	
9	4.484500	1.100000	4.020000	LZ-TF1	
10	3.206400	3.000000	3.190000	CAF2 P	
11	4.683200	16.000000	2.930000	AIR	
12	17.783600	2.600000	2.820000	CAF2 P	
13	7.546500	0.500000	2.670000	AIR	
14	4.521900	2.600000	2.750000	LZ-TF5 P	
15	3.608600	90.000000	2.130000	AIR	
16	214.800000	5.000000	7.660000	LZ-BK8	
17	-46.990000	0.300000	7.780000	AIR	
18	-46.880000	3.000000	7.760000	LZ-BF24	
19	-113.240000	157.811101 S	7.880000	AIR	
IMS	0.000000	0.000000	10.000000 S		

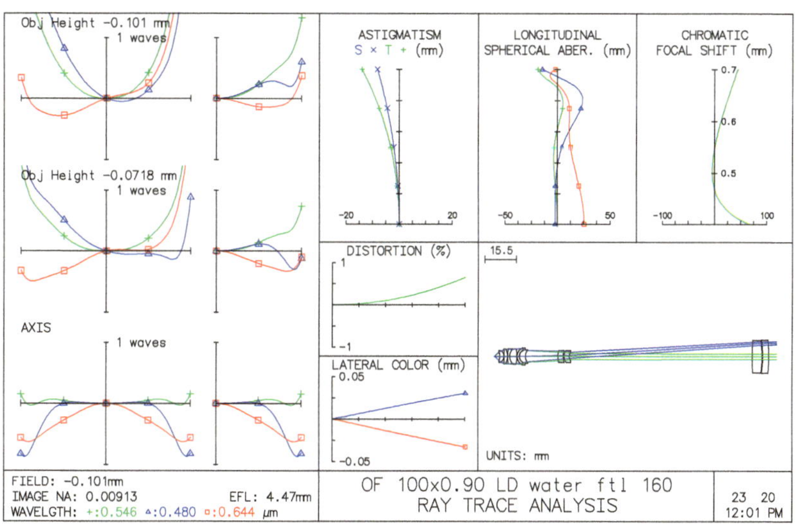

Figure 2.161 Optical design and graphs of aberration correction of a 100×/0.90 water immersion infinity LCR LD MicroFluar objective.

Table 2.11 Main technical parameters and basic optical layout of LWD water immersion (wi) objectives.

Magnification	*NA*	*WD* (mm)	*F′* (mm)	*R* (μm)	*DF* (μm)	*FOV* on object (mm)	*FOV* on image (mm)	Optical design at figure no.	The principal optical layout
10×	0.50 wi	2.1	16	0.67	1.33	2.0	20	2.158	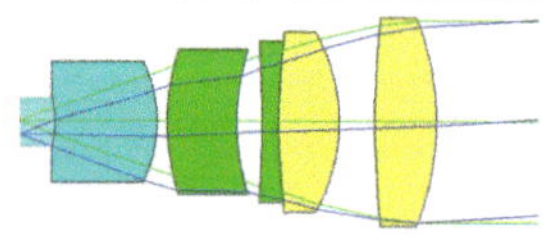
20×	0.60 wi	2.1	8	0.56	0.93	1.0	20	2.159	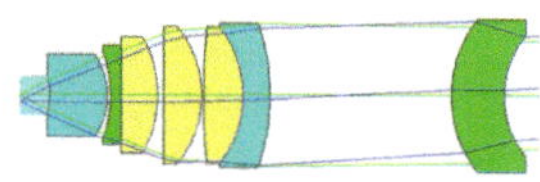
40×	0.85 wi	2.1	4	0.39	0.46	0.5	20	2.160	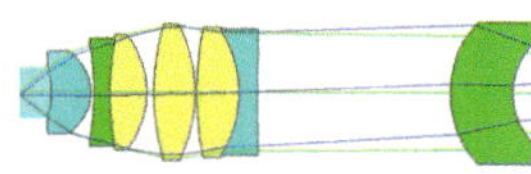
100×	0.90 wi	1.9	1.6	0.37	0.41	0.2	20	2.161	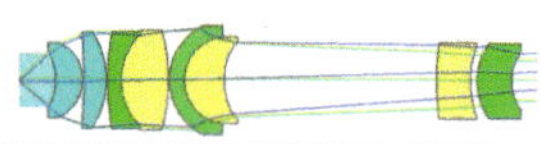

reach ±0.3 mm. In practice, an important characteristic of the objective is not only its degree of aberration correction (for example, to a glass thickness of 1.2 ± 0.3 mm), but also its potential to realize the largest possible value of the free working distance for viewing the object in depth to a considerable (by microscopic standards) distance. To achieve this, an *elongated* correction frame can be used in some objectives, which allows aberration correction when penetrating the sample thickness up to 1–1.5 mm. The optical design of an objective should allow for this elongated correction frame. Figures 2.167 and 2.168 show the design parameters and graphs of aberration correction of such objectives, and Figure 2.169 is a photograph of an objective with a linear magnification of 25× and a numerical aperture of 0.45.

One method for performing microscopic work similar to that done on an inverted biological microscope may be to use a conventional direct microscope and original objectives. Such objectives must have an original mechanical design that allows them to be waterproof while the objects under study are immersed in physiological substances. It is not too difficult to design and manufacture such objectives; in this case, the necessary depth of penetration into the thickness of physiological saline can always be accommodated, and the aberration correction does not need to take into account aberrations introduced by the laboratory glassware.

One such example are LWD water immersion objectives, the optical design of which is presented in Section 2.4.2.1. The body and the outer frame of the objective must be waterproof. The objective dips into the layer of physiological fluid in which the studied object is located; at the same time, it is possible to precisely study the upper part of the object (to investigate the

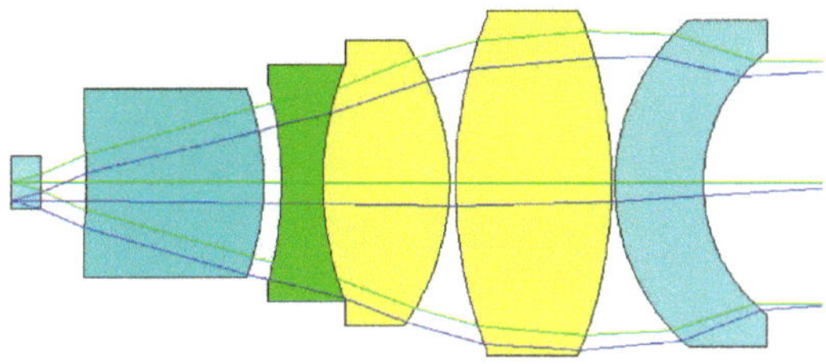

Lens:dualchromat 16x0.40 ftl160h=25mm

Object num aper 0.400000 Image height 10.000000 Primary wavln 1.480000

SRF	RADIUS	THICKNESS	APERTURE RADIUS	GLASS	SPECIAL
OBJ	0.000000	0.010000	0.616883	AIR	
AST	0.000000	1.000000	0.621247 AS	USERGLASS	
2	0.000000	1.531300	0.880000	AIR	
3	-16.482000	6.000000	1.520000	LZ_CTK12	
4	-8.632000	0.600000	3.110000	AIR	
5	-12.445000	1.400000	3.280000	LZ_TF3	
6	10.290000	4.300000	3.920000	CAF2	
7	-8.017000	0.200000	4.720000	AIR	
8	14.944000	5.300000	5.520000	CAF2 P	
9	-14.944000	0.100000	5.720000	AIR	
10	7.060000	3.000000	5.390000	LZ_K8	
11	5.546000	90.000000	4.340000	AIR	
12	214.800000	5.000000	9.210000	LZ_BK8	
13	-46.990000	0.300000	9.310000	AIR	
14	-46.880000	3.000000	9.290000	LZ_BF24	
15	-113.240000	160.757705 S	9.410000	AIR	
IMS	0.000000	0.000000	10.000000 S		

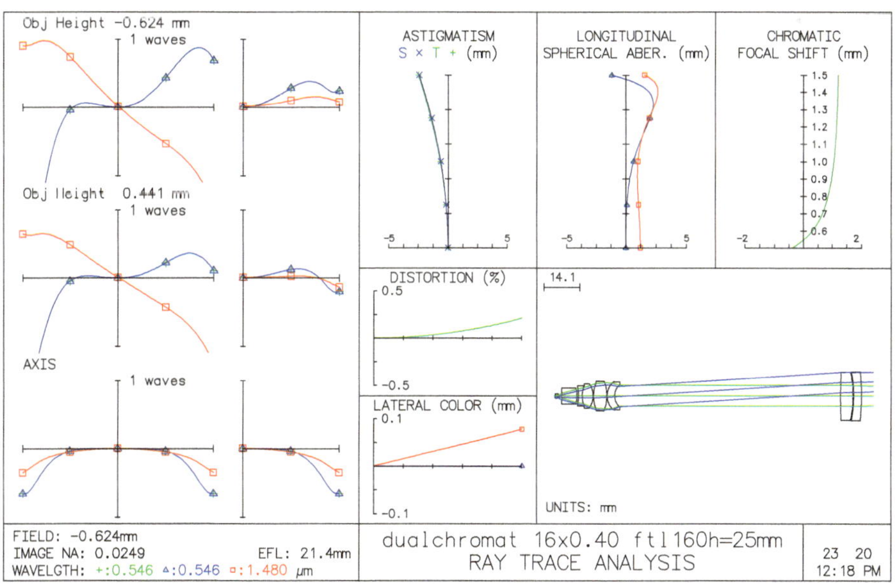

Figure 2.162 Optical design and graphs of aberration correction of a 16×/0.40 infinity dualchromat objective ($d = 1.0$ mm).

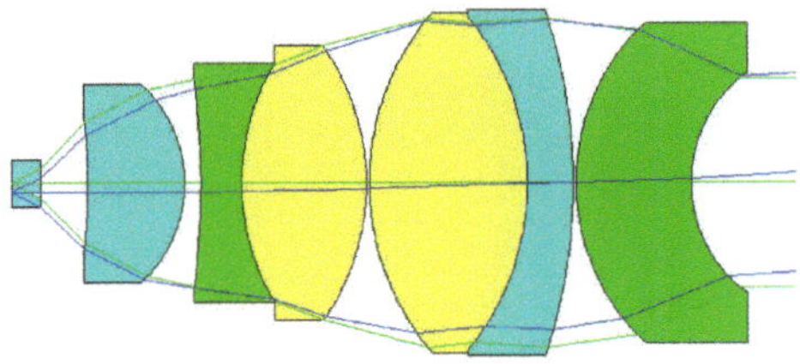

Lens: dualchromat 32x0.70 h25 ftl160

Object num aper 0.700000 Image height 10.000000 Primary wavln 1.480000

SRF	RADIUS	THICKNESS	APERTURE RADIUS	GLASS	SPECIAL
OBJ	0.000000	0.010000	0.309544	AIR	
AST	0.000000	1.000000	0.319345 AS	USERGLASS	
2	0.000000	1.675700	0.820000	AIR	
3	-22.438000	3.400000	2.340000	LZ_CTK19	
4	-4.407000	0.570000	3.350000	AIR	
5	-27.538300	1.400000	3.640000	LZ_TF5	
6	7.870000	4.300000	4.040000	CAF2	
7	-7.870000	0.100000	4.640000	AIR	
8	8.669000	5.500000	5.740000	CAF2 P	
9	-8.669000	1.600000	5.640000	LZ_CTK19	
10	-17.867800	0.100000	5.840000	AIR	
11	7.279000	4.000000	5.420000	LZ_TF5 P	
12	4.530000	90.000000	3.730000	AIR	
13	214.800000	5.000000	8.740000	LZ_BK8	
14	-46.990000	0.300000	8.830000	AIR	
15	-46.880000	3.000000	8.810000	LZ_BF24	
16	-113.240000	160.769139 S	8.920000	AIR	
IMS	0.000000	0.000000	10.000000 S		

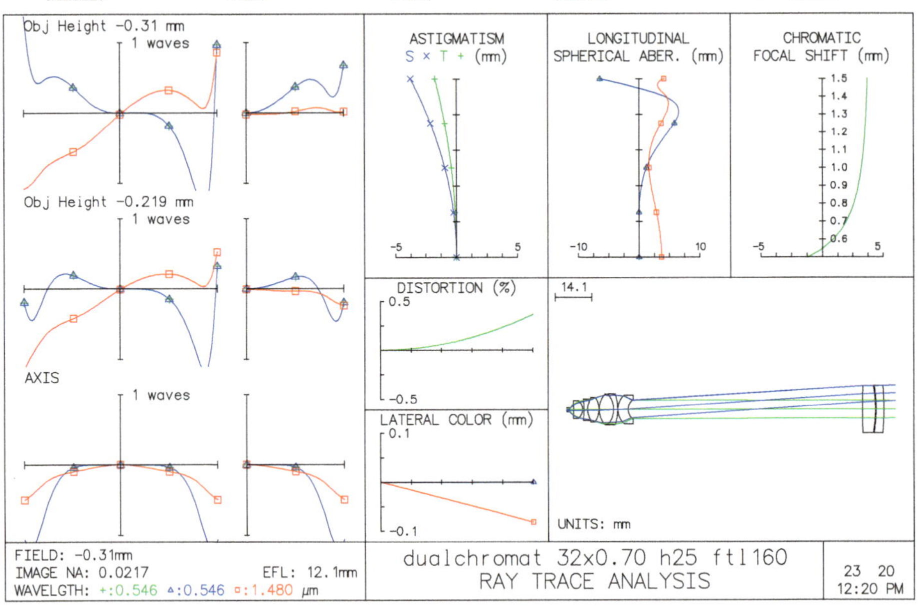

Figure 2.163 Optical design and graphs of aberration correction of a 32×/0.70 infinity dualchromat objective ($d = 1.0$ mm).

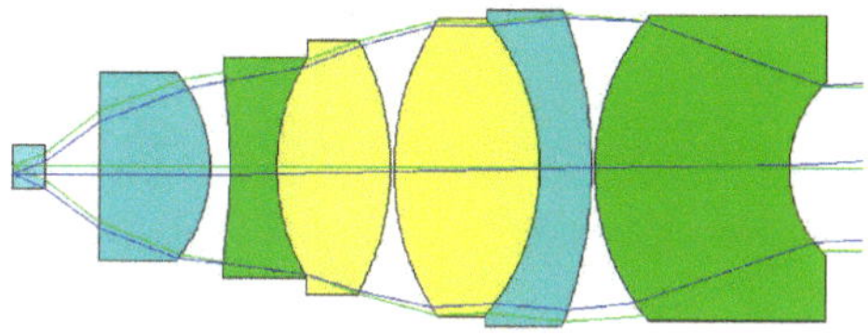

Lens:dualchromat 40x0.60 ftl160h=25mm

Object num aper 0.600000 Image height 10.000000 Primary wavln 1.480000

SRF	RADIUS		THICKNESS		APERTURE RADIUS		GLASS		SPECIAL
OBJ	0.000000		0.010000		0.249341		AIR		
AST	0.000000		1.000000		0.256841	AS	USERGLASS		
2	0.000000		1.673300		0.670000		AIR		
3	216.810000		3.400000		1.930000		LZ_CTK12		
4	-4.365000		0.600000		2.830000		AIR		
5	-20.320000		1.450000		2.990000		LZ_TF5		
6	6.400000		3.500000		3.320000		CAF2		
7	-7.656000		0.100000		3.810000		AIR		
8	8.017000		4.500000		4.470000		CAF2	P	
9	-6.730000		1.600000		4.440000		LZ_CTK19		
10	-13.100000		0.100000		4.740000		AIR		
11	7.060000		6.050000		4.590000		LZ_TF5	P	
12	3.597000		90.000000		2.600000		AIR		
13	214.800000		5.000000		7.890000		LZ_BK8		
14	-46.990000		0.300000		8.000000		AIR		
15	-46.880000		3.000000		7.990000		LZ_BF24		
16	-113.240000		160.715357	S	8.100000		AIR		
IMS	0.000000		0.000000		10.000000	S			

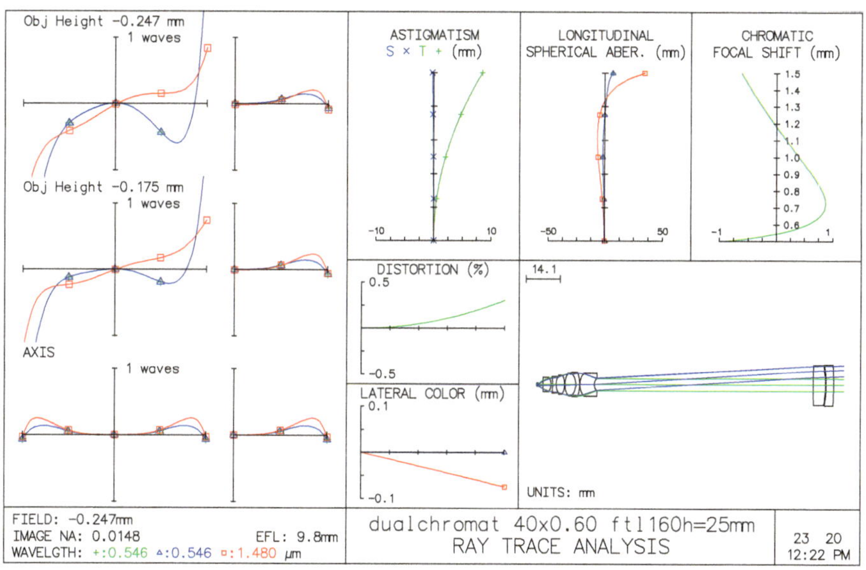

Figure 2.164 Optical design and graphs of aberration correction of a 40×/0.60 infinity dualchromat objective ($d = 1.0$ mm).

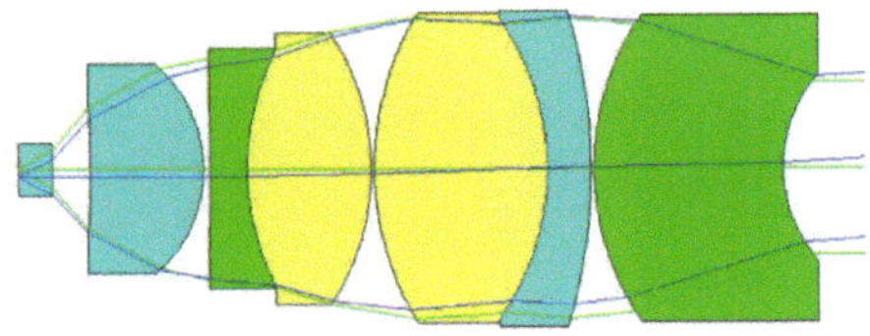

Lens: dualchromat 50x0.75 h25 ftl160

Object num aper	0.750000	Image height	10.000000	Primary wavln	1.480000

SRF	RADIUS	THICKNESS	APERTURE RADIUS	GLASS	SPECIAL
OBJ	0.000000	0.010000	0.208390	AIR	
AST	0.000000	1.000000	0.219729 AS	USERGLASS	
2	0.000000	1.130900	0.760000	AIR	
3	-25.940000	3.400000	1.960000	LZ_CTK19	
4	-3.908000	0.200000	3.050000	AIR	
5	-216.300000	1.100000	3.310000	LZ_TF10	
6	7.447000	3.700000	3.500000	CAF2	
7	-7.060000	0.100000	3.970000	AIR	
8	8.460000	5.200000	4.520000	CAF2 P	
9	-7.447000	1.300000	4.390000	LZ_TBF10	
10	-15.596000	0.100000	4.600000	AIR	
11	7.656000	5.640000	4.470000	LZ_TF10 P	
12	4.093000	90.000000	2.750000	AIR	
13	214.800000	5.000000	8.730000	LZ_BK8	
14	-46.990000	0.300000	8.850000	AIR	
15	-46.880000	3.000000	8.830000	LZ_BF24	
16	-113.240000	160.651723 S	8.970000	AIR	
IMS	0.000000	0.000000	10.000000 S		

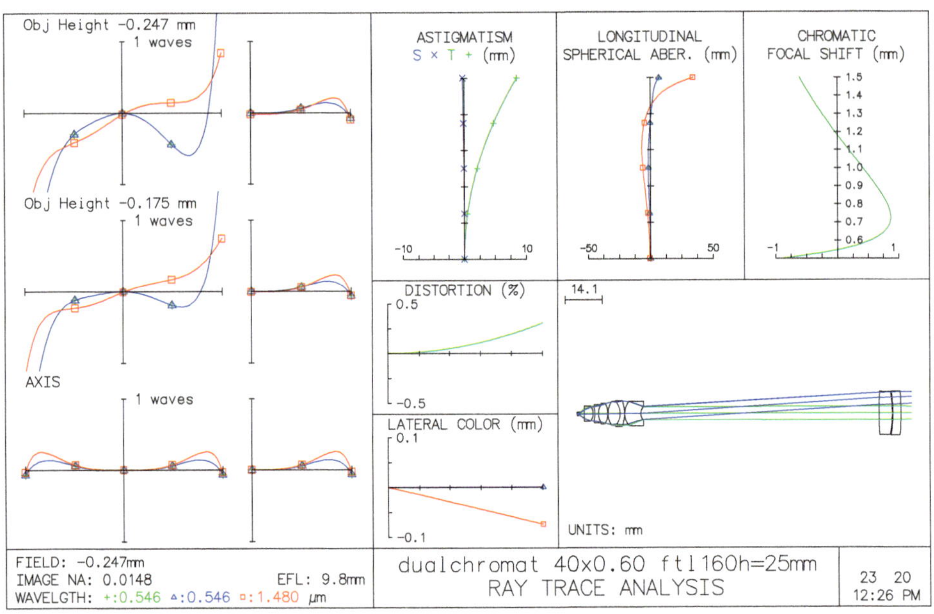

Figure 2.165 Optical design and graphs of aberration correction of a 50×/0.75 infinity dualchromat objective ($d = 1.0$ mm).

Table 2.12 Main technical parameters and basic optical layout of LWD dualchromat objectives.

Magnification	*NA*	*WD* (mm)	*F'* (mm)	*R* (μm)	*DF* (μm)	*FOV* on object (mm)	*FOV* on image (mm)	Optical design at figure no.	The principal optical layout
16×	0.40	1.2	10	2.26	5.64	1.25	20	2.162	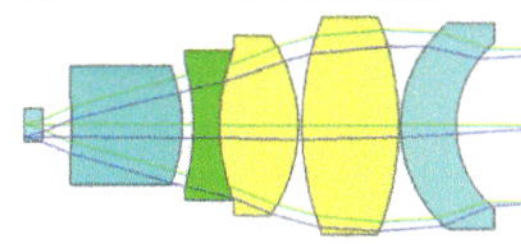
32×	0.70	1.2	58	1.29	1.84	0.625	20	2.163	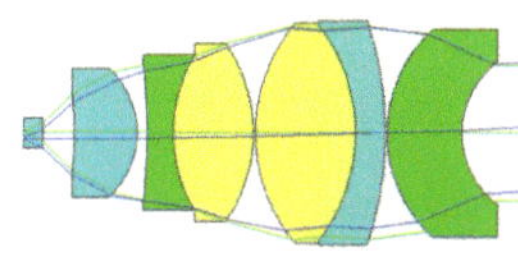
40×	0.60	1.6	4	1.5	2.5	0.5	20	2.164	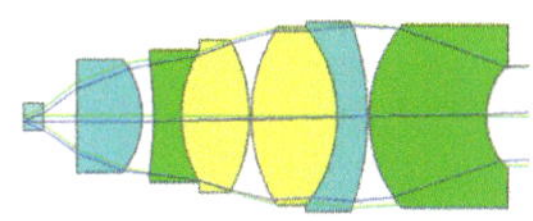
50×	0.75	0.82	3.2	1.2	1.6	0.4	20	2.165	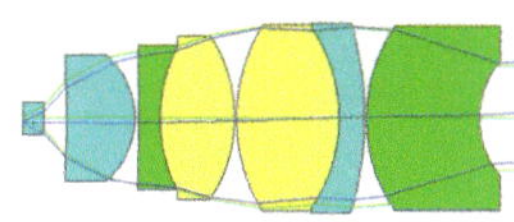

Figure 2.166 An example of a device, a ZILOS-tk, incorporating a 40× objective.

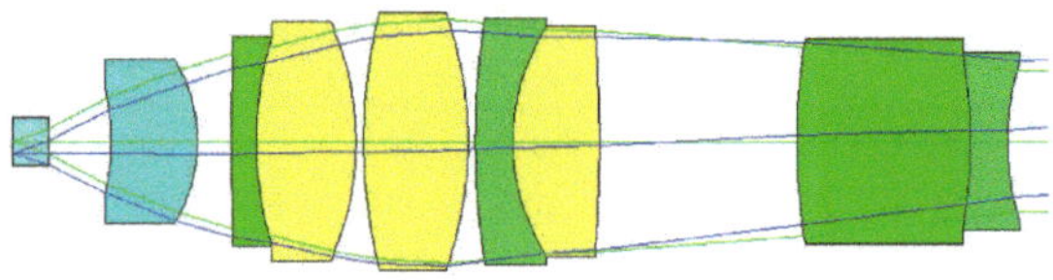

Lens: 25x0.45 d=1.0-2.0 Finity — Efl 6.53

Object num aper 0.450000 — Image height 11.000000 — Primary wavln 0.546070

SRF	RADIUS	THICKNESS	APERTURE RADIUS	GLASS	SPECIAL
OBJ	0.000000	0.010000	0.439791	PHIZIO	
AST	0.000000	1.500000	0.443368 AS	LZ_LK5	
2	0.000000	2.580000	0.930000	AIR	
3	-9.030000	3.500000	2.110000	LZ_TK14	
4	-6.194000	1.400000	3.240000	AIR	
5	200.400000	1.000000	3.890000	LZ_TF1	
6	13.335000	4.000000	4.150000	CAF2	
7	-11.668000	0.300000	4.700000	AIR	
8	20.420000	4.200000	5.000000	CAF2 P	
9	-14.655000	0.300000	5.070000	AIR	
10	31.120000	1.500000	4.870000	LZ_OF4	
11	8.204000	3.500000	4.530000	CAF2 P	
12	-65.610000	8.000000	4.410000	AIR	
13	24.890000	7.000000	4.060000	LZ_TF4	
14	-21.280000	1.500000	3.510000	LZ_OF4 P	
15	10.423000	151.800000	3.280000	AIR	
IMS	0.000000	0.000000	10.775929 S		

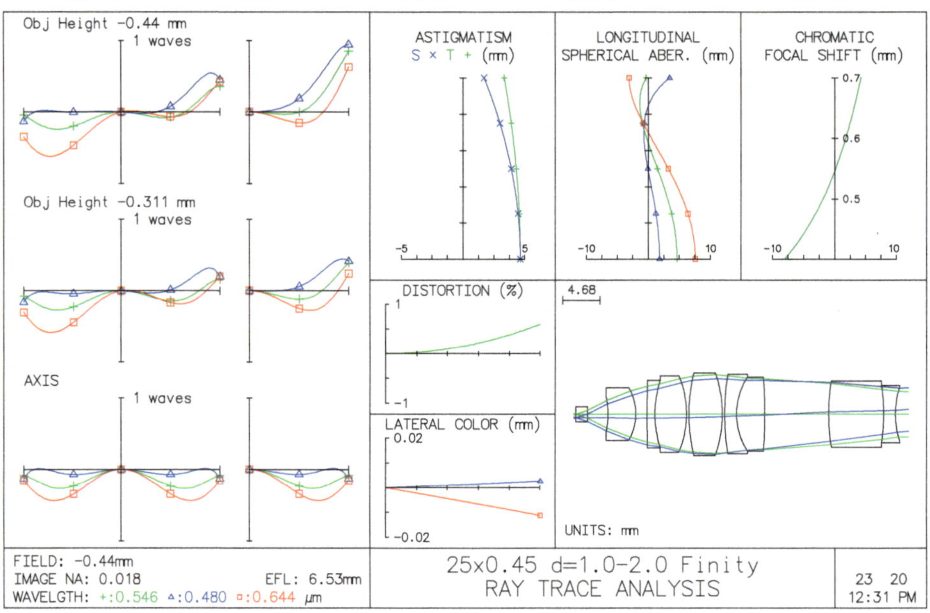

Figure 2.167 Optical design and graphs of aberration correction of a 25×/0.45 finite plan MicroFluar objective ($d = 1.5$ mm; nominal).

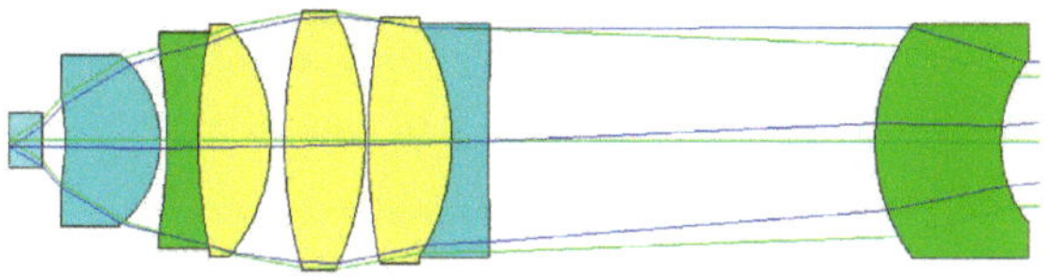

Lens: 40x0.75 d=1.5 Finity   Efl 3.97

Object num aper 0.750000 Image height 11.000000 Primary wavln 0.546070

SRF	RADIUS	THICKNESS	APERTURE RADIUS	GLASS	SPECIAL
OBJ	0.000000	0.010000	0.276213	PHIZIO	
AST	0.000000	1.500000	0.282995 AS	LZ_LK5	
2	0.000000	1.000000	1.160000	AIR	
3	-9.817000	4.200000	2.050000	LZ_CTK19	
4	-4.613000	0.300000	3.680000	AIR	
5	-23.230000	1.400000	4.140000	LZ_TF5	
6	23.230000	3.200000	4.650000	CAF2	
7	-7.656000	0.600000	4.950000	AIR	
8	19.588000	3.600000	5.470000	CAF2 P	
9	-12.474000	0.133800	5.540000	AIR	
10	27.425000	3.700000	5.280000	CAF2 P	
11	-9.550000	1.600000	5.010000	LZ_CTK19 P	
12	216.810000	17.000000	4.980000	AIR	
13	8.460000	5.570000	5.010000	LZ_TF5 P	
14	5.544000	148.149247 S	3.480000	AIR	
IMS	0.000000	0.000000	11.000000 S		

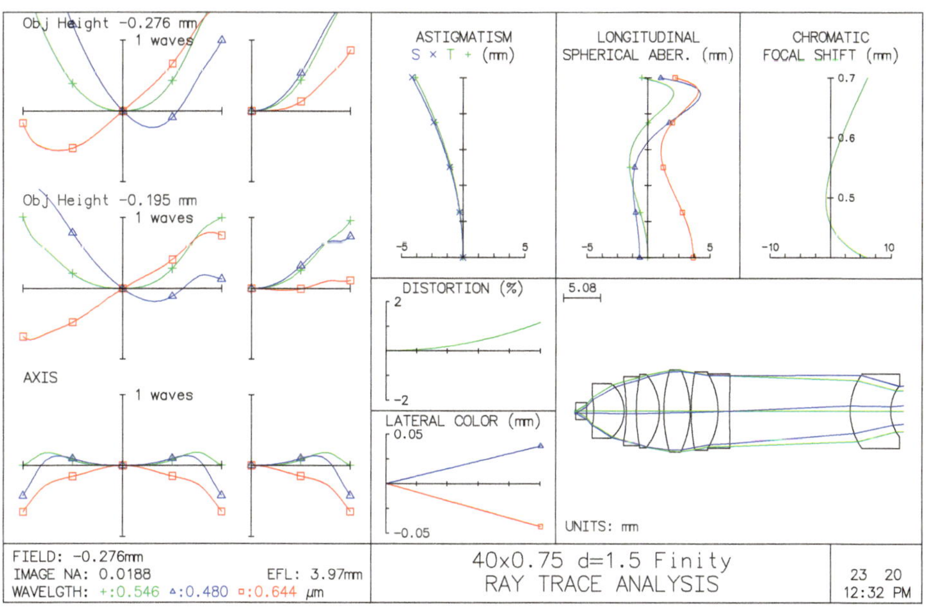

Figure 2.168 Optical design and graphs of aberration correction of a 40×/0.75 finite plan MicroFluar objective ($d = 1.5$ mm; nominal).

Figure 2.169 Photograph of a 25×/0.45 plan MicroFluar objective for an inverted biological microscope.

object's growth). Another example is the use of so-called *contact* objectives. A description of a device for research using contact microscopy is given in Volume 1 (Chapter 1, Section 1.2.6.6).

2.4.2.4 Objectives consisting of non-glued lenses

Let us consider glued components in the optical systems of lens objectives. Note that using glued components in these systems has both advantages and disadvantages. The motive for using glued components has always been to achieve the highest quality correction of chromatic aberrations. There are different opinions about and methods for choosing the optical materials (glass and/or crystal) that should be used in such components. For the optical design, the selected grades of glass in a double-glued component should have the maximum difference in dispersion values (the Abbe number) for the greatest impact on chromatic aberration. Additionally, to effectively correct chromatic aberrations such as the secondary spectrum, optical materials must be chosen with the closest possible relative partial dispersion values. When constructing the optical systems of objective lenses for microscopes, two-, three-, and even four-glued components are frequently used. An additional advantage of using several-glued components is a significant simplification of the mechanical design of the objective. In this case, the insertion of the frame is carried out for only one lens component. The remaining lenses are glued to the base lens. Modern bonding technologies can withstand the required accuracy of shaping the relative position of the lenses in the component and in the objective as a whole. Figure 2.170 presents example drawings of mounting

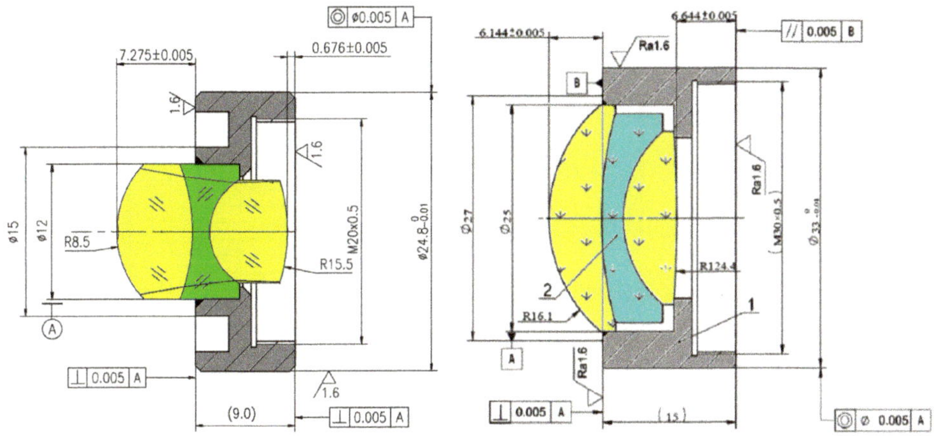

Figure 2.170 Examples of mechanical designs of three-glued components.

lenses in a frame with a mechanical design. Although these are not necessarily the most successful optical–mechanical constructions, they can be used in real objectives.

Objectives designed without glued components are required for some microscopy applications. Such objectives present a serious challenge for designing the optical system and greatly complicate the mechanical design. For these objectives, the mechanical rim of each lens must be designed separately because all the lenses are made individually. There may also be additional difficulties in attaching each of the lenses to the frame. Glue cannot necessarily be used for this process. In some cases, the *burnished lip* method must be used (a technique that is not used in typical microscope objectives). Figure 2.171 shows an assembly drawing of a Leica objective in which single lenses are used. We can see that this is an extremely complex optical and mechanical design.

Given their complexity, why would objectives without glued components be made? First, these objectives allow the microscope to be operated in conditions that are different from those in the laboratory. Indeed, in research laboratories where microscopes are most often used, the requirements for the atmosphere, including humidity, temperature, and other conditions, are quite stringent. If an objective is used in the field or in mobile laboratories, where there may be fluctuations in air temperature and atmospheric pressure, it should provide the effect of *image quality constant stability* if it (together with a microscope or separately) is transported, moving from one place to another. During vibrations and with long-term exposure to other external factors, the optical system, as well as the optical–mechanical design, must maintain the integrity of the components as well as the permanence of their mutual arrangement in the objective with high accuracy. One of these effects can be from an objective in conjunction with a laser, when it is required to focus a

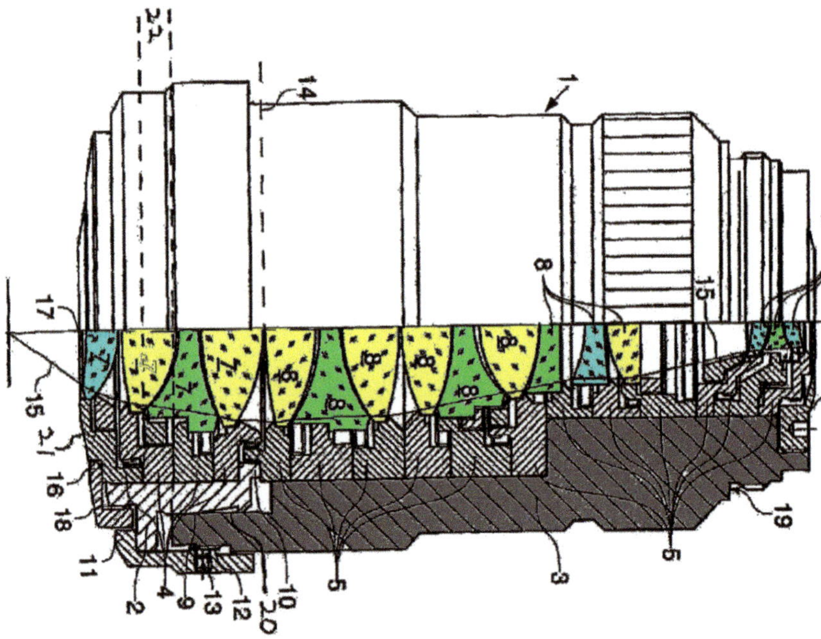

Figure 2.171 Assembly drawing of a Leica objective.

laser beam. A powerful laser radiation can damage the glued lenses. Another method of working on a microscope involves examining an object by using a heated table. The heating element has a temperature of ≥300°C; such a powerful thermal radiation can break an objective.

The complexity of the optical and the optical–mechanical designs explains why microscopes rarely use objectives with single lenses. This applies in particular to MicroFluar or apochromatic objectives in which fluorite crystals, which have unique technical parameters, are often used as crown lenses. This material has outstanding dispersive properties that make it possible to correct aberrations, such as the secondary spectrum and spherochromatism, in objectives. However, when gluing fluorite with other optical materials (e.g., flint or crown glass), a significant difference in the coefficients of thermal expansion occurs. In ordinary laboratory conditions, this phenomenon does not pose a significant danger to the objective because the ambient temperature is constant. However, if the objective is operating outside of typical laboratory conditions, it can be damaged if the air temperature drops, particularly in cases of *thermal shock*, when the temperature changes in a short time. In these cases, one or more components of the glued lenses becomes damaged (unglued). This objective almost completely loses the ability to build a high-quality image. To fix the problem, the damaged components almost always need to be replaced. Figure 2.172 shows photographs of the characteristic defects of unglued lenses.

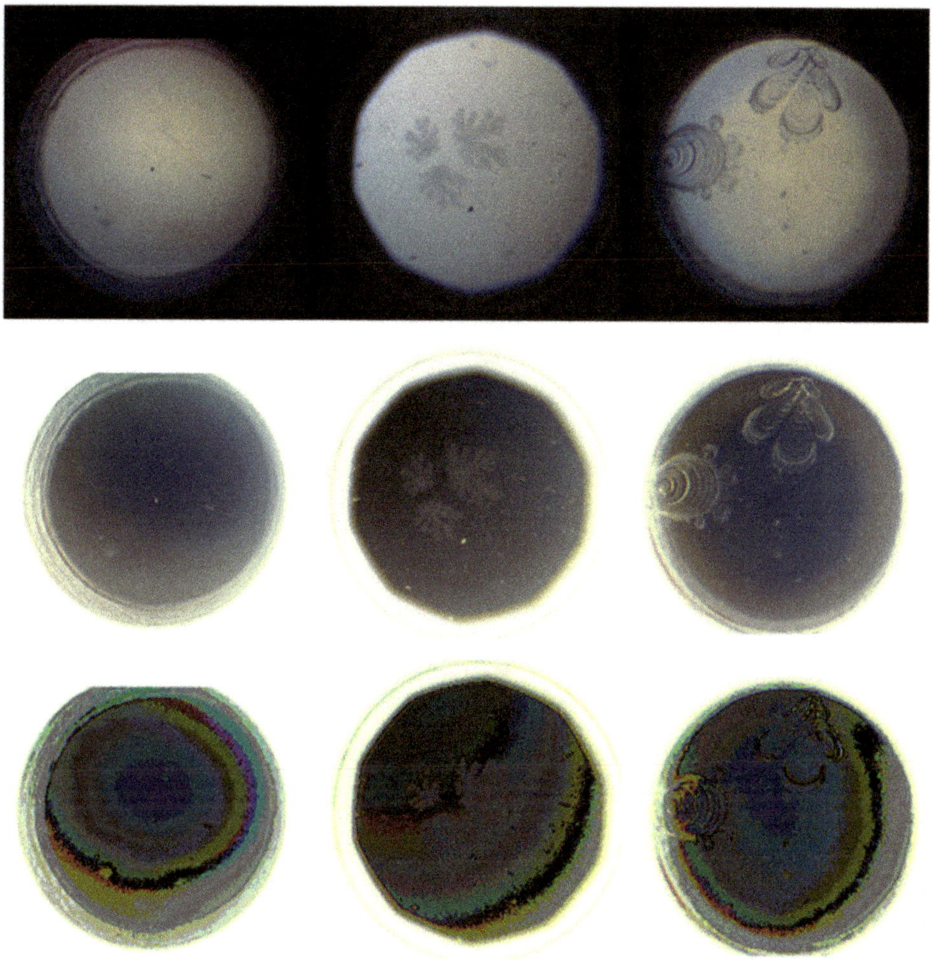

Figure 2.172 Photographs of typical defects of unglued lenses.

My colleagues and I studied the effects of residual deformations and inclusions in glass, including chips, stresses, and the appearance of so-called *local errors* on the surfaces of lenses, on a microscope's resulting image quality. It is no secret that in amateur photography, for example, some optical defects do not cause a critical deterioration in image quality. Although an objective can have reduced contrast (in the presence of residual negative spherical aberration, the image becomes "softened"), it may not have a significant impact on an image's sharpness or the resolution of the photographic lens. However, in light microscopy, many types of optical defects can cause various aberrations that significantly affect the resulting image quality.

We took some photographs with a polarizing microscope using the Bertrand system, which allowed us to obtain images of the glued surfaces of lenses. To achieve a higher contrast, the images were inverted. The first photograph (Fig. 2.172, top left) is of an unglued edge; the lens still works but with a smaller numerical aperture. The edge of the pupil (lens) of the objective is shielded. Unfortunately, both the aperture and the image contrast are reduced. The second photograph shows an example of thermal shock; although the drawing on this lens may resemble a work of art, the lens is not suitable for creating high-quality images and must be replaced in the objective. The third photograph illustrates a typical case of poor-quality lens installation; this lens has undergone tension, deformation, and destruction. Using a special algorithm and software for imaging lens surfaces with defects, we obtained so-called *Newton interference rings*, which can be used to assess the contribution of each aberration to the resulting image quality in an explicit numerical form. By the way, this technique can be used in the selection of both individual lenses (including those in frames) and objectives in general for polarizing microscopes. In this case, components and objectives with insignificant voltages (but nevertheless unsuitable for use in polarization) can be proposed for use, for example, in metallographic microscopes. In some cases, if the components or objectives as a whole have significant stresses, they can be used, for example, in biological or other microscopes. Here, we will not consider in detail the methodology for identifying the contribution of defective lenses to image quality (this could be an interesting research project for students); we will only conclude that, in almost all cases, broken unglued optical components are not suitable for a microscope objective.

The best solution might be to use a coupling technology for lenses by optical contact. In this case, such an objective might be repairable. However, this is a rather complicated and time-consuming technology that is inadvisable for the industrial manufacture of objectives. Moreover, there is no guarantee that the image quality would be improved.

To solve the problem of creating objectives that can be used outside of the laboratory environment, my colleagues and I designed two plan MicroFluar objectives. The first has a linear magnification of 20× and a numerical aperture in the space of objects of 0.75. The second objective has a linear magnification of 40× and a numerical aperture in the space of objects of 0.85. Both objectives contain single lenses arranged in a series along the optical axis. The optical systems of these objectives are built on the principle of a symmetrical design in which the system most fully meets the criteria of invariance. In the middle of the symmetrical design is a biconcave negative lens from which typical components—biconvex lenses—diverge to the left and right. Then the menisci follow, the first of which (from left to right along the ray) is turned with its concavity to the space of objects and the last to the space of images. By the way, this principle of constructing a symmetrical optical design allows the required level of aberration correction to be achieved using the minimum number of lenses in the objective. Figures 2.35 and 2.43 show

Table 2.13 Main technical parameters and basic optical layout of 20×/0.75 and 40×/0.85 objectives.

Magnification	*NA*	*WD* (mm)	*F′* (mm)	*R* (μm)	*DF* (μm)	*FOV* on object (mm)	*FOV* on image (mm)	Optical design at figure no.	The principal optical layout
20×	0.75	0.45	8.0	0.44	0.59	1.25	25	2.35	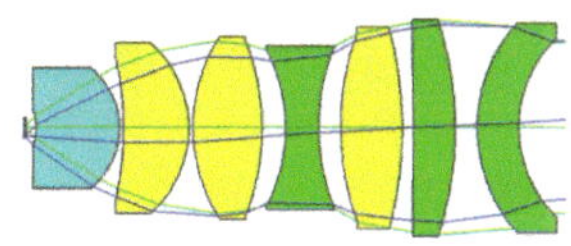
40×	0.85	0.20	4.0	0.39	0.46	0.625	25	2.43	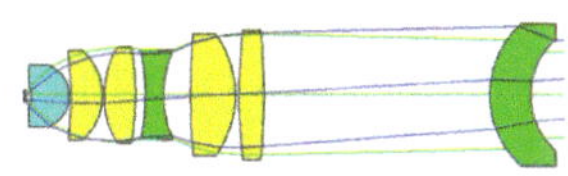

the design parameters and graphs of aberration correction of these objectives. Table 2.13 shows the main technical parameters and basic optical layout of these 20×/0.75 and 40×/0.85 objectives. The objectives are designed to work as part of microscopes with an infinite tube length; the focal length of the additional tube system is $F' = 160$ mm.

In the optical systems of both objectives, it is possible to implement specialized microscopy research techniques. For example, the choice of optical materials allows objectives for fluorescence analysis to be used; the transmission capacity of these objectives is high, in the near UV region of the spectrum, with the use of special antireflection coatings. It is also possible to implement the phase contrast method in the optical circuits of both objectives. For this purpose, a phase ring is applied to the 20× objective with a lens whose optical material has a low etch rate (if the ring is etched and a positive phase contrast is realized in the lens). In the 40× lens, an additional plane–parallel plate with applied phase rings can be used. In this objective, such a plate can be located near the penultimate lens.

It should be noted that these two objectives are designed in such a way that some aberrations are incompletely corrected in the objective itself but corrected in the system, which involves a combination of an objective and a tube system (tube lens). Such an engineering solution is quite common in many types of light microscopes, especially medium series versions. In some cases, the use of such a solution is an attempt to minimize the number of lenses in the objectives (when achieving enhanced technical parameters, as in the MicroFluar objectives discussed above). However, there is another reason to use this solution in the optical systems of microscopes: marketing. This is demonstrated when manufacturers of optical systems attempt to make their products original so that they cannot be interchangeable with products made by other manufacturers. In this case, not only LC but other types of optical system aberrations, such as spherical aberration, astigmatism, distortion, and others, individually or simultaneously, can be compensatory. Unfortunately, this trend creates inconvenience and uncertainty for microscope researchers when choosing the optimal equipment for their work; moreover, it presents a problem when they wish to upgrade their existing equipment (for example, with the best objectives). Engineers also do not like to create pseudo-original systems because, one way or another, they have limited versatility and can mainly be used for solving rather highly specialized tasks.

In our case, when we designed these two plan MicroFluar objectives with non-glued lenses, we considered the customer's wishes, creating an artificial limit to the universality of the equipment according to the technical task it was designed to perform.

2.4.2.5 Wide spectral transmitting objectives have monochromatic aberration correction

The same concept of non-glued lenses is found in silica/fluorite monochromatic aberration correction objectives. Although they are well-adjusted for the green region of the spectrum, they have very good transmission in a wide spectral range (from 200 to 2000 nm), which allows them to be used for dual- and multi-fluorescence methods (including the anti-Stock method) on a microscope. First, monochromatic correction does not require the use of glued lenses because there are no chromatic aberrations for such objectives. In such objectives, aberrations are corrected for a 546.07-nm wavelength for a linear field of 22 mm in image space. However, we tried to use various optical materials (crystals) for positive and negative lenses. Thus, in the entire working spectral range (in which good transmission is achieved), the aberrations will also not be huge. In any case, such objectives can be used to scan a sample from a spectrum at different wavelengths to identify the source of their own fluorescence. A very good correction of aberrations in the wavelengths of scanning is not required for this process. Using an interference green filter, for example, will allow the microscopist to get a sharp image (for green range) for a more detailed study of the selected sample area. Because these objectives allow researchers to work with live objects, the cover glass is not used.

Second, it is impossible to use glue in objectives that are designed to work in a wide spectral range, for example, in the wavelength range of 200 to 2000 nm. Modern types of optical glue are designed to work only in a relatively narrow spectral range. Third, the optical materials used in the optical systems of these objectives have a significant difference in their so-called *coefficient of thermal expansion* values. Therefore, bonding lenses from such materials is a very time-consuming process that does not guarantee that the glued component will adapt to temperature changes during operation or withstand thermal shock.

Figures 2.173–2.176 show the design parameters and graphs of aberration correction of these objectives. The set includes objectives with linear magnifications from 20× to 100×. All of them are designed to work as part of a microscope with an infinite tube length; the focal length of the additional focusing system is $F' = 160$ mm. The main technical parameters and optical schemes of these objectives are presented in Table 2.14.

Let us consider in more detail the basic principles of constructing optical systems for medium and high linear magnification objectives. The 20× objective uses the first (frontal) component in the form of a negative lens, which is made in the form of a biconcave lens. This solution allows a slight increase in the free working distance of the lens. In addition, it enables more

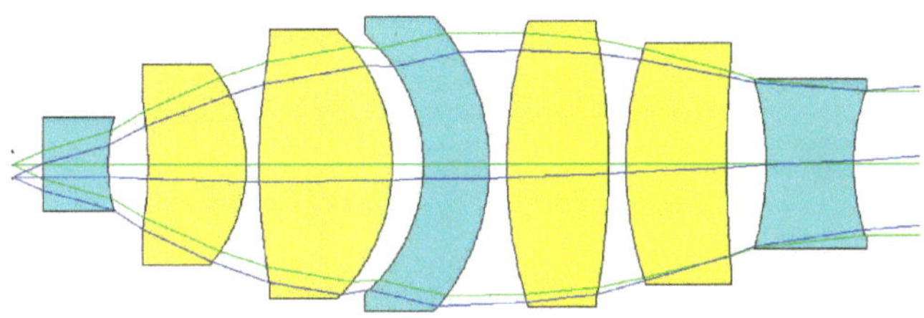

Lens: plan monochromat 20x0.40 ftl160

Object num aper 0.400000 Object height -0.550000 Primary wavln 0.546070

SRF	RADIUS	THICKNESS	APERTURE RADIUS	GLASS	SPECIAL
OBJ	0.000000	1.0000e-20	0.550000	AIR	
AST	0.000000	1.400000	0.550000 AS	AIR	
2	0.000000	3.000000	1.160000	SILICA	
3	8.164100	1.800000	2.090000	AIR	
4	-17.953900	4.500000	2.940000	CAF2	
5	-6.911900	0.600000	4.450000	AIR	
6	32.799900	6.000000	5.400000	CAF2 P	
7	-8.441400	1.400000	5.960000	AIR	
8	-7.546300	3.000000	5.750000	SILICA P	
9	-9.669900	0.800000	6.510000	AIR	
10	21.039600	4.600000	6.330000	CAF2 P	
11	-32.209100	0.800000	5.950000	AIR	
12	16.205300	4.600000	5.340000	CAF2 P	
13	52.664000	1.800000	4.700000	AIR	
14	-13.143500	4.000000	3.770000	SILICA P	
15	9.699300	100.000000	3.330000	AIR	
16	110.920000	4.000000	10.270000	LIF	
17	0.000000	1.000000	10.350000	AIR	
18	162.550000	4.000000	10.400000	LIF P	
19	-49.550000	2.000000	10.410000	SILICA P	
20	170.220000	4.000000	10.460000	LIF P	
21	-185.780000	152.119497 S	10.520000	AIR	
IMS	0.000000	0.000000	11.001339 S		

Figure 2.173 Optical design and graphs of aberration correction of a 20×/0.40 infinity plan monochromatic objective.

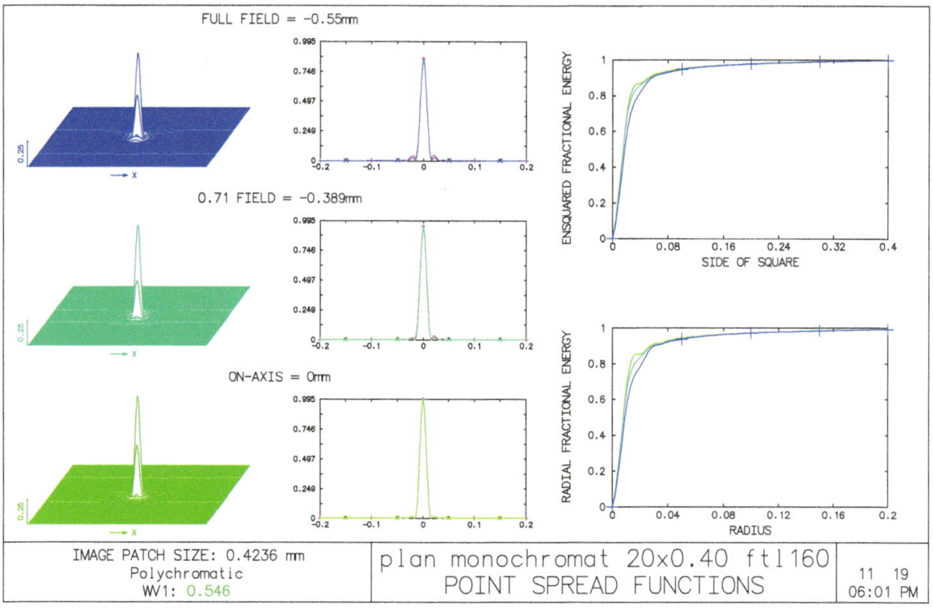

Figure 2.173 (*Continued*) Optical design and graphs of aberration correction of a 20×/0.40 infinity plan monochromatic objective.

careful aberration correction for a relatively large linear field. Further along the ray path (from left to the right), there are two positive components, a negative component and two more positive components, and one more negative component. The components are selected in such a way as to ensure a balance between positive and negative optical powers. In this case, an optimal aberration correction can be achieved with the minimum number of lenses in the objective. In 40× and 63× objectives, one positive component is set before the indicated negative component. It is made in the form of a meniscus, with a concavity to the space of the object. In the 100× objective, two such positive components are set before the indicated negative component. The above optical circuits can become the basis for the optical design of multi-apochromatic objectives that are equally well-adjusted throughout the indicated spectral range.

2.4.3 Plan apochromatic objectives for medium series microscopes

The idea of producing objectives with plan apochromatic aberration correction, in rather large quantities, deserves attention. The concept for such objectives may be as follows: objectives have moderate characteristics with respect to numerical apertures and linear fields. They can complete medium series light microscopes, which are distinguished by some versatility as compared, for example, with mass-market microscopes. At the same time, although medium series microscopes have obvious advantages in image quality due to the use of optics with plan apochromatic correction, they do not differ from other microscopes in terms of their general technical parameters.

In this regard, I would like to tell you a little about Professor Lev Nikolaevich Andreev. He has been teaching students for several decades,

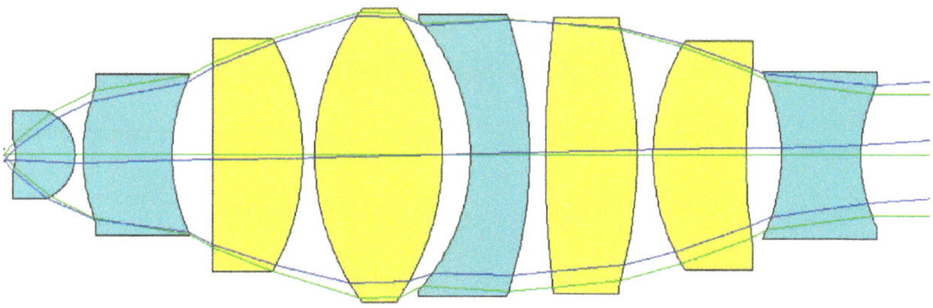

Lens: plan monochromat 40x0.75 ftl160

Object num aper	0.750000	Object height	-0.273000	Primary wavln	0.546070

SRF	RADIUS	THICKNESS		APERTURE RADIUS		GLASS		SPECIAL
OBJ	0.000000	1.0000e-20		0.273000		AIR		
AST	0.000000	0.600000		0.273000	AS	AIR		
2	-1.961800	3.000000		0.770000		SILICA		
3	-2.312000	0.400000		2.170000		AIR		
4	8.761000	4.500000		3.300000		SILICA	P	
5	8.925200	2.000000		3.920000		AIR		
6	855.167300	4.500000		4.540000		CAF2		
7	-11.568600	0.600000		5.680000		AIR		
8	11.898000	6.400000		7.160000		CAF2	P	
9	-12.280400	1.400000		7.100000		AIR		
10	-9.702000	3.000000		6.610000		SILICA	P	
11	-21.015000	0.800000		6.880000		AIR		
12	54.353700	4.600000		6.720000		CAF2	P	
13	-22.928000	0.800000		6.440000		AIR		
14	10.386300	4.600000		5.550000		CAF2	P	
15	28.428300	1.800000		4.310000		AIR		
16	-9.105100	4.000000		4.070000		SILICA	P	
17	7.831100	100.000000		3.450000		AIR		
18	110.920000	4.000000		10.150000		LIF		
19	0.000000	1.000000		10.230000		AIR		
20	162.550000	4.000000		10.270000		LIF	P	
21	-49.550000	2.000000		10.280000		SILICA	P	
22	170.220000	4.000000		10.330000		LIF	P	
23	-185.780000	152.092010	S	10.390000		AIR		
IMS	0.000000	0.000000		10.920658	S			

Figure 2.174 Optical design and graphs of aberration correction of a 40×/0.75 infinity plan monochromatic objective.

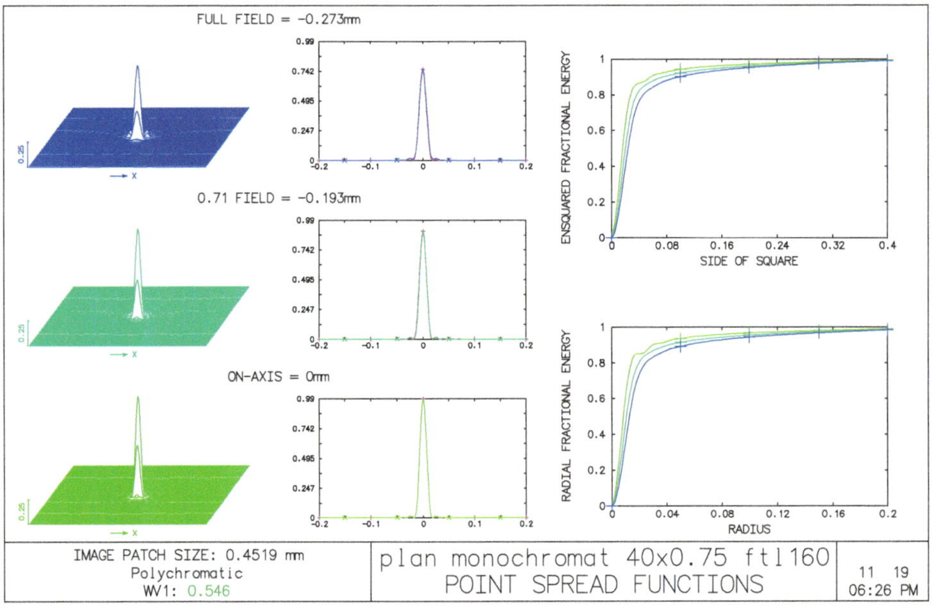

Figure 2.174 (*Continued*) Optical design and graphs of aberration correction of a 40×/0.75 infinity plan monochromatic objective.

sharing his experiences in the development of optical systems for microscopes. When he lectures, he tells his students that making an optical design of an objective for a microscope is like tasting a glass of good expensive wine. It was Professor Andreev who inspired and led the production of medium series plan apochromats; he performed most of the optical designs for such objectives. Now we can tell him that implementing the optical design of these objectives is more like a combination of a strong drink, sleepless nights, nervous breakdowns, etc. It is hard and exhausting work. But yes, the result is a taste that lasts a lifetime.

In Chapter 1, Fig. 1.251, we examined the optical system of an old Soviet Biolam-I microscope. This is an interesting device, primarily because it uses an original pancratical lighting system. The microscope has a large and heavy tripod, which makes it resistant to vibration. The use of original objectives with plan apochromatic aberration correction in this microscope also distinguishes it from such analogues as the Lumam-R8 microscope that was considered in Chapter 1, the stand of which is largely the same as the Biolam-I stand. In one of the configuration options in this microscope, the following plan apochromatic objectives are used: 10×/0.30 (called OPA-1), 16×/0.40 (OPA-2), 40×/0.65 (OPA-3), 60×/0.85 (OPA-4), and 100×/1.25 (OPA-5) oil immersion objectives. It could be said that such objectives have weak technical parameters (for plan apochromats), such as the numerical apertures and linear fields. However, if we consider the quantities of these objectives that were manufactured in the Soviet Union, we can agree that it was a significant achievement to produce medium series microscopes that had superior image quality compared with their analogues.

These objectives began to be developed around 1955–1957; by this time, engineers had already understood how to correct such aberrations in

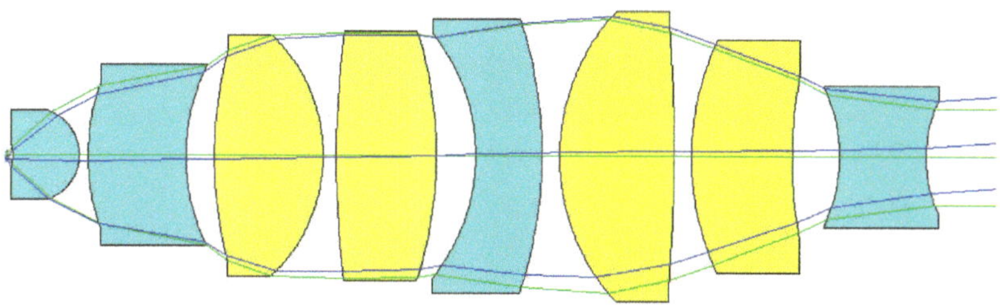

Lens: plan monochromat 63x0.85 ft1160

Object num aper 0.850000 Object height -0.174000 Primary wavln 0.546070

SRF	RADIUS	THICKNESS	APERTURE RADIUS	GLASS	SPECIAL
OBJ	0.000000	1.0000e-20	0.174000	AIR	
AST	0.000000	0.269870	0.174000 AS	AIR	
2	-1.899300	3.000000	0.500000	SILICA	
3	-2.082200	0.400000	1.990000	AIR	
4	9.240000	4.500000	3.200000	SILICA P	
5	8.846500	1.300000	4.020000	AIR	
6	18.161700	5.000000	4.690000	CAF2	
7	-7.237100	0.600000	5.350000	AIR	
8	42.795600	4.600000	5.530000	CAF2 P	
9	-17.691300	1.800000	5.530000	AIR	
10	-8.419400	3.000000	5.430000	SILICA P	
11	-18.907700	0.800000	6.090000	AIR	
12	9.403800	5.200000	6.460000	CAF2 P	
13	-77.450800	0.800000	5.950000	AIR	
14	11.703500	4.600000	5.160000	CAF2 P	
15	19.568700	2.200000	3.680000	AIR	
16	-6.888300	4.000000	3.130000	SILICA P	
17	5.785000	100.000000	2.540000	AIR	
18	110.920000	4.000000	8.700000	LIF	
19	0.000000	1.000000	8.780000	AIR	
20	162.550000	4.000000	8.810000	LIF P	
21	-49.550000	2.000000	8.830000	SILICA P	
22	170.220000	4.000000	8.880000	LIF P	
23	-185.780000	152.276192 S	8.940000	AIR	
IMS	0.000000	0.000000	10.965203 S		

Figure 2.175 Optical design and graphs of aberration correction of a 63×/0.85 infinity plan monochromatic objective.

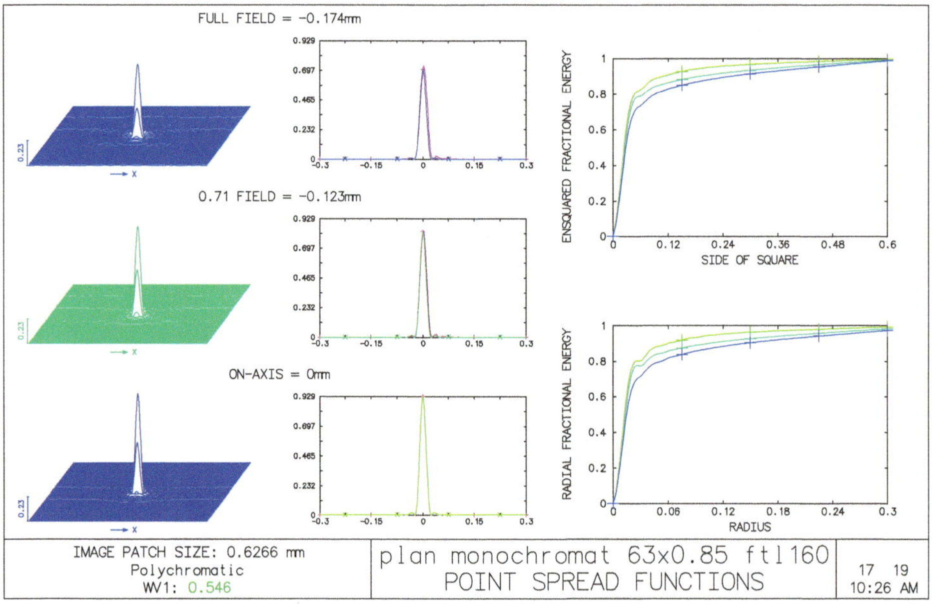

Figure 2.175 (*Continued*) Optical design and graphs of aberration correction of a 63×/0.85 infinity plan monochromatic objective.

objectives as image curvature and secondary spectrum. By then, the optical material fluorite, with its unique dispersion properties, was being mined on an industrial scale and the technology for processing and shaping lenses was already well developed. In addition, the economic advantages of a planned socialist economy, in which labor and materials did not have a specific price in the manufacture of a particular product by a particular factory, played a significant role. The slogan at that time—*"This is only possible under socialism"*—can be said to be embodied in these objectives. Indeed, if we look at the optical systems of these objectives (and understand that at that time the technology to grow artificial crystals was not yet well mastered in the optical industry), one wonders how producing just one of these objectives containing such a large number of fluorite lenses did not destroy the budget of the entire microscope. Recall that if we compare the optical systems of these microscopes, there were five such objectives with each one more complicated than the last. For this reason alone, it is interesting to talk about these lenses in more detail. Figure 2.177 is a photograph of a set of these plan apochromatic objectives, and Figs. 2.178–2.182 show the design parameters and graphs of aberration correction of each objective, respectively.

The engineers who developed these objectives did not confirm that nearly all of the objectives managed to achieve complete LC correction. When the optical design was implemented in the 40× and 60× objectives, the LC turned out to be completely corrected. This was most likely due to luck. Moreover, it is clear that for the 10× and 16× objectives, an LC value close to 0% was obtained automatically. It is important to note that the optical design was carried out using a theoretical framework for constructing optical systems with medium- and large-magnification objectives; however, a special algorithm for correcting the LC did not yet exist. The main unresolved problem with respect to the LC was the oil immersion objective with a linear

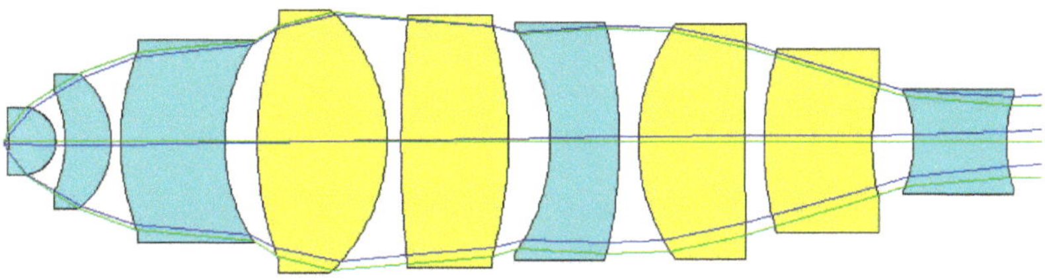

Lens:plan monochromat 100x0.95 ftl160

Object num aper 0.950000 Object height -0.110000 Primary wavln 0.546070

SRF	RADIUS		THICKNESS		APERTURE RADIUS		GLASS		SPECIAL
OBJ	0.000000		1.0000e-20		0.110000		AIR		
AST	0.000000		0.200000		0.110000	AS	AIR		
2	-1.228700		2.000000		0.450000		SILICA		
3	-1.425100		0.400000		1.420000		AIR		
4	-5.217300		2.000000		2.230000		SILICA	P	
5	-3.571600		0.400000		2.850000		AIR		
6	10.176400		4.500000		3.760000		SILICA	P	
7	7.626300		1.400000		4.270000		AIR		
8	13.884100		5.500000		4.980000		CAF2		
9	-7.532500		0.600000		5.530000		AIR		
10	45.384800		4.600000		5.290000		CAF2	P	
11	-18.108100		1.800000		4.990000		AIR		
12	-8.093700		3.000000		4.730000		SILICA	P	
13	-20.033800		0.800000		4.990000		AIR		
14	8.497800		4.600000		4.940000		CAF2	P	
15	233.754500		0.800000		4.270000		AIR		
16	13.601800		4.600000		3.860000		CAF2	P	
17	10.011000		1.800000		2.520000		AIR		
18	-5.319200		4.000000		2.190000		SILICA	P	
19	6.129500		100.000000		1.910000		AIR		
20	110.920000		4.000000		8.730000		LIF		
21	0.000000		1.000000		8.830000		AIR		
22	162.550000		4.000000		8.880000		LIF	P	
23	-49.550000		2.000000		8.910000		SILICA	P	
24	170.220000		4.000000		8.970000		LIF	P	
25	-185.780000		152.357320	S	9.050000		AIR		
IMS	0.000000		0.000000		11.006515	S			

Figure 2.176 Optical design and graphs of aberration correction of a 100×/0.95 infinity plan monochromatic objective.

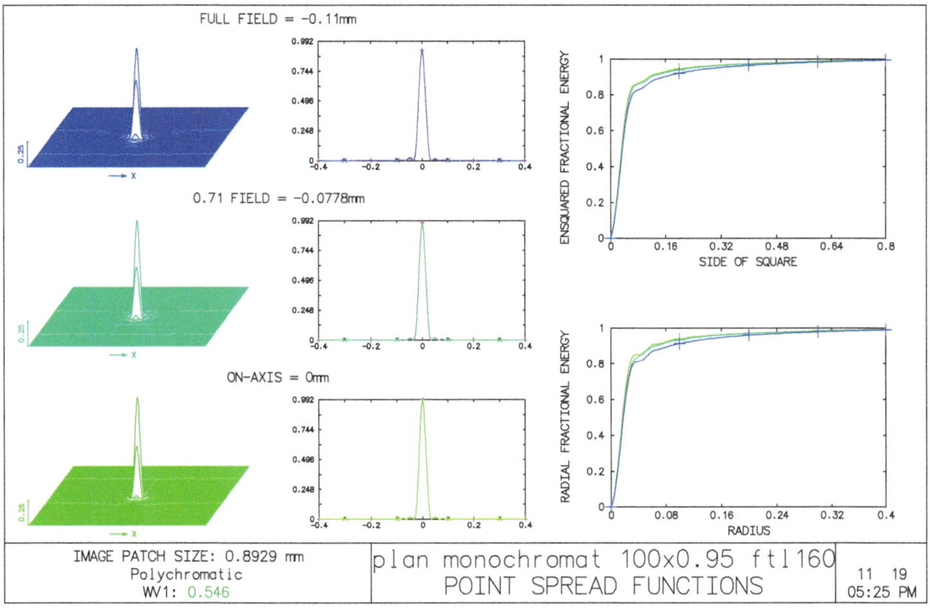

Figure 2.176 (*Continued*) Optical design and graphs of aberration correction of a 100×/0.95 infinity plan monochromatic objective.

Table 2.14 Main technical parameters and basic optical layout of silica/fluorite objectives.

Magnification	*NA*	*WD* (mm)	*F′* (mm)	*R* (μm)	*DF* (μm)	*FOV* on object (mm)	*FOV* on image (mm)	Optical design at figure no.	The principal optical layout
20×	0.40	1.0	8	0.83	2.08	1.1	22	2.173	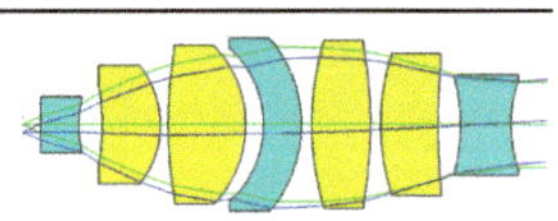
40×	0.75	0.33	4	0.44	0.59	0.55	22	2.174	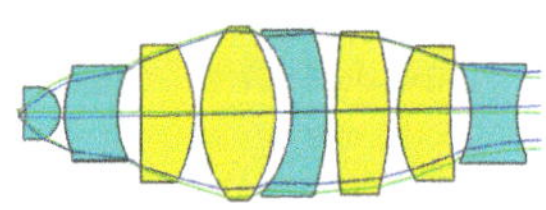
63×	0.85	0.18	2.54	0.39	0.46	0.35	22	2.175	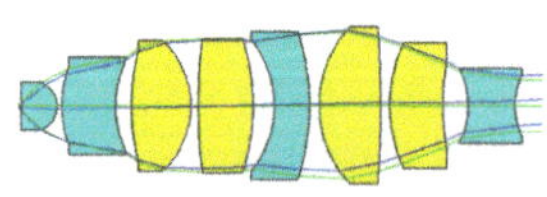
100×	0.95	0.09	1.6	0.35	0.37	0.22	22	2.176	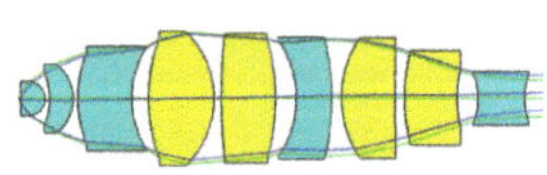

magnification of 100× and a numerical aperture of 1.25; its optical design was not as successful as the others in the kit and did not allow for LC correction. Of course, today we can explain that a more successful optical design for the 100× objective, the most complex objective in the kit, most likely could not

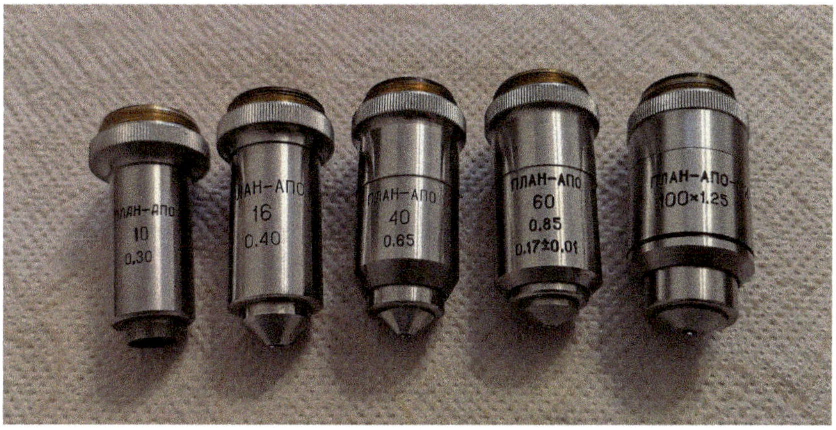

Figure 2.177 Photograph of a set of old Soviet plan apochromatic objectives.

have been achieved with the computing tools that were available then. At the time, ray tracing in optical systems was calculated manually using trigonometric tables, and aberrations in the system were also calculated manually using, for example, Feder's formulas [9]. A miracle did not happen; it was not possible to correct the LC aberration in the 100× objective. Thus, the LC of the objective was compensated by the eyepiece. Unfortunately, the aberration correction (LC) of the remaining lenses was consistent with that of the 100× objective; that is, the residual LC in all objectives was 2%.

Today, we can only simulate LC correction in the optical systems of these plan apochromats. Unfortunately, they were manufactured as is for more than 40 years; that is, they had their own LC of about 2%, which was compensated by an eyepiece in a microscope. As an example of the universality of the principal optical circuits of the 40×, 60×, and 100× objectives, we consider optical designs that can be LC corrected by only slightly changing the basic optical circuits and using other optical materials in separate components. Figures 2.183–2.185 show the design parameters and graphs of aberration correction of advanced lenses (upgraded to LC = 0%).

An analysis of these optical designs shows that when using modern computing tools and specialized software for calculating optical systems, it is quite simple to make an optical design that allows the LC to be fixed within the framework of the proposed basic optical schemes. Therefore, the engineering solutions proposed by Russian experts in 1960 to correct the LC in microscope objectives can most likely be accepted as relevant. Table 2.15 shows a comparison of one of the characteristics of optical image quality, referred to as the *point spread function*, for the initial and modernized optical design options.

A skeptic may ask, why are expensive objectives with a plan apochromatic aberration correction needed for medium series microscopes? After all, there

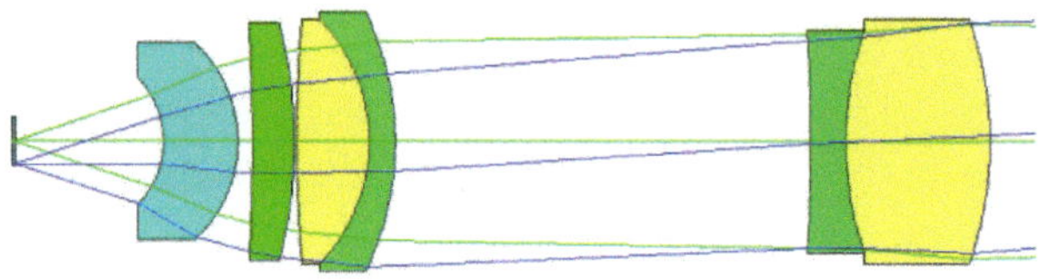

Lens: OPA-1 (10x0.30) finity | Efl 15.7

Object num aper 0.300000 | Image height 9.000000 | Primary wavln 0.546070

SRF	RADIUS	THICKNESS	APERTURE RADIUS	GLASS	SPECIAL
OBJ	0.000000	0.010000	0.894097	AIR	
AST	0.000000	0.170000	0.897242 AS	LZ_K14	
2	0.000000	6.044000	0.940000	AIR	
3	-3.499000	2.990000	2.500000	LZ_CTK9	
4	-5.152000	0.600000	3.850000	AIR	
5	-94.620000	1.700000	4.440000	LZ_TF3	
6	-18.113000	0.100000	4.650000	AIR	
7	51.760000	2.900000	4.750000	CAF2	
8	-6.761000	1.000000	4.790000	LZ_OF4	
9	-11.613000	16.780000	5.080000	AIR	
10	-69.980000	1.400000	4.380000	LZ_F8	
11	13.305000	5.800000	4.380000	CAF2 P	
12	-14.093000	152.510000	4.800000	AIR	
13	0.000000	0.000000	8.964540 S	AIR	
IMS	0.000000	0.000000	8.964540 S		

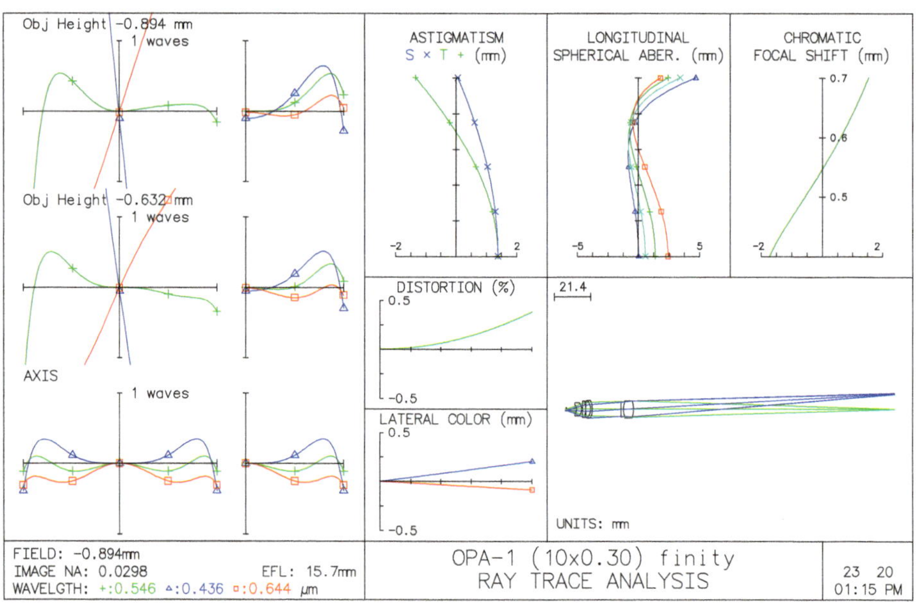

Figure 2.178 Optical design and graphs of aberration correction of a 10×/0.30 finite LCP plan apochromatic objective.

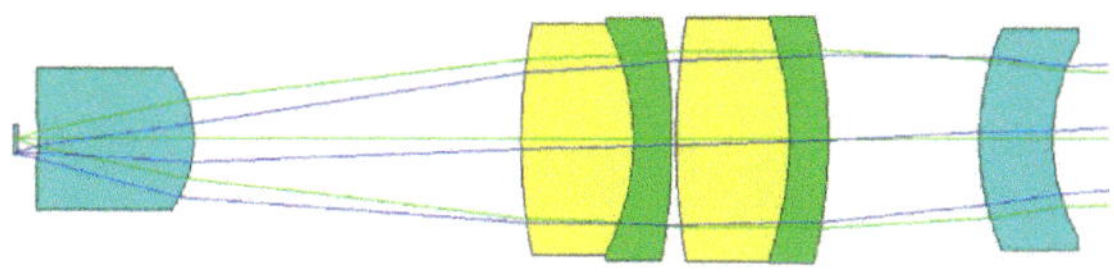

Lens: OPA-2 (16x0.40) finity — Efl 9.62

Object num aper 0.300000 — Image height 9.000000 — Primary wavln 0.546070

SRF	RADIUS	THICKNESS	APERTURE RADIUS	GLASS	SPECIAL
OBJ	0.000000	0.010000	0.556913	AIR	
AST	0.000000	0.170000	0.560058 AS	LZ_K14	
2	0.000000	0.880000	0.600000	AIR	
3	-4.055000	6.400000	0.940000	LZ_CTK9	
4	-5.012000	13.600000	2.890000	AIR	
5	22.650000	4.700000	4.630000	CAF2	
6	-10.069000	1.600000	4.700000	LZ_OF4	
7	-28.050000	0.200000	4.910000	AIR	
8	30.620000	4.700000	4.960000	CAF2 P	
9	-14.223000	1.600000	4.900000	LZ_OF4 P	
10	-21.180000	6.250000	4.980000	AIR	
11	11.561000	3.000000	4.510000	LZ_BK6	
12	7.980000	148.890000	3.980000	AIR	
13	0.000000	0.000000	8.965964 S	AIR	
IMS	0.000000	0.000000	8.965964 S		

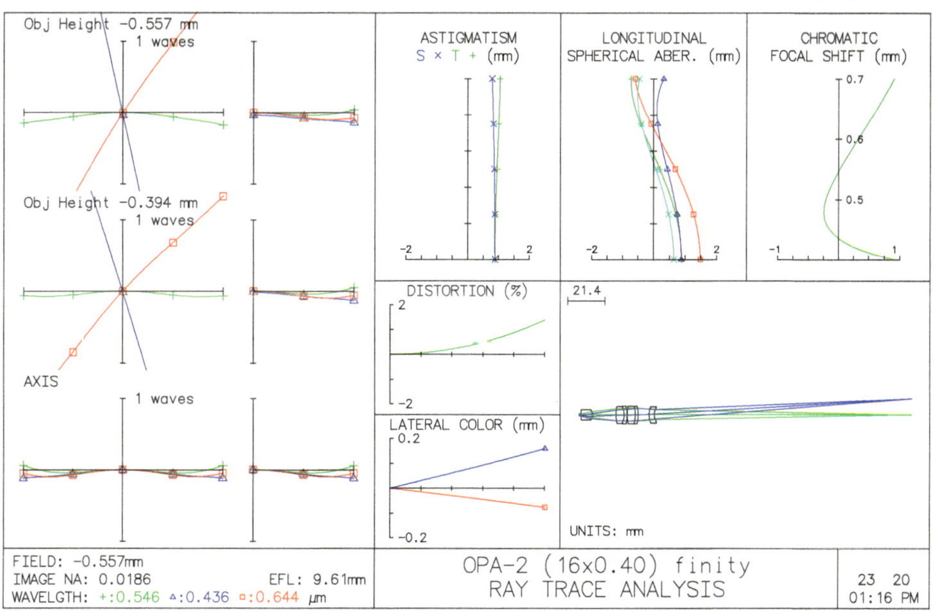

Figure 2.179 Optical design and graphs of aberration correction of a 16×/0.40 finite LCP plan apochromatic objective.

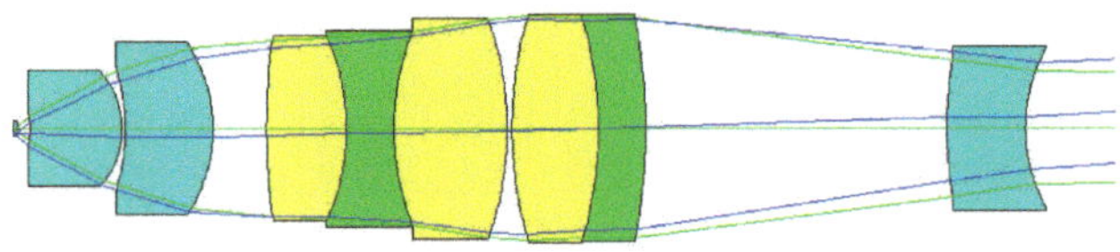

Lens: OPA-3 (40x0.65) finity			Efl	3.86	
Object num aper	0.650000	Image height	9.000000	Primary wavln	0.546070

SRF	RADIUS	THICKNESS	APERTURE RADIUS	GLASS	SPECIAL
OBJ	0.000000	0.010000	0.218071	AIR	
AST	0.000000	0.170000	0.226624 AS	LZ_K14	
2	0.000000	0.547000	0.310000	AIR	
3	-2.858000	4.000000	0.700000	LZ_CTK9	
4	-3.404000	0.150000	2.410000	AIR	
5	-8.128000	3.800000	2.620000	LZ_CTK9 P	
6	-6.397000	2.290000	3.600000	AIR	
7	26.920000	3.500000	3.850000	CAF2	
8	-8.995000	2.000000	3.860000	LZ_OF4	
9	11.169000	4.800000	4.100000	CAF2 P	
10	-13.062000	0.200000	4.610000	AIR	
11	15.996000	3.700000	4.770000	CAF2 P	
12	-15.996000	2.000000	4.710000	LZ_OF4 P	
13	-24.270000	12.950000	4.750000	AIR	
14	17.906000	3.360000	2.850000	LZ_BK6	
15	6.792000	148.510000	3.425940	AIR	
16	0.000000	0.000000	8.916001 S	AIR	
IMS	0.000000	0.000000	8.916001 S		

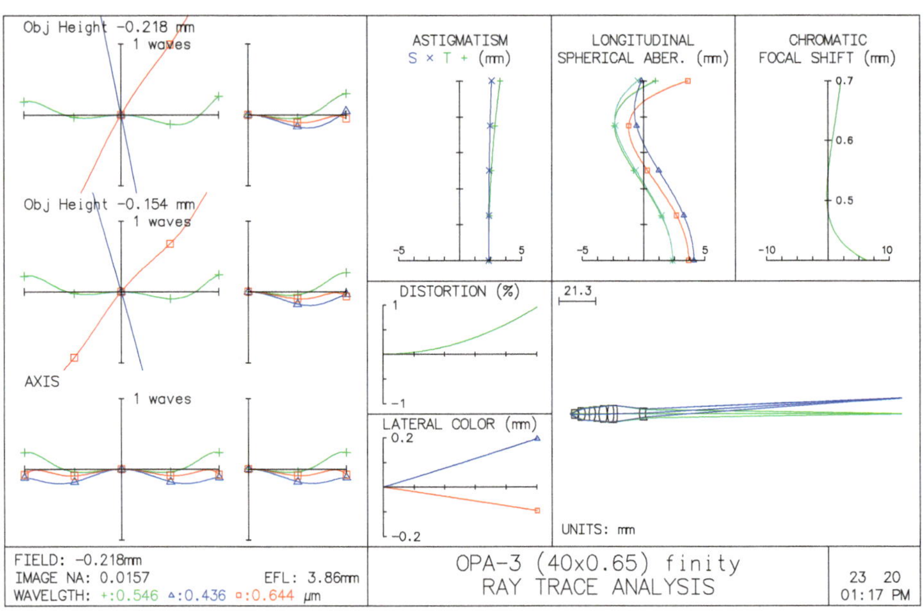

Figure 2.180 Optical design and graphs of aberration correction of a 40×/0.65 finite LCP plan apochromatic objective.

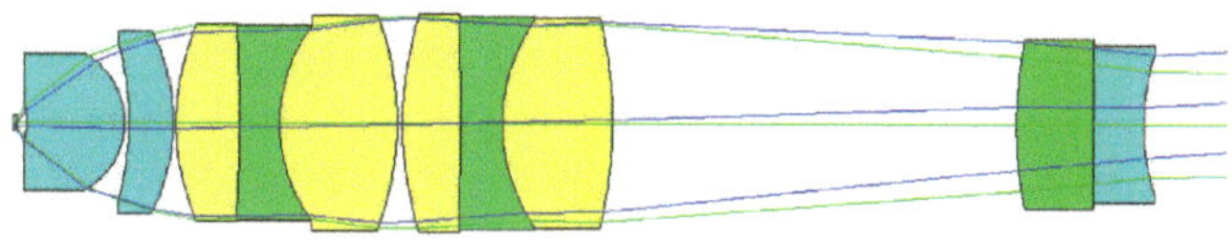

Lens: OPA-4 (60x0.85) finity			Efl	2.63	
Object num aper	0.850000	Image height	9.000000	Primary wavln	0.546070

SRF	RADIUS	THICKNESS	APERTURE RADIUS	GLASS	SPECIAL
OBJ	0.000000	0.010000	0.147394	AIR	
AST	0.000000	0.170000	0.163530 AS	LZ_K14	
2	0.000000	0.369200	0.280000	AIR	
3	-2.512000	3.880000	0.710000	LZ_CTK9	
4	-2.938000	0.200000	2.620000	AIR	
5	-10.069000	1.750000	3.050000	LZ_CTK9 P	
6	-7.379000	0.100000	3.480000	AIR	
7	9.120000	2.500000	3.780000	CAF2	
8	-56.360000	1.600000	3.740000	LZ_OF4	
9	6.109000	4.600000	3.740000	CAF2 P	
10	-11.429000	0.200000	4.100000	AIR	
11	13.646000	2.300000	4.180000	CAF2 P	
12	-162.930000	1.600000	4.070000	LZ_OF4 P	
13	6.855000	4.300000	3.920000	CAF2 P	
14	-16.827000	15.900000	4.030000	AIR	
15	16.827000	3.000000	3.260000	LZ_F1	
16	0.000000	2.000000	2.990000	LZ_TK16	
17	8.395000	147.530000	2.740000	AIR	
18	0.000000	0.000000	8.969697 S	AIR	
IMS	0.000000	0.000000	8.969697 S		

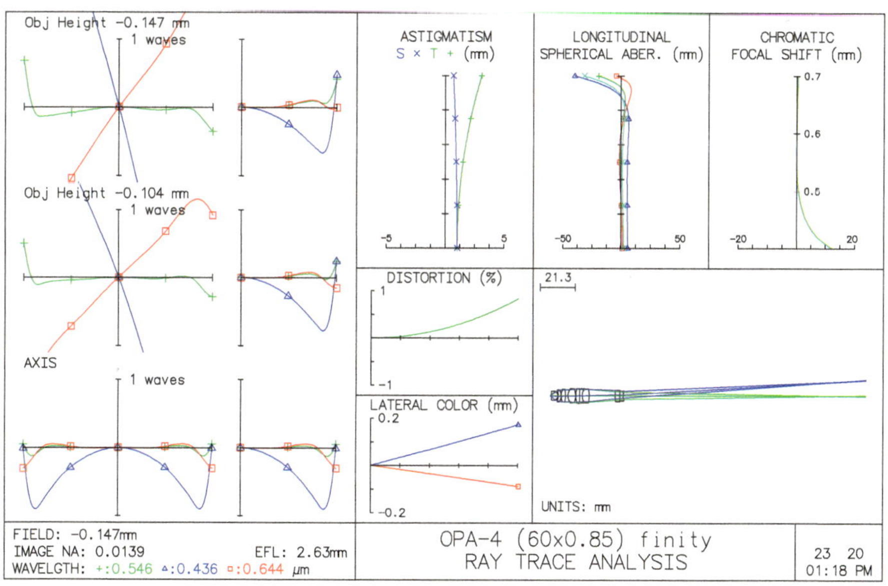

Figure 2.181 Optical design and graphs of aberration correction of a 60×/0.85 finite LCP plan apochromatic objective.

are plan achromats corrected for LC, for example, that have high image quality. However, several arguments can be given in favor of recommending objectives with apochromatic (plan apochromatic) aberration correction; for some types of studies that use a light microscope, plan achromats cannot give an adequate image or enable the correct identification of certain objects under study.

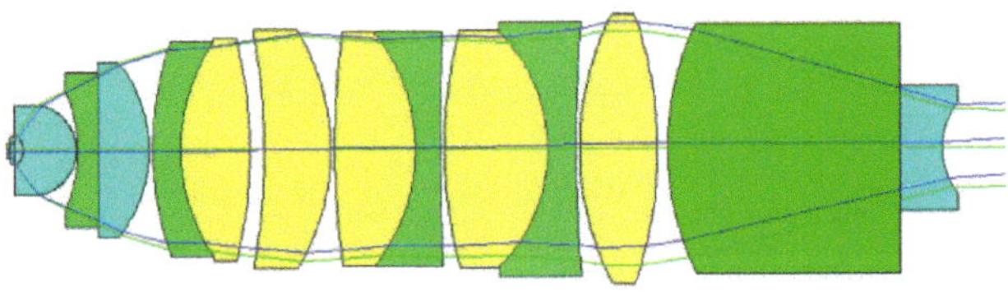

Lens: OPA-5 (100x1.25 oil) Finity — Efl 1.5

Object num aper 1.250000 Image height 9.000000 Primary wavln 0.546070

SRF	RADIUS	THICKNESS	APERTURE RADIUS		GLASS		SPECIAL
OBJ	0.000000	0.010000	0.083509		OIL_IMM		
AST	0.000000	0.170000	0.098036	AS	LZ_K14		
2	0.000000	0.192200	0.350000		OIL_IMM	P	
3	0.000000	0.420000	0.630000		LZ_BK6		
4	-0.816600	2.500000	0.690000		LZ_CTK9		
5	-2.089000	0.100000	2.080000		AIR		
6	-6.792000	1.000000	2.760000		LZ_OF4		
7	0.000000	2.450000	3.580000		LZ_CTK3		
8	-6.339000	0.200000	4.070000		AIR		
9	15.066000	1.300000	4.880000		LZ_OF4	P	
10	8.750000	3.300000	4.970000		CAF2		
11	-21.580000	0.700000	5.150000		AIR		
12	-29.040000	3.200000	5.260000		CAF2	P	
13	-10.568000	0.200000	5.560000		AIR		
14	38.990000	3.800000	5.450000		CAF2	P	
15	-8.590000	1.200000	5.320000		LZ_OF4	P	
16	125.890000	0.200000	5.430000		AIR		
17	20.750000	5.000000	5.500000		CAF2	P	
18	-7.482000	1.300000	5.450000		LZ_OF4	P	
19	61.240000	0.200000	5.890000		AIR		
20	13.305000	3.700000	6.270000		CAF2	P	
21	-18.493000	0.500000	6.270000		AIR		
22	13.032000	11.000000	5.790000		LZ_TF4		
23	0.000000	2.000000	2.910000		LZ_CTK9	P	
24	3.404000	147.350000	2.100000		AIR		
25	0.000000	0.000000	8.648896	S	AIR		
IMS	0.000000	0.000000	8.648896	S			

Figure 2.182 Optical design and graphs of aberration correction of a 100×/1.25 oil immersion finite LCP plan apochromatic objective.

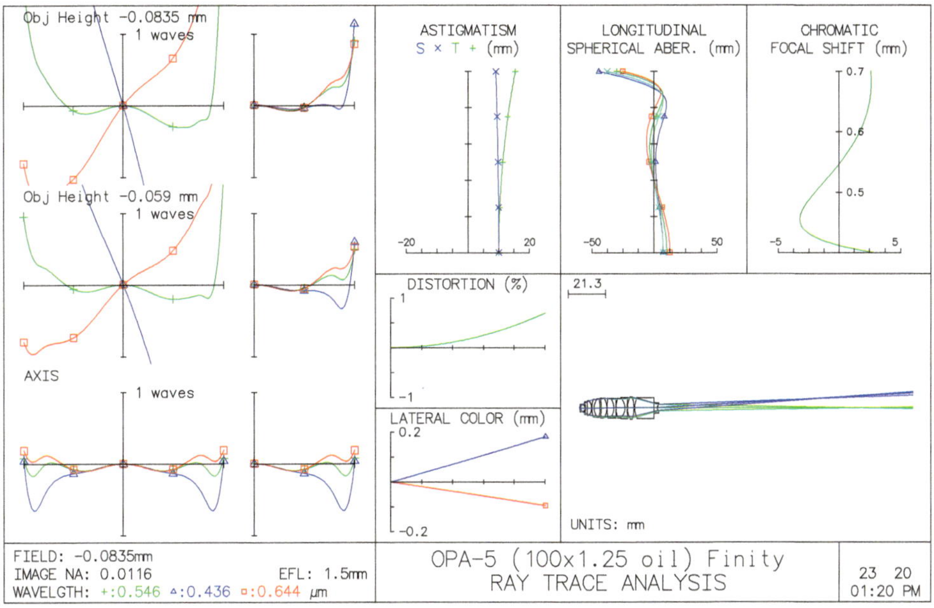

Figure 2.182 (*Continued*) Optical design and graphs of aberration correction of a 100×/1.25 oil immersion finite LCP plan apochromatic objective.

We should note that we are talking about lens objectives. Lens objectives for microscopes are the most complex elements among diverse optical systems to date. This is due to the high complexity of and use of special materials in their optical design, the high accuracy required for manufacturing lenses, and the technological difficulties of assembly. Objectives with plan apochromatic aberration correction are the highest achievement in the field of microscope optics. Compared with lenses with other types of correction, these objectives are distinguished by their spectral transmission and aberration correction range extended from the near UV region from 435 to 640 nm. Moreover, the very nature of the aberration correction in these objectives is somewhat different. For some types of microscope objectives, compromises are possible in correcting residual aberrations in the resulting image (for example, in the case discussed in the previous paragraphs of this chapter). This is due to the specific uses of such objectives, which are intended mainly for visual observation. There are no such compromises for plan apochromatic objectives; aberration correction must be performed carefully for the entire working spectral range.

Some skeptics argue that objectives with plan apochromatic aberration correction are extremely complex in design and therefore very expensive. For most routine microscopy techniques, such a high level of correction is not always necessary, which means that the use of plan apochromats is not always justified. For both visual observation and projecting an image onto a digital

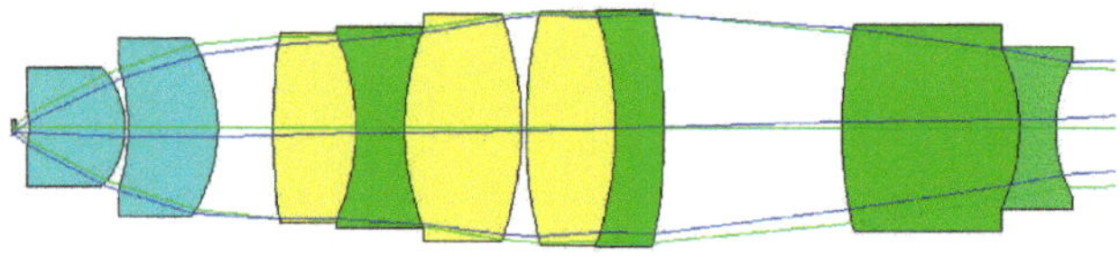

Lens: OPA-3 upgrade to LC=0% Efl 3.88

Object num aper 0.650000 Image height 9.000000 Primary wavln 0.546070

SRF	RADIUS	THICKNESS	APERTURE RADIUS	GLASS	SPECIAL
OBJ	0.000000	0.010000	0.224994	AIR	
AST	0.000000	0.170000	0.233547 AS	LZ_K14	
2	0.000000	0.548000	0.320000	AIR	
3	-2.858000	4.000000	0.710000	LZ_CTK9	
4	-3.404000	0.150000	2.420000	AIR	
5	-8.128000	3.800000	2.630000	LZ_CTK9 P	
6	-6.397000	2.290000	3.600000	AIR	
7	26.920000	3.500000	3.850000	CAF2	
8	-8.995000	2.000000	3.860000	LZ_TF1	
9	11.169000	4.800000	4.100000	CAF2 P	
10	-13.062000	0.200000	4.600000	AIR	
11	19.453500	3.700000	4.740000	CAF2 P	
12	-14.358900	2.000000	4.720000	LZ_TF2	
13	-23.601700	7.400000	4.800000	AIR	
14	17.283000	7.400000	4.200000	LZ_TF4	
15	-7.861000	1.500000	3.310000	LZ_OF4	
16	5.796000	148.503118 S	2.770000	AIR	
17	0.000000	0.000000 S	9.000000 S	AIR	
IMS	0.000000	0.000000	9.000000 S		

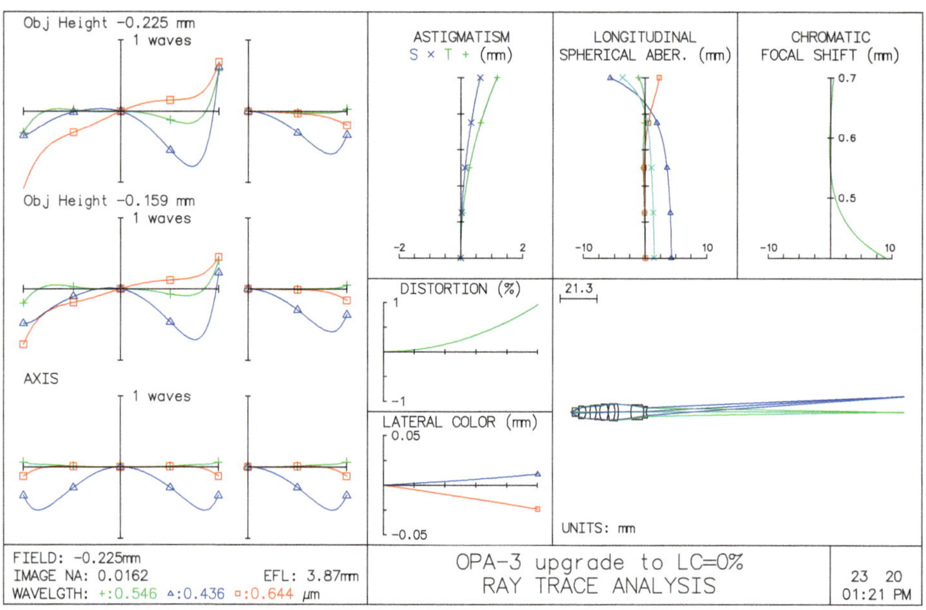

Figure 2.183 Optical design and graphs of aberration correction of a 40×/0.65 finite LCF plan apochromatic objective.

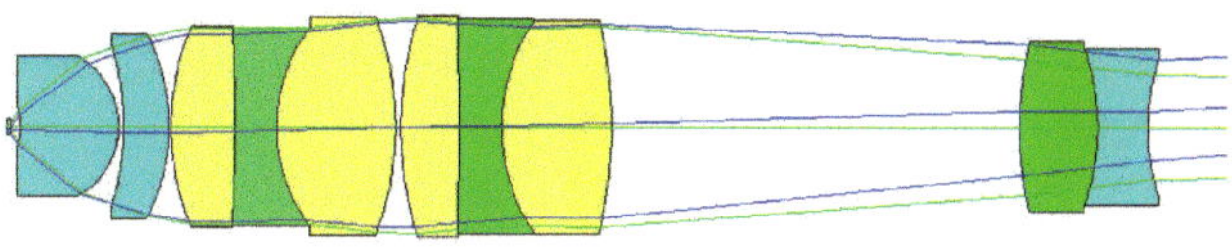

Lens: OPA-4 upgrade to LC=0% — Efl 2.51

Object num aper 0.850000 — Image height 9.000000 — Primary wavln 0.546070

SRF	RADIUS	THICKNESS	APERTURE RADIUS	GLASS	SPECIAL
OBJ	0.000000	0.010000	0.142439	AIR	
AST	0.000000	0.170000	0.158575 AS	LZ_K14	
2	0.000000	0.367500	0.280000	AIR	
3	-2.512000	3.880000	0.710000	LZ_CTK9	
4	-2.938000	0.200000	2.620000	AIR	
5	-10.069000	1.750000	3.050000	LZ_CTK9 P	
6	-7.379000	0.100000	3.480000	AIR	
7	9.120000	2.500000	3.780000	CAF2	
8	-56.360000	1.600000	3.740000	LZ_OF4	
9	6.109000	4.600000	3.740000	CAF2 P	
10	-11.429000	0.200000	4.100000	AIR	
11	13.646000	2.300000	4.180000	CAF2 P	
12	-162.930000	1.600000	4.070000	LZ_TF1	
13	6.855000	4.300000	3.920000	CAF2 P	
14	-16.827000	15.900000	4.030000	AIR	
15	14.622000	3.000000	3.200000	LZ_TF4	
16	-8.630000	2.000000	2.940000	LZ_CTK19	
17	7.870000	147.702206 S	2.580000	AIR	
18	0.000000	0.000000 S	9.000000 S	AIR	
IMS	0.000000	0.000000	9.000000 S		

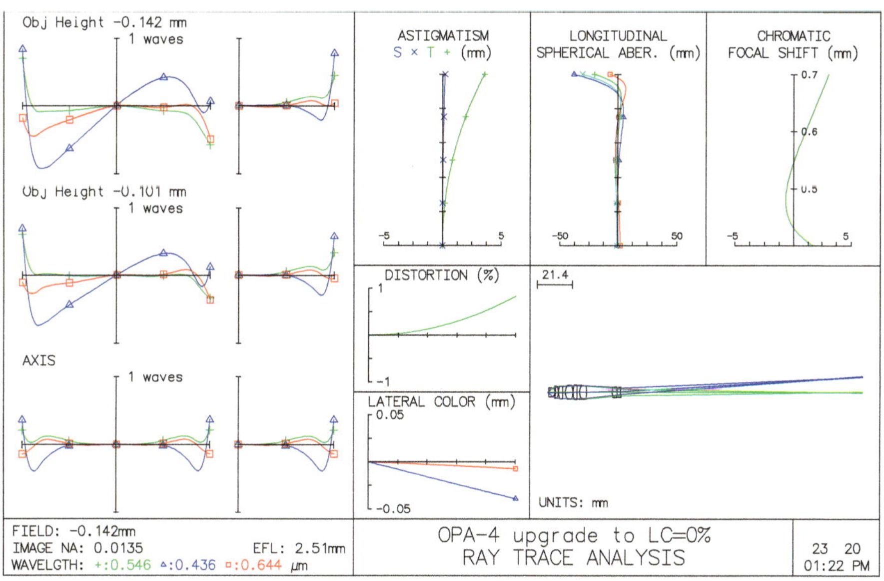

Figure 2.184 Optical design and graphs of aberration correction of a 60×/0.85 finite LCF plan apochromatic objective.

receiver, the advantages of plan apochromats are not obvious, as the spectral sensitivity decreases sharply in the near UV range for the microscopist's eye and for most electronic receivers.

By the way, not a skeptic, but a very positive-minded person, Alexander Grebyonkin, with whom it is interesting to have discussions on microscopy, suggested that I make some arguments here about the advantages of optics with apochromatic aberration correction. This young man is an active member of the Russian Cosmonaut Corps; microscopy is his favorite hobby.

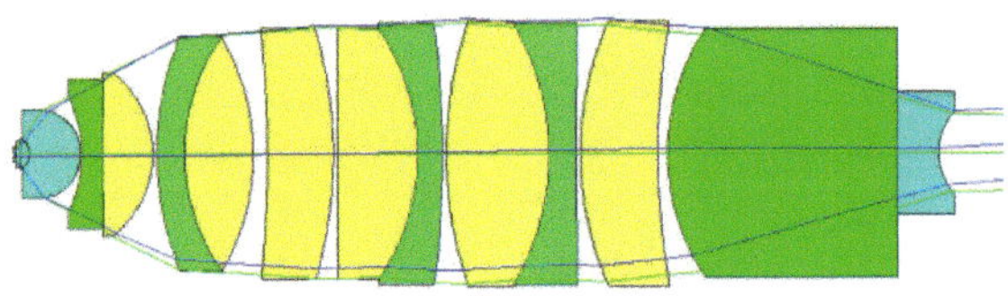

Lens: OPA-5 upgrade to LC=0% Efl 1.51

Object num aper 1.250000 Image height 9.000000 Primary wavln 0.546070

SRF	RADIUS		THICKNESS		APERTURE RADIUS		GLASS		SPECIAL
OBJ	0.000000		0.010000		0.090109		OIL_IMM		
AST	0.000000		0.170000		0.104635	AS	LZ_K14		
2	0.000000		0.193200		0.340000		OIL_IMM	P	
3	0.000000		0.420000		0.620000		LZ_BK6		
4	-0.816600		2.500000		0.680000		LZ_CTK9		
5	-2.089000		0.100000		2.080000		AIR		
6	-5.670600		1.000000		2.660000		LZ_TF5		
7	140.020000		2.450000		3.530000		CAF2		
8	-4.824800		0.200000		3.850000		AIR		
9	14.239000		1.300000		5.370000		LZ_OF4		
10	9.446000		3.300000		5.530000		CAF2	P	
11	-10.880400		0.700000		5.530000		AIR		
12	-55.960500		3.200000		5.690000		CAF2	P	
13	-20.485000		0.200000		5.930000		AIR		
14	692.206000		3.800000		5.930000		CAF2	P	
15	-10.274000		1.200000		5.920000		LZ_OF4	P	
16	-45.471200		0.200000		6.160000		AIR		
17	19.175300		5.000000		6.250000		CAF2	P	
18	-11.846500		1.300000		6.110000		LZ_OF4	P	
19	623.890000		0.200000		6.190000		AIR		
20	15.207900		3.700000		6.270000		CAF2	P	
21	34.317400		0.500000		5.980000		AIR		
22	10.267000		11.000000		5.860000		LZ_TF5	P	
23	420.080000		2.000000		2.890000		LZ_CTK9	P	
24	2.993400		147.161860	S	1.990000		AIR		
25	0.000000		0.000000	S	9.000000	S	AIR		
IMS	0.000000		0.000000		9.000000	S			

Figure 2.185 Optical design and graphs of aberration correction of a 100×/1.25 oil immersion finite LCF plan apochromatic objective.

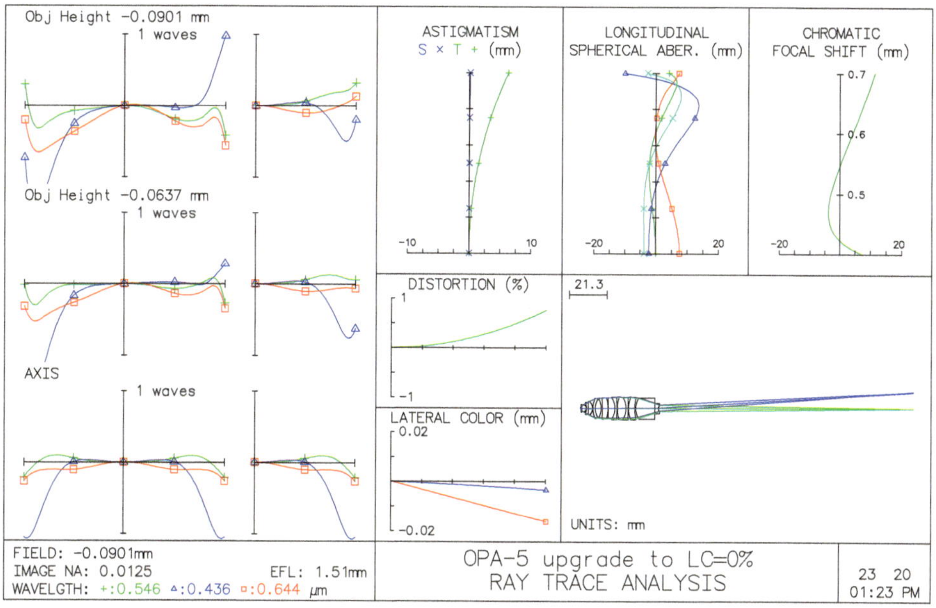

Figure 2.185 (*Continued*) Optical design and graphs of aberration correction of a 100×/1.25 oil immersion finite LCF plan apochromatic objective.

He is passionate about this applied science, has a small collection of light microscopes, and is happy to share his impressions of working with a microscope.

So, plan apochromats have significant advantages in the practice of microscopy over objectives with a different type of aberration correction, for example, plan achromats. These advantages are fully manifested in the study of low-contrast biological objects (for example, diatoms), for which the slightest nuances of the color palette are important. Figure 2.186 presents photographs of a diatom. To obtain the photograph on the left, the modern 40×/0.70 LCF plan achromat was used. On the right is an old Zeiss 40×/0.95 apochromat. Elements of the structure in the image in the photograph on the left have a color that is not likely present in the object. The appearance of such a color in the image is a consequence of the plan achromatic aberration correction. An objective with apochromatic correction gives a more adequate color reproduction in the image (photograph on the right). Images obtained using plan achromats or plan apochromats may not differ in terms of quality and adequacy of color reproduction in the study of stained preparations. For the study of unstained preparations, it is preferable to use plan apochromats.

Nevertheless, plan apochromats also have an advantage in the study of stained biological preparations because they are better able to correct certain types of aberrations. In some cases, there is no alternative to objectives with a plan apochromatic type of aberration correction. An objective with this type

Table 2.15 Comparison of the quality of the optical image between initial and modernized optical designs.

Initial optical design	Upgraded optical design
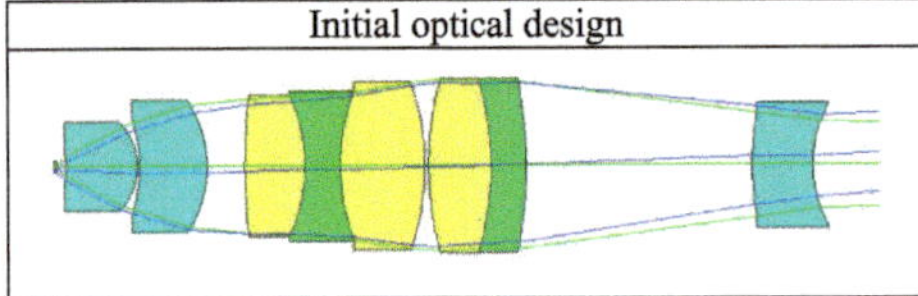	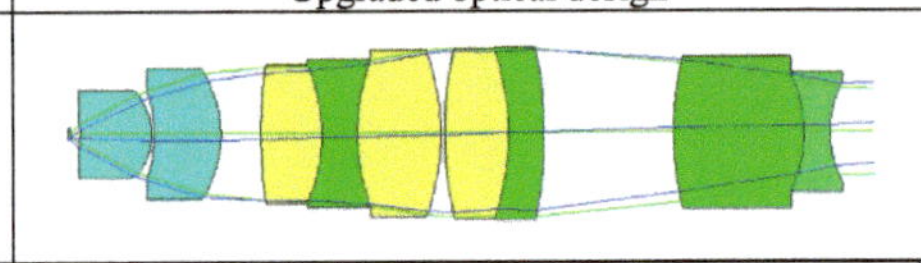
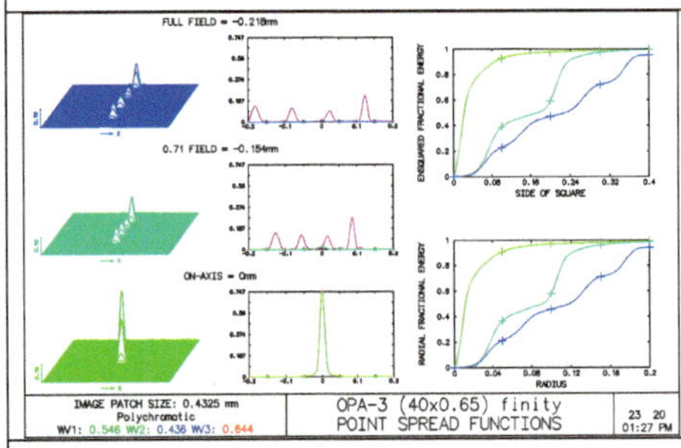	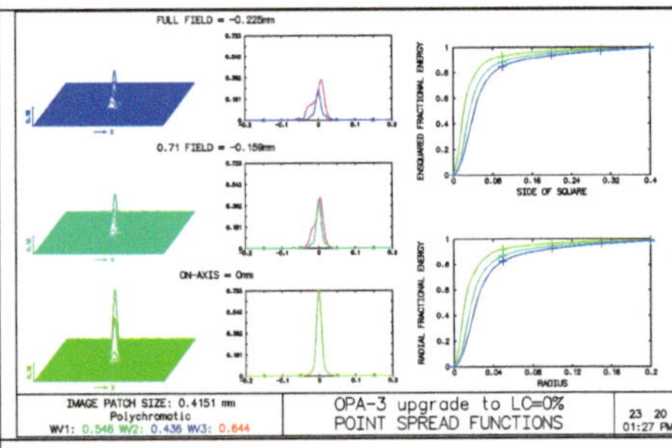
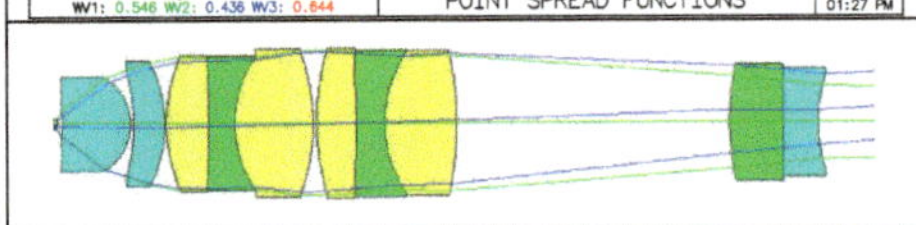	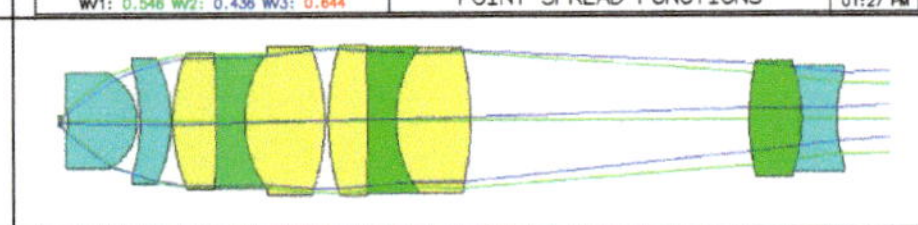
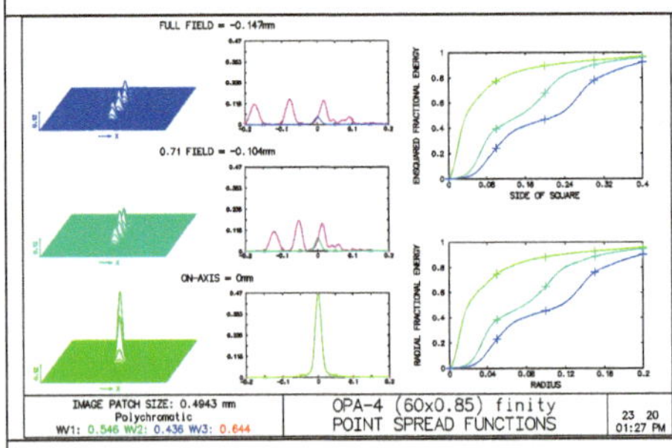	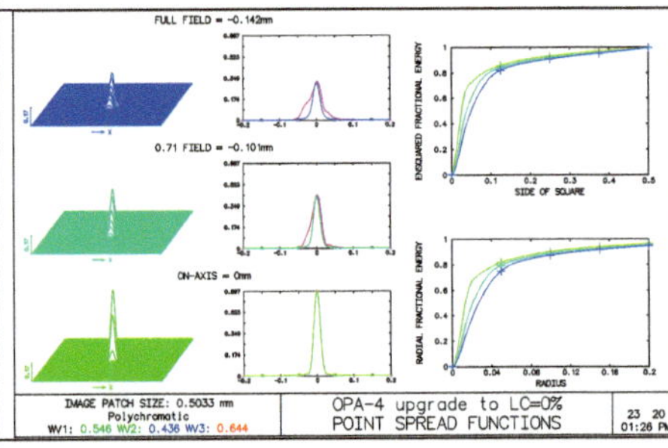
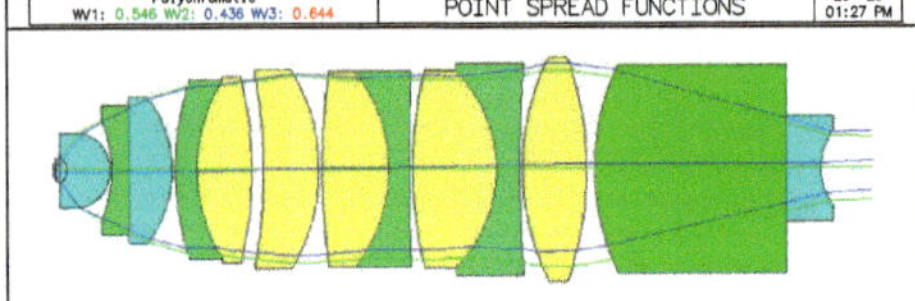	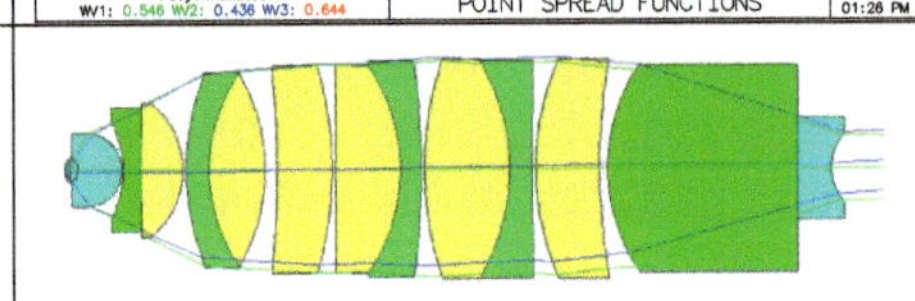
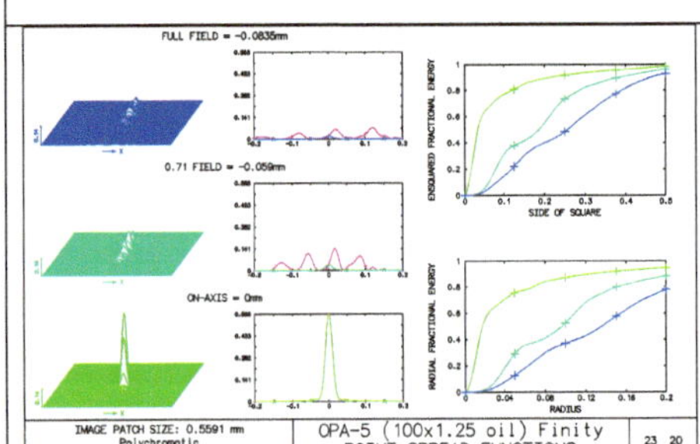	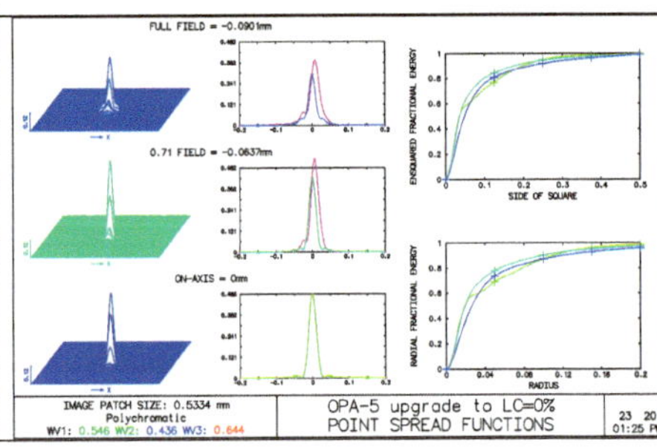

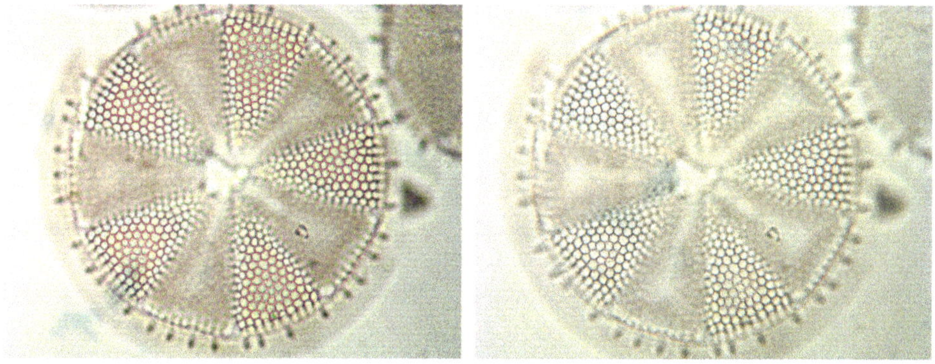

Figure 2.186 Comparison of images made by plan achromatic and apochromatic objectives.

of correction is especially valuable if it is used as a tool for obtaining contrasting methods using a biological microscope, such as polarization, phase, interference, or other contrast, as well as the dark field method. In microscopes for metallography and microelectronics, the nuances of color reproduction are also very important. Of particular importance is the correct color rendering of optics in the fields of applied science and technology, where polarization research techniques such as geology, crystallography, forensics, and others are used.

There is another crucial argument in favor of using plan apochromatic objectives in light microscopes. Let us turn again to the description of a standard biological sample (Volume 1, Chapter 1, Fig. 1.35). This sample is a combination of a glass slide, glue, substance of the object, and a cover glass. In at least 90% of such samples, it becomes impossible to achieve a high degree of reliable information about the investigated object because it is no longer alive. In addition, the parameters of the dye, glue, and cover glass contribute some information to the image on the microscope. A particular color (for example, red or green) may distort the information about the true color of the object under study. The thickness of the glue and its spectral characteristics can also distort information about the true state of the object. However, in diagnostic practice it is important to investigate the true properties of matter in its natural state, namely, as a living form in motion that undergoes constant development and modification. Thus, it is necessary to remove any obstacles that may prevent the object from being studied in its natural living form.

The main components in light microscopes are their optical systems, lighting, and projection and recording systems. The microscope illumination system should not affect the object under study (for example, it should not be too strong in its radiation and should not generate the spectral composition of the radiation, which kills matter). The projection system should provide the necessary resolution and image scale of the object under study. In this case, it is necessary to take into account the minimum size of the object structure that must be distinguished. The image registration system should also be designed in such a way as to ensure recognition of the minimum size of the image structure of the object. Naturally, the main element of a light microscope's optical system is its projection system, namely, the objective. The objective should allow the object to be studied in its vital state. Many types of biological objects are composed of 80–90% water or substances close to water. To design the optical system of a microscope, it is necessary to take into account such parameters as the values of the refractive index and spectral characteristics. Needless to say, only objectives have plan apochromatic aberration correction that can give true information about objects in their living state (i.e., not stained and not treated with glue). Moreover, there is every reason to assume that as production techniques for plan apochromatic objectives in medium series and mass-produced microscopes are mastered, the proportion of stained and pre-glued objects under the microscope will decrease. Ideally, all objects examined using a light microscope should be alive and in their natural state.

Using modern computing tools and specialized software for calculating optical systems, it is possible to significantly simplify the optical systems of plan apochromatic objectives designed to equip medium series microscopes, which means that they can be manufactured in rather significant quantities. Figures 2.187–2.195 show examples of the optical designs and graphs of aberration correction of some modern plan apochromatic objectives that have fairly simple optical structures.

To illustrate the possibilities of using plan apochromatic objectives, I would like to present some photographs taken by Anatoly Mikhaltsov. In Volume 1 of this book, I wrote about how Mikhaltsov devoted his professional career to the study of biological objects and to teaching schoolchildren the basics of light microscopy. The photographs he made using plan apochromatic objectives have amazing beauty, as seen below in Figs 2.196–2.202. Some of these photographs have been modified by the so-called *stacking* method.

2.5 Tube Systems with Fluorite Lenses

Like in plan MicroFluar objectives with aberration correction, optical materials with unusual dispersion properties must also be used in tube systems. This is due to the tube system parameters that need to be achieved, particularly the value of the output numerical aperture after the objective and the linear field in the image plane. In some cases, the use of conventional optical materials is not sufficient to achieve satisfactory correction of spherical and image curvature aberrations, as well as chromatic aberrations such as the secondary spectrum. The light diameters of the lenses of tube systems, as well

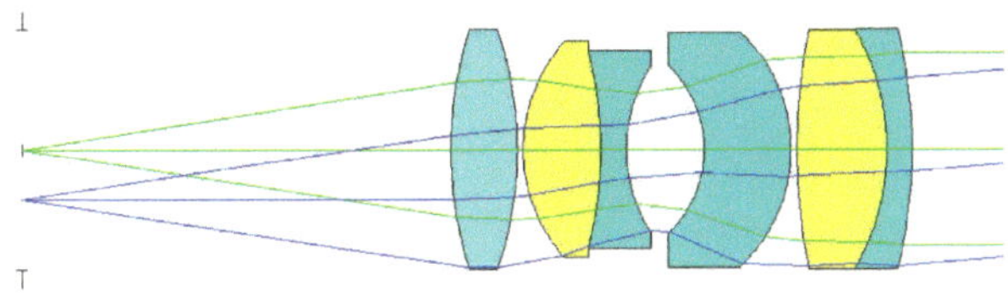

Lens: PlanApo 5x0.15 d=0 Ftl160

Object num aper 0.150000 Image height 12.500000 Primary wavln 0.546070

SRF	RADIUS	THICKNESS	APERTURE RADIUS	GLASS	SPECIAL
OBJ	0.000000	1.0000e-20	2.499986	AIR	
AST	0.000000	22.000000	5.990000 A	AIR	
2	19.408200	3.400000	5.990000	SILICA	
3	-18.172400	0.300000	5.990000	AIR	
4	7.891300	4.000000	5.400000	CAF2	
5	-19.147800	1.300000	4.950000	LZ_CTK19	
6	7.632500	4.000000	4.180000	AIR	
7	-5.676600	4.500000	4.220000	LZ_CTK19 P	
8	-7.767300	0.300000	5.870000	AIR	
9	28.818600	4.600000	5.960000	CAF2 P	
10	-11.618200	1.300000	5.850000	LZ_CTK19 P	
11	-24.042100	90.000000	5.980000	AIR	
12	214.800000	5.000000	10.850000	LZ_BK8	
13	-46.990000	0.300000	10.960000	AIR	
14	-46.880000	3.000000	10.930000	LZ_BF24	
15	-113.240000	157.836726 S	11.100000	AIR	
IMS	0.000000	0.000000	12.500000		

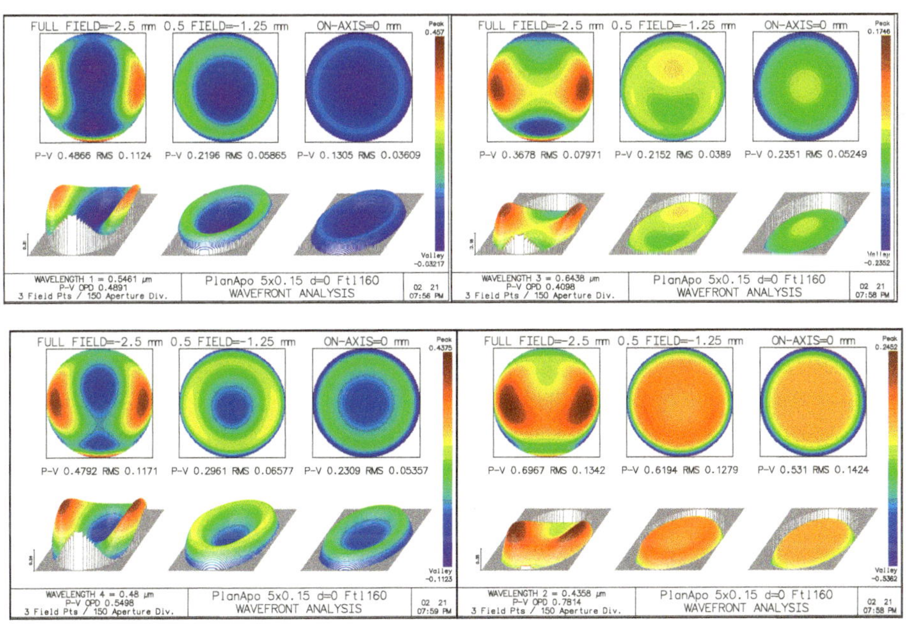

Figure 2.187 Optical design and graphs of aberration correction of a 5×/0.15 infinity CCF plan apochromatic objective ($F'_{ts} = 160$ mm).

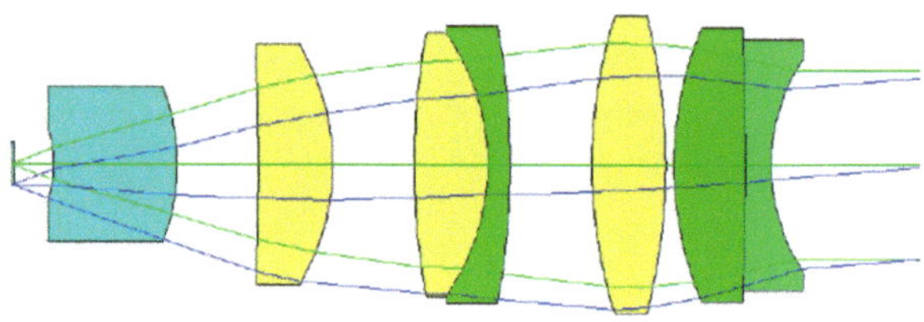

Lens: PLAPO 10x0.35 d=0.17 Ftl160

Object num aper	0.350000	Image height	12.500000	Primary wavln	0.546070

SRF	RADIUS	THICKNESS	APERTURE RADIUS	GLASS	SPECIAL
OBJ	0.000000	0.010000	1.250153	AIR	
AST	0.000000	0.170000	1.250000 A	LZ_K14	
2	0.000000	2.492000	1.290000	AIR	
3	-6.637000	7.500000	2.100000	LZ_CTK19	
4	-12.677000	5.000000	4.580000	AIR	
5	-212.300000	4.500000	6.490000	CAF2	
6	-13.740000	5.000000	7.110000	AIR	
7	40.360000	4.500000	7.800000	CAF2 P	
8	-13.305000	1.300000	7.800000	LZ_TF4	
9	-46.560000	5.000000	8.180000	AIR	
10	27.930000	4.500000	8.790000	CAF2 P	
11	-34.120000	0.500000	8.710000	AIR	
12	20.510000	4.300000	8.110000	LZ_TF4 P	
13	-145.880000	1.500000	7.420000	LZ_OF4	
14	11.066000	90.000000	6.300000	AIR	
15	214.800000	5.000000	11.740000	LZ_BK8	
16	-46.990000	0.300000	11.840000	AIR	
17	-46.880000	3.000000	11.820000	LZ_BF24	
18	-113.240000	157.854341 S	11.980000	AIR	
IMS	0.000000	0.000000	12.500000		

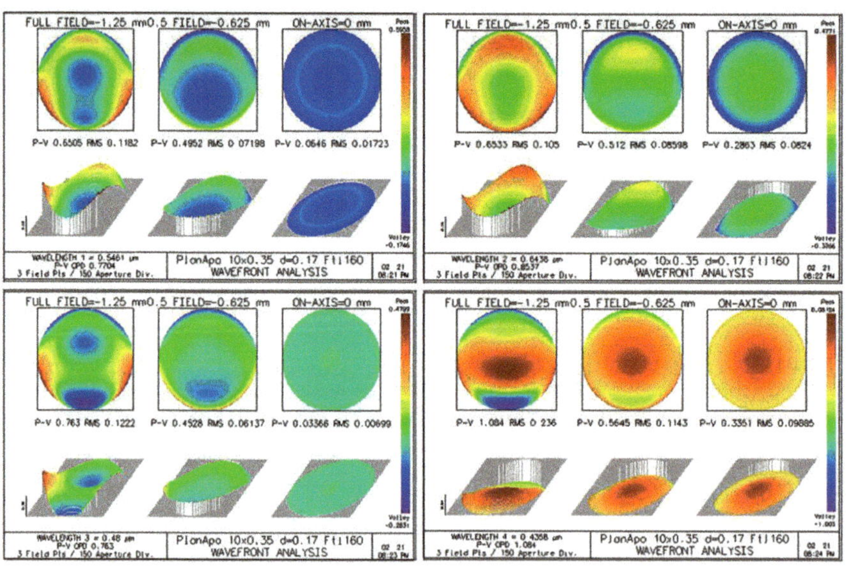

Figure 2.188 Optical design and graphs of aberration correction of a 10×/0.35 infinity CCF plan apochromatic objective ($F'_{ts} = 160$ mm).

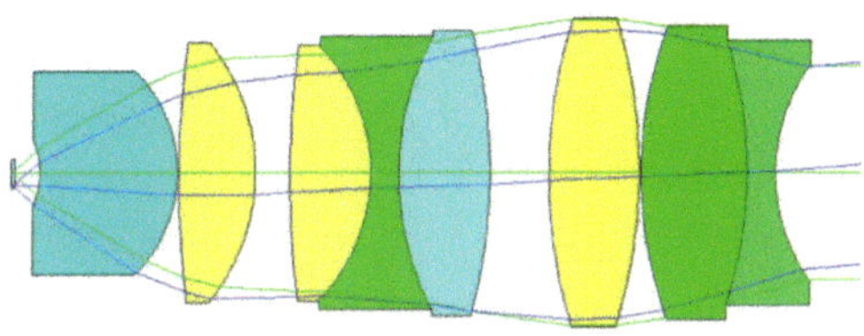

Lens: PlanApo 20x0.65 d=017 ftl160

Object num aper 0.650000 Image height 12.500000 Primary wavln 0.546070

SRF	RADIUS	THICKNESS	APERTURE RADIUS	GLASS	SPECIAL
OBJ	0.000000	0.010000	0.626375	AIR	
AST	0.000000	0.170000	0.634929 AS	LZ_K14	
2	0.000000	1.330000	0.710000	AIR	
3	-3.945000	7.000000	1.570000	LZ_CTK19	
4	-6.792000	0.100000	4.950000	AIR	
5	46.990000	3.900000	6.050000	CAF2	
6	-9.683000	1.750000	6.320000	AIR	
7	46.990000	4.100000	6.230000	CAF2 P	
8	-8.670000	1.400000	6.170000	LZ_TF1	
9	13.932000	4.600000	6.630000	LZ_LK3	
10	-26.060000	3.000000	6.940000	AIR	
11	24.550000	4.600000	7.510000	CAF2 P	
12	-24.550000	0.100000	7.490000	AIR	
13	19.815000	5.400000	7.140000	LZ_TF4	
14	-19.815000	1.400000	6.440000	LZ_OF4	
15	8.670000	90.000000	5.300000	AIR	
16	214.800000	5.000000	15.050000	LZ_BK8	
17	-46.990000	0.300000	15.100000	AIR	
18	-46.880000	3.000000	15.060000	LZ_BF24	
19	-113.240000	157.686534 S	15.240000	AIR	
IMS	0.000000	0.000000	12.500000 S		

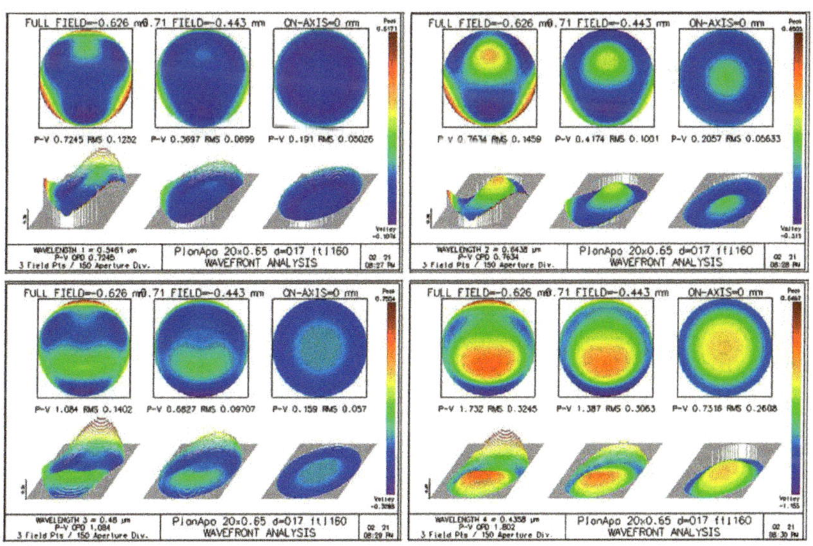

Figure 2.189 Optical design and graphs of aberration correction of a 20×/0.65 infinity CCF plan apochromatic objective ($F'_{ts} = 160$ mm).

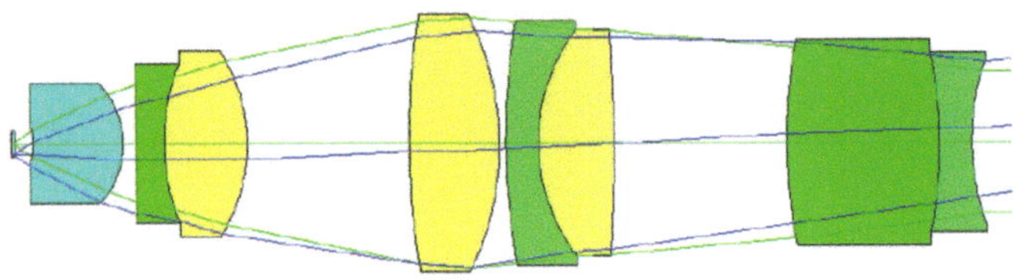

Lens: Plapo 25x0.55 tub160			Efl	6.11
Object num aper	0.550000	Image height 12.500000	Primary wavln	0.546070

SRF	RADIUS	THICKNESS	APERTURE RADIUS	GLASS	SPECIAL
OBJ	0.000000	0.010000	0.492717	AIR	
AST	0.000000	0.170000	0.520000 A	LZ_K14	
2	0.000000	0.840300	0.580000	AIR	
3	-2.911000	4.100000	1.000000	LZ_TK14	
4	-3.767000	0.600000	2.690000	AIR	
5	-146.220000	1.300000	3.200000	LZ_TF1	
6	9.290000	3.800000	3.590000	CAF2	
7	-7.656000	7.500000	4.150000	AIR	
8	34.670000	4.200000	5.620000	CAF2 P	
9	-12.560000	0.300000	5.760000	AIR	
10	31.120000	1.500000	5.460000	LZ_OF4	
11	8.204000	3.500000	5.060000	CAF2 P	
12	-65.610000	8.000000	4.990000	AIR	
13	24.890000	7.000000	4.620000	LZ_TF4	
14	-21.280000	1.500000	4.030000	LZ_OF4 P	
15	10.423000	147.674435 S	3.730000	AIR	
IMS	0.000000	0.000000	12.500000		

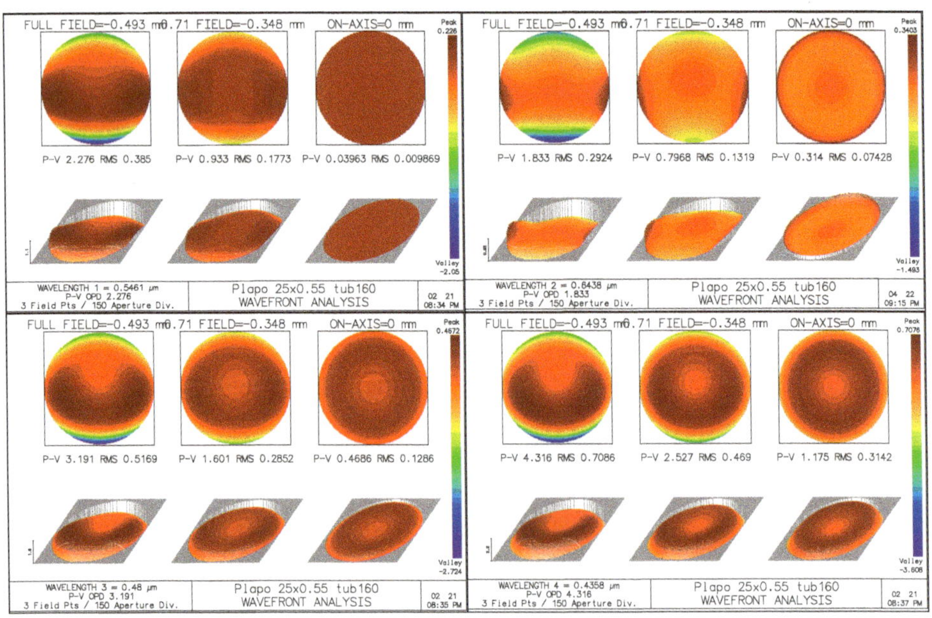

Figure 2.190 Optical design and graphs of aberration correction of a 25×/0.55 CCF plan apochromatic objective (finite tube length, 160 mm).

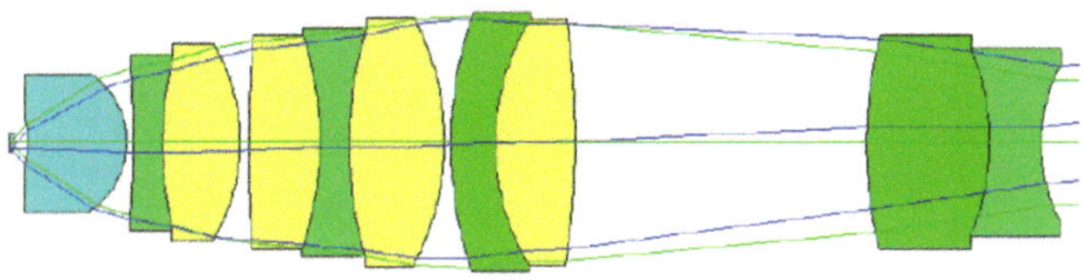

Lens: Plapo 40x0.75 tub160 Efl 3.87

Object num aper 0.750000 Image height 12.500000 Primary wavln 0.546070

SRF	RADIUS	THICKNESS	APERTURE RADIUS		GLASS		SPECIAL
OBJ	0.000000	0.010000	0.311692		AIR		
AST	0.000000	0.170000	0.320000	A	LZ_K14		
2	0.000000	0.641500	0.420000		AIR		
3	-3.404000	4.460000	0.980000		LZ_CTK19		
4	-3.597000	0.300000	3.010000		AIR		
5	-65.010000	1.300000	3.510000		LZ_OF4		
6	16.144000	3.500000	3.850000		CAF2		
7	-8.241000	0.400000	4.310000		AIR		
8	78.700000	3.200000	4.510000		CAF2	P	
9	-14.191000	1.300000	4.670000		LZ_OF4	P	
10	17.298000	4.300000	5.010000		CAF2	P	
11	-11.324000	0.300000	5.440000		AIR		
12	17.100000	2.000000	5.650000		LZ_TF4		
13	10.280000	3.600000	5.400000		CAF2	P	
14	-35.480000	13.100000	5.410000		AIR		
15	18.880000	5.500000	4.670000		LZ_TF4	P	
16	-10.471000	2.300000	4.110000		LZ_OF4	P	
17	7.145000	145.630000	3.410000		AIR		
IMS	0.000000	0.677219	12.500000				

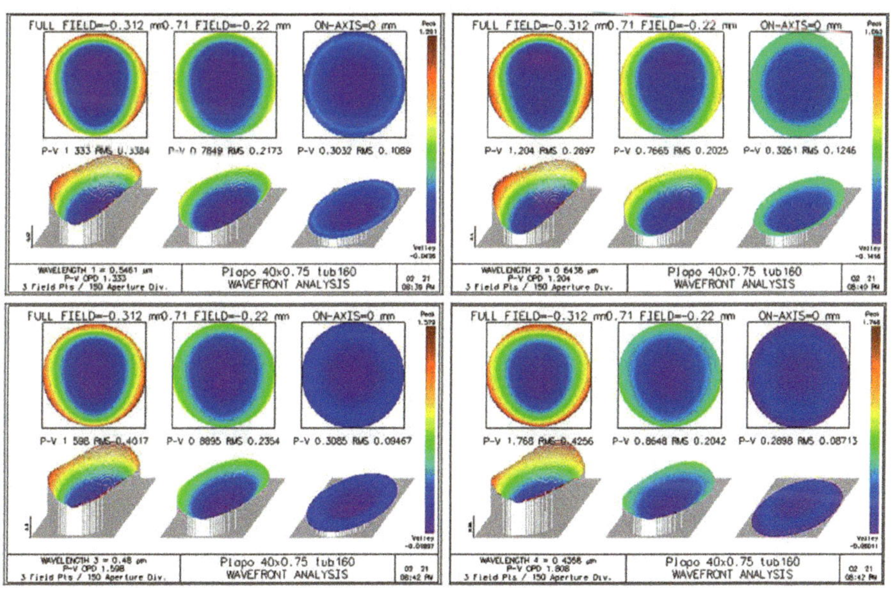

Figure 2.191 Optical design and graphs of aberration correction of a 40×/0.75 CCF plan apochromatic objective (finite tube length, 160 mm).

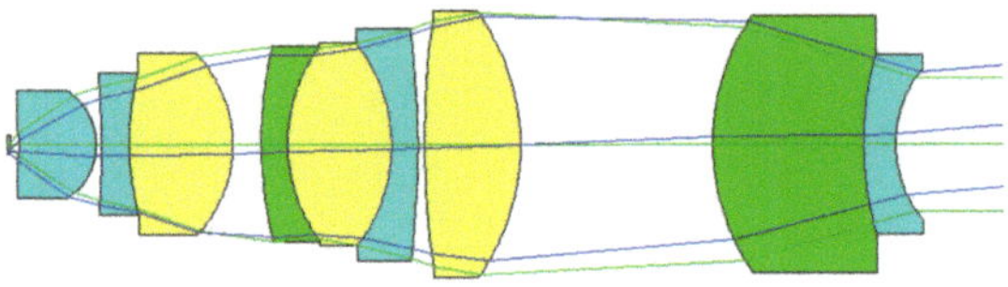

Lens: PLAPO 40x0.75 d=0.17 ftl160

Object num aper 0.750000 Image height 12.500000 Primary wavln 0.546070

SRF	RADIUS	THICKNESS	APERTURE RADIUS	GLASS	SPECIAL
OBJ	0.000000	0.010000	0.310217	AIR	
AST	0.000000	0.170000	0.320000 A	LZ_K14	
2	0.000000	0.400270	0.410000	AIR	
3	-2.270000	3.580000	0.730000	LZ_CTK19	
4	-2.831000	0.250000	2.410000	AIR	
5	-31.550000	1.300000	2.820000	LZ_CTK19 P	
6	11.830000	4.800000	3.190000	CAF2	
7	-5.834000	1.300000	4.050000	AIR	
8	18.408000	1.200000	4.380000	LZ_TF5	
9	7.145000	4.800000	4.310000	CAF2 P	
10	-7.145000	1.200000	4.540000	LZ_CTK19 P	
11	-38.280000	0.290000	5.180000	AIR	
12	38.280000	4.400000	5.560000	CAF2 P	
13	-9.954000	8.920000	5.950000	AIR	
14	9.954000	7.000000	5.740000	LZ_TF5 P	
15	13.868000	1.500000	4.010000	LZ_CTK19 P	
16	5.200000	90.000000	3.360000	AIR	
17	214.800000	5.000000	10.310000	LZ_BK8	
18	-46.990000	0.300000	10.430000	AIR	
19	-46.880000	3.000000	10.410000	LZ_BF24	
20	-113.240000	157.833700 S	10.570000	AIR	
IMS	0.000000	0.000000	12.500000		

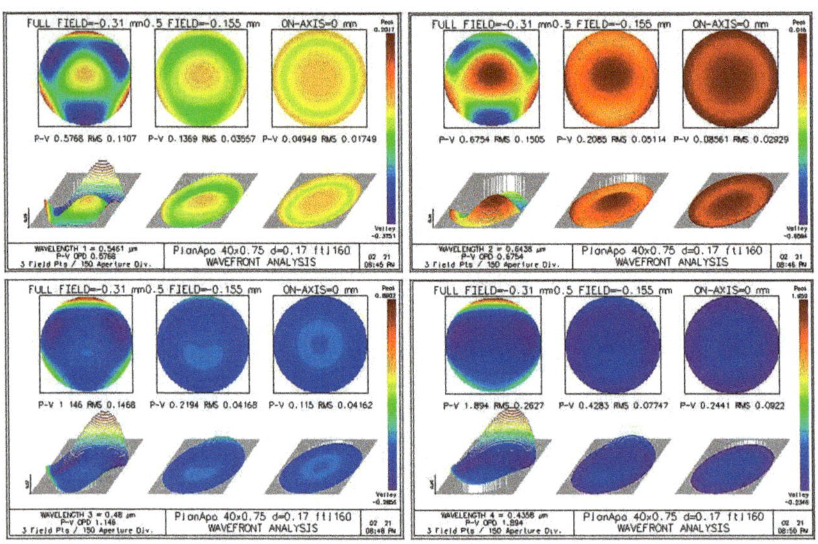

Figure 2.192 Optical design and graphs of aberration correction of a 40×/0.75 infinity CCF plan apochromatic objective ($F'_{ts} = 160$ mm).

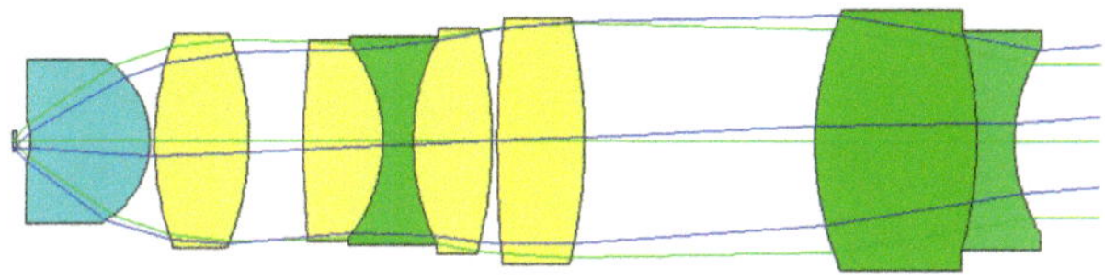

Lens: PlanApo 40x0.75 d=0.17 ftl160

Object num aper 0.750000 Image height 12.500000 Primary wavln 0.546070

SRF	RADIUS		THICKNESS		APERTURE RADIUS		GLASS		SPECIAL
OBJ	0.000000		0.010000		0.312671		AIR		
AST	0.000000		0.170000		0.320000	A	LZ_K14		
2	0.000000		0.550000		0.410000		AIR		
3	-2.280300		4.800000		0.850000		LZ_CTK19		
4	-3.672500		0.200000		3.190000		AIR		
5	11.396500		3.800000		4.020000		CAF2		
6	-10.551700		2.200000		4.180000		AIR		
7	42.133100		3.200000		3.940000		CAF2	P	
8	-5.975700		1.200000		3.830000		LZ_TF1		
9	8.899800		3.200000		4.090000		CAF2	P	
10	-17.118600		0.200000		4.390000		AIR		
11	44.505800		3.500000		4.560000		CAF2	P	
12	-20.798900		9.200000		4.790000		AIR		
13	12.817600		6.500000		5.070000		LZ_TF4		
14	-15.964500		1.500000		4.260000		LZ_OF4		
15	6.866700		160.000000		3.640000		AIR		
16	214.800000		5.000000		16.140000		LZ_BK8		
17	-46.990000		0.300000		16.180000		AIR		
18	-46.880000		3.000000		16.120000		LZ_BF24		
19	-113.240000		157.887810	S	16.320000		AIR		
IMS	0.000000		0.000000		12.500000				

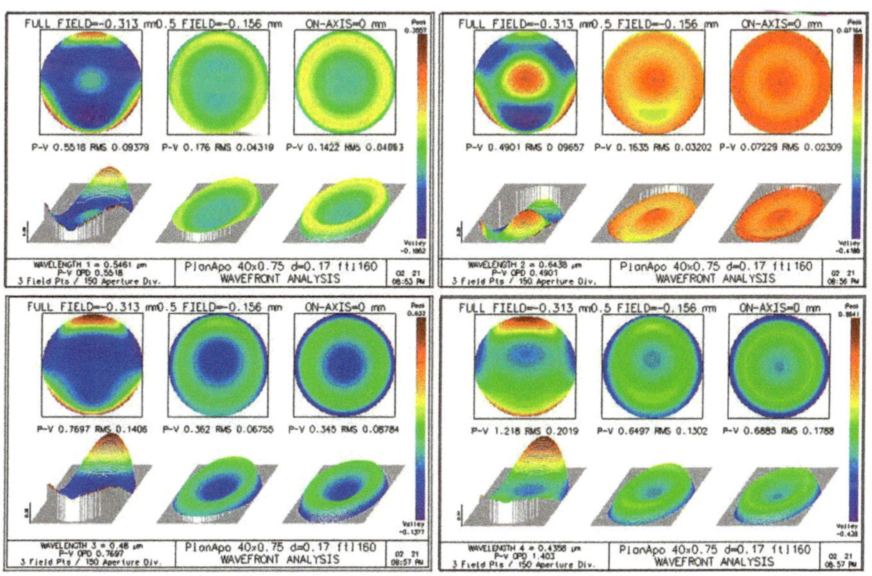

Figure 2.193 Optical design and graphs of aberration correction of a 40×/0.75 infinity CCF plan apochromatic objective ($F'_{ts} = 160$ mm).

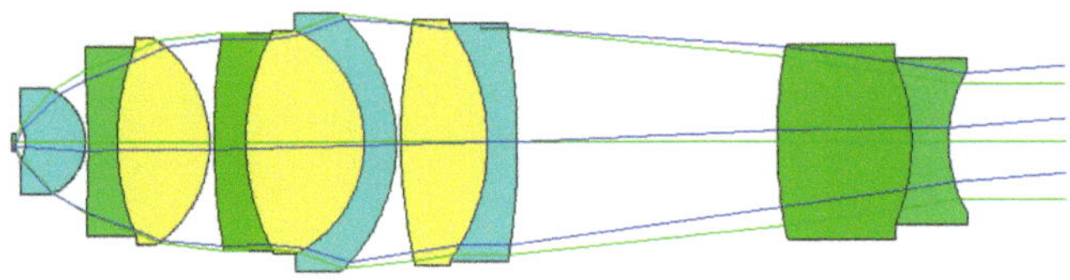

Lens: Plapo 63x0.90 d=0.17 Ftl160

Object num aper	0.900000	Image height	12.500000	Primary wavln	0.546070

SRF	RADIUS	THICKNESS	APERTURE RADIUS	GLASS	SPECIAL
OBJ	0.000000	0.010000	0.198927	AIR	
AST	0.000000	0.170000	0.220000 A	LZ_K14	
2	0.000000	0.400000	0.340000	AIR	
3	-1.481500	2.500000	0.750000	LZ_CTK19	
4	-2.233500	0.130000	2.120000	AIR	
5	-63.886000	1.200000	3.020000	LZ_KF4	
6	10.066000	3.800000	3.770000	CAF2	
7	-4.914500	0.200000	4.120000	AIR	
8	25.759000	1.300000	4.360000	LZ_TF5	
9	9.158700	4.800000	4.340000	CAF2 P	
10	-4.946000	1.300000	4.460000	LZ_CTK19 P	
11	-6.898600	0.200000	5.140000	AIR	
12	22.800000	3.500000	4.870000	CAF2 P	
13	-8.102300	1.200000	4.690000	LZ_CTK19 P	
14	-29.889000	10.540000	4.750000	AIR	
15	18.596700	5.500000	3.910000	LZ_TF5 P	
16	-8.265500	1.500000	3.320000	LZ_OF4	
17	5.647800	90.000000	2.820000	AIR	
18	214.800000	5.000000	9.960000	LZ_BK8	
19	-46.990000	0.300000	10.080000	AIR	
20	-46.880000	3.000000	10.060000	LZ_BF24	
21	-113.240000	157.838623 S	10.230000	AIR	
IMS	0.000000	0.000000	12.500000		

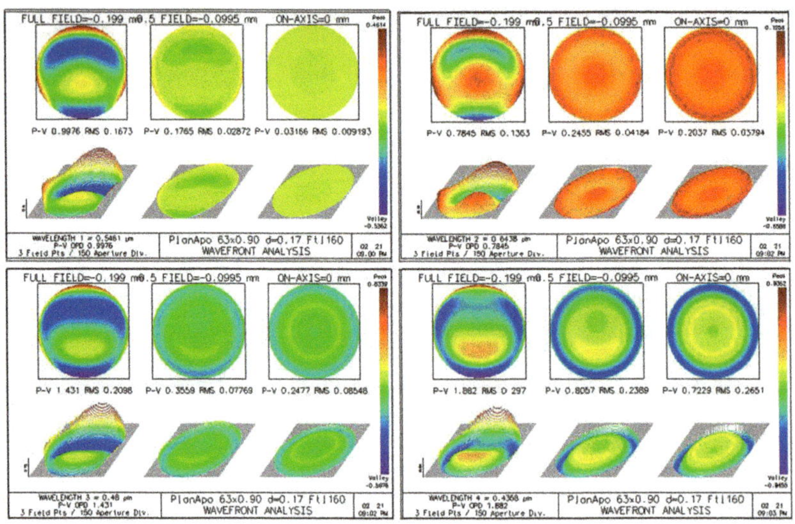

Figure 2.194 Optical design and graphs of aberration correction of a 63×/0.90 infinity CCF plan apochromatic objective ($F'_{ts} = 160$ mm).

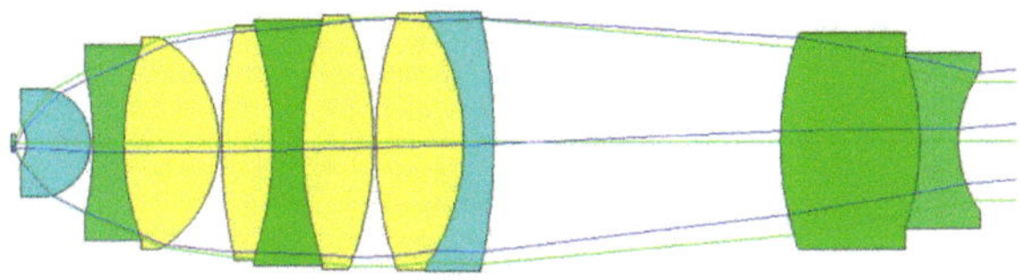

Lens: PlanApo 63x0.90 d=0.17 Ftl160

Object num aper 0.900000 Image height 12.500000 Primary wavln 0.546070

SRF	RADIUS	THICKNESS	APERTURE RADIUS		GLASS		SPECIAL
OBJ	0.000000	0.010000	0.198483		AIR		
AST	0.000000	0.170000	0.220000	A	LZ_K14		
2	0.000000	0.450000	0.340000		AIR		
3	-1.445400	2.500000	0.790000		LZ_CTK19		
4	-2.168000	0.130000	2.100000		AIR		
5	-12.359000	1.200000	2.840000		LZ_KF7		
6	10.209000	3.800000	3.770000		CAF2		
7	-4.742000	0.100000	4.110000		AIR		
8	19.454000	2.000000	4.520000		CAF2	P	
9	-14.588000	1.200000	4.550000		LZ_TF1		
10	14.588000	2.800000	4.760000		CAF2	P	
11	-12.794000	0.100000	4.910000		AIR		
12	14.588000	3.500000	4.980000		CAF2	P	
13	-8.590000	1.200000	4.890000		LZ_CTK19	P	
14	-23.770000	11.330000	5.000000		AIR		
15	12.359000	5.500000	4.190000		LZ_TF4		
16	-10.209000	1.500000	3.400000		LZ_OF4		
17	4.742000	90.000000	2.730000		AIR		
18	214.800000	5.000000	9.750000		LZ_BK8		
19	-46.990000	0.300000	9.880000		AIR		
20	-46.880000	3.000000	9.860000		LZ_BF24		
21	-113.240000	158.012507 S	10.020000		AIR		
IMS	0.000000	0.000000	12.500000				

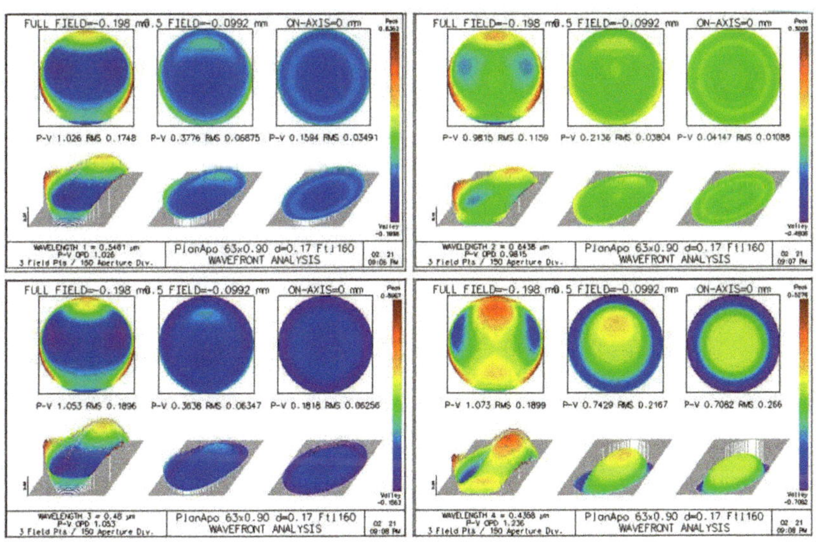

Figure 2.195 Optical design and graphs of aberration correction of a 63×/0.90 infinity CCF plan apochromatic objective ($F'_{ts} = 160$ mm).

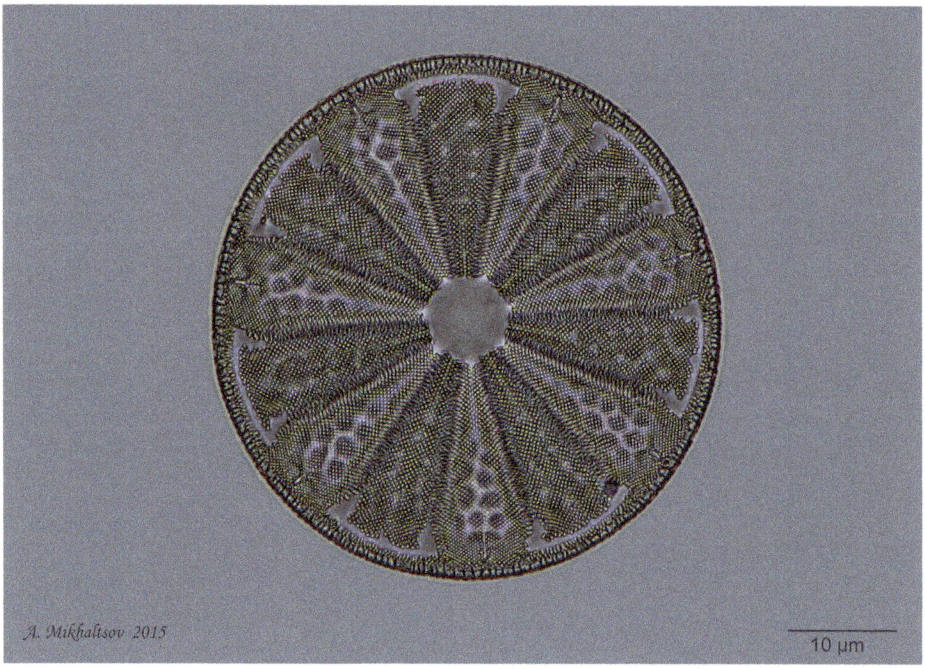

Figure 2.196 Microphotograph of *Actinoptychus maculatus* (taken by A. Mikhaltsov).

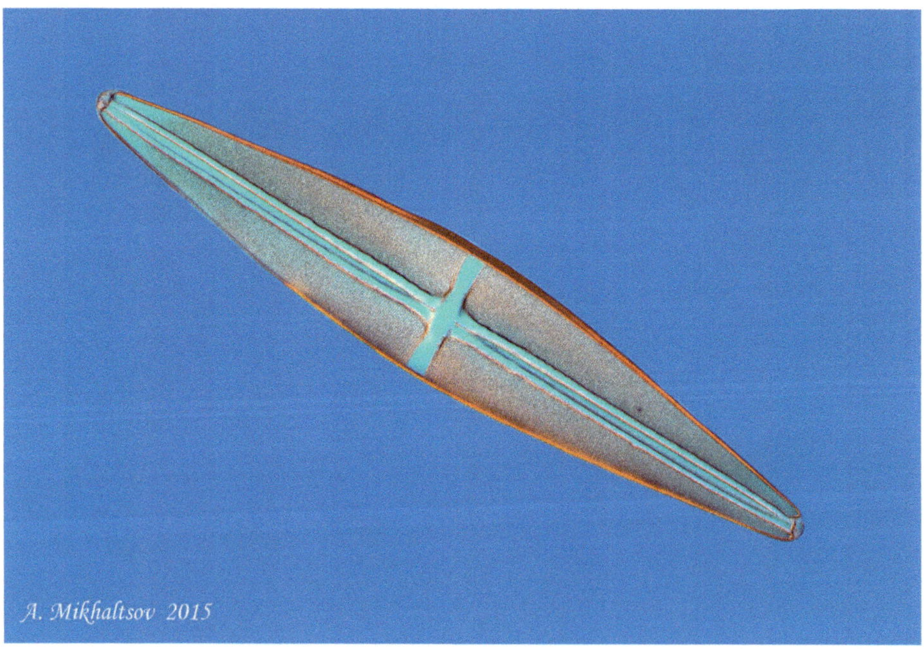

Figure 2.197 Microphotograph taken by A. Mikhaltsov.

Figure 2.198 Microphotograph of *Actinoptychus heterostrophus* (taken by A. Mikhaltsov).

Figure 2.199 Microphotograph of Xylem *Pteridium aquilinum* (taken by A. Mikhaltsov).

Figure 2.200 Microphotograph of *Auliscus oamaruensis* (taken by A. Mikhaltsov).

Figure 2.201 Microphotograph of a root of a fern (taken by A. Mikhaltsov).

Figure 2.202 Microphotograph of *Actinoptychus seductilis* (taken by A. Mikhaltsov).

as the values of image curvature, astigmatism, and LC aberrations, strongly depend on the calculated position of the entrance pupil (this issue is discussed in detail in Chapter 1, Section 1.8.2). Figures 2.203–2.208 show optical designs and graphs of aberration correction of some tube systems containing fluorite lenses; these tube systems have different input numerical apertures, linear fields in the image space, and entrance pupil positions.

These aberration correction graphs show that for all optical design options, a high degree of aberration correction has been achieved. However, one remark must be made about the technological capabilities of manufacturing fluorite lenses. Indeed, in all the optical designs, the fluorite lens diameters are large, and in one design, the diameter is especially large. The diameters of the lenses in these tube systems to a very large extent depend on the calculated position of the entrance pupil. It has already been emphasized that in the optical design of the lens, we adopted the so-called *telecentric ray path* when the tube system is located at a distance approximately equal to its focal length. The optical design of many tube systems was essentially based on this. Making fluorite lenses with such large diameters is a really big problem. However, in medium series microscopes, because the optical design of the microscope system generally does not adhere to the principle of telecentricity (the distance between the lens and the tube system is 40–90 mm), the tube systems required to complete such microscopes may contain fluorite lenses with diameters of a more acceptable size for manufacturing.

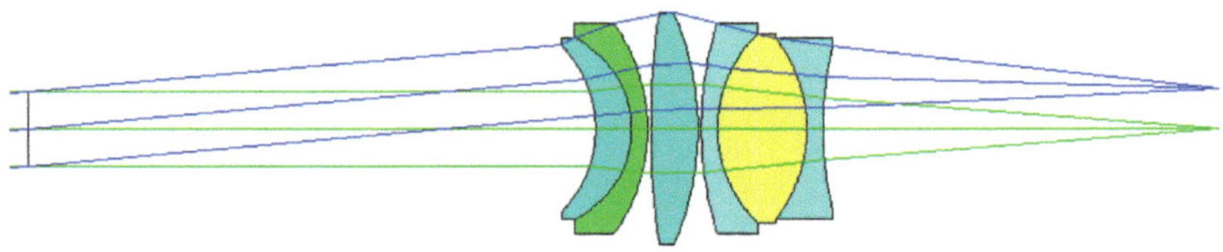

Lens: Ftl F130 NA=0.077 2y'=22					Efl 129.98
Image num aper	0.077000	Image height	11.000000	Primary wavln	0.587600

SRF	RADIUS		THICKNESS		APERTURE RADIUS		GLASS		SPECIAL
OBJ	0.000000		1.0000e+20		8.4627e+18		AIR		
AST	0.000000		0.001000		10.008662	AS	BK7	C	
2	0.000000		160.000000		10.008717	S	AIR		
3	-31.220000		10.000000		22.730000		H-LAK2		
4	-26.750000		4.500000		24.440000		CN_F6		
5	-47.896000		1.000000		28.680000		AIR		
6	218.830000		13.000000		31.190000		H-LAK2	P	
7	-78.500000		1.000000		31.420000		AIR		
8	94.340000		4.500000		28.660000		H-K9		
9	34.680000		25.000000		25.700000		CAF2		
10	-39.370000		4.500000		24.620000		H-K9	P	
11	95.550000		109.996249	S	22.250000		AIR		
IMS	0.000000		0.000000		11.000000	S			

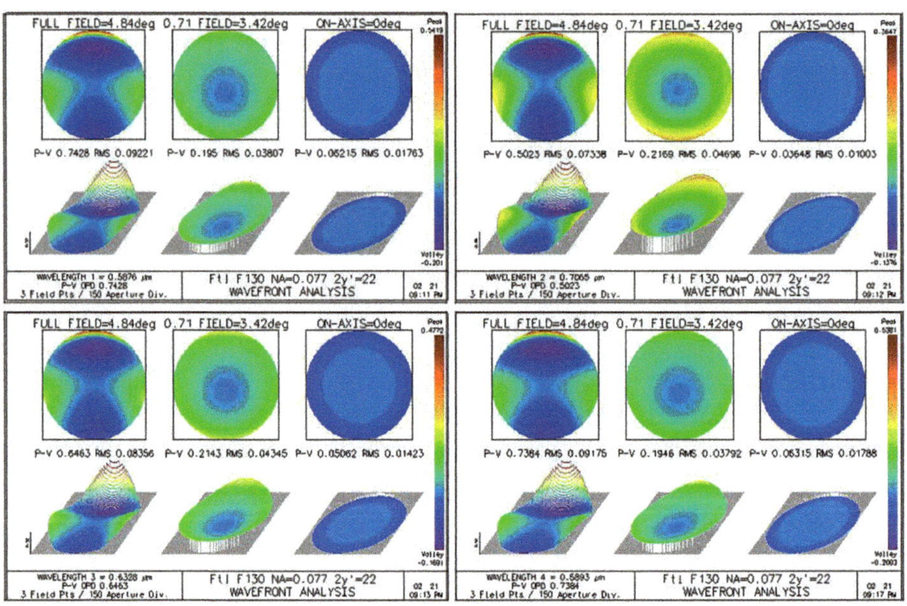

Figure 2.203 Design parameters and graphs of aberration correction of a tube system ($F'_{ts} = 130$ mm); the position of the entrance pupil is $S_p = -160$ mm ($NA' = 0.077$; $2y' = 22$ mm).

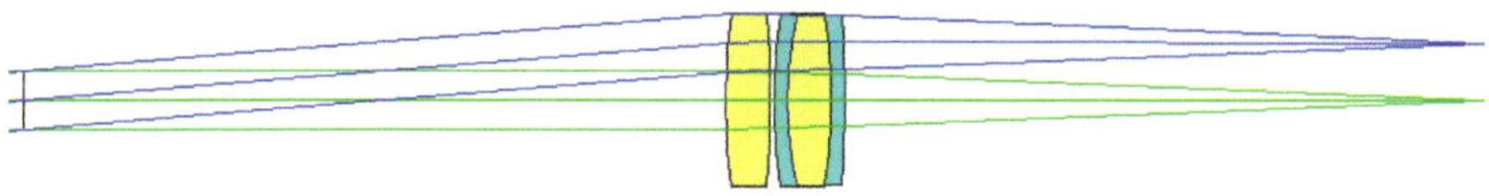

Lens: Ftl 160 NA=0.04 2y'=25mm			Efl 159.99		
Image num aper	0.040000	Image height	12.500000	Primary wavln	0.546074
SRF	**RADIUS**	**THICKNESS**	**APERTURE RADIUS**	**GLASS**	**SPECIAL**
OBJ	0.000000	1.0000e+20	7.8125e+18	AIR	
AST	0.000000	0.001000	6.399993 AS	BK7 C	
2	0.000000	160.000000	6.400045 S	AIR	
3	143.937600	10.000000	19.050000	CAF2	
4	-260.022900	1.000000	19.160000	AIR	
5	142.737100	3.000000	19.120000	LZ_CTK19	
6	78.119400	10.000000	18.860000	CAF2 P	
7	-121.937800	3.000000	18.710000	LZ_CTK19 P	
8	-254.880700	145.608176 S	18.730000	AIR	
IMS	0.000000	0.000000	12.500000 S		

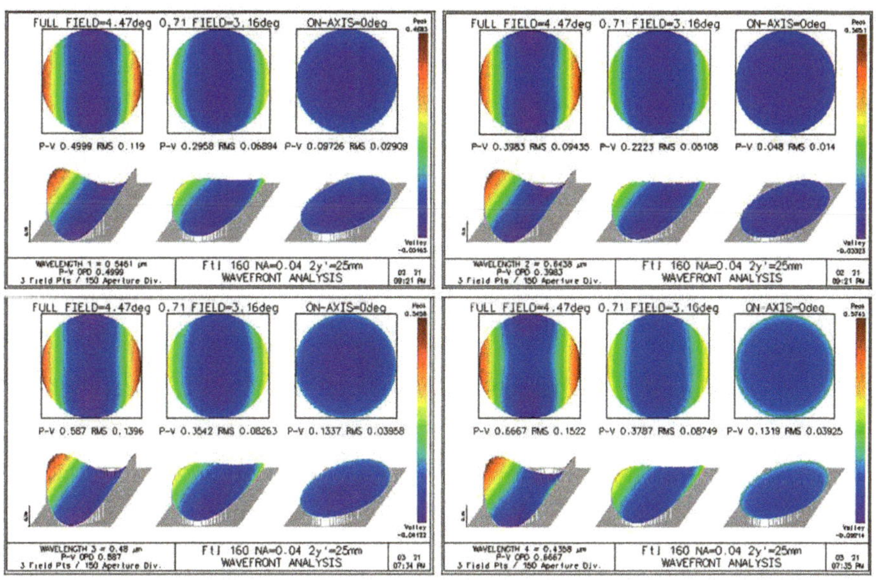

Figure 2.204 Design parameters and graphs of aberration correction of a tube system ($F'_{ts} = 160$ mm); the position of the entrance pupil is $S_p = -160$ mm ($NA' = 0.04$; $2y' = 25$ mm).

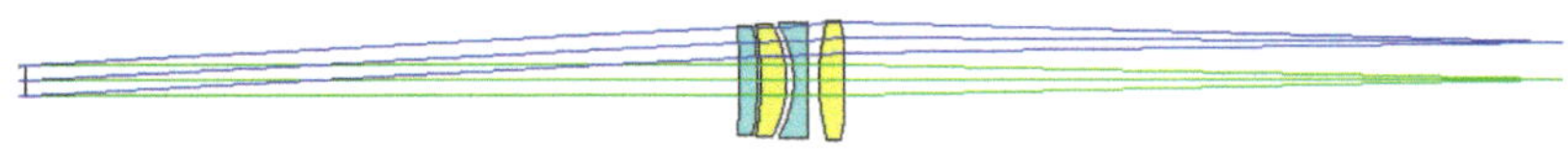

Lens: Ftl=200 NA=0.02 2y'=20 MultiApo — Efl 199.13

Image num aper 0.020000 Image height 10.000000 Primary wavln 0.546074

SRF	RADIUS	THICKNESS		APERTURE RADIUS		GLASS		SPECIAL
OBJ	0.000000	1.0000e+20		5.0218e+18		AIR		
AST	0.000000	0.001000		3.982621	AS	BK7	C	
2	0.000000	200.000000		3.982654	S	AIR		
3	-367.600000	5.000000		14.690000		SILICA	C	
4	-97.660000	1.000000		14.890000		AIR		
5	-78.686000	7.000000		14.880000		LIF	C	
6	-30.700000	2.000000		15.100000		AIR		
7	-28.980000	4.000000		14.750000		SILICA	C	
8	0.000000	3.000000		15.480000		AIR		
9	68.137000	7.000000		16.170000		CAF2	C	
10	-153.581000	199.011008	S	16.230000		AIR		
IMS	0.000000	0.000000		10.000000	S			

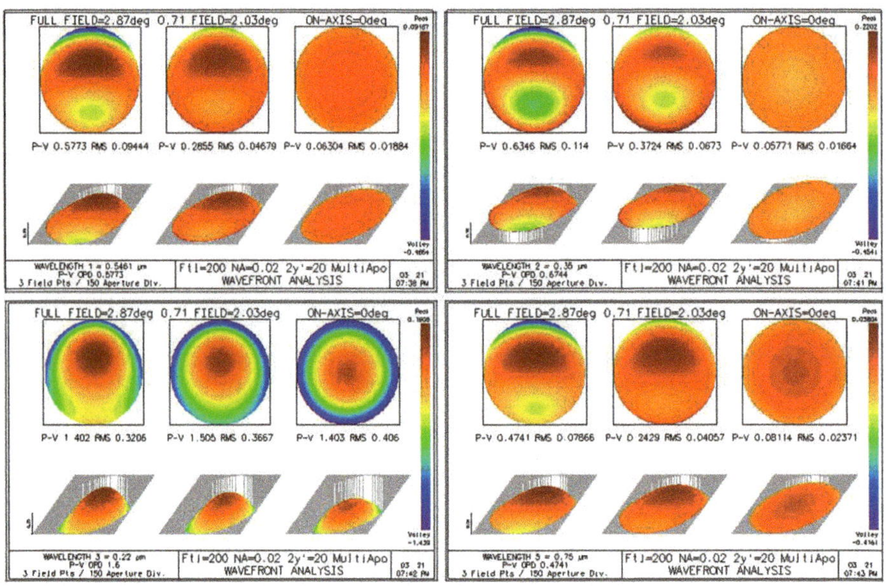

Figure 2.205 Design parameters and graphs of aberration correction of a tube system ($F'_{ts} = 200$ mm); the position of the entrance pupil is $S_p = -200$ mm ($NA' = 0.02$; $2y' = 20$ mm).

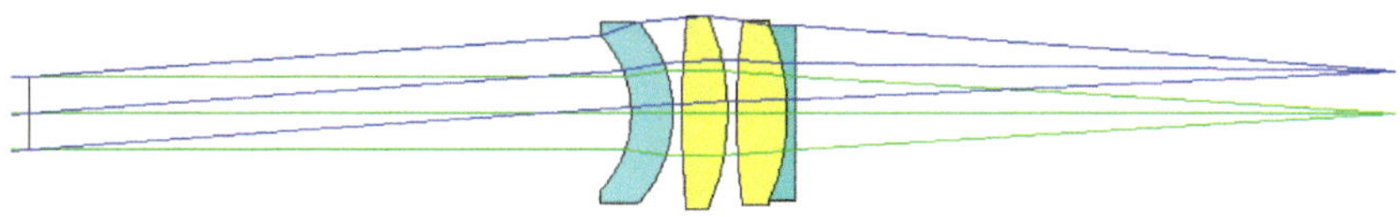

Lens: Ftl 200 NA=0.06 2y'=28mm Efl 200.0

Image num aper 0.060000 Image height 14.000000 Primary wavln 0.546074

SRF	RADIUS		THICKNESS		APERTURE RADIUS		GLASS		SPECIAL
OBJ	0.000000		1.0000e+20		6.9999e+18		AIR		
AST	0.000000		0.001000		12.000215	AS	BK7	C	
2	0.000000		200.000000		12.000261	S	AIR		
3	-36.100000		14.000000		25.290000		LZ_K8		
4	-43.560000		3.000000		29.720000		AIR		
5	377.100000		15.000000		31.170000		CAF2		
6	-81.500000		3.000000		31.440000		AIR		
7	221.300000		17.000000		29.940000		CAF2	P	
8	-72.000000		3.000000		28.750000		LZ_CTK12		
9	0.000000		199.995723	S	28.370000		AIR		
IMS	0.000000		0.000000		14.000000	S			

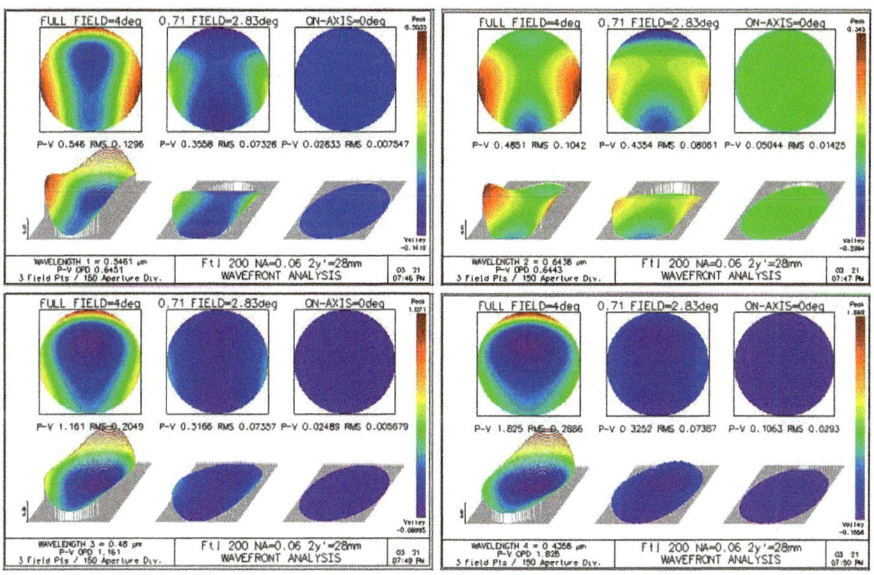

Figure 2.206 Design parameters and graphs of aberration correction of a tube system ($F'_{ts} = 200$ mm); the position of the entrance pupil is $S_p = -200$ mm ($NA' = 0.06$; $2y' = 28$ mm).

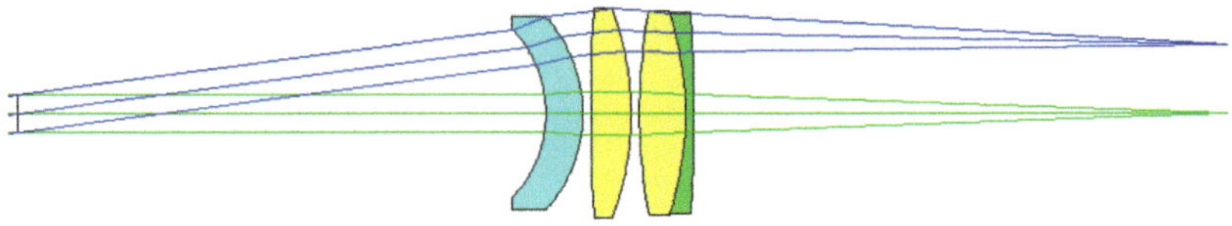

Lens: Ft1200 NA=0.035 2y'=50mm				Efl 199.99	
Image num aper	0.035000	Image height	25.000000	Primary wavln	0.546074

SRF	RADIUS	THICKNESS	APERTURE RADIUS	GLASS	SPECIAL
OBJ	0.000000	1.0000e+20	1.2500e+19	AIR	
AST	0.000000	0.001000	6.999919 AS	BK7 C	
2	0.000000	200.000000	7.000001 S	AIR	
3	-43.930000	14.000000	30.600000	H-QK3	
4	-51.200000	3.000000	35.220000	AIR	
5	667.000000	15.000000	37.650000	CAF2 C	
6	-105.900000	3.000000	38.160000	AIR	
7	194.900000	17.000000	37.360000	CAF2 C	
8	-118.000000	3.000000	36.610000	H-ZLAF50B	
9	-754.000000	200.023351 S	36.530000	AIR	
IMS	0.000000	0.000000	25.000000 S		

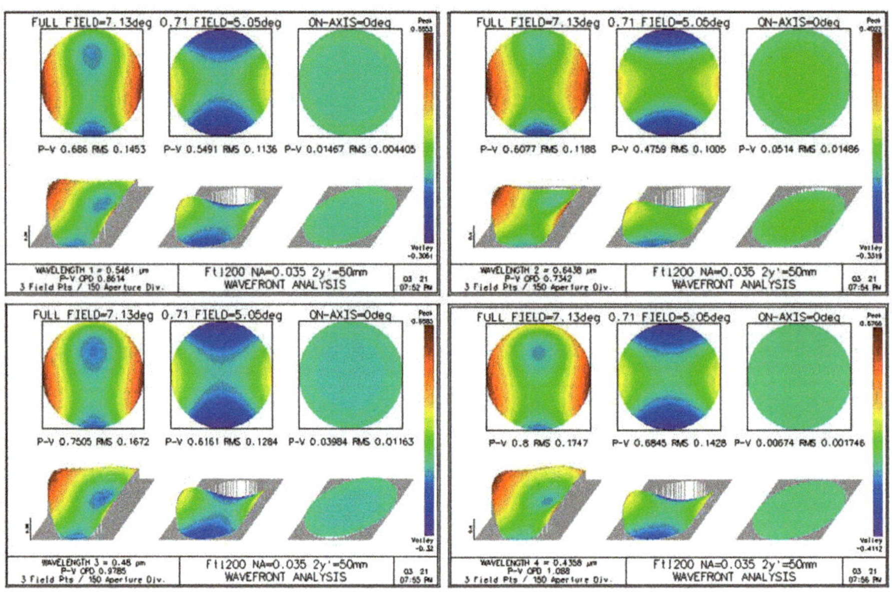

Figure 2.207 Design parameters and graphs of aberration correction of a tube system ($F'_{ts} = 200$ mm); the position of the entrance pupil is $S_p = -200$ mm ($NA' = 0.035$; $2y' = 50$ mm).

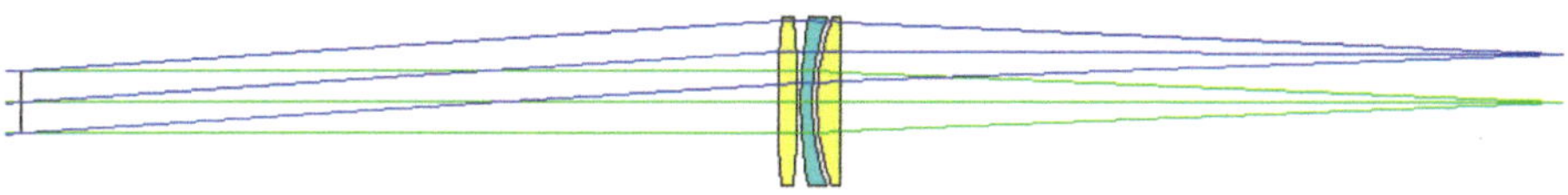

Lens: Ftl 200 NA=0.04 2y'=25mm Efl 199.62

Image num aper 0.040000 Image height 12.500000 Primary wavln 0.546074

SRF	RADIUS		THICKNESS		APERTURE RADIUS		GLASS		SPECIAL
OBJ	0.000000		1.0000e+20		6.2620e+18		AIR		
AST	0.000000		0.001000		7.984641	AS	BK7	C	
2	0.000000		200.000000		7.984682	S	AIR		
3	266.700000		5.000000		21.590000		CAF2		
4	-266.700000		1.000000		21.650000		AIR		
5	119.400000		3.000000		21.620000		LZ_CTK19		
6	62.810000		1.300000		21.240000		AIR		
7	59.430000		6.000000		21.430000		CAF2	P	
8	0.000000		190.966877	S	21.370000		AIR		
IMS	0.000000		0.000000		12.500000	S			

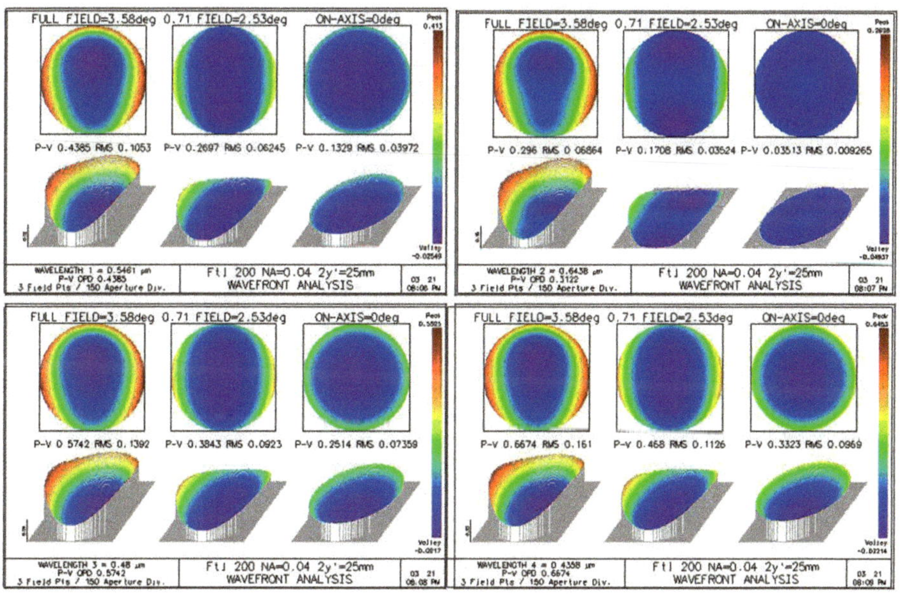

Figure 2.208 Design parameters and graphs of aberration correction of a tube system ($F'_{ts} = 200$ mm); the position of the entrance pupil is $S_p = -200$ mm ($NA' = 0.04$; $2y' = 25$ mm).

References

[1] D. N. Frolov, O. A. Vinogradova, V. N. Frolov, and P. S. Vakulov, "Optical design of objectives: LCF PlanFluor Pol for 27 mm observation in the polarizing microscope," *Proc. SPIE* **10590**, 105901J (2017) [https://doi.org/10.1117/12.2285227].

[2] O. A. Vinogradova, A. D. Frolov, and D. N. Frolov, "Optical lens design with forced performance for metallographic light microscope," Proc. SPIE **7849**, 78492N (2010) [https://doi.org/10.1117/12.870500].

[3] N. O. Tagiltseva, "Fluoroaluminate glasses containing some rare earth elements," Ph.D. dissertation, St.-Petersburg State Technological Institute, SPbGTI (TU), (2000).

[4] Optical Society of America, *Handbook of Optics, Volume 1: Fundamentals, Techniques, and Design*, 2nd ed., McGraw-Hill, New York (1994).

[5] H. Appelt, *Introduction to the Methods of Microscopic Research*, State Publishing House of Medical Literature, Moscow (1959).

[6] V. A. Panov and L. N. Andreev, *Optics of Microscopes*, Mashinostroenie, Leningrad (1976).

[7] H. Boegehold. Microscope objective. US Patent 2,206,155, 1940.

[8] D. N. Frolov, "Synthesis of the optical systems of lens objectives for microscopes," *J. Opt. Technol.* **69**(9), 614 (2002).

[9] D. P. Feder, "Optical calculations with automatic computing machinery," *J. Opt. Soc. Am.* **41**(9), 630–635 (1951).

Chapter 3
Synthesis of Optical Systems for Small Series Microscopes

3.1 Time-tested Optical Technologies in Use Today

We will begin by discussing the optical system of a classical inverted metallographic microscope (the Soviet microscope MMR-4) as an example. This is an interesting discussion because, first of all, the old and time-tested fundamental optical systems of light microscopes can almost always be used in new developments, that is, in modern microscopes. At the same time, light sources, image receivers, and, most importantly, lenses should be replaced with modern ones. Of course, the external design should also be changed (although there are many adherents of the classical style among microscope professionals, they emphasize that the microscope must be reliable and heavy), while nevertheless maintaining time-tested ergonomic characteristics. Figure 3.1 presents the principal optical system of the MMR-4, a metallographic microscope, which was previously manufactured at LOMO. This optical system includes a whole set of progressive engineering solutions that remain relevant today and can fully meet the requirements of the current level of technology.

The optical scheme of the microscope consists of three main systems: lighting, observation, and photographic. In the lighting system, the rays from the source 1 by the collector 2 and the prism 3 are projected into the plane of the aperture diaphragm 4. Further, the image of the source 1 and the aperture diaphragm 4 is projected onto the plane of the exit pupil of the objective by lens 5, mirror 6, lens 7, and translucent reflective plate 8. Field diaphragm 9 is placed in the front focal plane of the second illumination lens 7 and is projected by it into infinity and after the objective to the plane of the subject. In the observational system, the rays passing through the objective and reflected from the object again pass through the objective; the reflective plate 8 and the telephoto lens 10 are collected in the same intermediate plane, which is the plane of the object for the pancratical system 11. By the pancratical system 11, the image is transferred to a plane that coincides with the front focal plane of the first lens of the turning system 12, which projects the image

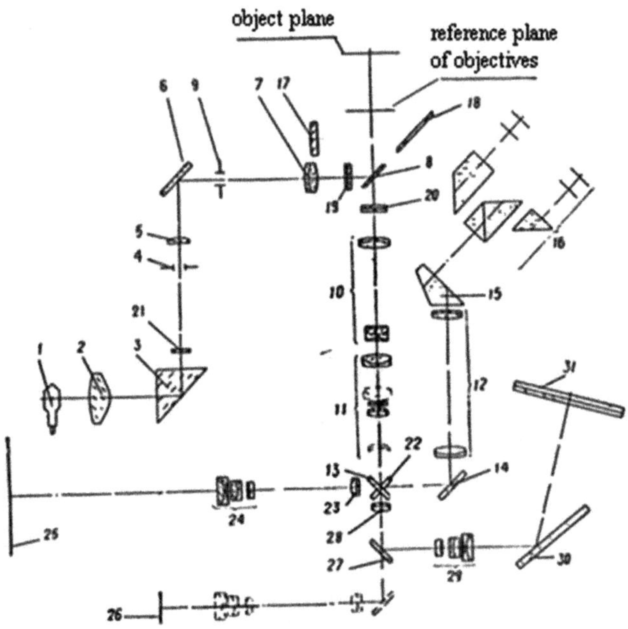

Figure 3.1 Principal optical layout of the MMR-4, an inverted metallurgical microscope.

to infinity. After the second lens of the turning system 12, the rays are collected in the front focal plane of the eyepiece and create a real inverse and enlarged image of the object. Using mirrors 13 and 14 and prism 15, the direction of the optical axis of the microscope is changed. The prism unit of the binocular head 16 divides the beam of rays and provides the possibility of binocular observation of the object.

When illuminated by the dark field method, instead of lens 7 and reflective plate 8, lens 17 (annular diaphragm) and annular mirror 18 are included in the ray path, which directs a parallel beam of rays to the parabolic reflective surface of the illumination path of the (dark field) objective. After the parabolic surface (a kind of condenser), illumination rays are collected on the object under study. The scattered light reflected from the object enters the objective. After that, the course of the rays is the same as when observed in a bright field.

When working in polarized light, a polarizer 19 and an analyzer 20 are simultaneously introduced into the beam path. Oblique illumination of the object is carried out by mixing the aperture diaphragm 9 perpendicular to the axis of the illuminator. To increase the contrast of the image or to weaken the luminous flux during all types of work, interchangeable light filters 21 can be included in the course of the rays in the lighting system.

When photographing, a mirror 22 is introduced into the rays and the mirror 13 is turned off. The image of the object with a lens 23 and a homal 24 is projected onto a 60×120 mm photographic plate or a 24×36 mm photographic film (digital receiver) 26. We can also observe the image of the investigated object on a special screen. For this, mirrors 13 and 22 are turned off from the path of the rays; instead, the mirror 27 is turned on. The image of the object by lens 28 and a homal 29 using the mirrors 27 and 30 is projected onto the special screen 31.

3.1.1 Research microscope Metam-LV

A good example that confirms the thesis of a continuity of generations of light microscopes, especially those that can be classified as small series, is the Metam-LV microscope. This is a microscope that has been manufactured at the LOMO factory for more than 30 years; it is still manufactured today. Its optical and mechanical design was completed in the beginning of the 1990s. The engineering solutions used in this microscope are still relevant today. The Metam-LV is a metallographic inverted microscope of the research class, sometimes referred to as a *small series*. It was designed with original engineering solutions such as telecentric optical systems, convenient ergonomics, and an increased linear field. It also allows techniques such as differential contrast to be implemented (with all the objectives in the kit) and has some advantages over analogues. Since its inception, the Metam-LV has not lost its originality and relevance, primarily due to its use of original objectives with plan apochromatic aberration correction. Figure 3.2 is a photograph of this microscope. Of course, it has many shortcomings, for example, its nozzle interferes with the location of extended objects; but who is without shortcomings? Even its appearance displays a certain standard of engineering excellence.

Figures 3.3–3.7 show the design parameters and graphs of aberration correction of the objectives used in this microscope.

Figure 3.2 Photograph of the Metam-LV, an inverted research microscope.

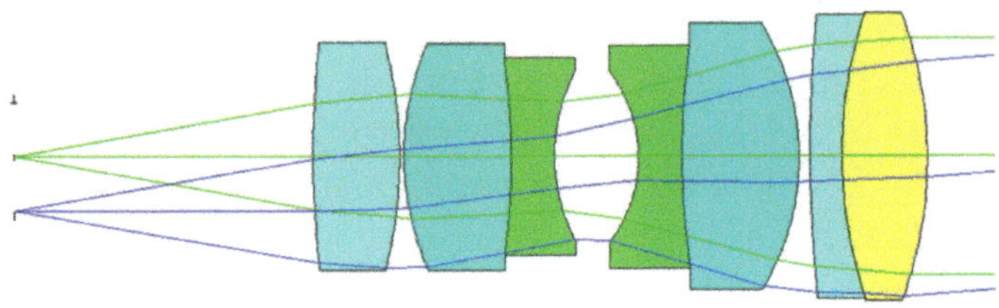

Lens: PlanApo 5x0.17 d=0 Ftl160

Object num aper 0.170000 Image height 12.500000 Primary wavln 0.546070

SRF	RADIUS	THICKNESS	APERTURE RADIUS	GLASS	SPECIAL
OBJ	0.000000	1.0000e-20	2.493029	AIR	
AST	0.000000	14.120000	2.493029 AS	AIR	
2	41.690000	4.000000	5.000000	LZ_LK3	
3	-20.990000	0.200000	5.220000	AIR	
4	12.417000	5.000000	5.190000	LZ_CTK3	
5	-33.810000	2.000000	4.530000	LZ_OF4	
6	8.318000	3.930000	3.910000	AIR	
7	-6.360000	2.000000	3.950000	LZ_OF4 P	
8	35.730000	5.500000	5.100000	LZ_CTK3 P	
9	-11.588000	0.500000	6.120000	AIR	
10	59.980000	1.500000	6.410000	LZ_K8	
11	20.320000	4.000000	6.500000	CAF2	
12	-18.880000	90.000000	6.600000	AIR	
13	214.800000	5.000000	11.630000	LZ_BK8	
14	-46.990000	0.300000	11.730000	AIR	
15	-46.880000	3.000000	11.700000	LZ_BF24	
16	-113.240000	158.862402 S	11.870000	AIR	
IMS	0.000000	0.000000	12.500000 S		

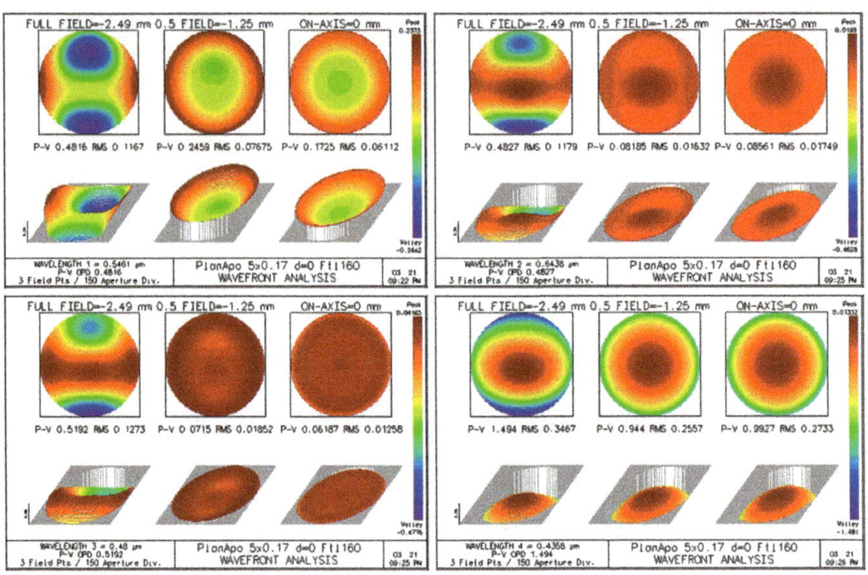

Figure 3.3 Design parameters and graphs of aberration correction of a $5\times/0.17$ completely color free (CCF) plan apochromatic objective ($F'_{ts} = 160$ mm).

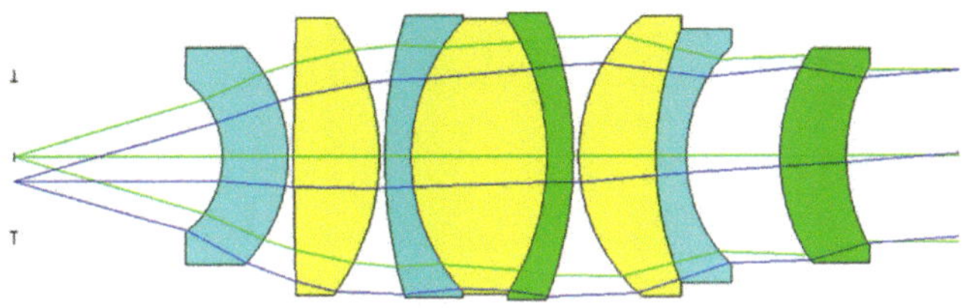

Lens: PlanApo 10x0.28 d=0 Ftl160

Object num aper 0.280000 Image height 12.500000 Primary wavln 0.546070

SRF	RADIUS		THICKNESS		APERTURE RADIUS		GLASS		SPECIAL
OBJ	0.000000		1.0000e-20		1.247967		AIR		
AST	0.000000		11.168500		3.940000	A	AIR		
2	-4.966000		3.500000		3.940000		LZ_LK3		
3	-8.017000		0.300000		5.670000		AIR		
4	158.120000		4.600000		6.740000		CAF2		
5	-11.967000		0.300000		7.230000		AIR		
6	26.180000		1.400000		7.380000		LZ_CTK12		
7	10.520000		7.200000		7.080000		CAF2	P	
8	-13.428000		1.400000		7.190000		LZ_TF5		
9	-20.140000		0.300000		7.500000		AIR		
10	9.727000		4.100000		7.320000		CAF2	P	
11	17.061000		1.600000		6.620000		LZ_K8		
12	7.943000		5.000000		5.720000		AIR		
13	10.139000		3.500000		5.620000		LZ_TF5	P	
14	9.376000		90.000000		4.720000		AIR		
15	214.800000		5.000000		11.180000		LZ_BK8		
16	-46.990000		0.300000		11.290000		AIR		
17	-46.880000		3.000000		11.260000		LZ_BF24		
18	-113.240000		157.826745	S	11.430000		AIR		
IMS	0.000000		0.000000		12.500000				

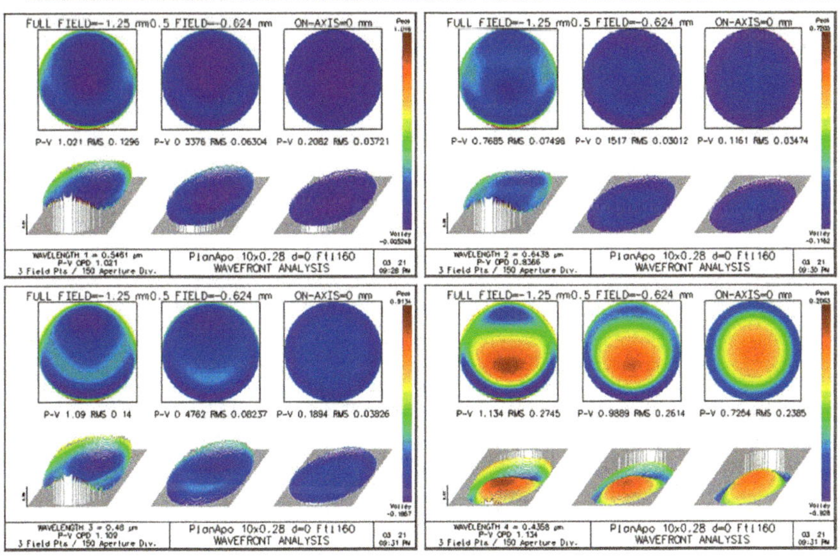

Figure 3.4 Design parameters and graphs of aberration correction of a 10×/0.28 CCF plan apochromatic objective ($F'_{ts} = 160$ mm).

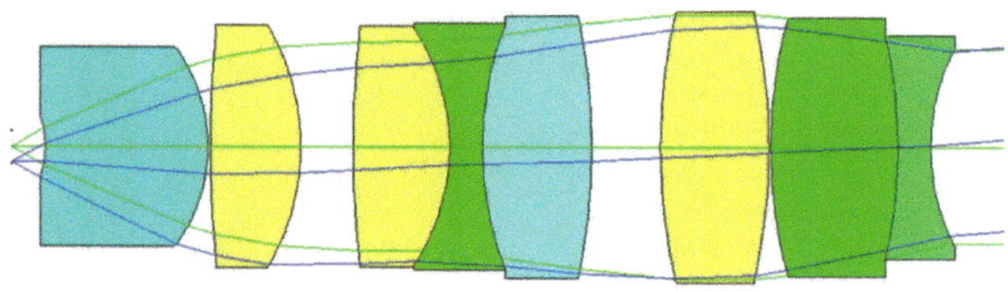

Lens: PlanApo 20x0.50 d=0 ft1160

Object num aper	0.500000	Image height	12.500000	Primary wavln	0.546070

SRF	RADIUS	THICKNESS		APERTURE RADIUS		GLASS		SPECIAL
OBJ	0.000000	-1.0000e-08		0.625644		AIR		
AST	0.000000	1.399800		0.625644	AS	AIR		
2	-3.767000	7.000000		1.300000		LZ_CTK19		
3	-6.699000	0.100000		4.110000		AIR		
4	46.990000	3.900000		4.620000		CAF2		
5	-9.683000	2.200000		5.010000		AIR		
6	46.990000	4.100000		4.970000		CAF2	P	
7	-8.670000	1.400000		4.880000		LZ_TF1		
8	13.932000	4.600000		5.080000		LZ_LK3		
9	-26.060000	2.900000		5.400000		AIR		
10	24.550000	4.600000		5.600000		CAF2	P	
11	-24.550000	0.100000		5.510000		AIR		
12	19.815000	5.400000		5.330000		LZ_TF4		
13	-19.815000	1.400000		4.600000		LZ_OF4		
14	8.670000	90.000000		4.060000		AIR		
15	214.800000	5.000000		11.060000		LZ_BK8		
16	-46.990000	0.300000		11.160000		AIR		
17	-46.880000	3.000000		11.130000		LZ_BF24		
18	-113.240000	157.868331	S	11.300000		AIR		
IMS	0.000000	0.000000		12.500000	S			

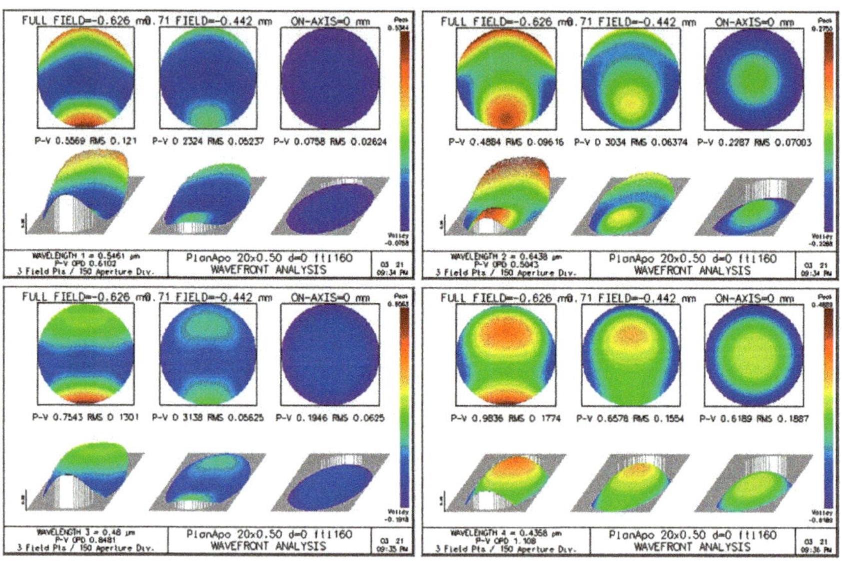

Figure 3.5 Design parameters and graphs of aberration correction of a 20×/0.50 CCF plan apochromatic objective ($F'_{ts} = 160$ mm).

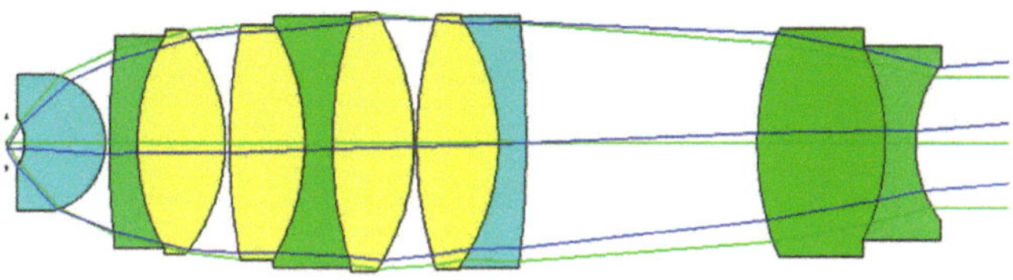

Lens: PlanApo 50x0.85 d=0 Ftl160

Object num aper 0.850000 Image height 12.500000 Primary wavln 0.546070

SRF	RADIUS	THICKNESS	APERTURE RADIUS	GLASS	SPECIAL
OBJ	0.000000	1.0000e-20	0.250838	AIR	
AST	0.000000	0.799980	0.250838 AS	AIR	
2	-1.737800	3.510000	1.010000	LZ_CTK19	
3	-2.938000	0.200000	2.850000	AIR	
4	41.020000	1.200000	3.940000	LZ_KF6	
5	9.036000	3.800000	4.440000	CAF2	
6	-7.278000	0.200000	4.700000	AIR	
7	27.040000	3.200000	4.910000	CAF2 P	
8	-9.817000	1.200000	4.930000	LZ_TF1	
9	17.579000	3.500000	5.250000	CAF2 P	
10	-9.817000	0.100000	5.430000	AIR	
11	17.579000	3.500000	5.320000	CAF2 P	
12	-9.036000	1.200000	5.190000	LZ_CTK19 P	
13	-41.020000	9.980000	5.260000	AIR	
14	12.503000	5.500000	4.760000	LZ_TF4	
15	-9.036000	1.300000	4.050000	LZ_OF4	
16	5.272000	90.000000	3.220000	AIR	
17	214.800000	5.000000	10.270000	LZ_BK8	
18	-46.990000	0.300000	10.390000	AIR	
19	-46.880000	3.000000	10.370000	LZ_BF24	
20	-113.240000	157.839557 S	10.530000	AIR	
IMS	0.000000	0.000000	12.500000 S		

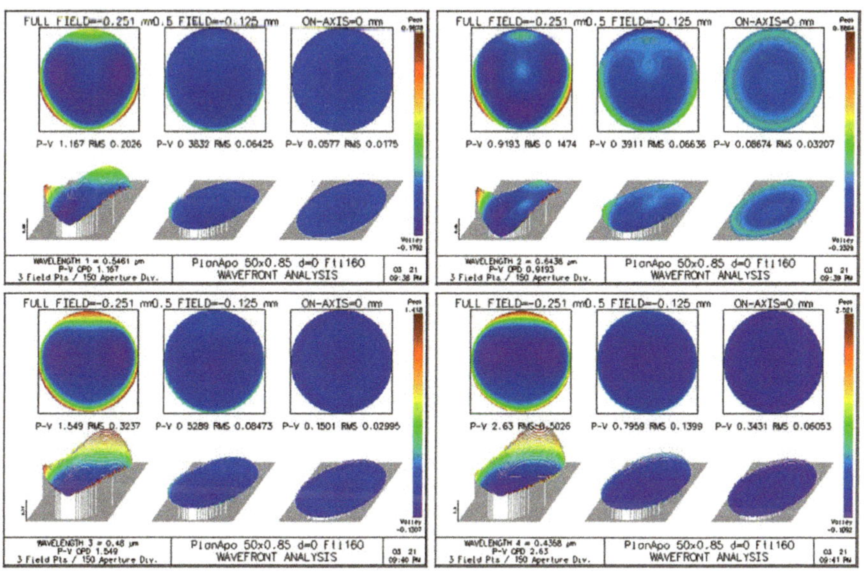

Figure 3.6 Design parameters and graphs of aberration correction of a 50×/0.85 CCF plan apochromatic objective ($F'_{ts} = 160$ mm).

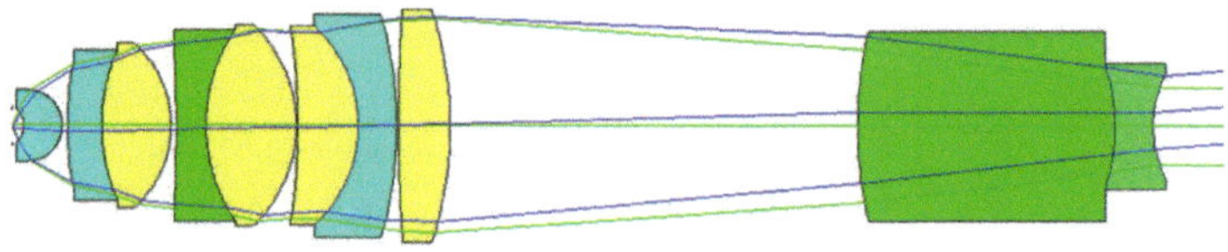

Lens: PlanApo 100x0.95 d=0 ftl160

Object num aper 0.950000 Image height 12.500000 Primary wavln 0.546070

SRF	RADIUS	THICKNESS	APERTURE RADIUS	GLASS	SPECIAL
OBJ	0.000000	-1.0000e-08	0.125306	AIR	
AST	0.000000	0.450000	0.125306 AS	AIR	
2	-0.895400	1.500000	0.650000	LZ_CTK19	
3	-1.367700	0.200000	1.370000	AIR	
4	13.213000	1.300000	2.480000	LZ_CTK19 P	
5	6.950000	2.800000	2.870000	CAF2	
6	-4.055000	0.200000	3.150000	AIR	
7	-74.300000	1.200000	3.410000	LZ_TF1	
8	6.457000	3.500000	3.680000	CAF2 P	
9	-5.346000	0.050000	3.880000	AIR	
10	-25.120000	2.400000	3.710000	CAF2 P	
11	-5.012000	1.500000	3.800000	LZ_CTK19 P	
12	-14.859000	0.050000	4.300000	AIR	
13	63.530000	2.000000	4.400000	CAF2 P	
14	-14.859000	15.970000	4.470000	AIR	
15	15.241000	10.000000	3.610000	LZ_TF4	
16	-7.551000	1.500000	2.410000	LZ_OF4	
17	4.285000	90.000000	2.030000	AIR	
18	214.800000	5.000000	9.410000	LZ_BK8	
19	-46.990000	0.300000	9.550000	AIR	
20	-46.880000	3.000000	9.540000	LZ_BF24	
21	-113.240000	157.808040 S	9.700000	AIR	
IMS	0.000000	0.000000	12.500000 S		

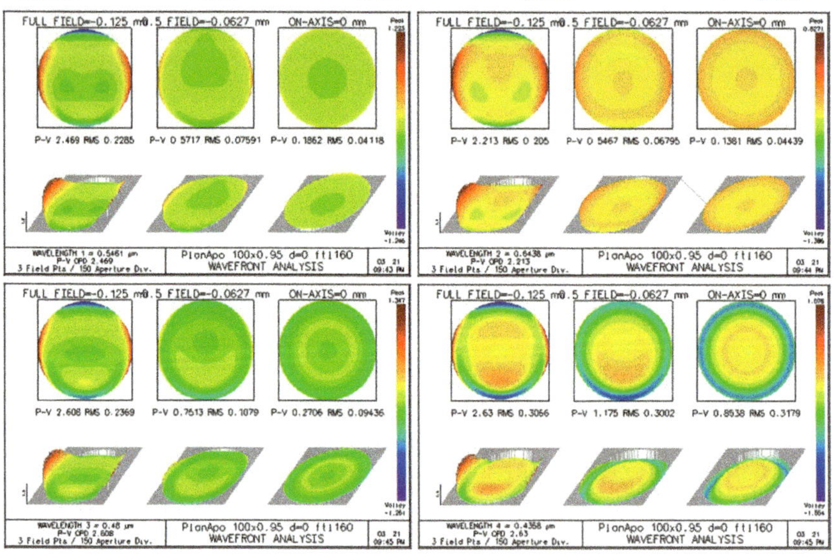

Figure 3.7 Design parameters and graphs of aberration correction of a 100×/0.95 CCF plan apochromatic objective ($F'_{ts} = 160$ mm).

Table 3.1 Main technical parameters and basic optical layout of the CCF plan apochromatic objectives from the Metam-LV microscope.

Magnification	*NA*	*WD* (mm)	*F′* (mm)	*R* (μm)	*DF* (μm)	*FOV* on object (mm)	*FOV* on image (mm)	Optical design at figure no.	The principal optical layout
5×	0.17	13.2	32.0	1.96	11.5	5.0	25	3.3	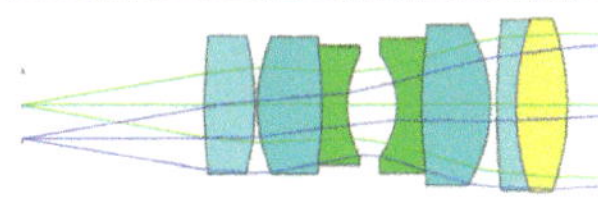
10×	0.28	9.80	16.0	1.19	4.25	2.5	25	3.4	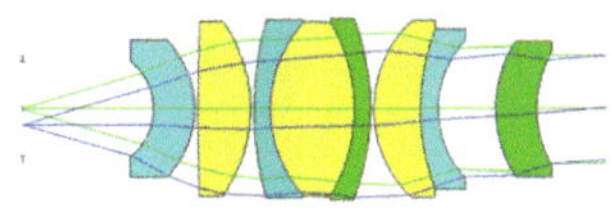
20×	0.50	0.86	8.0	0.67	1.33	1.25	25	3.5	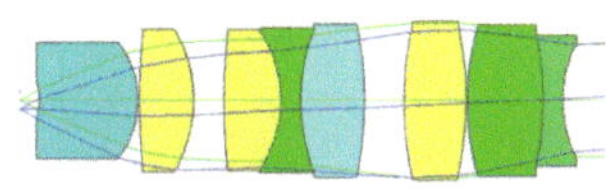
50×	0.85	0.45	3.2	0.42	0.39	0.50	25	3.6	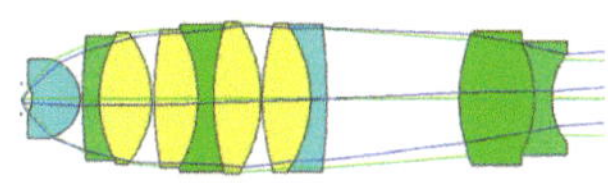
100×	0.95	0.09	1.6	0.35	0.37	0.25	25	3.7	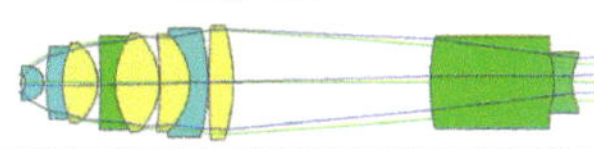

Table 3.1 shows the main technical parameters of the plan apochromatic objectives included in the Metam-LV research microscope kit, and Fig. 3.8 is a photograph of some of them.

These were the first domestic objectives for working in reflected light with completely color free (CCF) plan apochromatic aberration correction. Their optical and mechanical designs were based on proposed engineering solutions

Figure 3.8 Photograph of some CCF plan apochromatic objectives from the Metam-LV microscope.

from experienced scientists and engineers. However, many of these objectives are distinguished by their original designs, which developed and improved upon existing engineering ideas. Figures 3.9–3.11 are photographs of real objects taken using these objectives.

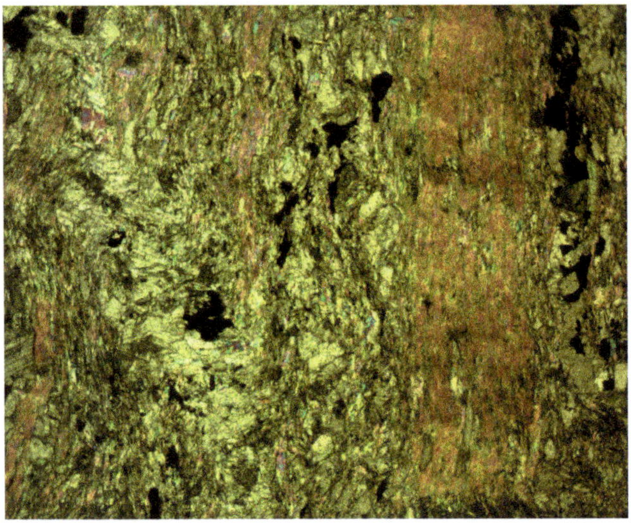

Figure 3.9 Photograph of a real object using a 10×/0.28 CCF plan apochromatic objective.

Figure 3.10 Photograph of a real object using a 20×/0.50 CCF plan apochromatic objective.

Figure 3.11 Photograph of a real object using a 50×/0.85 CCF plan apochromatic objective.

3.1.2 Phase contrast device KF-3

Although the KF-3 is a rather old device that is no longer manufactured, it may be useful for researchers working on a metallographic microscope, as it can significantly increase the information content of images of low-contrast objects. The KF-3 is intended to be an addition to an inverted metallographic microscope in order to study low-contrast thin section structures in reflected light by the phase contrast method. When using a phase contrast device on a microscope, it is possible to project images after the microscope onto a digital image receiver. The optical system of the KF-3 does not work as an independent optical–mechanical component; it is used only in conjunction with a metallographic inverted microscope system. It is designed to work with ordinary (not phase) objectives in which the so-called phase rings are not present. That is, the so-called remote phase contrast method is used when the plane of the exit pupil of the objective is projected using some additional optical components into an intermediate plane. Figure 3.12 shows the KF-3 device and its location on an inverted microscope.

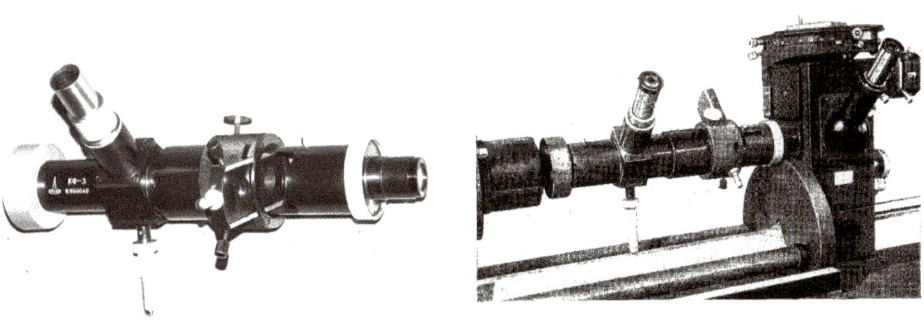

Figure 3.12 The KF-3 device and its location on an inverted microscope.

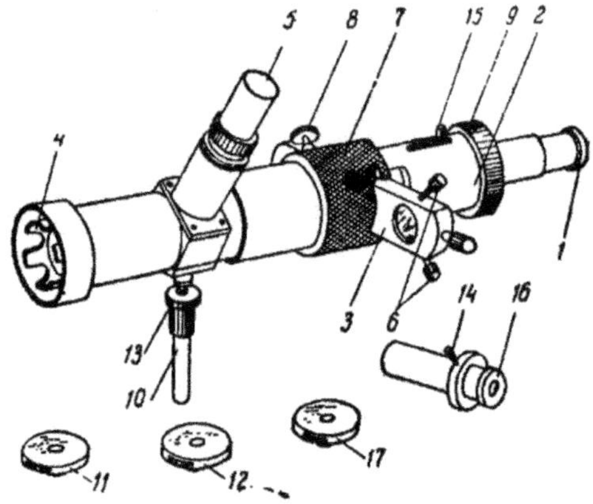

Figure 3.13 The KF-3 device and the principle of operation of its basic elements: 1, a tube with the first lens component of the system; 2, a tube; 3, a device with a diffraction plate; 4, a reference plane for attaching a photographic eyepiece; 5, a tube for visual observation; 6, adjusting screws for the diffraction plate; 7, a ring that moves with a diffraction plate and fixes with a screw; 8, 9, a ring for attaching the KF-3 device on an inverted microscope; 10, a stopper for centering the image of the field diaphragm in the eyepiece field; 11, 12, 17, annular diaphragms; 13, a locknut for fixing an emphasis; and 14, an auxiliary telescope.

Figure 3.13 is an illustration of the KF-3 device and the principle of operation of its main elements. The optical scheme of the KF-3, together with some elements of a microscope, is presented in Figure 3.14.

An annular diaphragm 1 is placed in the plane of the aperture diaphragm, which is projected by the lighting system into the exit pupil of the objective 2. Lenses 3 project the exit pupil, and therefore the image of the diaphragm 1, onto the phase plate 4. The image of the preparation by the objective 2

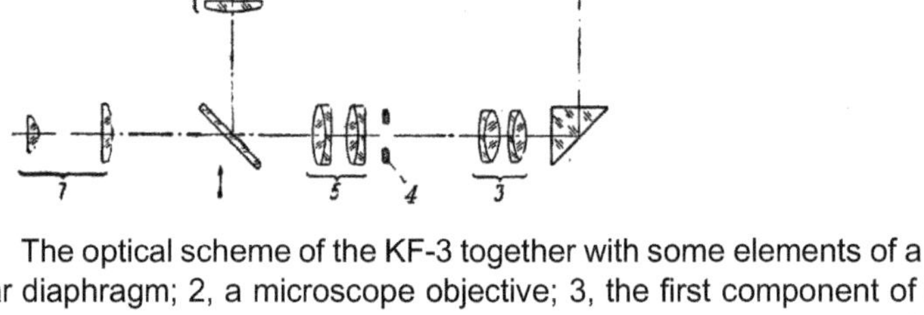

Figure 3.14 The optical scheme of the KF-3 together with some elements of a microscope: 1, an annular diaphragm; 2, a microscope objective; 3, the first component of the inverting system; 4, a centered diffraction (phase) plate; 5, the second component of the inverting system; 6, an eyepiece for visual observation; and 7, an eyepiece for photographing.

together with lenses 3 and 5 is projected into the focal plane of either the eyepiece 6 during visual observation or the eyepiece 7 when photographing. Behind the photographic eyepiece 7, a film camera or a digital image receiver can be installed. An annular diaphragm is placed in the plane of the aperture diaphragm of the microscope. To observe the phase ring when adjusting the lighting (i.e., when centering the diffraction plate ring with respect to the image of the annular diaphragm), an auxiliary telescope 3 is used that is inserted into the eyepiece tube 4 instead of the eyepiece.

The device uses three interchangeable annular diaphragms, 11, 12, and 17 (Fig. 3.13), which are glass plates enclosed in frames. The surface of the plates is varnished. A part of each plate in the form of a ring is freed from varnish, and light passes through this ring into the illuminating tube of the microscope. The choice of a specific aperture depends on the lens used. The phase plate has a transmittance of about 25% and introduces a phase shift of 90 deg into the light passing through it. The plate gives the effect of positive phase contrast. The phase plate of the device (one for all lenses) is centered by screws 6 (Fig. 3.13). Visual observation in this case is made through the eyepiece, which is replaced by an auxiliary microscope when the lighting is set up. To project the image onto a digital receiver, a special adapter can be mounted in the eyepiece tube.

Because the exit pupils of the objectives are located at different distances from the reference plane to combine the image plane of the pupil with the plane of the phase plate, the plate can move along the optical axis. Such a phase contrast device can be used in modern metallographic microscopes and can significantly increase the information content in images of low-contrast objects; in some cases, the use of phase contrast techniques in reflected light can be even more effective than, for example, the use of polarization or differential contrast techniques.

3.2 A Brief History of the Development of Reflected Light Objectives

Objects that are typically studied in reflected light in metallography, mineralogy, and petrography can be low or high contrast, multicolor or monochromatic, phase, anisotropic, and others. The specificity of research performed on a metallographic microscope in reflected light imposes specific requirements on the optical systems of objectives, including a flat field of the image of the object and the absence of scattered light. In microscopes that use a digital image pickup, there are additional requirements for aberration correction of reflected light objectives, such as a complete correction of astigmatism, distortion, and others.

The creation of microscope objectives with a flat image field was initiated in the 1930s by Boegehold, who developed the first set of plan achromats with the following magnifications and numerical apertures: 3.7×/0.11, 9×/0.20, 20×/0.45, and 40×/0.65 [1]. In contrast to the well-known designs of photographic objectives based on systems of a symmetric type, Boegehold proposed correcting the curvature of the image in small-magnification objectives by using the last meniscus-shaped element convex to the object. This marked the beginning of the design of the plan objective, which can be attributed to the *inverted telephoto lens* in the reverse direction of the rays.

With increasing magnifications and numerical apertures, the compensatory action of one meniscus is not enough, and Boegehold introduced a way to correct the field curvature in the frontal element: the frontal meniscus. Boegehold's set remained incomplete, as it did not contain objectives with 60× and 100× magnifications. The creation of a complete set of plan objectives with an achromatic and apochromatic type of aberration correction occurred at a later period, around the 1960s.

Today, the industry produces a large number of objectives for microscopes working in reflected light. However, it should be noted that some of these objectives do not meet modern technical requirements. The fact is that, traditionally, the calculation of such objectives was carried out on the principle of minimizing residual aberrations in the image. It is only recently that the question of how to provide specific features for this kind of research has arisen. Mineralogy, petrography, and metallography require microscopes with reduced image illumination because of the need to study poorly reflecting objects (ore, coal, etc.). In addition, some new microscopy methods are rapidly developing to address the impact of the presence of scattered light in the image, which changes the apparent reflectivity of the studied object and negatively affects the reliability of research and measurements. Therefore, further efforts by developers of complex optical systems aimed at increasing the contrast of images in flat field objectives are needed.

3.3 Scattered Light

Studies have found that the image quality of opaque objects is significantly influenced by the scattered light resulting from the reflection of the light flux from the working surfaces of an objective's lenses. Individual components of this reflected light flux, called *first-order reflexes*, can create significant scattering light of the illumination of the observation field, thereby reducing the contrast of the image of the object [2]. One method used to reduce the coefficient of illumination is the use of multilayer antireflection coatings on the surfaces of lenses bordering the air. In some cases, this reduces the scattered light by about two to three times; however, it does not have adequate efficiency. This is compounded by the fact that fluorite is widely used in modern plan MicroFluar and plan apochromatic objectives; fluorite does not have efficiently stable and effective antireflection coatings. From this we can conclude that, as much as possible, it is necessary to design another way to reduce the illumination coefficient.

To calculate the scattering light of the illumination created by first-order reflexes, the first and second paraxial rays from the aperture and field diaphragms of the microscope lighting system should be calculated. Using a special algorithm, it is possible to analyze the illumination of the illumination in the reflected light objectives included in the sets of manufactured metallographic and polarization microscopes. An analysis of the data allows us to make very important conclusions: objectives that have almost the same optical characteristics and one type of optical system and aberration correction can differ significantly in terms of the illumination coefficient and reflexes. Thus, if the illumination coefficient is largely determined by the shape of the lenses, it is possible to control its magnitude even within the same principal design of the optical system. Studying the characteristics of individual surfaces of optical systems with respect to the illumination of the image created by them allows one to choose certain types of surfaces suitable for creating objectives with increased image contrast. The complex implications of the predicted decrease in the illumination coefficient in reflected light microscopes, as well as the theoretical and practical tools related to this process, were studied by Ivanova and Kirillovsky [3]. So, for example, a study was conducted on the influence of the deflection of one surface—the shape of one thin lens and a thin double-glued component—on the illumination coefficient of the illumination; recommendations were given on the use of surfaces and single lenses of a certain shape.

All of these discussions can define the general concept of the design of reflected light objectives. However, it is not entirely clear whether one should use a particular optical design in order to achieve the maximum numerical aperture in the objective. It is most likely true that for working in reflected light, it is not always necessary to use high-aperture plan apochromatic objectives in which the requirements to minimize the illumination coefficient are quite difficult to fulfill. This is explained by the fact that when calculating an apochromatic objective with an aperture close to the limiting one, straightening radii that are associated with the correction of higher-order spherical and spherochromatic aberration prevail in the design. At the same time, one method for minimizing the illumination coefficient in the objective is the use of lenses that have rather significant optical powers (that is, lenses that have radii of significant curvature should prevail in the optical system of the objective). Therefore, for some objectives used in reflected light microscopes there is sometimes a choice: use a plan apochromatic objective or use an objective characterized by a low-illumination coefficient. This also determines the difficulties in the optical design of low-magnification objectives (long-focus) in whose optical systems it is not always possible to use short-focus lenses with significant optical powers.

3.4 Special Objectives with Super–High Linear Magnifications

In the microelectronic industry, when testing the topological structures of photomasks, as well as in *small photolithography*, objectives with a maximum numerical aperture in air with linear magnifications greater than 100×, reaching values of 300×, are required. The design parameters and graphs of aberration correction of objectives that meet these requirements are presented in Figs. 3.15–3.18; upon reaching a numerical aperture of 0.95, aberration correction of these LCF plan MicroFluar objectives was achieved.

3.5 Objectives with Long Working Distances and Standard Parfocal Height

To solve some of the problems that arise when researching samples in reflected light, objectives with so-called *increased free working distances*, i.e., the *long* space between the object under study and the frame of the first front lens of the objective, are required. These are necessary, for example, when examining hard-to-reach surfaces located in various ledges and cavities (e.g., when testing parts of automobile injection systems). An increased working distance may also be required, for example, when using various kinds of manipulators when testing elements of topological structures in the microelectronic industry. Some techniques require, for example, polarizing elements, various kinds of modulators, or other special optical elements that provide certain microscopic methods of research to be located in the free space between the object and the objective. Under the condition of increased working distance, the overall dimensions of such objectives should be standard, for example, the parfocal height should be 45 mm. The type of aberration correction for such objectives should be CCF plan apochromat (this is especially true when using such objectives for testing microelectronic products). Figures 3.19–3.23 show examples of optical designs of such objectives (principal optical schemes, design parameters, and graphs of aberration correction). These objectives are distinguished by increased linear fields.

I would like to note some characteristics for the design of high linear magnification objectives. As can be seen from the drawings of the principal optical schemes of the 20×, 50×, and 100× objectives, all of these objectives are constructed using some basic optical component; by adding single positive lenses in the front of the objective (their number, for example, may be $N = 1$, 2, 3), we can increase the value of the linear magnification in the objective. Figure 3.24 illustrates the implementation of this principle for building the optical systems of these objectives. By the way, this option is presented as the last one in Table 3.5 in Volume 1, Chapter 3, of this book.

Examples of the use of such objectives are shown in the photographs below, which were taken when testing a sample of an air filter used in

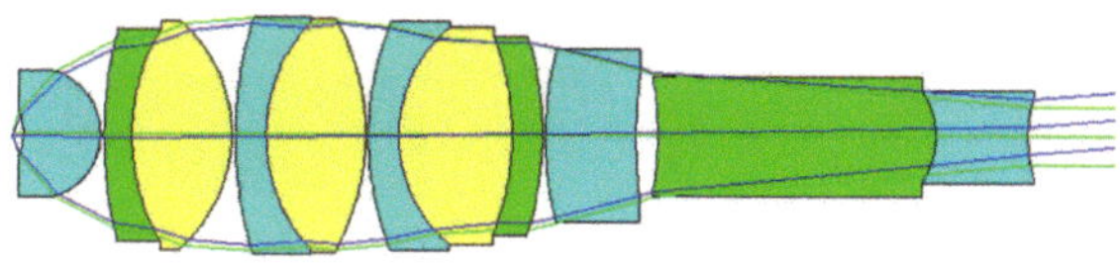

Lens: PlanFluar 150x0.95 d0 ftl160

Object num aper 0.950000 Image height 12.500000 Primary wavln 0.546070

SRF	RADIUS		THICKNESS		APERTURE RADIUS		GLASS		SPECIAL
OBJ	0.000000		1.0000e-20		0.083332		AIR		
AST	0.000000		0.389320		0.750000	A	AIR		
2	-1.757900		2.750000		0.750000		LZ_CTK19		
3	-2.290400		0.140000		2.230000		AIR		
4	14.159500		1.000000		3.540000		LZ_TF5		
5	7.318000		3.600000		3.750000		CAF2		
6	-5.270700		0.140000		4.020000		AIR		
7	14.119600		1.100000		4.190000		LZ_CTK19	P	
8	5.795500		3.600000		4.000000		CAF2	P	
9	-8.207100		0.100000		4.110000		AIR		
10	10.507500		1.100000		4.040000		LZ_CTK19	P	
11	4.907900		4.100000		3.820000		CAF2	P	
12	-9.639400		1.100000		3.490000		LZ_TF5	P	
13	-11.024300		0.100000		3.490000		AIR		
14	12.240200		3.300000		3.030000		LZ_CTK19	P	
15	15.522100		0.800000		2.120000		AIR		
16	-11.047000		10.000000		2.110000		LZ_TF5	P	
17	-2.887200		3.300000		1.630000		LZ_CTK19	P	
18	4.952500		90.000000		1.420000		AIR		
19	214.800000		5.000000		9.000000		LZ-BK8		
20	-46.990000		0.300000		9.160000		AIR		
21	-46.880000		3.000000		9.140000		LZ-BF24		
22	-113.240000		157.851597	S	9.310000		AIR		
IMS	0.000000		0.000000		12.500000				

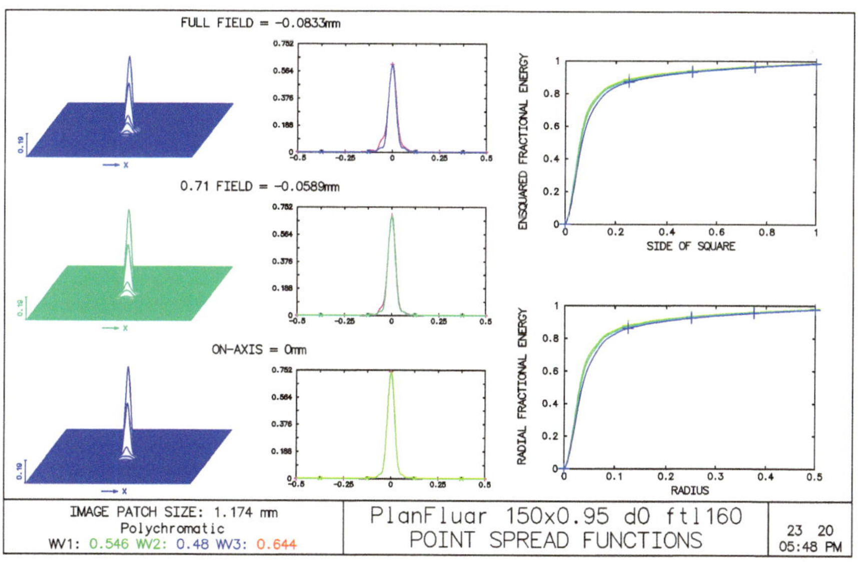

Figure 3.15 Optical design and graphs of aberration correction of a 150×/0.95 infinity LCF plan MicroFluar objective ($F'_{ts} = 160$ mm).

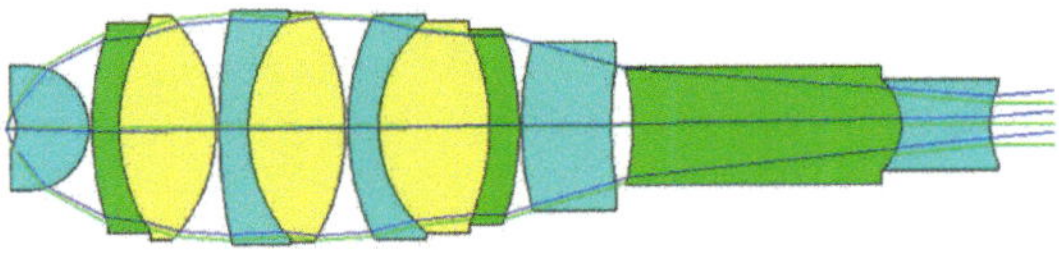

Lens: PlanFluar 200x0.95 d0 ftl160

Object num aper 0.950000 Image height 12.500000 Primary wavln 0.546070

SRF	RADIUS	THICKNESS	APERTURE RADIUS	GLASS	SPECIAL
OBJ	0.000000	1.0000e-20	0.062495	AIR	
AST	0.000000	0.389246	0.690000 A	AIR	
2	-1.370100	2.750000	0.690000	LZ_CTK19	
3	-2.211200	0.140000	2.170000	AIR	
4	11.065100	1.000000	3.500000	LZ_TF5	
5	6.751200	3.600000	3.660000	CAF2	
6	-5.669300	0.140000	3.940000	AIR	
7	19.199400	1.100000	4.100000	LZ_CTK19 P	
8	5.955200	3.600000	4.000000	CAF2 P	
9	-7.839600	0.100000	4.120000	AIR	
10	9.016800	1.100000	4.010000	LZ_CTK19 P	
11	5.118600	4.100000	3.620000	CAF2 P	
12	-8.172700	1.100000	3.450000	LZ_TF5 P	
13	-10.024200	0.100000	3.460000	AIR	
14	11.522200	3.300000	3.140000	LZ_CTK19 P	
15	15.271600	0.800000	2.280000	AIR	
16	-8.167600	10.000000	2.150000	LZ_TF5 P	
17	-2.212900	3.300000	1.370000	LZ_CTK19 P	
18	3.444600	90.000000	1.040000	AIR	
19	214.800000	5.000000	9.010000	LZ-BK8	
20	-46.990000	0.300000	9.180000	AIR	
21	-46.880000	3.000000	9.160000	LZ-BF24	
22	-113.240000	157.856493 S	9.340000	AIR	
IMS	0.000000	0.000000	12.500000		

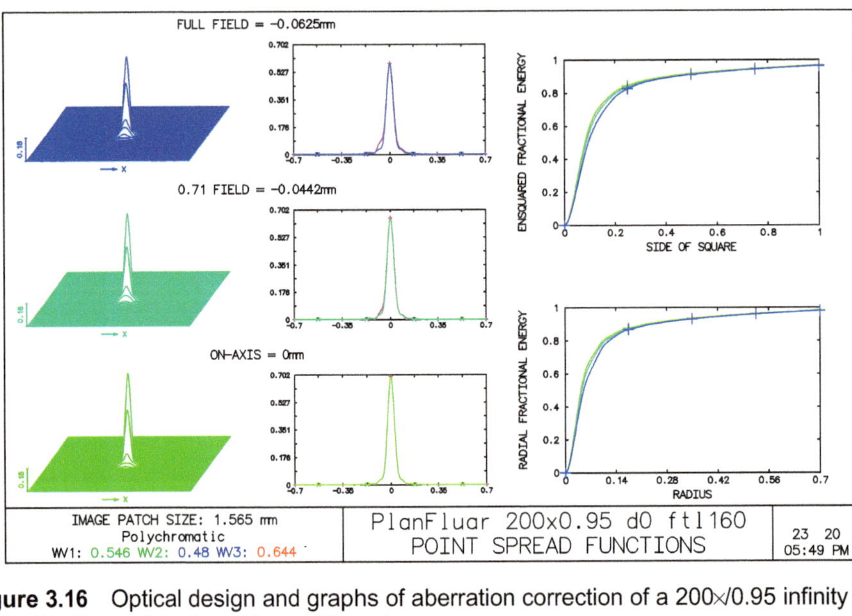

Figure 3.16 Optical design and graphs of aberration correction of a 200×/0.95 infinity LCF plan MicroFluar objective ($F'_{ts} = 160$ mm).

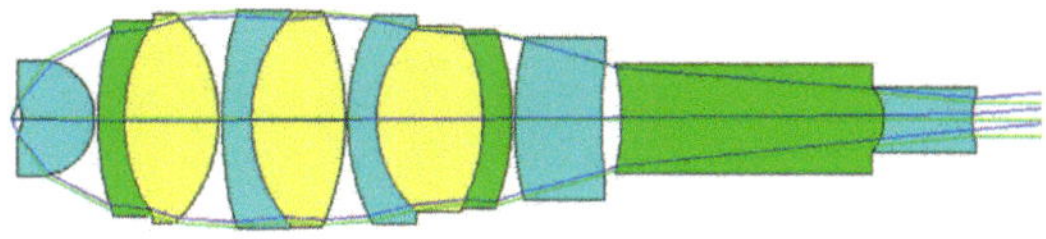

Lens: PlanFluar 250x0.95 d0 ftl160

Object num aper 0.950000 Image height 12.500000 Primary wavln 0.546070

SRF	RADIUS	THICKNESS	APERTURE RADIUS	GLASS	SPECIAL
OBJ	0.000000	1.0000e-20	0.049995	AIR	
AST	0.000000	0.389328	0.640000 A	AIR	
2	-1.169500	2.750000	0.640000	LZ_CTK19	
3	-2.160800	0.140000	2.140000	AIR	
4	10.913800	1.000000	3.500000	LZ_TF5	
5	6.497600	3.600000	3.650000	CAF2	
6	-5.461100	0.140000	3.910000	AIR	
7	14.791800	1.100000	4.070000	LZ_CTK19 P	
8	5.807400	3.600000	3.900000	CAF2 P	
9	-9.530200	0.100000	4.010000	AIR	
10	8.495700	1.100000	3.880000	LZ_CTK19 P	
11	4.673800	4.100000	3.460000	CAF2 P	
12	-6.983300	1.100000	3.300000	LZ_TF5 P	
13	-9.035700	0.100000	3.330000	AIR	
14	11.703100	3.300000	3.010000	LZ_CTK19 P	
15	14.206100	0.800000	2.170000	AIR	
16	-8.320000	10.000000	2.030000	LZ_TF5 P	
17	-1.820600	3.300000	1.210000	LZ_CTK19 P	
18	2.668000	90.000000	0.870000	AIR	
19	214.800000	5.000000	10.000000	LZ-BK8	
20	-46.990000	0.300000	10.180000	AIR	
21	-46.880000	3.000000	10.160000	LZ-BF24	
22	-113.240000	157.873965 S	10.380000	AIR	
IMS	0.000000	0.000000	12.500000		

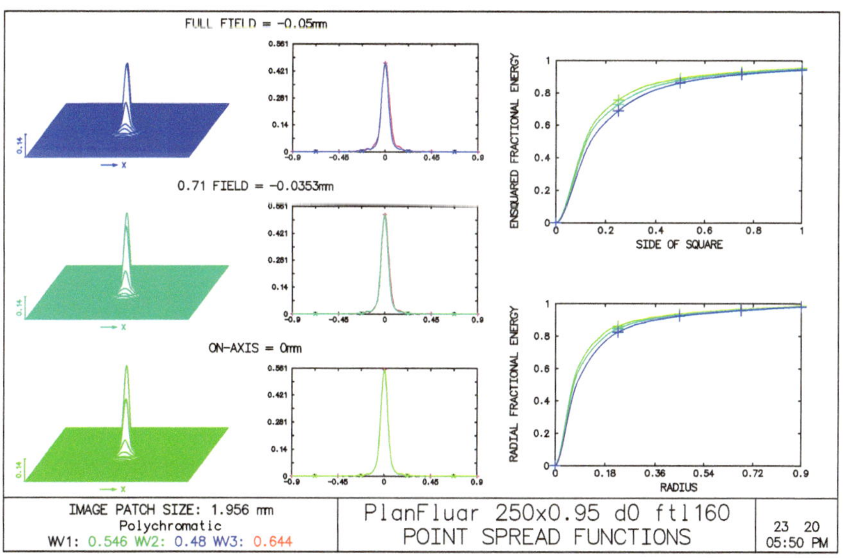

Figure 3.17 Optical design and graphs of aberration correction of a 250×/0.95 infinity LCF plan MicroFluar objective ($F'_{ts} = 160$ mm).

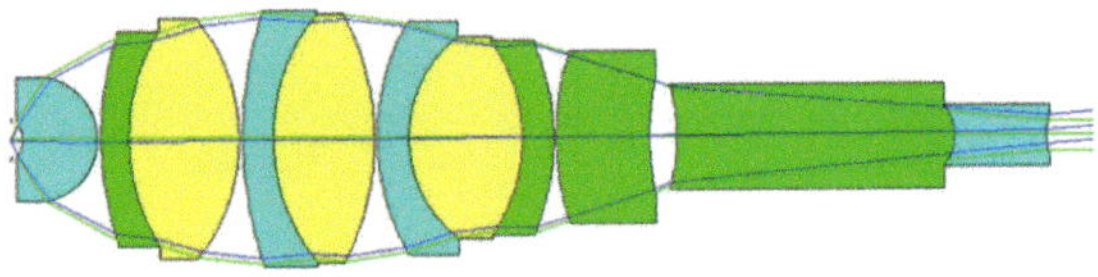

Lens: PlanFluar 300x0.95 d0 ftl160

Object num aper 0.950000 Image height 12.500000 Primary wavln 0.546070

SRF	RADIUS	THICKNESS	APERTURE RADIUS	GLASS	SPECIAL
OBJ	0.000000	1.0000e-20	0.041727	AIR	
AST	0.000000	0.398225	0.630000 A	AIR	
2	-1.068500	2.750000	0.630000	LZ_CTK19	
3	-2.147600	0.150000	2.140000	AIR	
4	11.169700	1.000000	3.560000	LZ_TF5	
5	6.886500	4.000000	3.720000	CAF2	
6	-6.400500	0.140000	4.160000	AIR	
7	13.187500	1.100000	4.430000	LZ_CTK19 P	
8	6.665100	3.600000	4.270000	CAF2 P	
9	-9.293700	0.100000	4.340000	AIR	
10	8.041300	1.100000	4.050000	LZ_CTK19 P	
11	4.301900	4.100000	3.490000	CAF2 P	
12	-5.671500	1.100000	3.330000	LZ_TF5 P	
13	-8.382600	0.100000	3.380000	AIR	
14	9.459300	3.300000	2.960000	LZ_TF5 P	
15	8.987000	1.000000	2.010000	AIR	
16	-6.605900	10.000000	1.800000	LZ_TF5 P	
17	-1.657800	3.300000	1.050000	LZ_CTK19 P	
18	2.404800	90.000000	0.750000	AIR	
19	214.800000	5.000000	10.050000	LZ-BK8	
20	-46.990000	0.300000	10.230000	AIR	
21	-46.880000	3.000000	10.220000	LZ-BF24	
22	-113.240000	157.702939 S	10.430000	AIR	
IMS	0.000000	0.000000	12.500000		

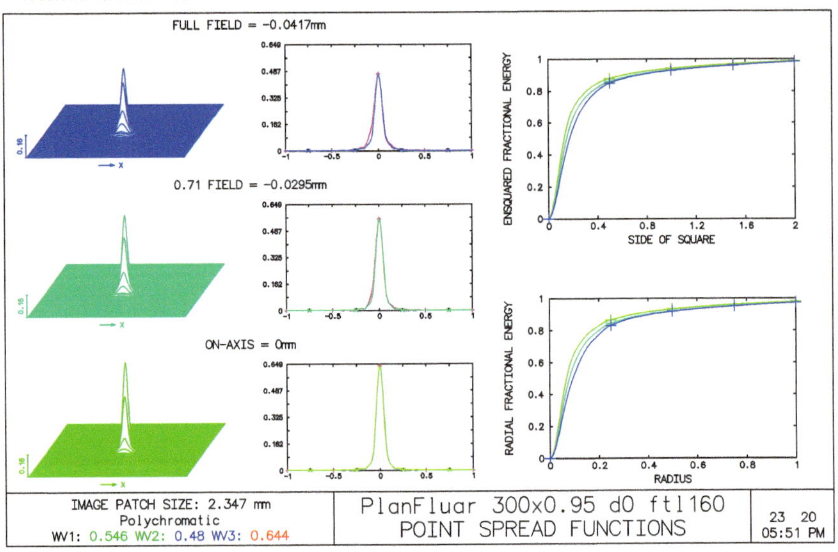

Figure 3.18 Optical design and graphs of aberration correction of a 300×/0.95 infinity LCF plan MicroFluar objective ($F'_{ts} = 160$ mm).

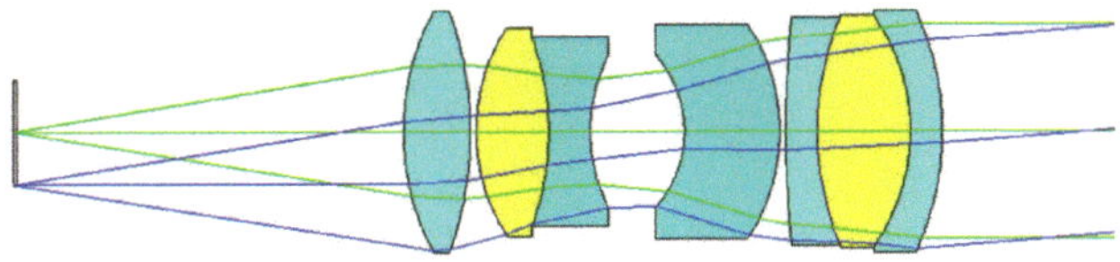

Lens: PLAPO 5x0.16 d=0.17 Ftl160

Object num aper	0.160000	Image height	12.500000	Primary wavln	0.546070

SRF	RADIUS		THICKNESS		APERTURE RADIUS		GLASS		SPECIAL
OBJ	0.000000		0.010000		2.499994		AIR		
AST	0.000000		0.170000		2.500000	A	LZ_K14		
2	0.000000		18.659000		2.520000		AIR		
3	12.174500		3.200000		5.780000		SILICA		
4	-17.374300		0.300000		5.730000		AIR		
5	9.041700		3.500000		4.980000		CAF2		
6	-13.125100		2.000000		4.520000		LZ_CTK19		
7	8.207100		4.500000		3.700000		AIR		
8	-5.503700		4.500000		3.650000		LZ_CTK19	P	
9	-9.206300		0.300000		5.100000		AIR		
10	53.700400		1.500000		5.350000		LZ_K8		
11	14.022500		4.500000		5.440000		CAF2	P	
12	-9.611800		1.500000		5.560000		LZ_K8	P	
13	-13.976000		90.000000		5.760000		AIR		
14	214.800000		5.000000		11.670000		LZ_BK8		
15	-46.990000		0.300000		11.770000		AIR		
16	-46.880000		3.000000		11.740000		LZ_BF24		
17	-113.240000		157.841255	S	11.910000		AIR		
IMS	0.000000		0.000000		12.500000				

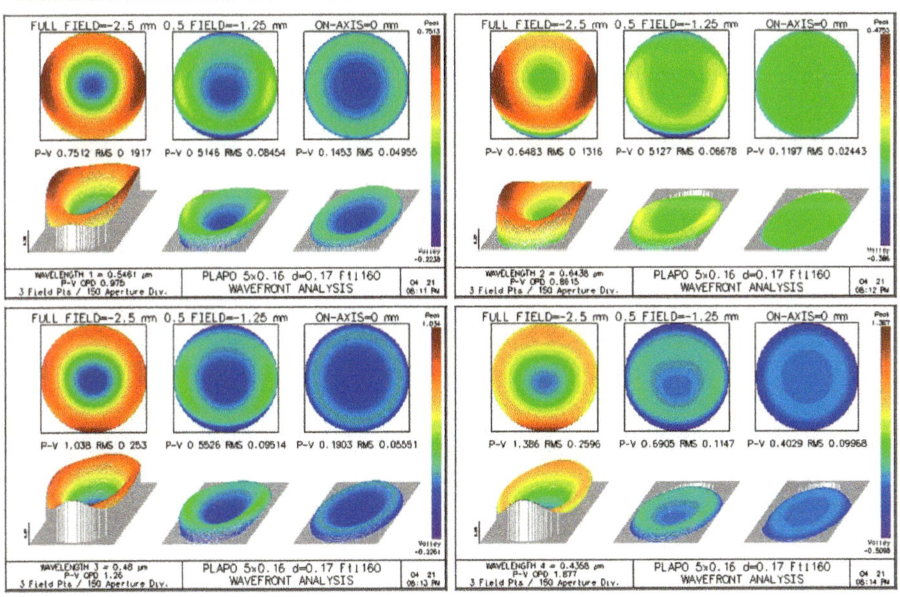

Figure 3.19 Design parameters and graphs of aberration correction of a 5×/0.16 CCF plan apochromatic objective ($F'_{ts} = 160$ mm).

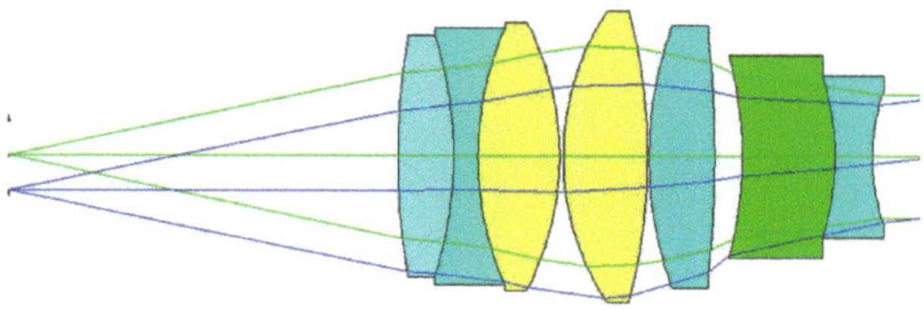

Lens: PlanApo LLD 10x0.2 d=0 Ft1160

Object num aper 0.200000 Image height 18.000000 Primary wavln 0.546070

SRF	RADIUS	THICKNESS	APERTURE RADIUS	GLASS	SPECIAL
OBJ	0.000000	1.0000e-20	1.800097	AIR	
AST	0.000000	21.000000	1.800097 AS	AIR	
2	36.244000	3.000000	6.190000	LZ_LK3	
3	-20.481700	1.300000	6.300000	LZ_CTK19	
4	16.279800	4.300000	6.720000	CAF2	
5	-15.061500	0.200000	7.000000	AIR	
6	13.594900	4.500000	7.630000	CAF2 P	
7	-29.781400	0.100000	7.500000	AIR	
8	19.917000	3.500000	6.840000	LZ_CTK19 P	
9	-56.072400	1.500000	6.220000	AIR	
10	-19.980000	5.000000	5.290000	LZ_TF5	
11	-13.433300	2.000000	4.200000	LZ_CTK19 P	
12	10.138000	100.000000	3.470000	AIR	
13	214.800000	5.000000	14.140000	LZ_BK8	
14	-46.990000	0.300000	14.250000	AIR	
15	-46.880000	3.000000	14.220000	LZ_BF24	
16	-113.240000	157.831394 S	14.500000	AIR	
IMS	0.000000	0.000000	18.000000 S		

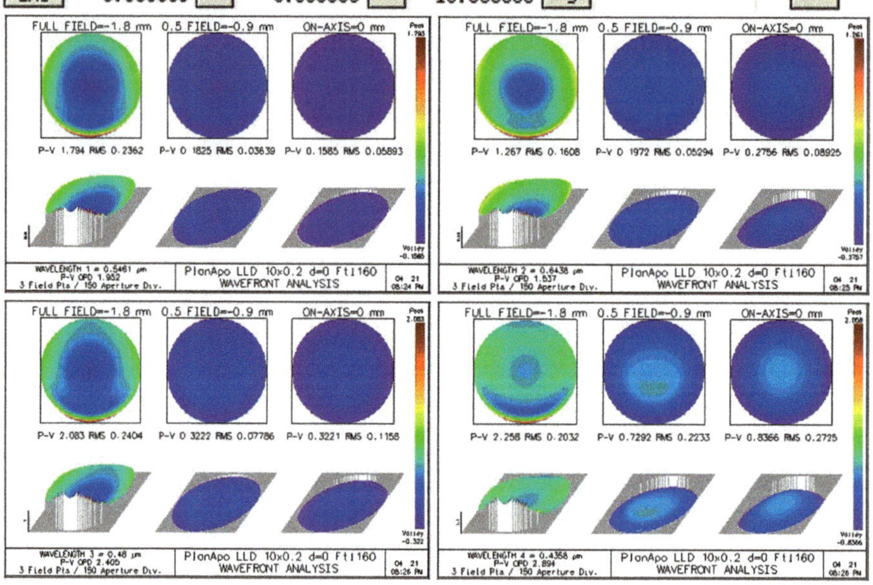

Figure 3.20 Design parameters and graphs of aberration correction of a 10×/0.20 CCF LLD plan apochromatic objective ($F'_{ts} = 160$ mm).

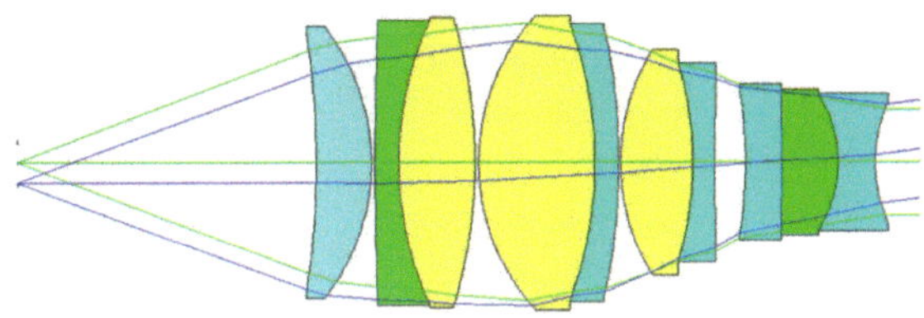

Lens: PlanApo LLD 20x0.35 d=0 Ftl160

Object num aper 0.350000 Image height 21.500000 Primary wavln 0.546070

SRF	RADIUS		THICKNESS		APERTURE RADIUS		GLASS		SPECIAL
OBJ	0.000000		1.0000e-20		1.075778		AIR		
AST	0.000000		16.300000		1.075778	AS	AIR		
2	-43.672200		3.000000		6.950000		LZ_LK3		
3	-11.574200		0.200000		7.250000		AIR		
4	190.982900		1.300000		7.500000		LZ_TF1		
5	17.596200		4.200000		7.610000		CAF2		
6	-24.851500		0.200000		7.750000		AIR		
7	11.450700		6.400000		7.870000		CAF2		
8	-20.389200		1.300000		7.410000		LZ_CTK19		
9	-31.861900		0.100000		7.170000		AIR		
10	11.135300		4.000000		6.010000		CAF2		
11	-17.692900		1.200000		5.310000		LZ_CTK19	P	
12	182.299600		1.700000		4.710000		AIR		
13	-23.826600		2.000000		4.150000		LZ_CTK19	P	
14	-155.959000		3.000000		3.870000		LZ_TF5		
15	-6.691900		2.000000		3.650000		LZ_CTK19	P	
16	7.551800		100.000000		3.190000		AIR		
17	214.800000		5.000000		16.720000		LZ_BK8		
18	-46.990000		0.300000		16.790000		AIR		
19	-46.880000		3.000000		16.750000		LZ_BF24		
20	-113.240000		157.822323	S	17.120000		AIR		
IMS	0.000000		0.000000		21.500000	S			

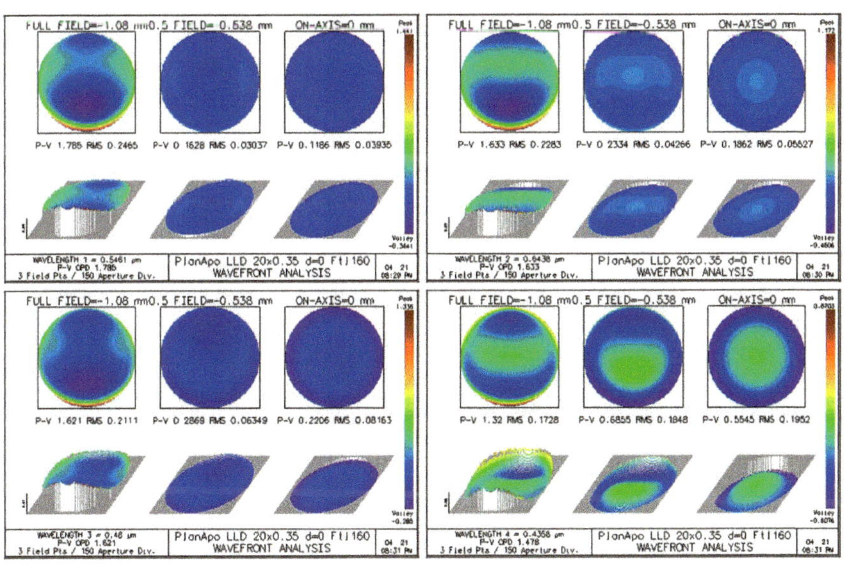

Figure 3.21 Design parameters and graphs of aberration correction of a 20×/0.35 CCF LLD plan apochromatic objective ($F'_{ts} = 160$ mm).

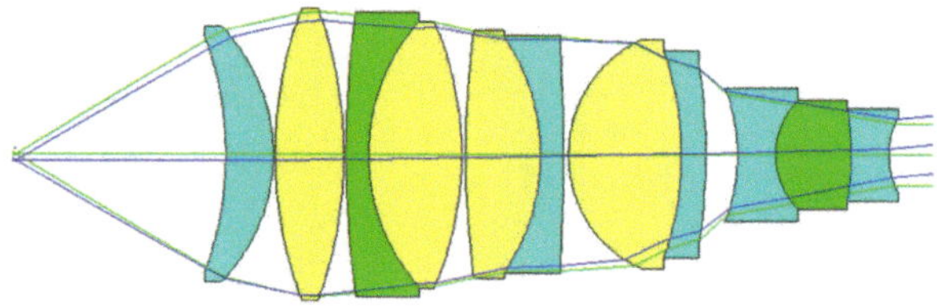

Lens: PlanApo LLD 50x0.50 d=0 Ftl160

Object num aper 0.500000 Image height 15.000000 Primary wavln 0.546070

SRF	RADIUS	THICKNESS		APERTURE RADIUS		GLASS		SPECIAL
OBJ	0.000000	1.0000e-20		0.300333		AIR		
AST	0.000000	11.300000		0.300333	AS	AIR		
2	-16.981600	2.500000		6.160000		LZ_CTK12		
3	-9.412900	0.100000		6.620000		AIR		
4	21.387700	3.600000		7.570000		CAF2		
5	-21.748100	0.100000		7.600000		AIR		
6	51.860400	1.200000		7.360000		LZ_TF1		
7	10.158600	5.000000		6.930000		CAF2	P	
8	-16.780900	0.200000		6.920000		AIR		
9	46.100900	4.000000		6.420000		CAF2	P	
10	-10.742600	1.200000		6.150000		LZ_CTK19		
11	-115.871200	0.300000		6.150000		AIR		
12	6.486000	6.000000		5.950000		CAF2	P	
13	-15.773800	1.300000		5.340000		LZ_CTK19	P	
14	-28.882700	1.700000		4.760000		AIR		
15	-9.145700	2.000000		3.440000		LZ_CTK19	P	
16	3.899600	4.000000		2.810000		LZ_TF5		
17	-13.943200	2.000000		2.380000		LZ_CTK19	P	
18	4.083900	100.000000		1.920000		AIR		
19	214.800000	5.000000		11.320000		LZ_BK8		
20	-46.990000	0.300000		11.460000		AIR		
21	-46.880000	3.000000		11.440000		LZ_BF24		
22	-113.240000	157.685274	S	11.640000		AIR		
IMS	0.000000	0.000000		15.000000	S			

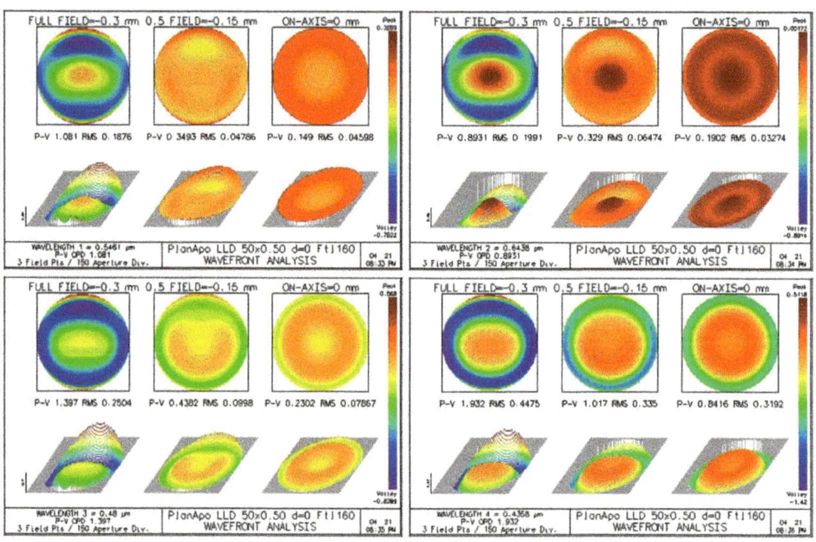

Figure 3.22 Design parameters and graphs of aberration correction of a 50×/0.50 CCF LLD plan apochromatic objective ($F'_{ts} = 160$ mm).

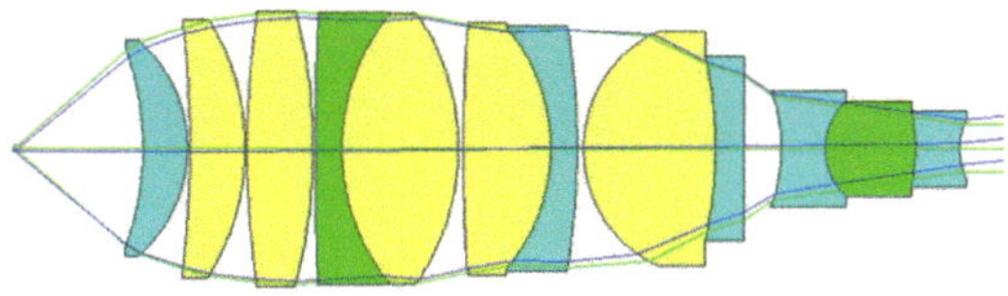

Lens: Plan Apo LLD 100x0.65 ftl160

Object num aper 0.650000 Image height 15.000000 Primary wavln 0.546070

SRF	RADIUS	THICKNESS	APERTURE RADIUS	GLASS	SPECIAL
OBJ	0.000000	1.0000e-20	0.150941	AIR	
AST	0.000000	5.800000	0.150941 AS	AIR	
2	-12.939200	2.000000	4.440000	LZ_CTK12	
3	-6.263300	0.100000	4.790000	AIR	
4	-46.413000	2.500000	5.470000	CAF2	
5	-10.761700	0.100000	5.770000	AIR	
6	45.766000	3.000000	6.050000	CAF2 P	
7	-22.939300	0.100000	6.130000	AIR	
8	140.035100	1.200000	6.070000	LZ_TF2	
9	9.435600	5.300000	5.930000	CAF2 P	
10	-10.295900	0.200000	6.050000	AIR	
11	74.742600	4.000000	5.610000	CAF2 P	
12	-8.104900	1.200000	5.350000	LZ_CTK19	
13	-32.430500	0.300000	5.450000	AIR	
14	5.721300	6.000000	5.220000	CAF2 P	
15	-17.825800	1.300000	4.100000	LZ_CTK19 P	
16	-226.324700	1.700000	3.490000	AIR	
17	-7.273300	2.000000	2.580000	LZ_CTK19 P	
18	2.865400	4.000000	2.110000	LZ_TF5	
19	-8.230300	2.000000	1.690000	LZ_CTK19 P	
20	3.080100	100.000000	1.330000	AIR	
21	214.800000	5.000000	11.060000	LZ-BK8	
22	-46.990000	0.300000	11.210000	AIR	
23	-46.880000	3.000000	11.190000	LZ-BF24	
24	-113.240000	157.294900 S	11.400000	AIR	
IMS	0.000000	0.000000	15.000000 S		

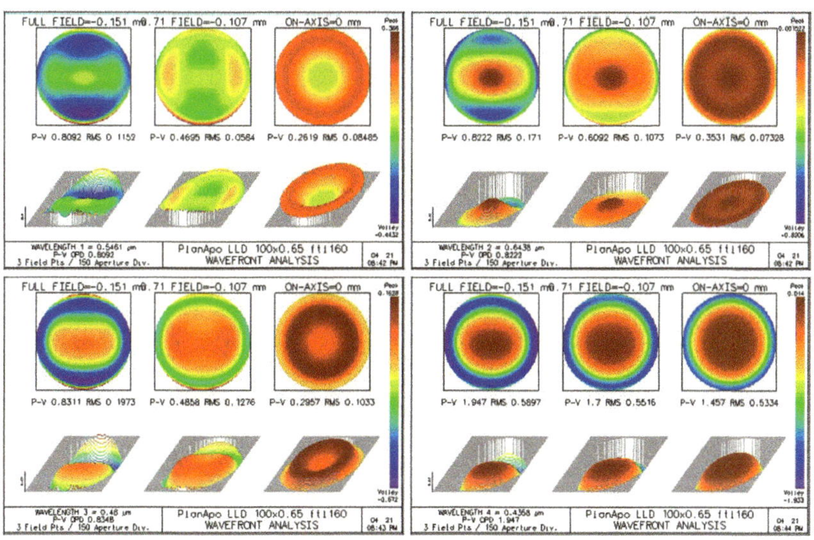

Figure 3.23 Design parameters and graphs of aberration correction of a 100×/0.65 CCF LLD plan apochromatic objective ($F'_{ts} = 160$ mm).

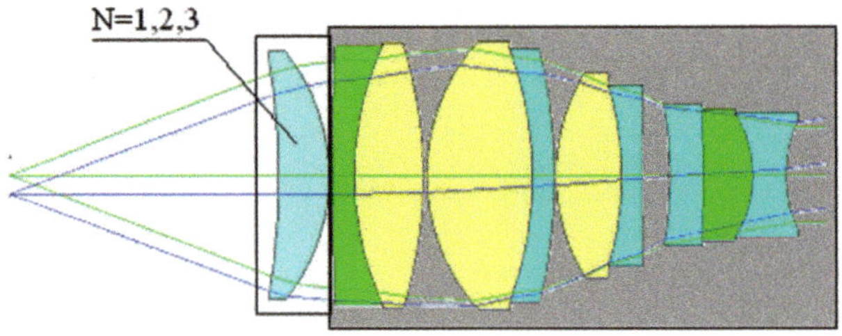

Figure 3.24 The principle for building the optical systems of objectives.

specialized equipment for air purification. Here are microscopic objects of both an organic and nonorganic origin. Such studies can be carried out using, for example, a light microscope of reflected light with the option of polarized light. At the same time, to obtain a certain *bulk* effect of the resulting image, we can apply an additional lighting device for oblique lighting. To do this, we need to have a sufficient distance between the object and the objective so that the lighting device (for example, the fiberglass of the illuminator) is brought closer to the object. We can try to make the testing process of such an object a little "creative" by using, for example, a combination of direct light through the lens with oblique light from an additional illuminator. If we add the polarized light technique to such *mixed* lighting, we can obtain satisfactory image quality in both the visual channel of the microscope (when observed through the eyepiece) and the microphotography by using, for example, a digital receiver. Figures 3.25–3.30 are examples of photographs of the described object using such objectives [often called long working distance (LWD) objectives].

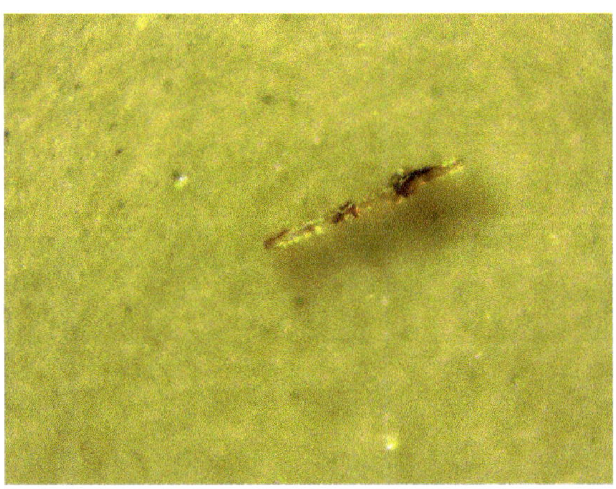

Figure 3.25 Photograph of a real object using an LWD objective.

Figure 3.26 Photograph of a real object using an LWD objective.

Figure 3.27 Photograph of a real object using an LWD objective.

Figure 3.28 Photograph of a real object using an LWD objective.

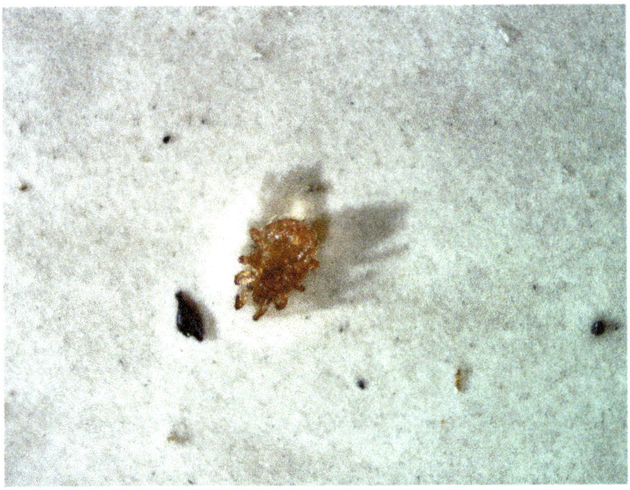

Figure 3.29 Photograph of a real object using an LWD objective.

Figure 3.30 Photograph of a real object using an LWD objective.

3.6 Synthesis of Plan Apochromatic Objective Structures

Among various designs of microscope objectives there is a large group of symmetric systems or systems built on the principle of partial symmetry. Such designs, often used in photo optics, are the most commonly encountered. Using such solutions, it is possible to construct low-powered magnification plan apochromatic objectives. Typically, such synthesis boils down to using optical materials with special dispersion properties in conventional camera objectives. In general, such synthesis and subsequent optical design of low-powered magnification plan apochromatic objectives will not increase the design's complexity; therefore, it does not require our consideration.

Synthesis of high-powered magnification plan apochromatic objective structures is a different matter. The process of constructing high-powered magnification plan apochromatic objective structures can be represented as a synthesis of optical components of the front part, middle part, and follow-on part. This approach to the synthesis and analysis of objectives was proposed in [4]. The synthesis problem is reduced to selection and arrangement of some optical components possessing a priori known dimensional and aberrational properties within the objective optical scheme.

Following the modern concept of image quality, it is necessary to choose the original optical scheme and, accordingly, individual elements of the scheme in order to affect the degree of off-axis beam aberration correction (mostly image curvature and astigmatism). In this regard, the design of the objective frontal part is defined by objective magnification (focal length), the front segment, and total dimensions (height). An example of typical optical designs of objectives with plan apochromatic aberration correction is given in [5].

In complex microscope objectives, the front part may be represented by one or more menisci made of high-refraction material and with the concave side facing the subject. Non-concentricity of the surfaces of these elements determines whether an objective's astigmatism can be corrected. Image curvature can be corrected not only by the front elements, but also by the follow-on elements of the scheme. The deflection of the first surface of the meniscus with the concave side facing the object determines, essentially, the degree of correction of the astigmatism meridional component, and the thickness, refractive index, and follow-on radius determine the degree of correction of the astigmatism sagittal component and, respectively, of the curvature. An additional frontal element can be executed as both single or glued, and this is determined by microobjective characteristics (mostly by aperture and degree of apochromaticism). When single frontal menisci are applied, it is advisable to use fluorite or *special crown* glass as an optical material for their production to reduce chromatic aberrations. Here the manufacturability of objectives decreases since using optical materials with such distinctive features for complex shaping to manufacture meniscus lenses is an outstanding engineering task.

In any case, when using single lenses as the frontal components, the correctional load on the follow-on components of the system increases and is distributed between them. When making calculations for a plan apochromatic objective, designing the second component as glued together pieces is not always feasible because zonal and spherochromatic aberrations can appear on the glued surfaces. However, such a constructive solution can help in correcting chromatic aberrations in the objective. It is known that minimal forces in glued lenses are ensured by selecting optical glasses with a maximum difference of dispersion coefficients. For glued lenses, the selection of optical materials with a minimum difference of relative partial dispersion may reduce secondary spectrum aberration in a lens. Refractive indices of the materials to be bonded must be close in order to reduce spherochromatic and zonal aberrations. The most rational decision is to choose a positive fluorite lens as an optical material. Choosing *short flint* or *light flint* glass as a negative lens optical material (for gluing with fluorite) is optimal. Designing the middle (compensatory) objective elements is determined by the lens magnification and aperture. A thrice-cemented element is often chosen as a main compensatory element in plan apochromats due to its advantages as compared with a double-cemented one. It has an additional compensation parameter, and it has a scope of solutions in which the minimum higher order of spherical and spherochromatic aberrations is provided, and other aberrations can vary in a fairly wide range. Besides, choosing *heavy flint* glass as an optical material for a negative glass lens enables secondary spectrum correction to be fixed at a sufficient level in an objective.

An effective tool for constructing plan apochromats is choosing certain glass brands as lens materials. For example, short flint, *special flint*, and *lanthanum crown* optical glass are used for negative glued lenses, and special crown optical glass or fluorite is used for positive glued lenses. In this case, better secondary spectrum correction is achieved due to the approximate equality of relative partial dispersion of bonded materials. Some developers conclude that it is not expedient to use heavy flint optical materials for plan apochromats because such materials possess excessively large relative partial dispersion. However, using only short flint or special flint optical materials results in significant optical power of both positive and negative optical components of an objective, which determines the steep radii of glued-surface curvature. In turn, greater curvature of lens-glued surfaces leads to the appearance of hard-to-fix spherochromatic aberrations. In fact, the use of heavy flint optical materials for plan apochromatic objectives is justified for certain types of negative lenses.

One (two) double-cemented component in the shape of a meniscus with the concave side facing the image space is used as a follow-on part. Each component is represented by positive and negative lenses glued together. It is also possible to arrange the follow-on part component as a single- or glued double-concave lens. This optical component can be made as a single piece or can be glued using two or three lenses. In all cases described above, this component works by the principle of reversible teleobjective lenses.

3.6.1 Main technological parameters for the manufacture of plan apochromatic objectives

Here are the main technological parameters for plan apochromatic objectives:

- These objectives complete 10–15% of all produced microscopes.
- As a rule, several lenses are made of CaF_2 crystals (or analogue).

- Objectives comprising chains of 7–20 lenses are primarily double, and one (two) of the three cemented lenses are made of glasses and crystals.

Optical parts have the following structural and technological parameters:

- Lens diameters: 1–5, 5–30 mm; accuracy on the diameter, ±0.005–0.010 mm
- Radius values: from ±0.5 mm to ∞; accuracy on the radius, ±0.01–0.05%
- Lens thickness, 0.7–15 mm; thickness accuracy: for 10% of the details, ±0.002 mm; for 30% of the details, ±0.005; others, ±0.01–0.03
- Accuracy in the sphere: $N = 0.5$–2.0; $N = 0.1$–0.3
- Lens centering: for 60% of the parts, 0.002–0.005 mm; for 40% of the details, 0.01–0.03 mm

These parameters demonstrate that it is quite difficult and time-consuming to manufacture plan apochromatic objectives. For further illustration, here are the drawings of some lenses and assembly components (framed lenses) that demonstrate the complexity of the assembly operations required for bonding lenses to the frames and adjusting the objective as a whole (see Figs. 3.94–3.147).

It is also complicated to assemble and adjust objectives with plan apochromatic correction. Lens components, which are single or glued lenses, should be placed on a single axis with no offsets or slewing for high accuracy, given that the aberrations resulting from misalignment are commensurate with fractions of the wavelength. Air gaps between components should be maintained with high precision and accuracy up to 0.002–0.005 mm. These conditions determine the complexity of the mechanical design of plan apochromats and explain the high complexity of integrating components in an objective body. The availability of a so-called *corrective frame* necessitates the use of a double body.

3.6.2 Aberration correction of plan apochromats: Criteria and degree

The criteria for evaluating the appropriate level of aberration correction are specified in [4]. The emphasis is on the correction of spherical aberration measured at a single wavelength and evaluated in terms of deformation of the wave surface emerging from the subject point. It is known that spherical aberration disturbs the beam homocentricity in the image domain; therefore, beams intersect in a curve forming a caustic surface instead of a point image. Light energy is dissipated when moved from the central spot to diffraction rings, which results in reduced resolving power and contrast in the image. For aperture rays depicting object axial points, the mean-square deviation of wavefront from an ideal sphere for the main wavelength $\lambda = 546$ nm should not exceed 0.10–0.15 λ; the spherical aberration for additional chromatic wavelengths of the operating spectral range within 480–644 nm should not exceed 0.25 λ. The aberration should not exceed 0.5 λ for the range of 480 to 435 nm. This means that compared with other types of microscope objectives, the spherical aberration in plan apochromats is fixed more thoroughly over the entire operating spectral range. In addition, degrees of chromatic aberration and secondary spectrum correction are substantially different in plan apochromats. Spherochromatic aberration correction is of higher difficulty.

Let us briefly discuss the necessity of secondary (lateral) chromatic aberration correction in a modern plan apochromatic objective. When developing the optical design of these complex and expensive objectives, this type of aberration must be corrected in the objective itself and should not be subjected to compensation (which is possible for some other types of microscope objectives) by ocular or projection systems. The same applies to correction of other aberrations that should be minimized in an objective's optical system. Off-axis beam aberration measured at a single wavelength within the entire linear field does not exceed 0.5–1.0 λ.

The practical aspect of aberration correction with negative impacts described in [6] is also noteworthy. This applies to the expected defocusing that is present in some optical designs. Optical designers often use a virtual technique of introducing a so-called *setup plane* in the object image, where the geometrical and wave aberrations are minimal. However, an objective's optical system will have top-of-the-line image quality only if a single image plane is used, in which the energy characteristics are optimal, and the values of geometrical and wave aberrations are minimal. The well-known Maréchal criterion implies that the brightness in the center of diffraction spots must be maximum [7]. In this sense, introducing in the calculations a certain abstraction represented by a shift of the actual image plane relative to the Gauss plane does not improve the image quality. Such a shift can be considered the first-order aberration: defocusing. Thus, it can be expected that it has no less significant an impact on the image quality compared with higher-order aberrations. Energy distribution in the defocusing spot is uniform (proportional to degree I), and it is not uniform in the aberration spot (proportional to the degree of aberration). For microscope objectives, this circumstance is especially important because many components are working at high relative apertures (there are aberrations, including spherochromatic). Of course, defocusing is unacceptable for objectives with plan apochromatic aberration correction.

Indeed, developers and designers of optics aim to achieve aberration correction for plan apochromats at the highest level by trying to fix all known aberrations that arise when an object is being imaged. This explains the complexity of their optical structure, the extremely high labor intensity of their manufacture, and, as a result, their high cost. But that is all justified for such objectives.

3.6.3 Unification of CCF plan apochromatic objectives

Microscopists use a wide variety of microscopes. Microscopes differ not only by their purpose and research functions, but also by their overall dimensions. The optical length of a microscope's body tube is one of these dimensions. There are microscopes with a finite optical tube length of 160 mm. Many microscopes operate by the *infinity* principle in which the observed subject is located in the front focal plane of an objective. This microscope provides images using an additional tube system; an image is created in the rear focal plane of this system. In this case, an objective optical design is developed for the infinite tube length, and a parametric series of objective focal lengths is determined by the formula $F'_{ob} = \frac{F'_{ts}}{V_{ob}}$, where F'_{ts} is the equivalent focal distance of the tube system, which creates a subject image in its rear focal plane, and V_{ob} is the linear magnification of the objective. Here in the optical calculations, the condition of ray telecentricity is fulfilled. But, as a rule, telecentricity is not actually present in the real microscopes.

Historically, different microscope manufacturers have chosen original dimensional standards for their microscopes. In particular, different values of equivalent focal length have been selected for tube systems. Accordingly, different parametric series of objectives are used. Obviously, it is possible to obtain the same magnification in the image plane by different combinations of objectives and tube systems. For example, if a microscope uses a tube system with an equivalent focal length of $F' = 200$ mm, then producing an image with 100× magnification would require an objective with a focal length of 2 mm. If a tube system has $F' = 164$ mm, the objective focal length should be 1.64 mm (that is about 1.25 times less). In [8], my colleagues and I consider the peculiarities of the optical systems of microscopes with different overall dimensions.

These arguments about the complexity of manufacturing separate components and plan apochromatic objectives in general are a prerequisite for developing unified designs that allow objectives of various parametric series to be built. The idea is to apply the most complex components of plan apochromatic objectives as unchangeable parts. For example, frontal components can remain unchanged. The elements of an objective's optical system with the highest aperture can also remain unchanged. We can synthesize an objective's optical system with plan apochromatic correction for different lengths of optical tube, such as a finite tube length of 160 mm and an infinite tube length. For an infinite tube length, we obtained objectives corresponding to a parametric series of tube systems: $F' = 164$ mm, $F' = 180$ mm, and $F' = 200$ mm. The optical design was performed for objectives with linear magnifications from 20× to 100×. Such unification does not result in an obvious advantage for small-magnification objectives. Different parts of plan apochromatic objectives can be changed when adapted to the selected parametric series. Table 3.2 illustrates different variations of permanent and variable parts of an objective, depending on the desired objective linear magnification.

Table 3.2 Synthesizing different F'_{ob} plan apochromatic objectives.

20×/0.70	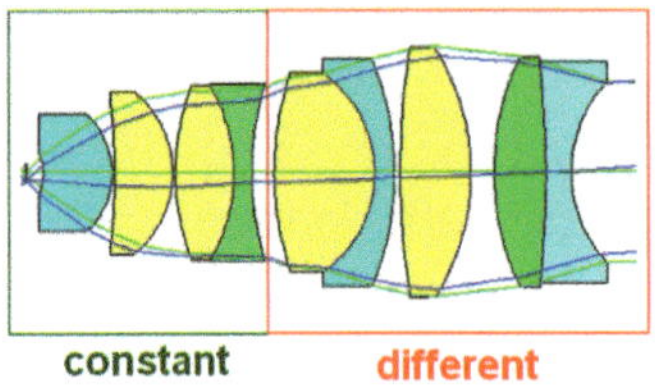	The first three components remain unchanged. The other three components are changed.
40×/0.90	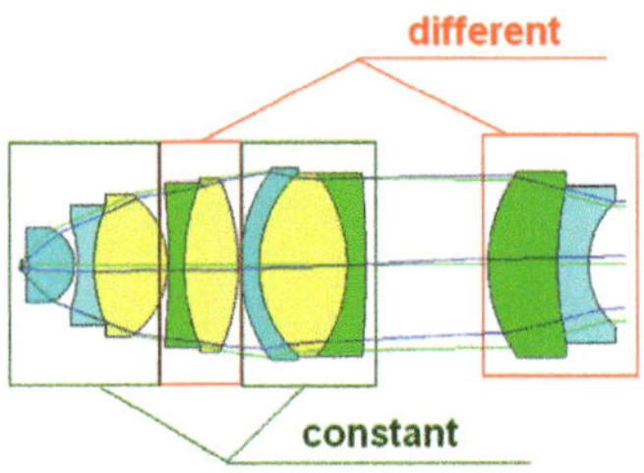	The first two components remain unchanged. The third component is changed. The fourth component is unchanged. The last fifth component is changed.
60×/0.92	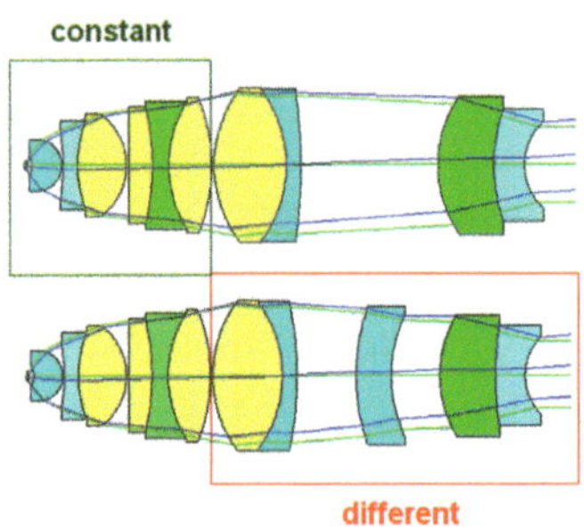	The first three components remain unchanged. Other components are changed. In addition, one more component is added.
100×/1.40 oil immersion	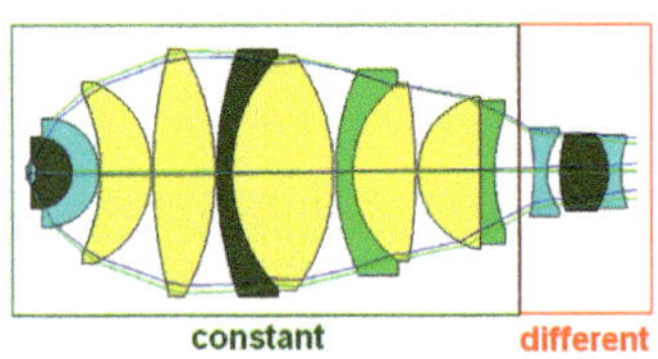	Only the last two components are changed.

Manufacturers of microscope objectives can use solutions for unifying the optical systems of plan apochromatic objectives. Often microscopists do not know the parametric series of their microscopes. However, an objective's manufacturer can identify the parametric series by a microscope's name and brand. Due to the high complexity of their manufacture, it is particularly important to use a general design approach, unified structures, group methods for processing designs, etc., for plan apochromatic objectives. In this case, the same equipment, technological techniques, and methods of assembly and adjusting are used.

Success in using any of the suggested plan apochromatic objectives in a specific microscope model requires independent aberration correction and alignment of the linear fields of the proposed objective and microscope.

Linear field values in modern optical microscopes reach 26.5 mm (diameter of the image in the intermediate plane of the microscope, or the so-called *ocular field*); however, many models implement linear fields of 25 mm, 22.5 mm, and even smaller. Therefore, our objectives must have a margin in regard to the implemented linear field for any microscope. We can choose the design linear field value of 28 mm for our plan apochromatic objectives. This is probably the maximum possible linear field for a visual microscope, which the human eye (a microscopist's visual apparatus) can still embrace. In addition, with this field it is still possible to use an eyepiece with a barrel diameter of 30 mm (a standard eyepiece diameter in an optical microscope).

Of course, achieving a design linear field of 28 mm, which is larger than previously known solutions, requires more careful aberration correction, especially for beams of non-axis rays. However, experience shows that these are only more precise optical designs; this does not lead to an increase in the complexity of the optical structures of plan apochromatic objectives.

Figures 3.31–3.70 show some examples of plan apochromatic objectives. They show the principal optical schemes, design parameters, and graphs of aberration correction. The design of some of the objectives shown is made in accordance with the indicated concept of unification, including their linear fields reaching 28 mm in the plane of the intermediate image of the microscope. Table 3.3 shows the main technical parameters of these objectives.

Using a basic optical design of a 100× water immersion plan apochromatic objective, we can demonstrate the options of using different unchangeable and changeable components in the optical design of an objective adapted to different parametric series. It is important that the fundamental optical scheme remain generally unchanged. That is, the arrangement of components—their aberrational and overall load—will remain the same within a single optical system. Changes are applied only to specific design parameters of individual components. Often, even the shape of components remains unchanged, enabling the use of group methods and techniques for manufacturing lenses. The resulting image quality (the aberration correction in particular) is almost identical for all variations of objective manufacturing.

Figure 3.71 shows the principal optical diagram of a water immersion plan apochromatic objective with a linear magnification of 100× and a numerical aperture of 1.20. The objective's optical design has 8 components. The first is two meniscus lenses glued together, with the concave side facing the subject space. The second, third, and fourth components are single positive lenses. According to the optical layout, the fifth component is negative and positive cemented lenses. The sixth component is positive and negative cemented lenses. Both the seventh and eighth components are biconvex and biconcave lenses glued together. The design parameters of different components are subject to changes during the design. Table 3.4 shows the parameters of 100×/1.20 water immersion finite (160 mm) and infinity

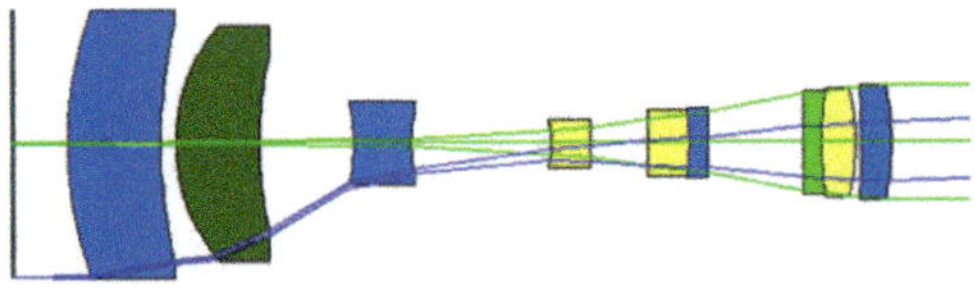

Lens: PlanApo 1x0.015 Ftl200

Object num aper 0.015000 Image height 7.000000 Primary wavln 0.546070

SRF	RADIUS	THICKNESS	APERTURE RADIUS	GLASS	SPECIAL
OBJ	0.000000	0.010000	6.997804	AIR	
AST	0.000000	0.170000	6.997954 AS	K9	
2	0.000000	2.959100	6.999634	AIR	
3	19.790000	5.000000	7.044025	H-ZLAF3	
4	27.040000	0.800000	6.260953	AIR	
5	9.461000	4.500000	6.187272	H-ZF62	
6	18.664000	5.000000	4.552269	AIR	
7	-10.800000	3.000000	2.178337	H-ZLAF3 P	
8	6.329000	7.500000	1.692907	AIR	
9	-3.448000	2.000000	1.161717	CAF2	
10	16.762000	3.000000	1.267249	AIR	
11	58.858000	2.200000	1.593047	CAF2 P	
12	-12.560000	1.000000	1.741499	H-ZLAF3 P	
13	0.000000	5.000000	1.825272	AIR	
14	52.060000	1.000000	2.604776	H-ZF1	
15	17.620000	1.800000	2.679386	CAF2 P	
16	-11.730000	0.300000	2.875456	AIR	
17	-41.800000	1.600000	2.890362	H-LAF10L	
18	-16.950000	200.000000	3.015671	AIR	
19	-36.100000	14.000000	11.550000	K9 P	
20	-43.560000	3.000000	13.620000	AIR	
21	377.100000	15.000000	13.780000	CAF2	
22	-81.500000	3.000000	14.090000	AIR	
23	221.300000	17.000000	13.890000	CAF2	
24	-72.000000	3.000000	13.120000	LZ_CTK12	
25	0.000000	200.111926 S	13.070000	AIR	
IMS	0.000000	0.000000	7.000000		

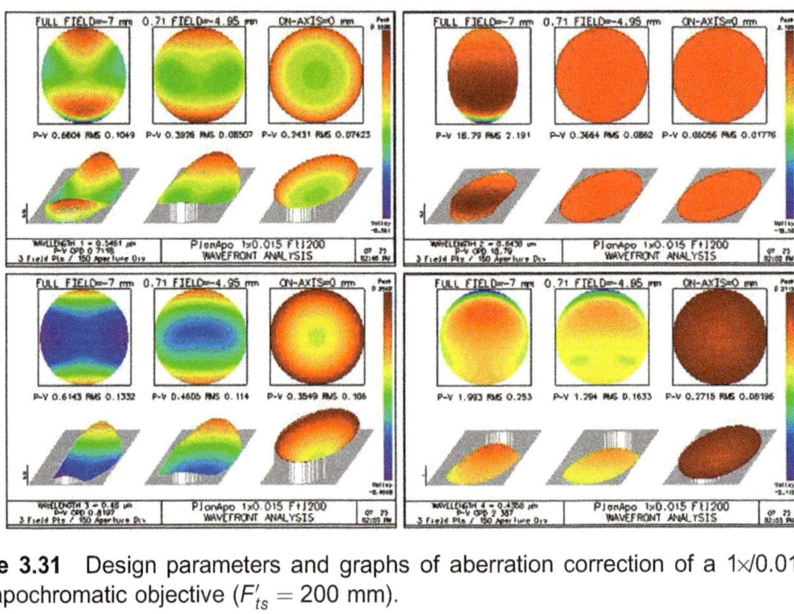

Figure 3.31 Design parameters and graphs of aberration correction of a 1×/0.015 CCF plan apochromatic objective ($F'_{ts} = 200$ mm).

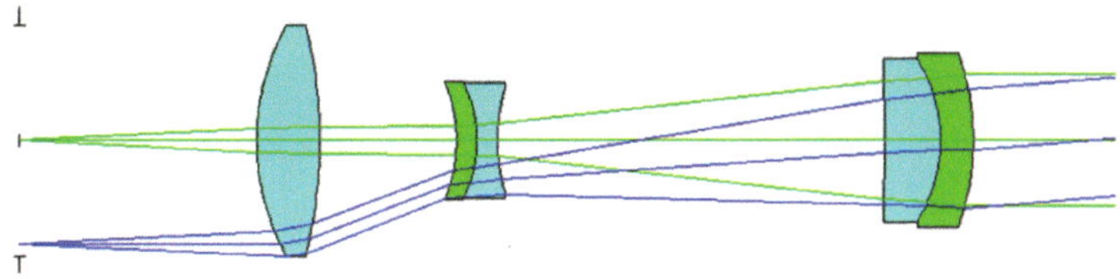

Lens: PlanApo 2.5x0.05 Ftl160

Object num aper 0.050000 Image height 12.500000 Primary wavln 0.546070

SRF	RADIUS	THICKNESS	APERTURE RADIUS	GLASS	SPECIAL
OBJ	0.000000	1.0000e-20	4.885138	AIR	
AST	0.000000	11.566000	5.490000 A	AIR	
2	11.117000	3.100000	5.490000	LZ_LK3	
3	-21.680000	6.700000	5.350000	AIR	
4	-7.047000	1.000000	2.780000	LZ_TF1	
5	-6.026000	1.000000	2.750000	LZ_K8	
6	8.790000	18.920000	2.590000	AIR	
7	-258.200000	2.800000	3.790000	LZ_K8 P	
8	-7.047000	1.600000	3.880000	LZ_TF1 P	
9	-12.246000	90.000000	4.100000	AIR	
10	214.800000	5.000000	9.750000	LZ_BK8	
11	-46.990000	0.300000	9.880000	AIR	
12	-46.880000	3.000000	9.860000	LZ_BF24	
13	-113.240000	157.835837 S	10.100000	AIR	
IMS	0.000000	0.000000	12.500000		

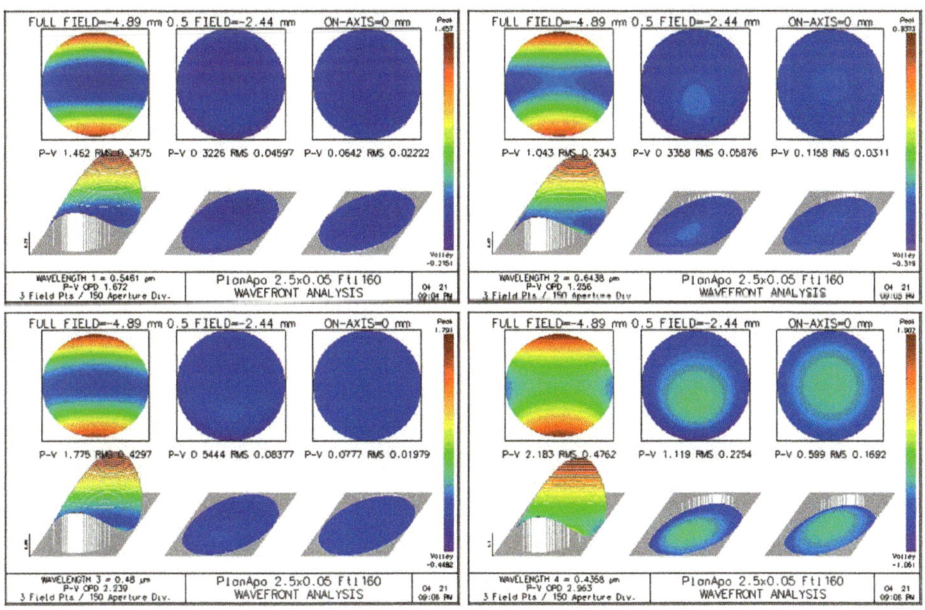

Figure 3.32 Design parameters and graphs of aberration correction of a 2.5×/0.05 CCF plan apochromatic objective ($F'_{ts} = 160$ mm).

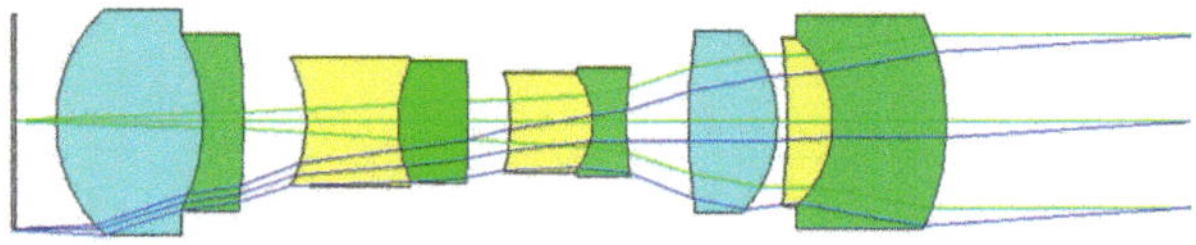

Lens: PlanApo 2.5x0.05 ft1200

Object num aper 0.050000 Image height 12.500000 Primary wavln 0.546070

SRF	RADIUS	THICKNESS	APERTURE RADIUS	GLASS	SPECIAL
OBJ	0.000000	0.010000	4.994994	AIR	
AST	0.000000	0.170000	4.960000 A	K9	
2	0.000000	1.976600	4.960000	AIR	
3	6.882000	7.200000	5.180000	SILICA	
4	-8.530000	2.000000	4.070000	H-LAF10L	
5	-32.617000	3.000000	3.740000	AIR	
6	-6.200000	4.500000	2.970000	CAF2	
7	7.060000	3.300000	2.820000	H-ZF1	
8	-42.270000	2.000000	2.620000	AIR	
9	-10.130000	4.000000	2.310000	CAF2 P	
10	-3.323000	1.600000	2.150000	H-LAF10L P	
11	14.670000	3.000000	2.490000	AIR	
12	23.230000	4.300000	3.710000	SILICA P	
13	-6.000000	0.500000	4.180000	AIR	
14	-30.430000	2.200000	3.910000	CAF2 P	
15	-5.129000	5.500000	3.880000	H-LAF10L P	
16	-11.355000	200.000000	4.930000	AIR	
17	-36.100000	14.000000	15.870000	K9	
18	-43.560000	3.000000	18.630000	AIR	
19	377.100000	15.000000	19.170000	CAF2	
20	-81.500000	3.000000	19.570000	AIR	
21	221.300000	17.000000	19.190000	CAF2	
22	-72.000000	3.000000	18.240000	LZ_CTK12	
23	0.000000	200.024666 S	18.150000	AIR	
IMS	0.000000	0.000000	12.500000		

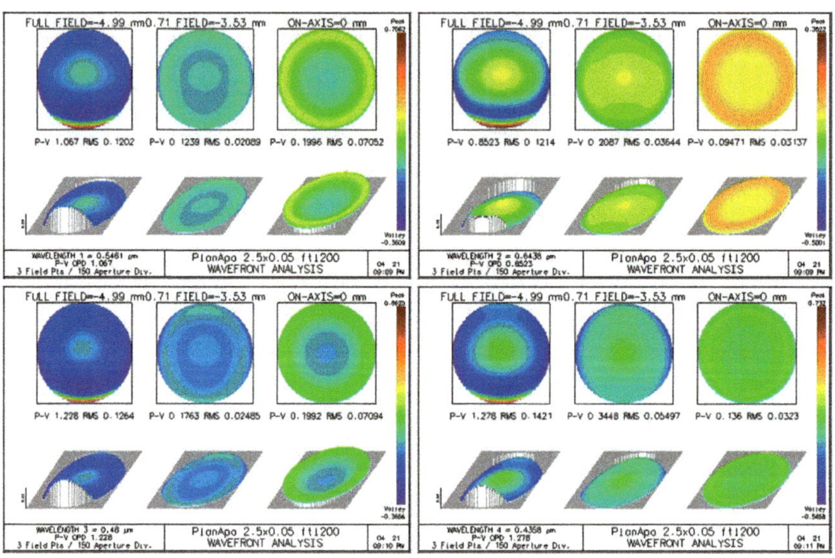

Figure 3.33 Design parameters and graphs of aberration correction of a 2.5×/0.05 CCF plan apochromatic objective ($F'_{ts} = 200$ mm).

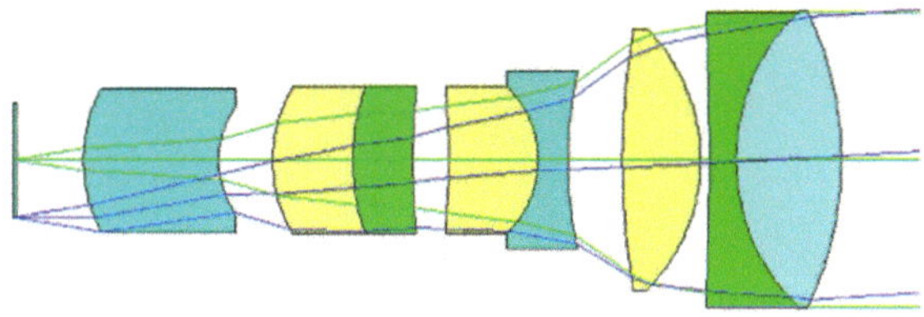

Lens: PlanApo 4x0.16 Ftl200

Object num aper 0.160000 Image height 14.000000 Primary wavln 0.546070

SRF	RADIUS	THICKNESS	APERTURE RADIUS	GLASS	SPECIAL
OBJ	0.000000	0.010000	3.499982	AIR	
AST	0.000000	0.170000	3.480000 A	K9	
2	0.000000	3.700000	3.500000	AIR	
3	8.000000	7.500000	4.300000	LZ_CTK12	
4	4.900000	3.000000	3.100000	AIR	
5	7.490000	4.500000	4.120000	CAF2	
6	11.660000	3.300000	4.100000	ZF1	
7	33.700000	2.000000	3.960000	AIR	
8	-46.900000	5.000000	3.940000	CAF2	
9	-5.401000	1.600000	4.000000	H-LAK3	
10	20.570000	3.000000	4.780000	AIR	
11	42.540000	4.300000	6.810000	CAF2	
12	-9.900000	0.500000	7.090000	AIR	
13	-200.000000	1.500000	7.500000	H-K10	
14	10.920000	5.800000	8.010000	H-QK3	
15	-18.500000	200.000000	8.100000	AIR	
16	-36.100000	14.000000	22.030000	K9	
17	-43.560000	3.000000	25.860000	AIR	
18	377.100000	15.000000	26.940000	CAF2	
19	-81.500000	3.000000	27.290000	AIR	
20	221.300000	17.000000	26.320000	CAF2	
21	-72.000000	3.000000	25.140000	LZ_CTK12 P	
22	0.000000	199.995452 S	24.910000	AIR	
IMS	0.000000	0.000000	14.000000		

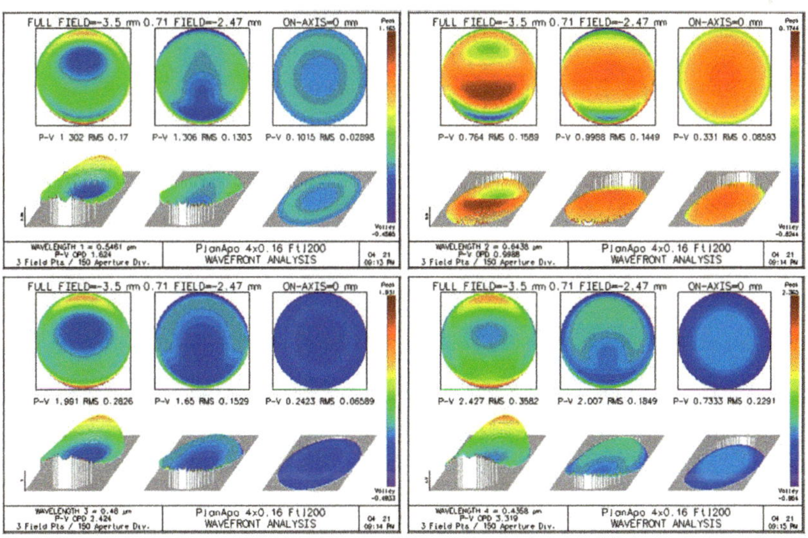

Figure 3.34 Design parameters and graphs of aberration correction of a 4×/0.16 CCF plan apochromatic objective ($F'_{ts} = 200$ mm).

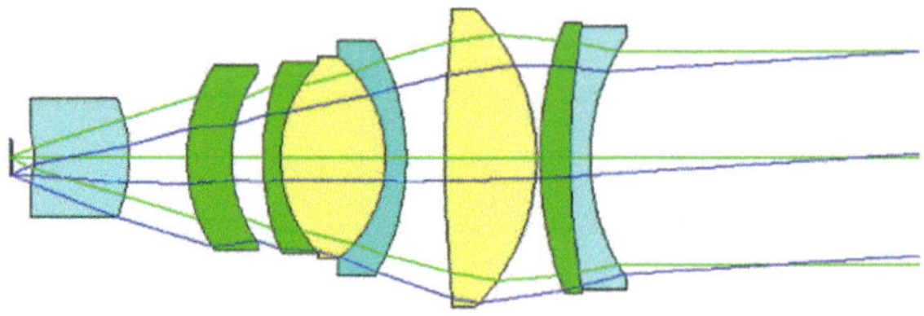

Lens: PlanApo 10x0.40 d=0.17 ft1200

Object num aper 0.400000 Image height 14.000000 Primary wavln 0.546070

SRF	RADIUS	THICKNESS		APERTURE RADIUS		GLASS		SPECIAL
OBJ	0.000000	0.010000		1.400351		AIR		
AST	0.000000	0.170000		1.400000	A	K9		
2	0.000000	1.700000		1.440000		AIR		
3	-6.680000	7.500000		2.040000		H-LAK3		
4	-10.850000	4.500000		4.660000		AIR		
5	12.210000	3.500000		7.210000		H-ZLAF7LA		
6	10.850000	2.500000		6.650000		AIR		
7	19.269000	1.300000		7.270000		H-ZLAF7LA	P	
8	10.281000	8.000000		7.280000		CAF2		
9	-9.720000	1.600000		7.670000		H-LAK3	P	
10	-17.390000	3.000000		8.980000		AIR		
11	97.700000	7.000000		11.180000		CAF2	P	
12	-15.810000	0.300000		11.470000		AIR		
13	29.620000	2.300000		10.350000		H-ZLAF7LA	P	
14	56.160000	1.500000		10.020000		K9	P	
15	16.120000	200.000000		9.110000		AIR		
16	-36.100000	14.000000		20.050000		K9		
17	-43.560000	3.000000		23.490000		AIR		
18	377.100000	15.000000		24.370000		CAF2		
19	-81.500000	3.000000		24.760000		AIR		
20	221.300000	17.000000		24.070000		CAF2		
21	-72.000000	3.000000		22.960000		LZ_CTK12		
22	0.000000	199.976694	S	22.790000		AIR		
IMS	0.000000	0.000000		14.000000				

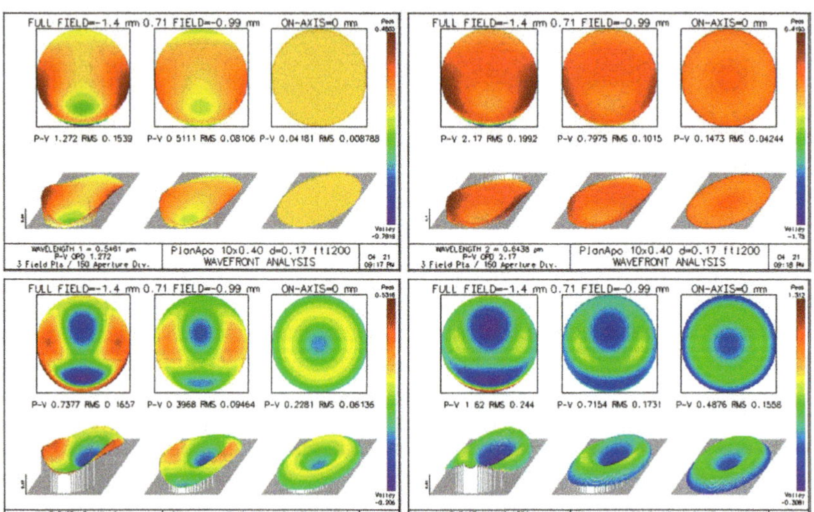

Figure 3.35 Design parameters and graphs of aberration correction of a 10×/0.40 CCF plan apochromatic objective ($F'_{ts} = 200$ mm).

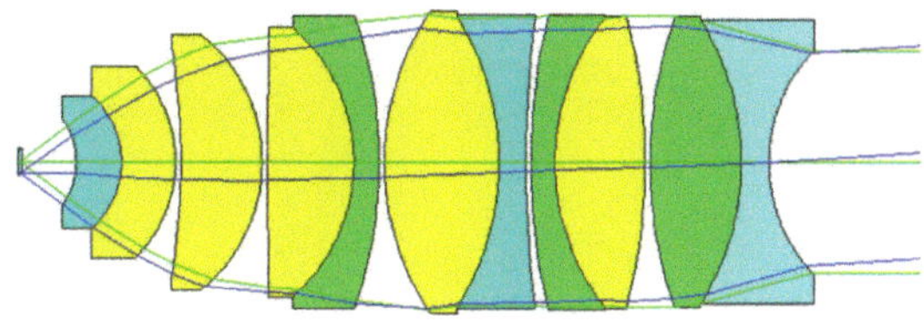

Lens: PlanApo 20x0.60 d=0.17 Ft1200

Object num aper 0.600000 Image height 12.500000 Primary wavln 0.546070

SRF	RADIUS		THICKNESS		APERTURE RADIUS		GLASS		SPECIAL
OBJ	0.000000		0.010000		0.623989		AIR		
AST	0.000000		0.170000		0.630000	A	K9		
2	0.000000		2.991800		0.700000		AIR		
3	-4.000000		2.500000		2.360000		H-LAK53A		
4	-4.720000		3.000000		3.580000		H-FK71		
5	-7.127000		0.300000		5.190000		AIR		
6	-38.950000		4.500000		6.270000		H-FK71	P	
7	-8.705000		0.300000		6.910000		AIR		
8	349.100000		4.800000		7.230000		H-FK71	P	
9	-9.480000		1.400000		7.300000		ZBAF4		
10	-23.800000		0.300000		7.920000		AIR		
11	14.510000		6.400000		8.200000		H-FK71	P	
12	-14.510000		1.500000		7.970000		H-LAK53A	P	
13	64.927000		0.300000		7.930000		AIR		
14	34.130000		1.300000		7.990000		ZBAF4	P	
15	12.560000		5.000000		7.850000		H-FK71	P	
16	-31.220000		0.300000		7.930000		AIR		
17	23.710000		5.000000		7.900000		ZBAF4	P	
18	-14.510000		1.500000		7.700000		SILICA		
19	8.705000		200.000000		6.060000		AIR		
20	-36.100000		14.000000		18.120000		K9		
21	-43.560000		3.000000		21.200000		AIR		
22	377.100000		15.000000		21.870000		CAF2		
23	-81.500000		3.000000		22.250000		AIR		
24	221.300000		17.000000		21.700000		CAF2		
25	-72.000000		3.000000		20.640000		LZ_CTK12		
26	0.000000		199.993795	S	20.500000		AIR		
IMS	0.000000		0.000000		12.500000				

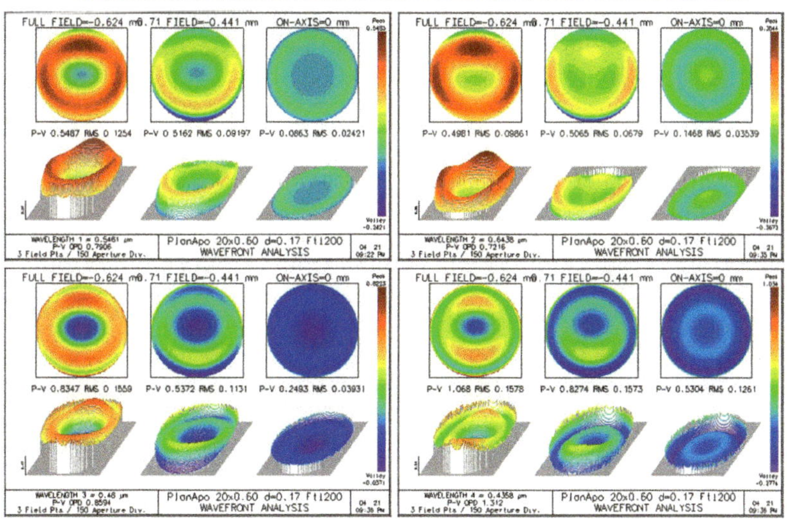

Figure 3.36 Design parameters and graphs of aberration correction of a 20×/0.60 CCF plan apochromatic objective ($F'_{ts} = 200$ mm).

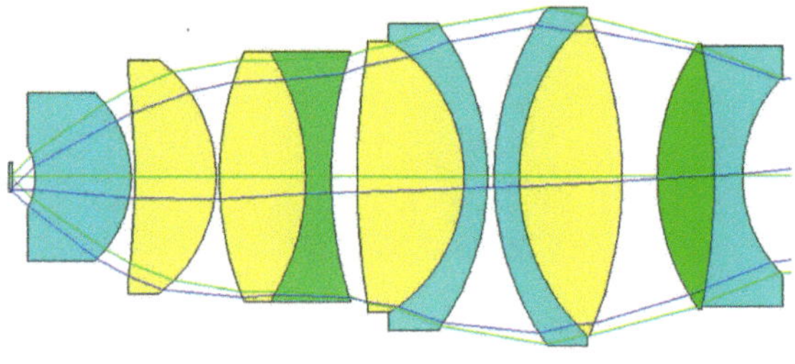

Lens: PLAPO 20x0.70 d017 finity					Efl	7.73		
Object num aper	0.700000	Image height		14.000000	Primary wavln	0.546070		

SRF	RADIUS	THICKNESS		APERTURE RADIUS		GLASS		SPECIAL
OBJ	0.000000	0.010000		0.700489		AIR		
AST	0.000000	0.170000		0.700000	A	K9		
2	0.000000	1.305400		0.790000		AIR		
3	-3.500000	5.510000		1.660000		H-LAK3		
4	-6.109000	0.200000		4.600000		AIR		
5	-41.350000	4.600000		5.640000		CAF2		
6	-7.900000	0.200000		6.370000		AIR		
7	17.320000	4.800000		6.810000		CAF2	P	
8	-12.886000	1.400000		6.720000		ZBAF4		
9	22.180000	1.450000		6.840000		AIR		
10	42.289500	6.000000		7.120000		CAF2	P	
11	-8.626100	1.300000		7.410000		H_LAF10LA		
12	-14.096200	0.400000		8.380000		AIR		
13	15.668500	1.400000		9.250000		H_LAF10LA	P	
14	11.720900	5.800000		8.700000		CAF2	P	
15	-21.500200	2.000000		8.690000		AIR		
16	12.471100	3.200000		7.230000		LZ_TF2		
17	-35.940000	1.600000		7.120000		LZ_CTK19		
18	7.690900	153.538155	S	5.430000		AIR		
IMS	0.000000	0.000000		14.000000				

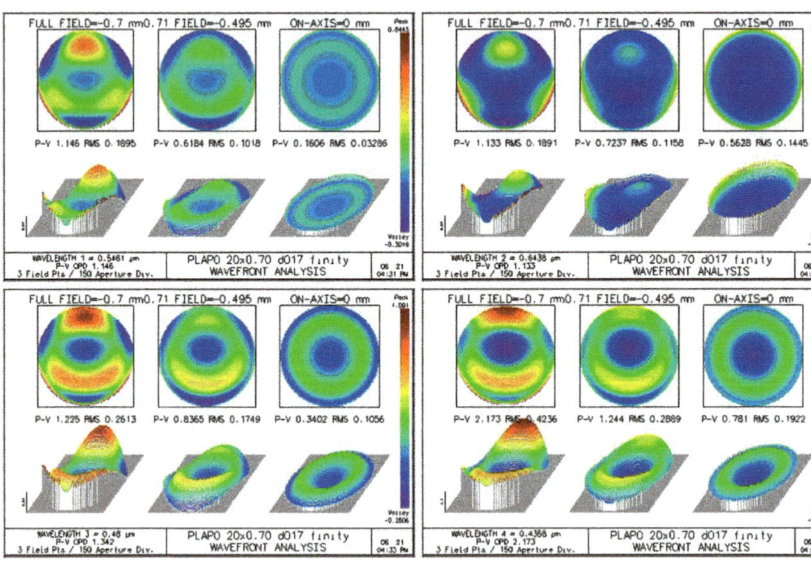

Figure 3.37 Design parameters and graphs of aberration correction of a 20×/0.70 CCF plan apochromatic objective with a finite tube length of 160 mm.

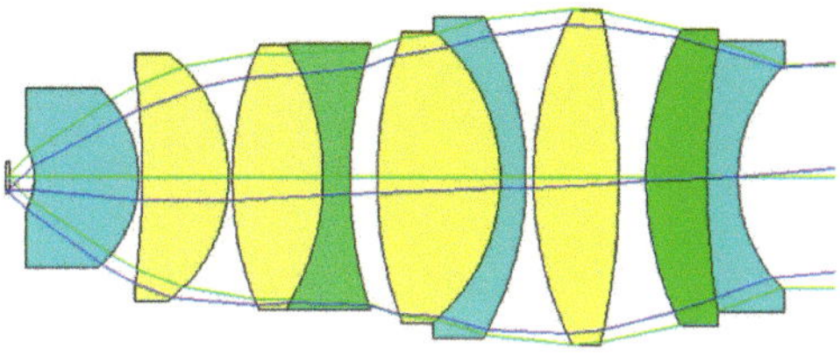

Lens: PlanApo 20x0.70 d=0.17 Ftl164

Object num aper 0.700000 Image height 14.000000 Primary wavln 0.546070

SRF	RADIUS	THICKNESS		APERTURE RADIUS		GLASS		SPECIAL
OBJ	0.000000	0.010000		0.700034		AIR		
AST	0.000000	0.170000		0.700000	A	K9		
2	0.000000	1.305400		0.790000		AIR		
3	-3.500000	5.510000		1.660000		H-LAK3		
4	-6.109000	0.200000		4.600000		AIR		
5	-41.350000	4.600000		5.640000		CAF2		
6	-7.900000	0.200000		6.370000		AIR		
7	17.320000	4.800000		6.810000		CAF2	P	
8	-12.886000	1.400000		6.720000		ZBAF4		
9	22.180000	1.450000		6.840000		AIR		
10	22.115600	6.500000		7.330000		CAF2	P	
11	-9.738200	1.300000		7.490000		H-ZLAF10LA		
12	-16.686700	0.400000		8.240000		AIR		
13	19.087000	4.500000		8.660000		CAF2	P	
14	-39.581000	1.450000		8.540000		AIR		
15	16.759500	3.200000		7.610000		LZ_TF5		
16	42.842900	1.600000		6.970000		H-LAK4L		
17	8.143300	164.000000		5.810000		AIR		
18	-29.602000	11.480000		19.020000		K9		
19	-35.719200	2.460000		22.370000		AIR		
20	309.222000	12.300000		23.490000		CAF2		
21	-66.830000	2.460000		23.820000		AIR		
22	181.466000	13.940000		22.990000		CAF2		
23	-59.040000	2.460000		22.100000		LZ_CTK12		
24	0.000000	163.918974	S	21.900000		AIR		
IMS	0.000000	0.000000		14.000000				

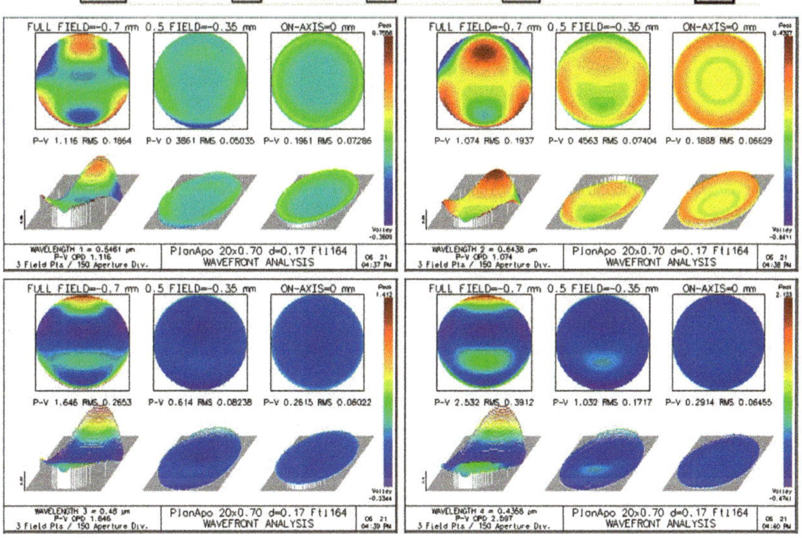

Figure 3.38 Design parameters and graphs of aberration correction of a 20×/0.70 CCF plan apochromatic objective ($F'_{ts} = 164$ mm).

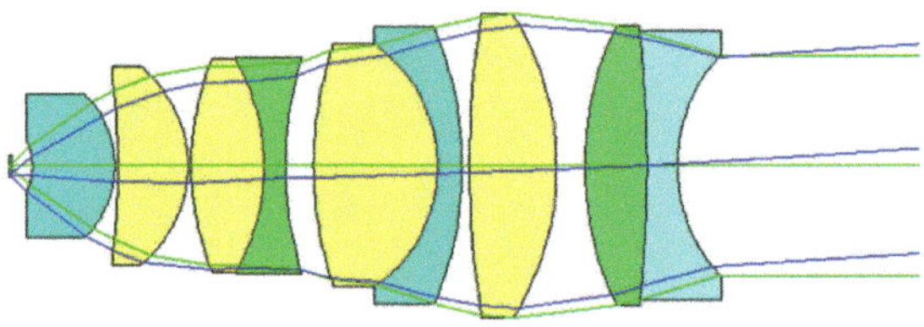

Lens: PlanApo 20x0.70 d=0.17 Ftl180

Object num aper 0.700000 Image height 14.000000 Primary wavln 0.546070

SRF	RADIUS	THICKNESS		APERTURE RADIUS		GLASS		SPECIAL
OBJ	0.000000	0.010000		0.699408		AIR		
AST	0.000000	0.170000		0.630000	A	K9		
2	0.000000	1.305400		0.720000		AIR		
3	-3.500000	5.510000		1.610000		H-LAK3		
4	-6.109000	0.200000		4.540000		AIR		
5	-41.350000	4.600000		5.550000		CAF2		
6	-7.900000	0.200000		6.300000		AIR		
7	17.320000	4.800000		6.790000		CAF2	P	
8	-12.886000	1.400000		6.700000		ZBAF4		
9	22.180000	1.800000		6.840000		AIR		
10	24.800000	8.000000		7.400000		CAF2	P	
11	-9.720000	1.600000		7.730000		H-ZLAF10LA		
12	-18.000000	0.500000		8.730000		AIR		
13	35.480000	5.600000		9.320000		CAF2	P	
14	-21.985000	1.800000		9.370000		AIR		
15	16.320000	4.000000		8.330000		H_F4		
16	-59.320000	2.000000		7.160000		H-LAK4L		
17	9.480000	180.000000		6.420000		AIR		
18	-32.490000	12.600000		17.130000		K9		
19	-39.204000	2.700000		20.020000		AIR		
20	339.390000	13.500000		20.680000		CAF2		
21	-73.350000	2.700000		21.000000		AIR		
22	199.170000	15.300000		20.430000		CAF2		
23	-64.800000	2.700000		19.430000		LZ_CTK12		
24	0.000000	179.977307	S	19.280000		AIR		
IMS	0.000000	0.000000		14.000000				

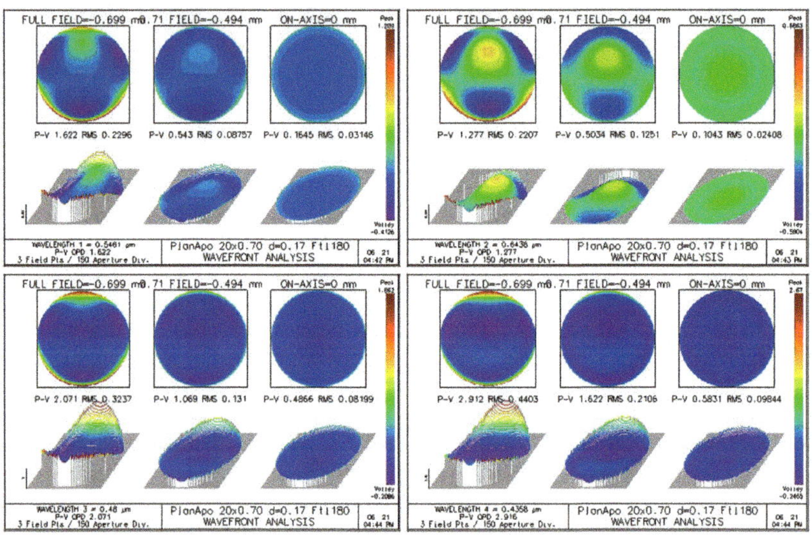

Figure 3.39 Design parameters and graphs of aberration correction of a 20×/0.70 CCF plan apochromatic objective ($F'_{ts} = 180$ mm).

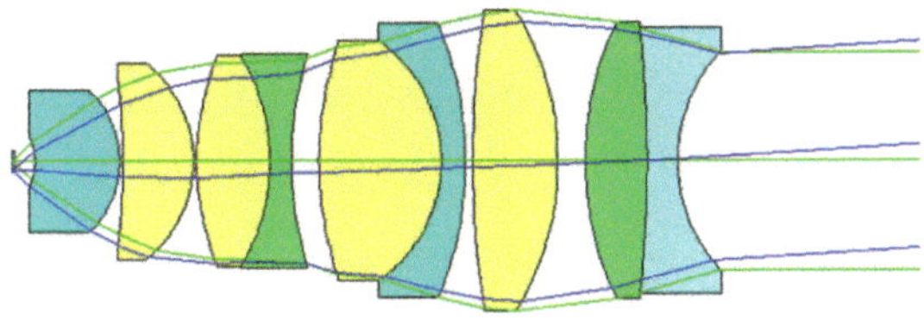

Lens: PlanApo 20x0.70 d=0.17 Ft1200

Object num aper 0.700000 Image height 14.000000 Primary wavln 0.546070

SRF	RADIUS	THICKNESS	APERTURE RADIUS	GLASS	SPECIAL
OBJ	0.000000	0.010000	0.699917	AIR	
AST	0.000000	0.170000	0.700000 A	K9	
2	0.000000	1.305400	0.790000	AIR	
3	-3.500000	5.510000	1.660000	H-LAK3	
4	-6.109000	0.200000	4.600000	AIR	
5	-41.350000	4.600000	5.640000	CAF2	
6	-7.900000	0.200000	6.370000	AIR	
7	17.320000	4.800000	6.810000	CAF2 P	
8	-12.886000	1.400000	6.720000	ZBAF4	
9	22.180000	1.800000	6.840000	AIR	
10	22.590000	8.000000	7.440000	CAF2 P	
11	-9.330000	1.600000	7.700000	H-LAK3 P	
12	-22.000000	0.500000	8.840000	AIR	
13	65.350000	5.600000	9.480000	CAF2 P	
14	-18.232000	1.800000	9.700000	AIR	
15	20.430000	4.000000	8.920000	H_F2	
16	-79.360000	2.000000	8.570000	H-QK3	
17	10.130000	200.000000	7.090000	AIR	
18	-36.100000	14.000000	20.420000	K9	
19	-43.560000	3.000000	23.930000	AIR	
20	377.100000	15.000000	24.840000	CAF2	
21	-81.500000	3.000000	25.220000	AIR	
22	221.300000	17.000000	24.480000	CAF2	
23	-72.000000	3.000000	23.360000	LZ_CTK12	
24	0.000000	199.963642 S	23.180000	AIR	
IMS	0.000000	0.000000	14.000000		

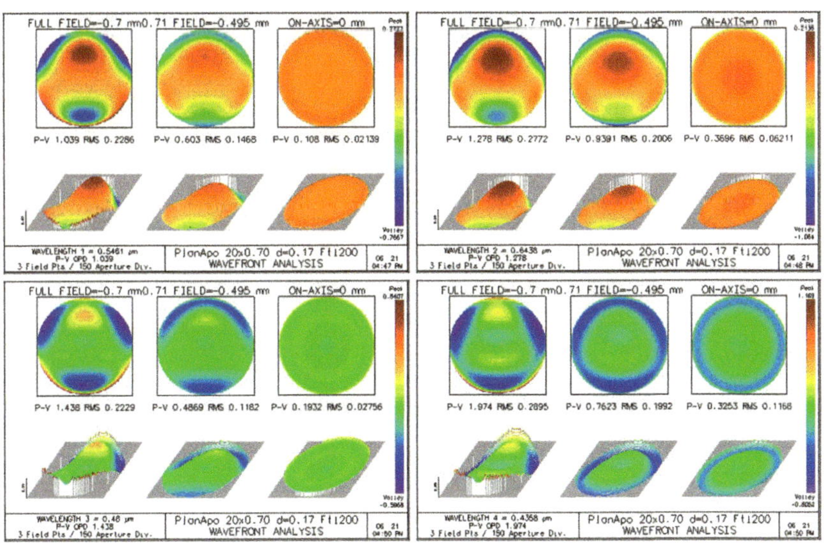

Figure 3.40 Design parameters and graphs of aberration correction of a 20×/0.70 CCF plan apochromatic objective ($F'_{ts} = 200$ mm).

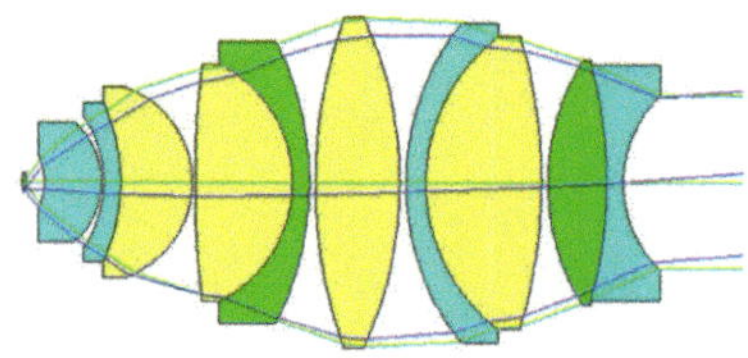

Lens: PlanApo 20x0.80 d=0.17 Ftl164

Object num aper 0.800000 Image height 12.500000 Primary wavln 0.546070

SRF	RADIUS		THICKNESS		APERTURE RADIUS		GLASS		SPECIAL
OBJ	0.000000		0.010000		0.625037		AIR		
AST	0.000000		0.170000		0.630000	A	K9		
2	0.000000		1.500000		0.740000		AIR		
3	-4.650000		4.500000		2.080000		H-ZLAF68B		
4	-5.890000		0.200000		4.520000		AIR		
5	-8.670000		1.300000		4.980000		H-ZLAF68B	P	
6	-14.150000		5.500000		5.990000		CAF2		
7	-7.600000		0.200000		7.200000		AIR		
8	75.600000		7.500000		8.760000		CAF2	P	
9	-9.900000		1.300000		8.950000		H-ZF6		
10	-23.140000		0.500000		10.690000		AIR		
11	37.224000		6.500000		12.400000		CAF2	P	
12	-29.470000		0.500000		12.560000		AIR		
13	19.538000		1.500000		12.150000		H-ZLAF68B	P	
14	13.820000		9.000000		11.100000		CAF2	P	
15	-37.535000		0.500000		10.860000		AIR		
16	17.030000		4.600000		9.230000		H-ZF7LA		
17	-33.600000		1.300000		8.910000		H-ZLAF68B	P	
18	9.160000		164.000000		6.640000		AIR		
19	-29.602000		11.480000		18.300000		K9		
20	-35.719200		2.460000		21.480000		AIR		
21	309.222000		12.300000		22.450000		CAF2		
22	-66.830000		2.460000		22.760000		AIR		
23	181.466000		13.940000		21.990000		CAF2		
24	-59.040000		2.460000		21.070000		LZ_CTK12		
25	0.000000		163.984983	S	20.080000		AIR		
IMS	0.000000		0.000000		12.500000				

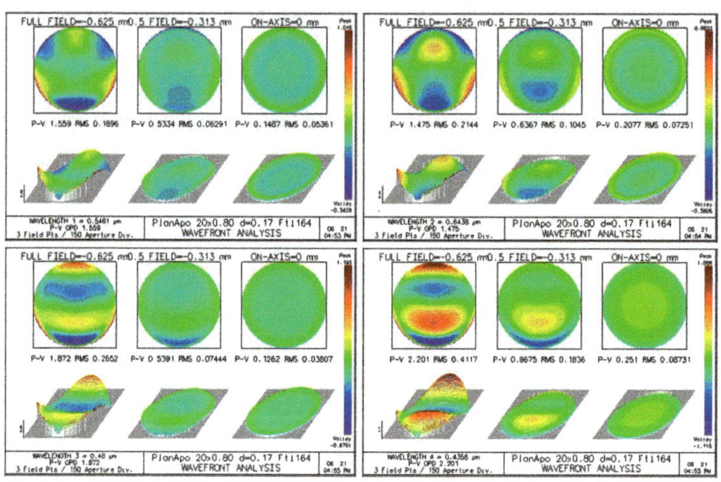

Figure 3.41 Design parameters and graphs of aberration correction of a 20×/0.80 CCF plan apochromatic objective ($F'_{ts} = 164$ mm).

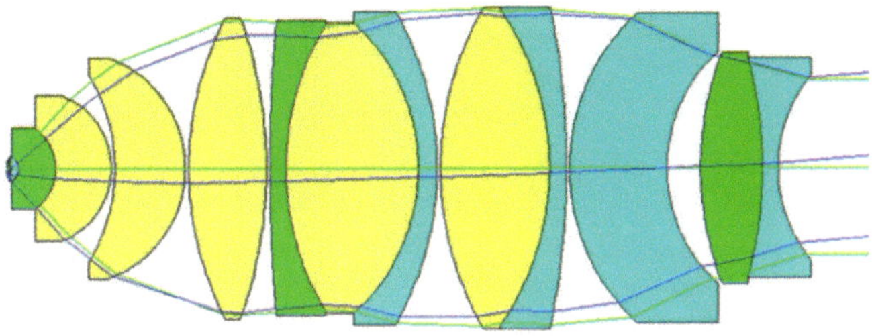

Lens:Plapo 25x1.10 water h60mm Ftl160

Object num aper 1.100000 Image height 12.500000 Primary wavln 0.546070

SRF	RADIUS	THICKNESS		APERTURE RADIUS		GLASS		SPECIAL
OBJ	0.000000	0.010000		0.499997		WATER		
AST	0.000000	0.170000		0.510000	A	K9		
2	0.000000	0.190000		0.690000		WATER		
3	-4.365000	0.600000		0.850000		BAK2		
4	-1.096500	3.000000		0.970000		H-ZF7LA		
5	-3.936000	4.600000		3.240000		CAF2		
6	-6.223100	0.300000		5.840000		AIR		
7	-14.218200	5.800000		7.620000		CAF2	P	
8	-9.357100	0.300000		8.870000		AIR		
9	25.931700	6.400000		12.050000		CAF2	P	
10	-35.289100	0.300000		12.100000		AIR		
11	144.246500	1.300000		11.830000		H-ZF7LA	P	
12	21.868700	11.000000		11.500000		CAF2	P	
13	-15.315600	1.500000		11.650000		H-ZLAF69		
14	-24.812900	0.300000		12.590000		AIR		
15	25.184200	9.000000		12.930000		CAF2	P	
16	-21.981400	1.300000		12.750000		H-ZLAF69	P	
17	-65.857700	0.300000		12.940000		AIR		
18	16.776200	8.000000		12.440000		H-ZLAF69	P	
19	12.457900	2.800000		9.300000		AIR		
20	25.629200	5.000000		9.280000		H-ZF7LA	P	
21	-33.527800	1.300000		8.860000		ZK6		
22	11.339300	90.000000		7.460000		AIR		
23	-28.880000	11.200000		14.040000		K9		
24	-34.848000	2.400000		16.530000		AIR		
25	301.680000	12.000000		17.150000		CAF2		
26	-65.200000	2.400000		17.540000		AIR		
27	177.040000	13.600000		17.200000		CAF2		
28	-57.600000	2.400000		16.470000		LZ_CTK12		
29	0.000000	160.001070	S	16.400000		AIR		
IMS	0.000000	0.000000		12.500000				

Figure 3.42 Design parameters and graphs of aberration correction of a 25×/1.10 water immersion CCF plan apochromatic objective ($F'_{ts} = 160$ mm; $h = 60$ mm).

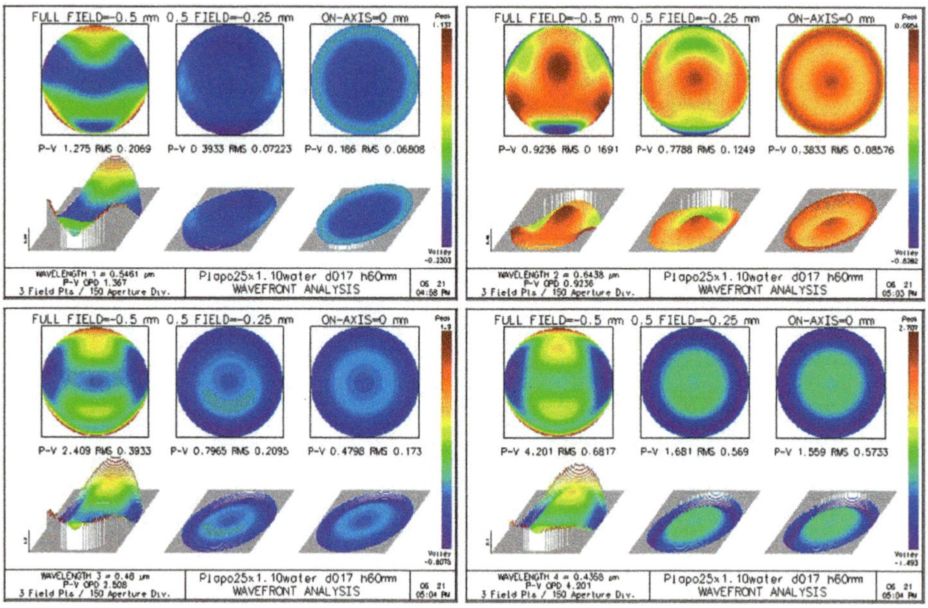

Figure 3.42 (*Continued*) Design parameters and graphs of aberration correction of a 25×/1.10 water immersion CCF plan apochromatic objective (F'_{ts} = 160 mm; h = 60 mm).

(F'_{ob} = 1.64 mm) objectives, and Table 3.5 shows the parameters of 100×/1.20 water immersion infinity (F'_{ob} = 1.8 mm and F'_{ob} = 2.0 mm) objectives.

In accordance with modern trends, we can eliminate the optical material H-ZF7LA, which has poor transmittance in the near UV region of the spectrum. For objectives with tube systems of F'_{ts} = 180 mm and F'_{ts} = 200 mm, we can use glass types ZF-5 and H-ZF6 for an improved transmittance wavelength of 365 nm.

Figure 3.72 shows a photograph and a technical drawing of a 60×/1.0 oil immersion objective (the design parameters and graphs of aberration correction are shown in Fig. 3.53). Figures 3.73 and 3.74 are photographs of real objects taken with it. Figure 3.73 shows the objective's ability to provide adequate color reproduction; Fig. 3.74 demonstrates the objective' resolution capability. These photographs were taken by Anatoly Mikhaltsov.

This objective is unique in that it uses the original design of the front component, made in the form of a triple-glued lens. Several mechanical frame-mounting options have been investigated for this component. This objective was chosen first in these studies because its input numerical aperture in oil is not very high. In this sense, this objective is a good prototype for designing the technology to manufacture a triple-glued lens and attach it to the frame.

Figure 3.75 shows a photograph and the principal mechanical design of a 100×/0.95 objective (the design parameters and graphs of aberration

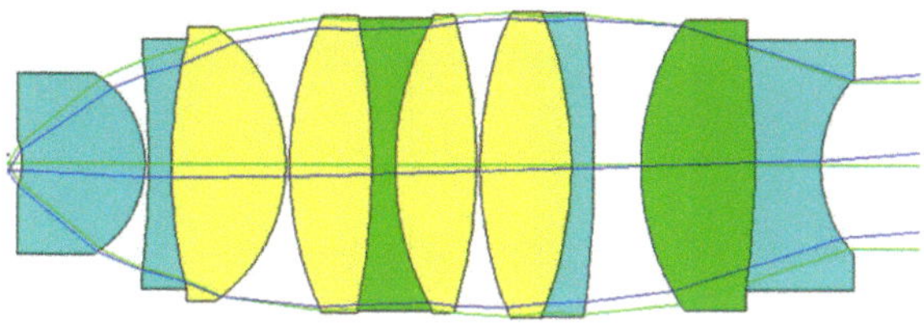

Lens: PlanApo 32x0.85 d0 Ftl160

Object num aper 0.850000 Image height 12.500000 Primary wavln 0.546070

SRF	RADIUS	THICKNESS	APERTURE RADIUS	GLASS	SPECIAL
OBJ	0.000000	-1.0000e-08	0.389837	AIR	
AST	0.000000	0.734400	0.387000 A	AIR	
2	-3.775600	6.500000	1.240000	LZ_CTK19	
3	-5.276000	0.200000	4.620000	AIR	
4	-51.766300	1.200000	5.730000	LZ_CTK19 P	
5	25.584900	6.000000	6.390000	CAF2	
6	-8.553500	0.200000	7.040000	AIR	
7	18.082200	4.300000	7.620000	CAF2 P	
8	-38.242400	1.200000	7.510000	LZ_TF5	
9	15.054400	4.300000	7.500000	CAF2 P	
10	-25.786200	0.200000	7.650000	AIR	
11	19.944100	4.800000	7.870000	CAF2 P	
12	-20.665900	1.200000	7.760000	LZ_CTK19 P	
13	-57.431500	2.500000	7.800000	AIR	
14	12.913900	6.000000	7.460000	LZ_TF5 P	
15	-48.093500	3.500000	6.420000	LZ_CTK19 P	
16	6.629500	90.000000	4.530000	AIR	
17	214.800000	5.000000	11.470000	LZ_BK8	
18	-46.990000	0.300000	11.580000	AIR	
19	-46.880000	3.000000	11.550000	LZ_BF24	
20	-113.240000	157.839802 S	11.710000	AIR	
IMS	0.000000	0.000000	12.500000		

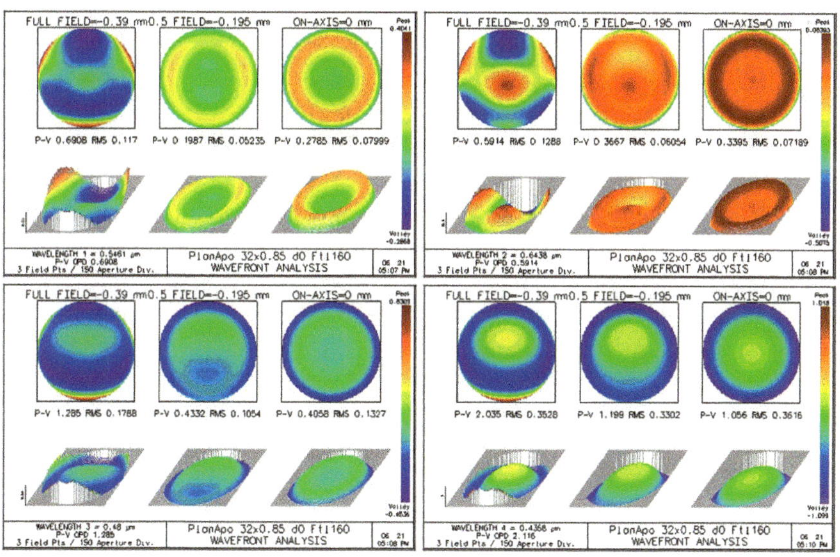

Figure 3.43 Design parameters and graphs of aberration correction of a 32×/0.85 CCF plan apochromatic objective ($F'_{ts} = 160$ mm).

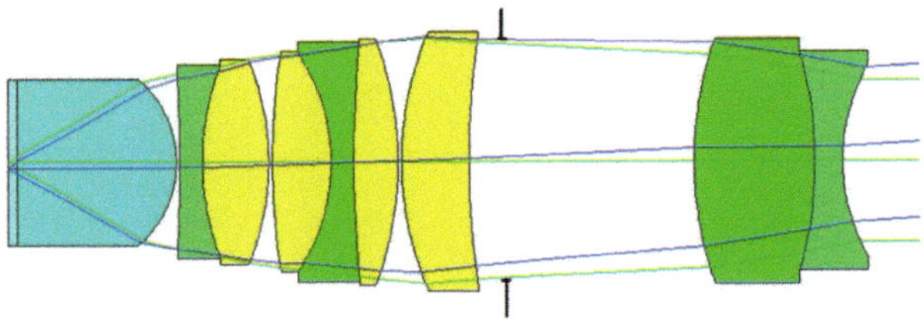

Lens: PLAPO 32x0.85 oil Ft1160

Object num aper 0.850000 Image height 10.000000 Primary wavln 0.546070

SRF	RADIUS	THICKNESS		APERTURE RADIUS		GLASS		SPECIAL
OBJ	0.000000	-1.0000e-08		0.312865		AIR		
AST	0.000000	0.500000		0.311000	A	OIL		
2	0.000000	8.700000		0.650000		LZ_CTK19		
3	-5.598000	0.200000		4.360000		AIR		
4	-80.540000	1.200000		4.760000		LZ_LF10		
5	13.709000	3.600000		5.090000		LZ_FFS2		
6	-13.709000	0.200000		5.400000		AIR		
7	36.310000	3.200000		5.710000		LZ_FFS2	P	
8	-10.093000	1.200000		5.790000		LZ_TF1		
9	58.210000	2.400000		6.320000		LZ_FFS2	P	
10	-18.030000	0.200000		6.490000		AIR		
11	16.482000	3.700000		6.830000		LZ_FFS2	P	
12	45.190000	12.200000		6.670000		AIR		
13	18.967000	6.500000		6.470000		LZ_TF4		
14	-20.700000	1.500000		5.740000		LZ_OF4		
15	9.954000	90.000000		5.030000		AIR		
16	-28.880000	11.200000		10.360000		LZ_K8		
17	-34.848000	2.400000		12.200000		AIR		
18	301.680000	12.000000		12.520000		CAF2		
19	-65.200000	2.400000		12.870000		AIR		
20	177.040000	13.600000		12.710000		CAF2	P	
21	-57.600000	2.400000		12.140000		LZ_CTK12		
22	0.000000	159.987786	S	12.110000		AIR		
IMS	0.000000	0.000000		10.000000				

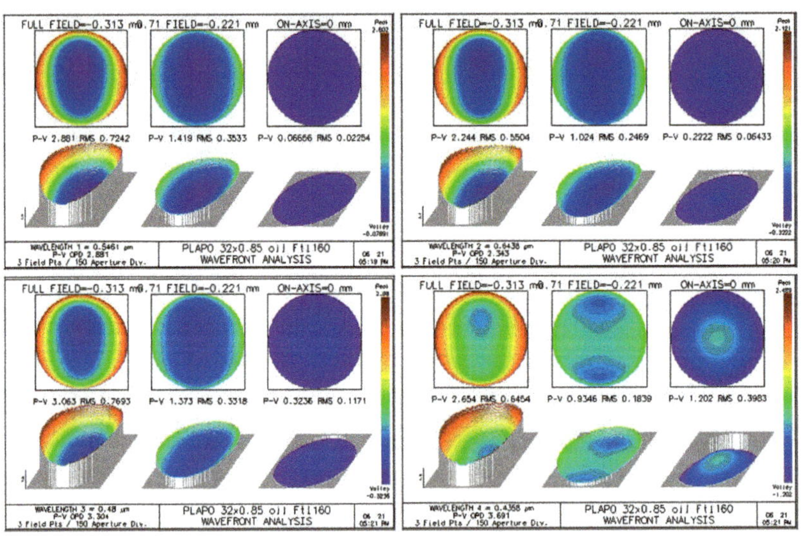

Figure 3.44 Design parameters and graphs of aberration correction of a 32×/0.85 oil immersion CCF plan apochromatic objective for polarization research ($F'_{ts} = 160$ mm).

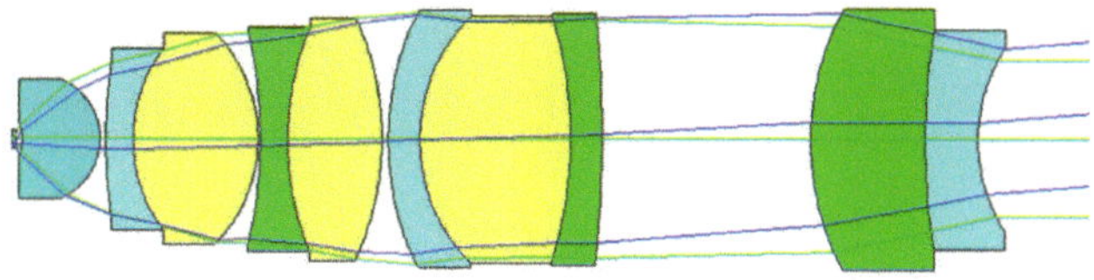

Lens: PLAPO 40x0.90 d=0.17 ftl160

Object num aper 0.900000 Image height 12.500000 Primary wavln 0.546070

SRF	RADIUS	THICKNESS	APERTURE RADIUS	GLASS	SPECIAL
OBJ	0.000000	0.010000	0.314170	AIR	
AST	0.000000	0.170000	0.330000 A	LZ_K14	
2	0.000000	0.300000	0.460000	AIR	
3	-2.270000	3.790000	0.790000	LZ_CTK19	
4	-2.965000	0.300000	2.760000	AIR	
5	27.860000	1.300000	3.730000	LZ_K8	
6	6.855000	6.000000	4.230000	CAF2	
7	-6.855000	0.100000	4.910000	AIR	
8	-21.090000	1.300000	4.910000	LZ_TF1	
9	13.274000	4.500000	5.230000	CAF2 P	
10	-13.274000	0.300000	5.680000	AIR	
11	13.274000	1.500000	6.040000	LZ_CTK12	
12	8.017000	7.200000	5.720000	CAF2 P	
13	-19.055000	1.500000	5.780000	LZ_TF1 P	
14	-49.430000	10.000000	5.900000	AIR	
15	12.474000	5.500000	6.100000	LZ_TF5	
16	31.120000	2.500000	5.110000	LZ_CTK12 P	
17	7.656000	90.000000	4.330000	AIR	
18	214.800000	5.000000	11.190000	LZ_BK8	
19	-46.990000	0.300000	11.300000	AIR	
20	-46.880000	3.000000	11.270000	LZ_BF24	
21	-113.240000	157.898965 S	11.430000	AIR	
IMS	0.000000	0.000000	12.500000		

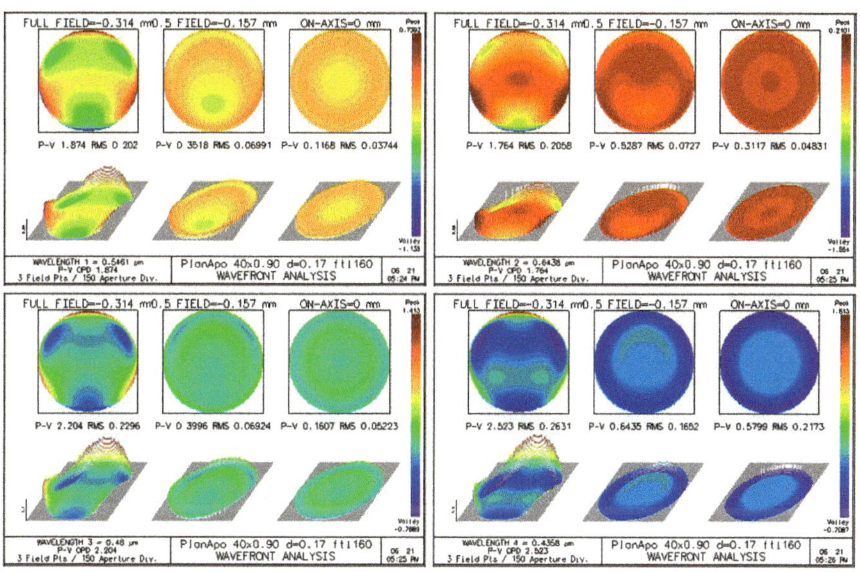

Figure 3.45 Design parameters and graphs of aberration correction of a 40×/0.90 CCF plan apochromatic objective ($F'_{ts} = 160$ mm).

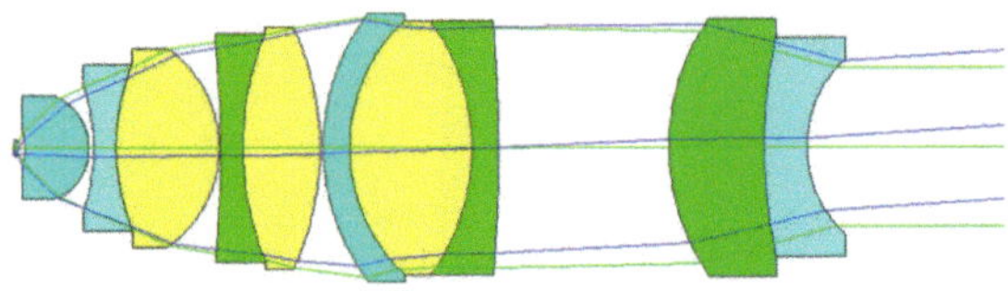

Lens: PlanApo 40x0.90 d=0.17 Ftl180

Object num aper	0.900000	Image height	14.000000	Primary wavln	0.546070

SRF	RADIUS	THICKNESS	APERTURE RADIUS	GLASS	SPECIAL
OBJ	0.000000	0.010000	0.350932	AIR	
AST	0.000000	0.170000	0.370000 A	K9	
2	0.000000	0.450000	0.490000	AIR	
3	-2.618000	3.800000	1.010000	H-LAK3	
4	-3.105000	0.300000	2.930000	AIR	
5	-10.700000	1.300000	3.620000	K9 P	
6	12.057000	6.000000	4.730000	CAF2	
7	-6.761000	0.100000	5.640000	AIR	
8	-54.700000	1.300000	6.050000	ZF1	
9	16.610000	4.200000	6.490000	CAF2	
10	-14.510000	0.200000	6.800000	AIR	
11	12.465000	1.500000	7.490000	LZ_CTK12	
12	9.817000	7.200000	7.080000	CAF2 P	
13	-11.885000	1.500000	6.930000	ZF1 P	
14	-88.830000	10.000000	7.030000	AIR	
15	11.885000	6.500000	7.020000	H-ZF7LA	
16	18.710000	1.500000	5.450000	ZK1	
17	6.300000	180.000000	4.610000	AIR	
18	-32.490000	12.600000	17.870000	K9	
19	-39.204000	2.700000	20.980000	AIR	
20	339.390000	13.500000	21.810000	CAF2	
21	-73.350000	2.700000	22.200000	AIR	
22	199.170000	15.300000	21.630000	CAF2	
23	-64.800000	2.700000	20.690000	LZ_CTK12	
24	0.000000	179.978659 S	20.550000	AIR	
IMS	0.000000	0.000000	14.000000		

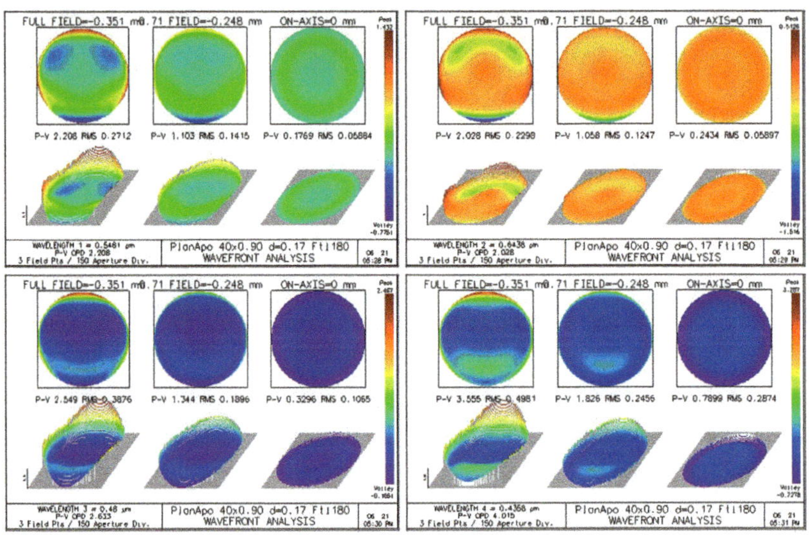

Figure 3.46 Design parameters and graphs of aberration correction of a 40×/0.90 CCF plan apochromatic objective ($F'_{ts} = 180$ mm).

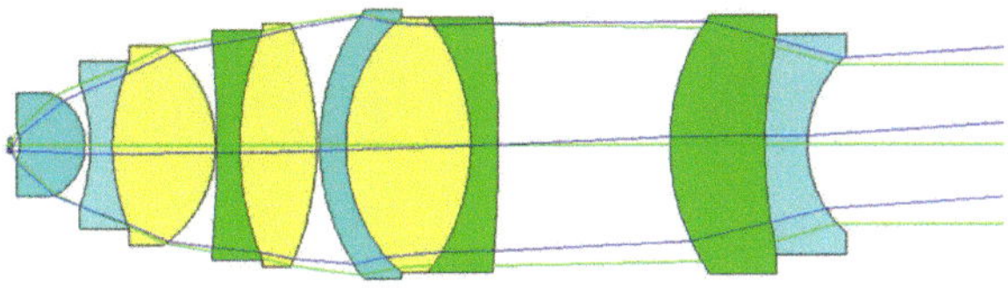

Lens: PlanApo 40x0.90 d=0.17 Ft1200

Object num aper 0.900000 Image height 14.000000 Primary wavln 0.546070

SRF	RADIUS	THICKNESS	APERTURE RADIUS	GLASS	SPECIAL
OBJ	0.000000	0.010000	0.350143	AIR	
AST	0.000000	0.170000	0.360000 A	K9	
2	0.000000	0.450000	0.490000	AIR	
3	-2.618000	3.800000	1.010000	H-LAK3	
4	-3.105000	0.300000	2.930000	AIR	
5	-10.700000	1.300000	3.620000	K9 P	
6	12.057000	6.000000	4.730000	CAF2	
7	-6.761000	0.100000	5.640000	AIR	
8	-61.800000	1.300000	6.060000	ZF1	
9	16.368000	4.500000	6.480000	CAF2	
10	-16.368000	0.200000	6.890000	AIR	
11	12.465000	1.500000	7.660000	LZ_CTK12	
12	9.817000	7.200000	7.240000	CAF2 P	
13	-11.885000	1.500000	7.120000	ZF1 P	
14	-88.830000	10.000000	7.250000	AIR	
15	12.700000	5.500000	7.390000	ZF6	
16	29.000000	2.500000	6.290000	H-QK3	
17	6.680000	200.000000	5.050000	AIR	
18	-36.100000	14.000000	18.730000	K9	
19	-43.560000	3.000000	21.970000	AIR	
20	377.100000	15.000000	22.760000	CAF2	
21	-81.500000	3.000000	23.180000	AIR	
22	221.300000	17.000000	22.630000	CAF2	
23	-72.000000	3.000000	21.590000	LZ_CTK12 P	
24	0.000000	199.985502 S	21.450000	AIR	
IMS	0.000000	0.000000	14.000000		

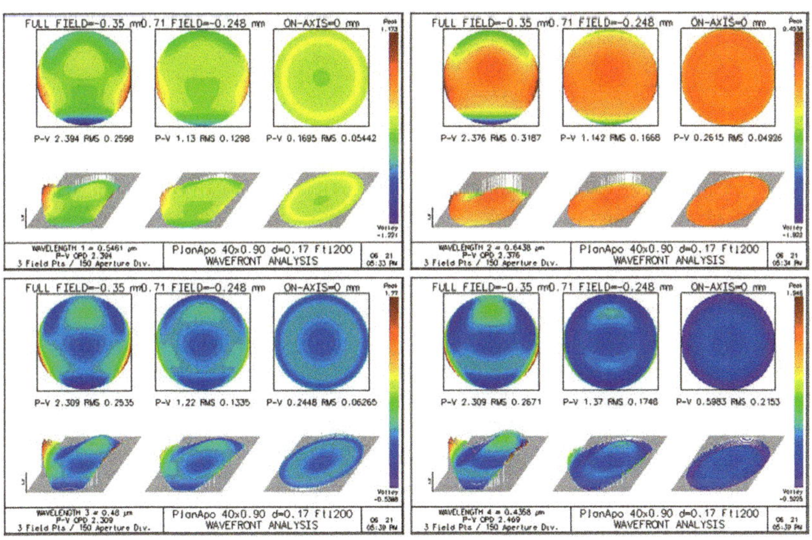

Figure 3.47 Design parameters and graphs of aberration correction of a 40×/0.90 CCF plan apochromatic objective ($F'_{ts} = 200$ mm).

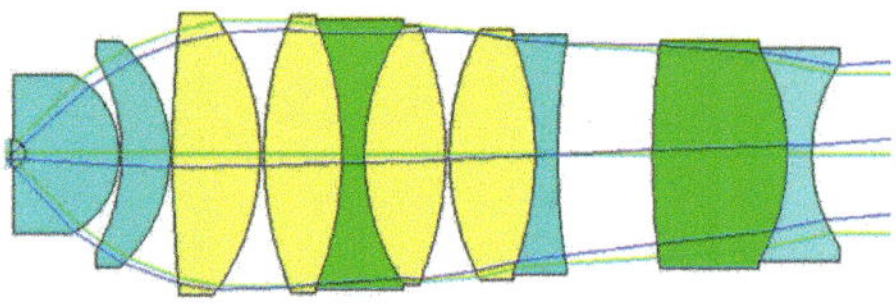

Lens: Plapo 40x1.0 oil Ftl164

Object num aper 1.000000 Image height 12.500000 Primary wavln 0.546070

SRF	RADIUS	THICKNESS		APERTURE RADIUS		GLASS		SPECIAL
OBJ	0.000000	1.0000e-20		0.312487		OIL		
AST	0.000000	0.350000		0.312487	AS	OIL		
2	0.000000	0.700000		0.620000		K9		
3	-0.700000	5.000000		0.690000		LZ_CTK19		
4	-4.512800	0.150000		4.060000		AIR		
5	-9.600900	2.600000		4.970000		LZ_CTK19	P	
6	-7.481100	0.150000		5.780000		AIR		
7	65.023300	4.600000		6.900000		CAF2		
8	-11.854800	0.200000		7.250000		AIR		
9	18.335300	4.100000		7.090000		CAF2	P	
10	-17.580200	1.200000		6.910000		LZ_TF2		
11	11.610100	4.300000		6.610000		CAF2	P	
12	-16.491600	0.200000		6.640000		AIR		
13	13.426700	4.500000		6.440000		CAF2	P	
14	-16.565400	1.200000		6.120000		LZ_CTK19	P	
15	37.996000	5.000000		5.900000		AIR		
16	41.838600	7.000000		5.800000		LZ_TF5		
17	-10.842900	1.300000		5.470000		K9	P	
18	8.092900	160.000000		4.670000		AIR		
19	-29.602000	11.480000		16.290000		K9	P	
20	-35.719200	2.460000		19.120000		AIR		
21	309.222000	12.300000		19.870000		CAF2	P	
22	-66.830000	2.460000		20.220000		AIR		
23	181.466000	13.940000		19.690000		CAF2	P	
24	-59.040000	2.460000		18.820000		LZ_CTK12		
25	0.000000	163.973168	S	18.700000		AIR		
IMS	0.000000	0.000000		12.500000	S			

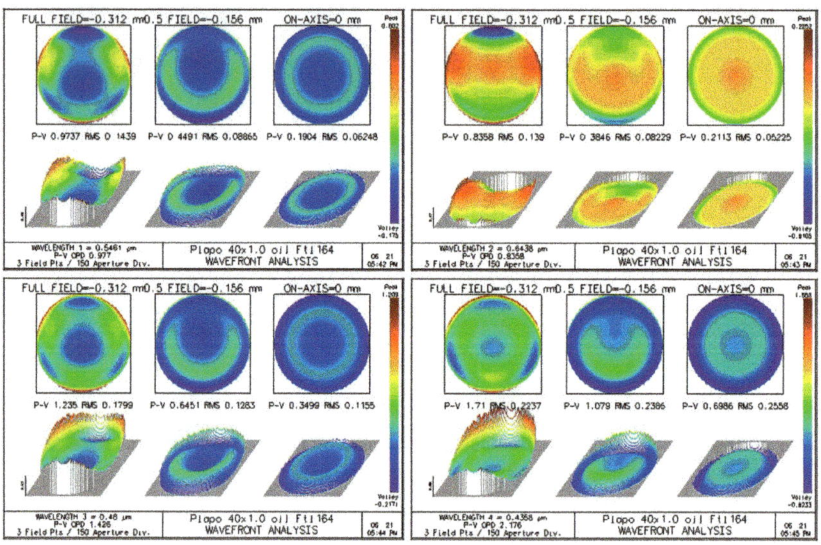

Figure 3.48 Design parameters and graphs of aberration correction of a 40×/1.0 oil immersion CCF plan apochromatic objective ($F'_{ts} = 164$ mm).

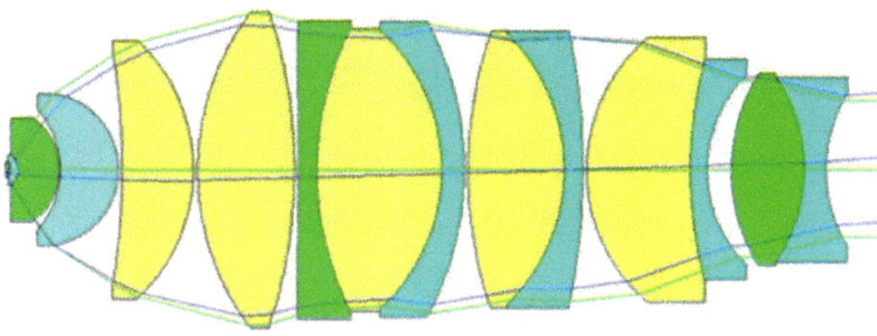

Lens: PlaPo 40x1.20water d017 Ftl160

Object num aper	1.200000	Image height	12.500000	Primary wavln	0.546070

SRF	RADIUS	THICKNESS	APERTURE RADIUS	GLASS	SPECIAL
OBJ	0.000000	0.010000	0.312518	WATER	
AST	0.000000	0.170000	0.330000 A	K9	
2	0.000000	0.190000	0.550000	WATER	
3	-1.054500	0.600000	0.580000	K9	
4	-1.341800	3.000000	1.050000	H-ZF6	
5	-4.321600	0.100000	3.620000	AIR	
6	-6.325500	4.200000	4.300000	K9 P	
7	-5.344100	0.300000	5.340000	AIR	
8	-49.563300	5.200000	8.250000	CAF2	
9	-12.106300	0.300000	9.080000	AIR	
10	18.128900	7.000000	11.130000	CAF2	
11	-34.063700	0.300000	11.030000	AIR	
12	-1.6705e+03	1.300000	10.470000	H-ZF7LA	
13	21.662200	9.000000	9.940000	CAF2	
14	-13.069200	1.600000	9.840000	LZ_CTK19	
15	-21.370000	0.300000	10.370000	AIR	
16	29.975600	7.000000	9.760000	CAF2 P	
17	-12.997000	1.300000	9.520000	LZ_CTK19 P	
18	-46.797300	0.300000	9.730000	AIR	
19	11.974600	7.600000	9.310000	CAF2 P	
20	29.182200	1.200000	7.740000	LZ_CTK19 P	
21	9.316400	1.800000	6.690000	AIR	
22	12.753900	5.200000	6.760000	H-ZF6 P	
23	-11.267800	1.600000	6.520000	LZ_CTK19 P	
24	9.924600	90.000000	5.360000	AIR	
25	-28.880000	11.200000	12.190000	K9	
26	-34.848000	2.400000	14.410000	AIR	
27	301.680000	12.000000	14.910000	CAF2	
28	-65.200000	2.400000	15.340000	AIR	
29	177.040000	13.600000	15.140000	CAF2	
30	-57.600000	2.400000	14.530000	LZ_CTK12	
31	0.000000	159.951330 S	14.500000	AIR	
IMS	0.000000	0.000000	12.500000		

Figure 3.49 Design parameters and graphs of aberration correction of a 40×/1.20 water immersion CCF plan apochromatic objective ($F'_{ts} = 160$ mm; $h = 60$ mm).

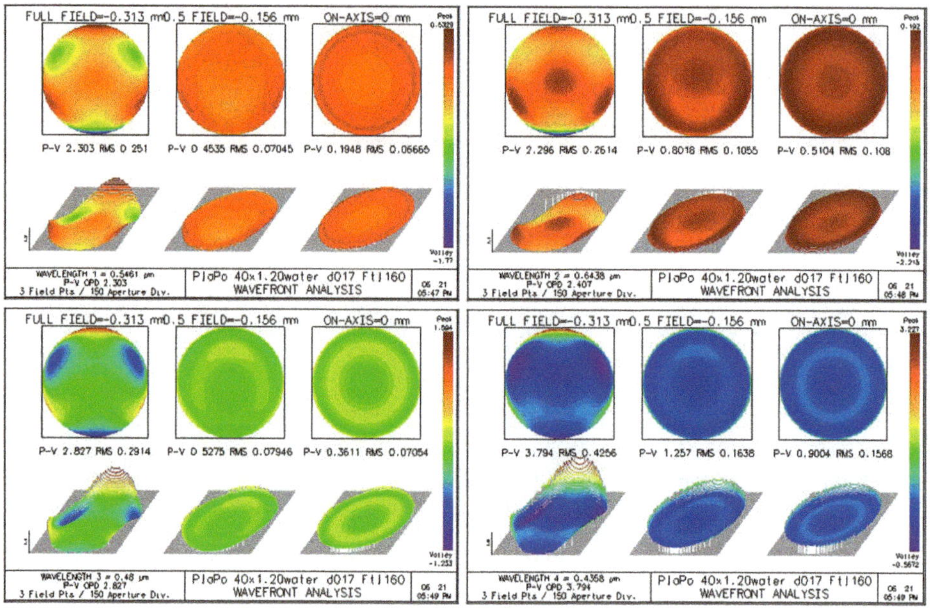

Figure 3.49 (*Continued*) Design parameters and graphs of aberration correction of a 40×/1.20 water immersion CCF plan apochromatic objective ($F'_{ts} = 160$ mm; $h = 60$ mm).

correction are shown in Fig. 3.7). Figures 3.76–3.79 are photographs of real objects taken with it.

Figure 3.80 shows a photograph of a 100×/1.40 oil immersion objective (the design parameters and graphs of aberration correction are shown in Fig. 3.66), and Figs. 3.81–3.85 are photographs of real objects taken with it.

The following photographs, taken by Anatoly Mikhaltsov using other plan apochromatic objectives, not only contain information about the objects being studied, but also have high artistic value (Figs. 3.86–3.90).

3.6.4 Unification of CCF plan apochromatic objectives for objects with and without a cover glass

Significant unification, or more precisely, reducing the *double* quantity of optical and mechanical parts used in objectives, can be achieved by using an original engineering solution. In old Soviet objectives, a plane–parallel plate was used for integration in the optical–mechanical system of an objective to eliminate, for example, scattering and spurious light in objectives with small linear magnifications. Such a mechanical design has also been known to be used for mounting a so-called *quarter-wave plate*, a polarizer, and other optical elements without their own optical power before the front lens of the objective. In the same way, it is possible to integrate a 0.17-mm-thick cover

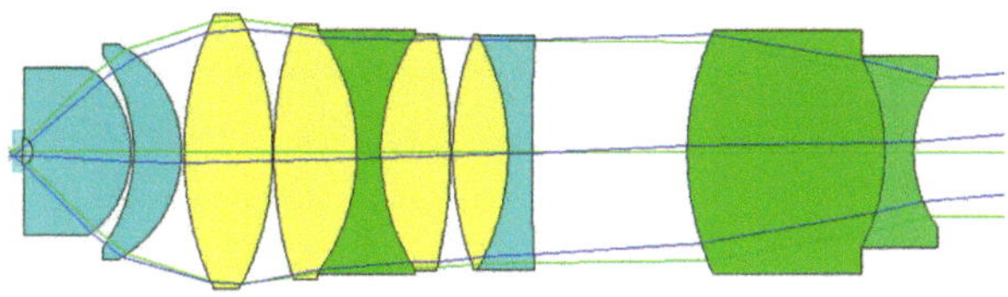

Lens: Plapo 50x1.0 oil Ftl160

Object num aper 1.000000 Image height 12.500000 Primary wavln 0.546070

SRF	RADIUS		THICKNESS		APERTURE RADIUS		GLASS		SPECIAL
OBJ	0.000000		1.0000e-20		0.250089		OIL		
AST	0.000000		0.480000		0.500000	A	OIL		
2	0.000000		0.500000		0.500000		K9		
3	-0.704700		5.000000		0.670000		LZ_CTK19		
4	-4.987500		0.150000		4.110000		AIR		
5	-7.818400		2.400000		4.700000		LZ_CTK19	P	
6	-6.004300		0.150000		5.300000		AIR		
7	15.823400		4.500000		6.740000		CAF2		
8	-14.437600		0.050000		6.790000		AIR		
9	20.100900		4.200000		6.300000		CAF2	P	
10	-10.094200		1.200000		6.030000		LZ_TF1		
11	10.507700		3.500000		5.770000		CAF2	P	
12	-24.584000		0.100000		5.830000		AIR		
13	15.839500		2.800000		5.790000		CAF2	P	
14	-11.761600		1.200000		5.770000		LZ_CTK19	P	
15	135.289800		7.822000		5.760000		AIR		
16	14.172000		10.000000		6.040000		LZ_TF5		
17	-9.737500		1.500000		4.700000		LZ_OF4		
18	6.254300		90.000000		3.740000		AIR		
19	214.800000		5.000000		10.570000		LZ_BK8		
20	-46.990000		0.300000		10.690000		AIR		
21	-46.880000		3.000000		10.660000		LZ_BF24		
22	-113.240000		157.601829	S	10.820000		AIR		
IMS	0.000000		0.000000		12.500000				

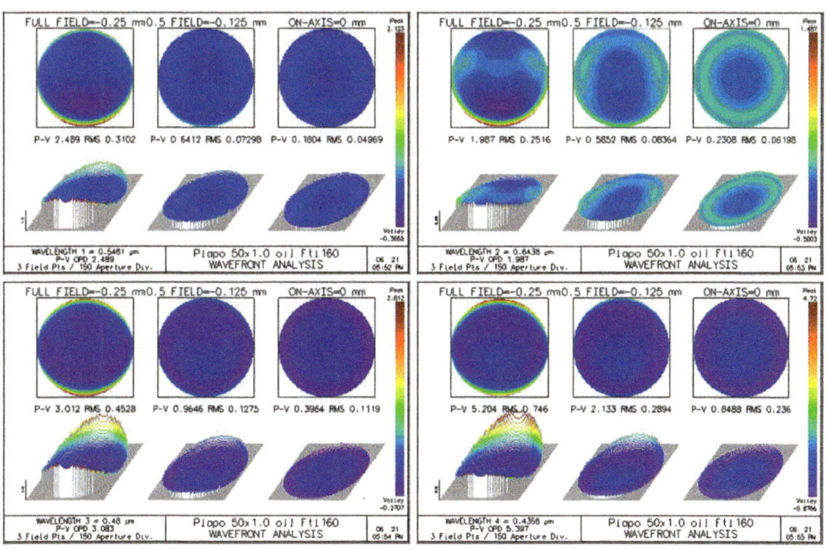

Figure 3.50 Design parameters and graphs of aberration correction of a 50×/1.0 oil immersion CCF plan apochromatic objective ($F'_{ts} = 160$ mm).

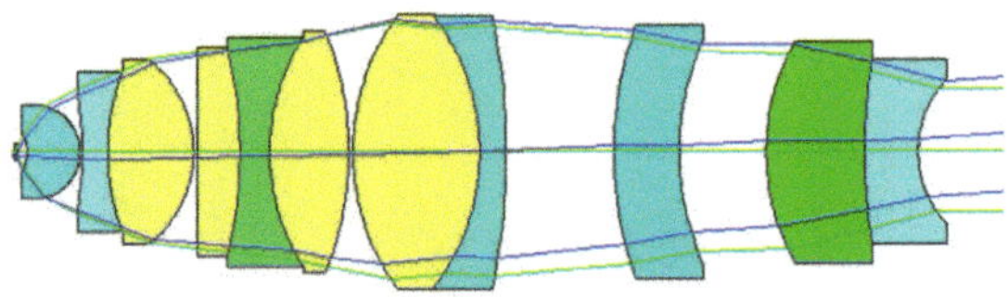

Lens: PlanApo 60x0.92 d017 Ft1180

Object num aper	0.920000	Image height	12.500000	Primary wavln	0.546070

SRF	RADIUS	THICKNESS	APERTURE RADIUS	GLASS	SPECIAL
OBJ	0.000000	0.010000	0.208327	AIR	
AST	0.000000	0.170000	0.230000 A	K9	
2	0.000000	0.450290	0.360000	AIR	
3	-1.519000	2.500000	0.830000	LZ_CTK12	
4	-2.073000	0.100000	2.060000	AIR	
5	-17.000000	1.200000	2.820000	H-K10	
6	10.000000	4.000000	3.590000	CAF2	
7	-5.082000	0.200000	4.130000	AIR	
8	0.000000	2.000000	4.520000	CAF2 P	
9	-18.923000	1.400000	4.680000	ZF1	
10	9.400000	3.600000	5.160000	CAF2 P	
11	-14.322000	0.200000	5.430000	AIR	
12	10.130000	5.900000	6.180000	CAF2 P	
13	-9.817000	1.200000	6.030000	LZ_CTK12 P	
14	-30.657000	5.000000	6.140000	AIR	
15	15.900000	3.000000	5.680000	H-ZLAF69	
16	12.210000	4.000000	5.050000	AIR	
17	9.900000	4.500000	4.990000	ZF6	
18	22.000000	2.500000	4.110000	H-QK3	
19	4.546000	180.000000	3.220000	AIR	
20	-32.490000	12.600000	15.730000	K9	
21	-39.204000	2.700000	18.480000	AIR	
22	339.390000	13.500000	19.110000	CAF2	
23	-73.350000	2.700000	19.510000	AIR	
24	199.170000	15.300000	19.110000	CAF2	
25	-64.800000	2.700000	18.240000	LZ_CTK12 P	
26	0.000000	180.005123 S	18.150000	AIR	
IMS	0.000000	0.000000	12.500000		

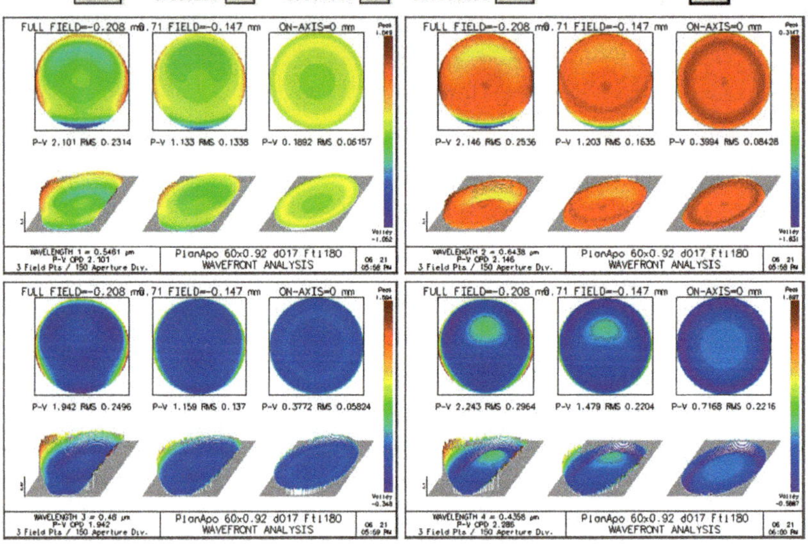

Figure 3.51 Design parameters and graphs of aberration correction of a 60×/0.92 CCF plan apochromatic objective ($F'_{ts} = 180$ mm).

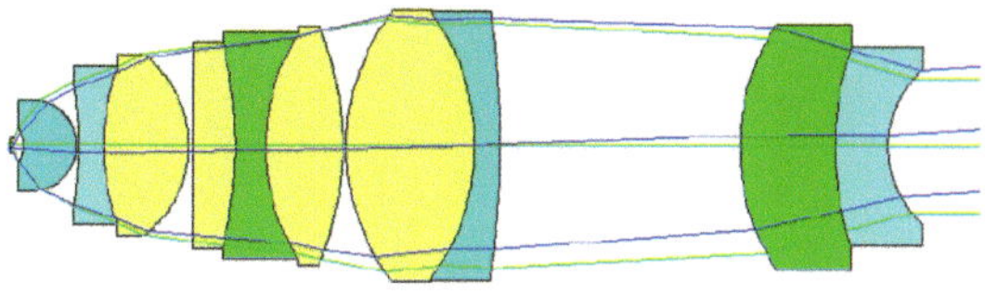

Lens: PlanApo 60x0.92 d=0.17 Ft1200

Object num aper 0.920000 Image height 14.000000 Primary wavln 0.546070

SRF	RADIUS	THICKNESS	APERTURE RADIUS	GLASS	SPECIAL
OBJ	0.000000	0.010000	0.233505	AIR	
AST	0.000000	0.170000	0.250000 A	K9	
2	0.000000	0.450200	0.380000	AIR	
3	-1.519000	2.500000	0.840000	LZ_CTK12	
4	-2.073000	0.100000	2.060000	AIR	
5	-17.000000	1.200000	2.820000	H-K10	
6	10.000000	4.000000	3.590000	CAF2	
7	-5.082000	0.200000	4.130000	AIR	
8	0.000000	2.000000	4.520000	CAF2 P	
9	-18.923000	1.400000	4.690000	ZF1	
10	9.400000	3.600000	5.170000	CAF2 P	
11	-14.322000	0.200000	5.450000	AIR	
12	10.000000	6.000000	6.230000	CAF2 P	
13	-10.000000	1.200000	6.060000	LZ_CTK12 P	
14	-43.330000	11.300000	6.160000	AIR	
15	9.738000	4.500000	5.650000	ZF6	
16	13.700000	2.500000	4.580000	H-QK3	
17	4.670000	200.000000	3.560000	AIR	
18	-36.100000	14.000000	18.000000	K9	
19	-43.560000	3.000000	21.150000	AIR	
20	377.100000	15.000000	21.910000	CAF2	
21	-81.500000	3.000000	22.360000	AIR	
22	221.300000	17.000000	21.890000	CAF2	
23	-72.000000	3.000000	20.920000	LZ_CTK12 P	
24	0.000000	200.106105 S	20.810000	AIR	
IMS	0.000000	0.000000	14.000000		

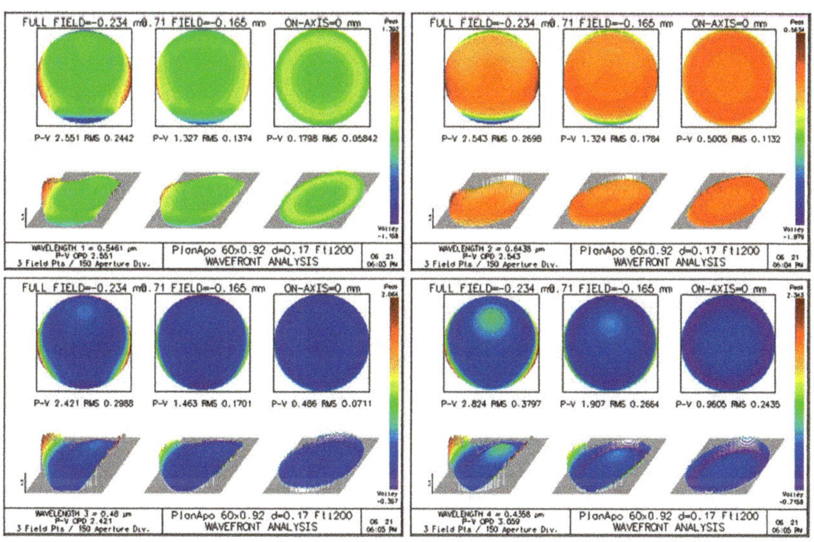

Figure 3.52 Design parameters and graphs of aberration correction of a 60×/0.92 CCF plan apochromatic objective ($F'_{ts} = 200$ mm).

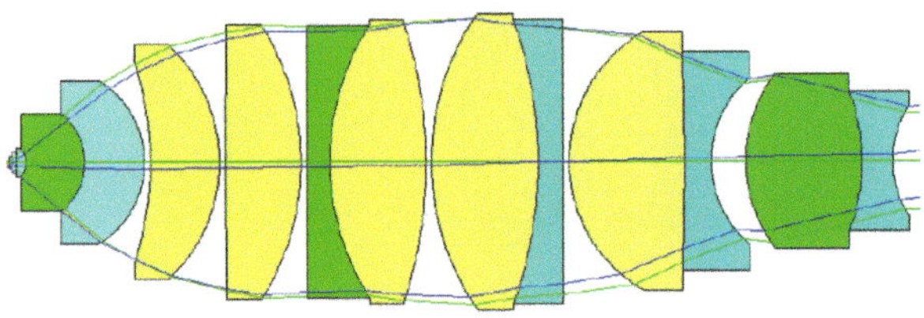

Lens: 60x1.0 oil tub160

Object num aper	1.000000	Image height	12.500000	Primary wavln	0.546070

SRF	RADIUS	THICKNESS	APERTURE RADIUS	GLASS	SPECIAL
OBJ	0.000000	0.010000	0.208491	OIL	
AST	0.000000	0.170000	0.170000 A	K9	
2	0.000000	0.200000	0.315000	OIL P	
3	0.000000	0.500000	0.500000	K9 P	
4	-1.050000	3.000000	0.700000	H-ZF6	
5	-3.133000	3.200000	2.490000	H-QK3	
6	-4.720000	0.300000	4.130000	AIR	
7	-16.200000	3.600000	5.130000	CAF2	
8	-8.050000	0.300000	5.960000	AIR	
9	438.300000	4.000000	6.710000	CAF2 P	
10	-12.750000	0.300000	7.015000	AIR	
11	0.000000	1.200000	6.950000	H-ZF6 P	
12	12.750000	5.000000	6.960000	CAF2 P	
13	-22.280000	0.300000	7.230000	AIR	
14	13.090000	5.600000	7.550000	CAF2 P	
15	-20.990000	1.300000	7.250000	LZ_CTK12	
16	0.000000	0.300000	7.070000	AIR	
17	7.551000	6.000000	6.590000	CAF2 P	
18	-176.710000	1.400000	5.570000	H-LAK3	
19	5.495000	1.800000	4.230000	AIR	
20	7.244000	6.000000	4.440000	H-ZF6 P	
21	-10.130000	1.600000	3.540000	H-LAK3 P	
22	5.105000	148.909808 S	2.790000	AIR	
IMS	0.000000	0.000000	12.500000		

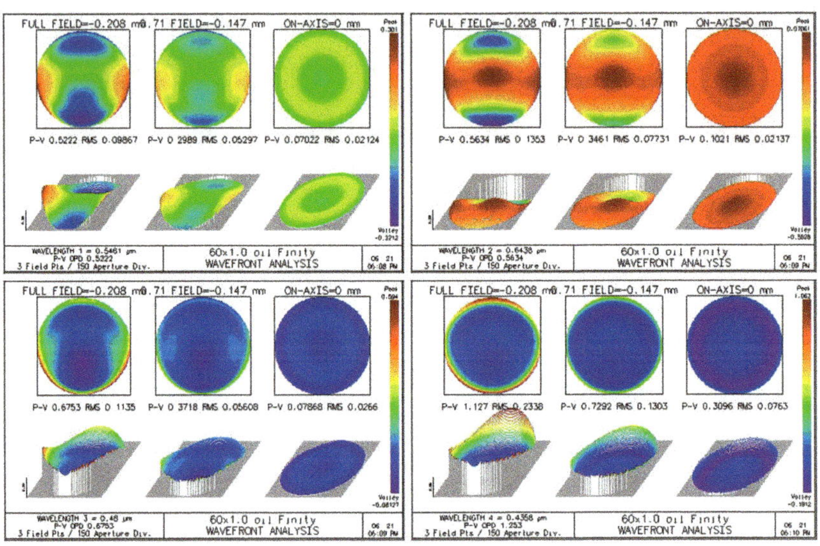

Figure 3.53 Design parameters and graphs of aberration correction of a 60×/1.0 oil immersion CCF plan apochromatic objective with a finite tube length of 160 mm.

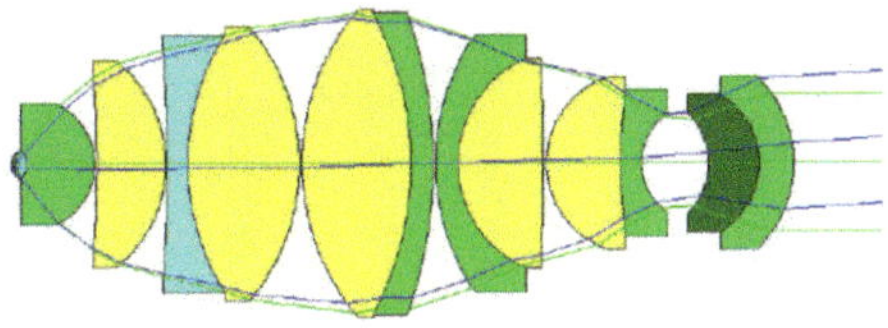

Lens: PlanApo 60x1.20oil Ft1200

Object num aper 1.200000 Image height 14.000000 Primary wavln 0.546070

SRF	RADIUS	THICKNESS	APERTURE RADIUS	GLASS	SPECIAL
OBJ	0.000000	0.010000	0.233238	OIL	
AST	0.000000	0.170000	0.240000 A	K9	
2	0.000000	0.150000	0.470000	OIL	
3	0.000000	0.500000	0.660000	K9	
4	-1.050000	4.000000	0.810000	ZLAF3	
5	-3.773000	0.120000	3.480000	AIR	
6	-64.450000	4.100000	5.230000	CAF2	
7	-7.420000	0.120000	5.930000	AIR	
8	-67.000000	1.200000	6.660000	ZK1	
9	13.550000	6.600000	7.430000	CAF2 P	
10	-11.970000	0.150000	7.880000	AIR	
11	13.550000	6.400000	8.850000	CAF2 P	
12	-17.950000	1.300000	8.720000	ZLAF3 P	
13	-25.500000	0.150000	8.750000	AIR	
14	12.953000	1.300000	7.470000	ZLAF3 P	
15	6.532000	5.000000	6.040000	CAF2	
16	-148.000000	0.200000	5.910000	AIR	
17	5.187000	4.400000	4.980000	CAF2	
18	36.000000	1.200000	4.320000	H-ZLAF3	
19	3.500000	4.000000	2.920000	AIR	
20	-3.820000	3.000000	2.940000	ZF62	
21	-4.546000	2.000000	4.000000	ZLAF3 P	
22	-6.761000	200.000000	5.070000	AIR	
23	-36.100000	14.000000	18.990000	K9	
24	-43.560000	3.000000	22.290000	AIR	
25	377.100000	15.000000	23.090000	CAF2	
26	-81.500000	3.000000	23.500000	AIR	
27	221.300000	17.000000	22.890000	CAF2	
28	-72.000000	3.000000	21.810000	LZ_CTK12	
29	0.000000	199.677565 S	21.670000	AIR	
IMS	0.000000	0.000000	14.000000		

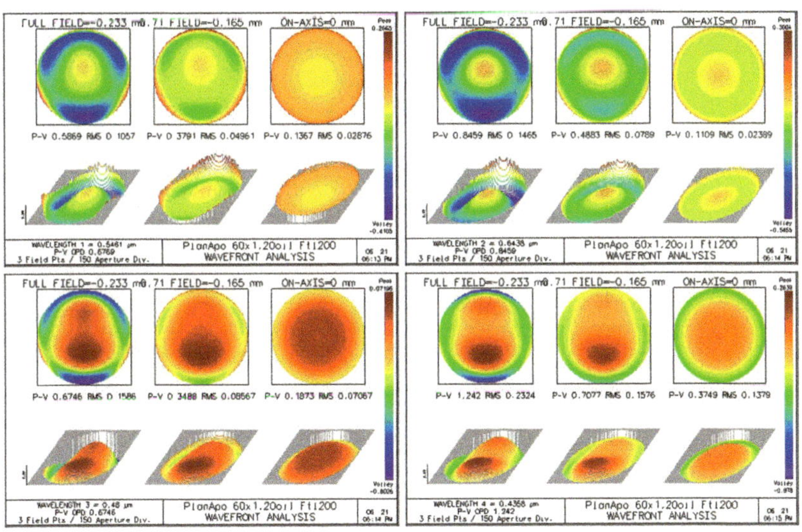

Figure 3.54 Design parameters and graphs of aberration correction of a 60×/1.20 oil immersion CCF plan apochromatic objective ($F'_{ts} = 200$ mm).

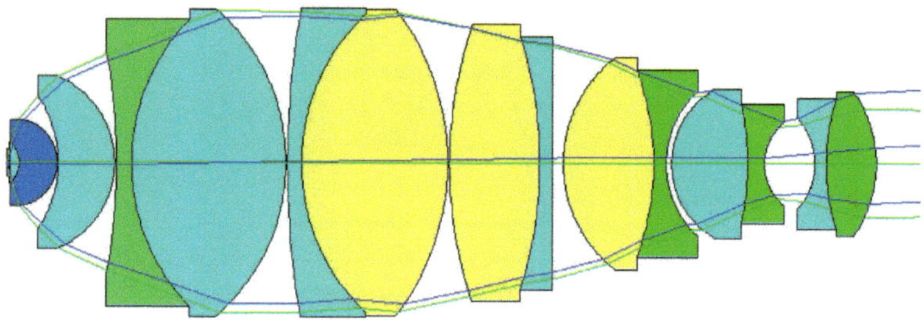

Lens:S-PlanApo 60x1.49 oil h75 Ftl200

Object num aper 1.490000 Image height 11.000000 Primary wavln 0.546070

SRF	RADIUS	THICKNESS		APERTURE RADIUS		GLASS		SPECIAL
OBJ	0.000000	0.010000		0.182809		OIL		
AST	0.000000	0.170000		0.240000	A	K9		
2	0.000000	0.100000		1.180000		OIL		
3	0.000000	0.880000		1.760000		BAK2		
4	-2.250000	3.550000		1.780000		S-LAH79		
5	-3.720000	0.100000		3.720000		AIR		
6	-9.950000	5.000000		5.820000		S-FPM2		
7	-8.000000	0.170000		7.460000		AIR		
8	-53.270000	1.300000		9.630000		H-F4A		
9	17.050000	14.000000		12.360000		S-FPM2	P	
10	-17.050000	0.150000		13.300000		AIR		
11	74.050000	1.300000		13.330000		LZ_CTK12		
12	17.430000	13.000000		12.920000		CAF2		
13	-21.180000	0.150000		13.230000		AIR		
14	36.130000	8.000000		11.950000		CAF2	P	
15	-36.130000	1.100000		10.940000		LZ_CTK12	P	
16	0.000000	1.000000		10.540000		AIR		
17	11.460000	8.000000		9.160000		CAF2	P	
18	-22.840000	1.100000		8.100000		H-ZF1		
19	8.365000	0.500000		6.280000		AIR		
20	7.330000	6.800000		6.410000		H-LAK4L		
21	-23.920000	1.500000		5.140000		H-ZF52		
22	4.670000	4.500000		3.620000		AIR		
23	-5.750000	1.000000		3.930000		S-PHM53		
24	18.000000	4.500000		5.690000		H-ZF52	P	
25	-10.516000	200.000000		6.220000		AIR		
26	-36.100000	14.000000		15.840000		K9	P	
27	-43.560000	3.000000		18.470000		AIR		
28	377.100000	15.000000		18.910000		CAF2	P	
29	-81.500000	3.000000		19.220000		AIR		
30	221.300000	17.000000		18.780000		CAF2	P	
31	-72.000000	3.000000		17.730000		LZ_CTK12	P	
32	0.000000	199.883672	S	17.610000		AIR		
IMS	0.000000	0.000000		11.000000				

Figure 3.55 Design parameters and graphs of aberration correction of a 60×/1.49 oil immersion CCF S-plan apochromatic objective ($F'_{ts} = 200$ mm).

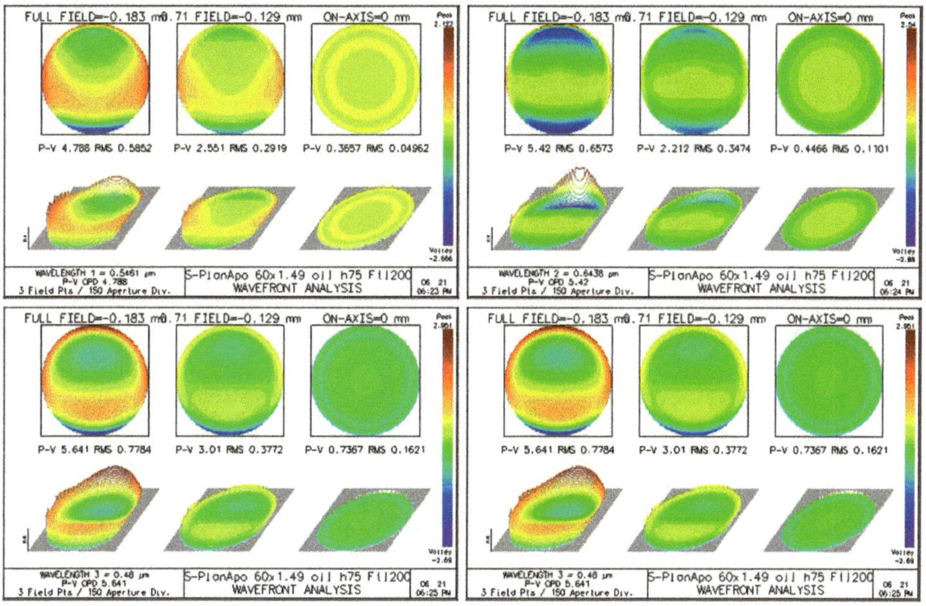

Figure 3.55 (*Continued*) Design parameters and graphs of aberration correction of a 60×/1.49 oil immersion CCF S-plan apochromatic objective ($F'_{ts} = 200$ mm).

glass into the optical–mechanical system of an objective. Figure 3.91 demonstrates the principle of such an engineering solution.

It turns out that if we design an objective to be used for biological research with a standard coverslip, we can use a mechanical design method (by integrating the coverslip into the objective design) to adapt the objective to work with objects without a coverslip, i.e., to be used for metallographic research. This engineering solution can actually work. However, we should discuss whether it is advisable to use it in real microscopic conditions. First, it makes sense to develop a more complex, universal mechanical design of the lens if the numerical aperture of the objective is (conditionally) more than 0.60. The objectives shown in Chapter 1, Fig. 1.189, work equally well for objects with and without a coverslip. Second, the use of such an engineering solution seems appropriate for small series objectives, i.e., plan MicroFluars and plan apochromats, which are expensive and have optical parts that are highly complex to manufacture. If such objectives are intended for use in metallography, their linear magnifications should not be 40× or 60×, as the so-called *scale estimate* methods include the use of a 50× objective, which is not used in biology. Therefore, the maximum benefit from such an engineering solution can be for 20× and 100× objectives that work without immersion. However, for these objectives, integrating the coverslip using the method shown in Fig. 3.91 is a big problem. For a 20× objective, the diameter of the

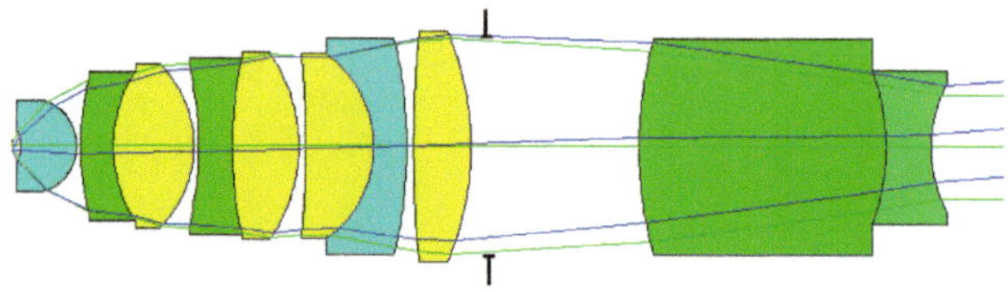

Lens: PlanApo 63x0.90 d=0 Ftl160 POL

Object num aper 0.900000 Image height 12.500000 Primary wavln 0.546070

SRF	RADIUS	THICKNESS	APERTURE RADIUS	GLASS	SPECIAL
OBJ	0.000000	-1.0000e-08	0.195586	AIR	
AST	0.000000	0.400000	0.192000 A	AIR	
2	-1.318300	2.500000	0.660000	LZ_CTK19	
3	-2.051000	0.200000	1.980000	AIR	
4	11.015000	1.500000	2.960000	LZ_TF1	
5	5.834000	3.600000	3.240000	LZ_FFS2	
6	-4.966000	0.200000	3.560000	AIR	
7	-15.524000	1.500000	3.580000	LZ_TF1 P	
8	17.298000	3.000000	3.850000	LZ_FFS2 P	
9	-6.982000	0.300000	4.080000	AIR	
10	-45.810000	3.000000	4.020000	LZ_FFS2 P	
11	-4.920000	1.500000	4.030000	LZ_CTK19 P	
12	-19.055000	0.300000	4.690000	AIR	
13	48.190000	2.600000	4.920000	LZ_FFS2 P	
14	-12.735000	7.520000	5.050000	AIR	
15	17.298000	11.000000	4.730000	LZ_TF4	
16	-9.419000	2.000000	3.370000	LZ_OF4	
17	5.675000	90.000000	2.790000	AIR	
18	-28.880000	11.200000	9.620000	LZ_K8	
19	-34.848000	2.400000	11.470000	AIR	
20	301.680000	12.000000	11.830000	CAF2	
21	-65.200000	2.400000	12.300000	AIR	
22	177.040000	13.600000	12.220000	CAF2 P	
23	-57.600000	2.400000	11.810000	LZ_CTK12	
24	0.000000	159.975525 S	11.810000	AIR	
IMS	0.000000	0.000000	12.500000		

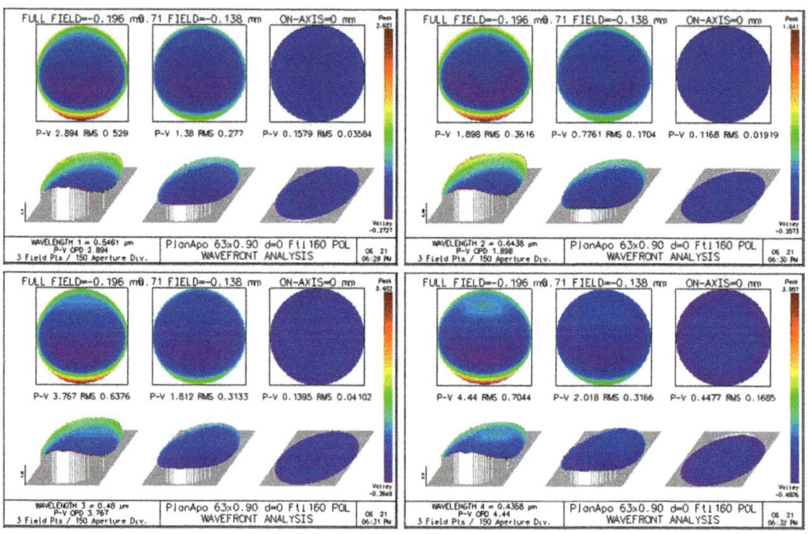

Figure 3.56 Design parameters and graphs of aberration correction of a 63×/0.90 CCF plan apochromatic objective for polarization research (F'_{ts} = 160 mm).

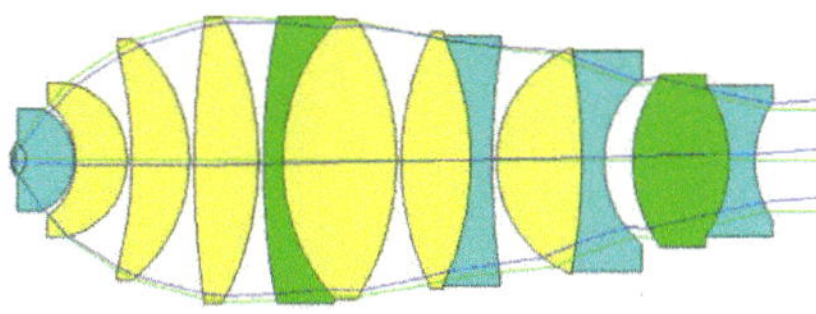

Lens: Plapo 63x1.20 water ftl=160

Object num aper 1.200000 Image height 12.500000 Primary wavln 0.546070

SRF	RADIUS	THICKNESS	APERTURE RADIUS	GLASS	SPECIAL
OBJ	0.000000	0.010000	0.199567	WATER	
AST	0.000000	0.170000	0.220000 A	LZ_K14	
2	0.000000	0.189530	0.440000	WATER P	
3	-4.365000	0.600000	0.710000	LZ_K8	
4	-1.096500	3.000000	0.930000	LZ_CTK19	
5	-3.353500	0.100000	3.090000	AIR	
6	-4.261800	3.200000	3.470000	CAF2	
7	-4.724800	0.300000	4.660000	AIR	
8	-24.359300	3.600000	6.770000	CAF2 P	
9	-9.073400	0.300000	7.260000	AIR	
10	57.809700	4.000000	8.500000	CAF2 P	
11	-17.842900	0.300000	8.680000	AIR	
12	43.211400	1.200000	8.660000	LZ_TF5	
13	12.552500	7.000000	8.350000	CAF2 P	
14	-15.609500	0.300000	8.460000	AIR	
15	17.208700	4.300000	7.720000	CAF2 P	
16	-17.431500	1.300000	7.520000	LZ_CTK19 P	
17	46.768100	0.300000	7.110000	AIR	
18	7.375000	5.200000	6.760000	CAF2 P	
19	-39.629800	1.400000	6.650000	LZ_CTK19 P	
20	6.515200	1.800000	5.000000	AIR	
21	8.654700	6.000000	5.180000	LZ_TF5 P	
22	-7.775800	1.600000	4.550000	LZ_CTK19 P	
23	6.186900	90.000000	3.530000	AIR	
24	214.800000	5.000000	11.100000	LZ-BK8	
25	-46.880000	0.300000	11.230000	AIR	
26	-46.990000	3.000000	11.200000	LZ-BF24	
27	-113.240000	157.814856 S	11.380000	AIR	
IMS	0.000000	0.000000	12.500000		

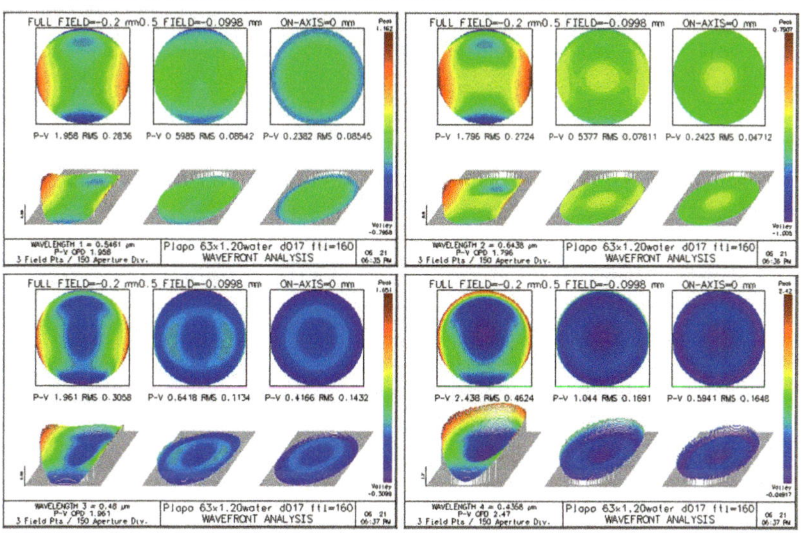

Figure 3.57 Design parameters and graphs of aberration correction of a 63×/1.20 water immersion CCF plan apochromatic objective ($F'_{ts} = 160$ mm).

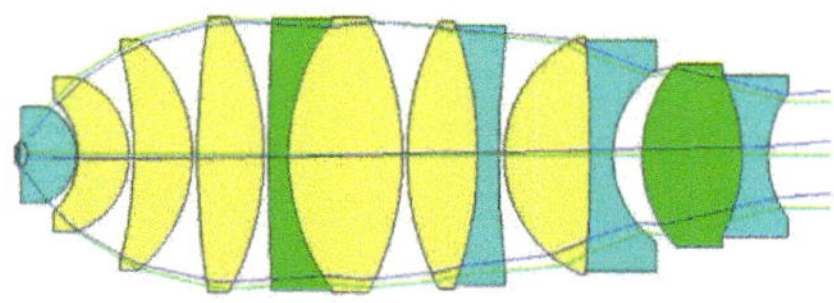

Lens: Plapo 63x1.25 oil f160

Object num aper	1.250000	Image height	12.500000	Primary wavln	0.546070

SRF	RADIUS		THICKNESS		APERTURE RADIUS		GLASS		SPECIAL
OBJ	0.000000		0.010000		0.198425		OIL		
AST	0.000000		0.170000		0.210000	A	LZ_K14		
2	0.000000		0.120000		0.460000		OIL		
3	0.000000		0.500000		0.640000		LZ_K8		
4	-1.096500		3.000000		0.820000		LZ_CTK19		
5	-3.080300		0.100000		2.860000		AIR		
6	-4.597100		3.200000		3.390000		CAF2		
7	-4.777900		0.300000		4.600000		AIR		
8	-23.883600		3.600000		6.320000		CAF2	P	
9	-8.735200		0.300000		6.860000		AIR		
10	49.863800		4.000000		8.040000		CAF2	P	
11	-16.745400		0.300000		8.200000		AIR		
12	146.768200		1.200000		8.100000		LZ_TF5		
13	13.125000		7.000000		7.930000		CAF2	P	
14	-17.171000		0.300000		8.190000		AIR		
15	17.125900		4.300000		7.910000		CAF2	P	
16	-22.460700		1.300000		7.710000		LZ_CTK19	P	
17	74.302400		0.300000		7.430000		AIR		
18	7.675100		5.200000		7.050000		CAF2	P	
19	-92.750800		1.400000		6.880000		LZ_CTK19	P	
20	6.320000		1.800000		5.120000		AIR		
21	8.184900		6.000000		5.450000		LZ_TF5	P	
22	-10.314500		1.600000		4.780000		LZ_CTK19	P	
23	6.259700		90.000000		3.730000		AIR		
24	214.800000		5.000000		11.490000		LZ_BK8		
25	-46.990000		0.300000		11.610000		AIR		
26	-46.880000		3.000000		11.580000		LZ_BF28		
27	-113.240000		157.767110	S	11.770000		AIR		
IMS	0.000000		0.000000		12.500000				

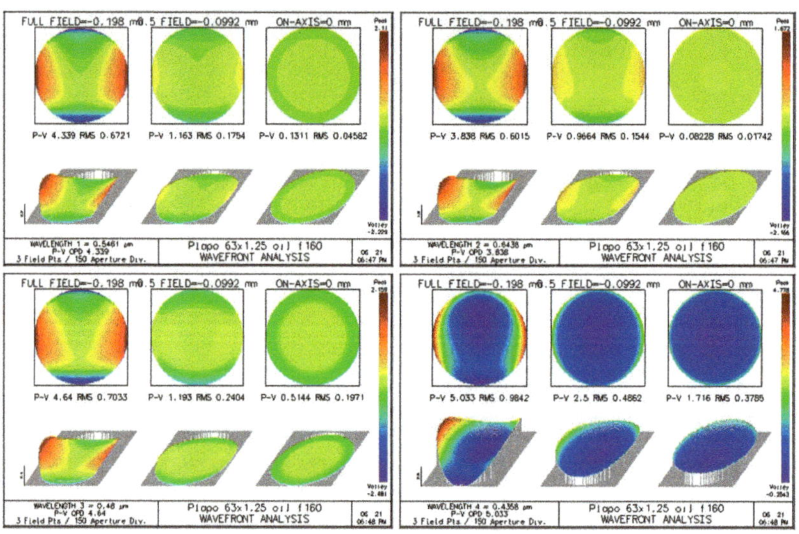

Figure 3.58 Design parameters and graphs of aberration correction of a 63×/1.25 oil immersion CCF plan apochromatic objective ($F'_{ts} = 160$ mm).

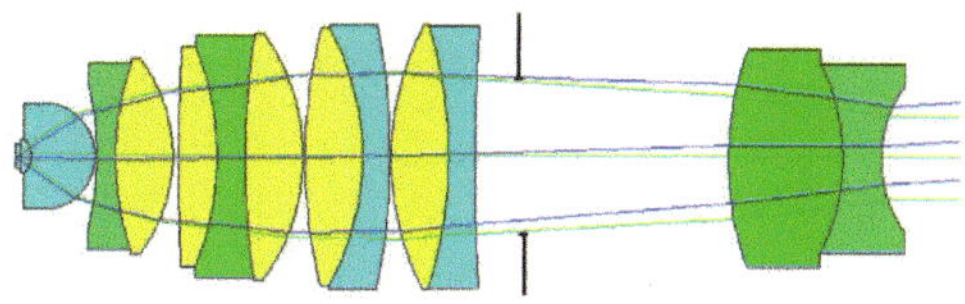

Lens: PlanApo 63x1.25 oil Ftl160

Object num aper 1.250000 Image height 10.000000 Primary wavln 0.546070

SRF	RADIUS		THICKNESS		APERTURE RADIUS		GLASS		SPECIAL
OBJ	0.000000		-1.0000e-08		0.155791		OIL		
AST	0.000000		0.310000		0.154000	A	OIL	P	
2	0.000000		0.550000		0.610000		LZ_K8		
3	-1.137600		3.350000		0.850000		LZ_CTK19		
4	-2.704000		0.100000		2.690000		AIR		
5	-15.596000		1.000000		3.740000		LZ_OF4		
6	14.555000		3.000000		4.730000		LZ_CTK3		
7	-8.017000		0.300000		4.950000		AIR		
8	100.000000		2.000000		5.470000		LZ_FFS2		
9	-14.622000		1.500000		5.580000		LZ_TF4		
10	41.590000		3.000000		6.180000		LZ_FFS2	P	
11	-10.544000		0.100000		6.300000		AIR		
12	26.550000		3.000000		6.560000		LZ_FFS2	P	
13	-13.305000		1.400000		6.550000		LZ_CTK19	P	
14	-35.650000		0.100000		6.750000		AIR		
15	14.521000		3.000000		6.750000		LZ_FFS2	P	
16	-21.730000		1.400000		6.710000		LZ_CTK19	P	
17	109.650000		13.330000		6.530000		AIR		
18	15.740000		6.000000		5.550000		LZ_TF4		
19	-10.423000		2.000000		4.810000		LZ_OF4		
20	6.223000		90.000000		3.710000		AIR		
21	-28.880000		11.200000		8.890000		LZ_K8		
22	-34.848000		2.400000		10.510000		AIR		
23	301.680000		12.000000		10.780000		CAF2		
24	-65.200000		2.400000		11.130000		AIR		
25	177.040000		13.600000		11.030000		CAF2	P	
26	-57.600000		2.400000		10.560000		LZ_CTK12		
27	0.000000		159.952242	S	10.550000		AIR		
IMS	0.000000		0.000000		10.000000				

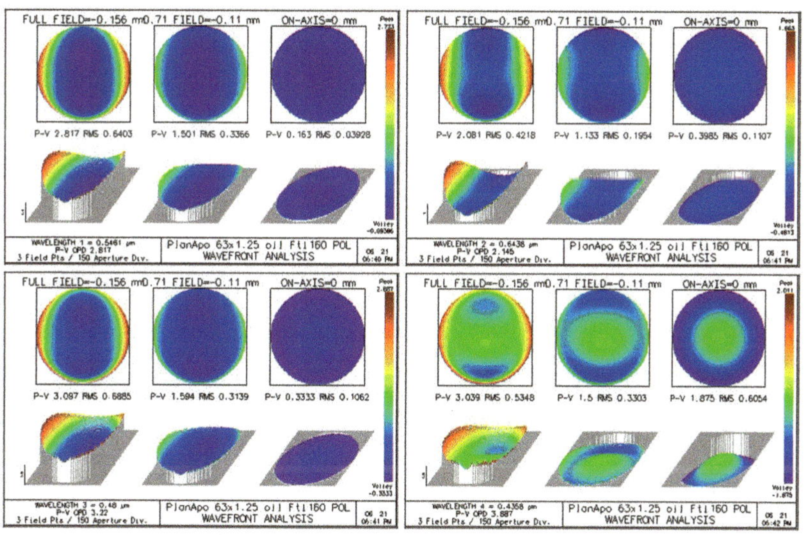

Figure 3.59 Design parameters and graphs of aberration correction of a 63×/1.25 oil immersion CCF plan apochromatic objective for polarization research ($F'_{ts} = 160$ mm).

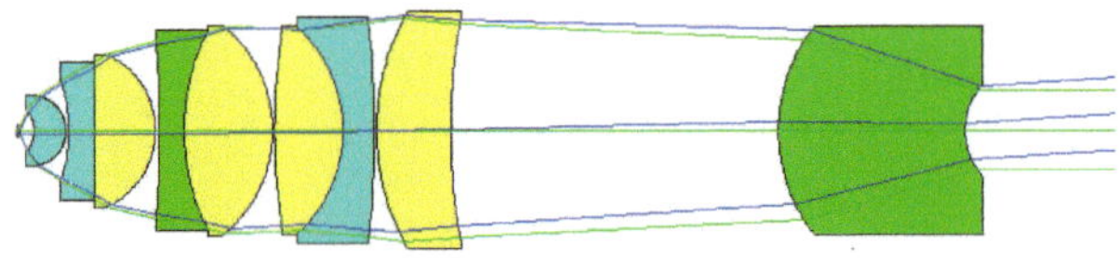

Lens: PlanApo 100x0.93 d=0.17 ft1200

Object num aper 0.930000 Image height 14.000000 Primary wavln 0.546070

SRF	RADIUS	THICKNESS	APERTURE RADIUS	GLASS	SPECIAL
OBJ	0.000000	0.010000	0.140038	AIR	
AST	0.000000	0.170000	0.160000 A	K9	
2	0.000000	0.650100	0.300000	AIR	
3	-1.500000	1.520000	0.990000	H-LAK3	
4	-1.670000	0.150000	1.660000	AIR	
5	-7.700000	1.300000	2.430000	H-LAK3 P	
6	0.000000	3.000000	3.270000	CAF2	
7	-3.900000	0.200000	3.620000	AIR	
8	-53.600000	1.200000	4.270000	ZF2	
9	10.189000	4.300000	4.760000	CAF2 P	
10	-6.281000	0.100000	4.980000	AIR	
11	32.000000	3.400000	4.960000	CAF2 P	
12	-6.399000	1.500000	4.930000	LZ_CTK12	
13	-49.200000	0.100000	5.390000	AIR	
14	11.220000	3.600000	5.670000	CAF2 P	
15	29.000000	16.000000	5.470000	AIR	
16	7.900000	9.000000	5.030000	ZF6	
17	3.200000	200.000000	2.200000	AIR	
18	-36.100000	14.000000	15.150000	K9	
19	-43.560000	3.000000	18.220000	AIR	
20	377.100000	15.000000	18.780000	CAF2	
21	-81.500000	3.000000	19.240000	AIR	
22	221.300000	17.000000	18.920000	CAF2	
23	-72.000000	3.000000	18.050000	LZ_CTK12 P	
24	0.000000	199.880059 S	17.980000	AIR	
IMS	0.000000	0.000000	14.000000		

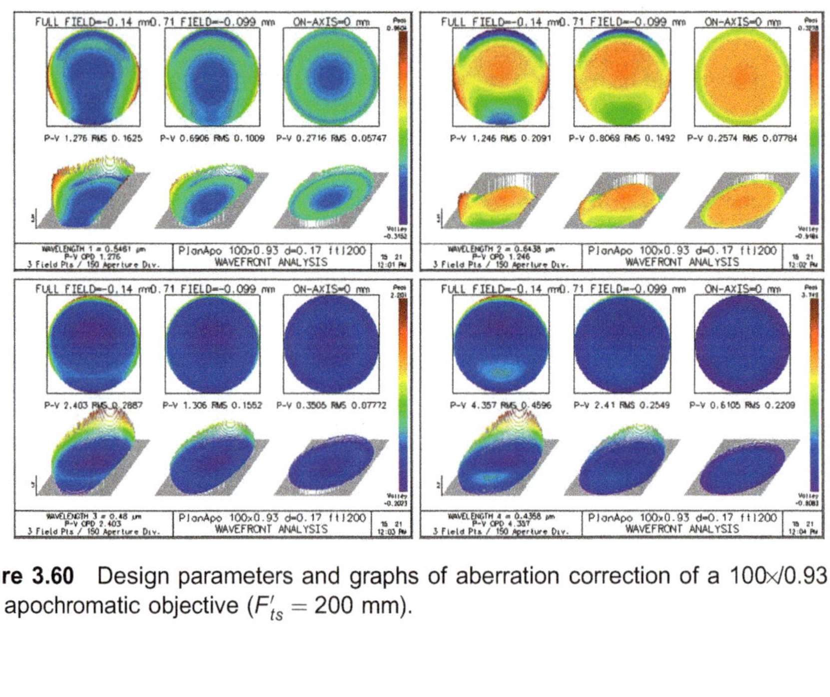

Figure 3.60 Design parameters and graphs of aberration correction of a 100×/0.93 CCF plan apochromatic objective ($F'_{ts} = 200$ mm).

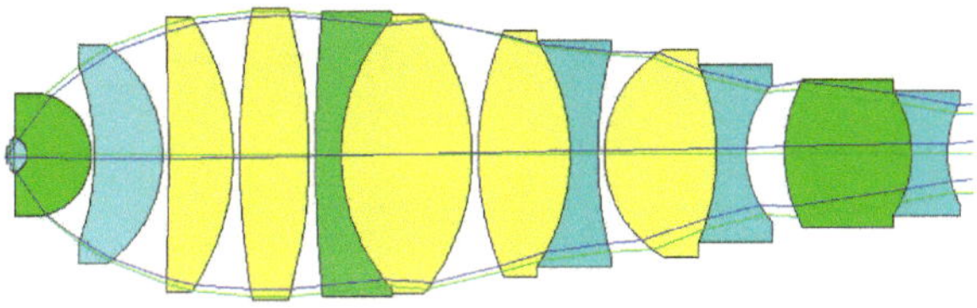

Lens: 100x1.20 water tub160

Object num aper 1.200000 Image height 12.500000 Primary wavln 0.546070

SRF	RADIUS	THICKNESS	APERTURE RADIUS	GLASS	SPECIAL
OBJ	0.000000	0.010000	0.125114	WATER	
AST	0.000000	0.170000	0.150000 A	K9	
2	0.000000	0.190000	0.370000	WATER P	
3	-4.365000	0.600000	0.660000	K9	
4	-0.850000	3.000000	0.800000	H-ZF6	
5	-2.816000	0.150000	2.770000	AIR	
6	-11.508000	3.200000	4.090000	H-QK3	
7	-5.998000	0.300000	4.950000	AIR	
8	-143.940000	3.000000	5.960000	CAF2	
9	-10.920000	0.300000	6.260000	AIR	
10	37.742000	3.200000	6.580000	CAF2 P	
11	-23.230000	0.300000	6.630000	AIR	
12	63.300000	1.200000	6.480000	H-ZF7LA	
13	9.330000	6.000000	6.180000	CAF2 P	
14	-10.233000	0.300000	6.330000	AIR	
15	13.550000	4.300000	5.580000	CAF2 P	
16	-9.330000	1.300000	5.150000	LZ_CTK12	
17	19.724000	0.300000	4.780000	AIR	
18	5.453000	5.200000	4.740000	CAF2 P	
19	-10.233000	1.400000	4.050000	LZ_CTK12 P	
20	4.600000	1.800000	3.130000	AIR	
21	7.170000	5.900000	3.380000	H-ZF6 P	
22	-4.897000	1.700000	2.830000	LZ_CTK12 P	
23	4.897000	151.125484 S	2.300000	AIR	
IMS	0.000000	0.000000	12.500000		

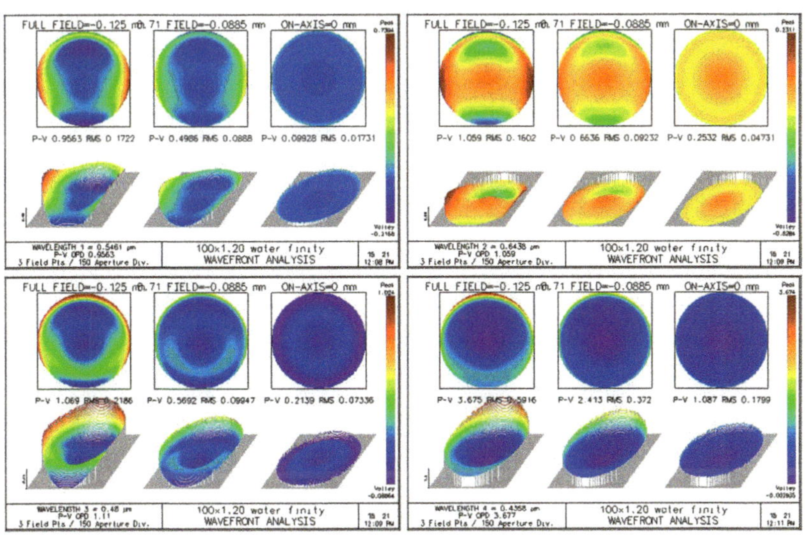

Figure 3.61 Design parameters and graphs of aberration correction of a 100×/1.20 water immersion CCF plan apochromatic objective with a finite tube length of 160 mm.

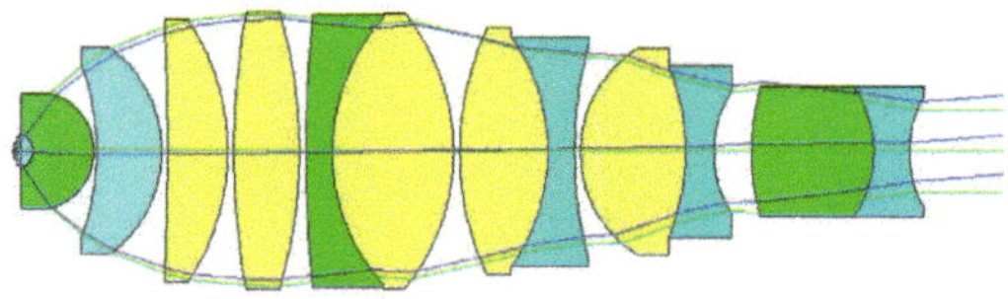

Lens: Plapo 100x1.20 water Ftl=164

Object num aper 1.200000 Image height 12.500000 Primary wavln 0.546070

SRF	RADIUS	THICKNESS		APERTURE RADIUS		GLASS		SPECIAL
OBJ	0.000000	0.010000		0.124322		WATER		
AST	0.000000	0.170000		0.150000	A	K9		
2	0.000000	0.188500		0.370000		WATER	P	
3	-4.365000	0.600000		0.650000		K9		
4	-0.850000	3.000000		0.800000		H-ZF6		
5	-2.816000	0.150000		2.760000		AIR		
6	-11.508000	3.200000		4.090000		H-QK3		
7	-5.998000	0.300000		4.950000		AIR		
8	-143.940000	3.000000		5.960000		CAF2		
9	-10.920000	0.300000		6.260000		AIR		
10	37.742000	3.200000		6.580000		CAF2	P	
11	-23.230000	0.300000		6.630000		AIR		
12	63.300000	1.300000		6.530000		H-ZF7LA		
13	9.330000	6.000000		6.380000		CAF2	P	
14	-10.233000	0.300000		6.560000		AIR		
15	13.550000	4.300000		5.910000		CAF2	P	
16	-9.330000	1.300000		5.470000		LZ_CTK12		
17	19.724000	0.300000		5.300000		AIR		
18	5.702000	5.190000		4.920000		CAF2	P	
19	-10.233000	1.400000		4.090000		LZ_CTK12	P	
20	5.495000	1.800000		2.940000		AIR		
21	10.233000	5.960000		3.000000		H-ZF7LA	P	
22	-5.557000	1.750000		3.110000		LZ_CTK12	P	
23	5.401000	164.000000		2.670000		AIR		
24	220.328800	5.128700		15.440000		LZ-BK8		
25	-48.199500	0.307700		15.500000		AIR		
26	-48.086700	3.077200		15.450000		LZ-BF24		
27	-116.154700	161.702893	S	15.650000		AIR		
IMS	0.000000	0.000000		12.500000				

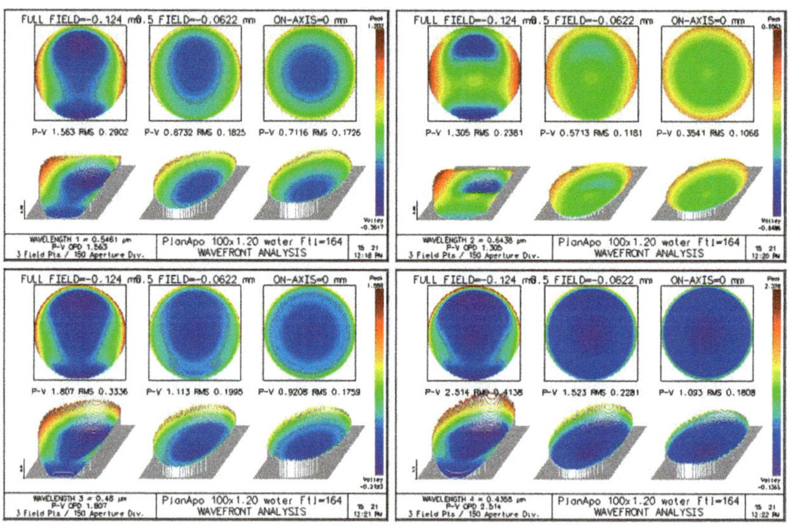

Figure 3.62 Design parameters and graphs of aberration correction of a 100×/1.20 water immersion CCF plan apochromatic objective ($F'_{ts} = 164$ mm).

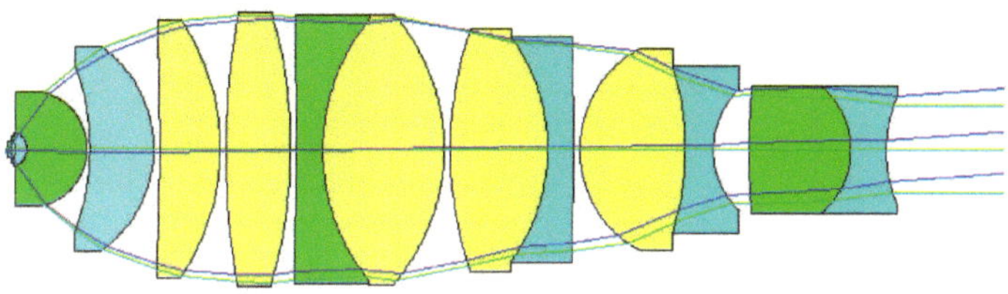

Lens: 100x1.20 water plapo tl=180

Object num aper 1.200000 Image height 14.000000 Primary wavln 0.546070

SRF	RADIUS	THICKNESS	APERTURE RADIUS	GLASS	SPECIAL
OBJ	0.000000	0.010000	0.139969	WATER	
AST	0.000000	0.170000	0.160000 A	K9	
2	0.000000	0.188000	0.380000	WATER P	
3	-4.365000	0.600000	0.670000	K9	
4	-0.850000	3.000000	0.800000	H-ZF6	
5	-2.816000	0.150000	2.770000	AIR	
6	-11.508000	3.200000	4.100000	H-QK3	
7	-5.998000	0.300000	4.950000	AIR	
8	-143.940000	3.000000	5.960000	CAF2	
9	-10.920000	0.300000	6.260000	AIR	
10	37.742000	3.200000	6.590000	CAF2 P	
11	-23.230000	0.300000	6.630000	AIR	
12	-214.370000	1.300000	6.530000	H-ZF6 P	
13	9.900000	6.000000	6.390000	CAF2 P	
14	-9.900000	0.300000	6.570000	AIR	
15	17.900000	4.800000	5.920000	CAF2 P	
16	-9.032000	1.300000	5.490000	LZ_CTK12	
17	-206.000000	0.300000	5.330000	AIR	
18	5.557000	5.200000	4.950000	CAF2 P	
19	-14.191000	1.400000	4.160000	LZ_CTK12 P	
20	4.000000	1.800000	2.980000	AIR	
21	44.700000	5.000000	3.050000	H-ZF6 P	
22	-4.093000	1.800000	3.170000	ZK1	
23	6.532000	180.000000	2.730000	AIR	
24	-32.490000	12.600000	17.820000	K9 P	
25	-39.204000	2.700000	20.990000	AIR	
26	339.390000	13.500000	21.900000	CAF2 P	
27	-73.350000	2.700000	22.350000	AIR	
28	199.170000	15.300000	21.820000	CAF2 P	
29	-64.800000	2.700000	20.940000	LZ_CTK12 P	
30	0.000000	179.783725 S	20.830000	AIR	
IMS	0.000000	0.000000	14.000000		

Figure 3.63 Design parameters and graphs of aberration correction of a 100×/1.20 water immersion CCF plan apochromatic objective ($F'_{ts} = 180$ mm).

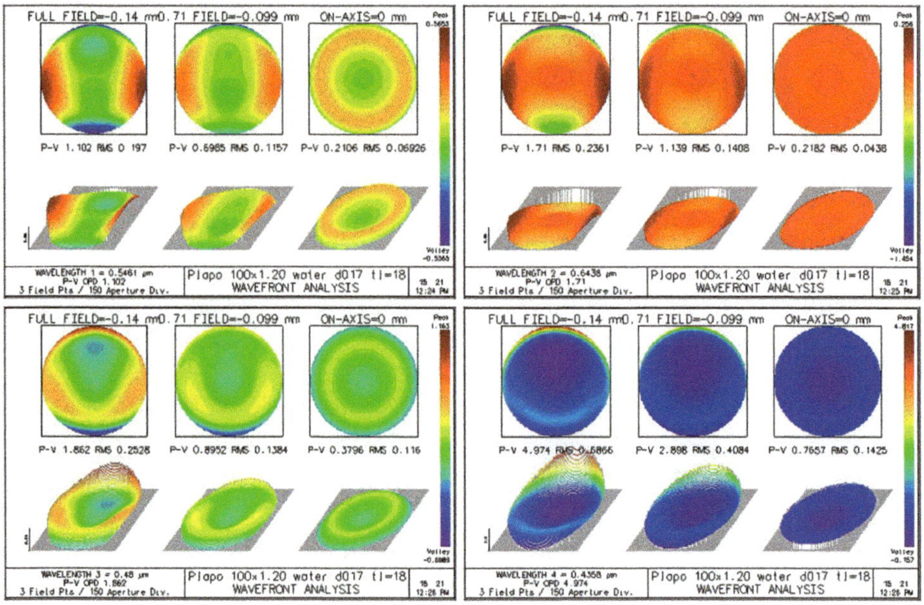

Figure 3.63 (*Continued*) Design parameters and graphs of aberration correction of a 100×/1.20 water immersion CCF plan apochromatic objective (F'_{ts} = 180 mm).

coverslip is significant with respect to its thickness, which makes the optical–mechanical component highly vulnerable to breakage. For a 100× objective, the free working distance is quite small, and the threaded connection of the optical–mechanical component cannot be stable in terms of the accuracy of positioning the coverslip with respect to the front lens of the objective. By the way, similar problems with integrating the coverslip with such an engineering method are also inherent in other high-magnification objectives, for example, 40× and 60×. Therefore, it makes sense to use such fundamental engineering, but with some adjustments. The optical–mechanical design will work much more reliably if the outer surface of the cover glass is integrated with the front lens of the objective as a single unit, as shown in Fig. 3.92 (some different designs are possible).

Of course, the cover glass in this case is not standard in diameter, but rather specially made; it can be connected to the front lens of the objective (which is also made in a special way) by bonding or optical contact. After that, such an optical component is mounted in the frame. In this case, when performing optical calculations of such a component, the amount of deflection of the first surface of the front lens of the objective is considered, and then the diameter of the cover glass integrated into the lens is selected. Thus, unified objectives for working with objects with and without a coverslip will be, as they say, factory-made; these will be two different objectives,

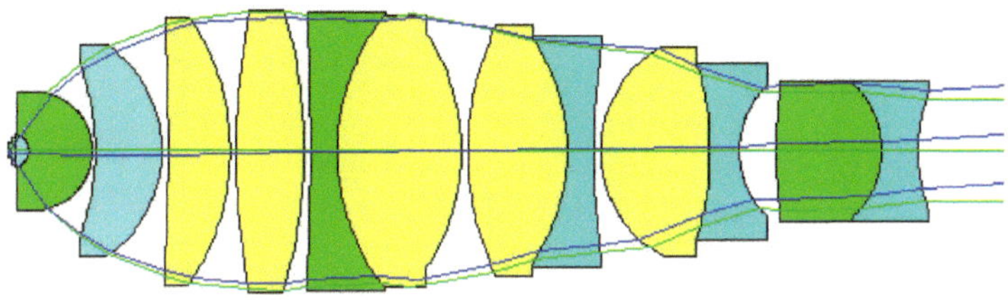

Lens:Plapo 100x1.20water d=0.17 tl200

Object num aper	1.200000	Image height	14.000000	Primary wavln	0.546070

SRF	RADIUS	THICKNESS	APERTURE RADIUS	GLASS	SPECIAL
OBJ	0.000000	0.010000	0.139945	WATER	
AST	0.000000	0.170000	0.160000 A	K9	
2	0.000000	0.188000	0.380000	WATER	
3	-4.365000	0.600000	0.670000	K9	
4	-0.850000	3.000000	0.800000	H-ZF6	
5	-2.816000	0.150000	2.770000	AIR	
6	-11.508000	3.200000	4.100000	H-QK3	
7	-5.998000	0.300000	4.950000	AIR	
8	-143.940000	3.000000	5.960000	CAF2	
9	-10.920000	0.300000	6.260000	AIR	
10	37.742000	3.200000	6.590000	CAF2	
11	-23.230000	0.300000	6.630000	AIR	
12	-214.370000	1.300000	6.530000	ZF6	
13	9.900000	6.000000	6.390000	CAF2	
14	-9.900000	0.300000	6.570000	AIR	
15	13.550000	4.800000	5.870000	CAF2	
16	-9.200000	1.300000	5.430000	LZ_CTK12	
17	42.270000	0.300000	5.140000	AIR	
18	5.540000	5.200000	4.960000	CAF2	
19	-13.550000	1.400000	4.240000	LZ_CTK12 P	
20	4.146000	1.800000	3.100000	AIR	
21	26.400000	5.000000	3.200000	ZF5	
22	-4.285000	1.800000	3.340000	ZK1	
23	8.336000	200.000000	2.980000	AIR	
24	-36.100000	14.000000	17.900000	K9	
25	-43.560000	3.000000	21.060000	AIR	
26	377.100000	15.000000	21.830000	CAF2	
27	-81.500000	3.000000	22.300000	AIR	
28	221.300000	17.000000	21.840000	CAF2	
29	-72.000000	3.000000	20.890000	LZ_CTK12 P	
30	0.000000	199.923939 S	20.790000	AIR	
IMS	0.000000	0.000000	14.000000		

Figure 3.64 Design parameters and graphs of aberration correction of a 100×/1.20 water immersion CCF plan apochromatic objective ($F'_{ts} = 200$ mm).

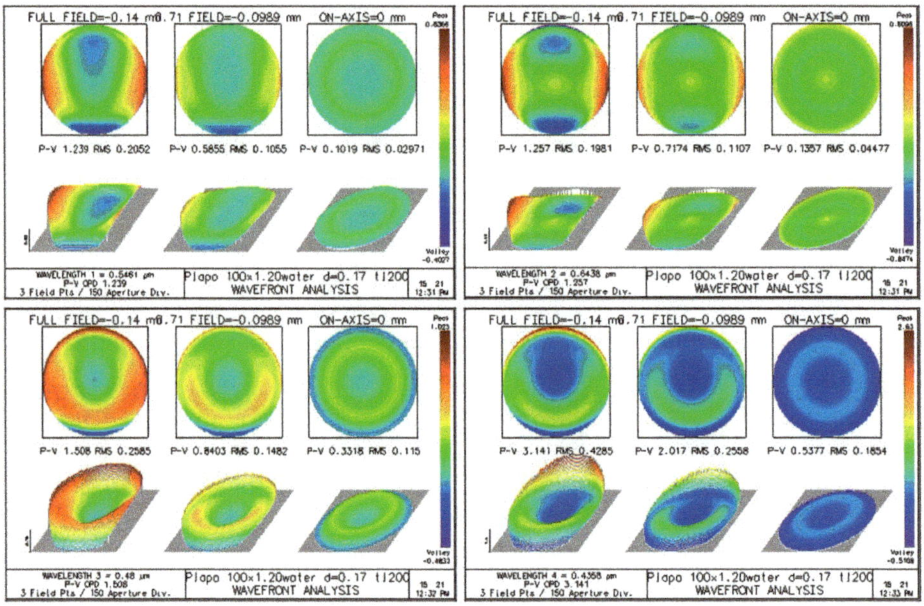

Figure 3.64 (*Continued*) Design parameters and graphs of aberration correction of a 100×/1.20 water immersion CCF plan apochromatic objective ($F'_{ts} = 200$ mm).

although a large number of optical and mechanical parts will be the same in them. Therefore, the optical design of the 100×/0.93 objective, for example, was first made to work with the coverslip (design parameters and graphs of aberration correction of this objective are shown in Fig. 3.60). Then, it can be easily transformed for work without a coverslip. Figure 3.93 shows this; the overall parameters of the objective, as well as the aberration correction, remain unchanged.

The free front working distance for an objective with a cover glass integrated in the design is approximately 0.28 mm, which is sufficient for stable operation for objects without a cover glass. For comparison, the free working distance of a 100×/0.95 objective is approximately 0.09 mm. The parameters of this objective are shown in Fig. 3.7, and photographs obtained with it are shown in Figs. 3.76–3.82.

3.6.5 Example drawings of plan apochromatic objective parts

Here are two example drawings of plan apochromatic objectives that show the typical requirements for manufacturing lenses, frames, and mounting lenses in frames. For optical design, the technological capabilities of the factory where the small series objectives (microscope) will be manufactured, the methods used for attaching lenses to frames, and the techniques for assembling and testing the finished products must be considered. This is very important

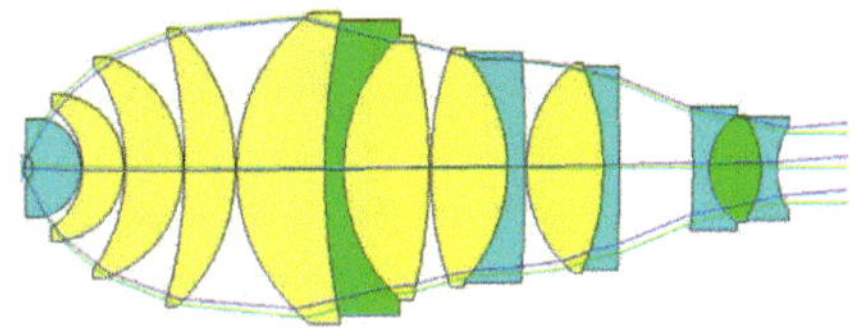

Lens: Plapo 100x1.35 oil Ftl160

Object num aper 1.350000 Image height 12.500000 Primary wavln 0.546070

SRF	RADIUS	THICKNESS	APERTURE RADIUS	GLASS	SPECIAL
OBJ	0.000000	1.0000e-20	0.124377	OIL	
AST	0.000000	0.279926	0.680000 A	OIL	
2	0.000000	0.500000	0.680000	LZ_K8	
3	-0.787000	3.000000	0.720000	LZ_CTK19	
4	-3.082800	0.100000	2.990000	AIR	
5	-4.444300	2.600000	3.680000	CAF2	
6	-4.500700	0.100000	4.470000	AIR	
7	-10.764700	3.600000	6.210000	CAF2 P	
8	-7.177900	0.100000	6.780000	AIR	
9	-30.795300	3.200000	8.270000	CAF2 P	
10	-11.759700	0.100000	8.520000	AIR	
11	12.111900	5.500000	9.520000	CAF2 P	
12	44.077900	1.200000	9.020000	LZ_TF5	
13	11.098600	5.300000	8.050000	CAF2 P	
14	-31.985100	0.100000	7.960000	AIR	
15	22.030200	4.800000	7.280000	CAF2 P	
16	-10.422500	1.100000	7.030000	LZ_CTK19	
17	-107.464200	0.100000	6.850000	AIR	
18	8.364800	4.800000	6.410000	CAF2 P	
19	-15.968300	1.000000	6.190000	LZ_CTK19 P	
20	-314.148700	4.600000	5.700000	AIR	
21	-54.466800	1.000000	3.760000	LZ_CTK19 P	
22	3.865400	3.300000	3.230000	LZ_TF5 P	
23	-5.486100	1.000000	3.120000	LZ_CTK19 P	
24	4.752100	90.000000	2.530000	AIR	
25	214.800000	5.000000	10.130000	LZ_BK8	
26	-46.990000	0.300000	10.270000	AIR	
27	-46.880000	3.000000	10.250000	LZ_BF24	
28	-113.240000	157.437922 S	10.430000	AIR	
IMS	0.000000	0.000000	12.500000		

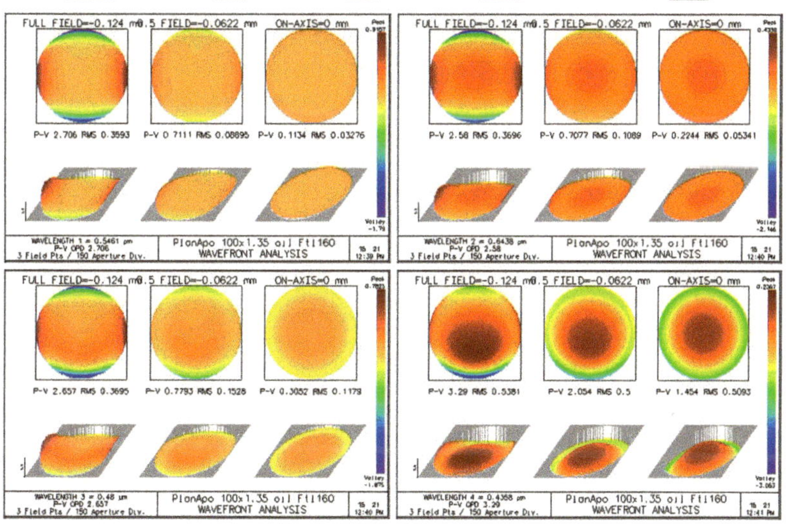

Figure 3.65 Design parameters and graphs of aberration correction of a 100×/1.35 oil immersion CCF plan apochromatic objective ($F'_{ts} = 160$ mm).

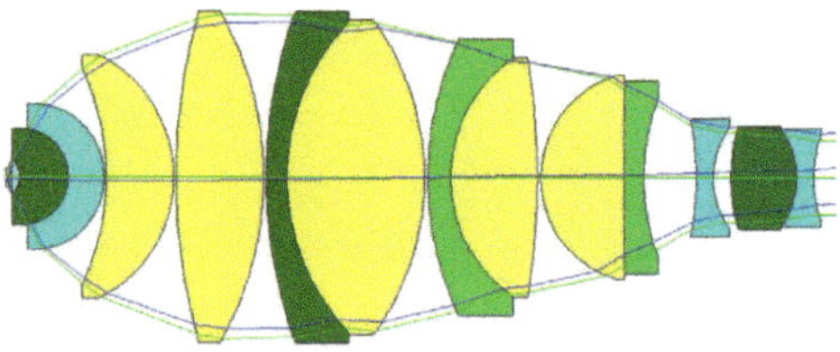

Lens: planApo 100x1.40 oil Finity Efl 1.539

Object num aper 1.400000 Image height 12.500000 Primary wavln 0.546070

SRF	RADIUS	THICKNESS	APERTURE RADIUS	GLASS	SPECIAL
OBJ	0.000000	0.010000	0.125612	OIL	
AST	0.000000	0.170000	0.120000 A	K9	
2	0.000000	0.100100	0.540000	OIL P	
3	0.000000	0.500000	0.780000	K9	
4	-1.300000	2.820000	0.960000	H-ZF62	
5	-2.710000	2.100000	2.690000	H-LAF10L	
6	-4.093000	0.120000	4.080000	AIR	
7	-13.900000	4.000000	6.150000	CAF2	
8	-7.503000	0.120000	6.860000	AIR	
9	33.700000	5.000000	9.130000	CAF2 P	
10	-18.923000	0.150000	9.340000	AIR	
11	26.410000	1.200000	9.320000	H-ZF62 P	
12	12.340000	8.000000	8.760000	CAF2 P	
13	-14.791000	0.200000	8.880000	AIR	
14	18.000000	1.300000	7.790000	H-ZLAF3	
15	8.128000	5.000000	6.740000	CAF2 P	
16	-41.980000	0.200000	6.680000	AIR	
17	5.960000	4.800000	5.750000	CAF2 P	
18	0.000000	1.100000	5.460000	H-ZLAF3 P	
19	12.340000	3.000000	4.470000	AIR	
20	-18.664000	1.000000	3.320000	H-LAK3	
21	4.760000	1.000000	2.840000	AIR	
22	10.000000	3.800000	2.920000	H-ZF62 P	
23	-4.720000	1.000000	2.760000	H-LAF10L P	
24	7.244000	147.640676 S	2.470000	AIR	
IMS	0.000000	0.000000	12.500000		

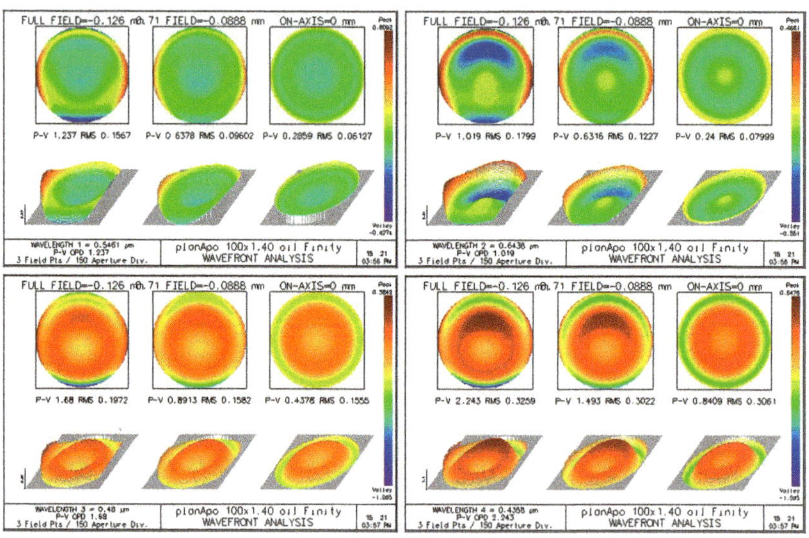

Figure 3.66 Design parameters and graphs of aberration correction of a 100×/1.40 oil CCF plan apochromatic objective with a finite tube length of 160 mm.

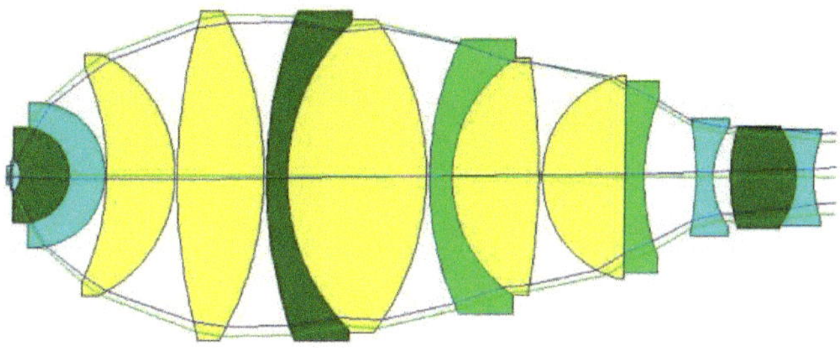

Lens: Plapo 100x1.40 oil Ftl164		Zoom	1 of 1	Efl	-8.056286
Object num aper	1.400000	Image height	14.000000	Primary wavln	0.546070

SRF	RADIUS	THICKNESS	APERTURE RADIUS	GLASS	SPECIAL
OBJ	0.000000	0.010000	0.140024	OIL	
AST	0.000000	0.170000	0.160000 A	K9	
2	0.000000	0.150000	0.580000	OIL	
3	0.000000	0.450000	0.940000	K9	
4	-1.300000	2.820000	0.970000	H-ZF62	
5	-2.710000	2.100000	2.690000	H-LAF10L	
6	-4.093000	0.120000	4.080000	AIR	
7	-13.900000	4.000000	6.160000	CAF2	
8	-7.503000	0.120000	6.860000	AIR	
9	33.700000	5.000000	9.140000	CAF2 P	
10	-18.923000	0.150000	9.350000	AIR	
11	26.410000	1.200000	9.350000	H-ZF62	
12	12.340000	8.000000	8.790000	CAF2	
13	-14.791000	0.200000	8.910000	AIR	
14	18.000000	1.300000	7.830000	H-ZLAF3	
15	8.128000	5.000000	6.780000	CAF2	
16	-41.980000	0.200000	6.720000	AIR	
17	5.960000	4.800000	5.810000	CAF2	
18	0.000000	1.100000	5.670000	ZLAF3	
19	12.340000	2.700000	4.620000	AIR	
20	-22.553800	1.000000	3.660000	H-LAK61	
21	4.854200	1.300000	3.090000	AIR	
22	12.687600	3.800000	3.200000	H-ZF62	
23	-4.641600	1.000000	3.130000	ZLAF1	
24	8.922400	164.000000	2.880000	AIR	
25	-29.602000	11.480000	18.670000	K9	
26	-35.719200	2.460000	22.050000	AIR	
27	309.222000	12.300000	23.280000	CAF2	
28	-66.830000	2.460000	23.670000	AIR	
29	181.466000	13.940000	22.970000	CAF2	
30	-59.040000	2.460000	22.160000	LZ_CTK12	
31	0.000000	164.148073 S	22.030000	AIR	
IMS	0.000000	0.000000	14.000000		

Figure 3.67 Design parameters and graphs of aberration correction of a 100×/1.40 oil immersion CCF plan apochromatic objective ($F'_{ts} = 164$ mm).

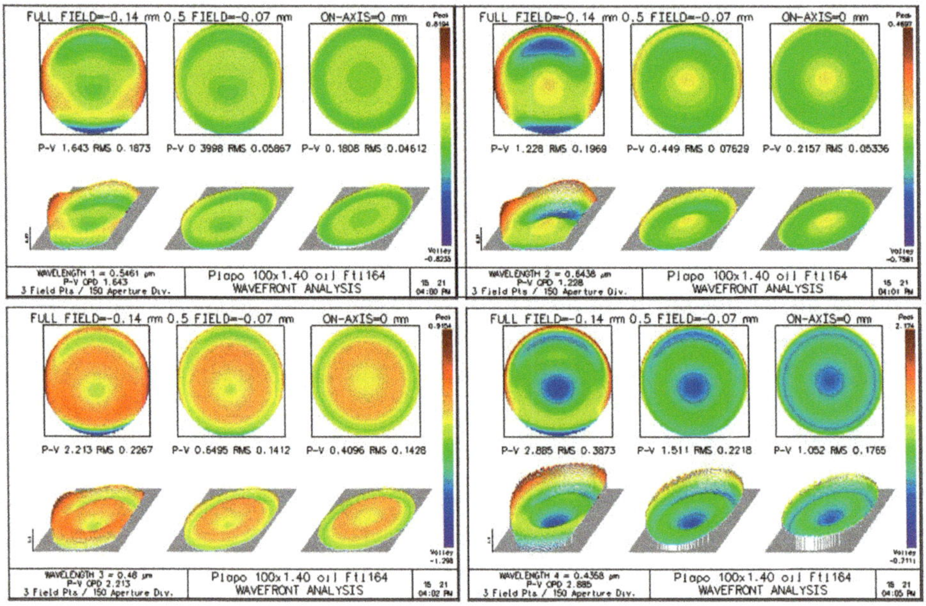

Figure 3.67 (*Continued*) Design parameters and graphs of aberration correction of a 100×/1.40 oil immersion CCF plan apochromatic objective ($F'_{ts} = 164$ mm).

because success in the manufacture of a complex optical–mechanical component, as well as the product as a whole, largely depends on understanding the capabilities and specific characteristics of a particular plant. Certain regulations and technical documents reflect a manufacturer's competence. The differences between manufacturers' methods can be very consequential; however, choosing the appropriate technique for the manufacture of each optical–mechanical product should guarantee the success and consistency of the finished product's image quality.

As an example, let us discuss two different methods for evaluating the decentering of lenses and optical–mechanical components. In some cases, the centering value is estimated in angular quantities (minutes, seconds), whereas in others, it is estimated in linear values (fractions of a millimeter). Scientific and engineering schools advocate for one or another concept, each of which determines the order of the original mechanical design. In these two versions, the principles for assigning tolerances, the technological equipment of the plant, and the assembly methods are completely different.

Figure 3.94 shows a technical pre-drawing of a general mechanical design of a 100×/0.93 CCF plan apochromatic objective. We have already discussed that the overall design of the external body of an objective can be different; it depends on the artistic preferences of the engineer and the ergonomic and stylistic features of the microscope in which the objective will be used. Thus,

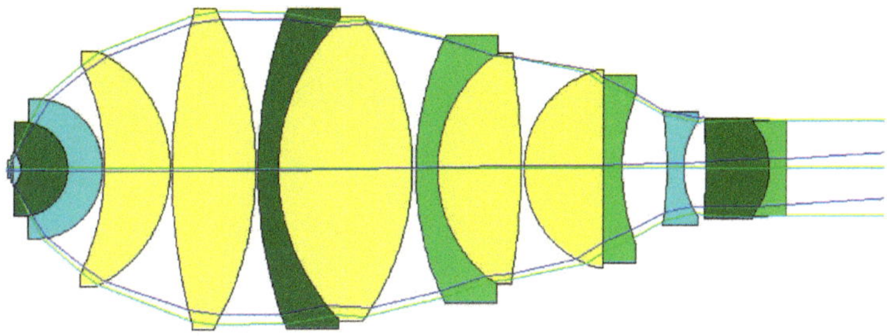

Lens: PlanApo 100x1.40 oil Ftl180

Object num aper 1.400000 Image height 14.000000 Primary wavln 0.546070

SRF	RADIUS	THICKNESS		APERTURE RADIUS		GLASS		SPECIAL
OBJ	0.000000	0.010000		0.139943		OIL		
AST	0.000000	0.170000		0.160000	A	K9		
2	0.000000	0.150000		0.580000		OIL		
3	0.000000	0.450000		0.820000		K9		
4	-1.300000	2.820000		0.970000		H-ZF62		
5	-2.710000	2.100000		2.690000		H-LAF10L		
6	-4.093000	0.120000		4.080000		AIR		
7	-13.900000	4.000000		6.160000		CAF2		
8	-7.503000	0.120000		6.860000		AIR		
9	33.700000	5.000000		9.140000		CAF2	P	
10	-18.923000	0.150000		9.350000		AIR		
11	26.410000	1.200000		9.350000		H-ZF62		
12	12.340000	8.000000		8.790000		CAF2		
13	-14.791000	0.200000		8.910000		AIR		
14	18.000000	1.300000		7.830000		H-ZLAF3		
15	8.128000	5.000000		6.780000		CAF2		
16	-41.980000	0.200000		6.720000		AIR		
17	5.960000	4.800000		5.810000		CAF2		
18	0.000000	1.100000		5.670000		H-ZLAF3		
19	12.340000	2.800000		4.620000		AIR		
20	-19.310000	1.000000		3.660000		LZ_CTK12		
21	4.680000	1.300000		3.100000		AIR		
22	36.130000	3.700000		3.180000		H-ZF62		
23	-4.234000	1.100000		3.250000		ZLAF1		
24	20.000000	180.000000		3.140000		AIR		
25	-32.490000	12.600000		19.550000		K9		
26	-39.204000	2.700000		23.060000		AIR		
27	339.390000	13.500000		24.230000		CAF2		
28	-73.350000	2.700000		24.660000		AIR		
29	199.170000	15.300000		23.980000		CAF2		
30	-64.800000	2.700000		23.090000		LZ_CTK12	P	
31	0.000000	179.921514	S	22.950000		AIR		
IMS	0.000000	0.000000		14.000000				

Figure 3.68 Design parameters and graphs of aberration correction of a 100×/1.40 oil immersion CCF plan apochromatic objective ($F'_{ts} = 180$ mm).

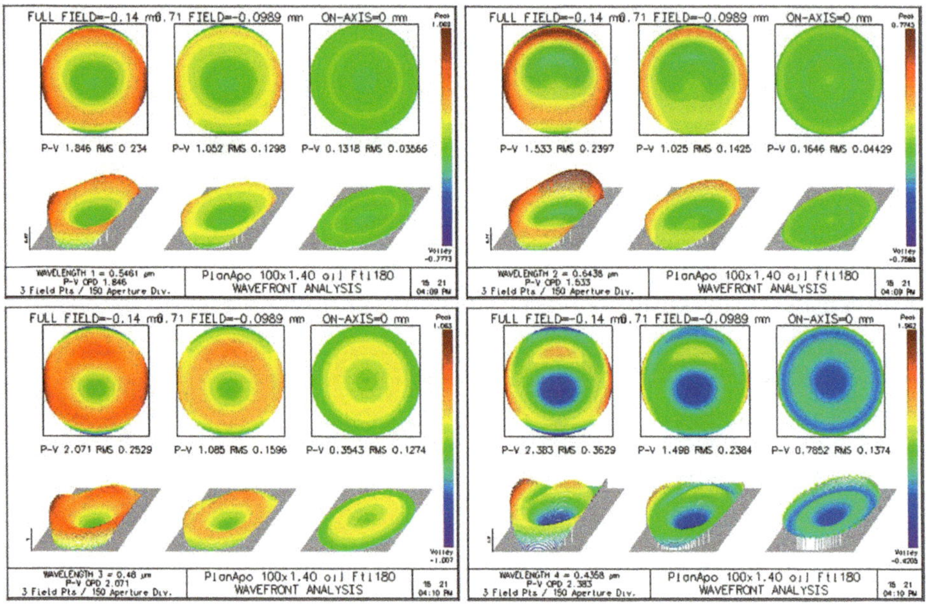

Figure 3.68 (*Continued*) Design parameters and graphs of aberration correction of a 100×/1.40 oil immersion CCF plan apochromatic objective (F'_{ts} = 180 mm).

the two components of mechanical design—drawings of lenses and drawings of lenses in frames—must be considered in detail. Both types of drawings can show which technologies are preferable for manufacturing a particular component and the objective as a whole. We can also see the assembly methods and the precision and tolerances for manufacturing parts. Figures 3.95–3.147 show drawings of optical parts, mechanical frames, and components, which are lenses integrated into the frames. These drawings are made for two objectives: a 100×/0.93 objective, the design parameters of which are presented in Fig. 3.60, and a 40×/0.90 objective, the design parameters of which are presented in Fig. 3.47.

3.7 Unifying Objectives by Changing Only the Frontal Lens

As the practice of designing light microscopes shows, if we consider some possible methods of working with a microscope in advance, we can quite significantly reduce the number of optical and mechanical components and parts used. This is especially true for microscope objectives, the nomenclature of which is very wide. The reduction in the number of some parts due to their unification, that is, the use of the same parts in different types of objectives, can significantly reduce production and manufacturing costs. Thus, it can optimize the manufacturing process of objectives.

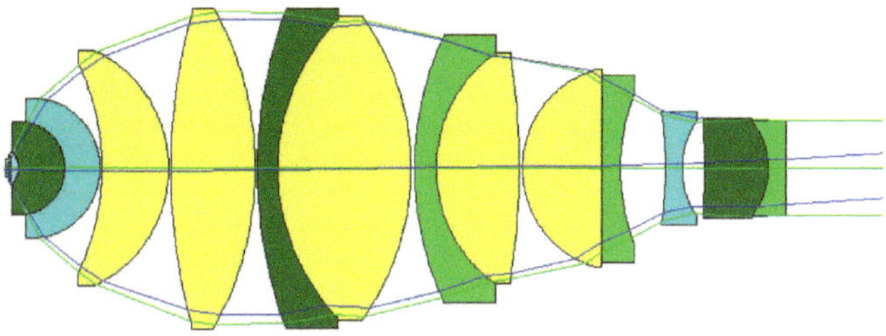

Lens: PlanApo 100x1.40 oil f200

Object num aper 1.400000 Image height 14.000000 Primary wavln 0.546070

SRF	RADIUS	THICKNESS	APERTURE RADIUS	GLASS	SPECIAL
OBJ	0.000000	0.010000	0.140014	OIL	
AST	0.000000	0.170000	0.160000 A	K9	
2	0.000000	0.150030	0.580000	OIL	
3	0.000000	0.450000	0.940000	K9	
4	-1.300000	2.820000	0.970000	H-ZF62	
5	-2.710000	2.100000	2.690000	H-LAF10L	
6	-4.093000	0.120000	4.080000	AIR	
7	-13.900000	4.000000	6.160000	CAF2	
8	-7.503000	0.120000	6.860000	AIR	
9	33.700000	5.000000	9.140000	CAF2 P	
10	-18.923000	0.150000	9.350000	AIR	
11	26.410000	1.200000	9.350000	H-ZF62	
12	12.340000	8.000000	8.790000	CAF2	
13	-14.791000	0.200000	8.910000	AIR	
14	18.000000	1.300000	7.830000	H-ZLAF3	
15	8.128000	5.000000	6.780000	CAF2	
16	-41.980000	0.200000	6.720000	AIR	
17	5.960000	4.800000	5.810000	CAF2	
18	0.000000	1.100000	5.670000	H-ZLAF3	
19	12.340000	2.700000	4.620000	AIR	
20	-21.480000	1.000000	3.670000	H-LAK61	
21	4.920000	1.300000	3.110000	AIR	
22	-440.300000	3.800000	3.140000	H-ZF62	
23	-4.285000	1.000000	3.360000	ZLAF1	
24	104.070000	200.000000	3.450000	AIR	
25	-36.100000	14.000000	20.040000	K9	
26	-43.560000	3.000000	23.600000	AIR	
27	377.100000	15.000000	24.630000	CAF2	
28	-81.500000	3.000000	25.110000	AIR	
29	221.300000	17.000000	24.490000	CAF2	
30	-72.000000	3.000000	23.500000	LZ_CTK12	
31	0.000000	199.979752 S	23.370000	AIR	
IMS	0.000000	0.000000	14.000000		

Figure 3.69 Design parameters and graphs of aberration correction of a 100×/1.40 oil immersion CCF plan apochromatic objective ($F'_{ts} = 200$ mm).

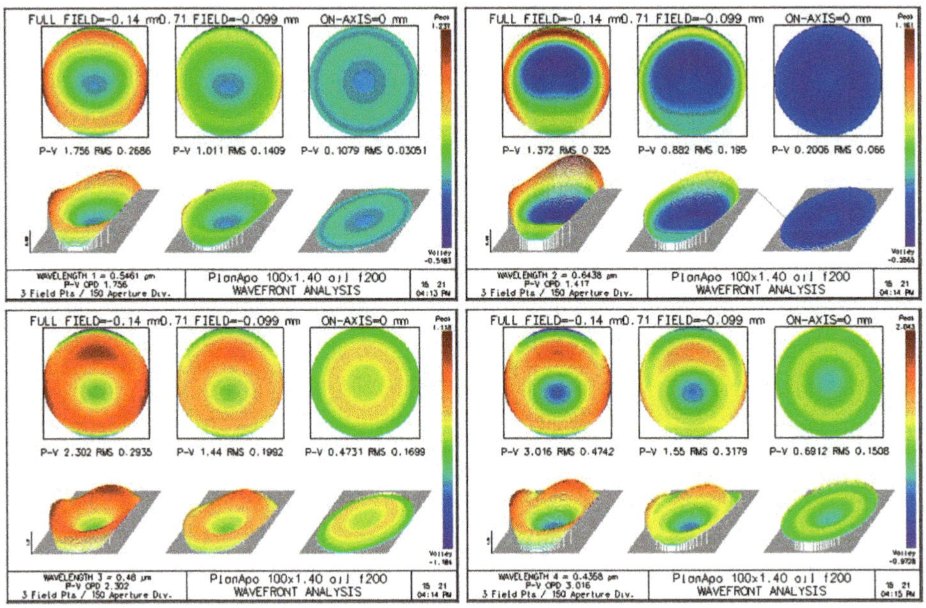

Figure 3.69 (*Continued*) Design parameters and graphs of aberration correction of a 100×/1.40 oil immersion CCF plan apochromatic objective ($F'_{ts} = 200$ mm).

One such method for optimizing the design of objectives, which reduces the number of parts used, is to change only the design parameters of the frontal lens, which is located first from the object under study along the rays. Such unification can be implemented in a rather simple way if the optical design initially provides for it. In this case, optical calculations are carried out in the usual way, and the level of aberration correction remains normal. However, the design of the front lens must be made in a particular way; namely, the first surface of the front lens from the object must be made in the form of a plane. In addition, the choice of the optical material of the frontal lens should correlate with the dispersion parameters, for example, with those of a coverslip or slide; in any case, the parameters of the refractive index and dispersion should not differ significantly. This is especially true for medium- and high-magnification objectives.

3.7.1 MIRO configuration plan MicroFluar objectives

Let us briefly describe the method of interference contrast using a light microscope and its advantage over other contrast methods, for example, phase contrast. An interference microscope is much more functional compared with, for example, a phase contrast microscope [9]. The theory of phase contrast is fully valid when applied to an ideal object, which is an *ideal phase lattice*; real microscopic objects differ not only from the ideal phase

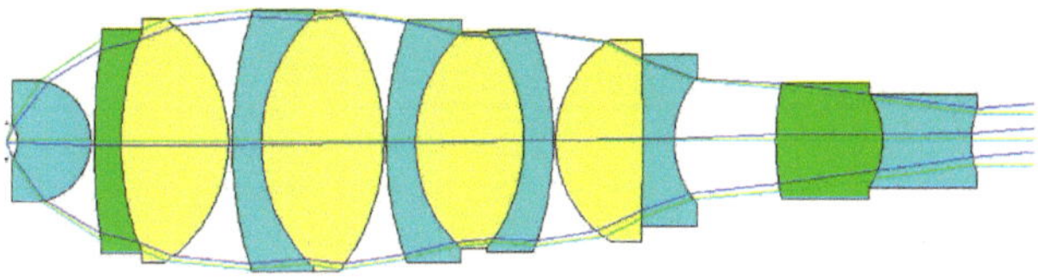

Lens: Plapo 160x0.95 f160

Object num aper 0.950000 Image height 12.500000 Primary wavln 0.546070

SRF	RADIUS	THICKNESS	APERTURE RADIUS	GLASS	SPECIAL
OBJ	0.000000	1.0000e-20	0.078188	AIR	
AST	0.000000	0.413845	0.650000 A	AIR	
2	-1.056800	2.750000	0.650000	LZ_CTK19	
3	-2.228000	0.140000	2.210000	AIR	
4	26.120000	1.000000	3.730000	LZ_TF5	
5	10.965000	4.000000	4.100000	CAF2	
6	-5.346000	0.140000	4.460000	AIR	
7	16.032000	1.100000	4.780000	LZ_CTK19 P	
8	6.457000	4.600000	4.620000	CAF2 P	
9	-7.727000	0.100000	4.770000	AIR	
10	13.032000	1.100000	4.440000	LZ_CTK19 P	
11	5.395000	4.100000	3.970000	CAF2 P	
12	-6.281000	1.100000	3.910000	LZ_CTK19 P	
13	-11.668000	0.100000	4.060000	AIR	
14	4.285000	3.300000	3.680000	CAF2 P	
15	-159.590000	1.200000	3.120000	LZ_CTK19 P	
16	3.499000	3.800000	2.310000	AIR	
17	10.666000	4.000000	2.110000	LZ_TF5 P	
18	-3.192000	3.300000	1.710000	LZ_CTK19 P	
19	3.192000	90.000000	1.190000	AIR	
20	214.800000	5.000000	7.810000	LZ-BK8	
21	-46.990000	0.300000	7.950000	AIR	
22	-46.880000	3.000000	7.940000	LZ-BF24	
23	-113.240000	157.839192 S	8.080000	AIR	
IMS	0.000000	0.000000	12.500000		

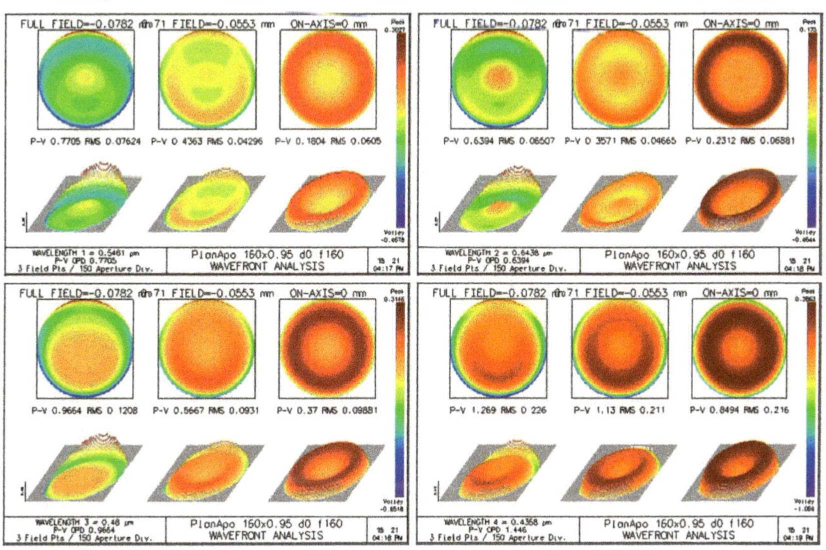

Figure 3.70 Design parameters and graphs of aberration correction of a 160×/0.95 CCF plan apochromatic objective ($F'_{ts} = 160$ mm).

Table 3.3 Main technical parameters and basic optical layout of some plan apochromatic objectives.

Magnification	*NA*	*WD* (mm)	*F'* (mm)	*R* (μm)	*DF* (μm)	*FOV* on object (mm)	*FOV* on image (mm)	Optical design at figure no.	The principal optical layout
1×	0.015	3.0	200	22	1480	14	14	3.31	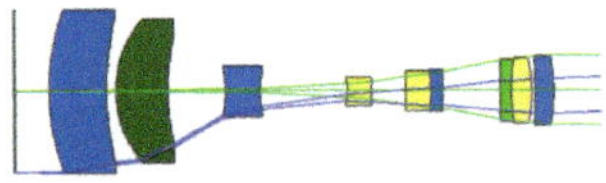
2.5×	0.05	11.5	64	6.7	134	10	25	3.32	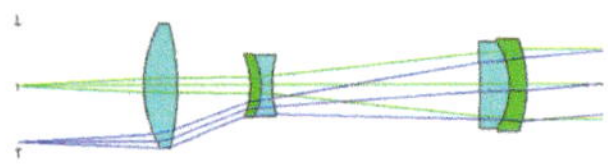
2.5×	0.05	2.0	80	6.7	134	10	25	3.33	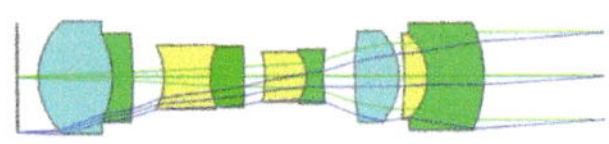
4×	0.16	3.7	50	2.1	13.1	7	28	3.34	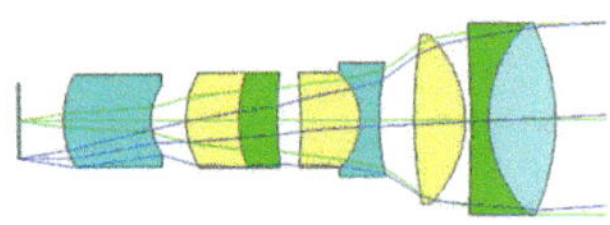
10×	0.40	1.3	20	0.84	2.1	2.8	28	3.35	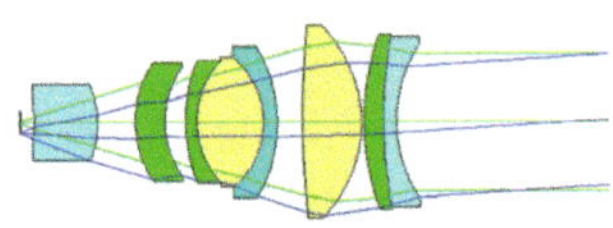
20×	0.60	1.9	10	0.56	0.93	1.25	25	3.36	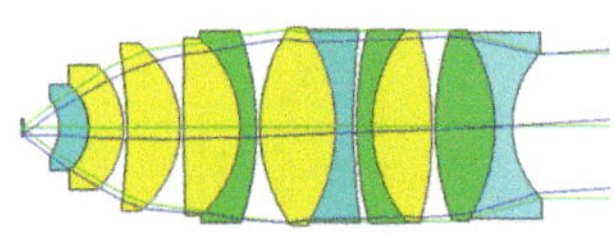
20×	0.70	0.8	7.73	0.48	0.68	1.4	28	3.37	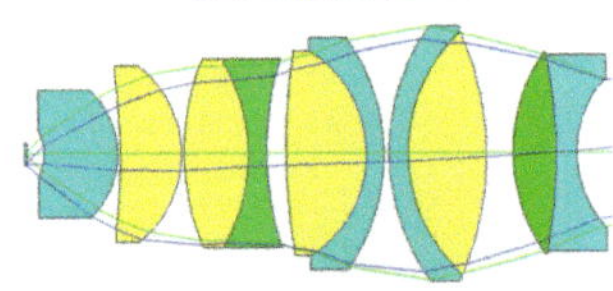
20×	0.70	0.8	8.2	0.48	0.68	1.4	28	3.38	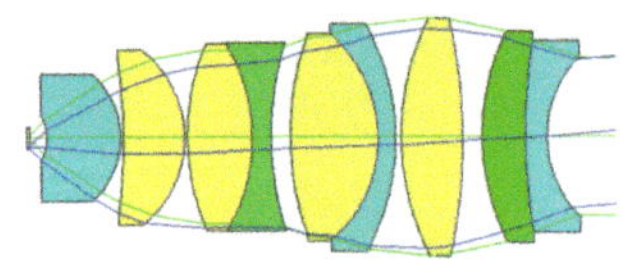
20×	0.70	0.8	9	0.48	0.68	1.4	28	3.39	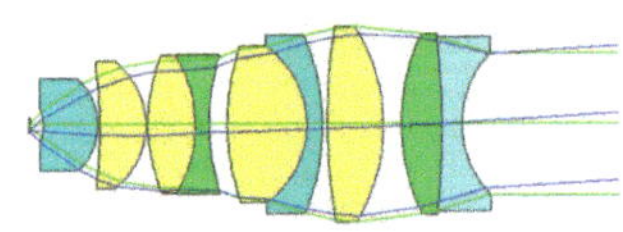
20×	0.70	0.8	10	0.48	0.68	1.4	28	3.40	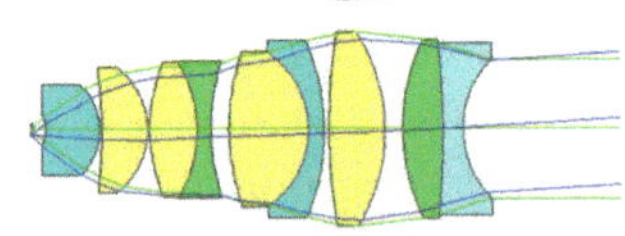
20×	0.80	1.1	8.2	0.42	0.52	1.25	25	3.41	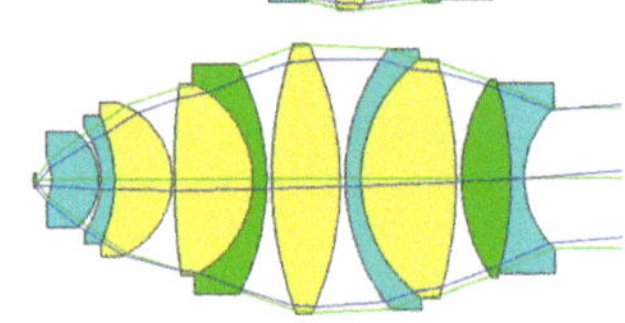

(*continued*)

Table 3.3 (*Continued*) Main technical parameters and basic optical layout of some plan apochromatic objectives.

Magnification	*NA*	*WD* (mm)	*F'* (mm)	*R* (μm)	*DF* (μm)	*FOV* on object (mm)	*FOV* on image (mm)	Optical design at figure no.	The principal optical layout
25×	1.0 water	0.12	6.4	0.33	0.33	1	25	3.42	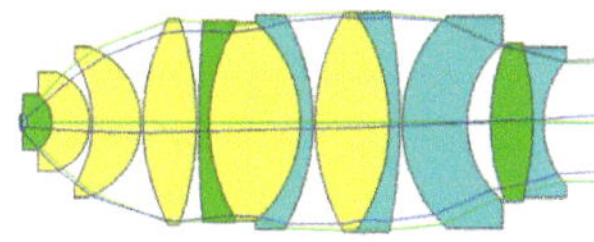
32×	0.85	0.47	5	0.39	0.46	0.78	25	3.43	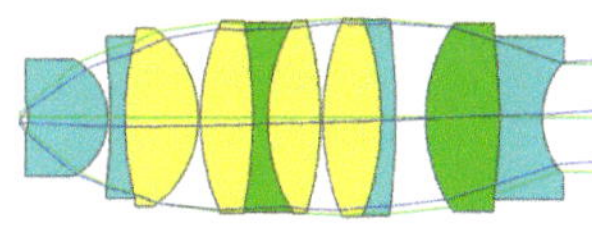
32×	0.85 oil	0.38	5	0.39	0.46	0.625	20	3.44	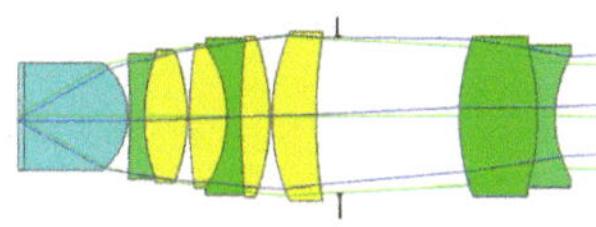
40×	0.90	0.17	4	0.38	0.41	0.625	25	3.45	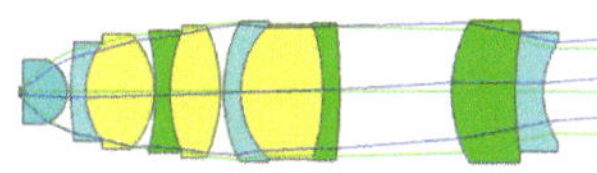
40×	0.90	0.3	4.5	0.38	0.41	0.7	28	3.46	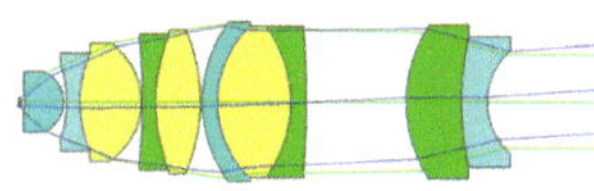
40×	0.90	0.3	5	0.38	0.41	0.7	28	3.47	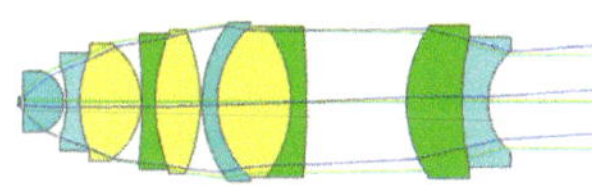
40×	1.0 oil	0.35	4.1	0.33	0.33	0.625	25	3.48	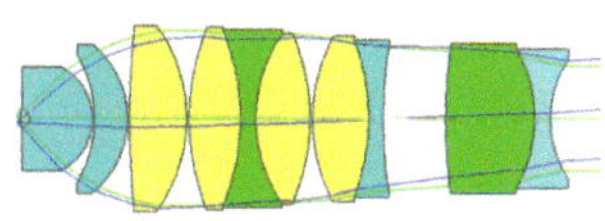
40×	1.2 water	0.1	4	0.28	0.23	0.625	25	3.49	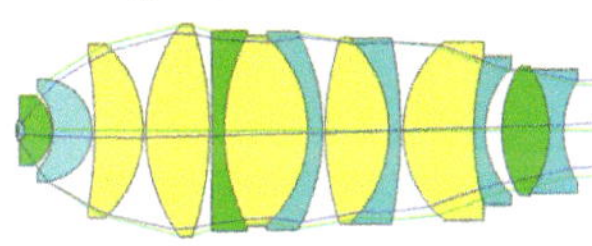
50×	1.0 oil	0.48	3.2	0.33	0.33	0.625	25	3.50	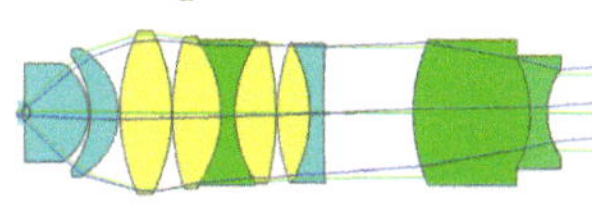
60×	0.92	0.18	3	0.37	0.40	0.42	25	3.51	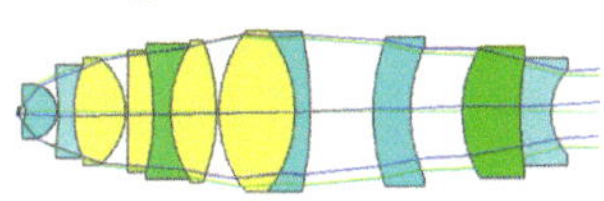
60×	0.92	0.18	3.33	0.37	0.40	0.47	28	3.52	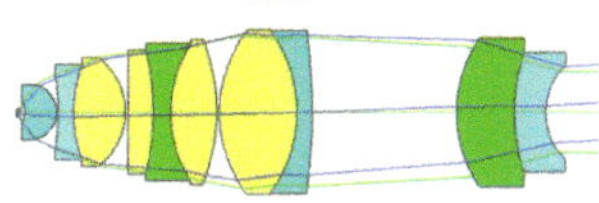
60×	1.0 oil	0.2	2.67	0.33	0.33	0.42	25	3.53	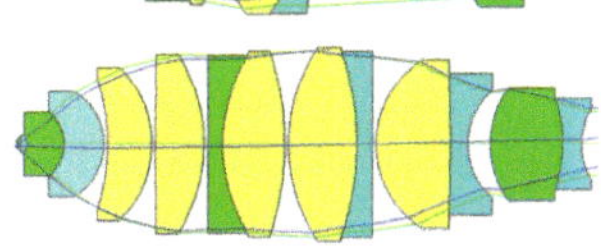

(*continued*)

Table 3.3 (*Continued*) Main technical parameters and basic optical layout of some plan apochromatic objectives.

Magnification	*NA*	*WD* (mm)	*F′* (mm)	*R* (μm)	*DF* (μm)	*FOV* on object (mm)	*FOV* on image (mm)	Optical design at figure no.	The principal optical layout
60×	1.2 oil	0.15	3.33	0.28	0.23	0.47	28	3.54	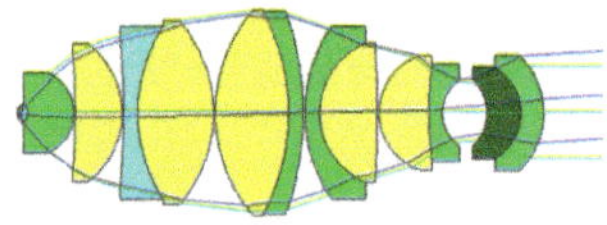
60×	1.49 oil	0.1	3.33	0.22	0.15	0.37	22	3.55	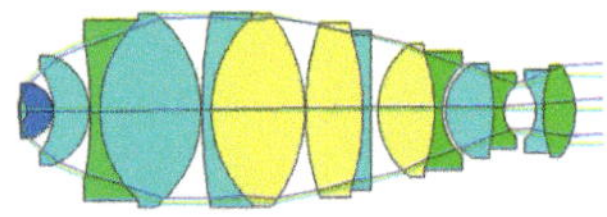
63×	0.9	0.16	2.54	0.38	0.41	0.4	25	3.56	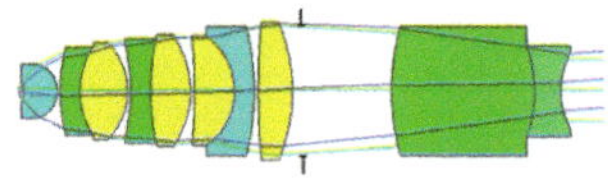
63×	1.2 water	0.11	2.54	0.28	0.23	0.4	25	3.57	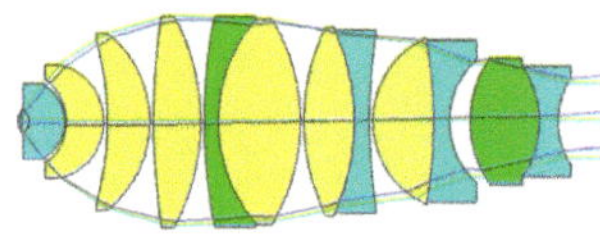
63×	1.25 oil	0.12	2.54	0.27	0.21	0.4	25	3.58	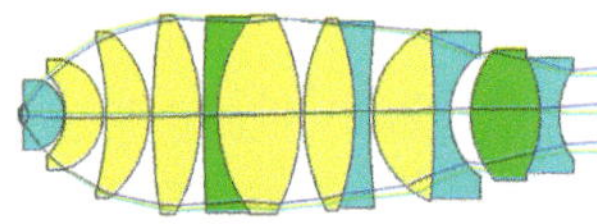
63×	1.25 oil	0.12	2.54	0.27	0.21	0.4	25	3.59	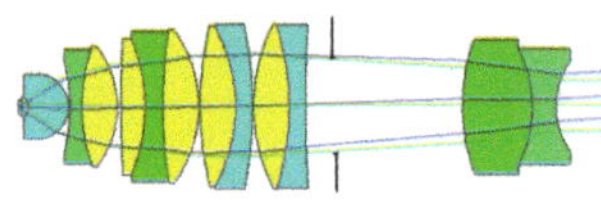
100×	0.93	0.23	2	0.36	0.39	0.28	28	3.60	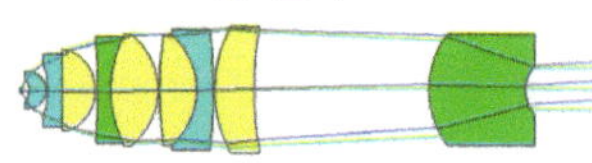
100×	1.2 water	0.1	1.57	0.28	0.23	0.25	25	3.61	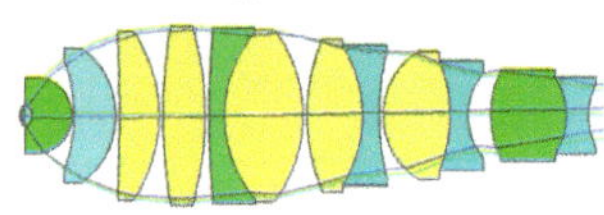
100×	1.2 water	0.1	1.64	0.28	0.23	0.25	25	3.62	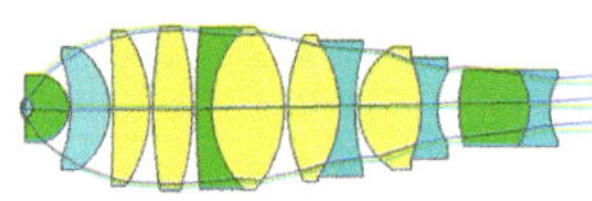
100×	1.2 water	0.1	1.8	0.28	0.23	0.25	25	3.63	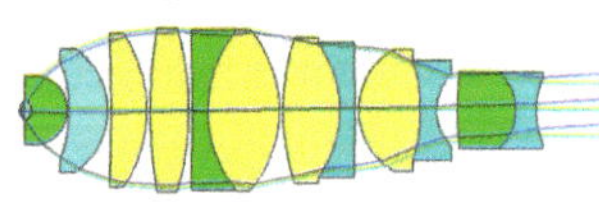
100×	1.2 water	0.1	2.0	0.28	0.23	0.25	25	3.64	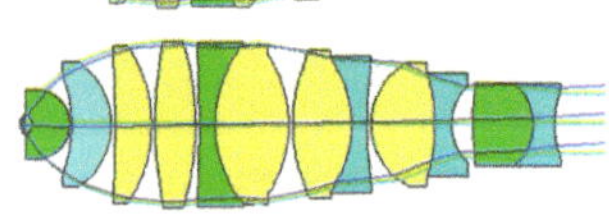
100×	1.35 oil	0.28	1.6	0.25	0.18	0.25	25	3.65	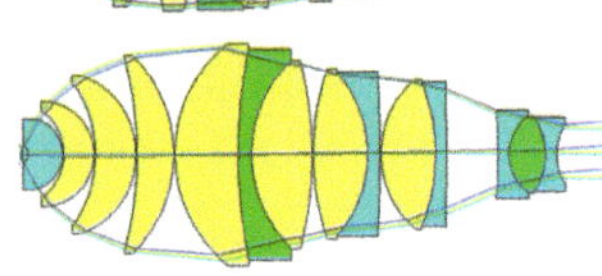

(*continued*)

Table 3.3 (*Continued*) Main technical parameters and basic optical layout of some plan apochromatic objectives.

Magnification	*NA*	*WD* (mm)	*F′* (mm)	*R* (μm)	*DF* (μm)	*FOV* on object (mm)	*FOV* on image (mm)	Optical design at figure no.	The principal optical layout
100×	1.40 oil	0.1	1.54	0.24	0.17	0.25	25	3.66	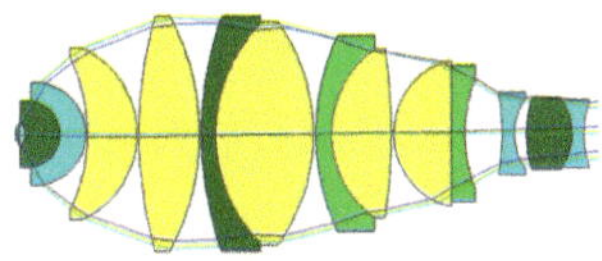
100×	1.40 oil	0.1	1.64	0.24	0.17	0.25	25	3.67	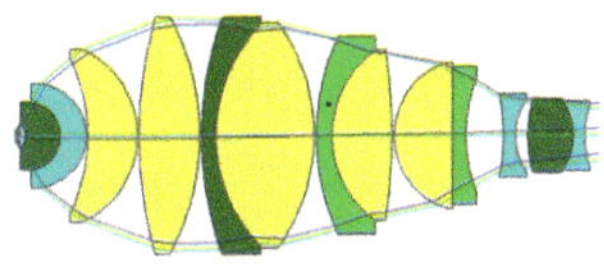
100×	1.40 oil	0.1	1.8	0.24	0.17	0.25	25	3.68	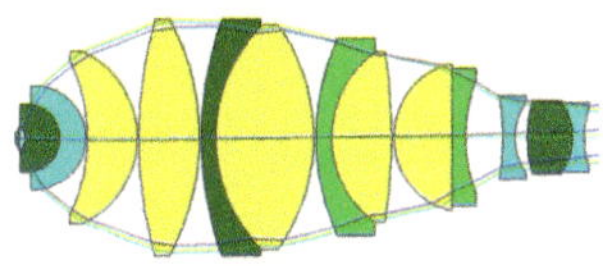
100×	1.40 oil	0.1	2.0	0.24	0.17	0.25	25	3.69	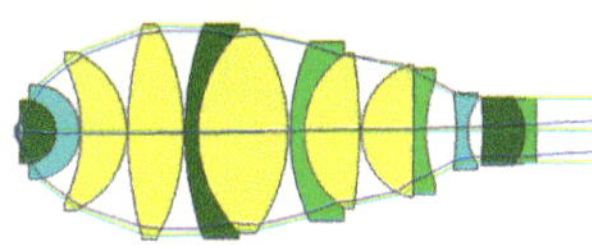
160×	0.95	0.18	1	0.36	0.40	0.16	25	3.70	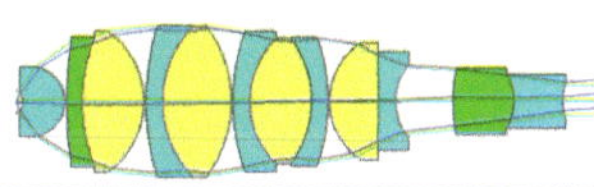

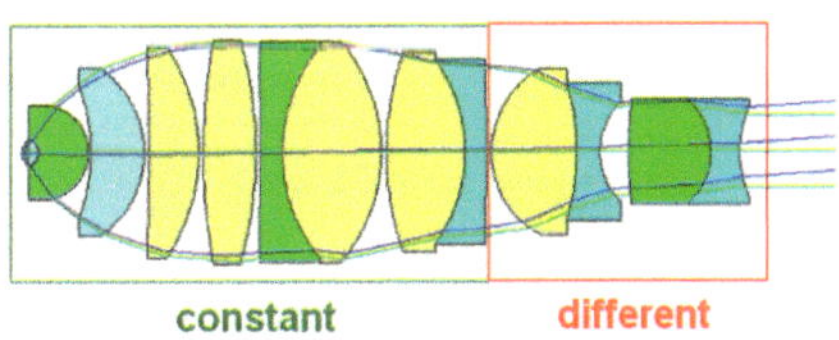

Figure 3.71 Comparison of different *F′* plan apochromatic 100× water immersion objectives.

lattice, but also from each other in terms of shape, size, thickness, refractive index, and other parameters. Because the phase contrast method is based on the separation of direct and diffracted light and their subsequent interference, it can be used to identify areas in the object where there are gradients of the refractive index or thickness. At small gradients, the diffraction angle is small, which does not allow a sufficient separation of direct and diffracted light; as a result, only fine details of the structure will be revealed in the phase contrast image. In the image of large coarse details, the contrast decreases rather quickly from the edge of the image to the center. Because the object does not absorb light, all the radiation energy passing through it is involved in the

Table 3.4 Parameters of 100x/1.20 water immersion finite (160 mm) and infinity (F'_{ob} = 1.64 mm) objectives.

Lens: 100x1.20 water Plapo tub160
Object num aper 1.200000 Image height 14.00

SRF	RADIUS	THICKNESS		
OBJ	0.000000	0.010000	WATER	
AST	0.000000	0.170000	K9	
2	0.000000	0.190000	WATER	P
3	-4.365000	0.600000	K9	
4	-0.850000	3.000000	H-ZF6	
5	-2.816000	0.150000	AIR	
6	-11.508000	3.200000	H-QK3	
7	-5.998000	0.300000	AIR	
8	-143.940000	3.000000	CAF2	
9	-10.920000	0.300000	AIR	
10	37.742000	3.200000	CAF2	P
11	-23.230000	0.300000	AIR	
12	63.300000	1.200000	H-ZF7LA	
13	9.330000	6.000000	CAF2	P
14	-10.233000	0.300000	AIR	
15	13.550000	4.300000	CAF2	P
16	-9.330000	1.300000	H-LAK2	
17	19.724000	0.300000	AIR	
18	5.453000	5.200000	CAF2	P
19	-10.233000	1.400000	H-LAK2	P
20	4.600000	1.800000	AIR	
21	7.170000	5.900000	H-ZF6	P
22	-4.897000	1.700000	H-LAK2	P
23	4.897000		AIR	

Lens: 100x1.20 water Plapo tl=164
Object num aper 1.200000 Image height 14.00

SRF	RADIUS	THICKNESS		
OBJ	0.000000	0.010000	WATER	
AST	0.000000	0.170000	K9	
2	0.000000	0.188500	WATER	P
3	-4.365000	0.600000	K9	
4	-0.850000	3.000000	H-ZF6	
5	-2.816000	0.150000	AIR	
6	-11.508000	3.200000	H-QK3	
7	-5.998000	0.300000	AIR	
8	-143.940000	3.000000	CAF2	
9	-10.920000	0.300000	AIR	
10	37.742000	3.200000	CAF2	P
11	-23.230000	0.300000	AIR	
12	63.300000	1.300000	H-ZF7LA	
13	9.330000	6.000000	CAF2	P
14	-10.233000	0.300000	AIR	
15	13.550000	4.300000	CAF2	P
16	-9.330000	1.300000	H-LAK2	
17	19.724000	0.300000	AIR	
18	5.702000	5.190000	CAF2	P
19	-10.233000	1.400000	H-LAK2	P
20	5.495000	1.800000	AIR	
21	10.233000	5.960000	H-ZF7LA	P
22	-5.557000	1.750000	H-LAK2	P
23	5.401000		AIR	

Table 3.5 Parameters of 100x/1.20 water immersion infinity (F'_{ob} = 1.8 mm and F'_{ob} = 2.0 mm) objectives.

Lens: 100x1.20 water plapo tl=180
Object num aper 1.200000 Image height 14.00

SRF	RADIUS	THICKNESS		
OBJ	0.000000	0.010000	WATER	
AST	0.000000	0.170000	K9	
2	0.000000	0.188000	WATER	P
3	-4.365000	0.600000	K9	
4	-0.850000	3.000000	H-ZF6	
5	-2.816000	0.150000	AIR	
6	-11.508000	3.200000	H-QK3	
7	-5.998000	0.300000	AIR	
8	-143.940000	3.000000	CAF2	
9	-10.920000	0.300000	AIR	
10	37.742000	3.200000	CAF2	P
11	-23.230000	0.300000	AIR	
12	-214.370000	1.300000	H-ZF6	P
13	9.900000	6.000000	CAF2	P
14	-9.900000	0.300000	AIR	
15	17.900000	4.800000	CAF2	P
16	-9.032000	1.300000	H-LAK2	
17	-206.000000	0.300000	AIR	
18	5.557000	5.200000	CAF2	P
19	-14.191000	1.400000	H-LAK2	P
20	4.000000	1.800000	AIR	
21	44.700000	5.000000	H-ZF6	P
22	-4.093000	1.800000	ZK1	
23	6.532000		AIR	

Lens: 100x1.20 water Plapo tl=200
Object num aper 1.200000 Image height 14.00

SRF	RADIUS	THICKNESS	
OBJ	0.000000	0.010000	WATER
AST	0.000000	0.170000	K9
2	0.000000	0.188000	WATER
3	-4.365000	0.600000	K9
4	-0.850000	3.000000	H-ZF6
5	-2.816000	0.150000	AIR
6	-11.508000	3.200000	H-QK3
7	-5.998000	0.300000	AIR
8	-143.940000	3.000000	CAF2
9	-10.920000	0.300000	AIR
10	37.742000	3.200000	CAF2
11	-23.230000	0.300000	AIR
12	-214.370000	1.300000	ZF6
13	9.900000	6.000000	CAF2
14	-9.900000	0.300000	AIR
15	13.550000	4.800000	CAF2
16	-9.200000	1.300000	H-LAK2
17	42.270000	0.300000	AIR
18	5.540000	5.200000	CAF2
19	-13.550000	1.400000	H-LAK2
20	4.146000	1.800000	AIR
21	26.400000	5.000000	ZF5
22	-4.285000	1.800000	ZK1
23	8.336000		AIR

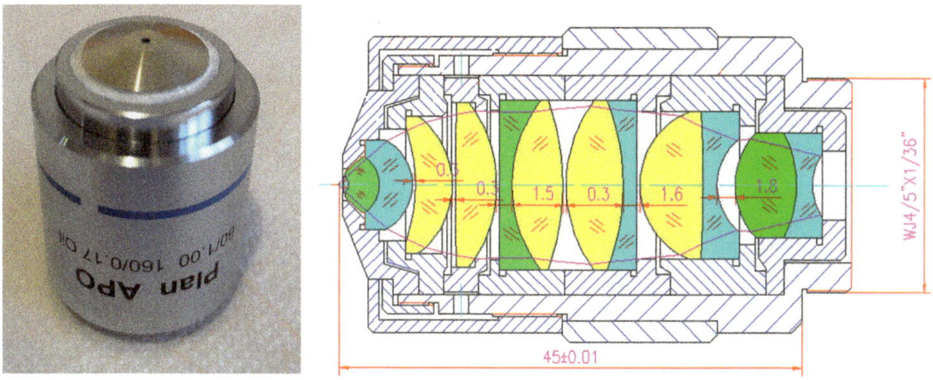

Figure 3.72 Photograph and technical drawing of a 60×/1.0 oil immersion CCF plan apochromatic objective.

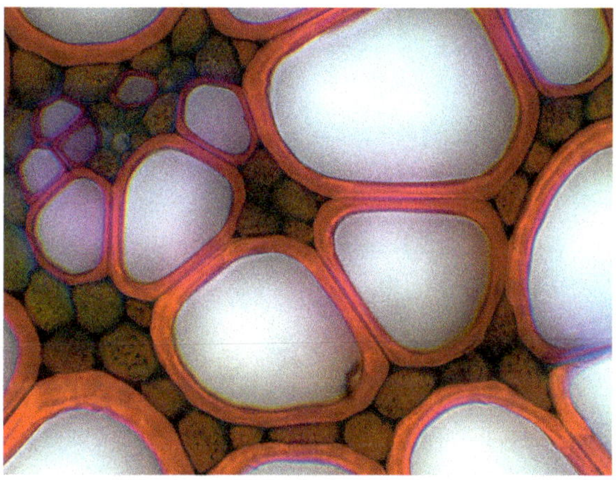

Figure 3.73 Photograph of a real object using a 60×/1.00 oil immersion CCF plan apochromatic objective (taken by A. Mikhaltsov).

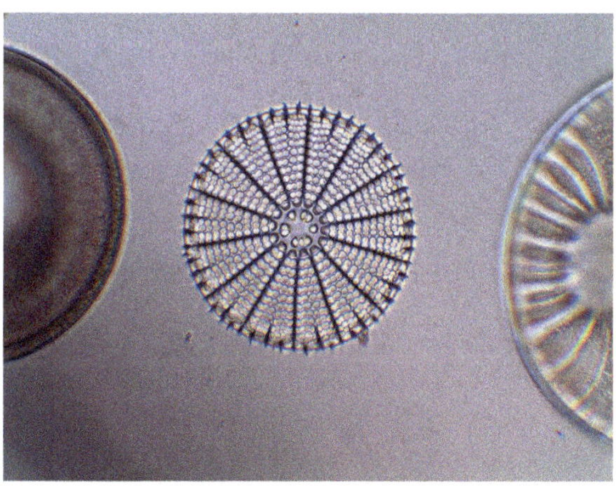

Figure 3.74 Photograph of a real object using a 60×/1.00 oil immersion CCF plan apochromatic objective (taken by A. Mikhaltsov).

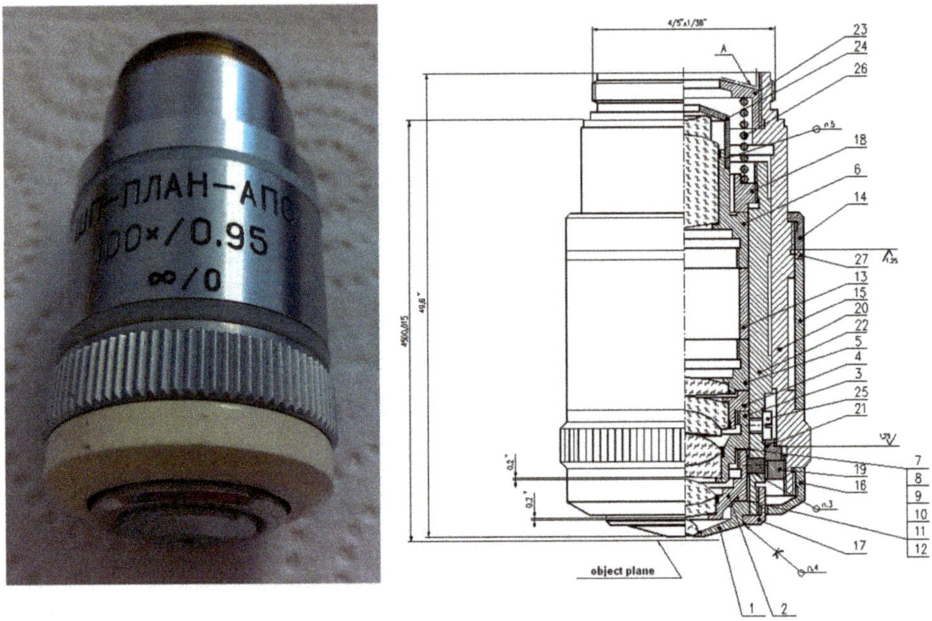

Figure 3.75 Photograph and principal mechanical design of a 100×/0.95 CCF plan apochromatic objective.

Figure 3.76 Photograph of a real object using a 100×/0.95 CCF plan apochromatic objective.

formation of the image. The phase ring redistributes energy only between different parts of the object (the absorption of the phase ring in this case does not play a role because it weakens the light equally for all points of the field of view). Therefore, the appearance of dark areas in the image should be

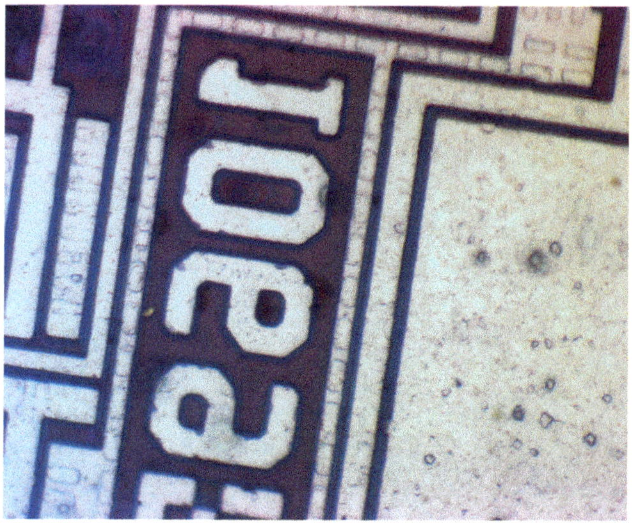

Figure 3.77 Photograph of a real object using a 100×/0.95 CCF plan apochromatic objective.

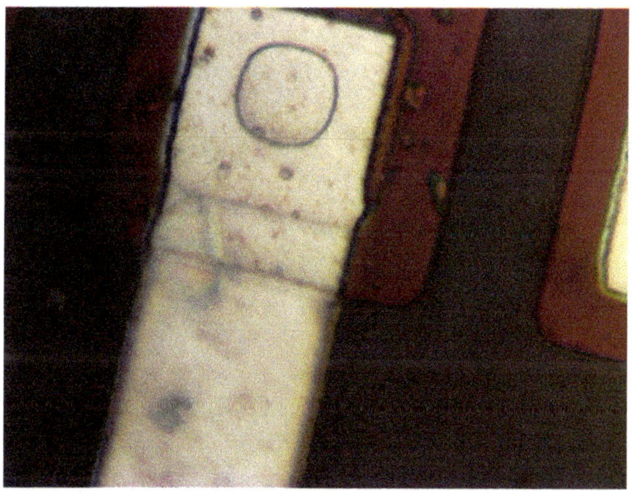

Figure 3.78 Photograph of a real object using a 100×/0.95 CCF plan apochromatic objective.

accompanied by an increase in illumination in neighboring areas, and vice versa. As a result, with a positive phase contrast, images of the dark elements of the structure are surrounded by light halos with a negative contrast; light images are accompanied by dark halos.

Interference contrast is free from such disadvantages. In white light, an interference microscope can produce both color and black and white images. With conventional phase microscopes, it is not possible to measure the path

Figure 3.79 Photograph of a real object using a 100×/0.95 CCF plan apochromatic objective.

Figure 3.80 Photograph of a 100×/1.40 oil immersion CCF plan apochromatic objective.

difference that occurs when light passes through a microscopic object. It is possible, in principle, to make such measurements with interference microscopy based on certain parameters of the object under study, e.g., concentration, weight, and others. That is, interference microscopy provides quantitative information about the object being examined with a microscope.

In 1931, Russian scientist A.A. Lebedev put into practice one of the first interference microscopes, the so-called "shearing" microscope. A device for studying opaque objects is implemented in the original interference objective

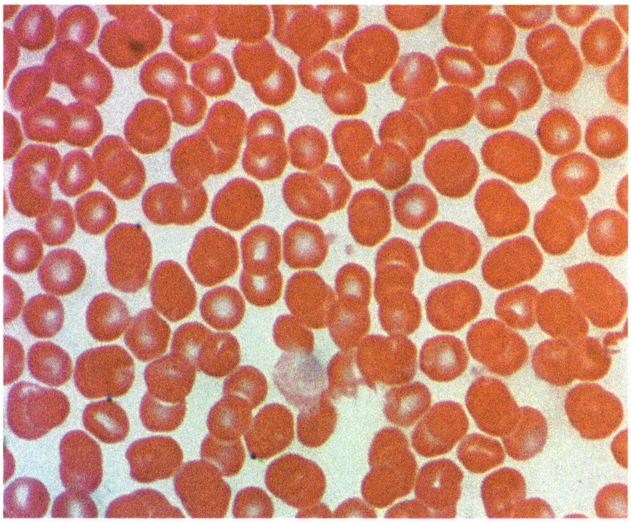

Figure 3.81 Photograph of a real object using a 100×/1.40 oil immersion CCF plan apochromatic objective.

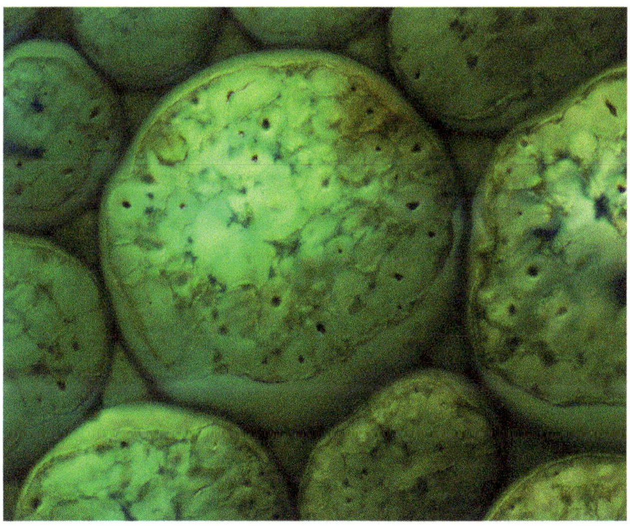

Figure 3.82 Photograph of a real object using a 100×/1.40 oil immersion CCF plan apochromatic objective.

designed by Professor Zakharyevsky (called the *MIRO configuration* in Western terminology) [10]. A principal diagram of such an objective is shown in Fig. 3.148.

A parallel beam of rays incident on the objective from the light source is focused by the lens system of the objective 1 on the studied surface of the object 5. The interferometric component of the objective consists of two

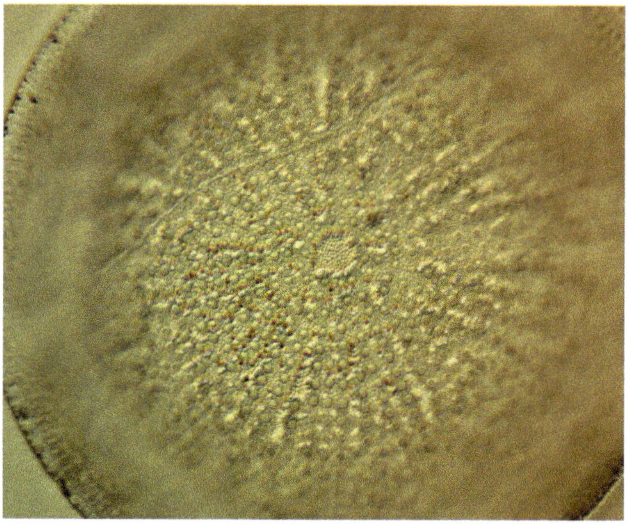

Figure 3.83 Photograph of a real object using a 100×/1.40 oil immersion CCF plan apochromatic objective.

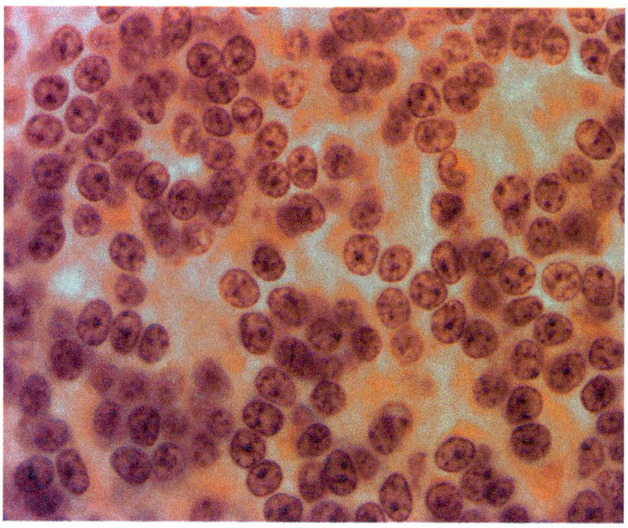

Figure 3.84 Photograph of a real object using a 100×/1.40 oil immersion CCF plan apochromatic objective.

plates, 3 and 4, of the same thickness. After applying to it a small mirror pad 2, plate 3 is glued to the front lens of the objective system 1. On the lens side, the mirror pad is coated with black opaque varnish. The lower surface of plate 4 is covered with a beam splitting layer and is located in the middle between mirror 2 and the surface of the investigated object 5. A ray of light from the lighting system passes through the objective and, having reached

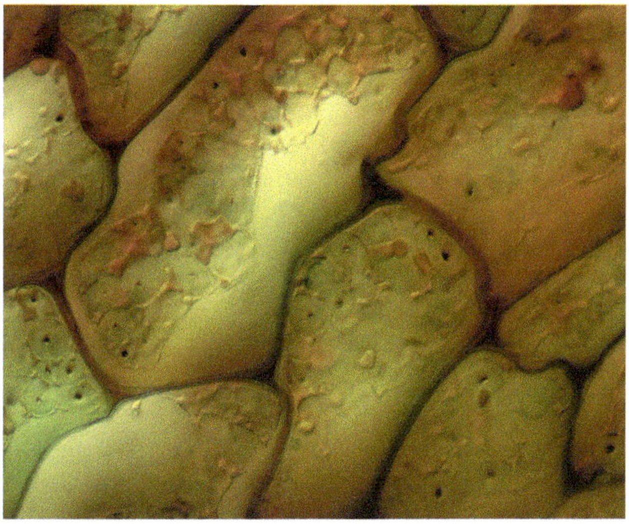

Figure 3.85 Photograph of a real object using a 100×/1.40 oil immersion CCF plan apochromatic objective.

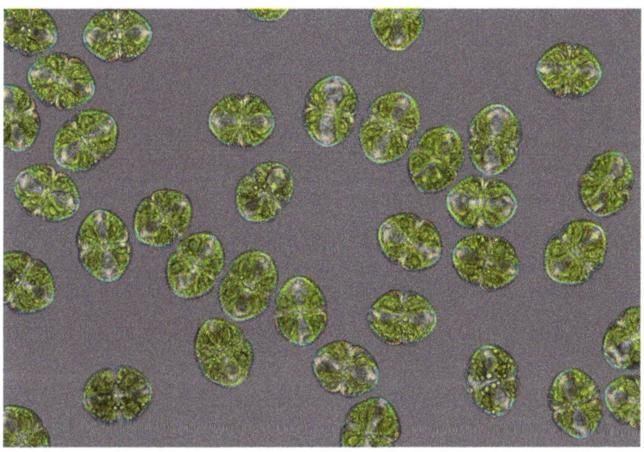

Figure 3.86 Photograph of a real object (taken by A. Mikhaltsov).

plate 4, bifurcates. One of the rays, without changing direction, falls on the object 5 and is reflected from the object to the plate 4. The second beam is reflected by the mirror 2 and also returns to the plate 4. Here, the rays are reunited and interfere, forming a system of interference bands of equal thickness localized on the surface of the sample; such an image is built in the plane of the intermediate image of the microscope.

Both interference beams should have equal intensities. Because the test surfaces have different reflections, the plates 4 are interchangeable. As a result of the addition (interference) of two wave systems, interference fringes will be

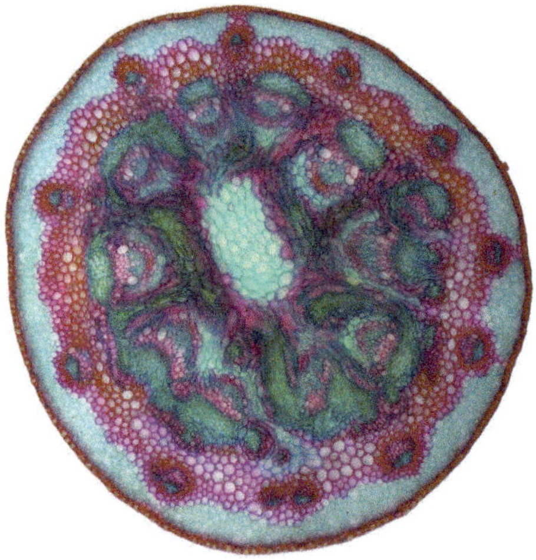

Figure 3.87 Photograph of a real object, *Hordeum vulgare* (taken by A. Mikhaltsov).

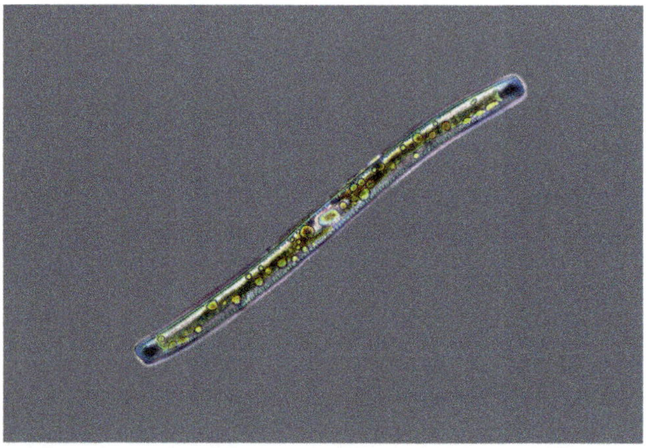

Figure 3.88 Photograph of a real object, *Nitzschia vermicularis* (taken by A. Mikhaltsov).

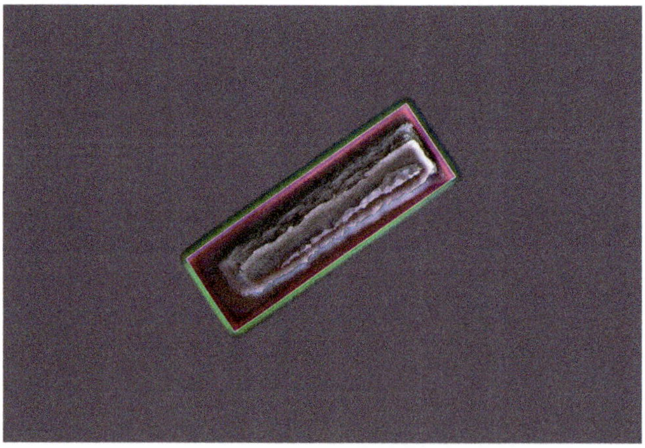

Figure 3.89 Photograph of a real object, crystal salt (taken by A. Mikhaltsov).

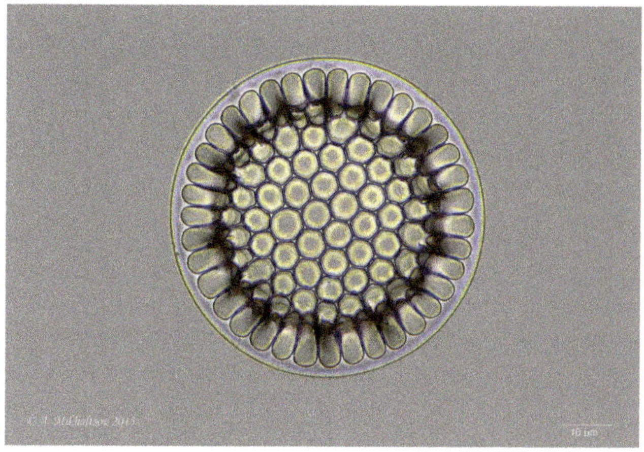

Figure 3.90 Photograph of a real object, *Stephanopyxis grunowii* (taken by A. Mikhaltsov).

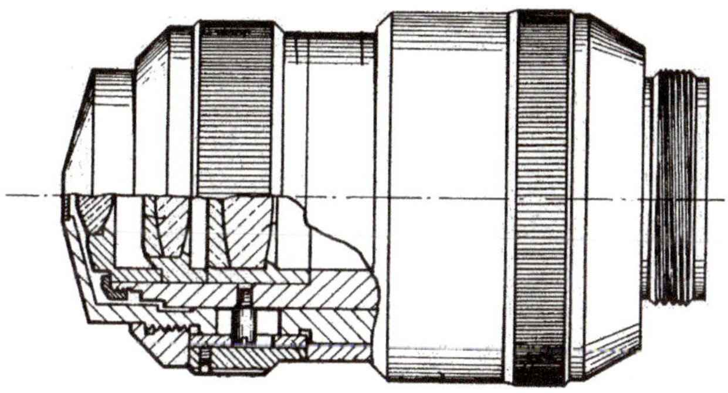

Figure 3.91 Cover glass in a frame as part of an objective.

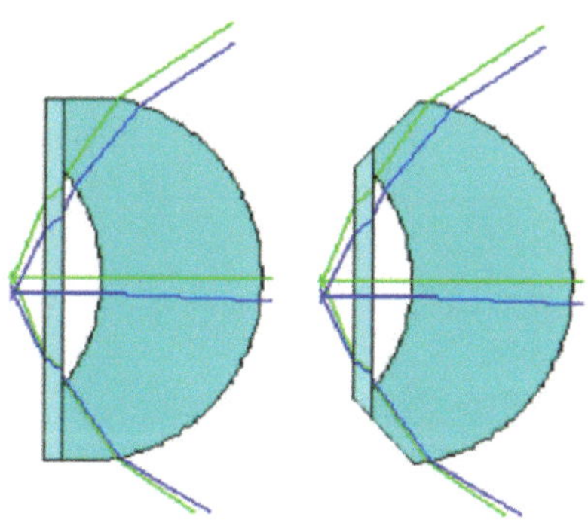

Figure 3.92 Integrating the cover glass into the objective design.

Lens: PlanApo 100x0.93 d=0 ft1200

Object num aper 0.930000 Image height 14.000000 Primary wavln 0.546070

SRF	RADIUS	THICKNESS	APERTURE RADIUS	GLASS	SPECIAL
OBJ	0.000000	0.280094	0.140046	AIR	
AST	0.000000	0.170000	0.160000 A	K9	
2	0.000000	0.380000	0.300000	AIR	
3	-1.500000	1.520000	0.990000	H-LAK3	
4	-1.670000	0.150000	1.660000	AIR	
5	-7.700000	1.300000	2.430000	H-LAK3 P	
6	0.000000	3.000000	3.270000	CAF2	
7	-3.900000	0.200000	3.620000	AIR	
8	-53.600000	1.200000	4.270000	ZF2	
9	10.189000	4.300000	4.760000	CAF2 P	
10	-6.281000	0.100000	4.980000	AIR	
11	32.000000	3.400000	4.960000	CAF2 P	
12	-6.399000	1.500000	4.930000	LZ_CTK12	
13	-49.200000	0.100000	5.390000	AIR	
14	11.220000	3.600000	5.670000	CAF2 P	
15	29.000000	16.000000	5.470000	AIR	
16	7.900000	9.000000	5.030000	ZF6	
17	3.200000	200.000000	2.200000	AIR	
18	-36.100000	14.000000	15.150000	K9	
19	-43.560000	3.000000	18.220000	AIR	
20	377.100000	15.000000	18.780000	CAF2	
21	-81.500000	3.000000	19.240000	AIR	
22	221.300000	17.000000	18.920000	CAF2	
23	-72.000000	3.000000	18.050000	LZ_CTK12 P	
24	0.000000	199.940023 S	17.980000	AIR	
IMS	0.000000	0.000000	14.000000		

Figure 3.93 Design parameters of a 100×/0.93 objective for work without a coverslip.

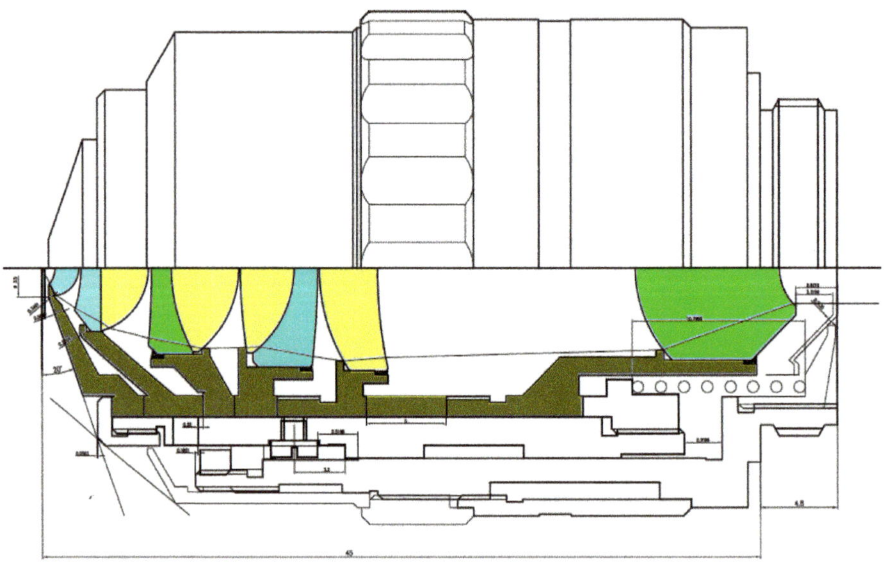

Figure 3.94 A technical pre-drawing of a general mechanical design of a 100×/0.93 CCF plan apochromatic objective.

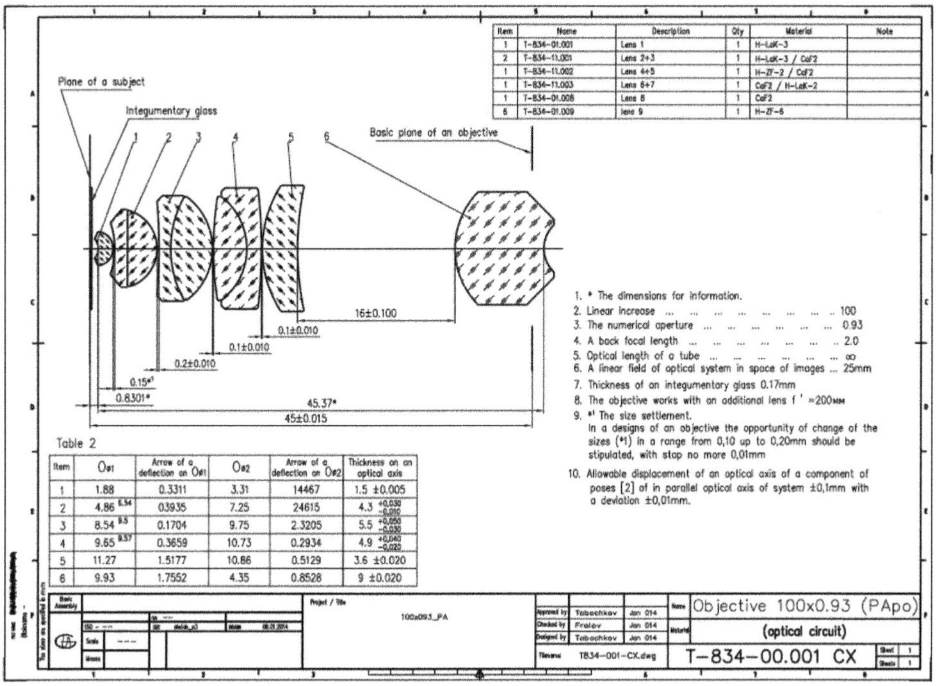

Figure 3.95 The principal optical layout of a 100×/0.93 infinity CCF plan apochromatic objective.

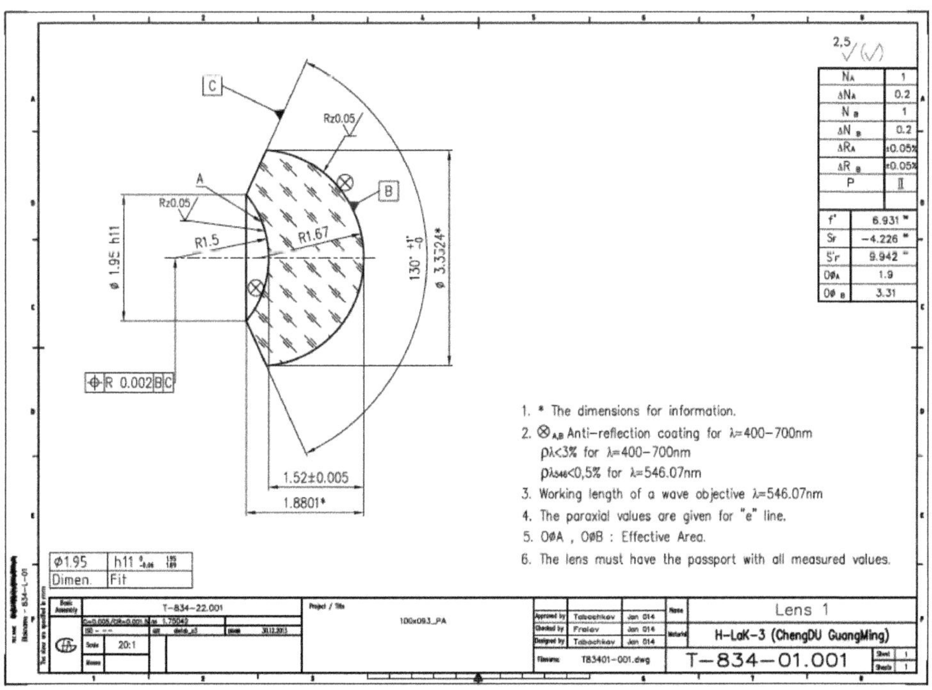

Figure 3.96 Drawing of the first lens.

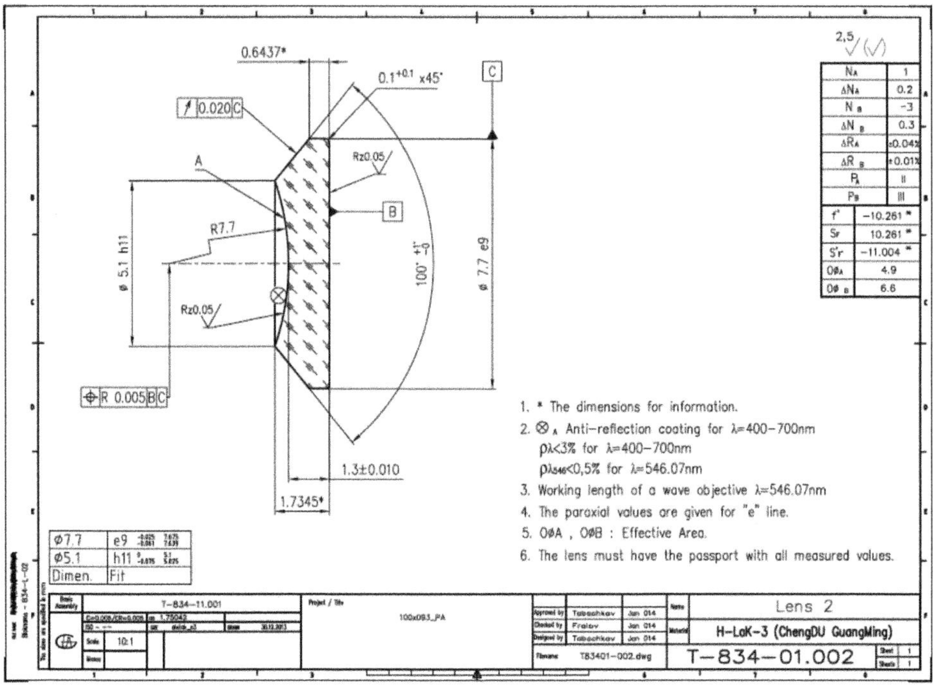

Figure 3.97 Drawing of the second lens.

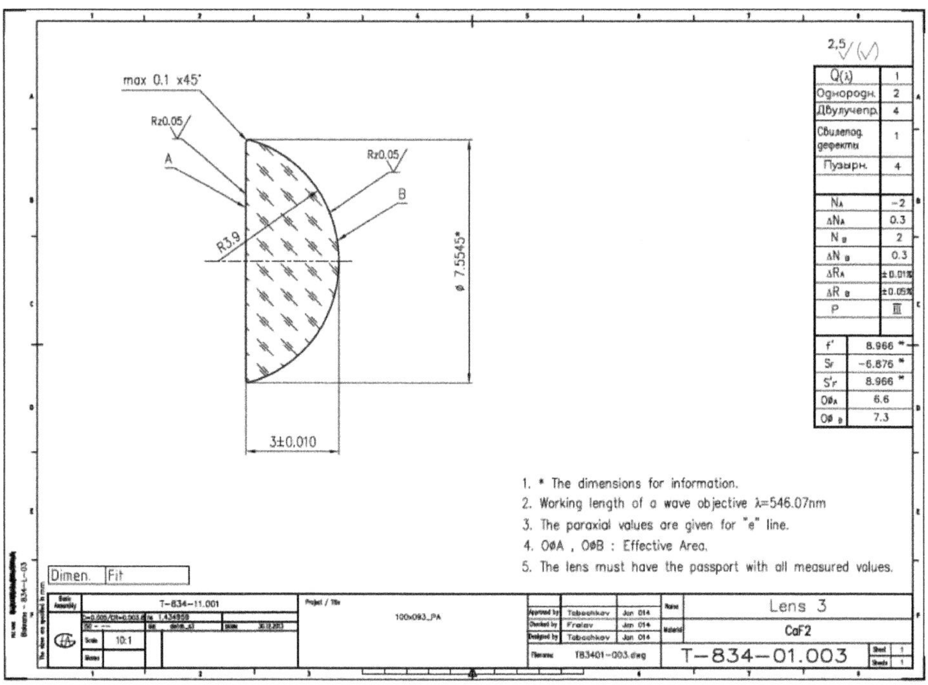

Figure 3.98 Drawing of the third lens.

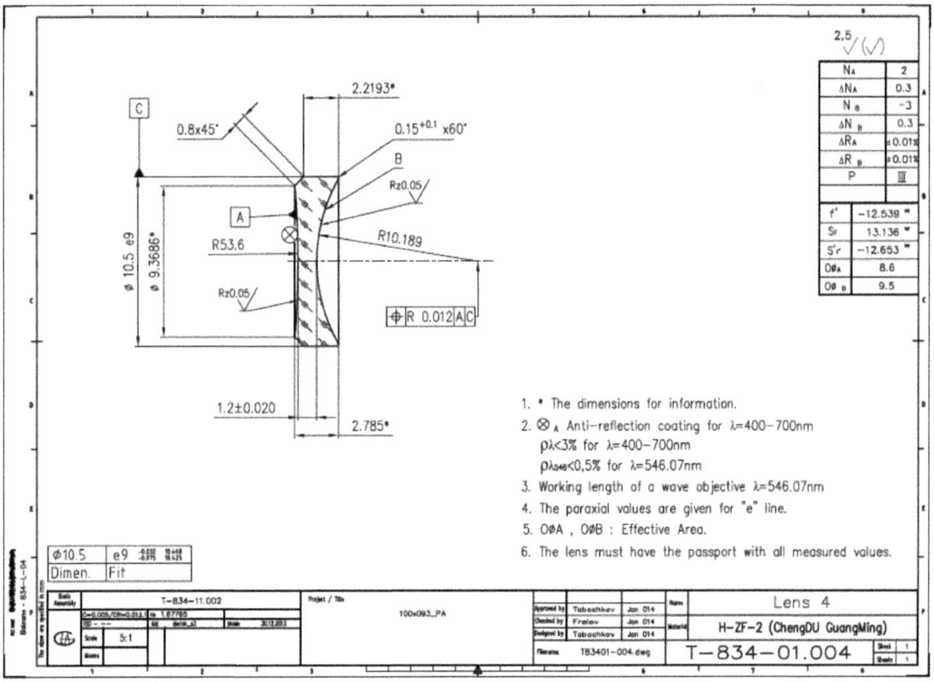

Figure 3.99 Drawing of the fourth lens.

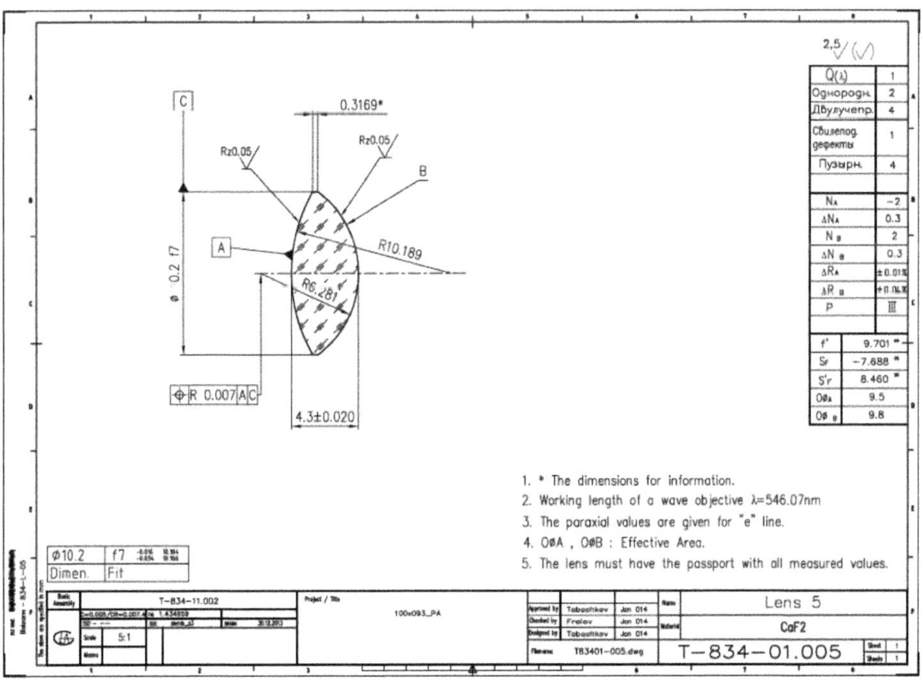

Figure 3.100 Drawing of the fifth lens.

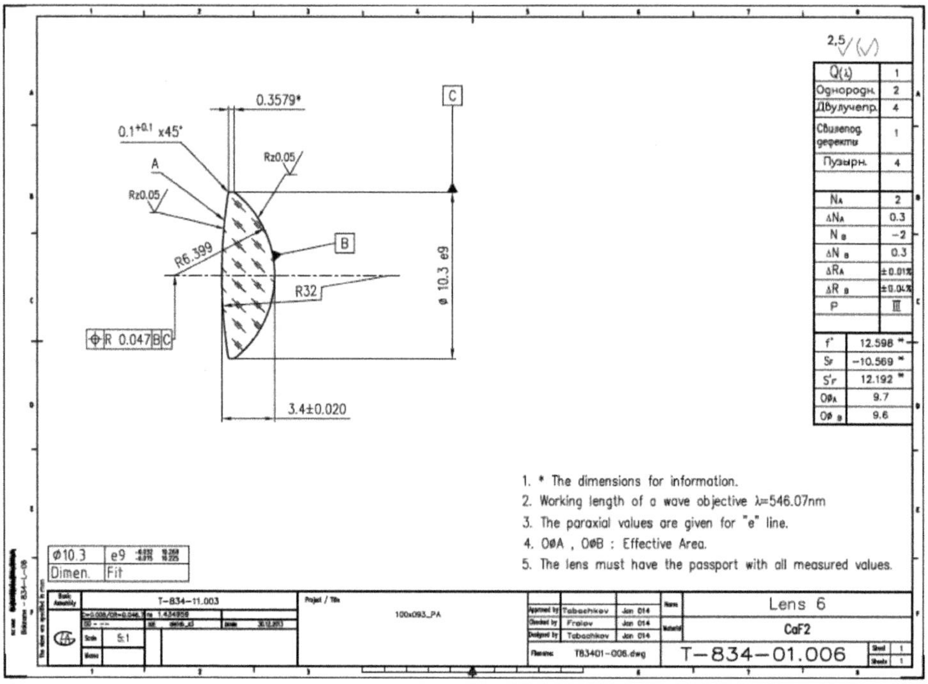

Figure 3.101 Drawing of the sixth lens.

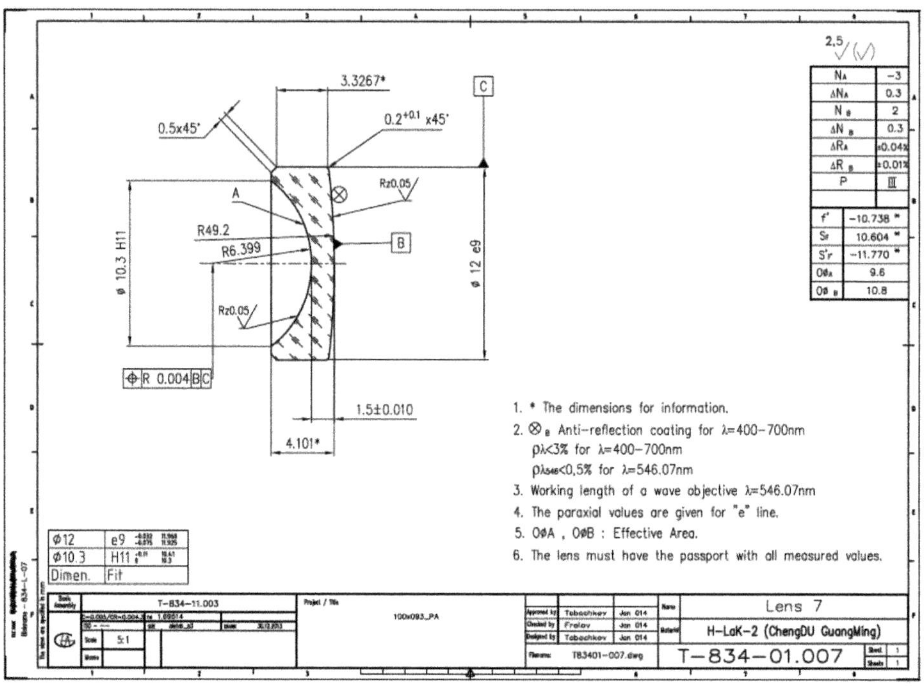

Figure 3.102 Drawing of the seventh lens.

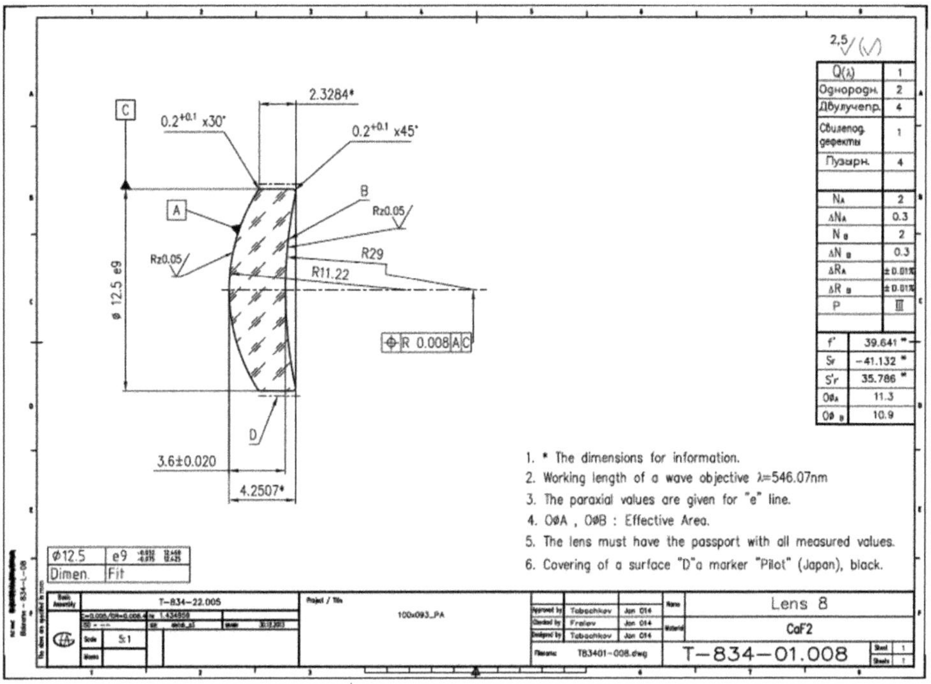

Figure 3.103 Drawing of the eighth lens.

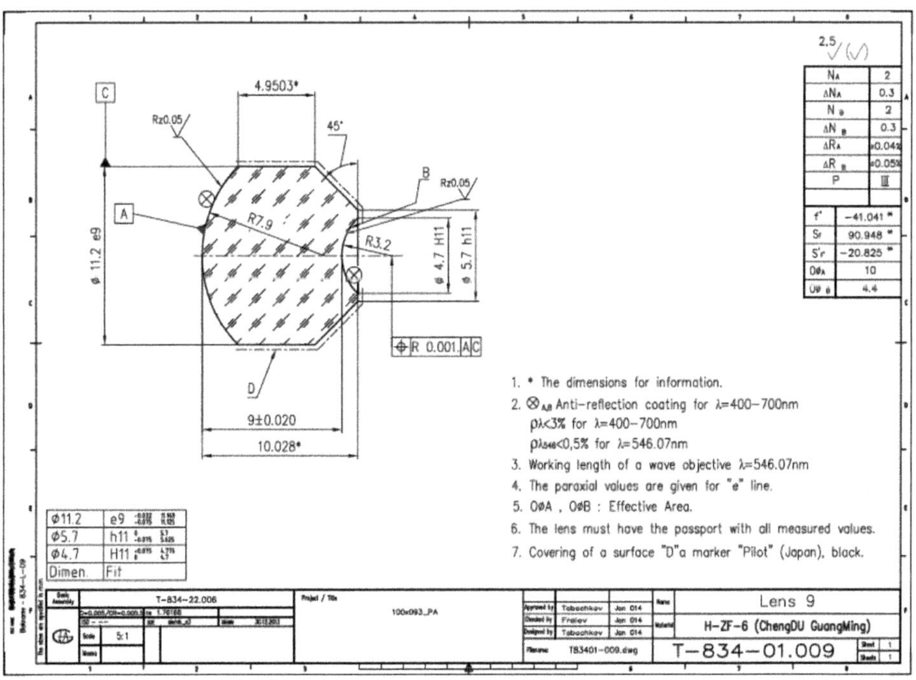

Figure 3.104 Drawing of the ninth lens.

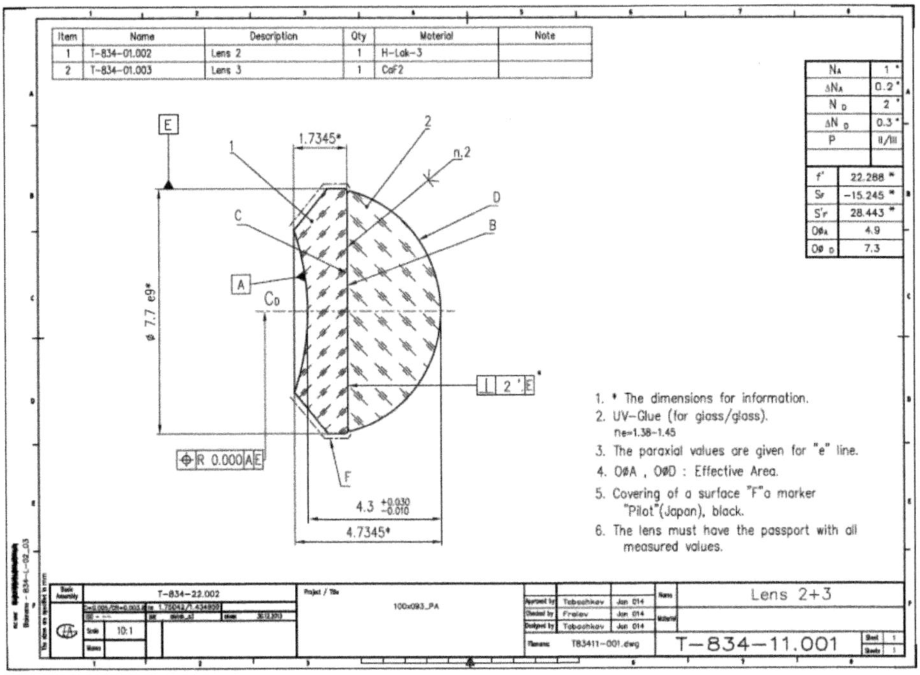

Figure 3.105 Drawing of the second and third lenses.

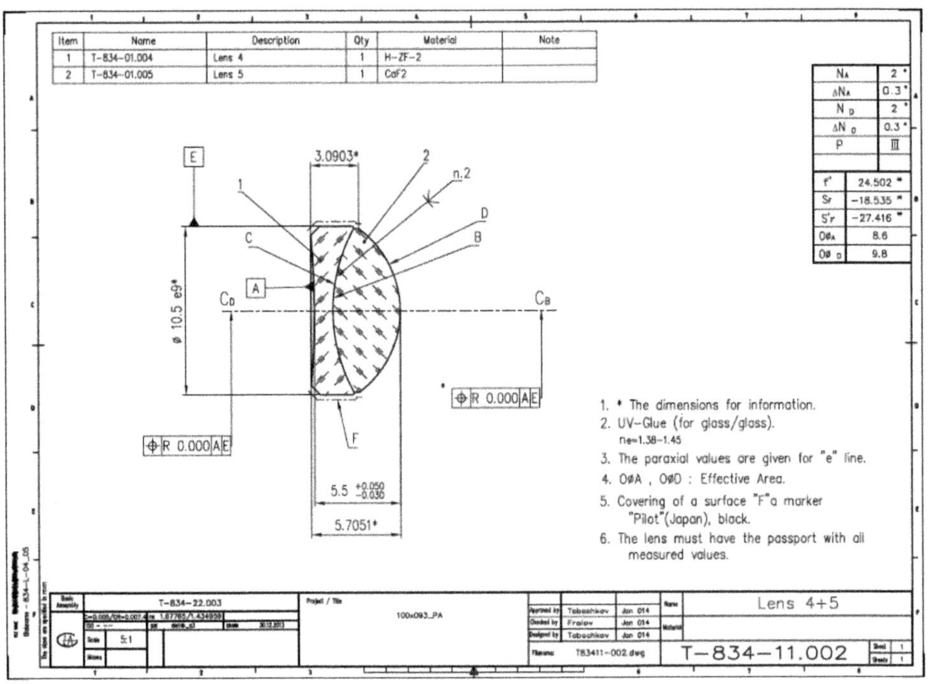

Figure 3.106 Drawing of the fourth and fifth lenses.

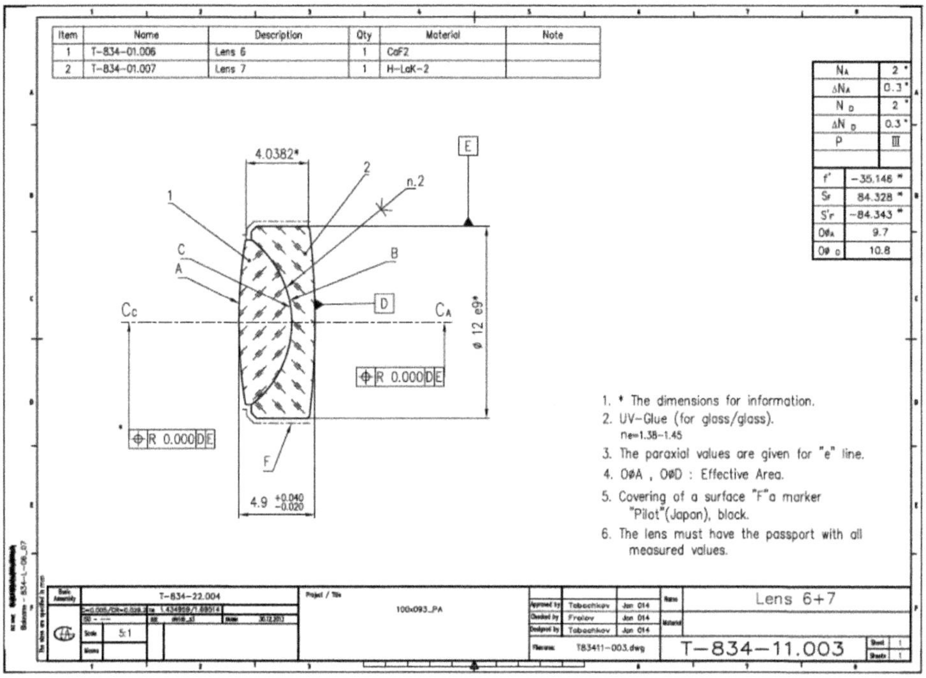

Figure 3.107 Drawing of the sixth and seventh lenses.

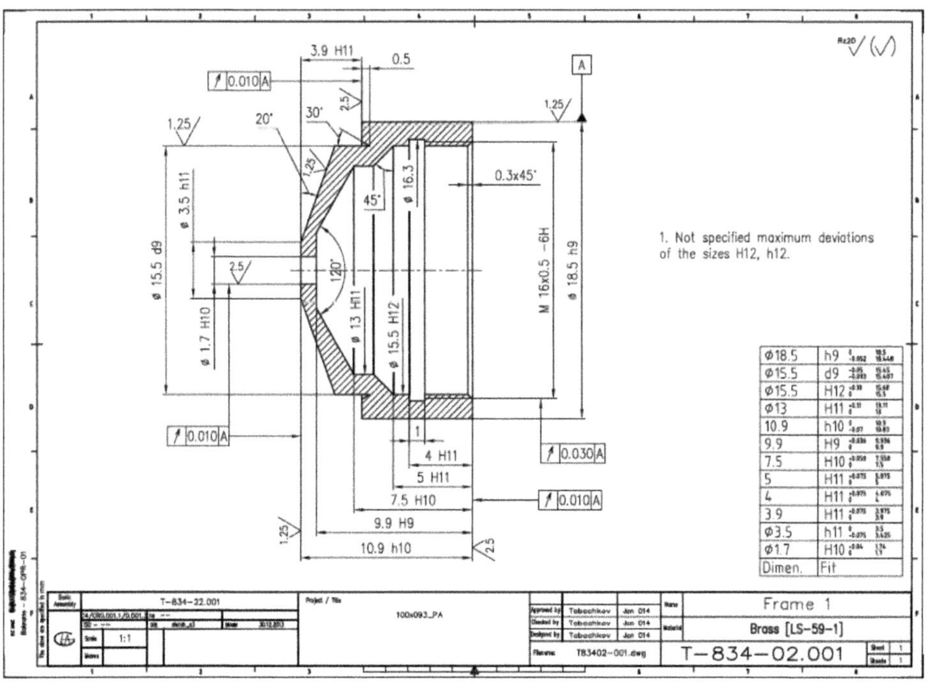

Figure 3.108 Drawing of the first frame.

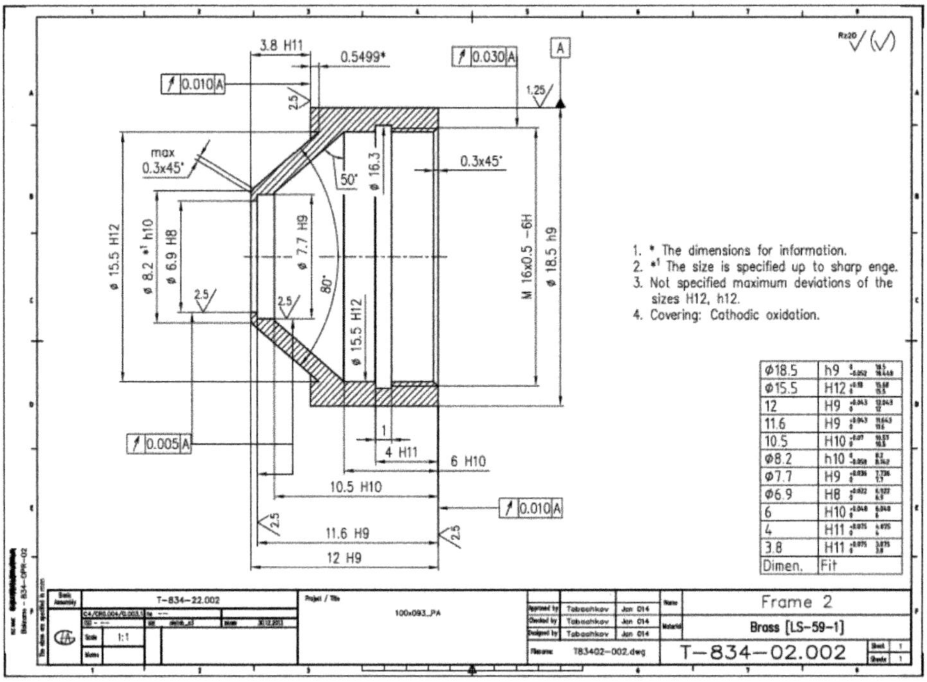

Figure 3.109 Drawing of the second frame.

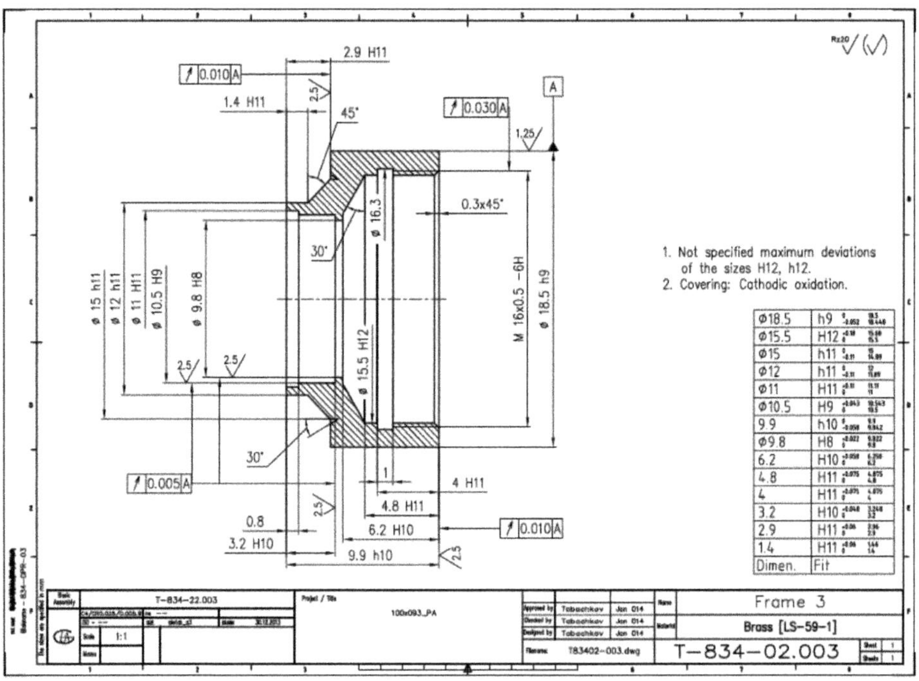

Figure 3.110 Drawing of the third frame.

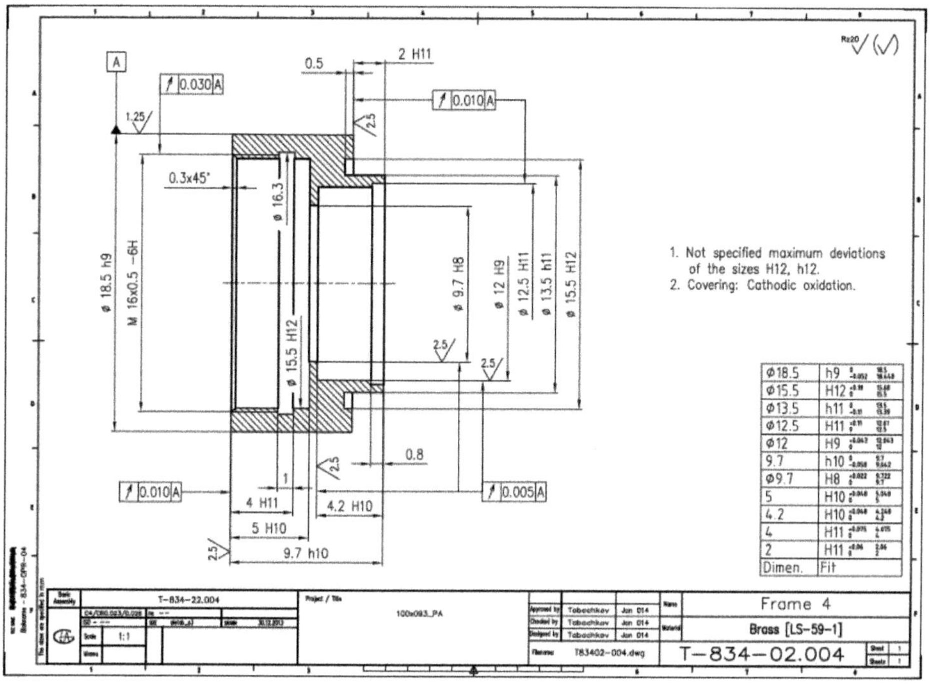

Figure 3.111 Drawing of the fourth frame.

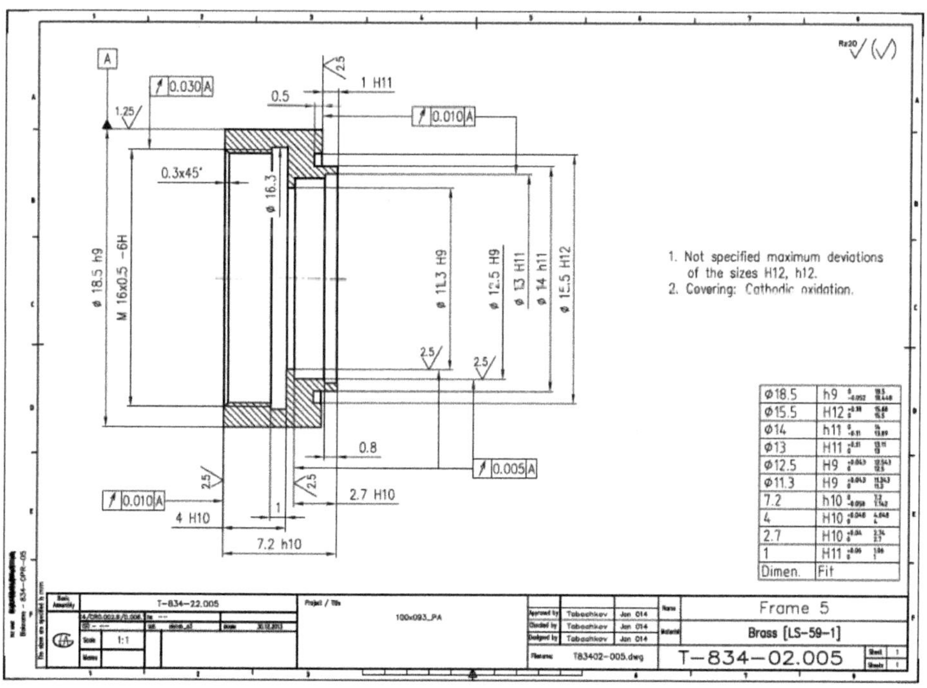

Figure 3.112 Drawing of the fifth frame.

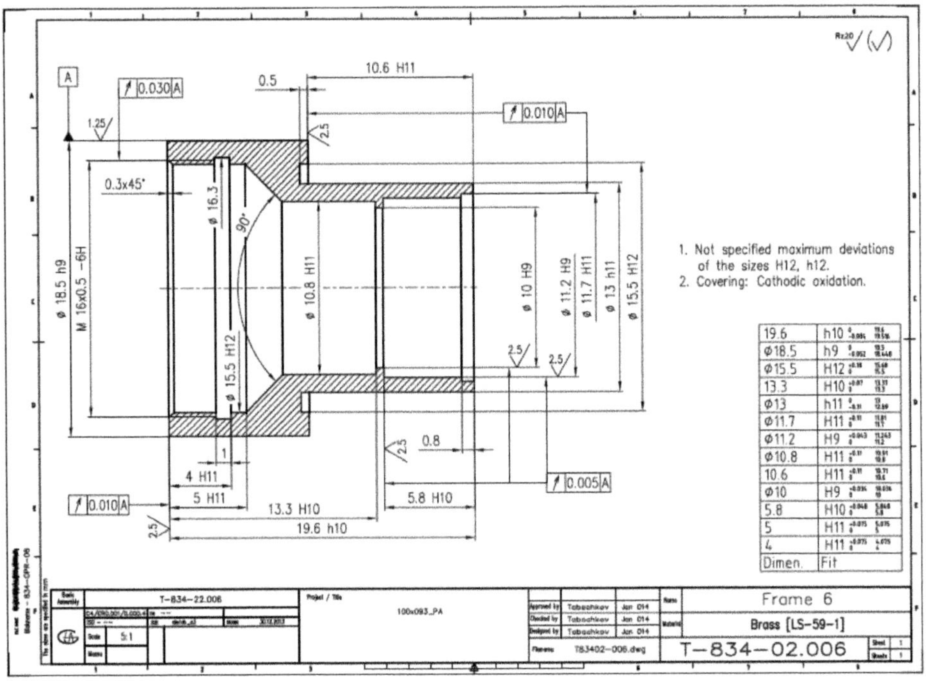

Figure 3.113 Drawing of the sixth frame.

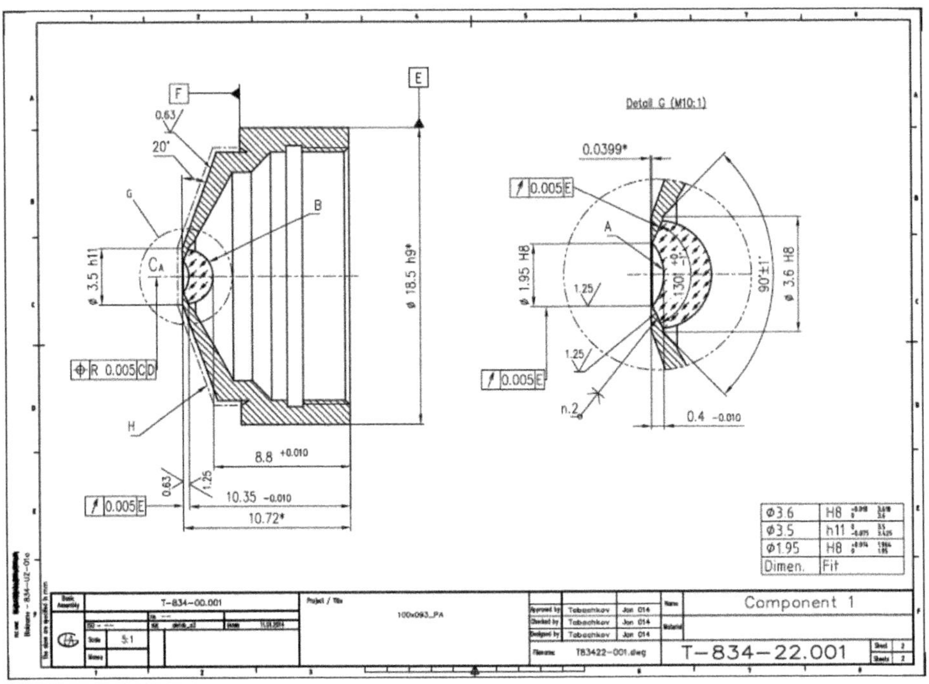

Figure 3.114 Drawing of the first component (no. 1).

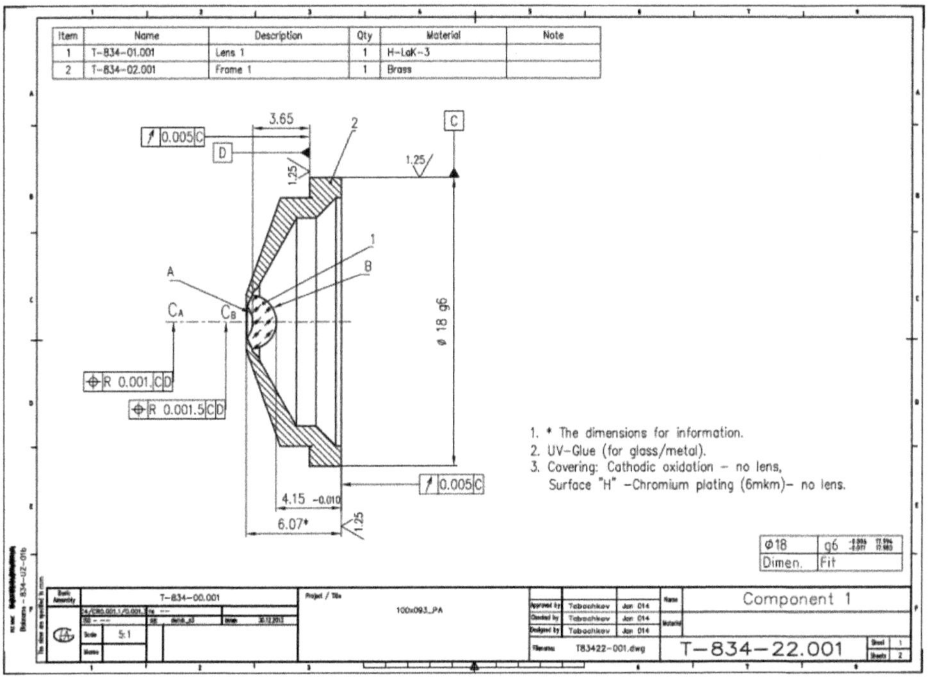

Figure 3.115 Drawing of the first component (no. 2).

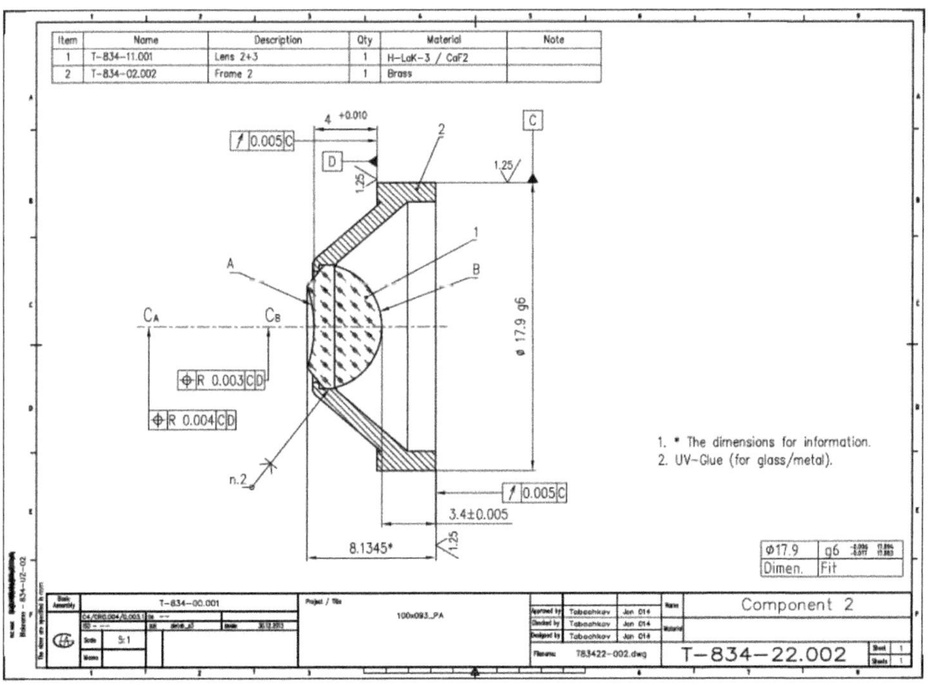

Figure 3.116 Drawing of the second component.

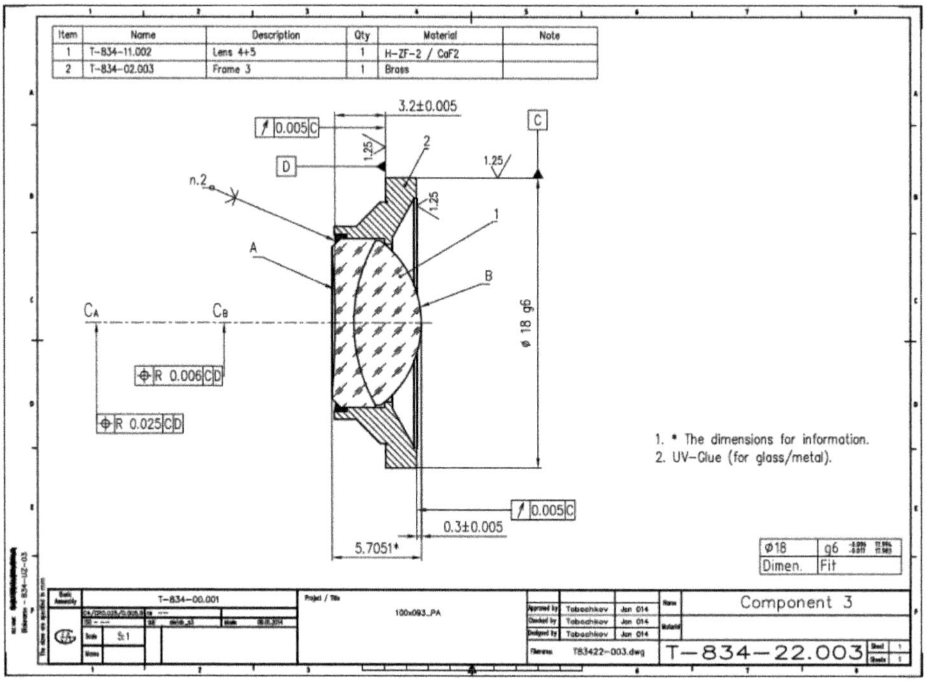

Figure 3.117 Drawing of the third component.

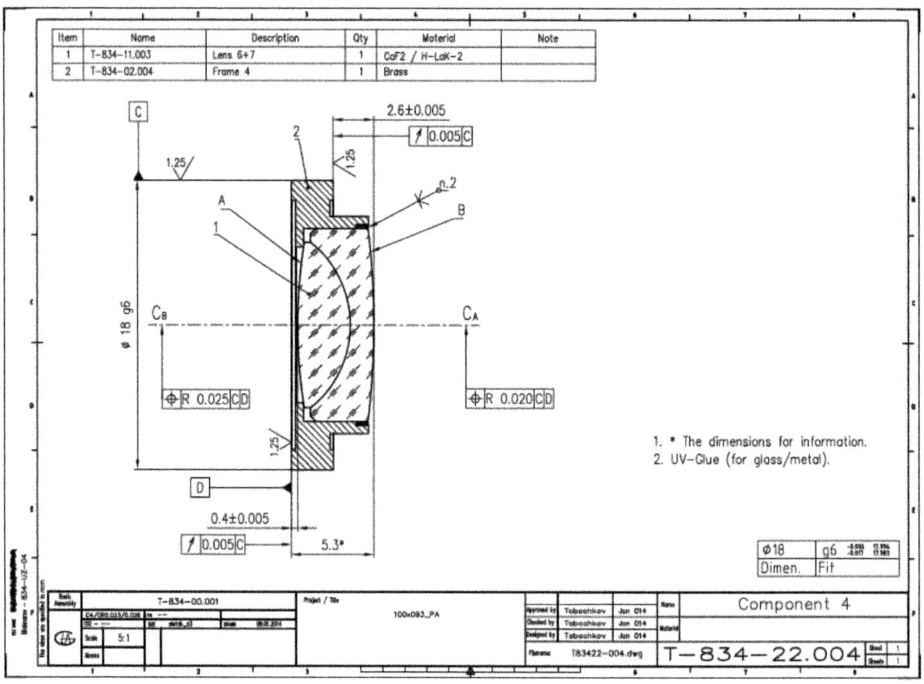

Figure 3.118 Drawing of the fourth component.

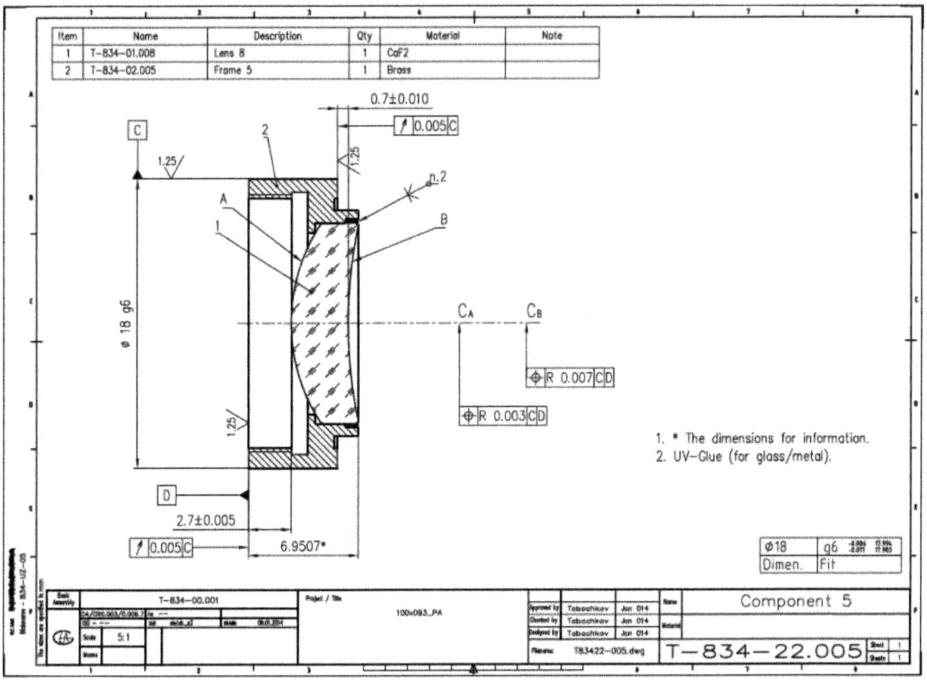

Figure 3.119 Drawing of the fifth component.

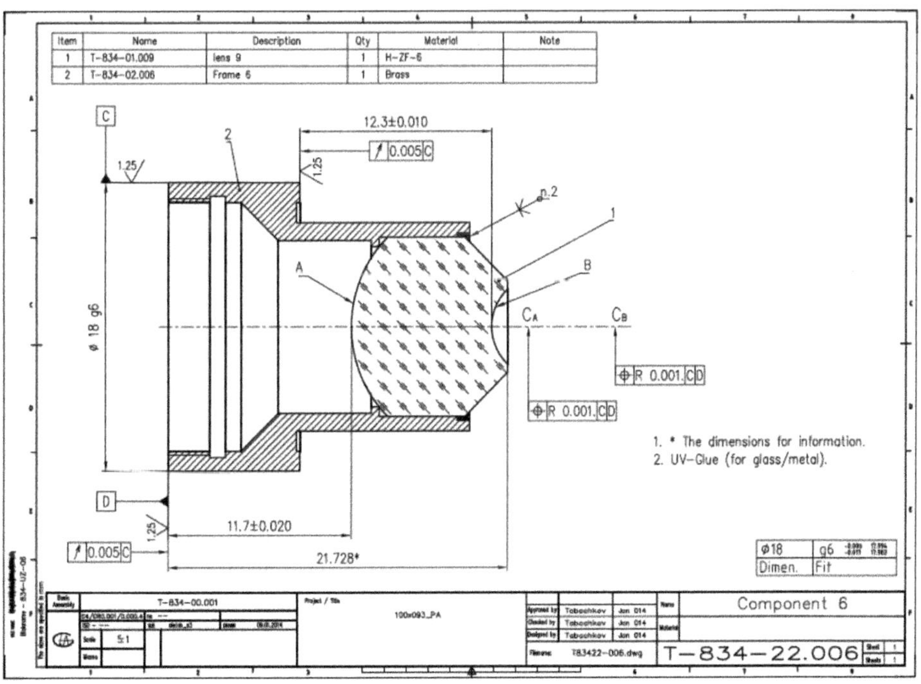

Figure 3.120 Drawing of the sixth component.

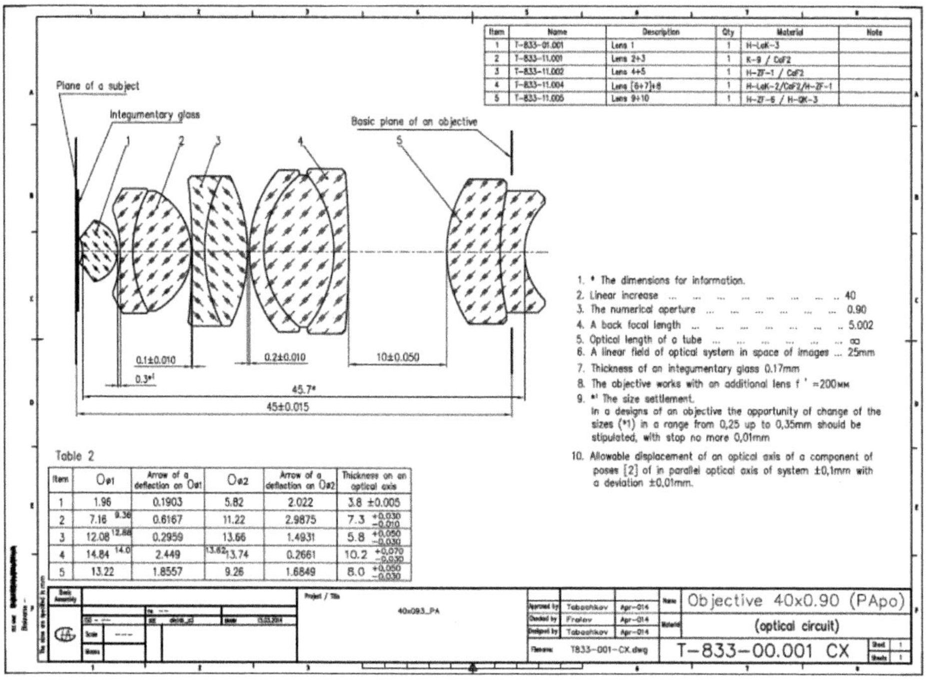

Figure 3.121 The principal optical layout of a 40×/0.90 infinity CCF plan apochromatic objective.

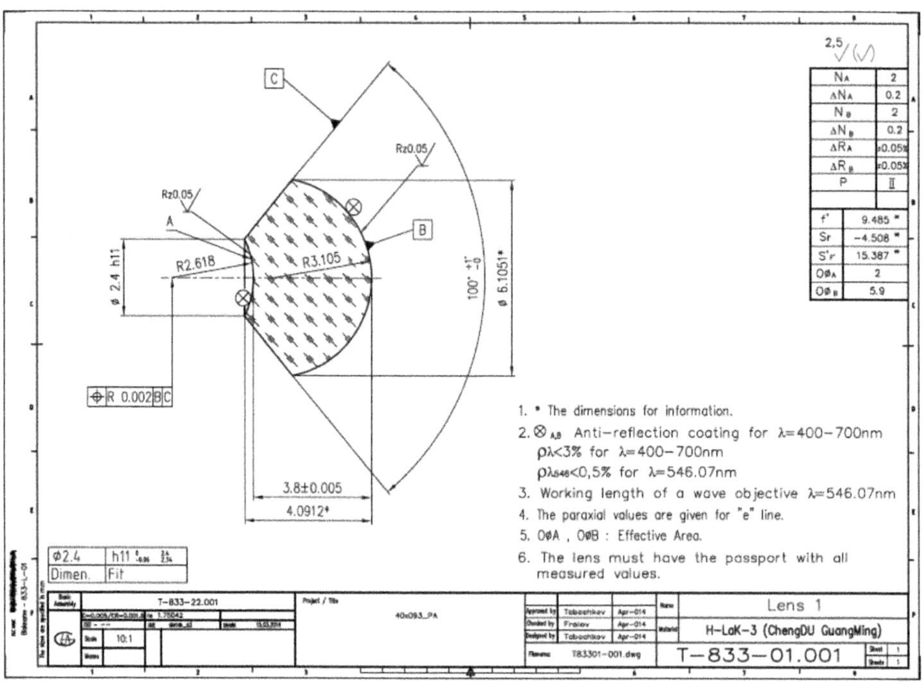

Figure 3.122 Drawing of the first lens.

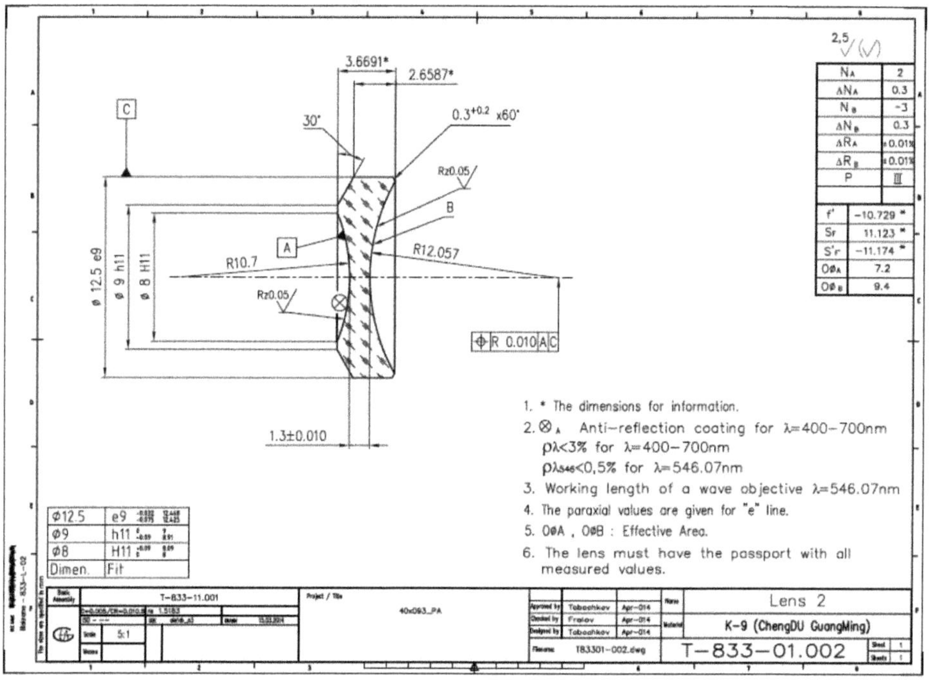

Figure 3.123 Drawing of the second lens.

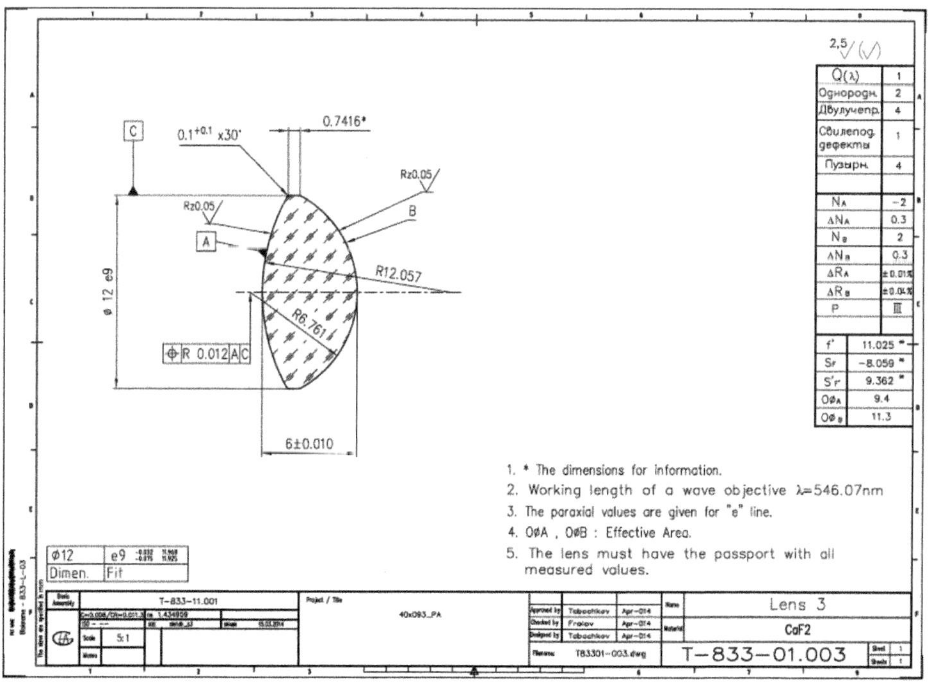

Figure 3.124 Drawing of the third lens.

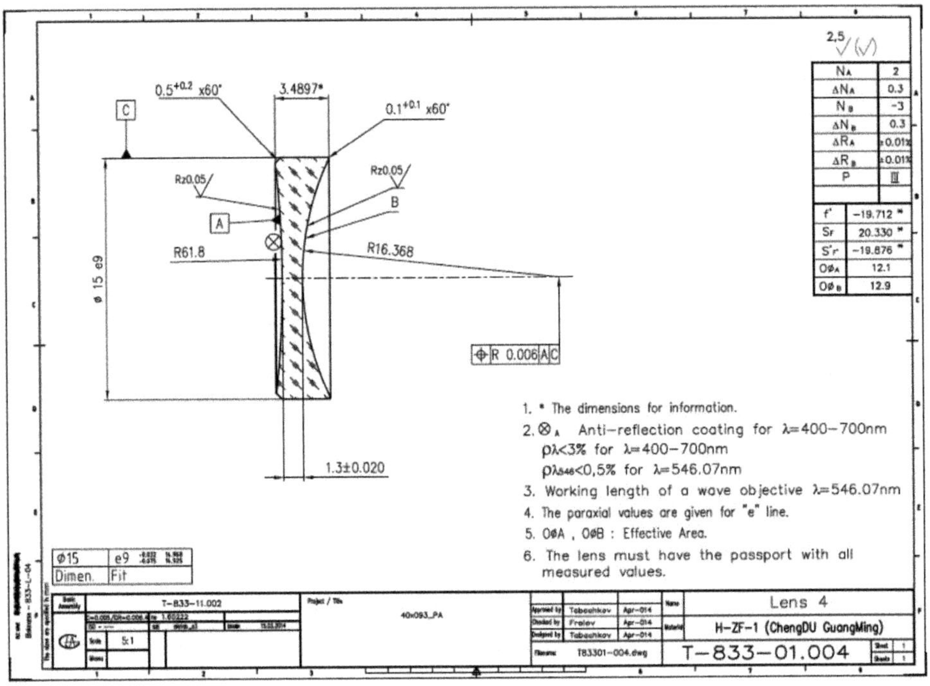

Figure 3.125 Drawing of the fourth lens.

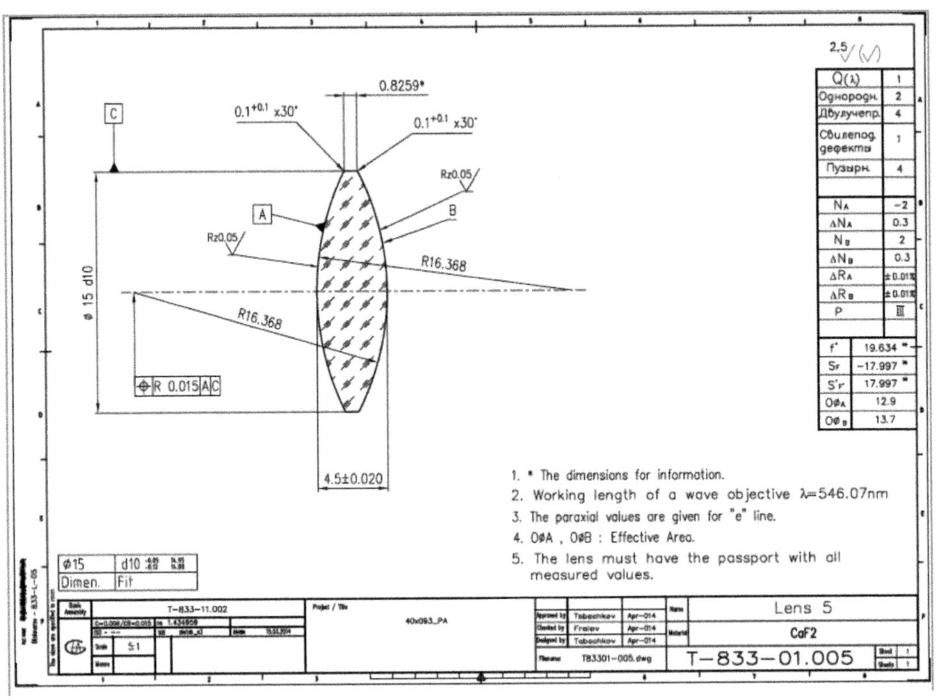

Figure 3.126 Drawing of the fifth lens.

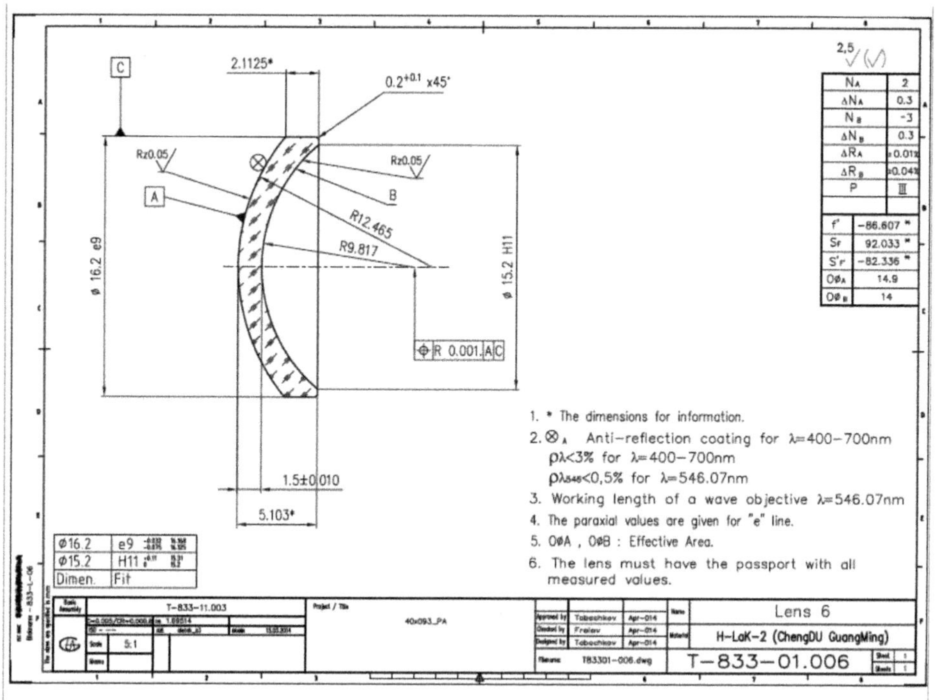

Figure 3.127 Drawing of the sixth lens.

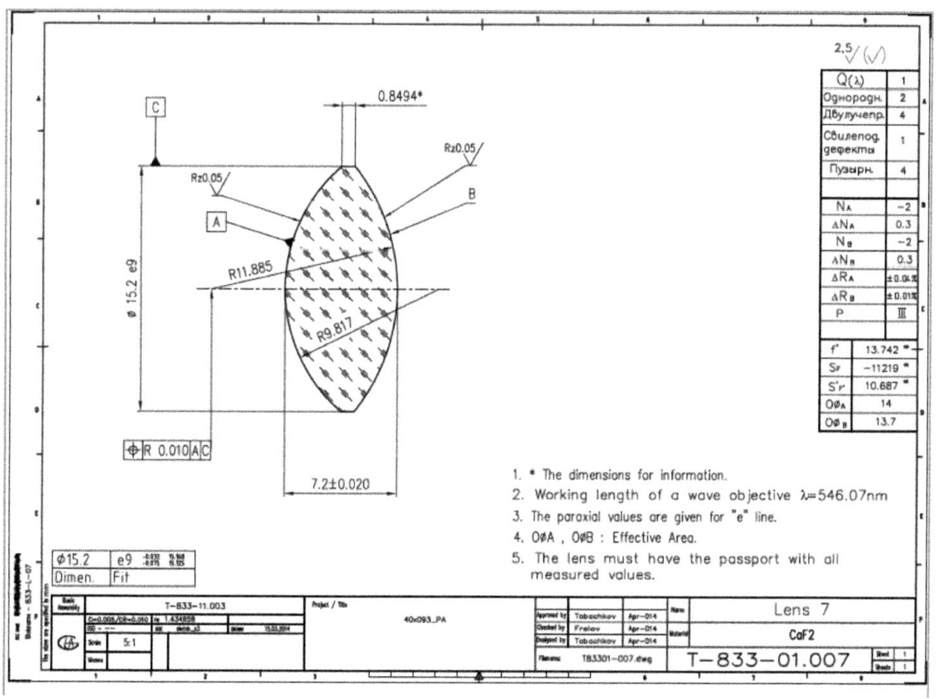

Figure 3.128 Drawing of the seventh lens.

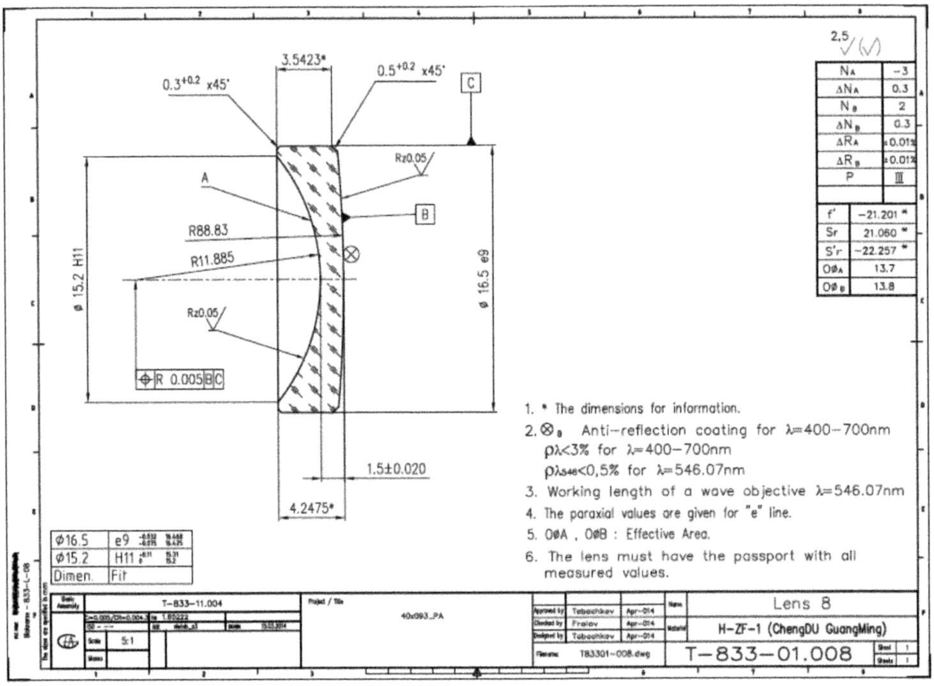

Figure 3.129 Drawing of the eighth lens.

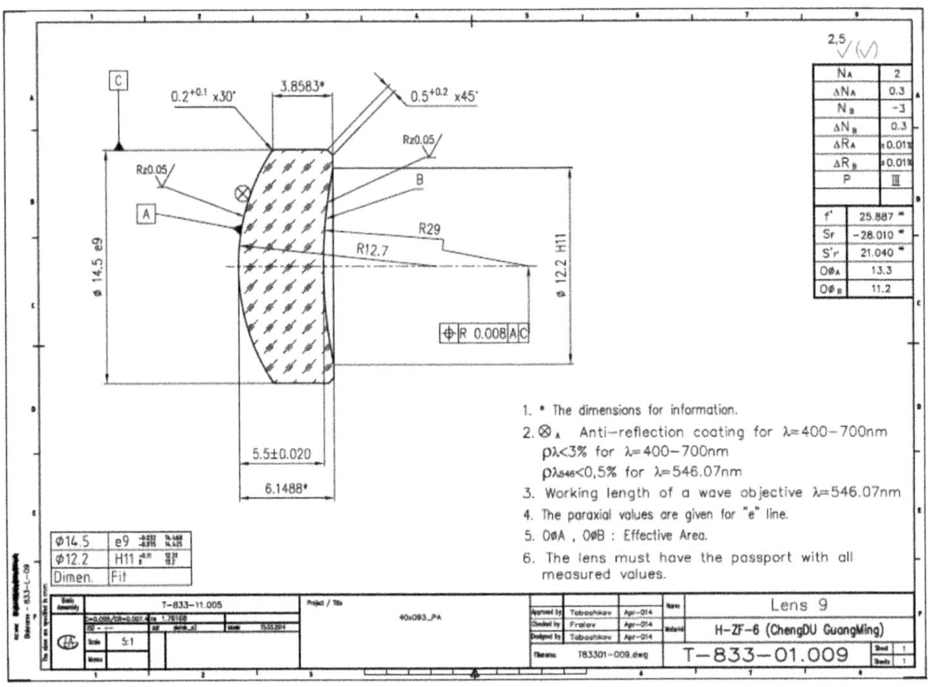

Figure 3.130 Drawing of the ninth lens.

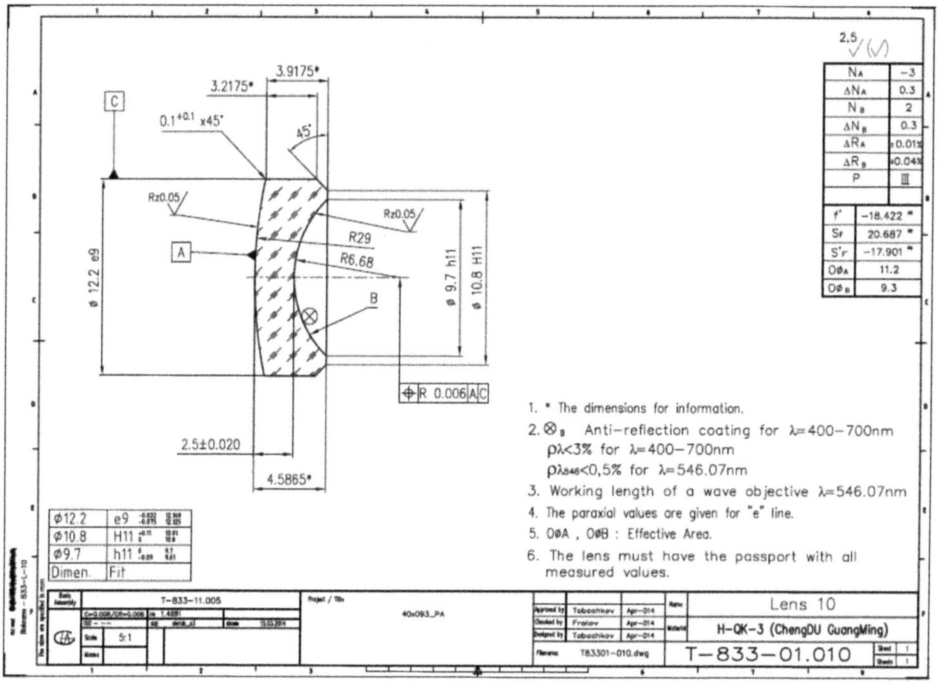

Figure 3.131 Drawing of the 10th lens.

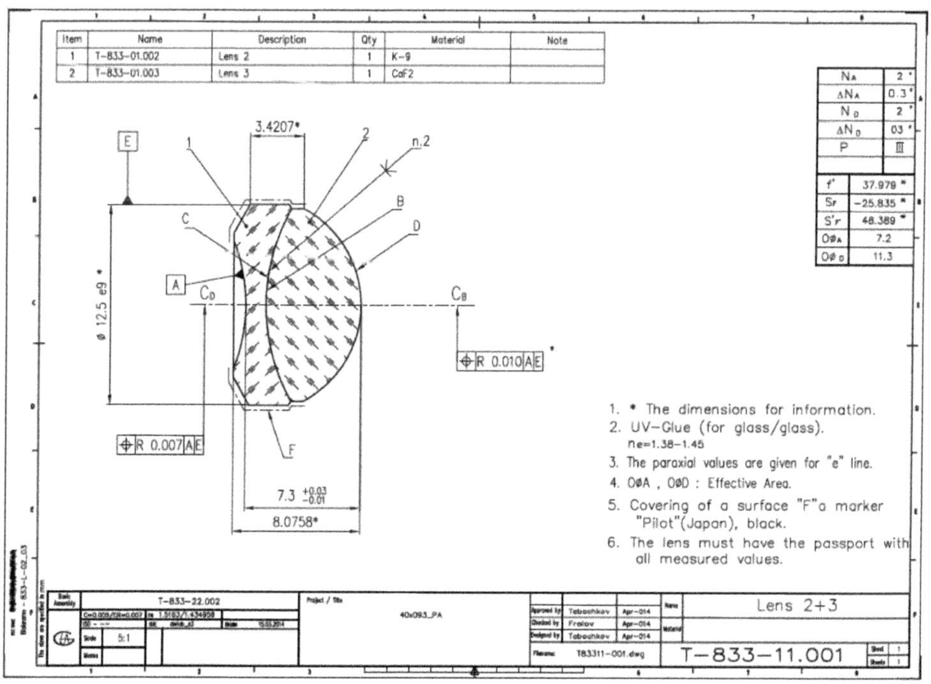

Figure 3.132 Drawing of the second and third lenses.

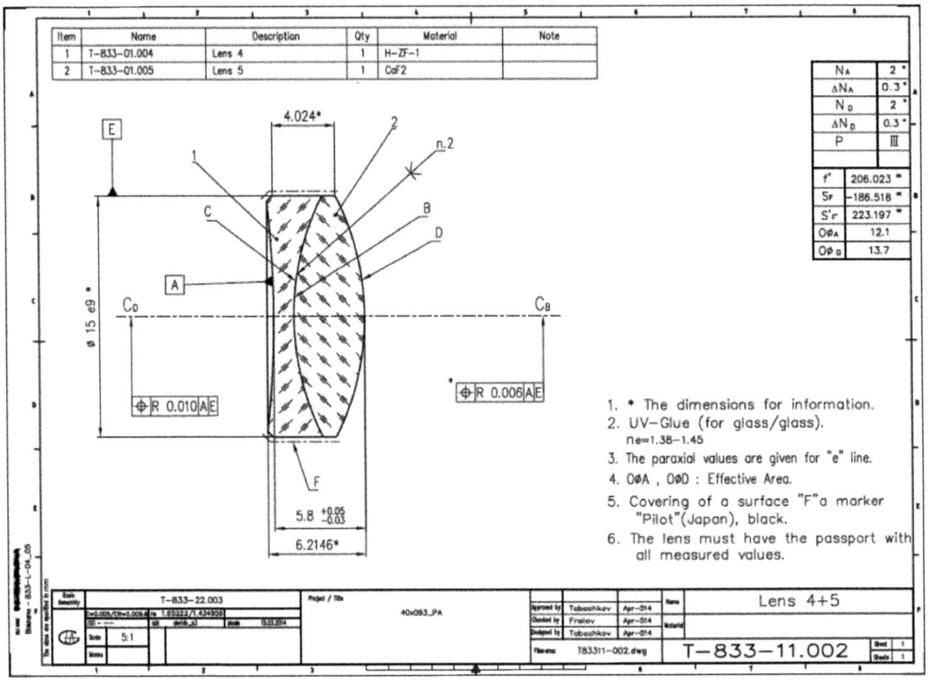

Figure 3.133 Drawing of the fourth and fifth lenses.

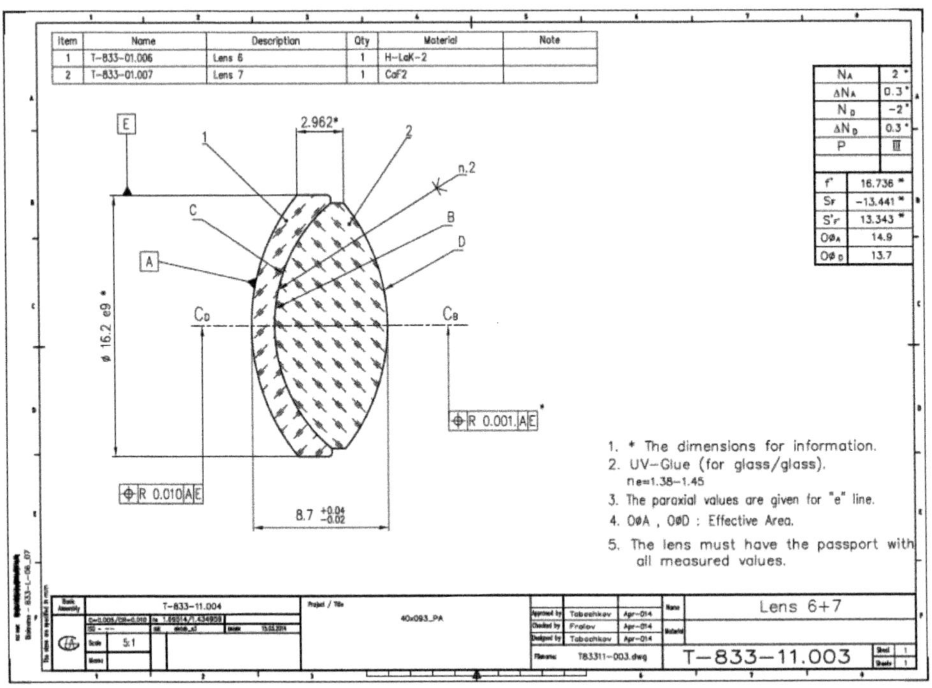

Figure 3.134 Drawing of the sixth and seventh lenses.

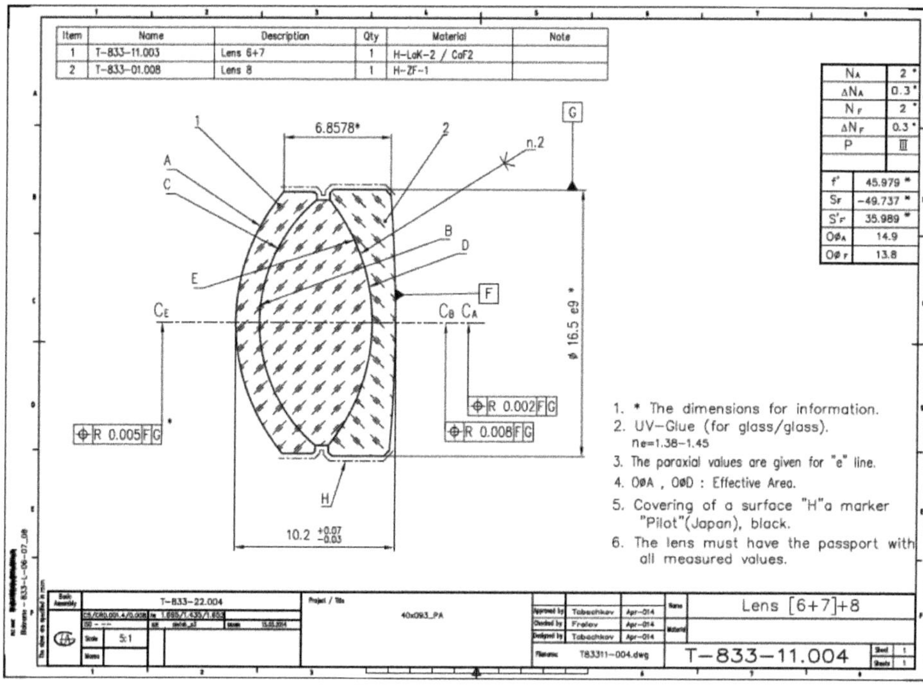

Figure 3.135 Drawing of the sixth, seventh, and eighth lenses.

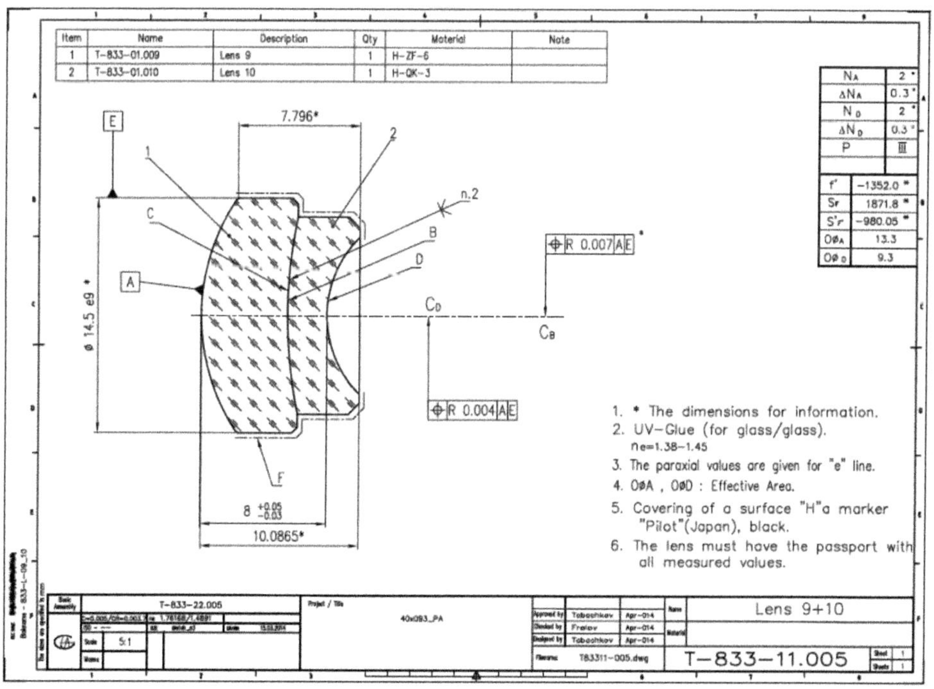

Figure 3.136 Drawing of the ninth and 10th lenses.

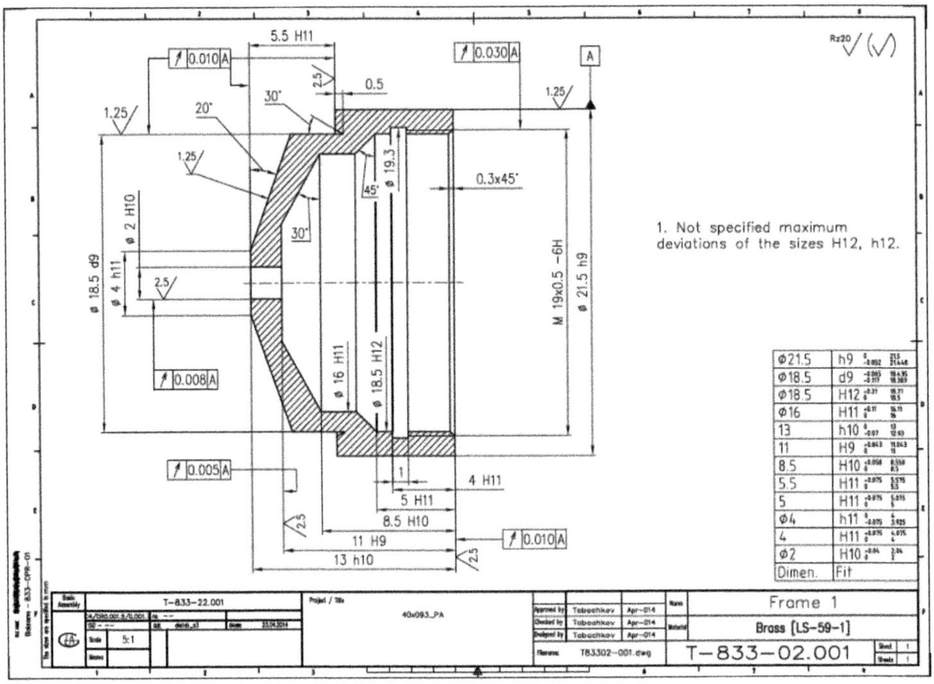

Figure 3.137 Drawing of the first frame.

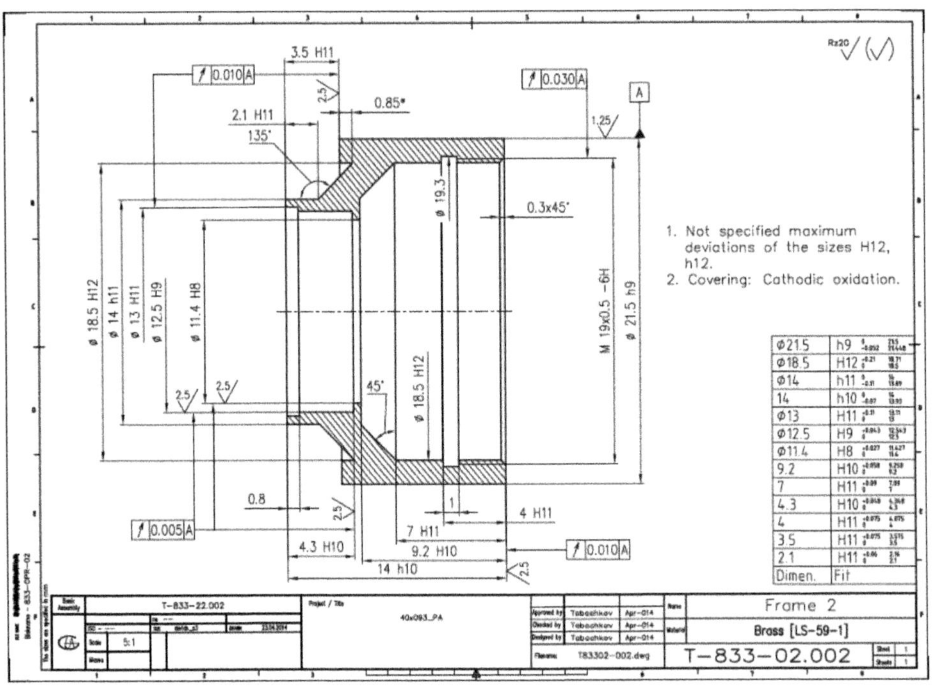

Figure 3.138 Drawing of the second frame.

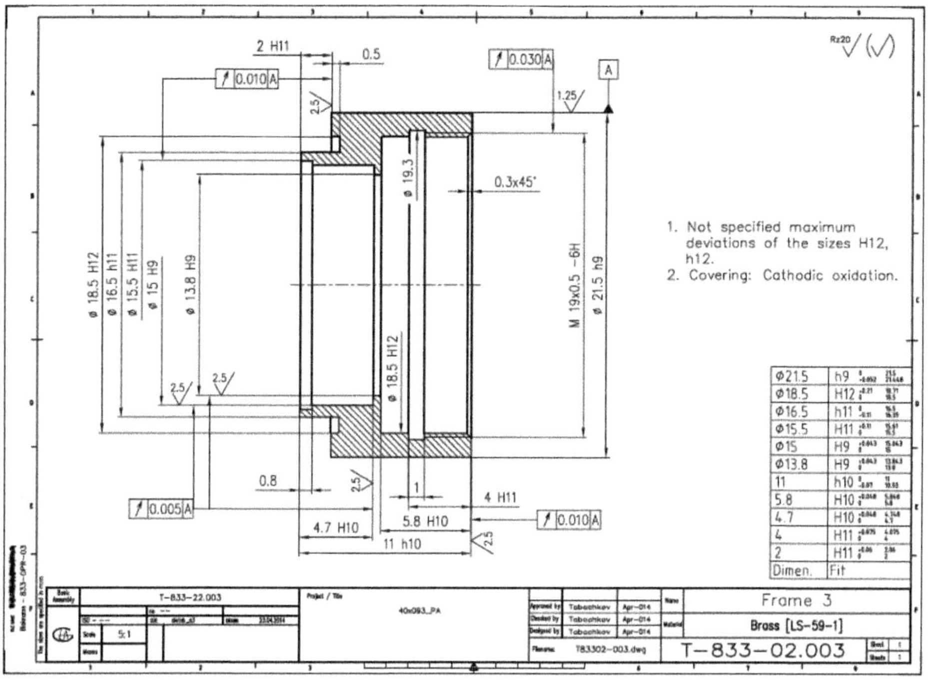

Figure 3.139 Drawing of the third frame.

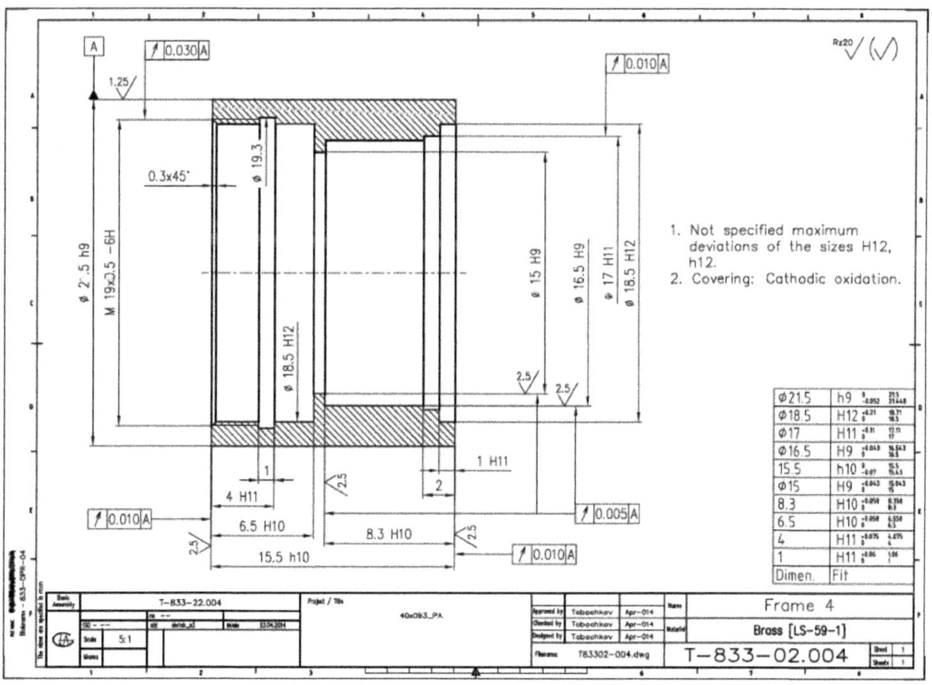

Figure 3.140 Drawing of the fourth frame.

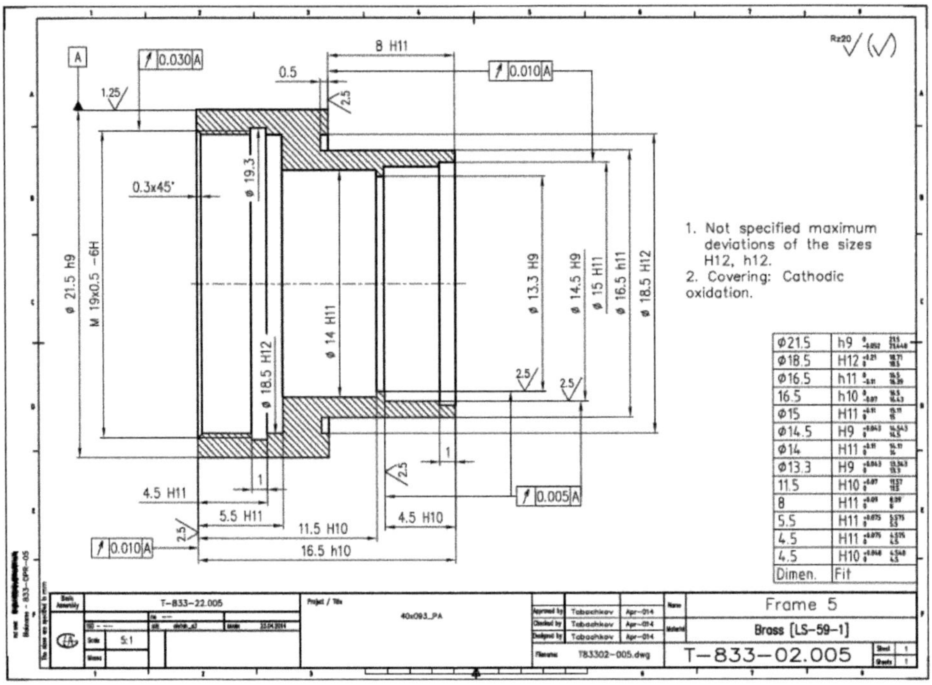

Figure 3.141 Drawing of the fifth frame.

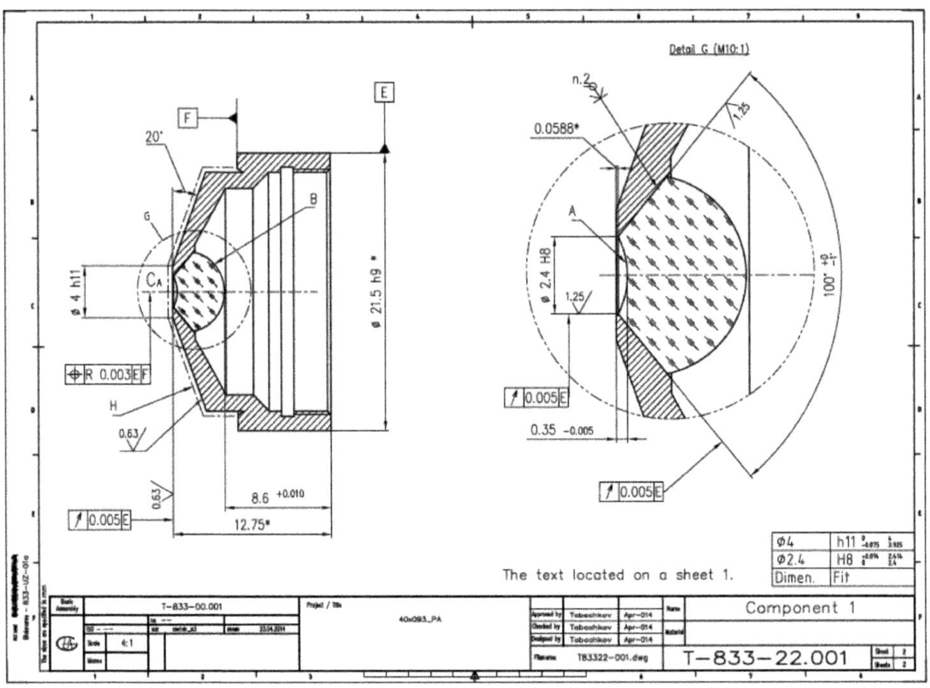

Figure 3.142 Drawing of the first component (no. 1).

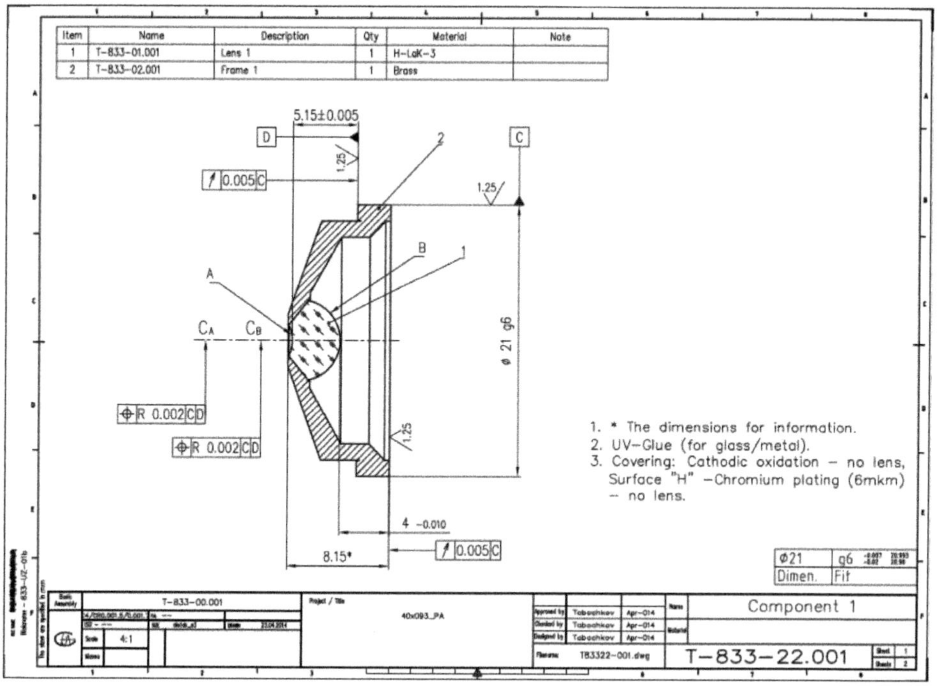

Figure 3.143 Drawing of the first component (no. 2).

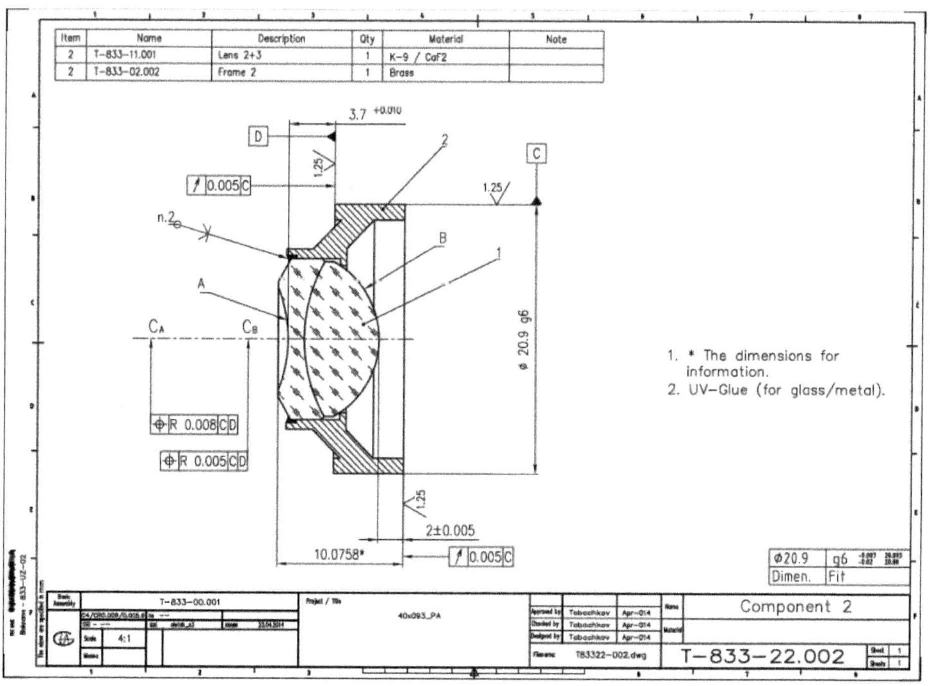

Figure 3.144 Drawing of the second component.

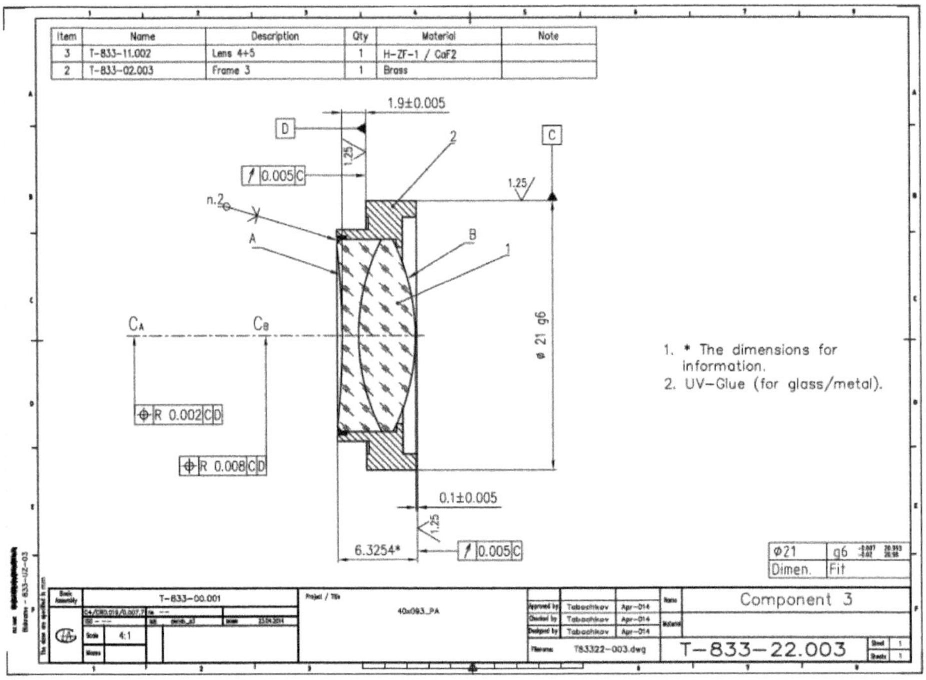

Figure 3.145 Drawing of the third component.

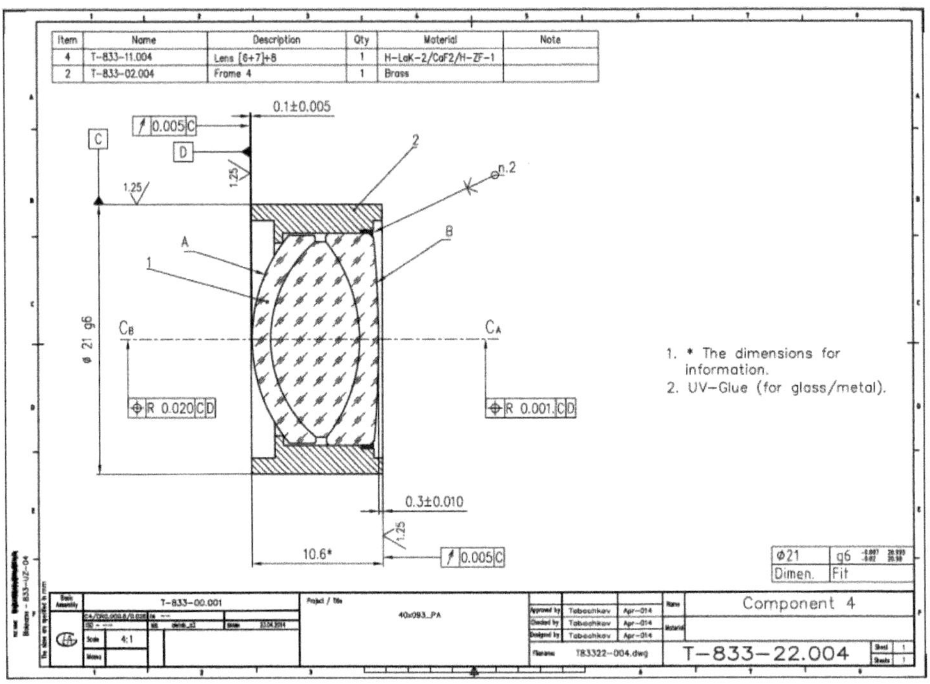

Figure 3.146 Drawing of the fourth component.

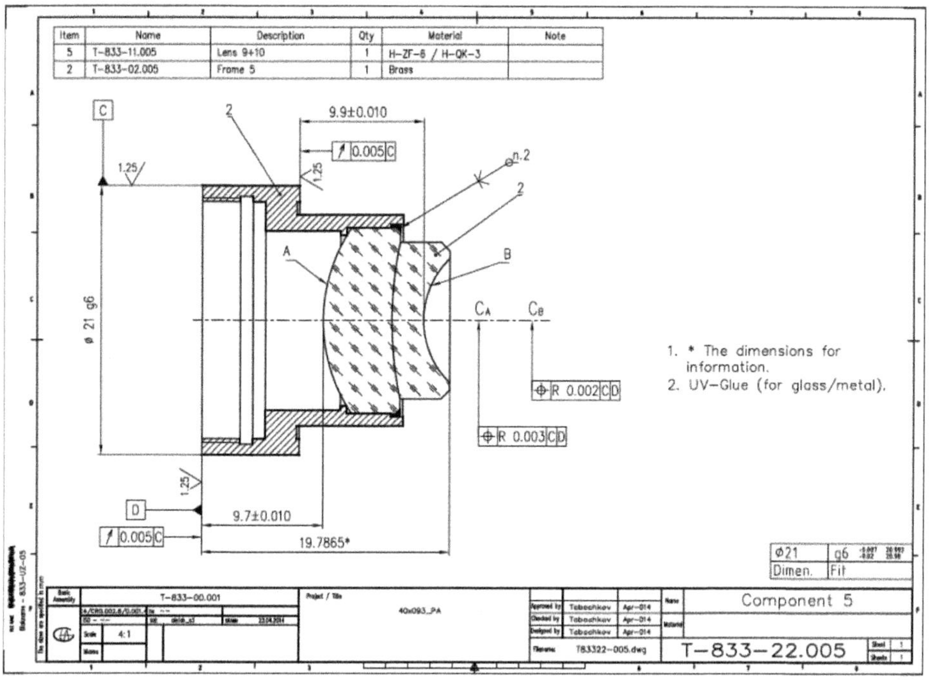

Figure 3.147 Drawing of the fifth component.

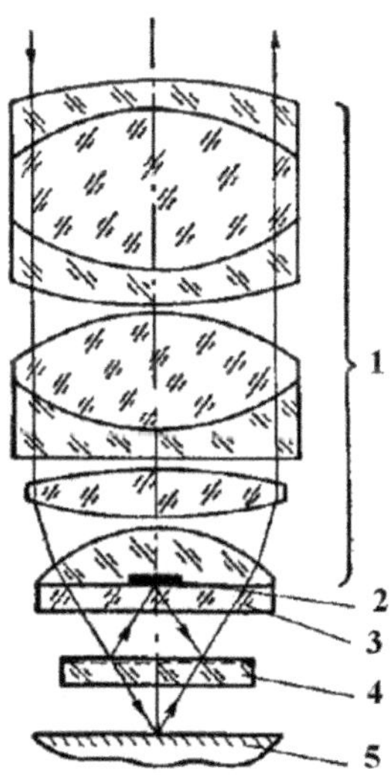

Figure 3.148 Principal diagram of a MIRO objective.

observed in the field of view of the microscope eyepiece. If the surface of the object is slightly tilted, the path difference between the interfering rays from the center of the field to the edges will increase and take all values: 0, $\lambda/2$, λ,

3/2λ, 2λ, etc., where λ is the wavelength. The interference pattern in white light will look like this: in the center there will be either a white achromatic strip, on both sides of which there will be two dark stripes and three to four colored stripes, or a dark strip, on both sides of which there will be three or four colored stripes. When working in monochromatic light, alternating black and light stripes should be visible in the field of view. The width and direction of the stripes is changed by the slope of the investigated object.

Figure 3.149 shows a photograph of the MIO-1 interference objective, which was manufactured at the LOMO factory and used in some models of metallographic microscopes. The MIO-1 interference device is intended for use in some models of metallographic microscopes; it is used to study the microstructure of metals (sections) and to visually assess, measure, and photograph the height of surface irregularities in the range of 0.8 to 0.05 μm. The objective is equipped with plates in special caps that are interchangeable; they can be screwed onto the main lens housing. The plates have different coatings and are used in the study of objects with different reflection coefficients. The plate number, ranging from 1 to 4, is engraved on the caps. Plates are selected depending on the reflection coefficient of the investigated surface. If the value of the surface reflection coefficient is in the range of 0.90 to 0.55, plate number 1 is selected. Plate number 2 corresponds to a reflection coefficient of 0.55 to 0.40. The third plate corresponds to a reflection coefficient of 0.40 to 0.20. When examining an object with a reflection coefficient of 0.20 to 0.04, plate number 4 is selected.

It should be noted that all components of such an objective should differ in the absence of internal stresses. This is why, in the optical designs of these objectives, the use of an optical material such as fluorite is limited.

Figure 3.149 Photograph of an MIO-1 interference objective.

3.7.2 Plan MicroFluars for inverted microscopes

The construction principle of the inverted microscope optical system is described in Chapter 1, Section 1.6.4.1. Depending on the class of the microscope, objectives with a certain degree of aberration correction can be used in it; however, the overall optical design is based on some general conditions. The special laboratory glassware used to locate the objects under study is almost always the same, regardless of the class of microscope. Nevertheless, the use of optics of a higher class provides new possibilities and makes the microscopic process better. This can explain the requirement of microscopists to develop inverted research class microscopes, including the creation of an appropriate element base for such microscopes, primarily objectives. In the principal optical–mechanical design of the microscope, little is changed. Perhaps the lighting system becomes better, for example, to achieve apochromatic aberration correction. Or modern positioning systems are applied based on high-precision stepper motors. Or perhaps specialized software is used to acquire a more advanced morphology of the studied objects.

Qualitative changes in the optics of inverted microscopes can be achieved by using objectives with higher consumer characteristics and/or a higher level of aberration correction. Therefore, in small series research inverted microscopes, the use of plan MicroFluar and plan apochromatic objectives is proposed.

3.7.3 Example optical designs of unified objectives

It is possible to design unified objectives that can be changed by only altering the front part so that they can be used without a cover glass. Then, by adding only some flat plates of a certain thickness (this is how it looks in the optical design) and by changing the thickness of the frontal lens, as well as the value of the free working distance, we can have different optical design options. We can create objectives that work with a standard cover glass ($d = 0.17$ mm), MIRO configurations (we assume that we use two plates, each of which is $d = 1.02$ mm thick), and the use of inverted microscopes (the thickness of the bottom of the laboratory plate may be $d = 1.1$–1.5 mm with the middle value of $d = 1.3$ mm). In some cases, we can use the objective to observe "through the window glass" with a thickness of $d = 5$ mm.

For an objective with a linear magnification of $V = -10\times$, the optical design shown in Chapter 1, Fig. 1.35, can be used as the initial one; to work as part of an inverted microscope, no changes are required (except for adjusting the parfocal height). Figure 3.150 shows the design parameters and graphs of aberration correction of a version of this objective designed to operate in the MIRO configuration. Figures 3.151–3.163 show the design parameters and

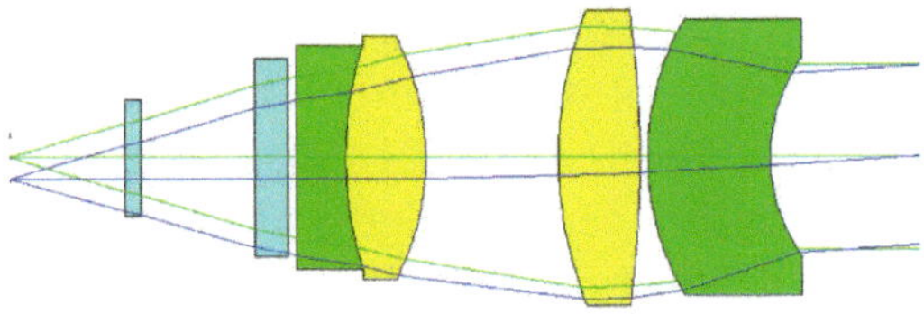

Lens: PlanFluor 10x0.28 MIRO Ft1200

Object num aper 0.280000 Image height 12.500000 Primary wavln 0.546070

SRF	RADIUS	THICKNESS	APERTURE RADIUS	GLASS	SPECIAL
OBJ	0.000000	1.0000e-20	1.247595	AIR	
AST	0.000000	7.070000	1.247595 AS	AIR	
2	0.000000	1.020000	3.310000	LZ_K8	
3	0.000000	7.070000	3.500000	AIR	
4	0.000000	1.020000	5.560000	LZ_K8 P	
5	0.000000	1.020000	5.750000	LZ_K8 P	
6	0.000000	0.500000	5.950000	AIR	
7	0.000000	3.000000	6.090000	LZ_TF5	
8	21.760000	5.000000	6.750000	FPL51	
9	-15.672000	8.200000	7.300000	AIR	
10	23.710000	5.000000	8.890000	FPL51 P	
11	-66.000000	0.500000	8.760000	AIR	
12	16.200000	7.500000	8.290000	LZ_TF5 P	
13	10.233000	200.000000	5.950000	AIR	
14	-36.100000	14.000000	17.460000	K9	
15	-43.560000	3.000000	20.440000	AIR	
16	377.100000	15.000000	21.070000	CAF2	
17	-81.500000	3.000000	21.460000	AIR	
18	221.300000	17.000000	20.980000	CAF2	
19	-72.000000	3.000000	19.960000	LZ_CTK12	
20	0.000000	199.012747 S	19.830000	AIR	
IMS	0.000000	0.000000	12.500000 S		

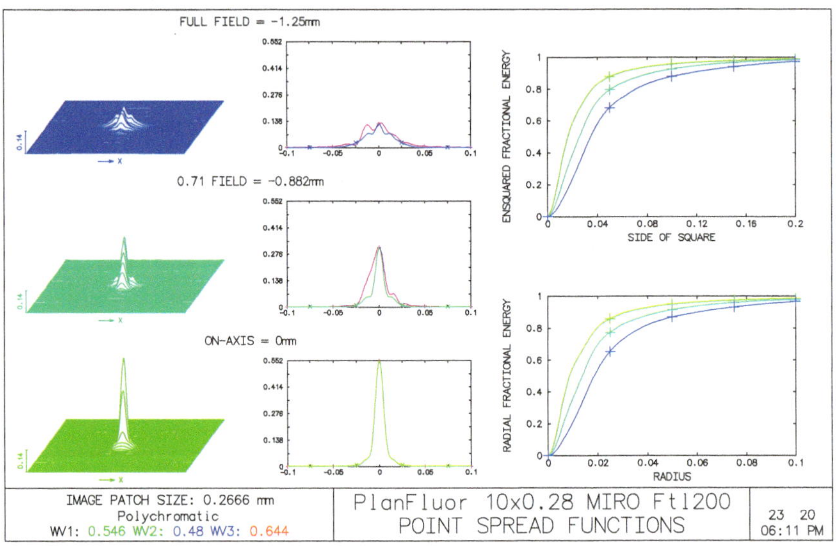

Figure 3.150 Optical design and graphs of aberration correction of a 10×/0.28 MIRO LCF plan MicroFluar objective ($F'_{ts} = 200$ mm).

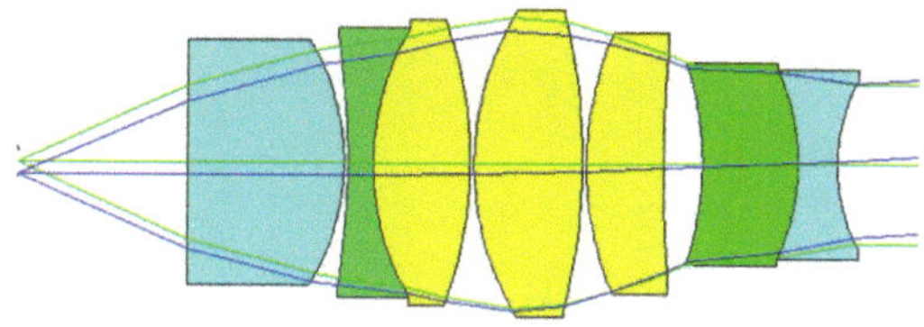

Lens: PlanFluor 20x0.40 d=0 Ft1200

Object num aper 0.400000 Image height 12.500000 Primary wavln 0.546070

SRF	RADIUS	THICKNESS	APERTURE RADIUS	GLASS	SPECIAL
OBJ	0.000000	1.0000e-20	0.625000	AIR	
AST	0.000000	8.793473	0.625000 AS	AIR	
2	0.000000	8.050000	4.460000	LZ_K8	
3	-11.529800	0.200000	6.170000	AIR	
4	-39.839400	1.300000	6.330000	LZ_BF28	
5	13.275200	5.000000	6.800000	FPL51	
6	-19.932300	0.200000	7.190000	AIR	
7	14.514300	5.600000	7.720000	FPL51 P	
8	-29.431800	0.200000	7.440000	AIR	
9	16.348800	4.000000	6.600000	FPL51	
10	87.791700	2.000000	5.580000	AIR	
11	-15.409700	5.000000	5.070000	TF5	
12	-10.919500	2.000000	4.760000	LZ_K8	
13	8.720800	200.000000	4.170000	AIR	
14	-36.100000	14.000000	16.340000	LZ_K8	
15	-43.560000	3.000000	19.140000	AIR	
16	377.100000	15.000000	19.700000	CAF2	
17	-81.500000	3.000000	20.100000	AIR	
18	221.300000	17.000000	19.700000	CAF2	
19	-72.000000	3.000000	18.740000	LZ_CTK12	
20	0.000000	199.996273 S	18.640000	AIR	
IMS	0.000000	0.000000	12.500000 S		

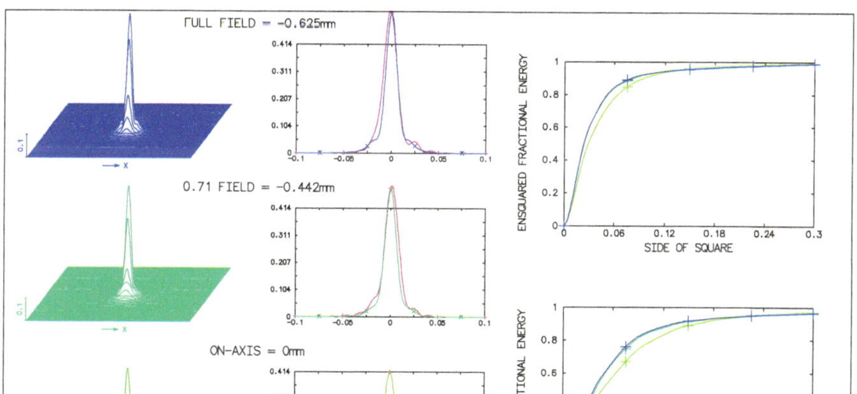

Figure 3.151 Optical design and graphs of aberration correction of a 20×/0.40 ($d = 0$) LCF plan MicroFluar objective ($F'_{ts} = 200$ mm).

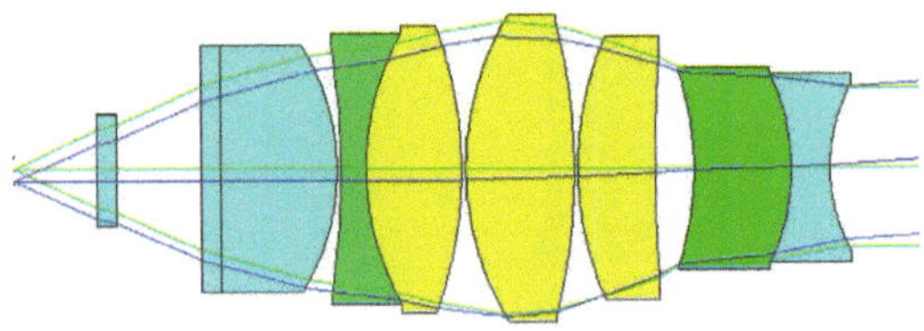

Lens: PlanFluor 20x0.40 MIRO Ftl200

Object num aper 0.400000 Image height 12.500000 Primary wavln 0.546070

SRF	RADIUS	THICKNESS		APERTURE RADIUS		GLASS		SPECIAL
OBJ	0.000000	1.0000e-20		0.625013		AIR		
AST	0.000000	4.400000		0.625013	AS	AIR		
2	0.000000	1.020000		2.540000		LZ_K8		
3	0.000000	4.400000		2.820000		AIR		
4	0.000000	1.020000		4.740000		LZ_K8	P	
5	0.000000	6.000000		5.020000		LZ_K8	P	
6	-11.529800	0.200000		6.170000		AIR		
7	-39.839400	1.300000		6.330000		LZ_BF28		
8	13.275200	5.000000		6.800000		FPL51		
9	-19.932300	0.200000		7.190000		AIR		
10	14.514300	5.600000		7.720000		FPL51	P	
11	-29.431800	0.200000		7.440000		AIR		
12	16.348800	4.000000		6.600000		FPL51		
13	87.791700	2.000000		5.580000		AIR		
14	-15.409700	5.000000		5.070000		TF5		
15	-10.919500	2.000000		4.760000		LZ_K8	P	
16	8.720800	200.000000		4.170000		AIR		
17	-36.100000	14.000000		16.340000		LZ_K8	P	
18	-43.560000	3.000000		19.140000		AIR		
19	377.100000	15.000000		19.700000		CAF2		
20	-81.500000	3.000000		20.100000		AIR		
21	221.300000	17.000000		19.700000		CAF2		
22	-72.000000	3.000000		18.740000		LZ_CTK12		
23	0.000000	200.020005	S	18.640000		AIR		
IMS	0.000000	0.000000		12.500000	S			

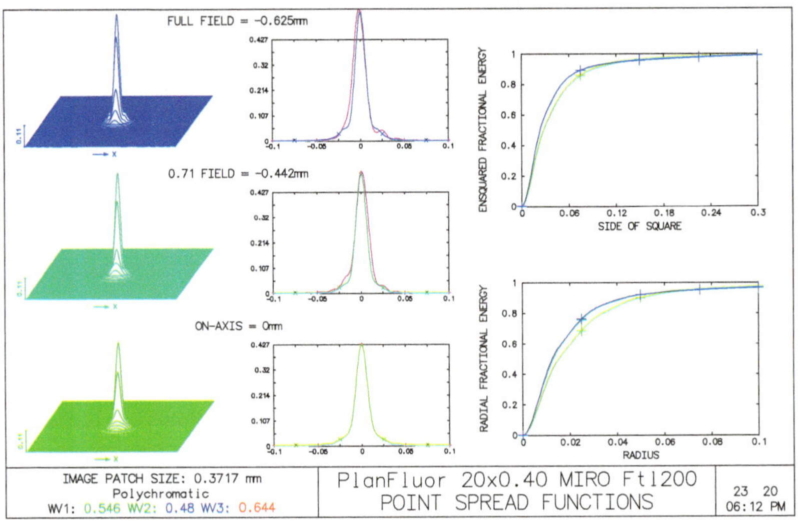

Figure 3.152 Optical design and graphs of aberration correction of a 20×/0.40 MIRO LCF plan MicroFluar objective ($F'_{ts} = 200$ mm).

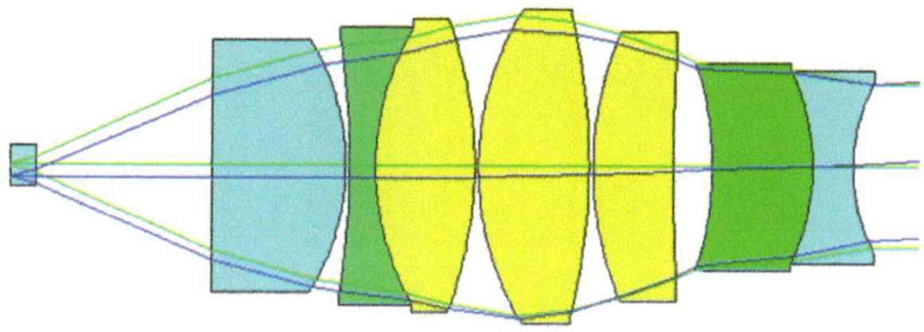

Lens:PlanFluor 20x0.40 d=1.3mm Ftl200

Object num aper 0.400000 Image height 12.500000 Primary wavln 0.546070

SRF	RADIUS	THICKNESS	APERTURE RADIUS	GLASS	SPECIAL
OBJ	0.000000	0.010000	0.625011	FIZIO	
AST	0.000000	1.300000	0.628148 AS	LZ_LK5	
2	0.000000	8.776900	0.990000	AIR	
3	0.000000	6.730000	4.820000	LZ_K8	
4	-11.529800	0.200000	6.180000	AIR	
5	-39.839400	1.300000	6.330000	LZ_BF28	
6	13.275200	5.000000	6.800000	FPL51	
7	-19.932300	0.200000	7.190000	AIR	
8	14.514300	5.600000	7.720000	FPL51 P	
9	-29.431800	0.200000	7.440000	AIR	
10	16.348800	4.000000	6.600000	FPL51	
11	87.791700	2.000000	5.580000	AIR	
12	-15.409700	5.000000	5.070000	TF5	
13	-10.919500	2.000000	4.760000	LZ_K8	
14	8.720800	200.000000	4.170000	AIR	
15	-36.100000	14.000000	16.340000	LZ_K8	
16	-43.560000	3.000000	19.140000	AIR	
17	377.100000	15.000000	19.700000	CAF2	
18	-81.500000	3.000000	20.100000	AIR	
19	221.300000	17.000000	19.700000	CAF2	
20	-72.000000	3.000000	18.740000	LZ_CTK12	
21	0.000000	200.015677 S	18.640000	AIR	
IMS	0.000000	0.000000	12.500000 S		

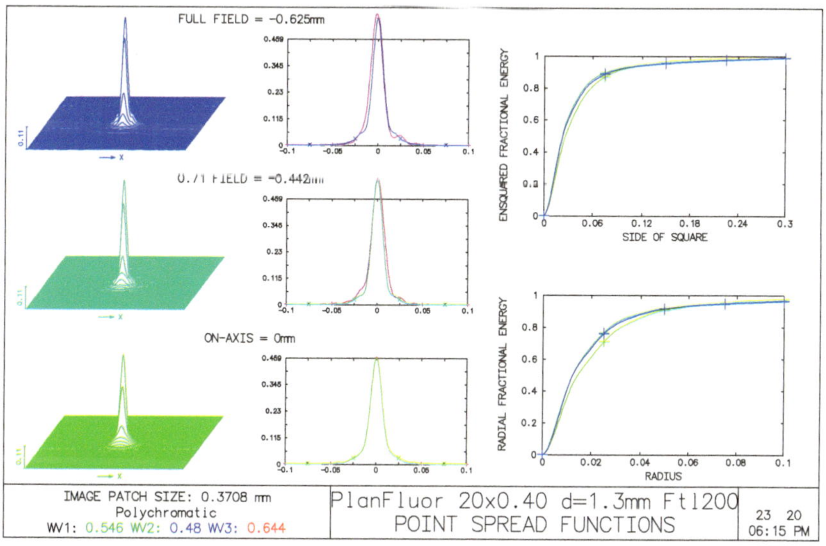

Figure 3.153 Optical parameters and graphs of aberration correction of a 20×/0.40 ($d = 1.3$ mm) LCF plan MicroFluar objective ($F'_{ts} = 200$ mm).

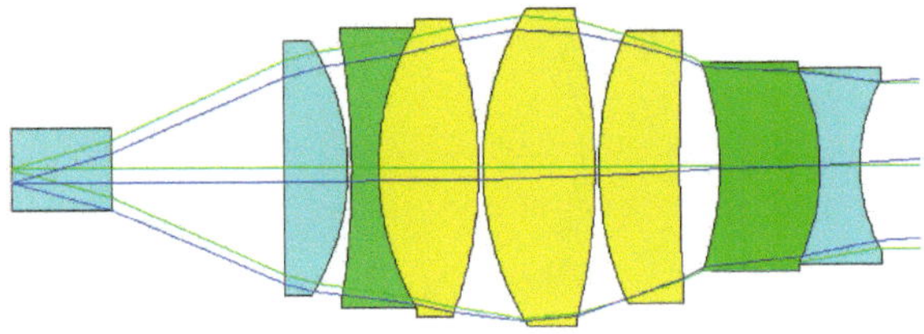

Lens:PlanFluor 20x0.40 d=5.0mm Ftl200

Object num aper	0.400000	Image height	12.500000	Primary wavln	0.546070

SRF	RADIUS	THICKNESS	APERTURE RADIUS	GLASS	SPECIAL
OBJ	0.000000	0.010000	0.625005	FIZIO	
AST	0.000000	5.000000	0.628143 AS	LZ_LK5	
2	0.000000	8.667600	2.030000	AIR	
3	0.000000	3.100000	5.820000	LZ_K8	
4	-11.529800	0.200000	6.180000	AIR	
5	-39.839400	1.300000	6.330000	LZ_BF28	
6	13.275200	5.000000	6.800000	FPL51	
7	-19.932300	0.200000	7.190000	AIR	
8	14.514300	5.600000	7.720000	FPL51 P	
9	-29.431800	0.200000	7.440000	AIR	
10	16.348800	4.000000	6.600000	FPL51	
11	87.791700	2.000000	5.580000	AIR	
12	-15.409700	5.000000	5.070000	TF5	
13	-10.919500	2.000000	4.760000	LZ_K8	
14	8.720800	200.000000	4.170000	AIR	
15	-36.100000	14.000000	16.340000	LZ_K8	
16	-43.560000	3.000000	19.140000	AIR	
17	377.100000	15.000000	19.700000	CAF2	
18	-81.500000	3.000000	20.100000	AIR	
19	221.300000	17.000000	19.700000	CAF2	
20	-72.000000	3.000000	18.740000	LZ_CTK12	
21	0.000000	200.005392 S	18.640000	AIR	
IMS	0.000000	0.000000	12.500000 S		

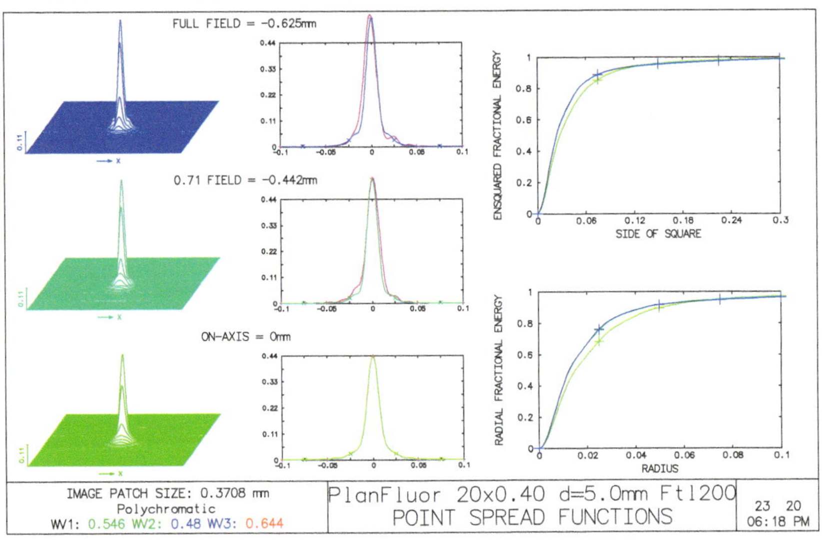

Figure 3.154 Optical design and graphs of aberration correction of a 20×/0.40 ($d = 5.0$ mm) LCF plan MicroFluar objective ($F'_{ts} = 200$ mm).

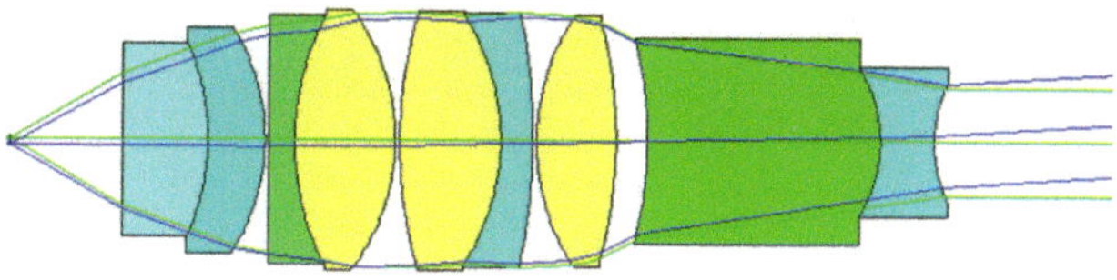

Lens: LD PlanApo 40x0.50 d=0 Ft1200

Object num aper 0.500000 Image height 12.500000 Primary wavln 0.546070

SRF	RADIUS		THICKNESS		APERTURE RADIUS		GLASS		SPECIAL
OBJ	0.000000		1.0000e-20		0.312649		AIR		
AST	0.000000		5.329000		0.312649	AS	AIR		
2	0.000000		4.040000		3.400000		H-K9L		
3	-10.516000		2.700000		4.470000		H-LAK61		
4	-9.480000		0.200000		5.240000		AIR		
5	0.000000		1.200000		5.540000		H-ZBAF5		
6	12.560000		4.800000		5.800000		CAF2		
7	-9.817000		0.200000		6.080000		AIR		
8	21.693000		4.800000		6.030000		CAF2	P	
9	-10.516000		1.500000		5.820000		H-LAK61	P	
10	-29.720000		0.200000		5.930000		AIR		
11	10.516000		3.800000		5.840000		CAF2	P	
12	-20.280000		1.500000		5.570000		AIR		
13	-18.400000		11.000000		4.760000		H-ZF6		
14	-6.516000		2.500000		3.450000		LZ_CTK12		
15	6.516000		200.000000		2.800000		AIR		
16	-36.100000		14.000000		15.220000		H-K9L	P	
17	-43.560000		3.000000		17.860000		AIR		
18	377.100000		15.000000		18.380000		CAF2		
19	-81.500000		3.000000		18.800000		AIR		
20	221.300000		17.000000		18.480000		CAF2		
21	-72.000000		3.000000		17.600000		LZ_CTK12	P	
22	0.000000		199.927672	S	17.530000		AIR		
IMS	0.000000		0.000000		12.500000	S			

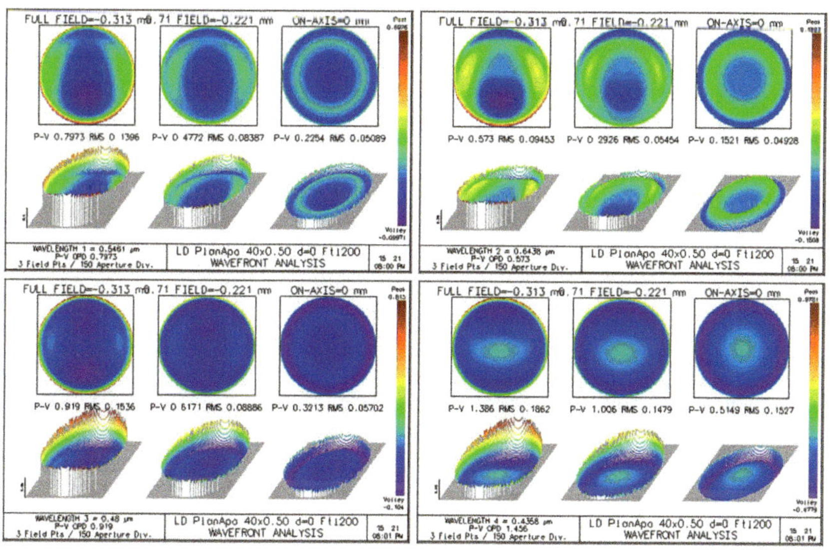

Figure 3.155 Optical design and graphs of aberration correction of a 40×/0.50 ($d = 0$) CCF plan apochromatic objective ($F'_{ts} = 200$ mm).

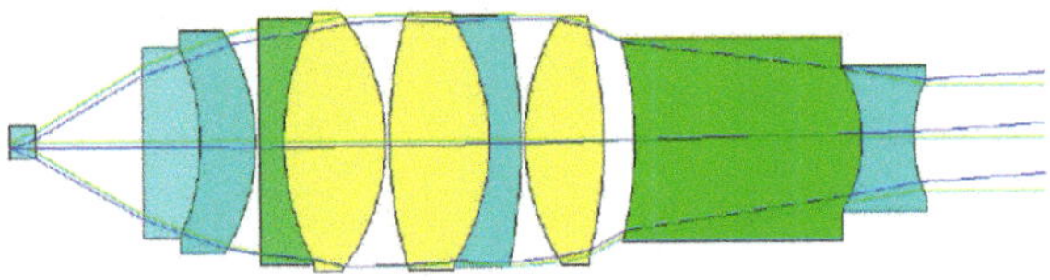

Lens: PlanApo 40x0.5 d=1.1-1.5 Ftl200

Object num aper 0.500000 Image height 12.500000 Primary wavln 0.546070

SRF	RADIUS	THICKNESS	APERTURE RADIUS	GLASS	SPECIAL
OBJ	0.000000	0.010000	0.312663	FIZIO	
AST	0.000000	1.300000	0.316698 AS	LZ_LK5	
2	0.000000	5.286100	0.780000	AIR	
3	0.000000	2.760000	3.840000	H-K9L	
4	-10.516000	2.700000	4.470000	H-LAK61	
5	-9.480000	0.200000	5.240000	AIR	
6	0.000000	1.200000	5.540000	H-ZBAF5	
7	12.560000	4.800000	5.800000	CAF2	
8	-9.817000	0.200000	6.080000	AIR	
9	21.693000	4.800000	6.030000	CAF2 P	
10	-10.516000	1.500000	5.820000	H-LAK61 P	
11	-29.720000	0.200000	5.930000	AIR	
12	10.516000	3.800000	5.840000	CAF2 P	
13	-20.280000	1.500000	5.570000	AIR	
14	-18.400000	11.000000	4.760000	H-ZF6	
15	-6.516000	2.500000	3.450000	LZ_CTK12	
16	6.516000	200.000000	2.800000	AIR	
17	-36.100000	14.000000	15.220000	H-K9L P	
18	-43.560000	3.000000	17.860000	AIR	
19	377.100000	15.000000	18.380000	CAF2	
20	-81.500000	3.000000	18.800000	AIR	
21	221.300000	17.000000	18.480000	CAF2	
22	-72.000000	3.000000	17.600000	LZ_CTK12 P	
23	0.000000	199.975922 S	17.530000	AIR	
IMS	0.000000	0.000000	12.500000 S		

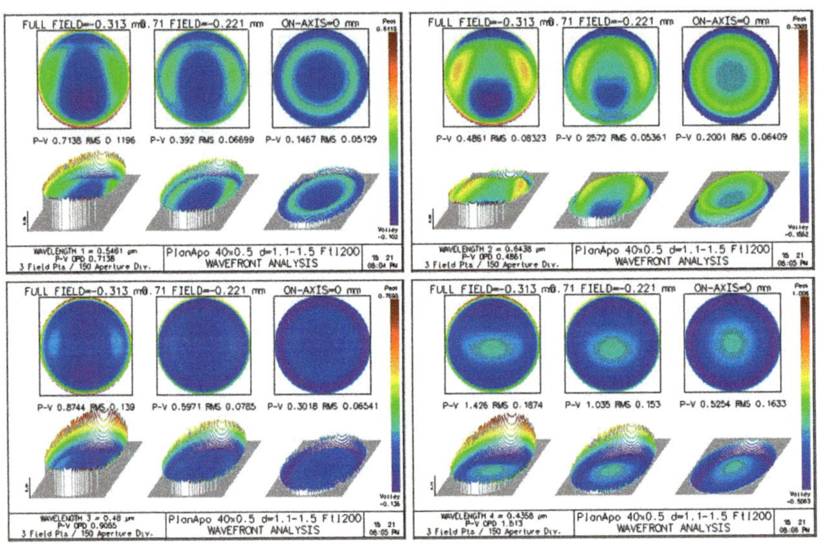

Figure 3.156 Optical design and graphs of aberration correction of a 40×/0.50 ($d = 1.3$ mm) CCF plan apochromatic objective ($F'_{ts} = 200$ mm).

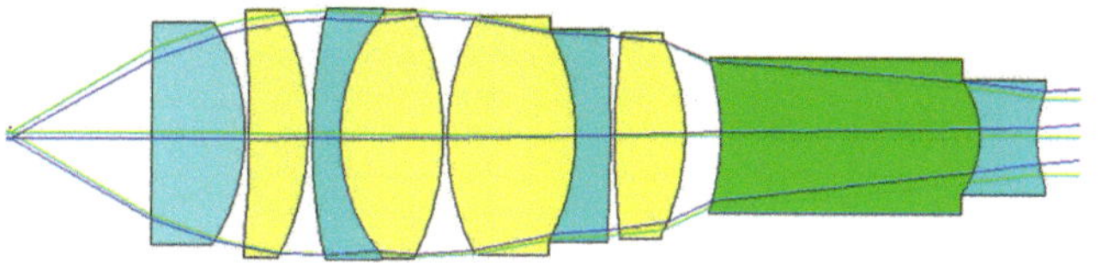

Lens: PlanApo 50x0.50 d0 Ftl160

Object num aper	0.500000	Image height	12.500000	Primary wavln	0.546070

SRF	RADIUS	THICKNESS	APERTURE RADIUS	GLASS	SPECIAL
OBJ	0.000000	1.0000e-20	0.249995	AIR	
AST	0.000000	6.000000	0.249995 AS	AIR	
2	0.000000	4.000000	3.710000	LZ_K8	
3	-8.395000	0.200000	4.630000	AIR	
4	-55.080000	2.500000	4.910000	CAF2	
5	-11.695000	0.200000	5.170000	AIR	
6	24.320000	1.200000	5.240000	LZ_CTK19	
7	8.017000	4.400000	5.070000	CAF2 P	
8	-11.695000	0.200000	5.190000	AIR	
9	10.715000	5.500000	4.980000	CAF2 P	
10	-9.204000	1.500000	4.420000	LZ_CTK19 P	
11	-112.720000	0.200000	4.330000	AIR	
12	34.750000	3.000000	4.270000	CAF2 P	
13	-9.204000	1.500000	4.090000	AIR	
14	-10.715000	11.000000	3.230000	LZ_TF5	
15	-4.093000	2.500000	2.380000	LZ_CTK12	
16	5.546000	90.000000	1.970000	AIR	
17	214.800000	5.000000	9.230000	LZ_BK8	
18	-46.990000	0.300000	9.350000	AIR	
19	-46.880000	3.000000	9.330000	LZ_BF24	
20	-113.240000	157.811339 S	9.480000	AIR	
IMS	0.000000	0.000000	12.500000 S		

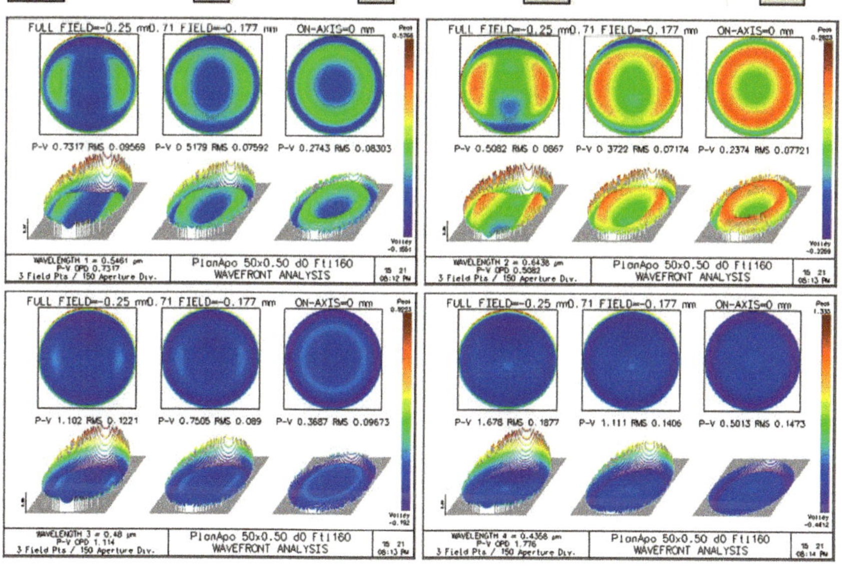

Figure 3.157 Optical design and graphs of aberration correction of a 50×/0.50 ($d = 0$) CCF plan apochromatic objective ($F'_{ts} = 160$ mm).

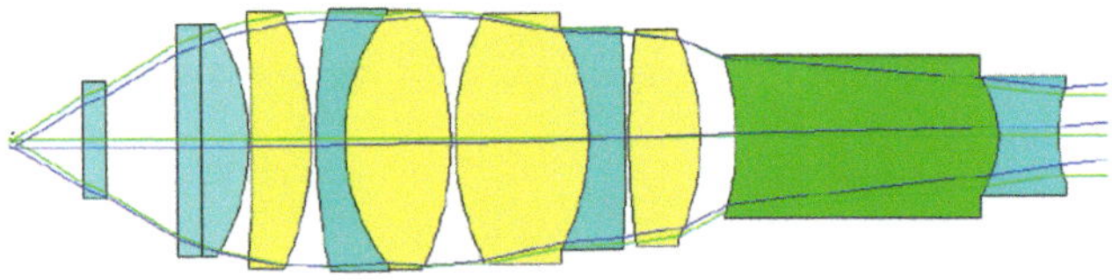

Lens: PlanApo 50x0.50 MIRO d0 Ftl160

Object num aper 0.500000 Image height 12.500000 Primary wavln 0.546070

SRF	RADIUS		THICKNESS		APERTURE RADIUS		GLASS		SPECIAL
OBJ	0.000000		1.0000e-20		0.249795		AIR		
AST	0.000000		2.990100		0.249795	AS	AIR		
2	0.000000		1.020000		1.970000		LZ_K8		
3	0.000000		2.990100		2.330000		AIR		
4	0.000000		1.020000		4.060000		LZ_K8	P	
5	0.000000		2.000000		4.410000		LZ_K8	P	
6	-8.395000		0.170000		4.630000		AIR		
7	-55.080000		2.500000		4.910000		CAF2		
8	-11.695000		0.200000		5.160000		AIR		
9	24.320000		1.200000		5.240000		LZ_CTK19		
10	8.017000		4.400000		5.070000		CAF2	P	
11	-11.695000		0.200000		5.190000		AIR		
12	10.715000		5.500000		4.980000		CAF2	P	
13	-9.204000		1.500000		4.420000		LZ_CTK19	P	
14	-112.720000		0.200000		4.330000		AIR		
15	34.750000		3.000000		4.270000		CAF2	P	
16	-9.204000		1.500000		4.090000		AIR		
17	-10.715000		11.000000		3.230000		LZ_TF5		
18	-4.093000		2.500000		2.380000		LZ_CTK12		
19	5.546000		90.000000		1.970000		AIR		
20	214.800000		5.000000		9.230000		LZ_BK8		
21	-46.990000		0.300000		9.350000		AIR		
22	-46.880000		3.000000		9.330000		LZ_BF24		
23	-113.240000		157.784619	S	9.480000		AIR		
IMS	0.000000		0.000000		12.500000	S			

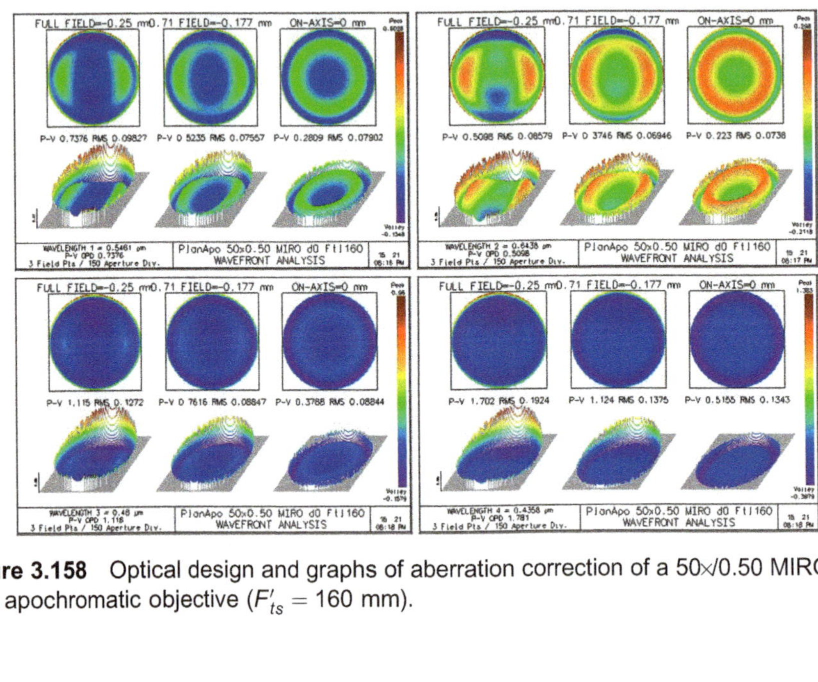

Figure 3.158 Optical design and graphs of aberration correction of a 50×/0.50 MIRO CCF plan apochromatic objective ($F'_{ts} = 160$ mm).

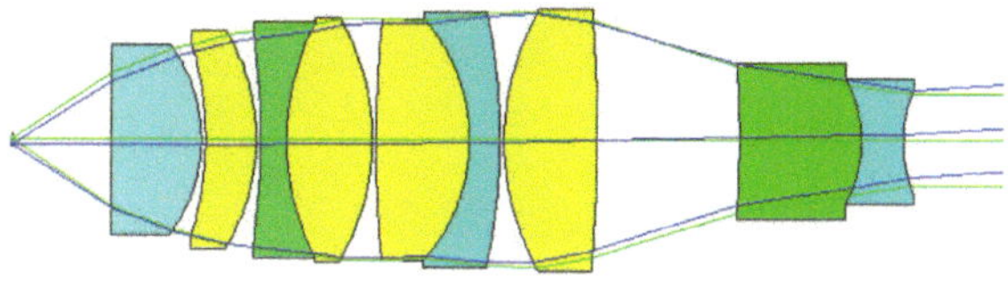

Lens: PlanFluor 50x0.55 d=0 Ftl200

Object num aper 0.550000 Image height 12.500000 Primary wavln 0.546070

SRF	RADIUS		THICKNESS		APERTURE RADIUS		GLASS		SPECIAL
OBJ	0.000000		1.0000e-20		0.249979		AIR		
AST	0.000000		4.793400		0.249979	AS	AIR		
2	0.000000		4.450000		3.410000		LZ_K8		
3	-7.531100		0.200000		4.550000		AIR		
4	-16.074200		2.500000		4.770000		FPL51		
5	-9.853900		0.200000		5.210000		AIR		
6	-42.148800		1.300000		5.330000		LZ_BF28		
7	12.113800		4.200000		5.640000		FPL51	P	
8	-11.424800		0.200000		5.850000		AIR		
9	70.190400		4.600000		5.810000		FPL51	P	
10	-8.483700		1.500000		5.760000		LZ_CTK19		
11	-28.461400		0.200000		6.130000		AIR		
12	12.363800		4.600000		6.260000		FPL51	P	
13	-64.747700		7.000000		5.840000		AIR		
14	-90.438000		6.000000		3.750000		TF5		
15	-6.120200		2.000000		3.000000		LZ_CTK19	P	
16	6.326200		200.000000		2.520000		AIR		
17	-36.100000		14.000000		14.680000		LZ_K8		
18	-43.560000		3.000000		17.230000		AIR		
19	377.100000		15.000000		17.710000		CAF2		
20	-81.500000		3.000000		18.140000		AIR		
21	221.300000		17.000000		17.840000		CAF2	P	
22	-72.000000		3.000000		16.990000		LZ_CTK12		
23	0.000000		199.912071	S	16.920000		AIR		
IMS	0.000000		0.000000		12.500000	S			

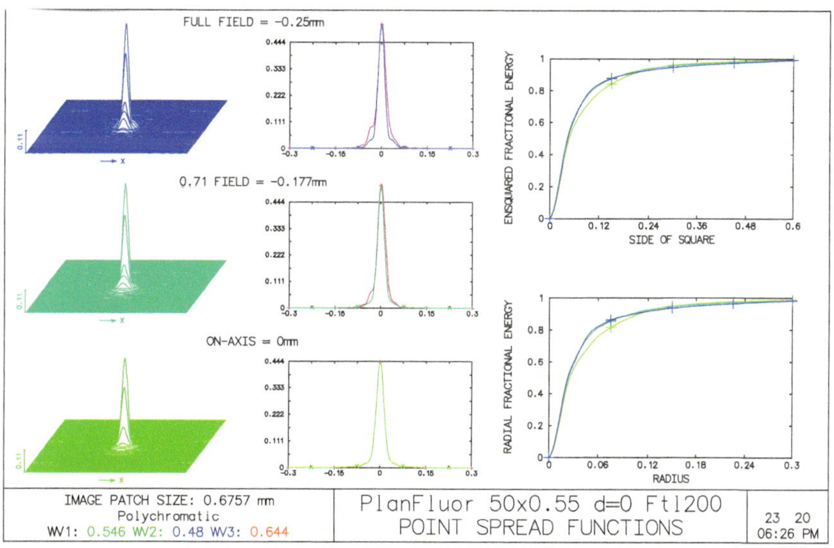

Figure 3.159 Optical design and graphs of aberration correction of a 50×/0.55 ($d = 0$) LCF plan MicroFluar objective ($F'_{ts} = 200$ mm).

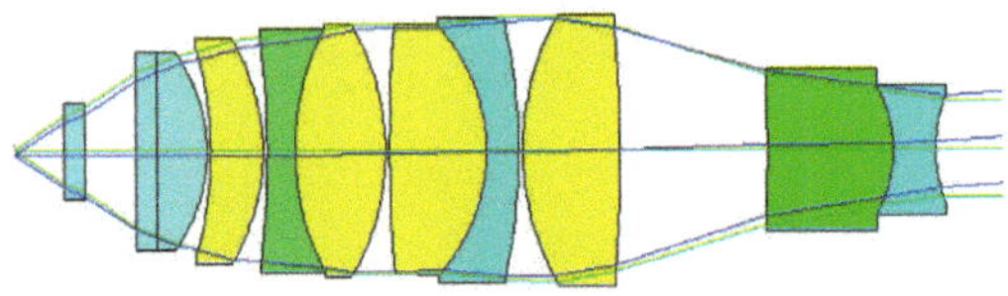

Lens: PlanFluor 50x0.55 d=0 MIRO

Object num aper 0.550000 Image height 12.500000 Primary wavln 0.546070

SRF	RADIUS	THICKNESS	APERTURE RADIUS	GLASS	SPECIAL
OBJ	0.000000	1.0000e-20	0.249971	AIR	
AST	0.000000	2.400000	0.249971 AS	AIR	
2	0.000000	1.020000	1.830000	LZ_K8	
3	0.000000	2.400000	2.230000	AIR	
4	0.000000	1.020000	3.810000	LZ_K8 P	
5	0.000000	2.400000	4.210000	LZ_K8 P	
6	-7.531100	0.200000	4.550000	AIR	
7	-16.074200	2.500000	4.770000	FPL51	
8	-9.853900	0.200000	5.210000	AIR	
9	-42.148800	1.300000	5.330000	LZ_BF28	
10	12.113800	4.200000	5.640000	FPL51 P	
11	-11.424800	0.200000	5.850000	AIR	
12	70.190400	4.600000	5.810000	FPL51 P	
13	-8.483700	1.500000	5.760000	LZ_CTK19	
14	-28.461400	0.200000	6.130000	AIR	
15	12.363800	4.600000	6.260000	FPL51 P	
16	-64.747700	7.000000	5.840000	AIR	
17	-90.438000	6.000000	3.750000	TF5	
18	-6.120200	2.000000	3.000000	LZ_CTK19 P	
19	6.326200	200.000000	2.520000	AIR	
20	-36.100000	14.000000	14.680000	LZ_K8 P	
21	-43.560000	3.000000	17.230000	AIR	
22	377.100000	15.000000	17.710000	CAF2	
23	-81.500000	3.000000	18.140000	AIR	
24	221.300000	17.000000	17.840000	CAF2 P	
25	-72.000000	3.000000	16.990000	LZ_CTK12	
26	0.000000	199.877892 S	16.920000	AIR	
IMS	0.000000	0.000000	12.500000 S		

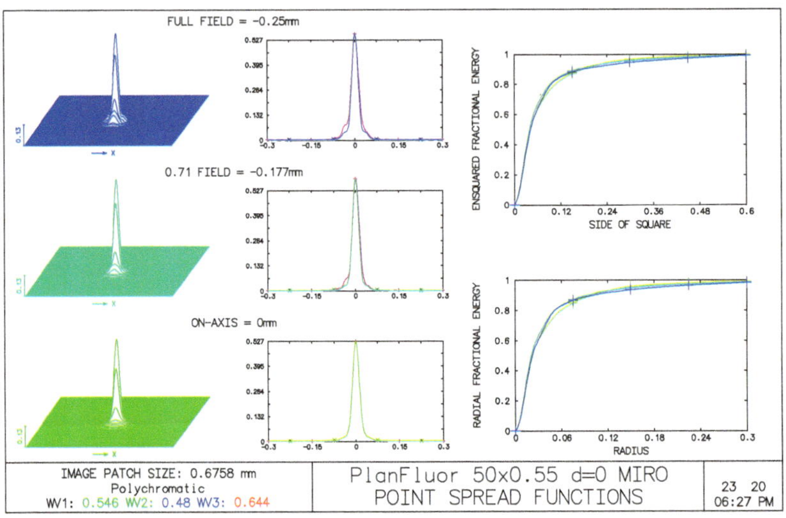

Figure 3.160 Optical design and graphs of aberration correction of a 50×/0.55 MIRO LCF plan MicroFluar objective ($F'_{ts} = 200$ mm).

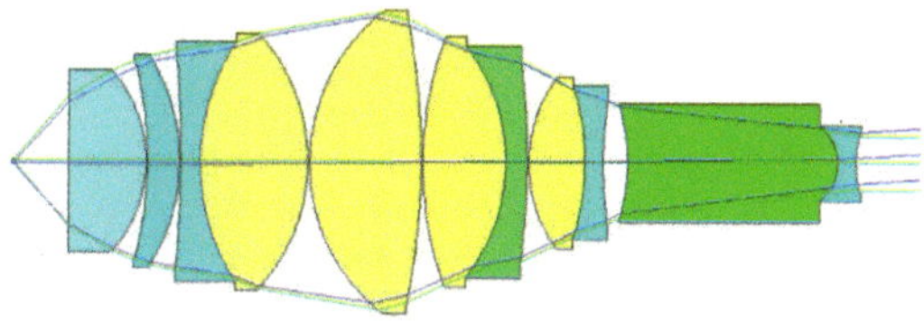

Lens:LD PlanFluar 100x0.75 d=0 ftl200

Object num aper 0.750000 Image height 12.500000 Primary wavln 0.546070

SRF	RADIUS	THICKNESS	APERTURE RADIUS	GLASS	SPECIAL
OBJ	0.000000	1.0000e-20	0.124893	AIR	
AST	0.000000	3.196000	0.124893 AS	AIR	
2	0.000000	4.440000	4.490000	H-K9L	
3	-7.503000	0.100000	5.130000	AIR	
4	-20.750000	1.800000	5.680000	H-LAK3	
5	-11.092000	0.100000	5.980000	AIR	
6	-60.900000	1.100000	6.240000	H-LAK3	
7	12.190000	6.200000	6.730000	CAF2	
8	-10.411100	0.100000	7.200000	AIR	
9	10.627000	6.400000	8.510000	CAF2 P	
10	-41.582000	0.100000	8.170000	AIR	
11	20.110000	4.700000	7.040000	CAF2 P	
12	-10.411100	1.200000	6.500000	ZBAF4	
13	-44.140000	0.100000	5.840000	AIR	
14	8.128000	3.200000	4.810000	CAF2 P	
15	-13.450000	1.000000	4.290000	H-LAK3 P	
16	23.230000	1.400000	3.600000	AIR	
17	-10.950000	12.000000	3.300000	H-ZF6	
18	-2.618000	1.000000	2.090000	H-LAK3 P	
19	5.401000	200.000000	1.820000	AIR	
20	-36.100000	14.000000	15.950000	H-K9L P	
21	-43.560000	3.000000	18.310000	AIR	
22	377.100000	15.000000	18.810000	CAF2	
23	-81.500000	3.000000	19.370000	AIR	
24	221.300000	17.000000	19.270000	CAF2	
25	-72.000000	3.000000	18.660000	LZ_CTK12	
26	0.000000	199.473986 S	18.650000	AIR	
IMS	0.000000	0.000000	12.500000		

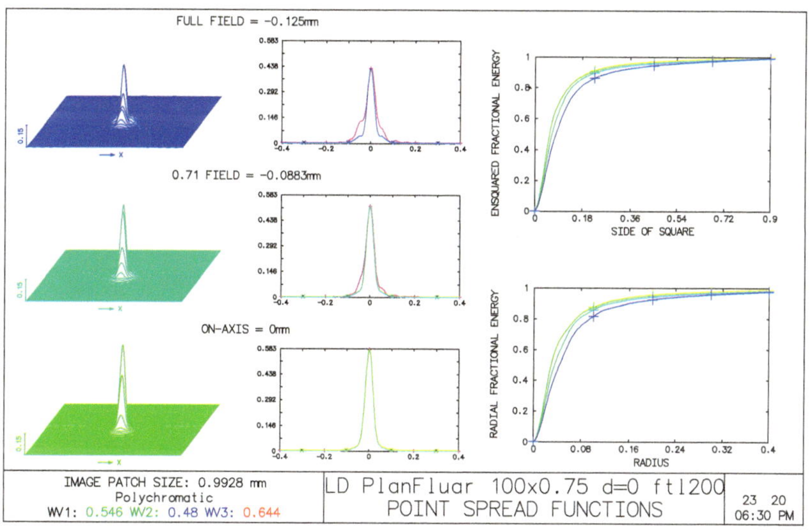

Figure 3.161 Optical design and graphs of aberration correction of a 100×/0.75 ($d = 0$) LCF plan MicroFluar objective ($F'_{ts} = 200$ mm).

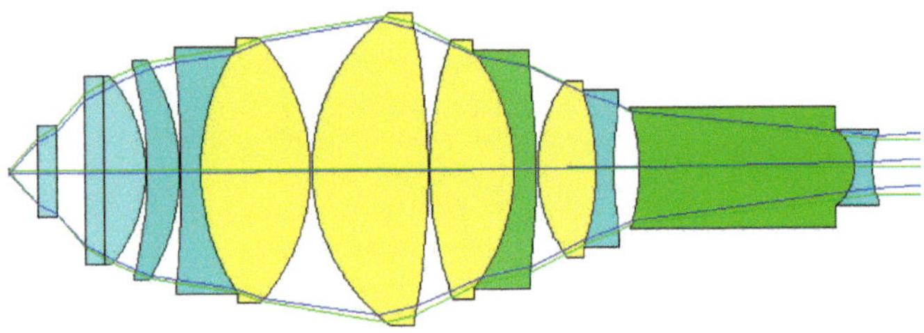

Lens: PlanFluar 100x0.75 MIRO ftl200

Object num aper	0.750000	Image height	12.500000	Primary wavln	0.546070

SRF	RADIUS	THICKNESS	APERTURE RADIUS	GLASS	SPECIAL
OBJ	0.000000	1.0000e-20	0.124893	AIR	
AST	0.000000	1.598000	0.124893 AS	AIR	
2	0.000000	1.020000	1.940000	H-K9L	
3	0.000000	1.598000	2.520000	AIR	
4	0.000000	1.020000	4.330000	H-K9L P	
5	0.000000	2.400000	4.910000	H-K9L	
6	-7.503000	0.100000	5.130000	AIR	
7	-20.750000	1.800000	5.680000	H-LAK3	
8	-11.092000	0.100000	5.980000	AIR	
9	-60.900000	1.100000	6.240000	H-LAK3	
10	12.190000	6.200000	6.730000	CAF2	
11	-10.411100	0.100000	7.200000	AIR	
12	10.627000	6.400000	8.510000	CAF2 P	
13	-41.582000	0.100000	8.170000	AIR	
14	20.110000	4.700000	7.040000	CAF2 P	
15	-10.411100	1.200000	6.500000	ZBAF4	
16	-44.140000	0.100000	5.840000	AIR	
17	8.128000	3.200000	4.810000	CAF2 P	
18	-13.450000	1.000000	4.290000	H-LAK3 P	
19	23.230000	1.400000	3.600000	AIR	
20	-10.950000	12.000000	3.300000	H-ZF6	
21	-2.618000	1.000000	2.090000	H-LAK3 P	
22	5.401000	200.000000	1.820000	AIR	
23	-36.100000	14.000000	15.950000	H-K9L P	
24	-43.560000	3.000000	18.310000	AIR	
25	377.100000	15.000000	18.810000	CAF2	
26	-81.500000	3.000000	19.370000	AIR	
27	221.300000	17.000000	19.270000	CAF2	
28	-72.000000	3.000000	18.660000	LZ_CTK12	
29	0.000000	199.473986 S	18.650000	AIR	
IMS	0.000000	0.000000	12.500000		

Figure 3.162 Optical design and graphs of aberration correction of a 100×/0.75 MIRO LCF plan MicroFluar objective ($F'_{ts} = 200$ mm).

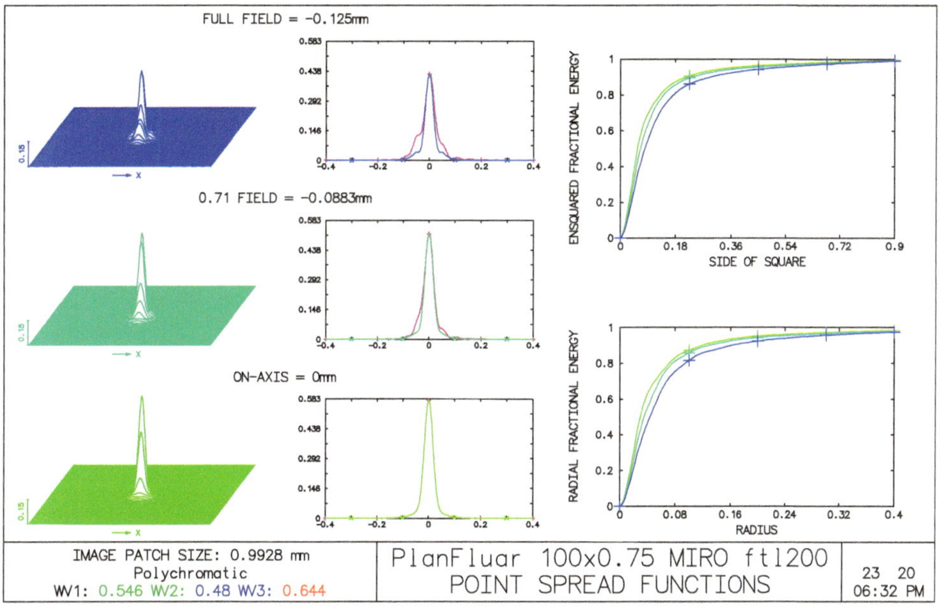

Figure 3.162 (*Continued*) Optical design and graphs of aberration correction of a 100×/0.75 MIRO LCF plan MicroFluar objective ($F'_{ts} = 200$ mm).

graphs of aberration correction of various objectives for working without a cover glass, in MIRO configurations, and as part of inverted microscopes.

3.8 Plan Apochromats for Inverted Biological Microscopes

Plan apochromatic objectives are still rarely used in inverted microscopes. Nevertheless, objects in inverted biological microscopes are examined in their intravital state, as a rule, without undergoing any coloring or cementing. Of course, objects may have some effect in the form of, for example, electric discharges or heating, but these techniques do not kill matter; the studied objects are preserved in their natural state. Therefore, for this type of research using an inverted light microscope, optics with a different plan apochromatic aberration correction are optimal.

By the way, it may make sense to make some improvements (while retaining the usual engineering solution) in the lighting system of an inverted biological microscope. Indeed, even in research models of inverted light microscopes, the objective focuses on the surface of the studied object from below, while the lighting system works from above the object. If the object is sufficiently long, or if the layer of physiological solution (in which the object is usually placed) is significant, then such a microscope does not fully correspond to the classical understanding of the theory of image formation

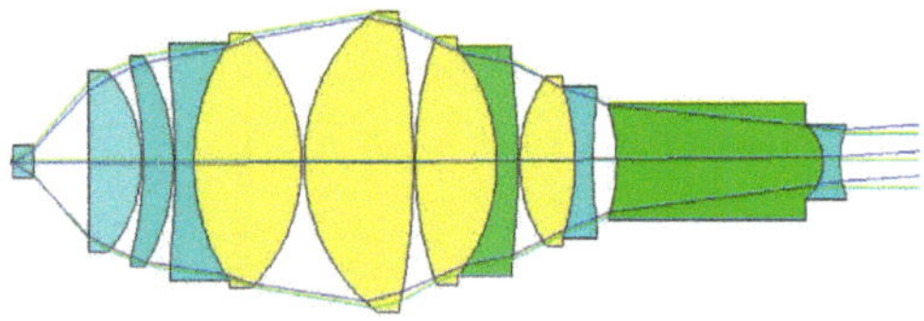

Lens: PlanFluor 100x0.75 d=1.3 ftl200

Object num aper 0.750000 Image height 12.500000 Primary wavln 0.546070

SRF	RADIUS	THICKNESS		APERTURE RADIUS		GLASS		SPECIAL
OBJ	0.000000	0.010000		0.125092		FIZIO		
AST	0.000000	1.300000		0.130000	A	LZ_LK5		
2	0.000000	3.167500		0.900000		AIR		
3	0.000000	3.140000		4.490000		H-K9L		
4	-7.503000	0.100000		5.130000		AIR		
5	-20.750000	1.800000		5.680000		H-LAK3		
6	-11.092000	0.100000		5.980000		AIR		
7	-60.900000	1.100000		6.240000		H-LAK3		
8	12.190000	6.200000		6.730000		CAF2		
9	-10.411100	0.150000		7.200000		AIR		
10	10.627000	6.400000		8.510000		CAF2	P	
11	-41.582000	0.100000		8.170000		AIR		
12	20.110000	4.700000		7.040000		CAF2	P	
13	-10.411100	1.200000		6.500000		ZBAF4		
14	-44.140000	0.100000		5.840000		AIR		
15	8.128000	3.200000		4.810000		CAF2	P	
16	-13.450000	1.000000		4.290000		H-LAK3	P	
17	23.230000	1.390000		3.600000		AIR		
18	-10.950000	12.000000		3.300000		H-ZF6		
19	-2.618000	1.000000		2.090000		H-LAK3	P	
20	5.401000	200.000000		1.820000		AIR		
21	-36.100000	14.000000		15.950000		H-K9L	P	
22	-43.560000	3.000000		18.310000		AIR		
23	377.100000	15.000000		18.810000		CAF2		
24	-81.500000	3.000000		19.370000		AIR		
25	221.300000	17.000000		19.270000		CAF2		
26	-72.000000	3.000000		18.660000		H-LAK2		
27	0.000000	199.502685	S	18.650000		AIR		
IMS	0.000000	0.000000		12.500000				

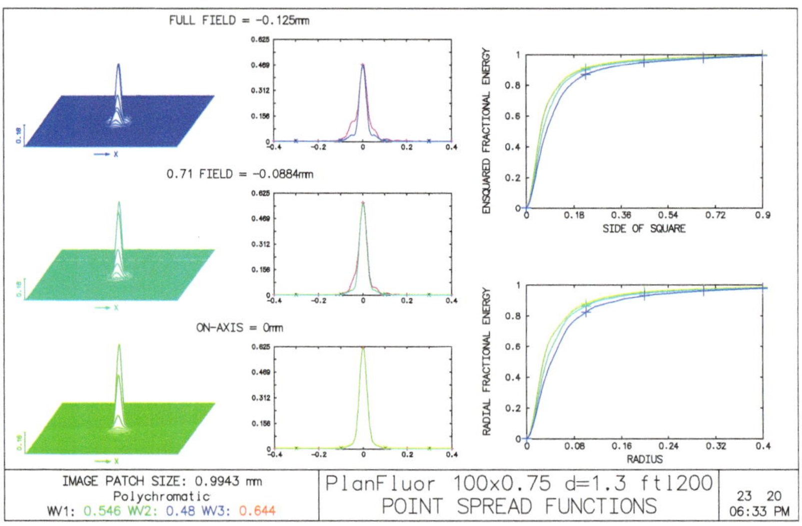

Figure 3.163 Optical design and graphs of aberration correction of a 100×/0.75 ($d = 1.3$ mm) LCF plan MicroFluar objective ($F'_{ts} = 200$ mm).

on a microscope. Although the studied surface of the object is between the objective and the condenser, the distance to the condenser is quite large, which does not allow for the creation of classical *high-aperture* lighting rays. Naturally, this reduces the resolution of the inverted microscope as compared with a conventional direct microscope, where the condenser is in close proximity to the object under study, and where the numerical apertures of the condenser and objective are close in value. In addition, the optical design of a high-aperture objective capable of working through the bottom of the glassware in an inverted microscope is challenging. Some improvement in resolution can be achieved by using a design in which the objective itself serves as a condenser, as in reflected light microscopes; however, objects examined with an inverted biological microscope are usually transparent, so the image will be corrupted by scattered light.

Significant progress can be achieved using the lighting system shown in Chapter 1, Fig. 1.306, or more precisely, an upgraded system in which the bottom of the laboratory glassware serves as a light guide layer. We will discuss the possibility of using conventional objectives designed to work with a standard coverslip in an inverted biological microscope. The aberration correction of ordinary objectives is designed for work with ($d = 0.17$ mm) or without ($d = 0$) a cover glass. But for small-magnification objectives, or rather, objectives with small numerical apertures, the aberration correction is not sensitive to the thickness of the cover glass; moreover, it is not sensitive to the thickness of the bottom of laboratory glassware. Therefore, a special optical design for such objectives is not required; they can be used "as is."

However, as practice shows, for objectives with a numerical aperture of 0.3–0.4, a slight change in the optical design from the original is required (Fig. 3.164 shows a comparison of the optical design with that shown in Fig. 3.35). For some objectives in which numerical aperture values of ≥ 0.5 are achieved, original optical designs are required. Figures 3.165–3.168, which show design parameters and graphs of aberration correction of some plan apochromatic objectives for working in inverted biological microscopes, confirm these theses.

Table 3.6 shows the main technical parameters of these objectives.

3.9 Objectives for 50-mm Observation and a Long Working Distance (95-mm Parfocal Distance)

The development of microscope objectives that have three advantages over other objectives is a complex engineering and technological task. What are the advantages?

The first is the attainment of plan apochromatic aberration correction. My colleagues and I have shown the advantage of using lens objectives with plan apochromatic correction for studies of biological objects by which the

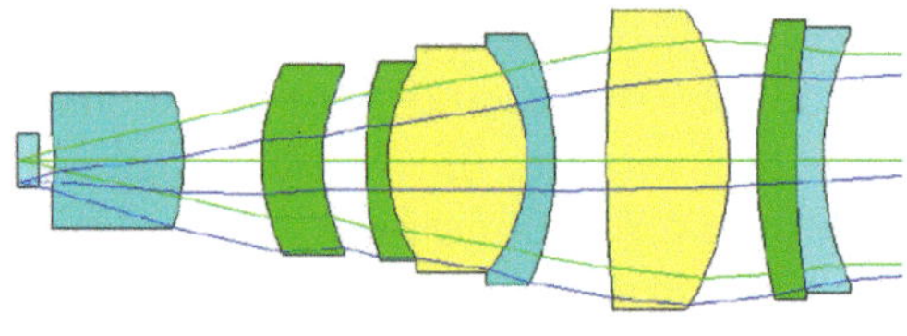

Lens: PlanApo 10x0.30 d=1.3 ftl200

Object num aper 0.300000 Image height 12.500000 Primary wavln 0.546070

SRF	RADIUS	THICKNESS		APERTURE RADIUS		GLASS		SPECIAL
OBJ	0.000000	0.010000		1.236252		FIZIO		
AST	0.000000	1.300000		1.238556	AS	LZ_LK5		
2	0.000000	0.903500		1.500000		AIR		
3	-6.680000	7.500000		1.720000		H-LAK3		
4	-10.850000	4.500000		3.780000		AIR		
5	12.210000	3.500000		5.300000		H-ZLAF7LA		
6	10.850000	2.500000		4.980000		AIR		
7	19.269000	1.300000		5.560000		H-ZLAF7LA	P	
8	10.281000	8.000000		5.570000		CAF2		
9	-9.720000	1.600000		6.320000		H-LAK3	P	
10	-17.390000	3.000000		7.040000		AIR		
11	97.700000	7.000000		7.950000		CAF2	P	
12	-15.810000	1.600000		8.410000		AIR		
13	29.620000	2.300000		7.800000		H-ZLAF7LA	P	
14	56.160000	1.500000		7.470000		LZ_K8		
15	16.120000	200.000000		6.950000		AIR		
16	-36.100000	14.000000		16.990000		K9		
17	-43.560000	3.000000		19.910000		AIR		
18	377.100000	15.000000		20.510000		CAF2		
19	-81.500000	3.000000		20.910000		AIR		
20	221.300000	17.000000		20.460000		CAF2		
21	-72.000000	3.000000		19.460000		LZ_CTK12		
22	0.000000	200.000494	S	19.340000		AIR		
IMS	0.000000	0.000000		12.500000	S			

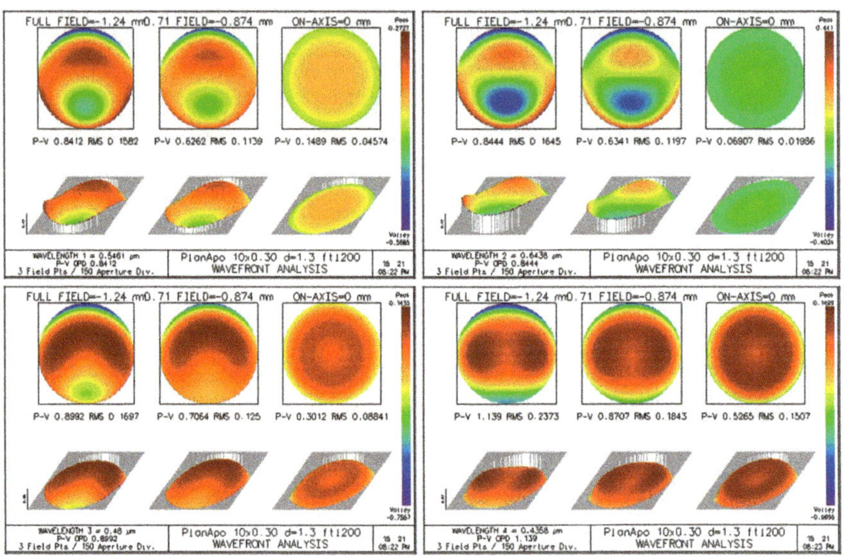

Figure 3.164 Optical design and graphs of aberration correction of a 10×/0.30 ($d = 1.3$ mm) CCF plan apochromatic objective ($F'_{ts} = 200$ mm).

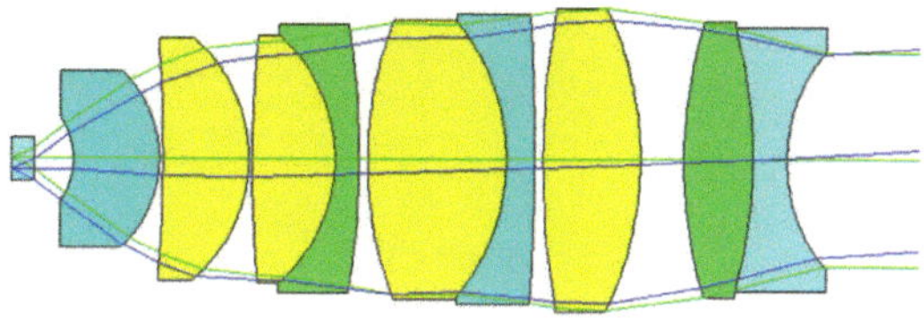

Lens: PlanApo 20x0.60 d=1.3 Ftl200

Object num aper 0.600000 Image height 12.500000 Primary wavln 0.546070

SRF	RADIUS	THICKNESS	APERTURE RADIUS	GLASS	SPECIAL
OBJ	0.000000	0.010000	0.625201	FIZIO	
AST	0.000000	1.300000	0.630226 AS	LZ_LK5	
2	0.000000	2.300000	1.200000	AIR	
3	-4.000000	5.000000	2.350000	H-LAK3	
4	-6.684000	0.200000	4.980000	AIR	
5	-82.310000	5.000000	6.130000	H-FK71	
6	-8.801000	0.200000	6.840000	AIR	
7	74.290000	4.800000	7.010000	H-FK71 P	
8	-9.160000	1.400000	7.010000	ZBAF4	
9	-46.260000	0.500000	7.570000	AIR	
10	21.580000	8.000000	7.880000	H-FK71 P	
11	-11.870000	1.600000	7.730000	H-LAK3 P	
12	-114.700000	0.500000	8.200000	AIR	
13	49.660000	5.600000	8.420000	H-FK71 P	
14	-19.790000	2.500000	8.550000	AIR	
15	26.410000	4.000000	7.760000	ZBAF4 P	
16	-26.410000	2.000000	7.420000	K9	
17	9.480000	200.000000	6.130000	AIR	
18	-36.100000	14.000000	18.460000	K9	
19	-43.560000	3.000000	21.590000	AIR	
20	377.100000	15.000000	22.290000	CAF2	
21	-81.500000	3.000000	22.660000	AIR	
22	221.300000	17.000000	22.080000	CAF2	
23	-72.000000	3.000000	21.000000	LZ_CTK12	
24	0.000000	200.017521 S	20.850000	AIR	
IMS	0.000000	0.000000	12.500000 S		

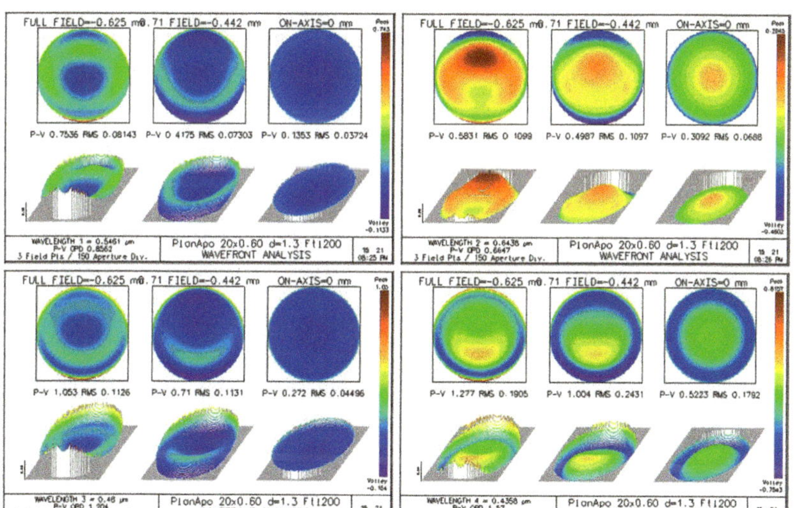

Figure 3.165 Optical design and graphs of aberration correction of a 20×/0.60 ($d = 1.3$ mm) CCF plan apochromatic objective ($F'_{ts} = 200$ mm).

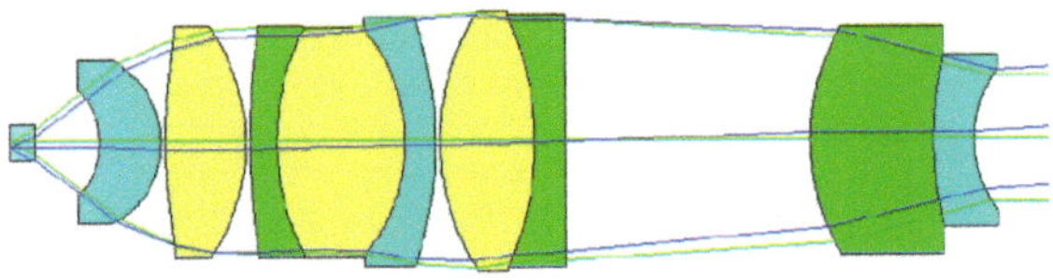

Lens: PlanApo 40x0.60 d=1.3 Ft1200

Object num aper 0.600000 Image height 12.500000 Primary wavln 0.546070

SRF	RADIUS	THICKNESS	APERTURE RADIUS	GLASS	SPECIAL
OBJ	0.000000	0.010000	0.312429	FIZIO	
AST	0.000000	1.300000	0.317454 AS	LZ_LK5	
2	0.000000	3.200000	0.890000	AIR	
3	-3.577800	3.000000	2.520000	LZ_CTK12	
4	-4.399100	0.200000	3.910000	AIR	
5	29.410400	4.000000	5.150000	CAF2	
6	-9.417400	0.200000	5.520000	AIR	
7	34.676700	1.300000	5.520000	LZ_TF5	
8	11.965900	6.200000	5.380000	CAF2	
9	-8.414100	1.500000	5.490000	LZ_CTK19	
10	-17.320200	0.200000	5.980000	AIR	
11	11.414300	4.600000	6.260000	CAF2	
12	-14.697200	1.500000	6.070000	LZ_CTK19 P	
13	0.000000	11.800000	6.000000	AIR	
14	10.828900	6.000000	5.460000	LZ_TF5 P	
15	18.640100	2.000000	4.130000	LZ_CTK19 P	
16	5.749300	200.000000	3.410000	AIR	
17	-36.100000	14.000000	15.690000	LZ_K8	
18	-43.560000	3.000000	18.400000	AIR	
19	377.100000	15.000000	18.940000	CAF2	
20	-81.500000	3.000000	19.350000	AIR	
21	221.300000	17.000000	19.000000	CAF2	
22	-72.000000	3.000000	18.080000	LZ_CTK12	
23	0.000000	199.813290 S	18.000000	AIR	
IMS	0.000000	0.000000	12.500000 S		

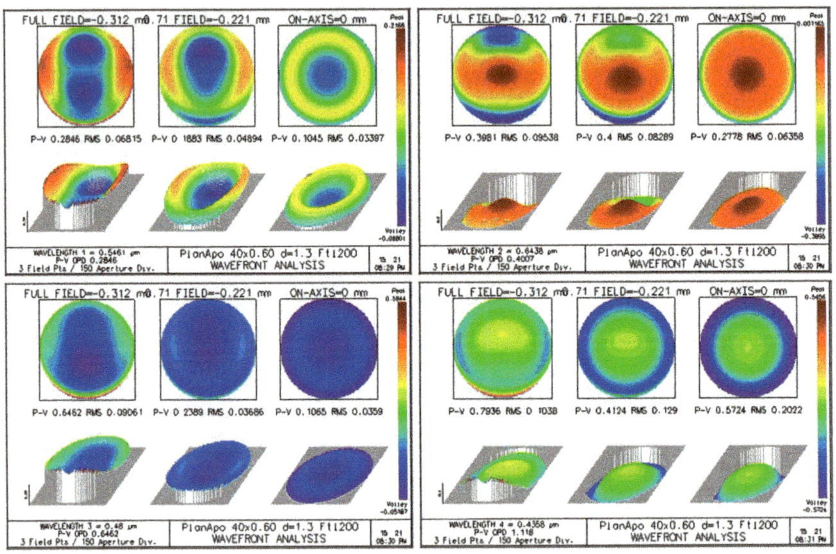

Figure 3.166 Optical design and graphs of aberration correction of a 40×/0.60 ($d = 1.3$ mm) CCF plan apochromatic objective ($F'_{ts} = 200$ mm).

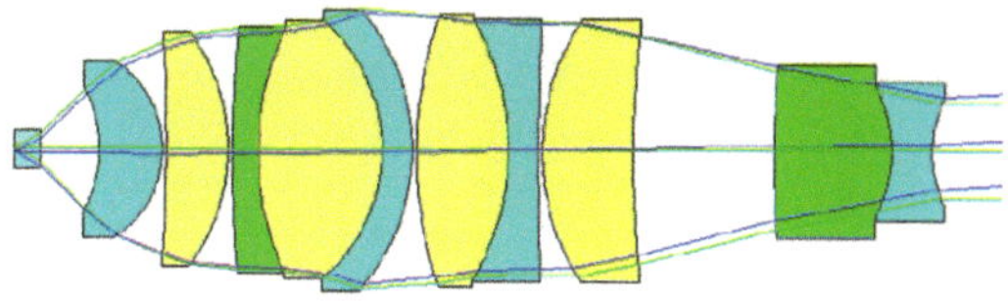

Lens: PlanApo 60x0.70 d=1.3 Ftl200

Object num aper 0.700000 Image height 12.500000 Primary wavln 0.546070

SRF	RADIUS	THICKNESS		APERTURE RADIUS		GLASS		SPECIAL
OBJ	0.000000	0.010000		0.208279		FIZIO		
AST	0.000000	1.300000		0.214430	AS	LZ_LK5		
2	0.000000	3.000000		0.910000		AIR		
3	-6.077400	3.200000		3.050000		LZ_CTK12		
4	-4.959500	0.200000		4.290000		AIR		
5	-92.507200	3.100000		5.300000		CAF2		
6	-9.283300	0.200000		5.680000		AIR		
7	69.772800	1.300000		5.940000		LZ_TF5		
8	13.948300	6.200000		5.990000		CAF2		
9	-8.145300	1.500000		6.280000		LZ_CTK19		
10	-10.239500	0.200000		6.860000		AIR		
11	20.107300	4.600000		6.630000		CAF2		
12	-12.319600	1.500000		6.390000		LZ_CTK19	P	
13	265.462000	0.200000		6.370000		AIR		
14	11.322300	4.600000		6.380000		CAF2	P	
15	58.741600	7.000000		5.830000		AIR		
16	37.941100	6.000000		4.210000		LZ_TF5	P	
17	-6.819700	2.000000		3.350000		LZ_CTK19	P	
18	5.719500	200.000000		2.680000		AIR		
19	-36.100000	14.000000		14.870000		LZ_K8		
20	-43.560000	3.000000		17.450000		AIR		
21	377.100000	15.000000		17.940000		CAF2		
22	-81.500000	3.000000		18.360000		AIR		
23	221.300000	17.000000		18.060000		CAF2		
24	-72.000000	3.000000		17.200000		LZ_CTK12		
25	0.000000	199.741664	S	17.120000		AIR		
IMS	0.000000	0.000000		12.500000	S			

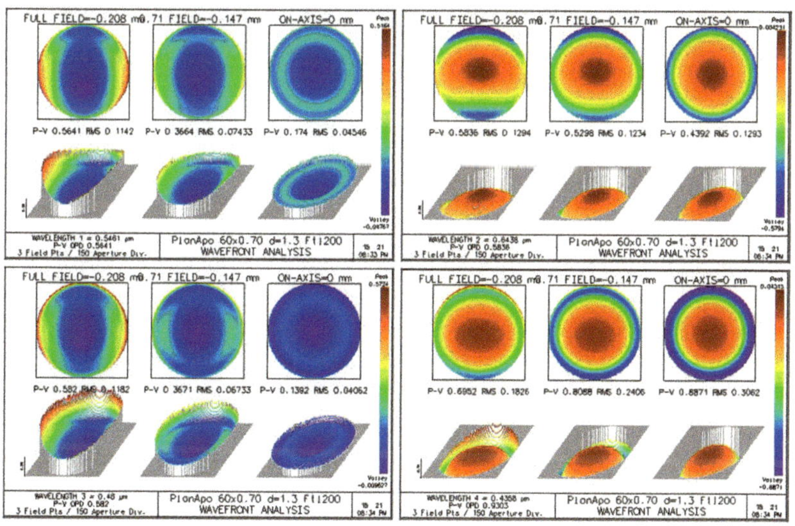

Figure 3.167 Optical design and graphs of aberration correction of a 60×/0.70 ($d = 1.3$ mm) CCF plan apochromatic objective ($F'_{ts} = 200$ mm).

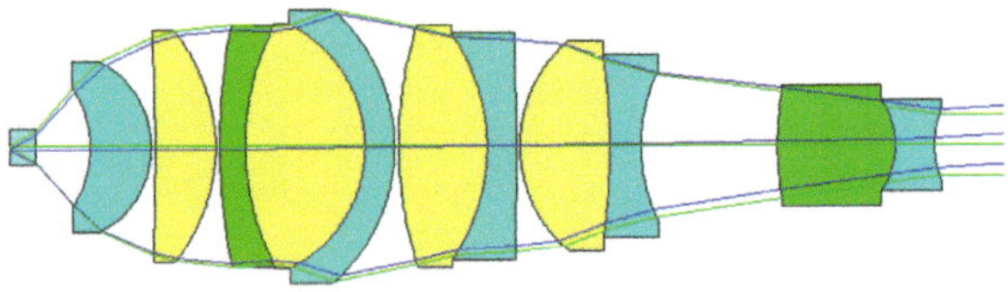

Lens: PlanApo 100x0.75 d=1.3 Ft1200

Object num aper 0.750000 Image height 12.500000 Primary wavln 0.546070

SRF	RADIUS	THICKNESS	APERTURE RADIUS	GLASS	SPECIAL
OBJ	0.000000	0.010000	0.124950	FIZIO	
AST	0.000000	1.300000	0.131733 AS	LZ_LK5	
2	0.000000	2.800000	0.900000	AIR	
3	-5.255600	3.200000	3.000000	LZ_CTK19	
4	-4.708000	0.200000	4.280000	AIR	
5	-177.607600	3.100000	5.510000	CAF2	
6	-9.020300	0.200000	5.820000	AIR	
7	30.589300	1.300000	6.080000	LZ_TF5	
8	12.600800	6.200000	6.030000	CAF2	
9	-6.764600	1.500000	6.110000	LZ_CTK19	
10	-9.051300	0.200000	6.890000	AIR	
11	19.576200	4.600000	6.070000	CAF2	
12	-9.607200	1.500000	5.660000	LZ_CTK19 P	
13	-175.042300	0.200000	5.530000	AIR	
14	6.731900	4.600000	5.270000	CAF2 P	
15	-31.402800	1.600000	4.590000	LZ_CTK19 P	
16	8.467500	7.000000	3.800000	AIR	
17	18.566100	6.000000	3.000000	LZ_TF5 P	
18	-3.960300	2.000000	2.280000	LZ_CTK19 P	
19	4.540900	200.000000	1.820000	AIR	
20	-36.100000	14.000000	14.450000	LZ_K8	
21	-43.560000	3.000000	16.990000	AIR	
22	377.100000	15.000000	17.470000	CAF2	
23	-81.500000	3.000000	17.910000	AIR	
24	221.300000	17.000000	17.640000	CAF2	
25	-72.000000	3.000000	16.830000	LZ_CTK12	
26	0.000000	199.613743 S	16.770000	AIR	
IMS	0.000000	0.000000	12.500000 S		

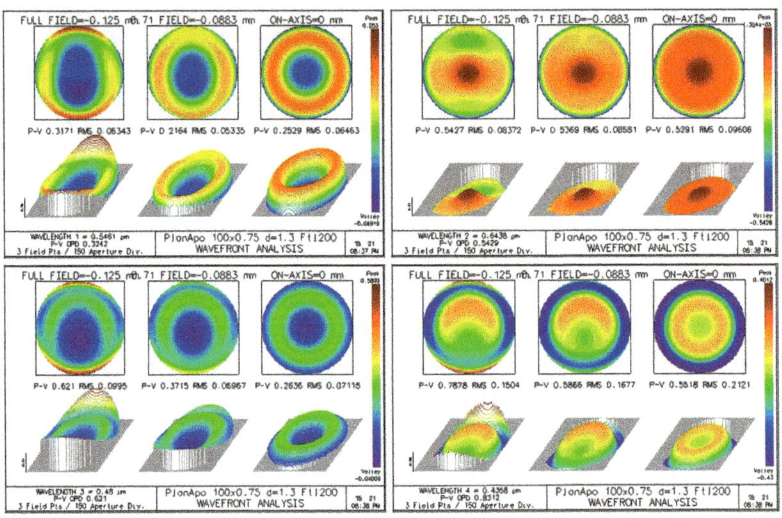

Figure 3.168 Optical design and graphs of aberration correction of a 100×/0.75 ($d = 1.3$ mm) CCF plan apochromatic objective ($F'_{ts} = 200$ mm).

Table 3.6 Main technical parameters and basic optical layout of plan apochromatic objectives for inverted biological microscopes.

Magnification	*NA*	*WD* (mm)	*F′* (mm)	*R* (μm)	*DF* (μm)	*FOV* on object (mm)	*FOV* on image (mm)	Optical design at figure no.	The principal optical layout
10×	0.3	0.5	20	1.11	3.7	2.5	25	3.164	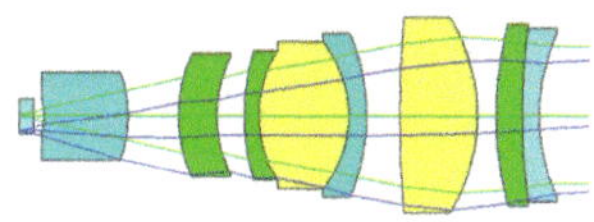
20×	0.6	0.9	10	0.56	0.93	1.25	25	3.165	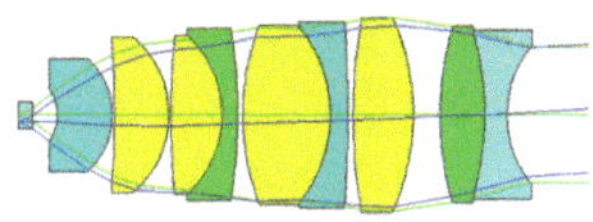
40×	0.6	1.6	5	0.56	0.93	0.625	25	3.166	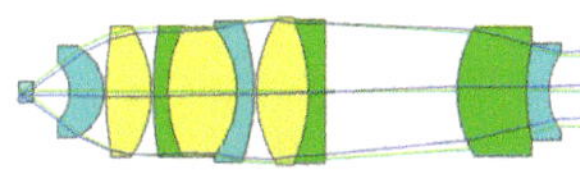
60×	0.7	1.9	3.33	0.48	0.68	0.42	25	3.167	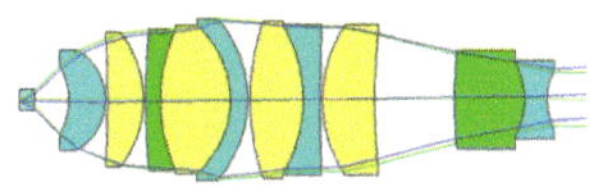
100×	0.75	1.7	2	0.44	0.59	0.25	25	3.168	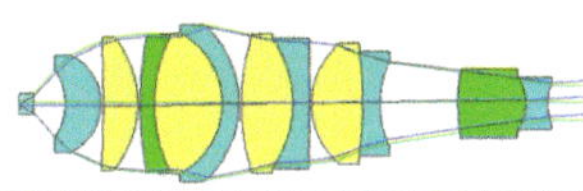

microscope works in transmitted light [11]. The same can be said for opaque objects with which the microscope works on reflection. First, these are microelectronic objects for which a light microscope is used. These objects can have different elements of structures, inclusions, and heterogeneities, which differ in color, shape, and composition from the surrounding fragments and structural elements. To identify and formalize such elements, the most perfect optics is required, including a plan apochromatic type of aberration correction. The structures of metals and alloys that are studied in a metallographic microscope can also contain original inclusions and inhomogeneities. The advantages of plan apochromatic aberration correction of optics are especially clear when using a polarizing microscope, including when working on reflection. The study of a spectra of multicolored inclusions in rocks, for example, allows their chemical composition to be identified and classified.

The second advantage is an increase (about twice) of the linear field in the image space. In modern models of light microscopes, this is no more than 25–26.5 mm; an improvement of this parameter by about two times can provide significant advantages and increase the productivity of work performed on a microscope. It can be taken into account that the image is obtained as a circle. The amount of information that is obtained in the image is proportional to the area of the circle.

The third advantage is an increase in the free working distance from the object plane to the first surface of the objective. This is a very important parameter of objective optics, particularly in the study of modern microelectronic objects. Microcircuits have several technological layers located at different depths from the surface of the sample. In addition, they are often "fenced off," with additional elements and structures that rise above the surface. A large working distance of the lens allows the researcher to "reach out" to any fragment of the sample, which is necessary for studying such objects. This is also true for manufacturing, for example, carburetors and other engineering products, where studying hard-to-reach and hidden cavities is required.

Today, optical components of microscopes and objectives are being developed to be increasingly informative. This process depends primarily on the main technical parameters of objectives, the quality of the images they create, and the degree of aberration correction. The search for optimal designs of objectives continues. Methods of designing, developing, and using new optical materials are being improved. The process of designing optical components and mechanical structures is improving. When analyzing the main elements in the optical–mechanical base of modern light microscopes, it is clear that they completely reflect the main modern trends in the development of microscopy. These tendencies include increasing the resolution, increasing the linear field, and achieving an increasingly higher degree of correction of aberrations.

We can proceed from the fact that researchers need to have an entire arsenal of objectives with different values of linear magnifications and numerical apertures. Assume that a microscope will be used in which there are two channels for projecting images. The first and main channel for such a microscope will be a projection onto a large-diameter digital receiver (a full-format camera receiver of 24×36 mm can be used). Having a second channel where the researcher can observe the image with their eyes is a good addition. But in this microscope, the second channel is not the main one, as it does not allow some advantages of the objective's ultrahigh linear magnifications to be realized (e.g., 150×, 200×, 250×, 500×). The full benefits of these objectives can be realized only when the image is projected onto a digital receiver.

The technique of working on such a microscope does not differ from the usual one. First, the researcher uses a low-magnification objective with the maximum linear field. In our case, this is, for example, a 1× objective with a linear field of about 34 mm or a 2× objective with a linear field of 25 mm. These are so-called *observer* objectives with which the researcher can select the site of interest for more detailed study (through an objective with a higher numerical aperture). When an interesting fragment of the object is selected, a medium- or high-magnification objective (depending on the parameters of the object) is included in the path of the rays. Linear fields of such objectives in the space of objects are smaller than in observer objectives; however, in the space of images they are approximately 50 mm. The ratio of the linear field in the object and in the image space is obviously connected through a linear magnification in the selected objective.

Based on the general requirements for the optical design of the objectives described above, we can design a set of objectives with linear magnifications from 1× to 500×. All the objectives in the kit are designed to work in the standard parfocal height of 95 mm. The focusing tube system is located 200 mm from the objective. The tube system is designed so that the focusing plane is 200 mm from its last surface. We used the same focusing system in the optical calculations of all the objectives; however, any focusing system using an independent aberration correction can be used. However, the level of correction should be very high. Therefore, the optical design of a new tube system for operation on a large linear field should be proposed. The proposed construction of the focusing tube system is rather complicated; it consists of three components. The first is a single lens made of a light crown. The second component is also a single positive lens made of a fluorite crystal. The third component is a double-glued lens made from a positive and negative lens; the positive lens is made from a fluorite crystal, and the negative lens is made from a high-refracting heavy lanthanum crown. Such a focusing tube system has a focal length of 200 mm, a good aberration for numerical apertures to 0.035, and a linear field of up to 50 mm.

Figures 3.169–3.181 show the optical design and graphs of aberration correction of some of these objectives.

Analyzing the optical designs of these objectives shows that they can be divided into four groups. The first group contains low linear magnification objectives (1×, 2×, and 5×), the designs of which are like those of classical or modernized photographic objectives. However, unlike standard photographic objectives, these have a plan apochromatic type of aberration correction and the telecentricity condition is strictly fulfilled. The second group is made up of medium magnification objectives (10× and 20×) in which a strong reversive component is already used at the end of the optical circuit of the objective. This is necessary for obtaining the largest possible working distance of the objective as well as successful compensation correction of the aberrations of its remaining part.

The third group of objectives have magnifications of 50× and 100× and more complex optical circuits. The fourth group contains ultrahigh linear magnification objectives (150×, 200×, 250×, and 500×). The concept of building these objectives is the same; however, 250× and 500× objectives have a triple-glued component, which makes them more difficult to manufacture. Here, too, to increase the working distance of the objective, a strong reverse component is used, which is the last in the circuit.

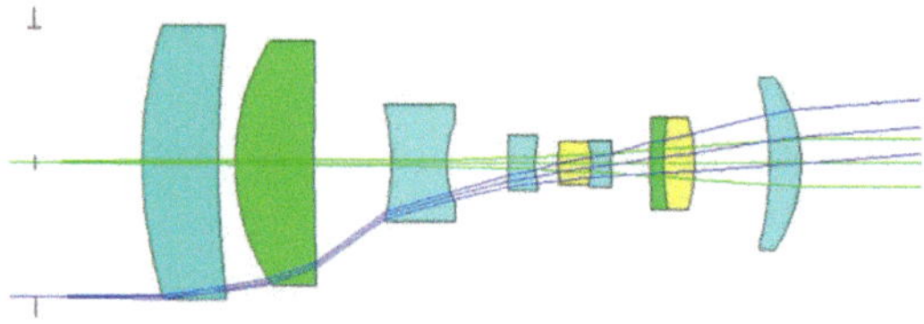

Lens: LD PlanApo 1x0.015 Ft1200

Object num aper 0.015000 Image height 17.000000 Primary wavln 0.546070

SRF	RADIUS	THICKNESS	APERTURE RADIUS	GLASS	SPECIAL
OBJ	0.000000	1.0000e-20	16.962072	AIR	
AST	0.000000	13.920000	16.962072 AS	AIR	
2	55.590000	9.700000	17.250000	H-LAK3	
3	152.600000	2.200000	16.240000	AIR	
4	28.689000	10.000000	15.400000	H-ZF7LA	
5	349.100000	10.000000	13.370000	AIR	
6	-30.760000	7.000000	7.380000	K9	
7	12.560000	8.000000	4.920000	AIR	
8	-135.790000	3.400000	3.490000	K9 P	
9	14.150000	3.000000	3.020000	AIR	
10	28.310000	4.200000	2.800000	CAF2 C	
11	-9.000000	2.500000	2.490000	H-LAK3 P	
12	71.430000	5.000000	2.940000	AIR	
13	318.980000	1.600000	5.350000	H-ZF7LA P	
14	33.700000	4.000000	5.830000	CAF2 P	
15	-19.538000	9.600000	5.690000	AIR	
16	-40.820000	4.000000	10.220000	H-LAK3 P	
17	-19.269000	200.000000	10.860000	AIR	
18	-43.930000	14.000000	25.730000	H-QK3	
19	-51.200000	3.000000	29.330000	AIR	
20	667.000000	15.000000	30.450000	CAF2 C	
21	-105.900000	3.000000	30.890000	AIR	
22	194.900000	17.000000	30.210000	CAF2 C	
23	-118.000000	3.000000	29.160000	H-ZLAF50B	
24	-754.000000	200.096936 S	29.040000	AIR	
IMS	0.000000	0.000000	17.000000		

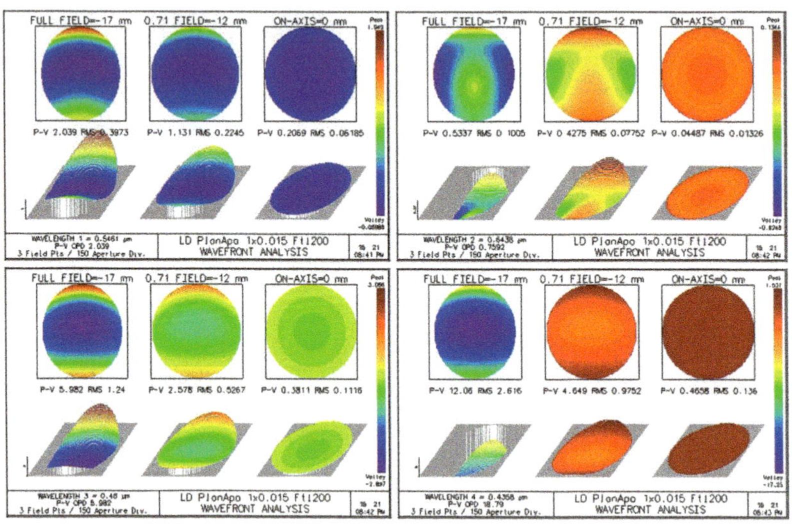

Figure 3.169 Optical design and graphs of aberration correction of a 1×/0.015 CCF plan apochromatic objective ($F'_{ts} = 200$ mm; $h = 95$ mm; $2y' = 34$ mm).

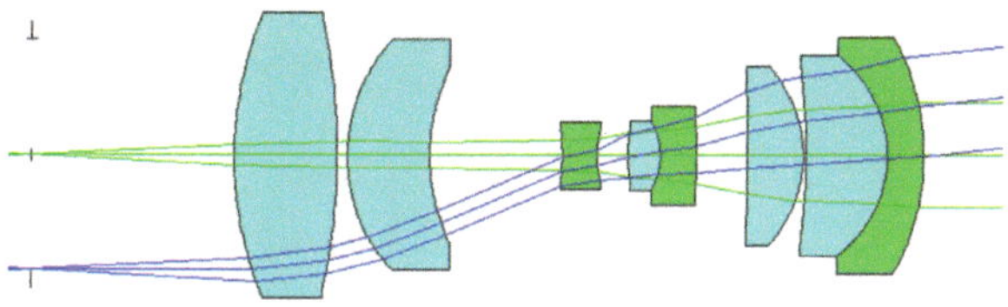

Lens: PlanApo LD 2x0.05 h95

Object num aper 0.050000 Image height 25.000000 Primary wavln 0.546070

SRF	RADIUS		THICKNESS		APERTURE RADIUS		GLASS		SPECIAL
OBJ	0.000000		1.0000e-20		12.489877		AIR		
AST	0.000000		20.000000		13.590000	A	AIR		
2	34.760000		10.000000		13.590000		K9		
3	-61.010000		1.000000		12.950000		AIR		
4	16.181000		8.000000		10.970000		H-QK3		
5	19.269000		13.000000		8.310000		AIR		
6	-21.180000		3.400000		3.170000		H-LAK3		
7	10.627000		3.000000		2.530000		AIR		
8	15.000000		3.200000		2.700000		H-QK3	P	
9	-6.300000		3.500000		3.270000		H-LAK3	P	
10	-139.830000		5.000000		4.670000		AIR		
11	-190.990000		5.500000		7.630000		H-QK3	P	
12	-12.340000		0.300000		8.500000		AIR		
13	-64.927000		8.000000		8.970000		H-QK3	P	
14	-11.660000		3.500000		9.510000		H-LAK3	P	
15	-23.140000		200.000000		11.190000		AIR		
16	-43.930000		14.000000		35.110000		H-QK3	P	
17	-51.200000		3.000000		40.440000		AIR		
18	667.000000		15.000000		43.890000		CAF2	C	
19	-105.900000		3.000000		44.200000		AIR		
20	194.900000		17.000000		42.620000		CAF2	C	
21	-118.000000		3.000000		42.030000		H-ZLAF50B		
22	-754.000000		200.021483	S	41.850000		AIR		
IMS	0.000000		0.000000		25.000000	S			

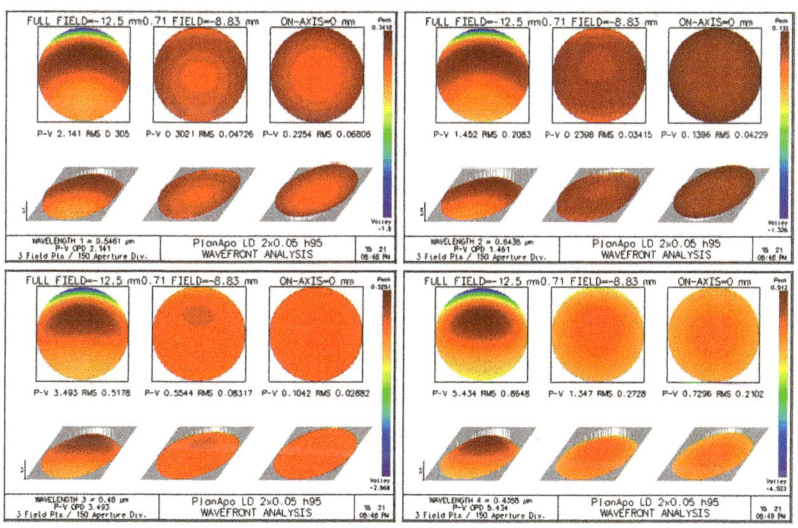

Figure 3.170 Optical design and graphs of aberration correction of a 2×/0.05 CCF plan apochromatic objective ($F'_{ts} = 200$ mm; $h = 95$ mm; $2y' = 50$ mm).

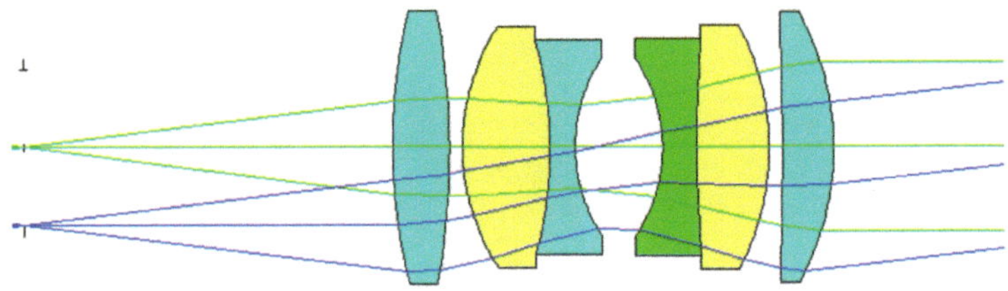

Lens: PlanApo LD 5x012 h95

Object num aper	0.120000	Image height	25.000000	Primary wavln	0.546070

SRF	RADIUS	THICKNESS	APERTURE RADIUS	GLASS	SPECIAL
OBJ	0.000000	1.0000e-20	5.015564	AIR	
AST	0.000000	21.929000	5.015564 AS	AIR	
2	30.270000	3.400000	7.830000	H-LAK3	
3	-42.482000	0.800000	7.760000	AIR	
4	12.340000	5.200000	6.920000	CAF2 C	
5	-21.580000	1.500000	6.150000	H-LAK4L	
6	8.801000	5.200000	5.120000	AIR	
7	-8.801000	2.000000	5.140000	H-ZF1	
8	75.000000	4.300000	6.200000	CAF2 C	
9	-14.510000	0.800000	6.960000	AIR	
10	-254.000000	3.000000	7.560000	LZ_CTK12	
11	-15.672000	200.000000	7.730000	AIR	
12	-43.930000	14.000000	26.560000	H-QK3	
13	-51.200000	3.000000	30.630000	AIR	
14	667.000000	15.000000	32.450000	CAF2 C	
15	-105.900000	3.000000	33.130000	AIR	
16	194.900000	17.000000	32.730000	CAF2 C	
17	-118.000000	3.000000	31.980000	H-ZLAF50B	
18	-754.000000	200.000819 S	31.990000	AIR	
IMS	0.000000	0.000000	25.000000 S		

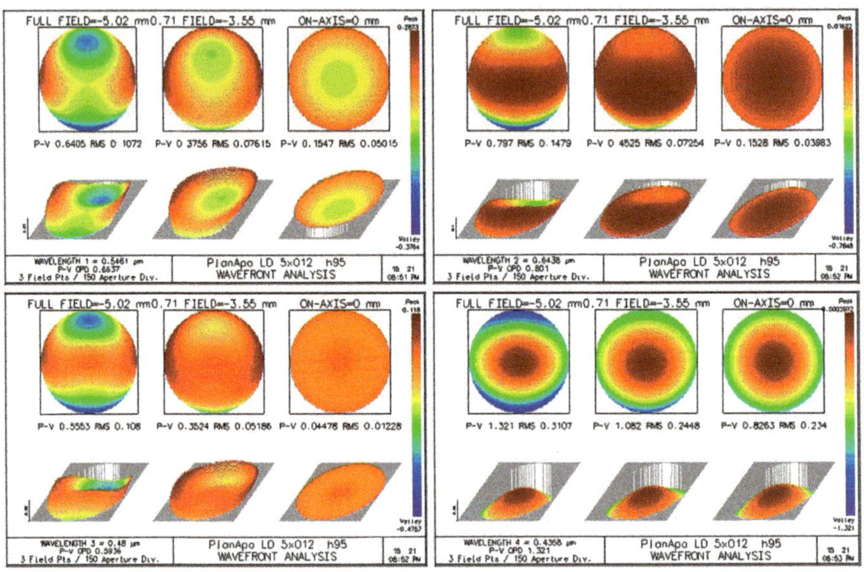

Figure 3.171 Optical design and graphs of aberration correction of a 5×/0.12 CCF plan apochromatic objective ($F'_{ts} = 200$ mm; $h = 95$ mm; $2y' = 50$ mm).

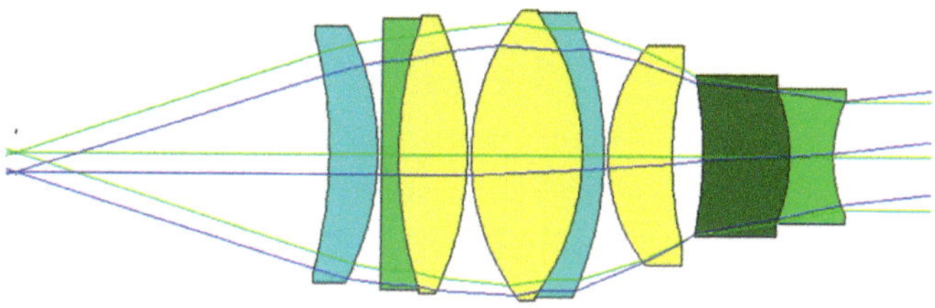

Lens: PlanApo LD 10x030 h95

Object num aper 0.300000 Image height 25.000000 Primary wavln 0.546070

SRF	RADIUS	THICKNESS		APERTURE RADIUS		GLASS		SPECIAL
OBJ	0.000000	1.0000e-20		2.501171		AIR		
AST	0.000000	36.485000		2.501171	AS	AIR		
2	-49.660000	5.600000		13.380000		H-LAK3		
3	-29.570000	0.500000		14.470000		AIR		
4	0.000000	2.000000		14.990000		H-ZLAF3		
5	47.500000	8.000000		15.280000		CAF2	C	
6	-34.760000	0.500000		15.620000		AIR		
7	25.500000	13.000000		16.260000		CAF2	C	
8	-25.500000	2.500000		15.960000		H-LAK3	P	
9	-37.070000	0.500000		15.950000		AIR		
10	18.883000	7.400000		12.230000		CAF2	C	
11	39.370000	3.600000		9.940000		AIR		
12	-44.240000	10.000000		8.960000		H-ZF62		
13	-19.724000	5.000000		7.490000		H-LAF10LA		
14	15.136000	200.000000		6.370000		AIR		
15	-43.930000	14.000000		29.930000		H-QK3		
16	-51.200000	3.000000		34.420000		AIR		
17	667.000000	15.000000		36.750000		CAF2	C	
18	-105.900000	3.000000		37.300000		AIR		
19	194.900000	17.000000		36.610000		CAF2	C	
20	-118.000000	3.000000		35.860000		H-ZLAF50B		
21	-754.000000	200.043121	S	35.790000		AIR		
IMS	0.000000	0.000000		25.000000	S			

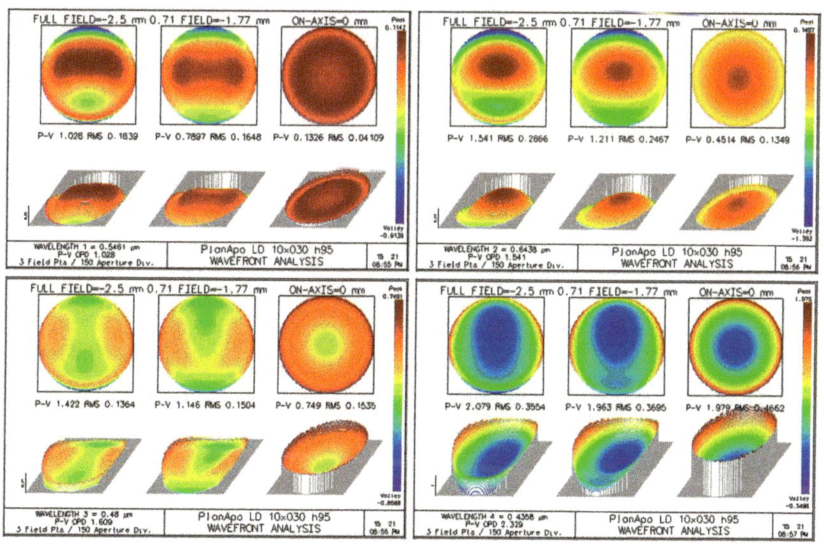

Figure 3.172 Optical design and graphs of aberration correction of a 10×/0.30 CCF plan apochromatic objective ($F'_{ts} = 200$ mm; $h = 95$ mm; $2y' = 50$ mm).

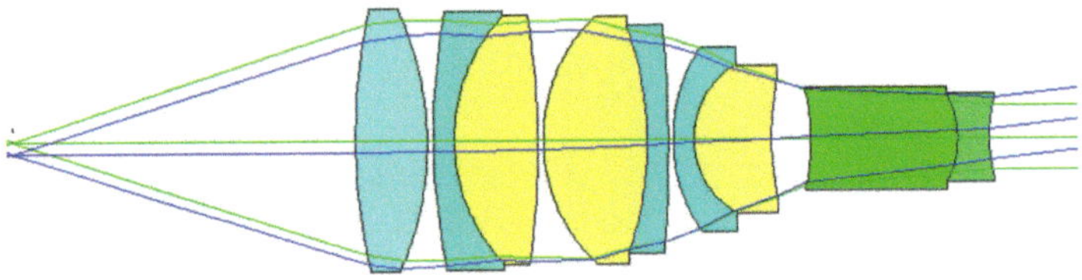

Lens: PlanApo LD 20x030 h95

Object num aper 0.300000 Image height 25.000000 Primary wavln 0.546070

SRF	RADIUS		THICKNESS		APERTURE RADIUS		GLASS		SPECIAL
OBJ	0.000000		1.0000e-20		1.249007		AIR		
AST	0.000000		32.967000		1.249007	AS	AIR		
2	53.680000		6.800000		12.030000		H-QK3		
3	-28.689000		0.600000		12.360000		AIR		
4	62.230000		2.000000		12.140000		LZ_CTK12		
5	16.530000		8.000000		11.540000		CAF2	C	
6	-74.290000		0.500000		11.640000		AIR		
7	15.672000		10.000000		11.580000		CAF2	C	
8	-30.760000		2.000000		10.680000		LZ_CTK12	P	
9	-119.500000		0.500000		10.030000		AIR		
10	14.791000		2.000000		8.570000		LZ_CTK12	P	
11	7.900000		7.400000		6.880000		CAF2	C	
12	25.800000		3.800000		5.570000		AIR		
13	-20.750000		14.000000		4.780000		H-ZF52A		
14	-8.336000		3.000000		4.210000		H-ZLAF50B		
15	14.510000		200.000000		3.930000		AIR		
16	-43.930000		14.000000		27.590000		H-QK3		
17	-51.200000		3.000000		31.760000		AIR		
18	667.000000		15.000000		33.710000		CAF2	C	
19	-105.900000		3.000000		34.350000		AIR		
20	194.900000		17.000000		33.870000		CAF2	C	
21	-118.000000		3.000000		33.120000		H-ZLAF50B		
22	-754.000000		200.027011	S	33.100000		AIR		
IMS	0.000000		0.000000		25.000000	S			

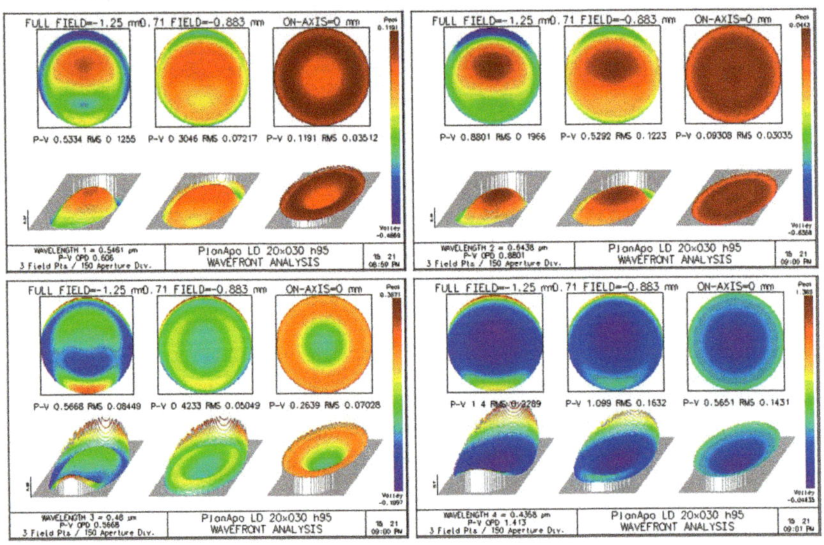

Figure 3.173 Optical design and graphs of aberration correction of a 20×/0.30 CCF plan apochromatic objective (F'_{ts} = 200 mm; h = 95 mm; $2y'$ = 50 mm).

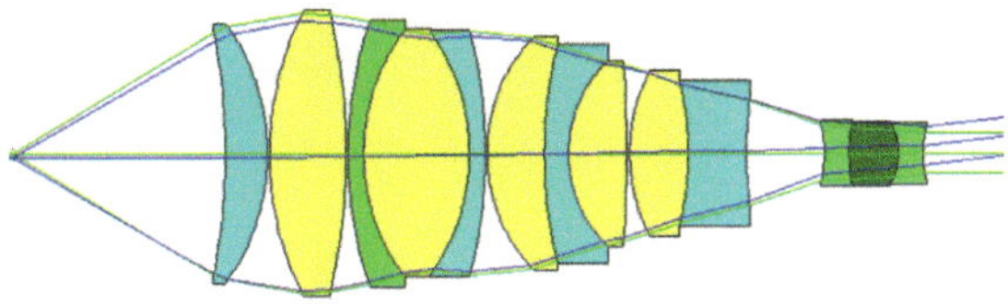

Lens: PlanApo LD 50x0.50 d0 h95

Object num aper 0.500000 Image height 25.000000 Primary wavln 0.546070

SRF	RADIUS	THICKNESS	APERTURE RADIUS	GLASS	SPECIAL
OBJ	0.000000	1.0000e-20	0.500292	AIR	
AST	0.000000	21.997000	0.500292 AS	AIR	
2	-59.320000	4.360000	12.430000	H-ZLAF1	
3	-21.580000	0.200000	12.970000	AIR	
4	30.270000	8.000000	14.300000	CAF2	
5	-57.800000	0.200000	14.150000	AIR	
6	40.190000	1.600000	13.330000	H-ZLAF3	
7	19.790000	11.000000	12.360000	CAF2 P	
8	-19.790000	1.600000	12.030000	H-ZLAF1 P	
9	-47.896000	0.200000	12.270000	AIR	
10	16.120000	6.000000	11.650000	CAF2 P	
11	42.657000	2.500000	10.850000	H-LAF10L	
12	12.465000	6.000000	9.220000	CAF2 P	
13	-124.360000	0.300000	8.960000	AIR	
14	17.760000	6.000000	8.260000	CAF2 P	
15	-27.534000	6.000000	7.240000	LZ_CTK12	
16	41.590000	8.000000	5.530000	AIR	
17	-13.305000	2.500000	3.350000	H-ZLAF3 P	
18	18.710000	5.100000	3.250000	H-ZF62	
19	-5.750000	2.500000	3.140000	H-ZLAF3 P	
20	10.850000	200.000000	2.890000	AIR	
21	-43.930000	14.000000	26.630000	H-QK3	
22	-51.200000	3.000000	30.680000	AIR	
23	667.000000	15.000000	32.500000	CAF2 C	
24	-105.900000	3.000000	33.170000	AIR	
25	194.900000	17.000000	32.780000	CAF2 C	
26	-118.000000	3.000000	32.040000	H-ZLAF50B	
27	-754.000000	199.674235 S	32.040000	AIR	
IMS	0.000000	0.000000	25.000000 S		

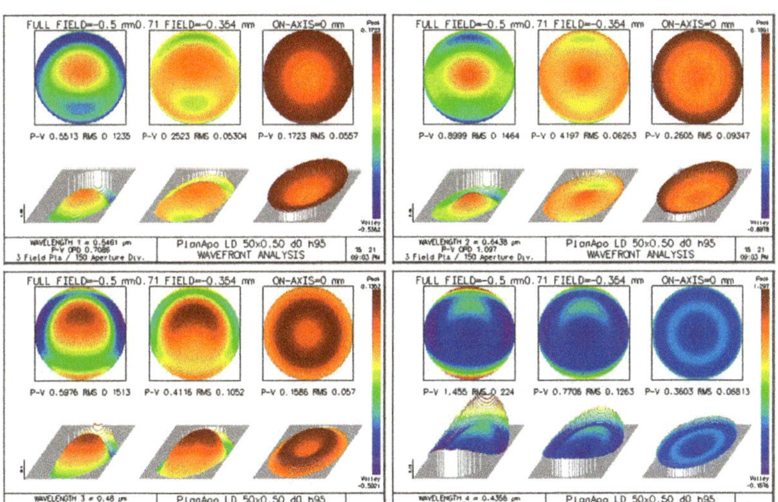

Figure 3.174 Optical design and graphs of aberration correction of a 50×/0.50 CCF plan apochromatic objective ($F'_{ts} = 200$ mm; $h = 95$ mm; $2y' = 50$ mm).

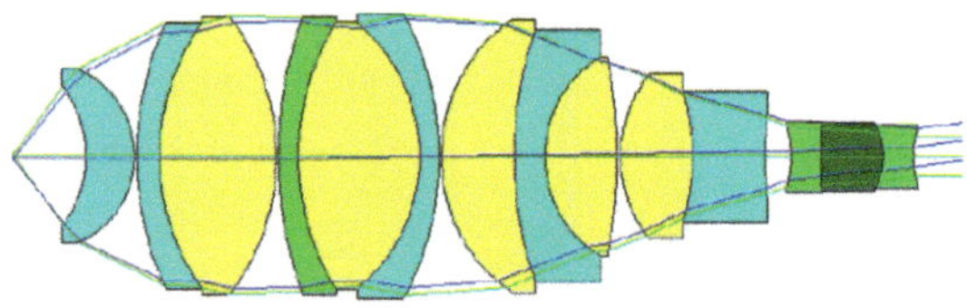

Lens: PlanApo LD 100x0.80 d0 h95

Object num aper 0.800000 Image height 25.000000 Primary wavln 0.546070

SRF	RADIUS	THICKNESS	APERTURE RADIUS		GLASS		SPECIAL
OBJ	0.000000	1.0000e-20	0.250206		AIR		
AST	0.000000	5.999236	0.250206	AS	AIR		
2	-9.000000	4.000000	5.620000		H-LAF10L		
3	-7.400000	0.200000	6.910000		AIR		
4	24.210000	1.900000	10.560000		H-LAF10L	P	
5	17.602000	9.700000	10.590000		CAF2	C	
6	-17.602000	0.200000	11.070000		AIR		
7	31.724000	1.600000	11.070000		H-ZLAF3		
8	19.790000	10.000000	10.630000		CAF2	P	
9	-13.305000	1.600000	10.570000		H-LAF10L	P	
10	-23.390000	0.200000	11.360000		AIR		
11	12.979000	6.000000	10.860000		CAF2	P	
12	30.000000	2.500000	10.080000		H-LAF10L	P	
13	9.032000	6.000000	7.890000		CAF2	P	
14	-38.950000	0.300000	7.800000		AIR		
15	10.068000	6.000000	6.560000		CAF2	P	
16	-17.701000	6.000000	5.200000		H-ZLAF3	P	
17	18.883000	2.200000	3.320000		AIR		
18	-9.900000	2.500000	2.810000		H-ZLAF3	P	
19	27.160000	5.200000	2.710000		H-ZF62		
20	-4.720000	2.500000	2.610000		H-LAF10L	P	
21	7.047000	200.000000	2.280000		AIR		
22	-43.930000	14.000000	24.910000		H-QK3		
23	-51.200000	3.000000	28.750000		AIR		
24	667.000000	15.000000	30.370000		CAF2	C	
25	-105.900000	3.000000	31.100000		AIR		
26	194.900000	17.000000	30.860000		CAF2	C	
27	-118.000000	3.000000	30.170000		H-ZLAF50B		
28	-754.000000	200.000716 S	30.190000		AIR		
IMS	0.000000	0.000000	25.000000	S			

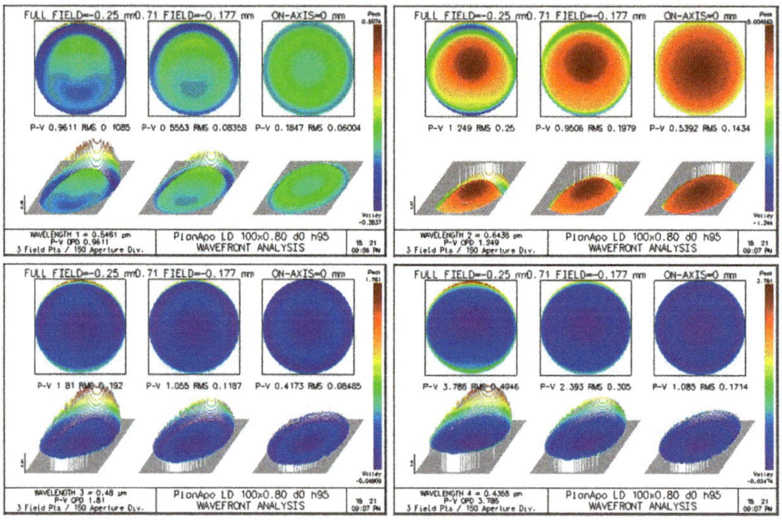

Figure 3.175 Optical design and graphs of aberration correction of a 100×/0.80 CCF plan apochromatic objective ($F'_{ts} = 200$ mm; $h = 95$ mm; $2y' = 50$ mm).

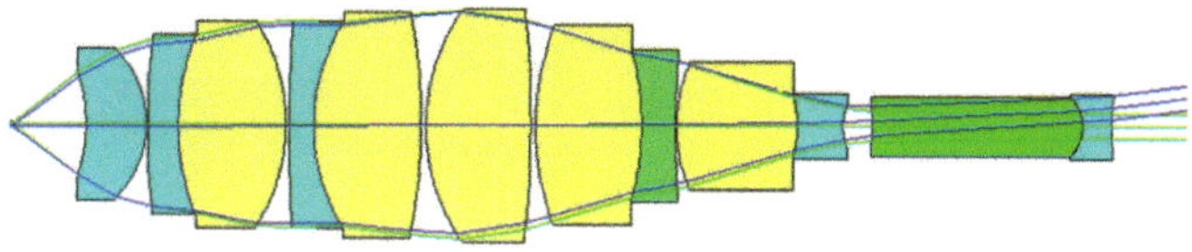

Lens: PlanFluor LD 150x0.60 d0 h95

Object num aper 0.600000 Image height 25.000000 Primary wavln 0.546070

SRF	RADIUS	THICKNESS		APERTURE RADIUS		GLASS		SPECIAL
OBJ	0.000000	1.0000e-20		0.166568		AIR		
AST	0.000000	5.012200		0.166568	AS	AIR		
2	-10.516000	4.000000		3.480000		H-LAF10L		
3	-6.780300	0.100000		4.880000		AIR		
4	48.540000	2.000000		5.530000		LZ_CTK12		
5	15.000000	7.200000		5.820000		CAF2	C	
6	-12.057000	0.250000		6.650000		AIR		
7	117.590000	1.600000		6.700000		LZ_CTK12	P	
8	12.340000	7.200000		6.670000		CAF2	P	
9	-35.480000	0.300000		7.250000		AIR		
10	12.840000	7.000000		7.580000		CAF2	C	
11	-64.450000	0.300000		6.960000		AIR		
12	18.000000	7.000000		6.450000		CAF2	C	
13	-18.000000	2.200000		4.880000		H-ZF7LA		
14	48.080000	0.100000		4.350000		AIR		
15	10.068000	8.000000		4.160000		CAF2	P	
16	-7.127000	3.000000		2.110000		LZ_CTK12	P	
17	3.448000	2.000000		1.400000		AIR		
18	-10.130000	14.000000		1.350000		H-ZF5		
19	-2.618000	2.000000		1.950000		H-LAK3		
20	29.990000	200.000000		2.080000		AIR		
21	-43.930000	14.000000		26.490000		H-QK3		
22	-51.200000	3.000000		30.560000		AIR		
23	667.000000	15.000000		32.410000		CAF2	C	
24	-105.900000	3.000000		33.090000		AIR		
25	194.900000	17.000000		32.720000		CAF2	C	
26	-118.000000	3.000000		31.980000		H-ZLAF50B		
27	-754.000000	200.204643	S	31.960000		AIR		
IMS	0.000000	0.000000		25.000000				

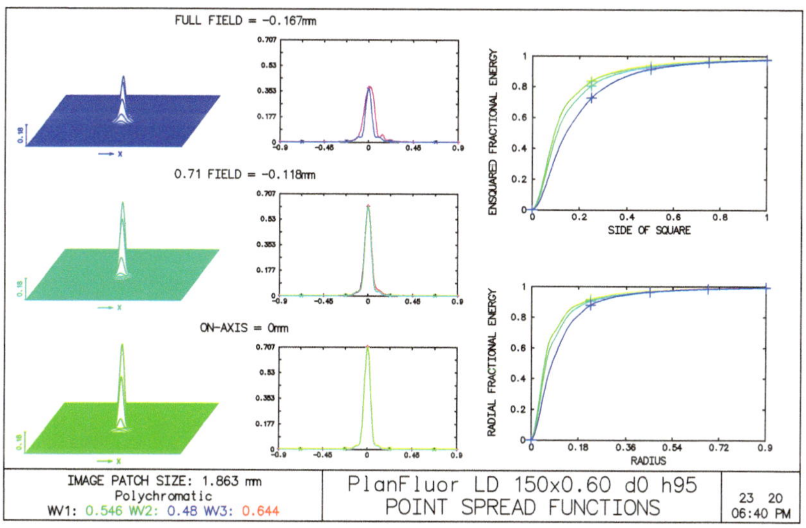

Figure 3.176 Optical design and graphs of aberration correction of a 150×/0.60 LCF plan MicroFluar objective ($F'_{ts} = 200$ mm; $h = 95$ mm; $2y' = 50$ mm).

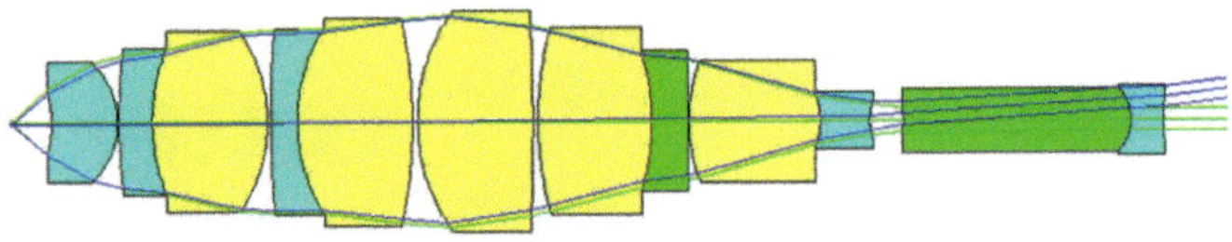

Lens: PlanFluor LD 200x0.65 d0 h95

Object num aper 0.650000 Image height 25.000000 Primary wavln 0.546070

SRF	RADIUS	THICKNESS	APERTURE RADIUS	GLASS	SPECIAL
OBJ	0.000000	1.0000e-20	0.125020	AIR	
AST	0.000000	2.498828	0.125020 AS	AIR	
2	-7.490000	4.000000	2.020000	H-LAF10L	
3	-4.950000	0.100000	3.570000	AIR	
4	43.250000	2.000000	4.060000	LZ_CTK12	
5	11.092000	7.200000	4.370000	CAF2 C	
6	-9.480000	0.250000	5.390000	AIR	
7	109.000000	1.600000	5.470000	LZ_CTK12 P	
8	10.068000	7.200000	5.520000	CAF2 P	
9	-27.160000	0.300000	6.220000	AIR	
10	11.092000	7.000000	6.560000	CAF2 C	
11	-103.020000	0.300000	5.890000	AIR	
12	19.538000	7.000000	5.580000	CAF2 C	
13	-13.900000	2.200000	4.170000	H-ZF7LA	
14	0.000000	0.100000	3.810000	AIR	
15	10.068000	8.000000	3.620000	CAF2 P	
16	-4.146000	3.000000	1.670000	LZ_CTK12 P	
17	3.569000	2.000000	1.130000	AIR	
18	-6.087000	14.000000	1.070000	H-ZF5	
19	-2.638000	2.000000	1.900000	H-LAK3	
20	75.860000	200.000000	2.070000	AIR	
21	-43.930000	14.000000	26.090000	H-QK3	
22	-51.200000	3.000000	30.090000	AIR	
23	667.000000	15.000000	31.850000	CAF2 C	
24	-105.900000	3.000000	32.540000	AIR	
25	194.900000	17.000000	32.180000	CAF2 C	
26	-118.000000	3.000000	31.440000	H-ZLAF50B	
27	-754.000000	199.999921 S	31.450000	AIR	
IMS	0.000000	0.000000	25.000000		

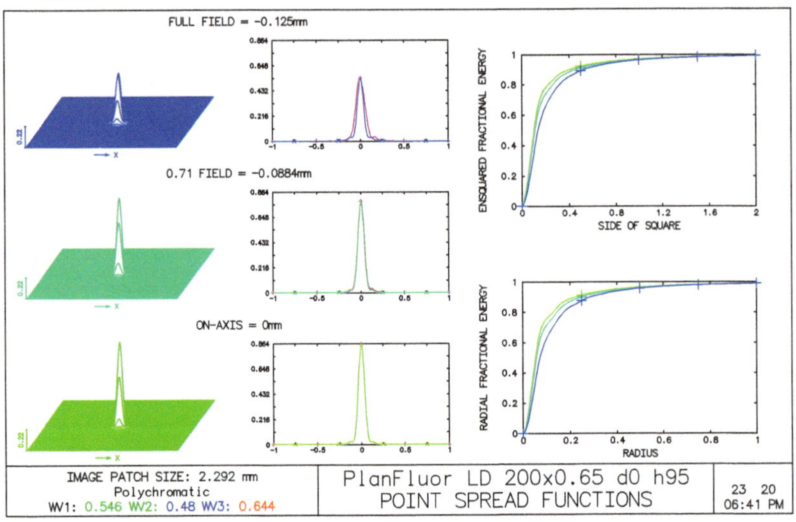

Figure 3.177 Optical design and graphs of aberration correction of a 200×/0.65 LCF plan MicroFluar objective ($F'_{ts} = 200$ mm; $h = 95$ mm; $2y' = 50$ mm).

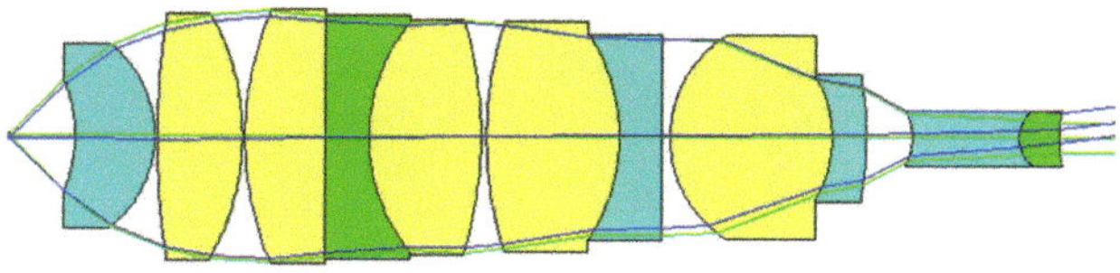

Lens: planFluor LD 250x0.70 d0 h95

Object num aper 0.700000 Image height 25.000000 Primary wavln 0.546070

SRF	RADIUS	THICKNESS	APERTURE RADIUS	GLASS	SPECIAL
OBJ	0.000000	1.0000e-20	0.099977	AIR	
AST	0.000000	2.502445	0.099977 AS	AIR	
2	-5.037000	3.000000	2.100000	H-LAK3	
3	-4.146000	0.100000	3.360000	AIR	
4	28.310000	3.200000	4.180000	CAF2 C	
5	-8.850000	0.100000	4.510000	AIR	
6	11.581000	3.000000	4.620000	CAF2 C	
7	0.000000	1.600000	4.430000	H-ZF1	
8	6.399000	4.200000	4.200000	CAF2 C	
9	-14.791000	0.200000	4.290000	AIR	
10	13.090000	5.000000	4.200000	CAF2 C	
11	-6.516000	1.600000	3.750000	LZ_CTK12	
12	0.000000	0.300000	3.730000	AIR	
13	4.338000	6.100000	3.690000	CAF2 C	
14	-4.338000	1.300000	2.310000	LZ_CTK12 P	
15	-9.817000	1.700000	1.950000	AIR	
16	-2.105000	4.000000	1.020000	H-LAK3 P	
17	1.260000	1.500000	1.000000	H-ZF7LA	
18	3.800000	200.000000	0.950000	AIR	
19	-43.930000	14.000000	26.130000	H-QK3	
20	-51.200000	3.000000	30.160000	AIR	
21	667.000000	15.000000	32.010000	CAF2 C	
22	-105.900000	3.000000	32.730000	AIR	
23	194.900000	17.000000	32.450000	CAF2 C	
24	-118.000000	3.000000	31.770000	H-ZLAF50B	
25	-754.000000	200.031019 S	31.790000	AIR	
IMS	0.000000	0.000000	25.000000		

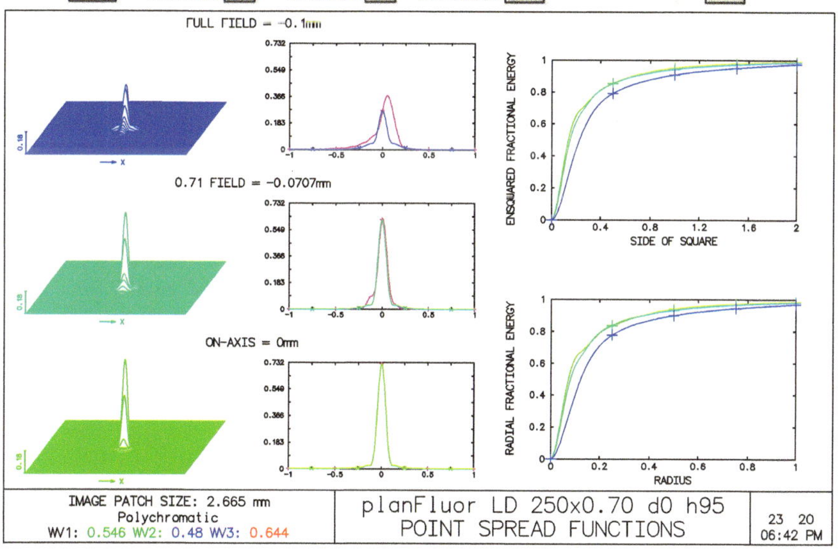

Figure 3.178 Optical design and graphs of aberration correction of a 250×/0.70 LCF plan MicroFluar objective (F'_{ts} = 200 mm; h = 45 mm; $2y'$ = 50 mm).

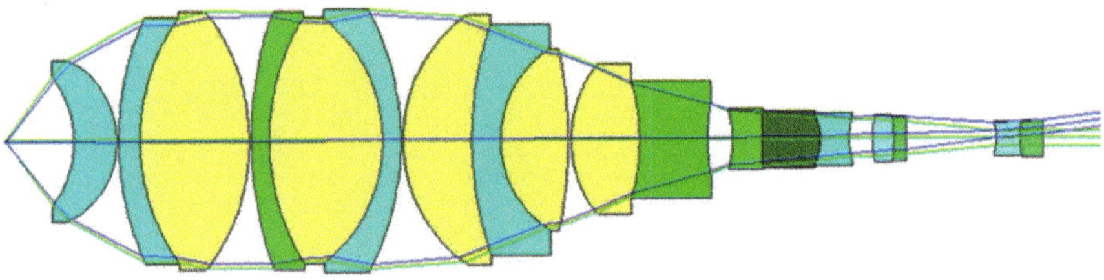

Lens: PlanApo LD 250x0.80 d0 h95

Object num aper 0.800000 Image height 25.000000 Primary wavln 0.546070

SRF	RADIUS	THICKNESS	APERTURE RADIUS	GLASS	SPECIAL
OBJ	0.000000	1.0000e-20	0.099984	AIR	
AST	0.000000	5.999148	0.099984 AS	AIR	
2	-9.000000	4.000000	5.550000	H-LAF10L	
3	-7.400000	0.200000	6.870000	AIR	
4	24.210000	1.900000	10.540000	H-LAF10L P	
5	17.602000	9.700000	10.570000	CAF2 C	
6	-17.602000	0.200000	11.070000	AIR	
7	31.724000	1.600000	11.040000	H-ZLAF3	
8	19.790000	10.000000	10.580000	CAF2 P	
9	-13.305000	1.600000	10.510000	H-LAF10L P	
10	-23.390000	0.200000	11.280000	AIR	
11	12.979000	6.000000	10.710000	CAF2 P	
12	30.000000	2.500000	9.880000	H-LAF10L P	
13	9.032000	6.000000	7.780000	CAF2 P	
14	-38.950000	0.300000	7.650000	AIR	
15	10.068000	6.000000	6.440000	CAF2 P	
16	-17.701000	6.000000	5.010000	H-ZLAF3 P	
17	18.883000	2.200000	3.140000	AIR	
18	-9.900000	2.500000	2.570000	H-ZLAF3 P	
19	27.160000	5.200000	2.420000	H-ZF62	
20	-4.720000	2.500000	2.190000	H-LAF10L P	
21	7.047000	2.000000	1.880000	AIR	
22	6.534100	1.900000	1.980000	LZ_LK3	
23	-15.447300	1.000000	1.870000	LZ_F4	
24	50.580300	8.000000	1.810000	AIR	
25	-3.694400	2.000000	1.290000	LZ_LK3 P	
26	4.990600	2.000000	1.470000	LZ_F4 P	
27	20.167400	200.000000	1.550000	AIR	
28	-43.930000	14.000000	26.510000	H-QK3	
29	-51.200000	3.000000	30.650000	AIR	
30	667.000000	15.000000	32.610000	CAF2 C	
31	-105.900000	3.000000	33.330000	AIR	
32	194.900000	17.000000	33.030000	CAF2 C	
33	-118.000000	3.000000	32.350000	H-ZLAF50B	
34	-754.000000	199.975315 S	32.380000	AIR	
IMS	0.000000	0.000000	25.000000 S		

Figure 3.179 Optical design and graphs of aberration correction of a 250×/0.80 CCF plan apochromatic objective ($F'_{ts} = 200$ mm; $h = 95$ mm; $2y' = 50$ mm).

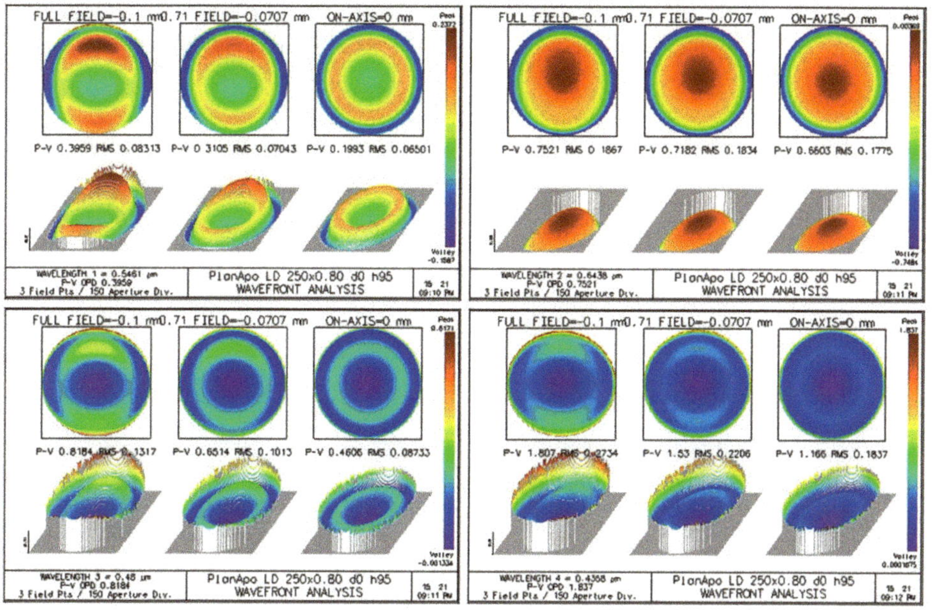

Figure 3.179 (*Continued*) Optical design and graphs of aberration correction of a 250×/0.80 CCF plan apochromatic objective ($F'_{ts} = 200$ mm; $h = 95$ mm; $2y' = 50$ mm).

Testing these objectives demonstrated their quality. Figure 3.182 shows photographs of these objectives. Naturally, the mechanical design using the same optical system can be different, which is shown in the example of a 20× objective. In addition, the mechanical design of the 250× and 500× objectives can be altered by shortening the parfocal distance to 45 mm and using standard 0.8” connection threads, respectively. Table 3.7 shows the main technical parameters of these objectives.

Figure 3.183 shows a variation of the standard equipment for using these objectives when the linear field corresponds to the microscope’s ocular field.

In other cases, when a larger linear field in the image space is required, the visual head must be dismantled and a special tube system must be used to realize the telecentric path of the rays. In these cases, the characteristics of these objectives can be realized most fully. Figures 3.184–3.208 show photographs of real objects taken using these objectives.

3.10 Expanding the Spectral Range of Lens Microscope Objectives

One of the main parameters of the objective, along with the numerical aperture, linear field, and aberration correction, is its—and hence the microscope’s—spectral range. Traditionally, light microscopes have been designed to operate in the visible range of the spectrum, as they are mainly

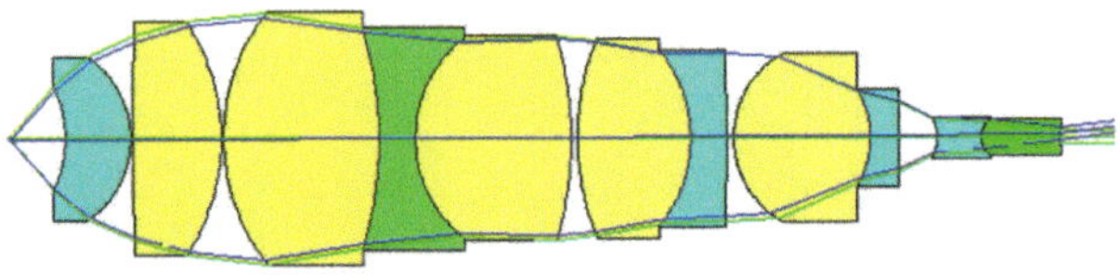

Lens: PlanFluor LD 500x0.75 d0 h95

Object num aper 0.750000 Image height 25.000000 Primary wavln 0.546070

SRF	RADIUS	THICKNESS		APERTURE RADIUS		GLASS		SPECIAL
OBJ	0.000000	1.0000e-20		0.049993		AIR		
AST	0.000000	2.503805		0.049993	AS	AIR		
2	-5.037000	3.000000		2.270000		H-LAK3		
3	-4.146000	0.100000		3.540000		AIR		
4	123.640000	4.000000		4.460000		CAF2	C	
5	-9.157000	0.100000		5.100000		AIR		
6	8.511000	7.000000		5.520000		CAF2	C	
7	-16.120000	1.600000		4.820000		H-ZBAF7		
8	5.401000	7.000000		4.330000		CAF2	C	
9	-15.489000	0.270000		4.470000		AIR		
10	11.041000	5.000000		4.320000		CAF2	C	
11	-6.000000	1.600000		3.820000		LZ_CTK12		
12	-176.120000	0.300000		3.780000		AIR		
13	4.120000	6.100000		3.670000		CAF2	C	
14	-4.120000	1.300000		2.110000		LZ_CTK12	P	
15	-38.000000	1.700000		1.640000		AIR		
16	-2.000000	2.000000		0.890000		LZ_CTK12	P	
17	1.010000	3.500000		0.810000		H-ZF7LA		
18	1.320000	200.000000		0.480000		AIR		
19	-43.930000	14.000000		26.130000		H-QK3		
20	-51.200000	3.000000		30.190000		AIR		
21	667.000000	15.000000		32.110000		CAF2	C	
22	-105.900000	3.000000		32.860000		AIR		
23	194.900000	17.000000		32.610000		CAF2	C	
24	-118.000000	3.000000		31.960000		H-ZLAF50B		
25	-754.000000	199.986799	S	32.000000		AIR		
IMS	0.000000	0.000000		25.000000				

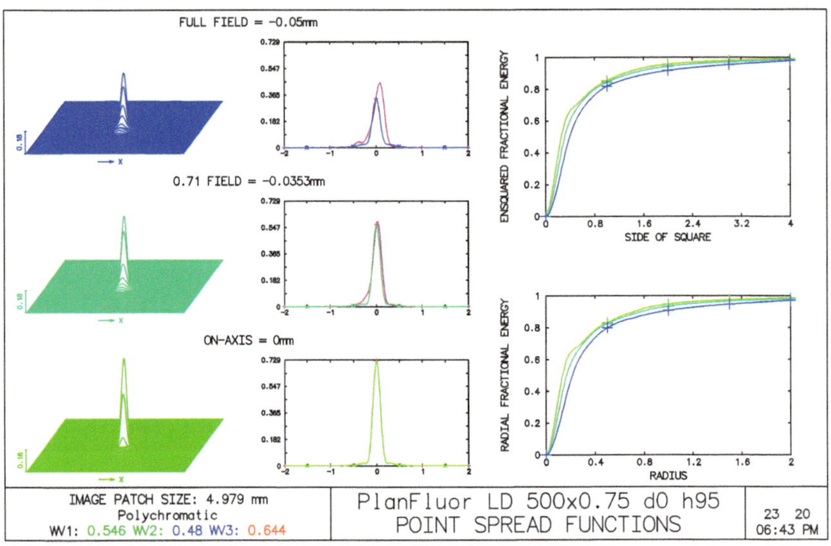

Figure 3.180 Optical design and graphs of aberration correction of a 500×/0.75 LCF plan MicroFluar objective ($F'_{ts} = 200$ mm; $h = 45$ mm; $2y' = 50$ mm).

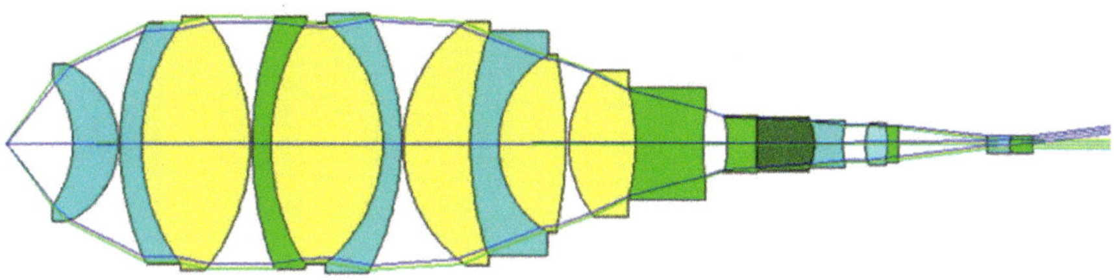

Lens: PlanApo LD 500x0.80 d0 h95

Object num aper 0.800000 Image height 25.000000 Primary wavln 0.546070

SRF	RADIUS		THICKNESS		APERTURE RADIUS		GLASS		SPECIAL
OBJ	0.000000		1.0000e-20		0.050301		AIR		
AST	0.000000		5.999148		0.050301	AS	AIR		
2	-9.000000		4.000000		5.520000		H-LAF10L		
3	-7.400000		0.200000		6.860000		AIR		
4	24.210000		1.900000		10.530000		H-LAF10L	P	
5	17.602000		9.700000		10.560000		CAF2	C	
6	-17.602000		0.200000		11.070000		AIR		
7	31.724000		1.600000		11.030000		H-ZLAF3		
8	19.790000		10.000000		10.570000		CAF2	P	
9	-13.305000		1.600000		10.490000		H-LAF10L	P	
10	-23.390000		0.200000		11.240000		AIR		
11	12.979000		6.000000		10.660000		CAF2	P	
12	30.000000		2.500000		9.810000		H-LAF10L	P	
13	9.032000		6.000000		7.740000		CAF2	P	
14	-38.950000		0.300000		7.590000		AIR		
15	10.068000		6.000000		6.400000		CAF2	P	
16	-17.701000		6.000000		4.940000		H-ZLAF3	P	
17	18.883000		2.200000		3.080000		AIR		
18	-9.900000		2.500000		2.490000		H-ZLAF3	P	
19	27.160000		5.200000		2.320000		H-ZF62		
20	-4.720000		2.500000		2.060000		H-LAF10L	P	
21	7.047000		2.000000		1.740000		AIR		
22	5.890000		1.910000		1.790000		LZ_K8		
23	-8.600000		1.000000		1.660000		LZ_TF2		
24	0.000000		7.990000		1.000000		AIR		
25	-2.630000		2.000000		0.720000		LZ_K8	P	
26	2.170000		2.000000		0.780000		LZ_TF5		
27	3.360000		200.000000		0.730000		AIR		
28	-43.930000		14.000000		28.150000		H-QK3		
29	-51.200000		3.000000		32.560000		AIR		
30	667.000000		15.000000		34.930000		CAF2	C	
31	-105.900000		3.000000		35.630000		AIR		
32	194.900000		17.000000		35.270000		CAF2	C	
33	-118.000000		3.000000		34.620000		H-ZLAF50B		
34	-754.000000		199.920989	S	34.660000		AIR		
IMS	0.000000		0.000000		25.000000				

Figure 3.181 Optical design and graphs of aberration correction of a 500×/0.80 CCF plan apochromatic objective ($F'_{ts} = 200$ mm; $h = 95$ mm; $2y' = 50$ mm).

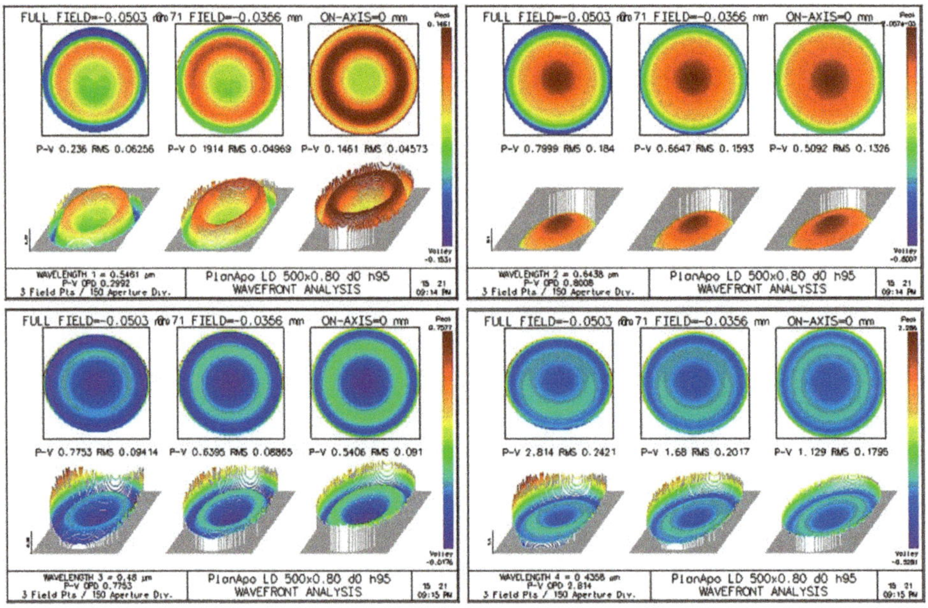

Figure 3.181 (*Continued*) Optical design and graphs of aberration correction of a 500×/0.80 CCF plan apochromatic objective ($F'_{ts} = 200$ mm; $h = 95$ mm; $2y' = 50$ mm).

Figure 3.182 Photographs of 50-mm observation and long working distance objectives (95-mm parfocal distance).

Table 3.7 Main technical parameters and the basic optical layouts of the objectives.

Magnification	*NA*	*WD* (mm)	*R* (μm)	*DF* (μm)	*FOV* object (mm)	*FOV* image (mm)	Aberration correction type/optical design at figure no.	The principal optical layout
1×	0.015	14	22	—	34	34	CCF plan apochromat/ 3.169	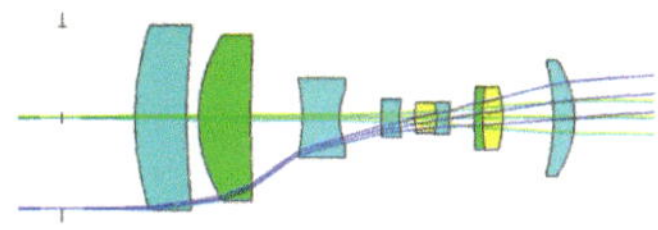
2×	0.05	20	6.7	134	25	50	CCF plan apochromat/ 3.170	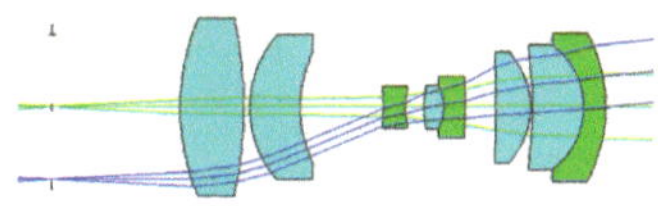
5×	0.12	22	2.7	23	10	50	CCF plan apochromat/ 3.171	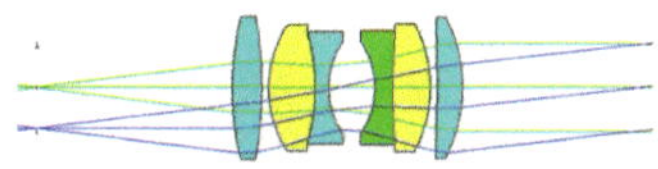
10×	0.30	34	1.1	3.7	5.0	50	CCF plan apochromat/ 3.172	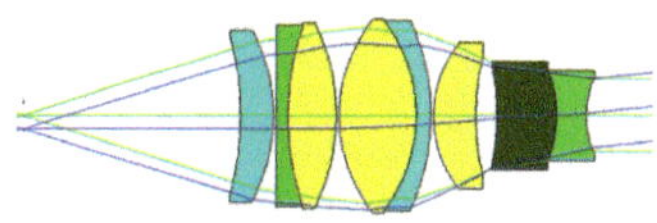
20×	0.30	33	1.1	3.7	2.5	50	CCF plan apochromat/ 3.173	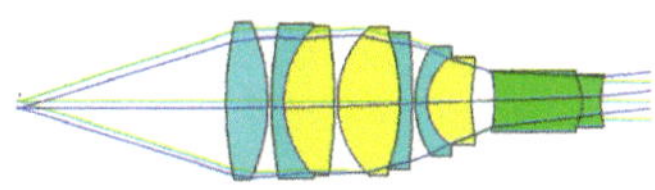
50×	0.50	18.5	0.67	1.33	1.0	50	CCF plan apochromat/ 3.174	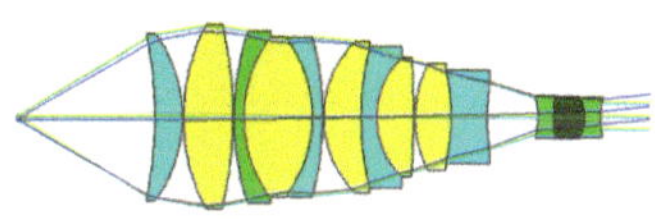
100×	0.80	4.0	0.42	0.52	0.5	50	CCF plan apochromat/ 3.175	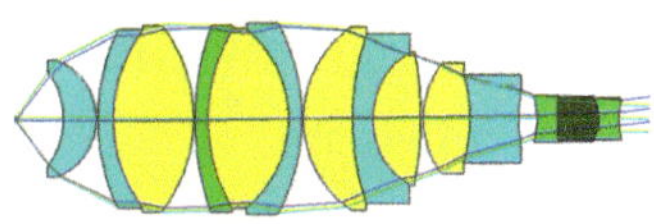
150×	0.60	4.2	0.56	0.93	0.33	50	LCF plan MicroFluar/ 3.176	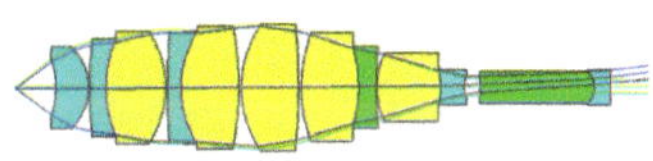
200×	0.65	2.0	0.51	0.79	0.25	50	LCF plan MicroFluar/ 3.177	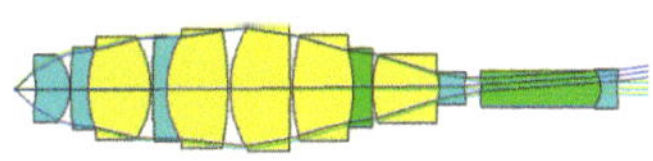
250×	0.70	2.0	0.48	0.68	0.2	50	LCF plan MicroFluar/ 3.178	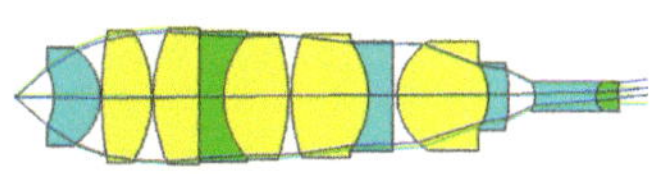
250×	0.80	4.0	0.42	0.52	0.2	50	CCF plan apochromat/ 3.179	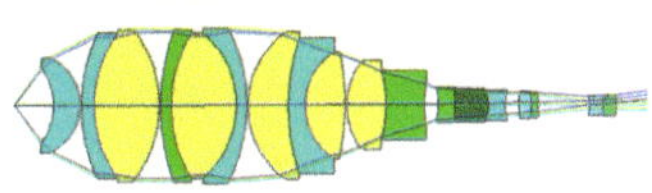
500×	0.75	1.6	0.44	0.59	0.1	50	LCF plan MicroFluar/ 3.180	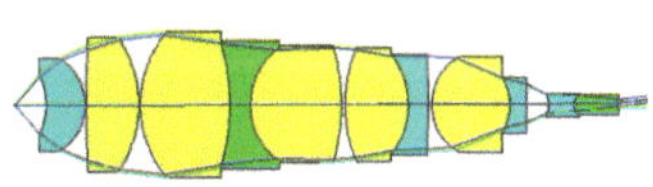
500×	0.80	4.0	0.42	0.52	0.1	50	CCF plan apochromat/ 3.181	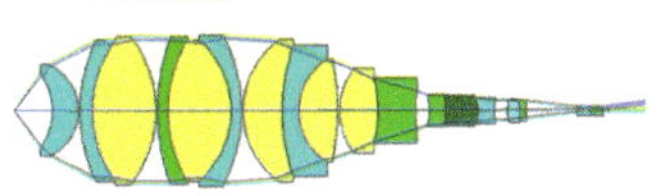

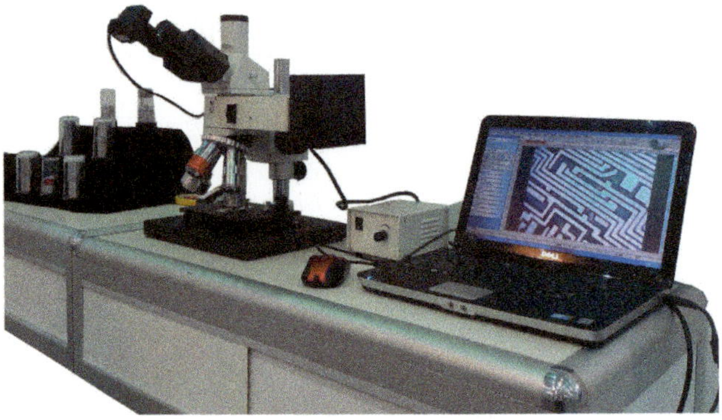

Figure 3.183 Variation of the standard equipment for using these objectives.

Figure 3.184 Photograph of a real object using a 1× CCF plan apochromatic objective ($h = 95$ mm).

Figure 3.185 Photograph of a real object using a 2× CCF plan apochromatic objective ($h = 95$ mm).

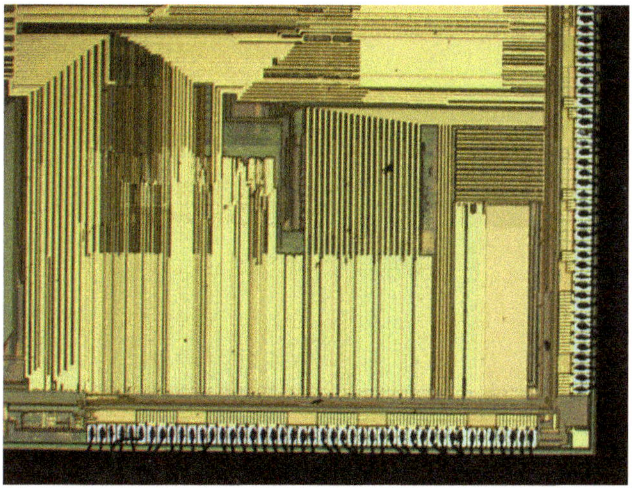

Figure 3.186 Photograph of a real object using a 2× CCF plan apochromatic objective ($h = 95$ mm).

Figure 3.187 Photograph of a real object using a 5× CCF plan apochromatic objective ($h = 95$ mm).

used by people looking through the eyepiece with their eyes. Accordingly, the most important parameters of the microscope—and first, the objective—must be correlated with the capabilities of the human visual apparatus. For example, the spectral range has been limited to the visible range (approximately 400–700 nm). In this spectral range, an aberration correction of the optical systems of the objective and the microscope as a whole was

Figure 3.188 Photograph of a real object using a 5× CCF plan apochromatic objective ($h = 95$ mm).

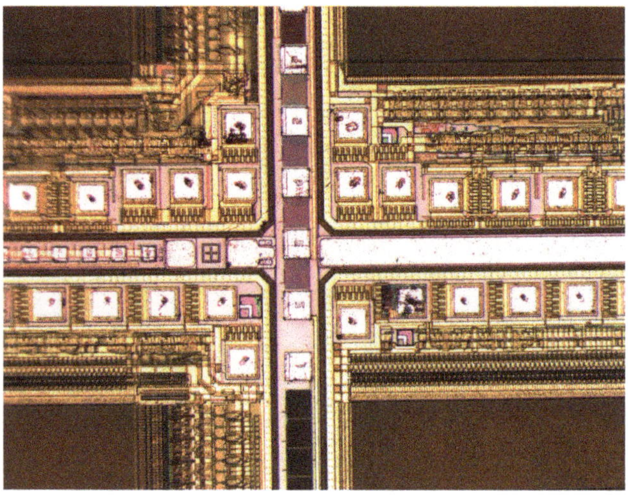

Figure 3.189 Photograph of a real object using a 5× CCF plan apochromatic objective ($h = 95$ mm).

carried out. Certain sources of lighting, including conventional tungsten or halogen lamps, have recently been successfully replaced with LED sources.

At the same time, alongside the usual methods for performing research on a microscope in the visible spectral range, special techniques that use a spectral range not visible by the human eye have been rapidly developing. An example is a fluorescence analysis technique, the so-called *Stokes* and *anti-Stokes* method, in which fluorescence excitation occurs outside the visible

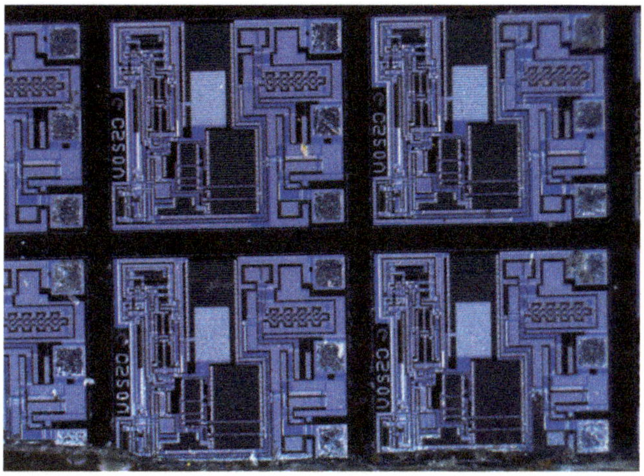

Figure 3.190 Photograph of a real object using a 5× CCF plan apochromatic objective (using the dark field method; $h = 95$ mm).

Figure 3.191 Photograph of a real object using a 10× CCF plan apochromatic objective ($h = 95$ mm).

spectral range. The observation of fluorescence in this case is possible in the visible region of the spectrum; however, this can substantially limit the effectiveness of the studies. A microscopist may wish to study fluorescence outside the visible spectral range. For example, some researchers would like to more effectively apply spectral ranges that extend into the shorter blue region of the spectrum [the near-UV (NUV) region in the wavelength range of 360 to

Figure 3.192 Photograph of a real object using a 10× CCF plan apochromatic objective ($h = 95$ mm).

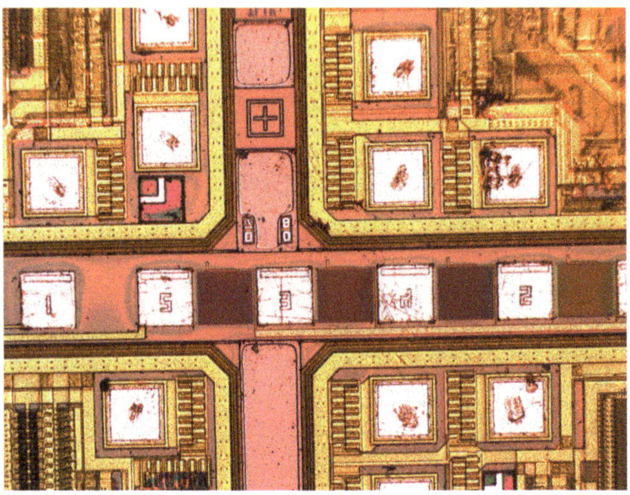

Figure 3.193 Photograph of a real object using a 10× CCF plan apochromatic objective ($h = 95$ mm).

400 nm]. Other researchers would like to use a microscope to work in a longer region [the near-IR (NIR) region in the wavelength range of 700 to 1700 nm].

It should be noted that to ensure operation in these spectral regions, the typical silicate glass grades can be used as optical materials for constructing the optical systems of objectives and microscopes, which in their main nomenclature have good throughput in the NIR region. There are also silicate

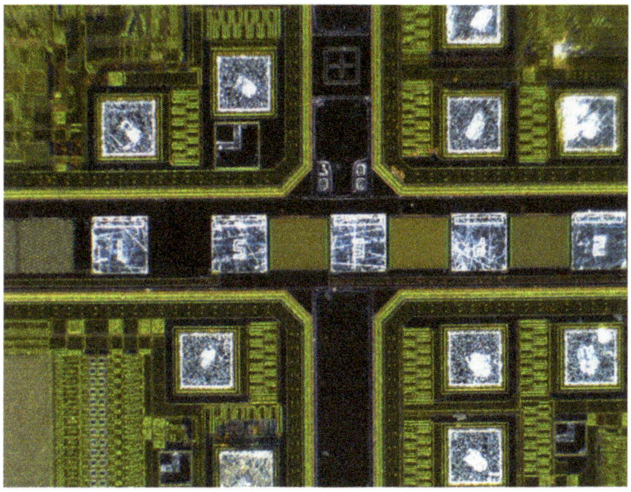

Figure 3.194 Photograph of a real object using a 10× CCF plan apochromatic objective (using the dark field method; $h = 95$ mm).

Figure 3.195 Photograph of a real object using a 10× CCF plan apochromatic objective ($h = 95$ mm).

glasses for working in the NUV region, although their nomenclature is not as broad. This also applies to antireflection coatings for lenses.

The throughput of the optical system of the objective and the microscope is a crucial characteristic that affects the quality of research. This is why, when developing the optical design, we should carefully choose the brands of optical materials and designate the types of antireflection coatings to be used. This also means considering the difference between the performance of the

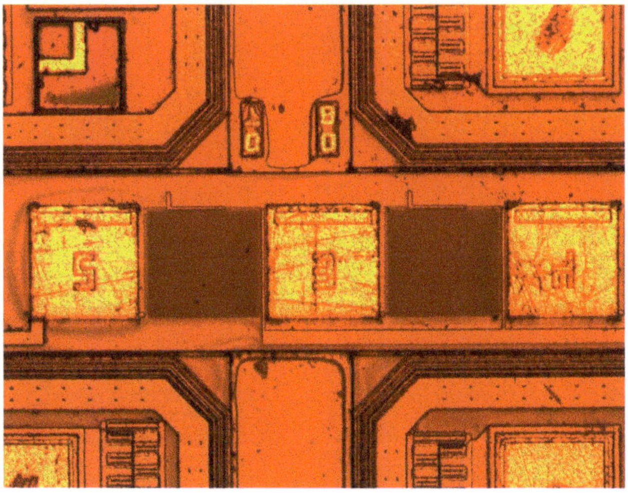

Figure 3.196 Photograph of a real object using a 20× CCF plan apochromatic objective ($h = 95$ mm).

Figure 3.197 Photograph of a real object using a 20× CCF plan apochromatic objective ($h = 95$ mm).

objective and the microscope in the visible spectral range (where the maximum sensitivity of the human eye corresponds to the green region of the spectrum) and in conditions outside the visible range, where the spectral sensitivity of different image receivers may be different.

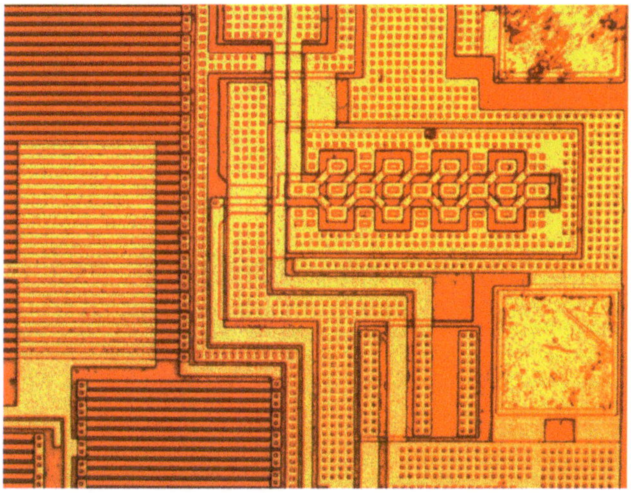

Figure 3.198 Photograph of a real object using a 20× CCF plan apochromatic objective ($h = 95$ mm).

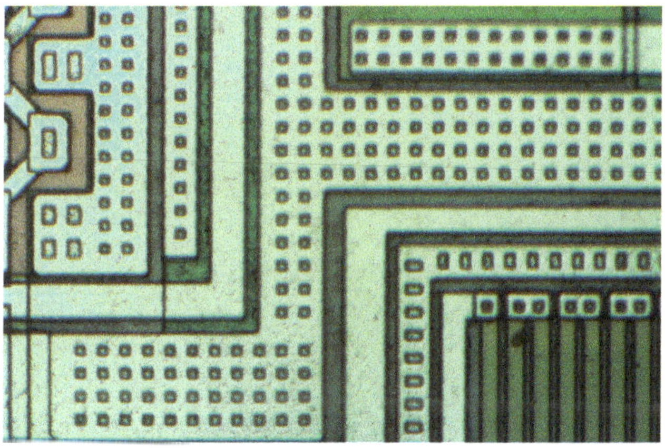

Figure 3.199 Photograph of a real object using a 50× CCF plan apochromatic objective ($h = 95$ mm).

3.10.1 A discussion about NUV and NIR microscopy research

NUV (NUV+VISIBLE)

In the spectral range of 355 to 430 nm, as a rule, excitation occurs under the so-called *Stokes fluorescence*, which is then observed in the visible range. This type of microscopy research is based on the effect of luminescence of many organic and inorganic substances under the influence of UV radiation. When a substance is irradiated, its molecules can go into an excited state by absorbing radiation of a certain wavelength. Then, returning to the initial (unexcited) state, they release some of the absorbed energy in the form of

Figure 3.200 Photograph of a real object using a 100× CCF plan apochromatic objective ($h = 95$ mm).

Figure 3.201 Photograph of a real object using a 150× LCF plan MicroFluar objective ($h = 95$ mm).

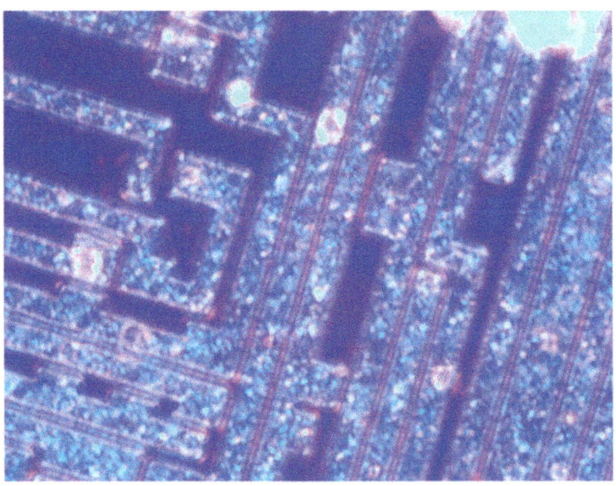

Figure 3.202 Photograph of a real object using a 150× LCF plan MicroFluar objective ($h = 95$ mm).

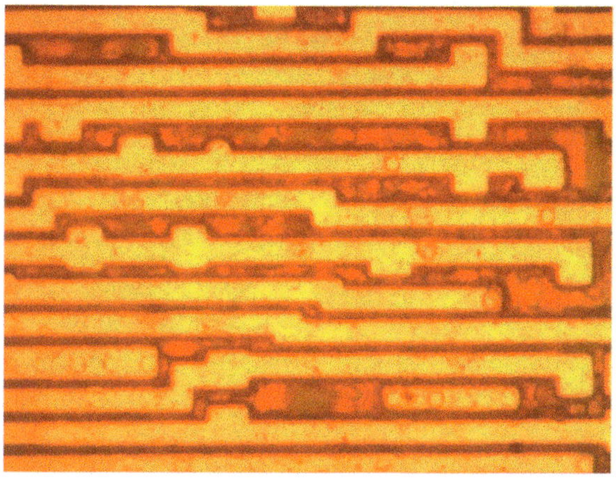

Figure 3.203 Photograph of a real object using a 150× LCF plan MicroFluar objective ($h = 95$ mm).

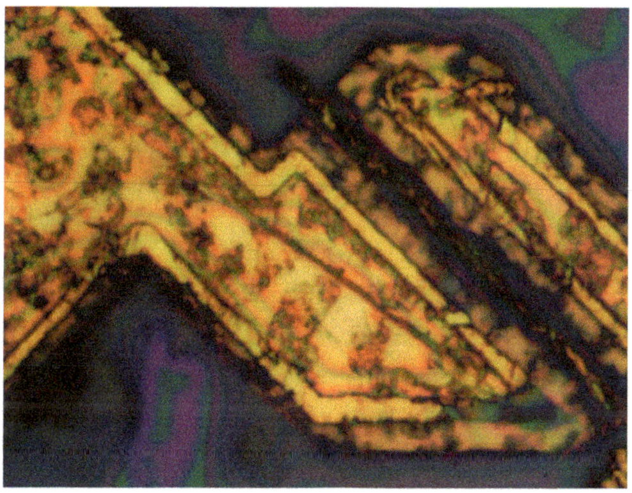

Figure 3.204 Photograph of a real object using a 150× LCF plan MicroFluar objective ($h = 95$ mm).

radiation, which usually has a longer wavelength. Thus, the area of achromatization of the microscope and the objective should correspond to the sum of these spectral ranges. We can make optical and mechanical designs of two sets of objectives that are made to work in such a spectral range. The aberration correction of the first set of objectives is in the range of 365 to 650 nm; however, the input numerical apertures are increased. The second set of objectives for working in both the NUV and visible ranges has an aberration correction in the spectral range of 355 to 800 nm; the numerical

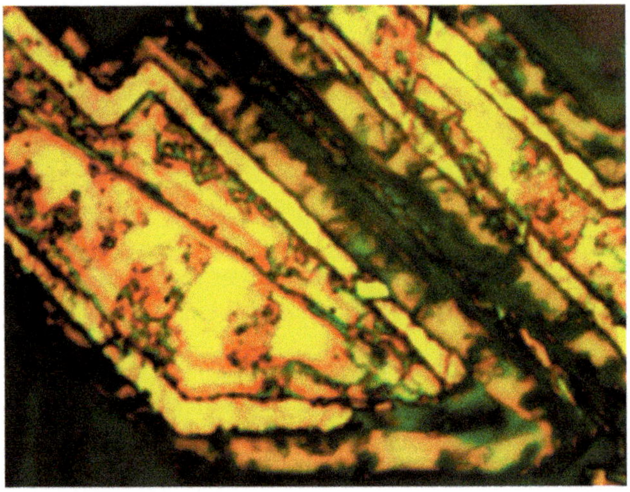

Figure 3.205 Photograph of a real object using a 200× LCF plan MicroFluar objective ($h = 95$ mm).

Figure 3.206 Photograph of a real object using a 250× LCF plan MicroFluar objective ($h = 95$ mm).

apertures of these objectives are not so high. However, the developed objectives of the second set can also be used in the implementation of the so-called *anti-Stokes fluorescence* when, for example, the excitation wavelength falls within a range of about 800 nm.

As a source of radiation in a microscope, a xenon lamp of ultrahigh pressure can be used, enabling the generation of a continuous spectrum of radiation. Such fluorescence methods are used to quantify the fluorescence spectra to research, for example, the selective properties of an object. Work on the microscope occurs in both transmitted and reflected light.

Figure 3.207 Photograph of a real object using a 250× LCF plan MicroFluar objective ($h = 95$ mm).

Figure 3.208 Photograph of a real object using a 500× LCF plan MicroFluar objective ($h = 95$ mm).

There are many techniques for performing research in the NUV range. Let us take as an example one of them, which is devoted to the identification of ancient works of art (paintings). Here are descriptions of the kinds of glow of different substances under the influence of UV [12]:

- Glues are revealed by bright whitish-bluish spots and sweats.
- Yolk, applied to the surface to strengthen the picturesque layer, gives a glow of grayish color.

- Drying oil becomes dark, and the image is not visible.
- Cinnabar under the influence of UV looks dark red-brown.
- Pollution usually looks darker than it looks in visible light.

By the heterogeneity of the luminescence of pigments, it is possible to detect restorative toning and recording. If a varnish has been applied to the surface of the product during the restoration, it can be detected by a glow of grayish, grayish-bluish color of different brightness. Lead white of a long-standing origin has a bright white glow, modern lead white has a grayish glow, and zinc white gives a yellowish-greenish glow of different shades. Restoration tinting applied over varnish will reveal a stronger, brighter lacquer glow. If the lacquer is applied over the tinting, then intensifying the glow of the lacquer, associated with the aging of the lacquer, will reduce the contrast between the tint and the artist's layer. Extinguished inscriptions painted with iron-gall inks containing iron compounds, chromium salts, lead, mercury, etc., are almost always well identified.

NIR (VISIBLE+NIR)

Traditionally, NIR research is conducted in botany, biophysics, medicine, forensics, alignment and testing of laser systems and optical fibers, observation of IR fluorescence, and others. A special place is occupied by research and attestation of products in the microelectronics industry.

One modern research method is the receiving and analysis of so-called *hyperspectral images*, which can be constructed by capture along adjacent spectral bands. In this case, the equipment includes at least two hyperspectral digital receivers of which one is tuned, for example, to a spectral range of 400 to 1000 nm, and the second to a spectral range of 900 to 1700 nm. According to the developers, when using a set of hyperspectral images from healthy and diseased tissue, a special algorithm was developed that makes it possible to detect differences in the spectral distribution. For example, more detailed cancer surveillance using hyperspectral instruments may eventually provide new information about the processes underlying tumor growth. Currently, the scientific community has several detailed hyperspectral medical images of the human body available, so an expanded database of hyperspectral images could be useful for studying cancers and other medical purposes.

Other studies [13] have already led to the creation of a measuring complex based on a light microscope, which allows graphical information about investigated objects to be analyzed, including localization of fluorescence foci, i.e., information on the spectral composition of fluorescent centers in the NIR region of the spectrum. In fact, it has been proposed that a microscope be equipped with a spectrum analyzer in which the localization of photosensitizer accumulation within a cell is determined from the distribution pattern of a fluorescent signal, and the spectrum of an NIR fluorescent signal in a required point is obtained. The developed device uses a laser source with a small angular divergence of the beam, which focuses the radiation in the irradiated region and allows a higher power density to be created, which enables hyper-thin operations to be performed on the cells.

Aberration correction for the objective and microscope should extend from the visible range to the near IR. We can make optical and mechanical designs of two sets of objectives to work in such a spectral range. The aberration correction of the first set of objectives is in the range of 480 to 1800 nm, and the 1064-nm line is chosen as the main wavelength. The numerical apertures of the objectives in this set are enlarged. In the second set of objectives, the operating spectral range is substantially expanded into a short spectral region (the aberration correction is from 430 to 1800 nm), and the 546.07-nm line is selected as the main wavelength. Although the numerical apertures of the objectives in the second set are not so high, these second-set objectives can be used to implement the modern microscopy techniques described above.

Lasers can be used as a source of radiation in a microscope. Figure 3.209 shows the main types of modern lasers operating in various regions of the spectrum, from NUV to NIR. Work on the microscope takes place in both transmitted and reflected light.

Microspectrophotometric heads can be used as image receivers. In addition, we can use modern image receivers, some of which work in the NUV and visible regions of the spectrum, while others work in the visible and IR range. Figure 3.210 shows the spectral sensitivity of some modern image receivers.

3.10.2 Example designs of NUV objectives (95-mm parfocal distance)

We can design some microscope objectives for performing NUV and DUV studies. Figures 3.211–3.213 show the design parameters and graphs of

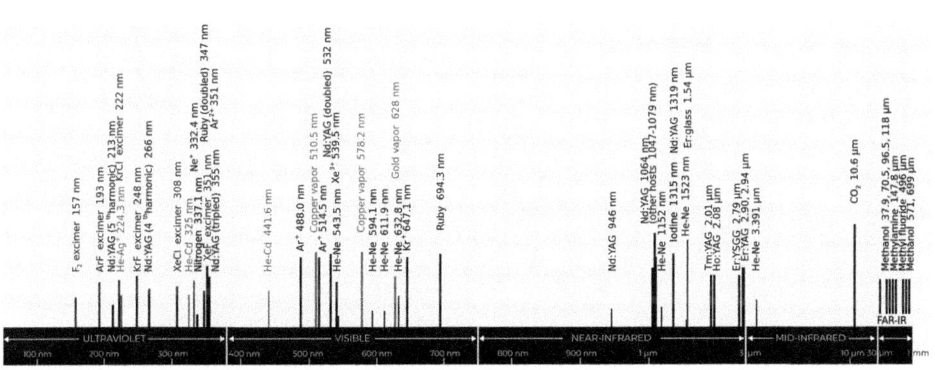

Figure 3.209 The main types of modern lasers.

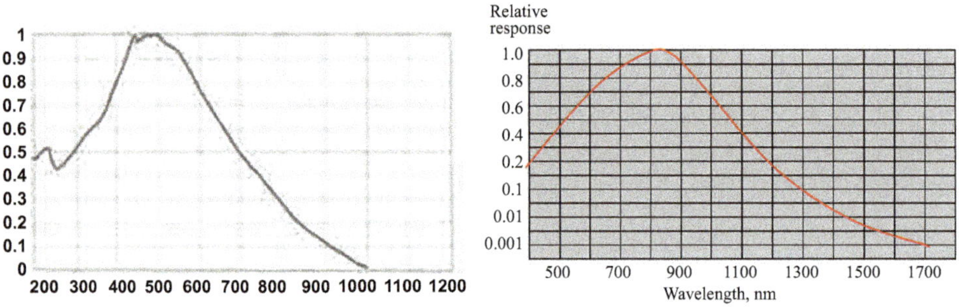

Figure 3.210 The spectral sensitivity of some modern image receivers.

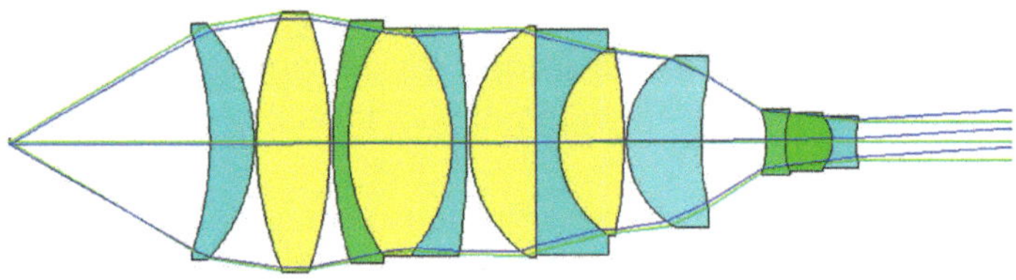

Lens: NUV PlanApo LD 50x0.50 d0 h95

Object num aper 0.500000 Image height 12.500000 Primary wavln 0.546070

SRF	RADIUS	THICKNESS	APERTURE RADIUS	GLASS	SPECIAL
OBJ	0.000000	1.0000e-20	0.250057	AIR	
AST	0.000000	21.800000	0.250057 AS	AIR	
2	-34.350000	4.700000	11.660000	H-LAF10LA	
3	-17.840000	0.300000	12.460000	AIR	
4	35.480000	8.000000	13.780000	CAF2	
5	-40.680000	0.300000	13.730000	AIR	
6	44.730000	1.600000	12.850000	H-ZLAF55C	
7	21.340000	11.000000	12.030000	CAF2 P	
8	-18.000000	1.600000	11.640000	H-LAF10LA P	
9	-81.000000	0.300000	12.020000	AIR	
10	15.076000	7.000000	12.210000	CAF2 P	
11	0.000000	2.500000	11.940000	H-LAF10LA P	
12	11.870000	7.000000	9.850000	CAF2 P	
13	-48.280000	0.300000	9.830000	AIR	
14	10.411100	8.000000	9.010000	SILICA	
15	31.720000	7.000000	7.180000	AIR	
16	-11.092000	2.000000	3.480000	H-ZLAF55C P	
17	10.068000	5.100000	3.100000	ZF6	
18	-4.093000	2.500000	2.740000	H-LAF10LA P	
19	8.395000	200.000000	2.350000	AIR	
20	-43.930000	14.000000	14.800000	H-QK3	
21	-51.200000	3.000000	16.970000	AIR	
22	667.000000	15.000000	17.370000	CAF2	
23	-105.900000	3.000000	17.880000	AIR	
24	194.900000	17.000000	17.770000	CAF2	
25	-118.000000	3.000000	17.150000	H-ZLAF50B	
26	-754.000000	200.004528 S	17.140000	AIR	
IMS	0.000000	0.000000	12.500000		

Figure 3.211 Optical design and graphs of aberration correction of a 50×/0.50 NUV CCF plan apochromatic objective ($F'_{ts} = 200$ mm; $h = 95$ mm).

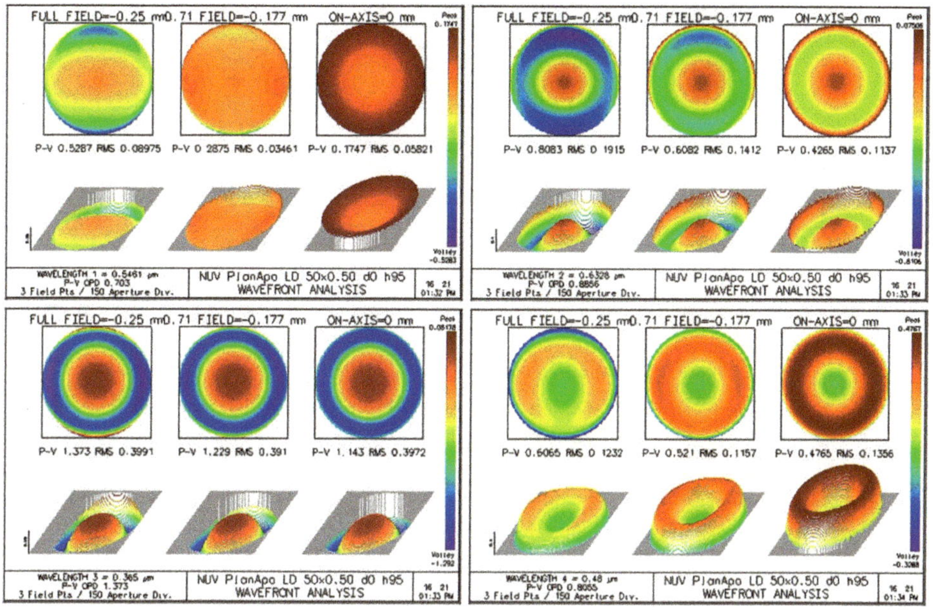

Figure 3.211 (*Continued*) Optical design and graphs of aberration correction of a 50×/0.50 NUV CCF plan apochromatic objective ($F'_{ts} = 200$ mm; $h = 95$ mm).

aberration correction of some of these objectives. These designs are made for the infinite length of the microscope tube using an additional focusing system ($F' = 200$ mm). The spectral line 546.07 nm was chosen as the so-called reference wavelength, and achromatization was performed for the spectral range of 365 to 650 nm. The parfocal height of the objectives is 95 mm.

Table 3.8 shows the main technical parameters of these objectives.

Figures 3.214–3.221 show the design parameters and graphs of aberration correction of another type of NUV objective. This objective is designed for an infinite length of the microscope tube using an additional focusing system ($F' = 200$ mm). The spectral line 546.07 nm was chosen as the so-called reference wavelength, and the achromatization was performed for a spectral range of 355 to 800 nm. This type of achromatization can be called a *poly-apochromatic* aberration correction because the achromatization region (more precisely, the apochromatization) is quite wide. For such objectives, we can use the designation NUV+VIS$_+$ because the working spectral range is slightly beyond the visible range, closer to the red region. In some cases, it is necessary to design an objective for use with a coverslip. In this case, when developing the objective's optical design, the technique of changing the parameters of the frontal lens (described above) is used.

Table 3.9 shows the main technical parameters of these objectives.

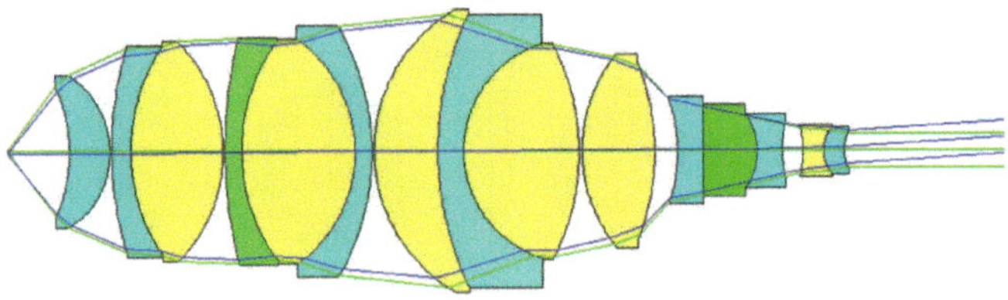

Lens: NUV PlanApo LD 100x0.80 d0 h95

Object num aper 0.800000 Image height 12.500000 Primary wavln 0.546070

SRF	RADIUS	THICKNESS	APERTURE RADIUS	GLASS	SPECIAL
OBJ	0.000000	1.0000e-20	0.124992	AIR	
AST	0.000000	5.950000	0.124992 AS	AIR	
2	-14.700000	4.000000	6.220000	H-LAF10LA	
3	-8.395000	0.200000	7.250000	AIR	
4	33.244000	1.900000	9.770000	H-LAF10LA P	
5	17.440000	9.000000	10.040000	CAF2	
6	-14.700000	0.200000	10.480000	AIR	
7	44.200000	1.600000	10.800000	H-ZLAF55C	
8	17.320000	11.000000	10.490000	CAF2 P	
9	-13.090000	1.600000	10.700000	H-LAF10LA P	
10	-23.540000	0.200000	11.940000	AIR	
11	15.360000	6.200000	13.410000	CAF2 P	
12	29.580000	2.500000	12.930000	H-LAF10LA P	
13	10.700000	11.000000	10.190000	CAF2 P	
14	-22.450000	0.300000	10.230000	AIR	
15	14.510000	7.000000	9.170000	CAF2 P	
16	-22.620000	2.200000	8.440000	AIR	
17	-14.894000	2.500000	5.100000	H-LAK53A	
18	0.000000	5.200000	4.370000	ZF6	
19	-5.960000	2.500000	3.530000	H-LAK53A P	
20	10.800000	2.000000	2.660000	AIR	
21	-9.600000	2.000000	2.410000	CAF2 P	
22	3.370000	2.000000	2.260000	H-LAK53A P	
23	4.093000	200.000000	1.950000	AIR	
24	-43.930000	14.000000	14.760000	H-QK3	
25	-51.200000	3.000000	16.930000	AIR	
26	667.000000	15.000000	17.340000	CAF2	
27	-105.900000	3.000000	17.870000	AIR	
28	194.900000	17.000000	17.770000	CAF2	
29	-118.000000	3.000000	17.180000	H-ZLAF50B	
30	-754.000000	199.415429 S	17.180000	AIR	
IMS	0.000000	0.000000	12.500000		

Figure 3.212 Optical design and graphs of aberration correction of a 100×/0.80 NUV CCF plan apochromatic objective ($F'_{ts} = 200$ mm; $h = 95$ mm).

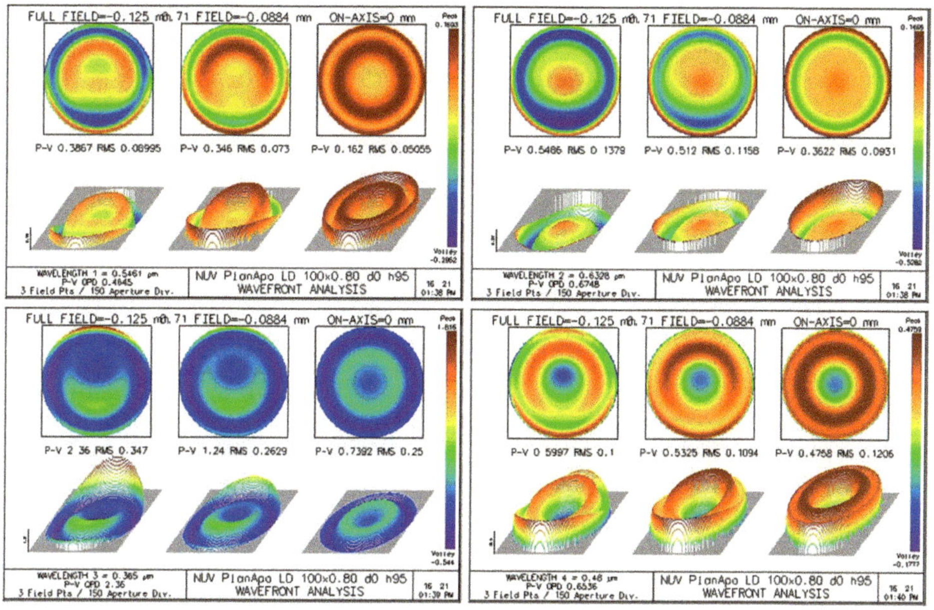

Figure 3.212 (*Continued*) Optical design and graphs of aberration correction of a 100×/0.80 NUV CCF plan apochromatic objective ($F'_{ts} = 200$ mm; $h = 95$ mm).

3.10.3 Example designs of NIR objectives (95-mm parfocal distance)

We can design some microscope objectives to perform NIR studies. Figures 3.222–3.227 show the design parameters and graphs of aberration correction of some of these objectives. These objectives are designed for the infinite length of the microscope tube using an additional focusing system ($F' = 200$ mm). The spectral line 1064 nm was chosen as the so-called reference wavelength, and achromatization was performed for the spectral range of 480 to 1800 nm. The parfocal height of the objectives is 95 mm.

Table 3.10 shows the main technical parameters and basic optical layout of these objectives.

Figure 3.228 is a photograph of VIS+NIR objectives that have minimum and maximum linear magnifications. The 10×/0.27 objective is designed for a parfocal distance of 95 mm and has a connecting thread of 26 mm, and the 500×/0.75 objective is made in the housing for a parfocal distance of 45 mm and has a 0.8" connecting thread.

Figures 3.229–3.232 show the design parameters and graphs of aberration correction of another type of NIR objective. These objectives are designed for an infinite length of the microscope tube using an additional focusing system ($F' = 200$ mm). The spectral line of 546.07 nm was chosen as the reference wavelength, and achromatization was performed for a spectral range of 430 to 1800 nm. This type of achromatization can be called a poly-apochromatic

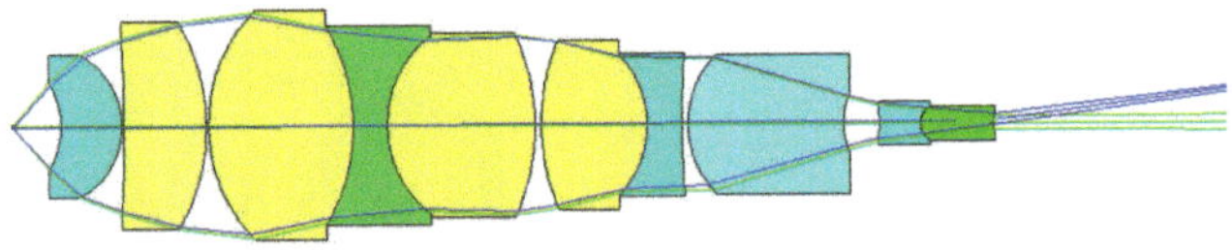

Lens: NUV PlanApo LD 500x0.75 d0 h95

Object num aper	0.750000	Image height	25.000000	Primary wavln	0.546070

SRF	RADIUS	THICKNESS	APERTURE RADIUS	GLASS	SPECIAL
OBJ	0.000000	1.0000e-20	0.050028	AIR	
AST	0.000000	2.299468	0.050028 AS	AIR	
2	-3.746000	3.000000	2.000000	H-LAF3A	
3	-3.600000	0.100000	3.290000	AIR	
4	-82.310000	4.000000	4.060000	CAF2	
5	-9.120000	0.100000	4.790000	AIR	
6	7.400000	7.000000	5.290000	CAF2 P	
7	-8.705000	1.600000	4.580000	BAF7	
8	5.160000	7.000000	4.080000	CAF2 P	
9	-10.000000	0.300000	4.250000	AIR	
10	9.720000	5.000000	3.930000	CAF2 P	
11	-4.650000	1.600000	3.270000	LZ_CTK12	
12	35.320000	0.300000	3.200000	AIR	
13	4.500000	7.400000	3.240000	H-K10	
14	3.200000	1.700000	1.310000	AIR	
15	-11.581000	2.000000	0.990000	H-LAK3	
16	0.910000	3.500000	0.780000	ZF6	
17	1.300000	200.000000	0.460000	AIR	
18	-43.930000	14.000000	26.060000	H-QK3	
19	-51.200000	3.000000	30.120000	AIR	
20	667.000000	15.000000	32.030000	CAF2	
21	-105.900000	3.000000	32.760000	AIR	
22	194.900000	17.000000	32.480000	CAF2	
23	-118.000000	3.000000	31.790000	H-ZLAF50B	
24	-754.000000	199.996545 S	31.830000	AIR	
IMS	0.000000	0.000000	25.000000		

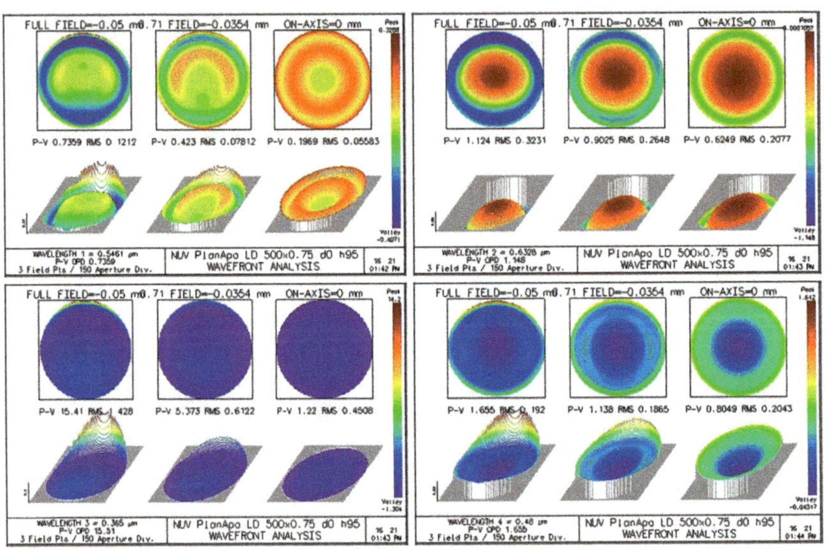

Figure 3.213 Optical design and graphs of aberration correction of a 500×/0.75 NUV CCF plan apochromatic objective ($F'_{ts} = 200$ mm; $h = 45$ mm).

Table 3.8 Main technical parameters and basic optical layout of NUV objectives ($\lambda = 546.07$ nm; 365–650 nm).

Magnification	*NA*	*WD* (mm)	*F′* (mm)	*R* (μm)	*DF* (μm)	*FOV* on object (mm)	*FOV* on image (mm)	Optical design at figure no.	The principal optical layout
50×	0.50	18	4	0.67	1.33	0.50	25	3.211	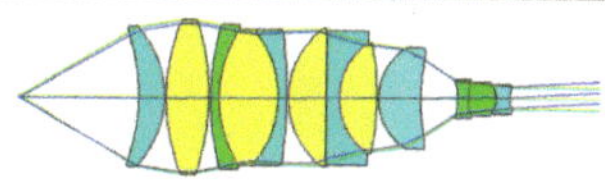
100×	0.80	3.2	2	0.42	0.52	0.25	25	3.212	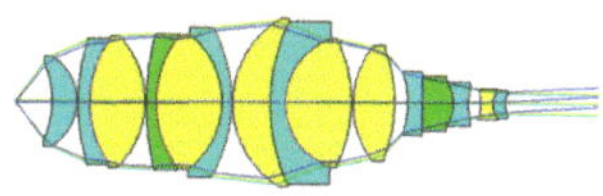
500×	0.75	1.5	0.4	0.44	0.59	0.1	50	3.213	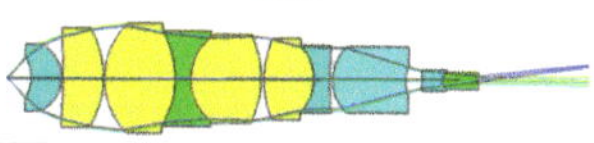

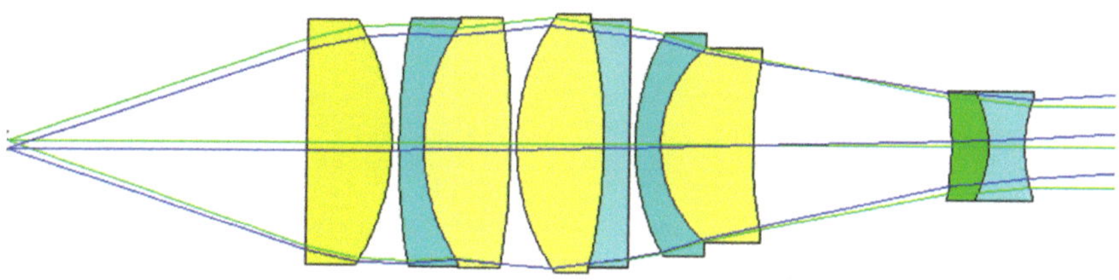

Lens: 20x0.32 355-800nm h95

Object num aper	0.320000	Image height	12.500000	Primary wavln	0.546070

SRF	RADIUS	THICKNESS	APERTURE RADIUS	GLASS	SPECIAL
OBJ	0.000000	1.0000e-20	0.624945	AIR	
AST	0.000000	23.997260	0.624945 AS	AIR	
2	0.000000	6.800000	8.730000	H-FK61	
3	-17.760000	0.600000	9.600000	AIR	
4	56.750000	2.000000	9.710000	H-ZLAF50E	
5	16.890000	7.000000	9.450000	CAF2	
6	-68.137000	0.500000	9.770000	AIR	
7	17.760000	7.000000	10.090000	CAF2 P	
8	-41.830000	2.000000	9.670000	H-K10	
9	0.000000	0.500000	9.270000	AIR	
10	17.760000	2.000000	8.710000	LZ_CTK12	
11	9.817000	7.400000	7.590000	CAF2 P	
12	34.130000	16.000000	6.680000	AIR	
13	-25.500000	3.000000	4.270000	ZF2	
14	-8.705000	3.000000	4.230000	H-K9	
15	10.627000	200.000000	3.790000	AIR	
16	-43.930000	14.000000	16.240000	H-QK3	
17	-51.200000	3.000000	18.560000	AIR	
18	667.000000	15.000000	19.020000	CAF2 P	
19	-105.900000	3.000000	19.510000	AIR	
20	194.900000	17.000000	19.360000	CAF2 P	
21	-118.000000	3.000000	18.650000	H-ZLAF50E P	
22	-754.000000	199.912259 S	18.630000	AIR	
IMS	0.000000	0.000000	12.500000		

Figure 3.214 Optical design and graphs of aberration correction of a 20×/0.32 NUV+VIS$_+$ CCF plan poly-apochromatic objective ($F'_{ts} = 200$ mm; $h = 95$ mm).

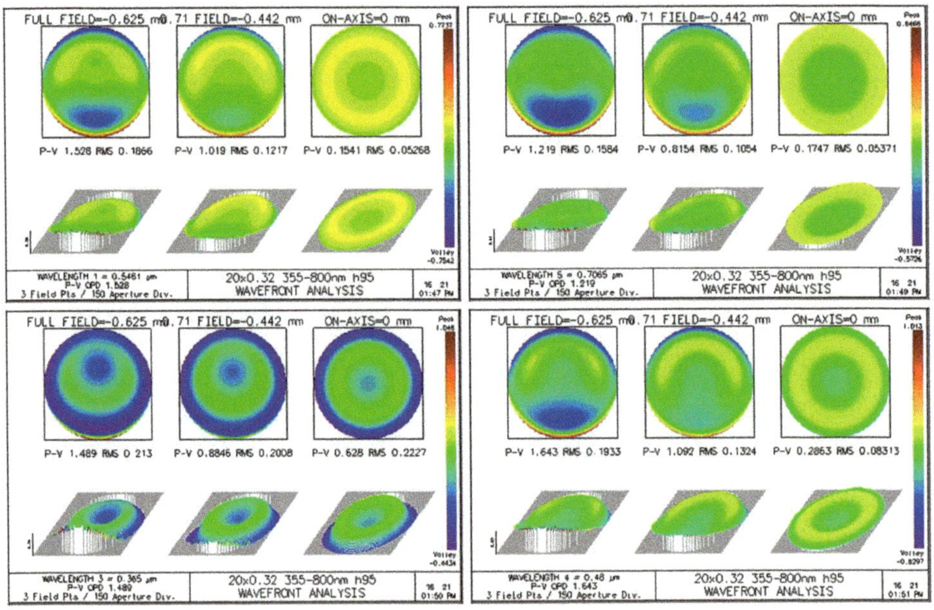

Figure 3.214 (*Continued*) Optical design and graphs of aberration correction of a 20×/0.32 NUV+VIS$_+$ CCF plan poly-apochromatic objective (F'_{ts} = 200 mm; h = 95 mm).

aberration correction because the achromatization region (more precisely, apochromatization) is quite wide. In some cases, it is necessary to design an objective for use with a coverslip. In this case, when developing the objective's optical design, the technique of changing the parameters of the frontal lens (described above) is used.

Table 3.11 shows the main technical parameters and basic optical layout of these objectives.

Figure 3.233 is a photograph of VIS+NIR objectives that have linear magnifications of 20× and 50×. Both objectives are designed for a parfocal distance of 95 mm and a connecting thread of 26 mm. Figure 3.234 shows photographs of a real object taken with these objectives.

3.10.4 Plan poly-apochromatic aberration correction objectives (45-mm parfocal distance)

Expanding the working spectral range in itself is not a difficult engineering task. In microscope optics, one should choose the grades of optical materials that can ensure maximum transmission within the entire selected spectral range. The parameters of antireflection coatings and glue must also be considered. In some cases, glue alone can provide equally good transmission at different wavelengths.

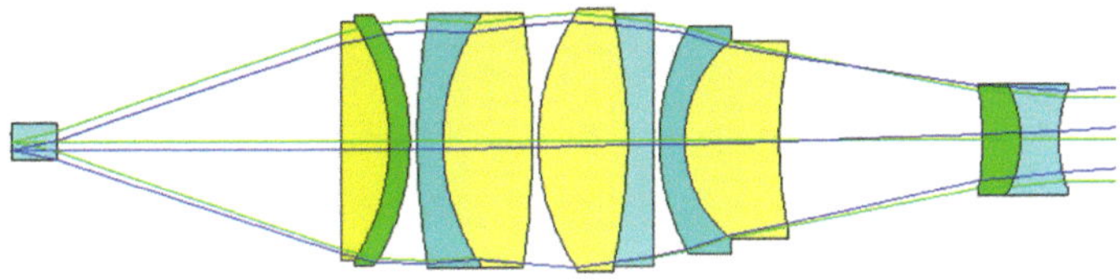

Lens: 20x032 d=3.5mm 355-800nm h95

Object num aper 0.320000 Image height 12.500000 Primary wavln 0.546070

SRF	RADIUS	THICKNESS	APERTURE RADIUS	GLASS	SPECIAL
OBJ	0.000000	1.0000e-20	0.624949	AIR	
AST	0.000000	0.050000	0.640000 A	AIR	
2	0.000000	3.500000	0.640000	H-K9L	
3	0.000000	22.275400	1.400000	AIR	
4	884.600000	3.800000	8.940000	H-FK61	
5	-16.630000	1.600000	9.150000	H-ZF7LA	
6	-17.760000	0.600000	9.590000	AIR	
7	56.750000	2.000000	9.710000	H-ZLAF50E	
8	16.890000	7.000000	9.450000	CAF2	
9	-68.137000	0.500000	9.770000	AIR	
10	17.760000	7.000000	10.090000	CAF2 P	
11	-41.830000	2.000000	9.670000	H-K10	
12	0.000000	0.500000	9.270000	AIR	
13	17.760000	2.000000	8.710000	LZ_CTK12	
14	9.817000	7.400000	7.590000	CAF2 P	
15	34.130000	15.850000	6.680000	AIR	
16	-25.500000	3.000000	4.270000	ZF2	
17	-8.705000	3.000000	4.230000	H-K9	
18	10.627000	200.000000	3.790000	AIR	
19	-43.930000	14.000000	16.240000	H-QK3	
20	-51.200000	3.000000	18.560000	AIR	
21	667.000000	15.000000	19.020000	CAF2 P	
22	-105.900000	3.000000	19.510000	AIR	
23	194.900000	17.000000	19.360000	CAF2 P	
24	-118.000000	3.000000	18.650000	H-ZLAF50E P	
25	-754.000000	199.943312 S	18.630000	AIR	
IMS	0.000000	0.000000	12.500000		

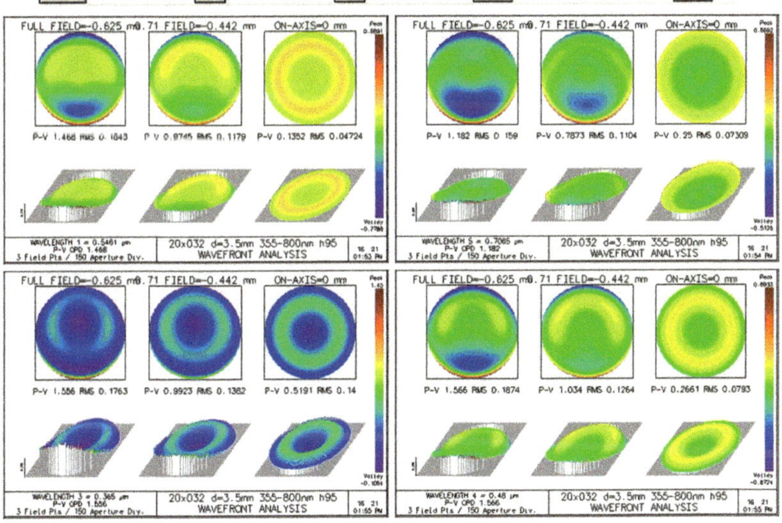

Figure 3.215 Optical design and graphs of aberration correction of a 20×/0.32 ($d = 3.5$ mm) NUV+VIS$_+$ CCF plan poly-apochromatic objective ($F'_{ts} = 200$ mm; $h = 95$ mm).

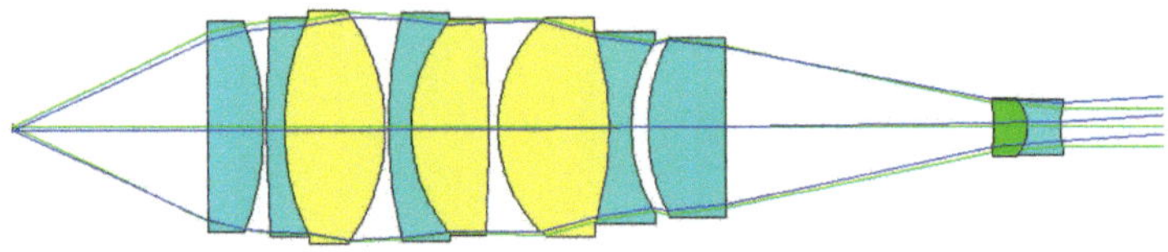

Lens: 50x0.42 d0 355-800nm h95

Object num aper	0.420000	Image height	12.500000	Primary wavln	0.546070

SRF	RADIUS		THICKNESS		APERTURE RADIUS		GLASS		SPECIAL
OBJ	0.000000		1.0000e-20		0.249999		AIR		
AST	0.000000		17.709545		0.249999	AS	AIR		
2	0.000000		5.000000		8.450000		H-ZLAF55D		
3	-25.350000		0.300000		9.220000		AIR		
4	142.120000		1.600000		9.490000		H-ZLAF50E		
5	21.985000		9.000000		9.610000		CAF2		
6	-17.320000		0.300000		10.210000		AIR		
7	40.130000		2.000000		9.950000		H-ZLAF50E	P	
8	14.370000		7.000000		9.360000		CAF2	P	
9	-110.280000		0.800000		9.460000		AIR		
10	12.840000		10.000000		9.600000		CAF2	P	
11	-27.160000		2.000000		8.370000		H-ZLAF50E	P	
12	14.370000		1.500000		7.560000		AIR		
13	16.762000		7.000000		7.880000		H-ZLAF50E	P	
14	0.000000		24.000000		7.330000		AIR		
15	-21.580000		3.000000		2.550000		ZF2		
16	-3.746000		3.000000		2.420000		LZ_CTK12		
17	9.612000		200.000000		2.160000		AIR		
18	-43.930000		14.000000		14.350000		H-QK3		
19	-51.200000		3.000000		16.450000		AIR		
20	667.000000		15.000000		16.840000		CAF2	P	
21	-105.900000		3.000000		17.340000		AIR		
22	194.900000		17.000000		17.240000		CAF2	P	
23	-118.000000		3.000000		16.630000		H-ZLAF50E	P	
24	-754.000000		199.951760	S	16.630000		AIR		
IMS	0.000000		0.000000		12.500000				

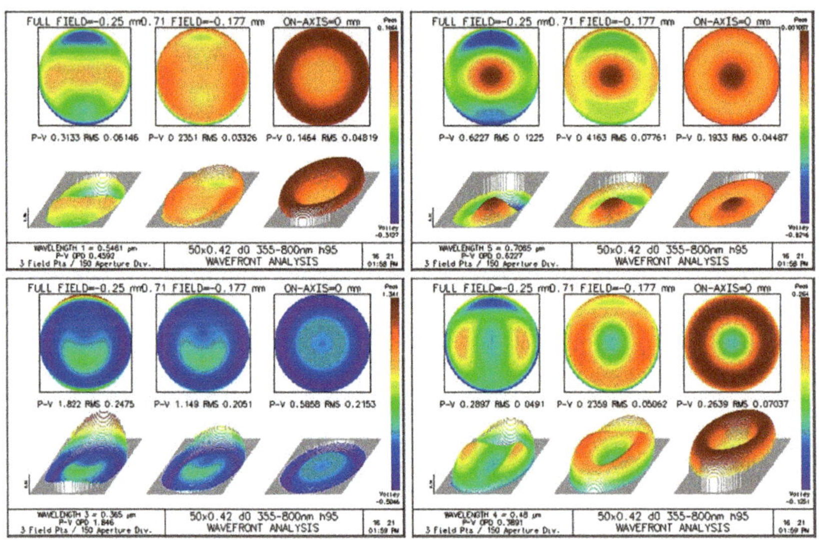

Figure 3.216 Optical design and graphs of aberration correction of a 50×/0.42 NUV+VIS$_+$ CCF plan poly-apochromatic objective (F'_{ts} = 200 mm; h = 95 mm).

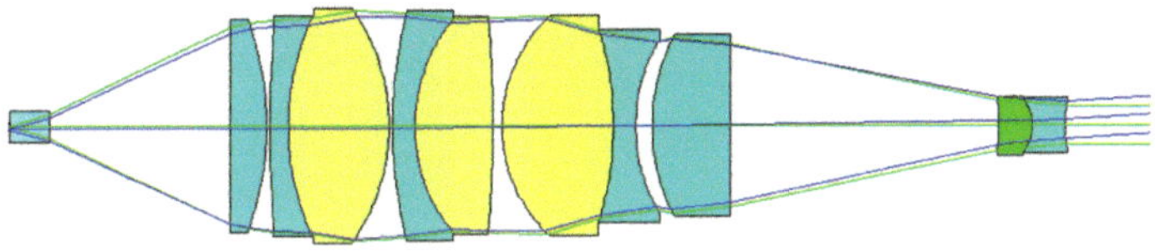

Lens: 50x0.42 d=3.5mm 355-800nm h95

Object num aper	0.420000	Image height	12.500000	Primary wavln	0.546070

SRF	RADIUS	THICKNESS	APERTURE RADIUS	GLASS	SPECIAL
OBJ	0.000000	1.0000e-20	0.249855	AIR	
AST	0.000000	0.050000	0.249855 AS	AIR	
2	0.000000	3.500000	0.270000	H-K9L	
3	0.000000	16.257700	1.280000	AIR	
4	1.1710e+03	3.300000	8.820000	H-ZLAF55D	
5	-25.730000	0.300000	9.190000	AIR	
6	142.120000	1.600000	9.490000	H-ZLAF50E	
7	21.985000	9.000000	9.610000	CAF2	
8	-17.320000	0.300000	10.210000	AIR	
9	40.130000	2.000000	9.950000	H-ZLAF50E P	
10	14.370000	7.000000	9.360000	CAF2 P	
11	-110.280000	0.800000	9.460000	AIR	
12	12.840000	10.000000	9.600000	CAF2 P	
13	-27.160000	2.000000	8.370000	H-ZLAF50E P	
14	14.370000	1.500000	7.560000	AIR	
15	16.762000	7.000000	7.880000	H-ZLAF50E P	
16	0.000000	23.890000	7.330000	AIR	
17	-21.580000	3.000000	2.550000	ZF2	
18	-3.746000	3.000000	2.420000	LZ_CTK12	
19	9.612000	200.000000	2.160000	AIR	
20	-43.930000	14.000000	14.350000	H-QK3	
21	-51.200000	3.000000	16.450000	AIR	
22	667.000000	15.000000	16.840000	CAF2 P	
23	-105.900000	3.000000	17.340000	AIR	
24	194.900000	17.000000	17.240000	CAF2 P	
25	-118.000000	3.000000	16.630000	H-ZLAF50E P	
26	-754.000000	199.977616 S	16.630000	AIR	
IMS	0.000000	0.000000	12.500000		

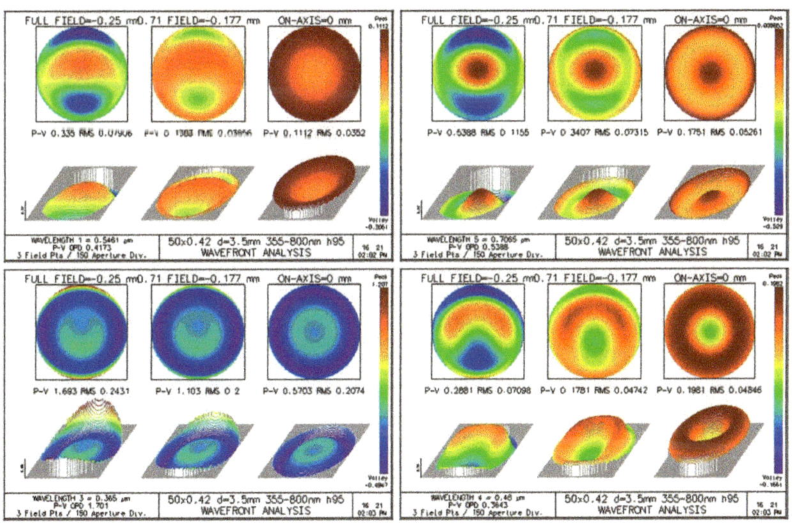

Figure 3.217 Optical design and graphs of aberration correction of a 50×/0.42 ($d = 3.5$ mm) NUV+VIS$_+$ CCF plan poly-apochromatic objective ($F'_{ts} = 200$ mm; $h = 95$ mm).

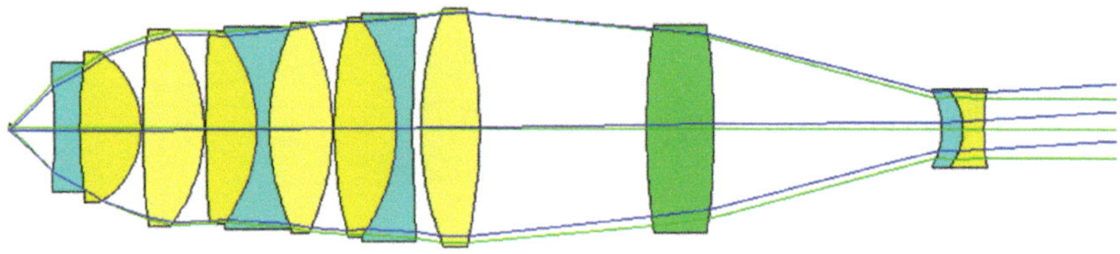

Lens: 50x0.70 d0 355-800nm h95

Object num aper 0.700000 Image height 12.500000 Primary wavln 0.546070

SRF	RADIUS	THICKNESS		APERTURE RADIUS		GLASS		SPECIAL
OBJ	0.000000	1.0000e-20		0.250119		AIR		
AST	0.000000	4.501798		0.250119	AS	AIR		
2	-39.370000	2.500000		4.420000		H-LAK61		
3	39.370000	6.000000		6.080000		H-FK61		
4	-8.240000	0.150000		7.090000		AIR		
5	78.880000	6.000000		8.800000		CAF2		
6	-14.400000	0.200000		9.250000		AIR		
7	145.400000	5.000000		9.070000		H-FK61	P	
8	-14.400000	1.200000		8.990000		H-LAK61	P	
9	22.390000	6.100000		9.510000		CAF2	P	
10	-22.390000	0.200000		9.870000		AIR		
11	39.370000	6.000000		10.280000		H-FK61	P	
12	-17.385000	1.500000		10.250000		H-LAK61	P	
13	162.500000	0.800000		10.670000		AIR		
14	30.270000	5.600000		11.170000		CAF2	P	
15	-51.720000	16.200000		11.180000		AIR		
16	71.430000	6.500000		9.750000		F_4		
17	-71.430000	22.000000		9.230000		AIR		
18	-7.244000	2.000000		3.480000		H-LAK61	P	
19	-5.750000	2.000000		3.640000		H-FK61	P	
20	21.180000	200.000000		3.580000		AIR		
21	-43.930000	14.000000		15.680000		H-QK3		
22	-51.200000	3.000000		17.920000		AIR		
23	667.000000	15.000000		18.340000		CAF2	P	
24	-105.900000	3.000000		18.830000		AIR		
25	194.900000	17.000000		18.690000		CAF2	P	
26	-118.000000	3.000000		18.010000		H-ZLAF50E		
27	-754.000000	199.917058	S	17.990000		AIR		
IMS	0.000000	0.000000		12.500000				

Figure 3.218 Optical design and graphs of aberration correction of a 50×/0.70 NUV+VIS$_+$ CCF plan poly-apochromatic objective (F'_{ts} = 200 mm; h = 95 mm).

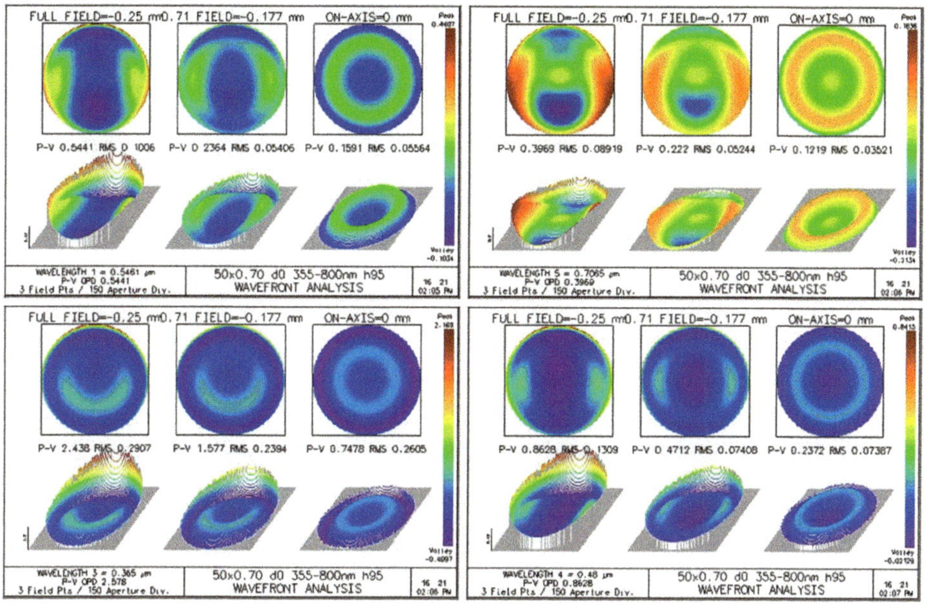

Figure 3.218 (*Continued*) Optical design and graphs of aberration correction of a 50×/0.70 NUV+VIS$_+$ CCF plan poly-apochromatic objective (F'_{ts} = 200 mm; h = 95 mm).

A difficult engineering task is to develop an optical design that provides aberration correction within a particular spectral range. In some cases, it is sufficient to develop an objective or other microscope optical component that has high transmittance in a fairly wide region of the spectrum but provides aberration correction only in a certain interval within this range. Of course, developing optical components that have an equally high level of aberration correction within an entire specific spectral range requires experience and the use of knowledge gained from the development of other optical components and devices. At the same time, optical calculation algorithms based on the designs of conventional optical components, for example, those used in mass produced and medium series microscopes, can be used to achieve this.

The use of fluorite (or optical materials with similar dispersion properties) is the basis for the development of optical components with aberration correction, such as MicroFluar, apochromatic, or poly-apochromatic (multi-apochromatic) objectives. However, fewer fluorite lenses (for example, for objectives) may be used in high linear magnification objectives than in medium and small linear magnification objectives. As an example, we can discuss oil immersion objectives with a 60× linear magnification and a numerical aperture in the space of objects equal to 1.49. Figure 3.235 displays a photograph of a Nikon objective and a drawing of its optical scheme [14], showing that only three fluorite lenses are used in the objective. A schematic optical diagram of a similar objective developed as part of the

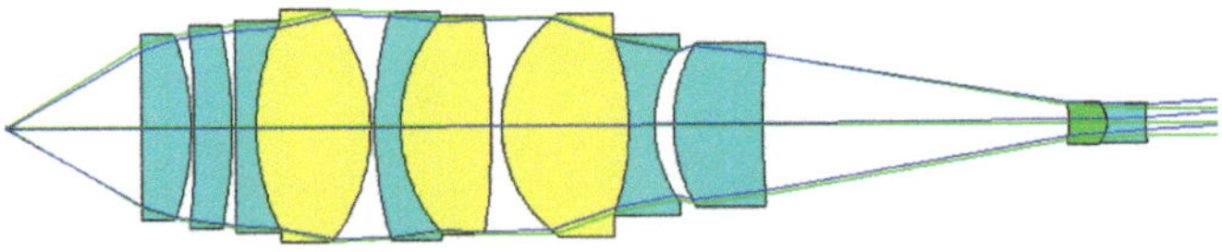

Lens: 100x0.50 d0 355-800nm h95

Object num aper	0.500000	Image height	12.500000	Primary wavln	0.546070

SRF	RADIUS	THICKNESS	APERTURE RADIUS	GLASS	SPECIAL
OBJ	0.000000	1.0000e-20	0.125040	AIR	
AST	0.000000	10.708590	0.125040 AS	AIR	
2	0.000000	4.000000	6.310000	H-ZLAF55D	
3	-19.470000	0.260000	7.070000	AIR	
4	-84.580000	3.000000	7.320000	H-ZLAF55D P	
5	-37.740000	0.300000	7.700000	AIR	
6	0.000000	1.600000	7.810000	H-ZLAF50E	
7	17.701000	9.060000	8.010000	CAF2	
8	-13.900000	0.300000	8.860000	AIR	
9	31.000000	2.000000	8.710000	H-ZLAF50E P	
10	12.057000	7.000000	8.140000	CAF2 P	
11	-61.650000	0.800000	8.320000	AIR	
12	10.233000	10.000000	8.440000	CAF2 P	
13	-20.750000	2.000000	6.850000	H-ZLAF50E P	
14	9.577000	1.500000	5.860000	AIR	
15	12.057000	7.000000	6.220000	H-ZLAF50E P	
16	87.970000	24.000000	5.570000	AIR	
17	-13.090000	3.000000	1.640000	ZF2	
18	-2.300000	3.000000	1.540000	LZ_CTK12	
19	8.179000	200.000000	1.410000	AIR	
20	-43.930000	14.000000	13.970000	H-QK3	
21	-51.200000	3.000000	16.050000	AIR	
22	667.000000	15.000000	16.430000	CAF2 P	
23	-105.900000	3.000000	16.950000	AIR	
24	194.900000	17.000000	16.870000	CAF2 P	
25	-118.000000	3.000000	16.300000	H-ZLAF50E P	
26	-754.000000	199.648161 S	16.300000	AIR	
IMS	0.000000	0.000000	12.500000		

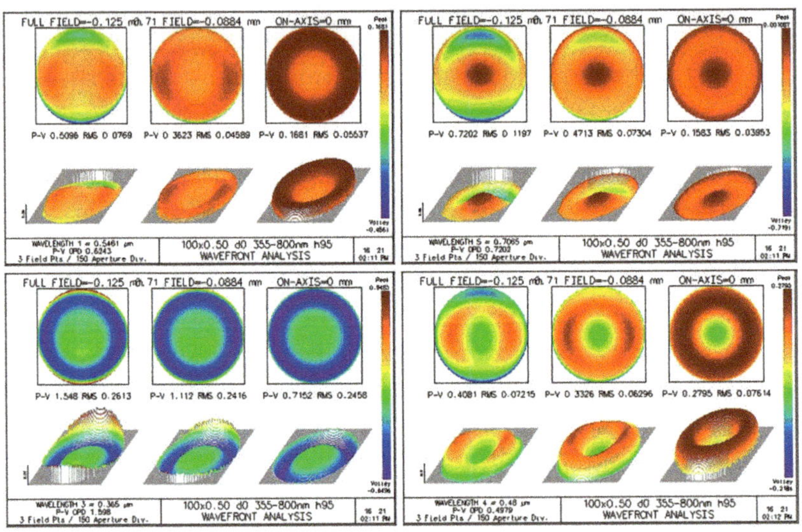

Figure 3.219 Optical design and graphs of aberration correction of a 100×/0.50 NUV+VIS$_+$ CCF plan poly-apochromatic objective (F'_{ts} = 200 mm; h = 95 mm).

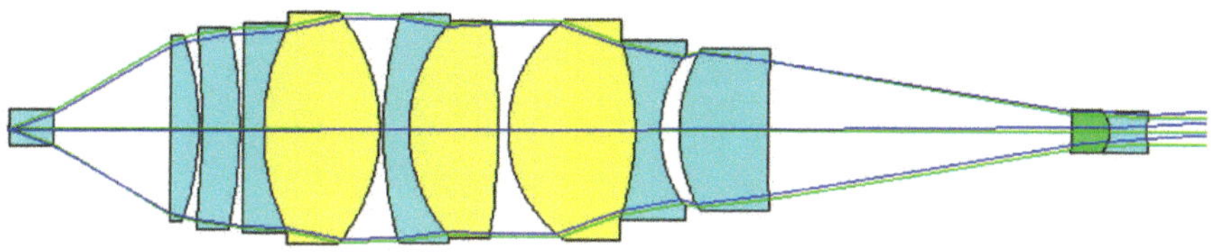

Lens: 100x0.50 d3.5 355-800nm h95

Object num aper	0.500000	Image height	12.500000	Primary wavln	0.546070

SRF	RADIUS	THICKNESS	APERTURE RADIUS	GLASS	SPECIAL
OBJ	0.000000	1.0000e-20	0.124926	AIR	
AST	0.000000	0.050000	0.124926 AS	AIR	
2	0.000000	3.500000	0.150000	H-K9L	
3	0.000000	9.279700	1.380000	AIR	
4	474.200000	2.300000	6.760000	H-ZLAF55D	
5	-20.110000	0.240000	7.070000	AIR	
6	-84.580000	3.000000	7.320000	H-ZLAF55D P	
7	-37.740000	0.300000	7.700000	AIR	
8	0.000000	1.600000	7.810000	H-ZLAF50E	
9	17.701000	9.060000	8.010000	CAF2	
10	-13.900000	0.300000	8.860000	AIR	
11	31.000000	2.000000	8.710000	H-ZLAF50E P	
12	12.057000	7.000000	8.140000	CAF2 P	
13	-61.650000	0.800000	8.320000	AIR	
14	10.233000	10.000000	8.440000	CAF2 P	
15	-20.750000	2.000000	6.850000	H-ZLAF50E P	
16	9.577000	1.500000	5.860000	AIR	
17	12.057000	7.000000	6.220000	H-ZLAF50E P	
18	87.970000	23.900000	5.570000	AIR	
19	-13.090000	3.000000	1.640000	ZF2	
20	-2.300000	3.000000	1.540000	LZ_CTK12	
21	8.179000	200.000000	1.410000	AIR	
22	-43.930000	14.000000	13.970000	H-QK3	
23	-51.200000	3.000000	16.050000	AIR	
24	667.000000	15.000000	16.430000	CAF2 P	
25	-105.900000	3.000000	16.950000	AIR	
26	194.900000	17.000000	16.870000	CAF2 P	
27	-118.000000	3.000000	16.300000	H-ZLAF50E P	
28	-754.000000	199.692714 S	16.300000	AIR	
IMS	0.000000	0.000000	12.500000		

Figure 3.220 Optical design and graphs of aberration correction of a 100×/0.50 ($d = 3.5$ mm) NUV+VIS$_+$ CCF plan poly-apochromatic objective ($F'_{ts} = 200$ mm; $h = 95$ mm).

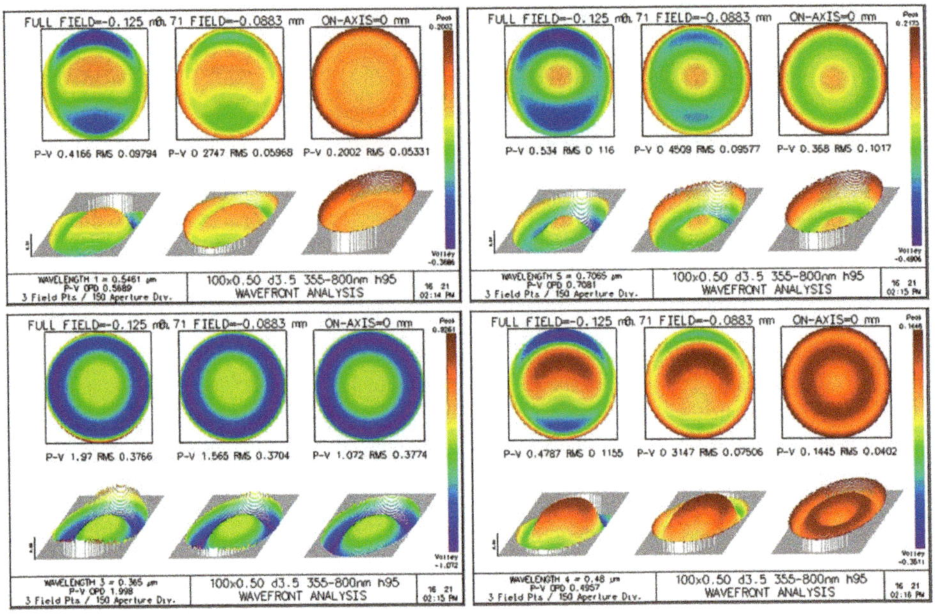

Figure 3.220 (*Continued*) Optical design and graphs of aberration correction of a 100×/0.50 ($d = 3.5$ mm) NUV+VIS$_+$ CCF plan poly-apochromatic objective ($F'_{ts} = 200$ mm; $h = 95$ mm).

Labor-Microscopes project is also shown (the parameters are shown in Fig. 3.55); this optical design also uses only three fluorite lenses. Apochromatic correction of chromatic aberrations was also achieved in both objectives.

At the same time, small and medium linear magnification objectives with apochromatic correction of chromatic aberrations have, as a rule, more fluorite lenses (or those made with a similar optical material). This can be explained by these objectives having higher output numerical aperture values after the lens. It is logical to assume that a similar situation exists with optical components that have poly-apochromatic aberration correction (multi-apochromatic).

3.10.4.1 Finite tube length objectives

The finite tube length for objectives with plan poly-apochromatic aberration correction may be preferable because there may be no additional elements in the optical design and in a real microscope, for focusing the image after the objective in the intermediate plane of the microscope. Otherwise, additional focusing (tube) systems should also be corrected for aberrations in the required spectral range. As examples, we can consider the optical designs of 10×/0.35 and 20×/0.70 objectives for operation in the spectral range of 400 to 1000 nm; chromatic aberrations are corrected within the limits of the entire specified spectral range. Unlike the CCF plan poly-apochromatic objectives

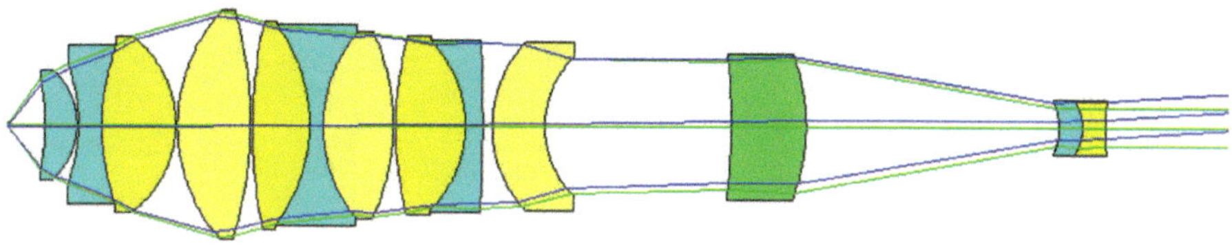

Lens: 100x0.80 ld h95 NUV+VIS

Object num aper 0.800000 Image height 12.500000 Primary wavln 0.546070

SRF	RADIUS	THICKNESS		APERTURE RADIUS		GLASS		SPECIAL
OBJ	0.000000	1.0000e-20		0.124984		AIR		
AST	0.000000	3.200100		0.124984	AS	AIR		
2	-23.390000	2.800000		3.950000		H-LAK61		
3	-6.121000	0.150000		4.620000		AIR		
4	-14.400000	2.000000		5.110000		H-LAK61		
5	18.270000	6.500000		6.700000		H-FK61		
6	-9.157000	0.150000		7.470000		AIR		
7	14.400000	6.400000		9.720000		CAF2		
8	-32.180000	0.200000		9.600000		AIR		
9	42.180000	5.000000		8.830000		H-FK61	P	
10	-13.808000	1.200000		8.520000		H-LAK61	P	
11	12.057000	6.100000		7.840000		CAF2	P	
12	-18.270000	0.200000		7.920000		AIR		
13	28.310000	6.000000		7.550000		H-FK61	P	
14	-10.068000	1.500000		7.210000		H-LAK61	P	
15	-130.000000	0.800000		7.250000		AIR		
16	10.233000	4.600000		7.140000		CAF2	P	
17	8.240000	16.200000		5.930000		AIR		
18	-49.000000	6.500000		5.910000		F_4		
19	-17.000000	22.000000		6.220000		AIR		
20	-5.401000	2.000000		2.170000		H-LAK61	P	
21	-4.093000	2.000000		2.320000		H-FK61	P	
22	17.000000	200.000000		2.310000		AIR		
23	-43.930000	14.000000		14.330000		H-QK3		
24	-51.200000	3.000000		16.420000		AIR		
25	667.000000	15.000000		16.800000		CAF2	P	
26	-105.900000	3.000000		17.290000		AIR		
27	194.900000	17.000000		17.190000		CAF2	P	
28	-118.000000	3.000000		16.590000		H-ZLAF50E		
29	-754.000000	199.907079	S	16.580000		AIR		
IMS	0.000000	0.000000		12.500000				

Figure 3.221 Optical design and graphs of aberration correction of a 100×/0.80 NUV+VIS$_+$ CCF plan poly-apochromatic objective ($F'_{ts} = 200$ mm; $h = 95$ mm).

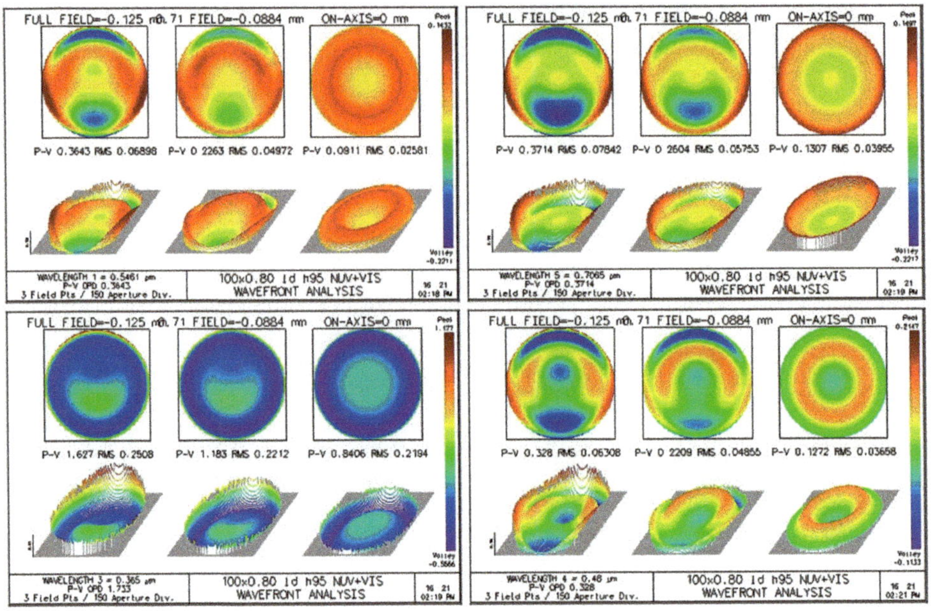

Figure 3.221 (*Continued*) Optical design and graphs of aberration correction of a 100×/0.80 NUV+VIS$_+$ CCF plan poly-apochromatic objective (F'_{ts} = 200 mm; h = 95 mm).

Table 3.9 Main technical parameters and basic optical layout of NUV+VIS$_+$ objectives (λ = 546.07 nm; 355–800 nm).

Magnification	*NA*	*WD* (mm)	*F'* (mm)	*R* (μm)	*DF* (μm)	*FOV* on object (mm)	*FOV* on image (mm)	Optical design at figure no.	The principal optical layout
20×	0.32	33	10	1.04	3.25	1.25	25	3.214	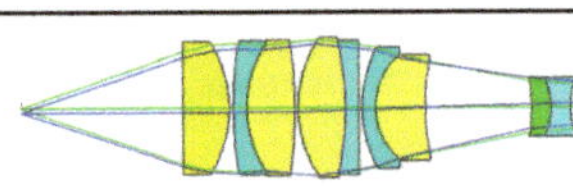
20×	0.32	21.5 d = 3.5	10	1.04	3.25	1.25	25	3.215	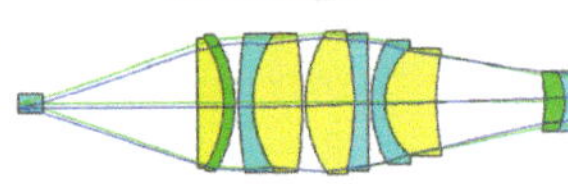
50×	0.42	17	4	0.79	1.89	0.50	25	3.216	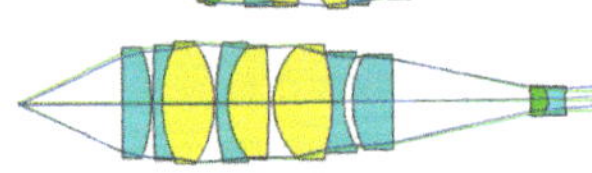
50×	0.42	17 d = 3.5	4	0.79	1.89	0.50	25	3.217	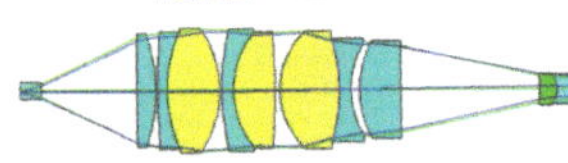
50×	0.70	3.6	4	0.48	0.68	0.50	25	3.218	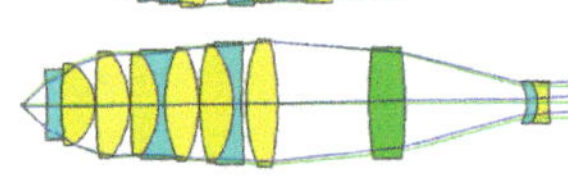
100×	0.50	10	2	0.67	1.33	0.25	25	3.219	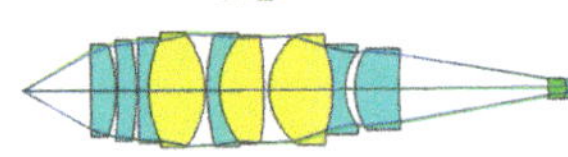
100×	0.50	8.5 d = 3.5	2	0.67	1.33	0.25	25	3.220	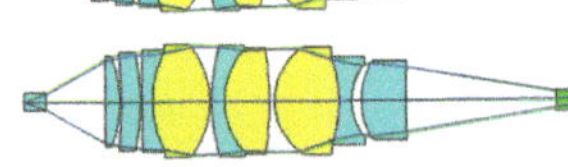
100×	0.80	1.8	2	0.42	0.52	0.25	25	3.221	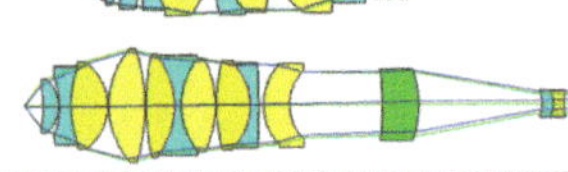

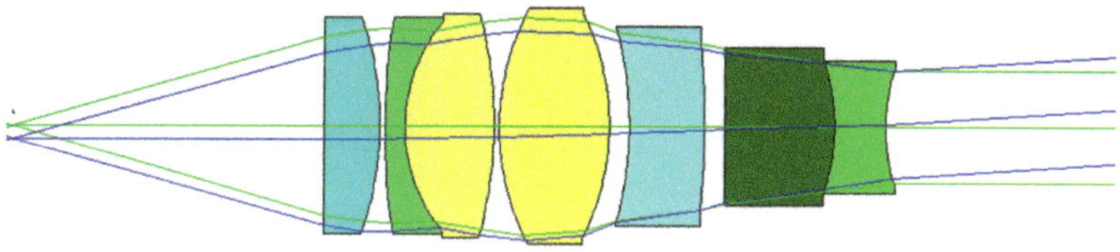

Lens: NIR PlanApo LD 10x0.27 h95

Object num aper 0.270000 Image height 12.500000 Primary wavln 1.060000

SRF	RADIUS	THICKNESS	APERTURE RADIUS	GLASS	SPECIAL
OBJ	0.000000	1.0000e-20	1.248771	AIR	
AST	0.000000	30.949090	1.248771 AS	AIR	
2	0.000000	5.600000	9.930000	H-LAK3	
3	-29.570000	0.500000	10.510000	AIR	
4	69.000000	2.000000	10.520000	H-ZLAF3	
5	15.394000	9.000000	10.180000	CAF2	
6	-39.800000	0.500000	10.810000	AIR	
7	23.710000	11.000000	11.420000	CAF2	
8	-23.710000	2.000000	11.000000	AIR	
9	-34.760000	7.400000	9.670000	SILICA	
10	-83.300000	2.000000	8.350000	AIR	
11	0.000000	11.000000	7.660000	H-ZF62	
12	-18.710000	5.000000	6.440000	H-ZLAF3 P	
13	18.710000	200.000000	5.660000	AIR	
14	-50.000000	14.000000	17.700000	H-QK3	
15	-63.800000	3.000000	19.960000	AIR	
16	0.000000	15.000000	20.400000	CAF2 C	
17	-87.500000	3.000000	21.130000	AIR	
18	137.000000	17.000000	20.160000	CAF2 C	
19	-145.000000	3.000000	20.100000	H-ZLAF50B	
20	0.000000	199.919527 S	20.000000	AIR	
IMS	0.000000	0.000000	12.500000		

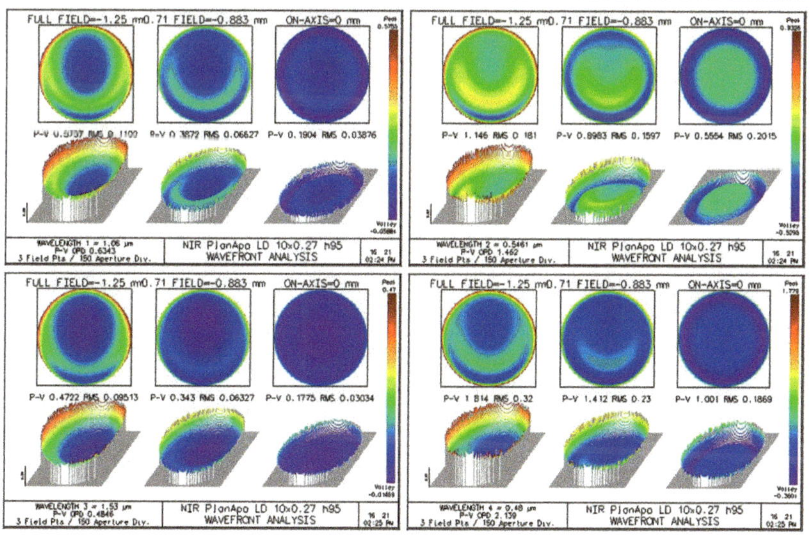

Figure 3.222 Optical design and graphs of aberration correction of a 10×/0.27 NIR CCF plan apochromatic objective ($F'_{ts} = 200$ mm; $h = 95$ mm).

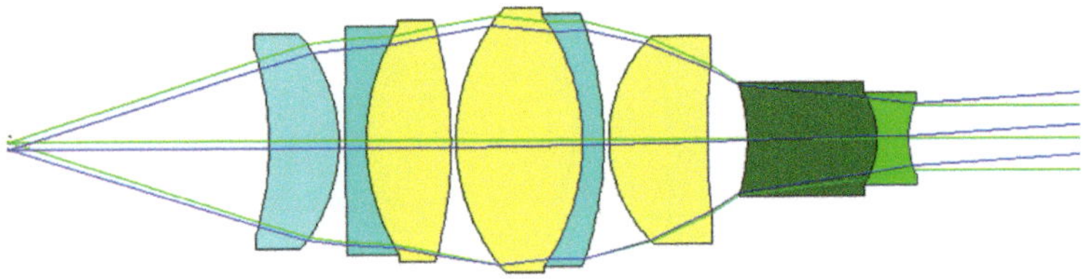

Lens: NIR PlanApo LD 20x0.30 h95

Object num aper	0.300000	Image height	12.500000	Primary wavln	1.060000

SRF	RADIUS	THICKNESS	APERTURE RADIUS	GLASS	SPECIAL
OBJ	0.000000	1.0000e-20	0.626071	AIR	
AST	0.000000	25.000000	0.626071 AS	AIR	
2	-29.700000	6.800000	8.700000	SILICA	
3	-15.000000	0.600000	10.030000	AIR	
4	567.240000	2.000000	10.420000	H-LAK3	
5	20.169000	8.000000	10.630000	CAF2	
6	-40.000000	0.500000	11.280000	AIR	
7	18.500000	12.000000	12.320000	CAF2	
8	-20.530000	2.000000	11.740000	H-LAK3 P	
9	-33.600000	0.500000	11.730000	AIR	
10	13.305000	9.400000	9.670000	CAF2 P	
11	62.400000	3.800000	6.920000	AIR	
12	-18.232000	12.500000	5.350000	H-ZF62	
13	-8.000000	3.000000	4.300000	H-ZLAF50B	
14	8.700000	200.000000	3.630000	AIR	
15	-50.000000	14.000000	15.650000	H-QK3	
16	-63.800000	3.000000	17.720000	AIR	
17	0.000000	15.000000	18.080000	CAF2 C	
18	-87.500000	3.000000	18.820000	AIR	
19	137.000000	17.000000	18.740000	CAF2 C	
20	-145.000000	3.000000	18.000000	H-ZLAF50B	
21	0.000000	199.879172 S	17.940000	AIR	
IMS	0.000000	0.000000	12.500000		

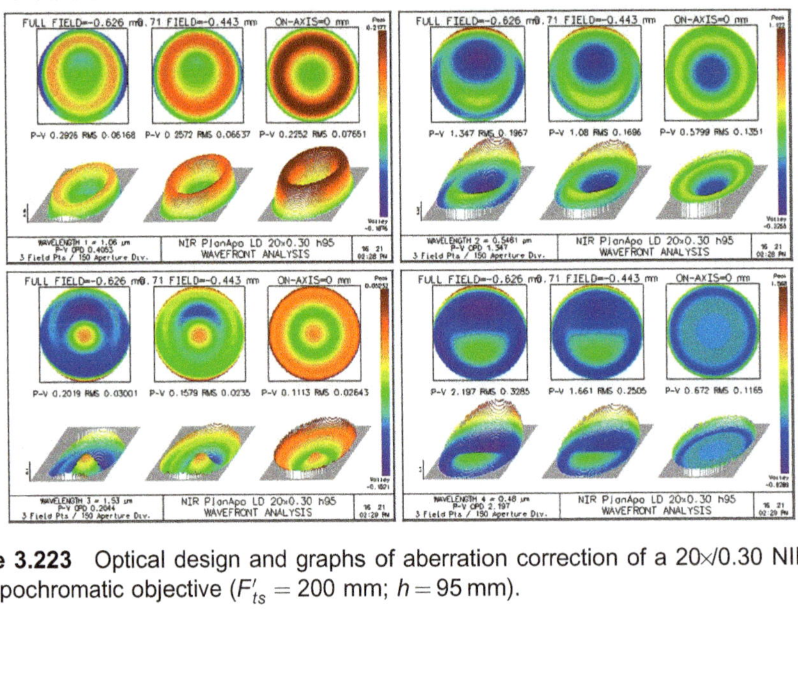

Figure 3.223 Optical design and graphs of aberration correction of a 20×/0.30 NIR CCF plan apochromatic objective ($F'_{ts} = 200$ mm; $h = 95$ mm).

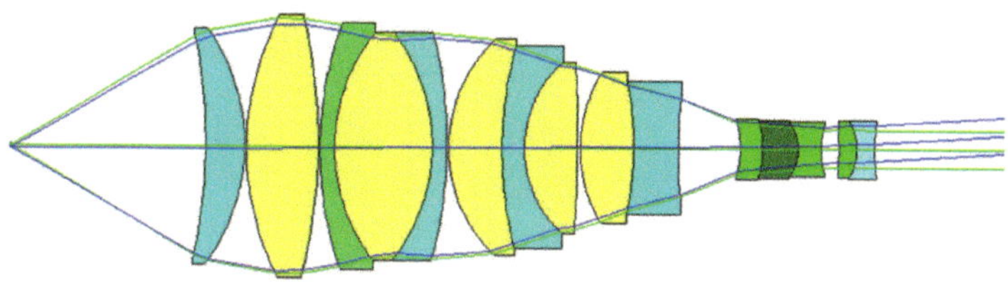

Lens: NIR PlanApo 50x0.50 d0 h95 Zoom 1 of 1 Efl -17.656353

Object num aper 0.500000 Image height 12.500000 Primary wavln 1.060000

SRF	RADIUS	THICKNESS	APERTURE RADIUS	GLASS	SPECIAL
OBJ	0.000000	1.0000e-20	0.250322	AIR	
AST	0.000000	22.001436	0.250322 AS	AIR	
2	-59.320000	4.360000	12.390000	ZLAF1	
3	-21.580000	0.200000	12.940000	AIR	
4	30.270000	8.000000	14.370000	CAF2	
5	-57.800000	0.200000	14.250000	AIR	
6	40.190000	1.600000	13.480000	H-ZLAF3	
7	19.790000	11.000000	12.520000	CAF2 P	
8	-19.790000	1.600000	12.240000	ZLAF1 P	
9	-47.896000	0.200000	12.500000	AIR	
10	16.120000	6.000000	11.860000	CAF2 P	
11	42.657000	2.500000	11.100000	H-ZLAF10LA	
12	12.465000	6.000000	9.400000	CAF2 P	
13	-124.360000	0.300000	9.170000	AIR	
14	15.672000	6.000000	8.330000	CAF2 P	
15	-35.900000	5.000000	7.270000	H-LAK3	
16	61.256000	7.000000	5.710000	AIR	
17	-11.495000	2.500000	3.290000	H-ZLAF3 P	
18	-11.495000	4.000000	3.160000	H-ZF62	
19	-4.950000	2.500000	3.010000	H-ZLAF3 P	
20	9.480000	2.000000	2.750000	AIR	
21	-57.800000	2.000000	2.970000	H-ZF6	
22	-5.890000	2.000000	3.100000	H-QK3	
23	18.000000	200.000000	3.140000	AIR	
24	-50.000000	14.000000	14.820000	H-QK3 P	
25	-63.800000	3.000000	16.790000	AIR	
26	0.000000	15.000000	17.130000	CAF2 C	
27	-87.500000	3.000000	17.860000	AIR	
28	137.000000	17.000000	17.790000	CAF2 C	
29	-145.000000	3.000000	17.110000	H-ZLAF50B	
30	0.000000	199.743529 S	17.050000	AIR	
IMS	0.000000	0.000000	12.500000		

Figure 3.224 Optical design and graphs of aberration correction of a 50×/0.50 NIR CCF plan apochromatic objective ($F'_{ts} = 200$ mm; $h = 95$ mm).

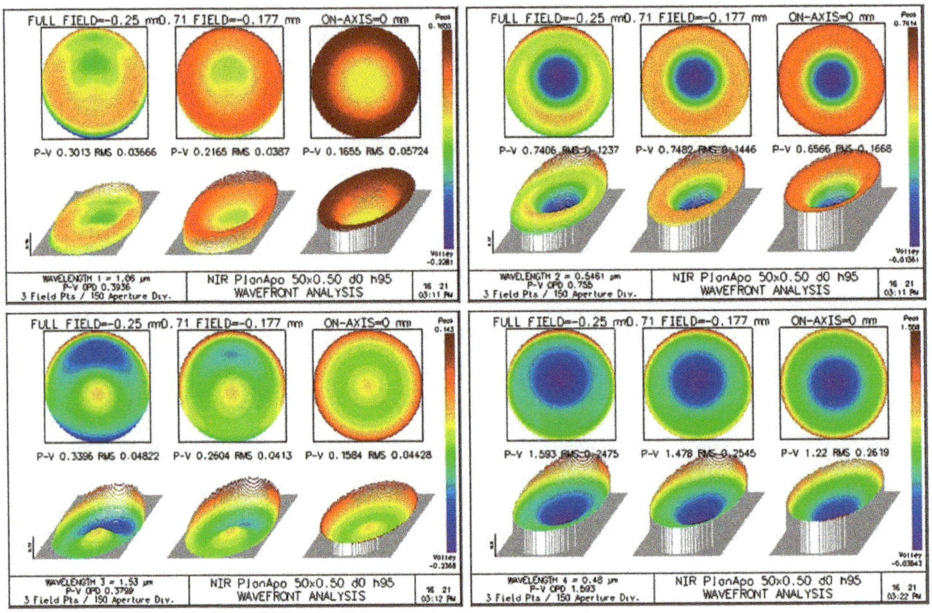

Figure 3.224 (*Continued*) Optical design and graphs of aberration correction of a 50×/0.50 NIR CCF plan apochromatic objective ($F'_{ts} = 200$ mm; $h = 95$ mm).

($h = 95$ mm) shown above, these objectives have significantly higher numerical apertures. Figures 3.236 and 3.237 show the design parameters and graphs of aberration correction of these objectives.

3.10.4.2 Example drawings of a 20×/0.70 plan poly-apochromatic objective

Figures 3.238–3.293 show a general mechanical design and drawings of optical parts, mechanical frames, and components, which are lenses integrated into the frames. These and other similar drawings are made by distinguished engineer Alexey Tabachkov of ITMO University in Russia.

Figure 3.294 is a photograph of a 20×/0.70 CCF plan poly-apochromatic objective, and Figures 3.295–3.304 show photographs of real objects taken with this objective.

Let us discuss these figures. An interesting feature of microelectronic objects is that they are not, as a rule, subjected to any additional processing before being examined using a light microscope. This is one of the main differences between them and biological objects, which, as a rule, are stained and exposed to mounting media. Obviously, adequate color rendering in the study of microelectronic objects is one of the main criteria that should characterize the level of aberration correction of microscope optical components. Another criterion may be the need to align the curvature in the image of the object since the objects in microelectronics are flat. Thus, to

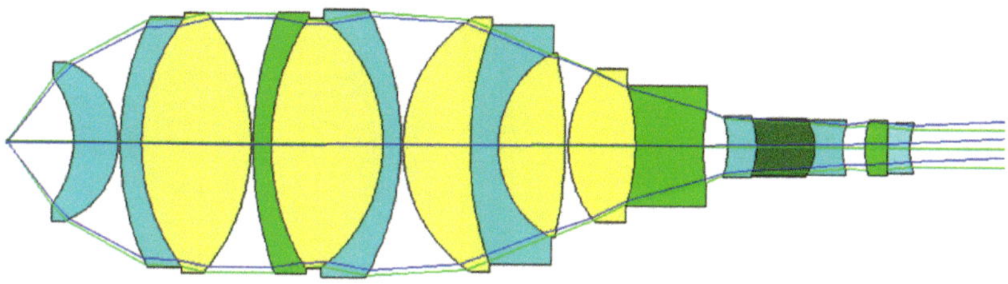

Lens: NIR PlanApo 100x0.80 d0 h95 Zoom 1 of 1 Efl -9.176924

Object num aper 0.800000 Image height 10.000000 Primary wavln 1.060000

SRF	RADIUS	THICKNESS	APERTURE RADIUS	GLASS	SPECIAL
OBJ	0.000000	1.0000e-20	0.100168	AIR	
AST	0.000000	6.000000	0.100168 AS	AIR	
2	-9.000000	4.000000	5.550000	H-LAF10L	
3	-7.400000	0.200000	6.870000	AIR	
4	24.210000	1.900000	10.730000	H-LAF10L P	
5	17.602000	9.700000	10.790000	CAF2	
6	-17.602000	0.200000	11.230000	AIR	
7	31.724000	1.600000	11.250000	H-ZLAF3	
8	19.790000	10.000000	10.800000	CAF2 P	
9	-13.305000	1.600000	10.750000	H-LAF10L P	
10	-23.390000	0.200000	11.580000	AIR	
11	12.979000	6.000000	11.030000	CAF2 P	
12	30.000000	2.500000	10.310000	H-LAF10L P	
13	9.032000	6.000000	8.010000	CAF2 P	
14	-38.950000	0.300000	7.950000	AIR	
15	10.068000	6.000000	6.610000	CAF2 P	
16	-17.701000	6.000000	5.230000	H-ZLAF3 P	
17	18.883000	2.200000	3.240000	AIR	
18	-9.200000	2.500000	2.670000	H-LAF10L P	
19	-7.300000	5.200000	2.520000	H-ZF62	
20	-4.000000	2.500000	2.360000	H-LAF10L P	
21	6.516000	2.000000	2.070000	AIR	
22	7.900000	2.000000	2.320000	H-ZF6	
23	-83.000000	2.000000	2.220000	H-QK3	
24	6.400000	200.000000	2.020000	AIR	
25	-50.000000	14.000000	10.790000	H-QK3 P	
26	-63.800000	3.000000	12.200000	AIR	
27	0.000000	15.000000	12.380000	CAF2 C	
28	-87.500000	3.000000	12.930000	AIR	
29	137.000000	17.000000	12.900000	CAF2 C	
30	-145.000000	3.000000	12.360000	H-ZLAF50B	
31	0.000000	199.163474 S	12.330000	AIR	
IMS	0.000000	0.000000	10.000000		

Figure 3.225 Optical design and graphs of aberration correction of a 100×/0.80 NIR CCF plan apochromatic objective ($F'_{ts} = 200$ mm; $h = 95$ mm).

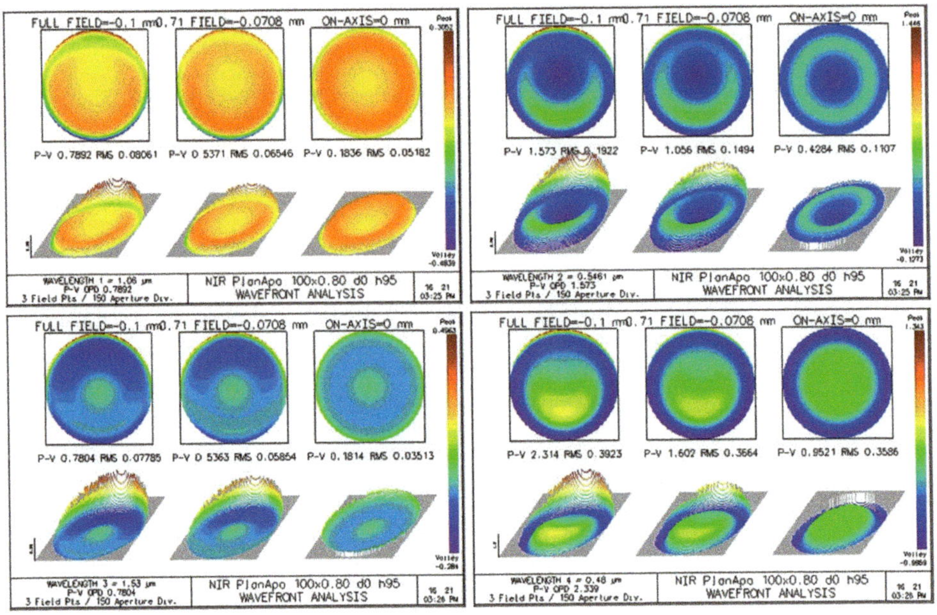

Figure 3.225 (*Continued*) Optical design and graphs of aberration correction of a 100×/0.80 NIR CCF plan apochromatic objective (F'_{ts} = 200 mm; h = 95 mm).

study microelectronic objects, it is necessary to use optical components (objectives, first and foremost) that have a plan apochromatic type of aberration correction. A better choice is a plan poly-apochromatic (plan multi-apochromatic) objective. Consider Fig. 3.304, which is a photograph of the surface of an ordinary household mirror. It is interesting that the objective used was able to "see" the interference fringes that occur from the coating that is sprayed onto the substrate. At the same time, in biological research, it is also necessary to not only use plan apochromatic objectives, but also objectives with a more advanced level of aberration correction, namely, plan poly-apochromatic (plan multi-apochromatic) objectives.

3.10.4.3 Infinite tube length objectives

Objectives used for equipping microscopes with an infinite optical tube length are considered more modern and (as many users often think) more advanced in terms of image quality. Of course, optical engineers may not agree with this; an interesting discussion can be devoted to this issue, and many arguments can be made in favor of finite microscope optical systems. However, the market dictates the need to develop an infinite optical system for *small series* light microscopes and objectives with an infinite tube length. Figures 3.305–3.312 show the design parameters and graphs of aberration

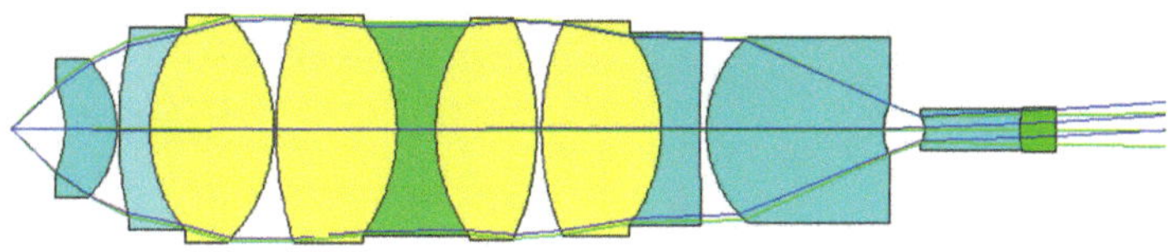

Lens: NIR PlanApo 250x0.70 d0 h95

Object num aper 0.700000 Image height 12.500000 Primary wavln 1.060000

SRF	RADIUS	THICKNESS	APERTURE RADIUS	GLASS	SPECIAL
OBJ	0.000000	1.0000e-20	0.050003	AIR	
AST	0.000000	2.197123	0.050003 AS	AIR	
2	-4.430000	2.200000	1.820000	H-ZLAF50B	
3	-3.773000	0.100000	2.790000	AIR	
4	16.030000	1.300000	3.630000	K9	
5	6.457000	5.200000	4.110000	CAF2	
6	-6.457000	0.100000	4.600000	AIR	
7	14.510000	5.000000	4.600000	CAF2 P	
8	-7.244000	1.600000	4.340000	BAF7	
9	7.244000	4.200000	4.340000	CAF2 P	
10	-10.850000	0.200000	4.500000	AIR	
11	10.850000	5.000000	4.360000	CAF2 P	
12	-6.399000	1.600000	3.880000	LZ-CTK12	
13	150.800000	0.300000	3.810000	AIR	
14	5.012000	7.400000	3.760000	H-ZLAF50B P	
15	2.230000	1.700000	1.070000	AIR	
16	-1.137600	4.000000	0.600000	LZ-CTK12 P	
17	3.400000	1.500000	0.840000	H-ZF7LA	
18	-128.800000	200.000000	0.900000	AIR	
19	-50.000000	14.000000	13.910000	H-QK3	
20	-63.800000	3.000000	15.820000	AIR	
21	0.000000	15.000000	16.150000	CAF2 C	
22	-87.500000	3.000000	16.900000	AIR	
23	137.000000	17.000000	16.890000	CAF2 C	
24	-145.000000	3.000000	16.300000	H-ZLAF50B	
25	0.000000	199.955076 S	16.270000	AIR	
IMS	0.000000	0.000000	12.500000		

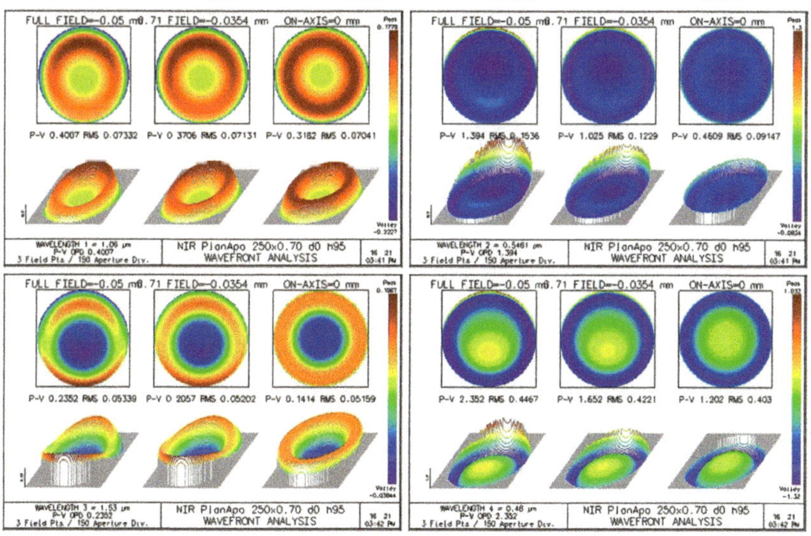

Figure 3.226 Optical design and graphs of aberration correction of a 250×/0.70 NIR CCF plan apochromatic objective ($F'_{ts} = 200$ mm; $h = 45$ mm).

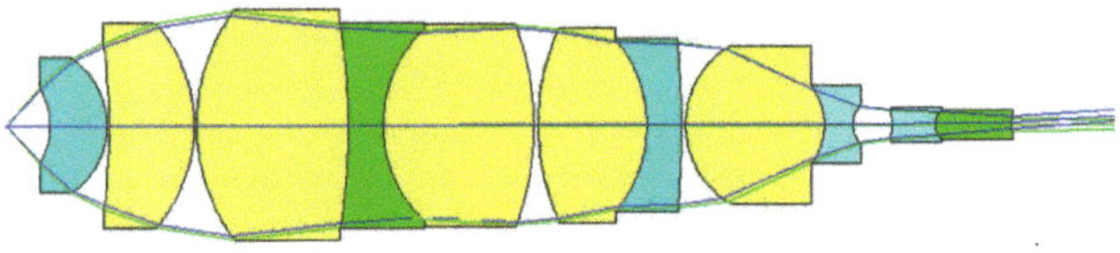

Lens: NIR PlanApo 500x0.75 d0 h95

Object num aper 0.750000 Image height 12.500000 Primary wavln 1.060000

SRF	RADIUS	THICKNESS	APERTURE RADIUS	GLASS	SPECIAL
OBJ	0.000000	1.0000e-20	0.024998	AIR	
AST	0.000000	1.997288	0.024998 AS	AIR	
2	-3.370000	2.700000	1.740000	H-ZLAF50B	
3	-3.492000	0.100000	3.010000	AIR	
4	-27.480000	4.000000	3.730000	CAF2	
5	-6.792000	0.100000	4.610000	AIR	
6	8.670000	7.000000	5.170000	CAF2 P	
7	-29.000000	1.600000	4.620000	ZF1	
8	5.890000	7.000000	4.310000	CAF2 P	
9	-12.840000	0.200000	4.540000	AIR	
10	10.233000	5.000000	4.370000	CAF2 P	
11	-5.890000	1.600000	3.870000	LZ_CTK12	
12	-44.372000	0.200000	3.800000	AIR	
13	4.040000	6.400000	3.550000	CAF2 P	
14	-2.864000	1.300000	1.750000	K9	
15	1.199000	1.800000	0.880000	AIR	
16	-30.000000	2.000000	0.790000	H-LAK53A	
17	0.828000	3.500000	0.660000	H-ZF7LA	
18	1.820000	200.000000	0.440000	AIR	
19	-50.000000	14.000000	16.350000	H-QK3	
20	-63.800000	3.000000	18.700000	AIR	
21	0.000000	15.000000	19.190000	CAF2 C	
22	-87.500000	3.000000	20.090000	AIR	
23	137.000000	17.000000	20.090000	CAF2 C	
24	-145.000000	3.000000	19.450000	H-ZLAF50B	
25	0.000000	199.908242 S	19.410000	AIR	
IMS	0.000000	0.000000	12.500000		

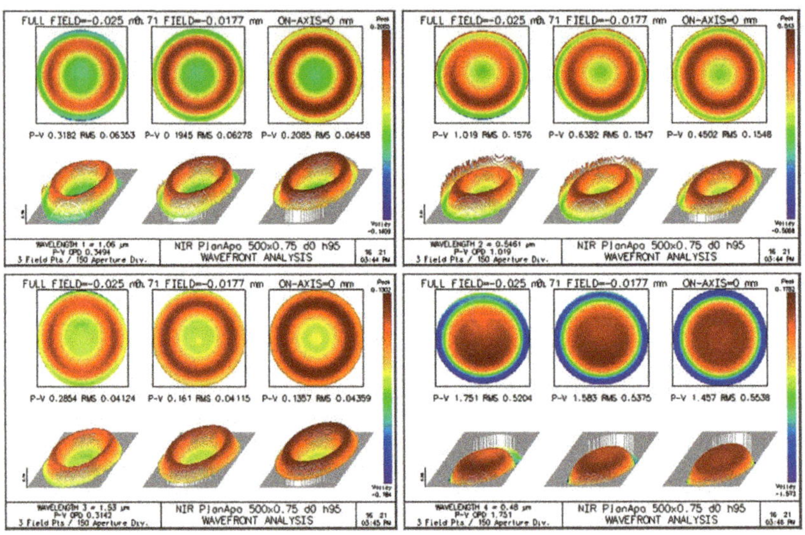

Figure 3.227 Optical design and graphs of aberration correction of a 500×/0.75 NIR CCF plan apochromatic objective ($F'_{ts} = 200$ mm; $h = 45$ mm).

Table 3.10 Main technical parameters and basic optical layout of VIS+NIR objectives ($\lambda = 1064$ nm; 480–1800 nm).

Magnification	*NA*	*WD* (mm)	*F′* (mm)	*R* (μm)	*DF* (μm)	*FOV* on object (mm)	*FOV* on image (mm)	Optical design at figure no.	The principal optical layout
10×	0.27	31	20	1.23	4.57	2.5	25	3.222	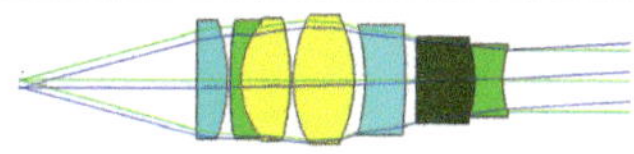
20×	0.30	22	10	1.11	3.70	1.25	25	3.223	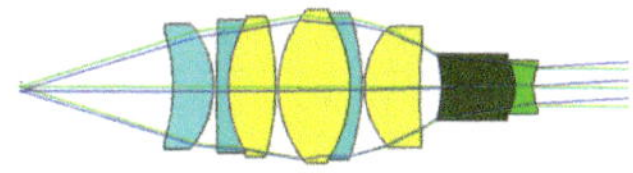
50×	0.50	18	4	0.67	1.33	0.50	25	3.224	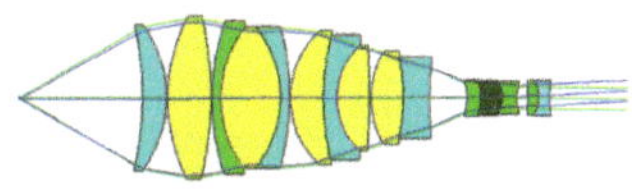
100×	0.80	3.2	2	0.42	0.52	0.25	25	3.225	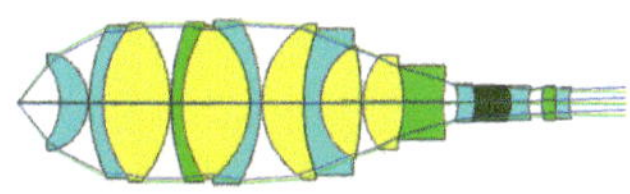
250×	0.70	1.3	0.8	0.48	0.68	0.10	25	3.226	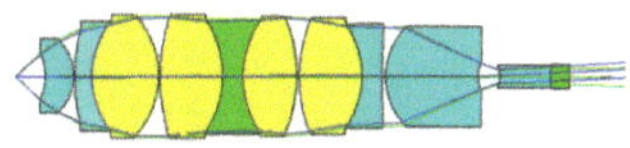
500×	0.75	1.1	0.4	0.44	0.59	0.05	25	3.227	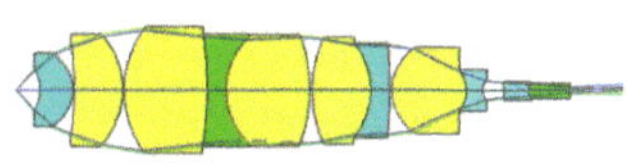

Figure 3.228 Two VIS+NIR objectives ($\lambda = 1064$ nm; 480–1800 nm).

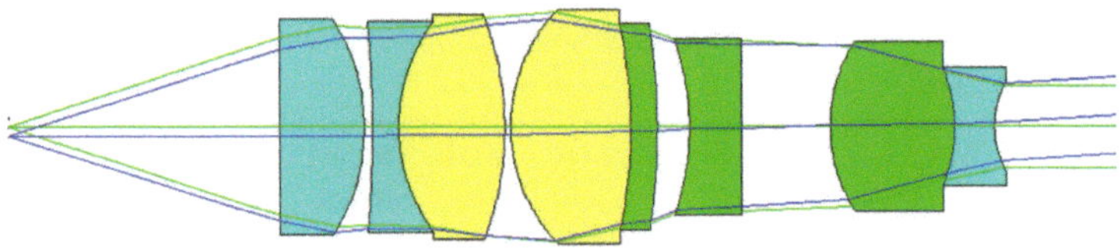

Lens: 20x0.30 430-1800nm h95

Object num aper 0.300000 Image height 12.500000 Primary wavln 0.546070

SRF	RADIUS	THICKNESS	APERTURE RADIUS	GLASS	SPECIAL
OBJ	0.000000	1.0000e-20	0.624239	AIR	
AST	0.000000	21.007200	0.624239 AS	AIR	
2	0.000000	6.500000	7.230000	H-ZK6	
3	-14.791000	0.600000	8.020000	AIR	
4	-105.900000	2.000000	7.840000	H-LAK3	
5	13.090000	8.000000	7.810000	CAF2	
6	-23.140000	0.500000	8.400000	AIR	
7	12.465000	9.200000	8.740000	CAF2 P	
8	-33.350000	2.000000	7.630000	H-ZF6	
9	-47.896000	2.500000	7.290000	AIR	
10	-18.923000	4.000000	6.570000	H-ZF6 P	
11	0.000000	6.800000	6.440000	AIR	
12	11.630000	9.500000	6.310000	H-ZF6 P	
13	-11.630000	3.000000	4.440000	H-ZLAF3	
14	6.516000	200.000000	3.310000	AIR	
15	-128.703000	3.700000	15.540000	H-ZLAF55D	
16	-152.600000	2.000000	15.880000	AIR	
17	148.130000	3.700000	16.160000	H-LAF50A	
18	89.840000	13.000000	16.140000	H-FK61	
19	-56.360000	3.700000	16.390000	LZ_CTK12	
20	-96.380000	199.993457 S	16.660000	AIR	
IMS	0.000000	0.000000	12.500000		

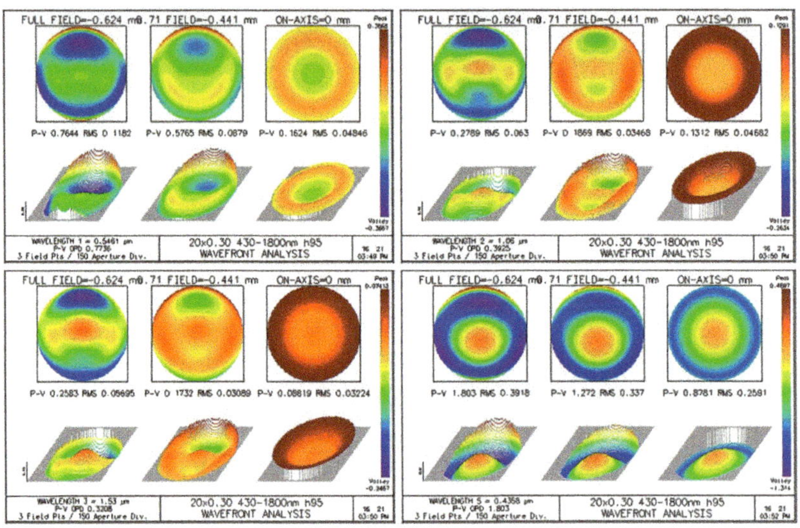

Figure 3.229 Optical design and graphs of aberration correction of a 20×/0.30 VIS+NIR CCF plan poly-apochromatic objective ($F'_{ts} = 200$ mm; $h = 95$ mm).

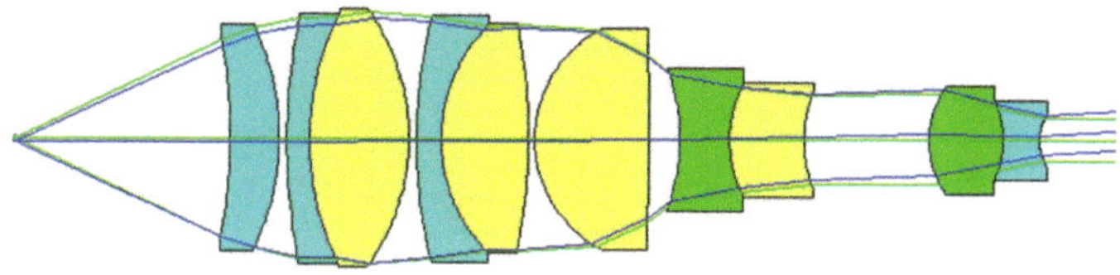

Lens: 50x0.42 d0 430-1800nm h95

Object num aper 0.420000 Image height 12.500000 Primary wavln 0.546070

SRF	RADIUS	THICKNESS	APERTURE RADIUS	GLASS	SPECIAL
OBJ	0.000000	1.0000e-20	0.249927	AIR	
AST	0.000000	17.501550	0.249927 AS	AIR	
2	-37.897100	4.000000	7.960000	H-ZLAF3	
3	-19.538000	0.600000	8.890000	AIR	
4	48.080000	1.800000	9.710000	H-LAK3	
5	21.580000	8.000000	9.820000	CAF2	
6	-17.120000	0.600000	10.180000	AIR	
7	39.500000	2.000000	9.740000	H-LAK3 P	
8	12.642000	7.000000	9.030000	CAF2 P	
9	-46.110000	0.500000	9.050000	AIR	
10	9.720000	9.200000	8.740000	CAF2 P	
11	-155.000000	2.500000	7.000000	AIR	
12	-16.465000	4.000000	5.580000	H-ZF6	
13	9.300000	6.000000	4.450000	CAF2 P	
14	9.300000	10.150000	3.800000	AIR	
15	7.047000	6.000000	4.180000	H-ZF6 P	
16	-7.047000	3.000000	3.080000	H-ZLAF3 P	
17	4.146000	200.000000	2.080000	AIR	
18	-128.703000	3.700000	14.270000	H-ZLAF55D	
19	-152.600000	2.000000	14.590000	AIR	
20	148.130000	3.700000	14.850000	H-LAF50A	
21	89.840000	13.000000	14.840000	H-FK61	
22	-56.360000	3.700000	15.110000	LZ_CTK12	
23	-96.380000	199.969148 S	15.370000	AIR	
IMS	0.000000	0.000000	12.500000		

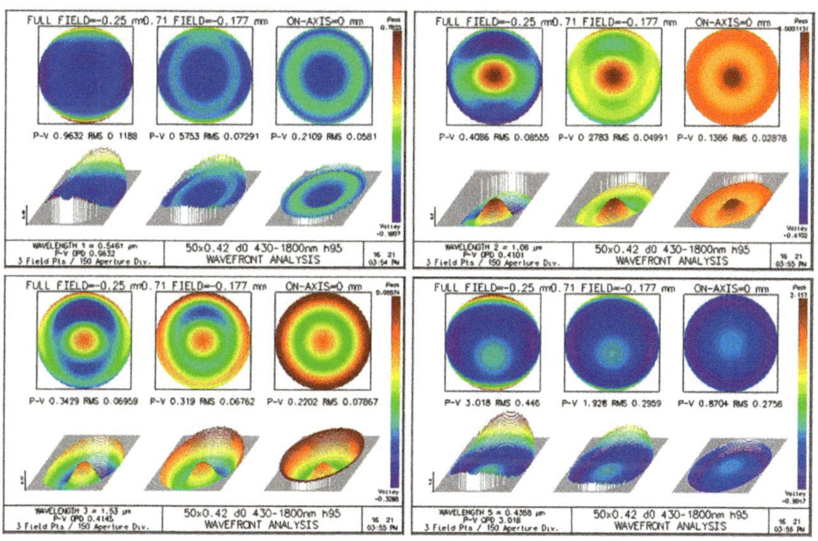

Figure 3.230 Optical design and graphs of aberration correction of a 50×/0.42 VIS+NIR CCF plan poly-apochromatic objective ($F'_{ts} = 200$ mm; $h = 95$ mm).

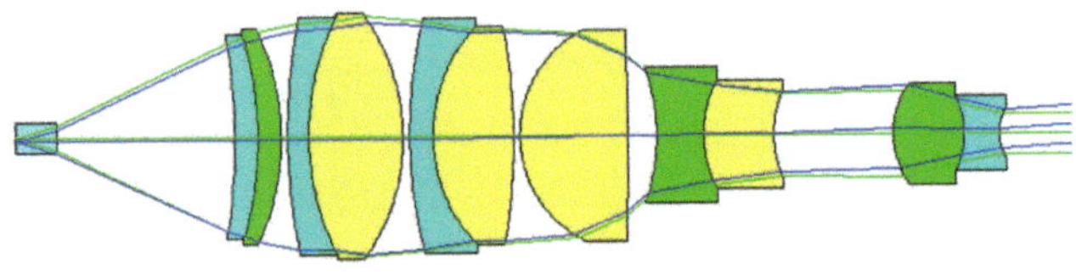

Lens: 50x0.42 d3.5 430-1800nm h95

Object num aper 0.420000 Image height 12.500000 Primary wavln 0.546070

SRF	RADIUS	THICKNESS		APERTURE RADIUS		GLASS		SPECIAL
OBJ	0.000000	0.010000		0.250012		AIR		
AST	0.000000	3.500000		0.250000	A	H-K9		
2	0.000000	15.293900		1.280000		AIR		
3	-48.710000	2.000000		8.040000		H-LAK3		
4	-27.930000	2.000000		8.490000		H-ZF6		
5	-19.910000	0.600000		8.890000		AIR		
6	48.080000	1.800000		9.710000		H-LAK3		
7	21.580000	8.000000		9.820000		CAF2		
8	-17.120000	0.600000		10.180000		AIR		
9	39.500000	2.000000		9.740000		H-LAK3	P	
10	12.642000	7.000000		9.030000		CAF2	P	
11	-46.110000	0.500000		9.050000		AIR		
12	9.720000	9.200000		8.740000		CAF2	P	
13	-155.000000	2.500000		7.000000		AIR		
14	-16.465000	4.000000		5.580000		H-ZF6		
15	9.300000	6.000000		4.450000		CAF2	P	
16	9.300000	10.150000		3.800000		AIR		
17	7.047000	6.000000		4.180000		H-ZF6	P	
18	-7.047000	3.000000		3.080000		H-ZLAF3		
19	4.146000	200.000000		2.080000		AIR		
20	-128.703000	3.700000		14.270000		H-ZLAF55D		
21	-152.600000	2.000000		14.590000		AIR		
22	148.130000	3.700000		14.850000		H-LAF50A		
23	89.840000	13.000000		14.840000		H-FK61		
24	-56.360000	3.700000		15.110000		LZ-CTK12		
25	-96.380000	199.876975	S	15.370000		AIR		
IMS	0.000000	0.000000		12.500000				

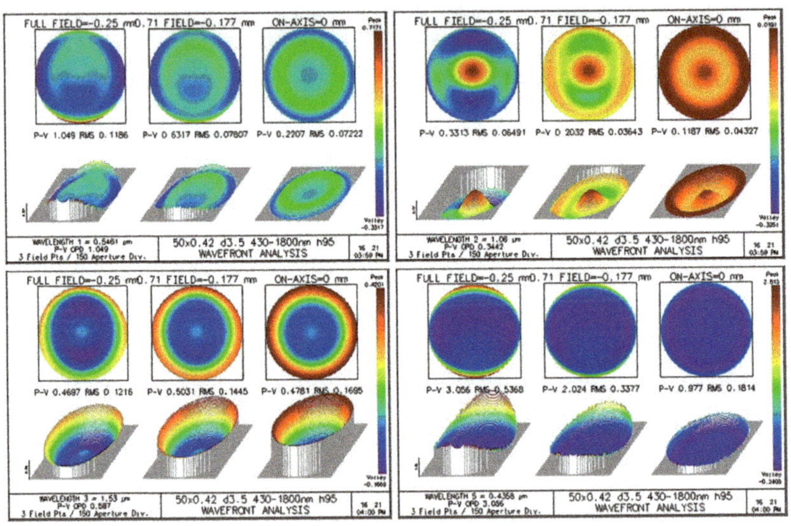

Figure 3.231 Optical design and graphs of aberration correction of a 50×/0.42 ($d = 3.5$ mm) VIS+NIR CCF plan poly-apochromatic objective ($F'_{ts} = 200$ mm; $h = 95$ mm).

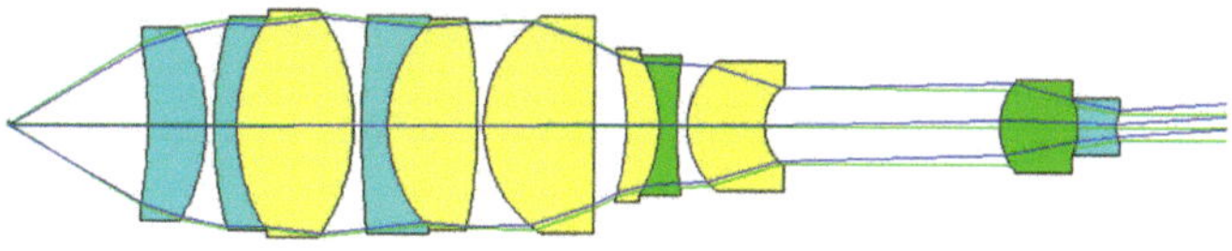

Lens: 100x0.50 d0 430-1800nm h95

Object num aper 0.500000 Image height 12.500000 Primary wavln 0.546070

SRF	RADIUS	THICKNESS	APERTURE RADIUS	GLASS	SPECIAL
OBJ	0.000000	1.0000e-20	0.124974	AIR	
AST	0.000000	11.000167	0.124974 AS	AIR	
2	-40.230000	4.500000	6.200000	H-ZLAF3	
3	-14.622000	0.600000	7.310000	AIR	
4	29.000000	1.800000	8.110000	H-LAK3	
5	15.000000	9.000000	8.100000	CAF2	
6	-15.000000	0.600000	8.620000	AIR	
7	81.800000	2.000000	8.230000	H-LAK3 P	
8	11.092000	7.000000	7.740000	CAF2 P	
9	-36.130000	0.500000	8.000000	AIR	
10	9.720000	8.500000	8.190000	CAF2 P	
11	0.000000	2.500000	6.700000	AIR	
12	-19.810000	2.600000	5.760000	CAF2 P	
13	-10.000000	1.400000	5.290000	ZF6	
14	42.482000	0.800000	5.040000	AIR	
15	6.399000	6.000000	4.900000	CAF2 P	
16	3.813000	18.300000	3.080000	AIR	
17	5.330000	6.000000	3.470000	ZF2	
18	-5.330000	3.000000	2.150000	H-ZLAF3 P	
19	2.710000	200.000000	1.310000	AIR	
20	-128.703000	3.700000	13.630000	H-ZLAF55D	
21	-152.600000	2.000000	13.940000	AIR	
22	148.130000	3.700000	14.190000	H-LAF50A	
23	89.840000	13.000000	14.190000	H-FK61	
24	-56.360000	3.700000	14.480000	LZ_CTK12	
25	-96.380000	199.991416 S	14.730000	AIR	
IMS	0.000000	0.000000	12.500000		

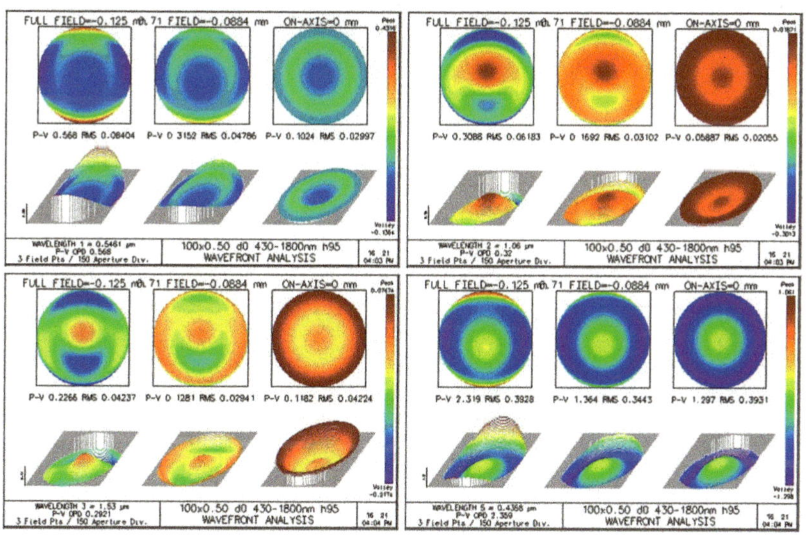

Figure 3.232 Optical design and graphs of aberration correction of a 100×/0.50 VIS+NIR CCF plan poly-apochromatic objective ($F'_{ts} = 200$ mm; $h = 95$ mm).

Table 3.11 Main technical parameters and basic optical layout of VIS+NIR objectives ($\lambda = 546.07$ nm; 430–1800 nm).

Magnification	*NA*	*WD* (mm)	*F′* (mm)	*R* (μm)	*DF* (μm)	*FOV* on object (mm)	*FOV* on image (mm)	Optical design at figure no.	The principal optical layout
20×	0.30	20	10	1.11	3.70	1.25	25	3.229	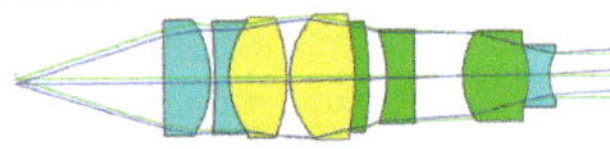
50×	0.42	16	4	0.79	1.89	0.50	25	3.230	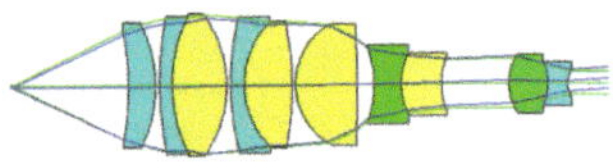
50×	0.42	15.3 $d = 3.5$	4	0.79	1.89	0.50	25	3.231	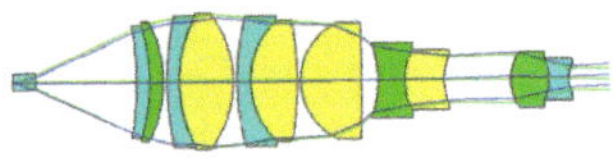
100×	0.50	8.7	2	0.67	1.33	0.25	25	3.232	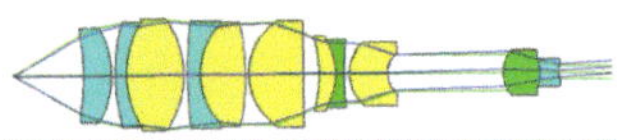

Figure 3.233 Two VIS+NIR objectives ($\lambda = 546.07$ nm; 430–1800 nm).

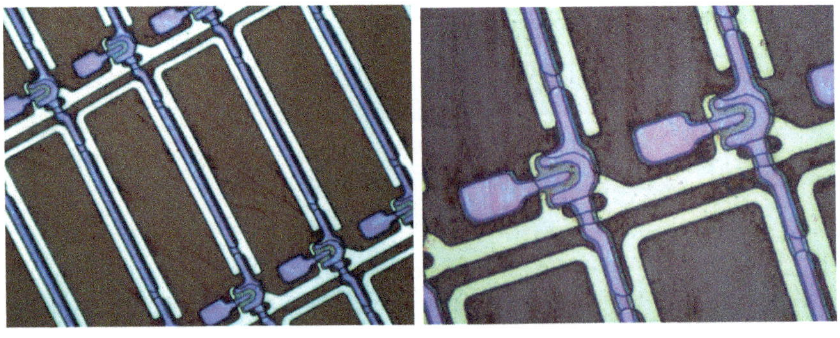

Figure 3.234 Photographs of a real object made by 20× and 50× objectives ($\lambda = 546.07$ nm; 430–1800 nm).

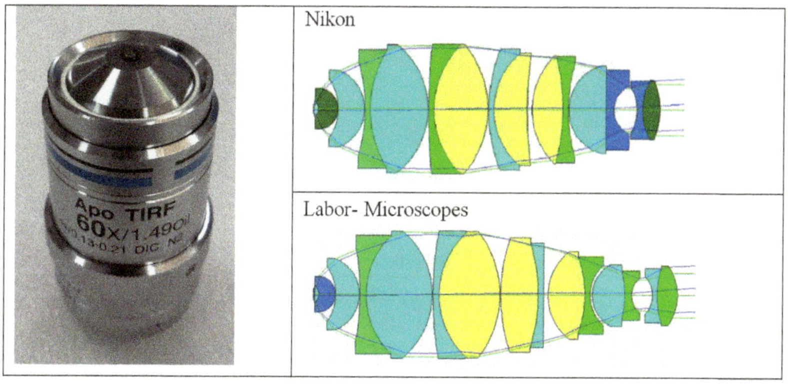

Figure 3.235 Picture and optical schemes of 60×/1.49 oil immersion objectives.

correction of some of these objectives, which have plan poly-apochromatic aberration correction in a spectral range of 365 to 1100 nm.

Table 3.12 shows the main technical parameters of these objectives.

3.10.5 Plan multi-apochromatic aberration correction objectives (45-mm parfocal distance)

If we include only crystals in the optical design, we can expand the pass band of the spectral transmission to the near UV region. However, achieving apochromatic aberration correction is rather difficult for engineers. Previously, old Soviet objectives were manufactured that had apochromatization in the spectral range of 250 to 800 nm [15]. But some researchers need optical components apochromatized in an even wider spectral range. For example, if we talk about the short-wavelength region of the NUV, in some cases, it is necessary to ensure focusing of laser radiation at wavelengths of 248, 222, 213, and even 193 nm. Or it is necessary to carry out photometric measurements within the wider range. However, there may be even wider requirements for the operating spectral range of optical components. Such components should also allow focusing for the NIR region, for example, for lasers operating at wavelengths of 1315, 1523, and even 2010 and 2080 nm. Thus, the corresponding apochromatization of optical components (the objective, first and foremost) can ensure operation in the extended spectral range from NUV to NIR inclusive without refocusing. To confirm the fundamental possibility of developing and manufacturing such components, optical designs of objectives that have plan multi-apochromatic aberration correction can be made. Figures 3.313–3.316 show the design parameters and graphs of aberration correction of some of these objectives.

Figure 3.317 is a photograph of the 10×/0.20 CCF plan multi-apochromatic objective.

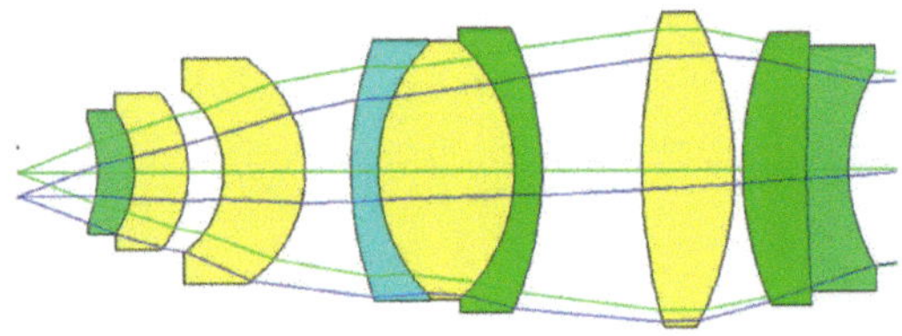

Lens: 10x0.35 PlanPolyApo d=0 Finity			Efl 14.46		
Object num aper 0.350000	Image height 12.500000		Primary wavln 0.546070		
SRF	**RADIUS**	**THICKNESS**	**APERTURE RADIUS**	**GLASS**	**SPECIAL**
OBJ	0.000000	1.0000e-20	1.248970	AIR	
AST	0.000000	4.501361	1.248970 AS	AIR	
2	-6.702200	2.000000	2.710000	H-ZLAF69	
3	-5.750000	3.000000	3.320000	CAF2	
4	-6.516000	2.000000	4.170000	AIR	
5	-5.297000	4.500000	4.350000	CAF2 P	
6	-7.244000	2.500000	6.000000	AIR	
7	19.724000	1.500000	6.970000	LZ_CTK12	
8	9.480000	7.500000	6.750000	CAF2 P	
9	-9.480000	1.500000	6.910000	H-ZF6	
10	-18.232000	5.500000	7.600000	AIR	
11	31.340000	5.000000	8.400000	CAF2 P	
12	-18.232000	0.480000	8.370000	AIR	
13	18.000000	3.500000	7.270000	H-ZF6 P	
14	112.910000	2.300000	6.520000	H-ZLAF69 P	
15	10.068000	149.217037 S	5.320000	AIR	
IMS	0.000000	0.000000	12.500000 S		

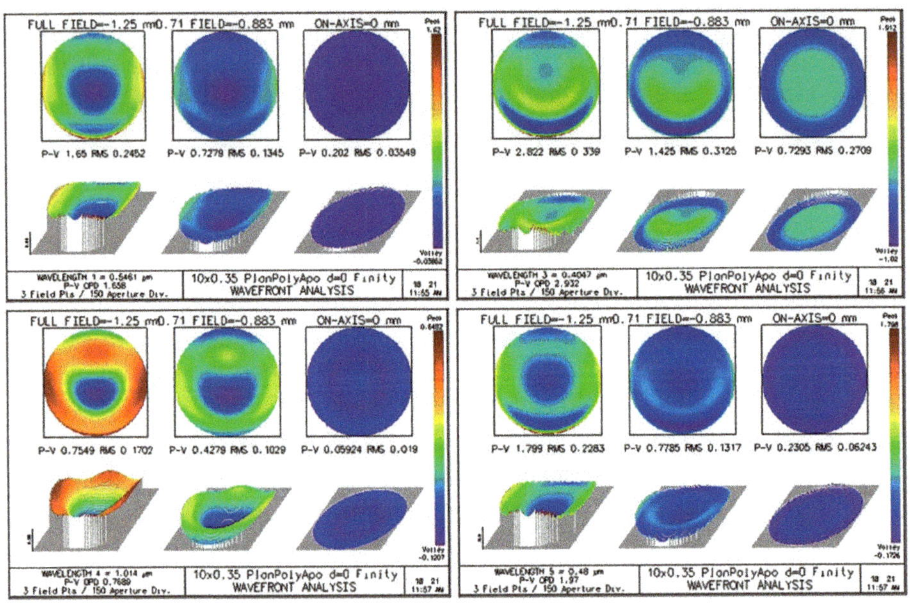

Figure 3.236 Optical design and graphs of aberration correction of a 10×/0.35 finite CCF plan poly-apochromatic objective ($h = 45$ mm).

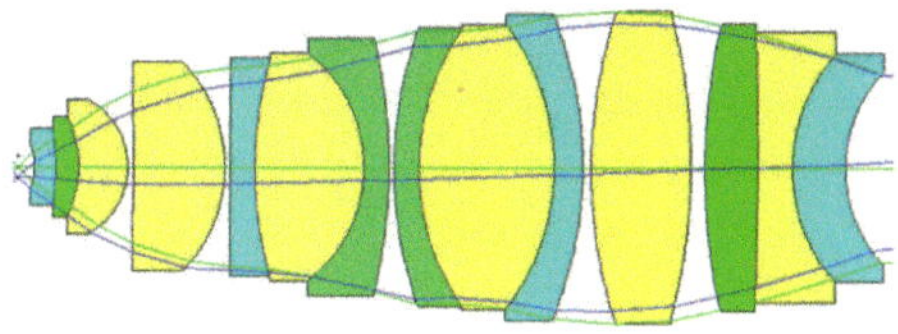

Lens: 20x0.70 PlanPolyApo d=0			Efl 7.52		
Object num aper	0.700000	Image height 11.000000	Primary wavln 0.546070		
SRF	**RADIUS**	**THICKNESS**	**APERTURE RADIUS**	**GLASS**	**SPECIAL**
OBJ	0.000000	1.0000e-20	0.550002	AIR	
AST	0.000000	0.920000	0.550002 AS	AIR	
2	-3.000000	1.200000	1.200000	LZ_CTK19	
3	-13.140000	1.300000	2.120000	LZ_TF5	
4	-6.329000	2.800000	2.730000	CAF2	
5	-4.093000	0.300000	3.640000	AIR	
6	-263.600000	5.200000	4.820000	CAF2 P	
7	-8.000000	0.300000	5.730000	AIR	
8	0.000000	1.400000	5.900000	LZ_CTK19 P	
9	20.600000	6.200000	5.990000	CAF2 P	
10	-7.686000	1.400000	6.220000	LZ_F1	
11	-28.310000	0.300000	7.060000	AIR	
12	22.180000	1.400000	7.620000	LZ_F1 P	
13	13.305000	7.600000	7.610000	CAF2 P	
14	-12.560000	1.600000	7.820000	LZ_CTK19 P	
15	-21.470000	0.500000	8.370000	AIR	
16	27.160000	5.600000	8.570000	CAF2 P	
17	-31.340000	0.800000	8.390000	AIR	
18	35.050000	3.000000	7.830000	LZ_TF5 P	
19	-188.600000	2.000000	7.430000	CAF2 P	
20	9.160000	3.000000	6.230000	LZ_CTK19 P	
21	7.600000	148.041293 S	5.260000	AIR	
IMS	0.000000	0.000000	11.000000 S		

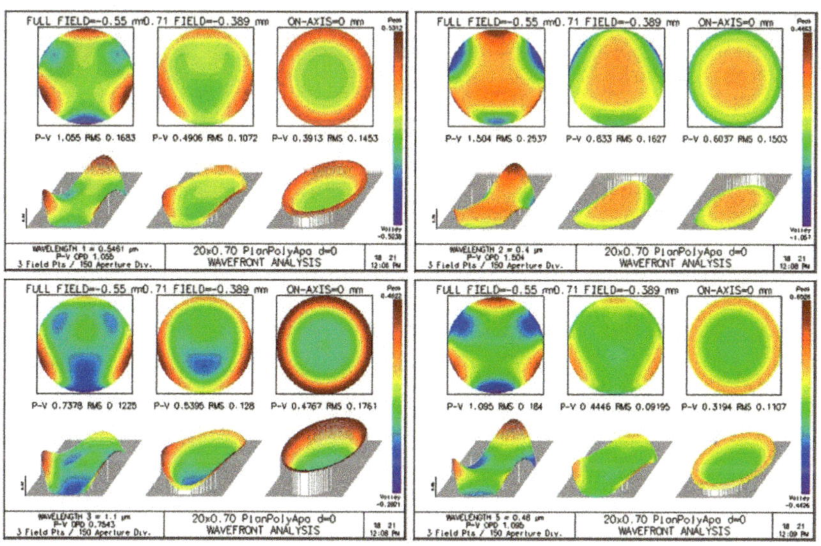

Figure 3.237 Optical design and graphs of aberration correction of a 20×/0.70 finite CCF plan poly-apochromatic objective ($h = 45$ mm).

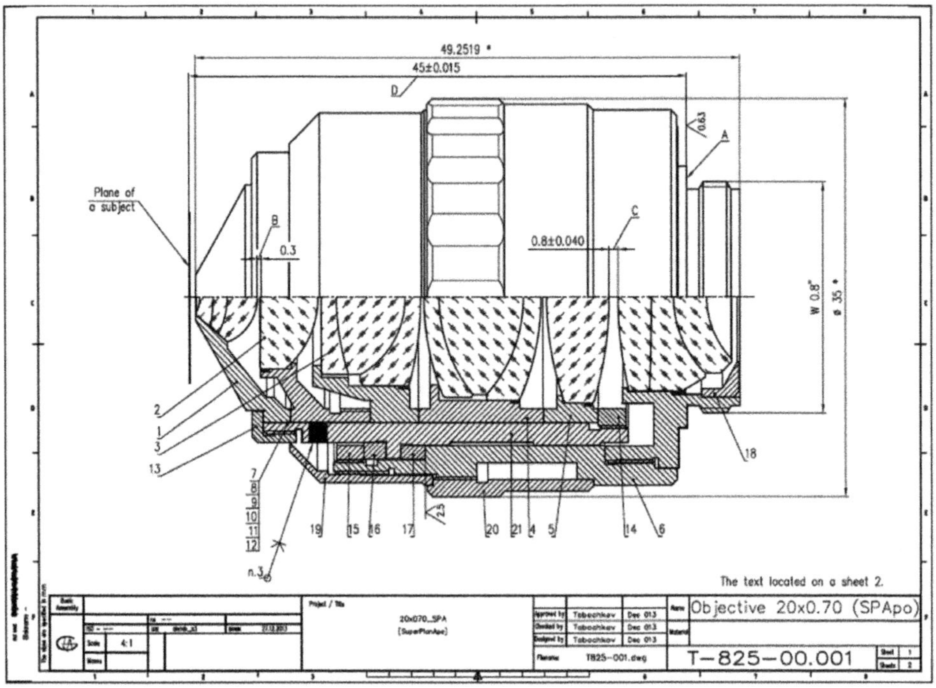

Figure 3.238 A drawing of a general mechanical design of a 20×/0.70 plan poly-apochromatic objective.

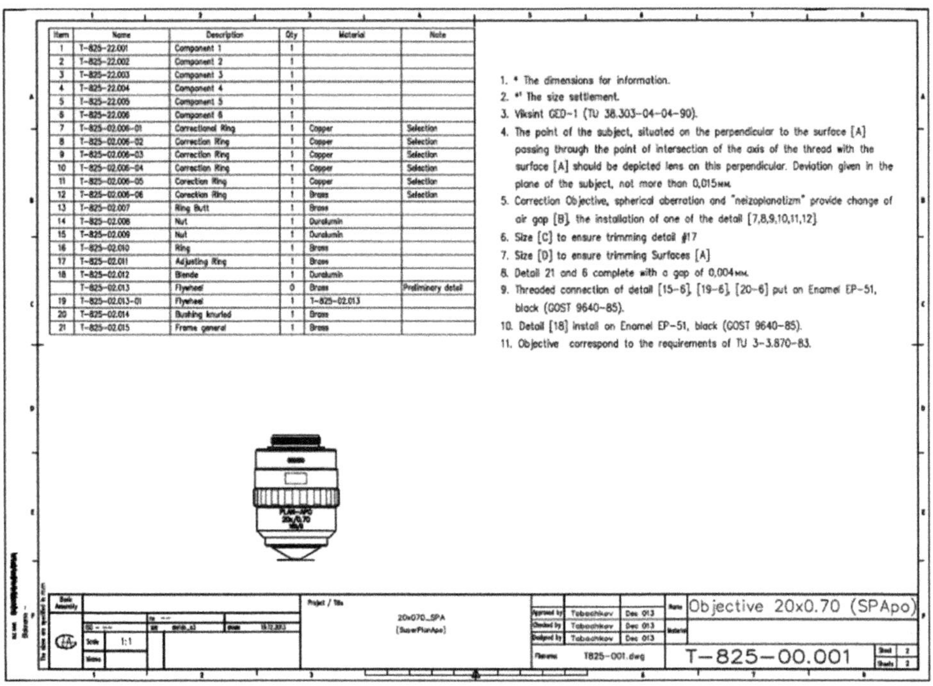

Item	Name	Description	Qty	Material	Note
1	T-825-22.001	Component 1	1		
2	T-825-22.002	Component 2	1		
3	T-825-22.003	Component 3	1		
4	T-825-22.004	Component 4	1		
5	T-825-22.005	Component 5	1		
6	T-825-22.006	Component 6	1		
7	T-825-02.006-01	Correctional Ring	1	Copper	Selection
8	T-825-02.006-02	Correction Ring	1	Copper	Selection
9	T-825-02.006-03	Correction Ring	1	Copper	Selection
10	T-825-02.006-04	Correction Ring	1	Copper	Selection
11	T-825-02.006-05	Corection Ring	1	Copper	Selection
12	T-825-02.006-06	Corection Ring	1	Brass	Selection
13	T-825-02.007	Ring Butt	1	Brass	
14	T-825-02.008	Nut	1	Duralumin	
15	T-825-02.009	Nut	1	Duralumin	
16	T-825-02.010	Ring	1	Brass	
17	T-825-02.011	Adjusting Ring	1	Brass	
18	T-825-02.012	Blende	1	Duralumin	
	T-825-02.013	Flywheel	0	Brass	Preliminary detail
19	T-825-02.013-01	Flywheel	1	T-825-02.013	
20	T-825-02.014	Bushing knurled	1	Brass	
21	T-825-02.015	Frame general	1	Brass	

1. * The dimensions for information.
2. *' The size settlement.
3. Viksint GED-1 (TU 38.303-04-04-90).
4. The point of the subject, situated on the perpendicular to the surface [A] passing through the point of intersection of the axis of the thread with the surface [A] should be depicted lens on this perpendicular. Deviation given in the plane of the subject, not more than 0,015мм.
5. Correction Objective, spherical aberration and "neizoplanatizm" provide change of air gap [B], the installation of one of the detail [7,8,9,10,11,12].
6. Size [C] to ensure trimming detail #17
7. Size [D] to ensure trimming Surfaces [A]
8. Detail 21 and 6 complete with a gap of 0,004мм.
9. Threaded connection of detail [15-6], [19-6], [20-6] put on Enamel EP-51, black (GOST 9640-85).
10. Detail [18] install on Enamel EP-51, black (GOST 9640-85).
11. Objective correspond to the requirements of TU 3-3.870-83.

Figure 3.239 Specifications.

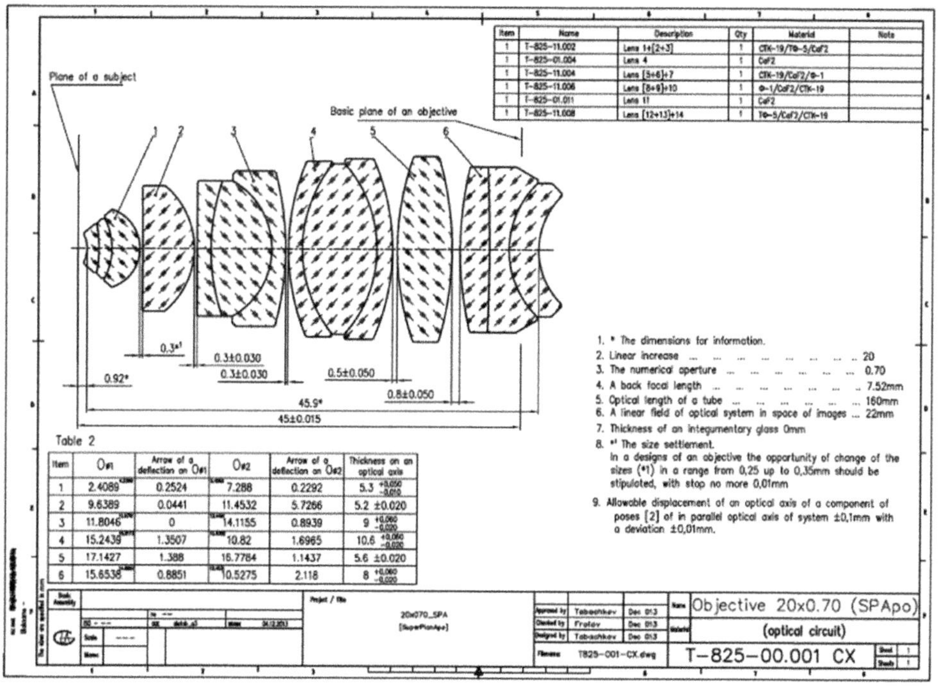

Figure 3.240 Principal optical layout of a 20×/0.70 plan poly-apochromatic objective.

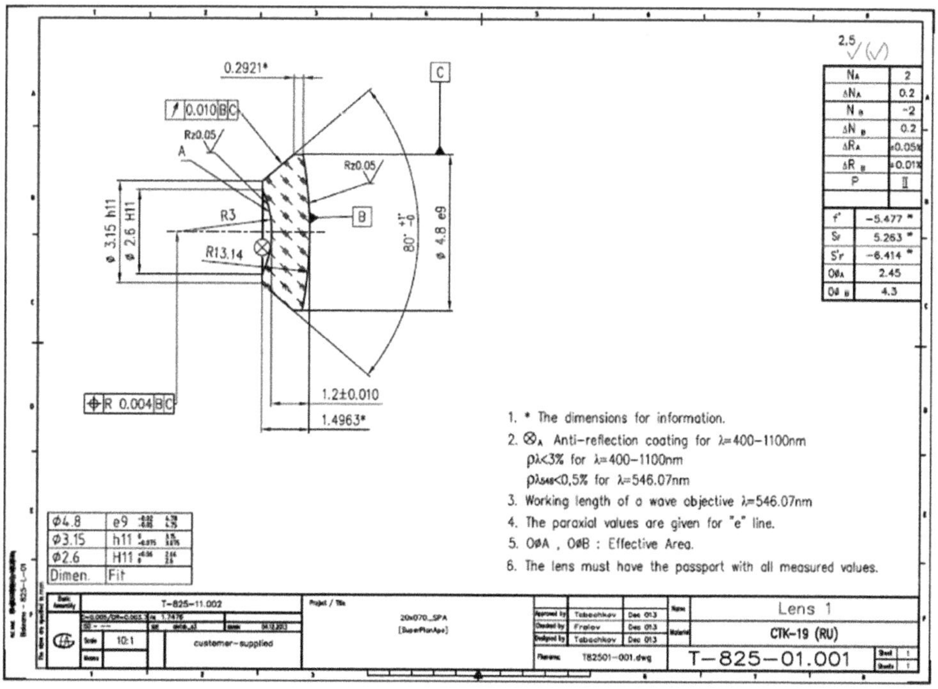

Figure 3.241 A drawing of the first lens.

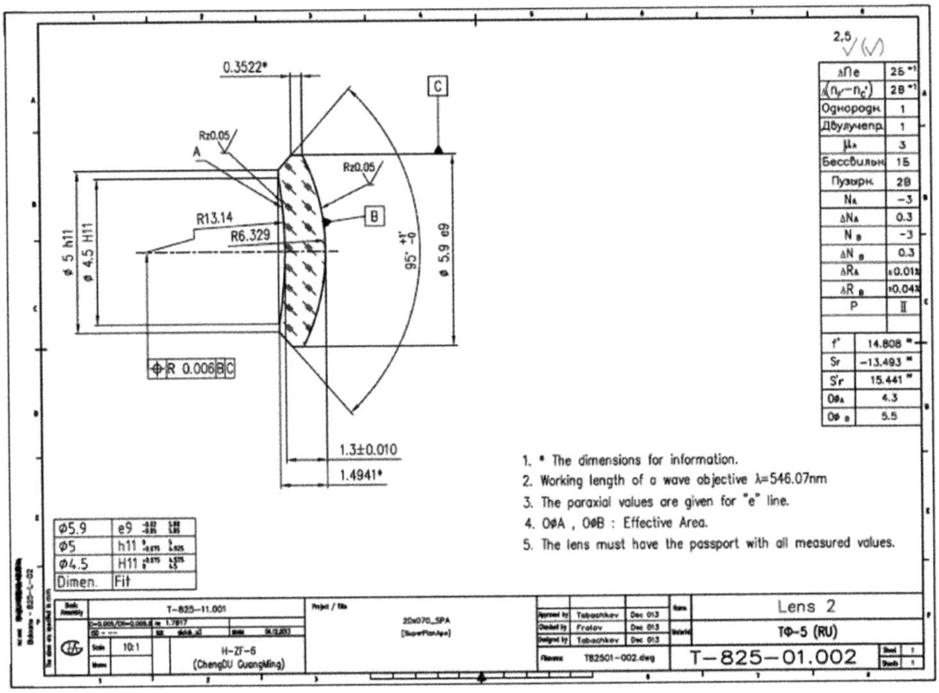

Figure 3.242 A drawing of the second lens.

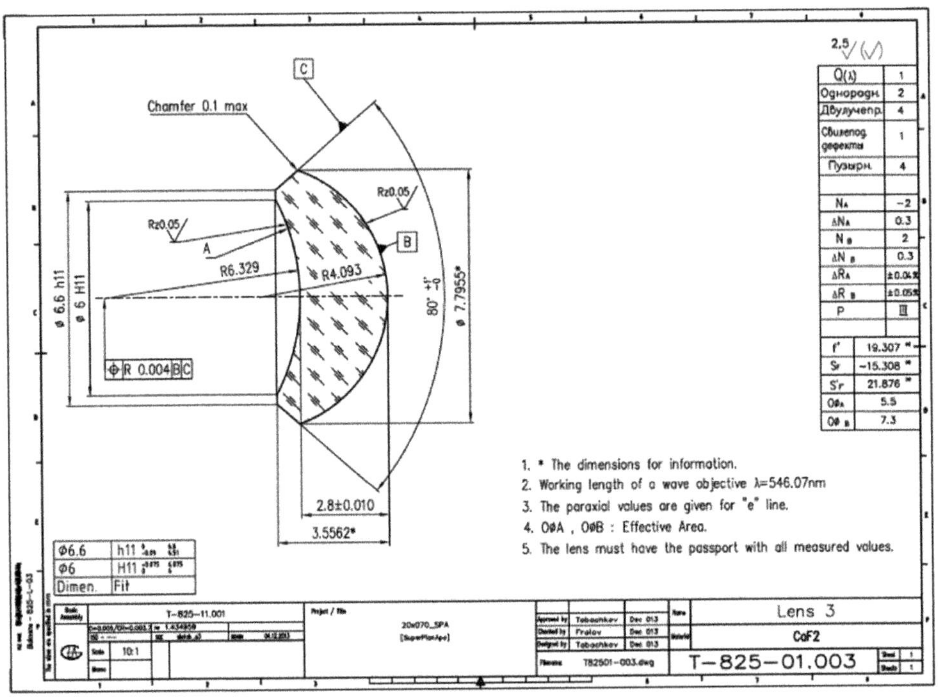

Figure 3.243 A drawing of the third lens.

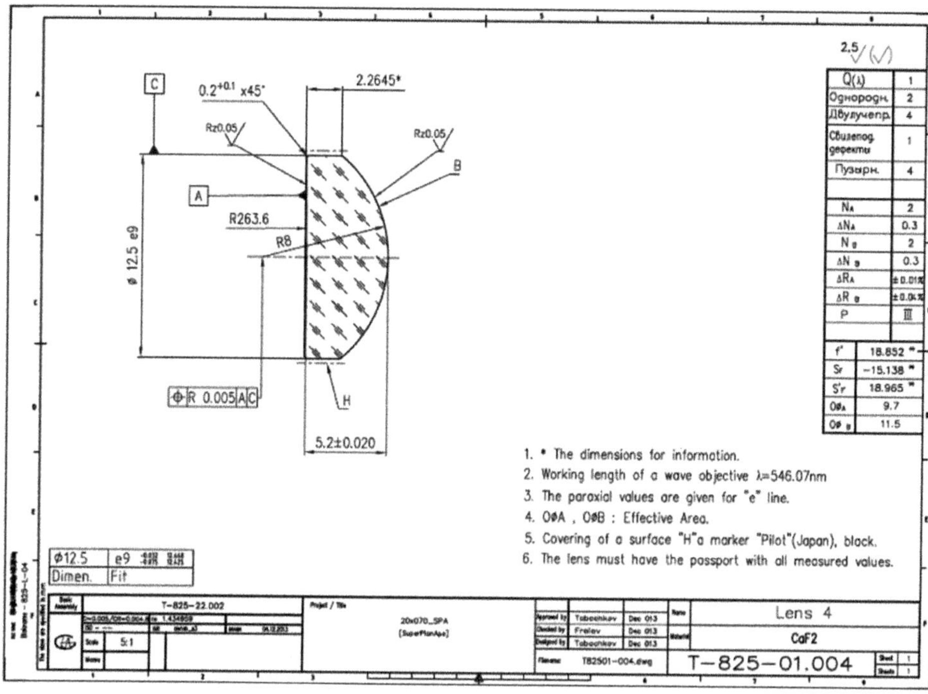

Figure 3.244 A drawing of the fourth lens.

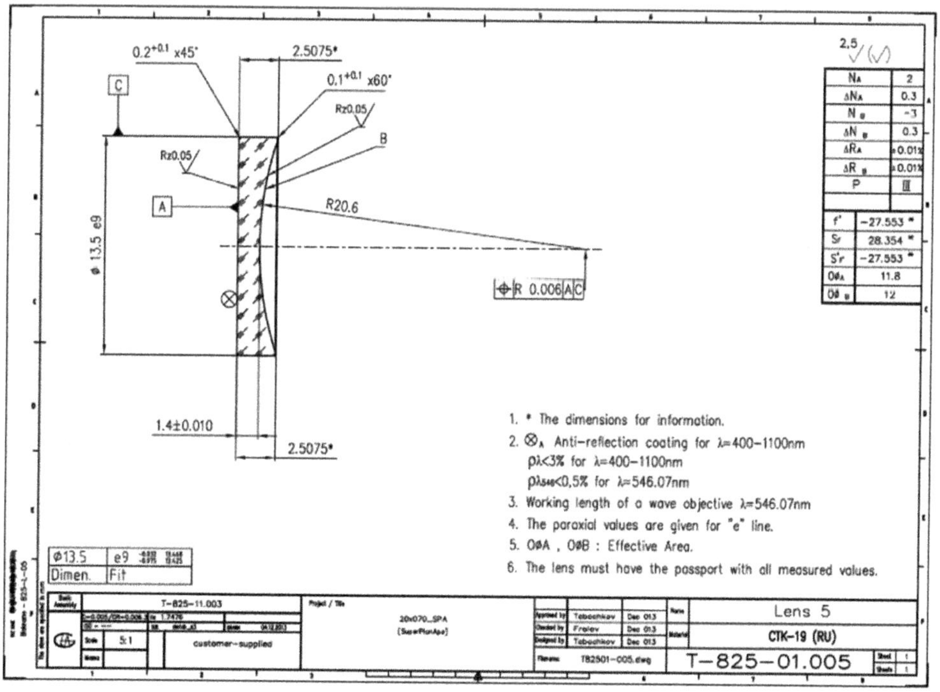

Figure 3.245 A drawing of the fifth lens.

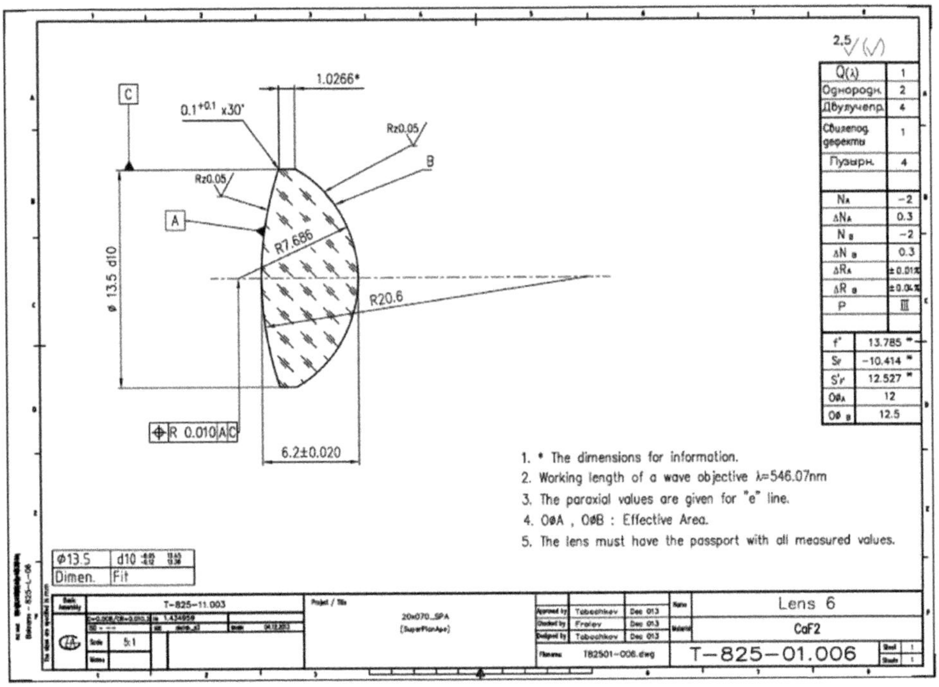

Figure 3.246 A drawing of the sixth lens.

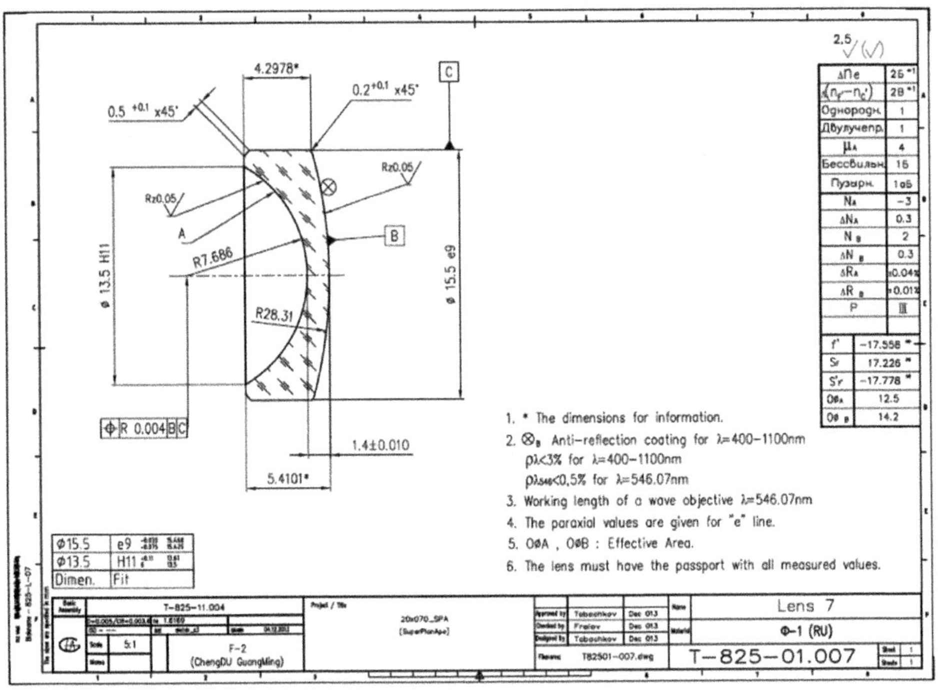

Figure 3.247 A drawing of the seventh lens.

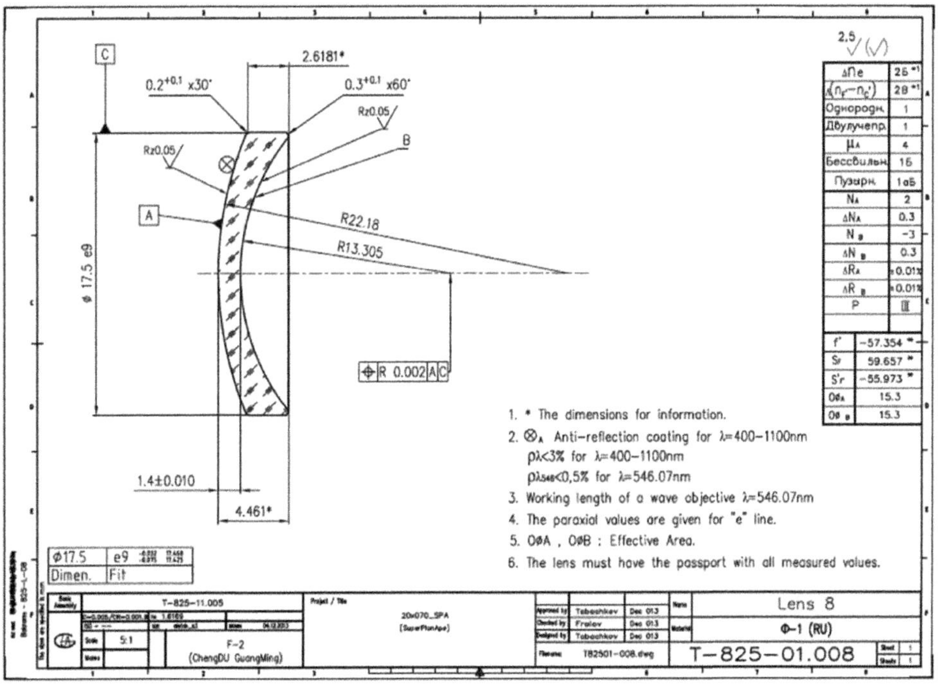

Figure 3.248 A drawing of the eighth lens.

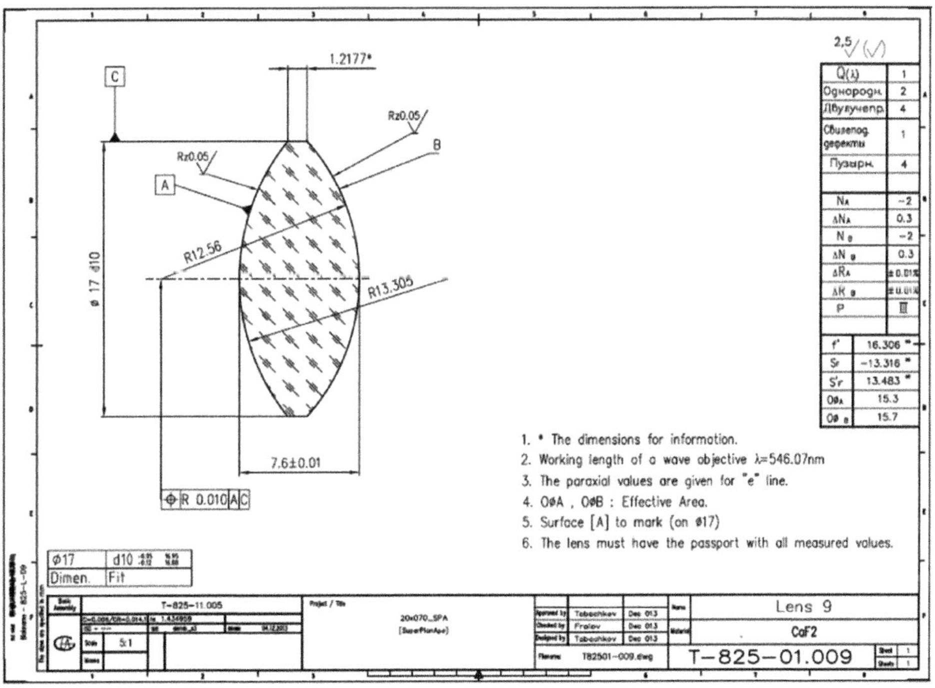

Figure 3.249 A drawing of the ninth lens.

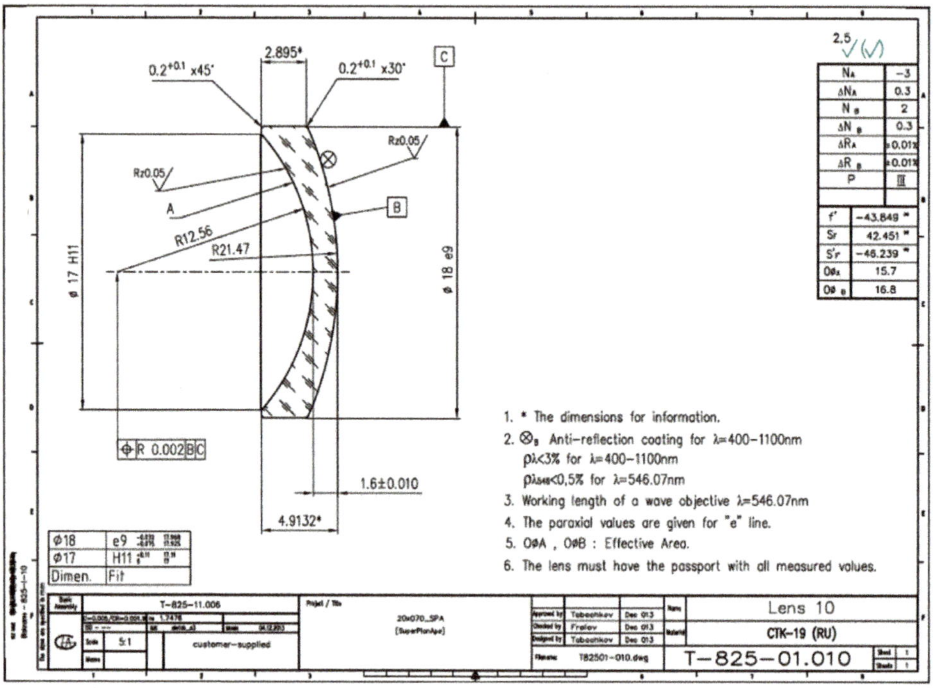

Figure 3.250 A drawing of the 10th lens.

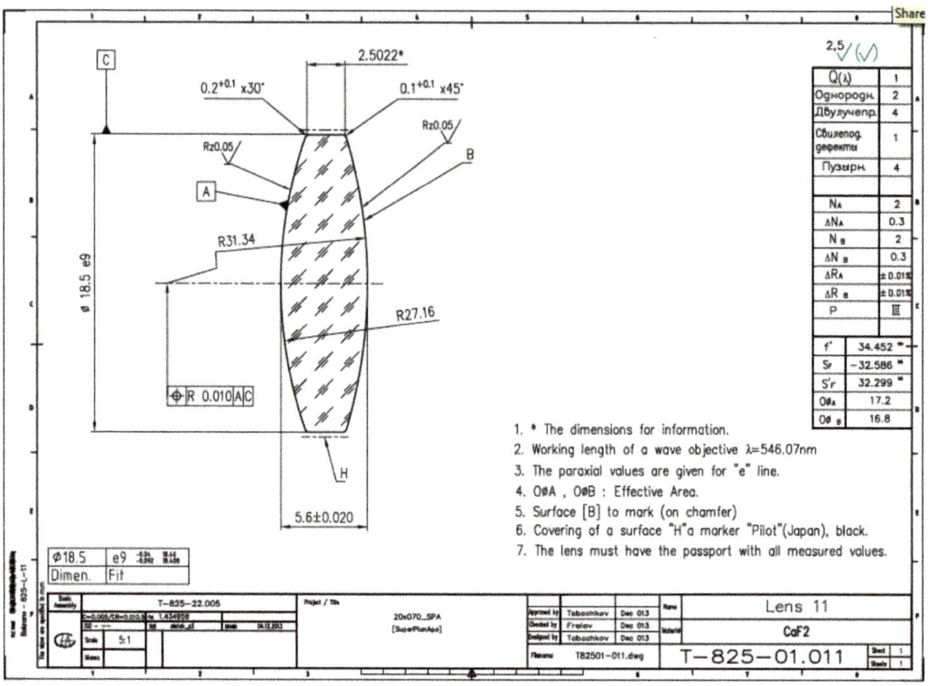

Figure 3.251 A drawing of the 11th lens.

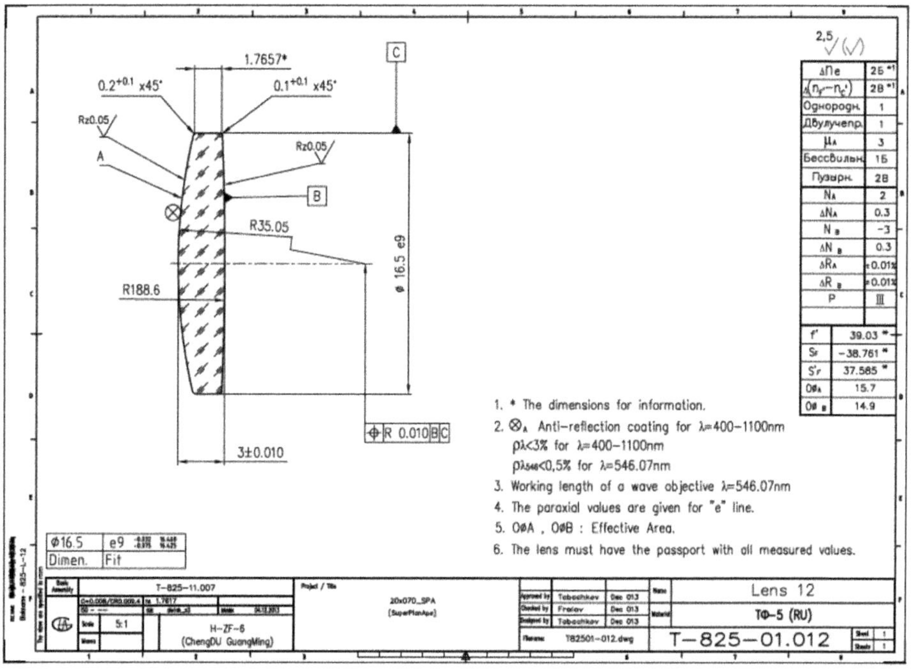

Figure 3.252 A drawing of the 12th lens.

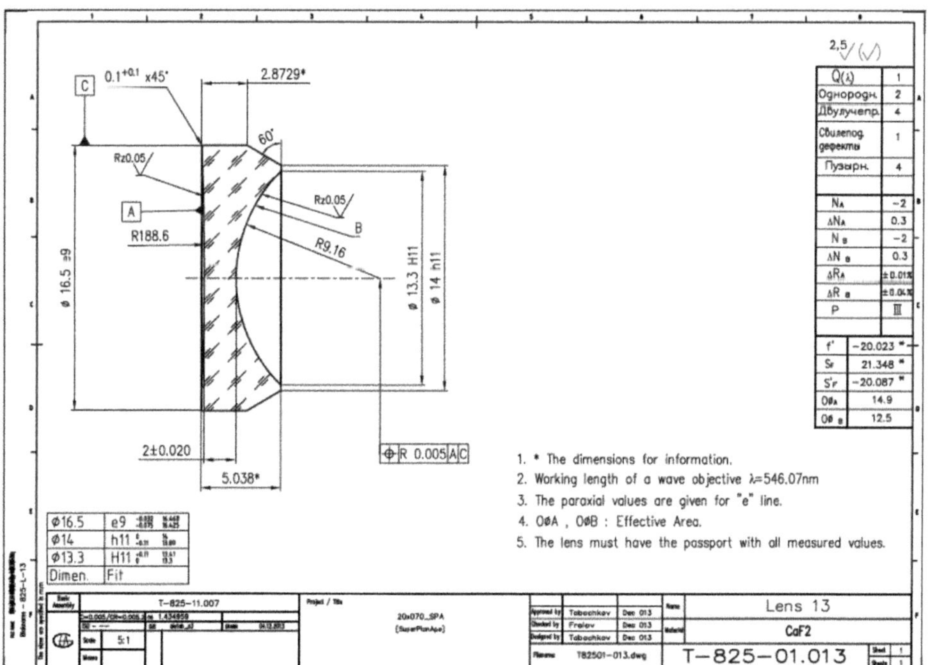

Figure 3.253 A drawing of the 13th lens.

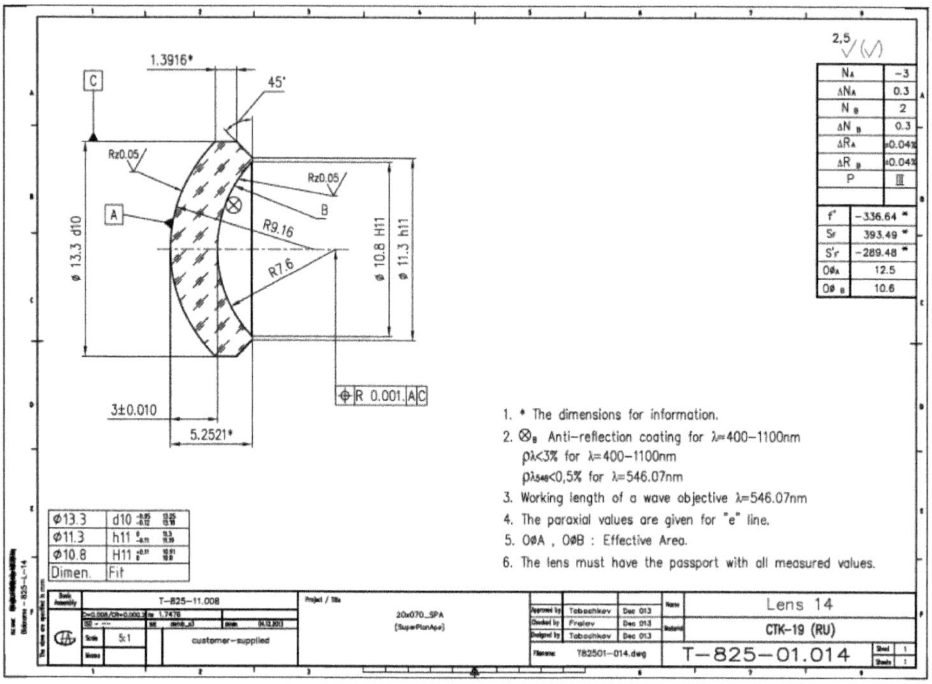

Figure 3.254 A drawing of the 14th lens.

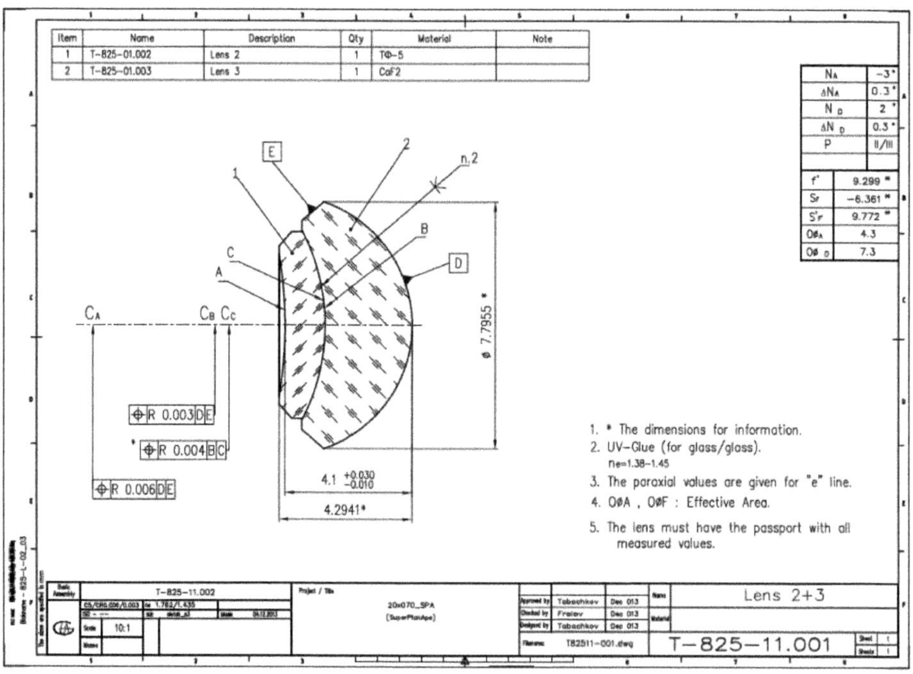

Figure 3.255 A drawing of the second and third lenses.

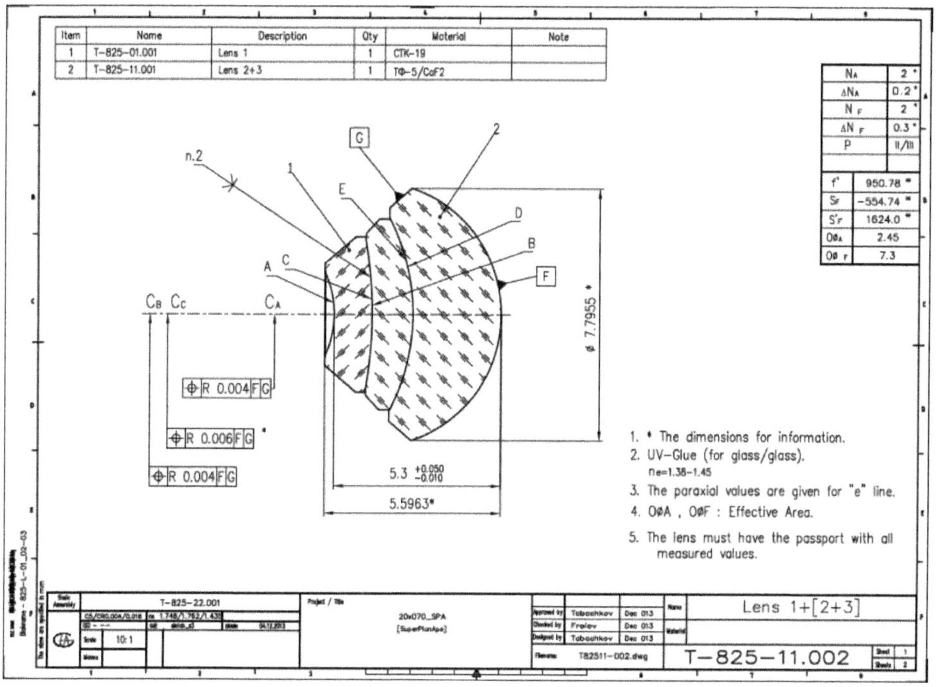

Figure 3.256 A drawing of the first, second, and third lenses (no. 1).

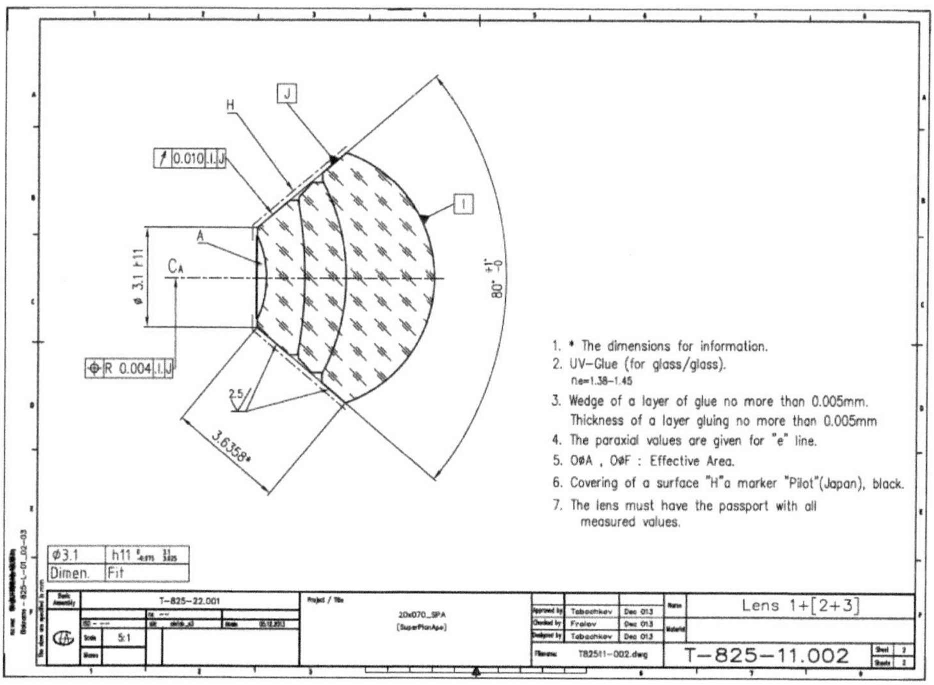

Figure 3.257 A drawing of the first, second, and third lenses (no. 2).

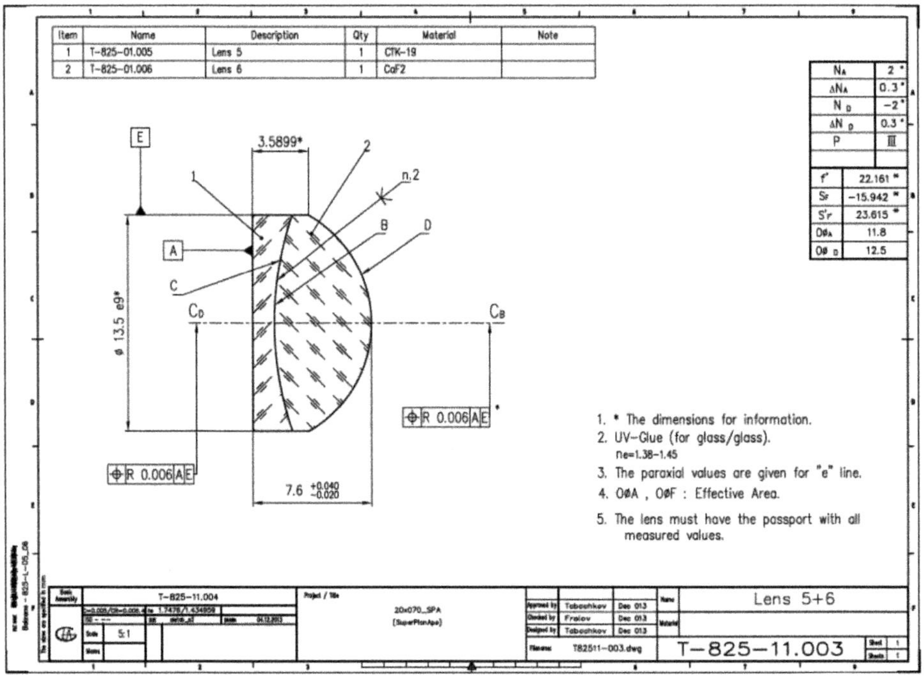

Figure 3.258 A drawing of the fifth and sixth lenses.

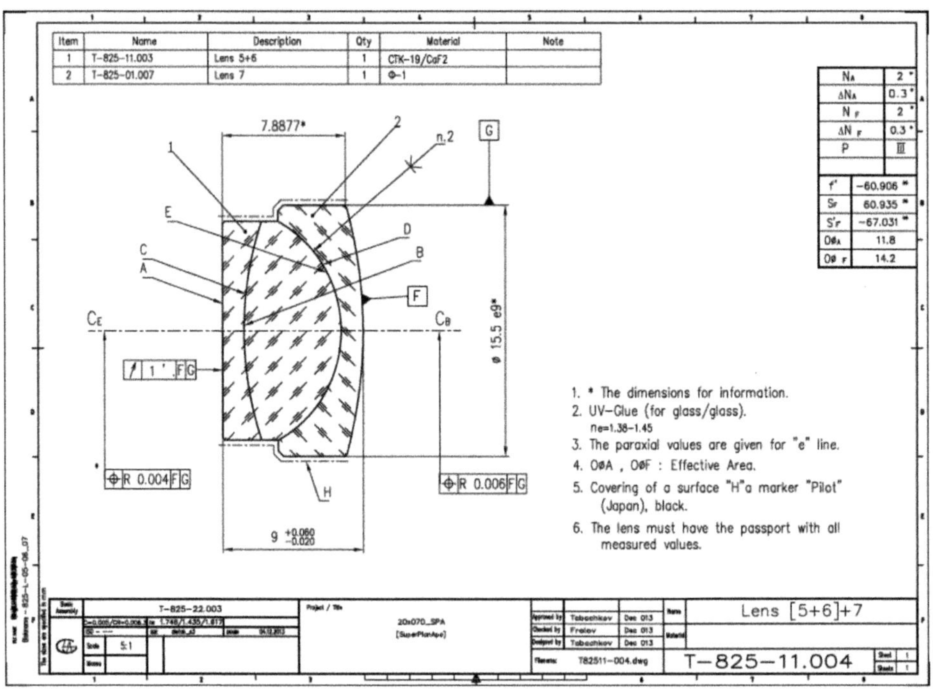

Figure 3.259 A drawing of the fifth, sixth, and seventh lenses.

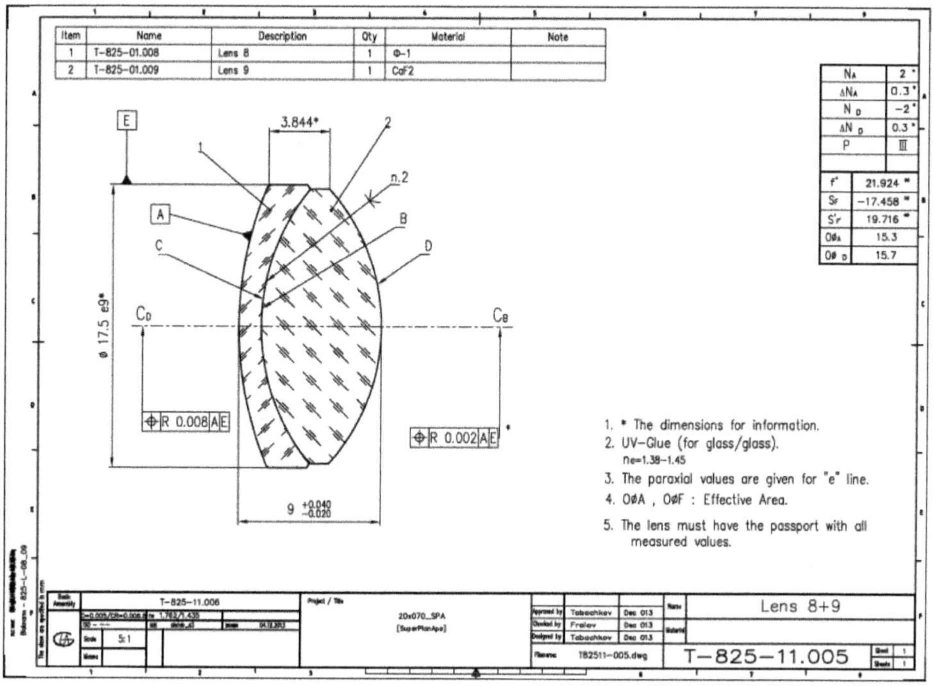

Figure 3.260 A drawing of the eighth and ninth lenses.

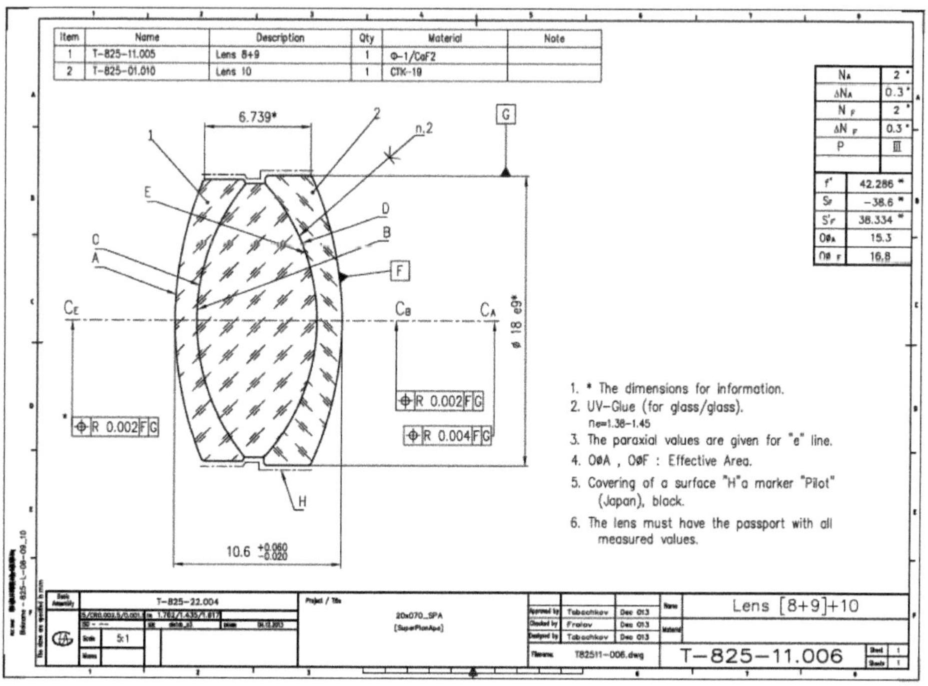

Figure 3.261 A drawing of the eighth, ninth, and 10th lenses.

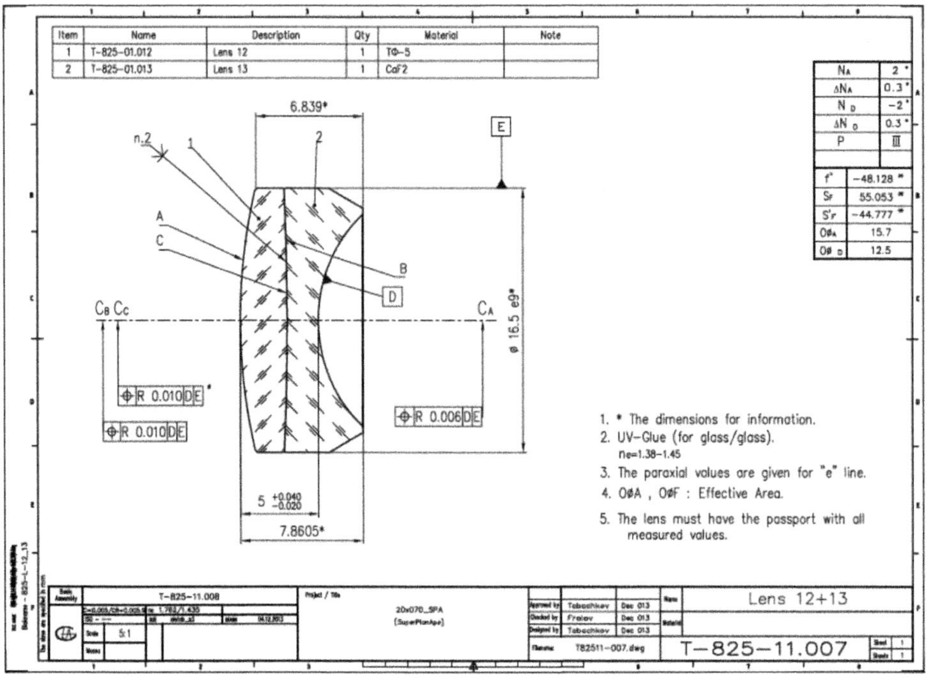

Figure 3.262 A drawing of the 12th and 13th lenses.

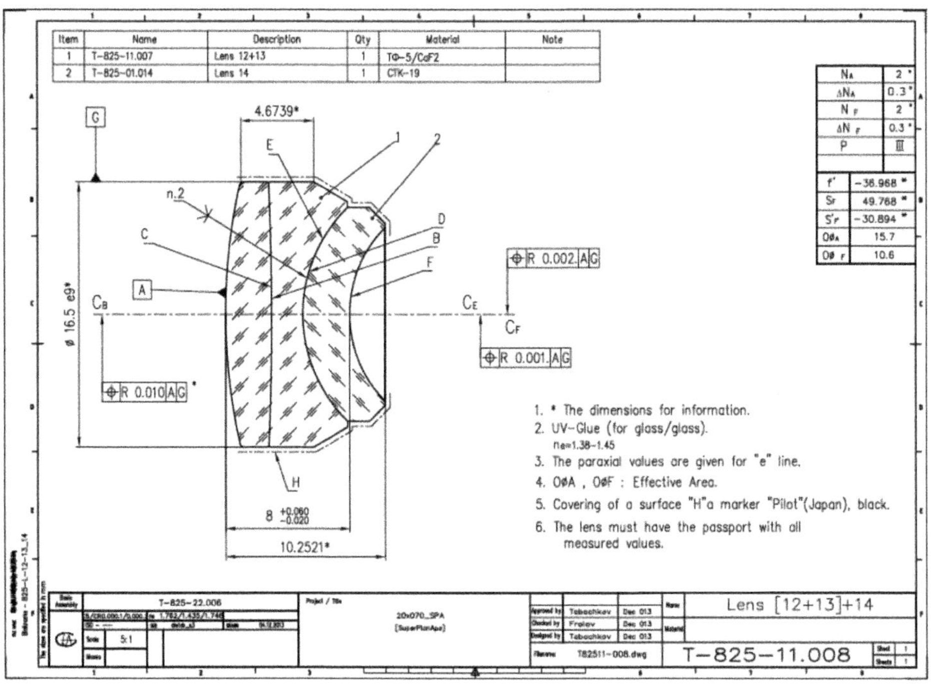

Figure 3.263 A drawing of the 12th, 13th, and 14th lenses.

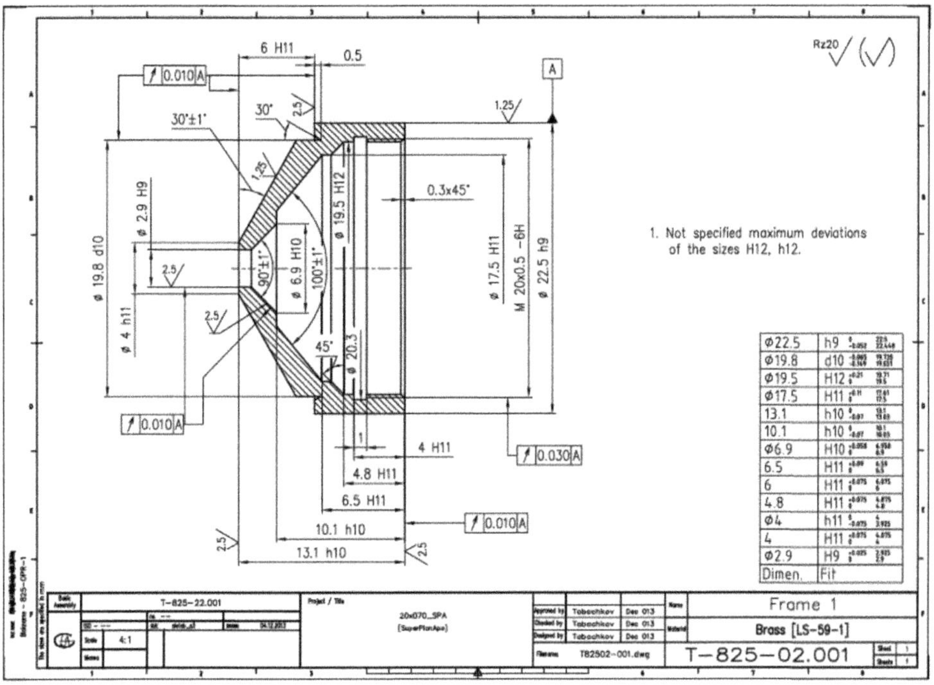

Figure 3.264 A drawing of the first frame.

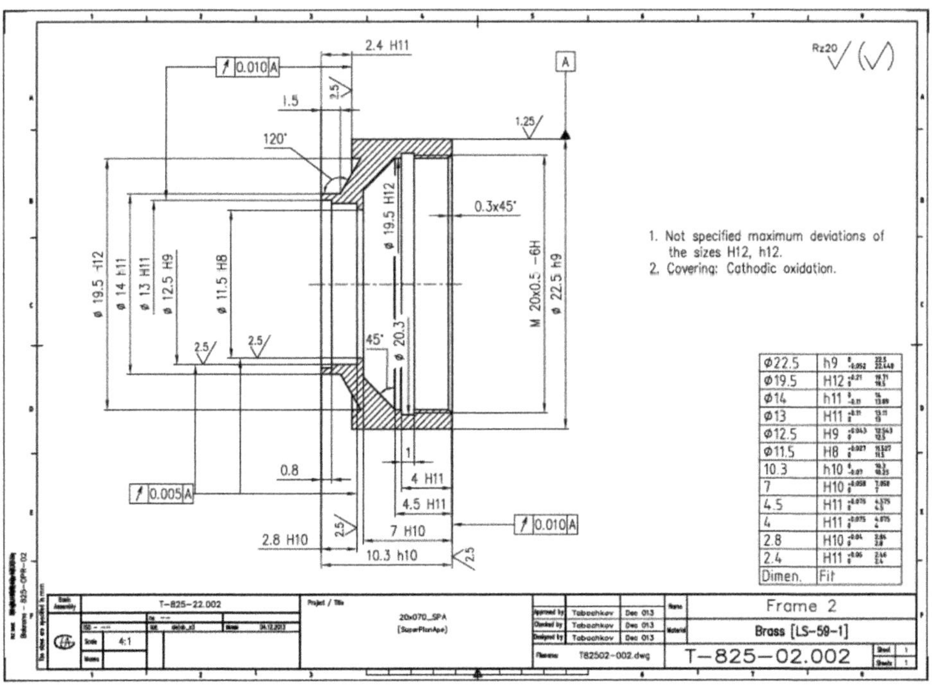

Figure 3.265 A drawing of the second frame.

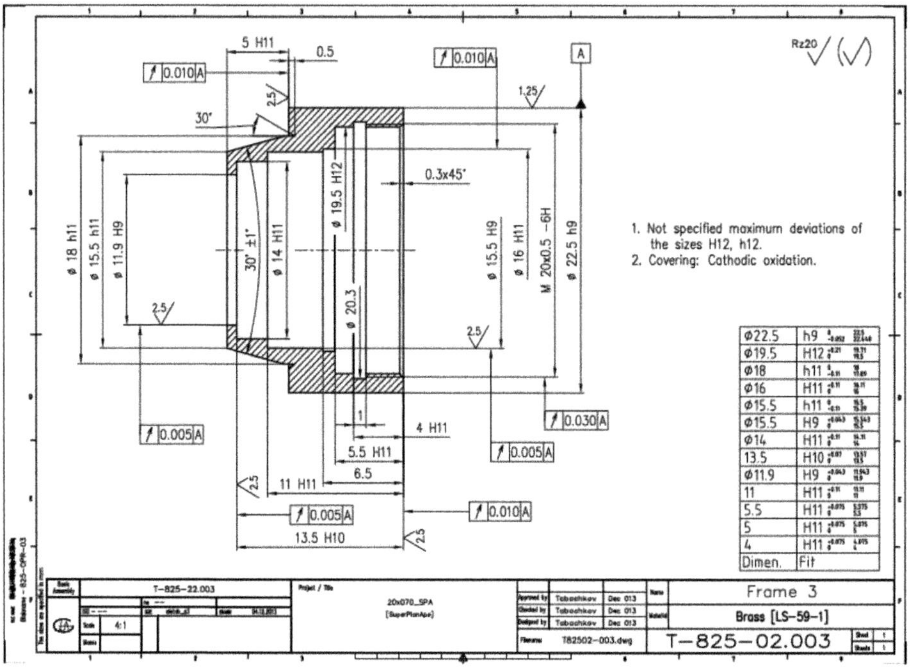

Figure 3.266 A drawing of the third frame.

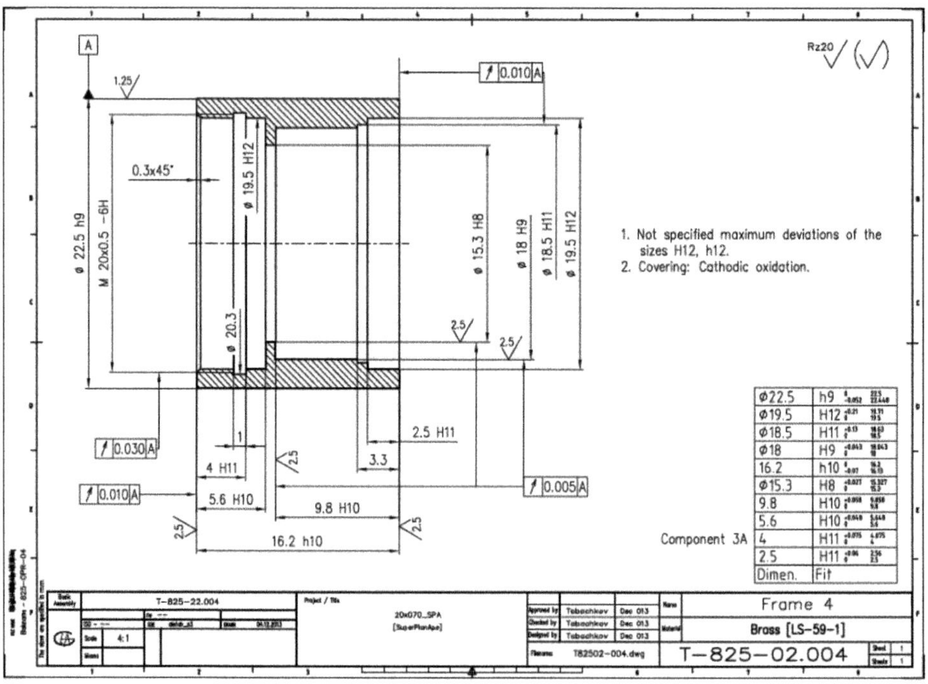

Figure 3.267 A drawing of the fourth frame (no. 1).

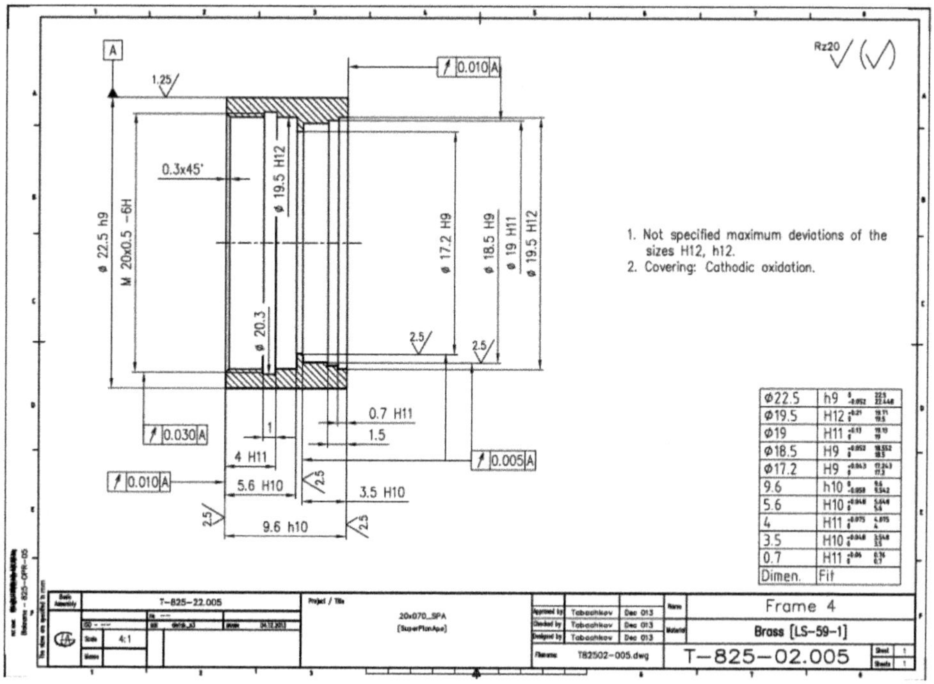

Figure 3.268 A drawing of the fourth frame (no. 2).

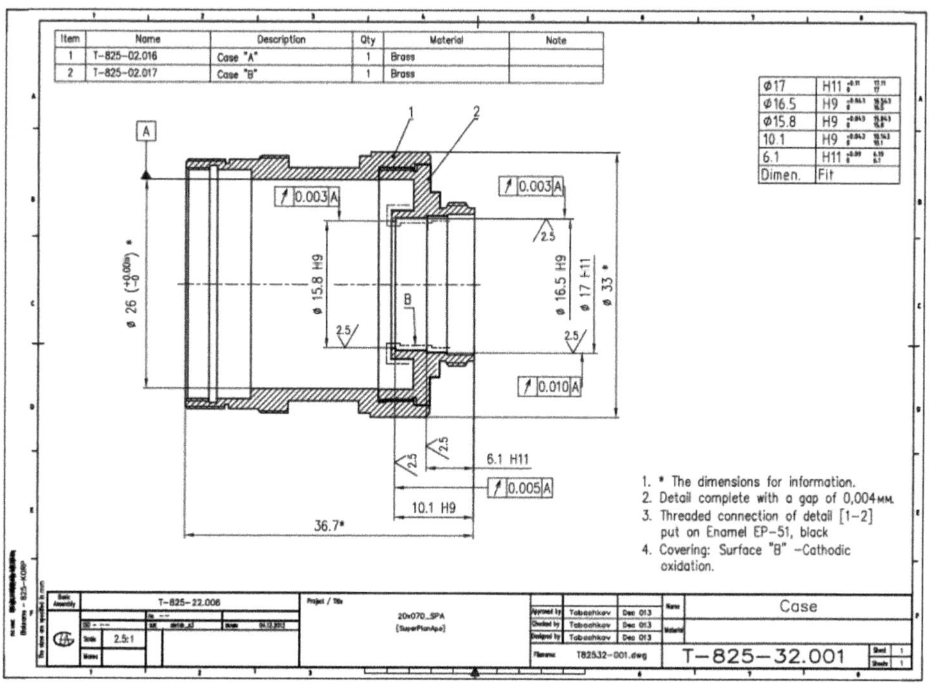

Figure 3.269 A drawing of the body (case).

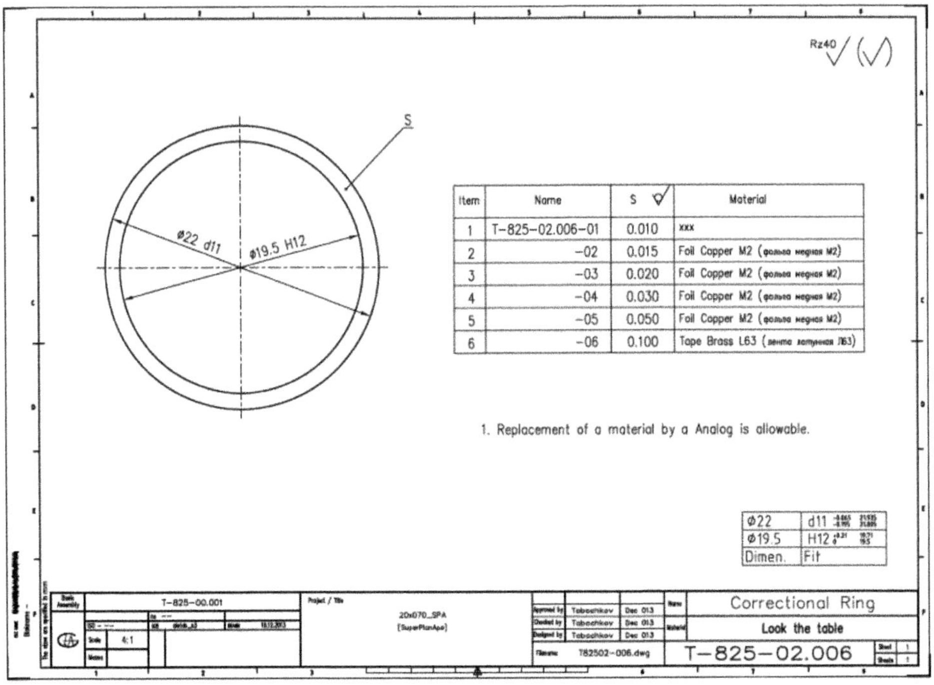

Item	Name	S	Material
1	T-825-02.006-01	0.010	xxx
2	-02	0.015	Foil Copper M2
3	-03	0.020	Foil Copper M2
4	-04	0.030	Foil Copper M2
5	-05	0.050	Foil Copper M2
6	-06	0.100	Tape Brass L63

Figure 3.270 A drawing of the correction ring.

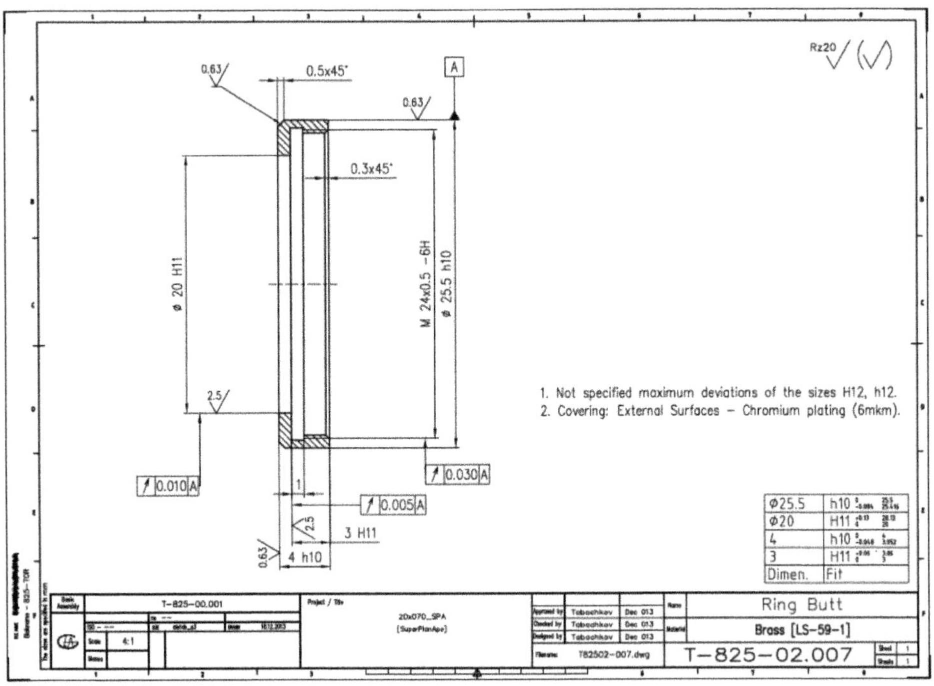

Figure 3.271 A drawing of the frame holder ring.

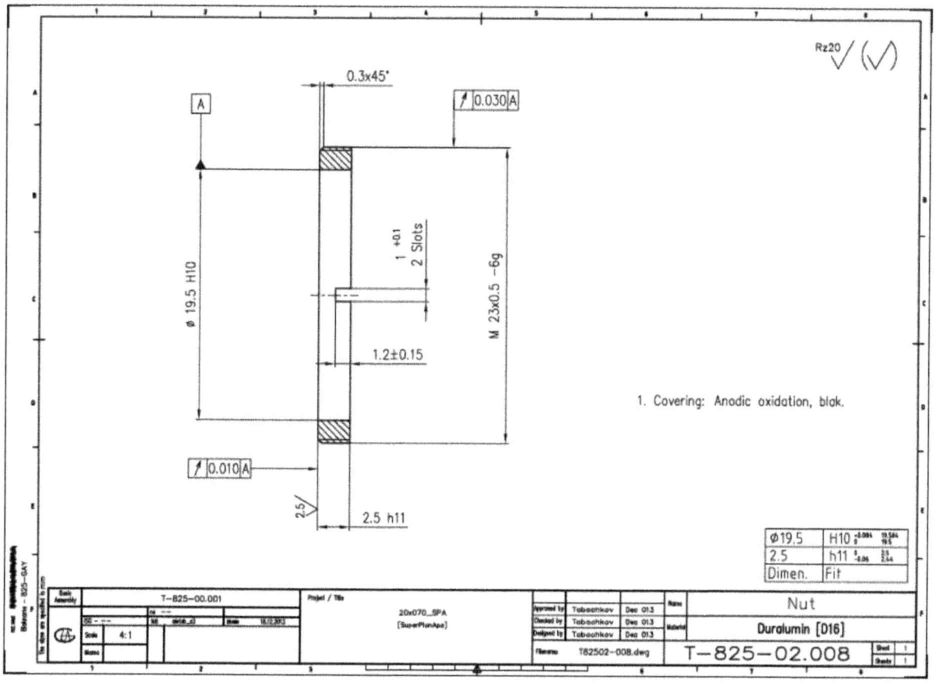

Figure 3.272 A drawing of the nut.

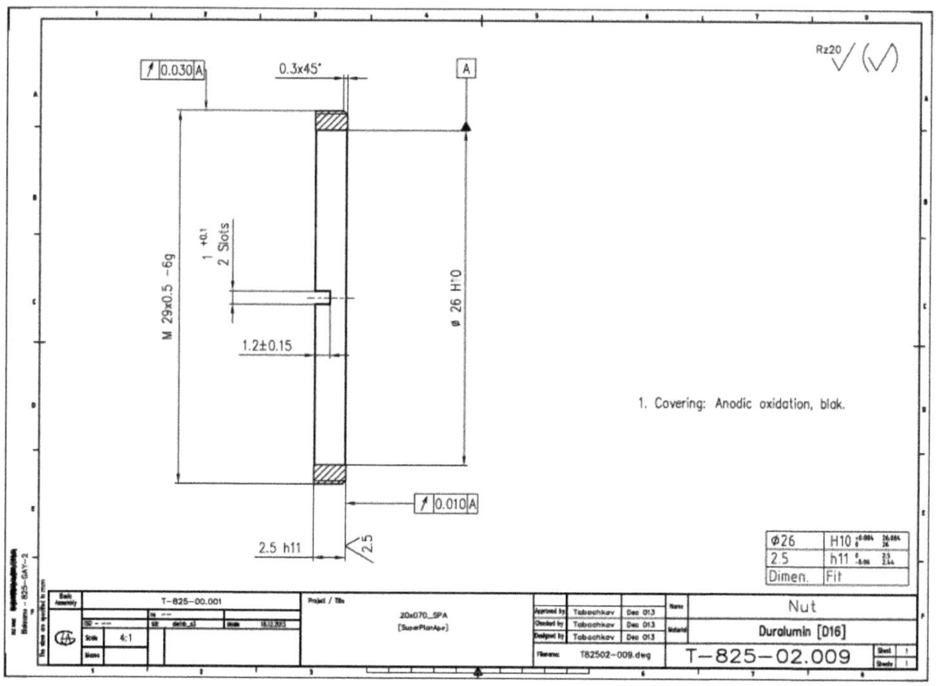

Figure 3.273 A drawing of the nut.

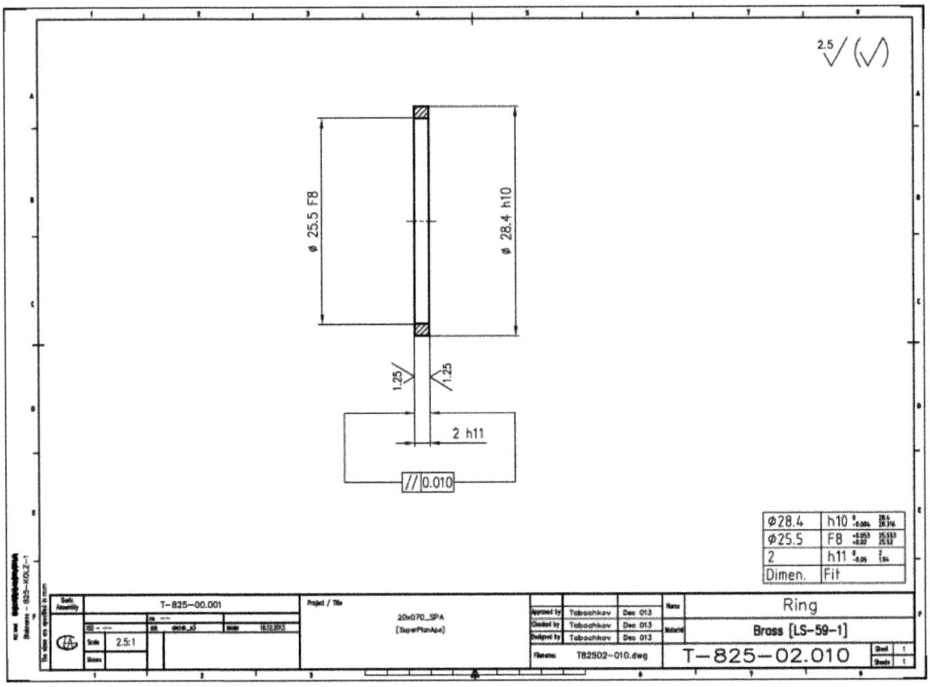

Figure 3.274 A drawing of the ring.

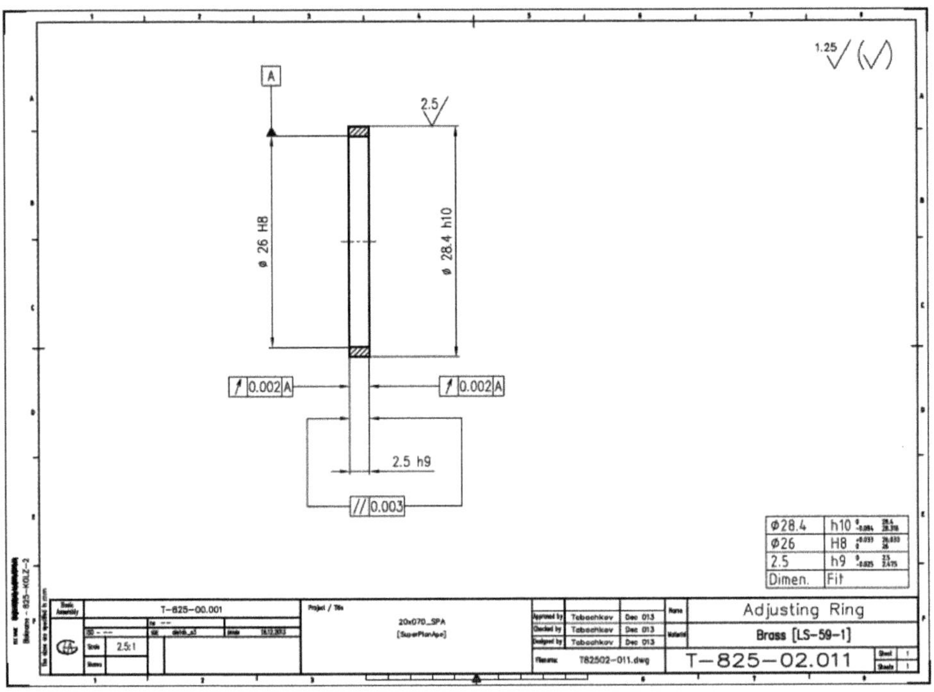

Figure 3.275 A drawing of the adjusting ring.

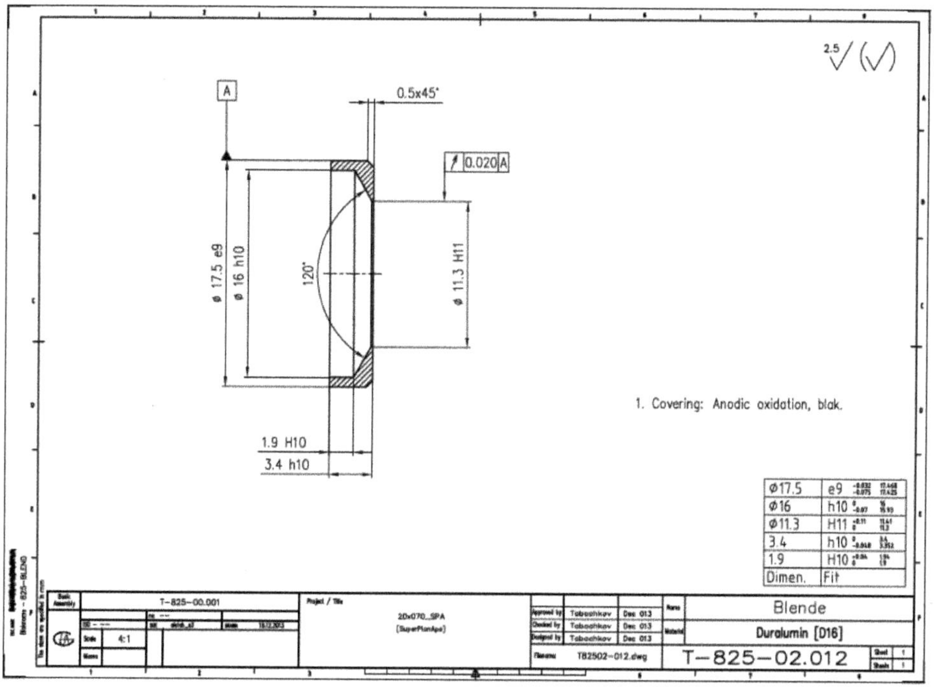

Figure 3.276 A drawing of the cover.

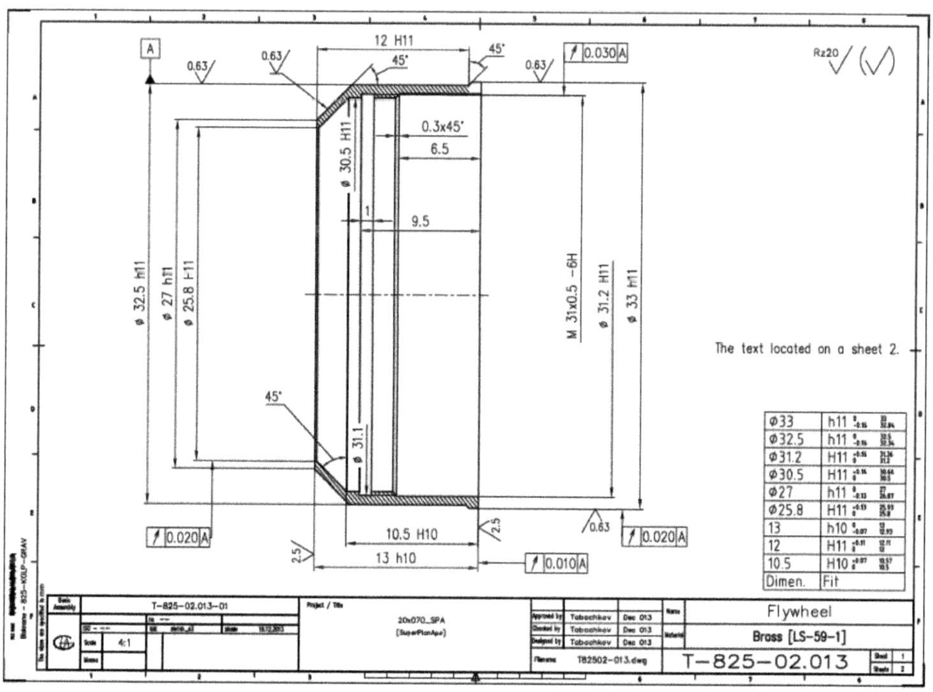

Figure 3.277 A drawing of the flywheel.

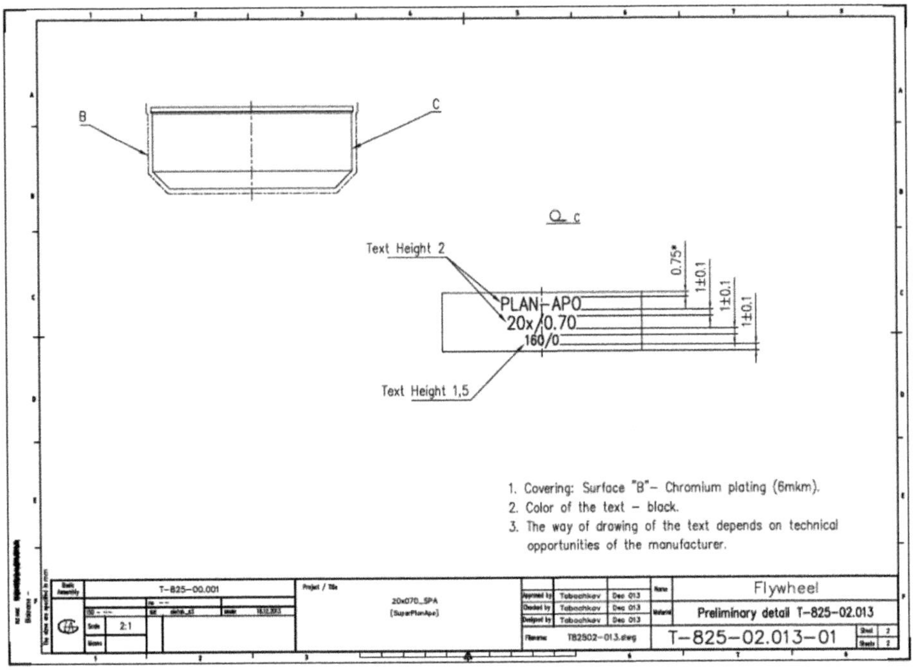

Figure 3.278 A drawing of the flywheel.

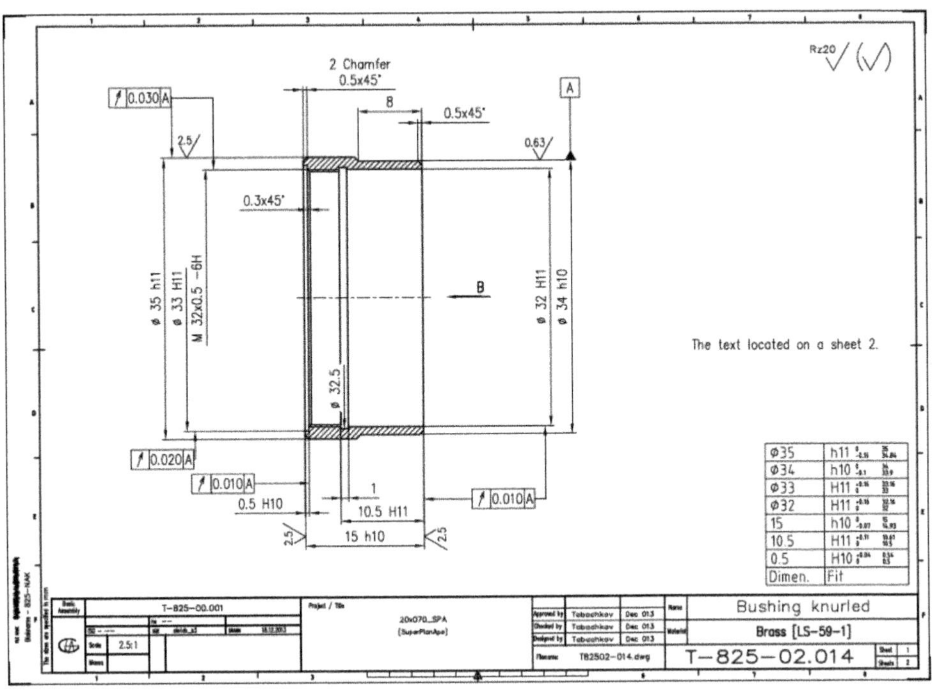

Figure 3.279 A drawing of the knurled bushing.

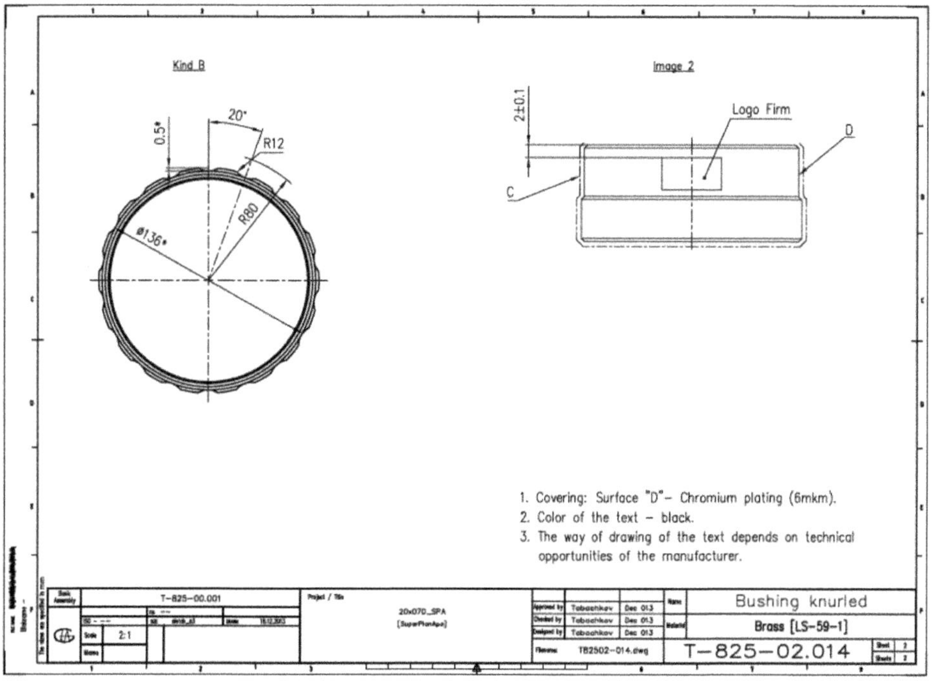

Figure 3.280 A drawing of the knurled bushing.

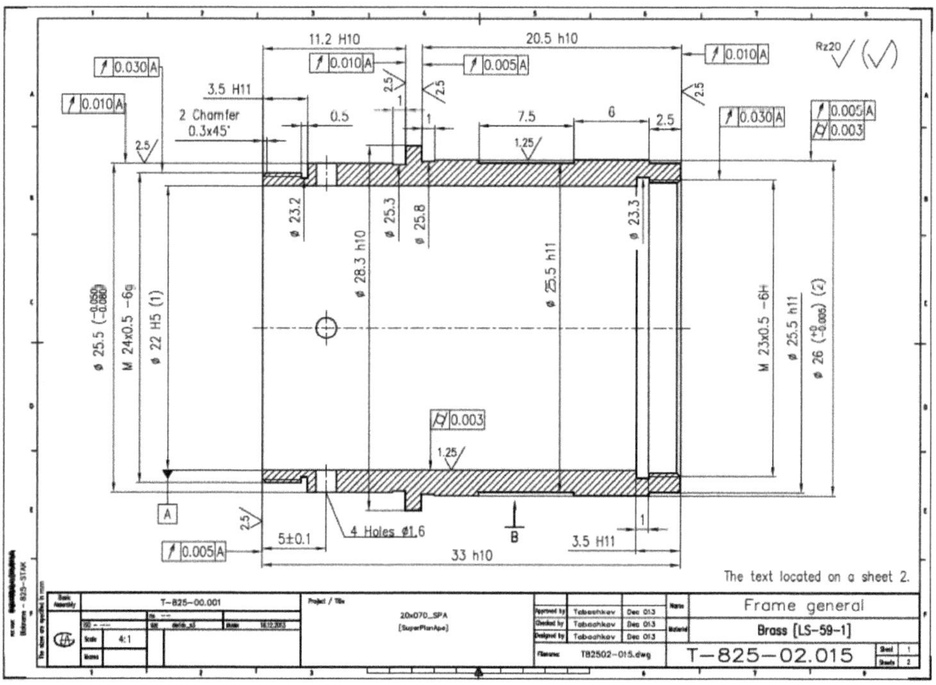

Figure 3.281 A drawing of the general frame (no. 1).

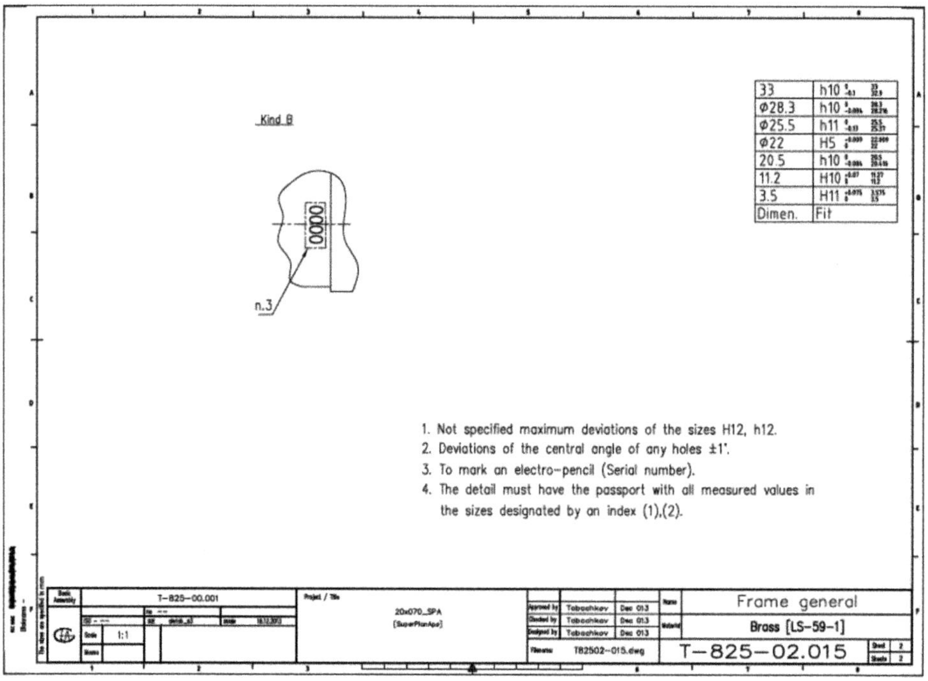

Figure 3.282 A drawing of the general frame (no. 2).

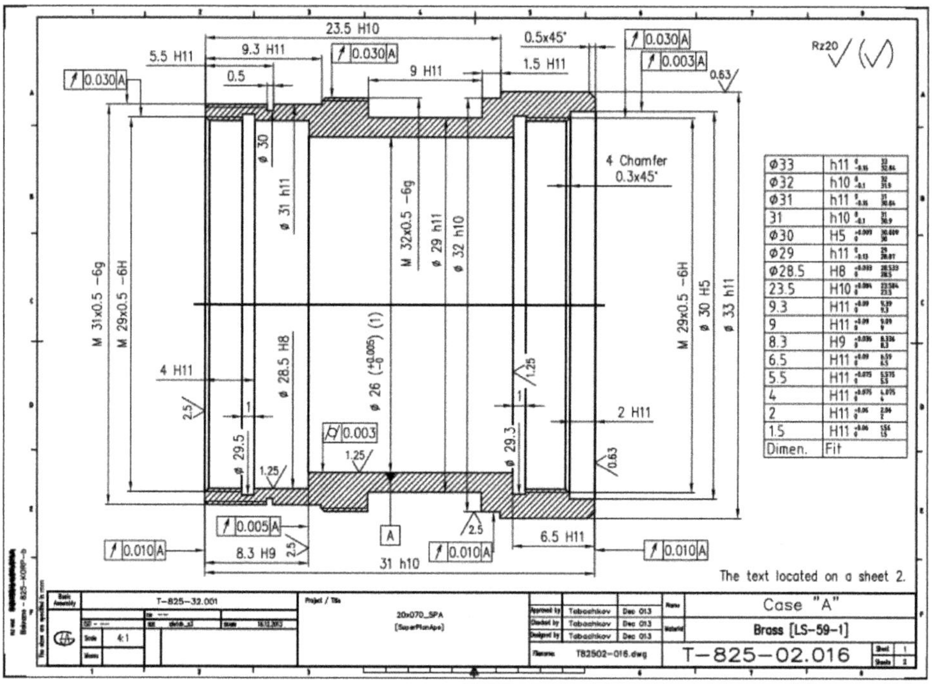

Figure 3.283 A drawing of the body (case) "A" (no. 1).

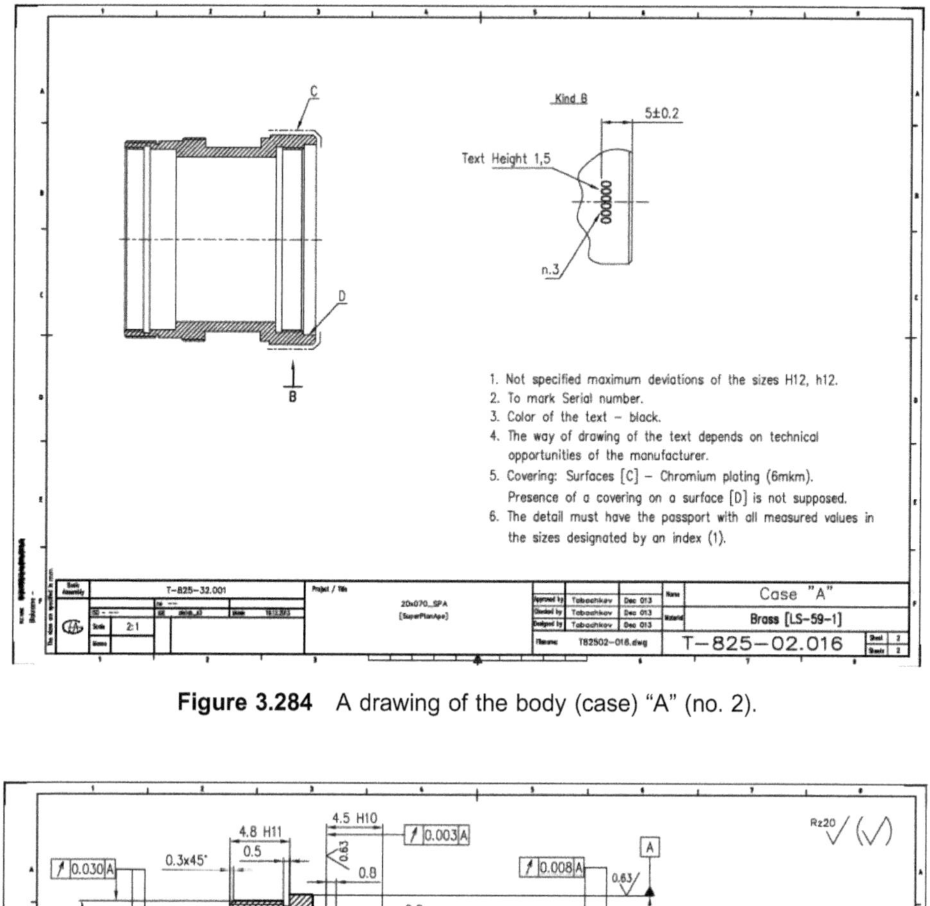

Figure 3.284 A drawing of the body (case) "A" (no. 2).

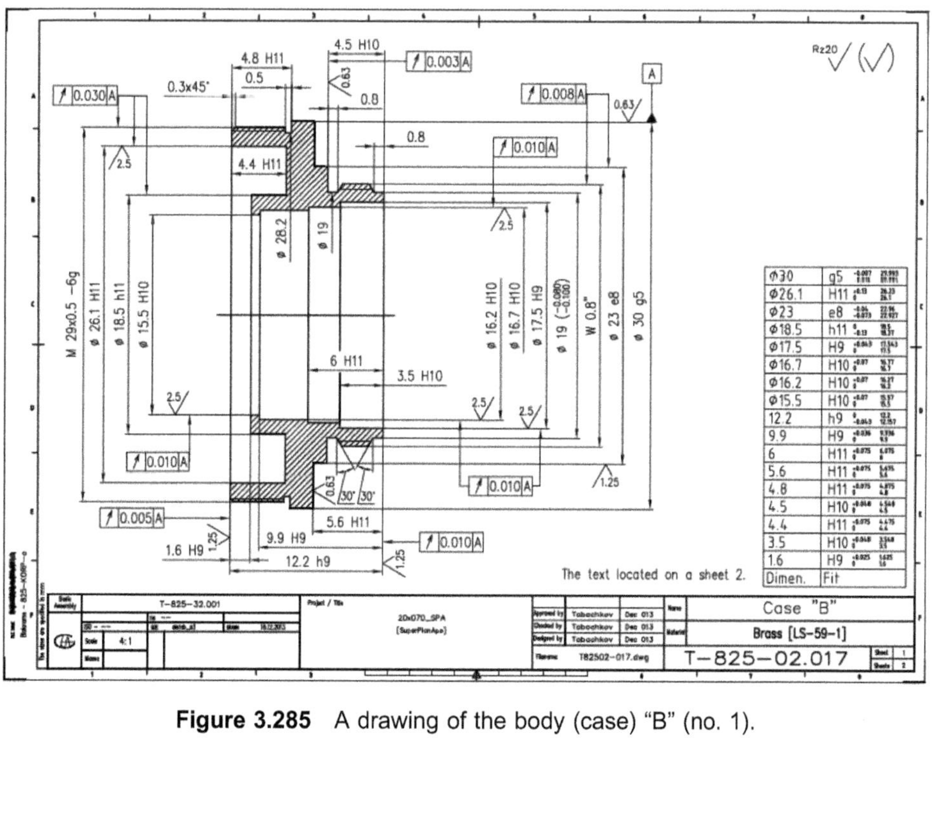

Figure 3.285 A drawing of the body (case) "B" (no. 1).

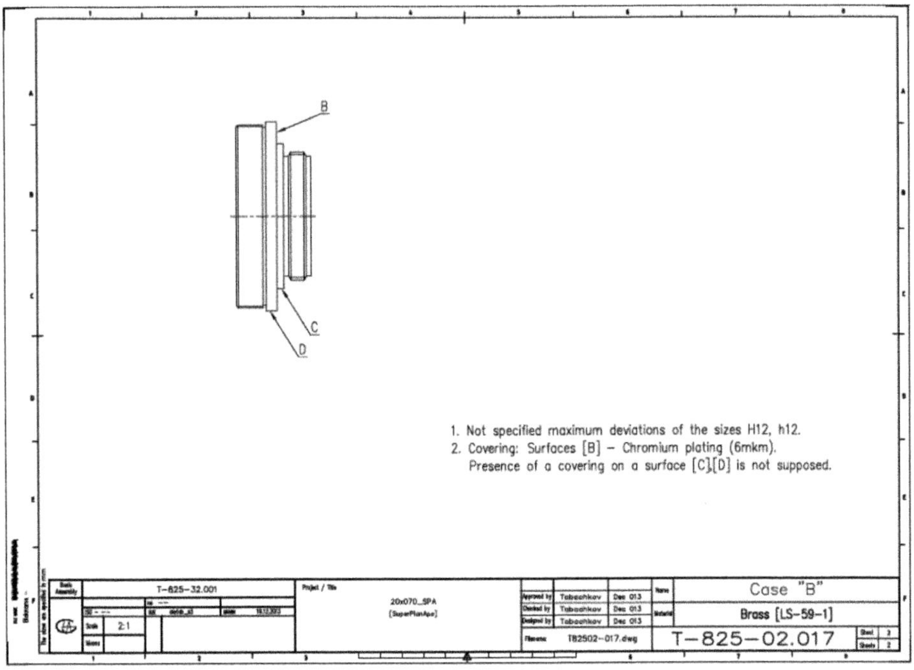

Figure 3.286 A drawing of the body (case) "B" (no. 2).

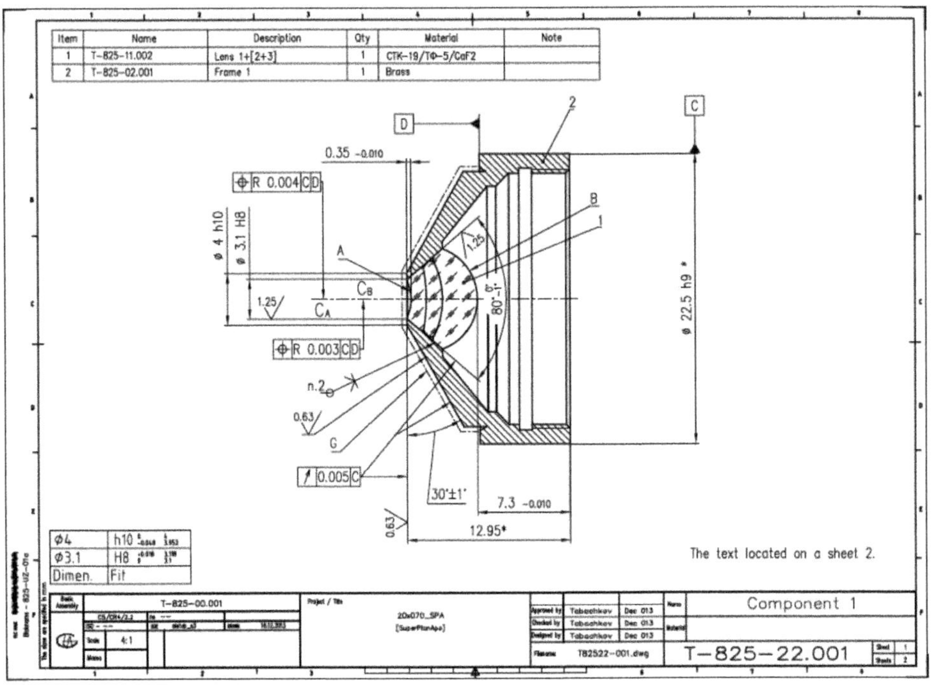

Figure 3.287 A drawing of the first component (no. 1).

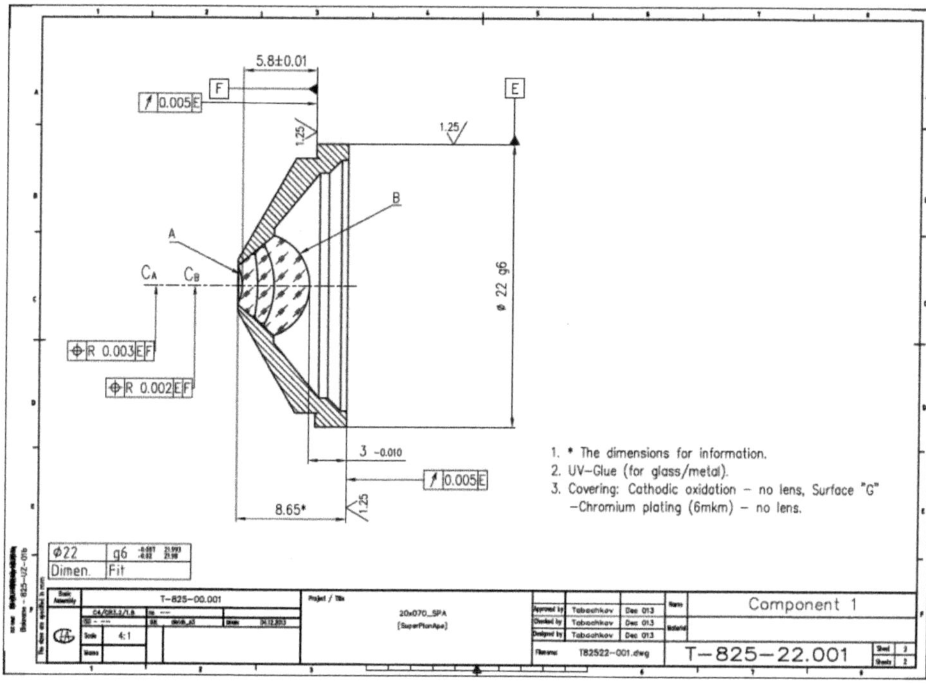

Figure 3.288 A drawing of the first component (no. 2).

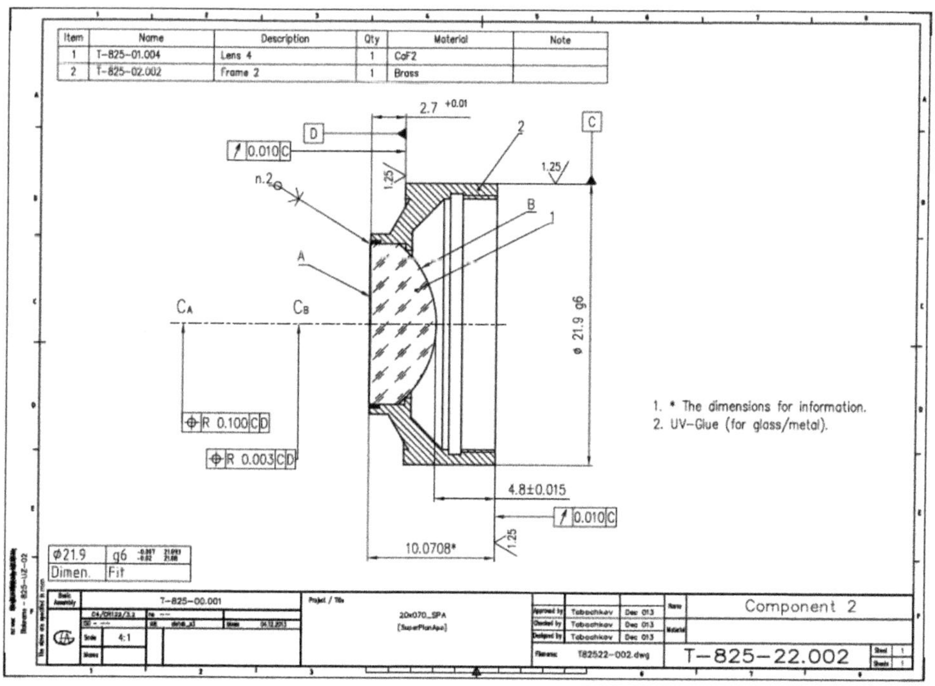

Figure 3.289 A drawing of the second component.

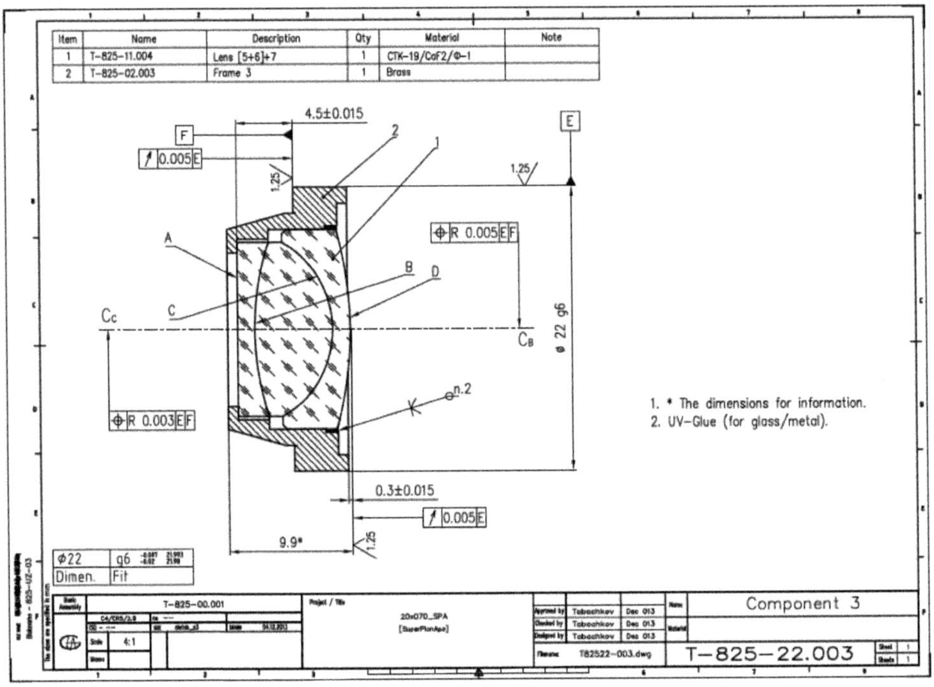

Figure 3.290 A drawing of the third component.

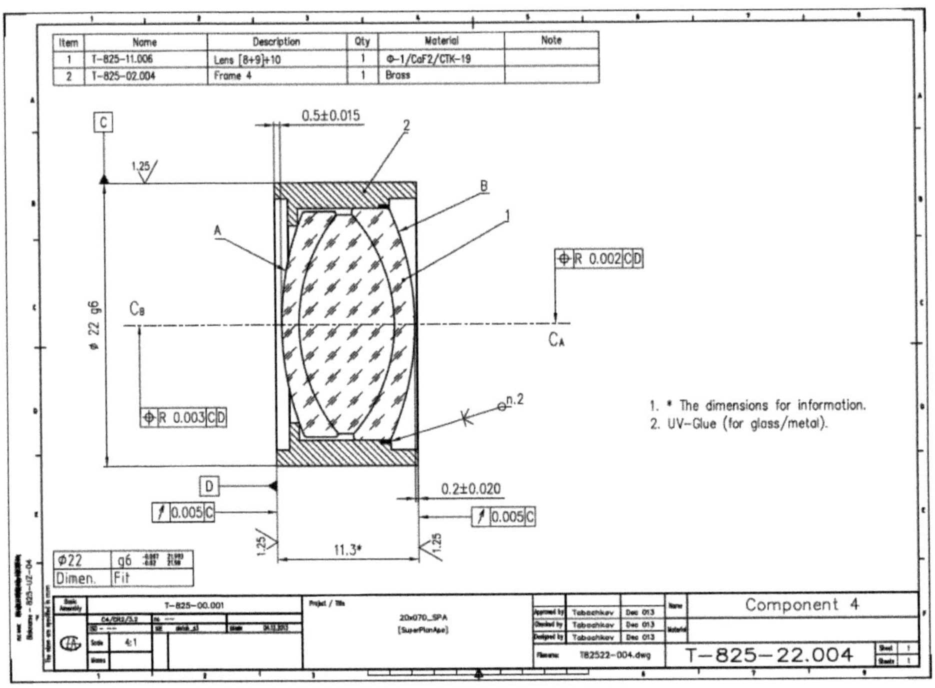

Figure 3.291 A drawing of the fourth component.

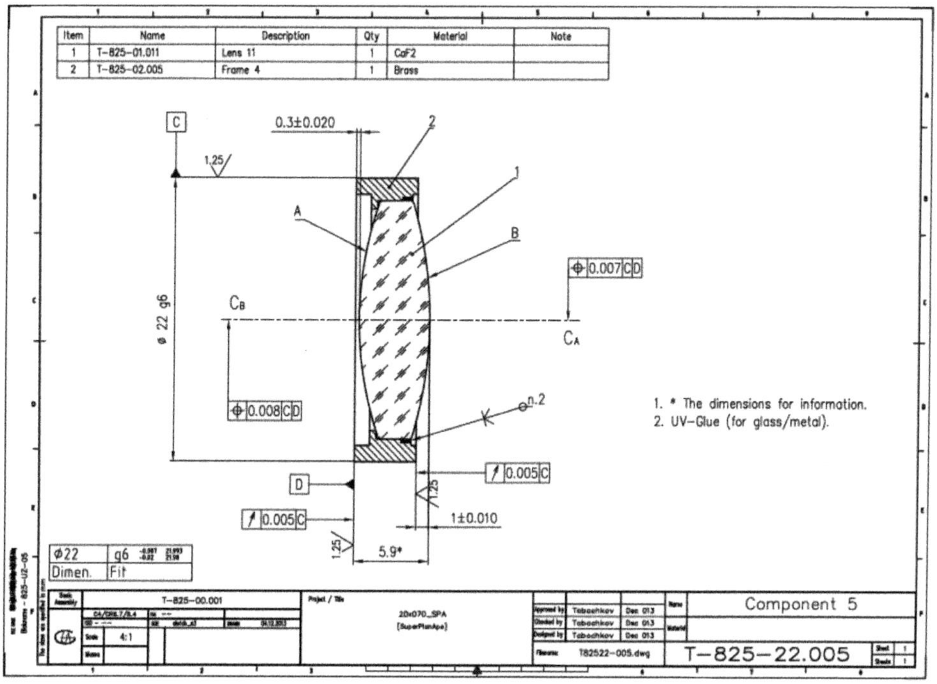

Figure 3.292 A drawing of the fifth component.

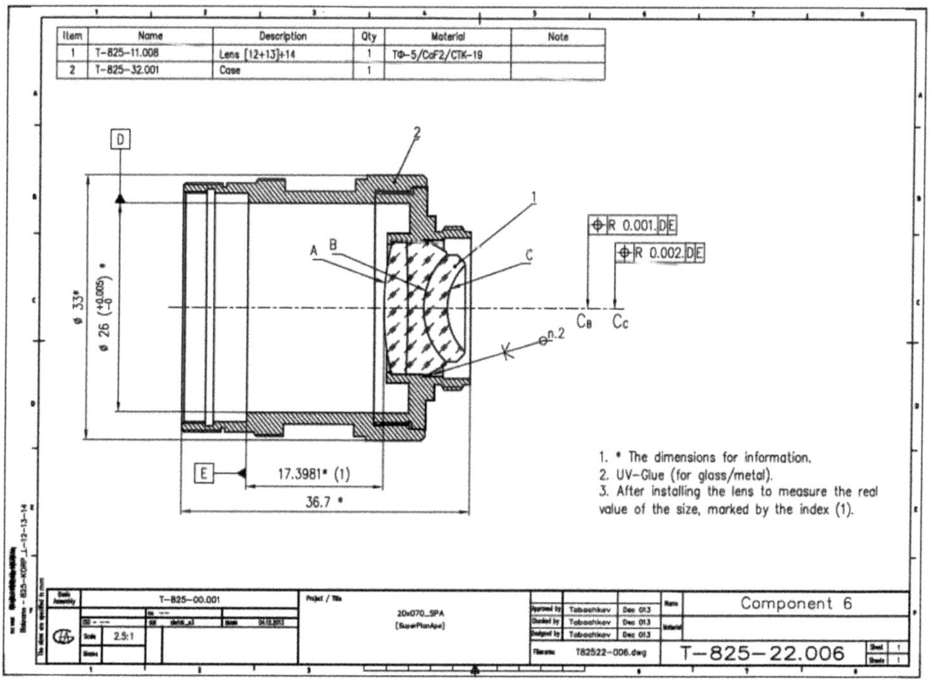

Figure 3.293 A drawing of the sixth component.

Figure 3.294 Photograph of a 20×/0.70 finite CCF plan poly-apochromatic objective.

Figure 3.295 Photograph of a real object using a 20×/0.70 CCF plan poly-apochromatic objective.

Table 3.13 shows the main technical parameters and basic optical layout of these objectives.

Using these objectives can significantly improve the quality of research. For example, we can focus on one plane of an image by using several lasers operating at different wavelengths simultaneously as a source of radiation. With the help of such objectives, it is also possible to perform other multispectral research.

Figure 3.296 Photograph of a real object using a 20×/0.70 CCF plan poly-apochromatic objective.

Figure 3.297 Photograph of a real object using a 20×/0.70 CCF plan poly-apochromatic objective.

3.11 Contact Objectives

So-called *contact objectives* can provide an original engineering solution in light microscopy. In biological studies, contact microscopy is not new (for example, Section 1.2.6.6 in Volume 1, Chapter 1, describes the OLK-2, a contact microscopy device). The main parameters of a standard biological (prepared) microscopic object and the basic requirements for the optical

Figure 3.298 Photograph of a real object using a 20×/0.70 CCF plan poly-apochromatic objective.

Figure 3.299 Photograph of a real object using a 20×/0.70 CCF plan poly-apochromatic objective.

system of light microscopes are given in Chapter 2, Section 2.4.3. In general, the reasoning behind these requirements also applies to contact microscopy; however, some may argue that "we must not do this." Let us analyze why using contact microscopy methodology allows one to reduce many factors that impair image quality and significantly simplify the optical and mechanical design of the microscope.

Figure 3.300 Photograph of a real object using a 20×/0.70 CCF plan poly-apochromatic objective.

Figure 3.301 Photograph of a real object using a 20×/0.70 CCF plan poly-apochromatic objective (polarized light).

First, I would like to describe contact objectives that can be used not for the study of biological objects, but rather for the study of metallographic, or nontransparent, objects. Contact objectives intended for use in so-called *nearfield microscopy* techniques, for example, for checking the quality of the magnetic layer of computer hard drives, are interesting. Such objectives work in contact with the studied objects and are aberration corrected for the visual

Figure 3.302 Photograph of a real object using a 20×/0.70 CCF plan poly-apochromatic objective.

Figure 3.303 Photograph of a real object using a 20×/0.70 CCF plan poly-apochromatic objective.

spectral range. The frontal (first) lenses of these epi plan achromats are made of a special material, gallium garnet gadolinium, and covered with a very thin ferromagnetic film. The optical constants for this optical material are as follows: $n_e = 1.963504$, $n_{F'-C'} = 0.047661$, and the Abbe number is $\nu_e = 20.22$. A feature of these objectives is an ultrahigh numerical aperture. Two

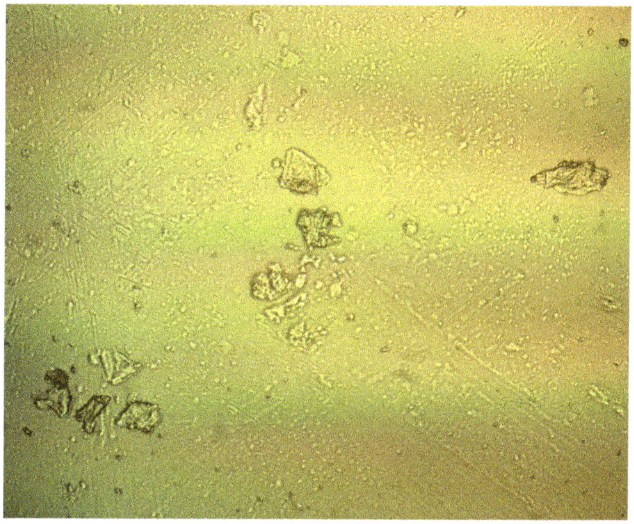

Figure 3.304 Photograph of a real object using a 20×/0.70 CCF plan poly-apochromatic objective.

objectives can be manufactured at the following magnifications and numerical apertures: 50×/1.40 and 125×/1.70. Additional requirements may be imposed on such objectives, namely, the complete absence of electrostatic electricity (by using special construction materials for the lens frames and housings), as well as the absence of intrinsic stresses in the lenses to enable polarization analysis (in another embodiment, using lenses from a special optical material, fluorophosphates glass type LZ_OK-4, instead of fluorite lenses). Figures 3.318 and 3.319 show the design parameters and graphs of aberration correction of these objectives.

In addition, the design of the 125× objective provides for an aperture diaphragm. However, unlike traditional iris mechanical diaphragms, the principle of a liquid crystal screen can be used here, the role of which is played by a plane–parallel plate, which is an integral part of the optical scheme of the lens. In another embodiment, a phase ring may be located on this plate; the phase contrast method allows you to increase the objective's depth of field.

To study the possibility of applying contact microscopy methodology in biology, we can use information about bacteria, viruses, and fungi, for example, given in Volume 1, Chapter 2, Section 2.3.1. This information allows us to formulate the basic requirements for the optical system of a microscope, which can be effectively used to study such objects that are very small in size and, as a rule, quite mobile. Living cells and tissues are also examined in cytology for various purposes, for example, to study changes in cells under various external influences; to determine patterns of metabolism in cells; and to study cellular structures, cytoplasm currents, cell permeability, and others. We can use the experience and technologies of such research. For example,

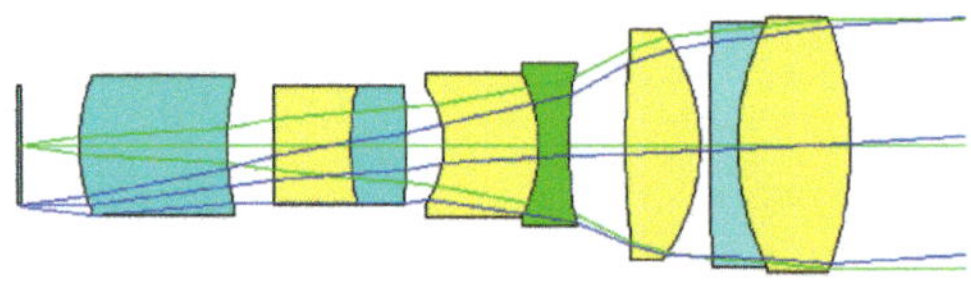

Lens: 4x0.13 365-1100nm d0-017

Object num aper 0.130000 Image height 12.500000 Primary wavln 0.546070

SRF	RADIUS	THICKNESS	APERTURE RADIUS	GLASS	SPECIAL
OBJ	0.000000	0.010000	3.130515	AIR	
AST	0.000000	0.170000	3.110000 A	K9	
2	0.000000	3.200000	3.120000	AIR	
3	9.120000	8.000000	3.640000	ZK6	
4	12.340000	2.500000	3.010000	AIR	
5	0.000000	4.000000	3.040000	CAF2	
6	13.900000	3.000000	3.100000	ZK6 P	
7	-29.140000	2.000000	3.060000	AIR	
8	-5.160000	5.000000	2.960000	CAF2 P	
9	-8.511000	1.600000	3.690000	ZBAF1	
10	20.990000	3.000000	4.210000	AIR	
11	49.000000	4.000000	5.580000	CAF2 P	
12	-9.817000	0.500000	6.010000	AIR	
13	133.000000	1.500000	6.240000	K9 P	
14	14.150000	6.000000	6.360000	CAF2 P	
15	-19.269000	200.000000	6.620000	AIR	
16	-36.100000	14.000000	18.820000	K9	
17	-43.560000	3.000000	22.060000	AIR	
18	377.100000	15.000000	22.780000	CAF2	
19	-81.500000	3.000000	23.130000	AIR	
20	221.300000	17.000000	22.480000	CAF2	
21	-72.000000	3.000000	21.330000	LZ_CTK12	
22	0.000000	199.993796 S	21.170000	AIR	
IMS	0.000000	0.000000	12.500000		

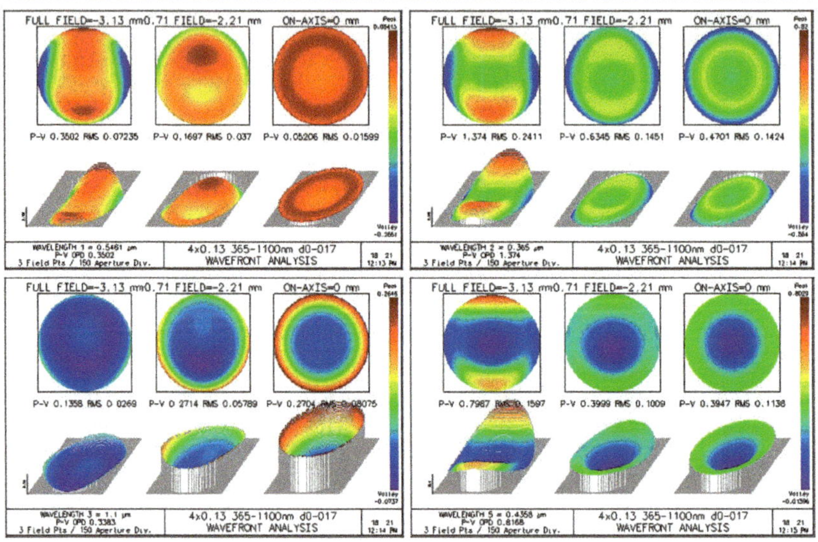

Figure 3.305 Optical design and graphs of aberration correction of a 4×/0.13 CCF plan poly-apochromatic objective (365–1100 nm; $F'_{ts} = 200$ mm).

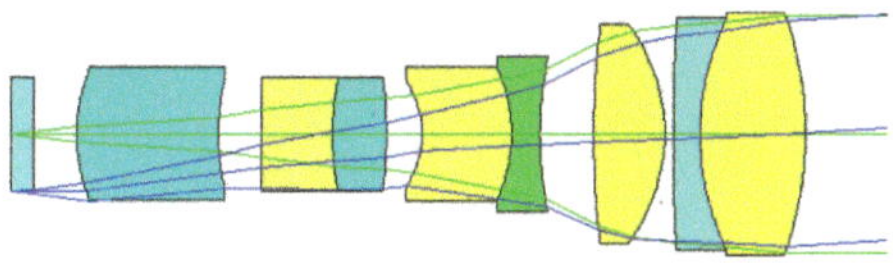

Lens:PlanPolyApo4x0.13 d1.1-1.5Ftl200

Object num aper 0.130000 Image height 12.500000 Primary wavln 0.546070

SRF	RADIUS	THICKNESS	APERTURE RADIUS	GLASS	SPECIAL
OBJ	0.000000	0.010000	3.130525	FISIO	
AST	0.000000	1.300000	3.110000 A	LZ_LK5	
2	0.000000	2.435800	3.120000	AIR	
3	9.120000	8.000000	3.640000	ZK6	
4	12.340000	2.500000	3.010000	AIR	
5	0.000000	4.000000	3.040000	CAF2	
6	13.900000	3.000000	3.100000	ZK6	
7	-29.140000	2.000000	3.060000	AIR	
8	-5.160000	5.000000	2.960000	CAF2	
9	-8.511000	1.600000	3.690000	ZBAF1	
10	20.990000	3.000000	4.210000	AIR	
11	49.000000	4.000000	5.580000	CAF2	
12	-9.817000	0.500000	6.010000	AIR	
13	133.000000	1.500000	6.240000	K9	
14	14.150000	6.000000	6.360000	CAF2	
15	-19.269000	200.000000	6.620000	AIR	
16	-36.100000	14.000000	18.820000	K9	
17	-43.560000	3.000000	22.060000	AIR	
18	377.100000	15.000000	22.780000	CAF2	
19	-81.500000	3.000000	23.130000	AIR	
20	221.300000	17.000000	22.480000	CAF2	
21	-72.000000	3.000000	21.330000	LZ_CTK12	
22	0.000000	199.997741 S	21.170000	AIR	
IMS	0.000000	0.000000	12.500000		

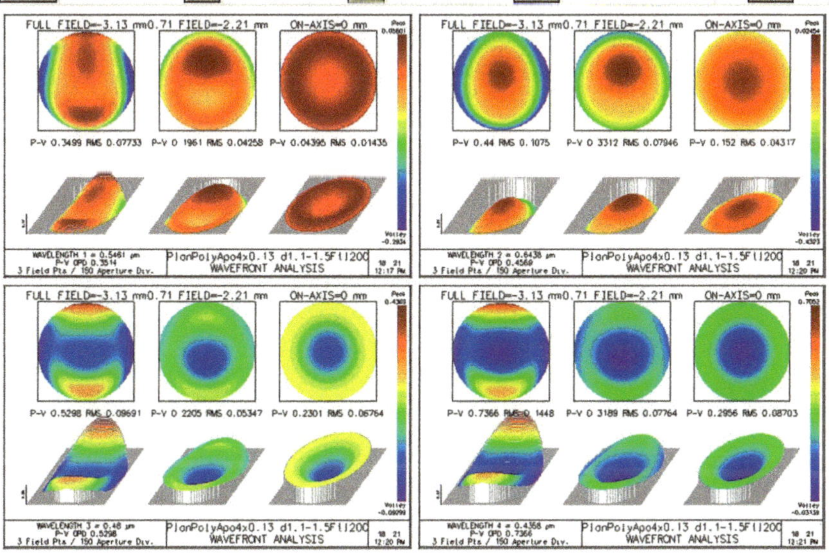

Figure 3.306 Optical design and graphs of aberration correction of a 4×/0.13 physiological immersion CCF plan poly-apochromatic objective for inverted microscopes (365–1100 nm; $F'_{ts} = 200$ mm).

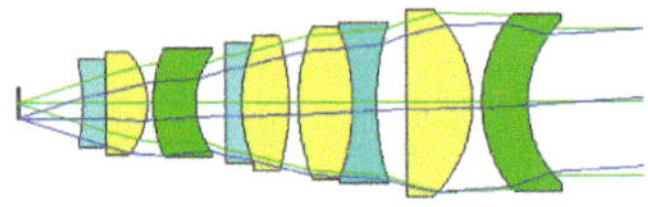

Lens: 10x0.30 365-1100nm d0-017

Object num aper 0.300000 Image height 12.500000 Primary wavln 0.546070

SRF	RADIUS	THICKNESS	APERTURE RADIUS	GLASS	SPECIAL
OBJ	0.000000	0.010000	1.250974	AIR	
AST	0.000000	0.170000	1.250000 A	K9	
2	0.000000	5.401100	1.290000	AIR	
3	-12.057000	2.000000	2.880000	K9 P	
4	0.000000	3.500000	3.560000	CAF2	
5	-6.300000	0.500000	4.160000	AIR	
6	11.092000	3.500000	4.450000	ZF2	
7	6.792000	2.500000	4.010000	AIR	
8	0.000000	1.300000	4.410000	K9 P	
9	12.840000	4.000000	4.910000	CAF2 P	
10	-19.350000	0.800000	5.480000	AIR	
11	16.465000	4.600000	6.160000	CAF2 P	
12	-18.000000	2.000000	6.240000	ZK6	
13	20.990000	2.500000	6.540000	AIR	
14	0.000000	5.500000	7.070000	CAF2 P	
15	-10.565000	0.800000	7.640000	AIR	
16	11.269000	3.800000	7.260000	ZF2 P	
17	9.214000	200.000000	6.650000	AIR	
18	-36.100000	14.000000	18.130000	K9	
19	-43.560000	3.000000	21.220000	AIR	
20	377.100000	15.000000	21.910000	CAF2	
21	-81.500000	3.000000	22.300000	AIR	
22	221.300000	17.000000	21.770000	CAF2	
23	-72.000000	3.000000	20.710000	LZ_CTK12	
24	0.000000	199.991411 S	20.580000	AIR	
IMS	0.000000	0.000000	12.500000		

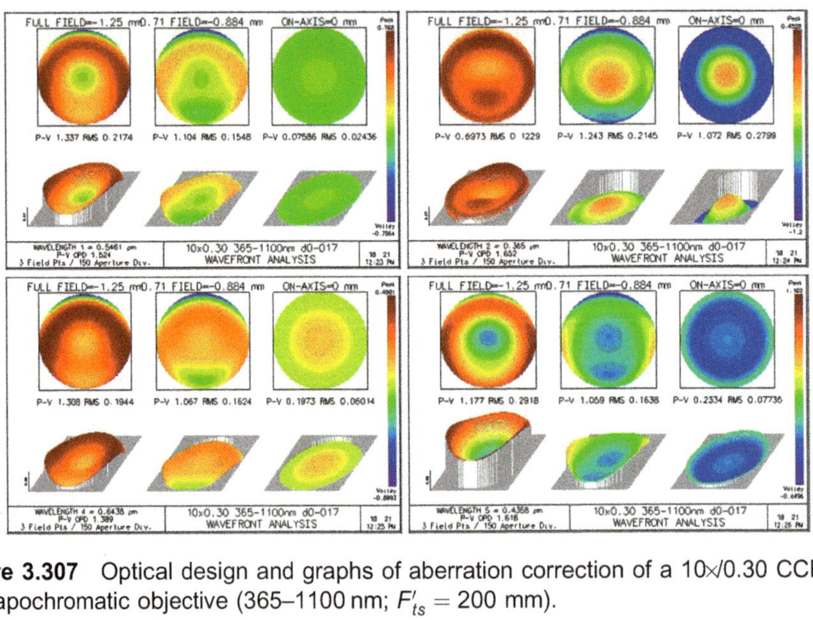

Figure 3.307 Optical design and graphs of aberration correction of a 10×/0.30 CCF plan poly-apochromatic objective (365–1100 nm; $F'_{ts} = 200$ mm).

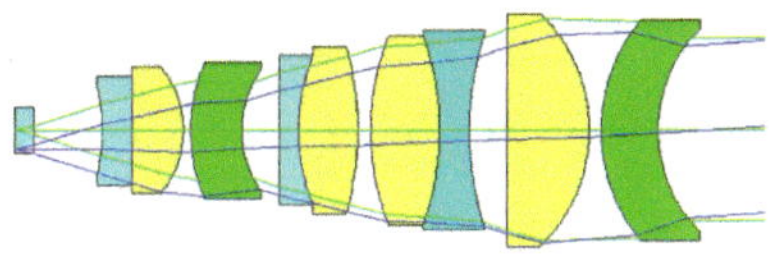

Lens:PlanPolyApo10x0.3d=1.1-1.5Ft1200

Object num aper 0.300000 Image height 12.500000 Primary wavln 0.546070

SRF	RADIUS	THICKNESS	APERTURE RADIUS	GLASS	SPECIAL
OBJ	0.000000	0.010000	1.251896	FISIO	
AST	0.000000	1.300000	1.254119 A	LK5	
2	0.000000	4.620600	1.510000	AIR	
3	-12.057000	2.000000	2.880000	K9	
4	0.000000	3.500000	3.560000	CAF2	
5	-6.300000	0.500000	4.160000	AIR	
6	11.092000	3.500000	4.450000	ZF2	
7	6.792000	2.500000	4.010000	AIR	
8	0.000000	1.300000	4.410000	K9	
9	12.840000	4.000000	4.910000	CAF2	
10	-19.350000	0.840000	5.480000	AIR	
11	16.465000	4.600000	6.160000	CAF2	
12	-18.000000	2.000000	6.240000	ZK6	
13	20.990000	2.500000	6.540000	AIR	
14	0.000000	5.500000	7.070000	CAF2	
15	-10.565000	0.800000	7.640000	AIR	
16	11.269000	3.800000	7.260000	ZF2	
17	9.214000	200.000000	6.650000	AIR	
18	-36.100000	14.000000	18.130000	K9	
19	-43.560000	3.000000	21.220000	AIR	
20	377.100000	15.000000	21.910000	CAF2	
21	-81.500000	3.000000	22.300000	AIR	
22	221.300000	17.000000	21.770000	CAF2	
23	-72.000000	3.000000	20.710000	LZ_CTK12	
24	0.000000	199.996473 S	20.580000	AIR	
IMS	0.000000	0.000000	12.500000		

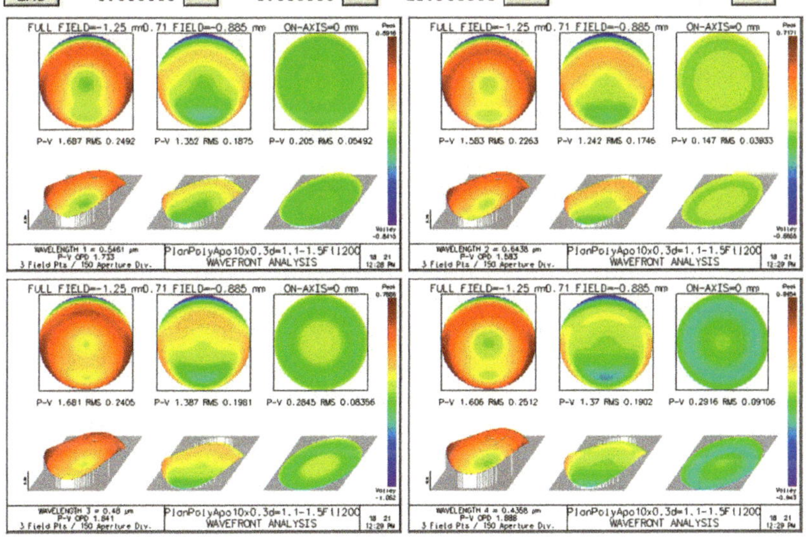

Figure 3.308 Optical design and graphs of aberration correction of a 10×/0.30 physiological immersion CCF plan poly-apochromatic objective for inverted microscopes (365–1100 nm; $F'_{ts} = 200$ mm).

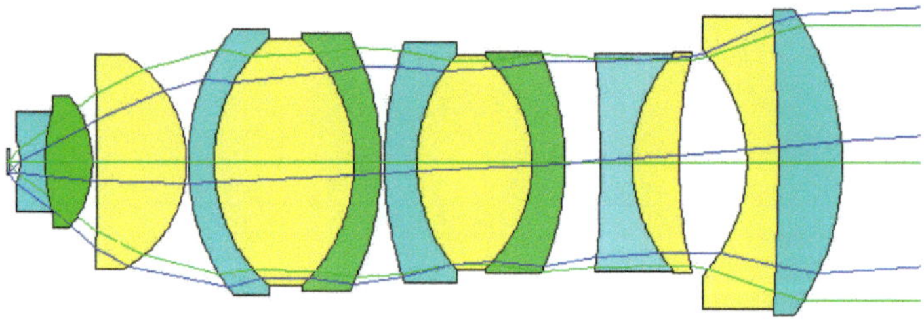

Lens: 20x0.75 365-1100nm d017

Object num aper 0.750000 Image height 12.500000 Primary wavln 0.546070

SRF	RADIUS	THICKNESS	APERTURE RADIUS	GLASS	SPECIAL
OBJ	0.000000	0.009706	0.625130	AIR	
AST	0.000000	0.170000	0.630000 A	K9	
2	0.000000	0.600000	0.730000	AIR	
3	-3.370000	1.500000	1.170000	H-ZLAF55C	
4	8.511000	2.700000	2.700000	ZF3	
5	-5.750000	0.300000	3.590000	AIR	
6	-92.000000	5.000000	4.940000	CAF2	
7	-6.608000	0.100000	5.860000	AIR	
8	11.440000	1.500000	7.250000	H-ZLAF55C P	
9	8.801000	8.000000	6.750000	CAF2 P	
10	-9.120000	1.500000	6.710000	ZBAF4	
11	-15.000000	0.200000	7.030000	AIR	
12	19.960000	1.800000	6.580000	H-ZLAF55C P	
13	8.801000	6.500000	5.880000	CAF2 P	
14	-7.686000	1.800000	5.750000	ZBAF4 P	
15	-14.850000	2.000000	6.050000	AIR	
16	-66.000000	1.800000	5.830000	K9 P	
17	8.801000	2.600000	5.920000	CAF2 P	
18	25.800000	3.800000	6.000000	AIR	
19	-8.801000	1.800000	6.140000	CAF2 P	
20	-115.500000	3.600000	7.970000	H-ZLAF55C P	
21	-14.150000	200.000000	8.380000	AIR	
22	-36.100000	14.000000	20.770000	K9 P	
23	-43.560000	3.000000	24.350000	AIR	
24	377.100000	15.000000	25.210000	CAF2 P	
25	-81.500000	3.000000	25.540000	AIR	
26	221.300000	17.000000	24.670000	CAF2 P	
27	-72.000000	3.000000	23.440000	LZ_CTK12	
28	0.000000	199.995248 S	23.230000	AIR	
IMS	0.000000	0.000000	12.500000		

Figure 3.309 Optical design and graphs of aberration correction of a 20×/0.75 CCF plan poly-apochromatic objective (365–1100 nm; $F'_{ts} = 200$ mm).

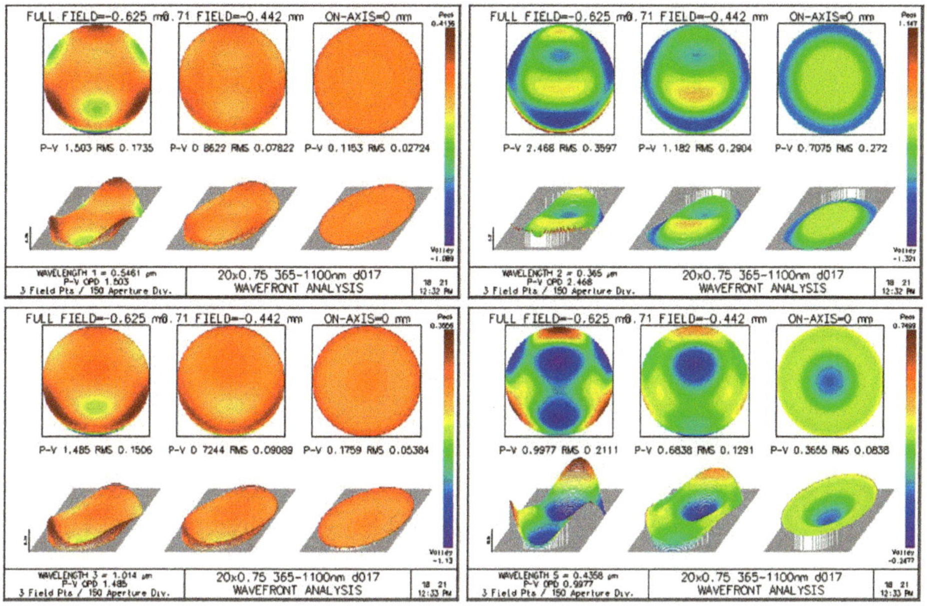

Figure 3.309 (*Continued*) Optical design and graphs of aberration correction of a 20×/0.75 CCF plan poly-apochromatic objective (365–1100 nm; $F'_{ts} = 200$ mm).

preparing a live specimen for study (in vivo) under a microscope. Small organisms, such as unicellular algae, protozoa, and bacteria, are transferred along with a drop of the medium in which they are cultivated onto a glass slide. Such a preparation is examined as is, that is, studied with a conventional objective that does not touch the medium drop (if the working distance of the objective is long enough) and without a cover glass. Living cells from the tissues of multicellular organisms are more difficult to investigate because sometimes cells need to be separated from the tissue. Cell isolation must be carried out in a drop of medium that does not chemically react with the preparation and is suitable for the organisms to be in for a rather long time. For example, plant cells are usually examined in water, and cells from various cold-blooded and warm-blooded animals are examined in saline or other liquid solutions. Some microscopic techniques do not involve the staining of living objects, while others do.

The requirements for dyes are the same as those for a drop of medium: they must not affect the cell under study or cause its mutation. Unlike basic and acidic dyes, living dyes are organic aromatic compounds with relatively little toxicity to living cells. Penetrating the cell, they are mainly connected with proteins. At first, the entire cytoplasm acquires a diffuse color; afterward, some dyes are deposited in the cytoplasm in the form of granules. Live staining can reveal many cell structures. Especially good results are obtained

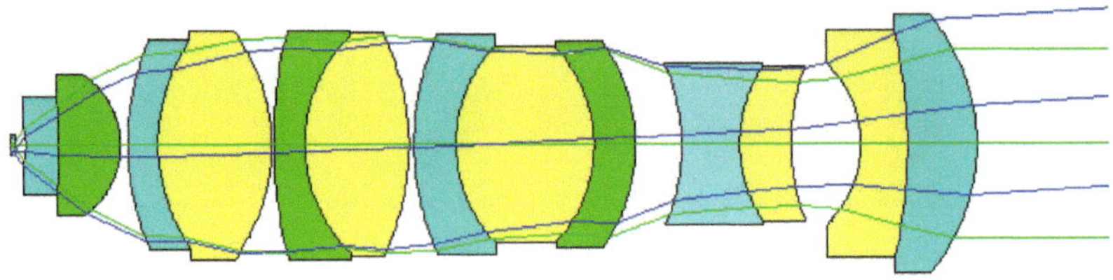

Lens: 40x0.80 365-1100nm d017

Object num aper	0.800000	Image height	12.500000	Primary wavln	0.546070

SRF	RADIUS	THICKNESS	APERTURE RADIUS	GLASS	SPECIAL
OBJ	0.000000	0.009706	0.312527	AIR	
AST	0.000000	0.170000	0.320000 A	K9	
2	0.000000	0.500000	0.430000	AIR	
3	-2.890000	1.500000	0.900000	H-ZLAF55C	
4	-31.340000	2.700000	2.070000	ZF4	
5	-3.569000	0.300000	3.000000	AIR	
6	10.600000	1.300000	4.340000	LZ_CTK19	
7	7.400000	5.000000	4.400000	CAF2	
8	-8.050000	0.100000	4.820000	AIR	
9	20.310000	1.300000	4.830000	ZF4 P	
10	6.274000	4.500000	4.570000	CAF2 P	
11	-10.700000	0.200000	4.700000	AIR	
12	10.950000	1.800000	4.620000	LZ_CTK19 P	
13	5.890000	6.000000	4.160000	CAF2 P	
14	-5.890000	1.800000	4.090000	ZF4 P	
15	-7.400000	2.000000	4.350000	AIR	
16	-8.670000	2.500000	3.410000	ZK1	
17	5.890000	2.200000	3.240000	CAF2 P	
18	8.670000	3.000000	3.270000	AIR	
19	-4.760000	2.000000	3.450000	CAF2 P	
20	-21.760000	3.000000	4.810000	H-ZLAF55C P	
21	-8.336000	200.000000	5.430000	AIR	
22	-36.100000	14.000000	17.830000	K9 P	
23	-43.560000	3.000000	20.860000	AIR	
24	377.100000	15.000000	21.520000	CAF2 P	
25	-81.500000	3.000000	21.910000	AIR	
26	221.300000	17.000000	21.410000	CAF2 P	
27	-72.000000	3.000000	20.380000	LZ_CTK12	
28	0.000000	199.987032 S	20.250000	AIR	
IMS	0.000000	0.000000	12.500000		

Figure 3.310 Optical design and graphs of aberration correction of a 40×/0.80 CCF plan poly-apochromatic objective (365–1100 nm; $F'_{ts} = 200$ mm).

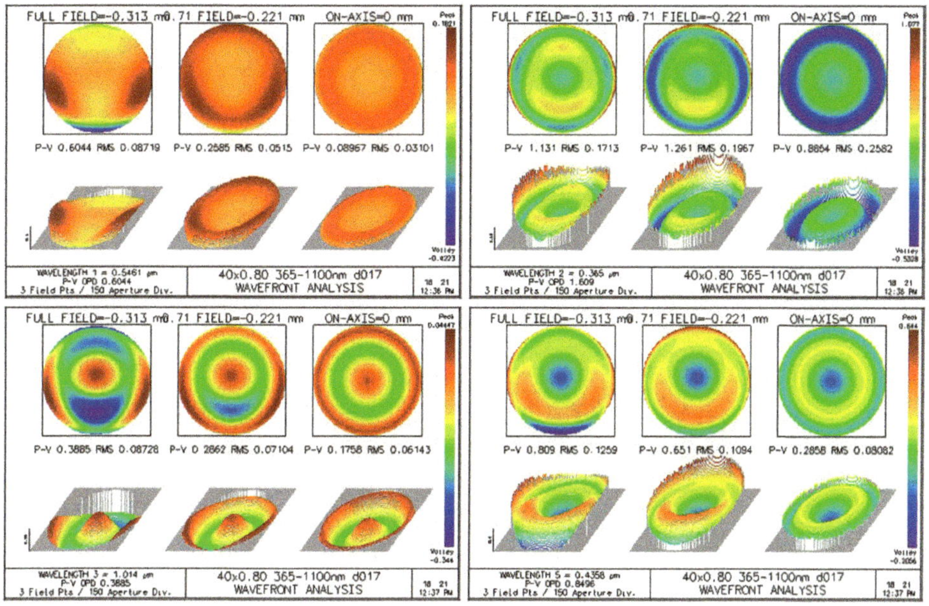

Figure 3.310 (*Continued*) Optical design and graphs of aberration correction of a 40×/0.80 CCF plan poly-apochromatic objective (365–1100 nm; $F'_{ts} = 200$ mm).

by using vital dyes as fluorochromes. The color of living cells makes it possible to identify changes occurring in cells and tissues under various external influences. In this case, it is extremely important that the amount of dye absorbed by undamaged and damaged cells (for example, by exposure to high temperature) can be accurately determined and quantified. The difference in the amount of dye absorbed by intact and damaged cells indicates the nature and extent of changes that they have undergone. Some staining methods make it possible to distinguish living cells from the dead. For example, eosin quickly stains dead cells in pink, and their very color turns out to be diffuse; eosin solution does not stain living cells.

However, a more objective method of studying living objects is to examine the sample as it is, that is, without staining; in this case, information about the object may not be complete, but it will be more reliable. For the appearance of a real image in the study of certain species of organisms, special methods of microscopic analysis are required, such as phase contrast, dark field, differential interference, polarization, or other so-called contrast methods. Fluorescence is probably the best known among the special methods of microscopy. This is because with fluorescence, the object is irradiated with light of different spectral compositions; the images obtained in the different spectral ranges are then integrated into a whole. But the task is the same here, namely, to not damage the preparation with irradiating light. Thus, the

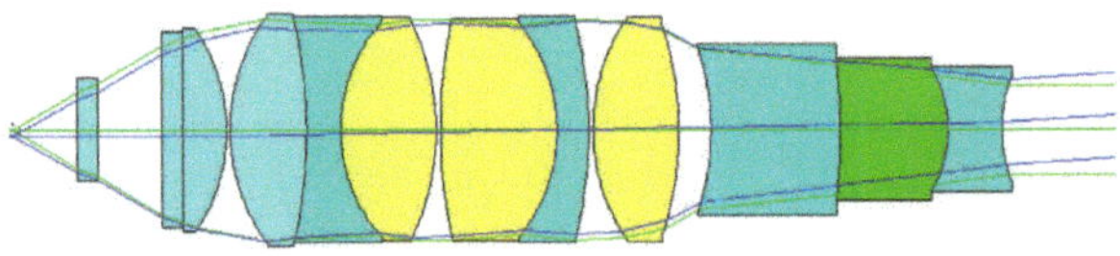

Lens: 50x0.50 interf MIRO 365-1100nm

Object num aper 0.500000 Image height 12.500000 Primary wavln 0.546070

SRF	RADIUS	THICKNESS		APERTURE RADIUS		GLASS		SPECIAL
OBJ	0.000000	1.0000e-20		0.249986		AIR		
AST	0.000000	2.990000		0.249986	AS	AIR		
2	0.000000	1.020000		1.950000		LZ_K8		
3	0.000000	2.990000		2.310000		AIR		
4	0.000000	1.020000		4.040000		LZ_K8	P	
5	0.000000	2.000000		4.400000		LZ_K8	P	
6	-7.959900	0.170000		4.590000		AIR		
7	8.616500	3.600000		5.250000		SILICA		
8	-21.809700	1.600000		5.110000		LZ_CTK19		
9	7.222800	4.400000		4.890000		CAF2		
10	-11.130600	0.200000		5.060000		AIR		
11	22.467500	5.500000		5.020000		CAF2	P	
12	-7.425200	1.500000		4.830000		LZ_CTK19	P	
13	-19.630200	0.200000		5.110000		AIR		
14	8.839600	4.000000		5.070000		CAF2	P	
15	-14.595800	1.500000		4.730000		AIR		
16	-11.537300	6.000000		3.870000		LZ_CTK19	P	
17	-35.650000	5.000000		3.220000		TF5		
18	-5.501800	2.500000		2.810000		LZ_CTK12		
19	6.435200	200.000000		2.320000		AIR		
20	-36.100000	14.000000		12.960000		LZ_K8	P	
21	-43.560000	3.000000		15.180000		AIR		
22	377.100000	15.000000		15.540000		CAF2		
23	-81.500000	3.000000		15.930000		AIR		
24	221.300000	17.000000		15.710000		CAF2		
25	-72.000000	3.000000		14.940000		LZ_CTK12	P	
26	0.000000	199.960019	S	14.890000		AIR		
IMS	0.000000	0.000000		12.500000	S			

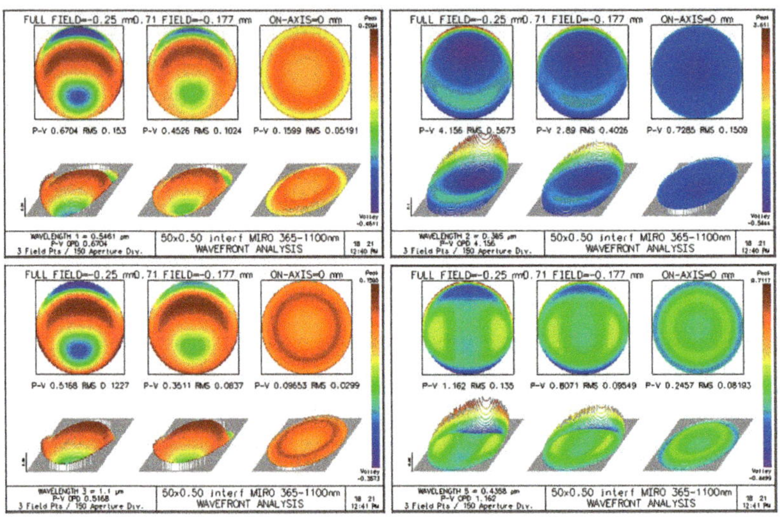

Figure 3.311 Optical design and graphs of aberration correction of a 50×/0.50 CCF plan poly-apochromatic objective (interference MIRO configuration; 365–1100 nm; $F'_{ts} = 200$ mm).

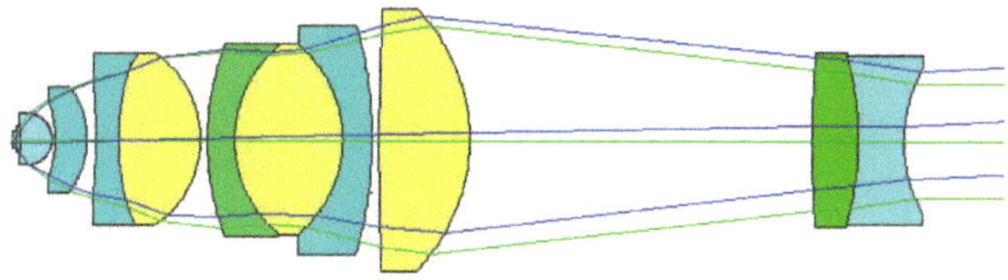

Lens: 100x1.20 water 365-1100nm d017

Object num aper 1.200000 Image height 10.000000 Primary wavln 0.546070

SRF	RADIUS		THICKNESS		APERTURE RADIUS		GLASS		SPECIAL
OBJ	0.000000		0.010000		0.102294		WATER		
AST	0.000000		0.170000		0.120000	A	K9		
2	0.000000		0.200000		0.350000		WATER	P	
3	-2.000000		1.400000		0.580000		BAK2		
4	-1.109000		0.130000		1.109000		AIR		
5	-5.037000		1.400000		1.730000		H-ZLAF69		
6	-3.746000		0.400000		2.270000		AIR		
7	-24.500000		1.000000		2.700000		H-ZLAF69	P	
8	10.850000		3.600000		3.150000		CAF2		
9	-4.368000		0.300000		3.680000		AIR		
10	11.426000		1.200000		4.140000		ZBAF4		
11	5.160000		4.800000		3.990000		CAF2	P	
12	-5.160000		1.300000		4.090000		H-ZLAF69	P	
13	-17.500000		0.300000		4.920000		AIR		
14	126.400000		4.000000		5.300000		CAF2	P	
15	-7.900000		15.000000		5.630000		AIR		
16	57.800000		2.000000		3.740000		H-ZF6		
17	-13.305000		2.000000		3.620000		H-QK3		
18	6.000000		200.000000		3.100000		AIR		
19	-36.100000		14.000000		13.060000		K9		
20	-43.560000		3.000000		15.270000		AIR		
21	377.100000		15.000000		15.620000		CAF2		
22	-81.500000		3.000000		15.980000		AIR		
23	221.300000		17.000000		15.740000		CAF2		
24	-72.000000		3.000000		14.940000		LZ_CTK12		
25	0.000000		199.962295	S	14.880000		ATR		
IMS	0.000000		0.000000		10.000000				

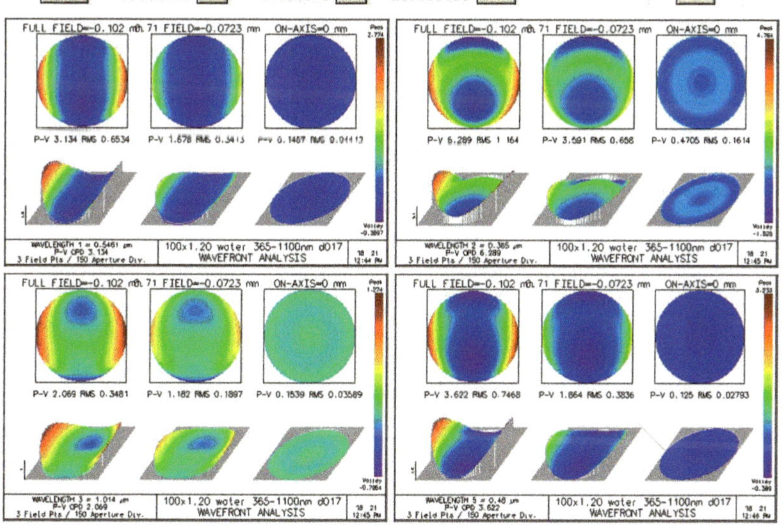

Figure 3.312 Optical design and graphs of aberration correction of a 100×/1.20 water immersion CCF plan poly-apochromatic objective (365–1100 nm; $F'_{ts} = 200$ mm).

Table 3.12 Main technical parameters and basic optical layout of plan poly-apochromatic objectives (365–1100 nm).

Magnification	*NA*	*WD* (mm)	*F′* (mm)	*R* (μm)	*DF* (μm)	*FOV* on object (mm)	*FOV* on image (mm)	Optical design at figure no.	The principal optical layout
4×	0.13	3.2	50	2.56	19.7	6.3	25	3.305	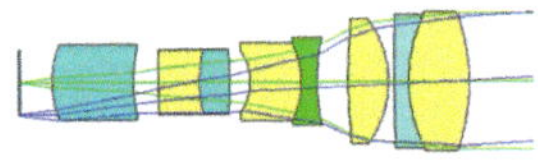
4×	0.13	2.4	50	2.56	19.7	6.3	25	3.306	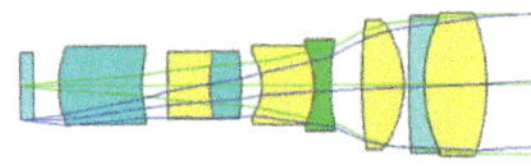
10×	0.30	4.1	20	1.11	3.7	2.5	25	3.307	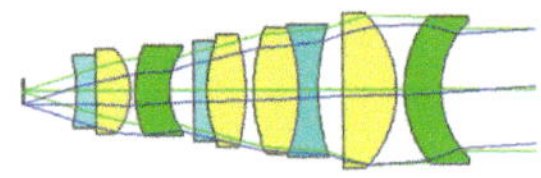
10×	0.30	3.2	20	1.11	3.7	2.5	25	3.308	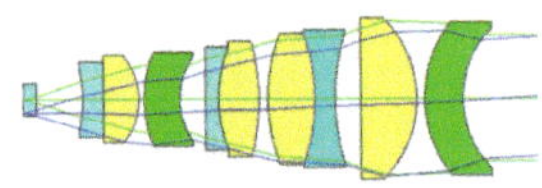
20×	0.75	0.42	10	0.44	0.59	1.25	25	3.309	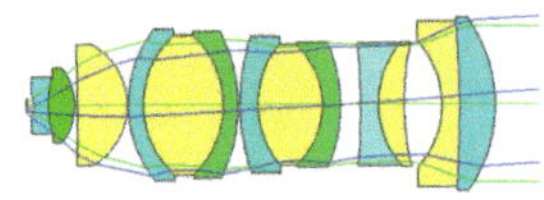
40×	0.80	0.33	5.0	0.42	0.52	0.63	25	3.310	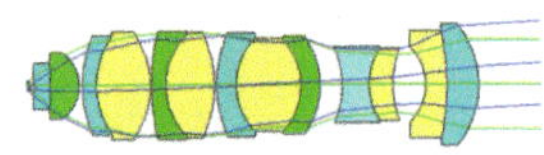
50×	0.50	3.0	4.0	0.67	1.33	0.50	25	3.311	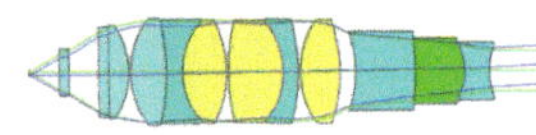
100×	1.2 water	0.13	2.0	0.28	0.23	0.20	20	3.312	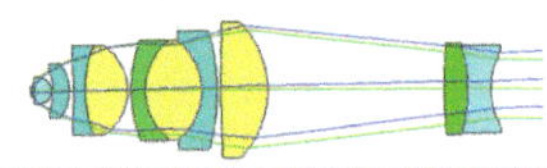

microscope should allow these contrast and fluorescent methods to be implemented. Implementing these methods should be possible since they mainly involve adapting the microscope's optical components.

Maximum numerical aperture and ultrahigh values of the linear magnification of the objective must also be achieved. This is due to at least two distinctive characteristics of the studied objects: their linear dimensions and their motor ability, that is, their speed of movement in the medium. Most likely, it will not be possible to use an ordinary microscope with which researchers only observe the obtained image with their eyes. To study the details of the most important cell processes, such as division, phagocytosis, cytoplasmic currents, and others, we will probably need a digital image receiver, which is quite versatile. Such a receiver should be able to record both fast and slow processes.

Thus, the usual block diagram of a light microscope, which contains such basic parts as the illumination, projection, and registration systems, can

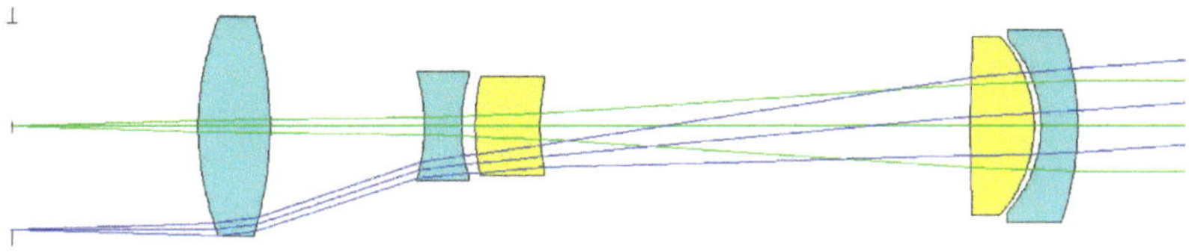

Lens:PlMultiApo2.5x0.03254-1100Ftl160

Object num aper 0.030000 Object height -4.360000 Primary wavln 0.546070

SRF	RADIUS	THICKNESS	APERTURE RADIUS	GLASS	SPECIAL
OBJ	0.000000	1.0000e-20	4.360000	AIR	
AST	0.000000	8.020300	4.360000 AS	AIR	
2	12.369400	3.100000	4.630000	SILICA	
3	-15.490700	6.700000	4.470000	AIR	
4	-7.636200	1.600000	2.310000	SILICA P	
5	7.331900	0.600000	2.090000	AIR	
6	9.460100	2.800000	2.080000	CAF2	
7	8.893200	18.500000	1.880000	AIR	
8	147.725100	2.800000	3.590000	CAF2 P	
9	-5.433500	0.300000	3.760000	AIR	
10	-5.256700	1.600000	3.670000	SILICA P	
11	-12.178700	100.000000	4.060000	AIR	
12	110.920000	4.000000	10.850000	LIF	
13	0.000000	1.000000	10.910000	AIR	
14	162.550000	4.000000	10.950000	LIF P	
15	-49.550000	2.000000	10.940000	SILICA P	
16	170.220000	4.000000	10.980000	LIF P	
17	-185.780000	152.184686 S	11.040000	AIR	
IMS	0.000000	0.000000	10.987449 S		

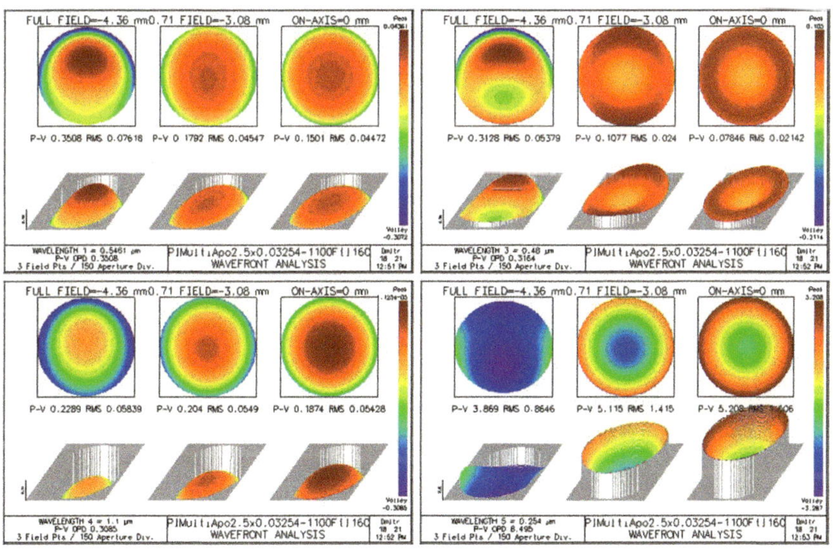

Figure 3.313 Optical design and graphs of aberration correction of a 2.5×/0.03 CCF plan multi-apochromatic objective (254–1100 nm; $F'_{ts} = 160$ mm).

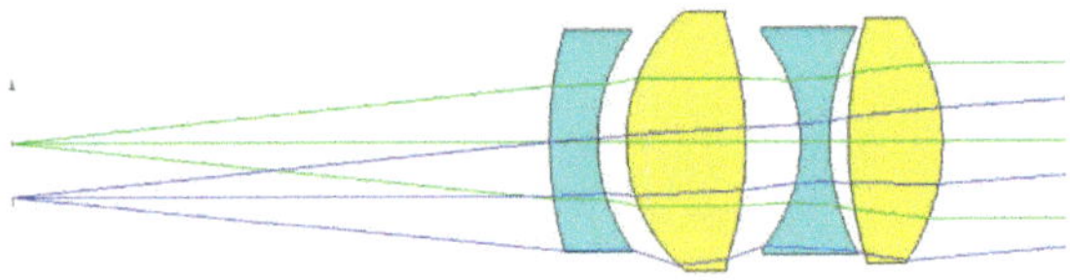

Lens:PlMultiApo 5x0.1 254-1100 Ftl160

Object num aper 0.100000 Object height -2.200000 Primary wavln 0.546070

SRF	RADIUS	THICKNESS	APERTURE RADIUS	GLASS	SPECIAL
OBJ	0.000000	1.0000e-20	2.200000	AIR	
AST	0.000000	22.760000	2.200000 AS	AIR	
2	17.017400	2.000000	4.550000	SILICA	
3	7.817200	1.200000	4.500000	AIR	
4	7.038500	5.000000	5.330000	CAF2	
5	-15.356300	2.300000	5.050000	AIR	
6	-6.748100	1.200000	4.410000	SILICA P	
7	9.860800	0.800000	4.640000	AIR	
8	16.303200	4.000000	4.780000	CAF2 P	
9	-8.704700	100.000000	5.030000	AIR	
10	110.920000	4.000000	8.370000	LIF	
11	0.000000	1.000000	8.470000	AIR	
12	162.550000	4.000000	8.520000	LIF P	
13	-49.550000	2.000000	8.560000	SILICA P	
14	170.220000	4.000000	8.620000	LIF P	
15	-185.780000	152.214142 S	8.700000	AIR	
IMS	0.000000	0.000000	10.990001 S		

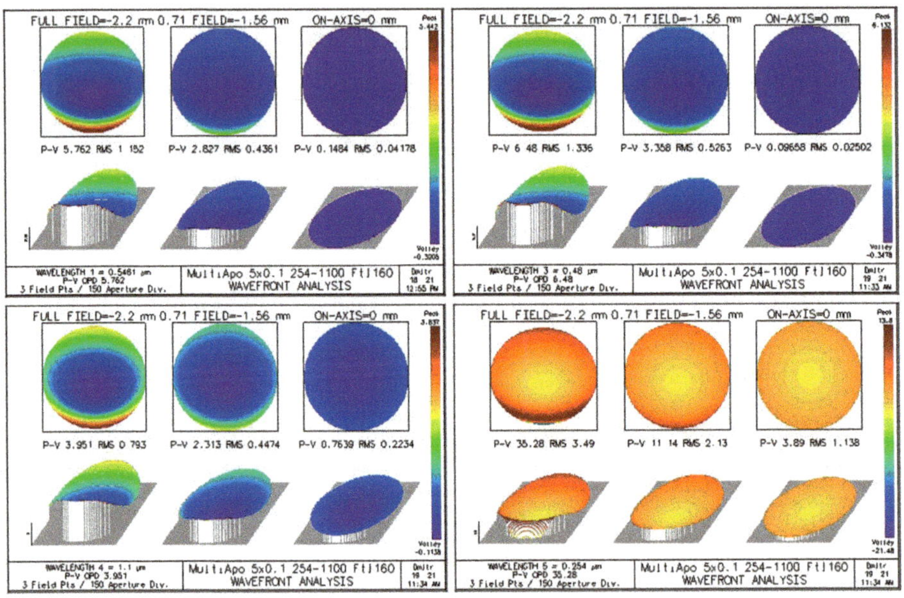

Figure 3.314 Optical design and graphs of aberration correction of a 5×/0.10 CCF S-plan multi-apochromatic objective (254–1100 nm; $F'_{ts} = 160$ mm).

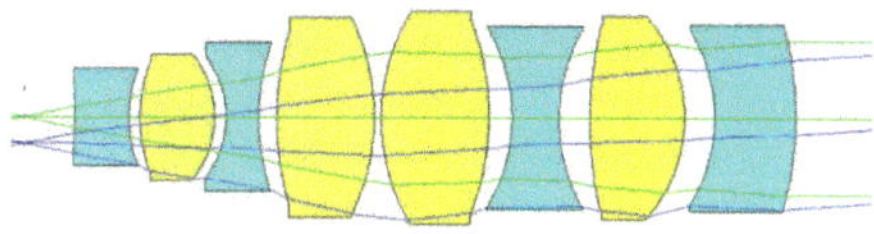

Lens:PlMultiApo 10x0.20 254-1100 f160

Object num aper 0.200000 Object height -1.100000 Primary wavln 0.546070

SRF	RADIUS	THICKNESS	APERTURE RADIUS	GLASS	SPECIAL
OBJ	0.000000	1.0000e-20	1.100000	AIR	
AST	0.000000	2.187200	1.100000 AS	AIR	
2	-15.334600	2.310000	1.530000	SILICA	
3	5.580700	0.480000	2.000000	AIR	
4	6.269200	3.200000	2.300000	CAF2	
5	-4.093000	0.500000	2.630000	AIR	
6	-4.090000	1.400000	2.560000	SILICA P	
7	8.928400	0.800000	3.090000	AIR	
8	12.252200	4.200000	3.540000	CAF2 P	
9	-8.903100	0.300000	4.160000	AIR	
10	8.415500	4.600000	4.430000	CAF2 P	
11	-10.748200	1.000000	4.120000	AIR	
12	-7.026300	2.000000	3.810000	SILICA P	
13	7.423300	1.300000	3.740000	AIR	
14	15.995900	4.000000	3.990000	CAF2 P	
15	-6.276800	1.300000	4.210000	AIR	
16	-6.781600	3.500000	3.680000	SILICA P	
17	-14.120400	100.000000	3.860000	AIR	
18	110.920000	4.000000	9.470000	LIF	
19	0.000000	1.000000	9.560000	AIR	
20	162.550000	4.000000	9.600000	LIF P	
21	-49.550000	2.000000	9.630000	SILICA P	
22	170.220000	4.000000	9.680000	LIF P	
23	-185.780000	152.149217 S	9.760000	AIR	
IMS	0.000000	0.000000	11.002019 S		

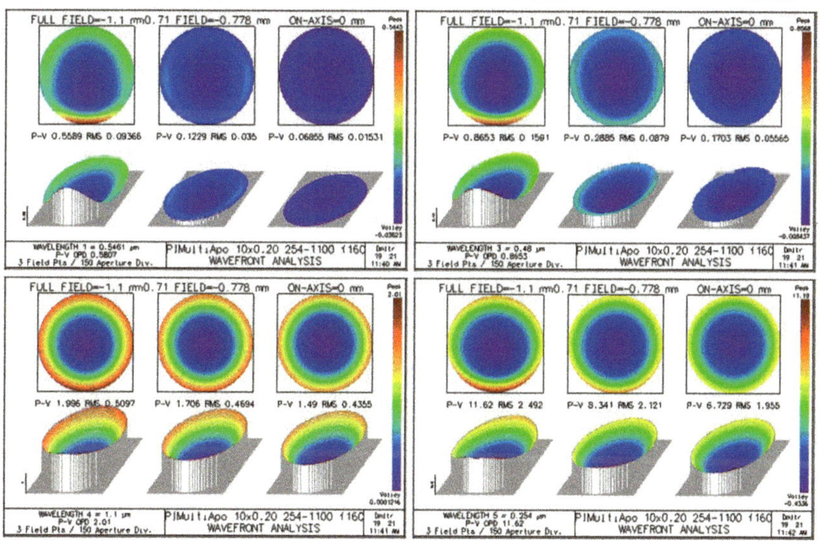

Figure 3.315 Optical design and graphs of aberration correction of a 10×/0.20 CCF plan multi-apochromatic objective (254–1100 nm; $F'_{ts} = 160$ mm).

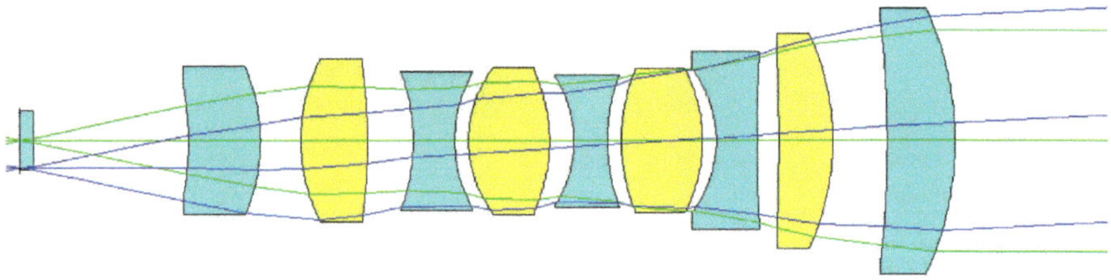

Lens:PlMultiApo10x0.20 200-1800Ftl200

Object num aper	0.200000	Image height	10.000000	Primary wavln	0.546070

SRF	RADIUS		THICKNESS		APERTURE RADIUS		GLASS		SPECIAL
OBJ	0.000000		0.001000		1.004563		AIR		
AST	0.000000		0.500000		1.010000	A	SILICA	C	
2	0.000000		5.800000		1.070000		AIR		
3	-11.809000		2.700000		2.210000		SILICA	C	
4	-6.516000		1.500000		2.680000		AIR		
5	6.220000		2.500000		2.950000		CAF2	C	
6	-19.538000		1.700000		2.800000		AIR		
7	-7.300000		1.500000		2.510000		SILICA	P	
8	4.950000		0.500000		2.460000		AIR		
9	4.260000		3.000000		2.650000		LIF	C	
10	-6.000000		0.900000		2.560000		AIR		
11	-4.260000		1.200000		2.330000		SILICA	C	
12	6.684000		0.500000		2.380000		AIR		
13	6.684000		3.000000		2.500000		LIF	C	
14	-5.495000		0.500000		2.610000		AIR		
15	-4.260000		1.500000		2.610000		SILICA	C	
16	68.137000		0.800000		3.220000		AIR		
17	-81.000000		2.000000		3.510000		CAF2	C	
18	-9.000000		2.000000		3.880000		AIR		
19	-37.535000		2.500000		4.460000		SILICA	C	
20	-11.370000		200.000000		4.790000		AIR		
21	-367.600000		5.000000		14.690000		SILICA	C	
22	-97.660000		1.000000		14.890000		AIR		
23	-78.686000		7.000000		14.880000		LIF	C	
24	-30.700000		2.000000		15.100000		AIR		
25	-28.980000		4.000000		14.750000		SILICA	C	
26	0.000000		3.000000		15.480000		AIR		
27	68.137000		7.000000		16.170000		CAF2	C	
28	-153.581000		198.792187	S	16.230000		AIR		
IMS	0.000000		0.000000		10.000000				

Figure 3.316 Optical design and graphs of aberration correction of a 10×/0.20 CCF plan multi-apochromatic objective (200–1800 nm; $F'_{ts} = 200$ mm).

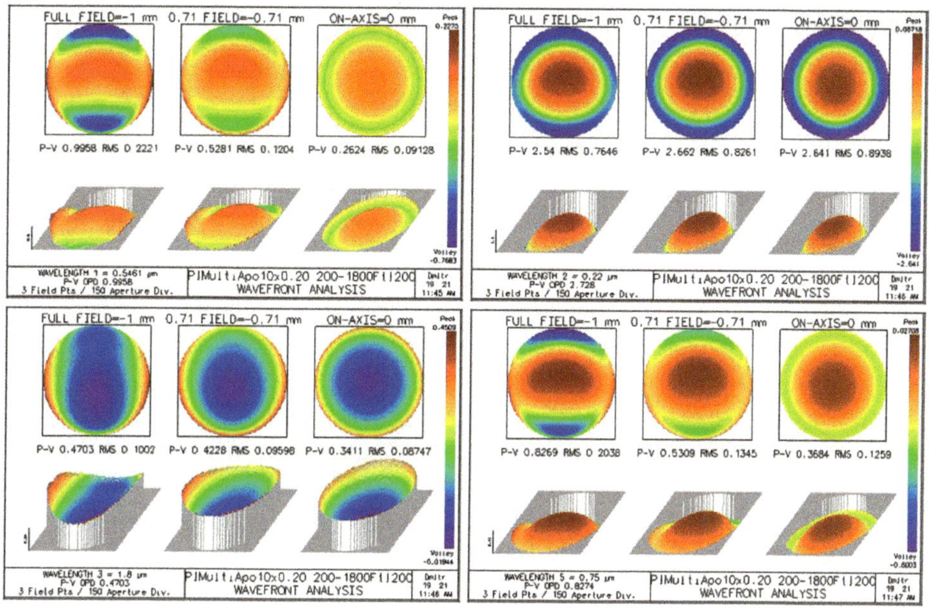

Figure 3.316 (*Continued*) Optical design and graphs of aberration correction of a 10×/0.20 CCF plan multi-apochromatic objective (200–1800 nm; $F'_{ts} = 200$ mm).

Figure 3.317 Photograph of a 10×/0.20 CCF plan multi-apochromatic objective ($F'_{tl} = 200$ mm).

Table 3.13 Main technical parameters and basic optical layout of CCF plan multi-apochromatic objectives.

Magnification	*NA*	*WD* (mm)	*F′* (mm)	*R* (μm)	*DF* (μm)	*FOV* on object (mm)	*FOV* on image (mm)	Effective spectral range (nm)	Optical design at figure no.	The principal optical layout
2.5×	0.03	8.0	64	11.1	370	8.8	22	254–1100	3.313	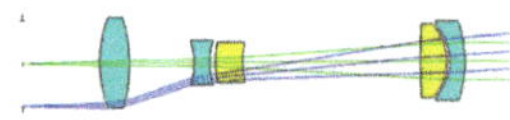
5×	0.10	23	32	3.32	33.3	4.4	22	254–1100	3.314	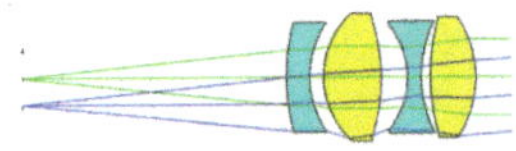
10×	0.20	1.8	16	1.66	8.32	2.2	22	254–1100	3.315	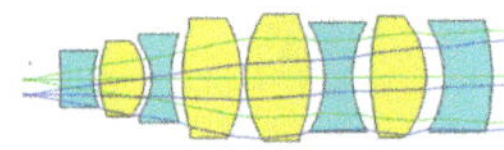
10×	0.20	5.5	20	1.66	8.32	2.2	22	200–1800	3.316	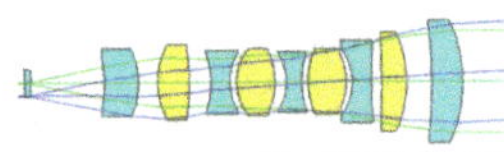

ensure the performance of the required task, namely, the study of bacteria and viruses. But if we compare them with devices widely used in conventional microscopy, these parts should be subjected to some modernization. However, there may be some simplifications in our microscope relative to a conventional microscope. If the so-called contact objective is used to study a living specimen, during operation it will come into contact with a drop of the medium in which the cell is located and integrate into it. In this case, there is no need to use a precision focusing mechanism in the microscope; the object under study will always be located in the front focal plane of the objective. In this case, focusing becomes a mechanism for moving the objective along the depth of a drop. Naturally, there is no need to use a cover glass; some microscopic errors will be missing. We can build an interesting microscope with two identical objectives, using one of them as a condenser; in this case, the achievable resolution value will be closest to the theoretical, corresponding to Abbe's theory. In this case, a drop will be enclosed between two objectives (or, more precisely, between their frontal lenses), which can be positioned at a very small distance from each other; perhaps, in this case, an effect close to a confocal near-field microscopy can be achieved. Finally, by using the technology of applying nanospheres (on the first surface of the frontal lens of the objective), one can also expect some increase in the resolution of the microscope.

A few words need to be said about the visualization system when examining living objects with a light microscope. If we are talking about a conventional microscope and a conventional objective used for visual observation with the eyes, it can be assumed that there is not a typical correlation between the linear magnification of the microscope and the resolution of the objective; the linear dimensions of the studied objects are too small and therefore may not be available in images enlarged within 500 to

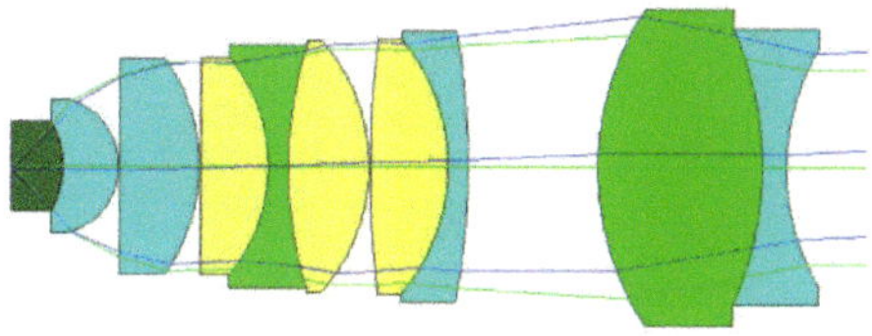

Lens: OF 50x1.40 contact ftl200

Object num aper	1.400000	Image height	10.000000	Primary wavln	0.546070

SRF	RADIUS		THICKNESS		APERTURE RADIUS		GLASS		SPECIAL
OBJ	0.000000		1.0000e-06		0.205669		GALI		
AST	0.000000		3.060000		0.205670	AS	GALI	P	
2	-5.544000		0.100000		2.670000		AIR		
3	-5.544000		3.400000		2.770000		LZ_CTK19		
4	-4.093000		0.100000		3.970000		AIR		
5	0.000000		4.840000		5.570000		LZ_CTK19	P	
6	-10.940000		0.100000		6.400000		AIR		
7	0.000000		4.200000		6.440000		CAF2		
8	-9.817000		1.200000		6.460000		LZ_TF5		
9	23.230000		5.000000		7.220000		CAF2	P	
10	-10.940000		0.100000		7.470000		AIR		
11	61.940000		4.800000		7.580000		CAF2	P	
12	-11.324000		1.200000		7.560000		LZ_CTK19	P	
13	-43.310000		7.980000		8.030000		AIR		
14	16.827000		9.970000		9.360000		LZ_TF5	P	
15	-19.588000		1.500000		8.200000		LZ_CTK19	P	
16	13.645000		100.000000		7.010000		AIR		
17	268.500000		5.000000		11.260000		LZ_BK8		
18	-58.737500		0.300000		11.310000		AIR		
19	-58.600000		3.000000		11.290000		LZ_BF24		
20	-141.550000		197.597544	S	11.370000		AIR		
IMS	0.000000		0.000000		10.000000	S			

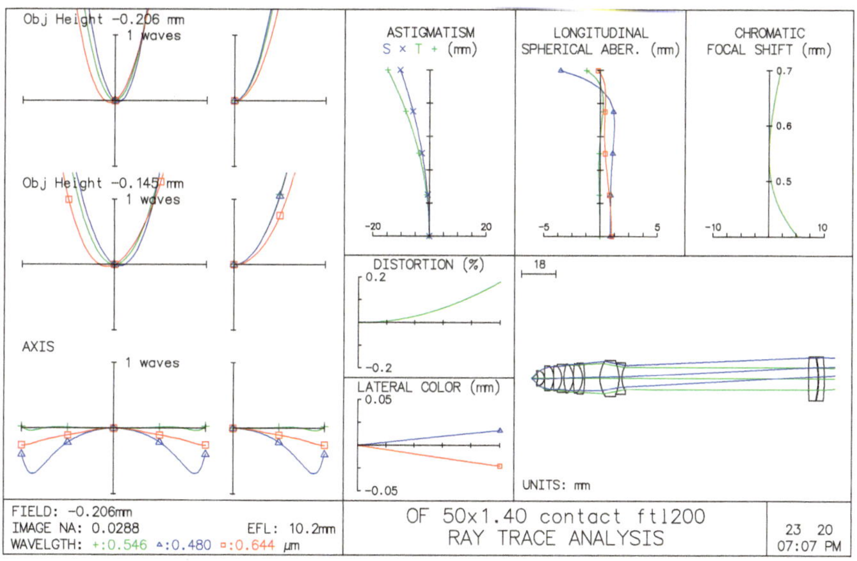

Figure 3.318 Optical design and graphs of aberration correction of a 50×/1.40 ($d = 0$) infinity S-plan MicroFluar contact objective ($F'_{ts} = 200$ mm).

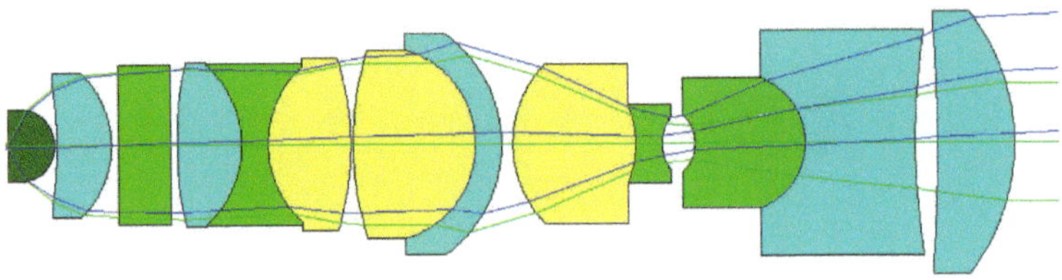

Lens: OF 125x1.70 contact ftl200

Object num aper	1.700000	Image height	10.000000	Primary wavln	0.546070

SRF	RADIUS	THICKNESS	APERTURE RADIUS	GLASS	SPECIAL
OBJ	0.000000	1.0000e-06	0.080008	GALI	
AST	0.000000	2.260000	0.080010 AS	GALI P	
2	-1.588800	0.100000	1.580000	AIR	
3	-20.993200	2.500000	2.620000	LZ_CTK19	
4	-4.375200	0.300000	3.210000	AIR	
5	0.000000	2.400000	3.380000	LZ_TF1	
6	0.000000	0.300000	3.530000	AIR	
7	16.241600	3.000000	3.610000	LZ_CTK19 P	
8	-4.700700	1.200000	3.570000	LZ_TF5	
9	4.624300	3.800000	3.430000	CAF2	
10	-12.321000	0.100000	3.830000	AIR	
11	13.380100	5.500000	4.110000	CAF2 P	
12	-4.344400	1.200000	4.180000	LZ_CTK19 P	
13	-5.923500	0.500000	4.900000	AIR	
14	5.068200	5.500000	3.520000	CAF2 P	
15	-4.545500	1.200000	1.740000	LZ_TF5 P	
16	1.993500	1.500000	1.220000	AIR	
17	-1.697800	5.000000	1.260000	LZ_TF5 P	
18	-3.044000	5.000000	2.880000	LZ_BK6	
19	31.547300	1.000000	4.990000	AIR	
20	-72.754500	3.500000	5.240000	LZ_CTK19 P	
21	-9.282200	100.000000	5.810000	AIR	
22	268.500000	5.000000	11.800000	LZ_BK8	
23	-58.737500	0.300000	11.910000	AIR	
24	-58.600000	3.000000	11.880000	LZ_BF24	
25	-141.550000	197.915550 S	12.020000	AIR	
IMS	0.000000	0.000000	10.000000 S		

Figure 3.319 Optical design and graphs of aberration correction of a 125×/1.70 ($d = 0$) infinity S-Plan MicroFluar contact objective ($F'_{ts} = 200$ mm).

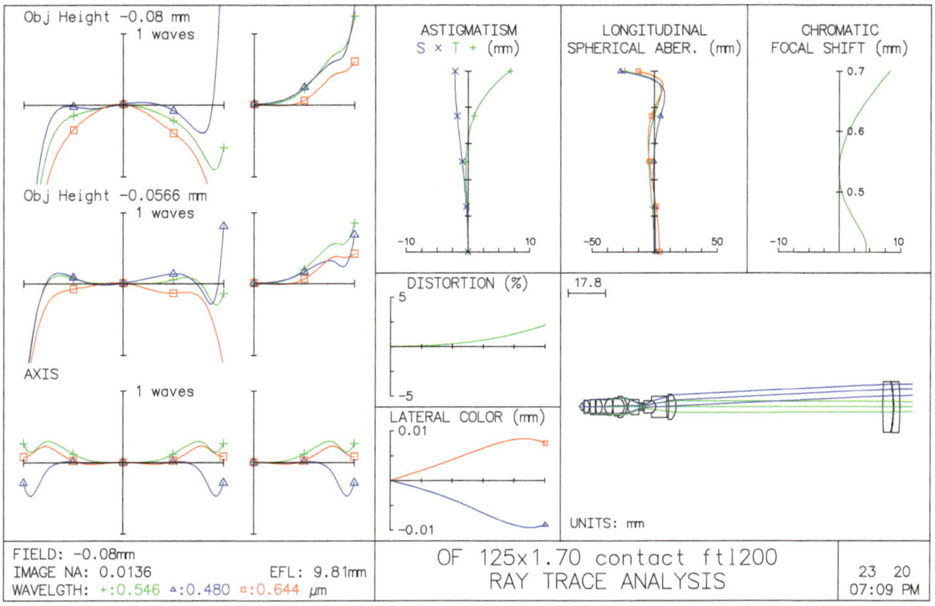

Figure 3.319 (*Continued*) Optical design and graphs of aberration correction of a 125×/1.70 ($d = 0$) infinity S-Plan MicroFluar contact objective ($F'_{ts} = 200$ mm).

1000 NA (NA is the numerical aperture of the objective). Large scale values of the magnifying projection system are required, that is, the values of linear magnifications, for example, of 250× or even 500× objectives. In this case, we are most likely only talking about the detection of small objects without the possibility of their detailed study at the limit of the microscope's resolution. As practice shows, we can detect such objects; the use of digital image receivers can significantly help in this process.

As an example, we can design three objectives that can help in the study of these kinds of living objects. These are the so-called contact objectives, which can provide an image of an object located directly on the first surface of the front lens. In this case, some part of the front lens of such an objective protrudes above its mechanical frame. In such objectives, this lens is very small (because these objectives are in excess of large linear magnifications); because its protrusion cannot be seen without a magnifying glass, photographs of some old similar contact objectives shown in Volume 1 (Chapter 1, Fig. 1.111) can serve as an example.

The method of contact microscopy is known; it was used in cytological, pathological, and other types of research under a microscope. At the same time, objectives were also able to implement fluorescence methods. These objectives were distinguished by an increased numerical aperture/linear magnification ratio. However, the parametric series of such objectives included ones with a maximum magnification of 60× and a numerical aperture of 1.15. However, the aberration correction of these objectives was not perfect; only achromatic correction was achieved (not "plan"), which made it impossible to use them effectively for micrographs.

The objectives we discuss now implement the concept of ultrahigh linear magnification and maximum numerical aperture; a plan MicroFluar type of aberration correction was achieved in the 250× and 500× objectives, and a plan poly-apochromatic type of aberration correction was achieved in the other 500× objective (the effective working spectral range is 355–800 nm), which significantly expands the functionality of the objective, especially when using the fluorescence method. The front numerical aperture of 1.7 is reached in all the presented objectives. Figures 3.320–3.322 show the design parameters and graphs of aberration correction of these objectives.

Table 3.14 shows the main technical parameters of some contact objectives.

Figure 3.323 shows a photograph of a 250×/1.70 objective and a drawing of its principal mechanical design. This objective achieved image quality close to the criterion of plan apochromatic correction; therefore, it can be designated as a plan apochromat.

3.11.1 Example drawings of 500× contact objectives

Figures 3.324–3.411 show a general mechanical design and drawings of optical parts, mechanical frames, and components, which are lenses integrated into the frames. These are for two objectives that have a linear magnification of 500× (the optical designs are shown in Figs. 3.321 and 3.322).

3.12 Microscope Objectives for Direct Projection onto a Digital Image Receiver

In a digital light microscope, observation through the eyepieces may be replaced by direct projection onto an electronic receiver to form an image on a monitor (a PC or TV screen). A special objective is the main element of such a digital light microscope. The extent of a conventional microscope's replacement with one in which a digital camera receiver and monitor screen are used depends on the technical and consumer characteristics of the digital microscope's objective. In the optical system of such a digital microscope, as well as in a conventional microscope, several objectives with different input numerical apertures and linear fields can be used.

If the object under study is non–self-luminous, a special calculation is required to correlate the parameters of the receiver, especially if it is non-monochromatic, with the parameters of the light source of the microscope. This calculation is a kind of convolution of two functions; the spectral characteristics of a specific receiver and the spectral composition of the radiation in a specific light microscope are used as initial data for the

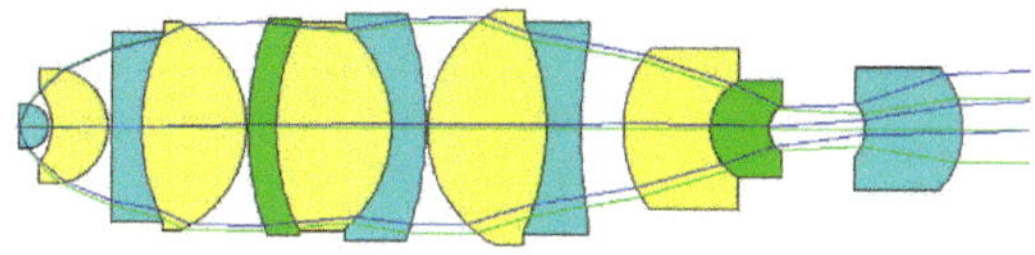

Lens: pl MFluar 250x1.70 cont t1200

Object num aper 1.700000 Image height 10.000000 Primary wavln 0.546070

SRF	RADIUS	THICKNESS	APERTURE RADIUS	GLASS	SPECIAL
OBJ	0.000000	0.000100	0.039990	H-ZLAF55D	
AST	0.000000	1.400000	0.040232 AS	H-ZLAF55D	
2	-1.010000	0.149720	1.010000	AIR	
3	-2.608000	2.600000	1.650000	CAF2	
4	-2.608000	0.200000	2.590000	AIR	
5	-300.000000	1.400000	3.690000	ZLAF1	
6	9.480000	4.900000	4.280000	CAF2 P	
7	-5.160000	0.100000	4.710000	AIR	
8	13.600000	1.300000	4.920000	H-ZF6	
9	10.758000	5.400000	4.720000	CAF2 P	
10	-5.916000	1.600000	4.600000	ZLAF1 P	
11	-14.900000	0.100000	5.140000	AIR	
12	6.066000	5.500000	5.340000	CAF2 P	
13	-11.290000	1.600000	4.820000	ZLAF1 P	
14	24.900000	2.000000	4.230000	AIR	
15	5.960000	4.000000	3.630000	CAF2 P	
16	2.500000	2.700000	2.170000	H-ZF6 P	
17	1.260000	4.500000	1.070000	AIR	
18	-1.820000	4.500000	1.220000	ZLAF1 P	
19	-4.178000	200.000000	2.770000	AIR	
20	-36.100000	14.000000	14.880000	K9	
21	-43.560000	3.000000	17.460000	AIR	
22	377.100000	15.000000	17.940000	CAF2	
23	-81.500000	3.000000	18.350000	AIR	
24	221.300000	17.000000	18.050000	CAF2	
25	-72.000000	3.000000	17.180000	LZ_CTK12	
26	0.000000	199.993929 S	17.100000	AIR	
IMS	0.000000	0.000000	10.000000 S		

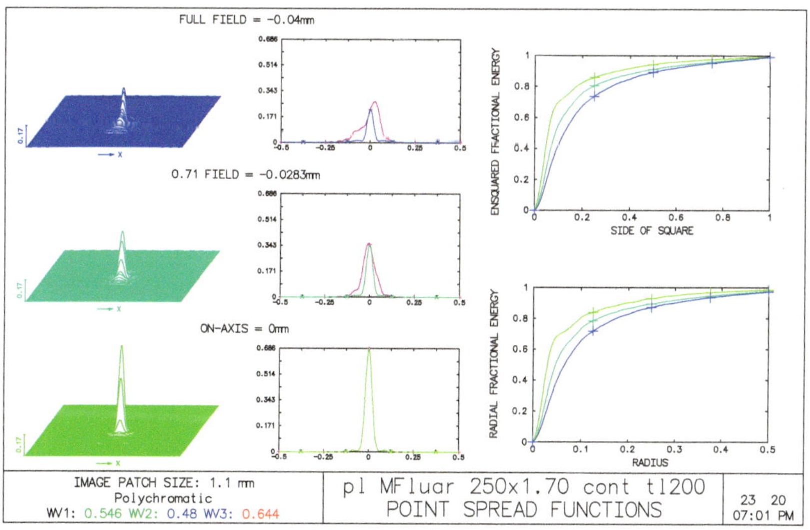

Figure 3.320 Optical design and graphs of aberration correction of a 250×/1.7 LCF plan MicroFluar contact objective ($F'_{ts} = 200$ mm).

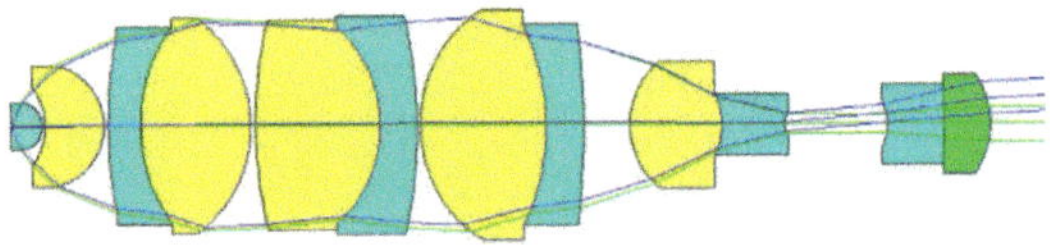

Lens: pl MFluar 500x1.70 cont tl200

Object num aper 1.700000 Image height 10.000000 Primary wavln 0.546070

SRF	RADIUS	THICKNESS	APERTURE RADIUS	GLASS	SPECIAL
OBJ	0.000000	0.000100	0.019995	H-ZLAF55D	
AST	0.000000	1.400000	0.020237 AS	H-ZLAF55D	
2	-1.010000	0.149706	1.010000	AIR	
3	-2.608000	2.600000	1.650000	CAF2	
4	-2.608000	0.200000	2.590000	AIR	
5	18.710000	1.400000	3.850000	LZ_CTK12	
6	7.060000	4.900000	4.190000	CAF2 P	
7	-5.569000	0.200000	4.590000	AIR	
8	18.270000	5.400000	4.500000	CAF2 P	
9	-5.633000	1.600000	4.210000	ZLAF1	
10	-20.945000	0.150000	4.650000	AIR	
11	5.842000	5.500000	4.930000	CAF2 P	
12	-9.817000	1.600000	4.290000	H-ZLAF55D P	
13	-38.950000	2.000000	3.960000	AIR	
14	3.492000	4.000000	2.700000	CAF2 P	
15	-3.070000	2.700000	1.310000	H-ZLAF55D P	
16	0.828000	4.500000	0.490000	AIR	
17	-2.105000	2.500000	0.970000	LZ_CTK12 P	
18	-28.590000	2.000000	1.700000	CDGM_F4	
19	-3.746000	200.000000	2.100000	AIR	
20	-36.100000	14.000000	14.620000	K9	
21	-43.560000	3.000000	17.170000	AIR	
22	377.100000	15.000000	17.660000	CAF2	
23	-81.500000	3.000000	18.090000	AIR	
24	221.300000	17.000000	17.810000	CAF2	
25	-72.000000	3.000000	16.970000	LZ_CTK12 P	
26	0.000000	199.994594 S	16.910000	AIR	
IMS	0.000000	0.000000	10.000000 S		

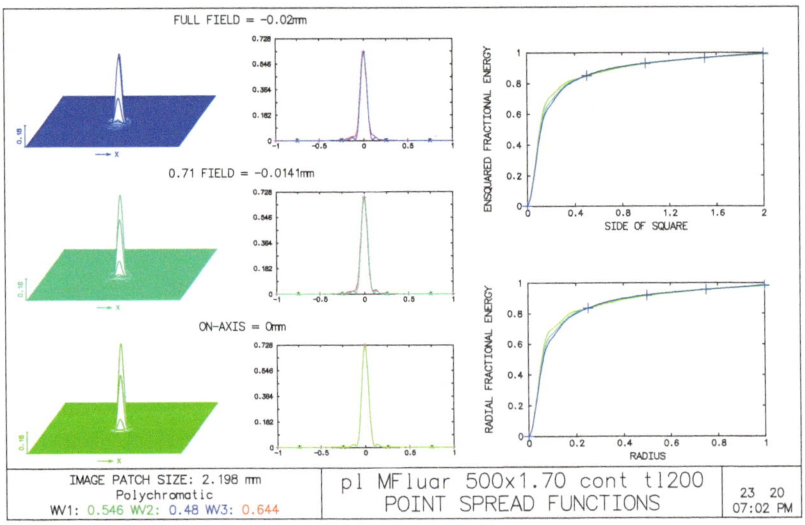

Figure 3.321 Optical design and graphs of aberration correction of a 500×/1.7 LCF plan MicroFluar contact objective ($F'_{ts} = 200$ mm).

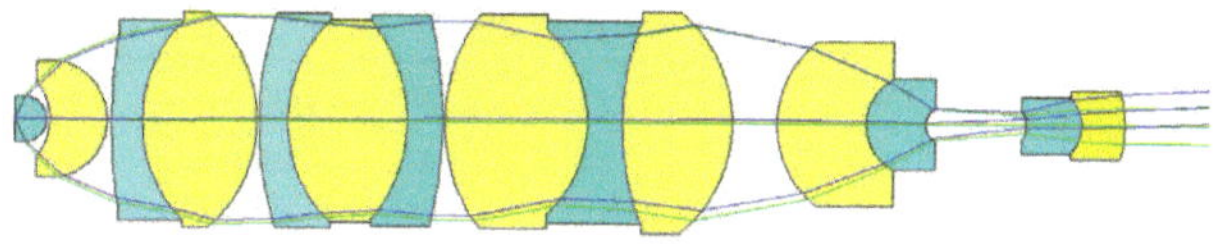

Lens:PlPolyApo500x1.70cont 355-800nm

Object num aper 1.700000 Image height 10.000000 Primary wavln 0.546070

SRF	RADIUS	THICKNESS	APERTURE RADIUS	GLASS	SPECIAL
OBJ	0.000000	0.000100	0.019986	H-ZLAF55D	
AST	0.000000	1.400000	0.020228 AS	H-ZLAF55D	
2	-1.010000	0.149607	1.010000	AIR	
3	-2.608000	2.600000	1.650000	CAF2	
4	-2.608000	0.193400	2.590000	AIR	
5	24.000000	1.400000	3.830000	H-K9L	
6	6.300000	5.200000	4.450000	CAF2 P	
7	-6.300000	0.100000	4.800000	AIR	
8	15.489000	1.300000	4.780000	H-K9L P	
9	6.260000	5.400000	4.510000	CAF2 P	
10	-6.748300	1.600000	4.390000	H-ZLAF69	
11	-19.000000	0.100000	4.710000	AIR	
12	7.900000	6.500000	4.740000	CAF2 P	
13	-5.495000	1.600000	4.120000	H-ZLAF69 P	
14	12.250000	5.000000	4.510000	CAF2 P	
15	-6.260000	2.000000	4.920000	AIR	
16	4.620000	4.000000	3.640000	CAF2 P	
17	2.260000	2.700000	2.020000	H-ZLAF69 P	
18	0.828000	4.500000	0.730000	AIR	
19	-1.248000	2.500000	0.660000	H-ZLAF55D P	
20	-1.755000	2.000000	1.240000	H-FK61	
21	-6.087000	200.000000	1.520000	AIR	
22	-36.100000	14.000000	12.780000	K9	
23	-43.560000	3.000000	15.110000	AIR	
24	377.100000	15.000000	15.520000	CAF2	
25	-81.500000	3.000000	15.960000	AIR	
26	221.300000	17.000000	15.760000	CAF2	
27	-72.000000	3.000000	15.030000	LZ_CTK12	
28	0.000000	200.013959 S	15.000000	AIR	
IMS	0.000000	0.000000	10.000000 S		

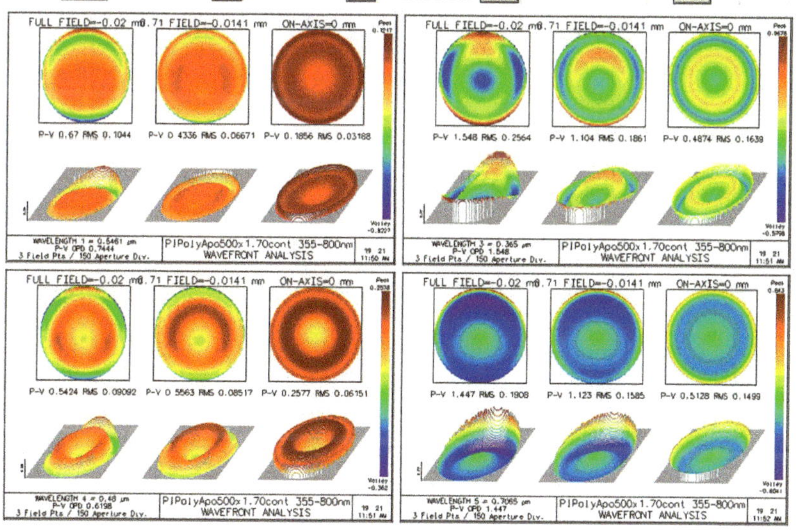

Figure 3.322 Optical design and graphs of aberration correction of a 500×/1.7 contact CCF plan poly-apochromatic objective (F'_{ts} = 200 mm; 355–800 nm).

Table 3.14 Main technical parameters and basic optical layout of contact objectives.

Magnification	*NA*	*WD* (mm)	*F′* (mm)	*R* (μm)	*DF* (μm)	*FOV* on object (mm)	*FOV* on image (mm)	Optical design at figure no.	The principal optical layout
50×	1.40	0	4.0	0.24	0.17	0.40	20	3.318	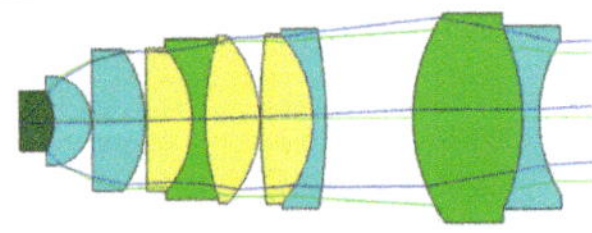
125×	1.70	0	1.6	0.20	0.12	0.16	20	3.319	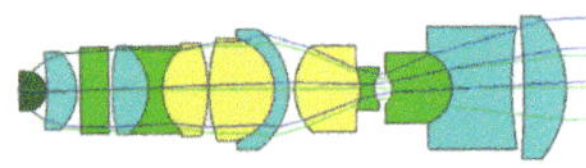
250×	1.70	0	0.8	0.20	0.12	0.08	20	3.320	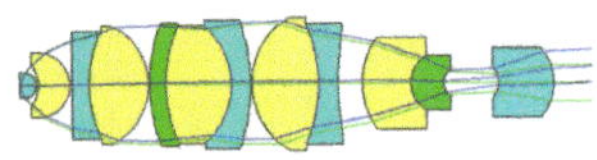
500×	1.70	0	0.4	0.20	0.12	0.04	20	3.321	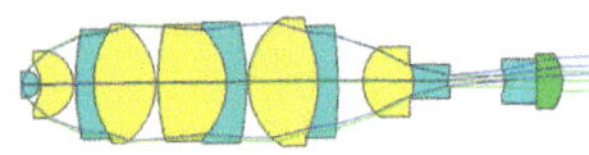
500×	1.70	0	0.4	0.20	0.12	0.04	20	3.322	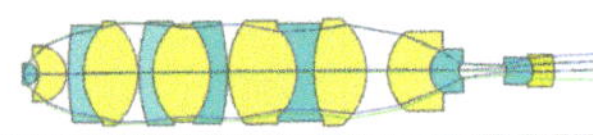

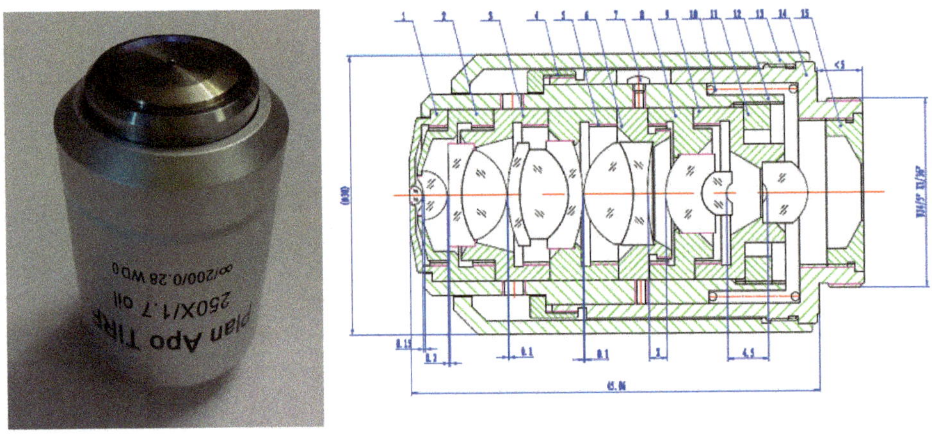

Figure 3.323 Photograph of a 250×/1.7 objective and a drawing of its principal mechanical design.

calculation. The calculated results can be used to design the optical circuit of the objective. Residual aberrations of the optical system in digital microscope objectives, such as curvature, astigmatism, and distortion, must be corrected with particular care. It is necessary to achieve a plan MicroFluar type of aberration correction and higher in the spectral range corresponding to the spectral range of the digital receiver.

Along with the traditional requirements for obtaining a congruent image of the studied object, specific conditions must be fulfilled, such as ensuring the

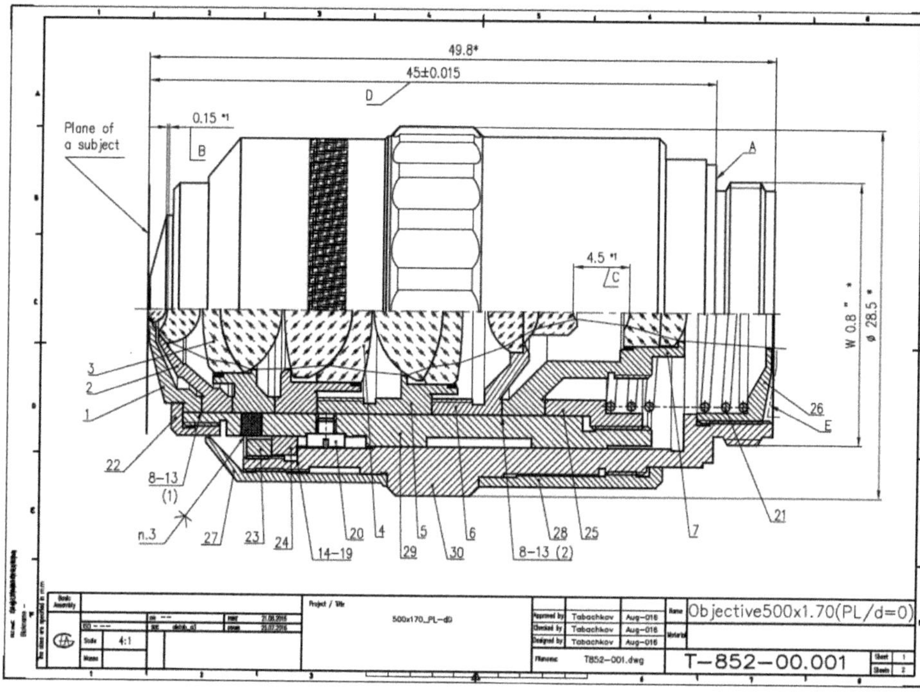

Figure 3.324 A drawing of a general mechanical design of a 500×/1.70 plan MicroFluar contact objective.

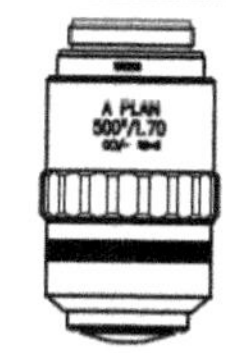

Item	Name	Description	Qty	Material	Note
1	T–852–22.001	Component 1	1		
2	T–852–22.002	Component 2	1		
3	T–852–22.003	Component 3	1		
4	T–852–22.004	Component 4	1		
5	T–852–22.005	Component 5	1		
6	T–852–22.007	Component 6B	1		
7	T–852–22.008	Component 7	1		
8	T–852–02.008–01	Correctional Ring	2	Foil Copper	Selection
9	T–852–02.008–02	Correctional Ring	2	Foil Copper	Selection
10	T–852–02.008–03	Correctional Ring	2	Foil Copper	Selection
11	T–852–02.008–04	Correctional Ring	2	Foil Copper	Selection
12	T–852–02.008–05	Correctional Ring	2	Foil Copper	Selection
13	T–852–02.008–06	Correctional Ring	2	Tape Brass	Selection
14	T–852–02.009–01	Correctional Ring	1	Foil Copper	Selection
15	T–852–02.009–02	Correctional Ring	1	Foil Copper	Selection
16	T–852–02.009–03	Correctional Ring	1	Foil Copper	Selection
17	T–852–02.009–04	Correctional Ring	1	Foil Copper	Selection
18	T–852–02.009–05	Correctional Ring	1	Foil Copper	Selection
19	T–852.02–009–06	Correctional Ring	1	Tape Brass	Selection
20	T–852–02.010	Screw	1	Brass	
21	T–852–02.011	Compression Spring	1	Spring Steel	
22	T–852–02.012	Ring Butt	1	Brass	
23	T–852–02.013	Nut	1	Duralumin	
24	T–852–02.014	Adjusting Ring	1	Brass	
25	T–852–02.015–01	Nut	1	Duralumin	
26	T–852–02.016–02	Blende	1	Duralumin	
27	T–852–02.018–17	Ring	1	T–852–02.017–01	
28	T–852–02.020–01	Engraving ring	1	T–852–02.019	
29	T–852–02.021	Frame general	1	Brass	
30	T–852–02.022	Case	1	Brass	

1. * The dimensions for information.
2. *1 The size settlement.
3. Viksint GED–1 (TU 38.303–04–04–90).
4. The point of the subject, situated on the perpendicular to the surface [A] passing through the point of intersection of the axis of the thread with the surface [A] should be depicted lens on this perpendiculur. Deviation given in the plane of the subject, not more than 0,015мм.
5. Correction Objective, spherical aberration provide change of air gap [B], the installation of one of the detail [8,9,10,11,12,13].
6. Correction of the focal segment Objective (the combination with the first surface of the front lens) to provide a change in the air gap [C] install one of the detail [8,9,10,11,12,13].
7. Size [D] to ensure the installation one of the detail [14–19]
8. Detail [29] and [30] complete with a gap of 0,004мм.
9. The movement of the detail [29] against the detail [30] to be smooth across the range. Details to put in oil [90–WA].
10. Threaded connection of detail [29/22, 29/25, 30/27, 30/28] put on Enamel EP–51, black (GOST9640–85).
11. Covering of a surface [E] – Enamel EP–51,black (GOST9640–85).
12. Objective correspond to the requirements of TU 3–3.870–83.

Project / Title: 500x170_PL–d0

Approved by	Tabachkov	Aug–016
Checked by	Tabachkov	Aug–016
Designed by	Tabachkov	Aug–016
Filename	T852–001.dwg	

Name: Objective500x1.70(PL/d=0)

T–852–00.001

Figure 3.325 Specifications.

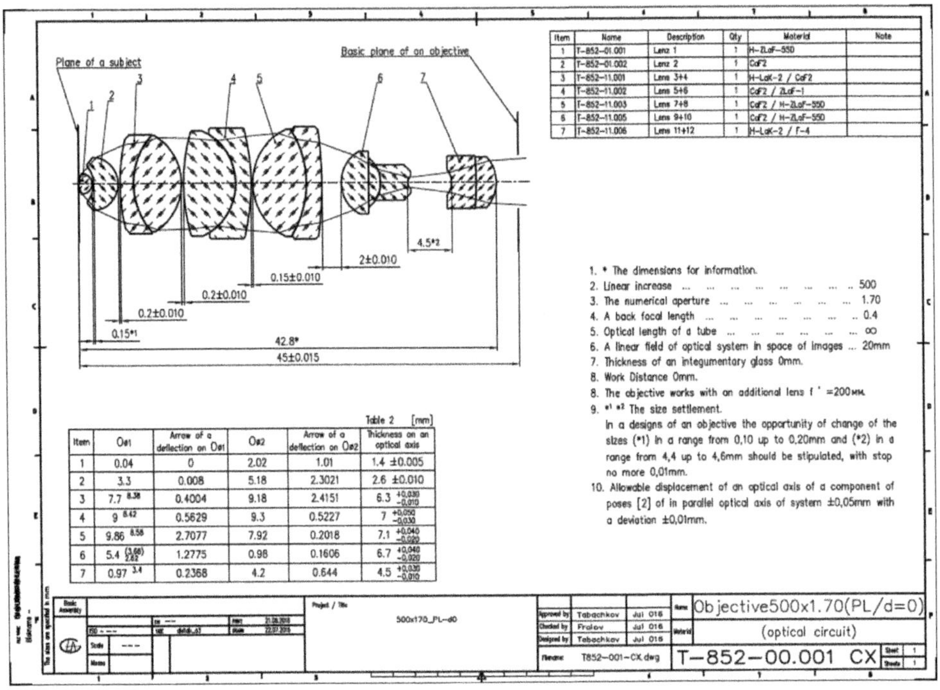

Figure 3.326 Principal optical layout of a 500×/1.70 plan MicroFluar contact objective.

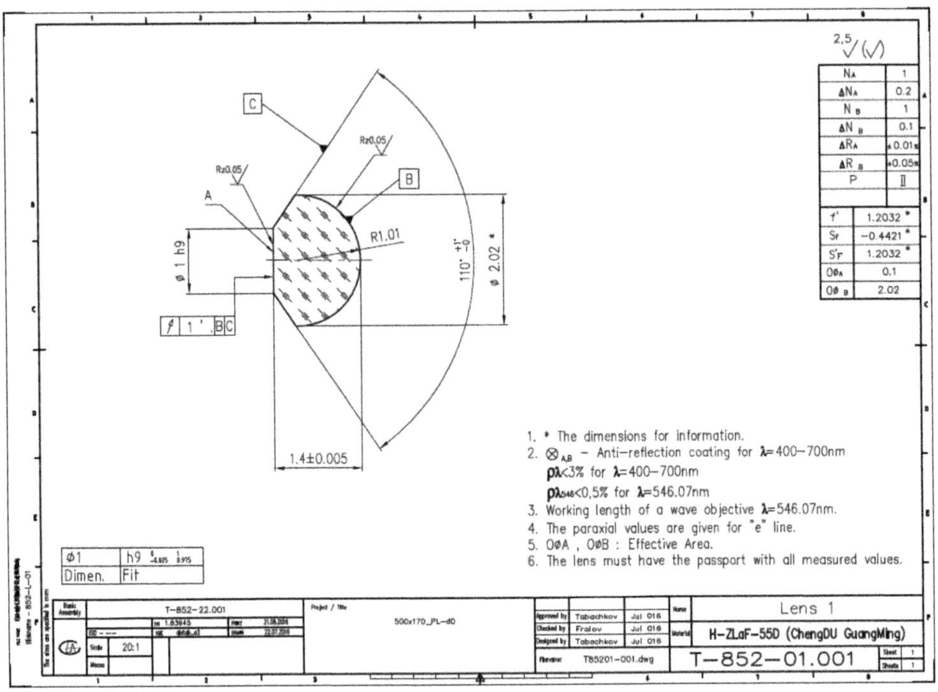

Figure 3.327 A drawing of the first lens.

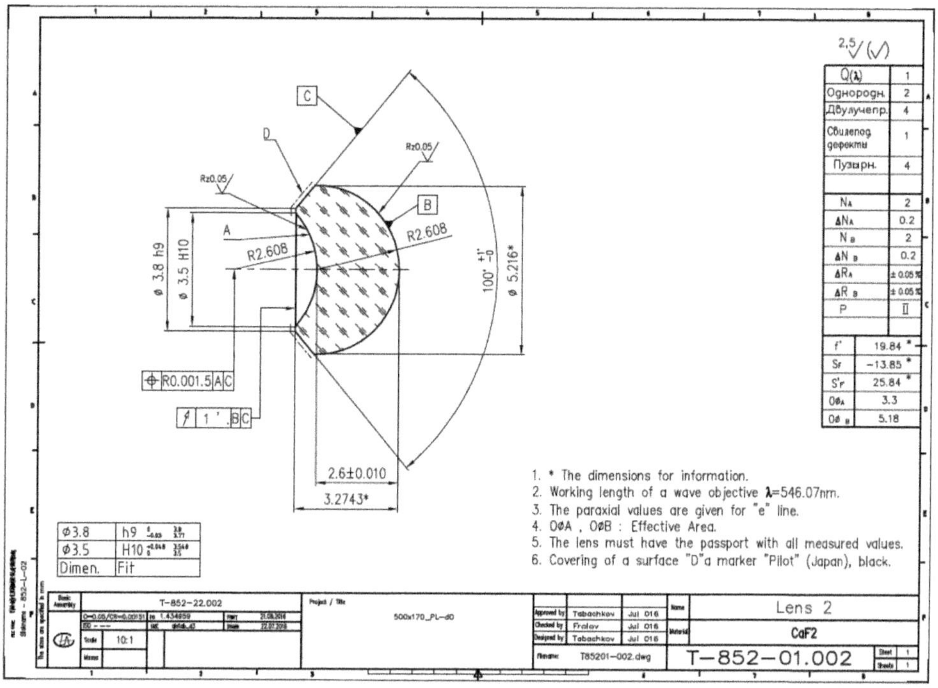

Figure 3.328 A drawing of the second lens.

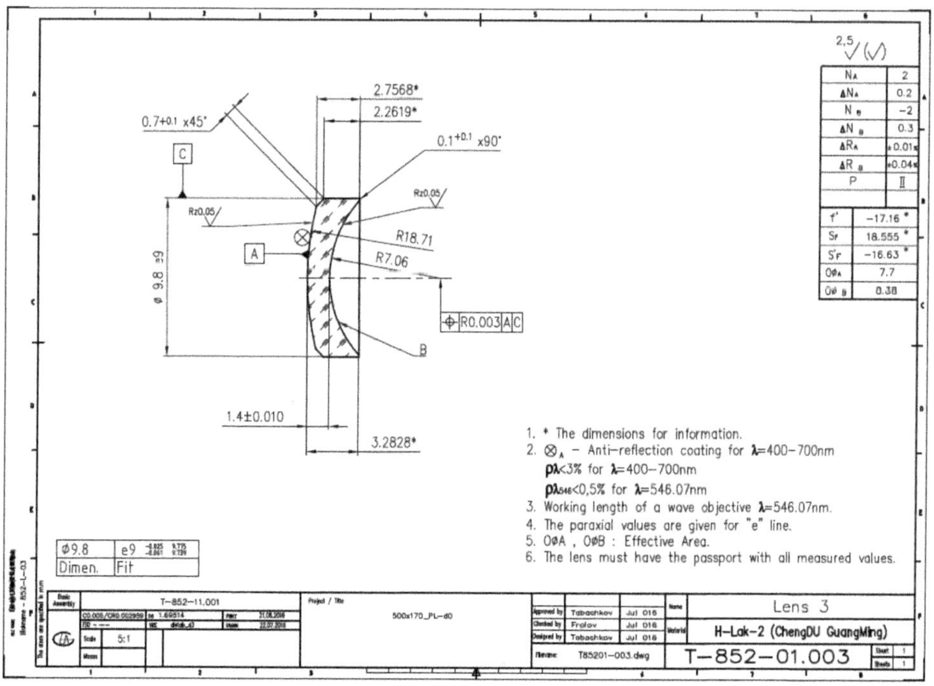

Figure 3.329 A drawing of the third lens.

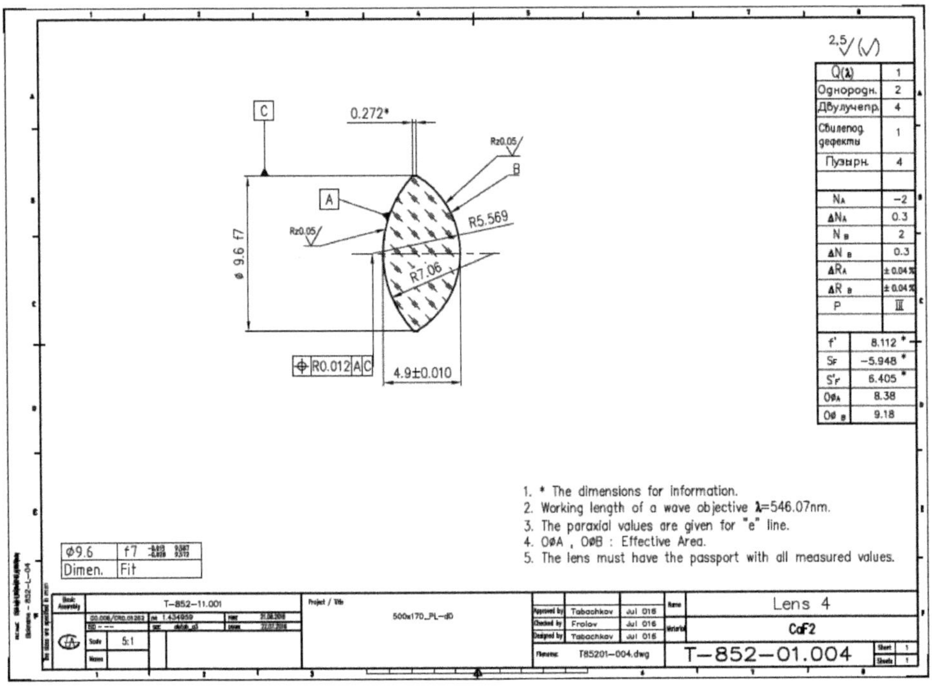

Figure 3.330 A drawing of the fourth lens.

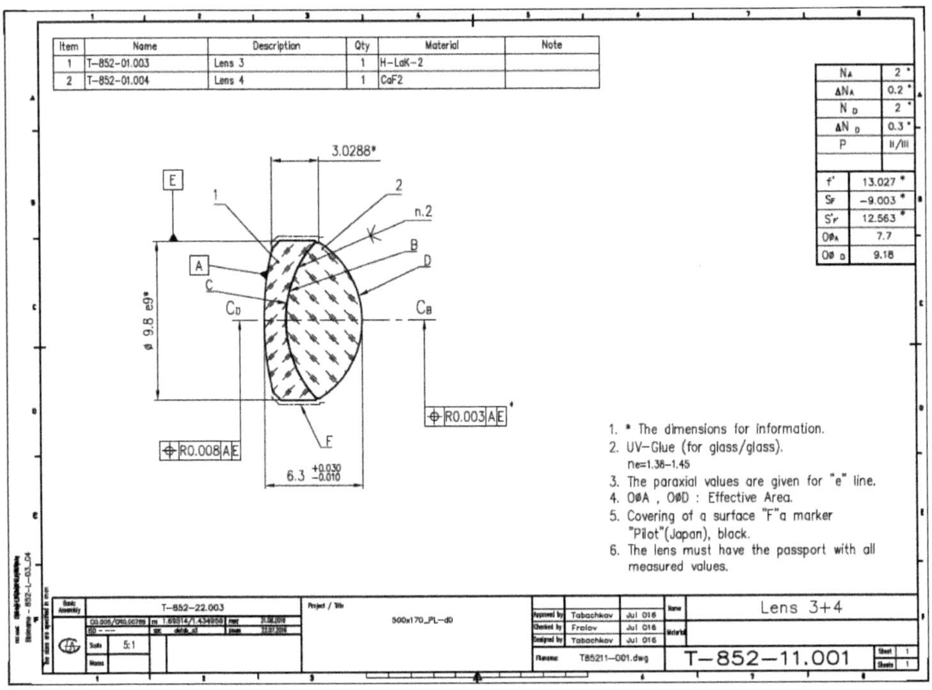

Figure 3.331 A drawing of the third and fourth lenses.

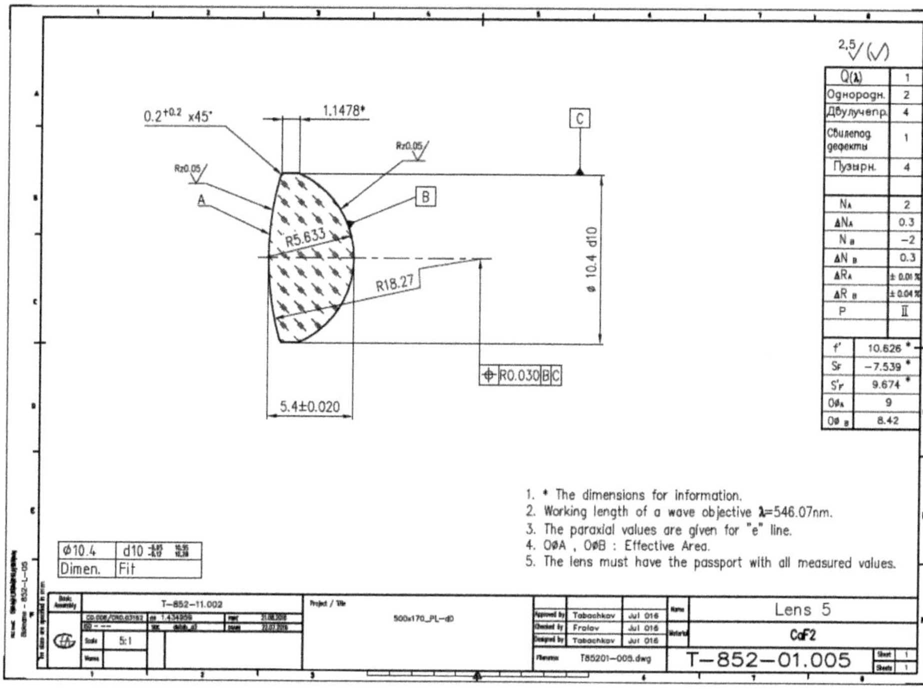

Figure 3.332 A drawing of the fifth lens.

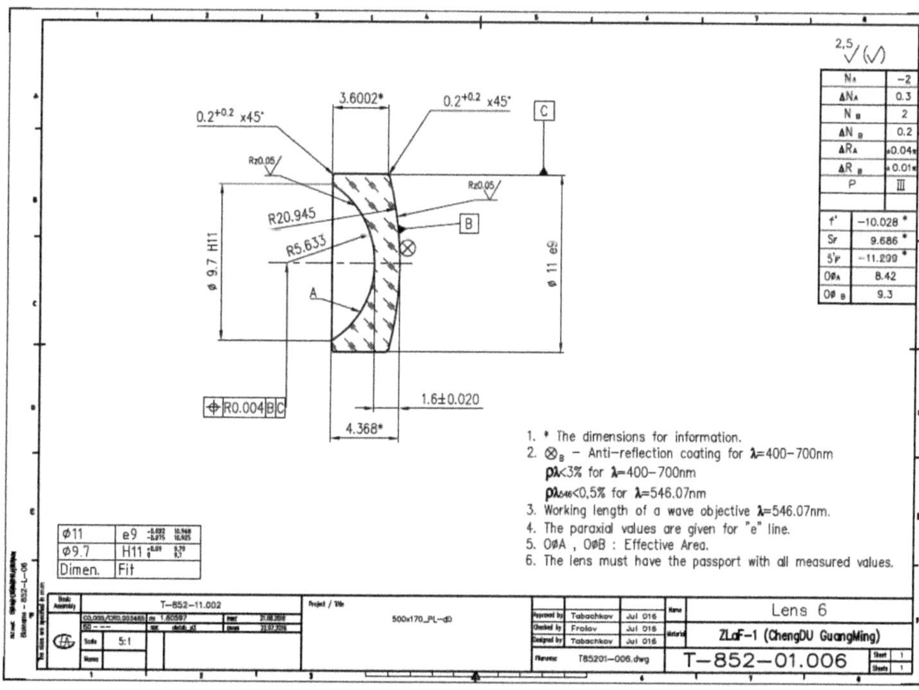

Figure 3.333 A drawing of the sixth lens.

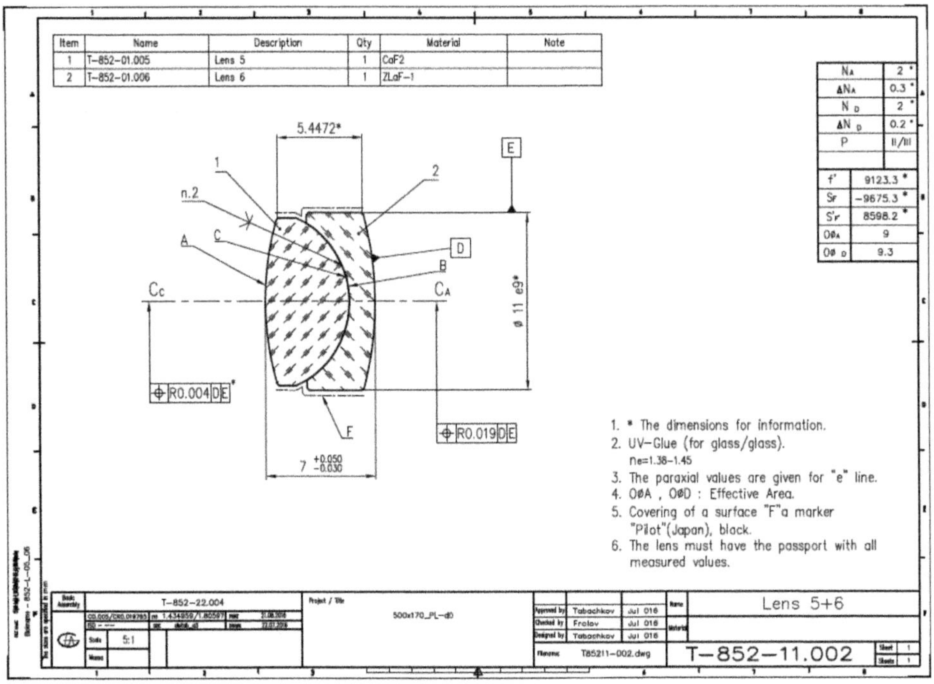

Figure 3.334 A drawing of the fifth and sixth lenses.

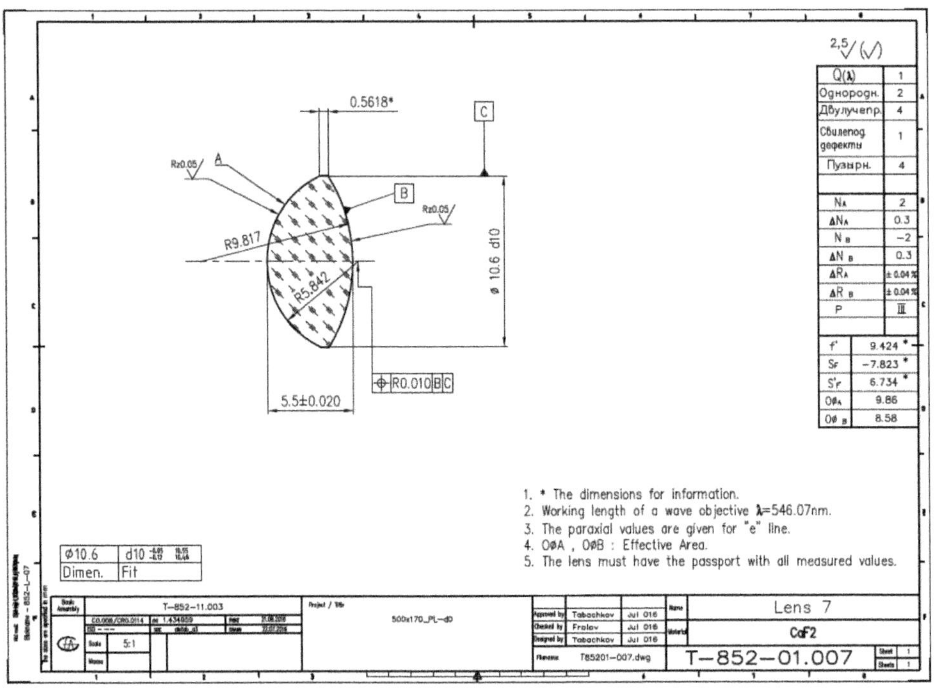

Figure 3.335 A drawing of the seventh lens.

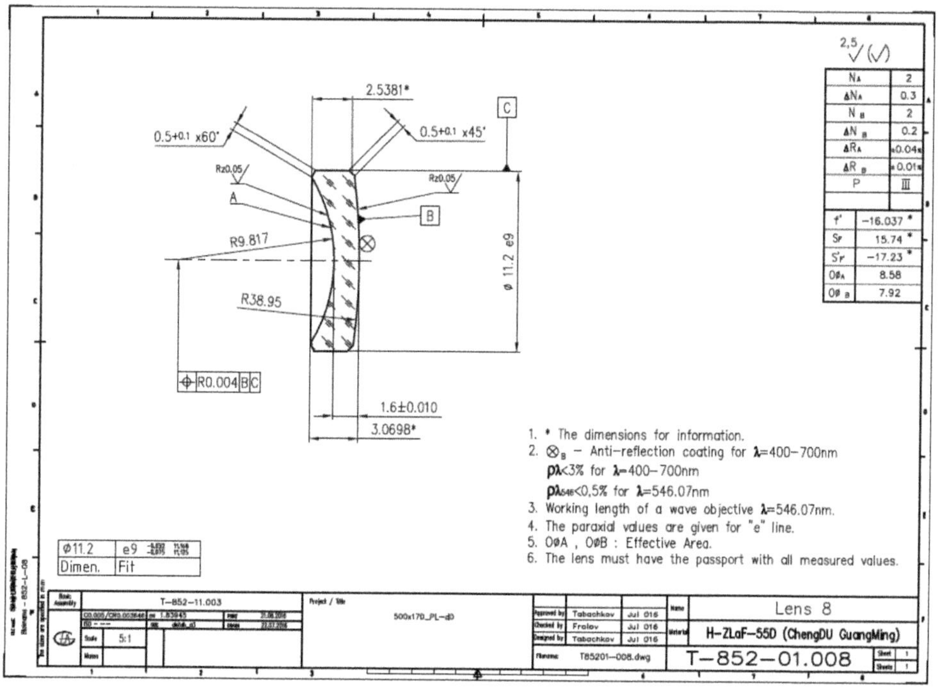

Figure 3.336 A drawing of the eighth lens.

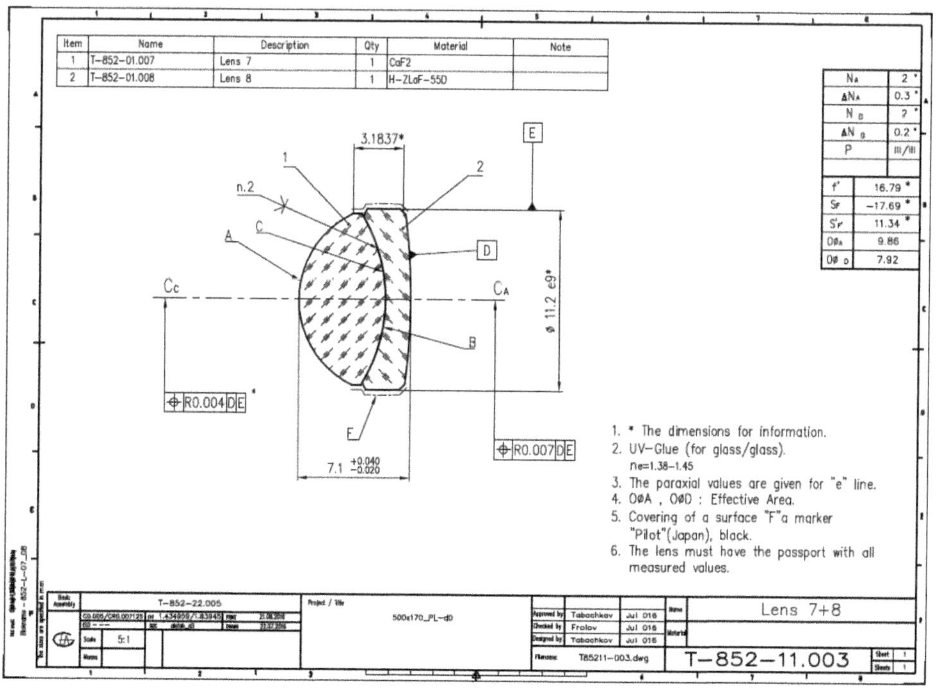

Figure 3.337 A drawing of the seventh and eighth lenses.

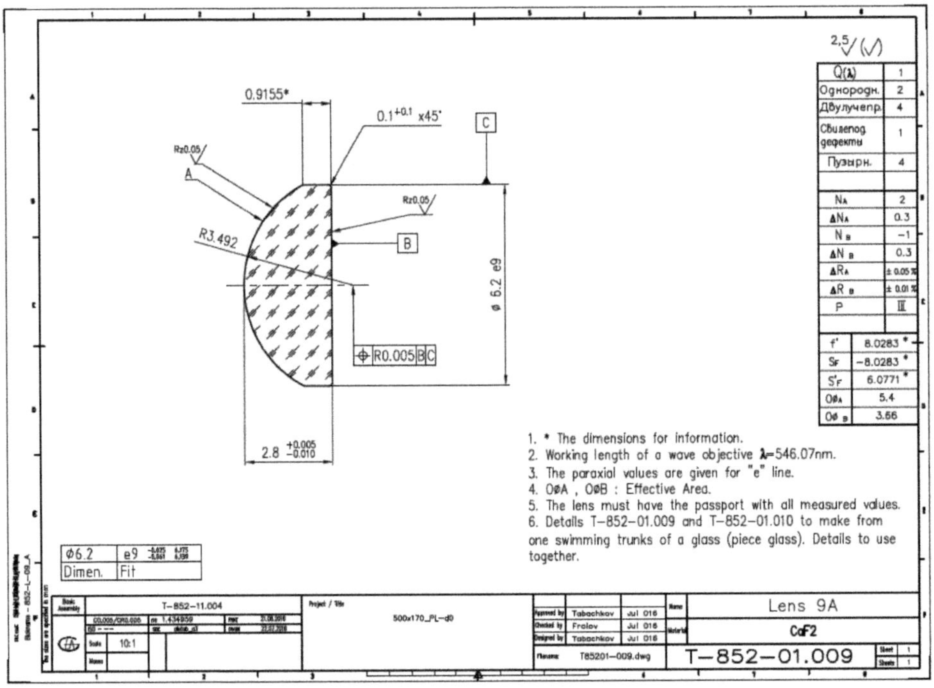

Figure 3.338 A drawing of the ninth "A" lens.

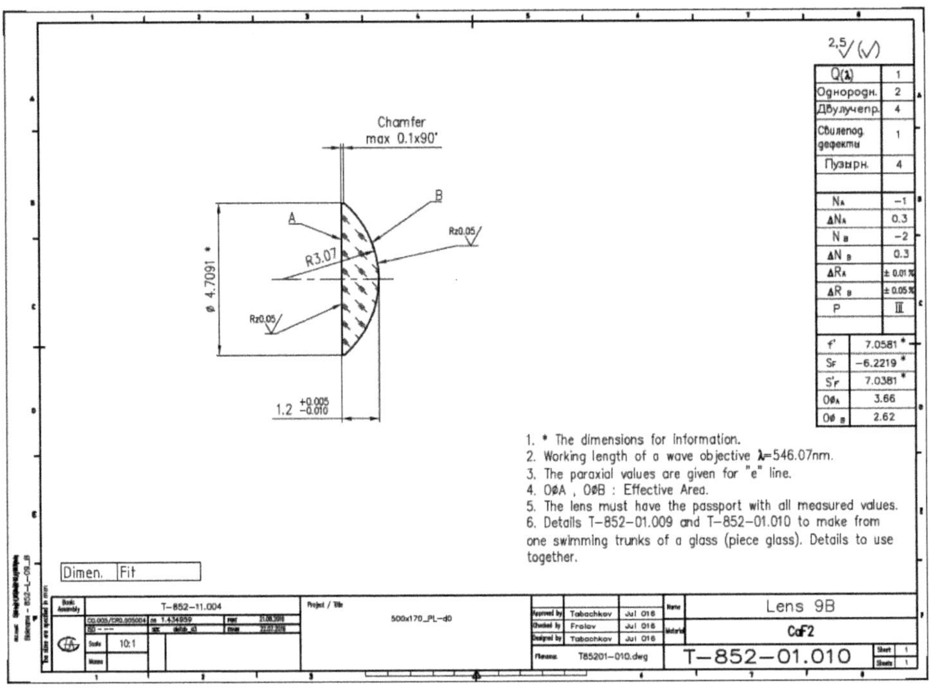

Figure 3.339 A drawing of the ninth "B" lens.

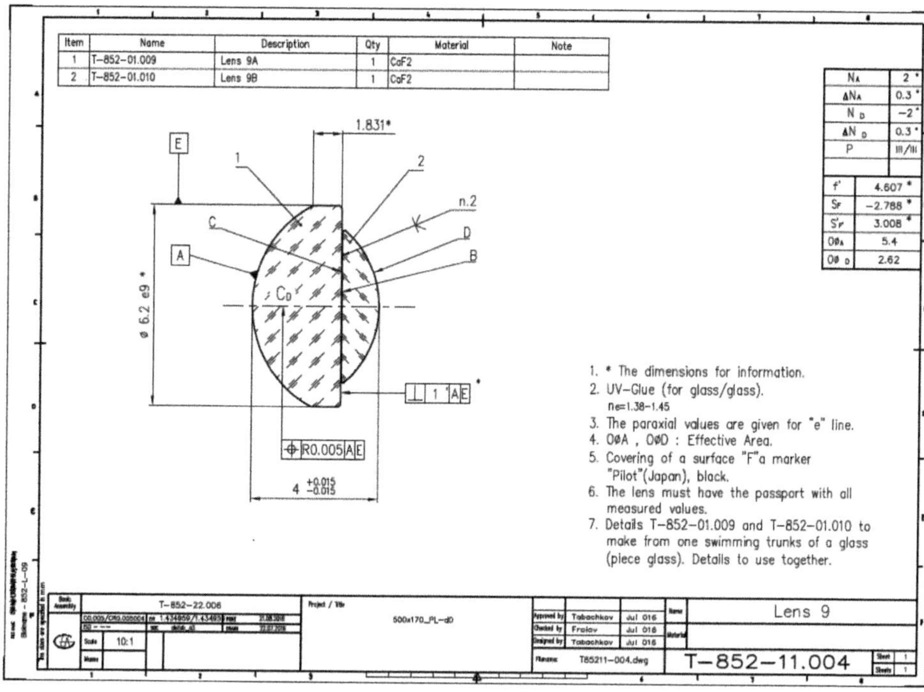

Figure 3.340 A drawing of the ninth lens.

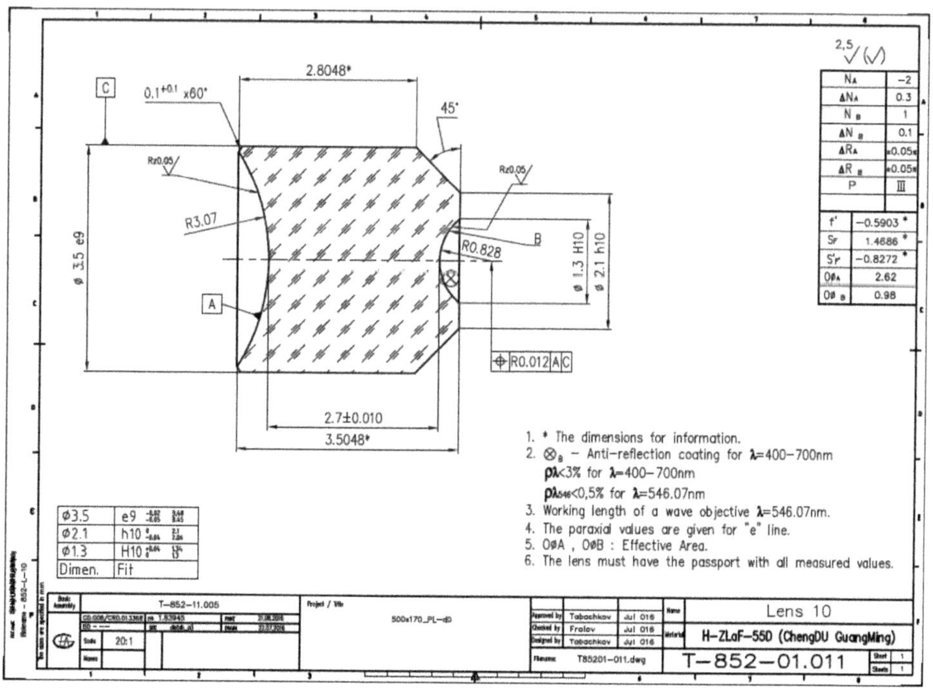

Figure 3.341 A drawing of the 10th lens.

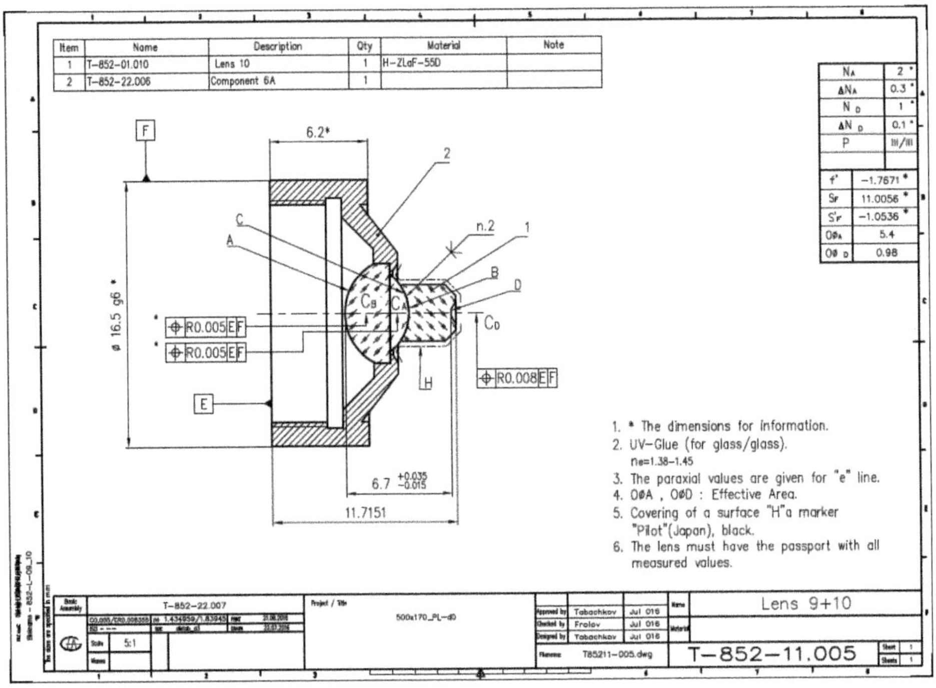

Figure 3.342 A drawing of the ninth and 10th lenses.

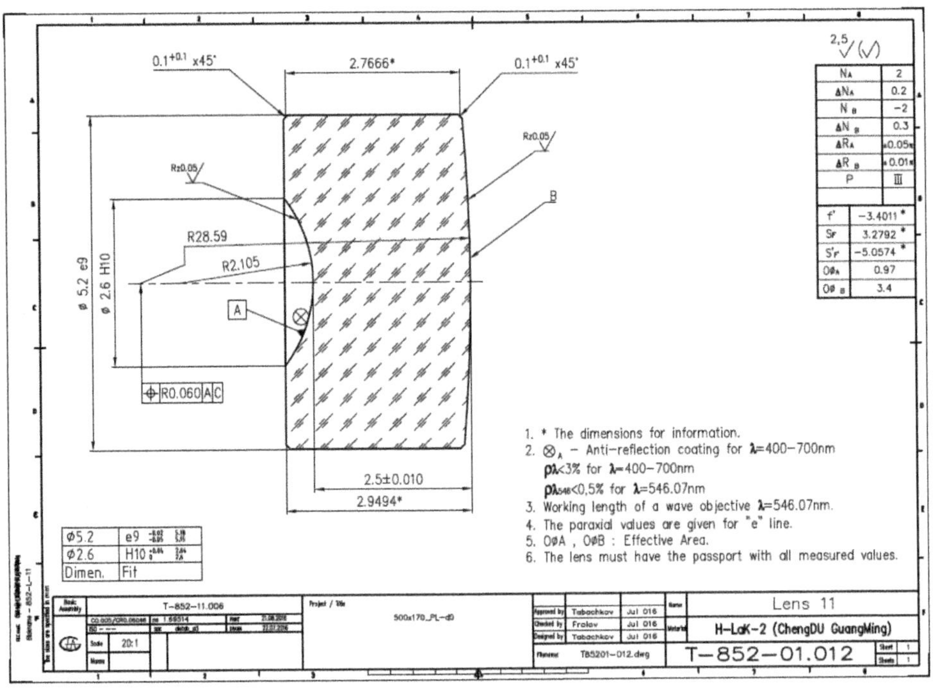

Figure 3.343 A drawing of the 11th lens.

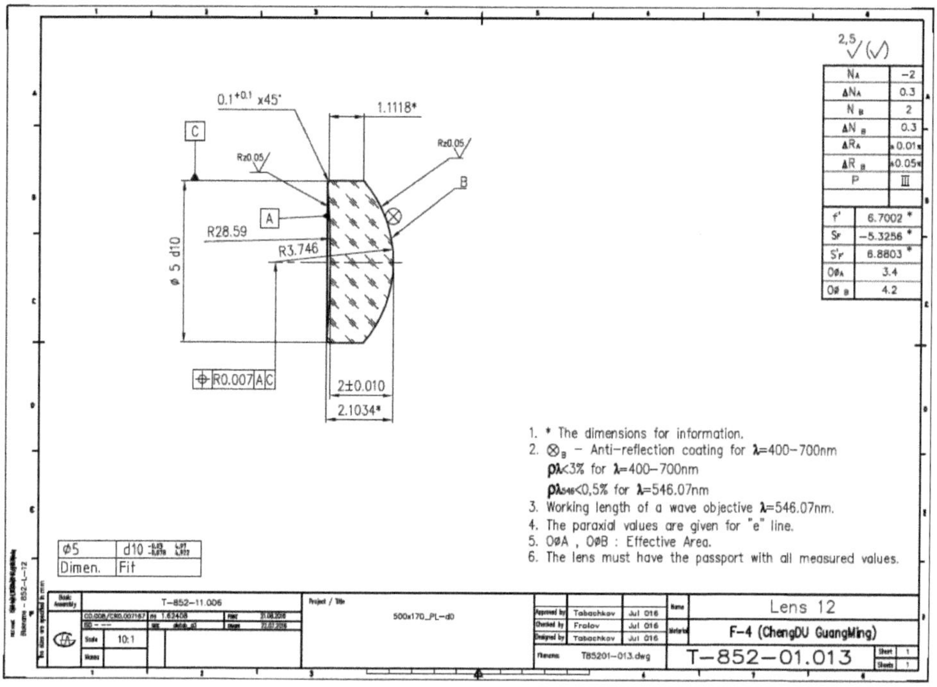

Figure 3.344 A drawing of the 12th lens.

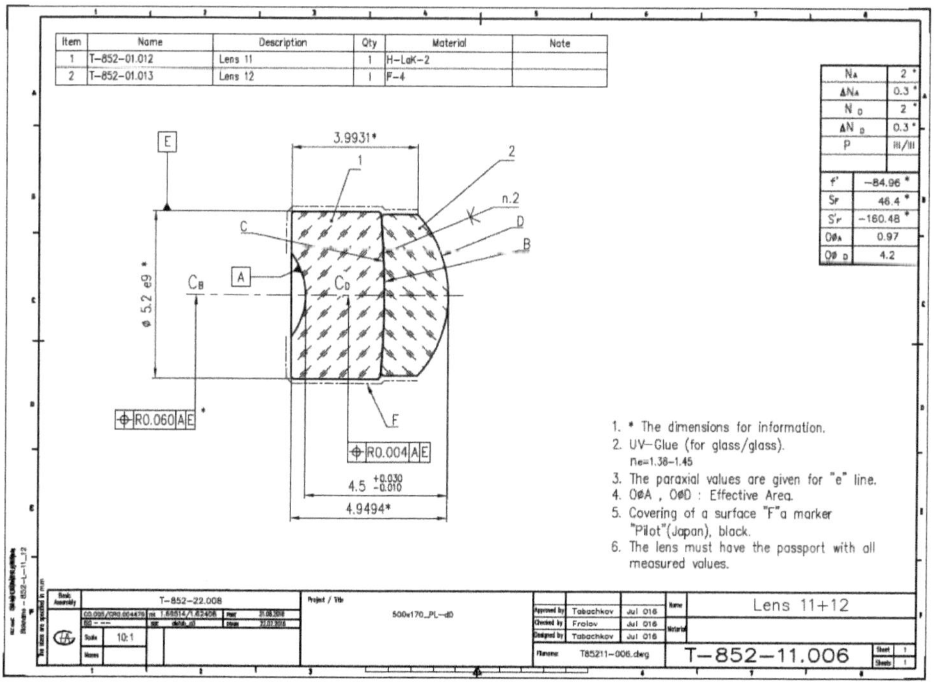

Figure 3.345 A drawing of the 11th and 12th lenses.

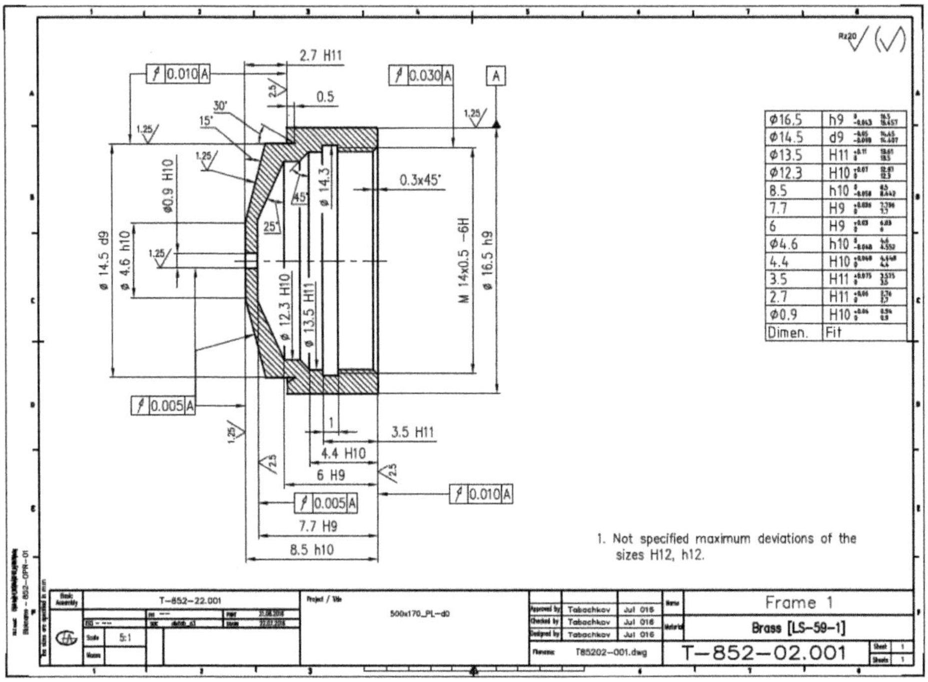

Figure 3.346 A drawing of the first frame.

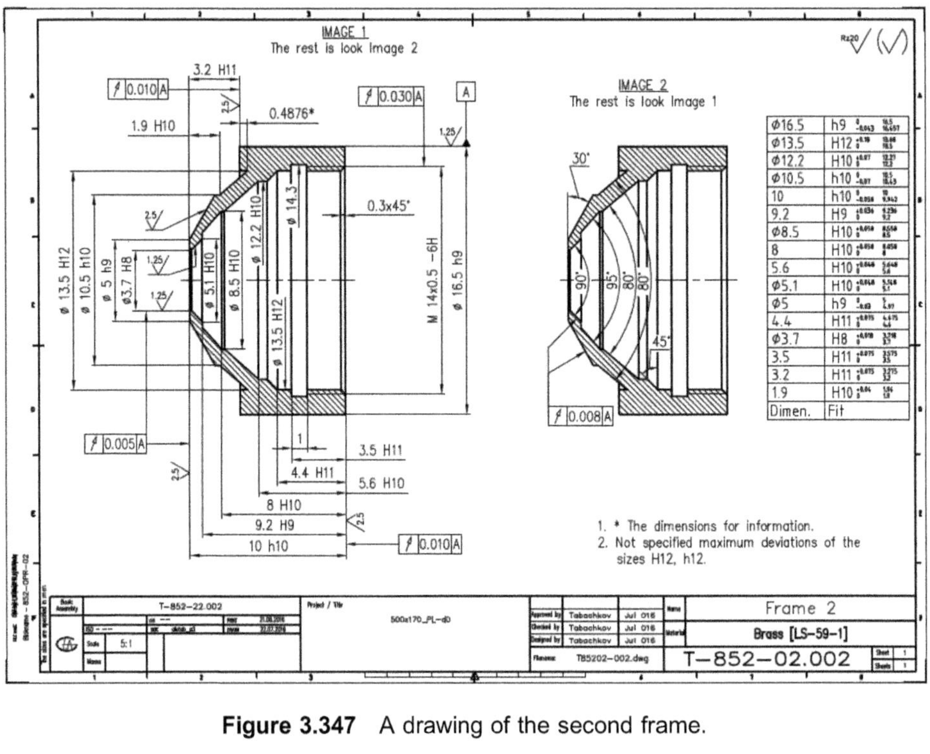

Figure 3.347 A drawing of the second frame.

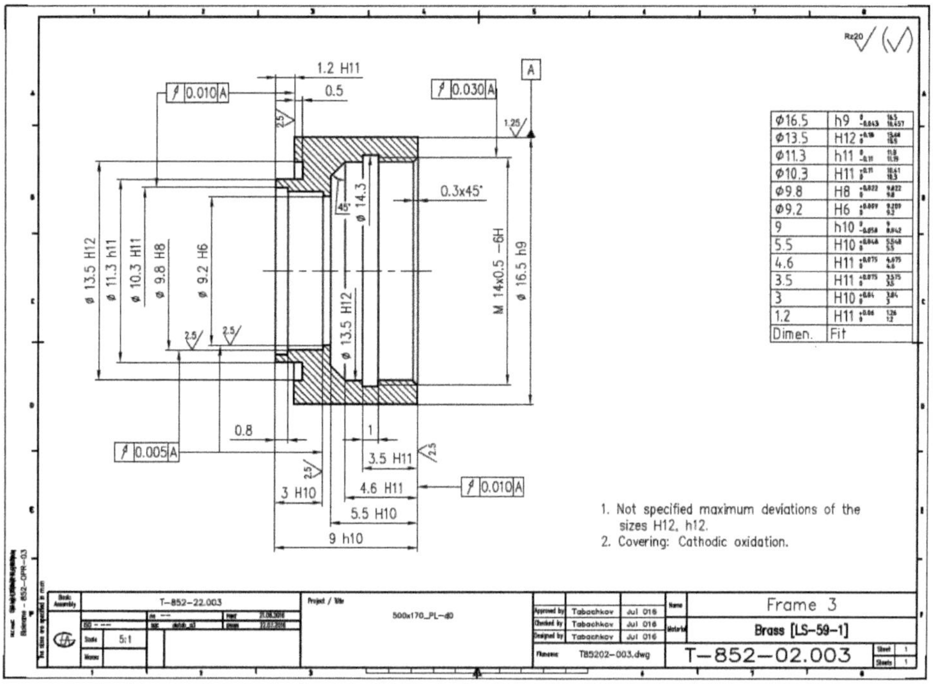

Figure 3.348 A drawing of the third frame.

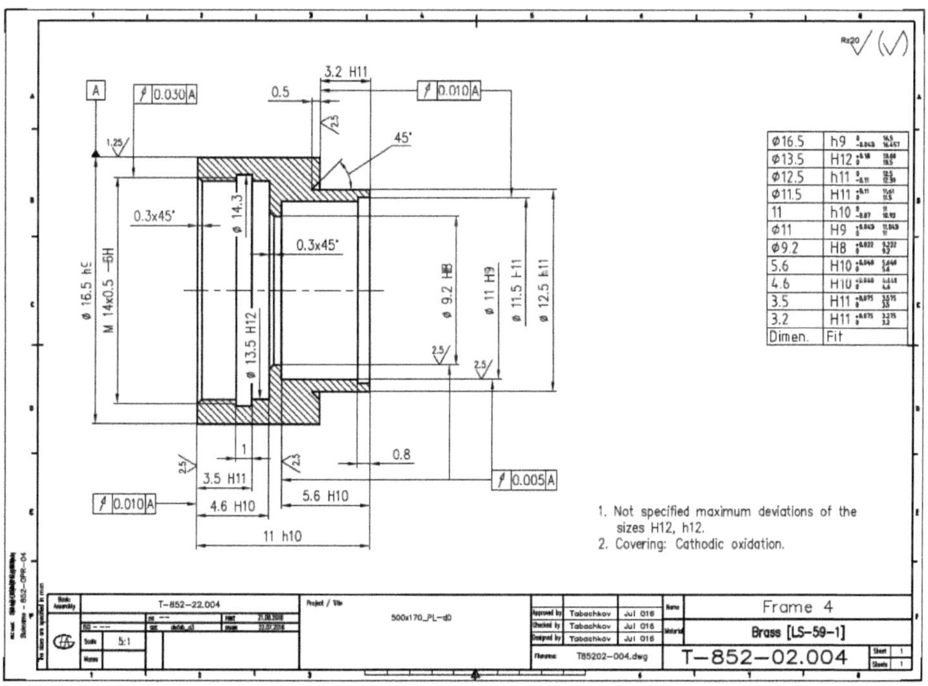

Figure 3.349 A drawing of the fourth frame.

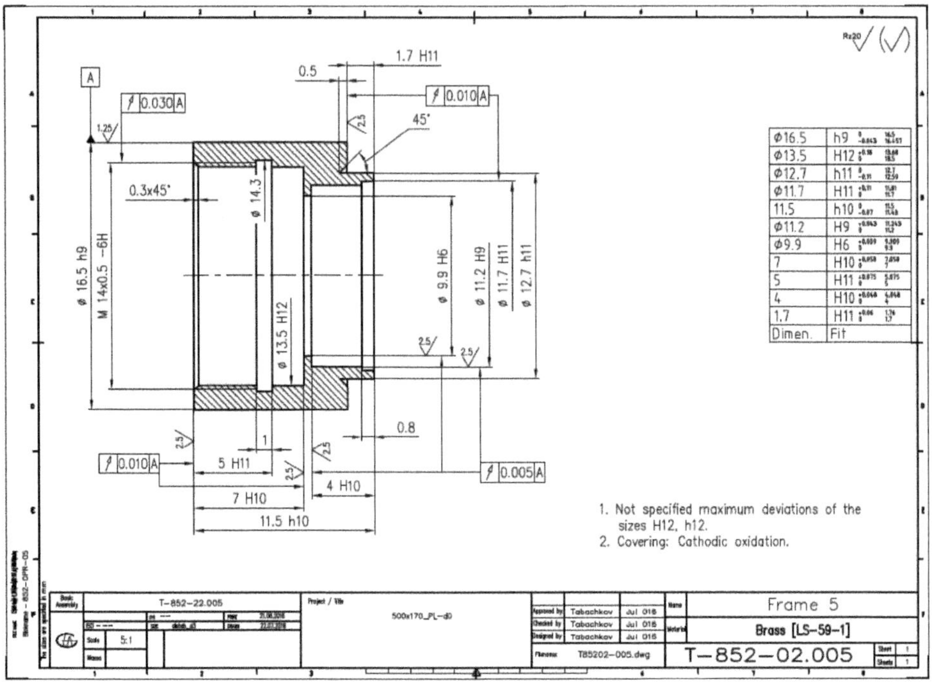

Figure 3.350 A drawing of the fifth frame.

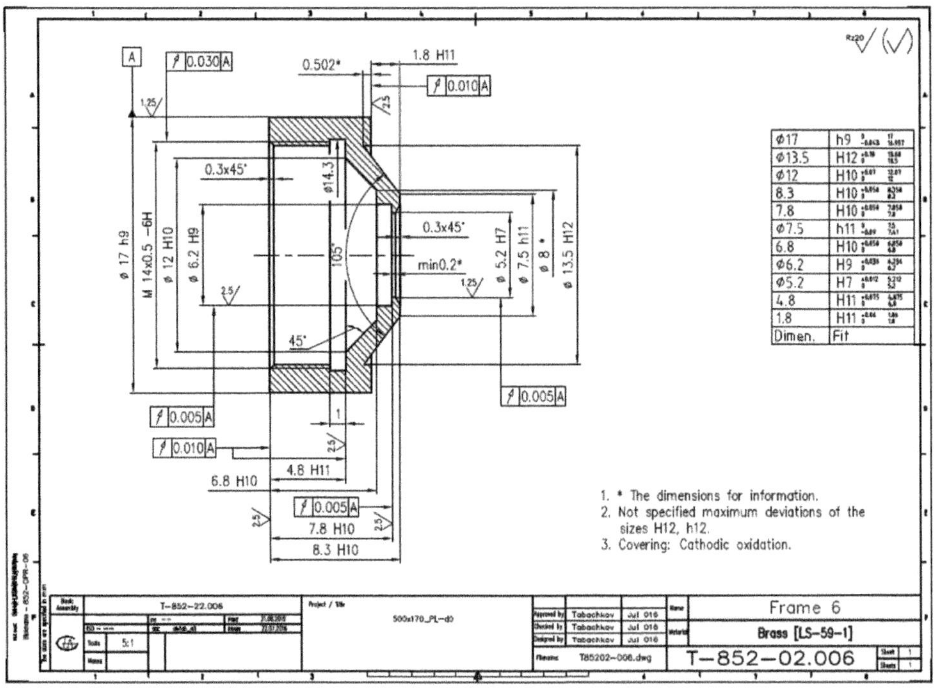

Figure 3.351 A drawing of the sixth frame.

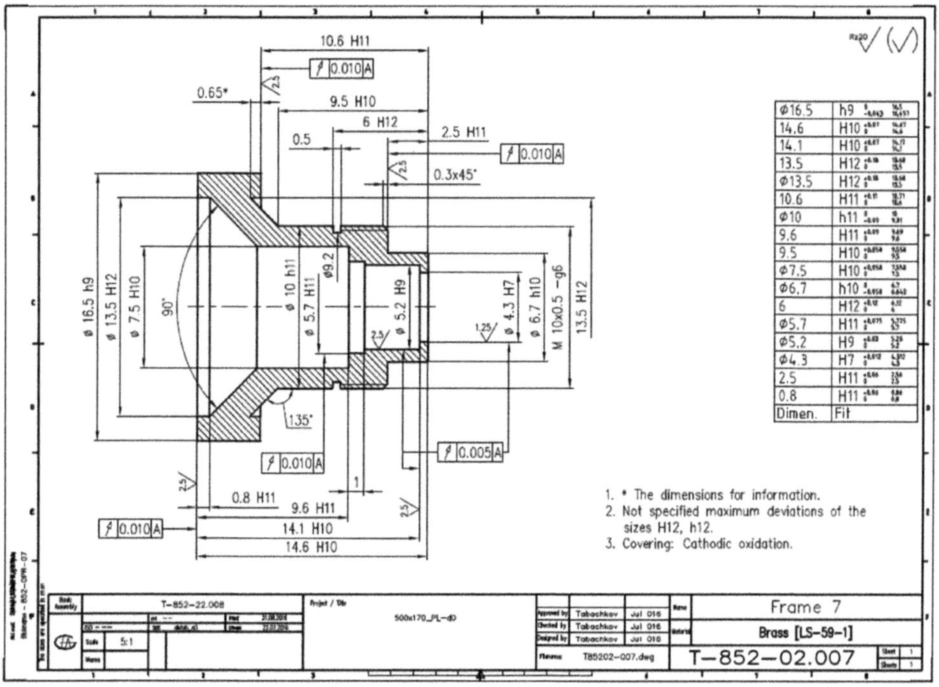

Figure 3.352 A drawing of the seventh frame.

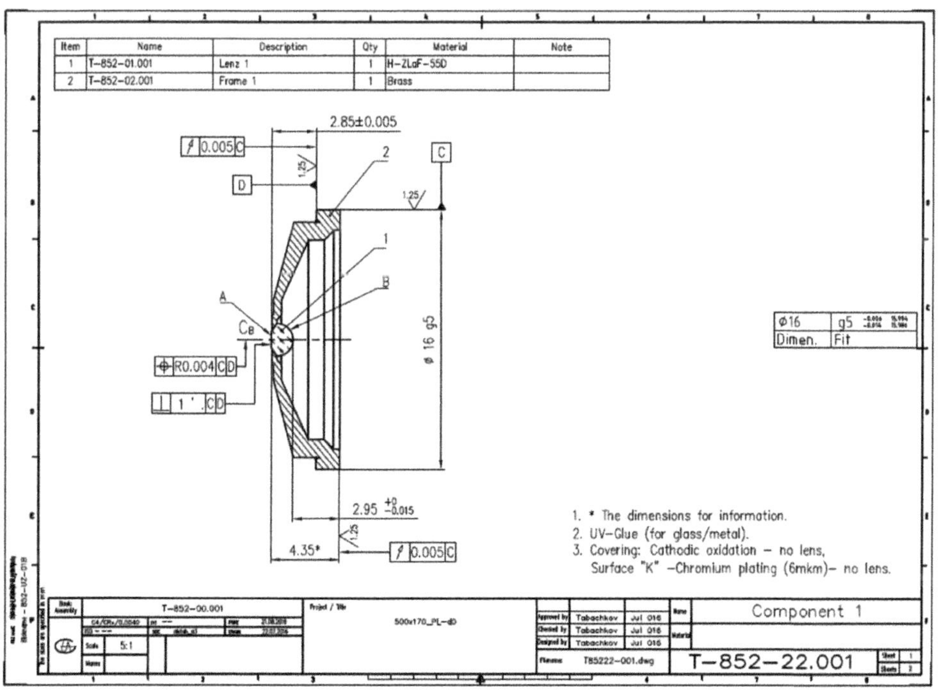

Figure 3.353 A drawing of the first component (no. 1).

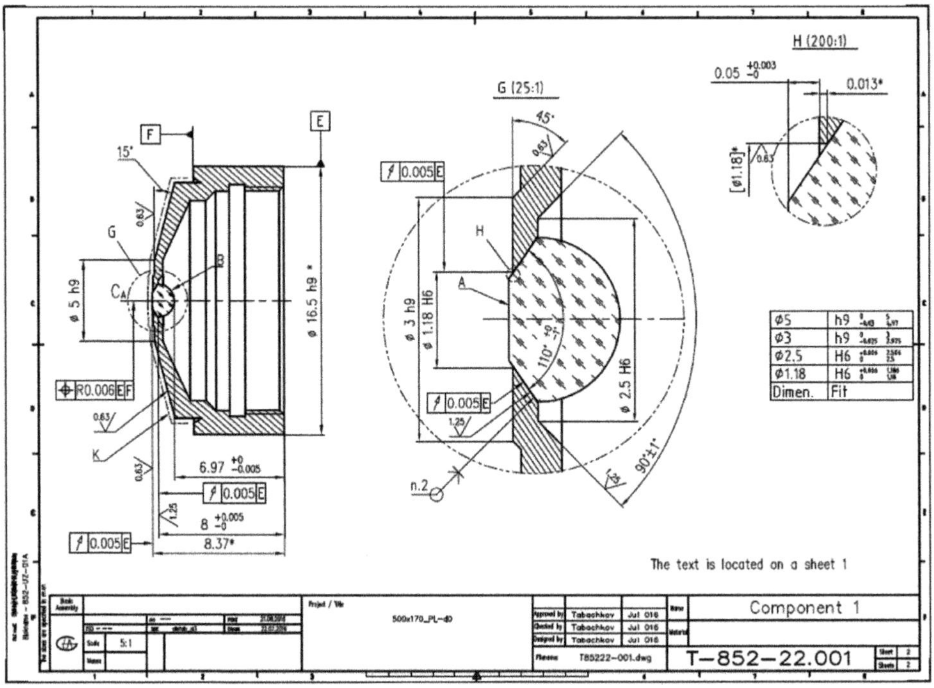

Figure 3.354 A drawing of the first component (no. 2).

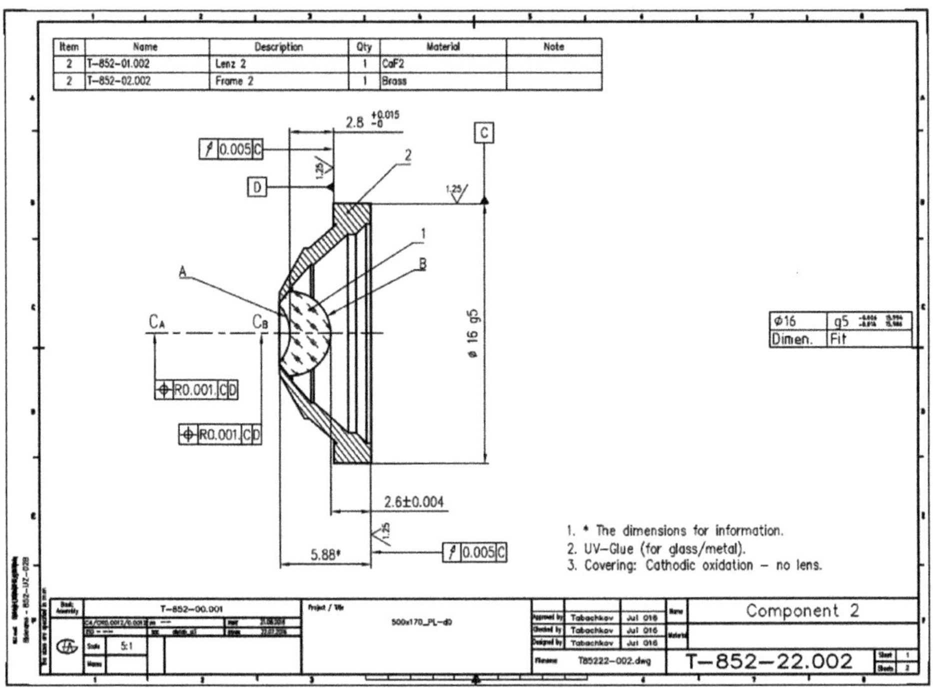

Figure 3.355 A drawing of the second component (no. 1).

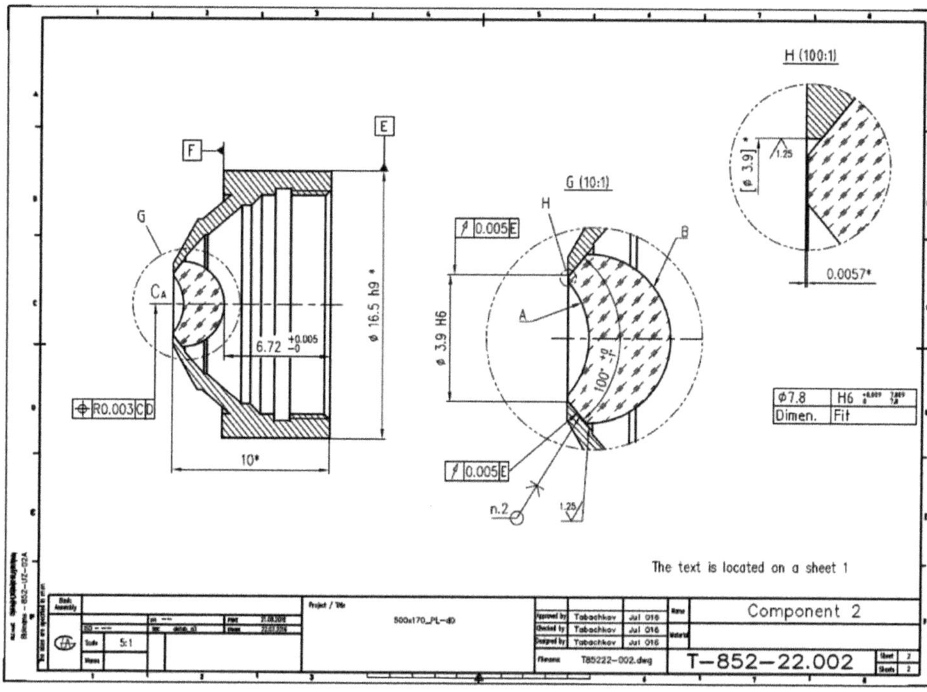

Figure 3.356 A drawing of the second component (no. 2).

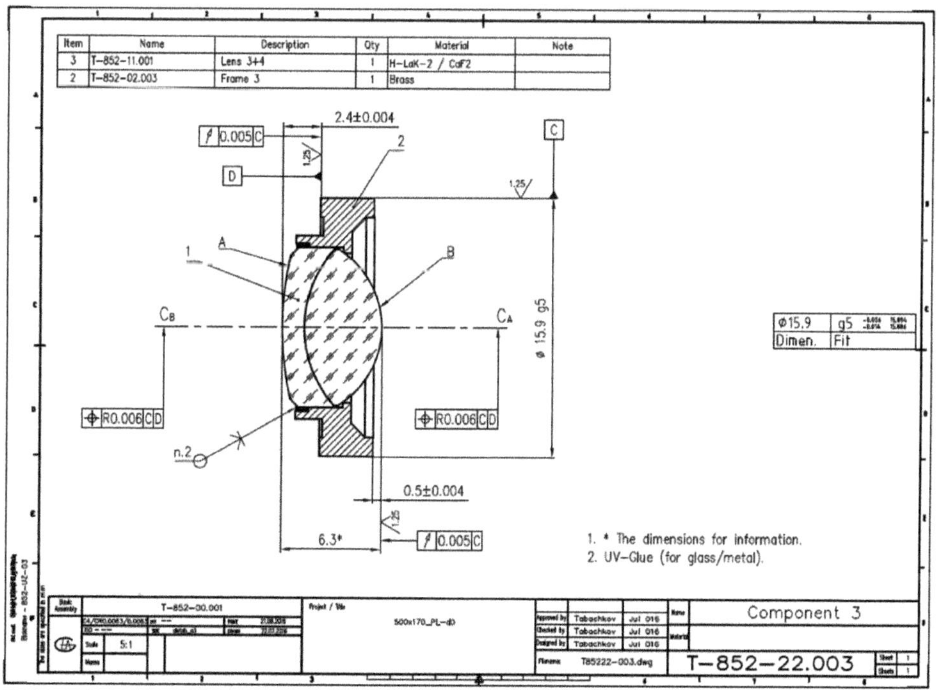

Figure 3.357 A drawing of the third component.

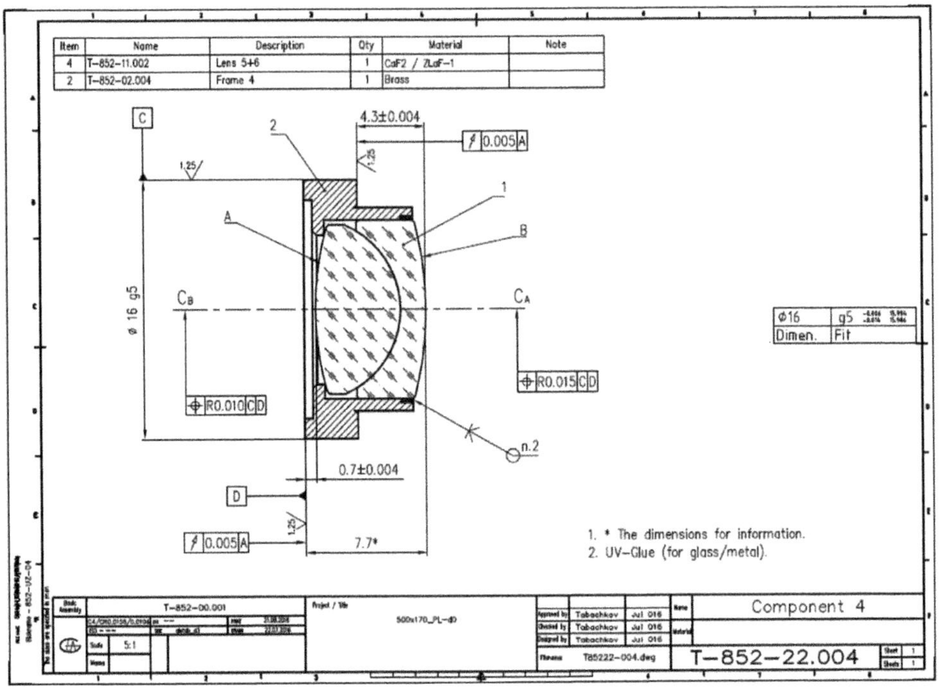

Figure 3.358 A drawing of the fourth component.

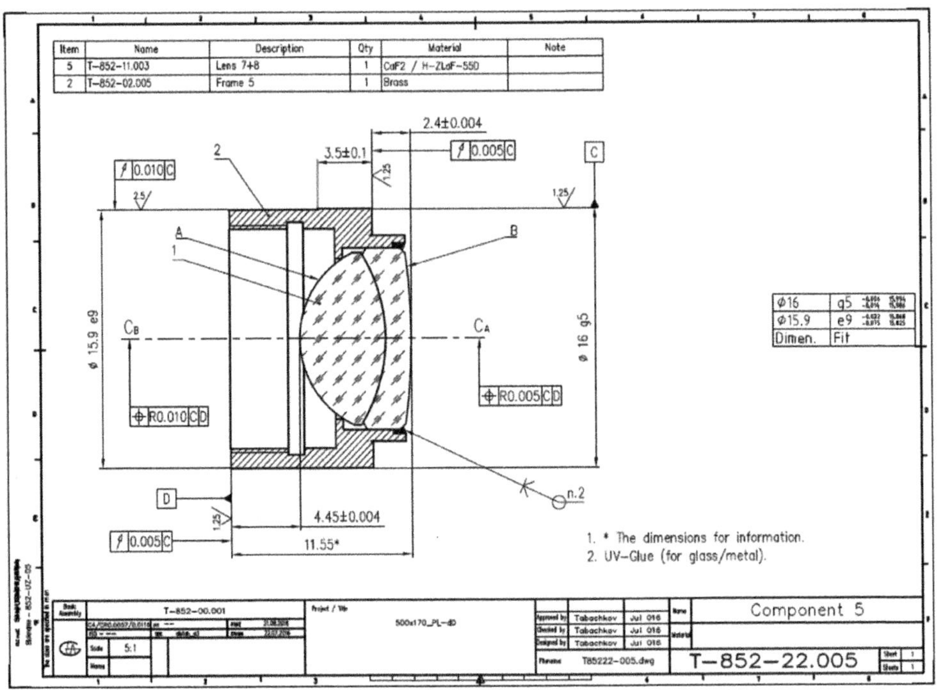

Figure 3.359 A drawing of the fifth component.

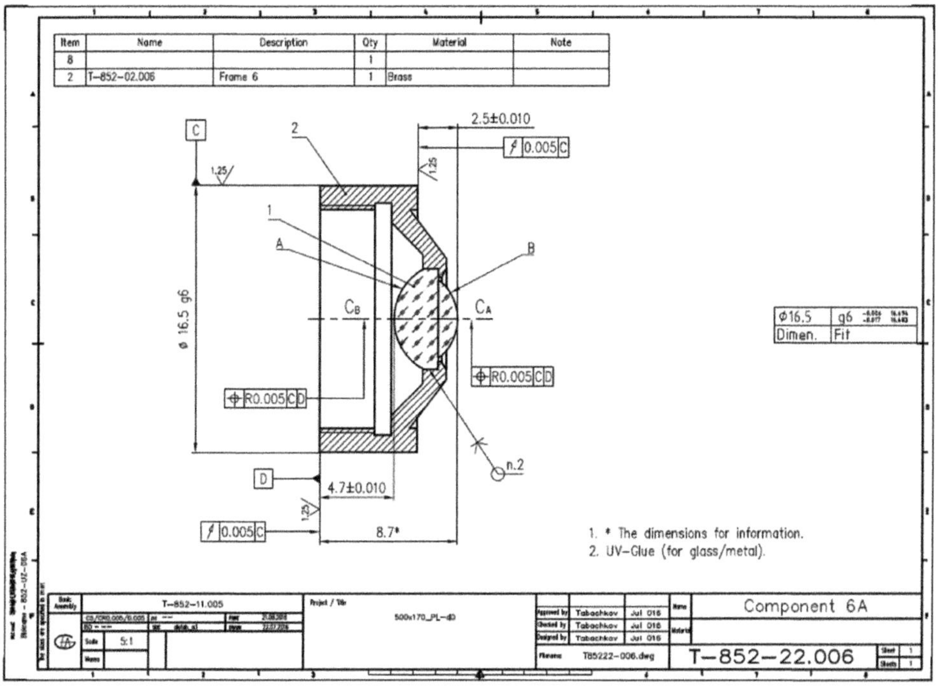

Figure 3.360 A drawing of the sixth "A" component.

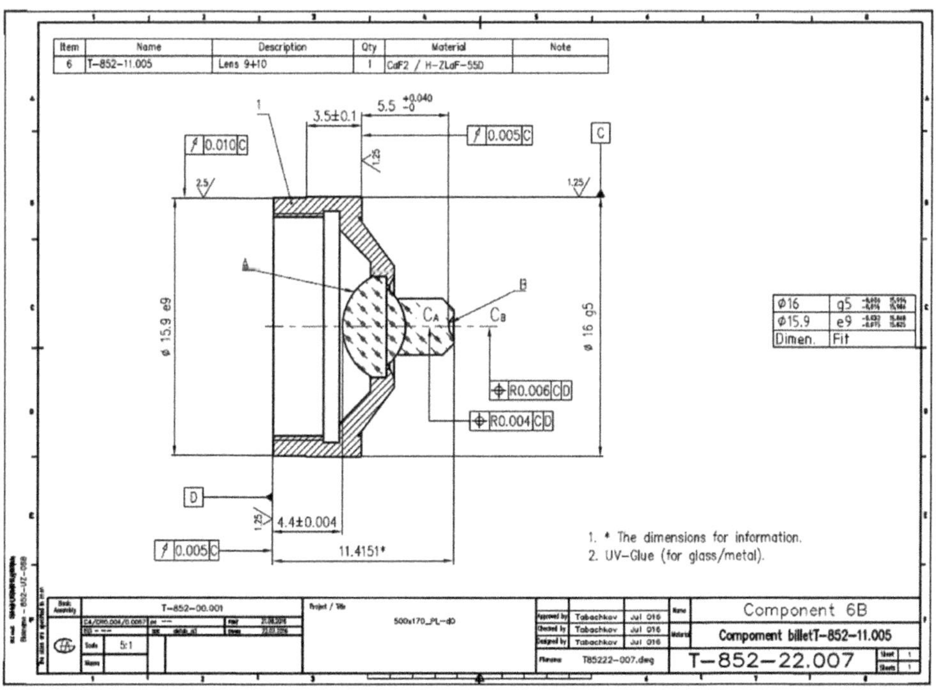

Figure 3.361 A drawing of the sixth "B" component.

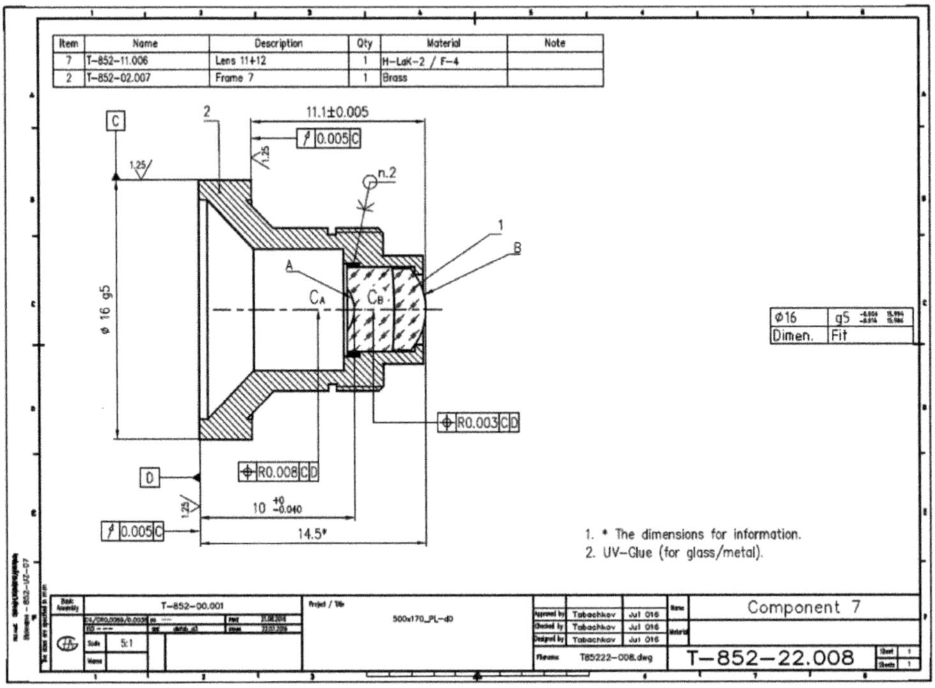

Figure 3.362 A drawing of the seventh component.

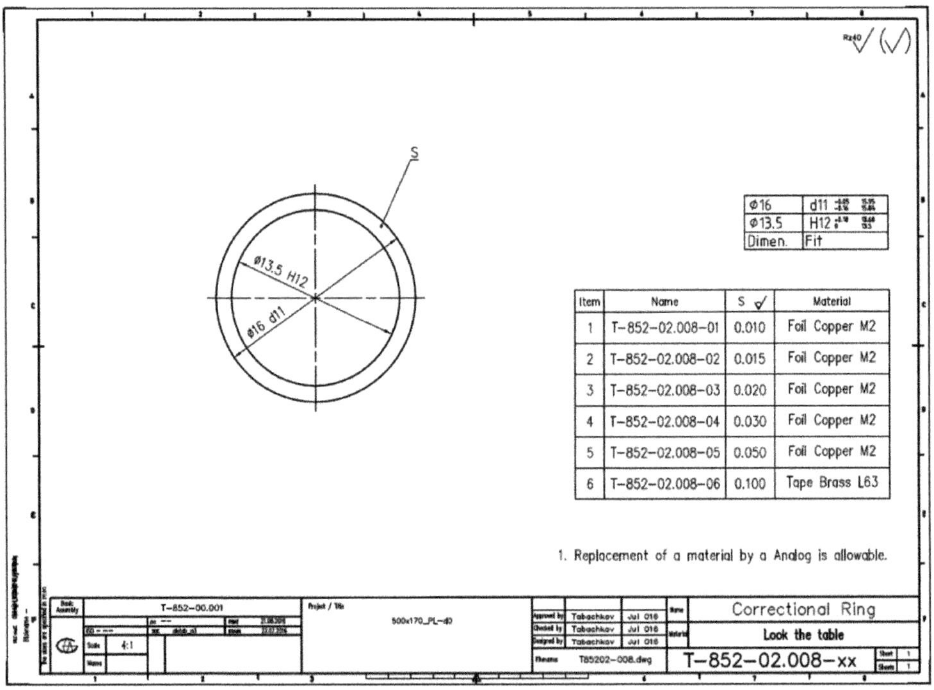

Figure 3.363 A drawing of the correction ring.

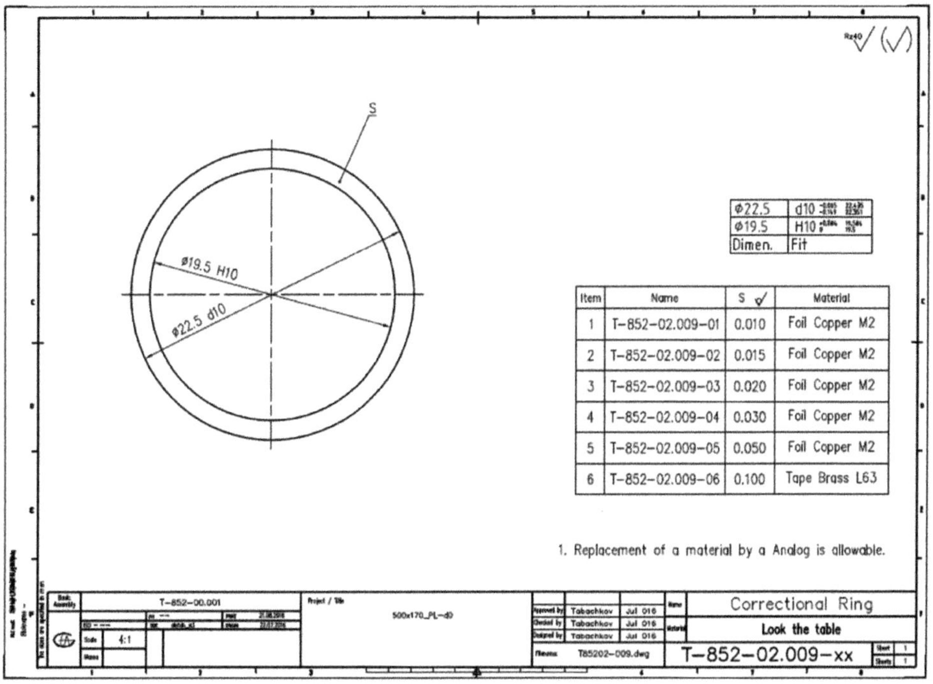

Figure 3.364 A drawing of the correction ring.

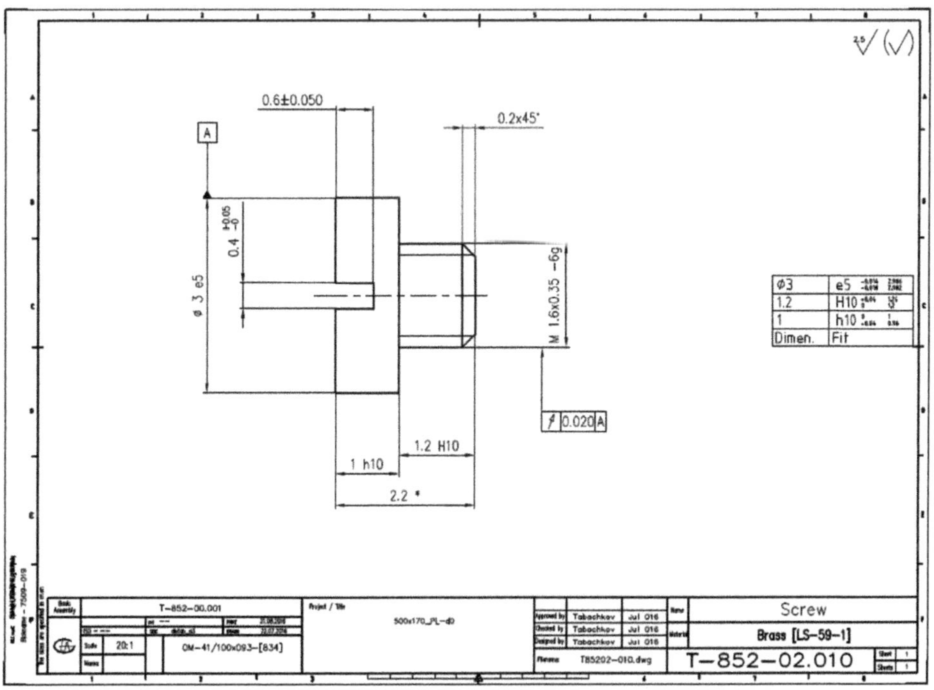

Figure 3.365 A drawing of the screw.

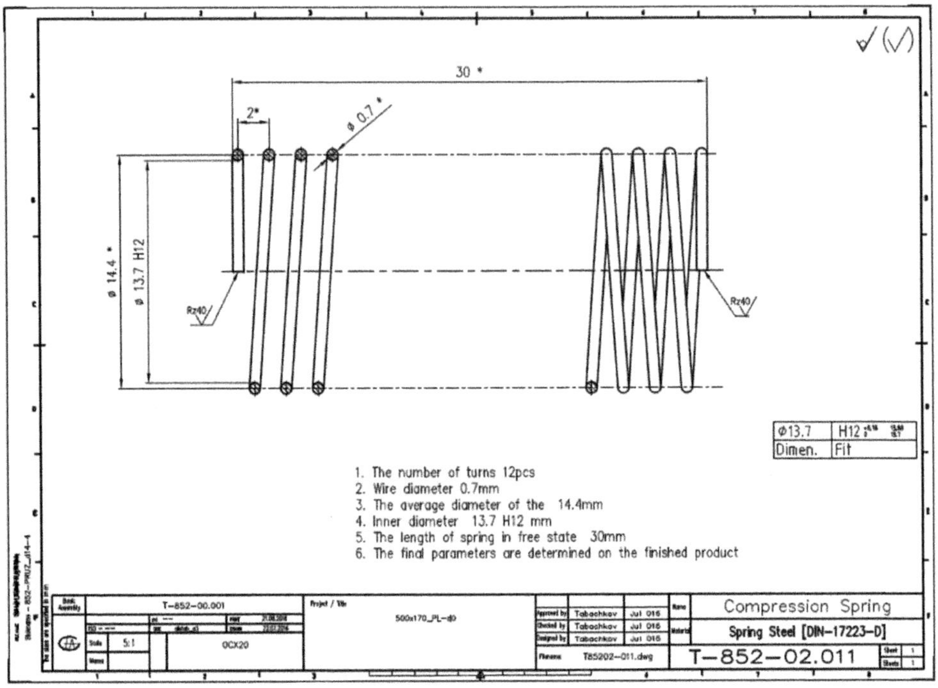

Figure 3.366 A drawing of the compression spring.

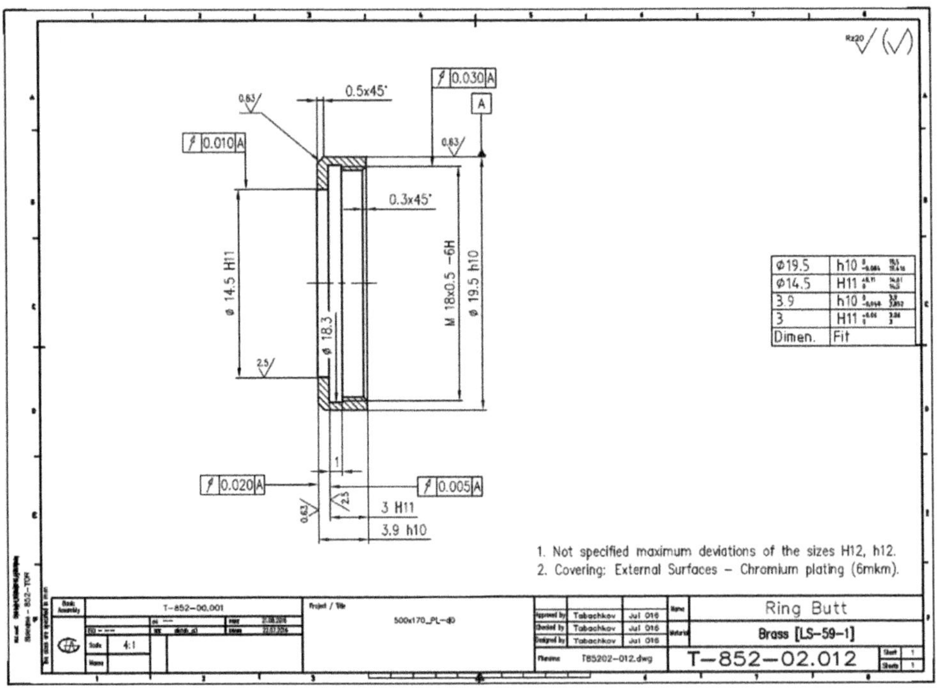

Figure 3.367 A drawing of the frame holder.

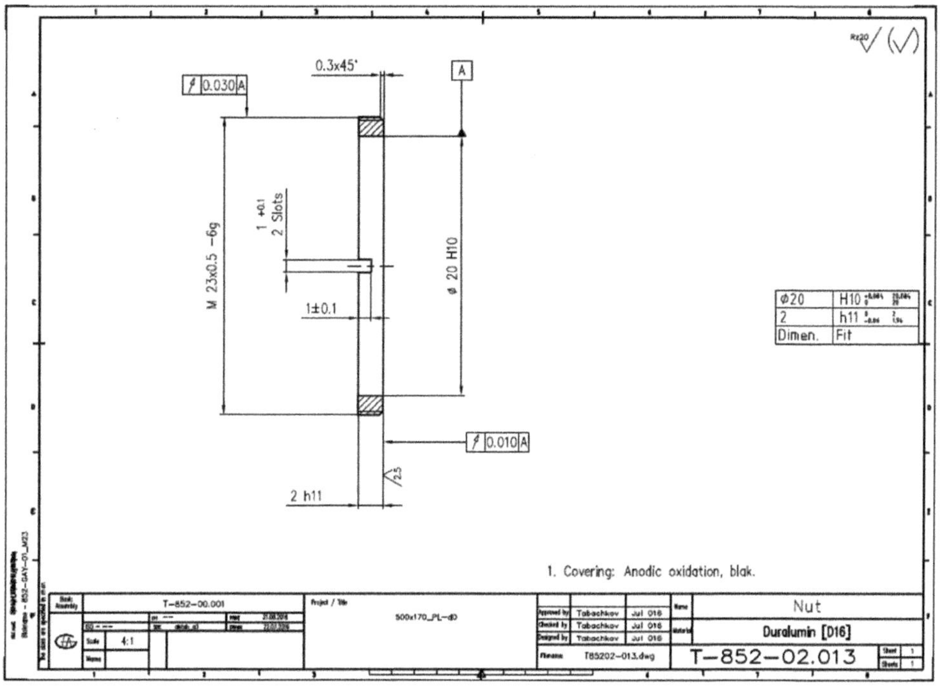

Figure 3.368 A drawing of the nut.

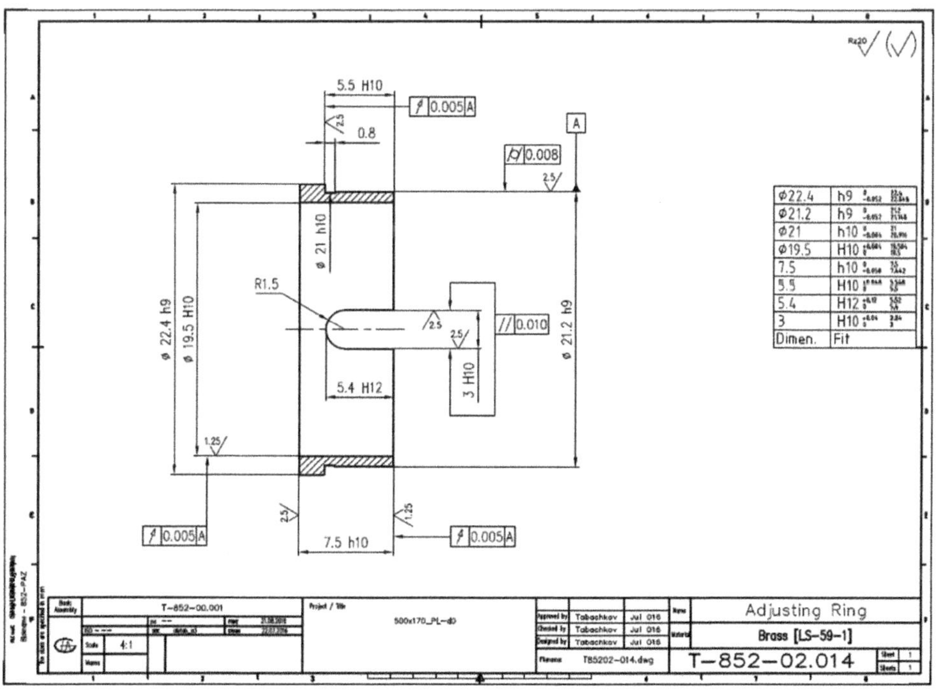

Figure 3.369 A drawing of the adjusting ring.

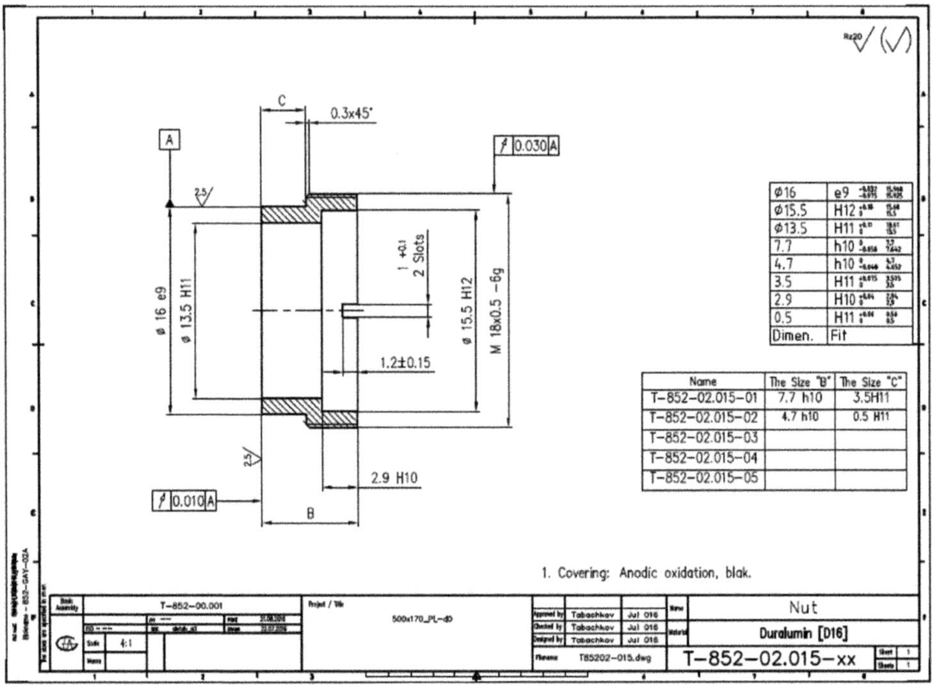

Figure 3.370 A drawing of the nut.

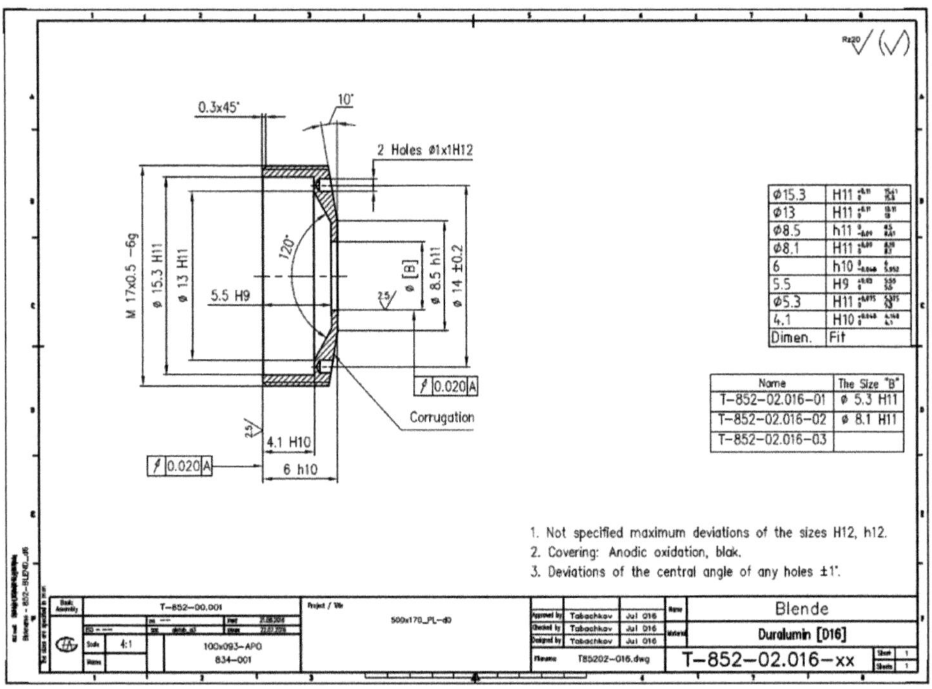

Figure 3.371 A drawing of the cover.

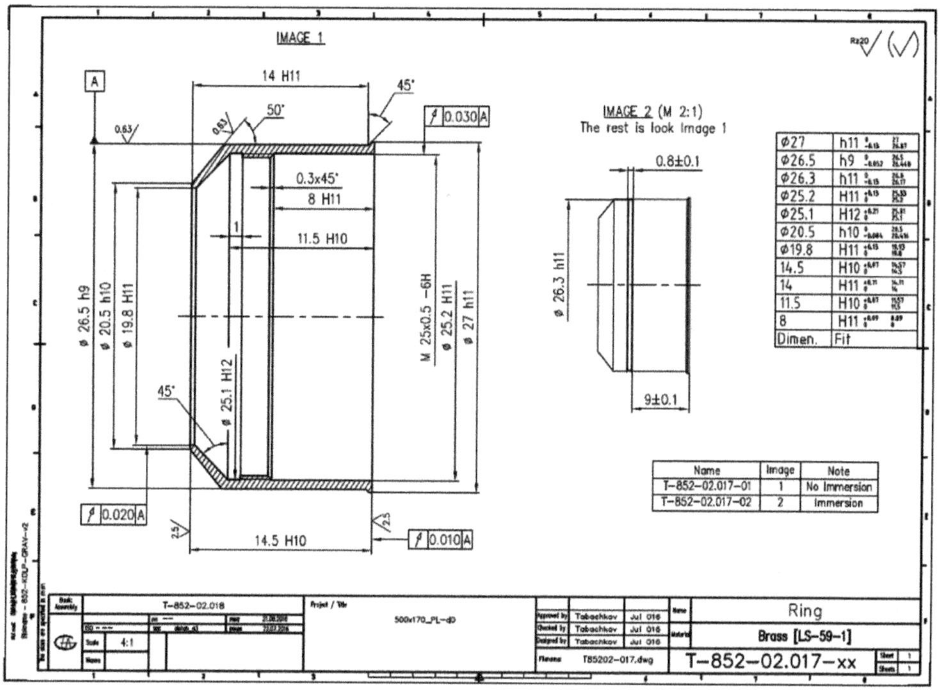

Figure 3.372 A drawing of the ring.

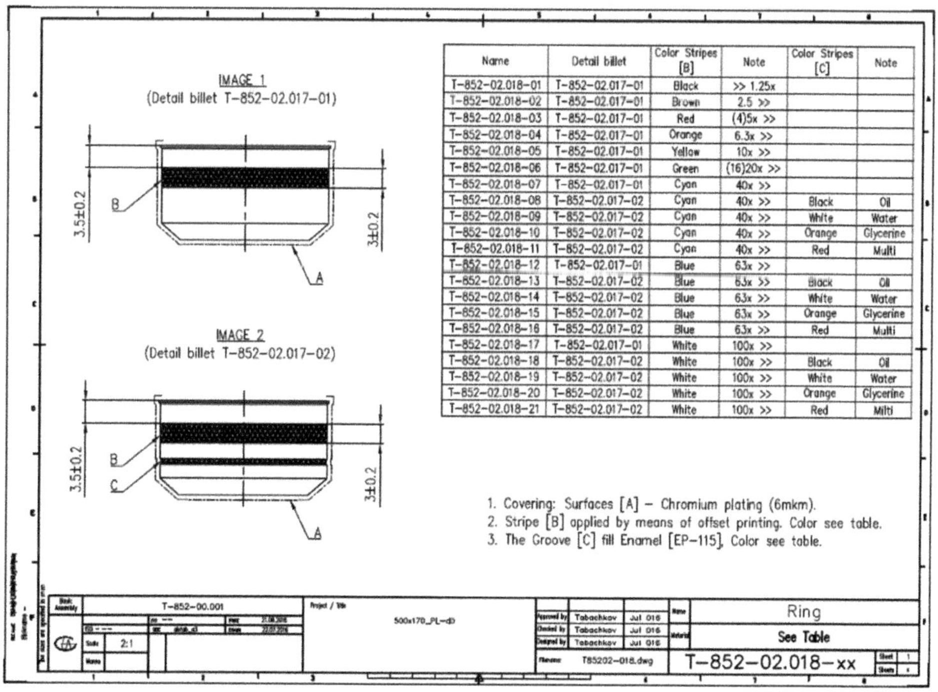

Figure 3.373 A drawing of the unified ring.

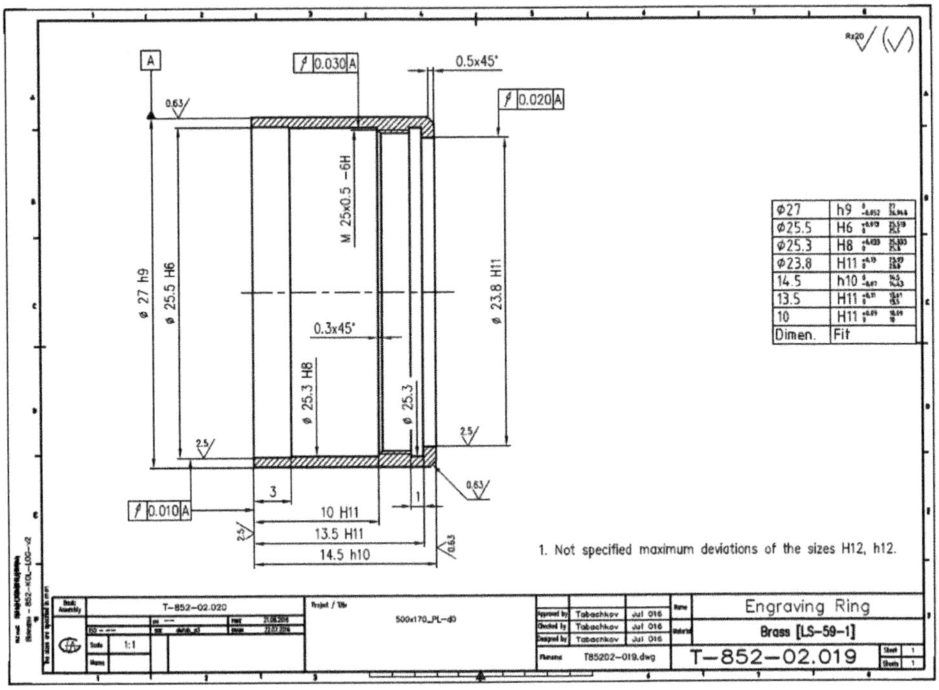

Figure 3.374 A drawing of the engraving ring.

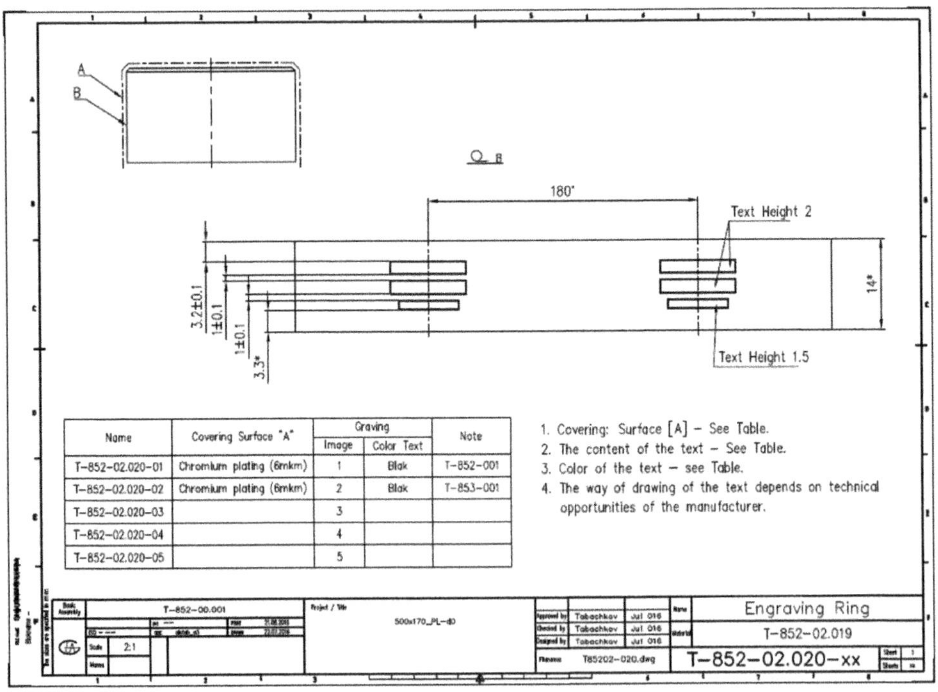

Figure 3.375 A drawing of the unified engraving ring.

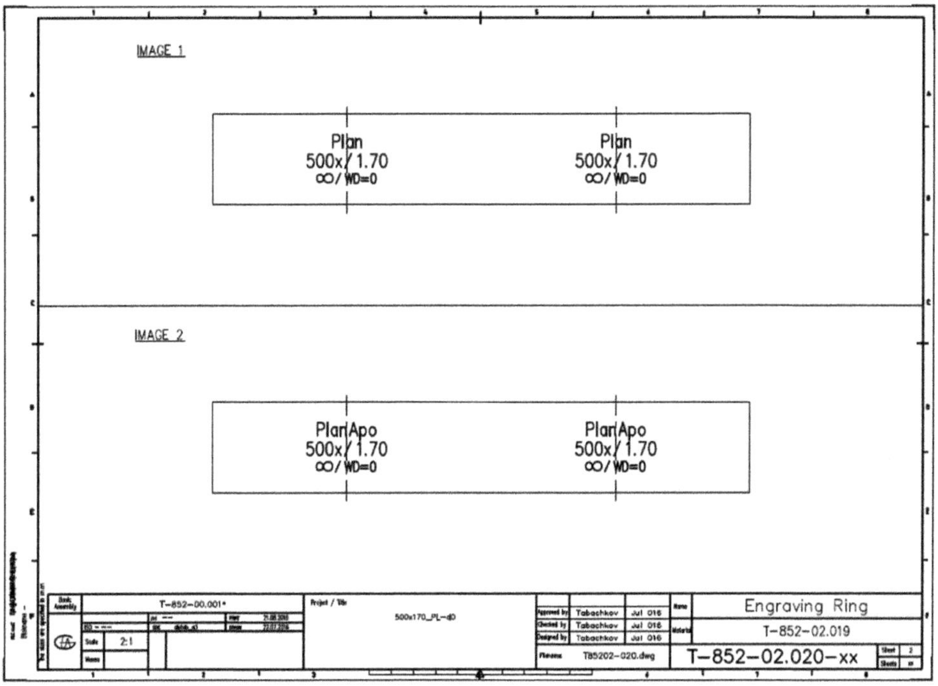

Figure 3.376 A drawing of the unified engraving ring.

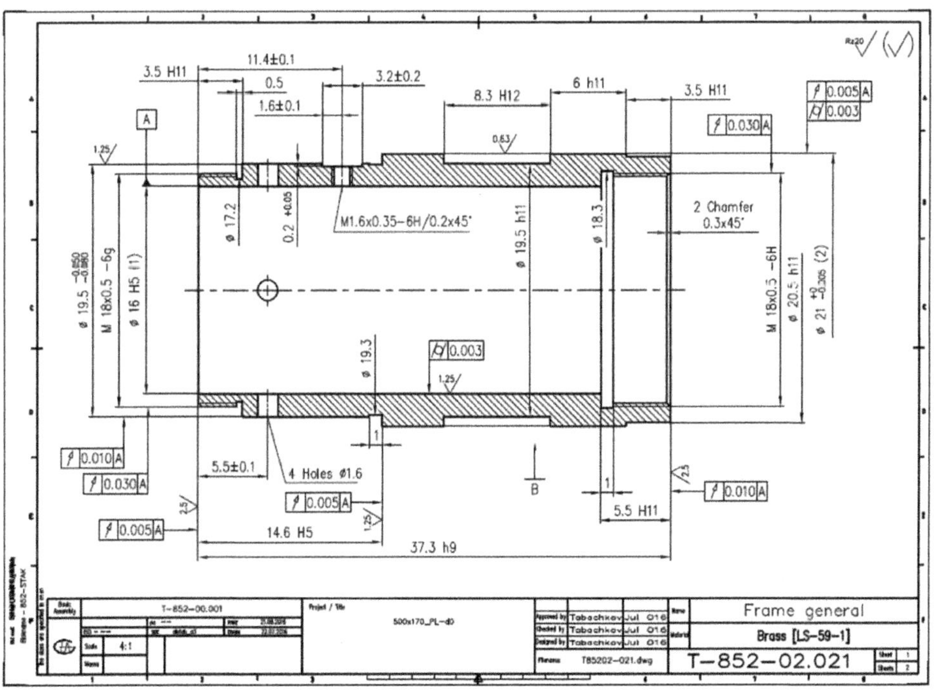

Figure 3.377 A drawing of the general frame (no. 1).

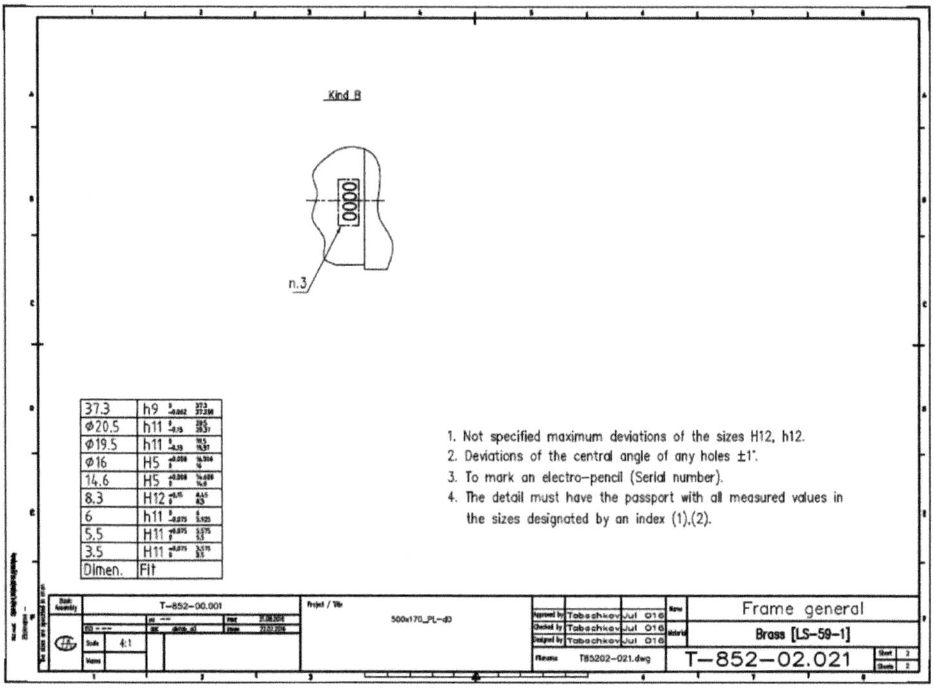

Figure 3.378 A drawing of the general frame (no. 2).

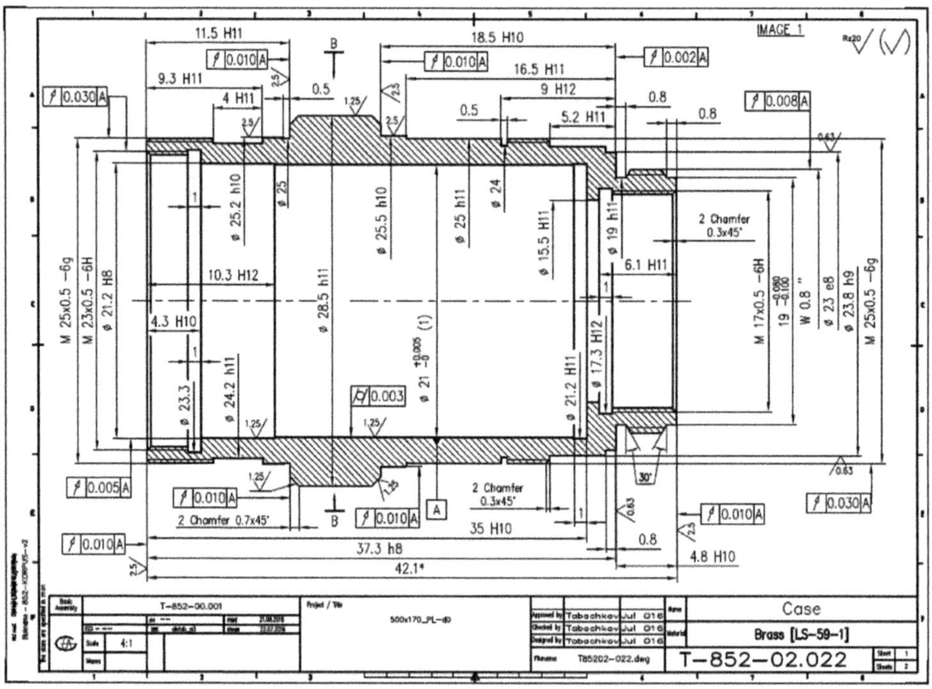

Figure 3.379 A drawing of the body (case) (no. 1).

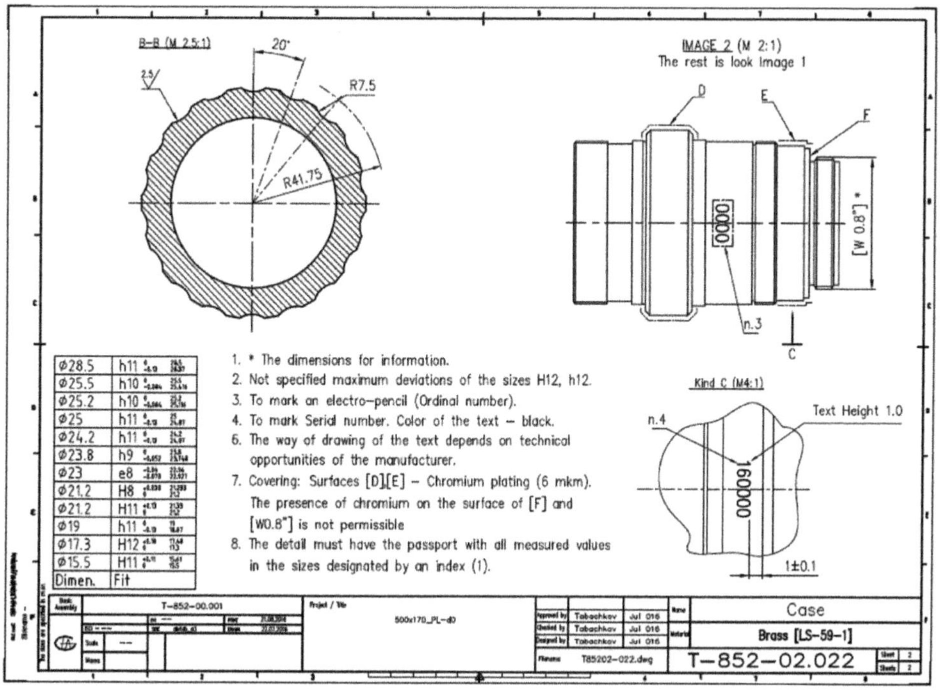

Figure 3.380 A drawing of the body (case) (no. 2).

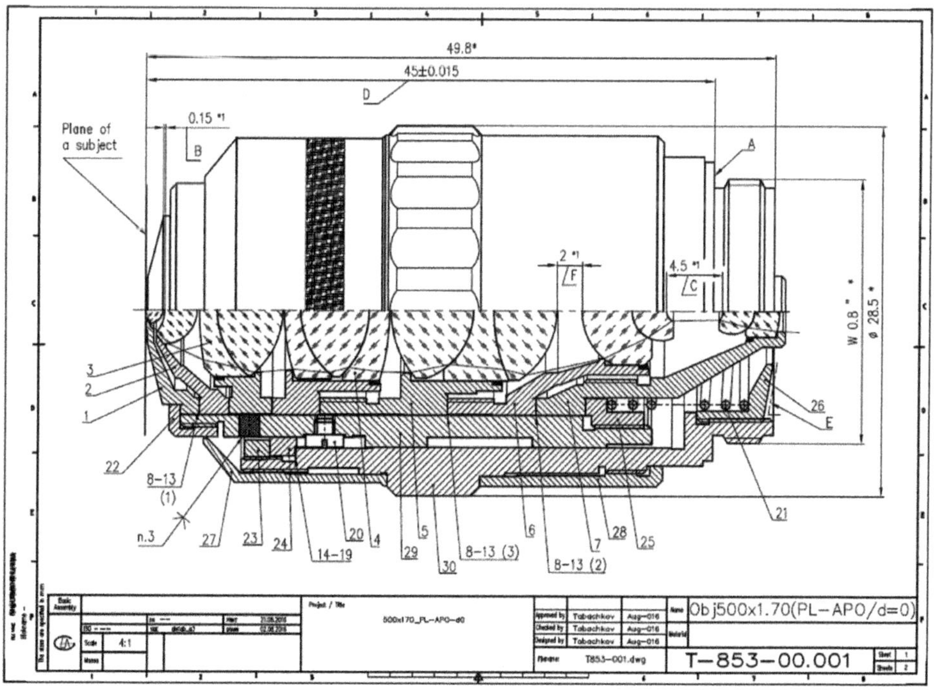

Figure 3.381 A drawing of a general mechanical design of a 500×/1.70 plan poly-apochromatic contact objective.

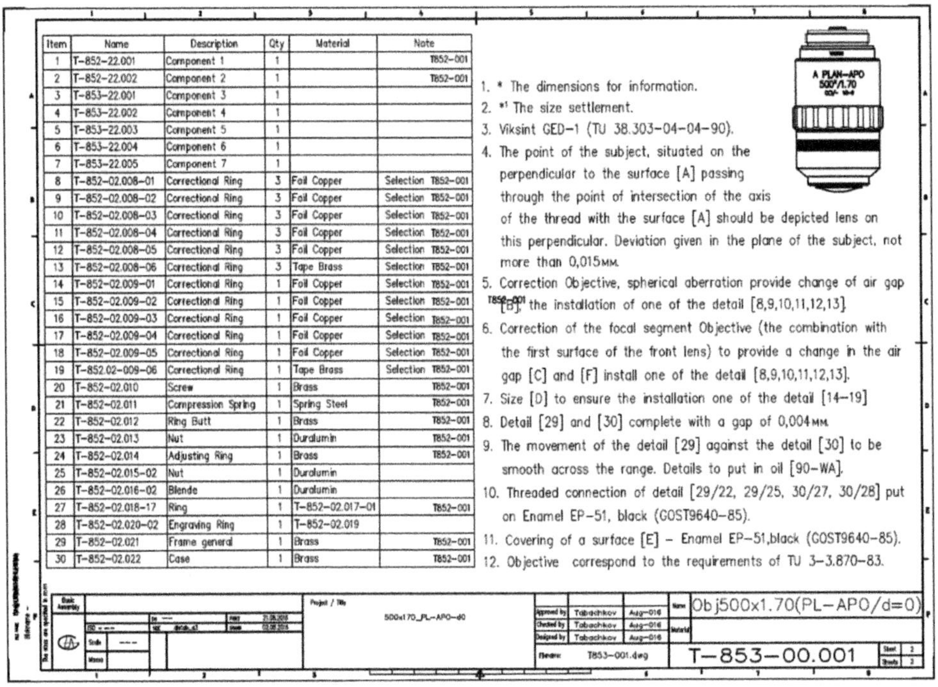

Figure 3.382 Specifications.

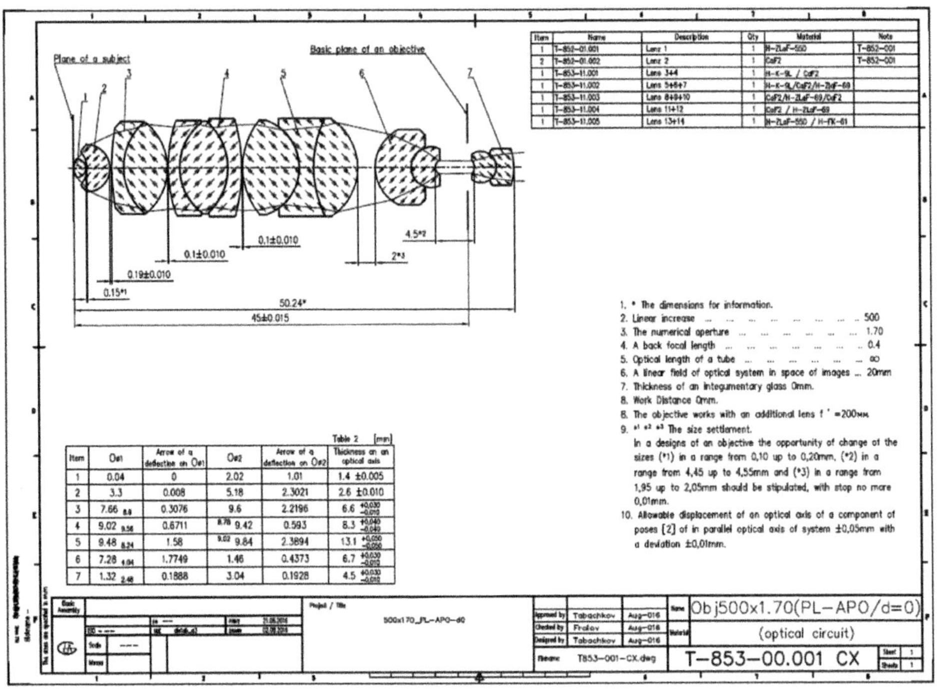

Figure 3.383 Principal optical layout of a 500×/1.70 plan poly-apochromatic contact objective.

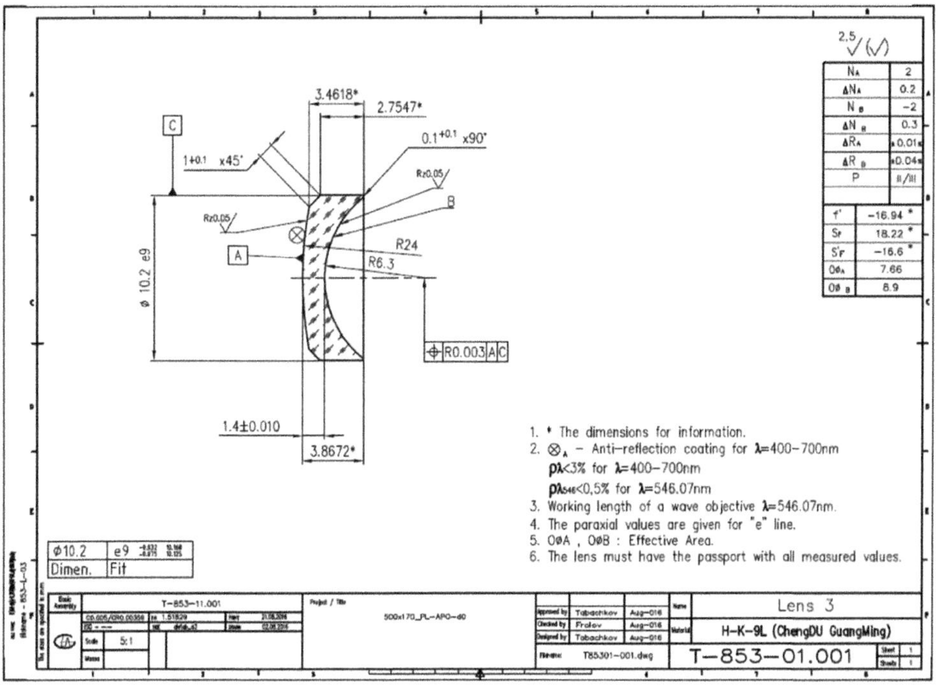

Figure 3.384 A drawing of the third lens.

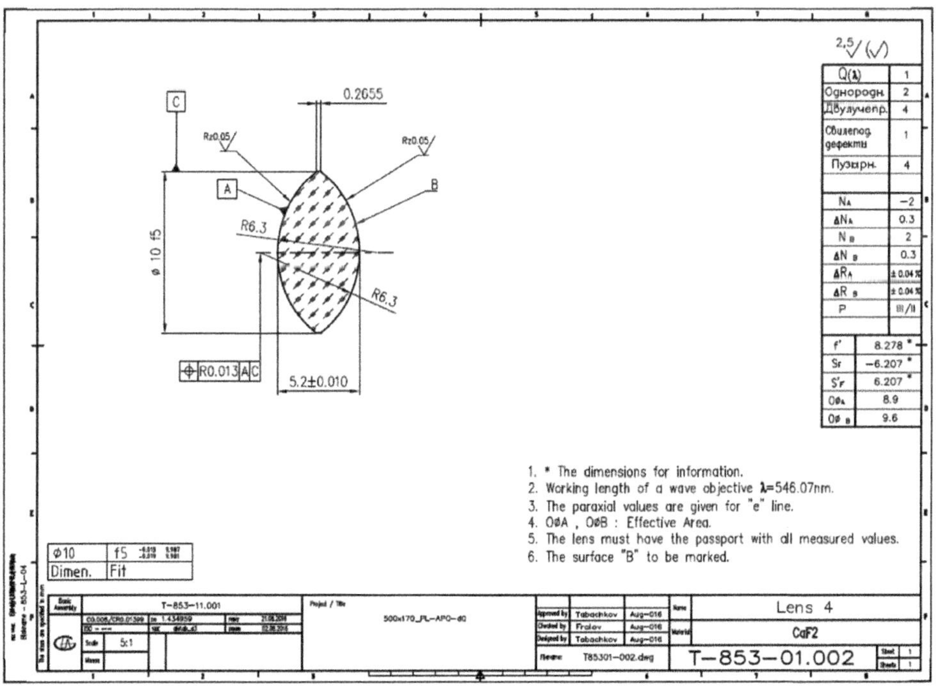

Figure 3.385 A drawing of the fourth lens.

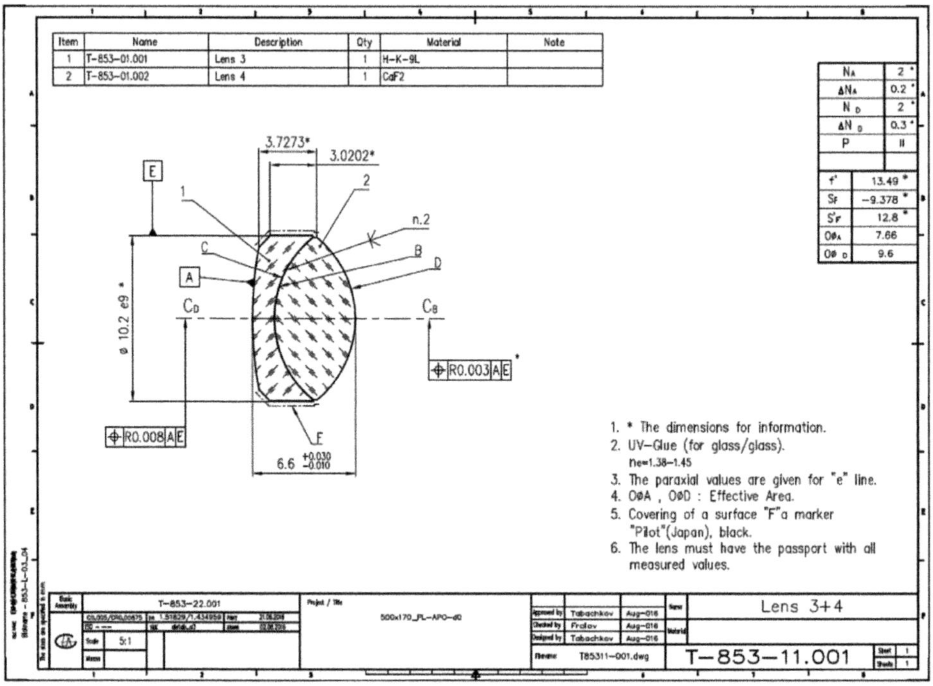

Figure 3.386 A drawing of the third and fourth lenses.

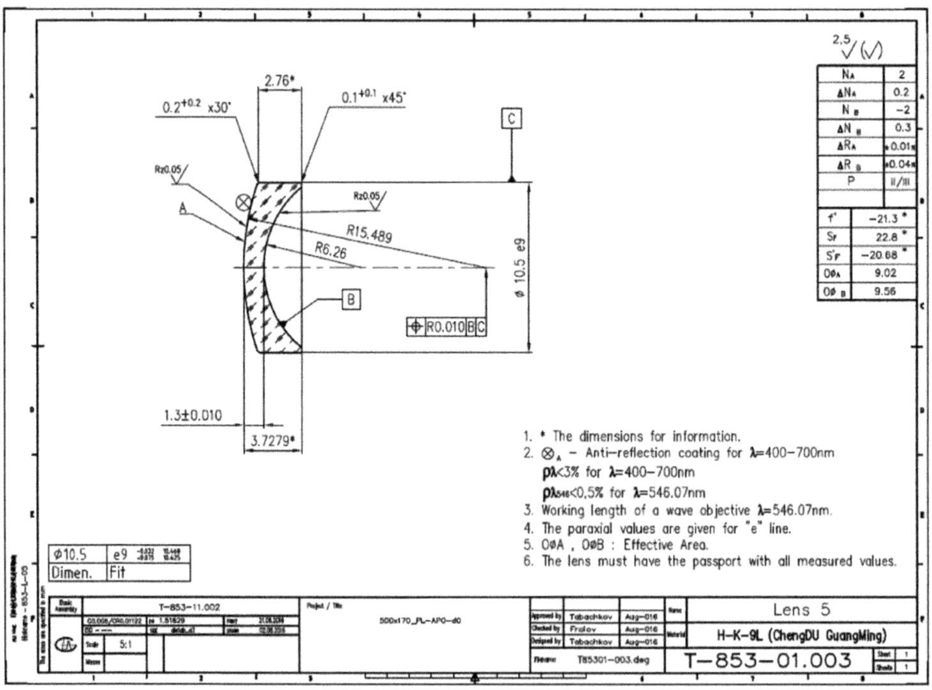

Figure 3.387 A drawing of the fifth lens.

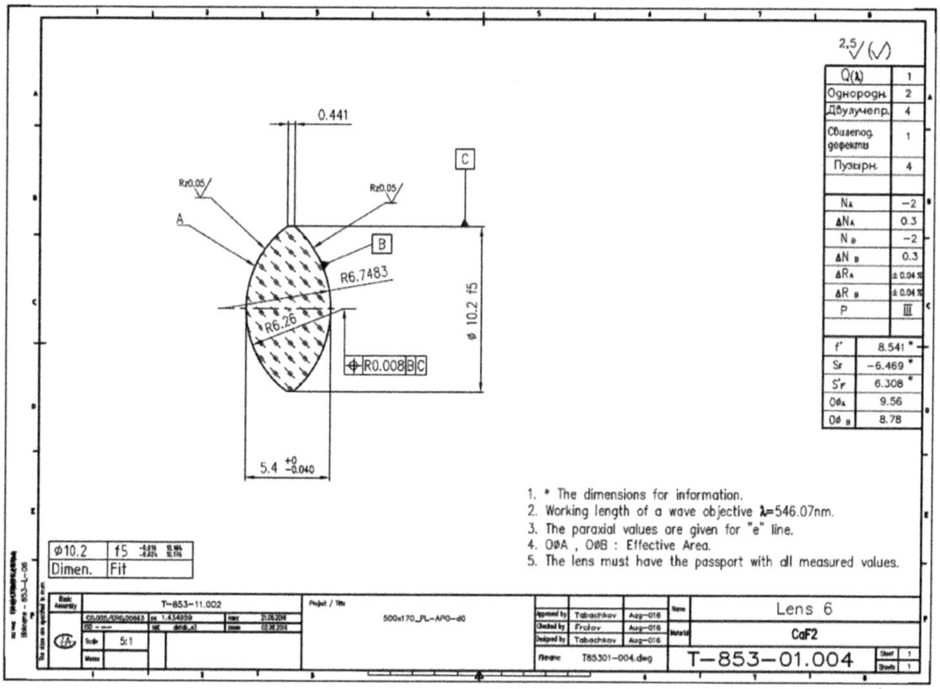

Figure 3.388 A drawing of the sixth lens.

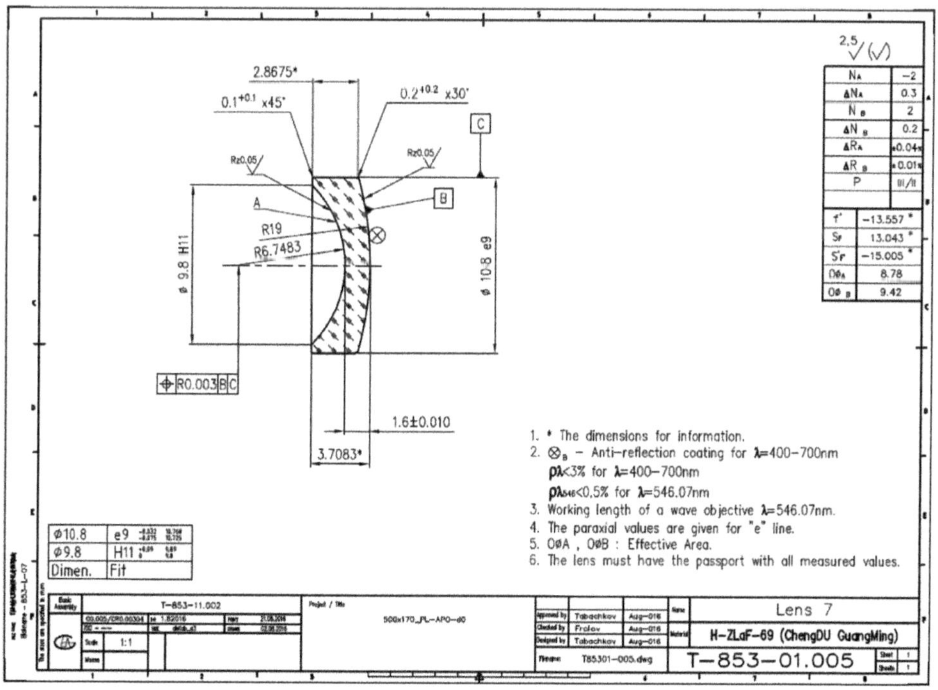

Figure 3.389 A drawing of the seventh lens.

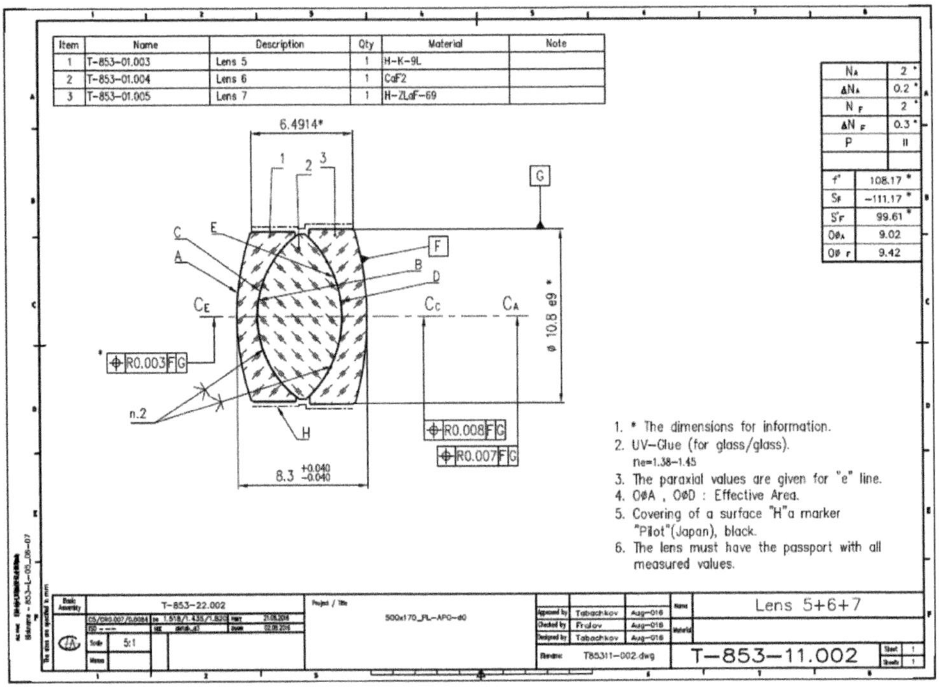

Figure 3.390 A drawing of the fifth, sixth, and seventh lenses.

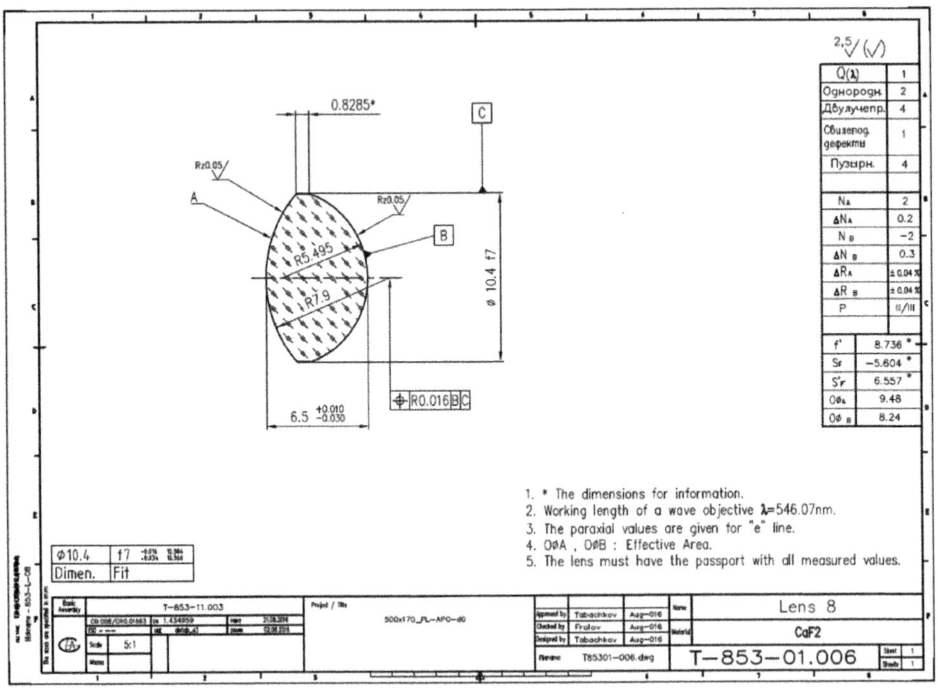

Figure 3.391 A drawing of the eighth lens.

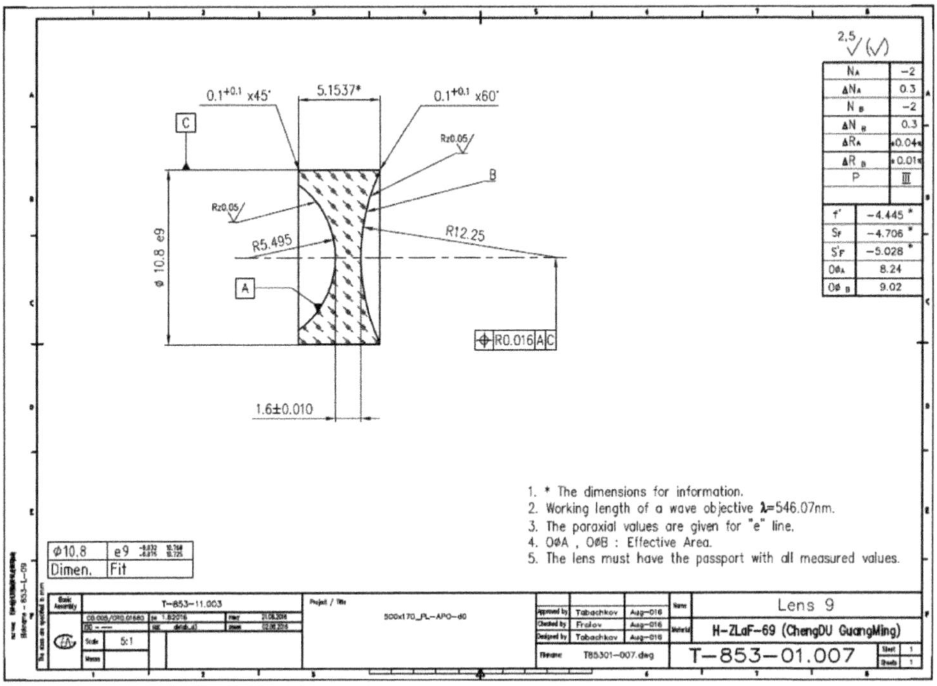

Figure 3.392 A drawing of the ninth lens.

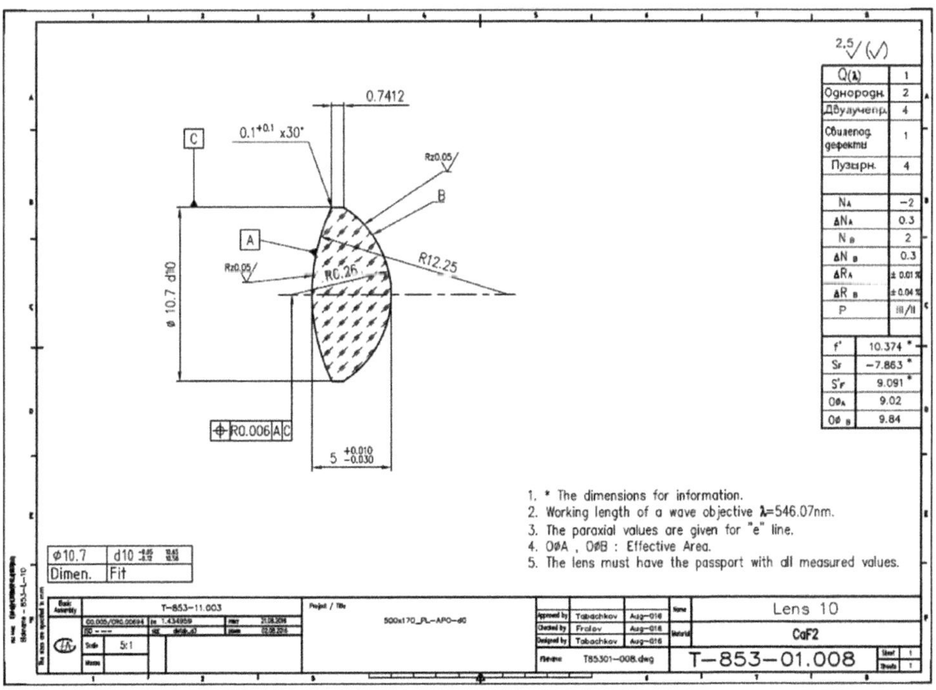

Figure 3.393 A drawing of the 10th lens.

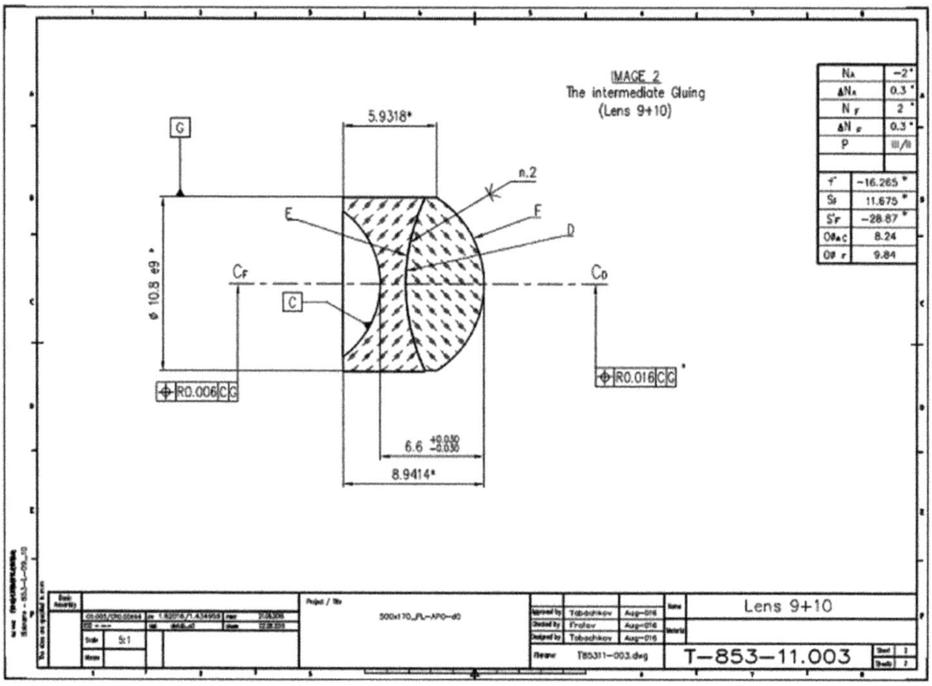

Figure 3.394 A drawing of the ninth and 10th lenses.

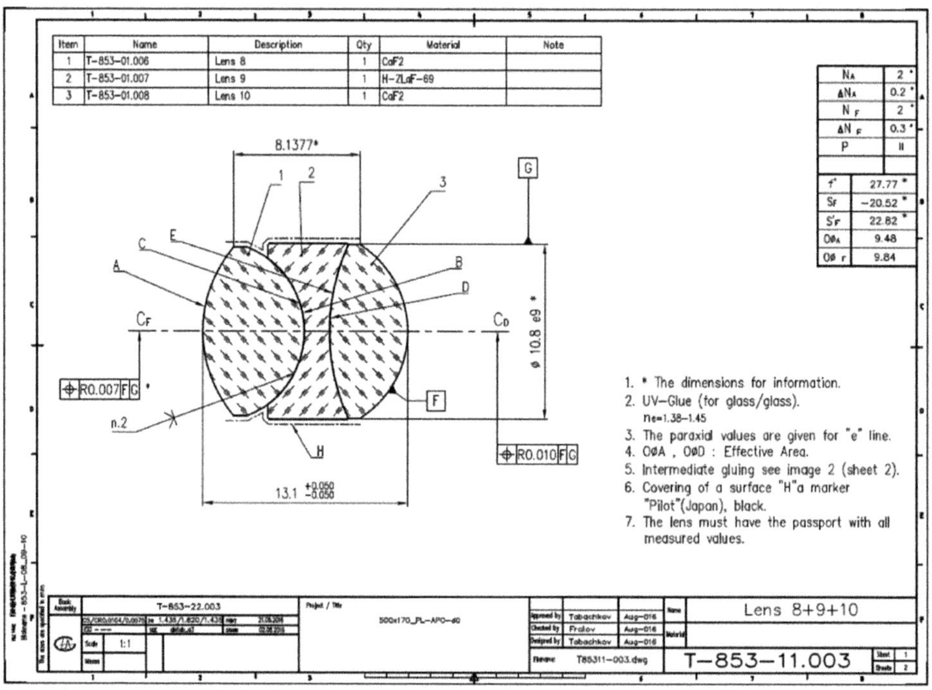

Figure 3.395 A drawing of the eighth, ninth, and 10th lenses.

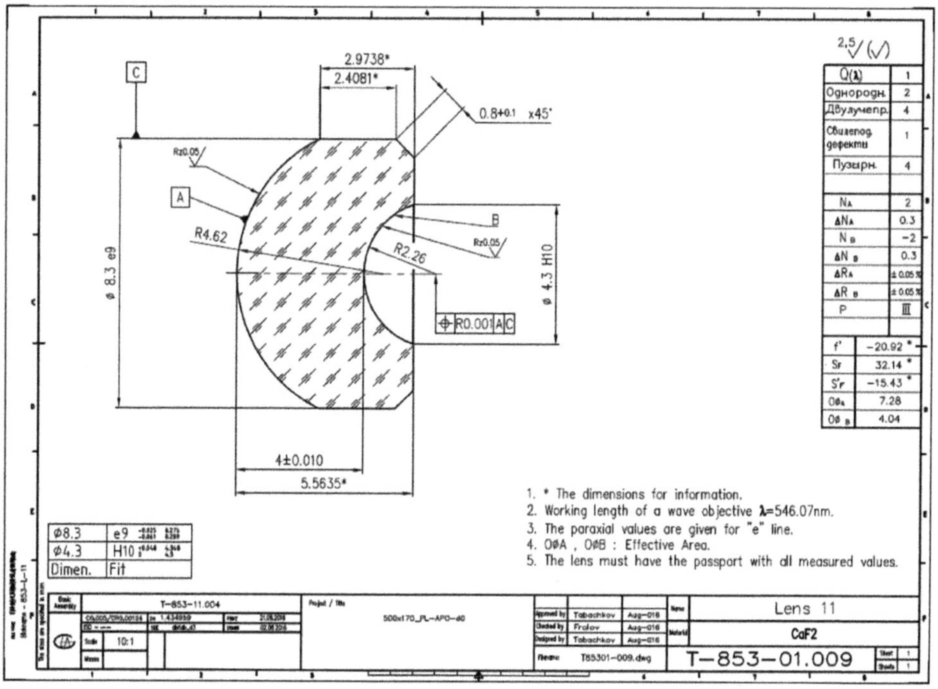

Figure 3.396 A drawing of the 11th lens.

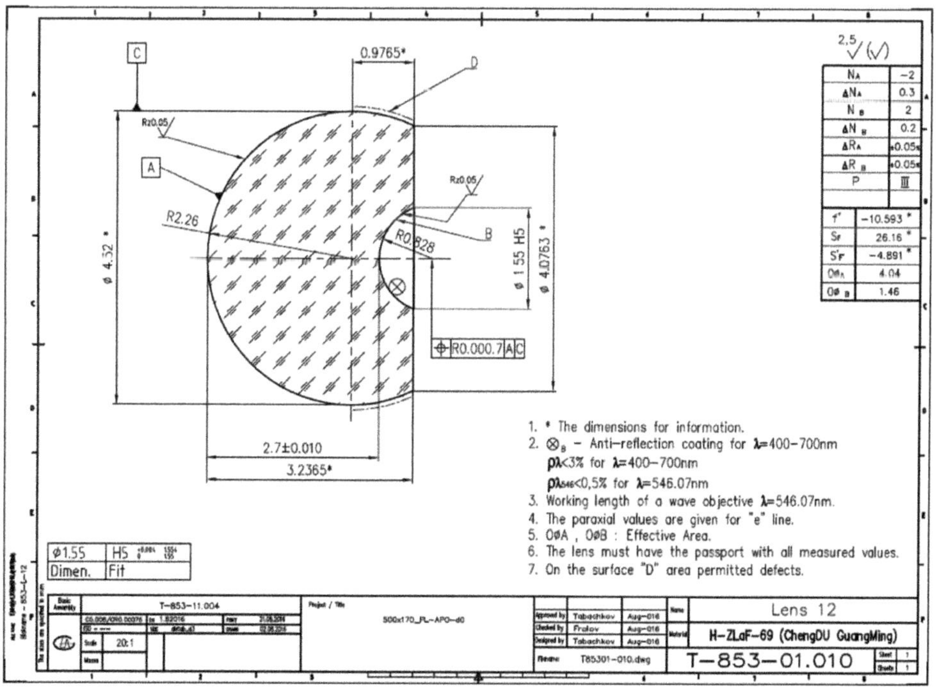

Figure 3.397 A drawing of the 12th lens.

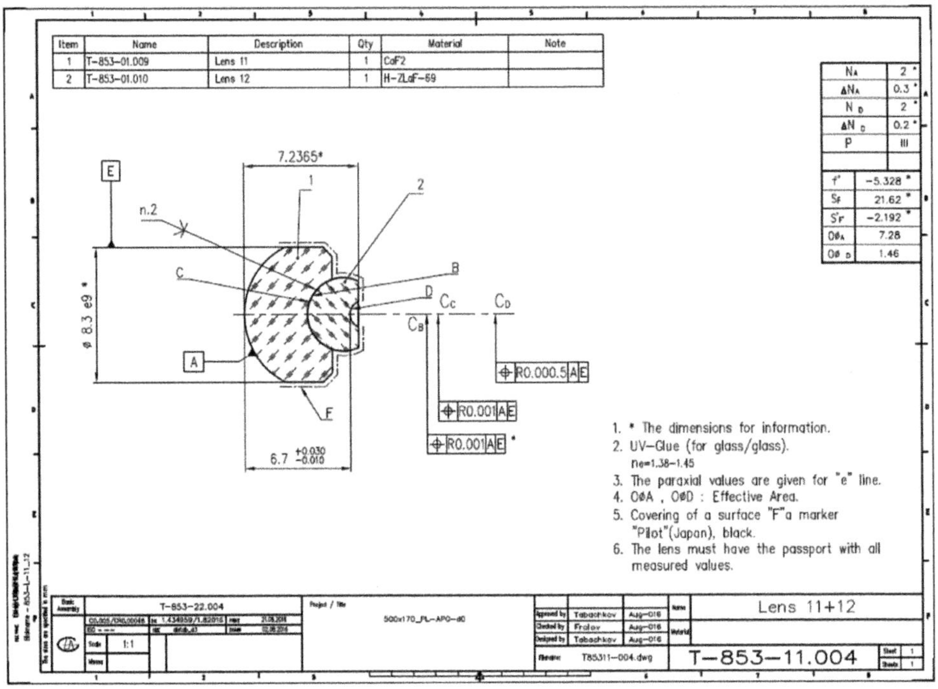

Figure 3.398 A drawing of the 11th and 12th lenses.

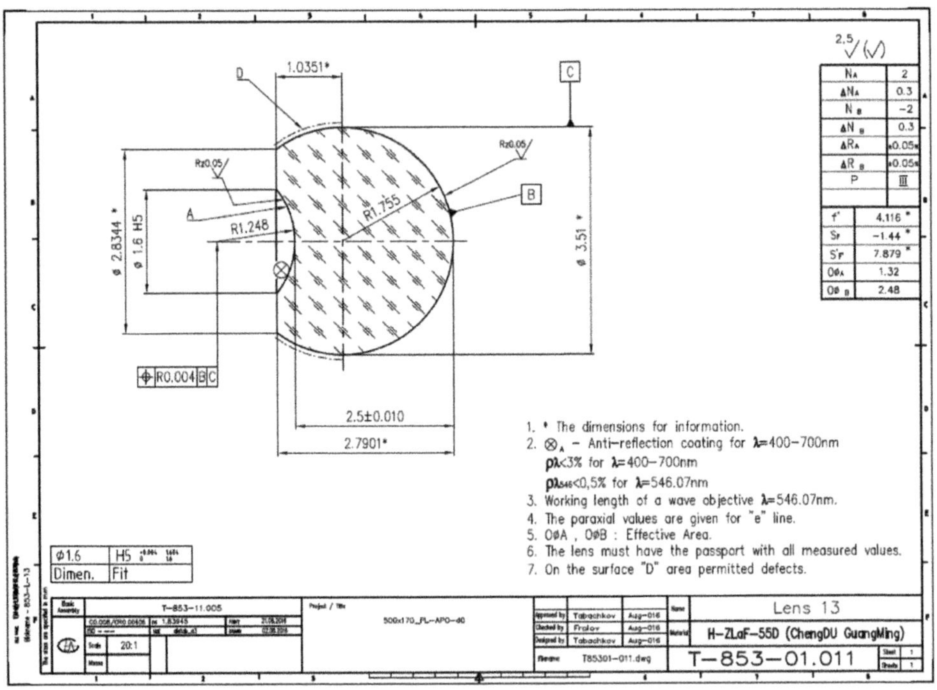

Figure 3.399 A drawing of the 13th lens.

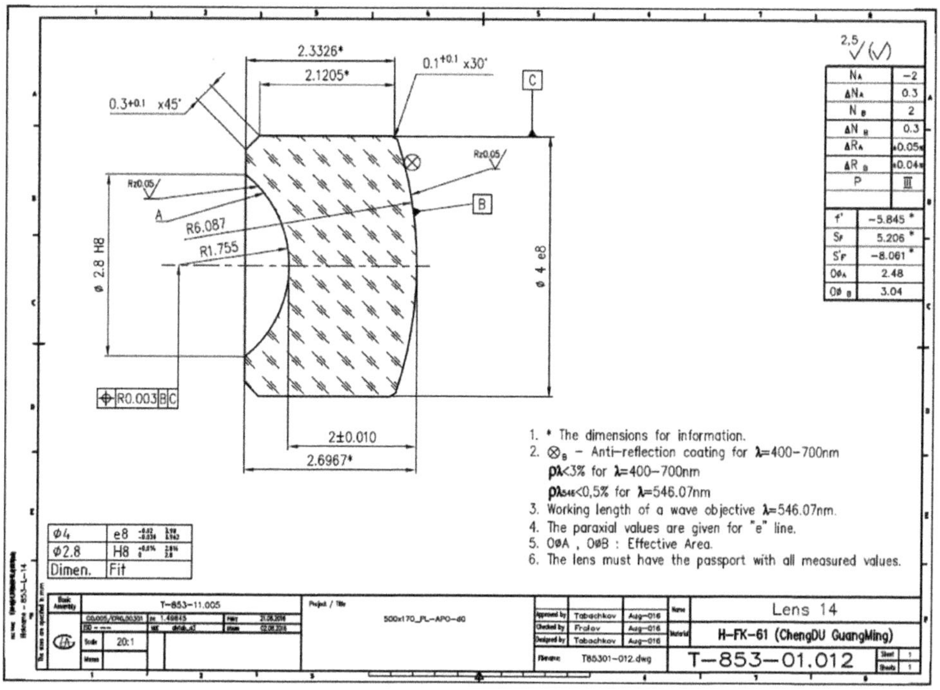

Figure 3.400 A drawing of the 14th lens.

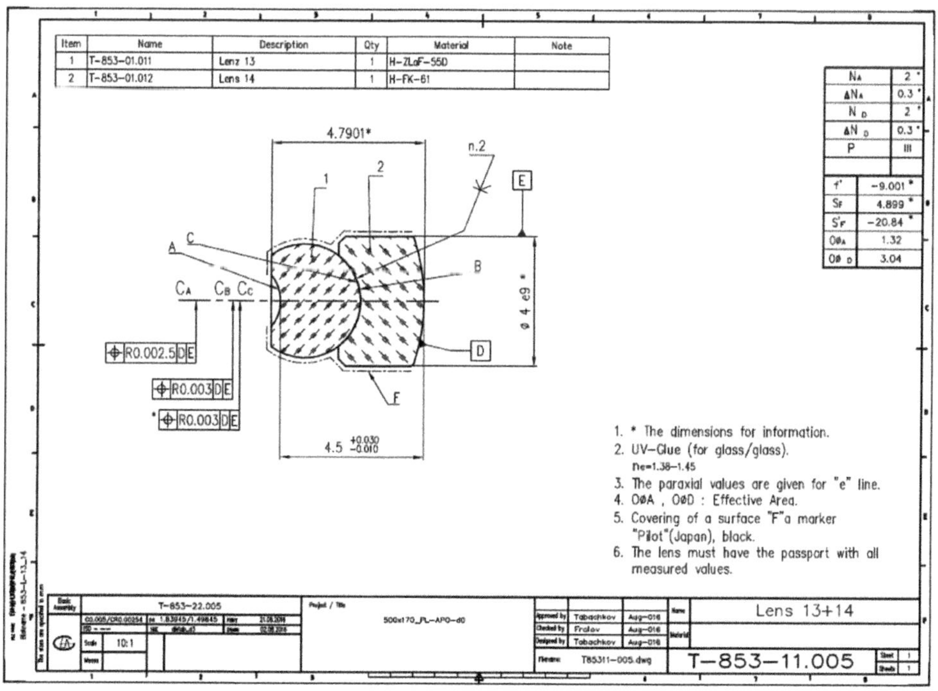

Figure 3.401 A drawing of the 13th and 14th lenses.

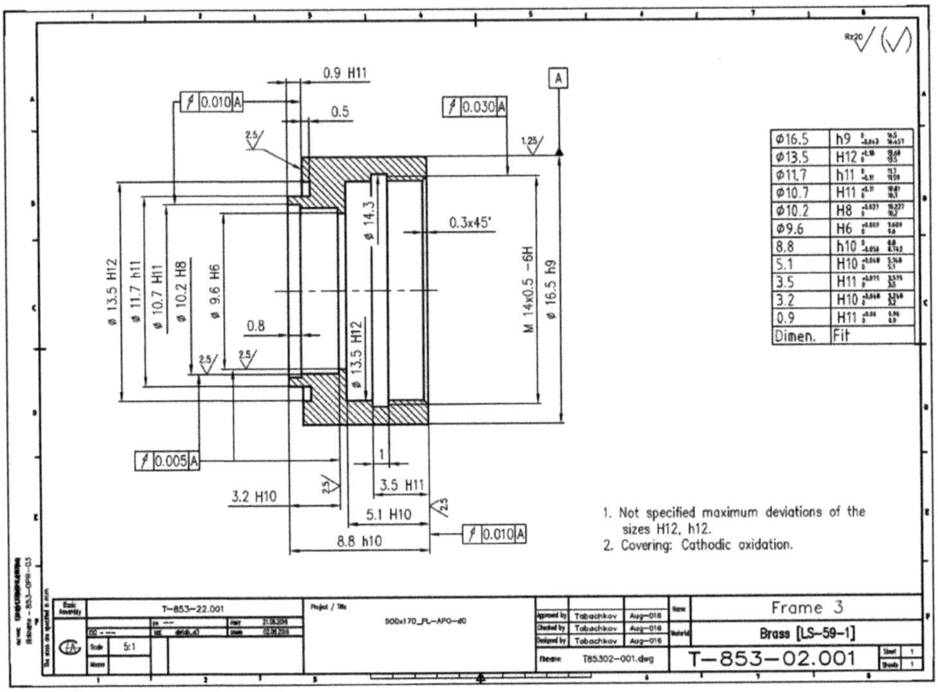

Figure 3.402 A drawing of the third frame.

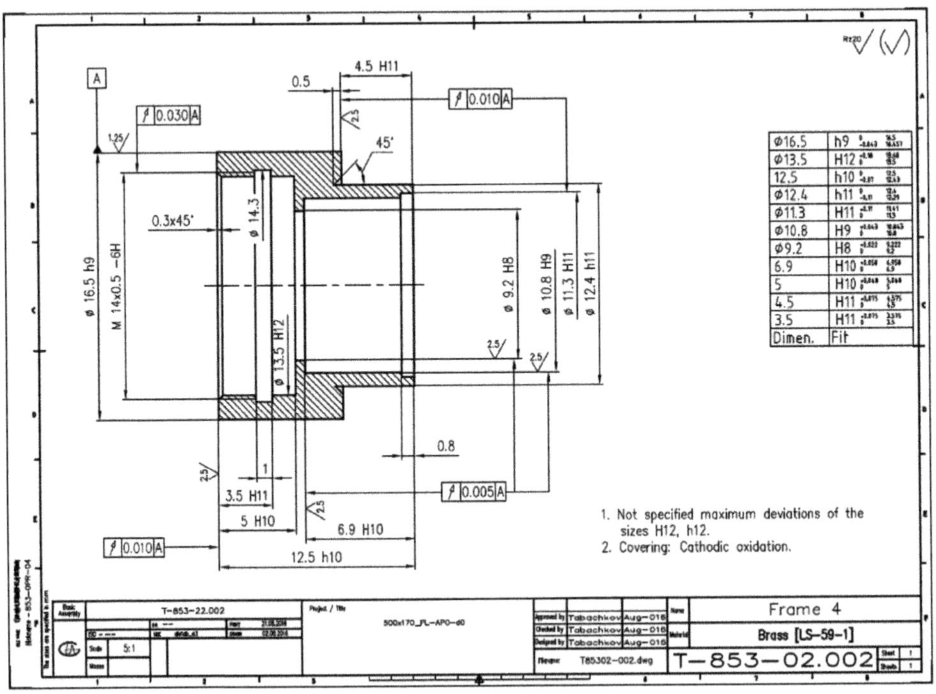

Figure 3.403 A drawing of the fourth frame.

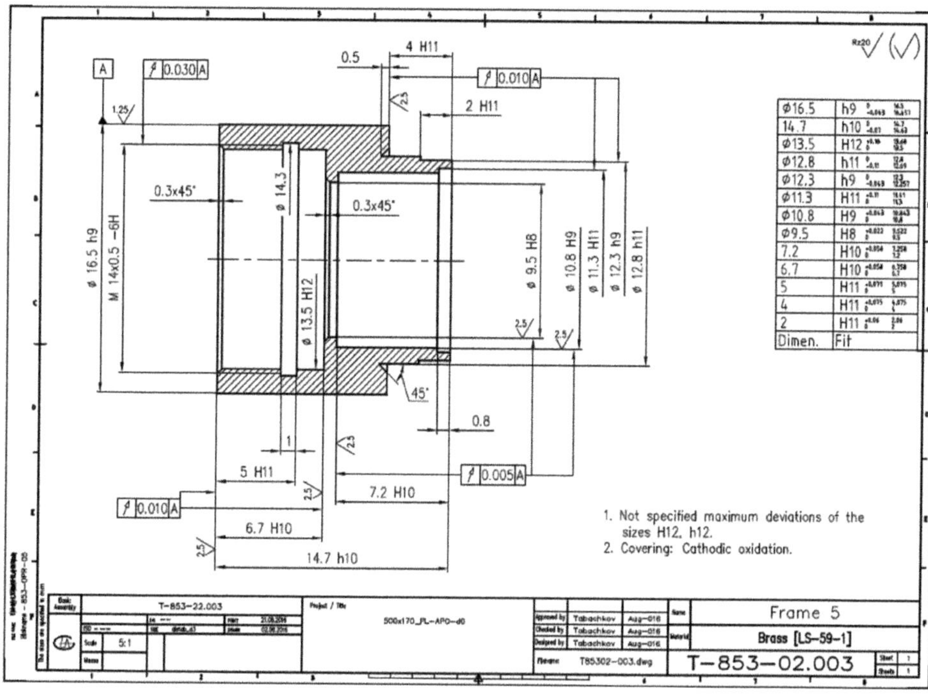

Figure 3.404 A drawing of the fifth frame.

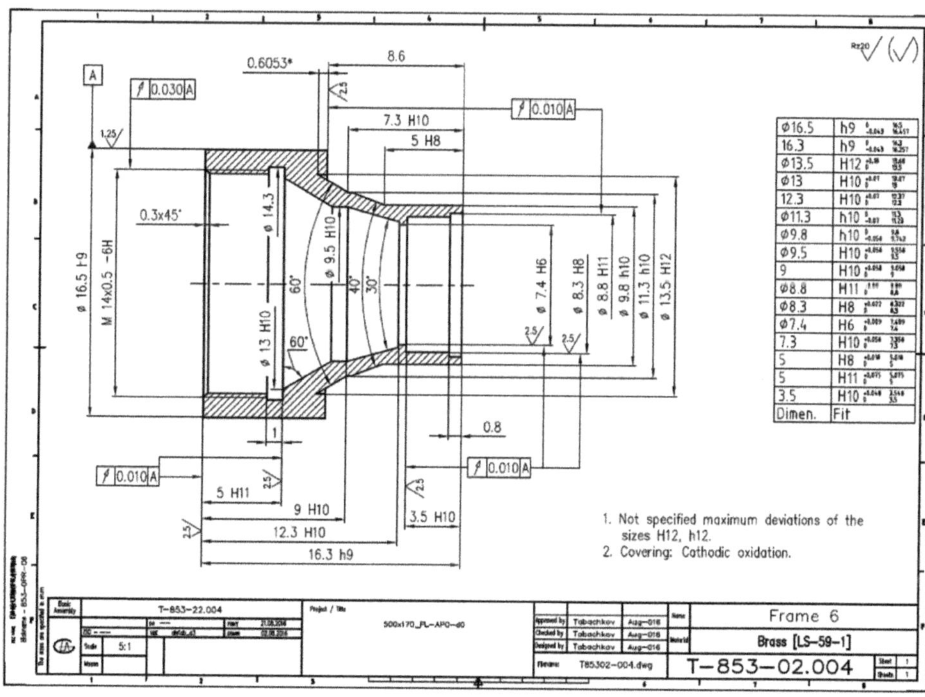

Figure 3.405 A drawing of the sixth frame.

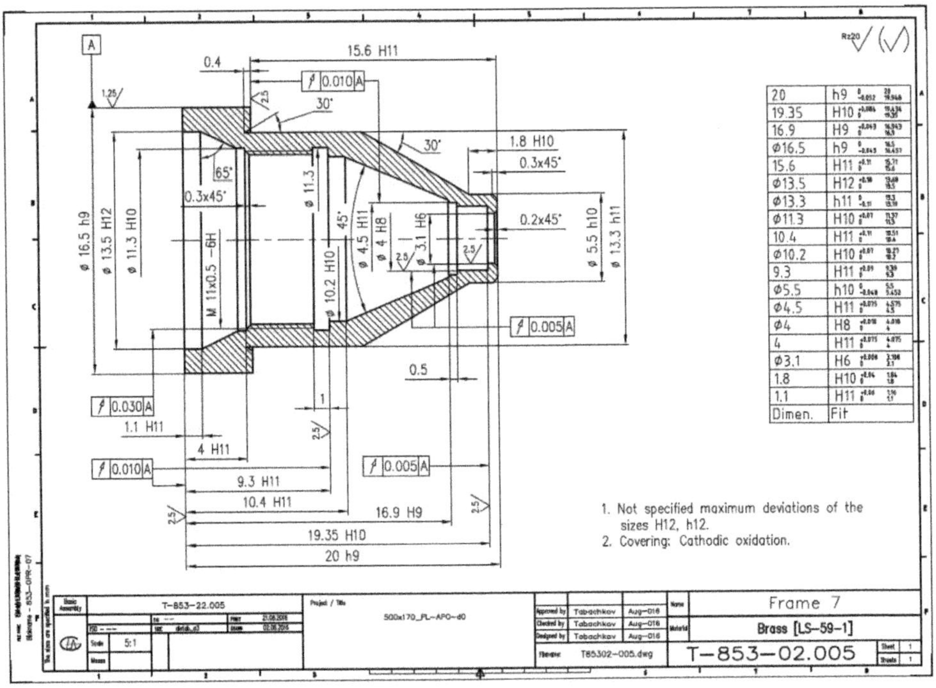

Figure 3.406 A drawing of the seventh frame.

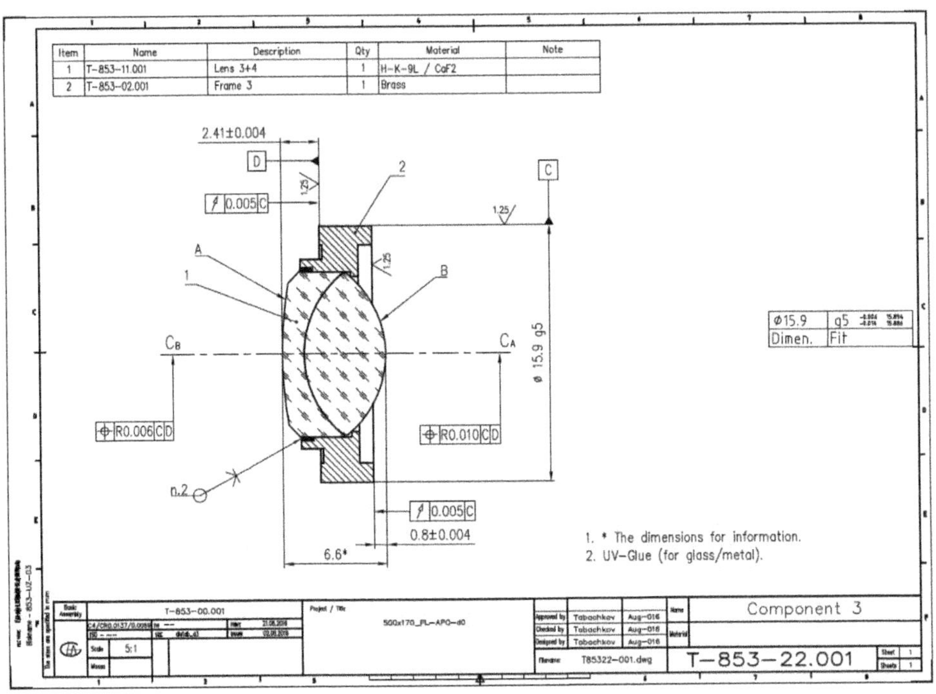

Figure 3.407 A drawing of the third component.

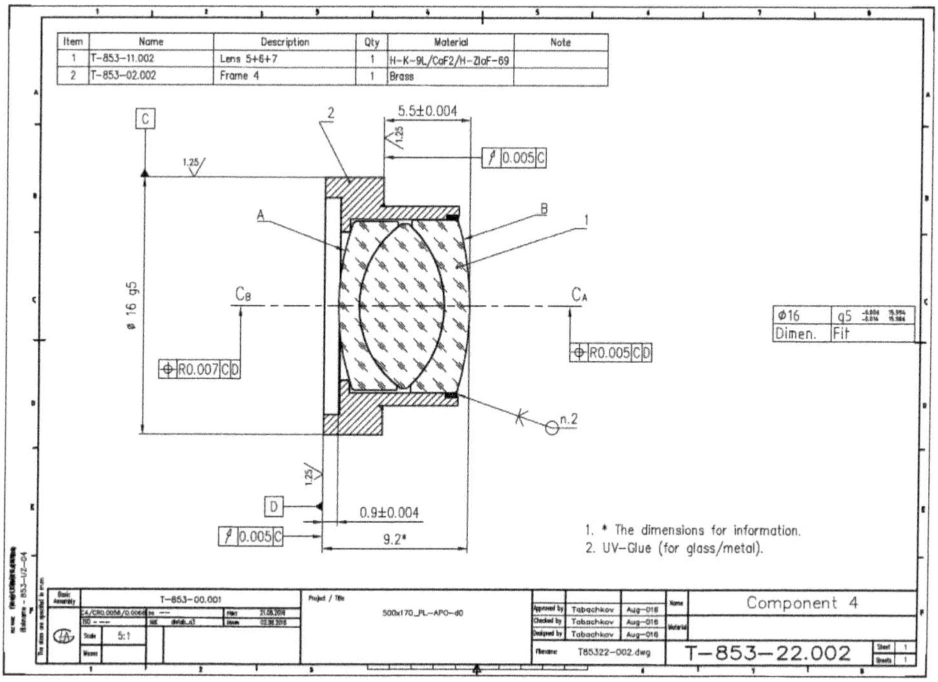

Figure 3.408 A drawing of the fourth component.

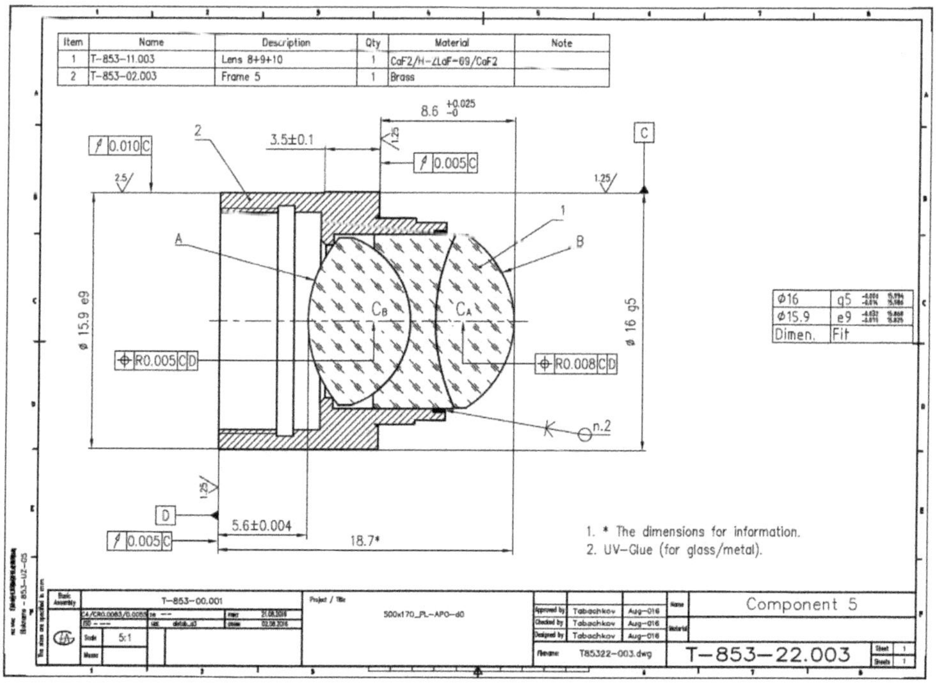

Figure 3.409 A drawing of the fifth component.

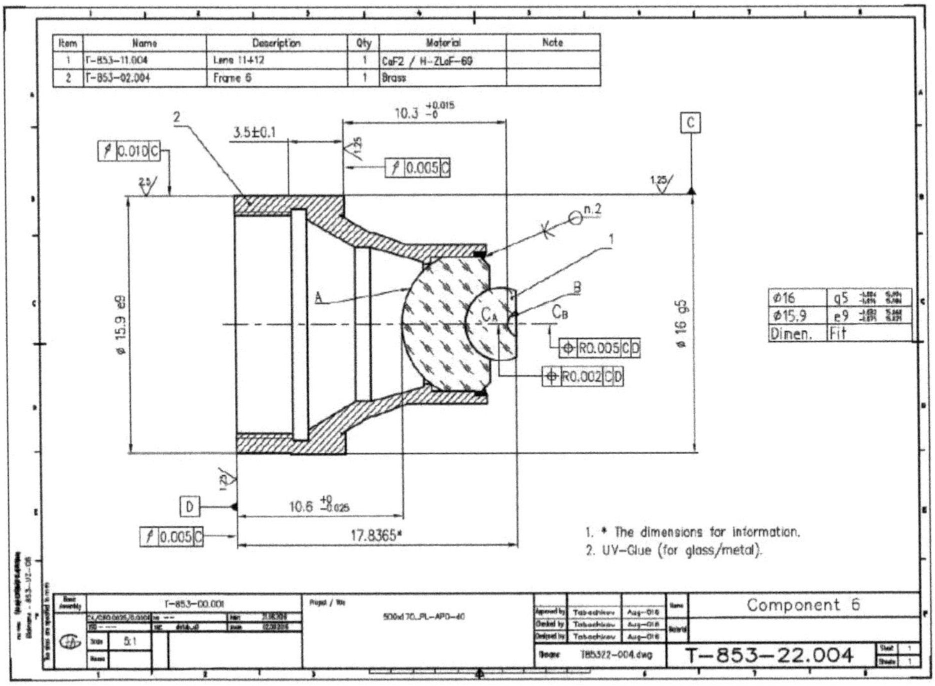

Figure 3.410 A drawing of the sixth component.

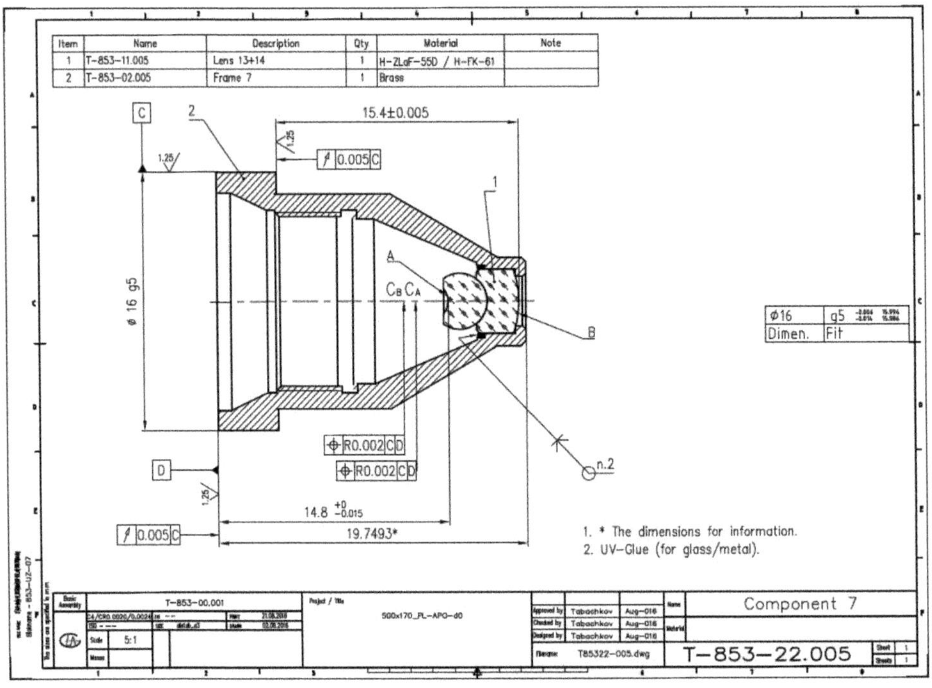

Figure 3.411 A drawing of the seventh component.

Table 3.15 Approximate values of digital microscope objective magnifications based on traditional requirements.

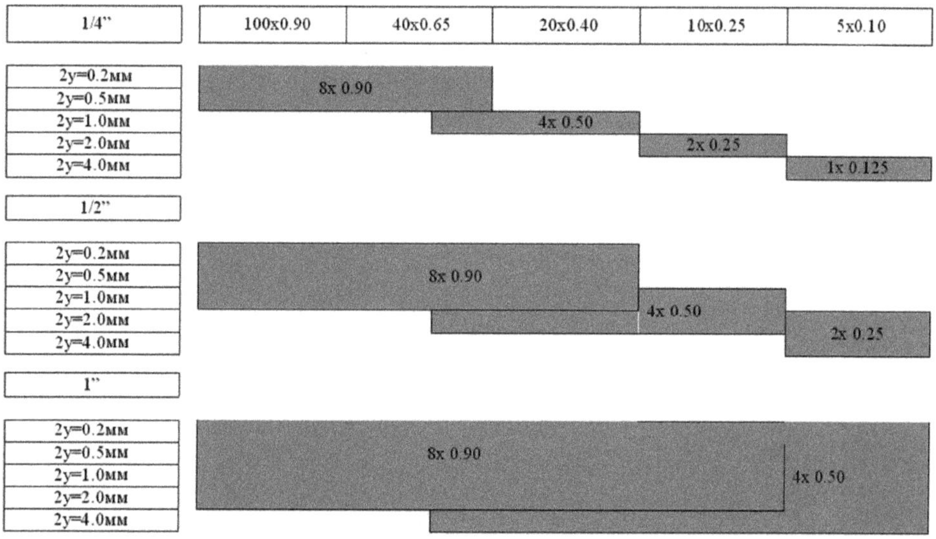

relationship between the parameters of the objective and the receiver. If we assume that the resolution of the receiver is limited by the size of a single (or two to three) elementary structure(s), i.e., a pixel, then we can determine the output numerical aperture of the objective. Table 3.15 presents approximate values of digital microscope objective magnifications based on traditional requirements. The linear field on the object is also taken into account.

This table shows that, depending on the linear size of the receiver, the standard set of conventional objectives (five items) can be replaced by four, three, or even just two kinds of special objectives for a digital microscope (depending on the size of the digital receiver). If we do not consider the presence of protective glass, which is usually located in front of the image receiver in a digital camera, then as examples, we can consider the optical designs of some special objectives. Figures 3.412–3.415 show the design parameters and graphs of aberration correction of some of these objectives.

Contact objectives with a linear magnification of 15× and a numerical aperture in the space of an object of 1.25 can be used for various studies of living objects. An example is the technique for working with such an objective when conducting a blood test. Today, a blood test performed in the clinic involves the traditional methodologies for killing cells, preparing the specimen, and using dyes, glass slides, cover glasses, and glue. The technique for using the specified contact objective can be as follows: a technician takes a blood sample from the patient, transfers it (in the form of a drop) to a special

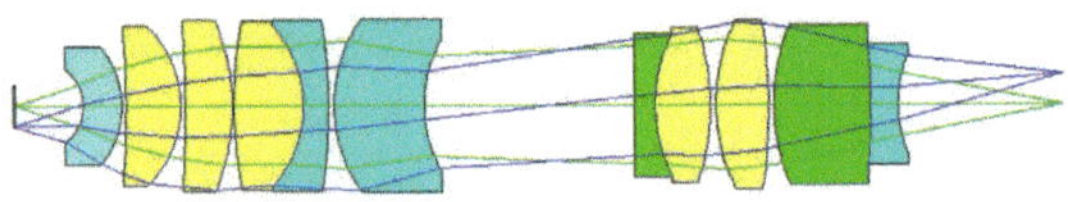

Lens: PlanApo 1.6x0.30 2y'=4.5

Object num aper	0.300000	Image height	2.250000	Primary wavln	0.546070

SRF	RADIUS	THICKNESS	APERTURE RADIUS	GLASS	SPECIAL
OBJ	0.000000	0.010000	1.407095	AIR	
AST	0.000000	0.170000	1.410240 AS	LZ_K14	
2	0.000000	5.000000	1.460000	AIR	
3	-3.162000	2.800000	2.600000	LZ_LK3	
4	-6.400000	0.300000	4.180000	AIR	
5	-54.200000	4.000000	4.980000	CAF2	
6	-7.656000	0.300000	5.610000	AIR	
7	-92.040000	3.500000	5.860000	CAF2 P	
8	-15.066000	0.100000	6.080000	AIR	
9	30.970000	5.000000	5.960000	CAF2 P	
10	-8.460000	1.800000	5.750000	LZ_CTK19	
11	-38.790000	0.400000	6.040000	AIR	
12	10.666000	6.500000	6.110000	LZ_CTK19 P	
13	9.120000	15.000000	4.560000	AIR	
14	342.800000	1.500000	4.840000	LZ_TF5	
15	11.320000	4.000000	5.050000	CAF2 P	
16	-16.827000	0.400000	5.470000	AIR	
17	13.243000	3.800000	5.960000	CAF2 P	
18	-40.786000	0.400000	5.900000	AIR	
19	14.191000	7.000000	5.670000	LZ_TF5 P	
20	-58.340000	2.000000	4.250000	LZ_CTK19 P	
21	8.460000	11.990000	3.510000	AIR	
22	0.000000	-5.9413e-05 S	2.250017 S	AIR	
IMS	0.000000	0.000000	2.250000 S		

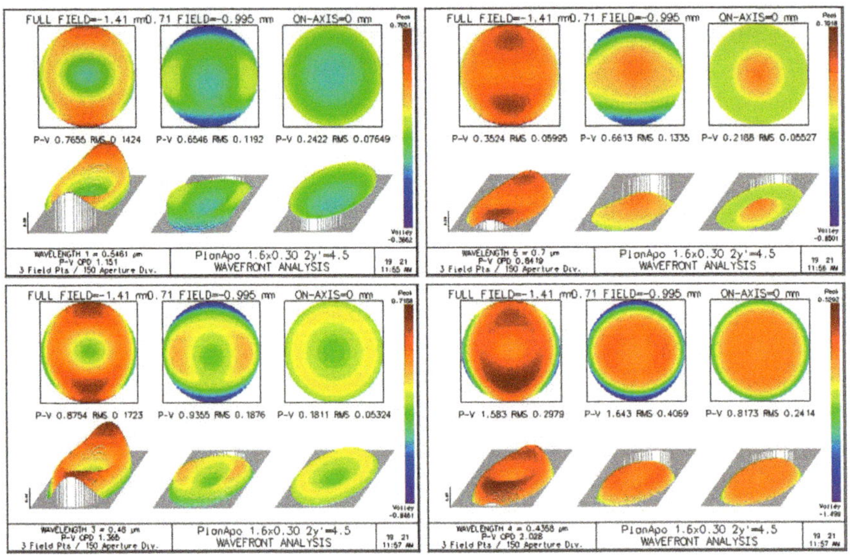

Figure 3.412 Optical design and graphs of aberration correction of a 1.6×/0.30 CCF plan apochromatic objective for direct digital receiver projection (435–700 nm).

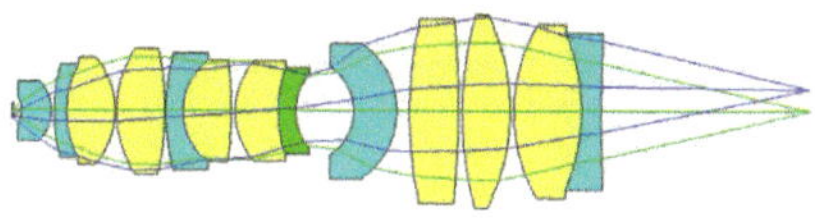

Lens: PlanM-Fluar 3.2x0.70 2y'=4.5

Object num aper 0.700000 Image height 2.250000 Primary wavln 0.546070

SRF	RADIUS		THICKNESS		APERTURE RADIUS		GLASS		SPECIAL
OBJ	0.000000		0.010000		0.703064		AIR		
AST	0.000000		0.170000		0.712866	AS	LZ_K14		
2	0.000000		0.842500		0.800000		AIR		
3	-2.611900		3.100000		1.290000		LZ_CTK19		
4	-4.208300		0.300000		3.000000		AIR		
5	18.389300		1.300000		4.330000		LZ_CTK19	P	
6	10.012100		5.000000		4.700000		CAF2		
7	-8.174300		0.150000		5.410000		AIR		
8	11.450900		5.000000		6.320000		CAF2	P	
9	-28.188700		0.100000		6.180000		AIR		
10	22.588200		1.800000		5.880000		LZ_CTK19	P	
11	6.314100		5.000000		5.090000		CAF2	P	
12	-34.704000		0.400000		5.060000		AIR		
13	7.405800		4.500000		5.010000		CAF2	P	
14	13.406100		1.800000		4.330000		LZ_TF5		
15	5.752700		6.000000		3.790000		AIR	P	
16	-5.245500		4.500000		4.440000		LZ_CTK19	P	
17	-8.083900		1.300000		6.810000		AIR		
18	40.809000		5.000000		8.790000		CAF2	P	
19	-108.917000		0.400000		9.320000		AIR		
20	35.862100		5.000000		9.700000		CAF2	P	
21	-21.522800		0.400000		9.740000		AIR		
22	17.130700		7.000000		8.760000		CAF2	P	
23	-19.892400		2.000000		7.850000		LZ_CTK19	P	
24	160.963500		21.470000		7.270000		AIR		
25	0.000000		0.003078	S	2.250972	S	AIR		
IMS	0.000000		0.000000		2.250000	S			

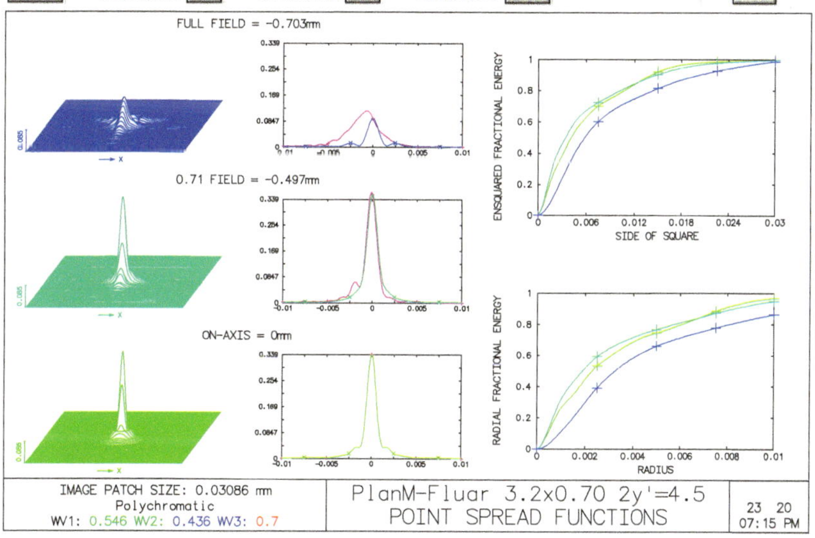

Figure 3.413 Optical design and graphs of aberration correction of a 3.2×/0.70 LCF plan MicroFluar objective for direct digital receiver projection (435–700 nm).

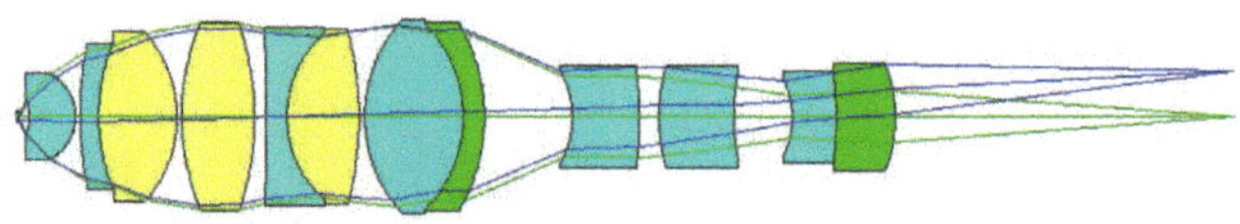

Lens: PlanM-Fluar 12.5x0.90 2y'=6.0

Object num aper 0.900000 Image height 3.000000 Primary wavln 0.546070

SRF	RADIUS	THICKNESS	APERTURE RADIUS	GLASS	SPECIAL
OBJ	0.000000	0.010000	0.240028	AIR	
AST	0.000000	0.170000	0.260675 AS	LZ_K14	
2	0.000000	0.700000	0.390000	AIR	
3	-2.995900	3.100000	1.260000	LZ_CTK19	
4	-3.171200	0.300000	2.870000	AIR	
5	22.183100	1.300000	4.420000	LZ_CTK19 P	
6	11.979300	5.500000	4.800000	CAF2	
7	-7.941700	0.300000	5.610000	AIR	
8	15.056000	5.000000	6.210000	CAF2 P	
9	-18.169500	0.800000	6.130000	AIR	
10	-138.265600	1.400000	5.800000	LZ_CTK19 P	
11	7.014100	5.000000	5.470000	CAF2 P	
12	-21.637800	0.400000	5.770000	AIR	
13	9.204000	6.600000	6.340000	CAF2 P	
14	-8.573600	1.400000	6.060000	LZ_TF5	
15	-12.109900	6.000000	6.170000	AIR P	
16	-6.012200	4.500000	3.120000	LZ_CTK19 P	
17	-23.962600	1.300000	3.390000	AIR	
18	11.860600	5.000000	3.360000	LZ_TF5 P	
19	14.110500	4.000000	2.730000	AIR	
20	-5.577700	3.000000	2.460000	LZ_CTK19 P	
21	-22.833700	4.000000	2.990000	LZ_TF5 P	
22	-9.502300	23.000000	3.540000	AIR	
23	0.000000	-0.006538 S	3.001408 S	AIR	
IMS	0.000000	0.000000	3.000000 S		

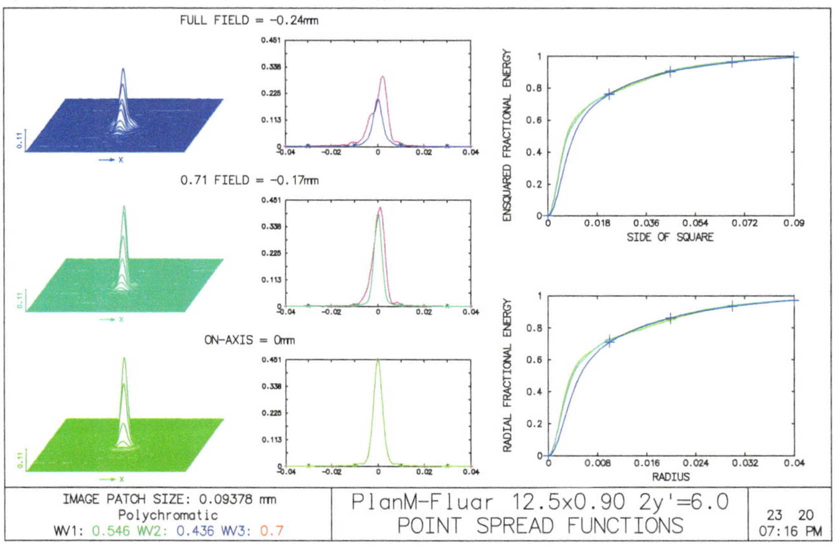

Figure 3.414 Optical design and graphs of aberration correction of a 12.5×/0.90 LCF plan MicroFluar objective for direct digital receiver projection (435–700 nm).

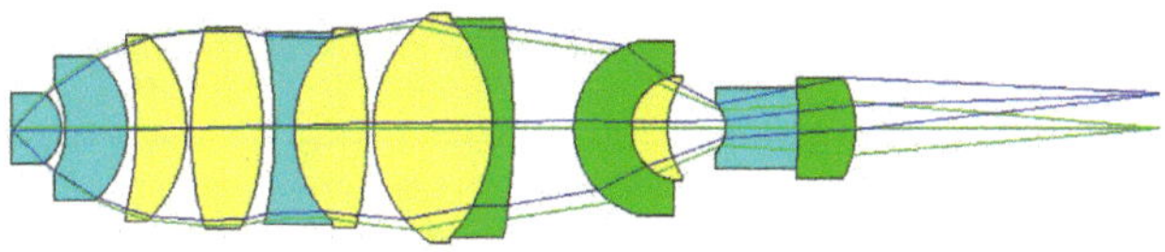

Lens: pl MFluar 15x1.25 cont finity				Efl	4.53
Object num aper	1.250000	Image height 2.250000		Primary wavln	0.546070
SRF	**RADIUS**	**THICKNESS**	**APERTURE RADIUS**	**GLASS**	**SPECIAL**
OBJ	0.000000	0.001000	0.150012	LZ_CTK19	
AST	0.000000	3.500000	0.151035 AS	LZ_CTK19	
2	-2.759200	0.300000	2.370000	AIR	
3	-5.786500	4.300000	2.970000	LZ_CTK19 P	
4	-6.308800	0.600000	4.850000	AIR	
5	-22.857900	3.500000	5.720000	CAF2	
6	-8.595400	0.300000	6.290000	AIR	
7	24.315800	5.000000	6.810000	CAF2 P	
8	-15.940000	0.800000	6.820000	AIR	
9	-30.130000	1.400000	6.470000	LZ_CTK19 P	
10	8.905600	5.000000	6.340000	CAF2 P	
11	-28.622100	0.400000	6.680000	AIR	
12	9.585300	8.200000	7.710000	CAF2 P	
13	-10.741300	1.400000	7.350000	LZ_TF5	
14	-40.261300	4.200000	7.320000	AIR	
15	6.048800	4.000000	5.840000	LZ_TF5 P	
16	3.564700	2.500000	3.510000	CAF2 P	
17	5.647100	4.000000	3.160000	AIR	
18	-2.275100	5.000000	1.620000	LZ_CTK19 P	
19	-28.545600	4.000000	2.710000	LZ_TF5 P	
20	-7.074700	21.000000	3.360000	AIR	
21	0.000000	-0.003635 S	2.250554 S	AIR	
IMS	0.000000	0.000000	2.250000 S		

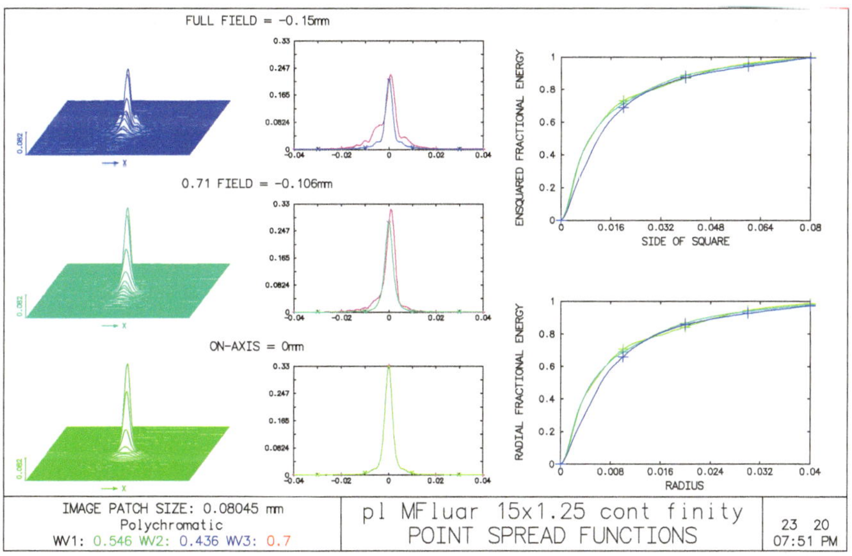

Figure 3.415 Optical design and graphs of aberration correction of a 15×/1.25 LCF plan MicroFluar contact objective for direct digital receiver projection (435–700 nm).

Table 3.16 Main technical parameters and basic optical layout of direct projection objectives.

Magnification	*NA*	*WD* (mm)	*R* (μm)	*DF* (μm)	*FOV* on object (mm)	*FOV* on image (mm)	Optical design at figure no.	The principal optical layout
1.6×	0.3	4.2	1.11	3.7	2.8	4.5	3.412	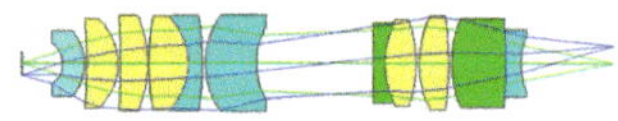
3.2×	0.70	0.48	0.48	0.68	1.4	4.5	3.413	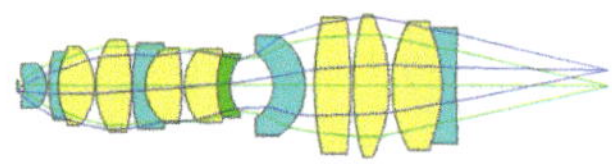
12.5×	0.9	0.37	0.37	0.41	0.48	6	3.414	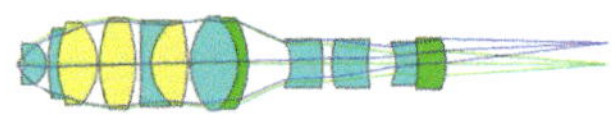
15×	1.25	0	0.27	0.21	0.3	4.5	3.415	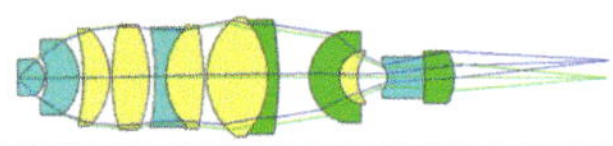

glass that has a backlight (similar to a glass slide; for an example, see the technical description of a glass slide given in Chapter 1, Section 1.8.6.2.2). The lab technician manually dips the lens into a drop, receiving images on the receiver of a digital camera. There is no need to use a microscope as such; there is no need for a focusing mechanism, a stage, and other usual parts of a light microscope. But most importantly, there is no need for a special preparation for the studied object (in this case, a drop of blood); the study can be carried out live and in real time.

Table 3.16 shows the main technical parameters of direct projection objectives.

We can also discuss the creation of special objectives: plan apochromats with extra-large working distances for research in metallography and microelectronics. These objectives are also designed for direct projection of images on a digital receiver; an explanation of the principle of obtaining such optical systems is shown in Chapter 1, Fig. 1.265. Such objectives can be constructed as a combination of two objectives; see the discussion of this in Section 3.9. A 5× objective (Fig. 3.171) is used as an adapter for projecting an image onto a digital receiver. The remaining objectives are used as the main interchangeable objectives; in one version, two identical 5× objectives are used, located as a mirror symmetry. A parallel course of the aperture beam is maintained between the objectives. Figures 3.416–3.422 show the design parameters and graphs of aberration correction of some objectives constructed in this way.

Table 3.17 shows the main technical parameters of direct projection objectives (plan apochromats with extra-large working distances).

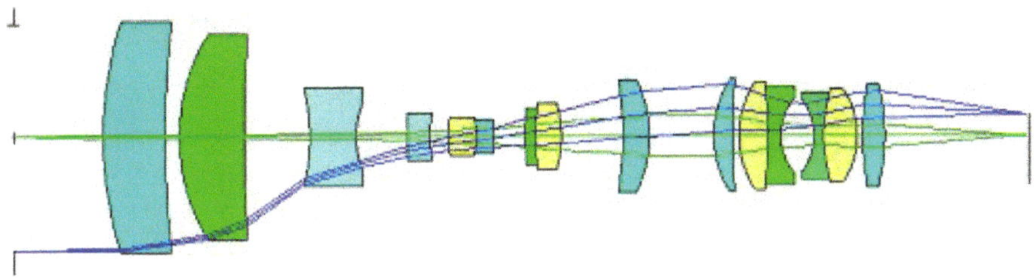

Lens: PlanApo 0.2x0.015 2y'=6.8 mm

Object num aper 0.015000 Image height 3.400000 Primary wavln 0.546070

SRF	RADIUS	THICKNESS	APERTURE RADIUS	GLASS	SPECIAL
OBJ	0.000000	1.0000e-20	16.908572	AIR	
AST	0.000000	13.920000	16.908572 AS	AIR	
2	55.590000	9.700000	17.250000	H-LAK3	
3	152.600000	2.200000	16.240000	AIR	
4	28.689000	10.000000	15.400000	H-ZF7LA	
5	349.100000	10.000000	13.370000	AIR	
6	-30.760000	7.000000	7.380000	K9	
7	12.560000	8.000000	4.920000	AIR	
8	-135.790000	3.400000	3.490000	K9 P	
9	14.150000	3.000000	3.020000	AIR	
10	28.310000	4.200000	2.800000	CAF2 C	
11	-9.000000	2.500000	2.490000	H-LAK3 P	
12	71.430000	5.000000	2.440000	AIR	
13	318.980000	1.600000	3.840000	H-ZF7LA P	
14	33.700000	4.000000	4.180000	CAF2 P	
15	-19.538000	9.600000	5.030000	AIR	
16	-40.820000	4.000000	7.720000	H-LAK3 P	
17	-19.269000	10.000000	8.410000	AIR	
18	15.672000	3.000000	9.450000	LZ_CTK12	
19	254.000000	0.800000	9.450000	AIR	
20	14.510000	4.300000	8.140000	CAF2	
21	-75.000000	2.000000	7.430000	H-ZF1	
22	8.801000	5.200000	5.670000	AIR	
23	-8.801000	1.500000	5.500000	H-LAK4L	
24	21.580000	5.200000	6.480000	CAF2	
25	-12.340000	0.800000	7.100000	AIR	
26	42.482000	3.400000	7.720000	H-LAK3	
27	-30.270000	21.926857 S	7.730000	AIR	
28	0.000000	0.000000	3.400000 S	AIR	
IMS	0.000000	0.000000	3.400000 S		

Figure 3.416 Optical design and graphs of aberration correction of a 0.2×/0.015 CCF plan apochromatic objective for direct digital receiver projection (435–650 nm).

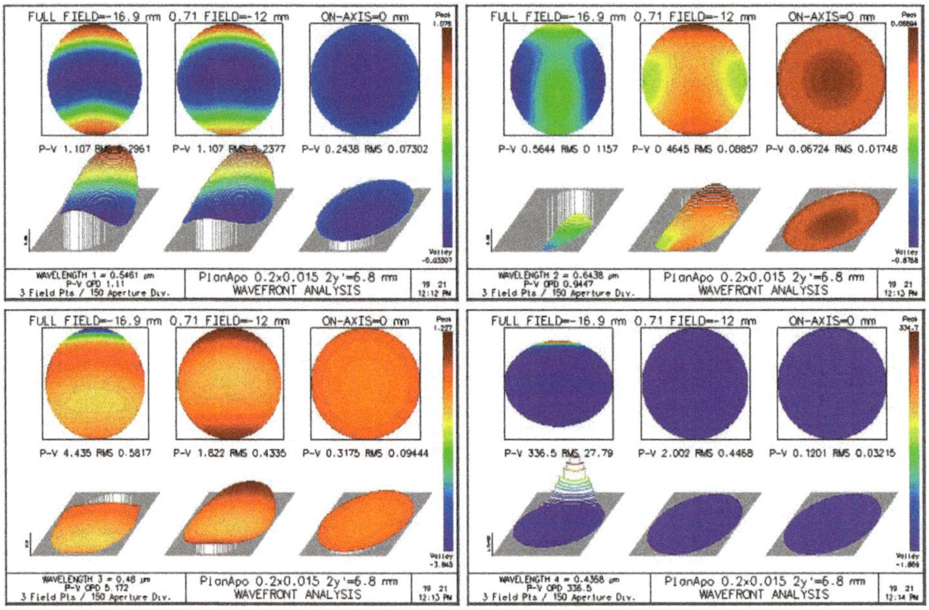

Figure 3.416 (*Continued*) Optical design and graphs of aberration correction of a 0.2×/0.015 CCF plan apochromatic objective for direct digital receiver projection (435–650 nm).

3.13 Telecentric Objectives

Telecentric objectives are in a separate class of microscope objectives. In the modern market, there are many telecentric objectives with different working distances, linear magnifications, and resolutions; various sized digital receivers are also used. Telecentric objectives are mainly used in so-called *machine vision* systems. However, they are essentially optical systems for projecting an image of an object under investigation onto a digital image receiver. We have already discussed the advantages of constructing optical systems using the telecentric path of rays; doing so has not only been proposed for objective systems, but also, for example, lighting and projection systems of microscopes. Thus, telecentric objectives are quite common in the practice of building optical components for microscopes. However, some features of telecentric objectives are noteworthy in terms of their optical design. These concern, first and foremost, the correction of distortion aberration; in ordinary mass market objectives for light microscopes, i.e., medium and small series, this aberration, as a rule, is not subject to special correction (exceptions include objectives used for measuring). On the contrary, distortion aberration is corrected with special care in telecentric objectives because they are mainly intended to be used for measuring. In fact, in a telecentric objective, the so-called *afocal system* consists of two independent objectives between which the parallel course of the aperture beam is preserved. The linear magnification of

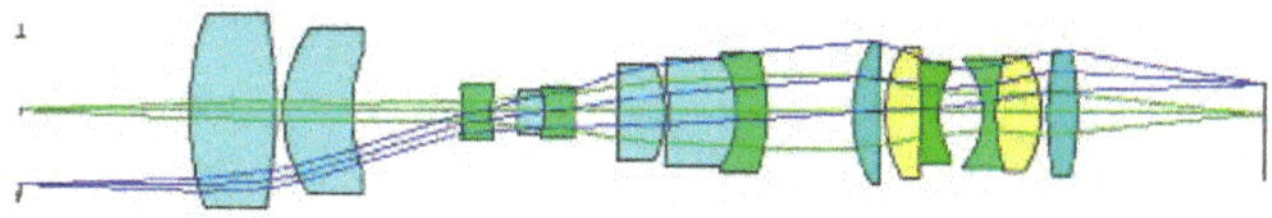

Lens: PlanApo 0.4x0.04 2y'=6.8 mm

Object num aper 0.040000 Image height 3.400000 Primary wavln 0.546070

SRF	RADIUS	THICKNESS	APERTURE RADIUS	GLASS	SPECIAL
OBJ	0.000000	1.0000e-20	8.467272	AIR	
AST	0.000000	20.000000	8.467272 AS	AIR	
2	34.760000	10.000000	10.990000	K9	
3	-61.010000	1.000000	10.410000	AIR	
4	16.181000	8.000000	9.350000	H-QK3	
5	19.269000	13.000000	7.030000	AIR	
6	-21.180000	3.400000	3.010000	H-LAK3	
7	10.627000	3.000000	2.510000	AIR	
8	15.000000	3.200000	2.450000	H-QK3 P	
9	-6.300000	3.500000	2.400000	H-LAK3 P	
10	-139.830000	5.000000	2.990000	AIR	
11	-190.990000	5.500000	4.540000	H-QK3 P	
12	-12.340000	0.300000	5.450000	AIR	
13	-64.927000	8.000000	5.560000	H-QK3 P	
14	-11.660000	3.500000	6.100000	H-LAK3 P	
15	-23.140000	10.000000	6.770000	AIR	
16	15.672000	3.000000	8.000000	LZ_CTK12	
17	254.000000	0.800000	7.180000	AIR	
18	14.510000	4.300000	7.400000	CAF2	
19	-75.000000	2.000000	5.800000	H-ZF1	
20	8.801000	5.200000	4.900000	AIR	
21	-8.801000	1.500000	6.000000	H-LAK4L	
22	21.580000	5.200000	6.300000	CAF2	
23	-12.340000	0.800000	6.600000	AIR	
24	42.482000	3.400000	7.000000	H-LAK3	
25	-30.270000	21.925351 S	7.000000	AIR	
26	0.000000	0.000000	3.400000	AIR	
IMS	0.000000	0.000000	3.400000 S		

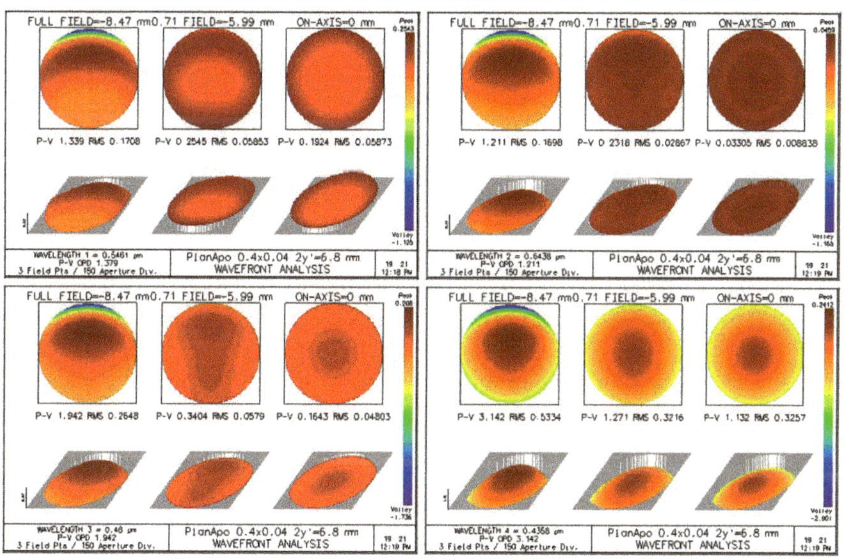

Figure 3.417 Optical design and graphs of aberration correction of a 0.4×/0.04 CCF plan apochromatic objective for direct digital receiver projection (435–650 nm).

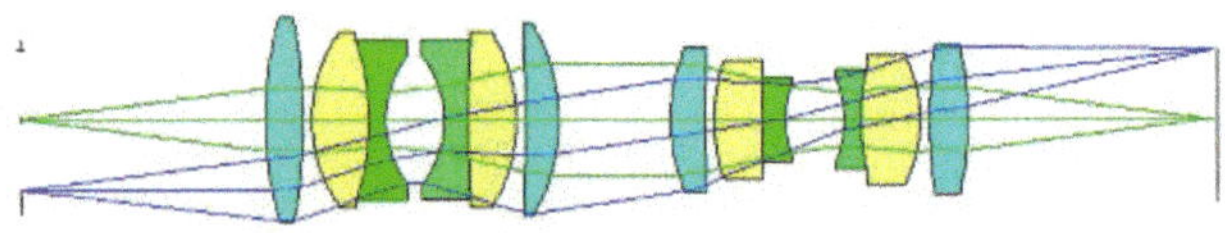

Lens: PlanApo 1x0.12 2y'=6.8 mm

Object num aper	0.120000	Image height	3.400000	Primary wavln	0.546070

SRF	RADIUS		THICKNESS		APERTURE RADIUS		GLASS		SPECIAL
OBJ	0.000000		1.0000e-20		3.400225		AIR		
AST	0.000000		21.929000		3.400225	AS	AIR		
2	30.270000		3.400000		6.120000		H-LAK3		
3	-42.482000		0.800000		6.080000		AIR		
4	12.340000		5.200000		5.700000		CAF2	C	
5	-21.580000		1.500000		4.870000		H-LAK4L		
6	8.801000		5.200000		4.280000		AIR		
7	-8.801000		2.000000		4.400000		H-ZF1		
8	75.000000		4.300000		5.210000		CAF2	C	
9	-14.510000		0.800000		6.020000		AIR		
10	-254.000000		3.000000		6.480000		LZ_CTK12		
11	-15.672000		10.000000		6.700000		AIR		
12	15.672000		3.000000		5.640000		LZ_CTK12	P	
13	254.000000		0.800000		5.230000		AIR		
14	14.510000		4.300000		4.710000		CAF2		
15	-75.000000		2.000000		3.530000		H-ZF1		
16	8.801000		5.200000		2.960000		AIR		
17	-8.801000		1.500000		3.090000		H-LAK4L		
18	21.580000		5.200000		3.470000		CAF2		
19	-12.340000		0.800000		4.370000		AIR		
20	42.482000		3.400000		4.640000		H-LAK3		
21	-30.270000		21.921842	S	4.740000		AIR		
22	0.000000		0.000000		3.400000	S	AIR		
IMS	0.000000		0.000000		3.400000	S			

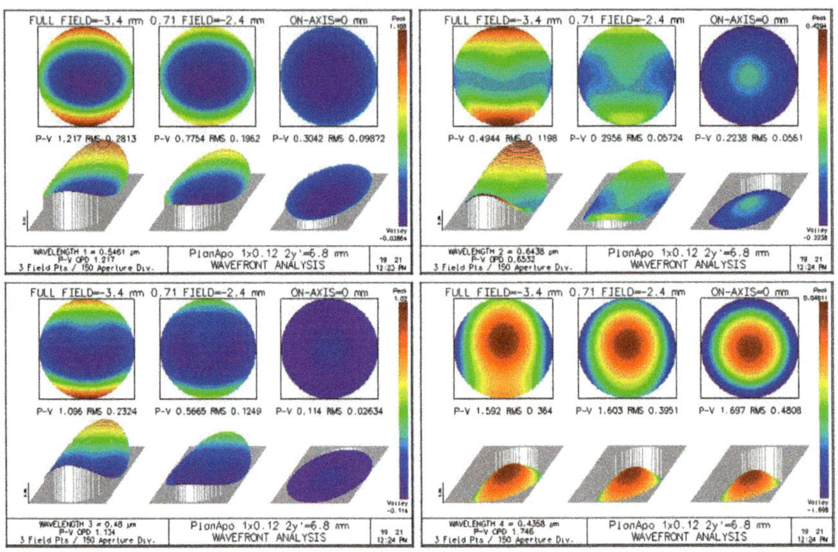

Figure 3.418 Optical design and graphs of aberration correction of a 1.0×/0.12 CCF plan apochromatic objective for direct digital receiver projection (435–650 nm).

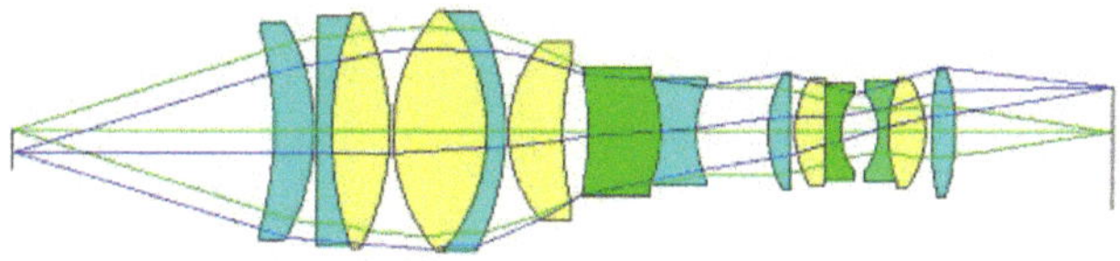

Lens: PlanApo 2x0.3 2y'=12 mm

Object num aper 0.300000 Object height -3.000000 Primary wavln 0.546070

SRF	RADIUS	THICKNESS	APERTURE RADIUS	GLASS	SPECIAL
OBJ	0.000000	1.0000e-20	3.000000	AIR	
AST	0.000000	36.485000	3.000000 AS	AIR	
2	-49.660000	5.600000	13.850000	H-LAK3	
3	-29.570000	0.500000	14.930000	AIR	
4	0.000000	2.000000	15.440000	H-ZLAF3	
5	47.500000	8.000000	15.730000	CAF2 C	
6	-34.760000	0.500000	16.000000	AIR	
7	25.500000	13.000000	16.580000	CAF2 C	
8	-25.500000	2.500000	16.390000	H-LAK3 P	
9	-37.070000	0.500000	16.380000	AIR	
10	18.883000	7.400000	12.310000	CAF2 C	
11	39.370000	3.600000	9.980000	AIR	
12	-44.240000	10.000000	8.960000	H-ZF62	
13	-19.724000	5.000000	7.480000	H-LAF10LA	
14	15.136000	10.000000	6.430000	AIR	
15	15.672000	3.000000	8.090000	LZ_CTK12	
16	254.000000	0.800000	7.960000	AIR	
17	14.510000	4.300000	7.310000	CAF2 C	
18	-75.000000	2.000000	6.620000	H-ZF1	
19	8.801000	5.200000	5.460000	AIR	
20	-8.801000	1.500000	5.480000	H-LAK4L	
21	21.580000	5.200000	6.850000	CAF2 P	
22	-12.340000	0.800000	7.500000	AIR	
23	42.482000	3.400000	8.730000	H-LAK3 P	
24	-30.270000	21.928899 S	8.810000	AIR	
25	0.000000	0.000000	6.016041 S	AIR	
IMS	0.000000	0.000000	6.016041 S		

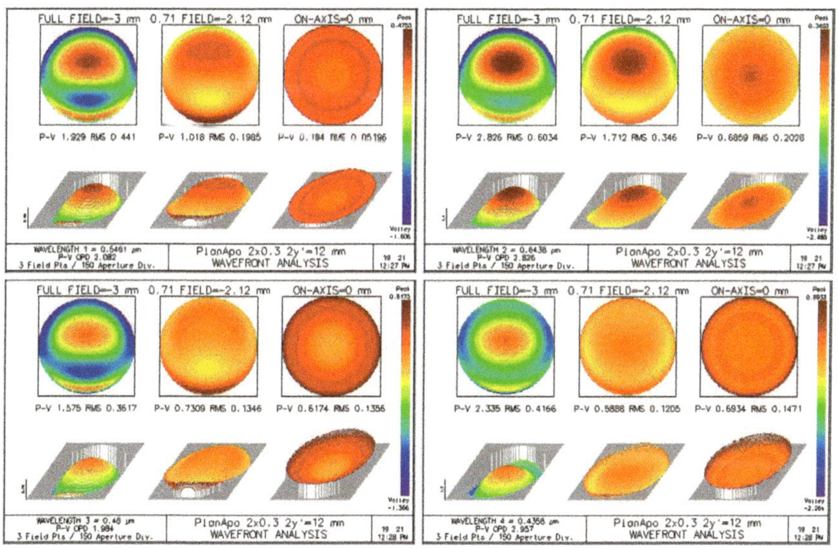

Figure 3.419 Optical design and graphs of aberration correction of a 2.0×/0.30 CCF plan apochromatic objective for direct digital receiver projection (435–650 nm).

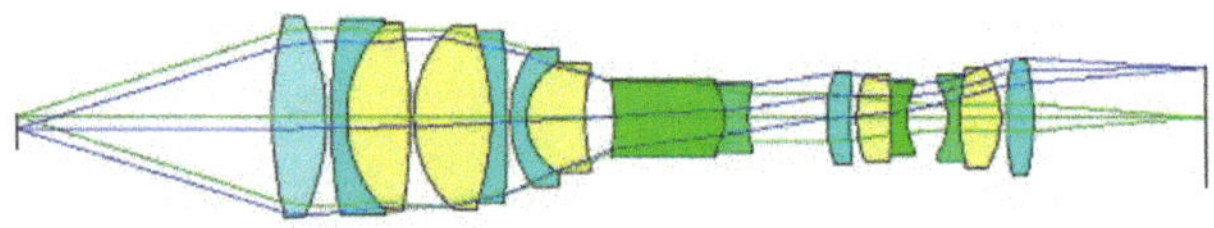

Lens: PlanApo 4x0.30 2y'=12 mm

Object num aper	0.300000	Image height	6.000000	Primary wavln	0.546070

SRF	RADIUS	THICKNESS	APERTURE RADIUS	GLASS	SPECIAL
OBJ	0.000000	1.0000e-20	1.494275	AIR	
AST	0.000000	32.967000	1.494275 AS	AIR	
2	53.680000	6.800000	12.270000	H-QK3	
3	-28.689000	0.600000	12.570000	AIR	
4	62.230000	2.000000	12.310000	LZ_CTK12	
5	16.530000	8.000000	11.670000	CAF2 C	
6	-74.290000	0.500000	11.760000	AIR	
7	15.672000	10.000000	11.650000	CAF2 C	
8	-30.760000	2.000000	10.770000	LZ_CTK12 P	
9	-119.500000	0.500000	10.090000	AIR	
10	14.791000	2.000000	8.590000	LZ_CTK12 P	
11	7.900000	7.400000	6.880000	CAF2 C	
12	25.800000	3.800000	5.570000	AIR	
13	-20.750000	14.000000	4.820000	H-ZF52A	
14	-8.336000	3.000000	4.350000	H-ZLAF50B	
15	14.510000	10.000000	4.090000	AIR	
16	15.672000	3.000000	5.640000	LZ_CTK12 P	
17	254.000000	0.800000	5.490000	AIR	
18	14.510000	4.300000	5.310000	CAF2	
19	-75.000000	2.000000	4.710000	H-ZF1	
20	8.801000	5.200000	4.200000	AIR	
21	-8.801000	1.500000	4.450000	H-LAK4L	
22	21.580000	5.200000	5.360000	CAF2	
23	-12.340000	0.800000	6.360000	AIR	
24	42.482000	3.400000	7.230000	H-LAK3	
25	-30.270000	21.841783 S	7.360000	AIR	
26	0.000000	0.000000	6.000000 S	AIR	
IMS	0.000000	0.000000	6.000000 S		

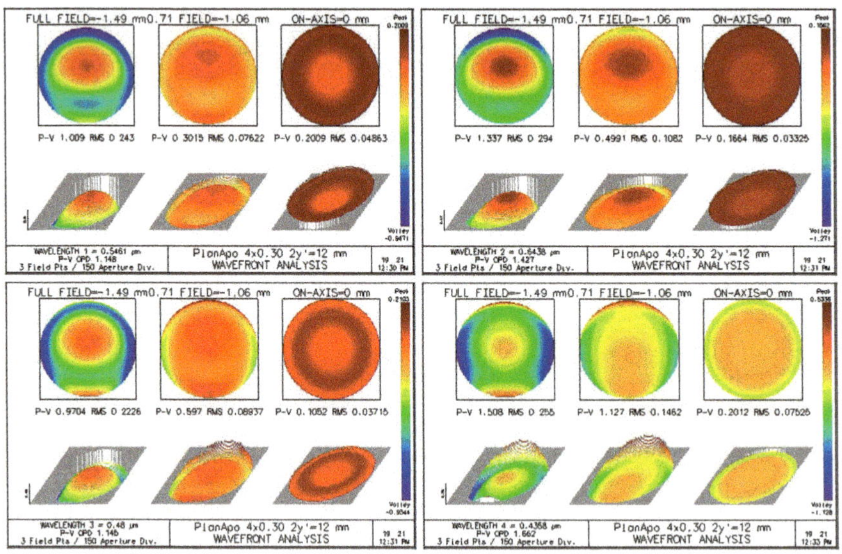

Figure 3.420 Optical design and graphs of aberration correction of a 4.0×/0.30 CCF plan apochromatic objective for direct digital receiver projection (435–650 nm).

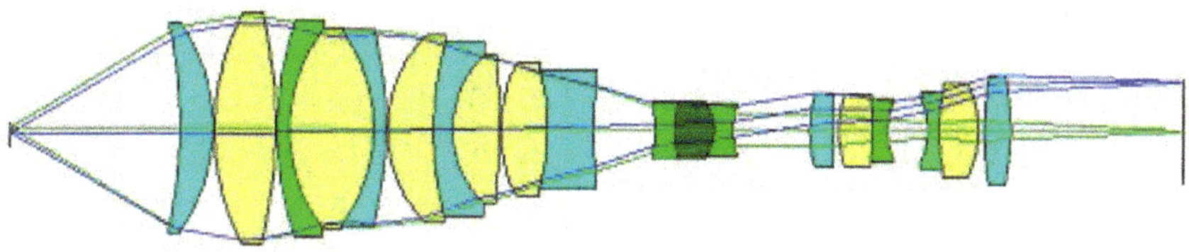

Lens: PlanApo 10x0.50 2y'=12 mm

Object num aper	0.500000	Image height	6.000000	Primary wavln	0.546070
SRF	**RADIUS**	**THICKNESS**	**APERTURE RADIUS**	**GLASS**	**SPECIAL**
OBJ	0.000000	1.0000e-20	0.598681	AIR	
AST	0.000000	21.997000	0.598681 AS	AIR	
2	-59.320000	4.360000	12.510000	H-ZLAF1	
3	-21.580000	0.200000	13.040000	AIR	
4	30.270000	8.000000	14.360000	CAF2	
5	-57.800000	0.200000	14.210000	AIR	
6	40.190000	1.600000	13.360000	H-ZLAF3	
7	19.790000	11.000000	12.380000	CAF2 P	
8	-19.790000	1.600000	12.050000	H-ZLAF1 P	
9	-47.896000	0.200000	12.280000	AIR	
10	16.120000	6.000000	11.660000	CAF2 P	
11	42.657000	2.500000	10.870000	H-LAF10L	
12	12.465000	6.000000	9.230000	CAF2 P	
13	-124.360000	0.300000	8.990000	AIR	
14	17.760000	6.000000	8.290000	CAF2 P	
15	-27.534000	6.000000	7.280000	H-LAK2	
16	41.590000	8.000000	5.590000	AIR	
17	-13.305000	2.500000	3.440000	H-ZLAF3 P	
18	18.710000	5.100000	3.380000	H-SF62	
19	-5.750000	2.500000	3.290000	H-ZLAF3 P	
20	10.850000	10.000000	3.040000	AIR	
21	15.672000	3.000000	4.540000	LZ_CTK12	
22	254.000000	0.800000	4.440000	AIR	
23	14.510000	4.300000	4.360000	CAF2	
24	-75.000000	2.000000	3.900000	H-ZF1	
25	8.801000	5.200000	3.590000	AIR	
26	-8.801000	1.500000	3.980000	H-LAK4L	
27	21.580000	5.200000	4.760000	CAF2	
28	-12.340000	0.800000	5.860000	AIR	
29	42.482000	3.400000	6.600000	H-LAK3	
30	-30.270000	21.912220 S	6.760000	AIR	
31	0.000000	0.000000	6.000000 S	AIR	
IMS	0.000000	0.000000	6.000000 S		

Figure 3.421 Optical design and graphs of aberration correction of a 10×/0.50 CCF plan apochromatic objective for direct digital receiver projection (435–650 nm).

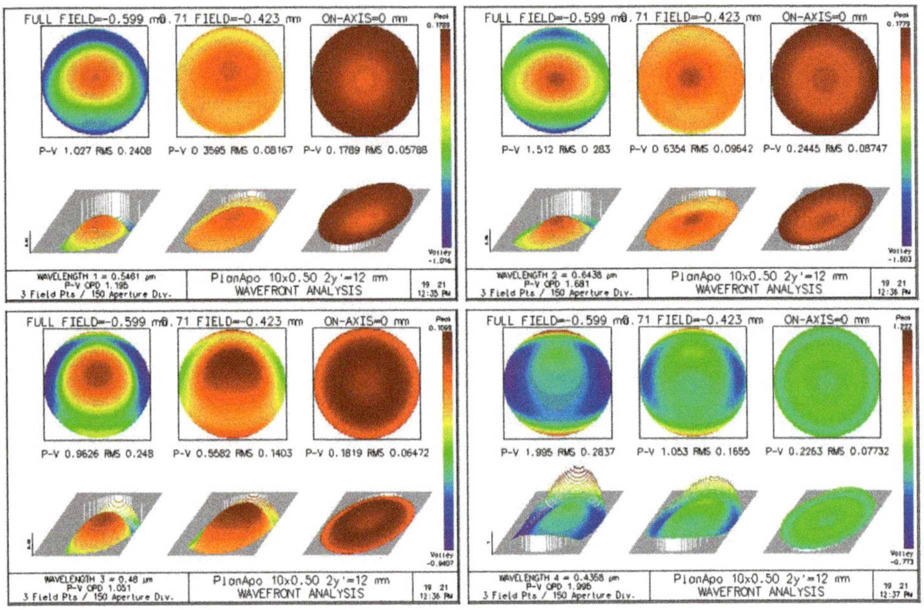

Figure 3.421 (*Continued*) Optical design and graphs of aberration correction of a 10×/0.50 CCF plan apochromatic objective for direct digital receiver projection (435–650 nm).

a telecentric objective is defined as the ratio of the focal length of the second objective to the focal length of the first objective. So, for example, if the focal length is $F' = 135.5$ mm in the first objective and $F' = 40$ mm in the second objective (shown in Fig. 3.171), then the resulting linear magnification in the telecentric objective will be $V = -0.3\times$. The optical design and graphs of aberration correction of such a telecentric objective, with a linear magnification $V = -0.3\times$, a front numerical aperture $NA = 0.035$, and a linear field in the image space approximately corresponding to the size of a 1/2-inch receiver, are shown in Fig. 3.423. This objective provides a resolution of about 10 μm on the object under study and a linear field in the object's space of 30 mm.

The objective consists of two parts. The first part is one objective. The studied object is in the front focal plane. It is important that the rear focal plane of this objective (the equivalent plane of the exit pupil) is aligned with the front focal plane of the second part, which is the other objective (the equivalent plane of the entrance pupil). The image of the studied object is built in the rear focal plane of the second objective. This ensures the principle of telecentricity of the second objective in which the paraxial parameters as well as the parameters of the entrance and exit pupils are coordinated in it. It is believed that the entrance pupil of the first objective and the telecentric objective as a whole are located at infinity; the exit pupil of the system is also

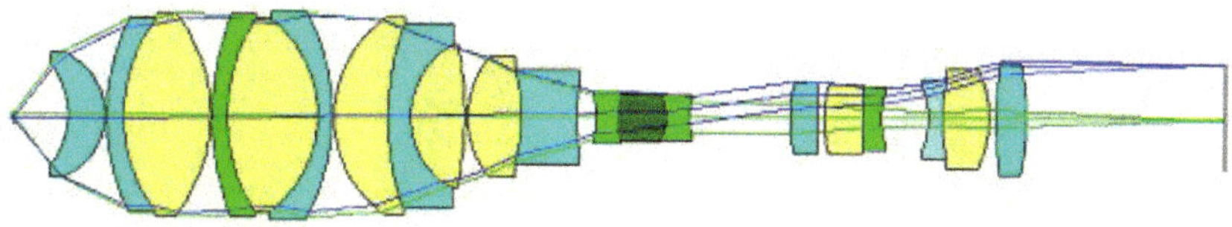

Lens: PlanApo 20x0.80 2y'=12 mm

Object num aper	0.800000	Image height	6.000000	Primary wavln	0.546070

SRF	RADIUS		THICKNESS		APERTURE RADIUS		GLASS		SPECIAL
OBJ	0.000000		1.0000e-20		0.299594		AIR		
AST	0.000000		5.999148		0.299594	AS	AIR		
2	-9.000000		4.000000		5.610000		H-LAF10L		
3	-7.400000		0.200000		6.910000		AIR		
4	24.210000		1.900000		10.560000		H-LAF10L	P	
5	17.602000		9.700000		10.590000		CAF2	C	
6	-17.602000		0.200000		11.070000		AIR		
7	31.724000		1.600000		11.070000		H-ZLAF3		
8	19.790000		10.000000		10.620000		CAF2	P	
9	-13.305000		1.600000		10.560000		H-LAF10L	P	
10	-23.390000		0.200000		11.350000		AIR		
11	12.979000		6.000000		10.830000		CAF2	P	
12	30.000000		2.500000		10.040000		H-LAF10L	P	
13	9.032000		6.000000		7.870000		CAF2	P	
14	-38.950000		0.300000		7.770000		AIR		
15	10.068000		6.000000		6.540000		CAF2	P	
16	-17.701000		6.000000		5.160000		H-ZLAF3	P	
17	18.883000		2.200000		3.290000		AIR		
18	-9.900000		2.500000		2.760000		H-ZLAF3	P	
19	27.160000		5.200000		2.660000		H-ZF62		
20	-4.720000		2.500000		2.530000		H-LAF10L	P	
21	7.047000		11.000000		2.200000		AIR		
22	15.672000		3.000000		3.900000		LZ_CTK12		
23	254.000000		0.800000		3.840000		AIR		
24	14.510000		4.300000		3.790000		CAF2		
25	-75.000000		2.000000		3.400000		H-ZF1		
26	8.801000		5.200000		3.180000		AIR		
27	-8.801000		1.500000		3.610000		H-LAK4L		
28	21.580000		5.200000		4.260000		CAF2		
29	-12.340000		0.800000		5.370000		AIR		
30	42.482000		3.400000		5.970000		H-LAK3		
31	-30.270000		21.495905	S	6.140000		AIR		
32	0.000000		0.000000		6.000000	S	AIR		
IMS	0.000000		0.000000		6.000000	S			

Figure 3.422 Optical design and graphs of aberration correction of a 20×/0.80 CCF plan apochromatic objective for direct digital receiver projection (435–650 nm).

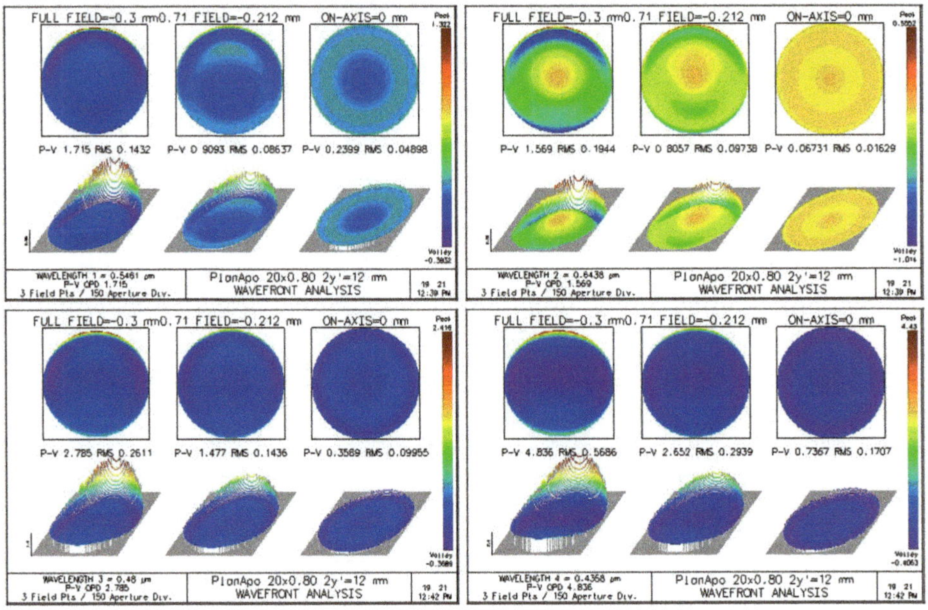

Figure 3.422 (*Continued*) Optical design and graphs of aberration correction of a 20×/0.80 CCF plan apochromatic objective for direct digital receiver projection (435–650 nm).

Table 3.17 Main technical parameters and basic optical layout of direct projection objectives (plan apochromats with extra-large working distances).

Magnification	*NA*	*WD* (mm)	*R* (μm)	*DF* (μm)	*FOV* on object (mm)	*FOV* on image (mm)	Optical design at figure no.	The principal optical layout
0.2×	0.015	14	22.2	1480	34	6.8	3.416	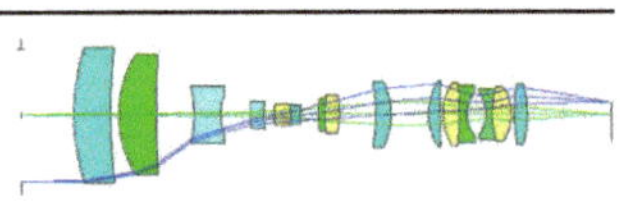
0.4×	0.04	20	8.33	208	17	6.8	3.417	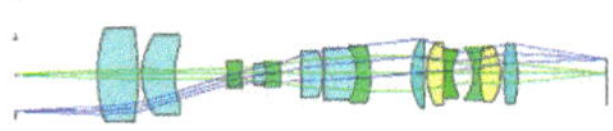
1×	0.12	22	2.78	23.13	6.8	6.8	3.418	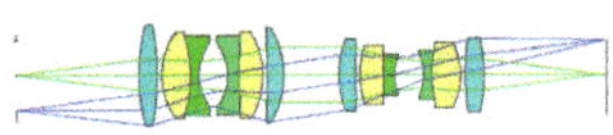
2×	0.3	34	1.11	3.7	6	12	3.419	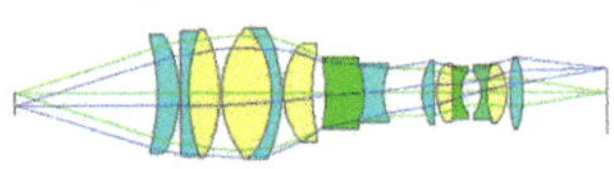
4×	0.3	33	1.11	3.7	3	12	3.420	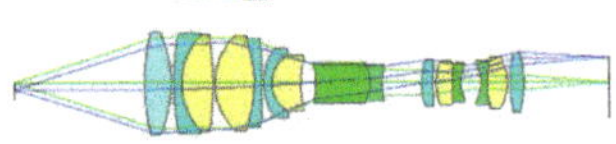
10×	0.5	18.5	0.67	1.33	1.2	12	3.421	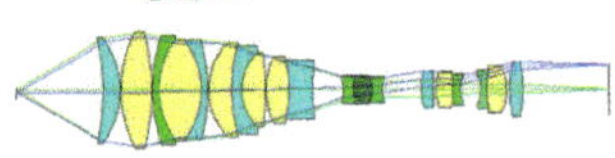
20×	0.8	4	0.42	0.52	0.6	12	3.422	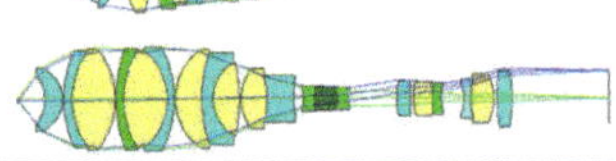

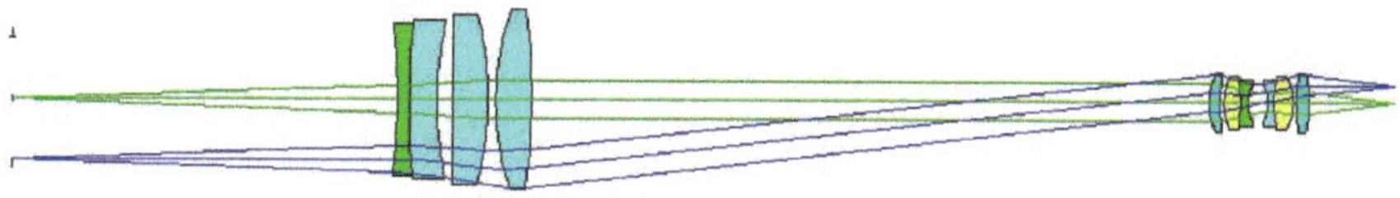

Lens: telecen 0.3x0.035 2y'=8.88 mm

Object num aper 0.035000 Image height 4.440000 Primary wavln 0.546070

SRF	RADIUS	THICKNESS	APERTURE RADIUS	GLASS	SPECIAL
OBJ	0.000000	1.0000e-20	14.997109	AIR	
AST	0.000000	99.180000	14.997109 AS	AIR	
2	-180.300000	3.000000	18.440000	TF5	
3	180.300000	7.000000	18.760000	K8	
4	89.130000	4.000000	19.540000	AIR	
5	0.000000	9.000000	20.050000	K8 P	
6	-81.660000	2.000000	21.160000	AIR	
7	71.120000	9.000000	22.270000	K8 P	
8	-180.300000	173.500000	22.160000	AIR	
9	15.672000	3.000000	7.240000	LZ_CTK12	
10	254.000000	0.800000	7.050000	AIR	
11	14.510000	4.300000	6.530000	CAF2	
12	-75.000000	2.000000	5.750000	H-ZF1	
13	8.801000	5.200000	4.820000	AIR	
14	-8.801000	1.500000	4.780000	H-LAK4L	
15	21.580000	5.200000	5.630000	CAF2	
16	-12.340000	0.800000	6.450000	AIR	
17	42.482000	3.400000	7.090000	H-LAK3	
18	-30.270000	21.924702 S	7.160000	AIR	
19	0.000000	0.000000	4.440000	AIR	
IMS	0.000000	0.000000	4.440000 S		

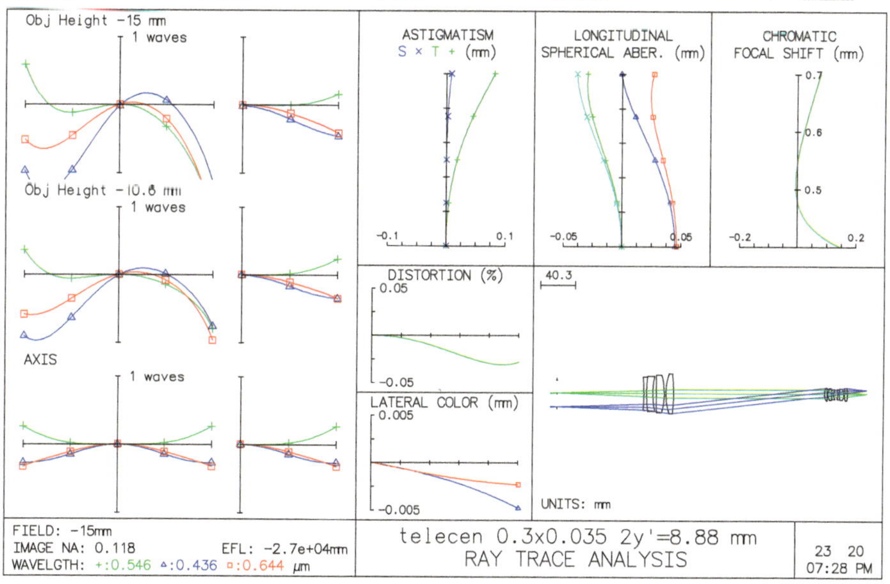

Figure 3.423 Optical design and graphs of aberration correction of a 0.3×/0.035 telecentric objective.

located at infinity. Both parts (both objectives) are separated by a significant air gap due to their rather long focus systems in which the so-called *pupil location planes* are relatively far removed from each objective's optical layout.

Telecentric objectives can be quite long; sometimes there is a need to reduce these sizes. We can use mirrors to change the direction of the rays. Figure 3.424 shows the optical design and graphs of aberration correction of a telecentric lens ($V = -0.25\times$ and NA = 0.025), and Fig. 3.425 shows photographs of a set of parts and a manufactured objective.

It can be convenient to place a beam-splitting cube where the focal planes of the two parts of the telecentric objectives coincide; see the example shown in Fig. 3.426. Then, in strict observance of the principle of telecentricity, we can use two or more optical image receivers. Note that the manufacture of such a beam-splitting cube can become a difficult or even unsolvable technological problem.

Let us discuss the possibility of using short-focus objectives as the first part of a telecentric objective. Doing so can exploit the significant advantages of telecentric objectives, primarily due to their increased resolution. However, there are several challenges. The plane of the exit pupil (when the entrance pupil is at infinity) of the first short-focus objective can be located inside the lens system of the objective. As an example, let us use a lens whose technical parameters are shown in Fig. 3.174 (50×/0.50 CCF plan apochromatic objective); Fig. 3.427 shows the location of the plane of the exit pupil of this objective.

Accordingly, the entrance pupil of the second objective must be located in the same place; that is, the front focal plane of the second objective must be located at a sufficient distance from the lens system. In addition, chromatic aberrations in a telecentric objective [lateral color (LC)] must be corrected in the first place (which is accomplished automatically if the first and second objectives are long-focus). Distortion aberration also needs to be corrected; this aberration is more difficult to correct in short-focus than in long-focus systems. To ensure that all of these conditions are met, it may be necessary to develop another objective as an image focusing system on a digital receiver. Figure 3.428 shows the optical design parameters and graphs of aberration correction of such an objective.

The development of some telecentric objectives with advanced technical parameters was achieved by using a combination of the specified objective (as a system for directly projecting an image onto a digital receiver) and some objectives described in Section 3.9. Figures 3.429–3.432 show the optical design parameters and graphs of aberration correction of some telecentric objectives.

Table 3.18 shows the main technical parameters of telecentric direct projection objectives.

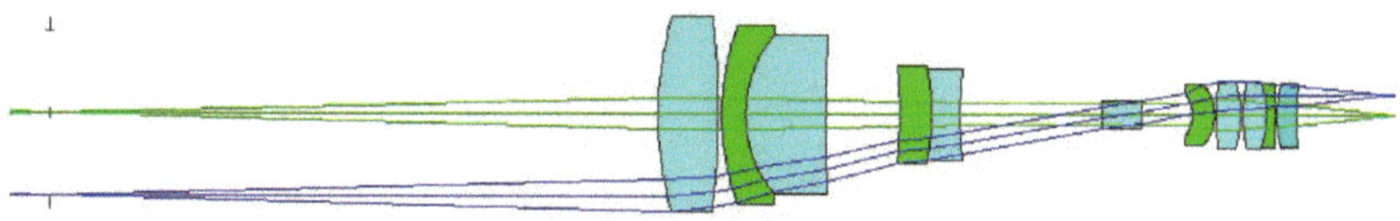

Lens:Telecent 0.25x NA=0.025 2y'=10mm

Object num aper 0.025000 Image height 5.000000 Primary wavln 0.546070

SRF	RADIUS	THICKNESS	APERTURE RADIUS	GLASS	SPECIAL
OBJ	0.000000	1.0000e-20	19.793543	AIR	
AST	0.000000	152.000000	19.793543 AS	AIR	
2	80.520000	15.000000	23.720000	LZ_K8	
3	-207.500000	1.000000	22.830000	AIR	
4	60.260000	6.000000	21.620000	LZ_TF5	
5	29.380000	20.000000	19.140000	LZ_K8 P	
6	250.000000	18.500000	16.520000	AIR	
7	-149.280000	8.000000	11.920000	LZ_TF5 P	
8	-52.480000	7.000000	11.120000	LZ_K8 P	
9	75.160000	35.000000	9.680000	AIR	
10	0.000000	10.000000	3.410000	LZ_K8 P	
11	0.000000	12.860000	3.340000	AIR	
12	-8.460000	5.500000	5.340000	LZ_TF5 P	
13	-11.320000	0.500000	7.350000	AIR	
14	79.430000	6.000000	7.930000	LZ_LK3	
15	-28.310000	0.700000	8.240000	AIR	
16	35.970000	6.000000	8.130000	LZ_LK3 P	
17	-21.090000	2.000000	7.770000	LZ_TF5 P	
18	136.460000	0.500000	7.680000	AIR	
19	30.410000	5.000000	7.680000	LZ_CTK12	
20	57.040000	24.902490 S	7.210000	AIR	
21	0.000000	0.000000	5.000000 S	AIR	
IMS	0.000000	0.000000	5.000000 S		

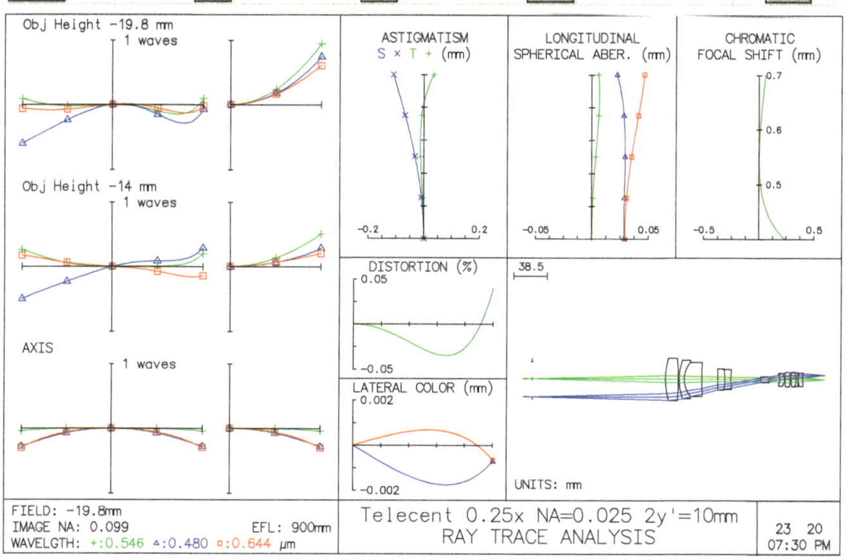

Figure 3.424 Optical design and graphs of aberration correction of a 0.25×/0.025 telecentric objective.

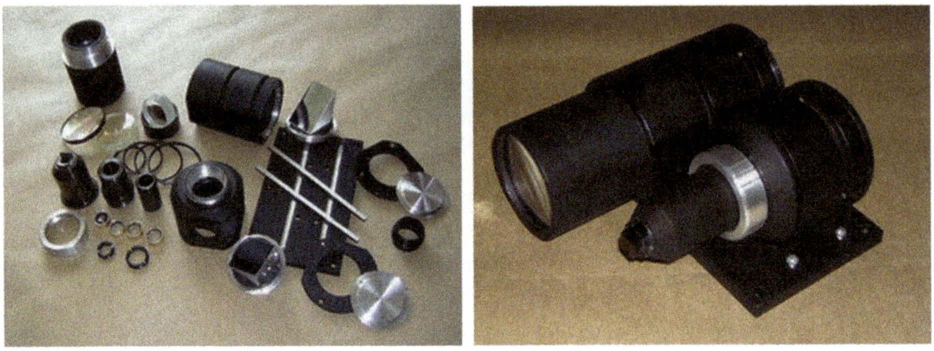

Figure 3.425 Photographs of a kit and a manufactured telecentric objective ($V = -0.25\times$; NA = 0.025).

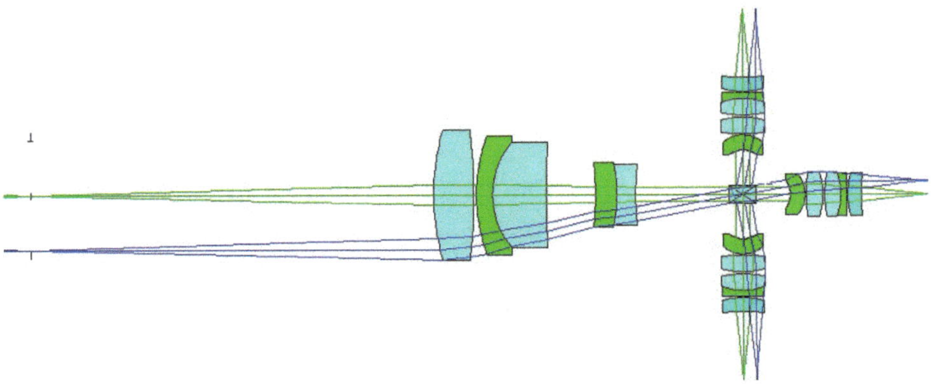

Figure 3.426 Schematic diagram of a telecentric objective for multiple image receivers.

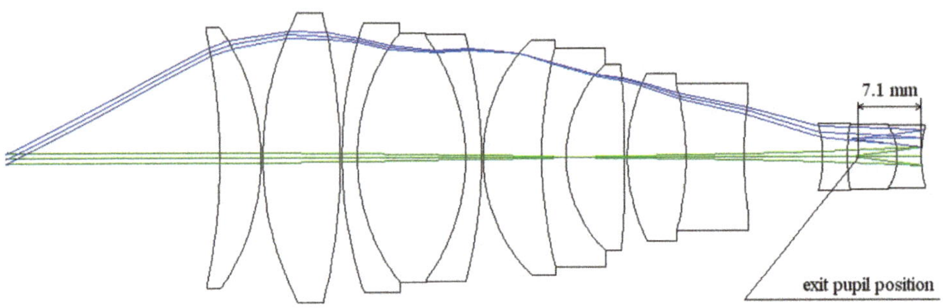

Figure 3.427 Exit pupil position for a 50×/0.50 CCF plan apochromatic objective.

3.14 Objectives Containing Mirrors

Many optical engineers have strong opinions about the advantages of using mirror and mirror–lens (catadioptric) systems for building optical systems of light microscopes. Other engineers find more convincing arguments for the

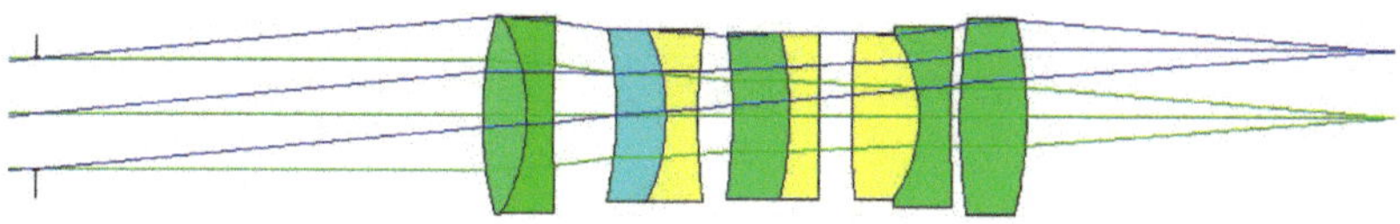

Lens:adapt F'=40 NA'=0.075 2y'=7.4 mm — Efl 40.06

Image num aper 0.075000 — Image height 3.700000 — Primary wavln 0.546074

SRF	RADIUS	THICKNESS	APERTURE RADIUS	GLASS	SPECIAL
OBJ	0.000000	1.0000e+20	9.2368e+18	AIR	
AST	0.000000	0.001000	3.004272 AS	BK7 C	
2	0.000000	25.320000	3.004332 S	AIR	
3	21.160000	2.500000	5.430000	H-ZLAF3	
4	-10.840000	1.500000	5.400000	LZ_TF10	
5	272.400000	3.600000	5.130000	AIR	
6	-20.280000	2.800000	4.680000	LZ_CTK19	
7	-11.530000	1.800000	4.730000	CAF2	
8	29.840000	2.000000	4.480000	AIR	
9	-29.840000	3.200000	4.420000	H-ZLAF3 P	
10	-22.500000	1.700000	4.560000	CAF2 P	
11	0.000000	1.800000	4.560000	AIR	
12	52.400000	3.800000	4.550000	CAF2 P	
13	-7.720000	1.800000	4.480000	H-ZLAF3 P	
14	225.500000	0.500000	4.940000	AIR	
15	35.550000	3.800000	5.190000	H-ZLAF3 P	
16	-21.900000	21.000000	5.400000	AIR	
17	0.000000	0.002223 S	3.700167 S	AIR	
IMS	0.000000	0.000000	3.700000 S		

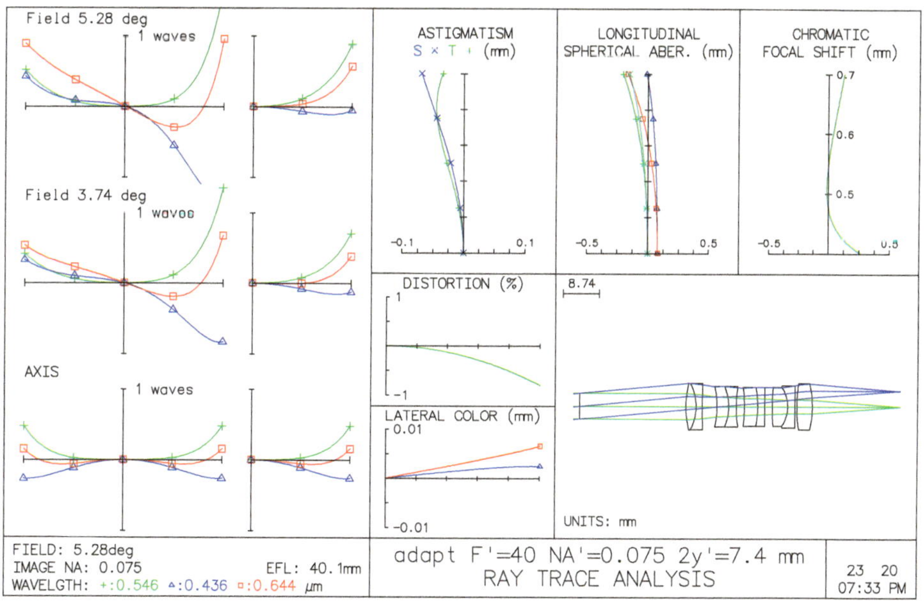

Figure 3.428 Optical design and graphs of aberration correction of a C-mount adapter (distortion compensate) for direct digital receiver projection ($F' = 40$ mm; $NA' = 0.075$).

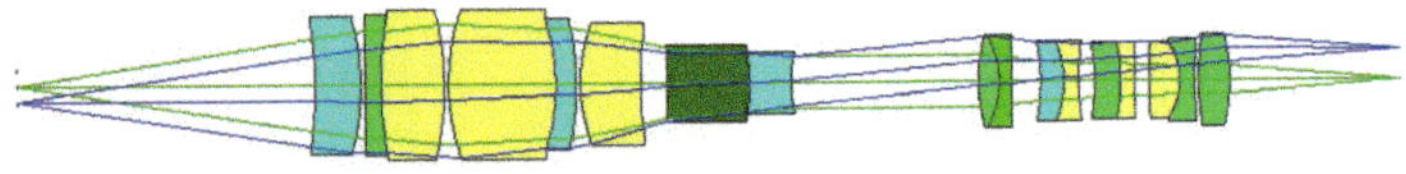

Lens: Telecen PlanApo 2x0.15 2y'=7.4

Object num aper 0.150000 Image height 3.700000 Primary wavln 0.546070

SRF	RADIUS	THICKNESS	APERTURE RADIUS	GLASS	SPECIAL
OBJ	0.000000	1.0000e-20	1.871707	AIR	
AST	0.000000	36.870000	1.871707 AS	AIR	
2	-49.660000	5.600000	7.380000	H-LAK3	
3	-29.570000	0.500000	8.140000	AIR	
4	0.000000	2.000000	8.230000	H-ZLAF3	
5	47.500000	8.000000	8.300000	CAF2	
6	-34.760000	0.500000	8.850000	AIR	
7	25.500000	13.000000	8.900000	CAF2 P	
8	-25.500000	2.500000	7.880000	H-LAK3 P	
9	-37.070000	0.500000	7.820000	AIR	
10	18.883000	7.400000	7.230000	CAF2 P	
11	39.370000	3.420000	5.520000	AIR	
12	-44.240000	10.000000	4.610000	H-ZF62	
13	-19.724000	5.000000	3.560000	H-LAF10LA	
14	15.136000	22.830000	3.260000	AIR	
15	21.160000	2.500000	5.430000	H-ZLAF3	
16	-10.840000	1.500000	5.400000	LZ_TF10	
17	272.400000	3.600000	5.130000	AIR	
18	-20.280000	2.800000	4.680000	LZ_CTK19	
19	-11.530000	1.800000	4.720000	CAF2	
20	29.840000	2.000000	4.480000	AIR	
21	-29.840000	3.200000	4.420000	H-ZLAF3 P	
22	-22.500000	1.700000	4.560000	CAF2 P	
23	0.000000	1.800000	4.550000	AIR	
24	52.400000	3.800000	4.540000	CAF2 P	
25	-7.720000	1.800000	4.480000	H-ZLAF3 P	
26	225.500000	0.500000	4.940000	AIR	
27	35.550000	3.800000	5.180000	H-ZLAF3 P	
28	-21.900000	21.000000	5.390000	AIR	
29	0.000000	-0.000520 S	3.700040 S	AIR	
IMS	0.000000	0.000000	3.700000 S		

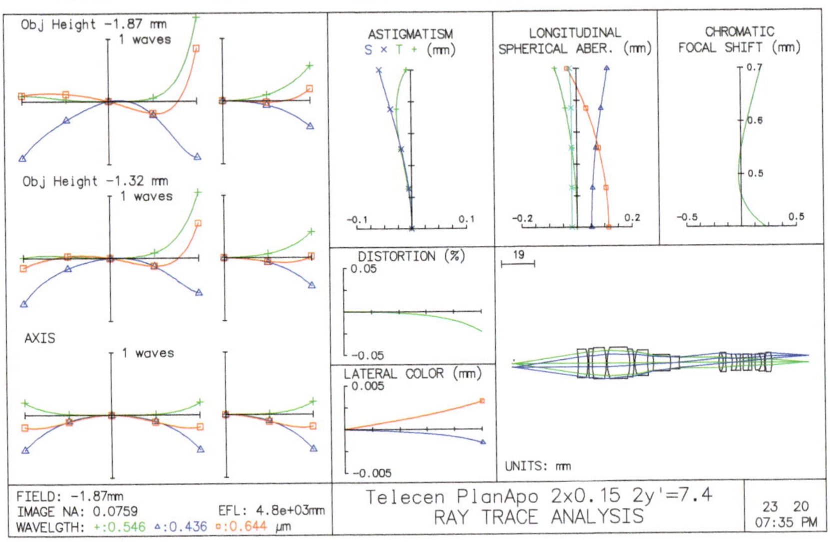

Figure 3.429 Optical design and graphs of aberration correction of a 2×/0.15 CCF plan apochromatic telecentric objective for direct digital receiver projection.

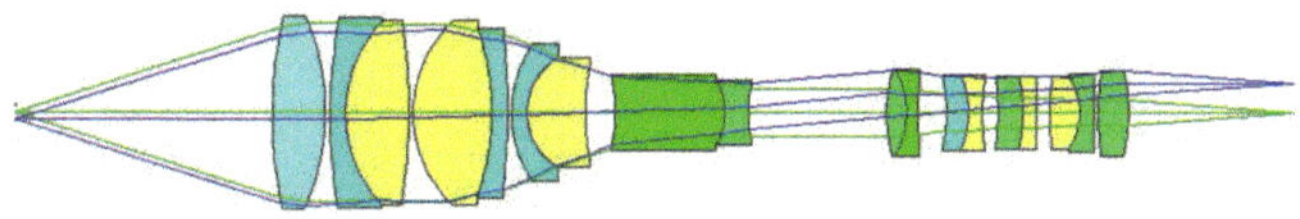

Lens: Telecen PlanApo 4x0.30 2y'=7.4

Object num aper 0.300000 Image height 3.700000 Primary wavln 0.546070

SRF	RADIUS		THICKNESS		APERTURE RADIUS		GLASS		SPECIAL
OBJ	0.000000		1.0000e-20		0.924480		AIR		
AST	0.000000		32.998378		0.924480	AS	AIR		
2	53.680000		6.800000		11.710000		H-QK3		
3	-28.689000		0.600000		12.060000		AIR		
4	62.230000		2.000000		11.900000		LZ_CTK12		
5	16.530000		8.000000		11.360000		CAF2	C	
6	-74.290000		0.500000		11.480000		AIR		
7	15.672000		10.000000		11.480000		CAF2	C	
8	-30.760000		2.000000		10.560000		LZ_CTK12	P	
9	-119.500000		0.500000		9.940000		AIR		
10	14.791000		2.000000		8.540000		LZ_CTK12	P	
11	7.900000		7.400000		6.870000		CAF2	C	
12	25.800000		3.770000		5.560000		AIR		
13	-20.750000		14.000000		4.720000		H-ZF52A		
14	-8.336000		3.000000		4.020000		H-ZLAF50B		
15	14.510000		17.760000		3.700000		AIR		
16	21.160000		2.500000		5.360000		H-ZLAF3		
17	-10.840000		1.500000		5.330000		LZ_TF10		
18	272.400000		3.600000		5.070000		AIR		
19	-20.280000		2.800000		4.620000		LZ_CTK19		
20	-11.530000		1.800000		4.660000		CAF2		
21	29.840000		2.000000		4.430000		AIR		
22	-29.840000		3.200000		4.370000		H-ZLAF3	P	
23	-22.500000		1.700000		4.500000		CAF2	P	
24	0.000000		1.800000		4.500000		AIR		
25	52.400000		3.800000		4.490000		CAF2	P	
26	-7.720000		1.800000		4.430000		H-ZLAF3	P	
27	225.500000		0.500000		4.870000		AIR		
28	35.550000		3.800000		5.110000		H-ZLAF3	P	
29	-21.900000		21.000000		5.320000		AIR		
30	0.000000		0.000000		3.700184	S	AIR		
IMS	0.000000		0.000000		3.700184	S			

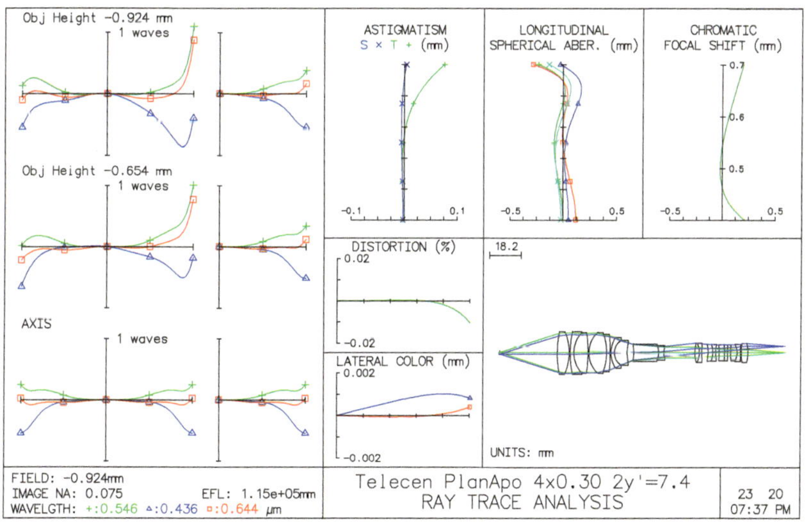

Figure 3.430 Optical design and graphs of aberration correction of a 4×/0.30 CCF plan apochromatic telecentric objective for direct digital receiver projection.

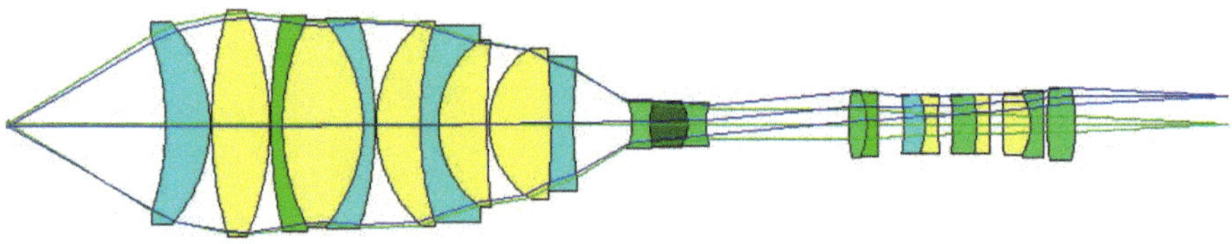

Lens:Telecen PlanApo 10x0.50 2y'=7.4

Object num aper 0.500000 Image height 3.700000 Primary wavln 0.546070

SRF	RADIUS	THICKNESS	APERTURE RADIUS	GLASS	SPECIAL
OBJ	0.000000	1.0000e-20	0.369936	AIR	
AST	0.000000	21.984361	0.369936 AS	AIR	
2	-59.320000	4.360000	12.320000	H-ZLAF1	
3	-21.580000	0.250000	12.880000	AIR	
4	30.270000	8.000000	14.210000	CAF2	
5	-57.800000	0.200000	14.080000	AIR	
6	40.190000	1.600000	13.280000	H-ZLAF3	
7	19.790000	11.000000	12.330000	CAF2 P	
8	-19.790000	1.600000	12.010000	H-ZLAF1 P	
9	-47.896000	0.200000	12.250000	AIR	
10	16.120000	6.000000	11.640000	CAF2 P	
11	42.657000	2.500000	10.840000	H-LAF10L	
12	12.465000	6.000000	9.200000	CAF2 P	
13	-124.360000	0.300000	8.930000	AIR	
14	17.760000	6.000000	8.220000	CAF2 P	
15	-27.534000	6.000000	7.180000	LZ_CTK12	
16	41.590000	8.000000	5.450000	AIR	
17	-13.305000	2.500000	3.200000	H-ZLAF3 P	
18	18.710000	5.100000	3.080000	H-SF62	
19	-5.750000	2.500000	2.930000	H-ZLAF3 P	
20	10.850000	18.214000	2.670000	AIR	
21	21.160000	2.500000	4.350000	H-ZLAF3 P	
22	-10.840000	1.500000	4.300000	LZ_TF10	
23	272.400000	3.600000	4.150000	AIR	
24	-20.280000	2.800000	3.840000	LZ_CTK19	
25	-11.530000	1.800000	3.900000	CAF2 P	
26	29.840000	2.000000	3.780000	AIR	
27	-29.840000	3.200000	3.760000	H-ZLAF3 P	
28	-22.500000	1.700000	3.930000	CAF2 P	
29	0.000000	1.800000	3.960000	AIR	
30	52.400000	3.800000	4.010000	CAF2 P	
31	-7.720000	1.800000	3.990000	H-ZLAF3 P	
32	225.500000	0.500000	4.370000	AIR	
33	35.550000	3.800000	4.570000	H-ZLAF3 P	
34	-21.900000	20.995202 S	4.790000	AIR	
35	0.000000	0.000000	3.700000 S	AIR	
IMS	0.000000	0.000000	3.700000 S		

Figure 3.431 Optical design and graphs of aberration correction of a 10×/0.50 CCF plan apochromatic telecentric objective for direct digital receiver projection (435–700 nm).

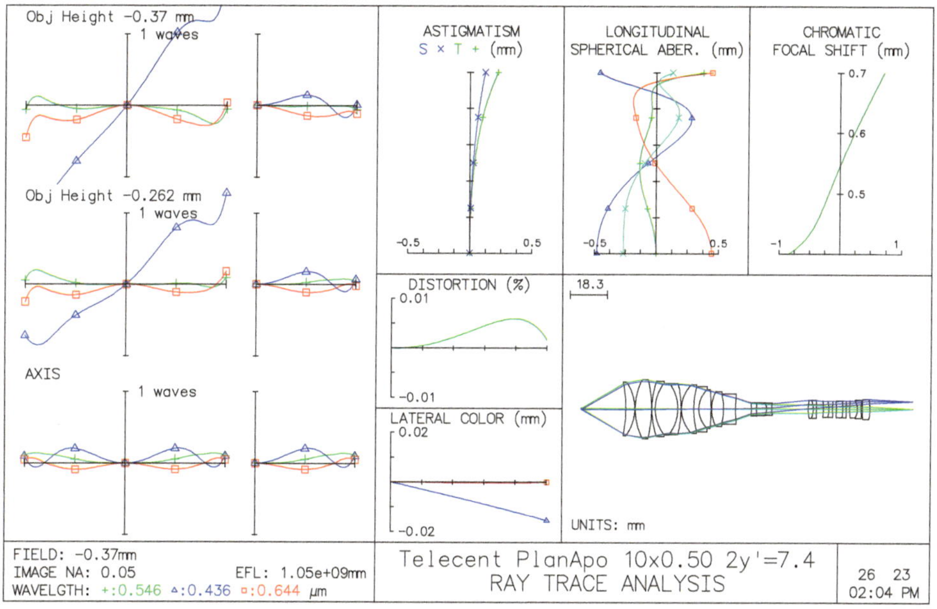

Figure 3.431 (*Continued*) Optical design and graphs of aberration correction of a 10×/0.50 CCF plan apochromatic telecentric objective for direct digital receiver projection (435–700 nm).

exclusive use of *dioptrical* lens components. The truth is most likely somewhere in between. Obvious advantages of mirror and mirror–lens optical systems can be found when creating, for example, so-called *photographic systems* due to the specifics of their use. Also, the development of telecentric mirror and mirror systems for *machine vision* could be an interesting engineering solution. The use of mirror optical systems for developing stereoscopic microscopes (for use in both the head and the projection system) can provide significant advantages not only in terms of the microscope's cost, but also in its consumer and technical parameters.

When creating the optical systems of light microscopes, it makes sense to use mirror optics, particularly if they can replace, for example, the glass prism systems in visual heads. Such optical components can provide many benefits when designing lighting systems. Less obvious advantages of mirror systems can be realized in the design of tube systems, all kinds of projectors, and adapters; mirrors can even be used in eyepieces. When designing objectives, especially high-aperture objectives, there are no obvious advantages to using only mirror surfaces instead of similar lens objectives; perhaps mastering the technologies of free-form optics will allow us to achieve a breakthrough in mirror system technology. There are also no obvious advantages to using mirror–lens systems over conventional lens systems when designing microscope objectives; the contrast in the image after the lens systems significantly

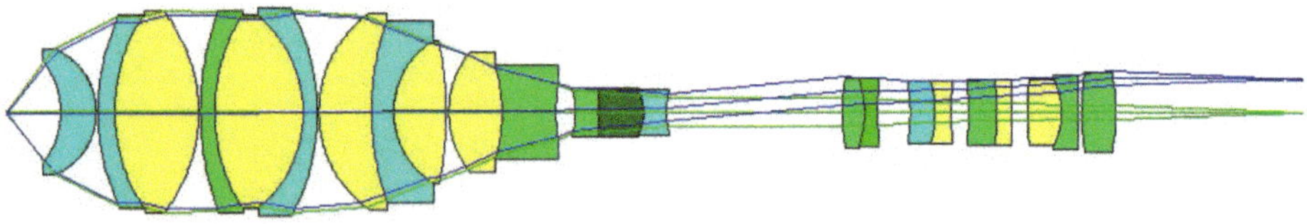

Lens: Telecen PlanApo 20x0.80 2y'=7.4

Object num aper 0.800000 Image height 3.700000 Primary wavln 0.546070

SRF	RADIUS	THICKNESS		APERTURE RADIUS		GLASS		SPECIAL
OBJ	0.000000	1.0000e-20		0.184891		AIR		
AST	0.000000	5.998101		0.184891	AS	AIR		
2	-9.000000	4.000000		5.590000		H-LAF10L		
3	-7.400000	0.200000		6.890000		AIR		
4	24.210000	1.900000		10.550000		H-LAF10L	P	
5	17.602000	9.700000		10.580000		CAF2		
6	-17.602000	0.200000		11.070000		AIR		
7	31.724000	1.600000		11.060000		H-ZLAF3		
8	19.790000	10.000000		10.610000		CAF2	P	
9	-13.305000	1.600000		10.540000		H-LAF10L	P	
10	-23.390000	0.200000		11.330000		AIR		
11	12.979000	6.000000		10.800000		CAF2	P	
12	30.000000	2.500000		9.990000		H-LAF10L	P	
13	9.032000	6.000000		7.850000		CAF2	P	
14	-38.950000	0.300000		7.740000		AIR		
15	10.068000	6.000000		6.510000		CAF2	P	
16	-17.701000	6.000000		5.120000		H-ZLAF3	P	
17	18.883000	2.200000		3.250000		AIR		
18	-9.900000	2.500000		2.710000		H-ZLAF3	P	
19	27.160000	5.200000		2.590000		H-ZF62		
20	-4.720000	2.500000		2.430000		H-LAF10L	P	
21	7.047000	19.630000		2.110000		AIR		
22	21.160000	2.500000		3.820000		H-ZLAF3		
23	-10.840000	1.500000		3.770000		LZ_TF10		
24	272.400000	3.600000		3.660000		AIR		
25	-20.280000	2.800000		3.400000		LZ_CTK19		
26	-11.530000	1.800000		3.470000		CAF2		
27	29.840000	2.000000		3.380000		AIR		
28	-29.840000	3.200000		3.380000		H-ZLAF3	P	
29	-22.500000	1.700000		3.550000		CAF2	P	
30	0.000000	1.800000		3.590000		AIR		
31	52.400000	3.800000		3.650000		CAF2	P	
32	-7.720000	1.800000		3.650000		H-ZLAF3	P	
33	225.500000	0.500000		3.980000		AIR		
34	35.550000	3.800000		4.150000		H-ZLAF3	P	
35	-21.900000	21.000000		4.370000		AIR		
36	0.000000	-0.008126	S	3.700541	S	AIR		
IMS	0.000000	0.000000		3.700000	S			

Figure 3.432 Optical design and graphs of aberration correction of a 20×/0.80 CCF plan apochromatic telecentric objective for direct digital receiver projection.

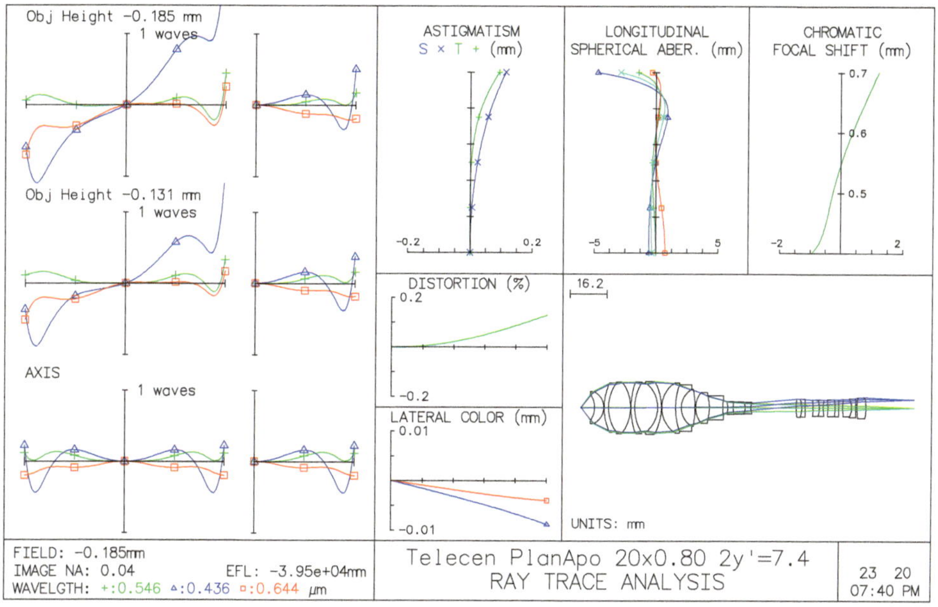

Figure 3.432 (*Continued*) Optical design and graphs of aberration correction of a 20×/0.80 CCF plan apochromatic telecentric objective for direct digital receiver projection.

Table 3.18 Main technical parameters and basic optical layout of telecentric direct projection objectives.

Magnification	*NA*	*WD* (mm)	*R* (μm)	*DF* (μm)	*FOV* on object (mm)	*FOV* on image (mm)	Optical design at figure no.	The principal optical layout
2×	0.15	34	2.22	14.8	3.7	7.4	3.429	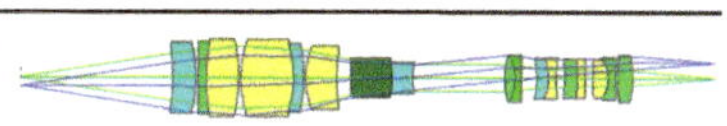
4×	0.3	33	1.11	3.7	1.85	7.4	3.430	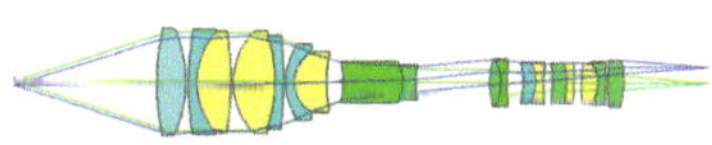
10×	0.5	18.5	0.67	1.33	0.74	7.4	3.431	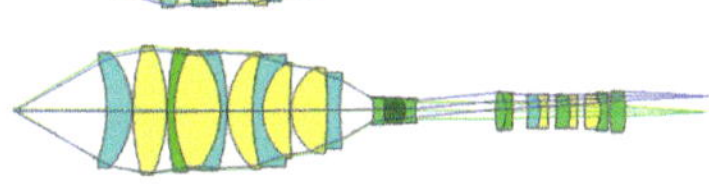
20×	0.8	4	0.42	0.52	0.37	7.4	3.432	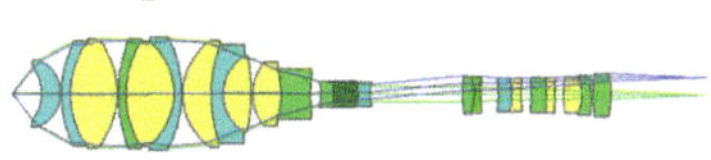

exceeds the contrast after the mirror and mirror–lens systems. At the same time, considering that low-contrast objects are often examined in microscopy, the advantage remains for using conventional lens objectives for the microscope.

Nevertheless, a discussion can be had about the possibility of designing mirror and mirror–lens microscope objectives. For mirror systems, the basis

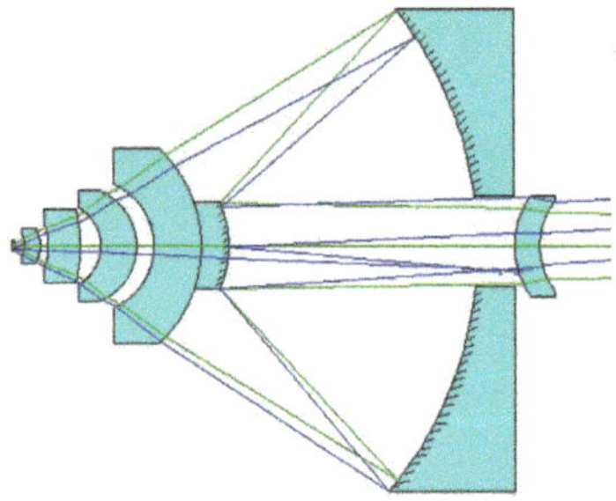

Lens:PlanMultiApo 40x0.50 tube160 h45			Efl	4.41	
Object num aper	0.500000	Image height 10.000000	Primary wavln	0.546070	
SRF	RADIUS	THICKNESS	APERTURE RADIUS	GLASS	SPECIAL
OBJ	0.000000	0.010000	0.249304	AIR	
AST	0.000000	0.170000	0.250000 A	SILICA	
2	0.000000	0.650000	0.320000	AIR	
3	-1.555000	1.000000	0.620000	SILICA P	
4	-1.947100	0.500000	1.050000	AIR	
5	-3.738500	2.000000	1.330000	SILICA P	
6	-8.476200	1.300000	2.210000	AIR	
7	-2.647600	2.300000	2.330000	SILICA P	
8	-3.588500	1.100000	3.380000	AIR	
9	-4.057600	3.000000	3.770000	SILICA P	
10	-7.949200	17.400000	5.990000	AIR	
11	-22.931900	-15.700000	15.360000	REFLECT	
12	-6.778100	18.000000	2.700000	REFLECT	
13	5.054600	1.600000	3.100000	SILICA P	
14	4.415600	163.718726 S	2.780000	AIR	
15	0.000000	0.000000 S	10.000000 S	AIR	
IMS	0.000000	0.000000	10.000000 S		

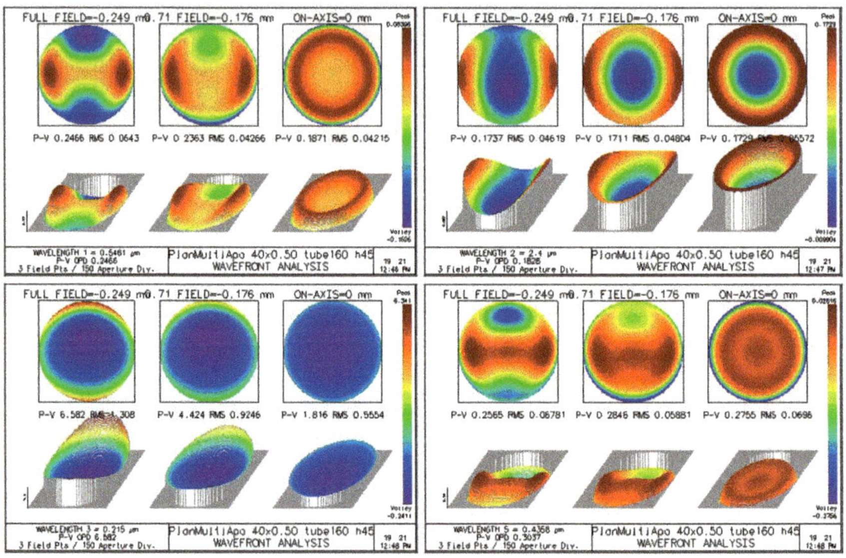

Figure 3.433 Optical design and graphs of aberration correction of a 40×/0.50 CCF plan multi-apochromatic catadioptric objective (200–2600 nm).

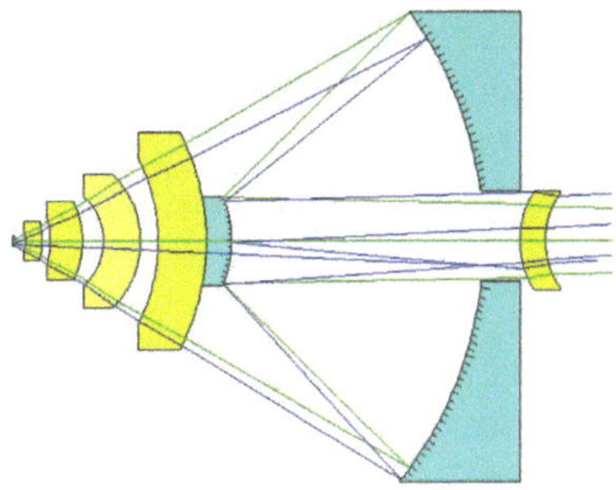

Lens:PlanMultiApo 40x0.50 tube160 h45 — Efl 4.42

Object num aper 0.500000 — Image height 10.000000 — Primary wavln 0.588000

SRF	RADIUS	THICKNESS		APERTURE RADIUS		GLASS		SPECIAL
OBJ	0.000000	0.010000		0.249964		AIR		
AST	0.000000	0.170000		0.255737	AS	MGF2		
2	0.000000	0.650000		0.320000		AIR		
3	-1.555000	1.000000		0.620000		MGF2	P	
4	-4.898700	0.500000		1.050000		AIR		
5	-5.164200	2.000000		1.330000		MGF2	P	
6	-5.340000	1.300000		2.210000		AIR		
7	-3.552900	2.300000		2.330000		LIF		
8	-5.617200	1.100000		3.380000		AIR		
9	-10.743800	3.000000		3.770000		MGF2	P	
10	-14.039000	17.287600		5.990000		AIR		
11	-23.155700	-15.718800		15.360000		REFLECT		
12	-7.167300	18.000000		2.700000		REFLECT		
13	5.538700	1.600000		3.100000		MGF2	P	
14	4.841400	163.402833	S	2.780000		AIR		
15	0.000000	3.0772e-14	S	10.000000	S	AIR		
IMS	0.000000	0.000000		10.000000	S			

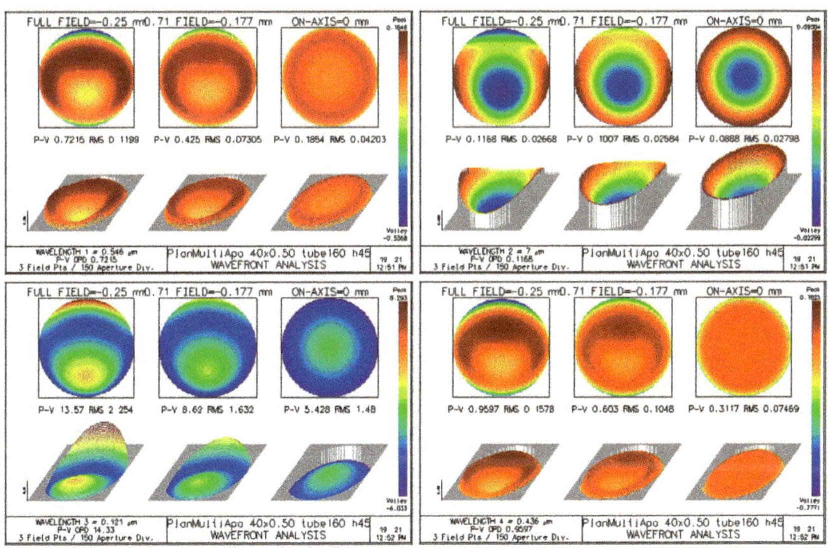

Figure 3.434 Optical design and graphs of aberration correction of a 40×/0.50 CCF plan multi-apochromatic catadioptric objective (120–7000 nm).

may be the Schwarzschild system [16], while the Burch objectives [17] represent one of the most successful attempts at implementing this idea. For mirror–lens systems, one can cite as examples the studies presented in publications [18, 19]. Some discussion regarding mirror systems is presented in Volume 1, Chapter 2, Sections 2.29 and 2.30.

Obviously, the possibility of expanding the spectral operating range for optics of a light microscope (without significantly complicating the system) is one of the main arguments for using mirror and mirror–lens optical elements. As examples, Figs. 3.433 and 3.434 show the design parameters and graphs of aberration correction of plan multi-apochromatic microscope objectives with a linear magnification of $V = -40\times$ and a numerical aperture of $NA = 0.50$. Both objectives are designed to work as part of a microscope for a finite optical tube length of 160 mm. Spherical mirrors are used here. The use of various types of crystals in the lens elements allows multi-apochromatic aberration correction for a wide spectral range—from NUV to NIR—to be implemented. It should be noted that creating lens objectives for working in such a wide spectral range is also possible; however, their optical systems (containing only crystals as the optical material for the lenses) can be extremely complex. It is also significant that for the objective to operate in a wide spectral range (both mirror–lens objectives and lens objectives), the lens components must consist exclusively of non-glued lenses.

References

[1] H. Boegehold, *Das Optische System des Mikroskops*, Verlag Technik, Berlin (1958).

[2] A. P. Grammatin, L. S. Agroskin, and R. M. Larina, "Reflected light microscopes," *Sov. J. Opt. Technol.* No. 3 (1969).

[3] T. A. Ivanova and V. K. Kirillovsky, *Design and Control in the Optics of Microscopes*, Mashinostroenie, Leningrad (1984).

[4] V. A. Panov and L. N. Andreev, *Optics of Microscopes*, Mashinostroenie, Leningrad (1976).

[5] D. N. Frolov, "Synthesis of the optical systems of lens objectives for microscopes," *J. Opt. Technol.* **69**(9), 614 (2002).

[6] I. L. Sakin, *Engineering Optics*, Mashinostroenie, Leningrad (1976).

[7] A. Maréchal, "Etude des effets combinés de la diffraction et des aberrations géométriques sur l'image d'un point lumineux," *Rev. Opt.* **2**, 257–277 (1947).

[8] D. N. Frolov, R. M. Raguzin, and V. A. Zverev, "Optical system of a modern microscope," *J. Opt. Technol.* **69**(9), 610 (2002).

[9] L. A. Fedin and I. Y. Barsky, *Microphotography*, Publishing House Science, Leningrad (1971).

[10] A. N. Zacharievsky, V. A. Panov, and A. F. Kuznetsova, "Interferential micro objective," *Optical and Mechanical Industry* **6**, 34–37 (1960).

[11] D. N. Frolov, O. A. Vinogradova, V. N. Frolov, and P. S. Vakulov, "Optical design and unification of objectives: CCF PlanApo for 28 mm observation in the microscope," *Proc. SPIE* **10679**, 106791G (2018) [https://doi.org/10.1117/12.2306504].
[12] I. E. Grabarya, *Restoration of Icons*, VNNRTS, Moscow (1993).
[13] Y. S. Maklygina, A. V. Borodkin, and V. B. Loschenov, "Laser microscope-spectrum analyzer for studying intracellular accumulation of near infrared emitting photosensitizers in vitro," *Biomedical Photonics* **4**(3):3–9 (2015) (In Russian).
[14] M. Mandai and K. Yamaguchi. Immersion microscope objective lens. US Patent 7046451B2, 2006.
[15] L. N. Andreev, S. V. Gribanova, M. M. Kravets-Kravchevskaya, and O. N. Fedorova, "Microscope objectives for the visible and UV spectral ranges," *GOI Proceedings* **49**(183) (1981).
[16] K. Schwarzschild, "Untersuchungen zur geometrischen optik. II: Theorie der spiegelteleskope," *Astronomische Mittheilungen der Koniglichen Sternwarte zu Gottingen* (1905).
[17] C. R. Burch, "Flat-fielded singlet aplanats," *Proc. Phys. Soc.* **57**, 567 (1945).
[18] H.-J. Mann and W. Ulrich, "Reflective high-NA projection lenses," *Proc. SPIE* **5962**, 596214 (2005) [https://doi.org/10.1117/12.625196].
[19] D. Shafer "Dennis Gabor's catadioptric design and some new variations," *Optical Design and Fabrication*, paper IW3A.1 (2017).

Chapter 4
Optical Systems for Single Exemplar Microscopes

Despite this being the shortest chapter—as it is impossible to discuss all or even some of the many current and developing engineering solutions and methods for working with a microscope—it should provide some of the necessary information. The most complex research microscopes, which implement modern microscopic concepts and advanced engineering solutions, are conditionally classified as *single exemplar*. Comprehensive information about every area of modern optical microscopy cannot be found in any one source; most of the information is scattered throughout the vast expanse of scientific publications and the internet. Some of it is presented on the official websites of the four optical giants: Leica, Nikon, Olympus, and Zeiss. Many engineers who work in the field of optical microscopy look to the engineers of these world-renowned companies for guidance and inspiration; the desire to somehow get closer to the achievements of these firms is one of the main motivations for everyone else.

But these company websites mainly provide general discussions on such topics as physical phenomena in light microscopes, the construction of optical and mechanical microscope systems, research methods, and the like. Corporate ethics and the usual competition in the market do not allow them to fully disclose information about the basics of their optical designs or the original engineering solutions that they have implemented in light microscopes. Moreover, it is not considered acceptable to publicly discuss the impact of not just theoretical image quality on a microscope, but also real image quality achieved from specific engineering projects. These considerations are especially relevant when discussing the so-called single exemplar devices in general and microscopes in particular. However, in most cases, to call these devices microscopes is not entirely correct; they are powerful research tools that are based not only on modern optical technologies, but also on the entire body of almost all human knowledge. Achievements in the fields of artificial intelligence, computer science, robotics—and much more—to one degree or another complement and serve OPTICS, the *queen of physics*, which realizes itself in modern microscopy in the most refined and magnificent way.

Therefore, we will not discuss here many of the original engineering solutions used to make single exemplar microscopes. There are many of these solutions; as a rule, they provide the required technical parameters for a particular device. However, optics and optical technologies can form a basis for constructing a wide range of microscopes that use original engineering solutions. We can also discuss some specific engineering solutions, which, for some reason, are not explicitly presented in many accessible sources.

4.1 Standard Cover Glass as Specimen Glass

It may seem strange, but light microscopy is still perhaps one of the most conservative areas of optical instrumentation; this is the case even if we are talking about mass-market, medium series, and small series microscopes. There are many examples of conservatism in engineering and research methods. One is the use of an extremely thick, so-called *specimen glass*. For many decades, a specimen glass of 1.1–1.2-mm thickness has been used for placing an object under study. Once again, we turn to Volume 1, Chapter 1, Fig. 1.35, which presents the model of an object and a picture of part of a transmitted light microscope. There may be some arguments that the use of such a specimen glass can be a serious obstacle to performing high-tech studies, especially for studies using single exemplar microscopes. Here, let us address several arguments confirming that the parameters of the specimen glass can and should be changed.

The first is that to achieve a high quality of research, not only do the numerical apertures of the condenser and the objective need to be consistent, but there must also be an equally high level of aberration correction. It is no secret that in the vast majority of cases, the aberration correction of the condenser is significantly inferior to that of the objective, especially if we choose condensers and objectives with the same numerical apertures. Many engineers assume that the aberration correction of a condenser absolutely does not affect the resolution of a light microscope if the necessary numerical aperture is achieved. We can agree or question this hypothesis regarding resolution, but the level of condenser aberration correction affects image quality; this is a fact. At the same time, it would be very difficult to make an optical design and develop a condenser with a high numerical aperture and a high level of aberration correction while also achieving a considerable working distance (to ensure that it would work with a 1.2-mm-thick specimen glass). Of course, this is possible, but the system of such a condenser would be extremely complex, which may make its use impractical.

Let us use a standard 0.17-mm-thick cover glass as a specimen glass. When preparing a specimen, microscopists often use a cover glass as the base, turning it over and gluing it to a specimen glass. Let us use two identical cover glasses in the model of the studied object that is illustrated in Volume 1, Chapter 1, Fig. 1.35. The use of such an engineering solution significantly expands the possibilities of the researcher's creativity; in this case, one of the objectives designed for working with a standard cover glass, which has the required aberration correction level and numerical aperture value, can be used as a condenser.

Russian microscope expert Anatoly Mikhaltsov, whose name is frequently mentioned in this book, made several of these samples of specimen glasses for testing by placing an object between two standard cover glasses; my colleagues and I have some experience working with such specimen glasses. Indeed, they are rather thin and fragile; they can break if they are handled carelessly (we have also had such a negative experience). In addition, these specimen glasses are quite difficult to position, for example, using a standard stage, which is often used in mass-market, medium series, and small series microscopes. A recommendation for these specimen glasses is to use the stage with a so-called *double layer*.

Applying this engineering solution can be useful in all classes of microscopes, particularly in single exemplar microscopes, where it can provide significant advantages when examining, for example, unstained and living objects (when, for example, achieving apochromatic aberration correction is required).

We took two photographs of such real objects located on a standard 0.17-mm cover glass used as a specimen glass; 20×/0.40 and 40×/0.65 standard plan achromatic objectives were used as condensers. Experiments were also conducted on the use of an original lighting device combined with a stage (shown in Chapter 1, Fig. 1.306, when a standard condenser is not required). Figures 4.1–4.7 are some of these photographs.

4.2 Microscope Design for NUV–DUV Ranges

Let us discuss the use of diopter optics for deep-UV (DUV) microscopy. Moving to the near-UV (NUV) and DUV ranges can significantly improve a microscope's resolution. Different kinds of crystals can be used as optical materials. The well-known Abbe and Rayleigh relationships describe the value of the resolving power of an objective, depending on its numerical aperture and the wavelength used. An optical design can allow objectives to be created whose maximum numerical aperture in air (vacuum) reaches 0.95. This is the technological limit in the creation of such systems. In this case, the use of a shorter wavelength is the only way to further enhance a microscope's resolution.

Currently, in light microscopy, the visible spectral range is mainly used as the working spectral range. In this spectral range, it is possible to obtain resolution values of about 270–470 nm in objectives without immersion. The

Figure 4.1 Photograph taken with a 40×/0.60 objective, a 0.17-mm-thick specimen glass, a 0.17-mm standard cover glass, and a 20×/0.40 objective used as a condenser.

use of immersion is a widely known and frequently used method of increasing resolution. In microscopes that use water as an immersion, it is possible to increase the resolving power by 1.3–1.4 times as compared with an air system. This is also true for systems designed to operate in NUV and DUV. Table 4.1 presents theoretical resolution values for numerical apertures of 0.95 and 1.20 in water immersion objectives at different NUV and DUV wavelengths.

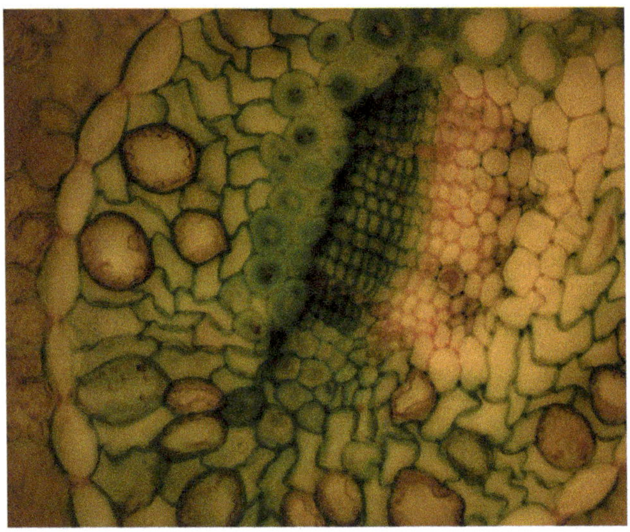

Figure 4.2 Photograph taken with a 40×/0.60 objective, a 0.17-mm-thick specimen glass, a 0.17-mm standard cover glass, and a 40×/0.65 objective used as a condenser.

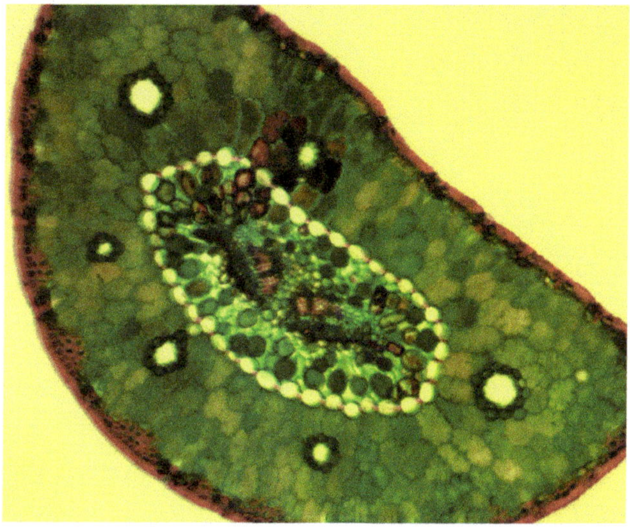

Figure 4.3 Photograph taken with a 10×/0.28 objective, a 0.17-mm-thick specimen glass, a 0.17-mm standard cover glass, and an LED panel used as a condenser.

Table 4.1 shows that when moving from the visible range to the shorter spectral range, the theoretical resolving power of a microscope can improve by a factor of 3–5. This is a strong argument for investigating the possibility of creating systems that operate in the NUV–DUV spectral ranges. Although some researchers' proposals for creating microscopes that operate in the DUV

Figure 4.4 Photograph taken with a 20×/0.40 objective, a 0.17-mm-thick specimen glass, a 0.17-mm standard cover glass, and an LED panel used as a condenser.

Figure 4.5 Photograph taken with a 20×/0.80 objective, a 0.17-mm-thick specimen glass, a 0.17-mm standard cover glass, and an LED panel used as a condenser.

range are known, these systems are catoptric and do not have high numerical apertures [1–3].

Dioptric objectives can be considered optimal optical systems because there is a possibility that total numerical aperture without obscuration can be obtained with their use. Such obscuration, inherent in mirror systems, changes the energy distribution at the center of the diffraction spot of an image.

Figure 4.6 Photograph taken with a 40×/0.60 objective, a 0.17-mm-thick specimen glass, a 0.17-mm standard cover glass, and an LED panel used as a condenser.

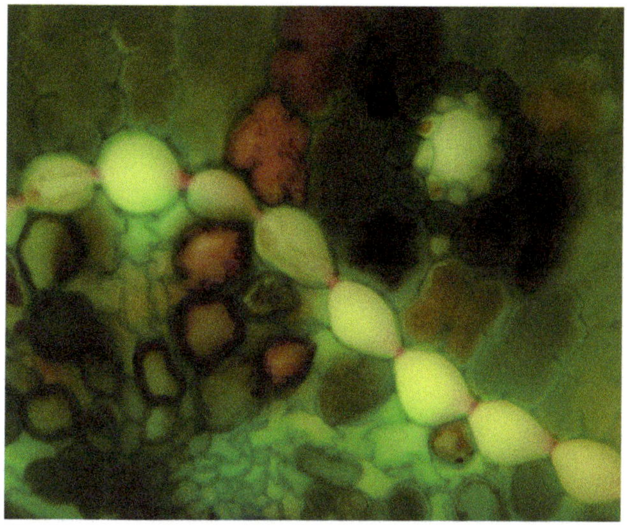

Figure 4.7 Photograph taken with a 40×/0.95 objective, a 0.17-mm-thick specimen glass, a 0.17-mm standard cover glass, and an LED panel used as a condenser.

Table 4.1 Theoretical resolution values for numerical apertures of 0.95 and 1.20 in water immersion objectives at different wavelengths.

Wavelengths (nm)	244	193.4	157	121.6	106
Resolution value (nm) NA = 0.95	157	124	101	78	68
Resolution value (nm) NA = 1.20 water	124	98	80	Not possible now	

This results in a significant reduction in an image's contrast and sharpness, and is not acceptable for high-aperture microscopy.

4.2.1 What objects can be investigated?

When exposed to it, living organisms absorb NUV and DUV radiation, which changes the chemical properties of their biological polymers. In microorganisms and cultured cells of plants and more complex animals, such radiation has a lethal and mutagenic effect, as the spectrum of the effects caused by ultraviolet radiation roughly coincides with the absorption spectrum of nucleic acids; in some cases, the spectrum of biological action is close to the absorption spectrum of proteins.

In this case, the optical properties of many substances in the ultraviolet range of the spectrum are significantly different from those, for example, in the visible region. Many substances can selectively absorb NUV and DUV; these substances can be used to detect and study various inclusions and impurities [4]. In ultraviolet microscopy, it is possible to investigate nonliving biological objects that are prepared using special methods. In addition, in a standard sample that includes a slide and cover glass, there should not be any glue, and the optical glass materials should provide high transmission of ultraviolet radiation; i.e., they must be made of crystal optics. The highest level of consumer satisfaction with an ultraviolet microscope is achieved when researching objects in the mining and ore industries. The study of such samples can be carried out in the fields of crystallography, petrography, and chemistry. Objects in the metallographic and semiconductor industries can be studied using NUV and DUV microscopy. To realize these methods, certain conditions are necessary for the optics of the microscope and its objectives.

4.2.2 Ultraviolet fluorescence method

For microscopic analysis of biological objects, the fluorescence analysis technique can be used successfully. Many substances have this property; namely, by absorbing irradiating light, they themselves start to emit light. The emitted light can be observed in a different wavelength than the irradiating one. Such fluorescent substances include, for example, proteins that begin to emit their own luminescence light when irradiated with ultraviolet light (NUV and DUV) in the spectral range.

However, different substances do not absorb light equally. Some absorb red light well but do not absorb blue; for others, the opposite is true. The proteins are colorless. They mainly absorb the light of the ultraviolet range of about 200 to 300 nm. In a protein, at least one of its constituent parts—aromatic amino acids—fluoresces. Light irradiation of a 240–245-nm wavelength, almost not absorbed by aromatic amino acids, is nevertheless able to cause a very strong fluorescence of the protein.

The fluorescence of proteins is a reliable and highly sensitive tool in the study of biological processes in matter. The excitation introduced into the peptide bonds due to irradiation of a protein molecule extends throughout the molecule. Interacting with aromatic amino acids (tryptophan, tyrosine, etc.), the energy obtained due to irradiation is emitted (fluoresces) in the ultraviolet range. This ability of proteins to fluoresce due to UV radiation makes it possible to quickly and accurately determine the amount of protein in various substances.

These aspects suggest that the fluorescent analysis technique in the ultraviolet spectral range (NUV and DUV) can be used to diagnose the presence of viruses in a substance. Issues related to such diagnostics and exposure to detected viruses have been described [5–7].

4.2.3 What optical materials can be used?

The spectral ranges indicated in Table 4.1 must be separated. Wavelengths within 244 and 193 nm can be said to be in the NUV range. Wavelengths within the range of 157 and 106 nm can be said to be in the DUV range.

The border between these spectral ranges can be air, the transmission of which ends at about 185 nm. The very popular and frequently used quartz crystal, which is used very effectively for the NUV range, ceases to transmit at about 156 nm. Fluorite, which is a universal and well-mastered optical material, can also be used for this range. When designing NUV optics, it is beneficial to use both of these materials in the same optical scheme. Here, quartz can be used as a material for so-called flint lenses, whereas crown lenses are made of fluorite. In some cases, this allows achromatization of the optical system in a given spectral range [8]. Nevertheless, creating the optical design, and especially the manufacture of such optical systems (objectives), is an extremely complex and time-consuming task. This particularly applies to high-aperture lenses (numerical aperture of 0.90–0.95). Achromatization is also theoretically possible in some areas of the DUV range, where sapphire can be used as a material for flint lenses. However, in practice, it is a very fragile material, so making high-aperture objective lenses poses many technological difficulties. Moreover, it is difficult to fix certain lenses in the frames. For this reason, and due to the extremely difficult optical design of dioptric lens objectives for the DUV range, the degree of optical correction is limited to correcting monochromatic aberrations for a selected fixed wavelength.

It should be noted that water, which can be considered a universal conductor of optical radiation, has a sufficiently high transparency at all wavelengths in the NUV range. The ultraviolet absorption limit of water vapor is approximately 140 nm. This means that water can be used as an immersion liquid in optical NUV microscopy systems (as well as at a 157-nm wavelength) to increase the numerical aperture of the objective and, accordingly, its resolving power. This quality is successfully exploited, for example, in photolithographic installations that operate at a 193-nm wavelength. In this case, the refractive index of water is about 1.44; the resolution can be increased approximately 1.5 times compared with objectives operating without immersion. In a wavelength range shorter than 185 nm, where air is not transparent, it is necessary to fill the working area of the microscope with a gaseous substance. Inert gases have the best transparency. Helium has the shortest wavelength transparency (approximately 50.4 nm). In terms of ultraviolet radiation receivers, the use of conventional photography materials in the NUV and DUV regions of the spectrum is restricted. This is because of the presence of gelatin in the photograph layer. We can use photovoltaic receivers, which use ultraviolet radiation to cause ionization and a photoelectric effect.

4.2.4 Optical design of high-aperture dioptric objectives

A serious consideration when designing a microscope that works in NUV and DUV spectral ranges is the need to achieve maximum numerical apertures. Achieving the maximum numerical aperture will ensure maximum resolution of the optics. In this case, it is not possible to use catoptric or catadioptric optics because they are characterized by the presence of a central obscuration.

The use of lens (dioptric) systems allows the maximum resolution to be realized. The main component in the optical system of such a microscope is the objective, which performs the optical projection of the image of the object into the plane of the image receiver. The objective must have a high transmission not only in the working wavelength but also in the specified spectral range, which allows the use of a fluorescence instrument. It has been argued that the transition to a shorter wavelength is justified only for extreme numerical apertures of objectives. For objectives without immersion, the theoretical values of the numerical aperture are 0.90 to 0.95; in practice, 0.90 is the most likely aperture value if it takes into account the technological errors in the manufacture of the lenses and the assembly of the objective. In this case, the optical system of the microscope containing the lighting system (collector, condenser, intermediate matching components) and the projection system (objective) should not only provide high transmission but also have a perfect aberration correction for the selected wavelength (spectral range).

For example, the condenser should also provide a numerical aperture of 0.90–0.95; its aberration correction should correspond to the aberration correction of the objective. It is logical to use two identical objectives, one of which serves as a condenser. It is believed that the intermediate optics of the illuminating microscope, as well as the collector, can be substantially inferior to the quality of the aberration correction of the condenser and the objective. However, to realize the maximum values of numerical apertures of 0.90, a very careful aberration correction of these microscope nodes is also required. From what has been stated above, it follows that the degree of correction of aberrations for a real NUV or DUV microscope can be a monochromatic correction for one selected working wavelength. Depending on the chosen wavelength, optical materials (crystals) are selected for optical design and fabrication of lens dioptric optics of the microscope. Naturally, the optical design of the microscope's optical system for a wavelength of about 106 nm is of the greatest interest. However, considering many factors (primarily technological aspects and availability of stable sources, image receivers, and others), we can choose several wavelengths of 106, 121.6, 193.4, and 244 nm as working ones for different objectives. We can make the optical designs of some types of NUV and DUV microscopes depending on the wavelength.

4.2.4.1 Cover glass and specimen glass (0.35-mm thick)

We can make optical designs for objectives that use a 0.35-mm-thick plane–parallel plate as a cover glass. We choose this thickness (twice as thick as the standard one) because crystal lithium fluoride (LiF) is used as an optical material for the cover glass. This material provides high transmission in these spectral ranges. However, it can be difficult to handle because it is soft when it undergoes grinding and polishing. The experience of manufacturing such a cover glass has shown that the chosen thickness is optimal for this optical material. This glass is a cover glass used for biological microscopy methods. For examining opaque objects in reflected light on a microscope, for which a cover glass is not required, it is possible to incorporate a specimen glass directly into the objective design, for example, as a removable part in its front decorative cap. Such a solution has long been known in ordinary microscopy, but it is not used often. In our case, a thickness of 0.35 mm is sufficient to maintain the integrity of the construction.

In addition, the use of such glass as specimen glass is also advisable. The same glass plate that we use as a cover glass ($d = 0.35$ mm; LiF) can be used as a specimen glass. This solution has significant advantages because we can use any of the proposed objectives as a condenser if transmitted light is used in a microscope. In addition, we can create a modernized version of a specimen glass that has side walls, which is especially important when working with live specimens; we can tightly seal the specimen by using a cover glass as a cover. The principal structures of two types of cover glass and specimen glass are shown in Fig. 4.8.

4.2.4.2 Example optical designs of some objectives

As examples, let us discuss the optical designs of several objectives designed for use in the NUV and DUV spectral ranges. All of them have monochromatic aberration correction when the image quality is optimized for only one wavelength. The optical designs of all objectives are based on the use of crystals as optical materials. However, optical materials, as well as the working spectral ranges of these objectives, are different. The logic of constructing the optical systems of these objectives is as follows. It is possible to arrange objectives in groups when each group is designed to be used for one of the wavelengths. Each group has objectives with different linear magnifications, from 100× to 500×, with the maximum possible numerical aperture. Wavelengths for the optical design can be selected in accordance with Table 4.1; depending on the optical transmission capabilities, various crystals can be selected as optical materials.

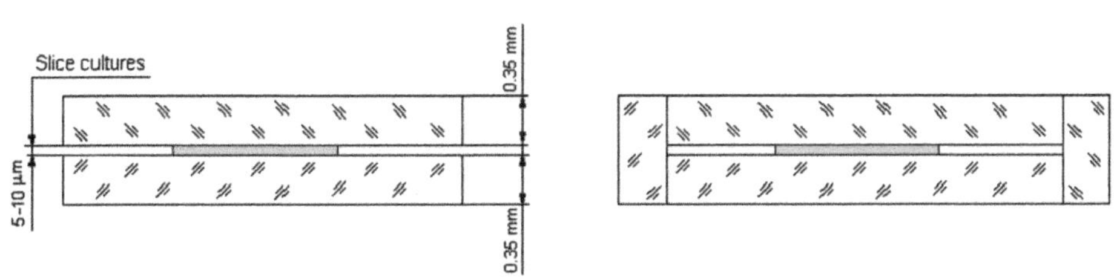

Figure 4.8 Two types of cover glass and specimen glass structures.

One of the objective groups is designed to operate at a wavelength of 244.08 nm, for which aberrations are optimized. Quartz glass is used as an optical material for the most technologically advanced and easiest to manufacture lenses. We can say that this is a universal optical material; it is well mastered in the manufacture of lenses and has practically no drawbacks. It is technologically advanced and hard enough to undergo the lens formation process (grinding and polishing), and it has low hygroscopicity and low thermal expansion ability. However, the disadvantage of this material is its limited (only for the NUV region) transmission. As a result, objectives made from this optical material are limited to the widespread use of ultraviolet fluorescence methods. To achieve the maximum numerical aperture for this spectral range, water immersion can be used; the numerical aperture achieved in the space of objects can be 1.20 in water immersion. An objective based on the use of such an optical material can be easily manufactured and have a relatively low cost.

The optical designs of the objectives in the other group also use quartz as an optical material; these objectives are designed for use at a wavelength of 193.4 nm. In this spectral region, quartz still has a fairly good optical transmission. A numerical aperture of 1.20 in water immersion is also implemented in these objectives. For the next group of objectives, designed to operate at a wavelength of 157 nm, CaF2 can be selected as an optical material for the manufacture of lenses. This crystal is well mastered by industry, but it is quite laborious to use to manufacture lenses that have a complex configuration. This material has good transmission in the NUV as well as DUV field. The disadvantages of this optical material include its low grinding hardness. In addition, this crystal is characterized by some instability of the coefficient of thermal expansion, which is reflected in the capabilities of its processing (grinding and polishing). Objectives based on the use of such an optical material can be manufactured under experimental conditions in accordance with the criteria for small-scale production. A numerical aperture of 0.95 can be achieved in these objectives.

Another group of objectives intended for use at a wavelength of 121.6 nm can use MgF2 as an optical material for the manufacture of lenses. The use of this optical material has already been discussed in Volume 1, Chapter 2, Section 2.28.5, where its use for the manufacture of large-diameter lenses for a photolithographic objective operating at the same wavelength is described. The last group of objectives discussed here are designed to operate at a wavelength of 106 nm, where maximum resolution can be achieved. LiF optical material is used in these objectives. This crystal is relatively rarely used as a material for the manufacture of lenses with complex configurations, since its technological properties are quite complex. This material is difficult for the lens-formation process (grinding and polishing), and its thermal expansion parameters often cause difficulties in manufacturing. Consequently, objectives whose lenses are made of this material will differ in relative complexity and high cost. However, the outstanding advantage of this optical material over others is its throughput and transparency, which allow the optics to operate in the DUV spectral range. In these objectives, the maximum numerical aperture reached is 0.95. Because air transmission in this spectral range is also limited, some gas is required (however, a vacuum is not required).

Figures 4.9–4.29 show the design parameters and graphs of aberration correction of all the discussed objectives.

If we analyze the design parameters of these objectives, we can see that two types of principal optical schemes are presented. As an example of constructing a basic optical system, we can give this description for the "dry" version of the objectives, where the numerical aperture is 0.95: an objective containing lenses sequentially located along the optical axis. The first lens is made in the form of a meniscus with its concave side facing the object. The second is a positive lens, and the third is a negative lens made in the form of a biconcave lens. The fourth lens is made in the form of a meniscus with its concave side facing the object. The fifth lens is positive. The sixth lens is biconcave. The seventh and eighth lenses are positive. The rest of the lenses may have different configurations.

Thus, for this type of principal optical scheme, a construction principle that uses the elements of unification is implemented. Moreover, the principle of unification can be extended to more specific objective parameters, namely, the design parameters of the first (along the rays from left to right) eight lenses. An analysis of optical designs shows that these same lenses are used in all objectives designed for use at a particular wavelength. Changing the linear magnification value is achieved in such objectives by changing the 9th, 10th, 11th, and 12th lenses. Figure 4.30 illustrates such a unification for dry objectives that have a numerical aperture of 0.95.

Figure 4.31 shows a similar unification principle for immersion objectives that have a numerical aperture of 1.20 in water immersion.

Such unification can be very convenient in the design of mechanical parts; interchangeable parts can be located, for example, in the so-called *revolving device*. Moreover, when changing the values of linear magnifications, additional focusing of the objective on the object is not required; the objectives remain parfocal. As an example of the operation of these types of objectives, we can describe the operation of lenses of immersion objectives in which the numerical aperture of 1.20 in water immersion is achieved. The optical system of such water immersion objectives is rather complicated. They contain (on the left in the course of the ray) the first lens in the form of a meniscus, with its concave side facing the object. The first surface, bordered by water, is made by taking into account the special hydrating properties of water. The second and third lenses are also made in the form of a meniscus, with its concave side facing the objects. For the most effective reduction of the

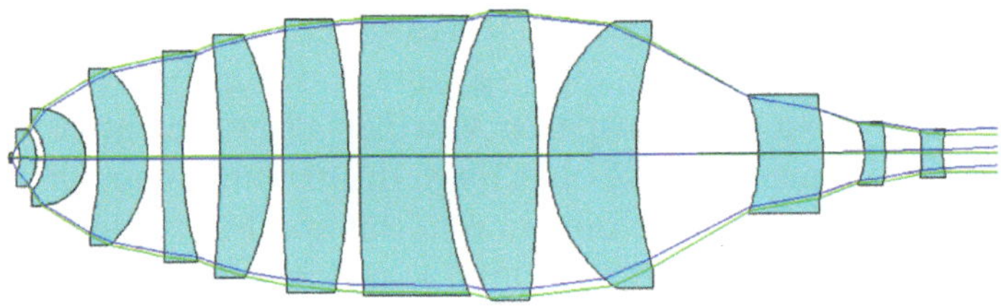

Lens: 100x0.90 mono L244 nm Fts200

Object num aper 0.900000 Image height 10.000000 Primary wavln 0.244080

SRF	RADIUS	THICKNESS	APERTURE RADIUS	GLASS	SPECIAL
OBJ	0.000000	0.010000	0.100002	AIR	
AST	0.000000	0.350000	0.120650 AS	LIF	
2	0.000000	0.756700	0.410000	AIR	
3	-2.770000	1.600000	1.300000	QUARTZ	
4	-4.530000	0.800000	2.630000	AIR	
5	-4.530000	4.000000	3.190000	QUARTZ P	
6	-4.530000	1.200000	4.530000	AIR	
7	-31.107200	4.800000	7.370000	QUARTZ P	
8	-10.925200	1.500000	8.270000	AIR	
9	406.841900	2.000000	9.440000	QUARTZ P	
10	34.540200	3.600000	9.890000	AIR	
11	-91.323200	5.000000	10.510000	QUARTZ P	
12	-25.723000	1.000000	11.390000	AIR	
13	261.031800	6.500000	12.200000	QUARTZ P	
14	-55.175000	1.000000	12.760000	AIR	
15	255.213600	8.000000	12.920000	QUARTZ P	
16	36.491000	1.000000	13.140000	AIR	
17	28.616300	8.000000	13.630000	QUARTZ P	
18	-105.983700	1.000000	13.510000	AIR	
19	15.345400	8.400000	12.510000	QUARTZ P	
20	37.823000	11.800000	10.850000	AIR	
21	-16.088600	6.200000	5.600000	QUARTZ P	
22	-28.842500	4.000000	4.370000	AIR	
23	-6.776700	2.000000	3.010000	QUARTZ P	
24	-15.290700	3.500000	2.900000	AIR	
25	-14.398100	2.000000	2.290000	QUARTZ P	
26	8.494300	200.000000	2.160000	AIR	
27	0.000000	8.500000	12.220000	QUARTZ P	
28	-135.084200	4.000000	12.480000	AIR	
29	-34.586500	5.000000	12.490000	QUARTZ P	
30	-36.706000	1.000000	13.130000	AIR	
31	0.000000	7.500000	13.150000	QUARTZ P	
32	-326.896600	200.048217 S	13.170000	AIR	
IMS	0.000000	0.000000	10.000000 S		

Figure 4.9 Optical design and graphs of aberration correction of a 100×/0.90 plan monochromatic objective (244.08 nm).

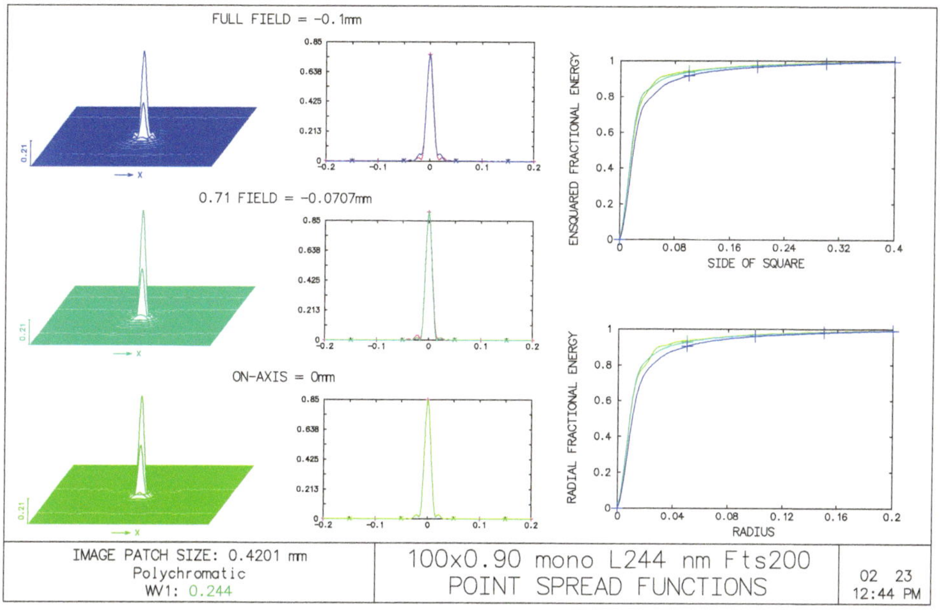

Figure 4.9 (*Continued*) Optical design and graphs of aberration correction of a 100×/0.90 plan monochromatic objective (244.08 nm).

input numerical aperture, the shape of the surfaces of these lenses is made with some deviation from aplanatics. The configuration of the fourth lens is interesting, as it is executed in the form of a biconcave shape. This negative lens has wide compensation capabilities for correcting aberrations in the objective as a whole. Although the numerical aperture after this lens increases somewhat, its use allows the number of lenses in the objective as a whole to be optimized.

The next several lenses serve to equalize the aberration correction of the objective lens while reducing the input numerical aperture. In addition, the image is warped and its projection is in the plane of the interface of the last group of lenses. The last group of lenses, in turn, makes an image in infinity (parallel beams of rays), compensating for residual aberrations (mainly image curvature and distortion).

Table 4.2 shows the main technical parameters and basic optical layout of some objectives for NUV and DUV research. These designs are for an infinite length of the microscope tube, using an additional focusing system ($F = 200$ mm).

4.2.5 Objectives as condensers

In a reflected light microscope, the objective simultaneously acts as a condenser and an objective because the light passes through it twice. In a

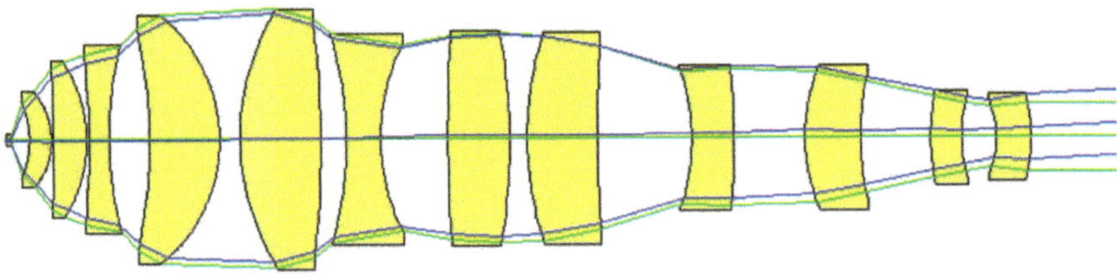

Lens: 100x0.95 mono L106 nm Fts200

Object num aper	0.950000	Image height	10.000000	Primary wavln	0.106000

SRF	RADIUS	THICKNESS	APERTURE RADIUS	GLASS	SPECIAL
OBJ	0.000000	0.010000	0.099977	AIR	
AST	0.000000	0.350000	0.130401 AS	LIF	
2	0.000000	0.969400	0.330000	AIR	
3	-4.801500	1.300000	1.980000	LIF P	
4	-3.911600	0.140000	2.690000	AIR	
5	-61.413000	2.000000	3.990000	LIF P	
6	-7.726500	0.140000	4.370000	AIR	
7	-43.778300	1.200000	4.770000	LIF P	
8	21.454700	2.300000	5.240000	AIR	
9	-28.078000	4.000000	6.200000	LIF P	
10	-8.664100	1.200000	6.900000	AIR	
11	13.284000	4.600000	7.220000	LIF P	
12	-56.708800	1.500000	6.660000	AIR	
13	-20.018000	2.000000	5.850000	LIF P	
14	11.347500	3.800000	5.230000	AIR	
15	90.005800	3.600000	5.760000	LIF P	
16	-32.727400	1.000000	5.950000	AIR	
17	18.039600	4.000000	5.860000	LIF P	
18	79.354200	5.400000	5.280000	AIR	
19	-11.634000	2.400000	3.960000	LIF P	
20	-52.876600	4.000000	4.060000	AIR	
21	9.652500	3.200000	4.030000	LIF P	
22	15.594600	4.000000	3.450000	AIR	
23	12.241800	1.800000	2.640000	LIF P	
24	7.656100	2.000000	2.240000	AIR	
25	-4.863000	2.000000	2.080000	LIF P	
26	-9.115200	200.000000	2.420000	AIR	
27	917.862600	8.500000	12.480000	LIF P	
28	-47.589000	0.800000	12.620000	AIR	
29	-53.960700	5.000000	12.380000	LIF P	
30	-211.141600	6.000000	12.380000	AIR	
31	-31.102400	7.500000	12.170000	LIF P	
32	-37.411500	199.685203 S	13.460000	AIR	
IMS	0.000000	0.000000	10.000000 S		

Figure 4.10 Optical design and graphs of aberration correction of a 100×/0.95 plan monochromatic objective (106 nm).

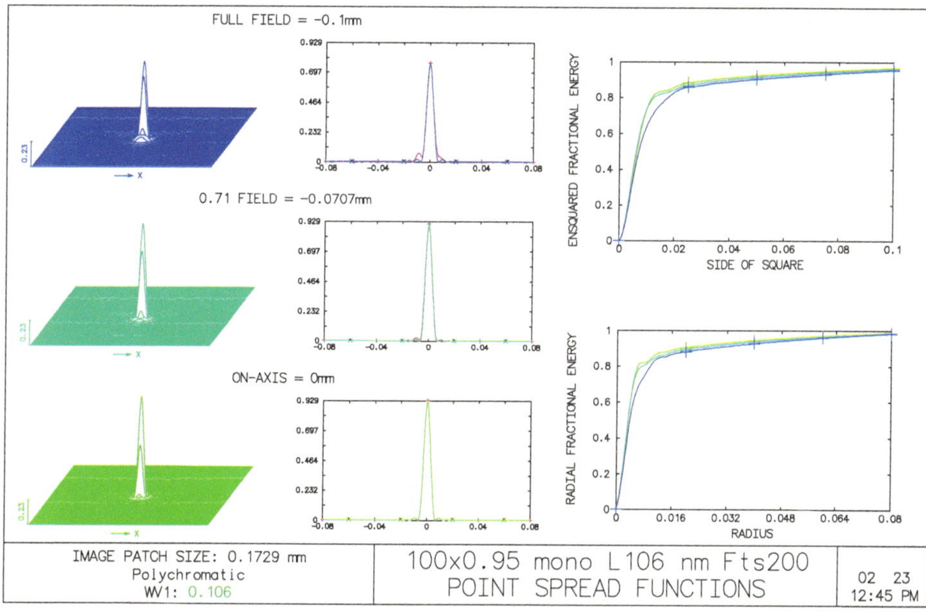

Figure 4.10 (*Continued*) Optical design and graphs of aberration correction of a 100×/0.95 plan monochromatic objective (106 nm).

transmitted light microscope, one device is used as an objective and the other as a condenser. According to Abbe's theory, the best resolution result can be achieved by using the same numerical apertures for the objective and condenser. An important addition may be the idea that the quality of the aberration correction of the condenser is also of great importance. Conventional microscopes, where standard condensers are used, do not have aberration correction, which is close to the aberration correction of objectives. In the lighting system, there is a redistribution of energy in the diffraction spot due to the presence of significant aberrations. In our case, when the whole microscope system works close to the resolution limit for dioptric microscopy, one cannot neglect such a powerful parameter for improving the quality of research as the most perfect aberration correction of a condenser.

In this case, to achieve maximum resolution, it is advisable to use two identical objectives, one of which serves as a condenser. This is very convenient because all four of our objectives are designed to work with a 0.35-mm LiF plate. The optical system of this microscope will use a "lighting" objective, which acts as a condenser; a 0.35-mm-thick specimen glass made of LiF; a 0.35-mm-thick cover glass made of LiF, and a "working for an image" objective.

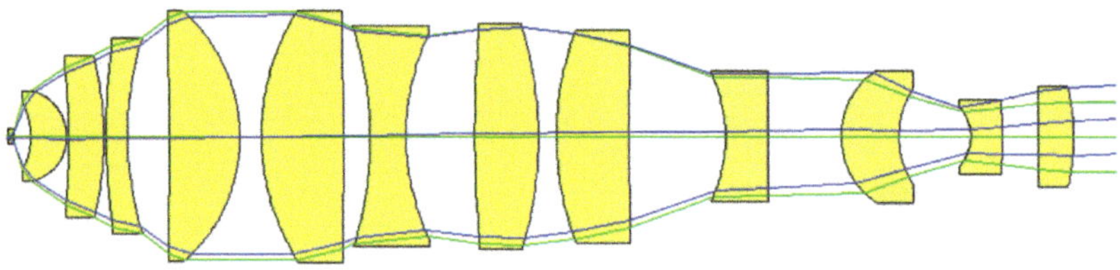

Lens: 100x0.95 mono L121.6 nm

Object num aper	0.950000	Image height	10.000000	Primary wavln	0.121600

SRF	RADIUS	THICKNESS	APERTURE RADIUS	GLASS	SPECIAL
OBJ	0.000000	0.010000	0.100003	AIR	
AST	0.000000	0.350000	0.130427 AS	LIF	
2	0.000000	0.969410	0.390000	AIR	
3	-3.693700	2.100000	1.840000	MGF2	
4	-2.463100	0.140000	2.440000	AIR	
5	-30.209200	2.000000	3.710000	MGF2 P	
6	-17.410500	0.140000	4.390000	AIR	
7	35.280200	1.200000	4.930000	MGF2 P	
8	17.595800	2.300000	5.310000	AIR	
9	-391.765400	4.000000	6.430000	MGF2 P	
10	-8.697900	1.200000	6.770000	AIR	
11	12.456900	4.600000	6.810000	MGF2 P	
12	-249.805800	1.500000	6.220000	AIR	
13	-18.503800	2.000000	5.970000	MGF2 P	
14	13.082000	3.800000	5.570000	AIR	
15	70.133100	3.600000	5.980000	MGF2 P	
16	-20.120200	1.000000	6.110000	AIR	
17	16.040900	4.000000	5.770000	MGF2 P	
18	388.504100	5.400000	5.130000	AIR	
19	-7.921400	2.400000	3.530000	MGF2 P	
20	0.000000	4.000000	3.540000	AIR	
21	4.274600	3.200000	3.580000	MGF2 P	
22	4.667200	4.000000	2.590000	AIR	
23	-2.362200	1.800000	1.620000	MGF2 P	
24	-26.681300	2.000000	2.030000	AIR	
25	94.763300	2.000000	2.530000	MGF2 P	
26	-10.665500	200.000000	2.750000	AIR	
27	0.000000	8.500000	12.670000	LIF	
28	-40.878200	0.800000	12.870000	AIR	
29	-46.593700	5.000000	12.690000	LIF P	
30	-116.493200	5.000000	12.760000	AIR	
31	-30.875900	7.500000	12.610000	LIF P	
32	-36.882100	200.029180 S	13.660000	AIR	
IMS	0.000000	0.000000	10.000000 S		

Figure 4.11 Optical design and graphs of aberration correction of a 100×/0.95 plan monochromatic objective (121.6 nm).

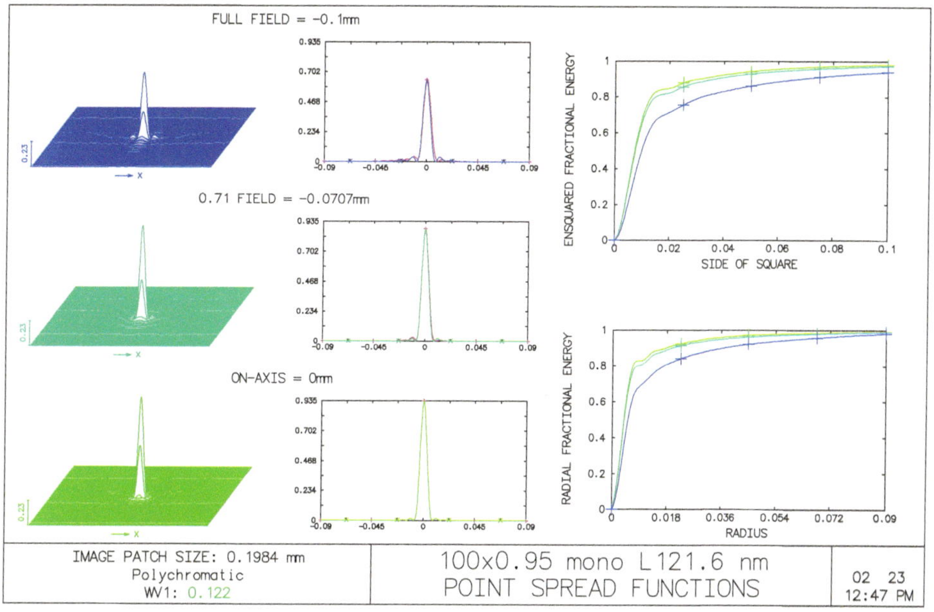

Figure 4.11 (*Continued*) Optical design and graphs of aberration correction of a 100×/0.95 plan monochromatic objective (121.6 nm).

4.2.6 Applications for these objectives

There is interest in designing a microscope for fluorescence analysis. As a tool for detecting fluorescence of a studied object, a special small- or medium-magnification objective can be used that has good transmission in the NUV to DUV spectral regions in which apochromatic aberration correction is achieved. For example, we can use a mirror-lens (catadioptric) objective; the design parameters and graphs of aberration correction of this objective are shown in Fig. 4.32.

The proposed objective is designed to operate in the spectral range of 106 to 1000 nm, that is, it has multi-apochromatic aberration correction. The requirements for the aberration correction of such a lens may be slightly different than usual. For example, uncorrected aberration of image curvature can contribute to the expansion of the objective's functional parameters. If it is mainly intended for the detection and registration of fluorescence from the object under study, this so-called *non-imaging* objective can work in depth at the periphery of the linear field and detect the fluorescence of objects out of the system's focus. This is a rather valuable and interesting property. Another interesting feature of mirror-lens objectives in general, and proposed here in particular, is their ability to accurately focus at any wavelength within the selected spectral range. In this case, the aberration correction at a fixed wavelength will be quite

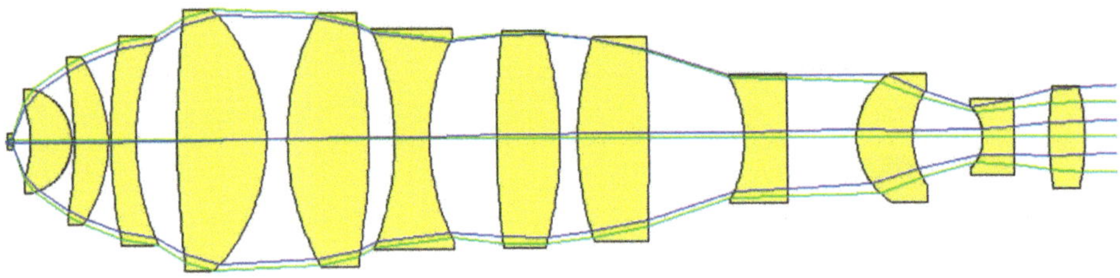

Lens: 100x0.95 mono L157 nm Fts200

Object num aper 0.950000 Image height 10.000000 Primary wavln 0.157000

SRF	RADIUS	THICKNESS	APERTURE RADIUS	GLASS	SPECIAL
OBJ	0.000000	0.010000	0.099991	AIR	
AST	0.000000	0.350000	0.130416 AS	LIF	
2	0.000000	0.969410	0.420000	AIR	
3	-6.225200	2.300000	2.180000	CAF2	
4	-2.946500	0.140000	2.840000	AIR	
5	-22.605400	2.000000	4.180000	CAF2 P	
6	-7.694100	0.140000	4.600000	AIR	
7	26.072000	1.500000	5.430000	CAF2 P	
8	13.787200	2.300000	5.730000	AIR	
9	56.698800	5.000000	6.690000	CAF2 P	
10	-9.902000	1.200000	7.180000	AIR	
11	13.966400	4.600000	6.960000	CAF2 P	
12	-34.271600	1.500000	6.490000	AIR	
13	-15.025900	2.000000	6.050000	CAF2 P	
14	12.292600	3.800000	5.550000	AIR	
15	49.883500	3.600000	5.870000	CAF2 P	
16	-21.471400	1.000000	5.940000	AIR	
17	17.449400	4.000000	5.620000	CAF2 P	
18	-336.692000	5.400000	5.030000	AIR	
19	-7.868700	2.400000	3.540000	CAF2 P	
20	0.000000	4.000000	3.540000	AIR	
21	4.198200	3.200000	3.530000	CAF2 P	
22	4.714500	4.000000	2.590000	AIR	
23	-2.403700	1.800000	1.680000	CAF2 P	
24	-86.126200	2.000000	2.080000	AIR	
25	21.697200	2.000000	2.610000	CAF2 P	
26	-13.594200	200.000000	2.770000	AIR	
27	0.000000	8.500000	12.770000	LIF P	
28	-40.966600	0.800000	12.980000	AIR	
29	-40.955400	5.000000	12.890000	LIF P	
30	-84.517300	5.000000	13.090000	AIR	
31	-34.552600	7.500000	12.990000	LIF P	
32	-38.247200	199.909455 S	13.830000	AIR	
IMS	0.000000	0.000000	10.000000 S		

Figure 4.12 Optical design and graphs of aberration correction of a 100×/0.95 plan monochromatic objective (157 nm).

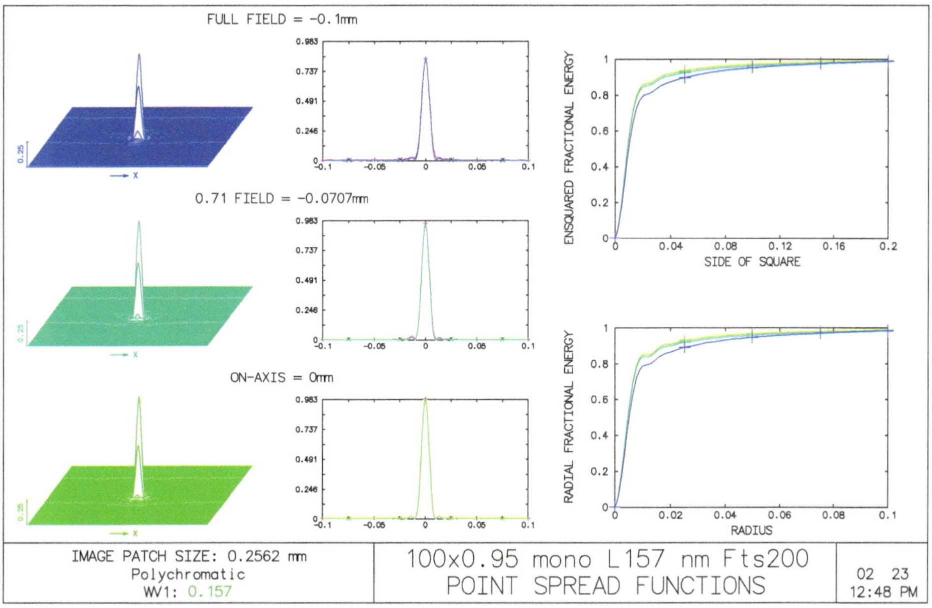

Figure 4.12 (*Continued*) Optical design and graphs of aberration correction of a 100×/0.95 plan monochromatic objective (157 nm).

satisfactory. This circumstance may allow some correlation between the wavelength of the radiation source and the objective; this may allow some scanning along several wavelengths (which are generated by the radiation source) using only one mirror-lens objective.

By the way, the use of a survey objective system, in which there are only mirrors in the optical system (catoptric system), has some advantages. You can try to exclude the so-called obscuration of the central zone of the pupil, for example, by using polarization optics. In this case, you can try to use translucent mirrors on which polarizing filters are applied. Because of the rotation of the plane of polarization, some rays are reflected from the mirrors and some pass through them without reflection. Note that manufacturing such mirrors can become a difficult or even unsolvable technological problem.

Such an objective can allow a sufficiently large linear field at the object to be observed, which helps locate the source of fluorescence. Because achieving the maximum resolution for such an objective is not required (it can be used as an overview), we can use an objective that has a mirror or mirror-lens design. After detecting the fluorescence of the sample under investigation, the wavelength (spectral range) of this fluorescence becomes known; after that, one or another objective with the maximum numerical aperture can be used for the detected wavelength for a more detailed study of the sample. Since the

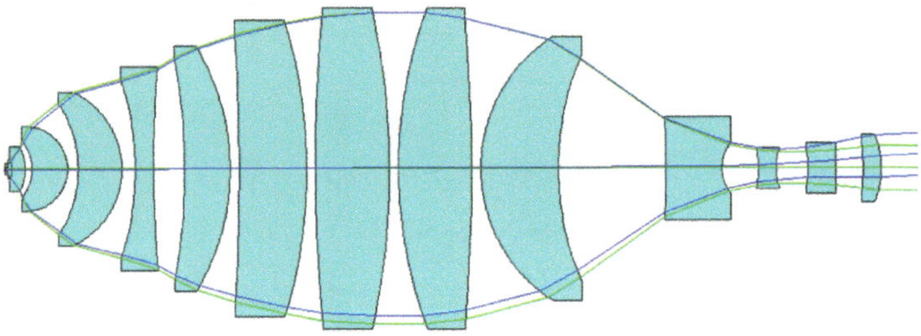

Lens: 100x1.20 water mono L193 nm

Object num aper 1.200000 Image height 10.000000 Primary wavln 0.193400

SRF	RADIUS	THICKNESS	APERTURE RADIUS	GLASS	SPECIAL
OBJ	0.000000	0.010000	0.100010	WATER	
AST	0.000000	0.350000	0.115086 AS	LIF	
2	0.000000	0.437200	0.640000	WATER P	
3	-2.770000	1.600000	1.010000	SILICA	
4	-4.530000	0.800000	2.450000	AIR	
5	-4.530000	4.000000	3.100000	SILICA P	
6	-4.530000	1.200000	4.530000	AIR	
7	-12.622800	4.800000	7.070000	SILICA P	
8	-8.961100	1.500000	8.230000	AIR	
9	-29.310000	2.000000	9.680000	SILICA P	
10	111.434900	3.600000	10.950000	AIR	
11	-66.168200	5.000000	12.110000	SILICA P	
12	-25.079400	1.000000	13.150000	AIR	
13	-238.518800	7.500000	14.610000	SILICA P	
14	-51.475200	1.000000	15.900000	AIR	
15	164.081500	8.000000	16.720000	SILICA P	
16	-81.148500	1.000000	17.150000	AIR	
17	48.902200	8.000000	17.180000	SILICA P	
18	-193.030900	1.000000	16.640000	AIR	
19	16.718700	8.400000	14.170000	SILICA P	
20	29.957800	11.800000	12.220000	AIR	
21	-131.461900	6.200000	5.590000	SILICA P	
22	4.363800	4.000000	2.760000	AIR	
23	-18.122900	2.000000	2.250000	SILICA P	
24	6.858500	3.500000	2.080000	AIR	
25	-9.455700	3.000000	2.290000	SILICA P	
26	228.992400	3.000000	2.740000	AIR	
27	-30.982000	2.000000	3.360000	SILICA P	
28	-9.275800	200.000000	3.620000	AIR	
29	0.000000	8.500000	13.680000	CAF2	
30	-35.576700	0.800000	13.870000	AIR	
31	-37.243300	5.000000	13.700000	CAF2 P	
32	-88.767700	6.000000	13.860000	AIR	
33	-28.355900	7.500000	13.700000	CAF2 P	
34	-32.528600	200.099595 S	14.830000	AIR	
IMS	0.000000	0.000000	10.000000 S		

Figure 4.13 Optical design and graphs of aberration correction of a 100×/1.20 water immersion plan monochromatic objective (193.4 nm).

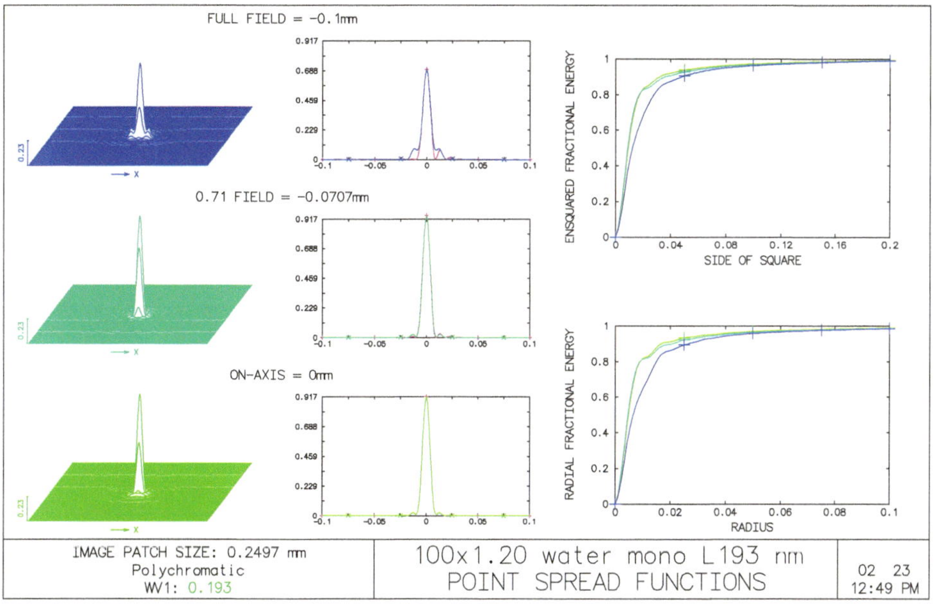

Figure 4.13 (*Continued*) Optical design and graphs of aberration correction of a 100×/1.20 water immersion plan monochromatic objective (193.4 nm).

linear dimensions of the studied objects can vary widely, we can propose the use of some objectives that have high and super–high linear magnifications (and maximum numerical apertures). In this case, the idea of using two identical objectives can be realized (the most effective is achieved for objectives with a numerical aperture of 0.95 for working in monochromatic light with a wavelength $\lambda = 106$ nm). However, if we consider this set of objectives to be a basic element of a microscope, we can assume that it will be applied in various ways to NUV and DUV microscopy.

A researcher working on a microscope with reflected light can use any objective in which a numerical aperture of 1.20 in water immersion is achieved. The aberrations of such a lens can be minimized, for example, for operation at $\lambda = 244$ nm; fluorescence can occur when the object under study is irradiated, for example, at a wavelength of 193 nm, since the optical material of this objective (and water immersion) are quite transparent in this area. If using a transmitted light microscope, the researcher can use an objective with a numerical aperture of 0.95, the aberrations of which are minimized at $\lambda = 106$ nm as a condenser, as well as any other objective to detect different fluorescence spectral ranges. In this case, a double- or multispectral range of research can be performed. Some results can be achieved using so-called *anti-Stokes* fluorescence.

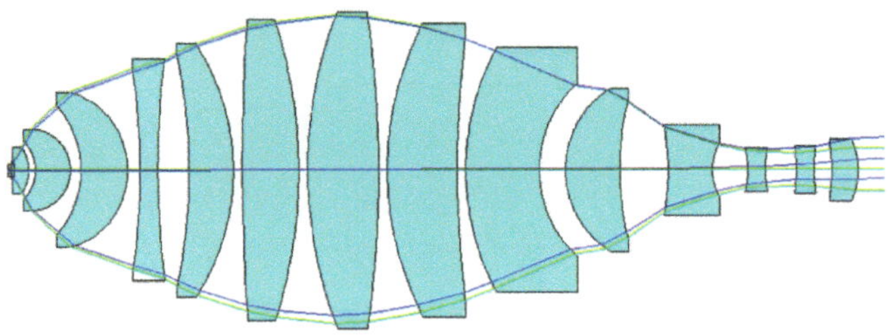

Lens: 100x1.20 water mono L244 nm

Object num aper 1.200000 Image height 10.000000 Primary wavln 0.244080

SRF	RADIUS	THICKNESS	APERTURE RADIUS	GLASS	SPECIAL
OBJ	0.000000	0.010000	0.099997	WATER	
AST	0.000000	0.350000	0.117559 AS	LIF	
2	0.000000	0.400042	0.670000	WATER P	
3	-2.770000	1.600000	1.030000	QUARTZ	
4	-4.530000	0.800000	2.550000	AIR	
5	-4.530000	4.000000	3.210000	QUARTZ P	
6	-4.530000	1.200000	4.530000	AIR	
7	-10.758300	5.400000	7.210000	QUARTZ P	
8	-8.851500	1.500000	8.580000	AIR	
9	-73.238800	2.000000	11.210000	QUARTZ P	
10	92.682600	3.600000	12.270000	AIR	
11	-59.747900	5.000000	13.040000	QUARTZ P	
12	-25.524300	1.000000	14.030000	AIR	
13	209.841600	6.500000	16.020000	QUARTZ P	
14	-52.686500	1.000000	16.630000	AIR	
15	47.416000	8.000000	17.470000	QUARTZ P	
16	-120.606700	1.000000	17.280000	AIR	
17	33.951600	8.000000	16.240000	QUARTZ P	
18	112.404200	1.000000	14.860000	AIR	
19	27.554400	8.400000	13.540000	QUARTZ P	
20	12.736400	3.000000	9.580000	AIR	
21	10.504200	6.200000	9.020000	QUARTZ P	
22	32.974200	5.500000	7.920000	AIR	
23	-24.260300	5.000000	5.000000	QUARTZ P	
24	6.093800	4.000000	3.010000	AIR	
25	-10.389500	2.000000	2.440000	QUARTZ P	
26	11.407900	3.500000	2.330000	AIR	
27	-15.398400	2.000000	2.460000	QUARTZ P	
28	17.946600	2.000000	2.650000	AIR	
29	-19.239900	3.000000	2.970000	QUARTZ P	
30	-8.036400	200.000000	3.440000	AIR	
31	0.000000	8.500000	13.590000	QUARTZ P	
32	-135.084200	4.000000	13.850000	AIR	
33	-34.586500	5.000000	13.840000	QUARTZ P	
34	-36.706000	1.000000	14.540000	AIR	
35	0.000000	7.500000	14.550000	QUARTZ P	
36	-326.896600	199.981423 S	14.550000	AIR	
IMS	0.000000	0.000000	10.000000 S		

Figure 4.14 Optical design and graphs of aberration correction of a 100×/1.20 water immersion plan monochromatic objective (244.08 nm).

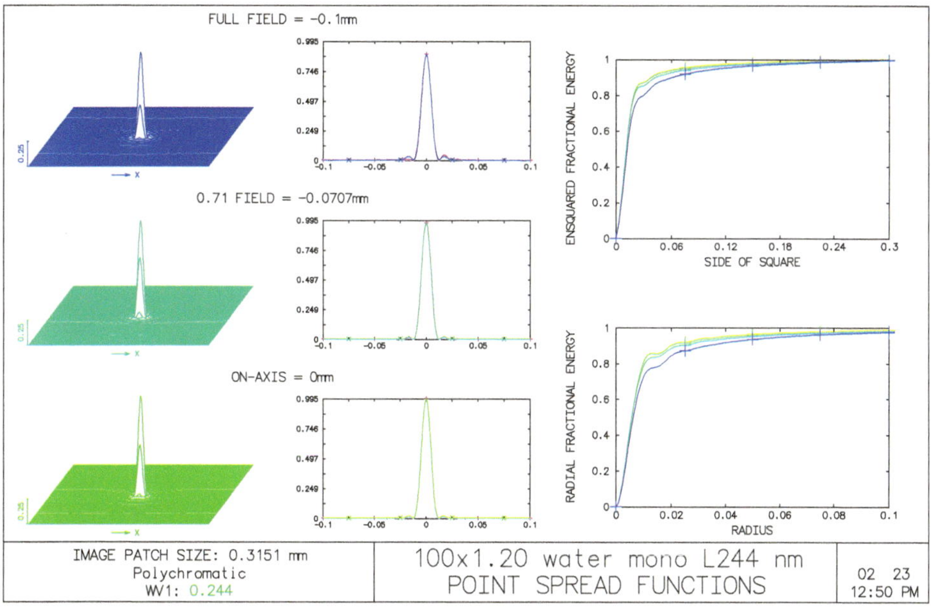

Figure 4.14 (*Continued*) Optical design and graphs of aberration correction of a 100×/1.20 water immersion plan monochromatic objective (244.08 nm).

4.2.7 Light source

When using the proposed objectives, it is necessary to provide illumination at different wavelengths of the NUV–DUV spectral ranges. On the one hand, such sources should be able to choose a narrow spectral band when obtaining monochromatic light. For example, an acceptable version of such a source could be a conventional monochromator supplemented with special optics for filling the pupil of the condenser for a microscope operating in transmitted light, or filling the pupil of the objective for a reflected light microscope. On the other hand, a problem with the detection of proteins and viruses on a transmitted light microscope is that the researcher does not know for sure when exactly the specimen is irradiated, i.e., at which wavelength the protein begins to fluoresce. In addition, different types of proteins and viruses fluoresce in different spectral regions. Therefore, at a minimum, the DUV lighting source and the optics of the lighting device must have high transparency and include scanning in the selected spectral range along the selected wavelengths, regulating them in turn.

4.2.8 Detectors

For the NUV range, use of a receiver is not difficult. However, the width of the registering spectrum is most likely somewhat limited. Therefore, it is better to use monochromatic receivers. For the DUV band, this is a rather difficult technological task for obtaining a broadband radiation receiver. Thus, a

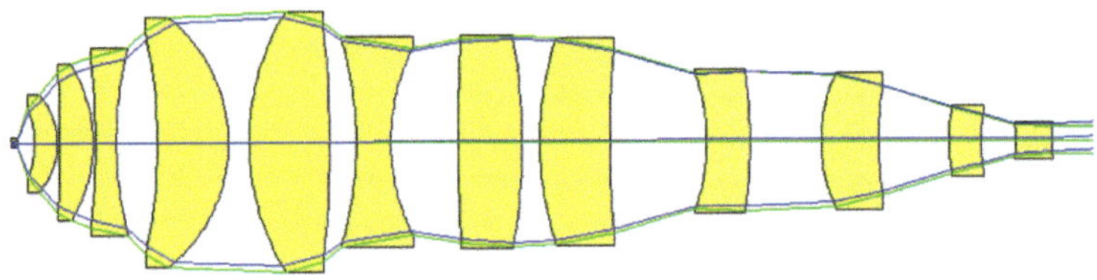

Lens: 250x0.95 mono L106 nm Fts200

Object num aper 0.950000 Image height 10.000000 Primary wavln 0.106000

SRF	RADIUS	THICKNESS	APERTURE RADIUS	GLASS	SPECIAL
OBJ	0.000000	0.010000	0.040002	AIR	
AST	0.000000	0.350000	0.070427 AS	LIF	
2	0.000000	0.969405	0.270000	AIR	
3	-4.801500	1.300000	1.950000	LIF P	
4	-3.911600	0.140000	2.670000	AIR	
5	-61.413000	2.000000	3.950000	LIF P	
6	-7.726500	0.140000	4.330000	AIR	
7	-43.778300	1.200000	4.740000	LIF P	
8	21.454700	2.300000	5.210000	AIR	
9	-28.078000	4.000000	6.180000	LIF P	
10	-8.664100	1.200000	6.890000	AIR	
11	13.284000	4.600000	7.200000	LIF P	
12	-56.708800	1.500000	6.630000	AIR	
13	-20.018000	2.000000	5.810000	LIF P	
14	11.347500	3.800000	5.190000	AIR	
15	90.005800	3.600000	5.680000	LIF P	
16	-32.727400	1.000000	5.870000	AIR	
17	18.039600	4.000000	5.760000	LIF P	
18	79.354200	5.400000	5.170000	AIR	
19	-10.886700	2.400000	3.840000	LIF P	
20	-33.145200	4.000000	3.950000	AIR	
21	8.733700	3.200000	3.780000	LIF P	
22	19.437000	4.000000	3.160000	AIR	
23	10.898700	1.800000	1.940000	LIF P	
24	7.492200	2.000000	1.500000	AIR	
25	-4.465100	2.000000	0.990000	LIF P	
26	7.678600	200.000000	0.930000	AIR	
27	917.862600	8.500000	11.010000	LIF P	
28	-47.589000	0.800000	11.160000	AIR	
29	-53.960700	5.000000	10.970000	LIF P	
30	-211.141600	6.000000	11.000000	AIR	
31	-31.102400	7.500000	10.820000	LIF P	
32	-37.411500	200.033668 S	11.980000	AIR	
IMS	0.000000	0.000000	10.000000 S		

Figure 4.15 Optical design and graphs of aberration correction of a 250×/0.95 plan monochromatic objective (106 nm).

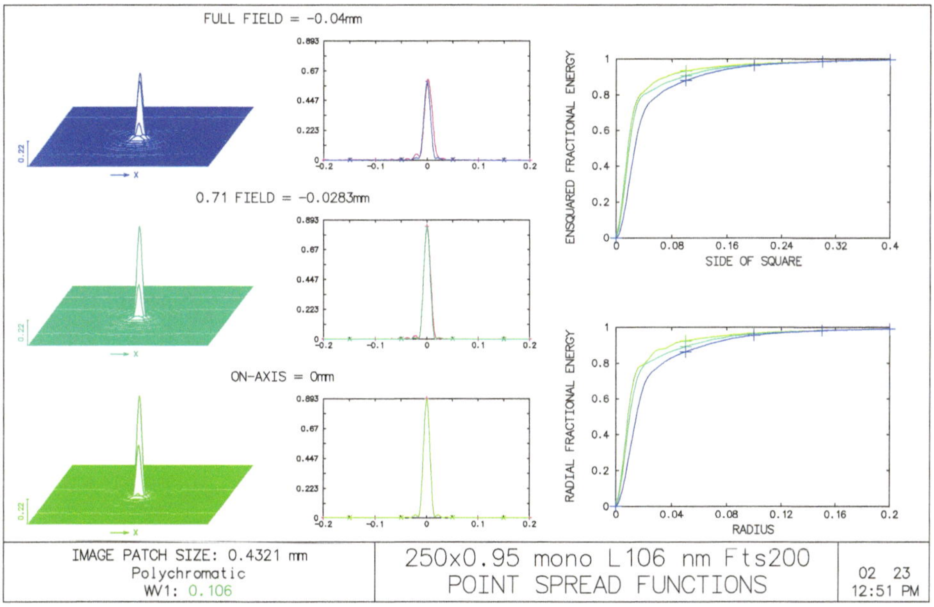

Figure 4.15 (*Continued*) Optical design and graphs of aberration correction of a 250×/0.95 plan monochromatic objective (106 nm).

monochromatic receiver is probably the only acceptable option in this case. The experience of creating experimental radiation receivers for a wavelength of 121.6 nm and some others has been reported [9].

4.2.9 Telecentric illuminator

The main principles of constructing the optical system of a modern light microscope have been described [10], including its lighting system. It has been shown that the principle of telecentric ray paths is the basis on which partially collimated light can be used to increase the local contrast of the illuminated elements of objects. The light source is located in the front focal plane of the collector, the front focal plane of each subsequent component is located in the back focal plane of the previous one, and the illuminated object is in the rear focal plane of the condenser. The advantage of such an optical illuminator circuit is the presence of intermediate images, which are interfaced with the planes of the aperture and field diaphragms of the illuminator. Thus, the material diaphragms can be located in the illuminator. Naturally, all the optical components of the lighting system should also be made of optical materials that have good transmittance in the NUV–DUV spectral region. For example, LiF can be used as the optical material for the collector and the lenses of the intermediate matching systems of the illuminator. The principal layout of an example telecentric illuminator is shown in Fig. 4.33.

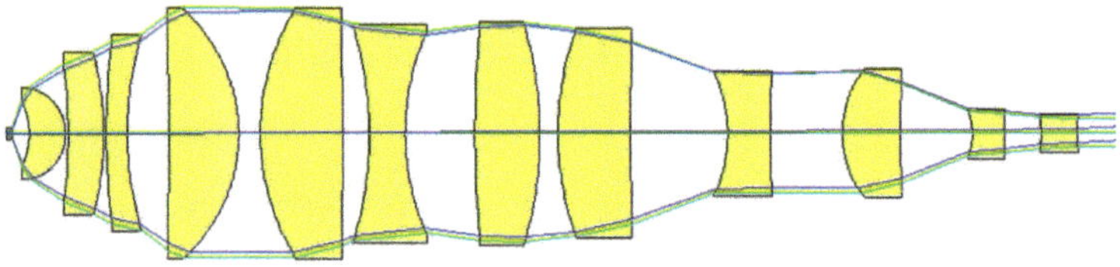

Lens: 250x0.95 mono L121.6 nm

Object num aper 0.950000 Image height 10.000000 Primary wavln 0.121600

SRF	RADIUS	THICKNESS	APERTURE RADIUS	GLASS	SPECIAL
OBJ	0.000000	0.010000	0.040000	AIR	
AST	0.000000	0.350000	0.070424 AS	LIF	
2	0.000000	0.969410	0.330000	AIR	
3	-3.693700	2.100000	1.820000	MGF2	
4	-2.463100	0.140000	2.440000	AIR	
5	-30.209200	2.000000	3.710000	MGF2 P	
6	-17.410500	0.140000	4.380000	AIR	
7	35.280200	1.200000	4.920000	MGF2 P	
8	17.595800	2.300000	5.300000	AIR	
9	-391.765400	4.000000	6.400000	MGF2 P	
10	-8.697900	1.200000	6.750000	AIR	
11	12.456900	4.600000	6.750000	MGF2 P	
12	-249.805800	1.500000	6.150000	AIR	
13	-18.503800	2.000000	5.890000	MGF2 P	
14	13.082000	3.800000	5.490000	AIR	
15	70.133100	3.600000	5.870000	MGF2 P	
16	-20.120200	1.000000	5.990000	AIR	
17	16.040900	4.000000	5.640000	MGF2 P	
18	388.504100	5.400000	5.000000	AIR	
19	-7.886500	2.400000	3.370000	MGF2 P	
20	66.113900	4.000000	3.350000	AIR	
21	5.496800	3.200000	3.460000	MGF2 P	
22	40.961200	4.000000	2.880000	AIR	
23	-3.413000	1.800000	1.330000	MGF2 P	
24	-174.971900	2.000000	1.210000	AIR	
25	-8.182300	2.000000	1.020000	MGF2 P	
26	9.224200	200.000000	1.000000	AIR	
27	0.000000	8.500000	11.110000	LIF	
28	-40.878200	0.800000	11.320000	AIR	
29	-46.593700	5.000000	11.190000	LIF P	
30	-116.493200	5.000000	11.280000	AIR	
31	-30.875900	7.500000	11.160000	LIF P	
32	-36.882100	199.923733 S	12.100000	AIR	
IMS	0.000000	0.000000	10.000000 S		

Figure 4.16 Optical design and graphs of aberration correction of a 250×/0.95 plan monochromatic objective (121.6 nm).

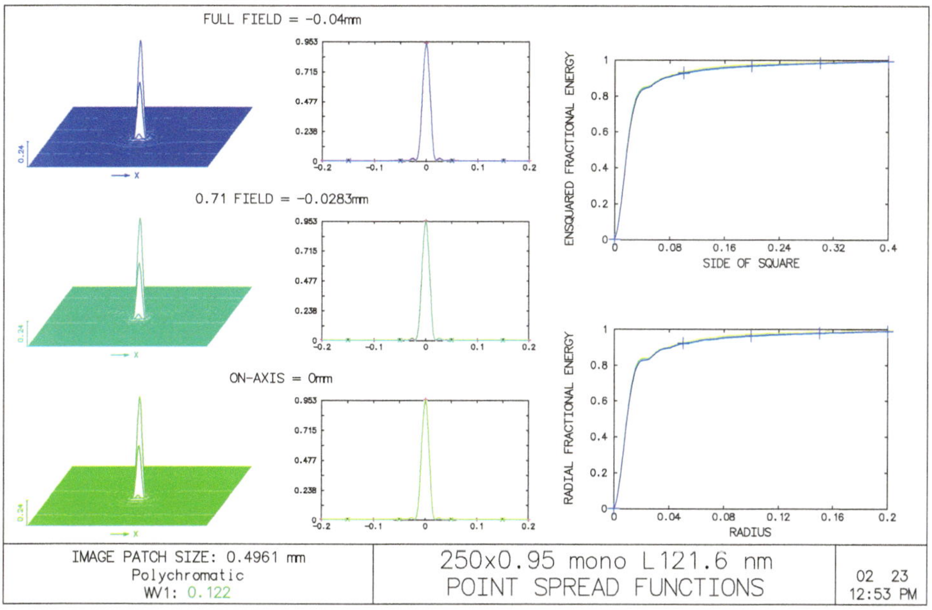

Figure 4.16 (*Continued*) Optical design and graphs of aberration correction of a 250×/0.95 plan monochromatic objective (121.6 nm).

4.2.10 Additional requirements for microscope parts

These are considered the main requirements because if they are not performed, work on the microscope will be impossible. Additional requirements include, for example, a way to provide some tightness to fill the internal space with gas (for the DUV range). The microscope must be protected from other external influences, such as vibration, humidity, and temperature. The microscope should use precision micro-mechanisms and motion and focus actuators. Optical filters (for conducting fluorescence analysis) should be of the highest quality.

4.3 Optics Achromatized in a Wide Spectral Range

4.3.1 Microscopy in NUV–VIS–NIR spectral ranges

The production and processing of images in the wide area of the spectrum (using simultaneously close ultraviolet, visible, and close infrared ranges) are two of the foundations for the development of modern research methods on a light microscope. Although it is impossible to describe every modern method, the trends of their development can be explained. For example, the methodology of broadband spectral visualization of an object under study is obviously an important method for obtaining images with a high spectral resolution [11]. Such studies are now called *hyperspectral* and *multispectral*. At

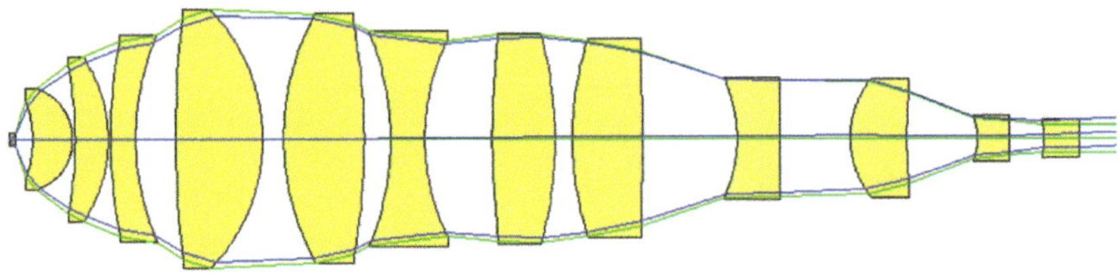

Lens: 250x0.95 mono L157 nm Fts200

Object num aper 0.950000 Image height 10.000000 Primary wavln 0.157000

SRF	RADIUS	THICKNESS	APERTURE RADIUS	GLASS	SPECIAL
OBJ	0.000000	0.010000	0.040001	AIR	
AST	0.000000	0.350000	0.070425 AS	LIF	
2	0.000000	0.969397	0.360000	AIR	
3	-6.225200	2.300000	2.150000	CAF2	
4	-2.946500	0.140000	2.830000	AIR	
5	-22.605400	2.000000	4.170000	CAF2 P	
6	-7.694100	0.140000	4.590000	AIR	
7	26.072000	1.500000	5.420000	CAF2 P	
8	13.787200	2.300000	5.730000	AIR	
9	56.698800	5.000000	6.680000	CAF2 P	
10	-9.902000	1.200000	7.170000	AIR	
11	13.966400	4.600000	6.920000	CAF2 P	
12	-34.271600	1.500000	6.440000	AIR	
13	-15.025900	2.000000	5.990000	CAF2 P	
14	12.292600	3.800000	5.480000	AIR	
15	49.883500	3.600000	5.780000	CAF2 P	
16	-21.471400	1.000000	5.840000	AIR	
17	17.449400	4.000000	5.510000	CAF2 P	
18	-336.692000	5.400000	4.900000	AIR	
19	-7.834300	2.400000	3.390000	CAF2 P	
20	198.419200	4.000000	3.350000	AIR	
21	5.095400	3.200000	3.310000	CAF2 P	
22	25.302800	4.000000	2.700000	AIR	
23	-3.249000	1.800000	1.250000	CAF2 P	
24	13.753500	2.000000	1.100000	AIR	
25	-4.927800	2.000000	0.970000	CAF2 P	
26	138.064100	200.000000	1.040000	AIR	
27	0.000000	8.500000	11.170000	LIF P	
28	-40.966600	0.800000	11.400000	AIR	
29	-40.955400	5.000000	11.330000	LIF P	
30	-84.517300	5.000000	11.520000	AIR	
31	-34.552600	7.500000	11.460000	LIF P	
32	-38.247200	200.006842 S	12.220000	AIR	
IMS	0.000000	0.000000	10.000000 S		

Figure 4.17 Optical design and graphs of aberration correction of a 250×/0.95 plan monochromatic objective (157 nm).

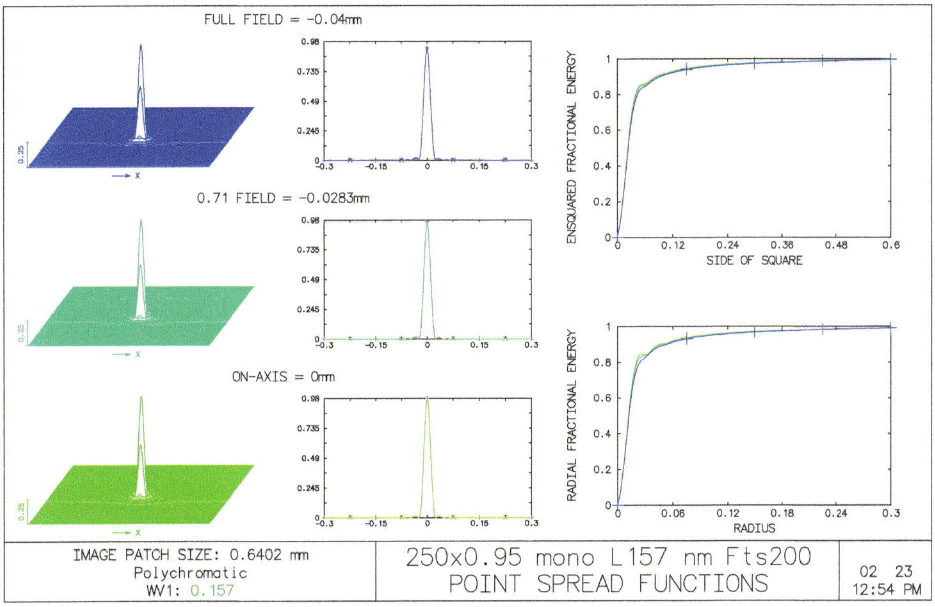

Figure 4.17 (*Continued*) Optical design and graphs of aberration correction of a 250×/0.95 plan monochromatic objective (157 nm).

the same time, spectral microscopic images allow the optical spectrum in a selected number of points of an object's image to be obtained and explored [12]. In biological applications, this methodology allows us to observe and diagnose the local properties of biological tissues. For example, in medicine, with the help of hyperspectral data, it is possible to distinguish normal, precancerous, and cancer cells in a specimen based on a combination of morphological and spectral characteristics of cells; this is the basis for the development of screening and diagnostic tests for the effective detection and diagnosis of serious diseases [13]. In addition, there are many advantages to using this methodology in microelectronics, as well as in material science, particularly when working in the near infrared range. However, the most information and reliable research can be achieved with complete spectral visualization of the object under study by combining visual and spectral microscopy methods. An example is the methodology of so-called *hybrid microscopy* [14].

Several factors should be considered when creating the appropriate equipment. A comprehensive analysis of the equipment's ability to study hyperspectral and multispectral images must be conducted. Such an analysis allows, for example, the original methodology of working on a microscope to be implemented [15]. This technique takes into account the features of the illumination of the object under study in monochrome radiation in the visible

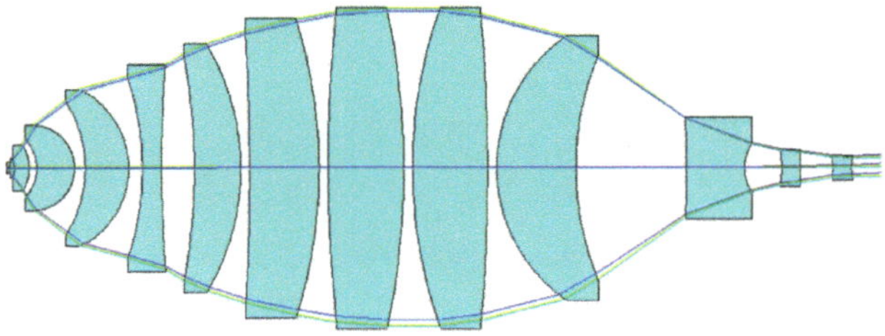

Lens: 250x1.20 water mono L193 nm

Object num aper	1.200000	Image height	10.000000	Primary wavln	0.193400

SRF	RADIUS	THICKNESS	APERTURE RADIUS	GLASS	SPECIAL
OBJ	0.000000	0.010000	0.040000	WATER	
AST	0.000000	0.350000	0.055076 AS	LIF	
2	0.000000	0.437222	0.580000	WATER P	
3	-2.770000	1.600000	0.970000	SILICA	
4	-4.530000	0.800000	2.420000	AIR	
5	-4.530000	4.000000	3.080000	SILICA P	
6	-4.530000	1.200000	4.520000	AIR	
7	-12.622800	4.800000	7.060000	SILICA P	
8	-8.961100	1.500000	8.220000	AIR	
9	-29.310000	2.000000	9.630000	SILICA P	
10	111.434900	3.600000	10.860000	AIR	
11	-66.168200	5.000000	12.010000	SILICA P	
12	-25.079400	1.000000	13.050000	AIR	
13	-238.518800	7.500000	14.440000	SILICA P	
14	-51.475200	1.000000	15.700000	AIR	
15	164.081500	8.000000	16.470000	SILICA P	
16	-81.148500	1.000000	16.890000	AIR	
17	48.902200	8.000000	16.890000	SILICA P	
18	-193.030900	1.000000	16.340000	AIR	
19	16.718700	8.400000	13.960000	SILICA P	
20	29.957800	11.800000	11.940000	AIR	
21	-135.885200	6.200000	5.370000	SILICA P	
22	4.606200	4.000000	2.650000	AIR	
23	-15.446000	2.000000	1.980000	SILICA P	
24	0.000000	3.500000	1.780000	AIR	
25	-10.584500	2.000000	1.300000	SILICA P	
26	5.653100	200.000000	1.190000	AIR	
27	0.000000	8.500000	11.110000	CAF2	
28	-35.576700	0.800000	11.330000	AIR	
29	-37.243300	5.000000	11.220000	CAF2 P	
30	-88.767700	6.000000	11.370000	AIR	
31	-28.355900	7.500000	11.260000	CAF2 P	
32	-32.528600	200.020695 S	12.220000	AIR	
IMS	0.000000	0.000000	10.000000 S		

Figure 4.18 Optical design and graphs of aberration correction of a 250×/1.20 water immersion plan monochromatic objective (193.4 nm).

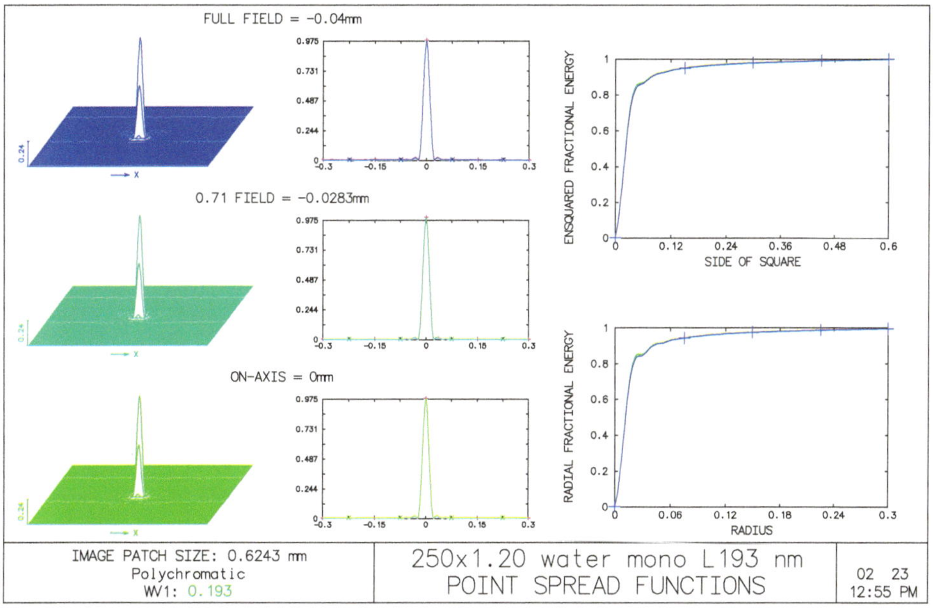

Figure 4.18 (*Continued*) Optical design and graphs of aberration correction of a 250x/1.20 water immersion plan monochromatic objective (193.4 nm).

and IR ranges and implements compensation due to the so-called *dark current* and the electrical noise of the matrix of the receiving device to correlate the parameters of the spectrum of radiation of the light source and the uneven sensitivity of the radiation receiver. The mathematical model of the proposed methodology allows us to implement color contrasting of the image based on the construction of a spectrum, which significantly increases the efficiency of research of low-contrast morphological objects. Also, the model for shifting and scaling the spectrum along the axis of wavelengths is useful for analyzing the spectral characteristics of the local properties of objects.

4.3.1.1 Plan multi-apochromatic lens objectives for the NUV–NIR (200–1800 nm) spectral range

Once again, we reiterate that it is the optical system, mainly the objective, that determines a microscope's functionality and power. Let us consider in more detail the issues related to creating objectives capable of ensuring that a microscope works in accordance with hyperspectral and multispectral research methods. The basics of the design of optical systems of such objectives are described in Chapter 3, Section 3.10.5; however, these are objectives with small linear magnifications and low numerical apertures, which are not always sufficient to solve the problems of microscopic research. Below are the optical designs of medium- and large-magnification multi-apochromatic objectives.

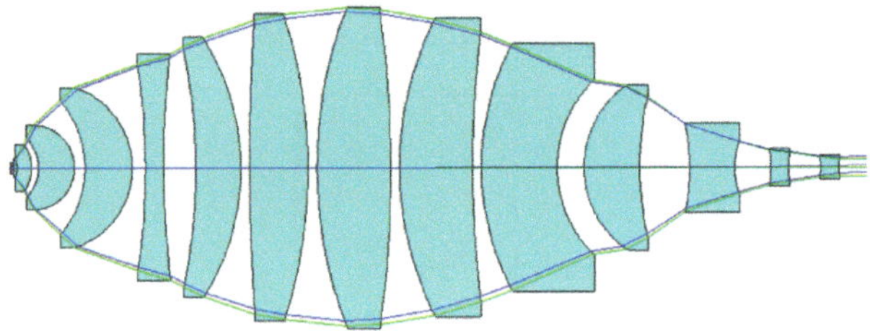

Lens: 250x1.20 water mono L244 nm

Object num aper 1.200000 Image height 10.000000 Primary wavln 0.244080

SRF	RADIUS		THICKNESS		APERTURE RADIUS		GLASS		SPECIAL
OBJ	0.000000		0.010000		0.040000		WATER		
AST	0.000000		0.350000		0.057562	AS	LIF		
2	0.000000		0.400042		0.610000		WATER	P	
3	-2.770000		1.600000		0.990000		QUARTZ		
4	-4.530000		0.800000		2.520000		AIR		
5	-4.530000		4.000000		3.190000		QUARTZ	P	
6	-4.530000		1.200000		4.530000		AIR		
7	-10.758300		5.400000		7.210000		QUARTZ	P	
8	-8.851500		1.500000		8.580000		AIR		
9	-73.238800		2.000000		11.150000		QUARTZ	P	
10	92.682600		3.600000		12.180000		AIR		
11	-59.747900		5.000000		12.950000		QUARTZ	P	
12	-25.524300		1.000000		13.940000		AIR		
13	209.841600		6.500000		15.850000		QUARTZ	P	
14	-52.686500		1.000000		16.460000		AIR		
15	47.416000		8.000000		17.240000		QUARTZ	P	
16	-120.606700		1.000000		17.040000		AIR		
17	33.951600		8.000000		16.020000		QUARTZ	P	
18	112.404200		1.000000		14.620000		AIR		
19	27.554400		8.400000		13.340000		QUARTZ	P	
20	12.736400		3.000000		9.440000		AIR		
21	10.504200		6.200000		8.870000		QUARTZ	P	
22	32.974200		5.500000		7.700000		AIR		
23	-24.276500		5.000000		4.800000		QUARTZ	P	
24	7.598700		4.000000		2.910000		AIR		
25	-16.832000		2.000000		2.040000		QUARTZ	P	
26	15.673100		3.500000		1.760000		AIR		
27	-13.097600		2.000000		1.310000		QUARTZ	P	
28	4.833000		200.000000		1.190000		AIR		
29	0.000000		8.500000		11.190000		QUARTZ	P	
30	-135.084200		4.000000		11.460000		AIR		
31	-34.586500		5.000000		11.480000		QUARTZ	P	
32	-36.706000		1.000000		12.070000		AIR		
33	0.000000		7.500000		12.100000		QUARTZ	P	
34	-326.896600		200.000729	S	12.140000		AIR		
IMS	0.000000		0.000000		10.000000	S			

Figure 4.19 Optical design and graphs of aberration correction of a 250×/1.20 water immersion plan monochromatic objective (244.08 nm).

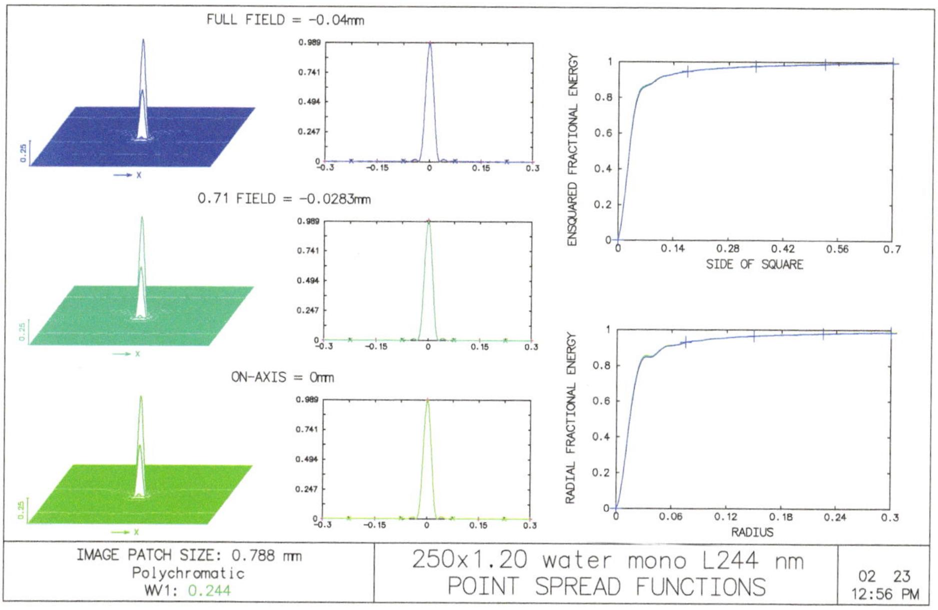

Figure 4.19 (*Continued*) Optical design and graphs of aberration correction of a 250×/1.20 water immersion plan monochromatic objective (244.08 nm).

Because of the extreme complexity of their optical systems, their design is a very difficult engineering task. These optical systems are tightly limited to include only lenses from quartz crystals, fluorite, and LiF located along the entire length of the objective. The parfocal height of these objectives differs from a standard value of 45 mm. Figures 4.34–4.36 show the design parameters and graphs of aberration correction of such objectives.

Calculating and optimizing the aberration characteristics of such objectives requires a certain heuristic algorithm; it is sometimes difficult to refer to a logical description and convincing formalization. If we try to formulate the sequence of actions when constructing the optical schemes of such objectives, we first need to admit that it can be arbitrary. This is also because the optical scheme of such an objective contains only single lenses. This, as practical experience shows, significantly complicates the entire process of optical design when choosing (or forming) the original optical system and during the mechanical process of achieving the required level of aberration correction. Much depends on the optical designer's intuition, experience, and even luck. Some will say that this resembles a modern trend in the use of so-called *freeform* optical surfaces, where the computer, namely, the optical designer's software, becomes a full co-author of the work. Moreover, the specialized software often becomes the main author, since the designer does not know what "free" form a particular optical surface will take in the calculations carried out by the

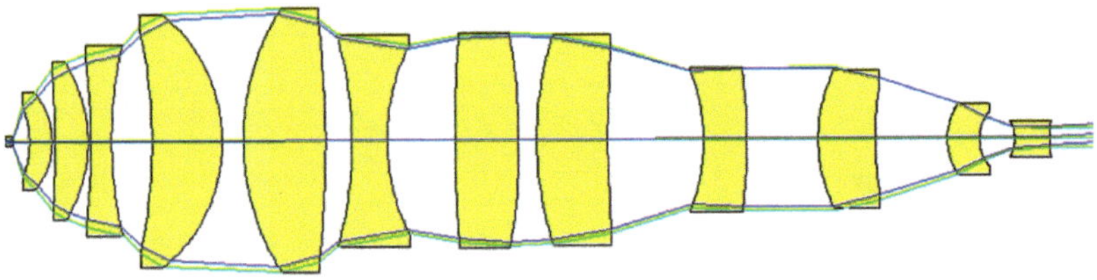

Lens: 350x0.95 mono L106 nm Fts200

Object num aper	0.950000	Image height	10.000000	Primary wavln	0.106000
SRF	**RADIUS**	**THICKNESS**	**APERTURE RADIUS**	**GLASS**	**SPECIAL**
OBJ	0.000000	0.010000	0.028570	AIR	
AST	0.000000	0.350000	0.058994 AS	LIF	
2	0.000000	0.969400	0.270000	AIR	
3	-4.801500	1.300000	1.950000	LIF P	
4	-3.911600	0.140000	2.670000	AIR	
5	-61.413000	2.000000	3.950000	LIF P	
6	-7.726500	0.140000	4.330000	AIR	
7	-43.778300	1.200000	4.740000	LIF P	
8	21.454700	2.300000	5.210000	AIR	
9	-28.078000	4.000000	6.180000	LIF P	
10	-8.664100	1.200000	6.890000	AIR	
11	13.284000	4.600000	7.200000	LIF P	
12	-56.708800	1.500000	6.630000	AIR	
13	-20.018000	2.000000	5.810000	LIF P	
14	11.347500	3.800000	5.190000	AIR	
15	90.005800	3.600000	5.680000	LIF P	
16	-32.727400	1.000000	5.870000	AIR	
17	18.039600	4.000000	5.760000	LIF P	
18	79.354200	5.400000	5.170000	AIR	
19	-10.400000	2.400000	3.840000	LIF P	
20	-44.602600	4.000000	3.950000	AIR	
21	8.441200	3.200000	3.780000	LIF P	
22	21.386900	4.000000	3.160000	AIR	
23	3.449400	1.800000	1.940000	LIF P	
24	2.290900	2.000000	1.500000	AIR	
25	-2.252100	2.000000	0.990000	LIF P	
26	7.428500	200.000000	0.930000	AIR	
27	917.862600	8.500000	11.010000	LIF P	
28	-47.589000	0.800000	11.160000	AIR	
29	-53.960700	5.000000	10.970000	LIF P	
30	-211.141600	6.000000	11.000000	AIR	
31	-31.102400	7.500000	10.820000	LIF P	
32	-37.411500	199.754858 S	11.980000	AIR	
IMS	0.000000	0.000000	10.000000 S		

Figure 4.20 Optical design and graphs of aberration correction of a 350×/0.95 plan monochromatic objective (106 nm).

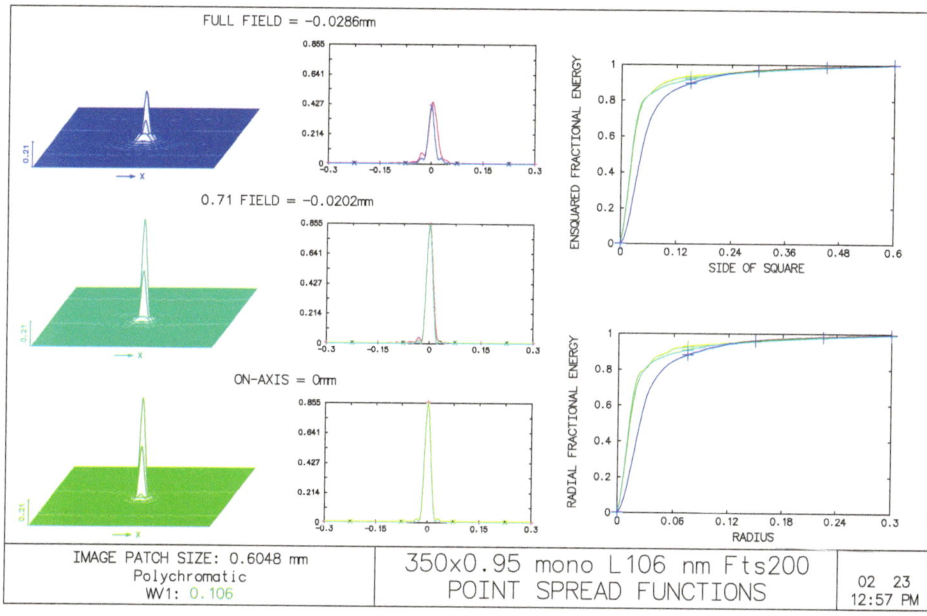

Figure 4.20 (*Continued*) Optical design and graphs of aberration correction of a 350×/0.95 plan monochromatic objective (106 nm).

software. Indeed, in our case, the layout of the optical system of the objective can be obtained by a similar method of determining the location of optical elements (lenses) within the framework of one optical scheme. However, when all the optical calculations have been completed and all the optical elements have taken their places, we find that the received optical systems turned out to be very logical and amenable to formal description. We can even talk about the peculiar beauty of such optical schemes.

Table 4.3 shows the main technical parameters and basic optical layout of some CCF plan multi-apochromatic objectives for the 200–1800-nm spectral range. This design is for an infinite length of the microscope tube, using an additional focusing system ($F_{ts} = 200$ mm). The values of the resolution of R and the depth of sharpness DF are presented for the main wavelength $\lambda = 546.07$ nm.

4.3.2 Optics of a microscope for working in the infrared range of the spectrum

A few decades ago, the infrared range was used quite rarely in studies using optical instruments, especially in microscopic applications. But modern photonics increasingly applies both classical and original methods of infrared visualization, particularly in applications such as data transmission, thermal visualization, and others. We will keep in mind that we are talking about a spectral range from approximately 0.75 to 15 μm. First, we must classify

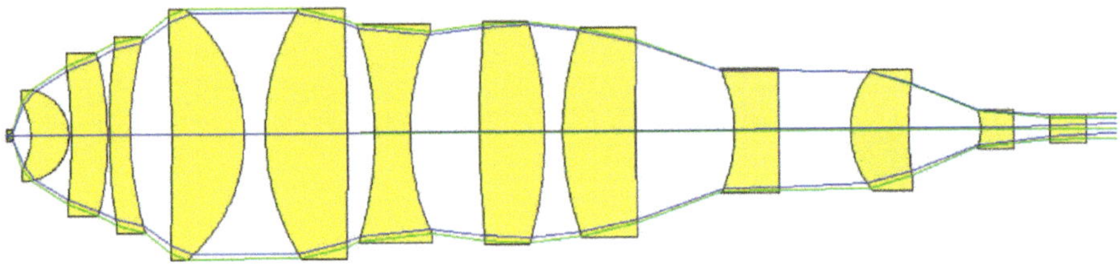

Lens: 350x0.95 mono L121.6 nm

Object num aper	0.950000	Image height	10.000000	Primary wavln	0.121600

SRF	RADIUS	THICKNESS	APERTURE RADIUS	GLASS	SPECIAL
OBJ	0.000000	0.010000	0.028573	AIR	
AST	0.000000	0.350000	0.058997 AS	LIF	
2	0.000000	0.969403	0.310000	AIR	
3	-3.693700	2.100000	1.820000	MGF2	
4	-2.463100	0.140000	2.440000	AIR	
5	-30.209200	2.000000	3.710000	MGF2 P	
6	-17.410500	0.140000	4.380000	AIR	
7	35.280200	1.200000	4.920000	MGF2 P	
8	17.595800	2.300000	5.290000	AIR	
9	-391.765400	4.000000	6.390000	MGF2 P	
10	-8.697900	1.200000	6.740000	AIR	
11	12.456900	4.600000	6.740000	MGF2 P	
12	-249.805800	1.500000	6.140000	AIR	
13	-18.503800	2.000000	5.870000	MGF2 P	
14	13.082000	3.800000	5.470000	AIR	
15	70.133100	3.600000	5.850000	MGF2 P	
16	-20.120200	1.000000	5.970000	AIR	
17	16.040900	4.000000	5.620000	MGF2 P	
18	388.504100	5.400000	4.970000	AIR	
19	-7.864900	2.400000	3.340000	MGF2 P	
20	0.000000	4.000000	3.310000	AIR	
21	5.077900	3.200000	3.250000	MGF2 P	
22	19.070300	4.000000	2.580000	AIR	
23	-2.903200	1.800000	1.040000	MGF2 P	
24	14.668400	2.000000	0.900000	AIR	
25	-5.896700	2.000000	0.740000	MGF2 P	
26	9.170300	200.000000	0.740000	AIR	
27	0.000000	8.500000	10.890000	LIF	
28	-40.878200	0.800000	11.110000	AIR	
29	-46.593700	5.000000	10.980000	LIF P	
30	-116.493200	5.000000	11.080000	AIR	
31	-30.875900	7.500000	10.960000	LIF P	
32	-36.882100	200.015456 S	11.890000	AIR	
IMS	0.000000	0.000000	10.000000 S		

Figure 4.21 Optical design and graphs of aberration correction of a 350×/0.95 plan monochromatic objective (121.6 nm).

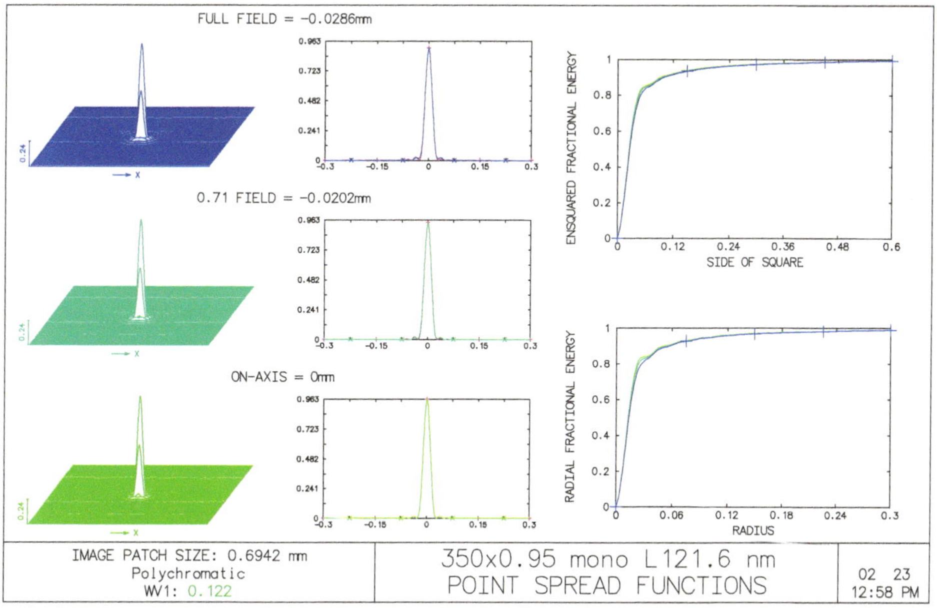

Figure 4.21 (*Continued*) Optical design and graphs of aberration correction of a 350×/0.95 plan monochromatic objective (121.6 nm).

spectral regions of the infrared spectrum. Although, of course, this is a rather conditional separation. Note that for biological microscopic studies, some restriction is associated with throughput in the IR range for water [16], as we keep in mind that many substances are 70–90% water. Perhaps the study of objects of inorganic origin in the infrared range is the most informative.

Thus, the first range, NIR, extends from about 750 nm to 1.1 μm. This range is well mastered by contemporary research methods under the microscope; many methods of such studies are described in the literature [17]. Today, even in a slightly wider range of up to 1.7 to 1.8 μm, microscopic techniques are represented, which, for example, allow blood vessels under a person's skin to be investigated and diagnosed (in medicine) or "crossed-out" typewritten or handwritten texts to be recognized (in forensics) [18]. The systems of objectives for research in metallography and microelectronics for working in the specified spectral range are discussed in Chapter 3, Section 3.10.

The spectral range with the conditional name short wave infrared (SWIR) extends from approximately 1 to 2.5 μm. Most of this range is not considered to be suitable for research (particularly in biological studies) because there is a rather intense absorption radiation of water. Nevertheless, studies have shown the presence of so-called *optical windows* in the ranges of 1100 to 1350 nm, 1600 to 1870 nm, and 2100 to 2350 nm [19]. This book contains many original discussions about the methods and advantages of research in the SWIR spectrum. Here, we are talking about the so-called *SWIR spectroscopy*

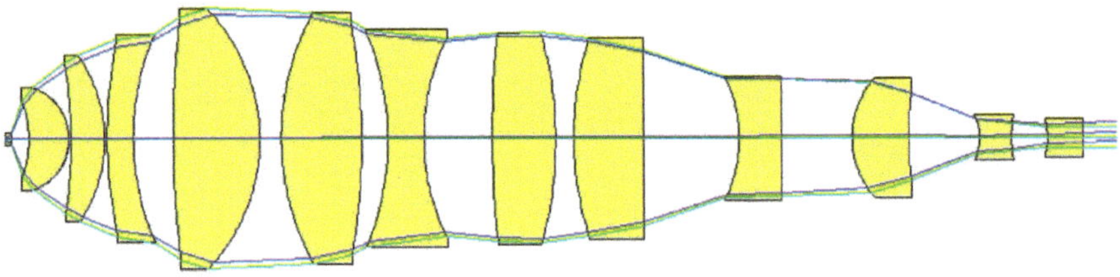

Lens: 350x0.95 mono L157 nm Fts200

Object num aper	0.950000	Image height	10.000000	Primary wavln	0.157000
SRF	**RADIUS**	**THICKNESS**	**APERTURE RADIUS**	**GLASS**	**SPECIAL**
OBJ	0.000000	0.010000	0.028572	AIR	
AST	0.000000	0.350000	0.058996 AS	LIF	
2	0.000000	0.969421	0.360000	AIR	
3	-6.225200	2.300000	2.150000	CAF2	
4	-2.946500	0.140000	2.830000	AIR	
5	-22.605400	2.000000	4.170000	CAF2 P	
6	-7.694100	0.140000	4.590000	AIR	
7	26.072000	1.500000	5.420000	CAF2 P	
8	13.787200	2.300000	5.730000	AIR	
9	56.698800	5.000000	6.680000	CAF2 P	
10	-9.902000	1.200000	7.170000	AIR	
11	13.966400	4.600000	6.920000	CAF2 P	
12	-34.271600	1.500000	6.440000	AIR	
13	-15.025900	2.000000	5.990000	CAF2 P	
14	12.292600	3.800000	5.480000	AIR	
15	49.883500	3.600000	5.780000	CAF2 P	
16	-21.471400	1.000000	5.840000	AIR	
17	17.449400	4.000000	5.510000	CAF2 P	
18	-336.692000	5.400000	4.900000	AIR	
19	-7.774700	2.400000	3.390000	CAF2 P	
20	-154.563200	4.000000	3.350000	AIR	
21	4.893100	3.200000	3.310000	CAF2 P	
22	18.463200	4.000000	2.700000	AIR	
23	-2.988500	1.800000	1.250000	CAF2 P	
24	7.639600	2.000000	1.100000	AIR	
25	-4.060500	2.000000	0.970000	CAF2 P	
26	17.172700	200.000000	1.040000	AIR	
27	0.000000	8.500000	11.170000	LIF P	
28	-40.966600	0.800000	11.400000	AIR	
29	-40.955400	5.000000	11.330000	LIF P	
30	-84.517300	5.000000	11.520000	AIR	
31	-34.552600	7.500000	11.460000	LIF P	
32	-38.247200	200.015577 S	12.220000	AIR	
IMS	0.000000	0.000000	10.000000 S		

Figure 4.22 Optical design and graphs of aberration correction of a 350×/0.95 plan monochromatic objective (157 nm).

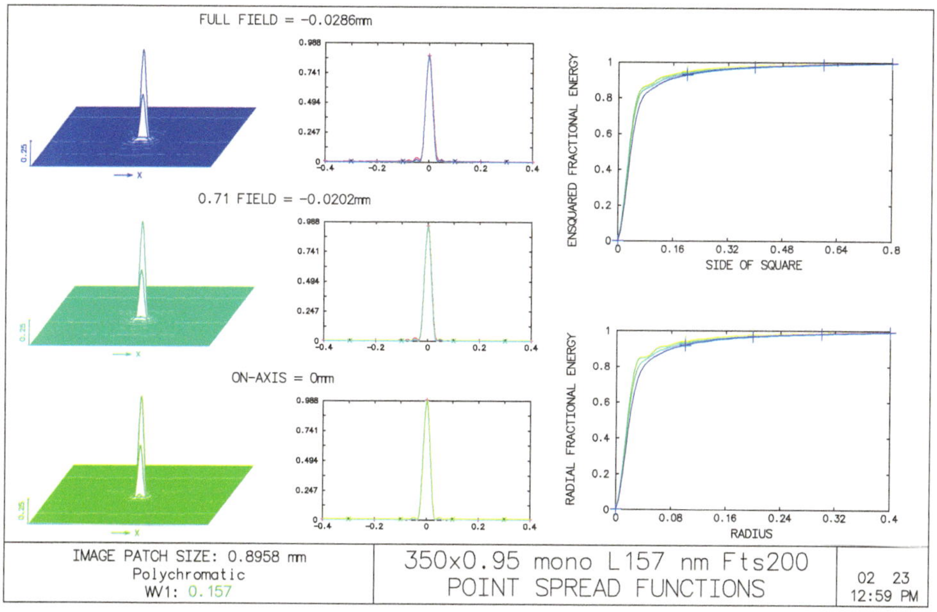

Figure 4.22 (*Continued*) Optical design and graphs of aberration correction of a 350×/0.95 plan monochromatic objective (157 nm).

methodology, which also has many modern applications, such as noninvasive and nondestructive control in the study of inorganic objects. Some methodologies in biological microscopy are based on the assumption that when a cell is exposed to infrared light, its temperature rises, which increases its geometric size. At the same time, various microscopic objects, as well as their areas, absorb light in different ways. In addition, areas of the object under study, warmer and colder, respond to infrared radiation in different ways. It is also true for the space inside the cell in which each type of molecule absorbs its infrared radiation. The study of absorption spectra allows researchers to accurately determine the internal structure of the studied object.

Similar reasoning can also be used by researchers when working on a microscope in the mid-wave infrared (MWIR) spectral range, which extends from about 3 to 8 μm. This thermal radiation can act as a catalyst for retaliatory radiation from the surface or internal structure of the object under study. If in normal practice the MWIR strip is used to check the temperature of the environment, then in the methodologies of microscopic studies, MWIR can become a tool for detecting temperature artifact and, accordingly, identification of the object. It must be kept in mind that the atmosphere (air) does not miss light in the range of 3 to 5 μm; however, it may be interesting here to look for any manifestations of unknown properties of the object under study (using, for example, some gas or special methods). In general, we can

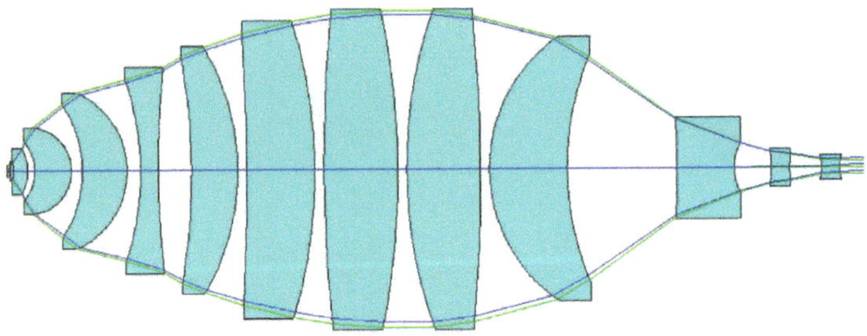

Lens: 350x1.20 water mono L193 nm

Object num aper	1.200000	Image height	10.000000	Primary wavln	0.193400

SRF	RADIUS	THICKNESS		APERTURE RADIUS		GLASS		SPECIAL
OBJ	0.000000	0.010000		0.028575		WATER		
AST	0.000000	0.350000		0.043651	AS	LIF		
2	0.000000	0.437222		0.580000		WATER	P	
3	-2.770000	1.600000		0.970000		SILICA		
4	-4.530000	0.800000		2.420000		AIR		
5	-4.530000	4.000000		3.080000		SILICA	P	
6	-4.530000	1.200000		4.520000		AIR		
7	-12.622800	4.800000		7.060000		SILICA	P	
8	-8.961100	1.500000		8.220000		AIR		
9	-29.310000	2.000000		9.630000		SILICA	P	
10	111.434900	3.600000		10.860000		AIR		
11	-66.168200	5.000000		12.010000		SILICA	P	
12	-25.079400	1.000000		13.050000		AIR		
13	-238.518800	7.500000		14.440000		SILICA	P	
14	-51.475200	1.000000		15.700000		AIR		
15	164.081500	8.000000		16.470000		SILICA	P	
16	-81.148500	1.000000		16.890000		AIR		
17	48.902200	8.000000		16.890000		SILICA	P	
18	-193.030900	1.000000		16.340000		AIR		
19	16.718700	8.400000		13.960000		SILICA	P	
20	29.957800	11.800000		11.940000		AIR		
21	-151.253600	6.200000		5.370000		SILICA	P	
22	5.240500	4.000000		2.650000		AIR		
23	-13.282500	2.000000		1.980000		SILICA	P	
24	37.442700	3.500000		1.780000		AIR		
25	-4.929400	2.000000		1.300000		SILICA	P	
26	4.738800	200.000000		1.190000		AIR		
27	0.000000	8.500000		11.110000		CAF2		
28	-35.576700	0.800000		11.330000		AIR		
29	-37.243300	5.000000		11.220000		CAF2	P	
30	-88.767700	6.000000		11.370000		AIR		
31	-28.355900	7.500000		11.260000		CAF2	P	
32	-32.528600	200.082368	S	12.220000		AIR		
IMS	0.000000	0.000000		10.000000	S			

Figure 4.23 Optical design and graphs of aberration correction of a 350×/1.20 water immersion plan monochromatic objective (193.4 nm).

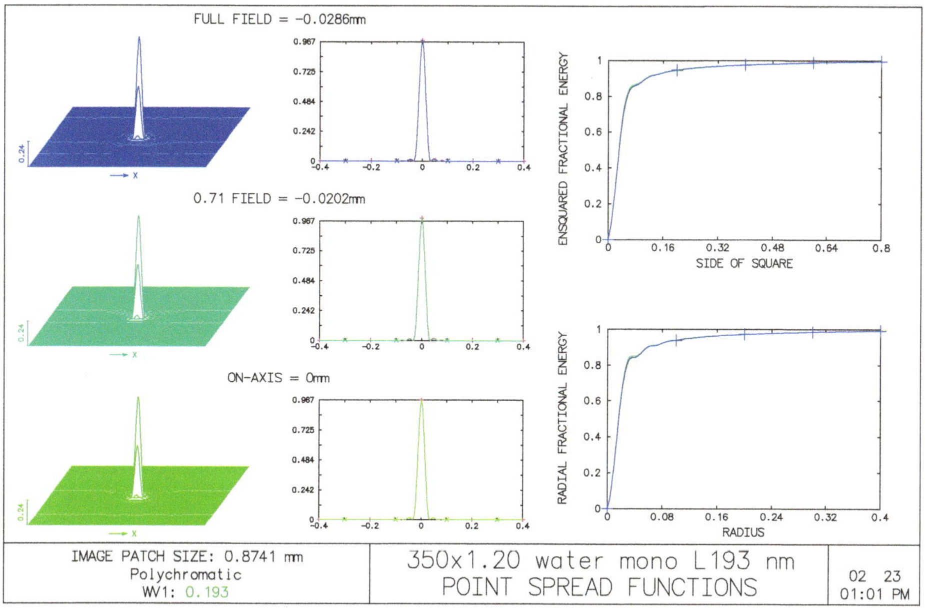

Figure 4.23 (*Continued*) Optical design and graphs of aberration correction of a 350×/1.20 water immersion plan monochromatic objective (193.4 nm).

say that this spectral range is successfully mastered in microscopy for biomedical and industrial applications [20].

The spectral range of long-wave infrared (LWIR) extends from 8 to 15 μm. One advantage when using this spectral range is that all kinds of "noises," interference, and other influences of the environment are the least prevalent here. Traditionally, the LWIR range is widely used in military applications. LWIR technologies are also used to view the earth's surface (trees) as well as detect moving objects (persons, animals). It can be assumed that some research methods in this spectral range will be able to successfully be used in light microscopy.

4.3.2.1 Plan poly-apochromatic objectives for the SWIR (1.1–2.4 μm) spectral range

Developing the optical design of such objectives does not deviate from the standard process. For the initial optical designs, we can choose fundamental solutions to the plan apochromatic objectives calculated for the visible spectral range, optimized with features for aberration correction for the SWIR range. We can use glued lenses, as well as fluorite lenses, which are necessary for correction of the secondary spectrum in objectives. Choosing optical materials that have good transmission in the SWIR region is not difficult. We can use conventional types of silicate glass, paying attention to the transmission characteristics, which are higher for some materials than for

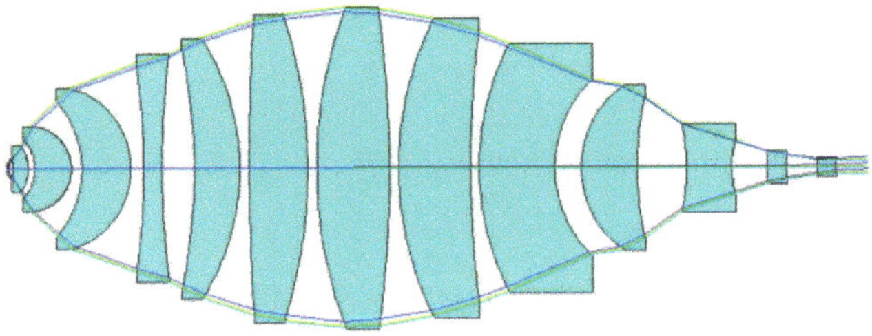

Lens: 350x1.20 water mono L244 nm

Object num aper	1.200000	Image height	10.000000	Primary wavln	0.244080

SRF	RADIUS	THICKNESS	APERTURE RADIUS	GLASS	SPECIAL
OBJ	0.000000	0.010000	0.028577	WATER	
AST	0.000000	0.350000	0.046139 AS	LIF	
2	0.000000	0.400036	0.600000	WATER P	
3	-2.770000	1.600000	0.980000	QUARTZ	
4	-4.530000	0.800000	2.520000	AIR	
5	-4.530000	4.000000	3.190000	QUARTZ P	
6	-4.530000	1.200000	4.530000	AIR	
7	-10.758300	5.400000	7.210000	QUARTZ P	
8	-8.851500	1.500000	8.580000	AIR	
9	-73.238800	2.000000	11.140000	QUARTZ P	
10	92.682600	3.600000	12.170000	AIR	
11	-59.747900	5.000000	12.930000	QUARTZ P	
12	-25.524300	1.000000	13.920000	AIR	
13	209.841600	6.500000	15.820000	QUARTZ P	
14	-52.686500	1.000000	16.420000	AIR	
15	47.416000	8.000000	17.200000	QUARTZ P	
16	-120.606700	1.000000	16.990000	AIR	
17	33.951600	8.000000	15.970000	QUARTZ P	
18	112.404200	1.000000	14.570000	AIR	
19	27.554400	8.400000	13.300000	QUARTZ P	
20	12.736400	3.000000	9.410000	AIR	
21	10.504200	6.200000	8.840000	QUARTZ P	
22	32.974200	5.500000	7.660000	AIR	
23	-24.205500	5.000000	4.760000	QUARTZ P	
24	10.426800	4.000000	2.920000	AIR	
25	-31.473200	2.000000	1.790000	QUARTZ P	
26	4.775800	3.500000	1.370000	AIR	
27	-6.677000	2.000000	0.950000	QUARTZ P	
28	4.379900	200.000000	0.870000	AIR	
29	0.000000	8.500000	10.900000	QUARTZ P	
30	-135.084200	4.000000	11.160000	AIR	
31	-34.586500	5.000000	11.180000	QUARTZ P	
32	-36.706000	1.000000	11.770000	AIR	
33	0.000000	7.500000	11.800000	QUARTZ P	
34	-326.896600	200.045276 S	11.840000	AIR	
IMS	0.000000	0.000000	10.000000 S		

Figure 4.24 Optical design and graphs of aberration correction of a 350x/1.20 water immersion plan monochromatic objective (244.08 nm).

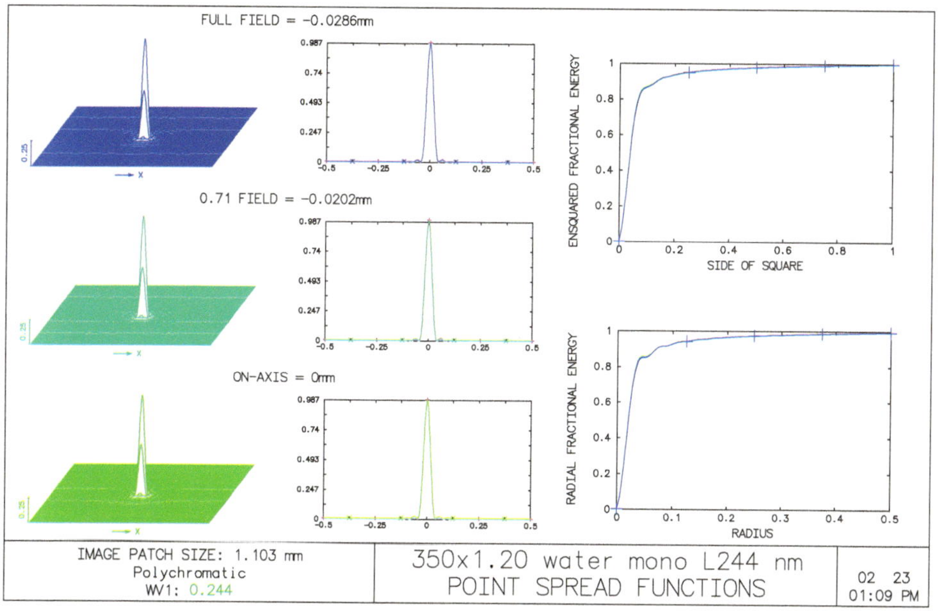

Figure 4.24 (*Continued*) Optical design and graphs of aberration correction of a 350×/1.20 water immersion plan monochromatic objective (244.08 nm).

others. Typically, such characteristics are presented in the appropriate catalogs of firms engaged in making glass. In the optical calculations of objectives given as examples, which demonstrate the fundamental possibility of creating such systems, optical materials from the Ohara Corporation were used. Figures 4.37–4.42 show example optical designs of such objectives (the principal optical schemes, design parameters, and graphs of aberration correction).

Table 4.4 shows the main technical parameters and basic optical layout of some SWIR CCF plan poly-apochromatic objectives. This design is for an infinite length of the microscope tube, using an additional focusing system ($F_{ts} = 200$ mm). The values of the resolution of R and the depth of sharpness DF are presented for the main wavelength $\lambda = 1.1$ μm.

4.3.2.2 Plan poly-apochromatic objectives for the MWIR (3–8 μm) spectral range

Figures 4.43–4.47 show example optical designs of such MWIR objectives (the principal optical schemes, design parameters, and graphs of aberration correction).

Table 4.5 shows the main technical parameters and basic optical layout of some MWIR CCF plan poly-apochromatic objectives. This design is for an infinite length of the microscope tube, using an additional focusing system ($F_{ts} = 200$ mm). The values of the resolution R and the depth of sharpness DF are presented for the main wavelength $\lambda = 5$ μm.

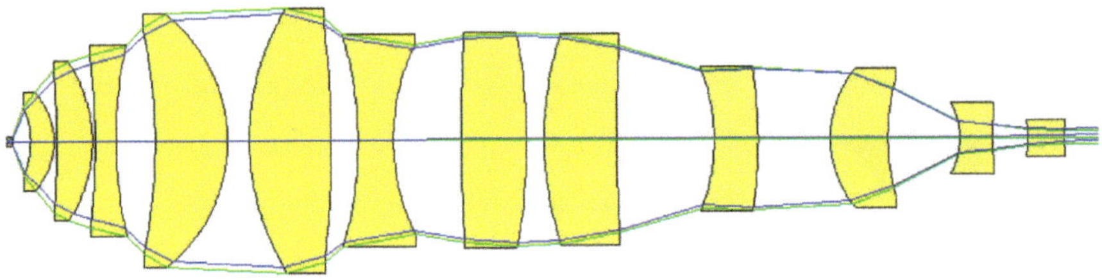

Lens: 500x0.95 mono L106 nm Fts200

Object num aper	0.950000	Image height	10.000000	Primary wavln	0.106000

SRF	RADIUS	THICKNESS	APERTURE RADIUS	GLASS	SPECIAL
OBJ	0.000000	0.010000	0.020012	AIR	
AST	0.000000	0.350000	0.050436 AS	LIF	
2	0.000000	0.969416	0.270000	AIR	
3	-4.801500	1.300000	1.950000	LIF P	
4	-3.911600	0.140000	2.670000	AIR	
5	-61.413000	2.000000	3.950000	LIF P	
6	-7.726500	0.140000	4.330000	AIR	
7	-43.778300	1.200000	4.740000	LIF P	
8	21.454700	2.300000	5.210000	AIR	
9	-28.078000	4.000000	6.180000	LIF P	
10	-8.664100	1.200000	6.890000	AIR	
11	13.284000	4.600000	7.200000	LIF P	
12	-56.708800	1.500000	6.630000	AIR	
13	-20.018000	2.000000	5.810000	LIF P	
14	11.347500	3.800000	5.190000	AIR	
15	90.005800	3.600000	5.680000	LIF P	
16	-32.727400	1.000000	5.870000	AIR	
17	18.039600	4.000000	5.760000	LIF P	
18	79.354200	5.400000	5.170000	AIR	
19	-10.496800	2.400000	3.840000	LIF P	
20	-26.888200	4.000000	3.950000	AIR	
21	5.859600	3.200000	3.780000	LIF P	
22	11.799200	4.000000	3.160000	AIR	
23	-3.896400	1.800000	1.940000	LIF P	
24	9.355000	2.000000	1.500000	AIR	
25	-4.805600	2.000000	0.990000	LIF P	
26	6.526100	200.000000	0.930000	AIR	
27	917.862600	8.500000	11.010000	LIF P	
28	-47.589000	0.800000	11.160000	AIR	
29	-53.960700	5.000000	10.970000	LIF P	
30	-211.141600	6.000000	11.000000	AIR	
31	-31.102400	7.500000	10.820000	LIF P	
32	-37.411500	200.063381 S	11.980000	AIR	
IMS	0.000000	0.000000	10.000000 S		

Figure 4.25 Optical design and graphs of aberration correction of a 500×/0.95 plan monochromatic objective (106 nm).

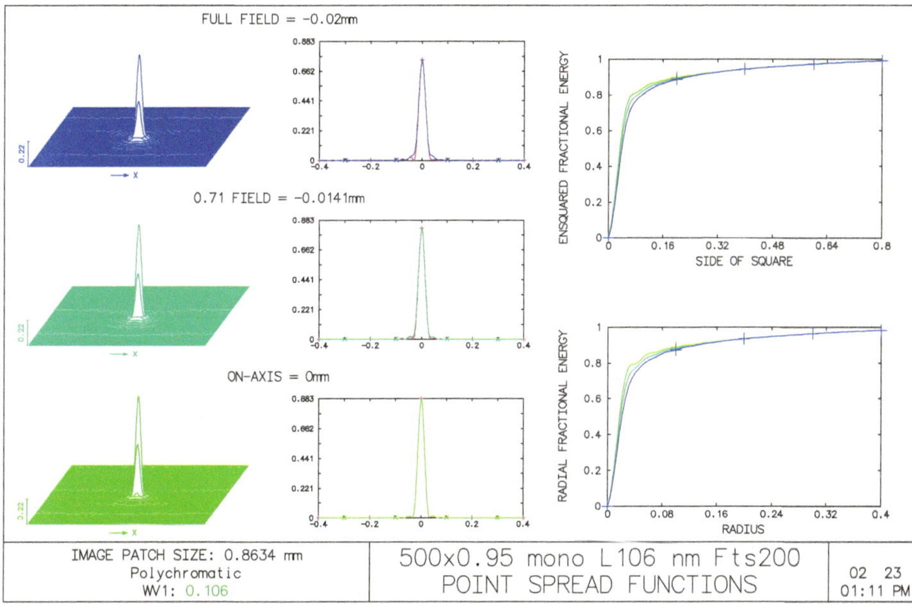

Figure 4.25 (*Continued*) Optical design and graphs of aberration correction of a 500×/0.95 plan monochromatic objective (106 nm).

4.3.2.3 Plan poly-apochromatic objectives for the LWIR (8–14 μm) spectral range

Figures 4.48–4.52 show example optical designs of such LWIR objectives (the principal optical schemes, design parameters, and graphs of aberration correction).

Table 4.6 shows the main technical parameters and basic optical layout of some LWIR CCF plan poly-apochromatic objectives. The parfocal distance is 95 mm. This design is for an infinite length of the microscope tube, using an additional focusing system ($F_{ts} = 200$ mm). The values of the resolution R and the depth of sharpness DF are presented for the main wavelength $\lambda = 10$ μm.

4.4 Concept and Design of an Underwater Microscope

One of the reasons my colleagues and I began developing a microscope for underwater research was a conversation we had with an interesting person. The person is *Natalia Ovodova*, a diver. She is a three-time world champion in deep diving, which she performs without the use of special equipment—the only method she uses is holding her breath. She talked about her diving experience, including how the illumination, water density, pressure, and more change. She described how a person changes when they are at sea depth.

The concept of developing a microscope for research in the aquatic environment has been proposed for situations in which studying living and inanimate, static, or moving microscopic objects in their natural environment

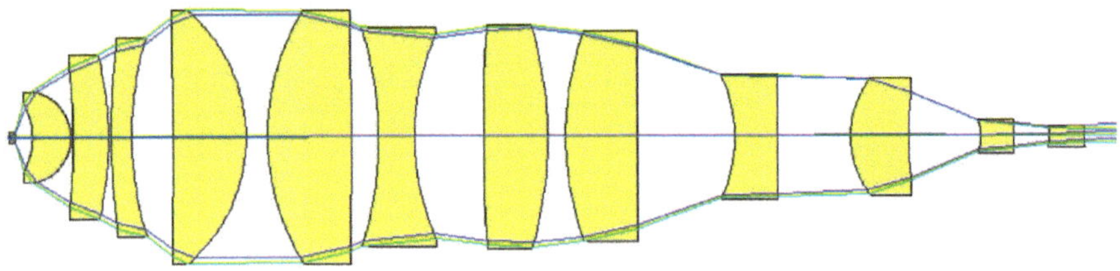

Lens: 500x0.95 mono L121.6 nm Fts200

Object num aper 0.950000 Image height 10.000000 Primary wavln 0.121600

SRF	RADIUS	THICKNESS	APERTURE RADIUS	GLASS	SPECIAL
OBJ	0.000000	0.010000	0.020001	AIR	
AST	0.000000	0.350000	0.050425 AS	LIF	
2	0.000000	0.969414	0.310000	AIR	
3	-3.693700	2.100000	1.810000	MGF2	
4	-2.463100	0.140000	2.440000	AIR	
5	-30.209200	2.000000	3.700000	MGF2 P	
6	-17.410500	0.140000	4.380000	AIR	
7	35.280200	1.200000	4.920000	MGF2 P	
8	17.595800	2.300000	5.290000	AIR	
9	-391.765400	4.000000	6.390000	MGF2 P	
10	-8.697900	1.200000	6.740000	AIR	
11	12.456900	4.600000	6.730000	MGF2 P	
12	-249.805800	1.500000	6.130000	AIR	
13	-18.503800	2.000000	5.860000	MGF2 P	
14	13.082000	3.800000	5.460000	AIR	
15	70.133100	3.600000	5.830000	MGF2 P	
16	-20.120200	1.000000	5.950000	AIR	
17	16.040900	4.000000	5.600000	MGF2 P	
18	388.504100	5.400000	4.950000	AIR	
19	-7.796100	2.400000	3.320000	MGF2 P	
20	-114.134600	4.000000	3.300000	AIR	
21	4.941300	3.200000	3.150000	MGF2 P	
22	16.546800	4.000000	2.440000	AIR	
23	-2.687000	1.800000	0.880000	MGF2 P	
24	6.581700	2.000000	0.690000	AIR	
25	-3.896600	2.000000	0.530000	MGF2 P	
26	7.243800	200.000000	0.540000	AIR	
27	0.000000	8.500000	10.640000	LIF	
28	-40.878200	0.800000	10.860000	AIR	
29	-46.593700	5.000000	10.740000	LIF P	
30	-116.493200	5.000000	10.840000	AIR	
31	-30.875900	7.500000	10.730000	LIF P	
32	-36.882100	199.999625 S	11.630000	AIR	
IMS	0.000000	0.000000	10.000000 S		

Figure 4.26 Optical design and graphs of aberration correction of a 500×/0.95 plan monochromatic objective (121.6 nm).

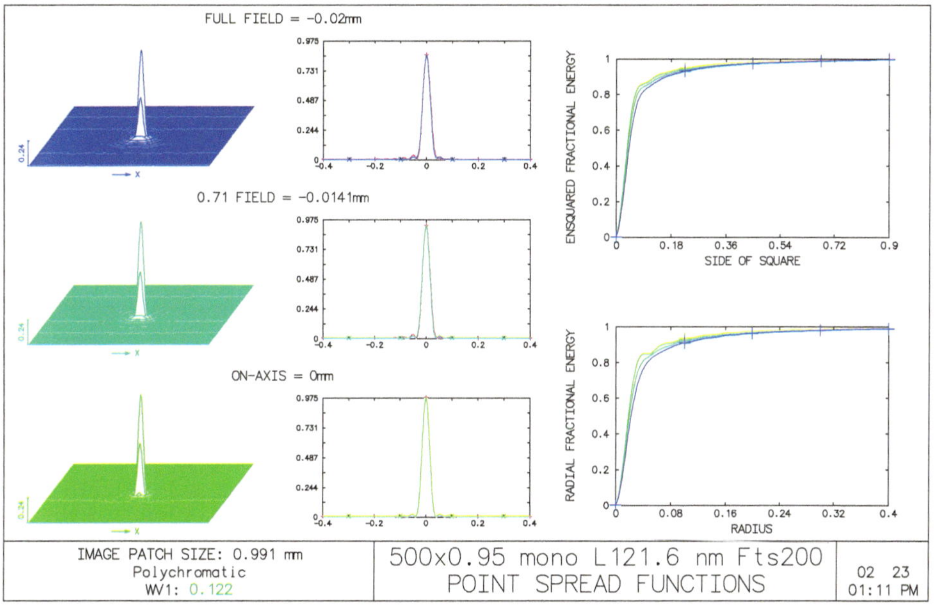

Figure 4.26 (*Continued*) Optical design and graphs of aberration correction of a 500×/0.95 plan monochromatic objective (121.6 nm).

is required. Like those of conventional microscopes, underwater microscopes use methods and technologies based on the theory of diffraction. The achievable resolution is up to 0.3–0.5 μm. The optical system of an underwater microscope should allow the study of microscopic objects in the water column (at different depths). For such microscopes, the only way to get images is to use a digital receiver. The objectives of these microscopes should work when immersed in water. The optical and mechanical design of the objective for an underwater microscope must be adapted to the conditions of immersion in the water column, including to a considerable depth.

My colleagues and I [21, 22] were among the first to propose the idea of creating an underwater microscope using modern optical and digital technologies. The goal was to study a wide range of microscopic objects at various depths underwater. We proposed using methods and technologies based on the theory of diffraction, with the achievable resolution values of 0.3–0.5 μm. Several research teams had the idea of an underwater microscope at the same time; for example, in 2016, results of a practical application of an underwater microscope to study coral properties were published [23].

For the study of microscopic objects in the aquatic environment, the principle of constructing an optical image as in a standard light microscope, where a congruent image of an object is constructed, can be used. In conventional microscopy, the objective and eyepiece are used as a projection system and as a surveillance system in conjunction with the human eye.

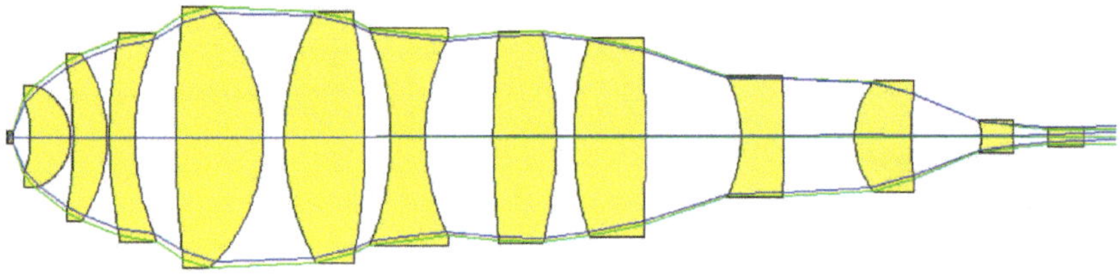

Lens: 500x0.95 mono L157 nm Fts200

Object num aper	0.950000	Image height	10.000000	Primary wavln	0.157000

SRF	RADIUS		THICKNESS		APERTURE RADIUS		GLASS		SPECIAL
OBJ	0.000000		0.010000		0.019999		AIR		
AST	0.000000		0.350000		0.050424	AS	LIF		
2	0.000000		0.969412		0.340000		AIR		
3	-6.225200		2.300000		2.140000		CAF2		
4	-2.946500		0.140000		2.820000		AIR		
5	-22.605400		2.000000		4.160000		CAF2	P	
6	-7.694100		0.140000		4.590000		AIR		
7	26.072000		1.500000		5.420000		CAF2	P	
8	13.787200		2.300000		5.730000		AIR		
9	56.698800		5.000000		6.670000		CAF2	P	
10	-9.902000		1.200000		7.160000		AIR		
11	13.966400		4.600000		6.900000		CAF2	P	
12	-34.271600		1.500000		6.420000		AIR		
13	-15.025900		2.000000		5.970000		CAF2	P	
14	12.292600		3.800000		5.460000		AIR		
15	49.883500		3.600000		5.750000		CAF2	P	
16	-21.471400		1.000000		5.800000		AIR		
17	17.449400		4.000000		5.470000		CAF2	P	
18	-336.692000		5.400000		4.860000		AIR		
19	-7.689200		2.400000		3.340000		CAF2	P	
20	-69.553800		4.000000		3.310000		AIR		
21	4.778200		3.200000		3.090000		CAF2	P	
22	16.793400		4.000000		2.400000		AIR		
23	-2.788600		1.800000		0.900000		CAF2	P	
24	4.760800		2.000000		0.690000		AIR		
25	-3.167900		2.000000		0.530000		CAF2	P	
26	7.851600		200.000000		0.540000		AIR		
27	0.000000		8.500000		10.600000		LIF	P	
28	-40.966600		0.800000		10.840000		AIR		
29	-40.955400		5.000000		10.770000		LIF	P	
30	-84.517300		5.000000		10.960000		AIR		
31	-34.552600		7.500000		10.900000		LIF	P	
32	-38.247200		199.908309	S	11.640000		AIR		
IMS	0.000000		0.000000		10.000000	S			

Figure 4.27 Optical design and graphs of aberration correction of a 500×/0.95 plan monochromatic objective (157 nm).

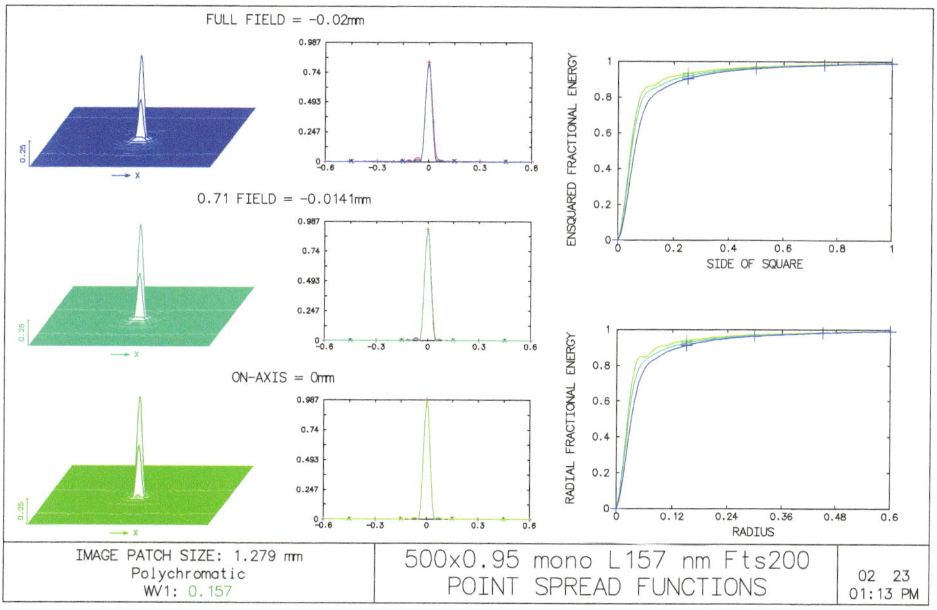

Figure 4.27 (*Continued*) Optical design and graphs of aberration correction of a 500×/0.95 plan monochromatic objective (157 nm).

However, it is impossible to use objectives and eyepieces from a standard light microscope when building a microscope for underwater research. Indeed, although a person can still be in relatively comfortable conditions at a depth of 5–10 m underwater, they cannot go below that depth without special diving equipment. The use of special diving equipment makes it almost impossible to visualize an image after a microscope with the help of a researcher's eye (even if a standard microscope could allow such an image to be obtained).

There can be some classification of underwater microscopes. One type could be an underwater microscope designed to work at depths of 5–10 m. In this case, the microscope could provide research with the participation of both a diver and autonomously. Another, more interesting type for scientific research could be a microscope that allows research to be conducted at considerable depths, from 100 m and below, including studies at a depth of several kilometers from the surface of the water.

There are fundamental differences between the way samples are examined under a microscope in the air (when the microscope is installed on a laboratory bench in a room where the temperature, humidity, atmospheric pressure, and other parameters are controlled) and in the water, especially when the microscope is immersed in water at a considerable depth. Under these conditions, the main factors determining the design of a microscope are the aquatic environment, parameters of water pressure at depth, spectral characteristics of light waves and scattering, and variable density of water.

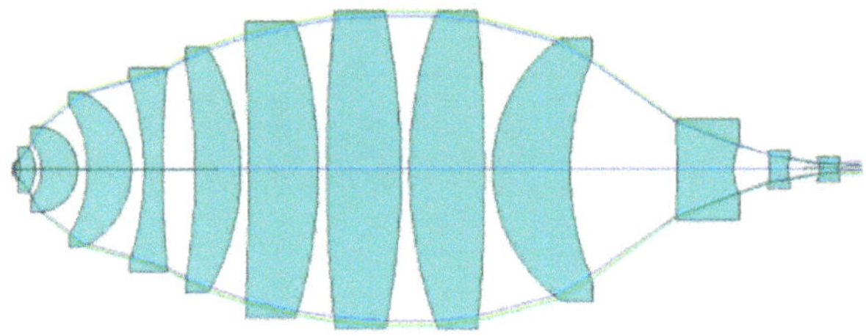

Lens: 500x1.20 water mono L193 nm

Object num aper	1.200000	Image height	10.000000	Primary wavln	0.193400

SRF	RADIUS	THICKNESS	APERTURE RADIUS	GLASS	SPECIAL
OBJ	0.000000	0.010000	0.020002	WATER	
AST	0.000000	0.350000	0.035077 AS	LIF	
2	0.000000	0.437219	0.580000	WATER P	
3	-2.770000	1.600000	0.970000	SILICA	
4	-4.530000	0.800000	2.420000	AIR	
5	-4.530000	4.000000	3.080000	SILICA P	
6	-4.530000	1.200000	4.520000	AIR	
7	-12.622800	4.800000	7.060000	SILICA P	
8	-8.961100	1.500000	8.220000	AIR	
9	-29.310000	2.000000	9.630000	SILICA P	
10	111.434900	3.600000	10.860000	AIR	
11	-66.168200	5.000000	12.010000	SILICA P	
12	-25.079400	1.000000	13.050000	AIR	
13	-238.518800	7.500000	14.440000	SILICA P	
14	-51.475200	1.000000	15.700000	AIR	
15	164.081500	8.000000	16.470000	SILICA P	
16	-81.148500	1.000000	16.890000	AIR	
17	48.902200	8.000000	16.890000	SILICA P	
18	-193.030900	1.000000	16.340000	AIR	
19	16.718700	8.400000	13.960000	SILICA P	
20	29.957800	11.800000	11.940000	AIR	
21	-158.474200	6.200000	5.370000	SILICA P	
22	6.156200	4.000000	2.650000	AIR	
23	-16.596700	2.000000	1.980000	SILICA P	
24	6.997100	3.500000	1.780000	AIR	
25	-2.857900	2.000000	1.300000	SILICA P	
26	4.771400	200.000000	1.190000	AIR	
27	0.000000	8.500000	11.110000	CAF2	
28	-35.576700	0.800000	11.330000	AIR	
29	-37.243300	5.000000	11.220000	CAF2 P	
30	-88.767700	6.000000	11.370000	AIR	
31	-28.355900	7.500000	11.260000	CAF2 P	
32	-32.528600	199.937041 S	12.220000	AIR	
IMS	0.000000	0.000000	10.000000 S		

Figure 4.28 Optical design and graphs of aberration correction of a 500×/1.20 water immersion plan monochromatic objective (193.4 nm).

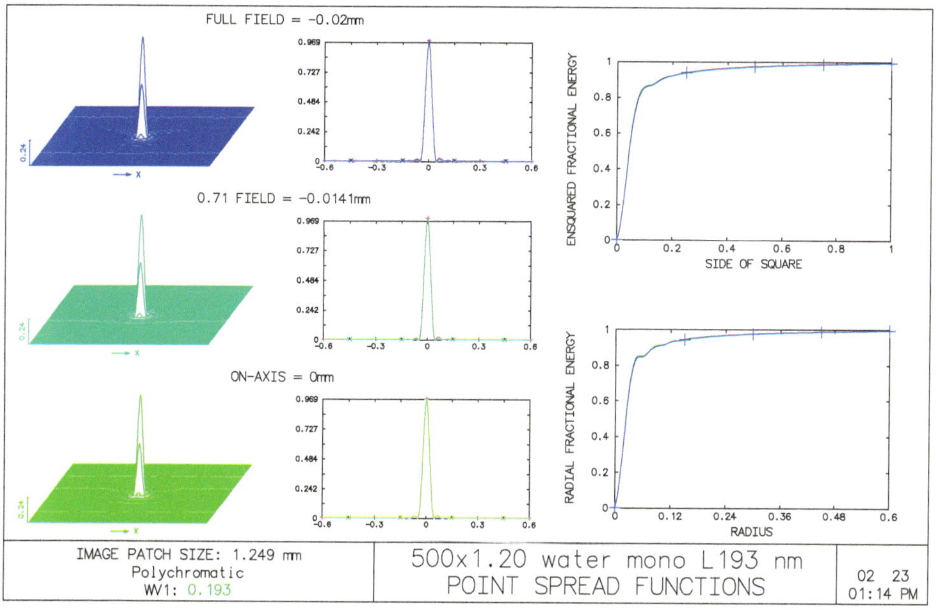

Figure 4.28 (*Continued*) Optical design and graphs of aberration correction of a 500×/1.20 water immersion plan monochromatic objective (193.4 nm).

The optical system of an underwater microscope should enable the study of microscopic objects in the water column at different depths. The only way to register images taken after the projection system (objective) in an underwater microscope is to use a digital receiver. At the same time, it is necessary to consider the parameters of the receiver used, such as the spectral sensitivity, the size of a single pixel, the transmission of contrast, and others. Of course, the parameters of the digital receiver are significantly different from the parameters of the human eye (with standard microscopy).

4.4.1 What parts of a conventional light microscope can be used to build an underwater microscope?

We will keep in mind that a functionally underwater microscope is no different from a conventional light microscope, which is usually located on a researcher's desk in a laboratory. When using an underwater microscope, as well as a conventional microscope, a researcher should be able to observe various microscopic objects whose images are formed by a projection system (an objective or an objective plus an additional projection tube system, using the so-called infinite optical length of the tube). The object to be examined must be located in the front focal plane of the microscope. More precisely, the objective should be focused on the object under study, and then the image is built in the so-called intermediate image plane, after which the resulting magnified image of

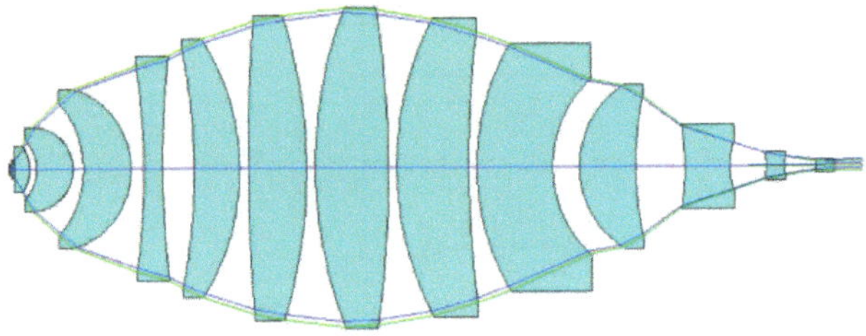

Lens: 500x1.20 water mono L244 nm

Object num aper 1.200000 Image height 10.000000 Primary wavln 0.244080

SRF	RADIUS	THICKNESS	APERTURE RADIUS	GLASS	SPECIAL
OBJ	0.000000	0.010000	0.020000	WATER	
AST	0.000000	0.350000	0.037562 AS	LIF	
2	0.000000	0.400039	0.590000	WATER P	
3	-2.770000	1.600000	0.980000	QUARTZ	
4	-4.530000	0.800000	2.510000	AIR	
5	-4.530000	4.000000	3.180000	QUARTZ P	
6	-4.530000	1.200000	4.530000	AIR	
7	-10.758300	5.400000	7.210000	QUARTZ P	
8	-8.851500	1.500000	8.580000	AIR	
9	-73.238800	2.000000	11.130000	QUARTZ P	
10	92.682600	3.600000	12.150000	AIR	
11	-59.747900	5.000000	12.920000	QUARTZ P	
12	-25.524300	1.000000	13.910000	AIR	
13	209.841600	6.500000	15.790000	QUARTZ P	
14	-52.686500	1.000000	16.400000	AIR	
15	47.416000	8.000000	17.170000	QUARTZ P	
16	-120.606700	1.000000	16.960000	AIR	
17	33.951600	8.000000	15.940000	QUARTZ P	
18	112.404200	1.000000	14.540000	AIR	
19	27.554400	8.400000	13.270000	QUARTZ P	
20	12.736400	3.000000	9.390000	AIR	
21	10.504200	6.200000	8.820000	QUARTZ P	
22	32.974200	5.500000	7.630000	AIR	
23	-24.218700	5.000000	4.730000	QUARTZ P	
24	9.757300	3.400000	2.890000	AIR	
25	-34.567000	2.000000	1.990000	QUARTZ P	
26	5.768000	7.000000	1.580000	AIR	
27	-4.957900	2.000000	0.720000	QUARTZ P	
28	3.314100	200.000000	0.640000	AIR	
29	0.000000	8.500000	10.720000	QUARTZ P	
30	-135.084200	4.000000	10.990000	AIR	
31	-34.586500	5.000000	11.010000	QUARTZ P	
32	-36.706000	1.000000	11.590000	AIR	
33	0.000000	7.500000	11.620000	QUARTZ P	
34	-326.896600	200.053414 S	11.670000	AIR	
IMS	0.000000	0.000000	10.000000 S		

Figure 4.29 Optical design and graphs of aberration correction of a 500×/1.20 water immersion plan monochromatic objective (244.08 nm).

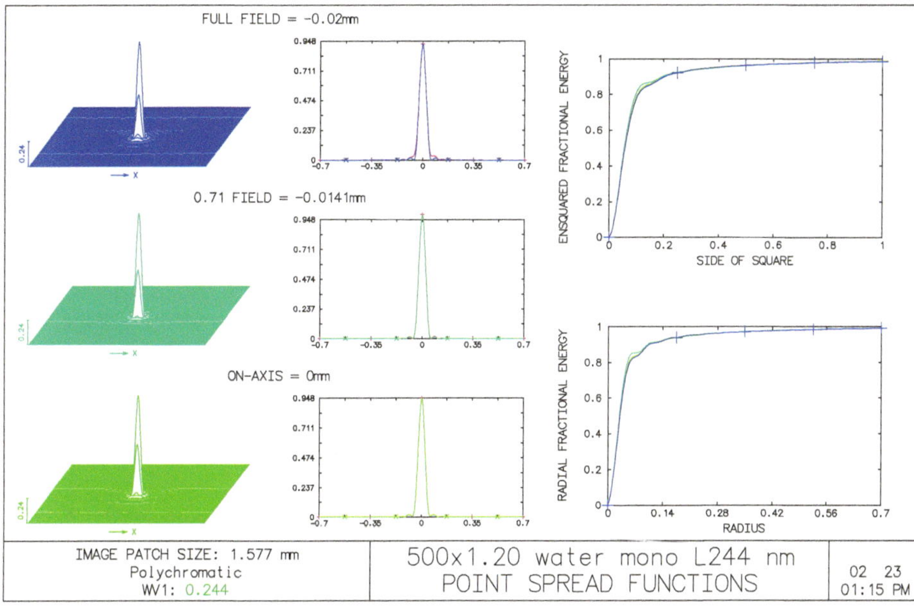

Figure 4.29 (*Continued*) Optical design and graphs of aberration correction of a 500×/1.20 water immersion plan monochromatic objective (244.08 nm).

the object under study is viewed with the help of the visualization system. Thus, functionally, the projection system of a conventional microscope and an underwater microscope may be the same. At the same time, the physical principles of obtaining images from a microscopic preparation remain unchanged, and the theory of diffraction determines the minimum size of the area under study; the optical invariant determines the relationship between the input and output parameters of the optical scheme of the projection system. The main technical parameters of the microscope are characteristics such as resolution (front numerical aperture of the objective), linear field on the object, and linear magnification of the system.

- In a conventional microscope, a digital image receiver is almost always used, although it is an alternative system; the main visualization system

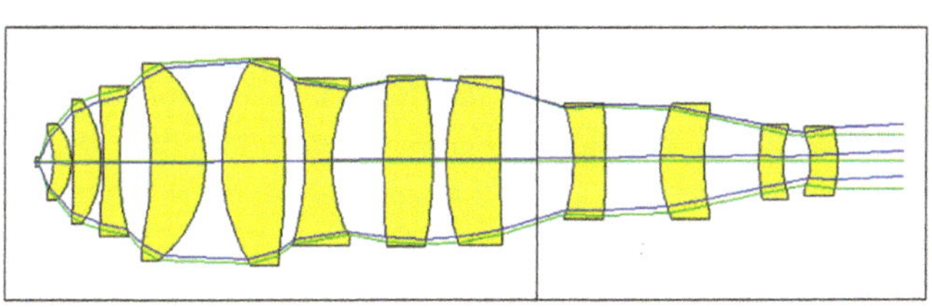

const var

Figure 4.30 Constant and variable parts of "dry" objectives (for working at the same wavelength).

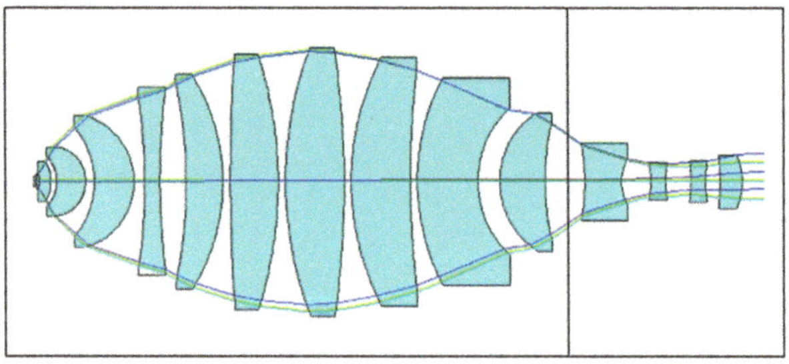

Figure 4.31 Constant and variable parts of immersion objectives (for working at the same wavelength).

remains the eyepiece and the eye of the researcher. For an underwater microscope, the use of a digital receiver as an element of the visualization system in most cases of underwater research of microscopic objects is the only option. In this sense, the technical parameters of the digital receiver are some of the main parameters when building an underwater microscope. First of all, these are such characteristics as the size of a single pixel, the effective spectral working range, and the linear size of the matrix. Let us keep in mind that the limiting resolution of the projection system of an underwater microscope must be 0.3–0.5 μm. To process such an image (taking into account the linear magnification of the projection system), a digital receiver is required that has a single pixel size of 1.3–1.5 μm. These are quite achievable parameters for modern digital cameras; however, such cameras are only suitable for operation in the spectral range of 400–700 nm. For spectral ranges such as the near UV and near IR ranges, the size of a single pixel (and, therefore, the values of the camera's own resolution) will be larger.

- In most cases, one of the elements of a conventional microscope's projection system is a lens objective that is designed to work in the air. Such objectives make it possible to obtain front numerical apertures up to 0.999 (theoretically and 0.95 practically), that is, resolution values on a microscope of 0.35 μm with a depth of field of 0.37 μm can be achieved. Similar objectives can theoretically be used when building an underwater microscope. In this case, the lens is located in a waterproof box (similar to boxes used for underwater photography). However, taking into account the thickness of the protective glass of the waterproof box, the quality of the aberration correction can be a serious obstacle when using objectives whose numerical aperture exceeds, for example, 0.15–0.20. Accordingly, the resolution for an underwater microscope using a standard microscope objective is limited to 1.67–2.2 μm. Using such standard objectives for standard microscopes, it is not necessary to talk about achieving the resolution of an underwater microscope of 0.3–0.5 μm.

Table 4.2 Main technical parameters and basic optical layout of NUV and DUV objectives.

Magnification/*F′*	*NA*	Optical material	λ (nm)	*WD* (mm)	*R* (nm)	*FOV* on object (mm)	*FOV* on image (mm)	Optical design at figure no.	The principal optical layout
100×/2.0	0.90	Quartz	244.08	0.22	165	0.20	20	4.16	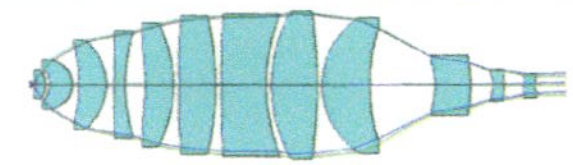
100×/2.0	0.95	LiF	106	0.40	68	0.20	20	4.17	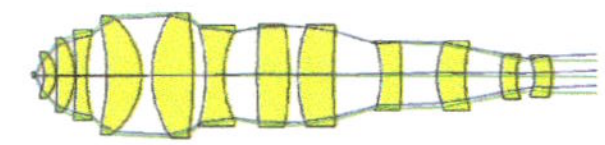
100×/2.0	0.95	MgF2	121.6	0.34	78	0.20	20	4.18	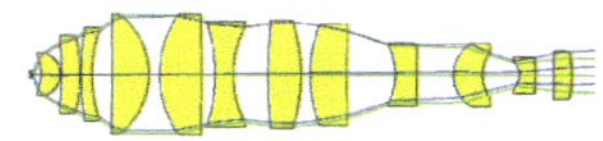
100×/2.0	0.95	CaF2	157	0.46	101	0.20	20	4.19	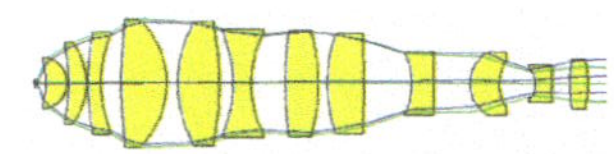
100×/2.0	1.20 water	Quartz	193.4	0.17	98	0.20	20	4.20	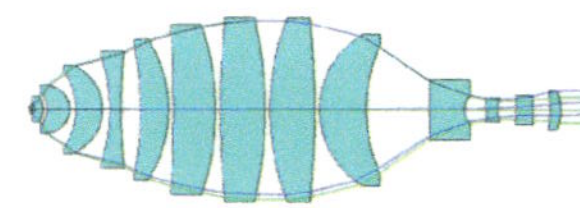
100×/2.0	1.20 water	Quartz	244.08	0.12	124	0.20	20	4.21	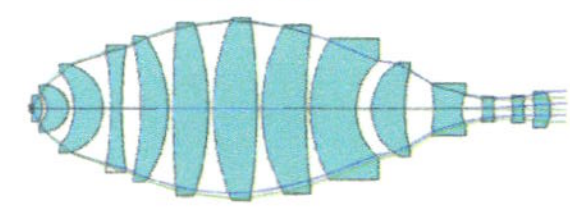
250×/0.8	0.95	LiF	106	0.40	68	0.08	20	4.22	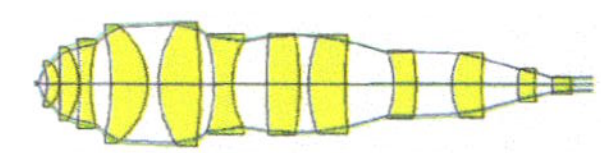
250×/0.8	0.95	MgF2	121.6	0.34	78	0.08	20	4.23	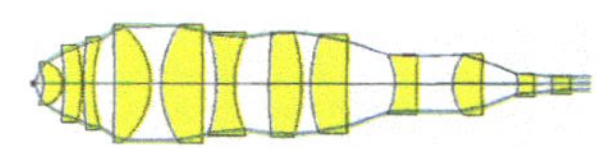

(*continued*)

Table 4.2 (*Continued*) Main technical parameters and basic optical layout of NUV and DUV objectives.

Magnification/F'	NA	Optical material	λ (nm)	WD (mm)	R (nm)	FOV on object (mm)	FOV on image (mm)	Optical design at figure no.	The principal optical layout
250×/0.8	0.95	CaF2	157	0.46	101	0.08	20	4.24	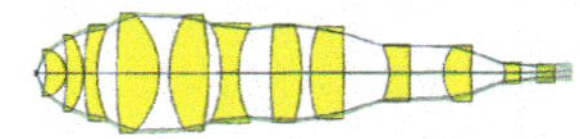
250×/0.8	1.20 water	Quartz	193.4	0.17	98	0.08	20	4.25	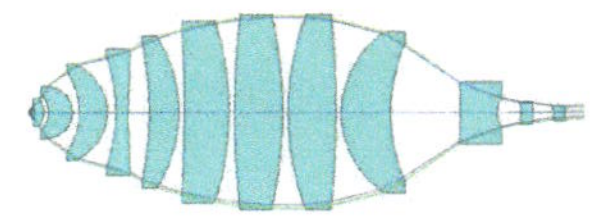
250×/0.80	1.20 water	Quartz	244.08	0.12	124	0.08	20	4.26	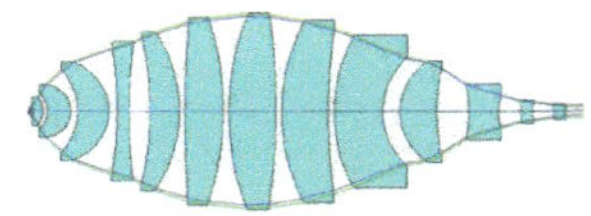
350×/0.57	0.95	LiF	106	0.40	68	0.057	20	4.27	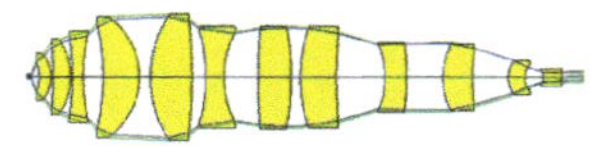
350×/0.57	0.95	MgF2	121.6	0.34	78	0.057	20	4.28	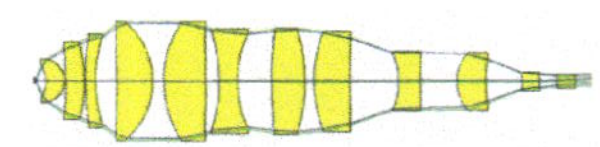
350×/0.57	0.95	CaF2	157	0.46	101	0.057	20	4.29	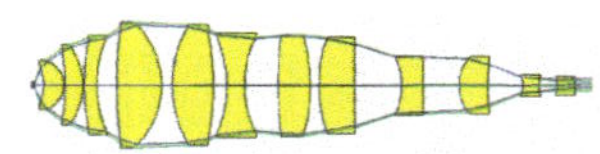
350×/0.57	1.20 water	Quartz	193.4	0.17	98	0.057	20	4.30	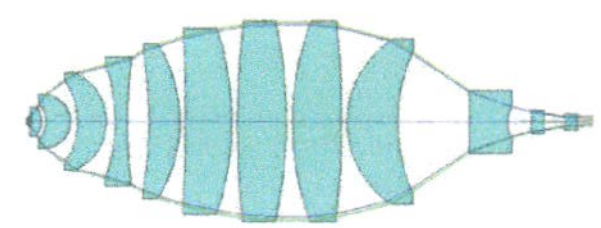
350×/0.57	1.20 water	Quartz	244.08	0.12	124	0.057	20	4.31	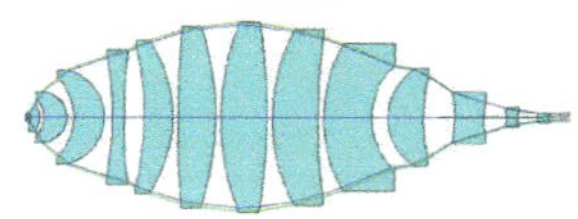

500×/0.4	0.95	LiF	106	0.40	68	0.04	20	4.32	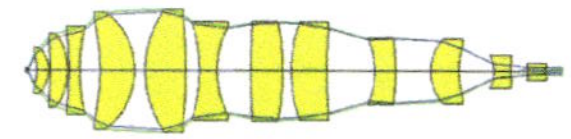
500×/0.40	0.95	MgF2	121.6	0.34	78	0.04	20	4.33	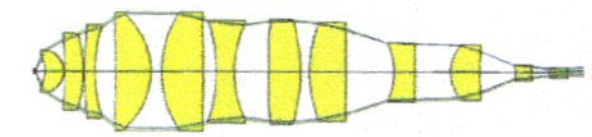
500×/0.40	0.95	CaF2	157	0.46	101	0.04	20	4.34	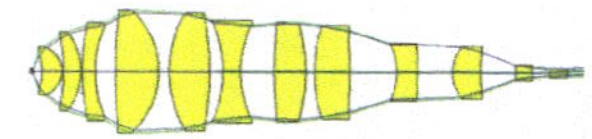
500×/0.4	1.20 water	Quartz	193.4	0.17	98	0.04	20	4.35	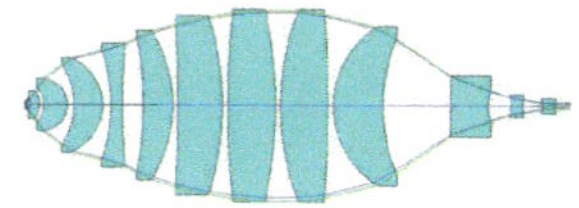
500×/0.4	1.20 water	Quartz	244.08	0.12	124	0.04	20	4.36	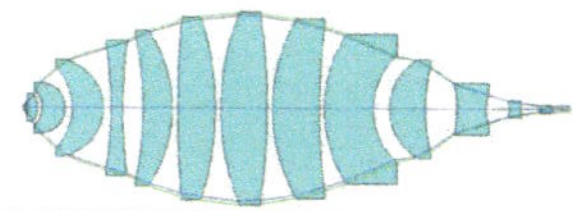

- A conventional microscope stand can be used as a base for an underwater microscope. If the underwater microscope does not need to reach the limiting values of numerical apertures and, accordingly, resolution, we can use the stand of a conventional microscope as a base. The essence and principle of operation of such a tripod comes down to its functional ability to ensure both the stability of the underwater microscope and the focusing of the projection system (objective) on the object under study. However, the practical feasibility of using such a tripod in an underwater microscope is limited to examining only fairly coarse objects, usually in a

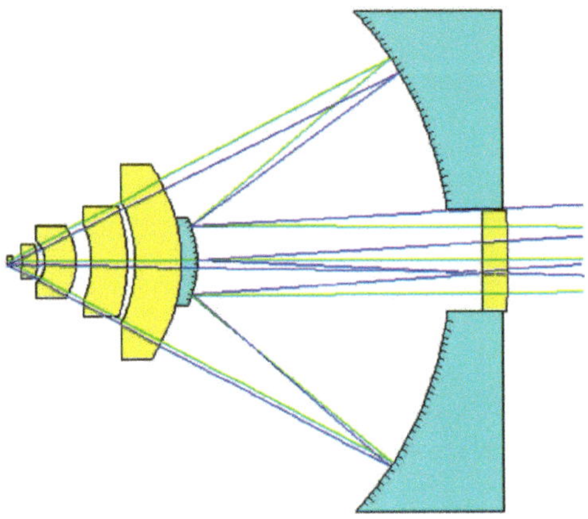

Lens:MultiApo 40x0.50 tube160 h45 LIF Efl 4.63

Object num aper 0.500000 Image height 10.000000 Primary wavln 0.546070

SRF	RADIUS	THICKNESS		APERTURE RADIUS		GLASS		SPECIAL
OBJ	0.000000	0.010000		0.249941		AIR		
AST	0.000000	0.350000		0.255714	AS	LIF		
2	0.000000	0.600000		0.320000		AIR		
3	-5.835300	1.000000		0.620000		LIF	P	
4	-2.865000	0.500000		1.050000		AIR		
5	-1.766900	2.000000		1.330000		LIF	P	
6	-4.015200	1.000000		2.210000		AIR		
7	-5.804000	2.300000		2.330000		LIF	P	
8	-10.254400	0.500000		3.380000		AIR		
9	-8.556400	3.000000		3.770000		LIF	P	
10	-10.154600	17.000000		5.990000		AIR		
11	-22.778500	-16.000000		15.360000		REFLECT		
12	-6.371200	18.000000		2.700000		REFLECT		
13	-104.288200	1.600000		3.100000		LIF	P	
14	-51.475000	163.405063	S	2.780000		AIR		
15	0.000000	0.000000	S	10.000000	S	AIR		
IMS	0.000000	0.000000		10.000000	S			

Figure 4.32 Optical design and graphs of aberration correction of a 40×/0.50 multi-apochromatic catadioptric objective (106–1000 nm).

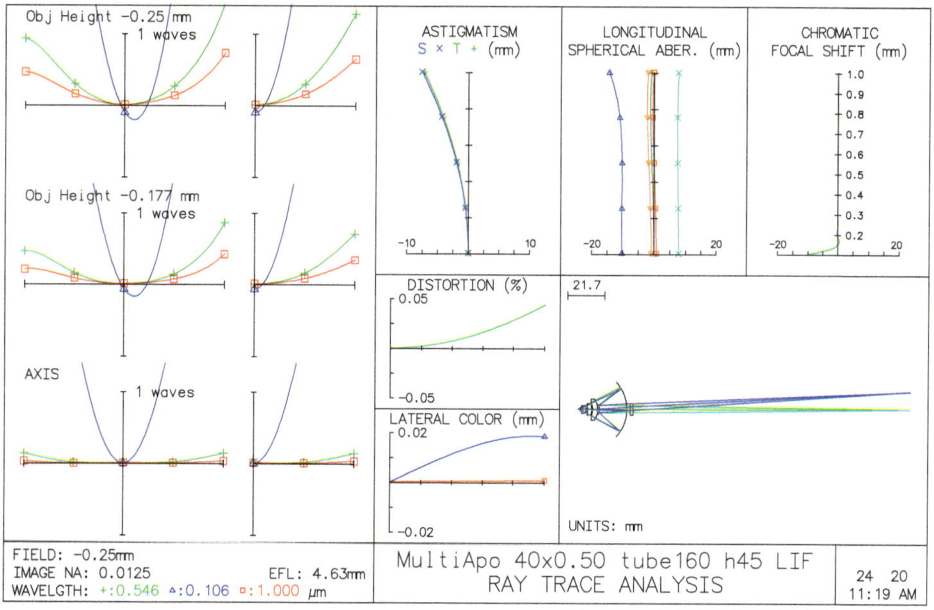

Figure 4.32 (*Continued*) Optical design and graphs of aberration correction of a 40×/0.50 multi-apochromatic catadioptric objective (106–1000 nm).

static state, which is quite rare among microscopic underwater objects. In fact, such a stand can be used in conjunction with lenses in which the numerical aperture values achieved are not more than 0.15–0.20.

- A conventional microscope illumination system can be used as a base system for an underwater microscope. However, among the types of lighting systems (the so-called *transmitted light*) of a conventional microscope, a critical lighting system should be chosen in which the light source is projected directly onto the object under study. Indeed, it is difficult to imagine an effective lighting system, for example, according to Köhler, when its implementation requires the passage of

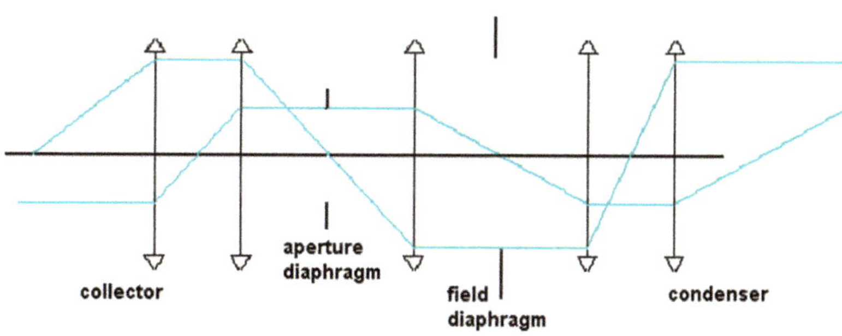

Figure 4.33 The principal layout of an example telecentric illuminator.

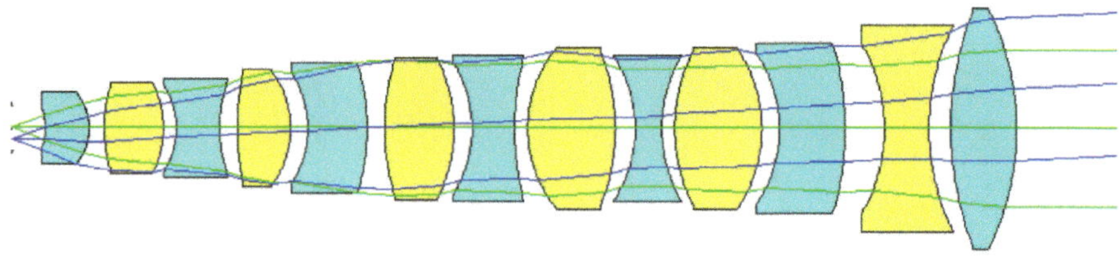

Lens:PlMultiApo20x0.35 200-1800Ftl200

Object num aper	0.350000	Image height	10.000000	Primary wavln	0.546070

SRF	RADIUS	THICKNESS	APERTURE RADIUS	GLASS	SPECIAL
OBJ	0.000000	0.001000	0.499699	AIR	
AST	0.000000	1.500000	0.990000 A	AIR	
2	-2.613000	2.000000	0.990000	SILICA	
3	-2.453800	0.700000	1.600000	AIR	
4	6.363100	2.800000	1.930000	CAF2	
5	-4.476200	0.600000	2.020000	AIR	
6	-3.556300	2.000000	1.930000	SILICA P	
7	6.502200	0.800000	2.200000	AIR	
8	11.375700	2.500000	2.420000	CAF2 P	
9	-4.162900	0.800000	2.610000	AIR	
10	-4.553900	2.700000	2.490000	SILICA P	
11	-9.535400	0.800000	2.920000	AIR	
12	9.749800	3.200000	3.140000	LIF	
13	-6.958500	0.800000	3.160000	AIR	
14	-5.647000	1.900000	3.030000	SILICA P	
15	12.434600	0.600000	3.240000	AIR	
16	5.467600	4.000000	3.590000	LIF P	
17	-9.106200	0.900000	3.410000	AIR	
18	-5.900500	1.200000	3.220000	SILICA P	
19	6.290800	0.600000	3.270000	AIR	
20	7.397800	4.000000	3.450000	LIF P	
21	-6.281200	0.800000	3.590000	AIR	
22	-6.066600	3.000000	3.430000	SILICA P	
23	-12.606900	1.800000	3.790000	AIR	
24	-7.477000	2.000000	3.820000	LIF P	
25	10.086100	1.000000	4.620000	AIR	
26	14.189000	3.000000	5.230000	SILICA P	
27	-11.888900	200.000000	5.370000	AIR	
28	-367.600000	5.000000	15.220000	SILICA P	
29	-97.660000	1.000000	15.410000	AIR	
30	-78.686000	7.000000	15.410000	LIF P	
31	-30.700000	2.000000	15.630000	AIR	
32	-28.980000	4.000000	15.300000	SILICA P	
33	0.000000	3.000000	16.040000	AIR	
34	68.137000	7.000000	16.710000	CAF2 P	
35	-153.581000	199.566473 S	16.760000	AIR	
IMS	0.000000	0.000000	10.000000		

Figure 4.34 Optical design and graphs of aberration correction of a 20×/0.35 CCF plan multi-apochromatic objective (200–1800 nm; $F'_{ts} = 200$ mm).

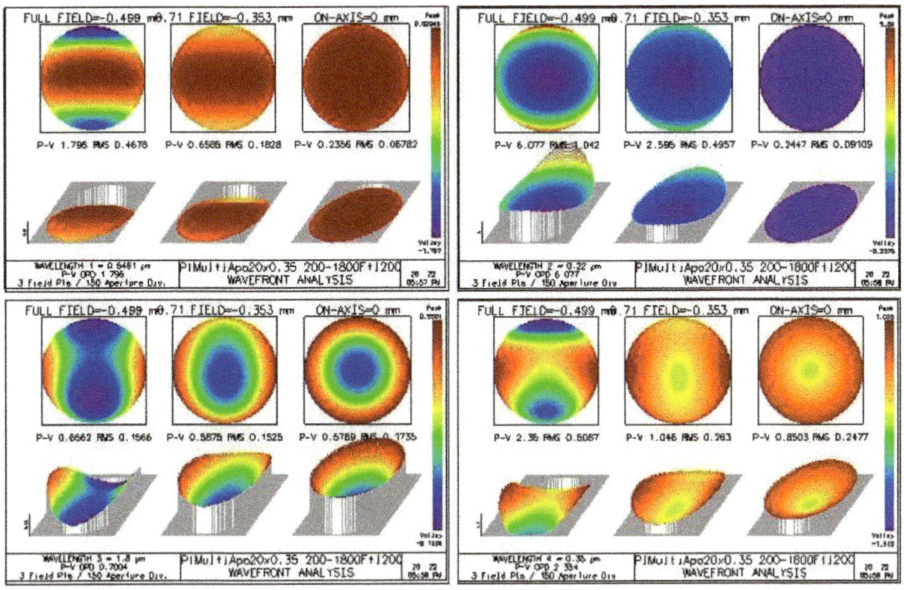

Figure 4.34 (*Continued*) Optical design and graphs of aberration correction of a 20×/0.35 CCF plan multi-apochromatic objective (200–1800 nm; $F'_{ts} = 200$ mm).

rays through a certain layer of water; most likely, a layer of water will cause the scattering of light rays and will not allow effective illumination of the object under study. To ensure the illumination of objects with a critical method, the use of a conventional microscope's illumination system may be sufficient, especially if the optical system of the underwater microscope uses objectives with a front numerical aperture of no more than 0.15–0.20. In the study of opaque underwater microscopic objects on reflection, simple lighting systems of sliding, oblique lighting (in the form of external spotlights), or systems built using the method of direct incident lighting through a lens, can be quite effective. There can also be an effective method of illumination, called *dark-field* in classical microscopy, in which lighting beams propagate parallel to the optical system of the objective and, like the objective, focus on the object under study using a mirror.

4.4.2 Some average water parameters

Certain properties of water must be considered when designing optical systems for the aquatic environment, especially ones meant to operate at a considerable depth. The study of microscopic objects at depths of more than 5–10 m is a complex scientific and technical problem whose solution is determined primarily by the current state and level of technological

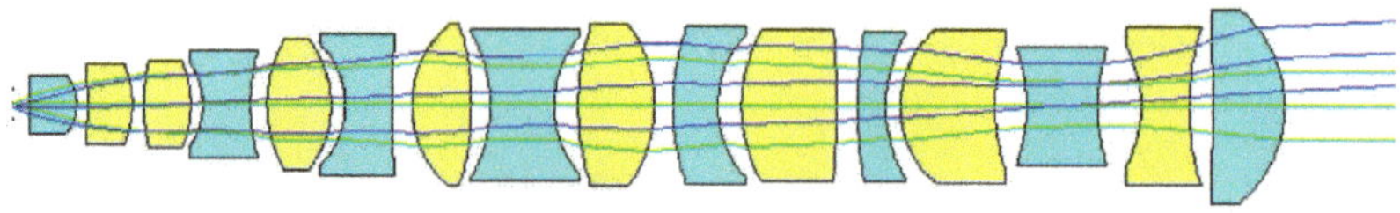

Lens:PlMultiApo50x0.50 200-1800Ftl200

Object num aper	0.500000	Image height	9.000000	Primary wavln	0.546070

SRF	RADIUS	THICKNESS	APERTURE RADIUS	GLASS	SPECIAL
OBJ	0.000000	1.0000e-05	0.179952	AIR	
AST	0.000000	0.700000	0.580000 A	AIR	
2	0.000000	2.000000	0.580000	SILICA	
3	-1.793200	0.500000	1.170000	AIR	
4	-10.911900	2.000000	1.330000	LIF	
5	-4.224700	0.500000	1.620000	AIR	
6	7.047700	2.000000	1.710000	LIF P	
7	-4.410500	0.500000	1.770000	AIR	
8	-3.036600	2.000000	1.740000	SILICA P	
9	7.681200	0.700000	2.180000	AIR	
10	5.036300	2.800000	2.610000	CAF2	
11	-4.123500	0.600000	2.660000	AIR	
12	-3.395300	2.000000	2.480000	SILICA P	
13	71.764000	0.800000	2.870000	AIR	
14	4.349900	2.500000	3.330000	LIF P	
15	-10.916300	0.800000	3.250000	AIR	
16	-5.831000	2.700000	3.070000	SILICA P	
17	4.370700	1.000000	2.830000	AIR	
18	10.477800	3.200000	3.010000	LIF	
19	-5.475100	0.800000	3.300000	AIR	
20	9.115600	1.900000	3.240000	SILICA P	
21	4.203100	1.000000	2.910000	AIR	
22	6.047800	4.000000	3.090000	LIF P	
23	-34.592400	0.900000	3.100000	AIR	
24	16.135800	1.200000	3.060000	SILICA P	
25	5.718000	0.600000	2.930000	AIR	
26	4.003300	4.000000	3.140000	LIF P	
27	8.518800	1.300000	2.500000	AIR	
28	-7.401300	3.000000	2.410000	SILICA P	
29	7.739100	1.800000	2.330000	AIR	
30	-5.157600	2.000000	2.430000	LIF P	
31	10.682600	1.000000	3.220000	AIR	
32	-106.145600	3.000000	3.520000	SILICA P	
33	-5.068600	200.000000	3.970000	AIR	
34	-367.600000	5.000000	12.370000	SILICA P	
35	-97.660000	1.000000	12.540000	AIR	
36	-78.686000	7.000000	12.530000	LIF P	
37	-30.700000	2.000000	12.750000	AIR	
38	-28.980000	4.000000	12.480000	SILICA P	
39	0.000000	3.000000	12.990000	AIR	
40	68.137000	7.000000	13.490000	CAF2 P	
41	-153.581000	199.823447 S	13.580000	AIR	
IMS	0.000000	0.000000	9.000000		

Figure 4.35 Optical design and graphs of aberration correction of a 50×/0.50 CCF plan multi-apochromatic objective (200–1800 nm; $F'_{ts} = 200$ mm).

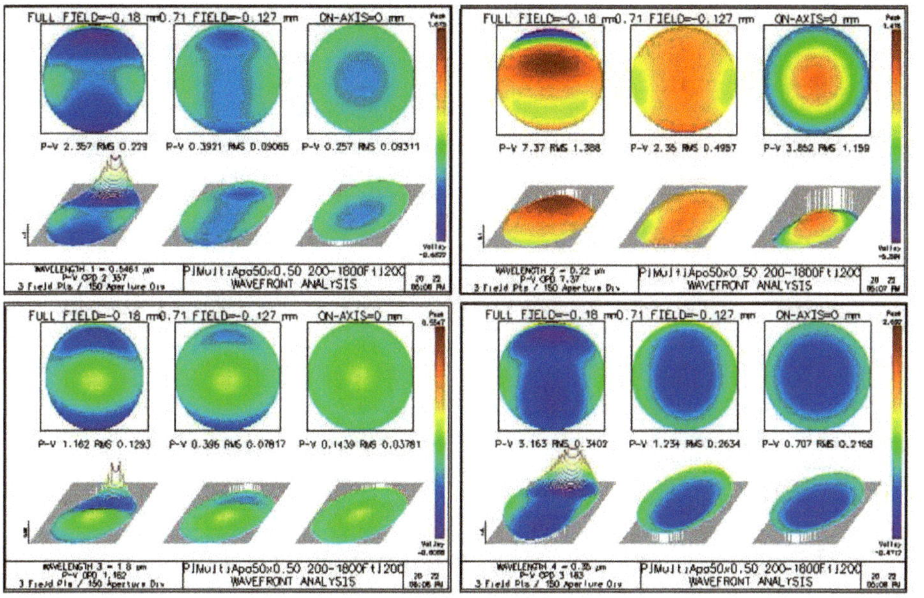

Figure 4.35 (*Continued*) Optical design and graphs of aberration correction of a 50×/0.50 CCF plan multi-apochromatic objective (200–1800 nm; $F'_{ts} = 200$ mm).

development. It is virtually unknown to a person what microscopic objects can exist at considerable depths underwater. A brief description of the most complex microscopic objects, such as bacteria, viruses, and others, is given in [24]. It would be unreasonable to assume that these and other types of microscopic objects are absent at depth; there are probably more of them there than on the earth's surface, and their diversity is much wider. That is why the proposed underwater microscope should allow the highest values of such parameters as resolution, contrast, and others to be reached; the accuracy of identifying the objects under study directly depends on these.

It should be taken into account that the physical properties of water are significantly different from those of the air [25]. For instance:

- The refractive index of water is much higher than that of the air; it may vary depending on the depth and temperature of the water.
- Water has a dispersion. The refractive index is different for different wavelengths, and it can vary depending on, for example, the degree of salinity of the water and its other chemical properties.
- It is believed that water is colorless.
- The density of water is more than 750 times higher than the density of air. An object immersed in water is affected by atmospheric pressure as

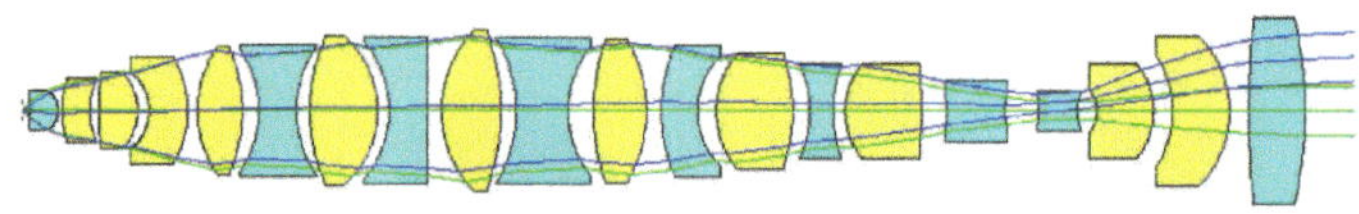

Lens:PlMultiApo100x0.65 200-2000Ftl20					
Object num aper	0.650000	Image height	9.000000	Primary wavln	0.546070
SRF	**RADIUS**	**THICKNESS**	**APERTURE RADIUS**	**GLASS**	**SPECIAL**
OBJ	0.000000	1.0000e-20	0.089888	AIR	
AST	0.000000	0.450000	0.420000 A	AIR	
2	-1.373000	1.500000	0.420000	SILICA	
3	-1.214900	0.200000	1.050000	AIR	
4	4.112400	1.500000	1.430000	CAF2	
5	4.042600	0.500000	1.620000	AIR	
6	33.661500	2.000000	1.700000	CAF2 P	
7	-2.702200	0.600000	1.980000	AIR	
8	-2.448100	2.000000	1.970000	LIF	
9	-5.297600	0.500000	2.720000	AIR	
10	5.548600	2.400000	3.280000	LIF P	
11	-7.834500	0.900000	3.250000	AIR	
12	-5.233300	2.000000	3.100000	SILICA P	
13	5.710800	0.700000	3.370000	AIR	
14	8.481500	3.600000	3.620000	CAF2 P	
15	-4.897100	0.600000	3.800000	AIR	
16	-5.357400	2.000000	3.470000	SILICA P	
17	322.263800	0.800000	3.750000	AIR	
18	5.620500	3.200000	4.080000	LIF P	
19	-11.143400	0.800000	3.940000	AIR	
20	-7.244800	2.700000	3.700000	SILICA P	
21	5.285800	1.300000	3.330000	AIR	
22	10.410800	2.800000	3.540000	LIF P	
23	-6.966800	0.800000	3.640000	AIR	
24	8.110800	1.900000	3.360000	SILICA P	
25	3.903900	1.000000	2.850000	AIR	
26	4.656000	4.000000	2.980000	LIF P	
27	-8.276800	0.900000	2.680000	AIR	
28	-5.461800	1.200000	2.430000	SILICA P	
29	7.388600	0.600000	2.330000	AIR	
30	3.664700	4.000000	2.420000	LIF P	
31	-25.588400	1.300000	1.890000	AIR	
32	-13.710300	3.000000	1.530000	SILICA P	
33	3.535700	1.800000	1.040000	AIR	
34	-3.134600	2.000000	0.920000	SILICA P	
35	2.719700	1.000000	1.030000	AIR	
36	-1.895400	3.000000	1.180000	CAF2 P	
37	-3.624100	1.000000	2.430000	AIR	
38	-4.923100	3.000000	2.800000	LIF P	
39	-5.340500	1.000000	3.820000	AIR	
40	40.363800	3.000000	4.520000	SILICA P	
41	-18.857800	200.000000	4.760000	AIR	
42	-367.600000	5.000000	13.650000	SILICA P	
43	-97.660000	1.000000	13.840000	AIR	
44	-78.686000	7.000000	13.840000	LIF P	
45	-30.700000	2.000000	14.090000	AIR	
46	-28.980000	4.000000	13.820000	SILICA P	
47	0.000000	3.000000	14.410000	AIR	
48	68.137000	7.000000	15.010000	CAF2 P	
49	-153.581000	139.406295 S	15.080000	AIR	
IMS	0.000000	0.000000	9.000000		

Figure 4.36 Optical design and graphs of aberration correction of a 100×/0.65 CCF plan multi-apochromatic objective (200–1800 nm; $F'_{ts} = 200$ mm).

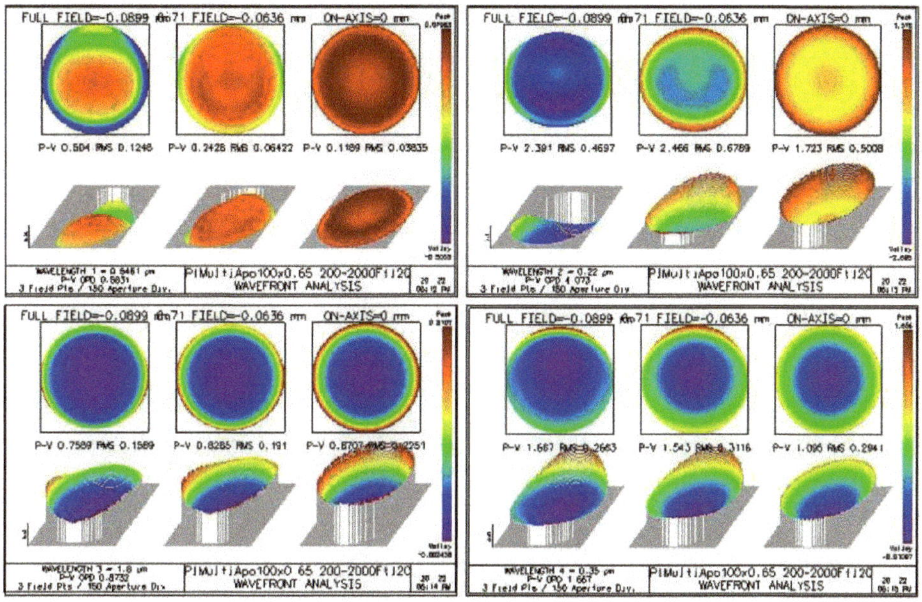

Figure 4.36 (*Continued*) Optical design and graphs of aberration correction of a 100×/0.65 CCF plan multi-apochromatic objective (200–1800 nm; $F'_{ts} = 200$ mm).

well as the hydrostatic pressure of water, which, as the object descends, increases proportionally by 1 kg/cm^2 every 10 m.

- A feature of water is the selectivity of its spectral transmittance. Water passes the blue part of the spectrum and displaces the red, which is practically absorbed at depths of more than 5–10 m. For large depths, the blue and purple parts of the spectrum are characteristic.
- Other properties of water cause strong scattering of the light entering it. Moreover, the light is scattered in all directions. This causes a

Table 4.3 Main technical parameters and basic optical layout of CCF plan multi-apochromatic objectives.

Magnification	*NA*	*WD* (mm)	*F'* (mm)	*R* (μm)	*DF* (μm)	*FOV* on object (mm)	*FOV* on image (mm)	Optical design at figure no.	The principal optical layout
20×	0.35	0.77	8	0.95	2.72	1.0	20	4.34	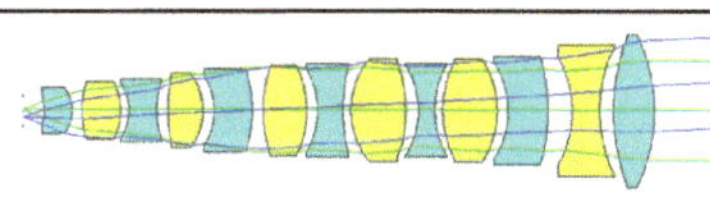
50×	0.50	0.45	4	0.67	1.33	0.36	18	4.35	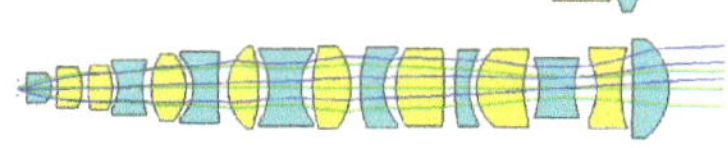
100×	0.65	0.2	2	0.51	0.79	0.18	18	4.36	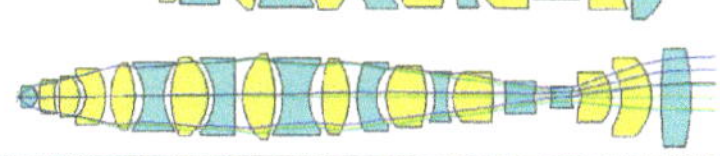

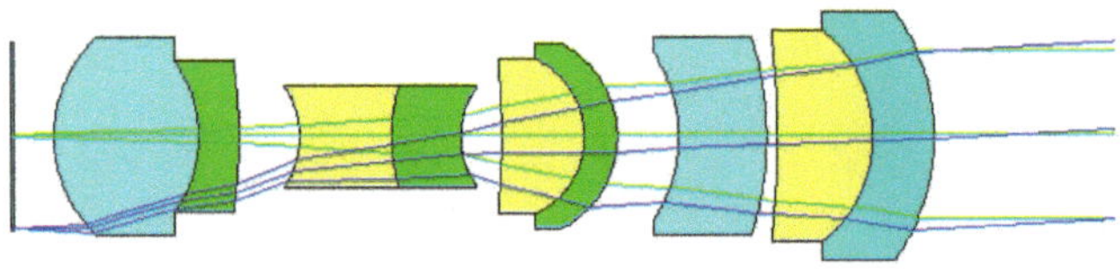

Lens: 2.5x0.05 1100-2400nm d017

Object num aper 0.050000 Image height 11.000000 Primary wavln 1.100000

SRF	RADIUS	THICKNESS	APERTURE RADIUS	GLASS	SPECIAL
OBJ	0.000000	0.010000	4.399681	AIR	
AST	0.000000	0.170000	4.480000 A	LZ_K8	
2	0.000000	2.000000	4.480000	AIR	
3	6.381600	7.200000	4.680000	SILICA C	
4	-6.246300	2.000000	3.630000	O_S-TIM22 C	
5	-19.719400	3.000000	3.230000	AIR	
6	-4.053800	4.500000	2.370000	CAF2 C	
7	8.581400	3.300000	2.440000	O_S-TIM22 P	
8	3.971500	2.000000	2.350000	AIR	
9	-38.547900	4.000000	3.010000	CAF2 P	
10	-4.077000	1.600000	3.720000	O_S-TIM22 P	
11	-4.906900	3.000000	4.400000	AIR	
12	-7.169800	4.300000	4.040000	SILICA P	
13	-15.350400	0.500000	4.750000	AIR	
14	-47.948800	4.700000	4.820000	CAF2 P	
15	-6.358900	3.000000	5.020000	O_S-LAM58 C	
16	-10.095100	200.000000	5.870000	AIR	
17	87.442400	10.000000	15.150000	LZ_LK3	
18	61.930600	5.000000	14.940000	AIR	
19	-99.657500	10.000000	15.130000	LZ_LK3 P	
20	-106.539600	3.000000	16.220000	AIR	
21	82.772900	10.000000	16.760000	CAF2	
22	-88.981200	3.000000	16.770000	LZ_TF1	
23	-156.432400	199.831035 S	16.850000	AIR	
IMS	0.000000	0.000000	11.000000		

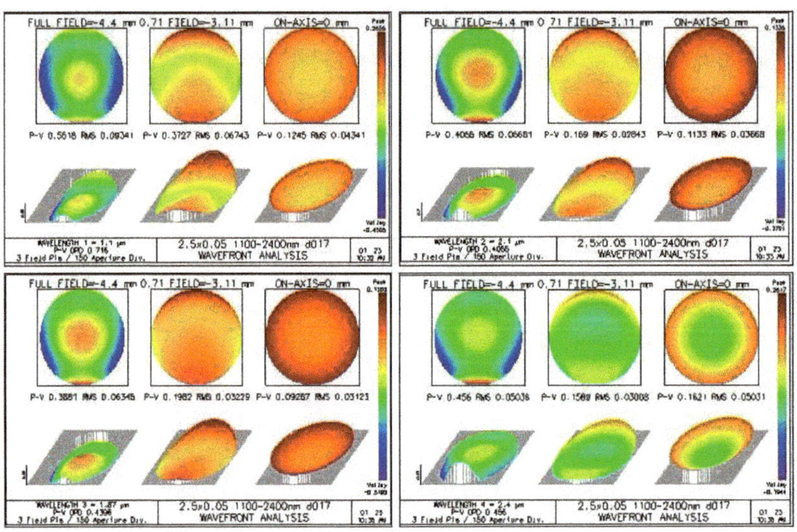

Figure 4.37 Optical design and graphs of aberration correction of a 2.5×/0.05 SWIR CCF plan poly-apochromatic objective ($F'_{ts} = 200$ mm).

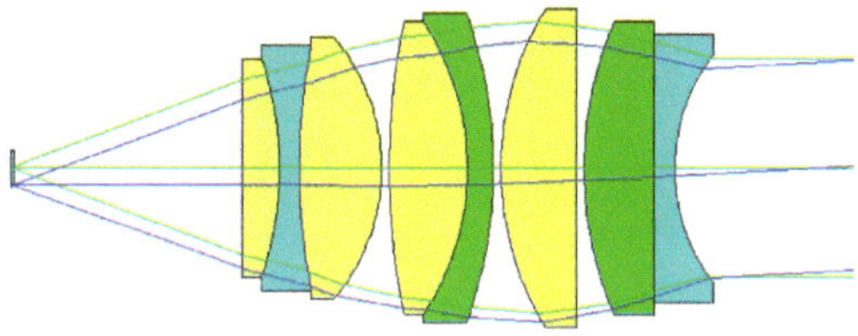

Lens: 10x0.35 1100-2400nm d017

Object num aper	0.350000	Image height	11.000000	Primary wavln	1.100000

SRF	RADIUS	THICKNESS	APERTURE RADIUS	GLASS	SPECIAL
OBJ	0.000000	0.010000	1.099943	AIR	
AST	0.000000	0.170000	1.100000 A	LZ_K8	
2	0.000000	15.000000	1.140000	AIR	
3	0.000000	2.400000	6.750000	CAF2 C	
4	-22.266800	1.300000	7.070000	O_S-LAM58 C	
5	40.097600	5.400000	7.930000	CAF2 C	
6	-12.920900	0.500000	8.460000	AIR	
7	45.422200	5.200000	9.360000	CAF2	
8	-16.741700	1.700000	9.440000	O_S-TIH53 C	
9	-29.200600	0.500000	9.980000	AIR	
10	19.517000	5.000000	10.260000	CAF2	
11	0.000000	0.500000	9.960000	AIR	
12	23.647000	4.600000	9.450000	O_S-TIH53 P	
13	0.000000	1.400000	8.640000	O_S-LAM58 C	
14	12.049200	200.000000	7.360000	AIR	
15	87.442400	10.000000	17.570000	LZ_LK3	
16	61.930600	5.000000	17.280000	AIR	
17	-99.657500	10.000000	17.370000	LZ_LK3 P	
18	-106.539600	3.000000	18.570000	AIR	
19	82.772900	10.000000	19.230000	CAF2	
20	-88.981200	3.000000	19.200000	LZ_TF1	
21	-156.432400	199.806109 S	19.290000	AIR	
IMS	0.000000	0.000000	11.000000		

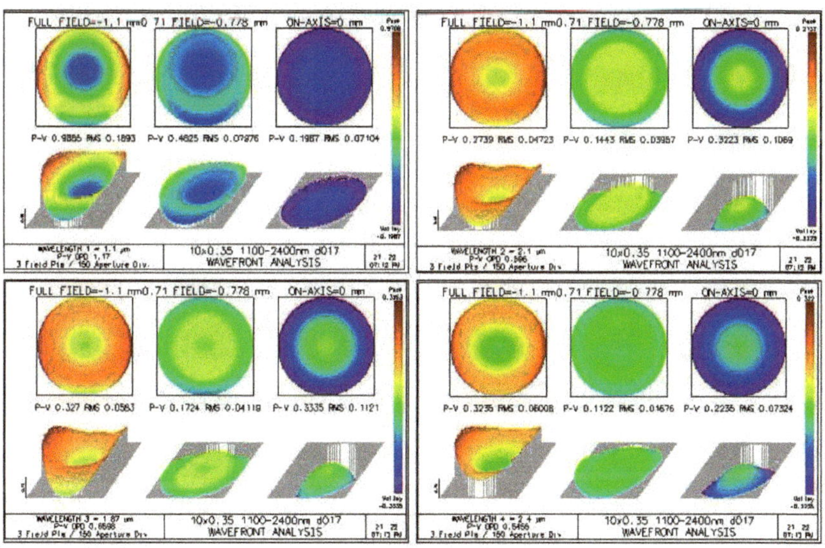

Figure 4.38 Optical design and graphs of aberration correction of a 10×/0.35 SWIR CCF plan poly-apochromatic objective ($F'_{ts} = 200$ mm).

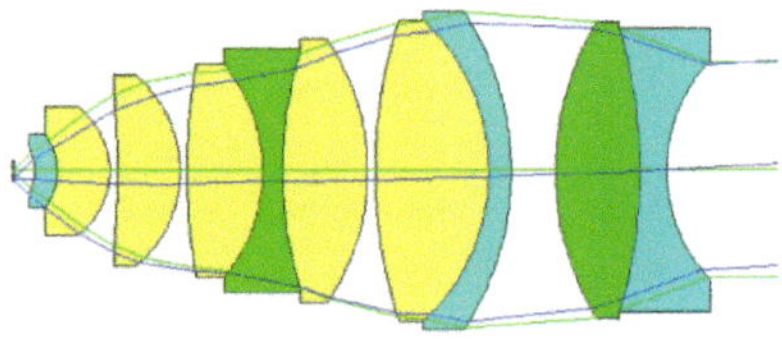

Lens: 20x0.70 1100-2400nm d017

Object num aper	0.700000	Image height	11.000000	Primary wavln	1.100000

SRF	RADIUS	THICKNESS	APERTURE RADIUS	GLASS	SPECIAL
OBJ	0.000000	0.010000	0.550012	AIR	
AST	0.000000	0.170000	0.560000 A	LZ_K8	
2	0.000000	1.350000	0.650000	AIR	
3	-3.261300	1.500000	1.570000	O_S-LAM58 C	
4	-3.839800	3.500000	2.400000	CAF2 C	
5	-5.195700	0.500000	4.160000	AIR	
6	-79.566100	4.200000	5.610000	CAF2	
7	-7.932600	0.500000	6.200000	AIR	
8	43.259100	5.000000	6.800000	CAF2 P	
9	-10.727900	1.300000	6.900000	O_S-TIH53 C	
10	27.031900	5.600000	7.920000	CAF2 P	
11	-14.898800	0.500000	8.540000	AIR	
12	30.271400	7.600000	9.670000	CAF2 P	
13	-13.169700	1.500000	9.780000	O_S-LAM58 C	
14	-18.399200	3.000000	10.340000	AIR	
15	18.258600	5.600000	9.630000	O_S-TIH53 P	
16	-31.298100	1.800000	9.180000	O_S-LAM58 C	
17	10.199100	200.000000	7.120000	AIR	
18	87.442400	10.000000	17.990000	LZ_LK3	
19	61.930600	5.000000	17.680000	AIR	
20	-99.657500	10.000000	17.760000	LZ_LK3 P	
21	-106.539600	3.000000	18.980000	AIR	
22	82.772900	10.000000	19.660000	CAF2 P	
23	-88.981200	3.000000	19.630000	LZ_TF1	
24	-156.432400	199.832752 S	19.710000	AIR	
IMS	0.000000	0.000000	11.000000		

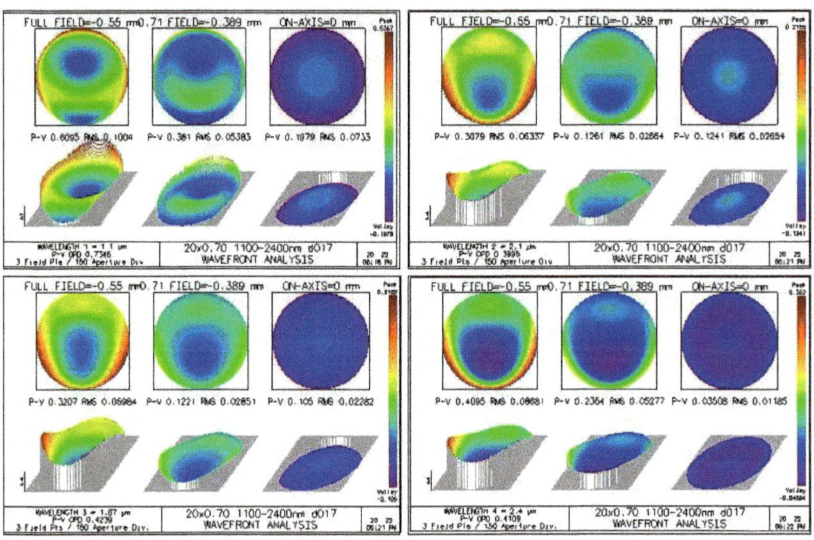

Figure 4.39 Optical design and graphs of aberration correction of a 20×/0.70 SWIR CCF plan poly-apochromatic objective ($F'_{ts} = 200$ mm).

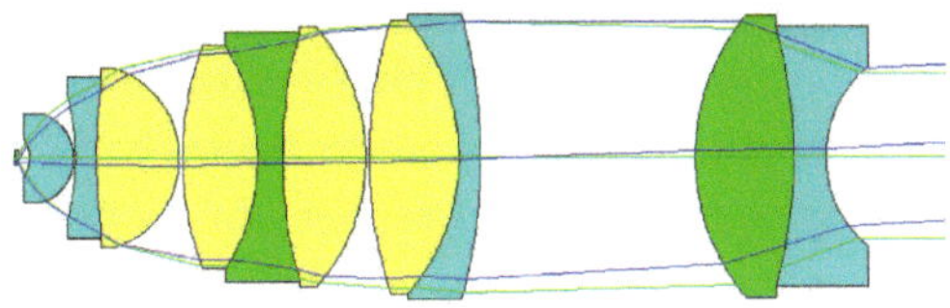

Lens: 40x0.90 1100-2400nm d017

Object num aper	0.900000	Image height	11.000000	Primary wavln	1.100000

SRF	RADIUS	THICKNESS	APERTURE RADIUS	GLASS	SPECIAL
OBJ	0.000000	0.010000	0.274500	AIR	
AST	0.000000	0.170000	0.290000 A	LZ_K8	
2	0.000000	0.619990	0.420000	AIR	
3	-1.840000	2.500000	1.040000	O_S-LAM58 C	
4	-2.522500	0.100000	2.400000	AIR	
5	-13.056800	1.200000	3.410000	O_S-LAM58 P	
6	35.290500	4.600000	4.410000	CAF2	
7	-5.103200	0.200000	4.890000	AIR	
8	16.758800	4.200000	6.000000	CAF2 P	
9	-11.178600	1.400000	6.070000	O_S-TIH53 C	
10	25.218800	4.600000	6.800000	CAF2 P	
11	-10.501500	0.200000	7.080000	AIR	
12	24.265900	5.000000	7.460000	CAF2 P	
13	-11.093100	1.200000	7.430000	O_S-LAM58 C	
14	-28.603600	12.000000	7.750000	AIR	
15	11.706900	5.600000	7.690000	O_S-TIH53 P	
16	-24.638100	1.800000	7.080000	O_S-LAM58 C	
17	6.358400	200.000000	4.930000	AIR	
18	87.442400	10.000000	15.180000	LZ_LK3	
19	61.930600	5.000000	14.960000	AIR	
20	-99.657500	10.000000	15.140000	LZ_LK3 P	
21	-106.539600	3.000000	16.200000	AIR	
22	82.772900	10.000000	16.710000	CAF2 P	
23	-88.981200	3.000000	16.700000	LZ_TF1	
24	-156.432400	199.818353 S	16.780000	AIR	
IMS	0.000000	0.000000	11.000000		

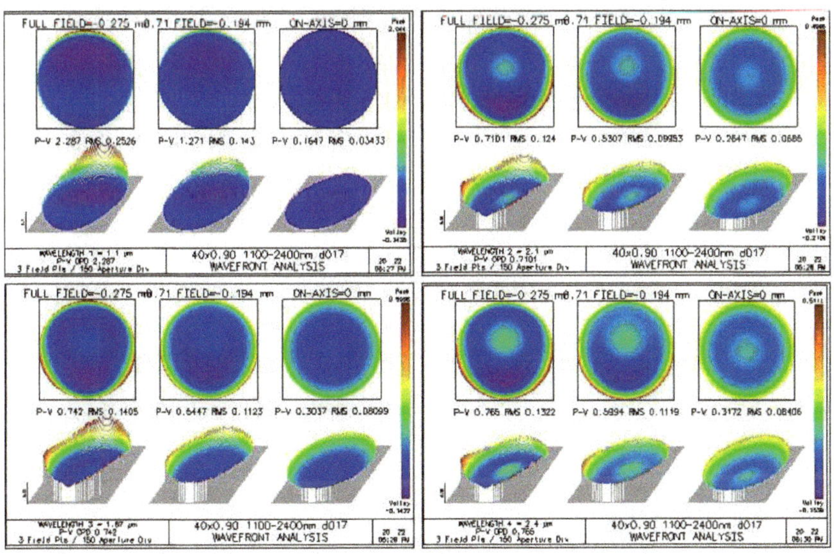

Figure 4.40 Optical design and graphs of aberration correction of a 40×/0.90 SWIR CCF plan poly-apochromatic objective ($F'_{ts} = 200$ mm).

significant reduction in light parameters such as brightness and color contrast. There is also the effect of *backscattering* of light in which some suspended particles of organic and inorganic origin scatter light in the direction opposite to the direction of its emission.

- Water absorbs light about 1000 times more than air. Depending on the degree of transparency, the use of an artificial illuminator is required for water at a depth of more than 5–10 m.

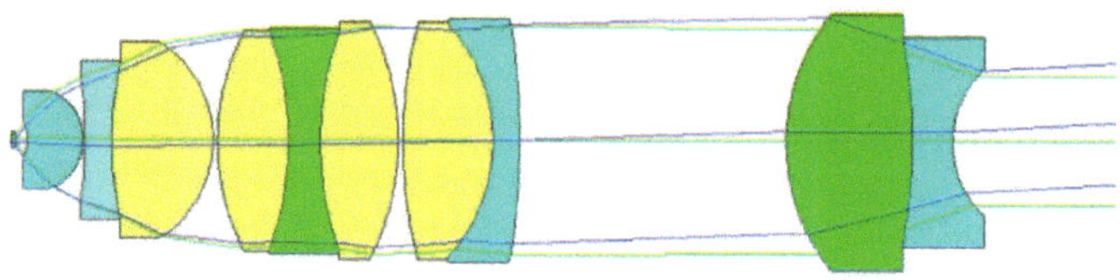

Lens: 60x0.85 1100-2400nm d017

Object num aper 0.850000 | Image height 11.000000 | Primary wavln 1.100000

SRF	RADIUS	THICKNESS	APERTURE RADIUS	GLASS	SPECIAL
OBJ	0.000000	0.010000	0.183372	AIR	
AST	0.000000	0.170000	0.200000 A	LZ_K8	
2	0.000000	0.549935	0.320000	AIR	
3	-1.688300	2.500000	0.840000	O_S-LAM58 C	
4	-2.260800	0.100000	2.130000	AIR	
5	-20.975100	1.200000	2.850000	O_S-LAM58 P	
6	16.792900	4.600000	3.460000	CAF2	
7	-4.974900	0.200000	4.310000	AIR	
8	10.185500	3.200000	4.870000	CAF2 P	
9	-13.650100	1.400000	4.830000	O_S-TIH53 C	
10	14.107000	3.600000	4.960000	CAF2 P	
11	-11.053600	0.200000	5.190000	AIR	
12	20.767400	4.000000	5.200000	CAF2 P	
13	-7.495700	1.200000	5.100000	O_S-LAM58 C	
14	-31.108100	12.000000	5.320000	AIR	
15	8.810000	5.600000	5.610000	O_S-TIH53 P	
16	-25.271300	1.800000	4.560000	O_S-LAM58 C	
17	4.395500	200.000000	3.230000	AIR	
18	87.442400	10.000000	14.500000	LZ_LK3	
19	61.930600	5.000000	14.350000	AIR	
20	-99.657500	10.000000	14.550000	LZ_LK3 P	
21	-106.539600	3.000000	15.620000	AIR	
22	82.772900	10.000000	16.130000	CAF2 P	
23	-88.981200	3.000000	16.160000	LZ_TF1	
24	-156.432400	200.005786 S	16.240000	AIR	
IMS	0.000000	0.000000	11.000000		

Figure 4.41 Optical design and graphs of aberration correction of a 60×/0.85 SWIR CCF plan poly-apochromatic objective ($F'_{ts} = 200$ mm).

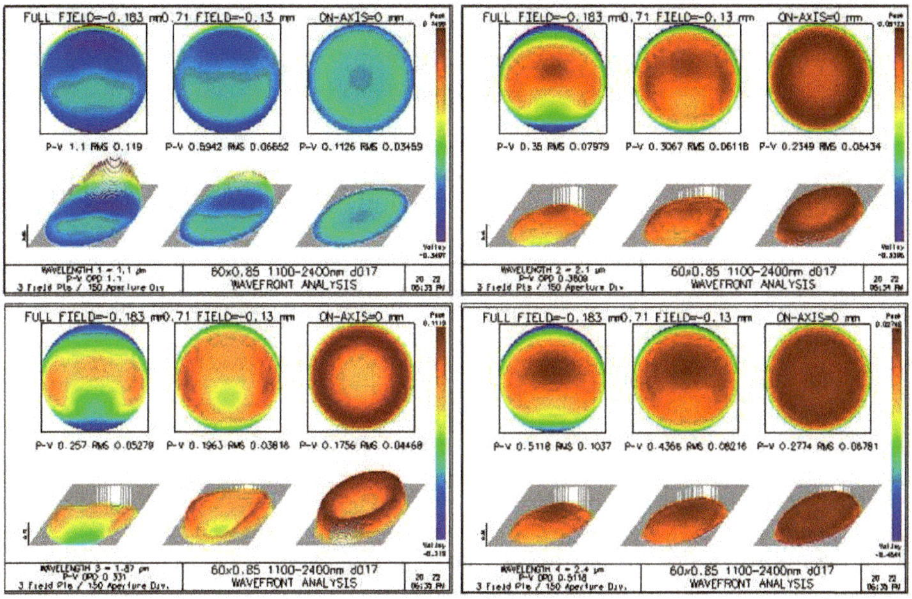

Figure 4.41 (*Continued*) Optical design and graphs of aberration correction of a 60×/0.85 SWIR CCF plan poly-apochromatic objective ($F'_{ts} = 200$ mm).

4.4.3 Parts of a light microscope for underwater research that must be modernized

If we proceed from the fact that a microscope for underwater research should become a full-fledged analogue of a conventional light microscope, it becomes obvious that it should reproduce all its technical parameters. Moreover, this not only concerns resolution values (and others, such as linear magnification, linear field on the object under study, etc.), but certain criteria for aberration correction adopted in classical microscopy must also be met. The criteria proposed by Abbe and Rayleigh to achieve aberration correction, in accordance with the criteria adopted in the modern concept of image quality on the microscope (for example, Strehl criterion or others), must be taken into account when designing an underwater microscope. The optical system of a microscopic projection system is a so-called *diffraction-limited* system, the optical system of which is geometrically limited, unlike, for example, a photographic objective.

Projection system: Objective

Using standard photographic objectives placed in a special waterproof box is considered acceptable for underwater photography in most cases [26]. A conventional photographic objective, the optical system of which works in air, focuses the image of an object on photographic film or a digital receiver, introducing some aberrations in the image; the level of aberration correction is considered acceptable for this objective and is characterized as a geometrically limited level. When such a photographic objective is placed inside a waterproof box with a protective glass (a window as a plane–parallel plate) and the box is immersed in water at some depth, although aberrations of the optical *window–objective* system vary compared with those of an objective

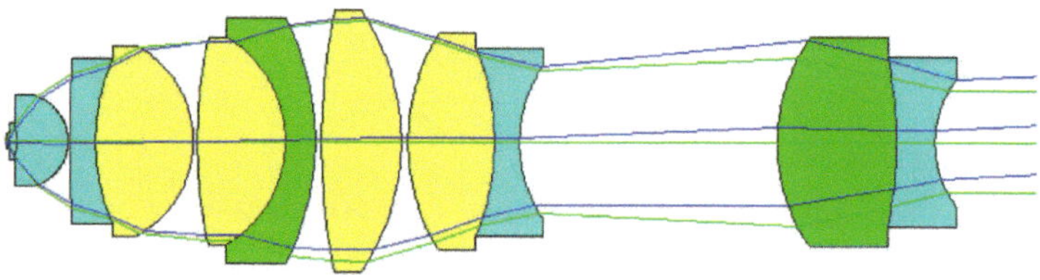

Lens: 100x1.20 1100-2400nm d017

Object num aper	1.200000	Image height	10.000000	Primary wavln	1.100000

SRF	RADIUS	THICKNESS	APERTURE RADIUS	GLASS	SPECIAL
OBJ	0.000000	0.010000	0.099958	WATER C	
AST	0.000000	0.170000	0.120000 A	LZ_K8	
2	0.000000	0.300517	0.360000	WATER C	
3	-3.840000	2.500000	0.850000	O_S-LAM58 C	
4	-2.120000	0.100000	2.070000	AIR	
5	48.960000	1.200000	3.200000	O_S-LAM58 P	
6	9.210000	4.600000	3.780000	CAF2 C	
7	-4.823000	0.200000	4.340000	AIR	
8	21.273100	4.200000	4.740000	CAF2 P	
9	-5.443000	1.400000	4.750000	O_S-TIH53 C	
10	-12.043700	0.300000	5.600000	AIR	
11	28.977000	3.800000	5.940000	CAF2 C	
12	-10.315500	0.300000	5.990000	AIR	
13	9.408900	4.000000	4.970000	CAF2 P	
14	-11.357000	1.200000	4.310000	O_S-LAM58 C	
15	6.117700	12.000000	3.580000	AIR	
16	8.306500	5.600000	4.780000	O_S-TIH53 P	
17	-17.807800	1.800000	3.870000	O_S-LAM58 C	
18	4.780000	200.000000	2.960000	AIR	
19	87.442400	10.000000	11.520000	LZ_LK3	
20	61.930600	5.000000	11.380000	AIR	
21	-99.657500	10.000000	11.610000	LZ_LK3 P	
22	-106.539600	3.000000	12.440000	AIR	
23	82.772900	10.000000	12.770000	CAF2 P	
24	-88.981200	3.000000	12.780000	LZ_TF1	
25	-156.432400	199.913873 S	12.840000	AIR	
IMS	0.000000	0.000000	10.000000		

Figure 4.42 Optical design and graphs of aberration correction of a 100×/1.20 water immersion SWIR CCF plan poly-apochromatic objective ($F'_{ts} = 200$ mm).

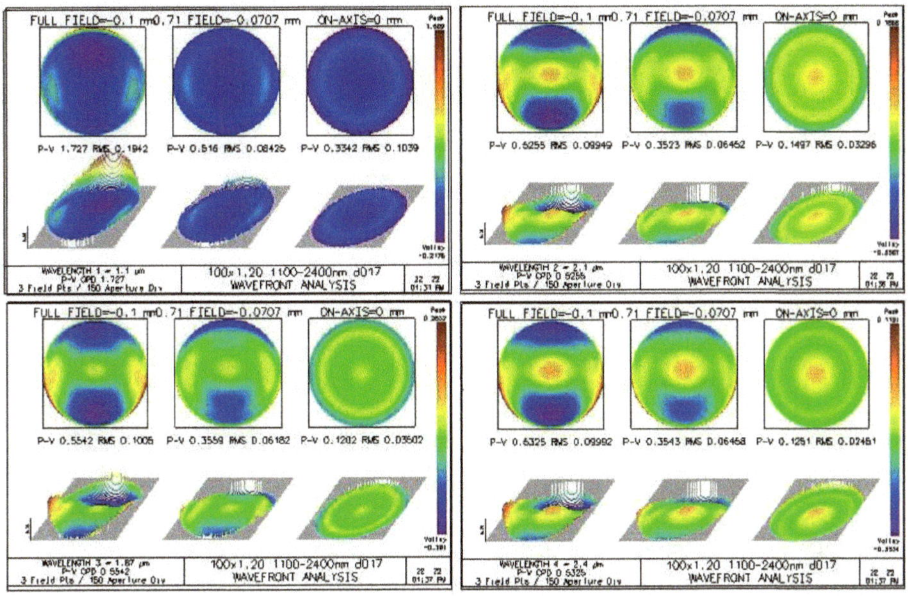

Figure 4.42 (*Continued*) Optical design and graphs of aberration correction of a 100×/1.20 water immersion SWIR CCF plan poly-apochromatic objective ($F'_{ts} = 200$ mm).

Table 4.4 Main technical parameters and basic optical layout of SWIR CCF plan poly-apochromatic objectives.

Magnification	*NA*	*WD* (mm)	*F'* (mm)	*R* (μm)	*DF* (μm)	*FOV* on object (mm)	*FOV* on image (mm)	Optical design at figure no.	The principal optical layout
2.5×	0.05	1.7	80	13.4	268	8.8	22	4.37	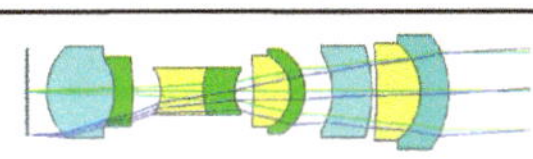
10×	0.35	14.3	20.0	1.9	5.5	2.2	22	4.38	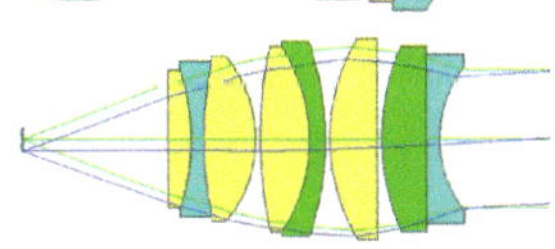
20×	0.70	0.88	8.0	0.96	1.37	1.1	22	4.39	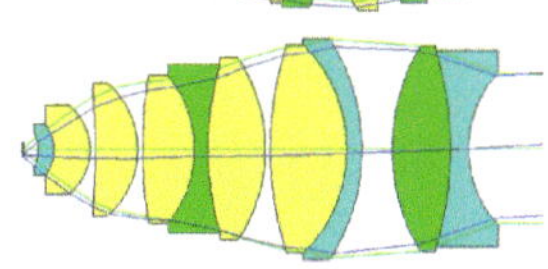
40×	0.90	0.38	4.0	0.75	0.83	0.55	22	4.40	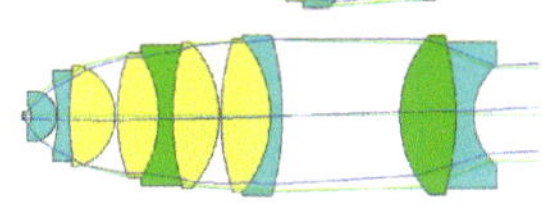
60×	0.85	0.33	3.33	0.79	0.93	0.37	22	4.41	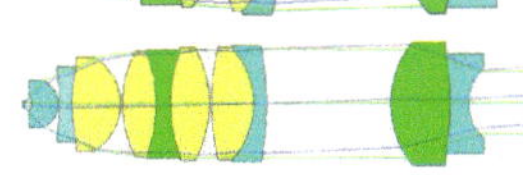
100×	1.20 water	0.12	2.0	0.56	0.47	0.20	20	4.42	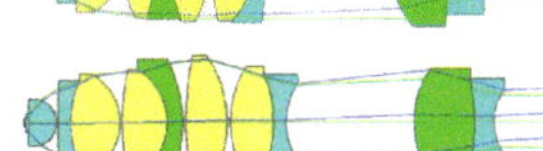

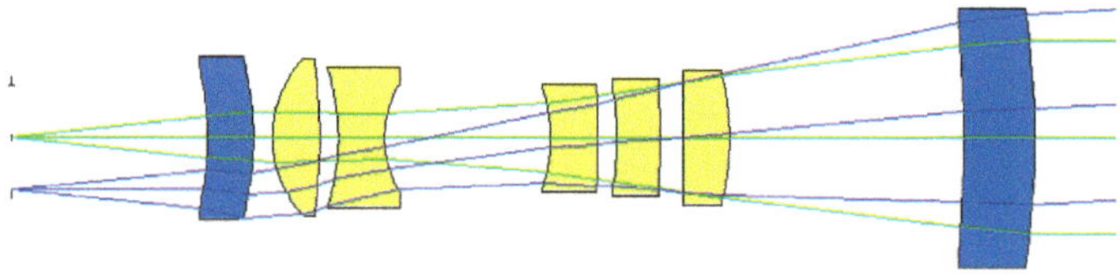

Lens: PlanPolyApo 5x010								
Object num aper	0.100000	Image height		11.000000		Primary wavln		5.000000
SRF	RADIUS	THICKNESS		APERTURE RADIUS		GLASS		SPECIAL
OBJ	0.000000	1.0000e-20		2.199992		AIR		
AST	0.000000	8.500000		2.199992	AS	AIR		
2	-13.042000	2.000000		3.020000		GERM	C	
3	-12.309700	0.800000		3.400000		AIR		
4	4.758800	2.000000		3.290000		CAF2	C	
5	-23.366600	0.800000		3.170000		AIR		
6	-9.141500	2.000000		2.940000		CAF2	P	
7	4.274200	7.000000		2.330000		AIR		
8	-6.512200	2.000000		2.100000		CAF2	P	
9	-33.279200	0.800000		2.230000		AIR		
10	-13.073400	1.800000		2.270000		CAF2	P	
11	-674.827500	1.000000		2.440000		AIR		
12	-177.085400	2.000000		2.560000		CAF2	P	
13	-11.653300	10.000000		2.830000		AIR		
14	-52.302400	3.000000		5.060000		GERM	P	
15	-36.156100	200.000000		5.420000		AIR		
16	1.5300e+03	5.000000		16.220000		GERM	P	
17	-500.000000	1.000000		16.240000		AIR		
18	-500.000000	2.000000		16.170000		GERM	P	
19	-1.0000e+03	197.683501	S	16.180000		AIR		
IMS	0.000000	0.000000		11.000000	S			

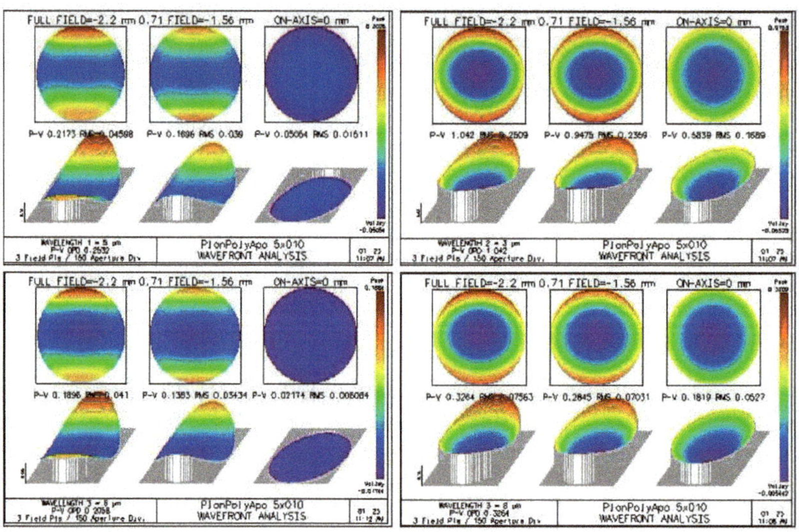

Figure 4.43 Optical design and graphs of aberration correction of a 5×/0.10 MWIR CCF plan poly-apochromatic objective ($F'_{ts} = 200$ mm).

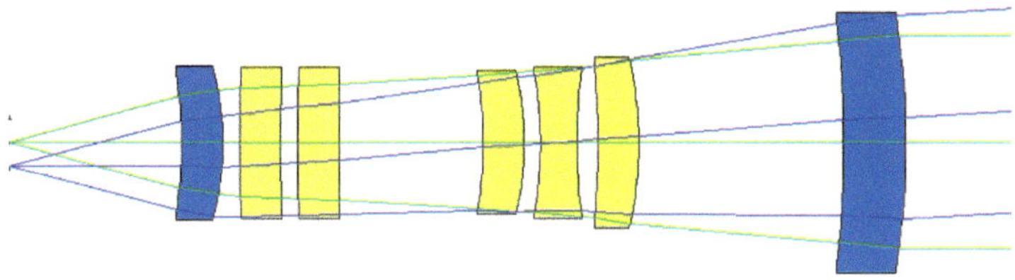

Lens: PlanPolyApo 10x025

Object num aper	0.250000	Image height	11.000000	Primary wavln	5.000000

SRF	RADIUS	THICKNESS	APERTURE RADIUS	GLASS	SPECIAL
OBJ	0.000000	1.0000e-20	1.099980	AIR	
AST	0.000000	8.500000	1.099980 AS	AIR	
2	-16.623400	2.000000	3.210000	GERM C	
3	-13.056800	0.800000	3.600000	AIR	
4	94.868800	2.000000	3.590000	CAF2 C	
5	288.672200	0.800000	3.560000	AIR	
6	100.571000	2.000000	3.550000	CAF2 P	
7	0.000000	7.000000	3.510000	AIR	
8	-17.782200	2.000000	3.320000	CAF2 P	
9	-14.019600	0.800000	3.370000	AIR	
10	-17.196400	1.800000	3.350000	CAF2 P	
11	33.499500	1.000000	3.570000	AIR	
12	-50.671100	2.000000	3.720000	CAF2 P	
13	-17.316700	10.000000	4.030000	AIR	
14	-47.579000	3.000000	5.760000	GERM P	
15	-39.021800	200.000000	6.140000	AIR	
16	1.5300e+03	5.000000	17.450000	GERM P	
17	-500.000000	1.000000	17.480000	AIR	
18	-500.000000	2.000000	17.390000	GERM P	
19	-1.0000e+03	197.681578 S	17.410000	AIR	
IMS	0.000000	0.000000	11.000000 S		

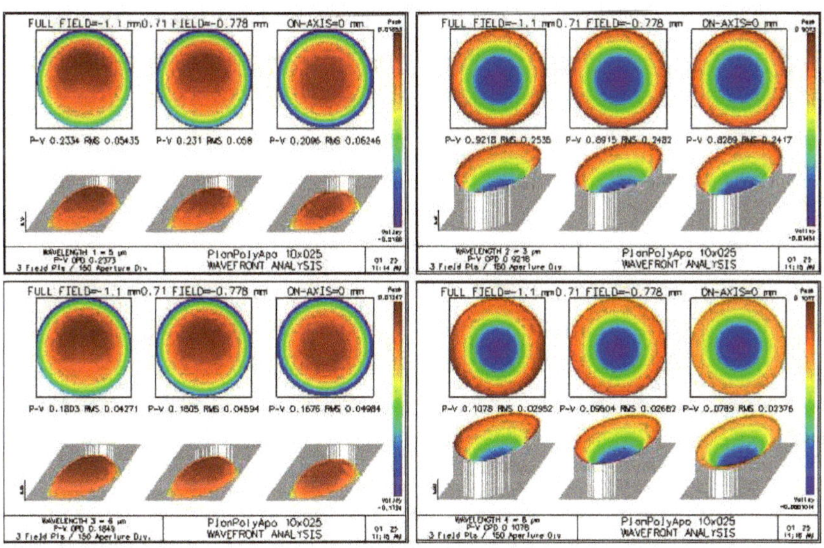

Figure 4.44 Optical design and graphs of aberration correction of a 10×/0.25 MWIR CCF plan poly-apochromatic objective ($F'_{ts} = 200$ mm).

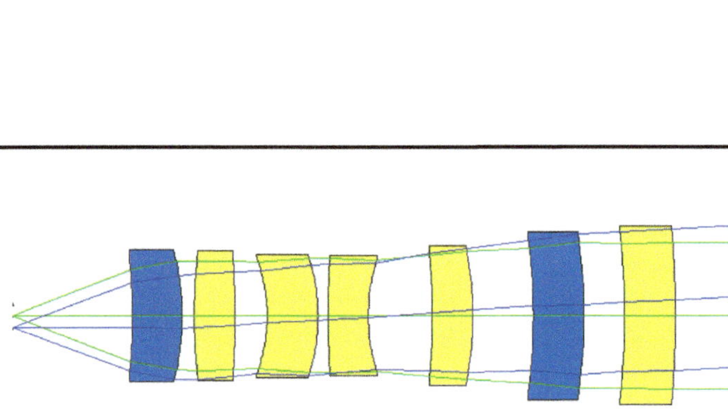

Lens: PlanPolyApo 20x035

Object num aper	0.350000	Image height	11.000000	Primary wavln	5.000000

SRF	RADIUS	THICKNESS		APERTURE RADIUS		GLASS		SPECIAL
OBJ	0.000000	1.0000e-20		0.549996		AIR		
AST	0.000000	6.000000		0.549996	AS	AIR		
2	-20.309500	2.500000		2.720000		GERM	C	
3	-12.064900	0.600000		3.160000		AIR		
4	26.120500	2.000000		3.110000		CAF2	C	
5	-48.298400	1.600000		3.000000		AIR		
6	-7.679900	2.500000		2.890000		CAF2	P	
7	-9.292000	0.500000		2.960000		AIR		
8	53.646700	2.000000		2.880000		CAF2	P	
9	8.742100	3.200000		2.710000		AIR		
10	-34.577800	2.000000		3.090000		CAF2	P	
11	-17.085600	3.000000		3.340000		AIR		
12	-27.499400	2.500000		3.700000		GERM	P	
13	-25.098700	2.000000		4.020000		AIR		
14	-34.074900	2.500000		4.110000		CAF2	P	
15	-52.587300	200.000000		4.270000		AIR		
16	1.5300e+03	5.000000		15.610000		GERM	P	
17	-500.000000	1.000000		15.640000		AIR		
18	-500.000000	2.000000		15.570000		GERM	P	
19	-1.0000e+03	197.660939	S	15.580000		AIR		
IMS	0.000000	0.000000		11.000000	S			

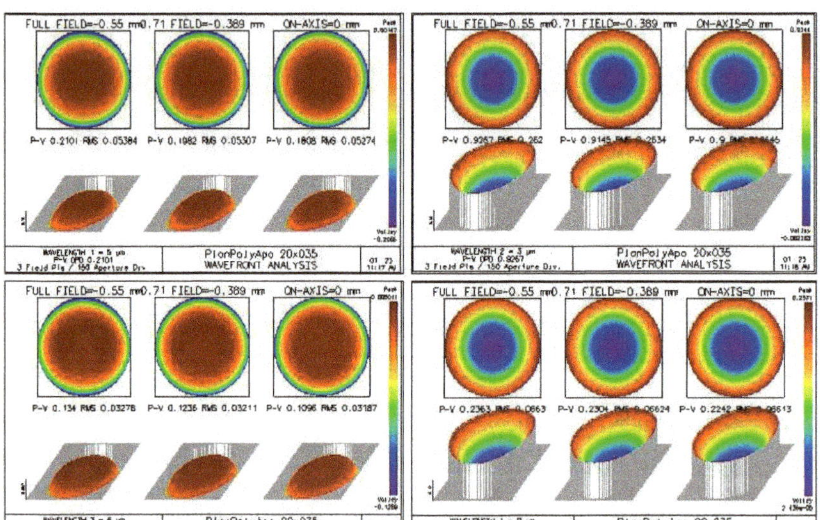

Figure 4.45 Optical design and graphs of aberration correction of a 20×/0.35 MWIR CCF plan poly-apochromatic objective ($F'_{ts} = 200$ mm).

operating in the air, they often remain acceptable for carrying out underwater photography. However, this approach does not always provide high-quality photographic images, especially if a high-aperture objective is used. Thus, special *hydro* photographic objectives have been created. The initial optical design of these objectives considered the effect of water on the resulting image quality of the window–objective system. In later developments, a lens was used as the window, significantly improving the image and consumer qualities of such photographic objectives for underwater use. Descriptions of the optical schemes of these so-called *GidroRussar* objectives, named after the outstanding Russian practical scientist Mikhail Mikhailovich Rusinov, are given in [27]. However, to date, even special photographic objectives for

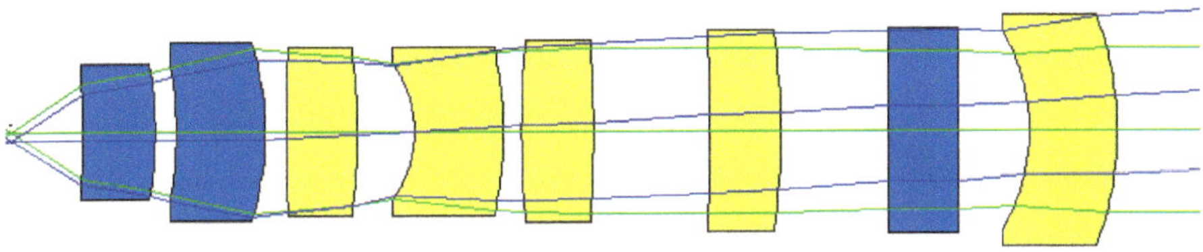

Lens: PlanPolyApo 50x055

Object num aper 0.550000 Image height 11.000000 Primary wavln 5.000000

SRF	RADIUS	THICKNESS	APERTURE RADIUS	GLASS	SPECIAL
OBJ	0.000000	1.0000e-20	0.219980	AIR	
AST	0.000000	2.000000	0.219980 AS	AIR	
2	-33.590500	2.000000	1.510000	GERM C	
3	-11.413600	0.600000	1.840000	AIR	
4	-9.823900	2.500000	1.960000	GERM C	
5	-7.933200	0.600000	2.430000	AIR	
6	41.525400	2.000000	2.280000	CAF2 C	
7	-16.485100	1.600000	2.070000	AIR	
8	-3.158000	2.500000	1.890000	CAF2 P	
9	-10.282700	0.500000	2.290000	AIR	
10	23.275100	2.000000	2.390000	CAF2 P	
11	-55.481700	3.200000	2.490000	AIR	
12	-71.091200	2.000000	2.660000	CAF2 P	
13	-18.227200	3.000000	2.750000	AIR	
14	-539.678400	2.000000	2.780000	GERM P	
15	-180.926200	2.000000	2.790000	AIR	
16	-5.312400	2.500000	2.760000	CAF2 P	
17	-8.730800	200.000000	3.170000	AIR	
18	1.5300e+03	5.000000	15.370000	GERM P	
19	-500.000000	1.000000	15.410000	AIR	
20	-500.000000	2.000000	15.350000	GERM P	
21	-1.0000e+03	197.621565 S	15.760000	AIR	
IMS	0.000000	0.000000	11.000000 S		

Figure 4.46 Optical design and graphs of aberration correction of a 50×/0.55 MWIR CCF plan poly-apochromatic objective ($F'_{ts} = 200$ mm).

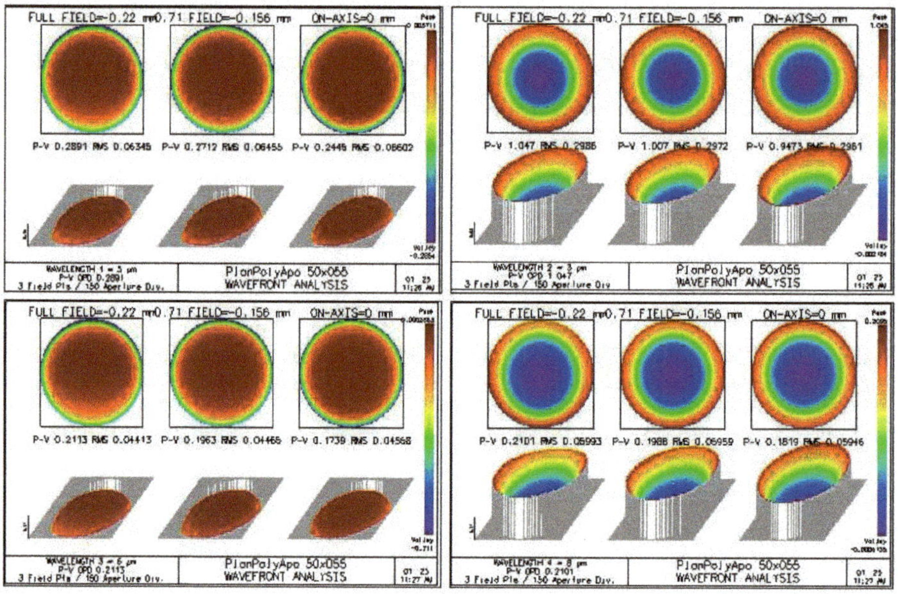

Figure 4.46 (*Continued*) Optical design and graphs of aberration correction of a 50×/0.55 MWIR CCF plan poly-apochromatic objective ($F'_{ts} = 200$ mm).

underwater shooting only meet the criteria for aberration correction of geometrically limited optical systems (not diffraction limited).

As an experiment, an approximate optical design can be performed to confirm that a standard objective (for a microscope) placed in a waterproof box underwater has significant variation in aberration correction compared with a standard microscope. Indeed, the level of the aberration correction of the window–objective optical system ceases to meet the requirements of the diffraction-limited optical system; does not meet the criteria of Abbe, Rayleigh, and others; and does not satisfy the Strehl relation. Optical calculations have confirmed that even for an objective with an aperture of 0.10–0.15, a special optical design is required that takes into account the impact of the water and the window on aberration, even if the window is made in the form of a plane–parallel plate. Such an optical design does not cause any difficulties. But even when obtaining a theoretically high image quality (matching the specified criteria of aberration correction for a microscope objective), the resulting system most likely will not have high functionality, as its resolution will not be better than 2 μm. Jaffe [28] provides some classification and separation of systems for underwater research, including macroscopes and microscopes; for example, a system that can achieve a resolution of 30 μm is macroscopic, while everything less than that is microscopic. Our opinion is that although a system with a resolution of 2 μm is microscopic, it has very weak parameters; a strong microscope should provide resolution, the value of which is determined by the diffraction limit. If we take into account that conventional light microscopes can achieve a resolution of 0.25–0.50 μm (using the visible spectral range), then these values should serve as a guide when creating an optical system as well as an underwater microscope. Thus, the optical system of the objective is the first and main part of the underwater microscope, which requires modernization. The aim is to achieve numerical apertures of 0.5–1.1 in objectives. At the same time, objectives with numerical apertures of 0.50–0.6 may differ in increased values of the free working distance (distance from the frame of the first objective lens to the object under study). Objectives of numerical apertures

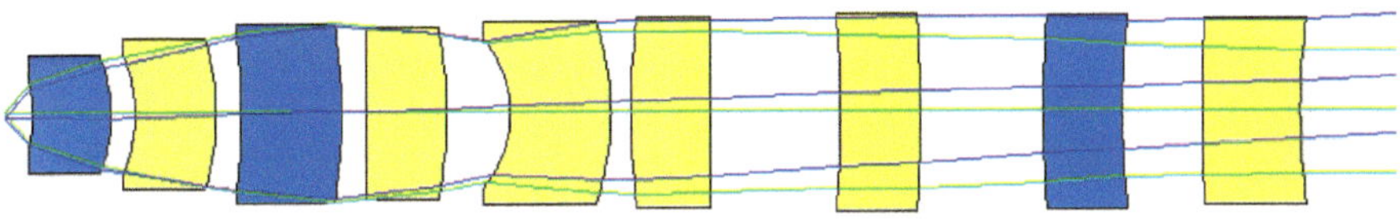

Lens: PlanPolyApo 100x075

Object num aper 0.750000 Image height 11.000000 Primary wavln 5.000000

SRF	RADIUS	THICKNESS	APERTURE RADIUS	GLASS	SPECIAL
OBJ	0.000000	1.0000e-20	0.109996	AIR	
AST	0.000000	0.700000	0.109996 AS	AIR	
2	-5.225100	2.000000	0.830000	GERM C	
3	-3.686600	0.600000	1.420000	AIR	
4	-3.704600	2.000000	1.470000	CAF2 C	
5	-6.190800	0.600000	1.830000	AIR	
6	-22.234900	2.500000	1.930000	GERM C	
7	-12.316200	0.600000	2.170000	AIR	
8	69.709700	2.000000	2.090000	CAF2 C	
9	-11.959400	1.600000	1.980000	AIR	
10	-2.683100	2.500000	1.810000	CAF2 P	
11	-6.815800	0.500000	2.210000	AIR	
12	17.203800	2.000000	2.280000	CAF2 P	
13	-140.492400	3.200000	2.300000	AIR	
14	-29.856700	2.000000	2.330000	CAF2 P	
15	-23.016300	3.000000	2.380000	AIR	
16	25.721000	2.000000	2.370000	GERM P	
17	22.441300	2.000000	2.230000	AIR	
18	48.832700	2.500000	2.270000	CAF2 P	
19	16.188000	200.000000	2.270000	AIR	
20	1.5300e+03	5.000000	13.870000	GERM P	
21	-500.000000	1.000000	13.900000	AIR	
22	-500.000000	2.000000	13.850000	GERM P	
23	-1.0000e+03	197.299055 S	13.860000	AIR	
IMS	0.000000	0.000000	11.000000 S		

Figure 4.47 Optical design and graphs of aberration correction of a 100×/0.75 MWIR CCF plan poly-apochromatic objective ($F'_{ts} = 200$ mm).

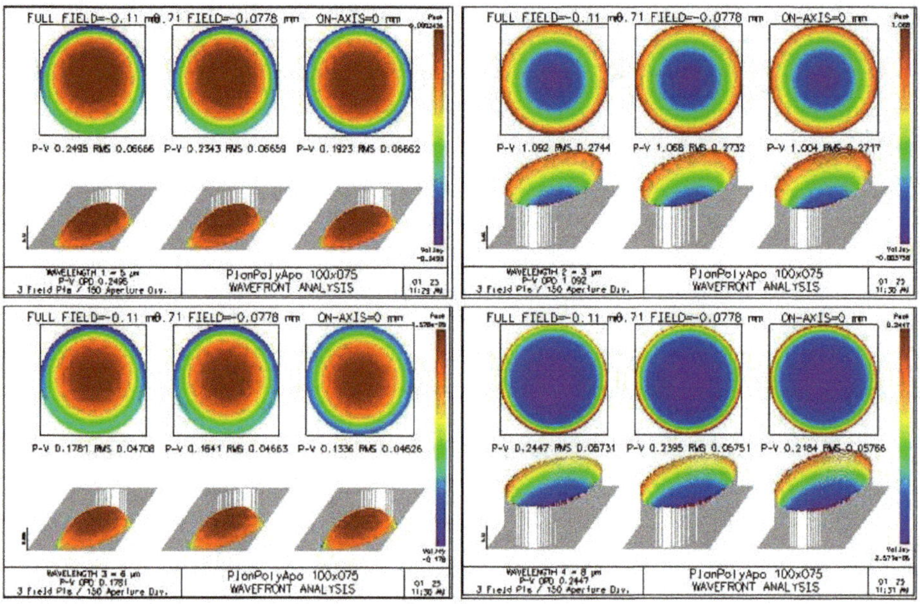

Figure 4.47 (*Continued*) Optical design and graphs of aberration correction of a 100×/0.75 MWIR CCF plan poly-apochromatic objective ($F'_{ts} = 200$ mm).

that exceed 1.0 (using water as an immersion) can work according to the principle of contact objectives, the advantages of which are described in Chapter 3, Section 3.11.

Table 4.5 Main technical parameters and basic optical layout of MWIR CCF plan poly-apochromatic objectives.

Magnification	*NA*	*WD* (mm)	*F'* (mm)	*R* (μm)	*DF* (μm)	*FOV* on object (mm)	*FOV* on image (mm)	Optical design at figure no.	The principal optical layout
5×	0.10	5.8	40.0	30.5	305	4.4	22	4.43	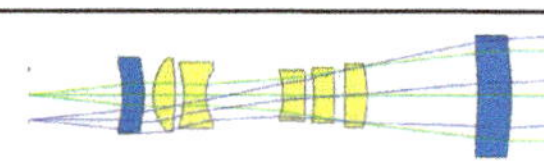
10×	0.25	6.2	20.0	12	49	2.2	22	4.44	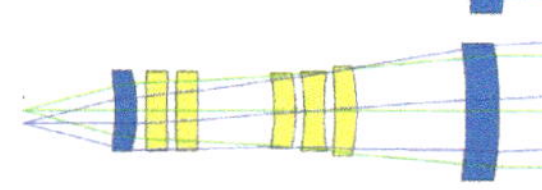
20×	0.35	4.2	8.0	8.7	25	1.1	22	4.45	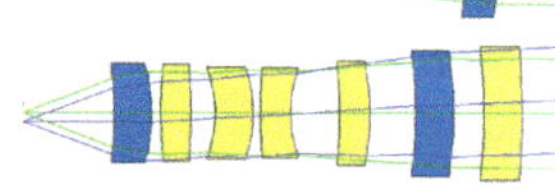
50×	0.55	1.2	4.0	5.6	10	0.44	22	4.46	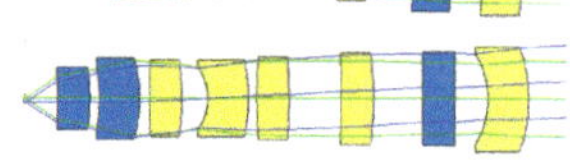
100×	0.75	0.4	2.0	4.1	5.4	0.22	22	4.47	

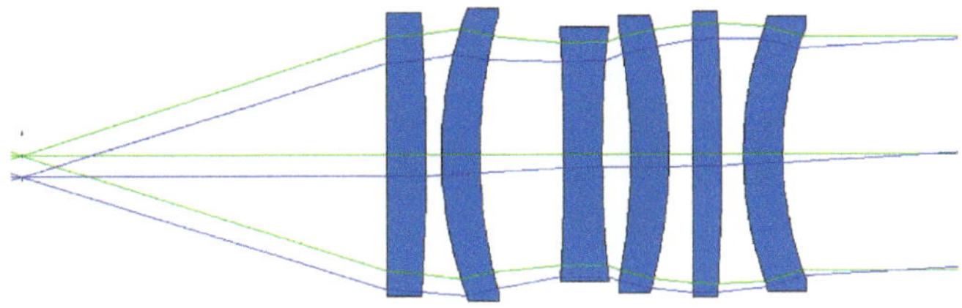

Lens: LD PlanApo 5x030 h95 GE

Object num aper	0.300000	Image height	11.000000	Primary wavln	10.000000

SRF	RADIUS	THICKNESS	APERTURE RADIUS	GLASS	SPECIAL
OBJ	0.000000	1.0000e-20	2.200004	AIR	
AST	0.000000	38.000000	2.200004 AS	AIR	
2	0.000000	4.000000	14.150000	GERMA	
3	-219.246800	1.600000	14.420000	AIR	
4	41.484800	4.000000	14.920000	GERMA P	
5	50.680000	8.800000	14.080000	AIR	
6	-213.680100	4.000000	12.970000	GERMA P	
7	121.361600	3.000000	12.980000	AIR	
8	-70.202800	4.000000	13.350000	GERMA P	
9	-51.700700	2.500000	14.130000	AIR	
10	0.000000	3.000000	14.530000	GERMA P	
11	-216.193000	2.200000	14.590000	AIR	
12	40.247100	4.000000	14.000000	GERMA P	
13	35.027600	200.000000	12.810000	AIR	
14	1.5000e+03	5.000000	22.120000	GERMA P	
15	-500.000000	1.000000	22.130000	AIR	
16	-500.000000	2.000000	22.010000	GERMA P	
17	-1.0000e+03	197.015744 S	22.020000	AIR	
IMS	0.000000	0.000000	11.000000 S		

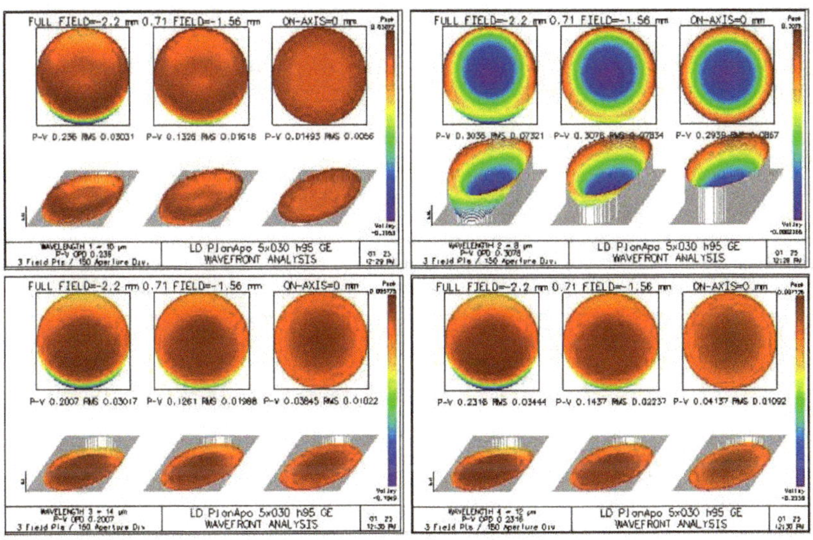

Figure 4.48 Optical design and graphs of aberration correction of a 5×/0.30 LWIR CCF plan poly-apochromatic objective ($F'_{ts} = 200$ mm).

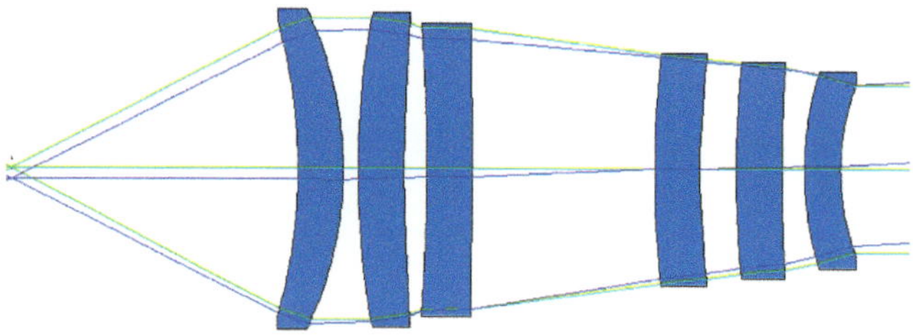

Lens: LD PlanApo 10x045 h95 GE

Object num aper	0.450000	Image height	11.000000	Primary wavln	10.000000

SRF	RADIUS	THICKNESS	APERTURE RADIUS	GLASS	SPECIAL
OBJ	0.000000	1.0000e-20	1.099996	AIR	
AST	0.000000	32.000000	1.099996 AS	AIR	
2	-55.765100	5.000000	16.040000	GERMA	
3	-39.223400	1.600000	17.210000	AIR	
4	87.974200	5.000000	16.920000	GERMA P	
5	192.560200	2.500000	16.280000	AIR	
6	-209.448400	5.000000	15.810000	GERMA P	
7	-1.5371e+03	20.300000	15.630000	AIR	
8	104.252100	5.000000	12.480000	GERMA P	
9	94.724500	4.000000	11.920000	AIR	
10	128.087500	5.000000	11.640000	GERMA P	
11	184.926700	2.500000	11.230000	AIR	
12	34.109900	4.000000	10.590000	GERMA P	
13	25.069600	200.000000	9.410000	AIR	
14	1.5000e+03	5.000000	20.670000	GERMA P	
15	-500.000000	1.000000	20.680000	AIR	
16	-500.000000	2.000000	20.570000	GERMA P	
17	-1.0000e+03	197.010673 S	20.580000	AIR	
IMS	0.000000	0.000000	11.000000 S		

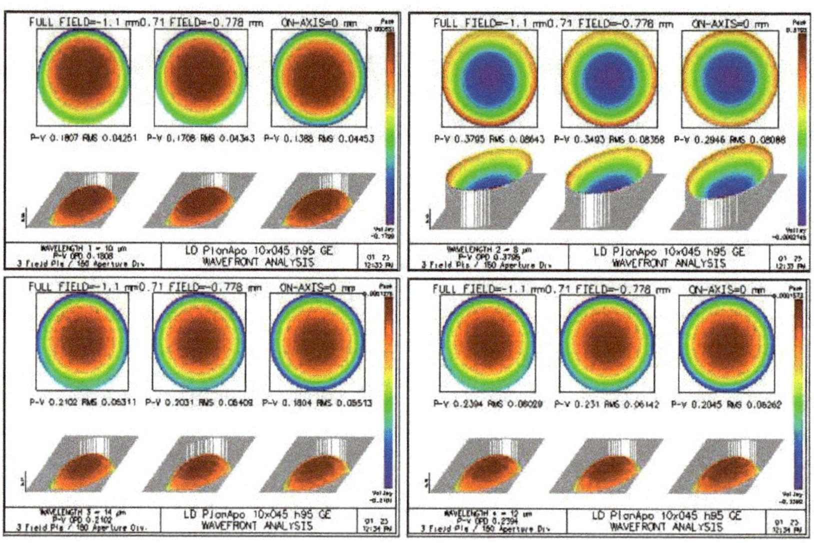

Figure 4.49 Optical design and graphs of aberration correction of a 10×/0.45 LWIR CCF plan poly-apochromatic objective ($F'_{ts} = 200$ mm).

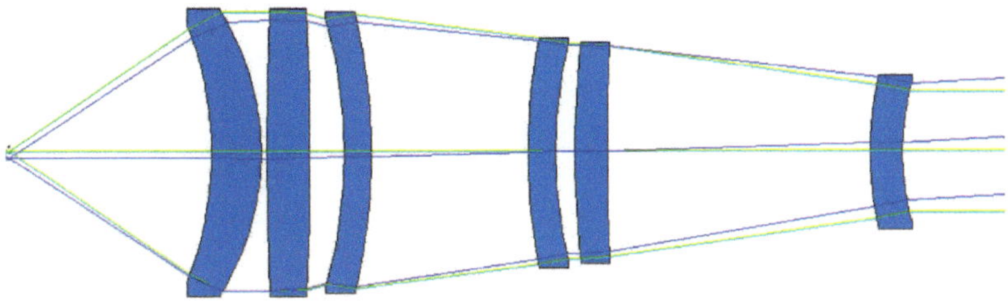

Lens: LD PlanApo 20x055 h95 GE

Object num aper	0.550000	Image height	11.000000	Primary wavln	10.000000

SRF	RADIUS	THICKNESS	APERTURE RADIUS	GLASS	SPECIAL
OBJ	0.000000	1.0000e-20	0.550016	AIR	
AST	0.000000	19.000000	0.550016 AS	AIR	
2	-30.228500	4.600000	11.550000	GERMA	
3	-23.773200	0.500000	12.980000	AIR	
4	254.102900	4.000000	12.950000	GERMA P	
5	-272.353300	3.200000	12.820000	AIR	
6	-44.309300	2.500000	12.250000	GERMA P	
7	-53.892400	14.500000	12.640000	AIR	
8	55.537100	2.500000	10.370000	GERMA P	
9	41.524900	1.800000	9.910000	AIR	
10	92.709500	3.000000	9.980000	GERMA P	
11	193.787100	24.500000	9.800000	AIR	
12	29.444200	3.000000	6.970000	GERMA P	
13	21.855800	200.000000	6.320000	AIR	
14	1.5000e+03	5.000000	17.660000	GERMA P	
15	-500.000000	1.000000	17.690000	AIR	
16	-500.000000	2.000000	17.600000	GERMA P	
17	-1.0000e+03	197.074176 S	17.610000	AIR	
IMS	0.000000	0.000000	11.000000 S		

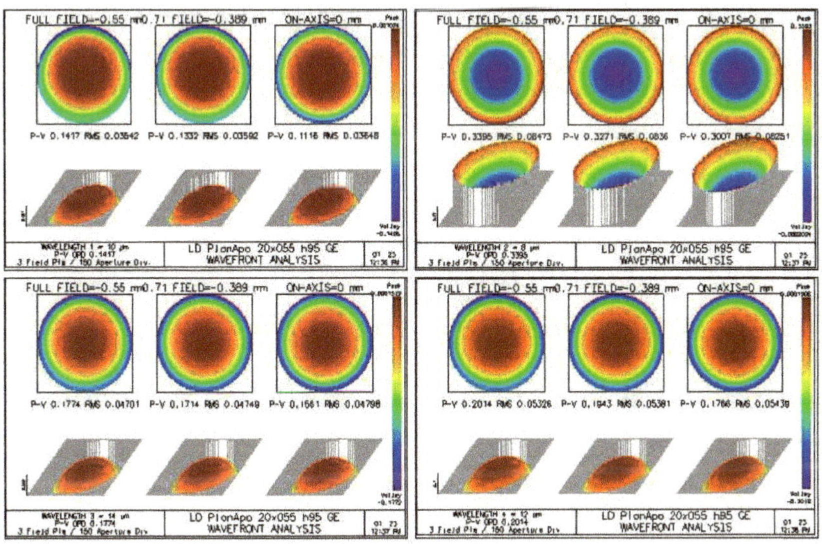

Figure 4.50 Optical design and graphs of aberration correction of a 20×/0.55 LWIR CCF plan poly-apochromatic objective ($F'_{ts} = 200$ mm).

Lighting system

The lighting system of the so-called *transmitted light* for an underwater microscope can differ from the classical one used in standard light microscopes. In a conventional microscope, a lighting system comprises several elements, such as a light source (halogen lamp or LED), a collector optical system, and a condenser optical system. In general, all of these systems can be used to build an underwater microscope if, for example, they are placed in a sealed box that withstands high pressure and allows these systems to be

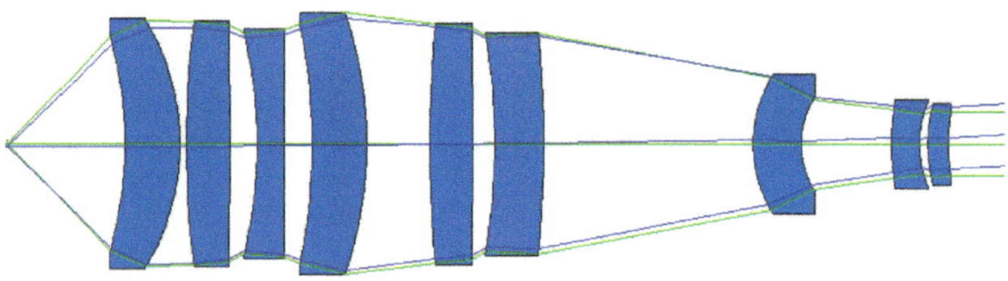

Lens: LD PlanApo 50x070 h95 GE

Object num aper 0.700000 | Image height 11.000000 | Primary wavln 10.000000

SRF	RADIUS	THICKNESS	APERTURE RADIUS	GLASS	SPECIAL
OBJ	0.000000	1.0000e-20	0.219995	AIR	
AST	0.000000	11.000000	0.219995 AS	AIR	
2	-32.378900	5.000000	9.580000	GERMA	
3	-20.964500	0.600000	11.040000	AIR	
4	85.559700	4.000000	10.880000	GERMA P	
5	-722.130900	2.500000	10.540000	AIR	
6	-40.529800	2.500000	9.830000	GERMA P	
7	2.9996e+03	2.500000	10.120000	AIR	
8	-52.317900	5.000000	10.590000	GERMA P	
9	-33.643800	5.800000	11.560000	AIR	
10	105.795400	3.900000	10.640000	GERMA P	
11	-623.862900	2.000000	10.300000	AIR	
12	-53.160600	4.500000	9.690000	GERMA P	
13	-103.423400	19.000000	9.830000	AIR	
14	11.350100	4.300000	6.190000	GERMA P	
15	7.533700	8.200000	4.240000	AIR	
16	20.997600	2.500000	3.520000	GERMA P	
17	8.957000	0.800000	3.900000	AIR	
18	12.908200	1.900000	3.600000	GERMA P	
19	24.634100	200.000000	3.450000	AIR	
20	1.5000e+03	5.000000	14.990000	GERMA P	
21	-500.000000	1.000000	15.020000	AIR	
22	-500.000000	2.000000	14.960000	GERMA P	
23	-1.0000e+03	196.815375 S	14.970000	AIR	
IMS	0.000000	0.000000	11.000000 S		

Figure 4.51 Optical design and graphs of aberration correction of a 50×/0.70 LWIR CCF plan poly-apochromatic objective ($F'_{ts} = 200$ mm).

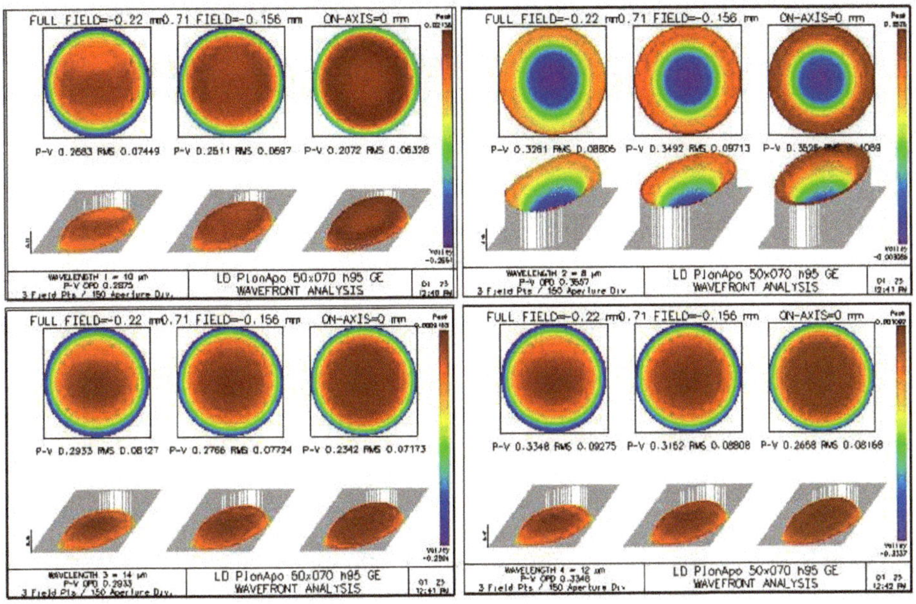

Figure 4.51 (*Continued*) Optical design and graphs of aberration correction of a 50×/0.70 LWIR CCF plan poly-apochromatic objective ($F'_{ts} = 200$ mm).

protected from the influence of temperature changes. However, the functionality of such an illumination system may not be high, and the dimensions and complexity of the hermetic box are considerable (especially given the need to descend to a great depth). Therefore, an alternative embodiment of the optical system of the transmitted light illuminator can be considered—one that is much smaller and uses the original optical system without requiring the use of additional equipment when immersed in water. An example of its principal optical system is shown in Chapter 1, Fig. 1.306.

4.4.4 Specifications of designed parts

The overall design of an underwater microscope can be determined by the tasks that the researcher solves by using it. Because the work of an underwater microscope without an illumination system is practically impossible at a depth of more than 5–10 m underwater, we will assume that the spectral range can be chosen by the researcher and ensured by the LEDs. The researcher may not know a priori which spectral range is preferable for certain observations under a microscope; it depends on many factors, including what kind of objects will be explored. Therefore, for the detection and possible primary identification of objects to be further studied in detail, the use of an auxiliary searching objective has been proposed. The optical calculation of this search objective is

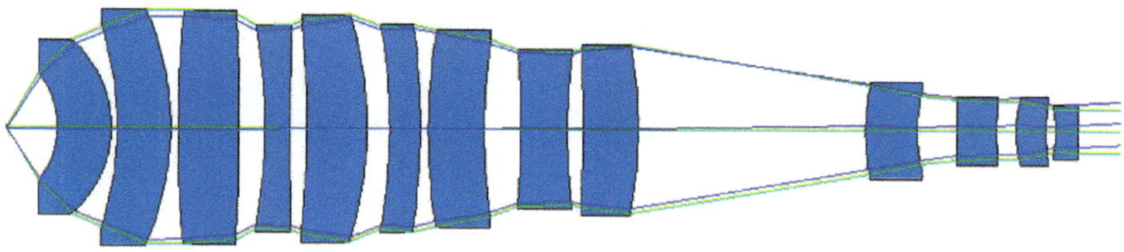

Lens: LD PlanApo 100x085 h95 GE

Object num aper	0.850000	Image height	11.000000	Primary wavln	10.000000

SRF	RADIUS	THICKNESS	APERTURE RADIUS	GLASS	SPECIAL
OBJ	0.000000	1.0000e-20	0.110009	AIR	
AST	0.000000	4.200000	0.110009 AS	AIR	
2	-6.615100	4.500000	4.310000	GERMA	
3	-8.859400	0.500000	6.990000	AIR	
4	-25.712600	4.600000	8.070000	GERMA P	
5	-21.596200	0.600000	9.390000	AIR	
6	94.766100	5.000000	9.330000	GERMA P	
7	-174.740300	2.200000	9.000000	AIR	
8	-41.543900	2.100000	8.130000	GERMA P	
9	255.609200	1.500000	8.240000	AIR	
10	-74.916800	5.000000	8.470000	GERMA P	
11	-30.222100	1.900000	9.070000	AIR	
12	-44.105800	2.600000	8.120000	GERMA P	
13	-59.033400	0.700000	8.280000	AIR	
14	46.920300	4.600000	7.870000	GERMA P	
15	53.713000	3.000000	7.080000	AIR	
16	-90.455800	3.900000	6.320000	GERMA P	
17	51.403200	1.300000	6.210000	AIR	
18	-280.029900	4.500000	6.450000	GERMA P	
19	-42.034000	18.800000	6.800000	AIR	
20	18.659700	4.300000	3.820000	GERMA P	
21	13.437900	3.300000	3.000000	AIR	
22	160.544900	3.300000	2.720000	GERMA P	
23	37.033500	1.600000	2.600000	AIR	
24	8.991700	2.500000	2.720000	GERMA P	
25	14.510600	0.800000	2.230000	AIR	
26	-10.565600	1.900000	2.050000	GERMA P	
27	-42.697000	200.000000	2.160000	AIR	
28	1.5000e+03	5.000000	14.650000	GERMA P	
29	-500.000000	1.000000	14.690000	AIR	
30	-500.000000	2.000000	14.630000	GERMA P	
31	-1.0000e+03	197.287031 S	14.650000	AIR	
IMS	0.000000	0.000000	11.000000 S		

Figure 4.52 Optical design and graphs of aberration correction of a 100×/0.85 LWIR CCF plan poly-apochromatic objective ($F'_{ts} = 200$ mm).

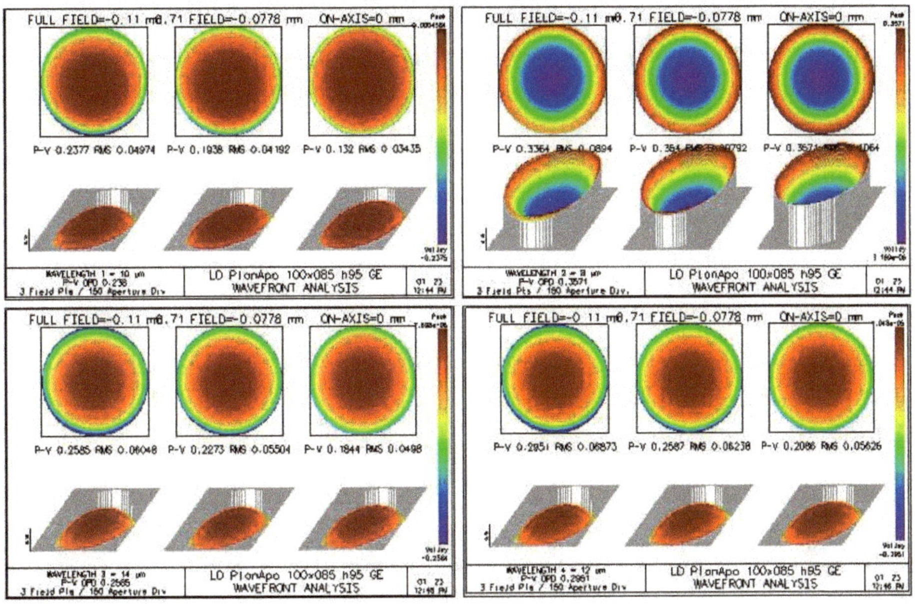

Figure 4.52 (*Continued*) Optical design and graphs of aberration correction of a 100×/0.85 LWIR CCF plan poly-apochromatic objective ($F'_{ts} = 200$ mm).

Table 4.6 Main technical parameters and basic optical layout of LWIR CCF plan poly-apochromatic objectives.

Magnification	*NA*	*WD* (mm)	*F′* (mm)	*R* (μm)	*DF* (μm)	*FOV* on object (mm)	*FOV* on image (mm)	Optical design at figure no.	The principal optical layout
5×	0.30	35	40.0	20	68	4.4	22	4.48	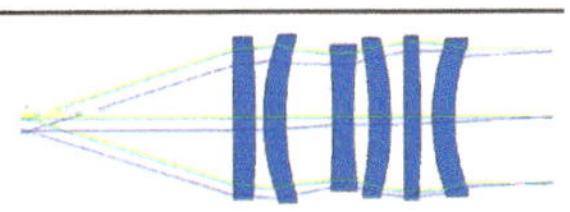
10×	0.45	26	20.0	14	30	2.2	22	4.49	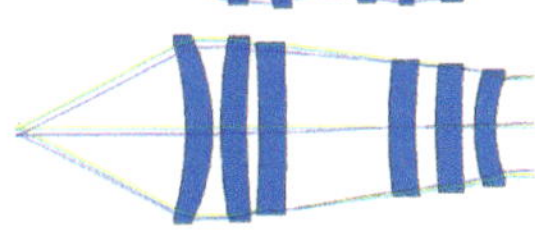
20×	0.55	15	8.0	11	20	1.1	22	4.50	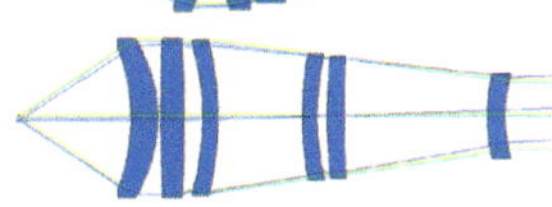
50×	0.70	7	4.0	8.7	12.5	0.44	22	4.51	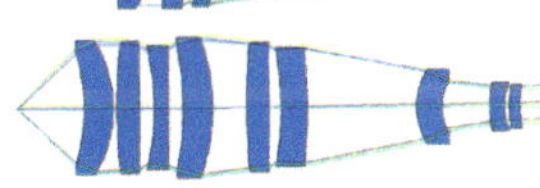
100×	0.85	2.5	2.0	7.2	8.4	0.22	22	4.52	

made to provide the possibility of scanning a certain area of water in a wide spectral range from 200 to 2000 nm; this is a long-focus objective, built like a photographic one, but with low parameters in terms of the angular field and aperture. However, its aberration correction performs equally well in the entire indicated spectral range due to the exclusive use of crystal lenses. Figure 4.53 shows the design parameters and graphs of aberration correction of such an auxiliary searching objective.

The equivalent focal length of this lens is 200 mm, and removal of the entrance pupil is 150 mm; it is designed to focus the image of an object located

Lens: tl water

Image num aper 0.025000 Image height 8.000000 Primary wavln 0.546074

SRF	RADIUS	THICKNESS		APERTURE RADIUS		GLASS		SPECIAL
OBJ	0.000000	1.0000e+20		4.0481e+18		WATER		
AST	0.000000	150.000000		9.800000	A	WATER	P	
2	395.013500	3.300000		9.800000		SILICA		
3	-38.654000	0.660000		9.870000		AIR		
4	-30.206200	4.650000		9.850000		LIF		
5	-18.470800	1.330000		10.000000		AIR		
6	-18.561800	2.660000		9.690000		SILICA	P	
7	46.443800	2.000000		10.100000		AIR		
8	48.649400	4.650000		10.520000		CAF2		
9	-40.210200	149.947311	S	10.670000		AIR		
IMS	0.000000	0.000000		8.000000				

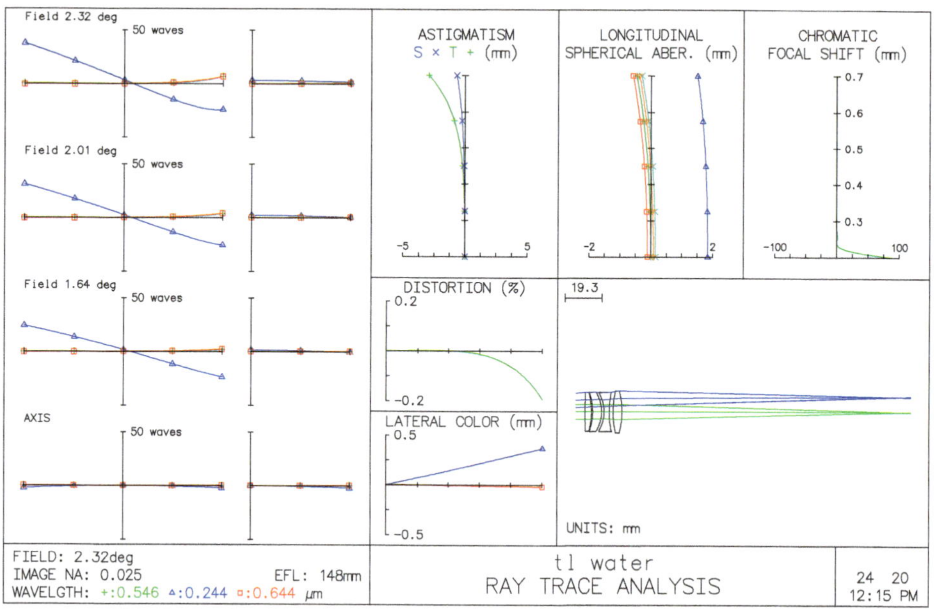

Figure 4.53 Design parameters and graphs of aberration correction of an auxiliary searching objective.

at infinity, i.e., in the plane of the digital image receiver, approximately 1" in size. The resolution in the image space for this lens (for a wavelength of $\lambda \approx$ 550 nm) is approximately 10 μm.

An example optical design of an underwater microscope objective is shown in Fig. 4.54. Such an objective can be used for microscopy at relatively shallow depths, perhaps up to 50–100 m. It is assumed that the first surface of the front lens is immersed in water; the mechanical design must ensure tightness for the second surface of the front lens, as well as for the other lenses.

The objective is designed for the projection of an image of an object located in water at a distance of about 7 mm from the rim of the front lens to a digital receiver up to 50 mm in size (diagonally). The linear magnification of this objective is 60×, and the numerical aperture in the space of objects is 0.55, which ensures that a resolution of 0.5–0.6 μm on the object under study is achieved. This objective is designed to work in the visible spectral range; its type of aberration correction is defined as plan achromatic. This objective can be used to study underwater objects in both transmitted and reflected light. To focus the objective on the object under study, a particular stand with a focusing mechanism is required.

The most interesting example of an objective for an underwater microscope is discussed below. Figure 4.55 shows the optical design and graphs of aberration correction of a high-aperture objective designed for deep-sea microscopic studies. It can be used at very high depths if the mechanical design allows for considerable stability of the optical system. The main distinctive feature of the optical design of this objective is the absence of glued lenses; all the lenses of this optical system are single. This is done because pressure and temperature drops can adversely affect optics at great depths, causing ungluing, stress, and other defects in glued lenses. Another feature is the implementation of the front lens in the form of a *thick lens* to ensure that it can be reliably attached and that its second surface remains tight under conditions of strong pressure. The third feature is the increased values of air gaps between the lenses to ensure that the original method of attaching lenses to the frames can be implemented.

As in the previous example, it is assumed that the first surface of the front lens is immersed in water; the mechanical design must ensure tightness for the second surface of the front lens, as well as for the other lenses. The objective is designed for the projection of an image of an object located in water, but practically on the first (frontal) lens to a 1" digital receiver. The linear magnification of this objective is 40×, and the numerical aperture in the space of objects is 1.1, which ensures that the object under study reaches a resolution of 0.25–0.30 μm, commensurate with the depth of field. This objective is designed to work in the visible spectral range; its type of aberration correction is defined as an LCF semi-plan fluor. Some curvature in the image at the periphery allows *orientation* during practical microscopy, when the planes of

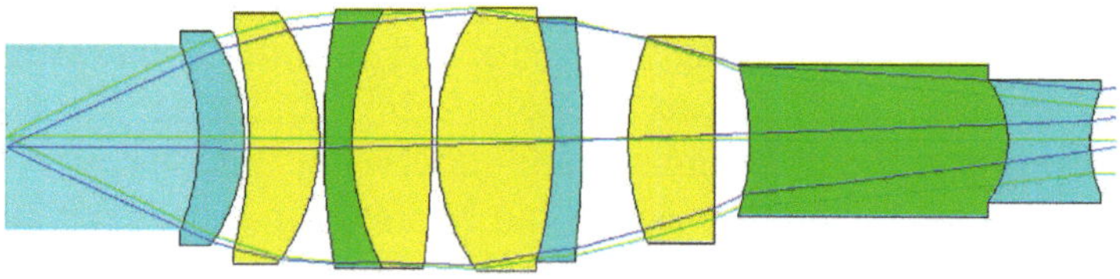

Lens: 60x/0.55 LD water finity

Object num aper	0.550000	Image height	25.000000	Primary wavln	0.546070

SRF	RADIUS	THICKNESS	APERTURE RADIUS	GLASS	SPECIAL
OBJ	0.000000	1.0000e-20	0.417083	WATER	
AST	0.000000	8.400000	0.550000 A	WATER P	
2	-9.865700	2.000000	3.860000	LZ_CTK19	
3	-7.459400	0.200000	4.450000	AIR	
4	-16.647400	3.000000	4.700000	LZ_OK4	
5	-8.503100	0.200000	5.210000	AIR	
6	29.198600	1.200000	5.370000	LZ_TF3	
7	11.326000	3.400000	5.300000	LZ_OK4 P	
8	-44.691400	0.200000	5.380000	AIR	
9	9.606800	5.000000	5.450000	LZ_OK4 P	
10	-17.820300	1.200000	5.060000	LZ_CTK12	
11	-44.344100	2.000000	4.940000	AIR	
12	11.382200	3.700000	4.310000	LZ_OK4 P	
13	211.753300	1.500000	3.540000	AIR	
14	-10.365900	11.000000	3.160000	LZ_TF5	
15	-4.275800	3.500000	2.550000	LZ_CTK12 P	
16	5.391900	149.811466 S	2.120000	AIR	
IMS	0.000000	0.000000	25.000000		

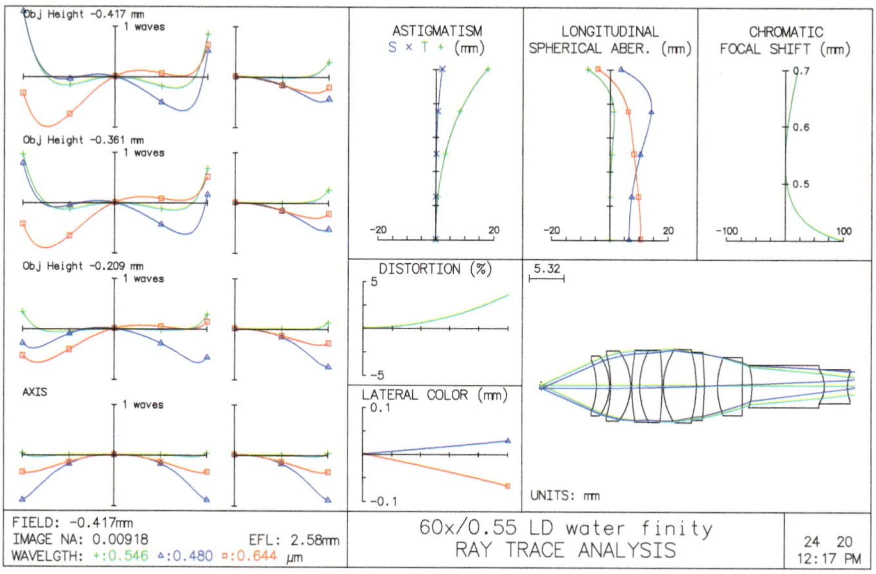

Figure 4.54 Design parameters and graphs of aberration correction of a 60×/0.55 LD water immersion LCF plan achromatic objective.

exact focusing do not coincide, allowing objects to be registered in different *monolayers*. It is possible to use this objective for the study of underwater objects in either transmitted or reflected light. Additional focusing for this objective is not required.

Let us discuss one option for the mechanical design of this lens, which is presented in Fig. 4.56. Some of the requirements determine the fundamental features of the mechanical design. The objective itself as an optical–mechanical device is not sealed, its frontal lens touches water, and its entire outer body is in the water. However, the lenses inside the objective must be

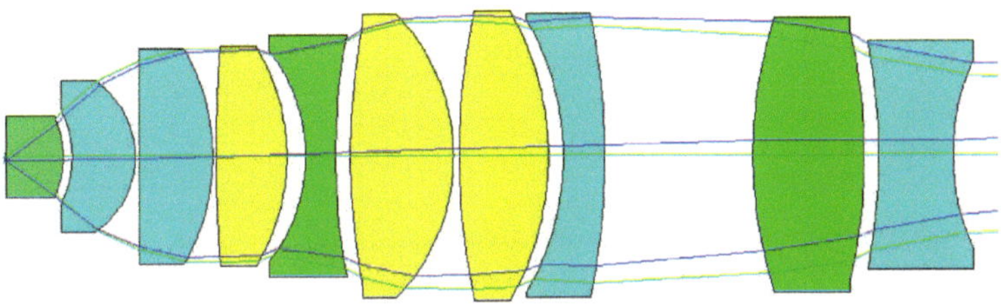

Lens: S-PlanFluor 40x1.10 cont finity — Efl 4.13

Object num aper 1.100000 — Image height 8.000000 — Primary wavln 0.546070

SRF	RADIUS		THICKNESS		APERTURE RADIUS		GLASS		SPECIAL
OBJ	0.000000		0.000100		0.199961		WATER		
AST	0.000000		3.100000		0.200107	AS	H-ZLAF68B		
2	-5.200000		0.500000		2.120000		AIR		
3	-5.200000		3.400000		2.460000		LZ_CTK19		
4	-4.662800		0.200000		3.940000		AIR		
5	709.651200		4.000000		4.970000		LZ_CTK19	P	
6	-10.801800		0.200000		5.590000		AIR		
7	77.998900		3.800000		5.680000		LZ_OK4		
8	-10.660800		1.000000		5.690000		AIR		
9	-8.717200		1.600000		5.470000		ZF7LGT		
10	32.776700		0.800000		6.290000		AIR		
11	34.096400		5.500000		6.850000		LZ_OK4	P	
12	-10.452000		0.300000		7.380000		AIR		
13	34.374500		4.800000		7.540000		LZ_OK4	P	
14	-14.751600		0.800000		7.460000		AIR		
15	-13.703400		2.200000		7.140000		LZ_CTK19	P	
16	-35.191900		8.000000		7.390000		AIR		
17	22.932000		6.000000		7.140000		ZF7LGT	P	
18	-28.295300		0.800000		6.500000		AIR		
19	-28.802800		4.000000		5.940000		LZ_CTK19	P	
20	11.390900		150.033842	S	4.870000		AIR		
IMS	0.000000		0.000000		8.000000	S			

Figure 4.55 Design parameters and graphs of aberration correction of a 40×/1.10 LCF semi-plan fluor contact objective.

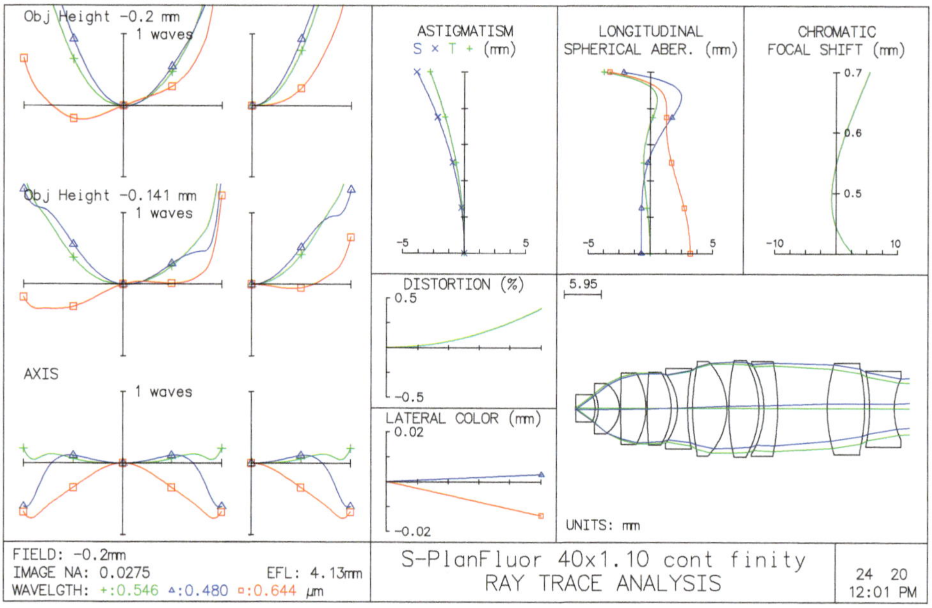

Figure 4.55 (*Continued*) Design parameters and graphs of aberration correction of a 40×/1.10 LCF semi-plan fluor contact objective.

sealed. In addition, it must be designed to withstand strong hydrostatic pressure. Therefore, all the mechanical parts are made thick.

The high-risk zone is the first (frontal) lens of the objective, which, like the entire objective, is under heavy pressure. The geometrical configuration of the front lens is not conventional; the mechanical frame is embedded in the part of the lens where there are no aperture rays. Such an engineering solution can prevent the lens from being "squeezed" by the water column; it also ensures the tightness of the inner side of this lens. The overall mechanical design of the objective is made in such a way that the front lens in the frame can be changed without disassembling the rest of the objective and without having to make further adjustments. The design takes into account that the front lens may be damaged as a result of exposure to water or other external factors; the researcher is able to replace it independently, without having to send the entire objective to a specialized workshop.

Although a lot of pressure affects the lenses and the frames inside the objective when it is underwater, the main influence on the structure's stability can be a temperature difference. Therefore, traditional methods of attaching lenses to the frames (such as fixing with a *cell lip* or using glue) must be abandoned. Instead, novel methods of installation and subsequent adjusting (centering) of the lenses into the frames should be used. For example, 3D printing technology could be used to install the lenses in the frames as follows: the printer prints half of the frame, and then a lens is installed; then the second half of the frame is printed.

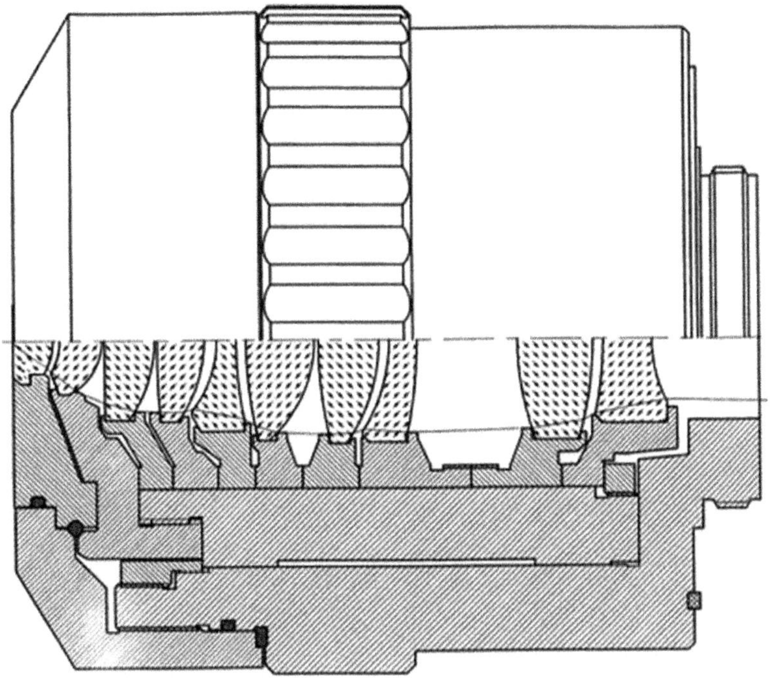

Figure 4.56 One possible option for the mechanical design of the objective.

Thus, the lens will be securely fixed in the frame from both sides. Using a threaded connection (or in some other way), the objective can be attached to another sealed body, which can contain, for example, a digital image receiver, a reflected light illuminator, and other accessories of an underwater microscope. By the way, 3D printing technologies for separate lenses, and even more so for lenses combined with frames, could become one of the main techniques for manufacturing a wide range of optical–mechanical devices.

The above reasoning is probably an attempt to fill in the content of the previously formulated concept of a microscope designed for real-time underwater studies of microscopic objects when a resolution of less than 1 μm is required. Theoretical and experimental studies have shown the fundamental possibility of creating such a microscope. Its main component is a projection system in the form of an objective, the numerical aperture of which is at least 0.5–0.55; the optical design of such an objective must take into account the dispersion properties of water, and its aberration correction must meet the image quality criteria adopted in microscopy. In the design of a visualization system, perhaps the only acceptable solution is to use a digital image receiver; however, the optics should allow working with relatively wide-format receivers. The illumination system used in classical microscopy can be significantly modified and simplified without an obvious loss of quality in such parameters as numerical aperture and illumination contrast. With a mechanical design, it seems appropriate to use the original method, which ensures that the lenses can be reliably fixed in the frames.

If we succeed in fulfilling all the listed conditions and innovations, we can obtain real microscopes, which are optical information systems that harmoniously combine some components, equipment, software, elements of systems engineering, artificial intelligence, and others. Such optical information systems can enable the study of animate and inanimate microscopic objects that cannot be removed from their natural habitat, namely, from the water column at various depths.

4.5 Reverse Microscope: Opportunities and Advantages

We have already discussed the main contradiction in the traditional method of constructing a classical light microscope. This contradiction lies in the impossibility of simultaneously observing a large linear field at the object under study and achieving the maximum possible resolution (determined by the numerical aperture). That is why for the classical light microscope, the study on which is carried out by the observer's eyes, a great many optical systems of condensers, objectives, and eyepieces that have different linear field and numerical aperture values were invented.

The prototype of the *reverse microscope* is described in Chapter 3, Section 3.12, where the linear magnification of the objective and its numerical aperture were selected based on the parameters of the digital image receiver. The optical designs of those objectives show that compared with conventional objectives, there is a clear imbalance between these values. For example, objectives are presented in which an input numerical aperture of 0.90 is achieved with a linear magnification of −12.5×. At the same time, in Chapter 1, Section 1.8.3, examples of optical designs for eyepieces, including those with their own linear magnification of 50× and even 100×, are given. We will also bear in mind the relationships presented in Table 1.15 in Chapter 1, as well as some discussion related to the analysis of this table.

In this case, an engineering idea can be proposed, namely, to functionally swap the objective and the eyepiece in the microscope's optical system. In this case, instead of using, for example, a set of objectives (with different linear magnification and numerical aperture values), only one objective that works with a maximum field and a maximum numerical aperture is used. The eyepieces can be replaceable parts of a wider nomenclature and can differ only in their linear field and linear magnification. Indeed, with the classical method of constructing a light microscope, for example, the value of a full linear magnification of 1000× in the microscope can be achieved by several combinations of objectives and eyepieces: a 100× objective + a 10× eyepiece, a 50× objective + a 20× eyepiece, or other combinations. For the proposed method of constructing a microscope, we can go further and get the value of the full magnification on a 1000× microscope by combining, for example, a 10× objective and a 100× eyepiece.

This objective should be designed so that, if we consider the space of the object under study, it can simultaneously realize the maximum linear field and the maximum numerical aperture. Figure 4.57 shows the design parameters and graphs of the aberration correction of such an objective.

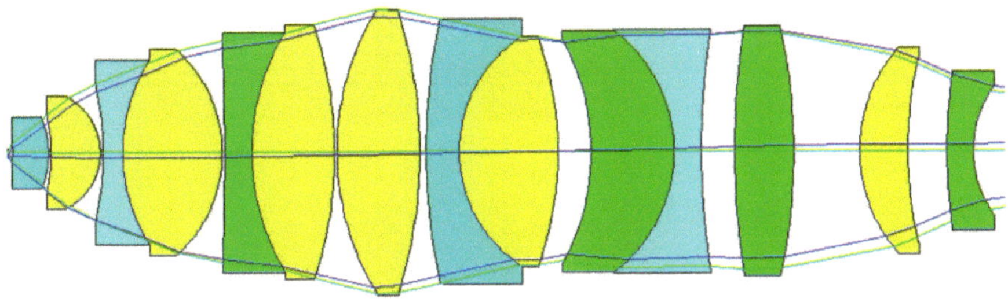

Lens: PlanApo 10x0.90 Finity 2y'=20 Efl 24.84

Object num aper 0.900000 Image height 10.000000 Primary wavln 0.546070

SRF	RADIUS	THICKNESS		APERTURE RADIUS		GLASS		SPECIAL
OBJ	0.000000	1.0000e-20		1.000000		AIR		
AST	0.000000	1.400000		1.000000	AS	AIR		
2	-15.074200	11.000000		3.180000		LZ_CTK19		
3	-27.830000	0.800000		10.670000		AIR		
4	-53.773300	15.000000		12.540000		H_FK71		
5	-18.540000	0.800000		16.950000		AIR		
6	-103.576200	6.000000		21.340000		LZ_K8		
7	50.258400	30.000000		27.340000		H_FK71	P	
8	-40.970000	0.800000		30.800000		AIR		
9	-823.117500	8.000000		33.350000		LZ_TF2		
10	67.528500	25.000000		35.700000		H_FK71	P	
11	-121.897600	0.800000		37.950000		AIR		
12	79.913000	25.000000		42.480000		H_FK71	P	
13	-137.746200	2.000000		42.200000		AIR		
14	192.047200	10.000000		39.390000		LZ_CTK19	P	
15	40.250200	30.000000		34.120000		H_FK71	P	
16	-98.397800	10.000000		34.280000		AIR		
17	-69.160800	25.000000		33.420000		LZ_TF2	P	
18	-44.314900	8.000000		35.850000		LZ_LK3		
19	219.728900	10.000000		36.120000		AIR		
20	243.566700	18.000000		37.070000		LZ_TF2	P	
21	-168.617700	20.000000		37.000000		AIR		
22	44.645100	15.000000		30.650000		H_FK71	P	
23	120.783400	12.000000		28.260000		AIR		
24	177.981800	8.000000		23.530000		LZ_TF2	P	
25	32.270900	199.999062	S	19.430000		AIR		
IMS	0.000000	0.000000		10.000000	S			

Figure 4.57 Design and graphs of aberration correction of a 10×/0.90 CCF plan apochromatic objective (finite).

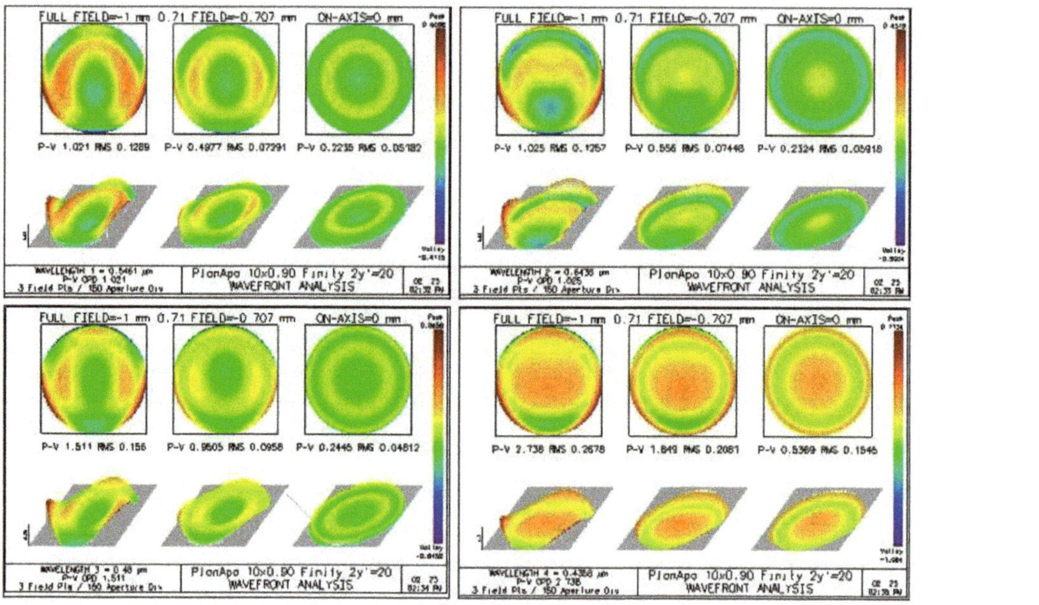

Figure 4.57 (*Continued*) Design and graphs of aberration correction of a 10×/0.90 CCF plan apochromatic objective (finite).

The optical design of this objective replaces several objectives that are commonly used in a microscope. The linear field on the object under investigation is 2 mm (which corresponds to a standard 10× objective). However, unlike a conventional 10× objective, the proposed objective implements a numerical aperture in the space of the object equal to 0.90. Thus, in the image space for such an objective, the linear field will be 20 mm (which corresponds to a standard 10× objective), the back (output) numerical aperture will be $NA' = 0.09$. This is significantly greater than for a conventional microscope. These are initial parameters for the design of optical eyepiece systems. Table 4.7 shows an example of the possible combinations of this objective and some eyepieces when constructing the reverse microscope optical system, as well as the values of the required technical parameters of the eyepieces.

It should be noted that there is some deviation from the classical method of constructing a light microscope, where observation is carried out with the eyes of a person, where intrinsic aberrations of the eye are not taken into account due to their insignificant contribution to the resulting image quality. On the contrary, with this construction, the observer's own aberrations of the eye should be taken into account when developing the optical design of the eyepiece and the combination system of the objective and eyepiece. Some examples of optical designs and

Table 4.7 Examples of possible combinations of an objective and different eyepieces.

Eyepiece	Objective	Total magnification on microscope	Linear field/numerical aperture of eyepiece
10×	10×	100×	20/0.09
12.5×	10×	125×	16/0.09
15×	10×	150×	13.3/0.09
16×	10×	160×	12.5/0.09
20×	10×	200×	10/0.09
25×	10×	250×	8/0.09
50×	10×	500×	4/0.09
100×	10×	1000×	2/0.09

aberration correction for the average human eye are given in the Optics Software for Layout and Optimization (OSLO®; Lambda Research Corporation) basic software package, which is used in all examples of optical designs in this book.

To complete the picture, we can also make an optical objective design that works in an immersion environment, specifically, in water immersion; the achieved numerical aperture is 1.2, and the linear magnification is −12.5×. Figure 4.58 shows the design parameters and graphs of aberration correction of such an objective.

Indeed, these two objectives are quite complex in design, especially considering the very large (for microscopy) lens diameters. That is why, as an optical material with special dispersion properties, optical glass type H-FK71 can be chosen, whose parameters are pretty close to those of CaF2; it is a bit easier to process lenses and gluing lenses containing this optical material. The manufacture of such objectives is, indeed, a rather complicated technical task, but the result of using a reverse microscope can be quite interesting due to a significant increase in the resolution at the microscope's output.

The position of the image relative to the last surface of the objective ($S' = 200$ mm) can be chosen on the basis of design considerations so that, between the objective and the image plane, it is possible to place a visual head that allows the eyes to observe (as in a conventional microscope). Implementing such an engineering solution causes large transverse dimensions of the objectives. Another example of the need to implement such a large distance to the image may be the need to place the image receiver, for example, in a vacuum chamber; at the same time, the thickness of the protective glass of such a chamber can reach 180–200 mm. In general, it would be interesting to view the resulting image using, for example, an electron microscope; an implementation of extra-large linear zoom may provide some additional information about the object. By the way, in such a system, we can explore objects that are in their natural state.

We can use a different layout of a light microscope for when the image of an investigated object is obtained and viewed by the observer on a small screen. This is a kind of analogue of virtual reality glasses; in this case, we can

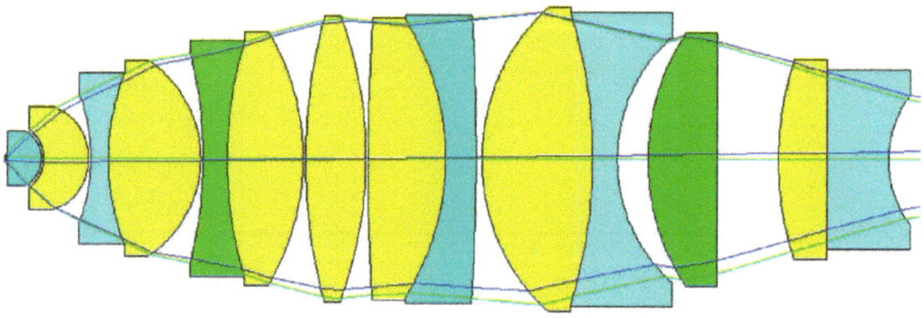

Lens: 12.5x1.20 water finity   Efl 20.2

Object num aper 1.200000 Image height 10.000000 Primary wavln 0.546070

SRF	RADIUS		THICKNESS		APERTURE RADIUS		GLASS		SPECIAL
OBJ	0.000000		0.010000		0.799973		WATER		
AST	0.000000		0.170000		0.820527	AS	K9		
2	0.000000		0.600000		1.040000		WATER	P	
3	-6.512100		11.000000		1.770000		H-ZLAF55D		
4	-9.510700		0.800000		8.540000		AIR		
5	-11.189300		15.000000		9.330000		H-FK71		
6	-17.055400		0.800000		16.370000		AIR		
7	-54.272400		6.000000		20.460000		K9	P	
8	69.726700		30.000000		27.240000		H-FK71	P	
9	-37.111300		0.800000		31.020000		AIR		
10	-132.571100		8.000000		33.330000		LZ_TF5		
11	124.667600		25.000000		37.280000		H-FK71	P	
12	-76.882300		0.800000		40.120000		AIR		
13	170.595100		20.000000		44.780000		H-FK71	P	
14	-123.697600		0.800000		45.240000		AIR		
15	694.296400		25.000000		44.620000		H-FK71	P	
16	-81.448300		10.000000		43.990000		H-ZLAF55D	P	
17	-780.789600		2.000000		45.790000		AIR		
18	64.879200		36.000000		48.380000		H-FK71	P	
19	-152.473700		8.000000		46.570000		K9	P	
20	51.892400		10.000000		39.030000		AIR		
21	72.345900		22.000000		40.230000		LZ_TF5	P	
22	0.000000		20.000000		38.710000		AIR		
23	98.805200		16.000000		31.590000		H-FK71	P	
24	0.000000		20.000000		28.350000		K9	P	
25	33.253600		200.001139	S	20.580000		AIR		
IMS	0.000000		0.000000		10.000000	S			

Figure 4.58 Design parameters and graphs of aberration correction of a 12.5×/1.20 water immersion LCF S-plan MicroFluar objective (finite).

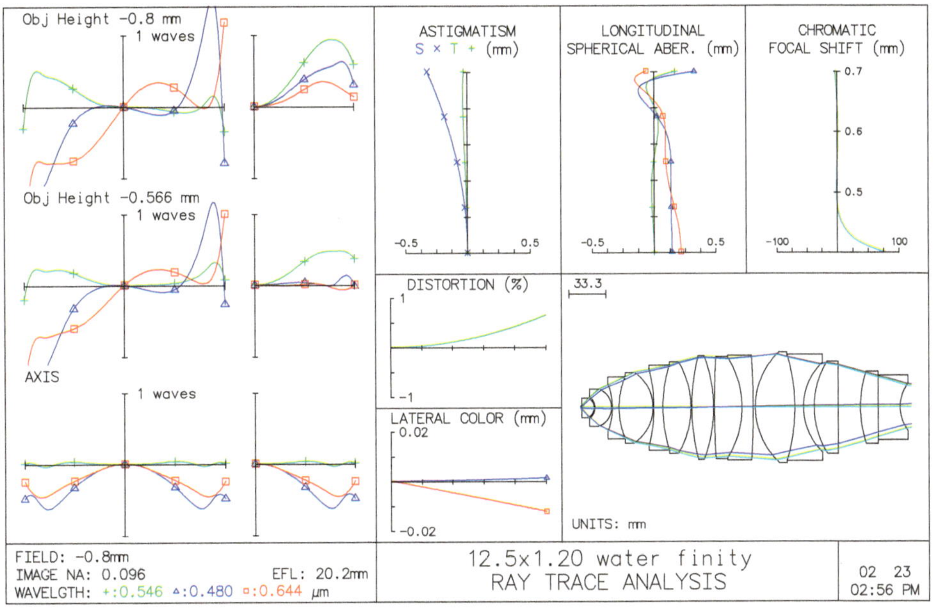

Figure 4.58 (*Continued*) Design parameters and graphs of aberration correction of a 12.5×/1.20 water immersion LCF S-plan MicroFluar objective (finite).

use an objective whose transverse dimensions differ little from ordinary ones. However, such a microscope may not be convenient and comfortable for the researcher, since a person, in this case, may lose their sense of reality and orientation in space.

4.6 Topical Reasoning about Single Exemplar Microscopes

To understand and draw adequate conclusions about the advantages of modern microscopic techniques is a rather difficult task for an ordinary engineer who designs microscopes. These advantages may include such microscopic technologies, for example, as methods for achieving ultrahigh resolution in light microscopy. Technologies such as extreme ultraviolet microscopy and even electron microscopy can be fully realized and accepted by people with special hands-on education and some experience in research in the field of physics. These are not technologies that can be widely discussed in society, even among engineers with a more sophisticated technical education and more experience designing optical instruments. In this context, high-profile optical engineers can contribute to the progress of modern microscopy only by offering highly engineered solutions for the design of various optical and optical–electronic systems. Such systems can represent some basic elements with which physicist researchers do their work.

Today, as a rule, researchers use the basic elements (microscopes, objectives, sources and image receivers, etc.) that are available. Of course, some progress can be achieved if researchers set tasks for engineers to develop narrowly specialized devices that will allow even the most minor thought or idea to be realized. The work of an engineer, as a rule, is inconspicuous. In the majority of cases, its result cannot be evaluated as an achievement; respect and recognition is for researchers. But engineers can drive progress by working hand-in-hand with researchers and offering technical solutions to achieve common goals. In this sense, it is informal communication between researchers and engineers that is the basis for the design of single exemplar microscopes, a necessary condition for achieving progress in microscopic studies.

References

[1] T. Jota, Y. Kim, D. Luepke, D. Kwok, and T. Milster, "Development of optical microscopy with a 121.6 nm source," *Optics InfoBase Conference Papers* (Optics InfoBase Conference Papers), Optical Society of America (2014).

[2] D. S. Keyes, T. S. Jota, W. Gao, D. Luepke, V. Densmore, Y.-S. Kim, G.-H. Kim, and T. D. Milster, "Optomechanical design and tolerance of a microscope objective at 121.6 nm," *Proc. SPIE* **9575**, 957506 (2015) [https://doi.org/10.1117/12.2188803]

[3] W. Gao, E. Finan, B. Jeong, G.-H. Kim, Y. Kim, and T. Milster, "An imaging microscope at 121.6 nm," *Optical Design and Fabrication (Freeform, IODC, OFT)*, OSA Technical Digest, paper JTu5A.6 (2017).

[4] A. Meyer and E. Seitz, *Ultraviolet Radiation*, Foreign Languages Publishing House, Moscow (1952).

[5] J. McGinty, N. P. Galletly, C. Dunsby, I. Munro, D. S. Elson, J. Requejo-Isidro, P. Cohen, R. Ahmad, A. Forsyth, A. V. Thillainayagam, M. A. A. Neil, P. M. W. French, and G. W Stamp, "Wide-field fluorescence lifetime imaging of cancer," *Biomed. Opt. Express.* **1**, 627–640 (2010).

[6] L. Echevarria, I. Orellana, I. Salazar, A. Hernandez, C. Alvarado-Castillo, and O. Estrada, "Spectroscopic evaluation of the interaction between gold nanoparticles and pomolic acid," *Boletin de la Academia de Ciencias Físicas, Matemáticas y Naturales.* **LXXI**(4), 9–17 (2011).

[7] C. J. de Jong, A. Lajevardipour, M. Gecevičius, M. Beresna, G. Gervinskas, P. G. Kazansky, Y. Bellouard, A. H. A. Clayton, and S. Juodkazis, "Deep-UV fluorescence lifetime imaging microscopy," *Photon Res.* **3**(5), 283–288 (2015).

[8] D.N. Frolov, "Synthesis of the optical systems of lens objectives for microscopes," *J. Opt. Technol.* **69**(9), 614 (2002).

[9] U. Schühle, J.-P. Halain, S. Meining, and L. Teriaca, "The Lyman-alpha telescope of the extreme ultraviolet imager on Solar Orbiter," *Proc. SPIE* **8148**, 81480K (2011) [https://doi.org/10.1117/12.893573].

[10] D. N. Frolov, R. M. Raguzin, and V. A. Zverev, "Optical system of a modern microscope," *J. Opt. Technol.* **69**(9), (2002).
[11] R. Schultz, T. Nielsen, J. Zavaleta, R. Ruch, R. Wyatt, and H. Garner, "Hyperspectral imaging: a novel approach for microscopic analysis," *Cytometry.* **43**, 239–247 (2001).
[12] R. B. Smith, *Introduction to Hyperspectral Imaging with TNTmips®*, MicroImages, Inc. (2012).
[13] J. A. Beach, "A richer view of bio structures," *Laser Focus World*, **2**, 68–71 (2009).
[14] M. Schnell, S. Mittal, K. Falahkheirkhah, A. Mittal, K. Yeh, S. Kenkel, A. Kajdacsy-Balla, P. S. Carney, and R. Bhargava, "All-digital histopathology by infrared-optical hybrid microscopy," *Proc. Natl. Acad. Sci. USA.* **117**(7), 3388–3396 (2020).
[15] A. Melnikov, "The formation and analysis of images in microscopy in visible and IR spectrum ranges by spectroscopy methods," Ph.D. dissertation, Saint Petersburg State University of Information Technologies, Mechanics and Optics (ITMO), Saint Petersburg, Russia (2012).
[16] G. M. Hale and M. R. Querry, "Optical constants of water in the 200-nm to 200-μm wavelength region," *Appl. Opt.* **12**(3), 555–563 (1973).
[17] L. V. Wang and H. Wu, *Biomedical Optics: Principles and Imaging*, Wiley, Hoboken, New Jersey, 2007.
[18] M. Vollmer, K.-P. Möllmann, and J. A. Shaw, "The optics and physics of near infrared imaging," *Proc. SPIE* **9793**, 97930Z (2015) [https://doi.org/10.1117/12.2223094].
[19] L. A. Sordillo and P. P. Sordillo, Eds., *Short-Wavelength Infrared Windows for Biomedical Applications*, SPIE Press, Bellingham, Washington, 2022.
[20] I. Kviatkovsky, H. M. Chrzanowski, E. G. Avery, H. Bartolomaeus, and S. Ramelow, "Microscopy with undetected photons in the mid-infrared," *Sci. Adv.* **6**(42), eabd0264 (2020).
[21] O. A. Vinogradova, A. D. Frolov, and D. N. Frolov, "Microscope for researching underwater objects," Theses: Paper 8372-47 of Conference 8372. SPIE Defense, Security, and Sensing, Baltimore, Maryland. 23–27 April 2012. [*Unpublished*]
[22] Dmitry N. Frolov, Olga A. Vinogradova, Alexey D. Frolov, and Alexey G. Tabachkov, "Concept and optical design of the underwater microscope," *Proc. SPIE* **11076**, 110761Y (2019) [https://doi.org/10.1117/12.2534738].
[23] A. D. Mullen, T. Treibitz, P. L. D. Roberts, E. L. A. Kelly, R. Horwitz, J. E. Smith, and J. S. Jaffe, "Underwater microscopy for in situ studies of benthic ecosystems," *Nat. Commun.* **7**, 12093 (2016).
[24] D. N. Frolov, O. A. Vinogradova, A. D. Frolov, and L. Wen, "Use of opportunities of contact microscopy: optical design," *Proc. SPIE* **11032**, 110320Y (2019) [https://doi.org/10.1117/12.2520606].

[25] I. G. Bronshtein, V. N. Vasiliev, I. L. Livshits, and M. B. Sergeev, *Optical Information Systems for Underwater Research*, Saint Petersburg State University of Information Technologies, Mechanics and Optics (ITMO), Saint Petersburg, Russia (2009).

[26] E. V. Babak and P. D. Ivanov, *Underwater Photography*, Mashinostroenie, Leningrad (1969).

[27] M. M. Rusinov, *Composition of Optical Systems*, Mashinostroenie, Saint Petersburg (1989).

[28] J. S. Jaffe, "Underwater optical imaging: The past, the present, and the prospects," *IEEE J. Ocean. Eng.* **40**(3), 683–700 (2015).

Dmitry Nikolayevich Frolov was born in 1961 in Saint Petersburg, Russia, where he was educated at the Institute of Precision Mechanics and Optics (today the University of Information Technology). He received his Ph.D. in technical sciences from the same institution. His specialty is the design of optical systems associated with optical–mechanical and optoelectronic devices. For more than 20 years, he worked for the Opto-Mechanical Association LOMO in Russia. He was promoted from an engineer to the head of the design bureau of complex instrument making. He has published more than 70 patents and more than 50 scientific articles. Currently, he works as a technical specialist for the project Labor-Microscopes. His primary field is microscopy, particularly the development of light microscopes and accessories; the design of optical systems of microscope objectives is his main professional skill. He has made more than 350 types of optical designs for microscope objectives, many of which were produced in different factories. He is a certified foreign expert in the field of optics and optical instrument making.